U	Internal energy, kJ
$\dot{U}$	Rate of u transport, $\dot{m}u$, kW
V	Velocity, m/s
V	Voltage, V
$\forall$	Volume, m^3
$\dot{\forall}$	Volume flow rate, m^3/s
$\forall_k$	Partial volume of component k, m^3
v	Specific volume, $\forall/m$, m^3/kg
$\bar{v}$	Molar v, vM, m^3/kmol
W	Total work including flow work, kJ
W	Variable representing a component
W_{ext}	External work transfer, kJ
$\dot{W}_{ext}$	Rate of external work, kW
w	Specific power kJ/kg
WinHip	Work in negative, heat in positive
x	Distance, m
x	Quality, m_v/m
x_k	Mass fraction of k, m_k/m
y_k	Mole fraction of k, n_k/n
Y	Yield of a gas liquefaction system
z	Elevation from a datum, m
Z	Compressibility factor
Z_h	Enthalpy departure factor
Z_s	Entropy departure factor

GREEK LETTERS

β	Volume expansivity, K^{-1}
Φ	Stored exergy, kJ
ϕ	Specific stored exergy, Φ/m, kJ/kg;
ϕ	Relative humidity, Equivalence ratio
η	Efficiency, mass based stoichiometric coefficient
κ	Compressibility, kPa^{-1}
λ	Percent theoretical air
μ_J	Joule Thomson coefficient
$\bar{\mu}$	Chemical potential, kJ/kmol
v	Molar stoichiometric coefficient
Ω	Number of microstates
θ	Angle, radian
ρ	Density, $1/v$, kg/m^3
ψ	Specific flow exergy, kJ/kg
$\bar{\psi}$	Molar ψ, kJ/kmol
$\dot{\Psi}$	Rate of transport of ψ, $\dot{m}\psi$, kW
ω	Specific humidity, kg of H$_2$O/kg d.a.

SUBSCRIPTS

0	Atmospheric condition
a	Dry air, Aircraft
af	Adiabatic flame
b	Beginning or initial, Back
B	Boundary
c	Clearance, Cut-off
C	Compressor, Cold, Cylinder
cr	Critical

d	
dp	Dew point
e	Exit
el	Electrical
ext	External
F	Flow, Fuel
f	Final
f	Saturated liquid ($x = 0$)
fg	Difference between g and f states
g	Saturated vapor ($y = 0$)
H	Hot
HP	Heat pump
i	Inlet
I	Related to first law
II	Related to second law
in	Transfer into the system
int	Internal
j	Jet
k	Component index, Reservoir index
M	Mechanical
net	Net transfer into or out of system
o	Overall
out	Transfer out of the system
p	Products
P	Pump, Weighted over products, Propulsive
$pd\forall$	Expansion or contraction
R	Refrigerator
ref	Reference value
rev	Reversible
r	Reduced, Reservoir, Relative, Reactants
R	weighted over reactants
reg	Regenerator
s	Isentropic, Stagnation, Specific
sh	Shaft
sat	Saturated
T	Turbine
t	total
th	Thermal, throat
tp	Triple point
u	Useful
univ	System and its immediate surroundings
v	Vapor
w	Water
wb	Wet bulb
*	Critical state

SUPERSCRIPTS

', ''	Phase identity
$^-$ (over bar)	Quantity per unit mole (per kmol)
$\cdot$ (over dot)	Quantity per unit time, rate of transport (per sec)
$^\circ$ (circle)	Standard condition
ch	Chemical

TEST: The Expert System for Thermodynamics

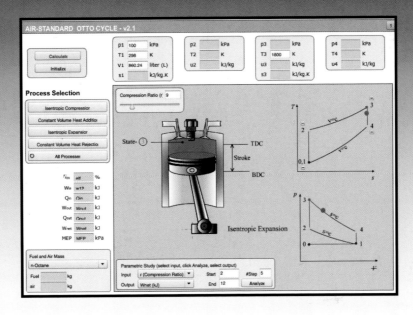

The Expert System for Thermodynamics, or **TEST**, is an interactive website offering tools that enable you to analyze thermofluids problems and verify your calculations. With **TESTcalc**, **Interactive Animations**, **Property Tables**, **Animations**, and **Test Code** examples, you can also pursue what-if scenarios, and visualize complex thermal systems. Throughout the text, you will find problems and examples with references to **TEST** tools.

Redeem access to TEST at:
www.pearsonhighered.com/bhattacharjee

Technical Support is available at www.247pearsoned.com

PEARSON One Lake Street, Upper Saddle River, NJ 07458

Thermodynamics: An Interactive Approach

Subrata Bhattacharjee
San Diego State University

New York Boston San Francisco
London Toronto Sydney Tokyo Singapore Madrid
Mexico City Munich Paris Cape Town Hong Kong Montreal

Vice President and Editorial Director, ECS: *Marcia Horton*
Executive Editor: *Norrin Dias*
Editorial Assistant: *Michelle Bayman*
Program and Project Management Team Lead: *Scott Disanno*
Program Manager: *Clare Romeo and Sandra Rodriguez*
Project Manager: *Camille Trentacoste*
Operations Specialist: *Maura Zaldivar-Garcia*
Product Marketing Manager: *Bram van Kempen*
Field Marketing Manager: *Demetrius Hall*
Cover Designer: *Black Horse Designs*
Cover Image: *The cover photo is from the BASS (Burning and Suppression of Solids) experiment conducted on the International Space Station in May, 2013. Principal Investigator: Paul Ferkul. Image courtesy of NASA.*
Media Project Manager: *Renata Butera*
Composition/Full Service Project Management: *Pavithra Jayapaul, Jouve North America*

10 9 8 7 6 5 4 3 2

Library of Congress Cataloging-in-Publication Data

Bhattacharjee, Subrata, 1961-
  Thermodynamics : an interactive approach : a text based on webware / Subrata (Sooby) Bhattacharjee, San Diego State University. — First edition.
     pages cm
  ISBN-13: 978-0-13-035117-3
  ISBN-10: 0-13-035117-2
  1. Thermodynamics—Textbooks. 2. Machinery, Dynamics of—Textbooks. 3. Thermodynamics—Computer-assisted instruction. I. Title.
  TJ265.B58 2014
  621.402'1—dc23   2013039655

PEARSON

ISBN-13: 978-0-13-035117-3
ISBN-10: 0-13-035117-2

TABLE OF CONTENTS

Thermodynamics: An Interactive Approach

With this new textbook, Subrata Bhattacharjee offers a new perspective on thermodynamic engineering by integrating his "layered approach" with online technological resources. Students are introduced to new terminology and integral applications, then called upon to apply these concepts in multiple scenarios throughout the text.

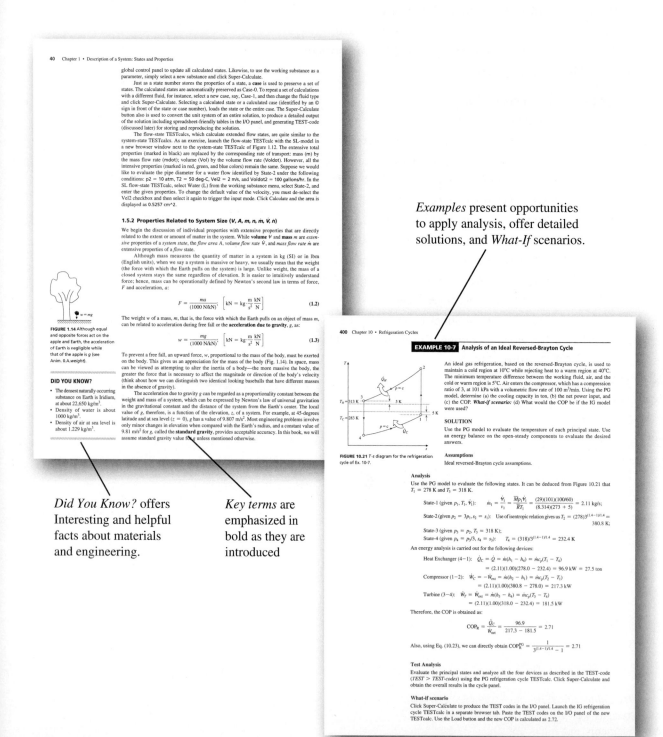

Examples present opportunities to apply analysis, offer detailed solutions, and *What-If* scenarios.

Did You Know? offers Interesting and helpful facts about materials and engineering.

Key terms are emphasized in bold as they are introduced

Thermodynamics: An Interactive Approach

Figures illustrate the textbook concepts and accompany examples.

Problems at the end of each chapter apply concepts and refine engineering analysis.

Property Tables allow for quick reference of material properties, as well as saturation and vapor tables, for many elements and compounds.

TEST: The Expert System for Thermodynamics

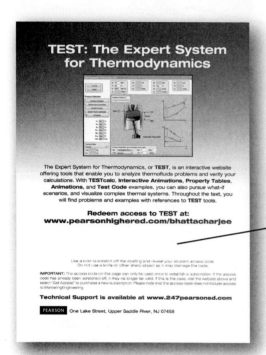

TEST is an online platform focusing on visualization and problem solving of complicated Thermodynamics topics. TEST was designed specifically to complement *Thermodynamics: An Interactive Approach* and using them together will greatly improve the learning experience.

TEST can be accessed at:
www.pearsonhighered.com/bhattacharjee

Throughout the text, examples, problems, and figures reference **TEST** to supplement and illustrate their content.

TEST: The Expert System for Thermodynamics

TESTcalc Map: A Clickable Map of Java Based Thermodynamic Calculators

TESTcalcs are thermodynamic calculators that correspond to specific states, systems, and processes in the book.

The *TESTcalc Map* allows for quick navigation to referenced TESTcalcs.

TESTcalcs contain many options, panels, and data entry points to verify manual solutions and check calculations.

Move the mouse over any widget or property object to see its definition appear here.

$$\eta_{th} = \frac{\dot{W}_{net}}{\dot{Q}_H} = \frac{\dot{Q}_H - \dot{Q}_C}{\dot{Q}_H} = 1 - \frac{\dot{Q}_C}{\dot{Q}_H} ; \quad \eta_{Carnot} = 1 - \frac{T_C}{T_H} ;$$

$$\dot{W}_{rev} = \eta_{Carnot} \dot{Q}_H ; \quad \eta_{II} = \frac{\eta_{th}}{\eta_{Carnot}} ;$$

$$\dot{S}_{gen} = \frac{\dot{Q}_C}{T_C} - \frac{\dot{Q}_H}{T_H} ; \quad \dot{I} = T_C \dot{S}_{gen} = \dot{W}_{rev} - \dot{W}_{net}$$

TEST: The Expert System for Thermodynamics

Interactives are presented in a map in which the animations are organized by types of process and system.

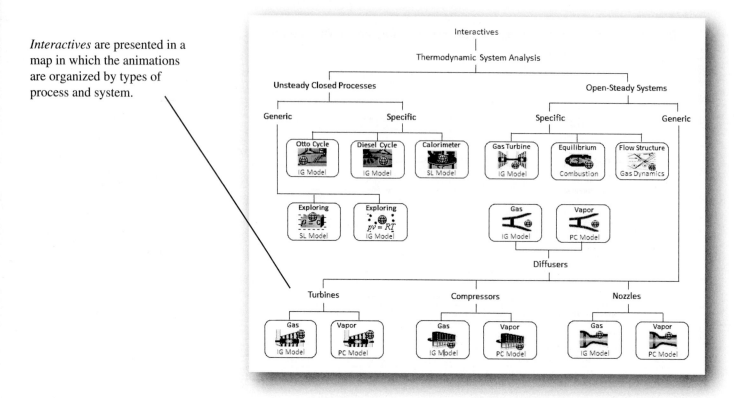

Each *Interactive* includes illustrative animated elements and graphs, while allowing for data entry to model a range of complex calculations.

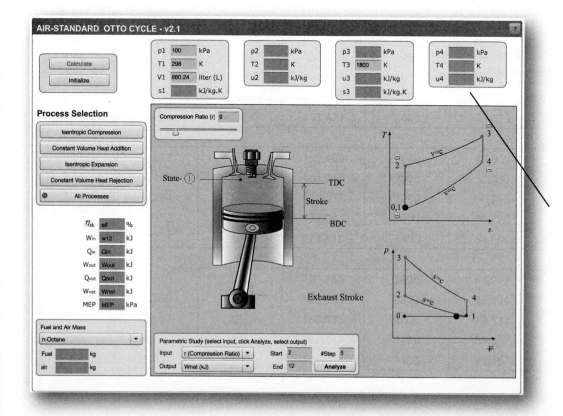

TEST: The Expert System for Thermodynamics

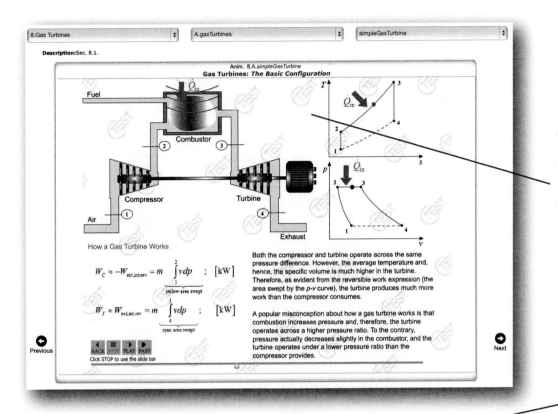

TEST contains more than 400 custom *Animation* assets that complement the figures and provide visual illustrations of many problems.

Property Tables are organized by model and feature links to corresponding *TESTcalc* assets

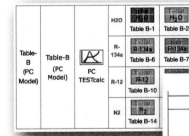

Saturation and superheated tables for phase-change (*PC*) fluids. In the PC TESTcalc, select the working fluid (from more than 60 fluids), enter two independent thermodynamic properties (say, *p* and *h*: all thermodynamic properties are colored

Property Tables display properties of elements and compounds, and allow the units to be toggled between SI Units and English Units.

☑ SI Units   ☐ English Units

Properties of common solids

Substance	Density kg/m³ ρ	Sp. heat kJ/kg·K $c_p = c_v$	Mol.Mass kg/kmol $\bar{M}$
Metals			
Aluminum	2,700	0.902	26.980
Bronze (76% Cu, 2% Zn, 2% Al)	8,280	0.400	50.120
Brass, yellow (65% Cu,	8,310	0.400	64.290
Copper	8,900	0.386	63.550
Iron	7,840	0.450	55.850
Lead	11,310	0.128	207.200
Magnesium	1,730	1.000	24.310
Nickel	8,890	0.440	58.690
Silver	10,470	0.235	107.870
Steel, mild	7,830	0.500	55.710
Tungsten	19,400	0.130	183.850
Nonmetals			
Asphalt	2,110	0.920	10.910
Brick, common	1,922	0.790	59.490
Brick, fireclay (500°C)	2,300	0.960	78.960
Concrete	2,300	0.653	270.100

Properties of common liquids

Substance	Temp. °C T	Density kg/m³ ρ	Sp. heat kJ/kg·K $c_p = c_v$	Mol.Mass kg/kmol $\bar{M}$
Ammonia	25	602	4.8	17.03
Argon	-185.6	1,394	1.14	39.95
Benzene	20	879	1.72	78.12
Brine(20% sodium cloride by mass)	20	1,150	3.11	26.1
n-Butane	-0.5	601	2.31	58.12
Carbon dioxide	0	298	0.59	44.01
Ethanol	25	783	2.46	46.07
Ethyl alcohol	20	789	2.84	46.07
Ethylene glycol	20	1,109	2.84	62.07
Glycerine	20	1,261	2.32	92.09
Helium	-268.9	146	22.8	4
Hydrogen	-252.8	71	10	1.01
Isobutane	-11.7	594	2.28	58.12
Kerosene	20	820	2	170.34
Mercury	25	13,560	0.139	200.59
Methane		423	3.49	16.04
	-161.5	423	3.49	16.04
	-100	301	5.79	16.04

PREFACE

It is hard to justify writing yet another textbook on Engineering Thermodynamics. Even when you compare the most popular textbooks in use today, you will notice striking similarities in content, organization, and the style of engaging students. Even after repeated attempts by different authors to integrate software (such as EES, IT, etc.) with thermodynamics, use of tables and charts are still the norm and parametric studies are still a thing of the future. The Internet revolution has not changed much in the way thermodynamics is taught or learned.

Based on a belief that if software is truly user friendly and easily accessible like a web page, and that students will welcome it is as a true learning tool, I started working on the web application TEST (www.pearsonhighered.com/bhattacharjee). TEST has proven to be a very popular resource with more than two thousand educators around the world using it in their classrooms, as well as among today's students, who enjoy learning from multiple sources, not just from a single textbook. This textbook is a result of melding the web-based resources of TEST with the traditional content of thermodynamics to create a rich learning environment. It differs from other textbooks in a number of ways:

1. *A Layered Approach:* Following the tradition of mechanics, most textbooks use a "spiral approach", where the concepts of *mass, energy,* and *entropy* are introduced sequentially. As a result, an analysis of a device—a turbine, for example—is carried out twice; once with the help of the energy equation, and later, more comprehensively, after the entropy balance equation is developed. By decoupling energy analysis from a comprehensive analysis in this manner, students are encouraged to ask the wrong question, "Is this an *energy* problem or is this an *entropy* problem?" Moreover, entropy is introduced so late in the semester that many students do not have sufficient time to fully appreciate the role entropy plays in practical problem solving, let alone the significance of entropy as a profound property. This textbook adopts a "layered approach", in which important concepts are introduced early based on physical arguments and numerical experiments, and progressively refined in subsequent chapters as the underlying theories unfold. The equation, for example, is first alluded to in Chapter 2, introduced in Chapter 3, derived in Chapter 5, and again revisited in Chapter 11 for a more complete discussion.

2. *Organization:* The layered approach requires a restructuring of the first few chapters, giving each chapter its distinctive theme. The first chapter, the Introduction, introduces a *system,* its *surroundings,* and their *interactions. Energy, heat,* and *work* are discussed in depth, borrowing concepts from mechanics. Chapter 1 is devoted to the description of a thermodynamic system through *system* and *flow states.* The concept of local thermodynamic equilibrium (LTE), a hypothesis that establishes thermodynamics as a practical subject, is discussed, and *properties* including *entropy* and *exergy* are introduced and classified, with the TESTcalcs serving as numerical laboratories. Chapter 2 consolidates the development of mass, energy, and entropy balance equations by exploiting their similarities. Closed steady systems, of which *heat engines, refrigerators,* and *heat pumps* are special cases, are analyzed. Chapter 3 centralizes evaluation of properties of all pure substances, divided into several *material models.* Emphasis is placed on thoughtful selection of the model before looking for the correct equation of state or the right chart in evaluating properties. TESTcalcs are used to verify manual calculations and compare competing models. It is a lengthy chapter and an instructor may cover just a single model and move on to Chapters 4 and 5 before coming back iteratively for additional models. Chapters 4 and 5 are dedicated to comprehensive mass, energy, and entropy analysis of *steady* and *unsteady* systems respectively. Every analysis, regardless the type of the system, begins with the same set of governing balance equations. TESTcalcs are used to verify manual solution and pursue "what-if" scenarios. The rest of the chapters, 6 through 15, follow a similar organization to that of most textbooks.

3. *Tables and Charts:* The Tables module of TEST organizes various property tables and charts into easily accessible web pages according to the underlying material model. Projecting a superheated table in the classroom alongside the constant-pressure lines produced by the PC state TESTcalc for example, can create a powerful visual connection between a series of dull numbers and coordinates in a thermodynamic plot.

4. *Animations:* TEST provides a huge library of animations, which are referenced throughout the book to explain almost every concept, device, and process discussed in this book. Whether it is the derivation of the energy balance equation, explaining the psychrometric chart, or the operation of a combined gas and vapor power cycle, a suitable animation can be used in the classroom to save time and visually connect operation of a device to its depiction in a thermodynamic plot, say, a *T–s* diagram. A slide bar under an animation can be used to go back and forth over a particular region of interest. A three-node address (chapter number followed by a section letter and a label— *9.B.CombinedCycle*—for example) is used to identify animations, which are organized according to the structure of this book.

5. *Interactives:* While an animation can be a good place to be introduced to the operation of a device, say, a turbine, an Interactive can go much further in establishing the parametric behavior. Simulating a system has never been easier. Simply launch an Interactive, for instance the IG turbine simulator, and you will notice that all the input parameters are already set to reasonable default values. To explore how the turbine power changes with inlet temperature, for

example, simply click the Parametric Study button, select inlet temperature T1 as the independent variable, and click Analyze. Once the plot appears, you can change the dependent variable and redo the plot instantly. Some of the advanced Interactives, such as the Combustion Chamber Simulator, can perform quite sophisticated equilibrium analyses of a large set of fuels and also, for example, plot how the equilibrium flame temperature and emissions would change with the equivalence ratio.

6. *TESTcalcs:* These thermodynamic calculators are the workhorses of TEST. The TESTcalc can be used to verify manual solutions, create thermodynamic plots, analyze a system or a process, and run "what-if" scenarios. But first and foremost, TESTcalc is a learning tool. To launch TESTcalc, one must make a successive series of assumptions and watch in real time how the governing equations simplify along with the system animation. Just hovering the pointer over a property brings up its definition and relation to other properties. It produces a comprehensive solution in a visual spreadsheet. For example, when calculating a property, it displays the entire state; when analyzing a device it calculates and displays each term of the energy, entropy, and exergy balance equations. Throughout the textbook, TESTcalcs are used to check manual calculations and gain further insight by performing "what-if" studies. Although using the TESTcalcs does not require any programming, a solution can be stored by generating what is called the TEST-code; a few statements about what is known about the analysis. TEST-code for a wide range of problems is posted in the TEST-code module. Parametric studies of complex systems, say, a modified Rankine cycle with *reheat* and *regeneration,* can be jump started by loading suitable TEST-code into the appropriate TESTcalc.

7. *Examples and Problems: Examples* are presented within the main text of every chapter. After presenting a complete manual solution, most Examples step students through a TEST solution for verification of manually calculated results. *What-if Scenarios* in the examples offer another opportunity to further explore the problem. At the end of each chapter, *Problems* are grouped by section to help students relate the questions to major sections of the chapter. Using the skills demonstrated in the Examples, students can now use TEST to verify the solutions their own. Many of the problems also include What-if Scenarios, which should be solved using TEST for further insight.

8. *Thermodynamic Diagrams:* When states are calculated by a state TESTcalc, they can be instantly visualized on a thermodynamic plot such as the *T–s* or the *p–v* diagram. Constant-property lines can be added by simply clicking the appropriate buttons. Another type of diagram called the flow diagram is introduced in this book, to graphically describe the inventory of energy, entropy, and exergy of a system or a process.

9. *"What-if" Studies:* Many problems have an extra section called *"What-if" Scenario,* which asks for evaluation of the effect of changing an input variable on the analysis. It is assumed that TESTcalcs will be used to explore such questions. Once a manual solution is verified through a TEST solution, a "What-if" study is almost effortless; simply change a parameter and click the Super-Calculate button to see its effect on the entire solution. The Interactives come with built-in parametric study tools. The effect of any input variable on the device can be studied with a few clicks.

10. *Consistent Terminologies and Symbols:* Thermodynamic textbooks are replete with conflicting use of symbols (for example, use of x as mass or mole fraction, V for velocity or volume, etc.). In Chapter 1, we develop a consistent notation: use of lowercase symbols for intensive properties and uppercase symbols for extensive properties, temperature (T), mass (m), and velocity (V) being the notable exceptions. Thus, pressure, an intensive property, is represented by p and not by P. Molar specific (per unit kmol) properties are expressed with a bar on top of a symbol, molar mass $\overline{M}$ included. The terms specific stored energy (e), stored exergy (ϕ), flow energy (j), and flow exergy (ψ) are consistently used in a standardized manner. Some textbooks use terms such as closed system exergy or fixed mass exergy to mean stored exergy while using the term total energy to mean stored energy. A few new terminologies, such as the *SL model* for incompressible solids and liquids and *PC model* to mean possible phase change by a pure substance, are introduced to parallel other standard models such as the IG or the RG model. Similarly, description of a system and description of a flow are distinguished through the terms *system* state and *flow* state.

11. *Benchmarking:* Ask a student of mechanics to estimate the kinetic energy of a truck moving at freeway speed, and you will probably see him or her reaching for the calculator. Throughout this book, topics selected in the "Did you know?" boxes are intended not only to grow curiosity but also to instill a sense of benchmarking. Students will not only learn to estimate how much fuel will be needed to accelerate a truck to its freeway kinetic energy, but to compare entropy of water vapor with liquid water without consulting any tables or charts.

12. *Physics Based Arguments:* Throughout this book physical arguments are used to simplify concepts, support mathematical arguments, and explain numerically derived conclusions. The lake analogy to explain the relations between energy, heat, and work in the Introduction, discussion of the mechanism of entropy generation in Chapter 2, explanation of the meaning of partial derivative using animations in Chapter 3, or understanding why compressing a vapor is more costly than compressing a liquid in Chapter 4 are examples of such physics based argument. To understand the effect of different modifications to a Rankine or Brayton cycle, the concept of an equivalent Carnot cycle with *effective temperatures* of heat addition and rejection is introduced in Chapter 7 and used in the subsequent chapters. Results of parametric studies are explained by evaluating the effect of the parameter change on the effective temperatures. In Chapter 14, the equilibrium criterion is derived not by mathematical manipulation, but from

physical arguments that a system in equilibrium has no internal differences to exploit to extract useful work. The discussion in Chapter 14 to demystify the chemical potential and understand it as a special thermodynamic property that determines the direction of movement of mass, just as temperature gradient decides the direction of heat transfer, is another case in point.

This book is written in such a manner that an instructor is free to decide to what extent TEST should be integrated. For instance, someone following the traditional approach may not use TEST much, but students may find this resource quite complementary, using some of the animations or on-line tables and charts in the classroom, or asking students to explore some of the devices simulated by the Interactives are a few ways to gradually incorporate TEST resources. Using the TESTcalc to verify homework solutions could be the next logical step. This will build confidence as students spend less time tracking down petty errors in interpolating properties from tables. Some educators have completely dispensed with manual evaluation of state properties in favor of TEST solution. Finally, parametric studies can be made part of the homework where students display their curiosity by pursuing different "what-if" scenarios on a given analysis.

I would greatly appreciate any suggestion for increasing the synergy between the textbook and the courseware TEST. Educators can access the professional TEST site and other instructor resources via the Pearson Instructor Resource Center at www.pearsonhighered.com. The professional site allows the instructor to use the full featured TESTcalcs, to quickly develop or customize new problems, leave a comment for the author, and to see what is next for TEST.

Finally, I must acknowledge the contribution from my students, who have helped me throughout the development of TEST and the manuscript. Even the mnemonic WinHip for the sign convention is their creation. I would be remiss if I do not mention at least a few names—Christopher Paolini, Grayson Lange, Luca Carmignani, Wynn Tran, Gaurav Patel, Shan Liang, Crosby Jhonson, Matt Patterson, Tommy Lin, Chris Ederer, Matt Smiley, Etanto Wijayanto, Jyoti Bhattacharjee, Kushagra Gupta, Deepa Gopal, Animesh Agrawal, and Jin Xing as the representatives of a large group I have been fortunate enough to befriend. I must also thank my children, Robi, Sarah, and Neil, who have helped me with the accuracy check and have been my biggest cheerleaders while sacrificing so much of their time with me.

— Subrata Bhattacharjee

your work...

State 1 (given P_1, T_1, V_1, m):

$\Rightarrow v_1 = 0.4744 \frac{m^3}{kg}$, $h_1 = 2960.6 \frac{kJ}{kg}$, $S_1 = 7.271 \frac{kJ}{kg \cdot K}$

$j_1 = h_1 + ke_1 = 2960.6 + \frac{30^2}{2000} = 2961.1 \frac{kJ}{kg}$

$A_1 = \frac{m}{V_1/v_1} = \frac{5}{30/0.4744} = 791 \, cm^2$

State 2 (given p_2, T_2, m):

$\Rightarrow v_2 = 0.7163 \frac{m^3}{kg}$, $h_2 = 2865 \frac{kJ}{kg}$

Using energy equation for nozzle:

$V_2 = \sqrt{2000 \, ke_2} = \sqrt{2000(j_1 - h_2)}$

$= \sqrt{2000(96.1)} = 438 \, m/s$

$A_2 = \frac{m}{V_2/v_2} = \frac{5}{438/0.4744} = 54 \, cm^2$

your answer specific feedback

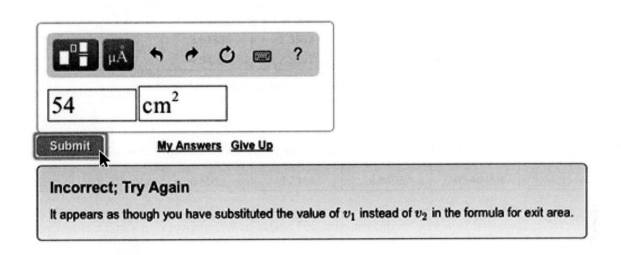

Incorrect; Try Again

It appears as though you have substituted the value of v_1 instead of v_2 in the formula for exit area.

Resources for Instructors

MasteringEngineering. This online Tutorial Homework program allows you to integrate dynamic homework with automatic grading and adaptive tutoring. MasteringEngineering allows you to easily track the performance of your entire class on an assignment-by-assignment basis, or the detailed work of an individual student.

Instructor's Resource. Visual resources to accompany the text are located on the Pearson Higher Education website: www.pearsonhighered.com. If you are in need of a login and password for this site, please contact your local Pearson representative. Visual resources include:

- All art from the text, available in PowerPoint slide and JPEG format.
- **Instructor's Solutions Manual.** This supplement provides complete solutions.

Video Solutions. Located on the TEST website, at www.pearsonhighered.com/bhattacharjee, as well as, MasteringEngineering, the Video Solutions offer step-by-step solution walkthroughs of select homework problems from the text. Make efficient use of class time and office hours by showing students the complete and concise problem-solving approaches that they can access any time and view at their own pace. The videos are designed to be a flexible resource to be used however each instructor and student prefers. A valuable tutorial resource, the videos are also helpful for student self-evaluation as students can pause the videos to check their understanding and work alongside the video.

Resources for Students

MasteringEngineering. Tutorial homework problems emulate the instructor's office-hour environment, guiding students through engineering concepts with self-paced individualized coaching. These in-depth tutorial homework problems are designed to coach students with feedback specific to their errors and optional hints that break problems down into simpler steps.

TEST Website. TEST, The Expert System for Thermodynamics, the Companion Website, located at www.pearsonhighered .com/bhattacharjee, includes:

- **TESTcalc:** A first of its kind thermodynamics calculator for students and professors to check their work, help create new problems, and analyze thermofluids problems.
- **Interactives:** Interactive models in which students can manipulate the inputs to help visualize the content while learning the theory.
- **Animations:** Models to help students visualize and conceptualize the theory they are learning.
- **Online Thermodynamic Tables:** Easily accessible and searchable thermodynamics tables. Tables are in both US and SI units and supplement TESTcalcs to verify interpolated data.

Video Solutions. Complete, step-by-step solution walkthroughs of select homework problems. Video Solutions offer fully worked solutions that show every step of how to solve and verify hand calculations using TESTcalcs—this helps students make vital connections between concepts.

0 INTRODUCTION

THERMODYNAMIC SYSTEM AND ITS INTERACTIONS WITH THE SURROUNDINGS

Thermodynamics is a word derived from the Greek words *thermo* (meaning energy or temperature) and *dunamikos* (meaning movement). Its origin began as the study of converting *heat* to *work,* that is, energy into movement. Today, scientists use the principles of thermodynamics to study the physical and chemical *properties* of matter. Engineers, however, apply thermodynamic principles to understand how the *state* of a practical *system* responds to interactions—transfer of *mass, heat,* and *work*—between the system and its *surroundings*. This understanding allows for more efficient designs of thermal systems that include steam-power plants, gas turbines, rocket engines, internal combustion engines, refrigeration plants, and air-conditioning units.

This chapter lays down the foundation by introducing a system, its surroundings, and all possible interactions between them in the form of mass, heat, and work transfer. A thorough understanding of these interactions is necessary for any thermodynamic analysis, whose major goal is to predict how a system responds to such interactions or, conversely, to predict the interactions necessary to bring about certain changes in the system. Properties will be discussed in Chapter 1 to keep the focus on mass, heat, and work interactions.

Throughout this book, we will adhere to *Système International d'Unités* (SI units), in developing theories and understanding basic concepts, while using mixed units—a combination of SI and English system units—for problem solving.

The accompanying courseware, The Expert System for Thermodynamics (TEST), accessible from *www.pearsonhighered.com/bhattacharjee,* will be used throughout this book for several purposes. The online video tutorials are the best resource to get current information on frequently used modules: (i) animations that are used to supplement discussions throughout this textbook; (ii) web-based thermodynamic calculators called *TESTcalcs* that are used as a numerical laboratory to develop a quantitative understanding of thermodynamic properties, step-by-step verification of manual solutions, and occasional "what-if" studies that provide deeper insights; (iii) Interactives that simulate thermodynamic systems, such as compressors, turbines, nozzles, internal combustion engines, etc., for exploring system behavior; and (iv) traditional and interactive tables and charts.

0.1 THERMODYNAMIC SYSTEMS

We are all familiar with the concept of a free-body diagram from our study of mechanics. To analyze the force balance on a body, or a portion of it, we isolate the region of interest with real or fictitious boundaries, call it a **free body**, and identify all relevant external forces that act on the surface and interior of the free body. For example, to determine the net reaction force, R, necessary to keep the book stationary (or moving without any acceleration) in Figure 0.1, we isolate the book and draw all the vertical forces acting on it. The ambient air applies uniform pressure all over the exposed surfaces. As a result there is no net contribution from the atmospheric pressure. Although pressure will be formally discussed in Sec. 1.5.5, it is sufficient for now to define **pressure** as the intensity of perpendicular compressive forces exerted by a fluid on a surface. In SI units, pressure is measured in kN/m^2 or kPa and in English units it is measured in lb/in^2 psi. To maintain mechanical equilibrium (no unbalanced force), the reaction force R in Figure 0.1, therefore, must be equal to the book's weight.

Applying the same process, we can obtain the pressure inside the piston-cylinder device of Figure 0.2 by drawing all the vertical forces acting on the piston after it has been

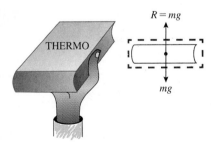

FIGURE 0.1 Free-body diagram of a textbook (see Anim. 0.A.*weight*).

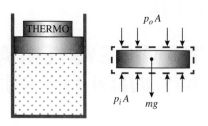

$$p_i A = mg + p_o A \ [\text{N}]$$

FIGURE 0.2 Force balance on a piston (see Anim. 0.A.*pressure*).

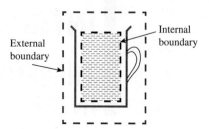

FIGURE 0.3 Each boundary can be used in analyzing the system—the coffee in the mug. In most analyses, we will use the external boundary (see Anim. 0.B.*systemBoundary*).

isolated in a free-body diagram. The steps involved in such an analysis are illustrated in Anim. 0.A.*pressure* and in Example 0-1.

In thermodynamics, a **system** is broadly defined as *any entity of interest within a well-defined boundary*. Just as a free-body diagram helps us analyze the force balance on a body, a thermodynamic system helps us analyze the interactions between a system and its surroundings.

Unlike a system in mechanics, a thermodynamic system does not need to have a fixed mass; it can be a practical device such as a pump or a turbine with all its possible interactions with its **surroundings**—*whatever lies outside its boundary*. Even a complete vacuum can constitute a valid (and interesting) system. A system's boundary is carefully drawn with the objective of separating what is of interest from its surroundings. For example, if the hot coffee within the black boundary in Figure 0.3 constitutes the system, then everything else—the mug, the desk, and the rest of the world for that matter—make up the surroundings. Collectively, the *system* and its *surroundings* form the **thermodynamic universe**.

A boundary can be real or imaginary, rigid or non-rigid, stationary or mobile, and internal or external with respect to a wall. The physical wall of a system, such as a pump's casing or the mug in Figure 0.3, is often considered a non-participant within the interactions between the system and its surroundings. As a result, the boundary can be placed *internally* or *externally* (outlined in black and red, respectively, in Fig. 0.3) without affecting the solution. The term **internal system** is sometimes used to identify the system bordered within the internal boundary. In this book, the external boundary passing through the ambient atmospheric air (as in Fig. 0.2) will be our default choice for a system boundary unless an analysis requires consideration of the internal system.

The material that constitutes or flows through a system is called the **working substance**. While the gasoline-air mixture in the cylinder of an automobile engine and steam flowing through a turbine are the working fluids of their respective systems, internal hardware such as the spark plug in the cylinder or blades inside the turbine can be considered non-participants and excluded from a thermodynamic analysis.

EXAMPLE 0-1   **Free-Body Diagram**

In Figure 0.4, the area of the piston is 25 cm^2, the mass of the hanging weight is 10 kg, the atmospheric pressure is 100 kPa, and the acceleration due to gravity is 9.81 m/s^2. Determine (a) the pressure inside the cylinder in kN/m^2. **What-if scenario:** (b) What is the maximum possible mass that this configuration can support?

SOLUTION

Draw a free-body diagram of the piston and balance the horizontal forces.

Assumptions

Neglect friction, if any, between the piston and the cylinder.

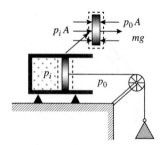

FIGURE 0.4 Schematic of an arrangement creating a sub-atmospheric pressure in Ex. 0-1 (see Anim. 0.A.*vacuumPressure*).

Analysis

Figure 0.4 shows the free-body diagram of the piston and all the horizontal forces that act on it. Because the piston has no acceleration (it is in *mechanical equilibrium*), a horizontal force balance yields:

$$p_i A_{\text{piston}} + \frac{mg}{(1000 \text{ N/kN})} = p_0 A_{\text{piston}}; \quad \left[\frac{\text{kN}}{\text{m}^2} \text{m}^2 = \text{kg} \frac{\text{m}}{\text{s}^2} \frac{\text{kN}}{\text{N}} = \text{kN} \right]$$

$$\Rightarrow \quad p_i = p_0 - \frac{mg}{(1000 \text{ N/kN}) A_{\text{piston}}} = 100 - \frac{(10)(9.81)}{(1000 \text{ N/kN})(25 \times 10^{-4})} = 60.76 \frac{\text{kN}}{\text{m}^2}$$

What-if scenario

As m increases, the piston moves to the right and new equilibrium positions are established. With p_0 and A_{piston} remaining constant, p_i will decrease according to the horizontal force balance equation. The minimum value of p_i is zero (a negative absolute pressure is impossible since pressure is always compressive). Therefore, the maximum mass that can be supported by the atmospheric pressure is:

$$p_i^0 = p_0 - \frac{m_{\text{max}} g}{(1000 \text{ N/kN}) A_{\text{piston}}}$$

$$\Rightarrow \quad m_{\text{max}} = \left(1000 \frac{\text{N}}{\text{kN}} \right) \frac{p_0 A_{\text{piston}}}{g} = \left(1000 \frac{\text{N}}{\text{kN}} \right) \frac{(100)(25 \times 10^{-4})}{9.81} = 25.48 \text{ kg}$$

Discussion

Notice the use of SI units in this problem. The unit of force used in thermodynamics is kN, as opposed to N in mechanics. The familiar expression for weight, mg, therefore must be divided by a unit conversion factor of 1000 N/kN 1000 to express the force in kN. View Anim. 0.A.*vacuumPressure* and similar animations for further insight on the application of free-body diagrams.

0.2 TEST AND ANIMATIONS

As stated earlier, we will make frequent references to different modules of TEST. A task bar at the top of the home page provides access to all the modules.

The structure of this book closely follows the organization of animations in TEST. They are referenced by a standard format—a short title for the animation is preceded by the section number and chapter number. For example, Anim. 0.D.*heatingValue* can be accessed by launching TEST, and clicking the Animation tab, selecting Chapter 0, Section D, and *heatingValue* from the drop-down menu located in the control panel of the Animation tab. Many animations have radio-buttons with interactive features. In this animation, for example, you can toggle among heat release from cookies, wine, and gasoline by selecting the corresponding radio-buttons. The Interactives, which can be used for advanced analysis and thermal system design, have the same look and feel as the animations, but take the concept further when a complete thermal system, such as a gas turbine or a combustion chamber, is simulated.

0.3 EXAMPLES OF THERMODYNAMIC SYSTEMS

The definition of a thermodynamic system—any entity inside a well-defined boundary—can make the scope of thermodynamic analysis mind-boggling. Through suitable placement of a boundary, *systems* can be identified in applications ranging from power plants, internal combustion engines, rockets, and jet engines to household appliances, such as air conditioners, gas ranges, pressure cookers, refrigerators, water heaters, propane tanks, and even hair dryers.

An overview of the thermodynamic systems we are going to analyze across this textbook are presented in Sec. 0.B (System Tour) of the Animations module. As you browse through these systems, there seems to be little in common among this diverse range of devices. Yet, when

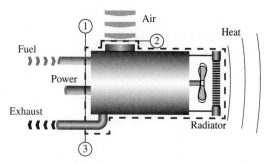

FIGURE 0.5(a) See Anim. 0.C.*carEngine*.

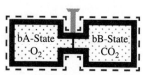

FIGURE 0.5(b) See Anim.
0.C.*closedMixing*.

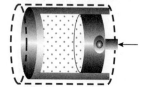

FIGURE 0.5(c) See Anim.
0.C.*compression*.

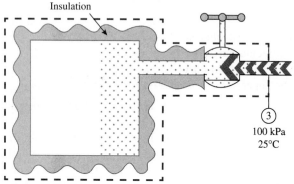

FIGURE 0.5(d) See Anim. 0.C.*charging*.

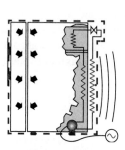

FIGURE 0.5(e) See
Anim. 0.C.*refrigerator*.

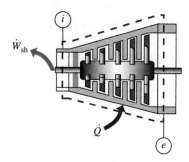

FIGURE 0.5(f) See Anim.
0.C.*turbine*.

some of these systems are examined more closely through animations in Sec. 0.C (interactions) and in Figure 0.5, they reveal a remarkably similar pattern in how they interact with their surroundings. Let's consider a few specific examples and explore these interactions throughout this chapter, qualitatively at first:

1. When we lift a car's hood, most of us are amazed by the complexity of the modern automobile engine. However, if we familiarize ourselves with how an engine works, we can understand the simplified system diagram shown in Figure 0.5(a). While the transfer of mass (in the form of air, fuel, and exhaust gases) and work (through the crankshaft) are obvious, to detect the heat radiating from the hot engine we have to get close enough to feel the heat.

2. Two rigid tanks containing two different gases are connected by a valve in Figure 0.5(b). As the valve is opened, the two working fluids flow and diffuse into each other, eventually forming a uniform mixture. In analyzing this mixing process, we can avoid the complexity associated with the transfer of mass between the two tanks if we draw the boundary to encompass both tanks. Furthermore, if the system is insulated, there can be no mass, heat, or work transfer during the mixing process.

3. The piston-cylinder device of Figure 0.5(c) is the heart of all internal combustion engines and also is found in reciprocating pumps and compressors. If the working fluid trapped inside is chosen as the system, there is no mass transfer. Furthermore, if the compression takes place rapidly, there is little time for any significant transfer of heat. The transfer of work between the system and its surroundings requires analyzing the displacement of the piston (boundary) due to the internal and external forces present.

4. Now, consider the completely evacuated rigid tank of Figure 0.5(d). As the valve is opened, outside air rushes in to fill up the tank and equalize the inside and outside pressures. What is not trivial about this process is that the air that enters becomes hot—hotter than the boiling temperature of water at atmospheric pressure. Mass and work transfer occur as the outside atmosphere pushes air in, but heat transfer may be negligible if the tank is insulated or the process takes place rapidly.

5. Let's now consider the household refrigerator shown in Figure 0.5(e). Energy is transferred into the system through the electric cord (to run the compressor), which constitutes work transfer in the form of electricity. Although a refrigerator is insulated, some amount of heat

leaks in through the seals and walls into the cold space maintained by the refrigerator. To keep the refrigerator temperature from going up, heat must be "pumped" out of the system. If we locate the condenser, a coil of narrow-finned tube placed behind or under the refrigerator, we will find it to be warm. Heat, therefore, must be rejected into the cooler atmosphere, thereby removing energy from the refrigerator. For this system, heat and electrical work transfer are the only interactions.

6. Finally, the steam turbine of Figure 0.5(f) extracts part of the useful energy transported by the working fluid through the turbine and delivers it as external work to the shaft. Although the boundary of the extended system may enclose all the physical hardware—casing, blades, nozzles, shaft, etc.—the actual analysis only involves mass, heat, and work transfer across the boundary and the presence of the hardware can be ignored without any significant effect on the solution.

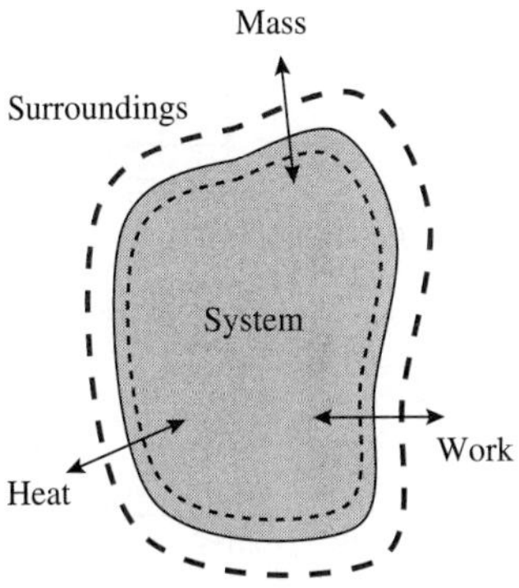

Here, our focus has been on the interactions between the system and the surroundings at the boundary. There is no need to complicate a system diagram with the complexities of non-participating hardware. The abstract or *generic* system shown in Figure 0.6 can represent each system discussed in this section adequately, as it incorporates all possible interactions between a system and its surroundings. The choice of an external (red) or internal (black) boundary cannot change the nature or the degree of these interactions.

FIGURE 0.6 Mass and energy interactions between a system and its surroundings are independent of whether an internal (black) or external (red) boundary is chosen (see Anim. 0.B.*genericSystem* and 0.B *systemBoundary*).

0.4 INTERACTIONS BETWEEN THE SYSTEM AND ITS SURROUNDINGS

A careful examination of the interactions discussed previously reveals that the interactions between a system and its surroundings fall into one of the three fundamental categories: transfer of *mass*, *heat*, or *work* (Fig. 0.6). A quantitative understanding of heat and work as *energy in transit* is essential in developing further concepts in the chapters that follow.

The simplest type of interaction is no interaction. A system segregated from its surroundings is called an **isolated system**. An isolated system can be complex and worth studying. The isolated system shown in Figure 0.7 contains oxygen and hydrogen, separated by a membrane; this system can undergo many changes if the membrane ruptures, triggering for example an exothermic reaction. Despite the heat released during this oxidation reaction, leading to a sharp rise in temperature and pressure, the system remains isolated as long as there are no interactions between the system and its surroundings. As we will discuss later, sometimes interactions among different subsystems can be internalized by drawing a large boundary encompassing the subsystems so that the combined system is isolated (as in Fig. 0.5b). Carefully choosing a boundary can sometimes simplify complex analysis.

FIGURE 0.7 An exothermic reaction may occur if the membrane ruptures (see Anim. 0.c.*isolatedSystem*).

0.5 MASS INTERACTION

Mass interactions between a system and its surroundings are the easiest to recognize. **Mass** is the measure of amount of matter in a system. *Mass cannot be created or destroyed*, only transported. Mass interactions between a system and its surroundings are the easiest to recognize. Usually ducts, pipes, or tubes connected to a system transport mass across the system's boundary. Depending on whether they carry mass in or out of the system, they are called either **inlet** or **exit** ports and are identified by the generic indices i and e. (Note: the term outlet is avoided in favor of exit so that the symbol o can be reserved to indicate ambient properties.) The **mass flow rate**, defined as the amount of fluid that passes through a cross-section per-unit time and measured in kg/s, is always represented by the symbol $\dot{m}$ (mdot in TEST). Thus, $\dot{m}_i$ and $\dot{m}_e$ symbolize the mass flow rates at the inlet i and exit e of the turbine in Figure 0.5(f). The dot on a thermodynamic symbol represents the *rate of transport* of a property as opposed to a time derivative in calculus. Thus, a system's time rate of change of mass is represented by dm/dt, not $\dot{m}$. The **volume flow rate**, the volume of fluid that passes through a cross-section, per-unit time measured in m³/s, can be regarded as the rate of transport of fluid volume and represented by the symbol $\dot{V}$ (Voldot in TEST).

To derive formulas for $\dot{V}$ and $\dot{m}$ at a given cross-section in a variable-area duct, consider Figure 0.8 in which the shaded differential element crosses the surface of interest (shown by the red line) in time Δt. The volume and mass of

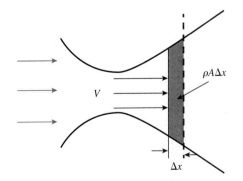

FIGURE 0.8 The volume of the shaded region is $A\Delta x$, where A is the area of cross section (see Anim. 0.C.*massTransport*).

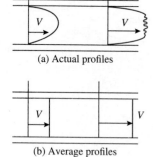

(a) Actual profiles

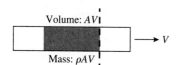

(b) Average profiles

FIGURE 0.9 Different types of actual profiles (a) (parabolic and top-hat profiles) and the corresponding average velocity profiles (b) at two different locations in a channel flow.

Volume: AV

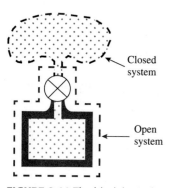

Mass: ρAV

FIGURE 0.10 Every second $\dot{V} = AV$ m³ and $\dot{m} = \rho AV$ kg cross the red mark.

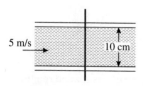

Closed system

Open system

FIGURE 0.11 The black boundary tracks a closed system while the red boundary defines an open system.

that element are given by $A\Delta x$ and $\rho A\Delta x$, respectively, where A is the cross-sectional area and Δx is the element's length. The corresponding flow rates, therefore, are given by the following transport equations:

$$\dot{V} = \lim_{\Delta t \to 0} \frac{A\Delta x}{\Delta t} = A \lim_{\Delta t \to 0} \frac{\Delta x}{\Delta t} = AV; \quad \left[\frac{m^3}{s} = m^2 \frac{m}{s}\right] \tag{0.1}$$

$$\dot{m} = \lim_{\Delta t \to 0} \frac{\rho(A\Delta x)}{\Delta t}; \quad \Rightarrow \quad \dot{m} = \rho AV = \rho \dot{V}; \quad \left[\frac{kg}{s} = \frac{kg}{m^3} \frac{m^3}{s}\right] \tag{0.2}$$

These equations express the instantaneous values of volume and mass flow rates, $\dot{V}$ and $\dot{m}$, at a given cross section in terms of flow properties A, V, and ρ. Implicit in this derivation is the assumption that V and ρ do not change across the flow area, and are allowed to vary only along the axial direction. This assumption is called the **bulk flow** or **one-dimensional flow** approximation, which restricts any change in flow properties only in the direction of the flow. In situations where the flow is not uniform, the average values (see Fig. 0.9) of V and ρ can be used in Eqs. (0.1) and (0.2) without much sacrifice in accuracy.

One way to remember these important formulas is to visualize a solid rod moving with a velocity V past a reference mark as shown in Figure 0.10. Every second AV m³ of solid volume and ρAV kg of solid mass moves past that mark, which are the volume flow rate $\dot{V}$ and mass flow rate $\dot{m}$ of the *solid flow* respectively.

Mass transfer, or the lack of it, introduces the most basic classification of thermodynamic systems: those with no significant mass interactions are called **closed systems** and those with significant mass transfer with their surroundings are called **open systems**. Always assume a system is open unless established otherwise. The advantage to this assumption is that any equation derived for a general open system can be simplified for a closed system by setting terms involving mass transfer, called the *transport* terms, to zero.

A simple inspection of the system's boundary can reveal if a system is open or closed. Open systems usually have inlet and/or exit ports carrying the mass in or out of the system. As a simple exercise, classify each system shown in Figure 0.5 as an open or closed system. Also, see animations in Sec. 0.C again, this time inspecting the system boundaries for any possible mass transfer. Select the Mass, Heat, or Work radio-button to identify the locations of a specific interaction. Sometimes the same physical system can be treated as an open or closed system depending on how its boundary is drawn. The system shown in Figure 0.11, in which air is charged into an empty cylinder, can be analyzed based on the open system, marked by the red boundary, or the closed system marked by the black boundary constructed around the fixed mass of air that passes through the valve into the cylinder.

EXAMPLE 0-2 Mass Flow Rate

A pipe of diameter 10 cm carries water at a velocity of 5 m/s. Determine (a) the volume flow rate in m³/min and (b) the mass flow rate in kg/min. Assume the density of water to be 997 kg/m³.

SOLUTION
Apply the volume and mass transport equations: Eqs. (0.1) and (0.2).

Assumptions
Assume the flow to be uniform across the cross-sectional area of the pipe with a uniform velocity of 5 m/s (see Fig. 0.12).

Analysis
The volume flow rate is calculated using Eq. (0.1):

5 m/s

10 cm

FIGURE 0.12 Schematic used in Ex. 0-2.

$$\dot{V} = AV = \frac{\pi(0.1^2)}{4}(5) = 0.0393 \frac{m^3}{s} = 2.356 \frac{m^3}{min}$$

The mass flow rate is calculated using Eq. (0.2):

$$\dot{m} = \rho A V = (997)\frac{\pi(0.1^2)}{4}(5) = 39.15\,\frac{\text{kg}}{\text{s}} = 2349\,\frac{\text{kg}}{\text{min}}$$

TEST Analysis

Although the manual solution is simple, a TEST analysis still can be useful in verifying results. To calculate the flow rates:

1. Navigate to the TESTcalcs > States > Flow page;
2. Select the SL-Model (representing a pure solid or liquid working substance) to launch the SL flow-state TESTcalc;
3. Choose Water (L) from the working substance menu, enter the velocity (click the check box to activate input mode) and area (use the expression '=PI*0.1^2/4' with the appropriate units), and press *Enter*. The mass flow rate (mdot1) and volume flow rate (Voldot1) are displayed along with other flow variables;
4. Now select a different working substance and click Calculate to observe how the flow rate adjusts according to the new material's density.

Discussion

Densities of solids and liquids often are assumed constant in thermodynamic analysis and are listed in Tables A-1 and A-2. Density and many other properties of working substances are discussed in Chapter 1.

0.6 TEST AND THE TESTcalcs

TESTcalcs, such as the one used in Example 0-2, are dedicated thermodynamic calculators that can help us verify a solution, visualize calculations in thermodynamic plots, and pursue "what-if" studies. Go through My First Solution in the Tutorial for a step-by-step introduction to a TESTcalc. Although there are a large number of TESTcalcs, they are organized in a tree-like structure (visit the TESTcalcs module) much like thermodynamic systems (e.g., open, closed, etc.). Each TESTcalc is labeled with its hierarchical location (see Fig. 0.13), which is a sequence of simplifying assumptions. For example, a page address $x > y > z$ means that assumptions $x, y,$ and z in sequence lead one into that particular branch of TESTcalcs to analyze the corresponding thermodynamic systems. To launch the SL flow-state TESTcalc click the Flow State node of the TESTcalcs, states, flow state branch to display all available models, and then the SL model icon.

   Launch a few TESTcalcs and you will realize that they look strikingly similar, sometimes making it hard to distinguish one from another. Once you learn how to use one TESTcalc, you can use any other TESTcalc, without much of a learning curve. The I/O **panel** of each TESTcalc also doubles as a built-in calculator that recognizes property symbols. To evaluate any arithmetic expression, simply type it into the I/O panel beginning with an equal sign— use the syntax as in: exp(-2)*sin(30) = PI*(15/100)^2/4, etc.—and press *Enter* to evaluate the expression. In the TEST solution of Example 0-2, you can use the expression '= rho1*Vel1*A1' to calculate the mass flow rate at State-1 in the I/O panel.

TESTcalcs (Java Applets) · States · Flow · SL Model

FIGURE 0.13 Each page in TEST has a hierarchical address.

0.7 ENERGY, WORK, AND HEAT

Recall that there are only three types of possible interactions between a system and its surroundings: mass, heat, and work. In physics, heat and work are treated as different forms of energy, but in engineering thermodynamics, an important distinction is made between energy stored in a system and energy in transit. *Heat and work are energies in transit*—they lose their individual identity and become part of stored energy as soon as they enter or leave a system (see Fig. 0.14). Therefore, an understanding of energy is crucial.

   Like mass, energy is difficult to define. In mechanics, **energy** is defined as *the measure of a system's capacity to do work,* that is, how much work a system is capable of delivering. Then again, we need a definition for work, a precise definition

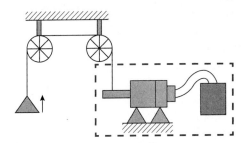

FIGURE 0.14 The battery possesses stored energy as evident from its ability to raise a weight (also see Anim. 0.D.*uConstructive*).

of which will be given in the next section. In qualitative terms, **work** is said to be performed *whenever a weight is lifted against the pull of gravity*. A system that is capable of lifting weight, therefore, must possess stored energy. To understand energy in a more direct manner, it is better to start with **kinetic energy** (KE) and relate all other forms of energy storage to this familiar form. A system's KE can be used to lift a weight as shown in Anim. 0.D.*keConstructive*. Yet another way to appreciate the energy stored as KE in a system is to appreciate the destructive potential of a system due to its motion. A projectile moving with a higher KE has the capacity to do more damage than one moving with a lower KE (see Anim. 0.D.*keDestructive*). Compare a slow moving bowling ball and a fast moving baseball, both having the same momentum. To bring these objects to halt in a given amount of time, an agent must exert the same amount of force in the opposite direction of motion. However, the baseball with its higher kinetic energy will push the agent further or cause more damage.

From our daily life experience, we know that gravitational **potential energy** (PE) can be easily converted to KE through a free fall of the system. Hence a system with higher PE must possess higher stored energy (see Anim. 0.D.*pe*). In fact, whenever a system is pushed or pulled by a force field (interconnected springs, electrical or magnetic fields, etc.), different modes of potential energy may arise. In this textbook, we will consider gravitational potential energy, represented by the symbol PE, as the only mode of potential energy (all other modes of PE are excluded in the kind of systems we generally study in engineering thermodynamics). The **mechanical energy** of a system is defined as the sum of its KE and PE (see Anim.0.D.*mechanicalEnergy*). Mechanical energy involves behavior (speed or position) of a system observable with naked eyes; KE and PE are, therefore, called **macroscopic energy**.

Besides mechanical energy, there are other modes of energy storage. After all, systems with little or no KE and PE can be used to lift a weight as shown in Anim. 0.D.*uConstructive*. Likewise, the battery of Figure 0.14, with no appreciable KE or PE, can be used to raise a weight. Similarly, fossil fuels with relatively little KE and PE possess enormous capacity for doing work or causing destruction (see Anim. 0.D.*uDestructive*). To appreciate how energy is stored in a system besides the familiar macroscopic modes (KE and PE), we must examine things at the microscopic level. Molecules or microscopic particles that comprise a system can also possess KE due to random or disorganized motion that is not captured in the macroscopic KE. For example, in a stationary solid crystal with zero macroscopic KE, a significant amount of energy can be stored in the vibrational KE of the molecules. Although molecular vibrations inside a solid cannot be seen with the naked eye, their effect can be directly felt as the solid's temperature, which is proportional to the average microscopic KE of the molecules (see Anim. 0.D.*uVibrationKE*). For gases, molecular KE can have different modes such as translation, rotation, and vibration as illustrated in Anim. 0.D.*uGasMoleculeKE*. Temperature, for gases, however, is directly proportional to the average translational KE. Microscopic particles can also have PE, energy that can be easily converted to KE, which arises out of inter-particle forces. At the microscopic level, gravitational forces are very weak; but several much stronger forces such as molecular binding forces, Coulomb forces between electrons and their nuclei, nuclear binding forces among protons and neutrons, etc., contribute to various modes of microscopic PE. When there is a change of phase or a chemical composition, the molecular PE can change drastically even if molecular KE (reflected by temperature) remain unchanged.

The aggregate of the various modes of kinetic and potential energies of the microscopic particles is a significant repository of a system's stored energy. For thermodynamic analysis, all are lumped into a single property called U, the **internal energy** of the system. Terms, such as microscopic energy, chemical energy, electrical energy, electronic energy, thermal energy, nuclear energy, used in other fields, are redundant in thermodynamics since U incorporates them all. Given the numerous types of energies that it encapsulates, U is difficult to measure in absolute terms. All we can claim is that it must be positive for all systems and zero for a perfect vacuum. Fortunately, it is not necessary to know the absolute internal energy of a system. In most analyses, it is the change in internal energy that matters.

While a change in mechanical energy of a system can be associated with changes in velocity and elevation, a change in

Table 0-1 Contribution from various sources to world energy consumption of 400 Quad (1 Quad = 1015 BTU or 1.055 x 1015 kJ) in year 2000 (see Anim. 0.D.*energyStats*).

Oil	40%
Natural Gas	23%
Coal	23%
Nuclear	6.5%
Hydroelectric	7%
Others	0.5%

U can be associated with a change in temperature, transformation of phase (as in boiling), or a change of composition through chemical reactions. However, for a large class of solids, liquids, and gases, a change in U often can be related to a change in temperature only.

Having explored all its components, the total **stored energy**, E, of a system can be defined as the sum of the microscopic and macroscopic contributions: $E \equiv U + \text{KE} + \text{PE}$ (Fig. 0.15). The term "stored" is used to emphasize the fact that E resides within the system as opposed to heat and work, which are always in transit. Clearly, the concept of E is much more general than mechanical energy. By definition, energy is stored not only in wind or water in a reservoir at high altitude, but also in stagnant air and water at sea level. Classical thermodynamics does not allow conversion of mass into energy. Hence the symbol for stored energy, E, should not be confused with Einstein's $E = mc^2$ formula that relates energy release to mass annihilation.

Although the unit J (joule) is used in mechanics for KE and PE, in thermodynamics the standard SI unit for stored energy E (and its components) is kJ. It is a good idea at this point to estimate the kinetic energy of some familiar objects as benchmarks and use the converter TESTcalc (located in basic tools branch) to relate some common units of energy, such as MJ, kWh, Btu, Therm, and Calorie, to kJ. We will discuss stored energy and all its components quantitatively in Sec. 1.5.7.

Now that we have defined stored energy, it is easier to define *heat* and *work* as two fundamental ways in which energy can penetrate the boundary of a system. When there is a temperature difference between two objects, nature finds a way to transfer energy from the hotter object to the cooler object through **heat transfer**. The amount of heat transfer depends on the temperature difference between a system and its surroundings, the duration, area of contact, and insulation. The best way to deduce if a particular energy interaction qualifies as heat transfer is to mentally eliminate the temperature difference between the system and its surroundings; the energy transfer will come to a halt if it is heat transfer. The E of the fluid in Figure 0.16 can be raised, as evident from an increase in temperature, by bringing the system in contact with a hotter body, such as placing the system on top of a flame or under focused solar radiation. In each case, energy crosses the boundary of the system through heat transfer driven by the temperature difference between the system and its surroundings. We will discuss heat transfer further in Sec. 0.7.1.

When a net external horizontal force acts on a rigid system (Fig. 0.17), Newton's second law of motion can explain the increase in velocity. However, in thermodynamic terms, we realize that the KE and, hence, the *stored energy* of the system has increased. In the absence of any heat transfer, there must be another fundamental way in which energy must have crossed the boundary of this closed system. Through application of Newton's law, it will be shown in Sec. 1.5.7 that the increase in the system's energy exactly equals *the integral of force times distance,* which is the operational definition of **work** in mechanics. Likewise, when a system is raised, the PE component of E increases as work (force times distance) is transferred in lifting the system upward with a force equal to its weight. We will discuss in detail different modes of work transfer associated with various types of force displacing its point of application in Sec. 0.8. The shaft in Fig. 0.18, for example, turns a paddle wheel and raises the KE and U of the system by transferring shaft work.

Once heat or work enters a system and becomes part of the system's stored energy, there is no way of knowing how the energy was originally transferred into the system. *Like mass, energy cannot be destroyed or created, but only transferred.* Terms such as *heat storage* or *work storage* have no place in thermodynamics, and are replaced by a more appropriate term: **energy storage**.

In a closed system, heat and work are the only ways in which energy can be transferred across a boundary. In an open system energy can also be transported across the boundary by mass. For example, when a pipeline carries oil, cold milk is added to hot coffee, or superheated vapor enters a steam turbine (Fig. 0.19), energy is *transported* by mass regardless of how hot or cold the flow is. Precisely how much energy is transported by a flow, of course, depends on the condition (state) of the flow; however, the direction of the transport is always coincident with the flow direction. We will discuss **energy transport** by mass in Sec. 1.5.8. Commonly used phrases such as *heat flow* or *heat coming out of an exhaust pipe* should be avoided in favor of the more precise term *energy transport* when energy is carried by mass.

To summarize the discussion in this section, energy is stored in a system as mechanical (KE and PE) and internal energy (U). Energy can be *transported* by mass and *transferred* across

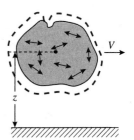

FIGURE 0.15 Stored energy in a system consists of its kinetic, potential, and internal energies (see Anim. 0.D.*storedEnergy*).

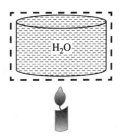

FIGURE 0.16 The stored energy of water in the tank increases due to transfer of heat (see Anim. 0.D.*heatTransfer*).

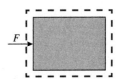

FIGURE 0.17 As the rigid body accelerates, acted upon by a net force F, the work done by the force is transferred into the system and stored as KE, one of the components of the stored energy of a system (see Anim. 0.D.*mechanicalWork*).

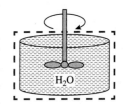

FIGURE 0.18 Transfer of work through the shaft raises the temperature of water in the tank (see Anim. 0.D.*workTransfer*).

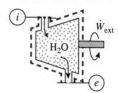

FIGURE 0.19 Energy is transported in at the inlet and out at the exit by mass. It is also transferred out of the system by the shaft (see Anim. 0.D.*energyTransport*).

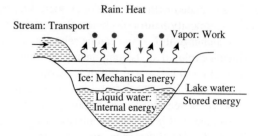

FIGURE 0.20 The lake analogy illustrates the distinction among stored energy, energy transport by mass, energy transfer by heat, and work (see Anim. 0.D. *lakeAnalogy*).

the boundary through heat and work. An analogy—we will call this the **lake analogy** illustrated in Figure 0.20—may be helpful to distinguish energy from heat and work. This figure shows a partially frozen lake that represents an open system; the total amount of water represents the stored energy. Just as stored energy consists of internal and mechanical energies, water in the lake consists of liquid water and ice. The water in the stream (analogous to transport of energy by mass), rain (analogous to heat), and evaporation (analogous to work) are all different ways of affecting the amount of stored water (stored energy) in the lake. Just as rain or vapor is different from water in the lake, heat and work are different from the stored energy of a system. The lake cannot hold rain or vapor just as a system cannot hold heat or work. Right after a rainfall, the rain water loses its identity and becomes part of the stored water in the lake; heat or work added to a system, similarly, becomes indistinguishable from the stored energy of the system once they are assimilated.

Carrying this analogy further, it is difficult to determine the exact amount of stored water in the lake, and the same is true about the absolute value of stored energy in a system. However, it is much easier to determine the change in the stored water by monitoring the water level. The change in stored energy also can be determined by monitoring quantities such as velocity, elevation, temperature, and phase and chemical composition of the working substance in a system. We will call upon this analogy again in later sections.

The discussion above is meant to emphasize the careful use of the terms *heat, work,* and *stored energy* in thermodynamics. In our daily lives and in many industries, the term energy continues to be loosely used. Thus, energy production in Table 0-1 actually means heat or work delivered from various sources. Another misuse of the term energy will be discussed when a property called *exergy* is introduced in Chapter 1.

0.7.1 Heat and Heating Rate (Q, $\dot{Q}$)

The symbol used for heat is Q. In SI units, stored energy and all its components as well as heat and work have the unit of kJ. An estimate for a kJ is the amount of heat necessary to raise the temperature by 1°C of approximately 0.24 kg of water or 1 kg of air. In the English system, the unit of heat is a Btu, the amount of heat required to raise the temperature of 1 lbm of water by 1°F. The symbol used for the rate of heat transfer is $\dot{Q}$ (Qdot in TEST), which has the unit of kJ/s or kW in SI and Btu/s in the English system. If $\dot{Q}$ is known as a function of time, the total amount of heat transferred, Q, in a process that begins at time $t = t_b$ and finishes at time $t = t_f$, can be obtained from:

$$Q = \int_{t_b}^{t_f} \dot{Q}dt; \quad [\text{kJ} = \text{kW} \cdot \text{s}] \tag{0.3}$$

For a constant value of $\dot{Q}$ during the entire duration $\Delta t = t_f - t_b$, Eq. (0.3) simplifies to:

$$Q = \dot{Q}\Delta t; \quad [\text{kJ} = \text{kW} \cdot \text{s}] \tag{0.4}$$

In this book, the phrase **heat transfer** means Q or $\dot{Q}$, depending on its context. For example, the heat transfer, Q, necessary to raise the temperature of 1 kg of water from 20°C (room temperature) to 100°C (boiling point) is about 335 kJ. Whereas the heat transfer, $\dot{Q}$, necessary to vaporize water at 100°C at a rate of 1 kg/s is about 2,260 kW or 2.26 MW under atmospheric pressure. The second heat transfer refers to the rate of heat transfer, which is evident from its units. A perfectly insulated system for which Q or $\dot{Q}$ are zero is called an **adiabatic system**. Classification of systems based on different kinds of interactions is illustrated in Anim. 0.C.*systemTypes*.

We are all familiar with the heat released from the combustion of fossil fuels or calories released from food during metabolism. A fuel's **heating value** (which, illustrated in Anim. 0.D.*heatingValue*, will be more thoroughly discussed in Chapter 13), is the magnitude of the maximum amount of heat that can be extracted by burning a unit mass of fuel with air when the products leave at atmospheric temperature. The heating value of gasoline can be looked up from Table G-2 (Tables link in TEST) as 44 MJ/kg. This means that to supply heat at the rate of 44 MW, at least 1 kg of gasoline has to be burned every second. If the entire amount of heat released is used to vaporize liquid water at 100°C under atmospheric pressure, 44/2.26 = 19.5 kg of water vapor will be produced for every kilogram of gasoline burned. In our daily lives, we see calorific values printed on food products. For example, a can of soda can release a maximum of 140 calories (or whatever amount is printed on the nutrition information) or 586 kJ (1 calorie = 4.187 kJ) of heat when metabolized.

Heat transfer can add or remove energy from a system. Hence it is bi-directional from the point of view of a system. A non-standard set of phrases such as heat gain, heat addition, heating rate, heat loss, heat rejection, and cooling rate, and symbols with special suffixes such as Q_{in}, Q_{out}, and $\dot{Q}_{loss}$, often are used to represent heat transfer in specific systems. With the subscripts specifying the direction of the transfer, symbols such as Q_{in}, Q_{out}, $\dot{Q}_{loss}$, etc., represent only the magnitude of heat transfer, which makes it difficult to formulate equations involving heat transfer. A more mathematical approach is to treat Q and $\dot{Q}$ as algebraic quantities and use algebraic signs to indicate the direction of heat transfer (Anim. 0.D.*WinHip*). For this, we need a standard definition of positive heat transfer.

*The **sign convention** that dates back to the days of early steam engines attributes heat added to a system with a positive sign.*

To illustrate this sign convention, suppose two bodies A and B have two different temperatures, say $T_A = 200°C$ and $T_B = 300°C$, and are placed in thermal contact as shown in Figure 0.21. Now suppose at a given instant the rate of heat transfer from B to A is 1 kW. In drawing the systems A and B separately, the heat transfer arrows are pointed in the positive directions irrespective of the actual directions of the transfers. This is a standard practice unless suffixes are used to indicate directions. Algebraically, we therefore express the heat transfers for the two systems as $\dot{Q}_A = 1$ kW and $\dot{Q}_B = -1$ kW. Similarly, if heat is lost from a system at a rate of 1 kW, it is expressed as either $\dot{Q}_{loss} = 1$ kW or $\dot{Q} = -1$ kW. If a system transfers heat with multiple external reservoirs, the net heat transfer can be calculated by summing the components, provided that each component has the correct sign.

The *lake analogy* (Sec. 0.7) in which we associated heat with rain also holds for the sign of heat transfer. A *positive* heat transfer adds energy to a system just as rainfall tends to increase the amount of water in the lake (Fig. 0.22).

The details of how heat is transferred can be found in any heat transfer textbook and only the overall mechanisms are illustrated in Anim. 0.D.*heatTransfer*. Regardless of the mechanism, the magnitude of heat transfer depends on the temperature difference, exposed surface area, thermal resistance (insulation), and exposure time. A system, therefore, tends to be *adiabatic* not

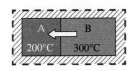

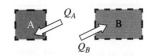

FIGURE 0.21 Two bodies, A and B, in thermal contact are treated as two separate systems (see Anim. 5.B.*blocksInContact*).

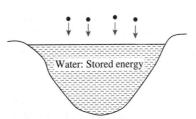

Direction of rain: Positive heat transfer

Water: Stored energy

FIGURE 0.22 Rain adds water to a lake just as positive heat transfer adds energy to a system (see Anim. 0.D.*lakeAnalogy*).

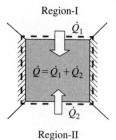

FIGURE 0.23 The net heat transfer is an algebraic sum of heat crossing the entire boundary.

only when it is well insulated, but also when the duration of a thermodynamic event is small. Quick compression of a gas in a piston-cylinder device, thus, can be considered adiabatic even if the cylinder is water-cooled.

In open systems, the inlet and exit ports usually have relatively small cross-sectional areas compared with the rest of the system's boundary area. Moreover, the flow of mass through the ports moderates the temperature gradient in the direction of the flow. Therefore, the rate of heat transfer $\dot{Q}$ through the ports of an open system can be neglected even as the flow transports significant amounts of energy. '*Heat flows out of an exhaust pipe,*' thus, is an incorrect statement for several reasons: first, it is mass that flows out transporting energy and second, heat transfer across an opening is generally negligible.

Sometimes the rate of heat transfer is not uniform over the entire boundary of a system. For example, a system may be in thermal contact with multiple thermal reservoirs (Fig. 0.23). The boundary in such situations is divided into different segments and the net heat transfer is obtained by summing the contribution from each segment:

$$Q = \sum_k Q_k; \quad [\text{kJ}]; \quad \dot{Q} = \sum_k \dot{Q}_k; \quad [\text{kW}] \tag{0.5}$$

Various suffixes such as *net, in, out*, etc., often are used in conjunction with particular system configurations, as clarified in Example 0-3 and Anim. 0.C.*WinHip*, to indicate the direction of heat transfer.

EXAMPLE 0-3   Heat Transfer Sign Convention

The net heat transfer rate between a system and its three surrounding reservoirs (Fig. 0.24) is $\dot{Q} = -1$ kW. Heat transfer rate from reservoir A to the system is 2 kW and $\dot{Q}_{out}$ is 4 kW. (a) Determine the heat transfer (including its sign) between the system and reservoir C. (b) Assuming heat transfer to be the only interaction between the system and its surroundings, determine the rate of change of the stored energy of the system in kJ/min .

SOLUTION

Let $\dot{Q}_A$, $\dot{Q}_B$, and $\dot{Q}_C$ represent the heat transfer between the system and the reservoirs. Using the sign convention, we can write $\dot{Q}_A = 2$ kW and $\dot{Q}_B = -\dot{Q}_{out} = -4$ kW. Therefore,

$$\dot{Q} = \dot{Q}_A + \dot{Q}_B + \dot{Q}_C; \quad [\text{kW}]$$
$$\Rightarrow \dot{Q}_C = \dot{Q} - \dot{Q}_A - \dot{Q}_B = (-1) - (2) - (-4) = 1 \text{ kW}$$

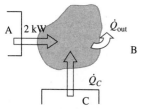
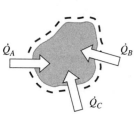

FIGURE 0.24 Schematic for Ex. 0-3.

The positive sign of $\dot{Q}_C$ means heat must be added at a rate of 1 kW from reservoir C to the system.

With no other means of energy transfer, the net flow of heat into the system, $\dot{Q} = -1$ kW, must be the rate at which the stored energy E accumulates (or depletes in this case) in a system. Therefore,

$$\frac{dE}{dt} = \dot{Q} = -1 \text{ kW} = -1\frac{\text{kJ}}{\text{s}} = -60\frac{\text{kJ}}{\text{min}}$$

Discussion

In practical systems, in which the directions of heat transfer are known or fixed by convention, use of algebraic signs is avoided in favor of subscripts, such as *in, out*, etc.

0.7.2 Work and Power (*W, Ẇ*)

The symbol used for work is W, and like heat or stored energy, it has the unit of kJ. In English units, however, work is expressed in ft · lbf while heat is expressed in Btu. The rate of work transfer $\dot{W}$ is called **power**. (Note the consistent use of a dot to indicate a time rate.) In SI units, $\dot{W}$ has the same unit as $\dot{Q}$: kJ/s or kW, but in English units, it has the unit of ft · lbf/s. If $\dot{W}$ is a function of time, the total amount of work W transferred in a process that begins at time $t = t_b$ and finishes at time $t = t_f$ can be obtained from:

$$W = \int_{t_b}^{t_f} \dot{W}dt; \quad [\text{kJ} = \text{kW} \cdot \text{s}] \qquad \qquad \textbf{(0.6)}$$

For a constant $\dot{W}$ during the interval $\Delta t = t_f - t_b$, Eq. (0.6) simplifies to:

$$W = \dot{W}\Delta t; \quad [\text{kJ} = \text{kW} \cdot \text{s}] \qquad \qquad \textbf{(0.7)}$$

The phrase **work transfer** will be used interchangeably to mean both W and $\dot{W}$ depending on the context. A 1-kN external force applied to a body transfers 1 kJ of work when the body (the system) translates by 1 m. An engine pulling a 1-kN load at a speed of 1 m/s supplies 1 kW of work transfer. The latter is actually the rate of work transfer or power, which is evident from the unit.

To add directionality to work, like heat, a sign convention is required. *By convention, work done by the system, i.e., the work delivered by a system, is considered positive.* The opposite signs used for heat and work addition to a system sometimes cause confusion; after all, heat and work are both energy in transit. The convention originated from the desire to label heat and work transfer as positive in a heat engine (a system whose sole purpose is to produce work while consuming heat). The mnemonic WinHip (work in negative, heat in positive) may be helpful to remember this important sign convention which is illustrated in Anim. 0.D.*WinHip* with several examples. The *lake analogy*, introduced earlier, is consistent with the sign convention (Fig. 0.25)—evaporation from the lake being analogous to work transfer, the positive direction of work can be associated with the natural direction of evaporation. Just as positive evaporation causes water loss from the lake, positive work causes loss of stored energy from a system.

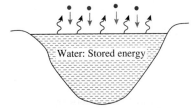

Direction of evaporation: Positive work transfer

Water: Stored energy

FIGURE 0.25 Direction of rain and evaporation can be connected with the sign convention of heat and work transfer (see Anim. 0.D.*lakeAnalogy*). Another way to remember the sign convention is to remember the mnemonic WinHip (Work in negative, heat in positive).

0.8 WORK TRANSFER MECHANISMS

While overall heat transfer without the details of heat transfer calculations often is adequate for thermodynamic analysis, a thorough understanding of different modes of work transfer is necessary. Although displacement by a force is at the root of all work transfer (upon close examination, any form of work transfer can be traced to the fundamental definition of force times distance), it is advantageous to classify work based on specific types of work interactions. Various mechanisms of work transfer are illustrated in Anim. 0.D.*workTransfer* and discussed below.

0.8.1 Mechanical Work (W_M, $\dot{W}_M$)

Mechanical work is the work introduced in mechanics (see Anim. 0.D.*mechanicalWork*). If F is the component of a force in the direction of displacement x of the rigid body shown in Figure 0.26, then the energy transfer due to mechanical work is given as

$$W_M = -\int_{x_b}^{x_f} Fdx; \quad [\text{kJ} = \text{kN} \cdot \text{m}] \qquad \qquad \textbf{(0.8)}$$

where x_b and x_f mark the beginning and final positions of the point of application. For a constant F, the integral reduces to:

$$W_M = -F(x_f - x_b) = -F\Delta x; \quad [\text{kJ} = \text{kN} \cdot \text{m}] \qquad \qquad \textbf{(0.9)}$$

The negative sign in this definition arises due to the WinHip sign convention (Sec. 0.7.2), indicating work is going into the system. A force, by itself, cannot transfer any energy into a system unless it succeeds in moving the point of application. When a force of 1 kN (approximately the weight of 50 textbooks) displaces a system (its point of application) by 1 m, then 1 kJ of mechanical work is done by the force. In thermodynamic terms, 1 kJ of work is transferred into the system or, in algebraic terms, $W = W_M = -1$ kJ. Suppose a body moves in the positive x direction from x_b to x_f, and is acted on by two constant opposing forces F_1 and F_2, as shown in Figure 0.27. If $F_1 > F_2$, the body accelerates in the x direction, which results in an increase in its kinetic energy and, hence, stored energy. $F_1 > F_2$ also implies that the net work $W_M = -(F_1 - F_2)(x_f - x_b)$ is negative, that is, energy is transferred into the system, which is stored as kinetic energy. Now

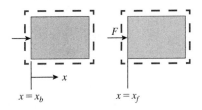

$x = x_b$

F

$x = x_f$

FIGURE 0.26 A body acted upon by a force F is displaced. Work is done by the force and transferred into the system.

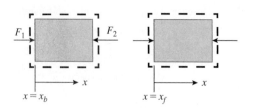

FIGURE 0.27 Work is done by F_1 against F_2.

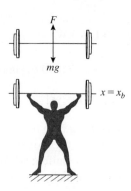

FIGURE 0.28 Benchmarking work: The work done in lifting a 100-kg mass through a height of one meter is approximately 1 kJ (see Anim. 0.D.*mechanicalWork*).

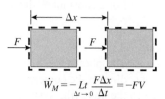

$$\dot{W}_M = -\lim_{\Delta t \to 0} \frac{F\Delta x}{\Delta t} = -FV$$

FIGURE 0.29 The rate of work done by F is the mechanical power transferred to the system (click Horizontal in Anim. 0.D.*mechanicalWork*).

suppose F_1 is only momentarily greater than F_2 so that the body starts moving in the direction of F_1. Thereafter, $F_1 = F_2$ will keep the body moving in the same direction at a constant velocity. In this case, there is no net work transfer as the work transferred into the system equals the work done by the system. As another benchmark for 1 kJ of work, consider the weightlifter in Figure 0.28 exerting an upward force of 0.981 kN to keep the 100-kg weight from falling ($g = 9.81 \, \text{m/s}^2$). Suppose the weightlifter succeeds in moving the weights up a distance of 1.02 m. In that case, the minimum work transferred to the weights is 1 kJ, that is, $W = W_M = -1$ kJ, which is stored as the weights' (system's) PE.

The time rate of work transfer, $\dot{W}_M$, is called **mechanical power** and can be expressed in the instantaneous velocity of the system as follows (see Fig. 0.29):

$$\dot{W}_M = -\lim_{\Delta t \to 0} \frac{F\Delta x}{\Delta t} = -FV; \quad \left[kN \cdot \frac{m}{s} = \frac{kJ}{s} = kW \right] \qquad (0.10)$$

The negative sign is necessary to comply with the WinHip sign convention. Using Eq. (0.10), it can be shown that a 100-kW engine can lift a 1000-kg weight vertically at a velocity of 36.7 km/h.

If the external force is in the opposite direction of the velocity, the rate of work transfer becomes positive, which is shown in Example 0-4.

EXAMPLE 0-4   **Power in Mechanics**

The **aerodynamic drag** force in kN on an automobile (see Fig. 0.30) is given as

$$F_d = \frac{1}{2000} c_d A \rho V^2$$

where c_d is the non-dimensional drag coefficient, A is the frontal area in m², ρ is the density of the surrounding air in kg/m³, and V is the velocity of air with respect to the automobile in m/s. (a) Determine the power required to overcome the aerodynamic drag for a car with $c_d = 0.8$ and $A = 5 \, \text{m}^2$ traveling at a velocity of 100 km/h. Assume the density of air to be $\rho = 1.15$ kg/m³. **What-if scenario:** (b) What would be the percent increase in power requirement if the car is traveling 20% faster?

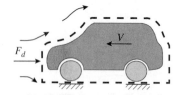

FIGURE 0.30 Aerodynamic drag force on a car is in the opposite direction of the car's velocity.

SOLUTION

The car must impart a force equal to F_d on its surroundings in the opposite direction of the drag to overcome the drag force. The power required is the rate at which it has to do (transfer) work.

Assumptions

The drag force remains constant over time.

Analysis

Use the converter TESTcalc in TEST to verify that 100 km/h is 27.78 m/s. Using Eq. (0.10), the rate of mechanical work transfer can be calculated as

$$\dot{W}_M = F_d V = \frac{1}{2000} c_d A \rho V^3 \left[kN \frac{m}{s} = \frac{kJ}{s} = kW \right]$$

$$= \frac{(0.8)(5)(1.15)(27.78^3)}{2000} = 49.3 \text{ kW}$$

What-if scenario

The power to overcome the drag is proportional to the cube of the automobile's speed. A 20% increase in V, therefore, will cause the power requirement for overcoming drag to go up by a factor of $1.2^3 = 1.728$ or 72.8%.

Discussion

The sign of the work transfer is positive, meaning work is done by the system (the car engine). A gasoline (heating value: 44 MJ/kg) powered engine with an overall efficiency of 30% will require heat release at the rate of 49.31/0.3 = 164.4 kW to supply 49.31 kW of shaft power. This translates to a consumption of 13.45 kg of fuel every hour, 7.43 km/kg, 5.57 km/L, or 13.1 mpg (assuming a gasoline density of 750 kg/m³) if the entire engine power is used to overcome drag only. Additional power is required for accelerating the car, raising it against a slope, and overcoming rolling resistance. At highway speed, the majority of the power, however, goes into overcoming aerodynamic drag.

0.8.2 Shaft Work (W_{sh}, $\dot{W}_{sh}$)

Torque acting through an angle (in radians) is the rotational counterpart of force acting through a distance. Work transfer through rotation of shafts is called **shaft work** and quite common in many practical systems such as automobile engines, turbines, compressors, and gearboxes.

The work done by a torque T in rotating a shaft through an angle $\Delta\theta$ in radians is given by $F\Delta s = Fr\Delta\theta = T\Delta\theta$ (see Fig. 0.31). The power transfer through a shaft, therefore, can be expressed as:

$$\dot{W}_{sh} = \lim_{\Delta t \to 0} \frac{T\Delta\theta}{\Delta t} = T\omega = 2\pi \frac{N}{60} T; \quad \left[\frac{kN \cdot m}{s} = kW \right] \qquad (0.11)$$

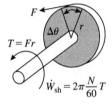

$T = Fr$

$\dot{W}_{sh} = 2\pi \dfrac{N}{60} T$

FIGURE 0.31 Power transfer by a rotating shaft is common in many engineering devices (click $\dot{W}_{sh}$ in Anim. 0.D.workTransfer).

where ω is the rotational speed in *radians/s* and N is the rotational speed measured in **rpm** (revolutions per minute). At 3000 rpm, the torque in a shaft carrying 50 kW of power can be calculated from Eq. (0.11) as 0.159 kN · m. Work transfer over a certain period from $t = t_b$ to $t = t_f$ can be obtained by integrating $\dot{W}_{sh}$ over time. For a constant torque, W_{sh} is given as:

$$W_{sh} = \int_{t_b}^{t_f} \dot{W}_{sh} dt = 2\pi \frac{N}{60} T\Delta t; \quad [kW \cdot s = kJ] \qquad (0.12)$$

Appropriate signs must be given to these expressions based on the direction of work transfer. For a paddle-wheel stirring a liquid in a container (Anim. 2.E.paddleWheel), the sign of shaft work transfer must be negative.

0.8.3 Electrical Work (W_{el}, $\dot{W}_{el}$)

When electrons cross a boundary, mechanical work done by the electromotive force in pushing the charged particles is called **electrical work**. For a potential difference of V (in volts) driving a current I (in amps) across a resistance R (in ohms), the magnitude of the electrical work can be expressed as:

$$\dot{W}_{el} = \frac{VI}{(1000 \text{ W/kW})} = \frac{V^2}{(1000 \text{ W/kW})R} = \frac{I^2 R}{(1000 \text{ W/kW})}; \quad [kW] \qquad (0.13)$$

$$W_{el} = \int_{t_b}^{t_f} \dot{W}_{el} dt = \dot{W}_{el}\Delta t; \quad [kJ] \qquad (0.14)$$

Like *shaft* work, *electrical* work is easy to identify and evaluate. Appropriate signs must be added to these expressions depending on the direction of the energy transfer in accordance with the WinHip convention. For example, suppose the electric heater in Figure 0.32 operates at 110 V and draws a current of 10 amps. For the heater as a system, the electrical work transfer rate can be evaluated as $\dot{W}_{el} = -1.1$ kW.

Sometimes an energy interaction can be worded in such a way that it is not immediately clear whether it is a heat or work interaction. For example, in an electrical water heater (Fig. 0.32), the water is said to be *heated* by electricity. To determine if this is a heat or work interaction, we must remember to examine the boundary rather than the system's interior. Accordingly, it is $\dot{W}_{el} = -1.1$ kW for the system within the red boundary and $\dot{Q} = 1.1$ kW for the system defined by the black boundary. These algebraic values also can be expressed as magnitudes only if appropriate subscripts are used: $\dot{W}_{in} = 1.1$ kW and $\dot{Q}_{in} = 1.1$ kW in Figure 0.32. However, we will prefer the algebraic quantities which can be directly substituted into the balance equations to be developed in Chapter 2.

0.8.4 Boundary Work (W_B, $\dot{W}_B$)

Boundary work is a general term that includes all types of work that involve displacement of any part of a system boundary. *Mechanical work*, work transfer during rigid-body motion, clearly qualifies as boundary work. However, boundary work is more general in that it can also account for distortion of the system. For example, work transferred while compressing or elongating a spring is a case in point. For a linear spring with a spring constant k (kN/m), the boundary work transfer in pulling the spring from an initial position $x = x_b$ to a final position $x = x_f$ (see Fig. 0.33) can be expressed as:

$$W_B = -\int_{x_b}^{x_f} F dx = -\int_{x_b}^{x_f} kx dx = -k \left[\frac{x^2}{2} \right]_{x_b}^{x_f} = -\frac{k}{2}(x_f^2 - x_b^2); \quad [kJ] \qquad \textbf{(0.15)}$$

where $x = 0$ is the undisturbed position of the spring. The negative sign indicates that work has been transferred into the spring (the system). For a linear spring with a k of 200 kN/m, the work transferred to the spring by stretching it by 10 cm from its rest position is 1 kJ.

The most prevalent mode of boundary work in thermal systems, however, accompanies expansion or contraction of a fluid. Consider the trapped gas in the piston-cylinder device of Figure 0.34 as the system. Heated by an external source, the system (gas) expands and lifts the piston carrying a load. Positive boundary work is transferred during the process from the system (which is stored in the PE of the load). If the heating process is slow, the piston is assumed to be in *quasi-equilibrium* (quasi meaning almost) at all times, and a free-body diagram of the piston (revisit Anim. 0.A.*pressure*) can be used to establish that the internal pressure, p, remains constant during the expansion process. Now suppose instead of heating the gas, the weight on top of the piston is increased incrementally as shown in Figure 0.35. As expected, the pressure inside will increase as weights are added. However, if the weights are differentially small, the piston can be assumed to be in quasi-equilibrium (equilibrium of the system is discussed in Sec. 1.2), allowing a free-body diagram of the piston to express p in the external forces on the piston.

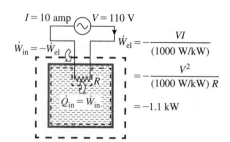

FIGURE 0.32 Electrical heating of water involves work or heat transfer depending on which boundary (the red or the black) defines the system (click $\dot{W}_{el}$ in Anim. 0.D.*workTransfer*).

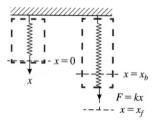

FIGURE 0.33 To elongate a linear spring, the applied force has to be only differentially greater than kx.

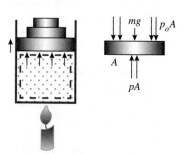

FIGURE 0.34 The system expands due to heat transfer and raises the weights (click the heating option in Anim. 0.A.*pressure*).

Whether it is compression or expansion, as the piston moves from a beginning position x_b to a final position x_f (Fig. 0.36), the boundary work transfer is related to the pressure and volume of the system as follows:

$$W_{pd\Psi} = \int_b^f F dx = \int_b^f pA dx = \int_b^f pd\Psi; \quad \left[\text{kJ} = \frac{\text{kN}}{\text{m}^2}\cdot\text{m}^3\right] \qquad (0.16)$$

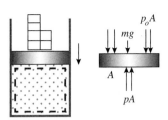

FIGURE 0.35 Weights on the piston are increased in small increments (try different modes of compression in Anim. 5.A.*pTs ConstCompressionWorkIG*).

Although the force on the piston can vary, the piston's internal pressure will adjust to a variable external force as long as the system can be assumed to be in quasi-equilibrium. Due to its frequent use, this type of boundary work is also called the $pd\Psi$ (pronounced p-d-V) work.

Equation (0.16) can be interpreted as the area under a p-Ψ diagram as the system volume goes from an initial volume Ψ_b to a final volume Ψ_f (Fig. 0.36). Instead of evaluating the integral, it is easier to calculate the area under the p-Ψ diagram and add the appropriate sign: positive for expansion and negative for compression. For a constant pressure or **isobaric** process, the boundary work equation simplifies to: $W_{pd\Psi} = p(\Psi_f - \Psi_b)$. Absolute pressure cannot be negative, so the boundary work must be positive when a system expands and negative when it is compressed.

The rate of $pd\Psi$ work transfer is deduced from Eq. (0.16) as follows:

$$\dot{W}_{pd\Psi} = \frac{dW_{pd\Psi}}{dt} = p\frac{d\Psi}{dt}; \quad \left[\text{kW} = \frac{\text{kN}}{\text{m}^2}\frac{\text{m}^3}{\text{s}} = \frac{\text{kJ}}{\text{s}}\right] \qquad (0.17)$$

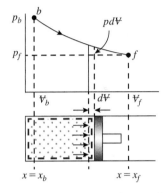

FIGURE 0.36 The boundary work during a resisted expansion or contraction can be obtained from a p-Ψ diagram (try different options in Anim. 5.A.*pTsConst ExpansionWorkIG*).

In the absence of mechanical work, $pd\Psi$ work is the only type of boundary work that is present in stationary systems and it is common to use the symbols W_B or $W_{pd\Psi}$ interchangeably (see Anim. 0.D.*externalWork*). In the analysis of reciprocating devices, such as automobile engines and some pumps and compressors, the $pd\Psi$ work is the primary mode of work transfer. Despite the high-speed operation of these devices, the quasi-equilibrium assumption and the resulting $pd\Psi$ formula produce acceptable accuracy when evaluating boundary work transfer.

EXAMPLE 0-5   Boundary Work During Compression

A gas is compressed in a horizontal piston-cylinder device (see Fig. 0.37). At the start of compression, the pressure inside is 100 kPa and the volume is 0.1 m³. Assuming that the pressure is inversely proportional to the volume, determine the boundary work in kJ when the final volume is 0.02 m³.

SOLUTION

Evaluate the $pd\Psi$ work using Eq. (0.16).

Assumptions

The gas is in quasi-equilibrium during the process.

Analysis

The pressure can be expressed as a function of volume and the conditions at the beginning of the process ($p_b = 100$ kPa, $\Psi_b = 0.1$ m³):

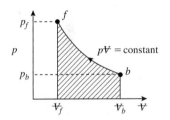

p-Ψ = constant

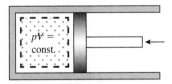

FIGURE 0.37 Schematic for Ex. 0-5 (Anim. 5.A.*pTsConstCompressionIG*).

$$p\Psi = \text{constant} = p_b\Psi_b = (100)(0.1) = 10; \quad \left[\text{kPa}\cdot\text{m}^3 = \frac{\text{kN}}{\text{m}^2}\cdot\text{m}^3 = \text{kN}\cdot\text{m} = \text{kJ}\right]$$

The boundary work, now, can be obtained by using Eq. (0.16):

$$W_B = \int_b^f pd\Psi = \int_b^f \frac{10}{\Psi}d\Psi = 10\left[\ln\Psi\right]_{0.1}^{0.02} = 10\ln\left(\frac{0.02}{0.1}\right) = -16.09 \text{ kJ}$$

Discussion

The key to direct evaluation of the $pd\Psi$ work is to examine how the internal pressure p varies with respect to Ψ. To find a relationship between p and Ψ, a free-body diagram of the piston (see Anim. 0.A.*pressure*) is helpful. In this problem, we arrive at the answer without any consideration of outside agents responsible for this work transfer. However, if we could calculate the exact amount of work done by the external link pushing the piston, it would be significantly less than 16.09 kJ. What other agent, do you think, is responsible for the difference? The outside atmosphere, of course.

EXAMPLE 0-6 Boundary Work During Expansion

A 9-mm pistol is test fired with sensors attached inside the barrel that measure the pressure of the explosive gases with respect to the projectile's position. The sample data is as follows:

Position of projectile, x, mm	0	10	20	30	40	50	60	70	80
Chamber pressure, p, MPa	250	220	200	180	150	120	100	80	40

Determine the boundary work transfer between (a) the gases and the projectile, and (b) the projectile and the outside air whose pressure is 101 kPa.

SOLUTION

Evaluate the boundary work transfer into the projectile by estimating the area under the $p - \Psi$ diagram.

Assumptions

Assume the measured pressure to be uniform. Represent the pressure at x as the mean of the two nearest measured values.

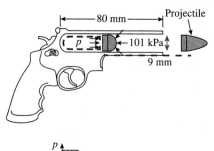

Analysis

The area under the $p - \Psi$ diagram can be divided into a number of adjacent rectangles (Figure 0.38) and approximated as follows:

$$W_B = \int_b^f p \, d\Psi = A \int_b^f p \, dx = A \sum_{i=1}^{8} p_i \Delta x_i$$

$$= \frac{\pi (9 \times 10^{-3})^2}{4} (235 \times 0.01 + 210 \times 0.01 + \cdots + 60 \times 0.01) \left(10^6 \frac{J}{MJ} \right)$$

$$= 760 \text{ J}$$

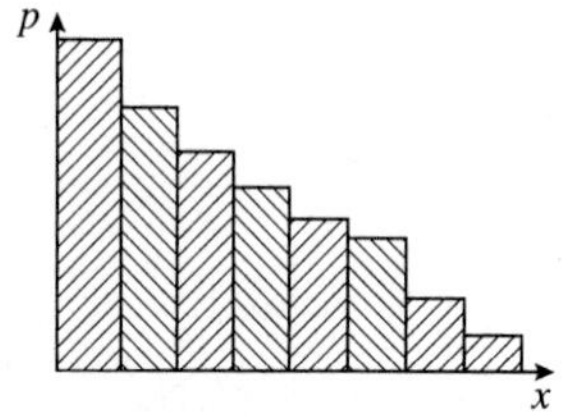

FIGURE 0.38 The area under the $p - x$ diagram is approximated as the sum of the areas of the rectangles.

If the projectile is treated as the system, the work transfer from the gases is −760 J. The work transfer from the projectile into the outside atmosphere is:

$$W_{atm} = p_{atm}\Delta \Psi = (101) \frac{\pi (9 \times 10^{-3})^2}{4} (0.08) \left(10^3 \frac{j}{kj} \right) = 0.514 \text{ J}$$

Discussion

The net boundary work transferred into the projectile goes KE and overcoming friction. If friction is neglected, the projectile's velocity at the end of the barrel can be calculated by equating its kinetic energy to the net work transferred. In calculating the atmospheric work done by the outer surface of the projectile, it can be shown that the resultant force over the curved surface comes out as if the outer surface were flat.

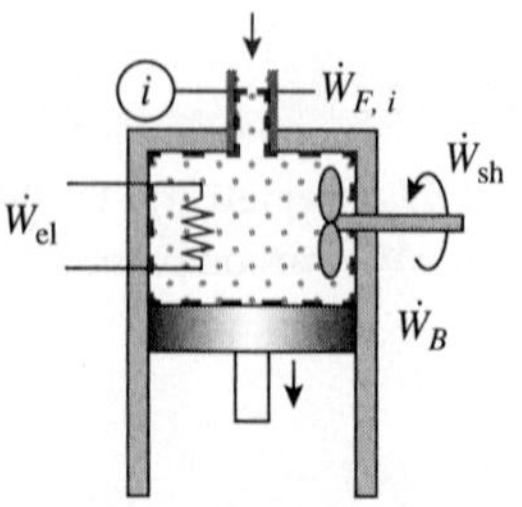

0.8.5 Flow Work ($\dot{W}_F$)

FIGURE 0.39 Although the shaft, electrical, and boundary work transfer are apparent from this system schematic, the flow work transfer is more subtle (see Anim. 0.D.*workTransfer*).

If we examine the boundary of the system shown in Figure 0.39, it is easy to detect the *electrical* and *shaft* work transfer. The *boundary* work also becomes evident once we realize that the piston is moving, but what about the flow that is being pushed into the cylinder at the inlet port? Work must be done by the external push-force to introduce fluid into the cylinder against the resistance of the internal pressure. To obtain an expression for such work, let's consider the force balance on a thin element of fluid at a particular port, say, the exit port e of the open system shown in

Figure 0.40. The element, which can be modeled as an imaginary piston, is subjected to the tremendous forces on its left and right sides. For the fluid in that element to be forced out of the system, the force exerted from the inside should be only differentially greater than the force from the outside (to overcome the small frictional force as the element rubs against the wall). By imagining the element to be sufficiently thin (wall friction becomes vanishingly small), the force on both faces of the element can be approximated as $p_e A_e$, where p_e is the pressure at that port. The element being forced out with a velocity V_e, the rate at which work is done on the element can be obtained from Eq. (0.10) as:

$$\dot{W}_{F,e} = F_e V_e = p_e A_e V_e; \quad \left[\frac{\text{kN}}{\text{m}^2} \cdot \text{m}^2 \cdot \frac{\text{m}}{\text{s}} = \frac{\text{kJ}}{\text{s}} = \text{kW} \right] \tag{0.18}$$

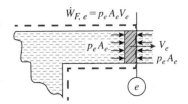

FIGURE 0.40 In pushing the mass out, the system does positive flow work at a rate $\dot{W}_{F,e} = p_e A_e V_e$ (click $\dot{W}_F$ in Anim. 0.D.*workTransfer*).

This is called the **flow work** or, more precisely, the rate of flow work transfer at the port. In Chapter 1 (Sec. 1.5.8), we will explore energy transfer by mass in full detail and realize that flow work is only part of the total energy transported by mass. Flow work always takes place in the direction of the flow and hence is bundled with all other modes of energy transferred by mass. To be consistent with the sign convention, flow work transfer at the exit port e of a system, $p_e A_e V_e$, must be assigned a positive sign and the flow work transfer at the inlet i, $p_i A_i V_i$, a negative sign.

In Sec. 0.7.1, we discussed why heat transfer through the port openings can be neglected in open systems. Can we make the same assumption about the work transfer through the inlets and exits of an open system? The rate of the flow work, $\dot{W}_F$, depends on the product of the pressure, velocity, and flow area. A small area, alone, cannot guarantee negligible flow work as the pressure can be quite high. If atmospheric air at 100 kPa enters a room through a 3 m × 1 m door at 3.33 m/s, the flow work transfer can be calculated as –1 MW, a power that is equivalent to turning on 1000 1-kW heaters. When the air leaves the room through a window or another door (normally air breezes through a room only when there is a cross flow), an almost equivalent amount of work is transferred from the room to the outside surroundings; therefore, no net amount of energy is stored in the room. Conversely, when air enters an evacuated insulated tank (Fig. 0.5d), the flow work transferred by the incoming flow causes the stored energy to increase, which raises the air's temperature inside the tank. Unlike shaft, electric, and boundary work, flow work is invisible (Fig. 0.40). We will exploit flow work's invisibility in Chapter 1 by combining it with the stored energy transported by a flow and defining a new property called *flow energy*.

0.8.6 Net Work Transfer ($\dot{W}$, $\dot{W}_{ext}$)

There are many other minor modes of work transfer, such as work transfer due to polarization or magnetism and work transfer in stretching a liquid film. Fortunately, an exhaustive knowledge of all these modes is not necessary for analyzing most thermodynamic systems.

Different modes of work transfer discussed so far can be algebraically added to produce an expression for the net work transfer, which can be split into various groups (see Fig. 0.41):

$$\dot{W} = \dot{W}_F + \underbrace{(\dot{W}_M + \dot{W}_{pd\forall})}_{\dot{W}_B:\ \text{Boundary Work}} + \underbrace{(\dot{W}_{sh} + \dot{W}_{el} + \ldots)}_{\dot{W}_O:\ \text{Other Types}} = \dot{W}_F + \underbrace{(\dot{W}_B + \dot{W}_O)}_{\dot{W}_{ext}:\ \text{External Work}}$$

$$= \dot{W}_F + \dot{W}_{ext}; \quad [\text{kW}] \tag{0.19}$$

where $\dot{W}_{ext} = \dot{W}_B + \dot{W}_{sh} + \dot{W}_{el}$ and $\dot{W}_F = \sum_e p_e A_e V_e - \sum_i p_i A_i V_i$; [kW] (0.20)

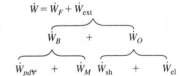

FIGURE 0.41 Relationship among different modes of work as expressed by Eq. (0.19) (see Anim. 0.D.*externalWork*).

In these equations, $\dot{W}_F$, the net flow work transferred out of a system, is expressed by summing the flow work given by Eq. (0.18) over all exit ports (positive work) and all inlet ports (negative work). Essentially, Eq. (0.19) divides total work transfer into two major components: Flow work, which takes place internally (and invisibly) within the working substance at the ports, and external work, which must cross the physical boundary of a system through a shaft, electrical cables, or movement of the boundary.

External work, a component of Eq. (0.19), consists of boundary work, which is the sum of mechanical and $pd\forall$ work, and all **other work**—the sum of the shaft, electrical, and any other type of work present. In most situations, however, external work is the sum of *boundary, shaft,* and *electrical* work as expressed by Eq. (0.20). In Chapters 1 and 2, flow work can be absorbed into other terms in the energy balance equation leaving external work as the only relevant work transfer in an energy analysis.

To identify a specific type of work transfer we must inspect the boundary and look for (a) displacement of any part of the boundary for boundary work, (b) electric cables for electrical work, (c) rotating shafts for shaft work, and (d) mass transfers for flow work. To determine the sign of work transfer, we must determine how the work transfer affects the stored energy of the system. In practical systems, only one or two of these modes may be present simultaneously. In closed systems the flow work, by definition, is absent so that $\dot{W} = \dot{W}_{ext}$. In the system shown in Figure 0.42, the external work consists of electrical, shaft, and boundary work. In Figure 0.39, despite the presence of flow work, the constituents of the external work remain unchanged. In an energy analysis of open systems, flow work is absorbed in other terms (bundled with energy transport by mass) so that work transfer usually means *external work* transfer, even though $\dot{W} \neq \dot{W}_{ext}$ according to Eq. (0.19).

EXAMPLE 0-7   **Different Types of Energy Interactions**

A gas trapped in a piston-cylinder device (Fig. 0.42) is subjected to the following energy interactions for 30 seconds: The electric resistance draws 0.1 amps from a 100-V source, the paddle wheel turns at 60 rpm with the shaft transmitting a torque of 5 N·m, and 1 kJ of heat is transferred into the gas from the candle. The volume of the gas increases by 6 L during the process. If the atmospheric pressure is 100 kPa and the piston is weightless, (a) determine the net transfer of energy into the system. (b) Air is now injected into the cylinder using a needle. Calculate the flow work transfer as 10 L of air is injected into the cylinder at a constant pressure of 100 kPa.

SOLUTION

Evaluate the different types of energy transfers during the process and add them algebraically to find the net energy transfer.

Assumptions

The piston is at mechanical equilibrium (no force imbalance) at all times.

Analysis

Evaluate different modes of energy transfer by treating the gas as the system.

Boundary Work:

During the expansion of the gas, boundary work transferred is positive. To obtain a relationship between p_i and Ψ, a vertical force balance on the piston (see the free-body in Fig. 0.42) yields the following:

$$p_i A_{piston} = \frac{\cancel{m^0}g}{(1000\ \text{N/kN})} + p_0 A_{piston}; \quad \Rightarrow \quad p_i = p_0$$

From Eq. (0.16), $W_B = \displaystyle\int_b^f pd\Psi = p_i(\Psi_f - \Psi_b) = p_0(\Psi_f - \Psi_b)$

$$= (100\ \text{kPa})(6 \times 10^{-3}\ \text{m}^3) = 0.6\ \text{kJ}$$

Shaft Work: From Eq. (0.12):

$$W_{sh} = -2\pi\frac{N}{60}T\Delta t = -2\pi\frac{60}{60}\left(\frac{5}{1000}\right)30 = -0.94\ \text{kJ}$$

Electrical Work: From Eq. (0.14):

$$W_{el} = -\frac{VI}{1000}\Delta t = \left(-\frac{100 \times 0.1}{1000}\ \text{kW}\right)(30\ \text{s}) = -0.3\ \text{kJ}$$

Heat Transfer: $Q = 1$ kJ (given)

The net work transfer is summed up as:

$$W = W_{ext} = W_B + W_{sh} + W_{el} = 0.6 - 0.94 - 0.3 = -0.64\ \text{kJ}$$

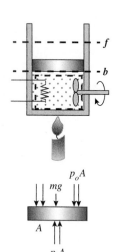

FIGURE 0.42 Different types of work transfer in Ex. 0-7.

DID YOU KNOW?

• To transfer enough sound energy to warm a cup of coffee, you would have to yell continuously for 8 years, 7 months, and 6 days.

Thus, 0.64 kJ of work and 1 kJ of heat are transferred into the system. Therefore, the net energy transfer into the system during the process is 1.64 kJ.

Flow Work:

Noting that there is a single inlet port, Eq. (0.20) can be simplified with the help of Eqs. (0.1) and (0.6) to produce:

$$\dot{W}_F = \sum_e p_e A_e V_e - \sum_i p_i A_i V_i = -p_i A_i V_i = -p_i \dot{V}_i; \quad [\text{kW}]$$

$$\Rightarrow \quad W_F = \int_{t=t_b}^{t_f} \dot{W}_F dt = -\int_{t=t_b}^{t_f} p_i \dot{V}_i dt = -p_i \int_{t=t_b}^{t_f} \dot{V}_i dt = -p_i V_i; \quad [\text{kJ}]$$

$$\Rightarrow \quad W_F = -(100)\left[\frac{10}{1000 \text{ L/m}^3}\right] = -1 \text{ kJ}; \quad [\text{kPa} \cdot \text{m}^3 = \text{kN} \cdot \text{m} = \text{kJ}]$$

The pressure at which the air is injected remains constant because the piston is free to move. This also simplifies the evaluation of the integral in the expression for W_F.

Discussion

When evaluating a particular mode of work or heat transfer, it is an acceptable practice to determine the magnitude before attaching an appropriate sign that specifies the energy transfer's direction. The net energy transferred into the system must give rise to an increase in the stored energy. Kinetic and potential energies of the system remaining unchanged, the transferred energy must be stored in the internal energy of the system. The energy balance equation, to be developed in Chapter 2, will precisely relate the energy transfer through mass, heat, and work with the inventory of stored energy of a system.

0.8.7 Other Interactions

Are there other interactions between a system and its surroundings besides mass, heat, and work interactions? What about microwaves, radio waves, lasers, sound waves, nuclear radiation, or, for that matter, the scent of a perfume? Actually, none of these interactions is unique; each is a special case of heat or mass transfer. We can therefore conclude that mass, heat, and work interactions, summarized in Anim. 0.C.*genericSystem,* are sufficient to capture all thermodynamic activities between a system and its surroundings. The lack of mass transfer makes a system *closed,* the lack of heat transfer makes it *adiabatic,* and the lack of all interactions makes a system *isolated* (see Anim. 0.C.*systemTypes*). Classification of systems based on interactions will have an important role in later chapters.

FIGURE 0.43 How do you classify this delightful interaction? Mass transfer, of course.

0.9 CLOSURE

In this Introduction we have introduced the basic vocabulary of thermodynamics by defining a system, its surroundings, and their interactions. Aided by a series of animations, an overview of diverse thermodynamic systems was presented with particular attention to different types of interactions. Mass interaction is analyzed producing the mass flow rate formula. TESTcalcs are introduced for calculating flow properties, including mass flow rate. Stored energy of a system was presented as a generalization of the familiar mechanical energy used in mechanics. Heat and work were introduced as energy in transit, as fundamental mechanisms of energy interactions. Different types of work transfer were analyzed, leading to a formula for the net work for its components: boundary work, shaft work, electrical work, and flow work. A lake analogy was formulated to explain the connection among heat, work, and stored energy and the sign convention for heat and work transfer.

PROBLEMS

SECTION 0-1: FREE-BODY DIAGRAM

0-1-1 Two thermodynamics books, each with a mass of 1 kg, are stacked one on top of another. Neglecting the presence of atmosphere, draw the free-body diagram of the book at the bottom to determine the vertical force on its (a) top and (b) bottom faces in kN.

0-1-2 Determine (a) the pressure felt on your palm to hold a textbook of mass 1 kg in equilibrium. Assume the distribution of pressure over the palm to be uniform and the area of contact to be $25\,\text{cm}^2$. (b) **What-if scenario:** How would a change in atmospheric pressure affect your answer (0: No change; 1: increase; −1: decrease)?

0-1-3 The lift-off mass of a Space Shuttle is 2 million kg. If the lift-off thrust (the net force upward) is 10% greater than the minimum amount required for a lift off, determine the acceleration.

0-1-4 A body weighs 0.05 kN on Earth where $g = 9.81\,\text{m/s}^2$. Determine its weight on (a) the moon, and (b) on Mars with $g = 1.67\,\text{m/s}^2$ and $g = 3.92\,\text{m/s}^2$, respectively.

0-1-5 Calculate the weight of an object of mass 50 kg at the bottom and top of a mountain with (a) $g = 9.8\,\text{m/s}^2$ and (b) $9.78\,\text{m/s}^2$ respectively.

0-1-6 According to Newton's law of gravity, the value of g at a given location is inversely proportional to the square of the distance of the location from the center of the Earth. Determine the weight of a textbook of mass 1 kg at (a) sea level and in (b) an airplane cruising at an altitude of 45,000 ft. Assume Earth to be a sphere of diameter 12,756 km.

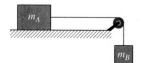

0-1-7 The maximum possible frictional force on a block of mass m_A resting on a table (see accompanying figure) is given as $F = \mu_s N$, where N is the normal reaction force from the table. Determine the maximum value for m_B that can be supported by friction. Assume the pulley to be frictionless.

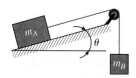

0-1-8 If the block A in Problem 0-1-7 sits on a wedge with an angle θ with the horizontal, how would the answer change?

0-1-9 A block with a mass of 10 kg is at rest on a plane inclined at 25° to the horizontal. If $\mu_s = 0.6$, determine the range of the horizontal push force F if the block is (a) about to slide down, and (b) about to slide up.

100 kPa

20 kg

0-1-10 A vertical piston cylinder device contains a gas at an unknown pressure. If the outside pressure is 100 kPa, determine (a) the pressure of the gas if the piston has an area of $0.2\,\text{m}^2$ and a mass of 20 kg. Assume $g = 9.81\,\text{m/s}^2$. (b) **What-if scenario:** What would the pressure be if the orientation of the device were changed and it were now upside down?

0-1-11 Determine the mass of the weight necessary to increase the pressure of the liquid trapped inside a piston-cylinder device like the one in Figure 0.2 to 120 kPa. Assume the piston to be weightless with an area of $0.1\,\text{m}^2$, the outside pressure to be 100 kPa and $g = 9.81\,\text{m/s}^2$.

0-1-12 A mass of 100 kg is placed on the piston of a vertical piston-cylinder device containing nitrogen. The piston is weightless and has an area of $1\,m^2$. The outside pressure is 100 kPa. Determine (a) the pressure inside the cylinder. The mass placed on the piston is now doubled to 200 kg. Also, additional nitrogen is injected into the cylinder to double the mass of nitrogen. (b) Determine the pressure inside under these changed conditions. Assume temperature to remain unchanged at 300 K and R for nitrogen to be 0.296 kJ/kg·K.

0-1-13 A piston with a diameter of 50 cm and a thickness of 5 cm is made of a composite material with a density of $4000\,kg/m^3$. (a) If the outside pressure is 101 kPa, determine the pressure inside the piston-cylinder assembly if the cylinder contains air. *What-if scenario:* (b)What would the inside pressure be if the piston diameter were 100 cm instead? (c) Would the answers change if the cylinder contained liquid water instead?

0-1-14 A piston-cylinder device contains $0.02\,m^3$ of hydrogen at 300 K. It has a diameter of 10 cm. The piston (assumed weightless) is pulled by a connecting rod perpendicular to the piston surface. If the outside conditions are 100 kPa, 300 K, (a) determine the pull force necessary in kN to create a pressure of 50 kPa inside. (b) The piston is now released; as it oscillates back and forth and finally comes to equilibrium, the temperature inside is measured as 600 K (reasons unknown). What is the pressure of hydrogen at equilibrium? *What-if scenario:* (c) What would be the answer in part (a) if the gas were oxygen instead?

0-1-15 Air in the piston-cylinder device shown is in equilibrium at 200°C. If the mass of the hanging weight is 10 kg, atmospheric pressure is 100 kPa, and the piston diameter is 10 cm, (a) determine the pressure of air inside. Assume $g = 9.81\,m/s^2$. (b) *What-if scenario:* What would the pressure be if the gas were hydrogen instead? The molar mass of air is 29 kg/kmol and that of hydrogen is 2 kg/kmol. Neglect piston mass and friction.

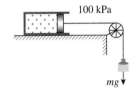

0-1-16 A vertical hydraulic cylinder has a piston with a diameter of 100 mm. If the ambient pressure is 100 kPa, determine the mass of the piston if the pressure inside is 1000 kPa.

0-1-17 Determine the pull force necessary on the rope to reduce the pressure of the liquid trapped inside the piston cylinder device to 80 kPa. Assume the piston to be weightless with a diameter of 0.1 m, the outside pressure to be 100 kPa, and $g = 9.81\,m/s^2$.

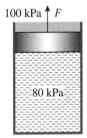

0-1-18 A piston-cylinder device contains $0.17\,m^3$ of hydrogen at 450°C. It has a diameter of 50 cm. The piston (assumed weightless) is pulled by a connecting rod perpendicular to the piston surface. If the outside conditions are 100 kPa, 25°C, (a) determine the pull force necessary in kN to create a pressure of 60 kPa inside. (b) The piston is now released. It oscillates back and forth and finally comes to equilibrium. Determine the pressure at that state.

0-1-19 A 5-cm diameter piston-cylinder device contains 0.04 kg of an ideal gas at equilibrium at 100 kPa, 300 K occupying a volume of $0.5\,m^3$. Determine (a) the gas density. (b) A weight is now hung from the piston (see figure) so that the piston moves down to a new equilibrium position. Assuming the piston to be weightless, determine the mass of the hanging weight necessary to reduce the gas pressure to 50 kPa. (Data supplied: $g = 9.81\,m/s^2$; outside pressure: 100 kPa.)

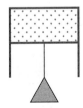

0-1-20 In Problem 0-1-19, the gas is now cooled so that the piston moves upward toward the original position with the weight still hanging. Determine the pressure when the gas shrinks to its original volume of $0.5\,m^3$.

SECTION 0-2: INTERACTIONS BETWEEN A SYSTEM AND ITS SURROUNDINGS

0-2-1 What do you call a system that has (a) no mass interaction, (b) no heat interaction, (c) no mass and energy interaction?

0-2-2 As shown in the figure below, electric current from the photo-voltaic (PV) cells runs an electric motor. The shaft of the motor turns the paddle wheel inside the water tank. Identify the interactions (mass, heat, work) for the following systems: (a) PV cells, (b) motor, (c) tank, and (d) the combined system that includes all these three subsystems.

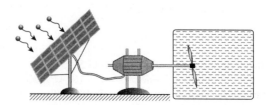

0-2-3 On a hot day, a student turns on the fan and keeps the refrigerator door open in a closed kitchen room, thinking that it would cool down the hot kitchen. Treating the room as a closed insulated system, identify the possible energy interactions between the room and the surroundings. Also determine the sign of (a) Q and (b) W_{ext}, if any.

0-2-4 An external force drags and accelerates a rigid body over a surface. Treating the body as a thermodynamic system, determine the sign of (a) W_{ext}, and (b) Q across its boundary. Assume friction to be present.

0-2-5 During the free fall of a rigid body (system), identify the interactions between the system and its surroundings.

0-2-6 An electric adaptor for a notebook computer (converting 110 V to 19 V) operates 10°C warmer than the surrounding temperature. Determine the sign of (a) W_{ext} and (b) Q interactions.

0-2-7 A block of ice dropped into a tank of water as shown begins melting. Identify the interactions for the (a) ice as a system, (b) water in the tank as a system, and (c) water and ice together as a system.

0-2-8 A gas trapped in an insulated piston-cylinder assembly expands as it is heated by an electrical resistance heater placed inside the cylinder. Treating the gas and the heater as the system, identify the interactions with its surroundings and the sign of (a) Q and (b) W_{ext}.

0-2-9 A piston-cylinder device contains superheated vapor at atmospheric pressure. The piston is pulled by an external force until the pressure inside drops by 50%. Determine the sign of W_B, treating the vapor trapped inside the cylinder as the system. Where does the work done in pulling the piston go?

0-2-10 A warm cup of coffee gradually cools down to room temperature. Treating the coffee as the system, determine the sign of Q during the cooling process.

0-2-11 A hot block of solid is dropped in an insulated tank of water at the temperature of the surroundings. Determine the sign of Q treating (a) the block as the system, (b) the water as the system and (c) the entire tank (with the block and water).

0-2-12 An insulated tank containing high-pressure nitrogen is connected to another insulated tank containing oxygen at low pressure. Determine the possible interactions as the valve is opened and the two gases are allowed to form a mixture by treating (a) one of the tanks as a system and (b) two tanks together as a single system.

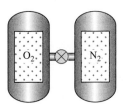

0-2-13 A fluid is accelerated by an insulated nozzle attached at the end of a pipe. Identify the interactions, treating the nozzle as an open system.

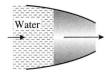

0-2-14 An insulated steam turbine produces Q and W_{sh} as steam flows through it, entering at a high pressure and a high temperature and leaving at a relatively low pressure. Identify the interactions between the turbine (as an open system) and its surroundings and determine the sign of (a) Q and (b) W_{ext}.

0-2-15 Identify the possible interactions of a steam turbine with poor insulation with its surroundings and determine the sign of (a) Q and (b) W_{ext}.

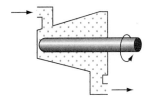

0-2-16 The pressure of a liquid flow is raised by a pump driven by an electrical motor. Identify the interactions treating (a) the pump as an open system and (b) the pump and the motor as a combined system.

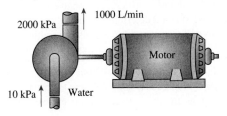

0-2-17 An insulated compressor raises the pressure of a gas flow. The temperature of the gas also is increased as a result. Identify the possible interactions between the compressor and its surroundings and determine the sign of (a) Q and (b) W_{ext}.

0-2-18 In a heat exchanger a flow of hot air is cooled by a flow of water. Identify the interactions treating (a) the entire heat exchanger as the system and (b) one of the streams as the system.

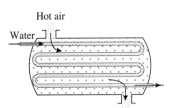

0-2-19 A pressure cooker containing water is heated on a stove. Determine the interactions and signs of (a) Q and (b) W_{ext}, if any, as steam is released.

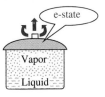

0-2-20 As you blow up a balloon, what are the interactions and the sign (positive: 1; negative: −1; none: 0) of (a) Q and (b) W_{ext}, if any, between the balloon as a system and its surroundings?

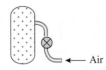

0-2-21   Air rushes in to fill an evacuated insulated tank as the valve is opened. Determine the interactions and the sign of Q and W_{ext}, if any, treating (a) the tank as the system and (b) the tank and the outside air that eventually enters as the system.

SECTION 0-3: MASS TRANSFER

0-3-1   Air with a density of 1 kg/m³ flows through a pipe of diameter 20 cm at a velocity of 10 m/s. Determine (a) the volume flow rate ($\dot{V}$) in L/min and (b) mass flow rate ($\dot{m}$) in kg/min. Use the PG flow state TESTcalc to verify your answer.

0-3-2   Steam flows through a pipe of diameter 5 cm with a velocity of 50 m/s at 500 kPa. If the mass flow rate ($\dot{m}$) of steam is measured at 0.2 kg/s, determine (a) the specific volume (v) of steam in m³/kg, and (b) the volume flow rate ($\dot{V}$) in m³/s.

0-3-3   Water flows through a variable-area pipe with a mass flow rate ($\dot{m}$) of 10,000 kg/min. Determine the minimum diameter of the pipe if the flow velocity is not to exceed 5 m/s. Assume density (ρ) of water to be 1000 kg/m³. Use the SL flow state TESTcalc to verify your answer.

0-3-4   A mixture of water ($\rho = 1000$ kg/m³) and oil ($\rho = 800$ kg/m³) is flowing through a tube of diameter 2 cm with a velocity of 4 m/s. The mass flow rate ($\dot{m}$) is measured to be 1.068 kg/s. Determine (a) the density (ρ) of the liquid mixture, (b) the percentage of oil in the mixture by mass. Assume liquids to be incompressible.

0-3-5   Air flows through a pipe of diameter 10 cm with an average velocity of 20 m/s. If the mass flow rate ($\dot{m}$) is measured to be 1 kg/s, (a) determine the density (ρ) of air in kg/m³. (b) *What-if scenario:* What would be the answer if CO_2 were flowing instead (with all the readings unchanged)?

0-3-6   Air flows steadily through a constant-area duct. At the entrance the velocity is 5 m/s and temperature is 300 K. The duct is heated such that at the exit the temperature is 600 K. (a) If the specific volume (v) of air is proportional to the absolute temperature (in K) and the mass flow rate ($\dot{m}$) remains constant throughout the duct, determine the exit velocity. (b) How would heating affect the pressure?

0-3-7   Hydrogen flows through a nozzle exit of diameter 10 cm with an average velocity of 200 m/s. If the mass flow rate ($\dot{m}$) of air is measured as 1 kg/s, determine (a) the density (ρ) of hydrogen at the exit in kg/m³. (b) Determine the specific volume (v) of hydrogen at the exit in m³/kg. (c) If hydrogen is supplied from a tank, determine the loss of mass of this tank (in kg) in 1 hour.

0-3-8   A horizontal-axis wind turbine has a diameter of 50 m and faces air coming at it at 20 miles per hour. If the density (ρ) of air is estimated as 1.1 kg/m³, determine the mass flow rate ($\dot{m}$) of air through the turbine in kg/s. Assume the flow of air is not disturbed by the rotation of the turbine. How would your answer change if the turbine is slightly angled with a yaw angle of 10 degrees (yaw angle is the angle between the turbine axis and the wind direction)?

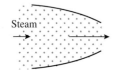

0-3-9   Steam at 400°C enters a nozzle with an average velocity of 20 m/s. If the specific volume and the flow area at the inlet are measured as 0.1 m³/kg and 0.01 m² respectively, determine (a) the volume flow rate ($\dot{V}$) in m³/s, and (b) the mass flow rate ($\dot{m}$) in kg/s. Use the PC flow state TESTcalc to verify your answers.

SECTION 0-4: WORK TRANSFER

0-4-1   A bucket of concrete with a mass of 5000 kg is raised without any acceleration by a crane through a height of 20 m. (a) Determine the work transferred into the bucket. (b) Also determine the power delivered to the bucket if it is raised at a constant speed of 1 m/s. (c) What happens to the energy after it is transferred into the bucket?

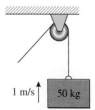

0-4-2   The accompanying figure shows a body of mass 50 kg being lifted at a constant velocity of 1 m/s by the rope and pulley arrangement. Determine power delivered by the rope.

0-4-3 An elevator with a total mass of 1500 kg is pulled upward using a cable at a velocity of 5 m/s through a height of 300 m. (a) Determine the rate at which is work is transferred into the elevator (magnitude only, no sign). (b) What is the sign of W? (c) The change in stored E of the elevator assuming the KE and U to remain unchanged.

0-4-4 In Problem 0-4-3, assume the energetic efficiency (work transfer to the elevator [desired] to the electrical work transfer [required] to the motor) of the system to be 80%, (a) determine the power consumption rate by the motor (magnitude only). (b) Assuming electricity costs 20 cents per kWh, determine the energy cost of operation in cents.

0-4-5 (a) Determine the constant force necessary to accelerate a car of mass 1000 kg from 0 to 100 km/h in 6 seconds. (b) Also calculate the work done by the force. (c) Verify that the work done by the force equals the change in KE of the car. Neglect friction. (d) **What-if scenario:** What would the work be if the acceleration were achieved in 5 seconds?

0-4-6 A driver locks the brake of a car traveling at 140 km/h. Without anti-lock brakes, the tires immediately start skidding. If the total mass of the car, including the driver, is 1200 kg, determine (a) the deceleration, (b) the stopping distance for the car and (c) the work transfer (include sign) treating the car as the system. Assume the friction coefficient between rubber and pavement to be 0.9. Neglect viscous drag.

0-4-7 A car delivers 200 hp to a winch used to raise a load of 1000 kg. Determine the maximum speed of lift.

0-4-8 A block of mass 100 kg is dragged on a horizontal surface with static and kinetic friction coefficients of 0.15 and 0.09 respectively. Determine (a) the pull force necessary to initiate motion (b) the work done by the pull force and (c) the work done against the frictional force as the block is dragged over a distance of 5 m. (d) What is the net work transfer between the block and its surroundings?

0-4-9 In the accompanying figure, determine (a) the work done by a force of 100 N acting at an angle $20°$ in moving the block of mass 10 kg by a distance of 3 m if the friction coefficient is 0.5. (b) What is the net work transfer if the block is treated as the system?

0-4-10 Twenty 50 kg suitcases are carried by a horizontal conveyor belt at a velocity of 0.5 m/s without any slippage. If $\mu_s = 0.9$, (a) determine the power required to drive the conveyor. Assume no friction loss on the pulleys. (b) **What-if scenario:** What would the power required be if the belt were inclined upward at an angle of $10°$?

0-4-11 A person with a mass of 70 kg climbs the stairs of a 50 m tall building. (a) What is the minimum work transfer if you treat the person as a system? Assume standard gravity. (b) If the energetic efficiency (work output/heat released by food) of the body is 30%, how many calories are burned during this climbing process?

0-4-12 A person with a mass of 50 kg and an energetic efficiency of 35% decides to burn all the calories consumed from a can of soda (140 calories) by climbing stairs of a tall building. Determine the maximum height of the building necessary to ensure that all the calories from the soda can is expended in the work performed in climbing.

0-4-13 The aerodynamic drag force F_d in kN on an automobile is given as $F_d = (1/2000)$ $C_d A \rho V^2$ [kN], where C_d is the non-dimensional drag coefficient, A is the frontal area in m^2, ρ is the density of the surrounding air in kg/m^3, and V is the velocity of air with respect to the automobile in m/s. Determine the power required to overcome the aerodynamic drag for a car with $C_d = 0.4$ and $A = 7\,m^2$, traveling at a velocity of 100 km/h. Assume the density of air to be $p = 1.2\,kg/m^3$.

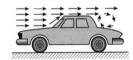

0-4-14 The rolling resistance of the tires is the second major opposing force (next to aerodynamic drag) on a moving vehicle and is given by $F_r = fW$[kN] where f is the rolling resistance coefficient and W is the weight of the vehicle in kN. Determine the power required to overcome the rolling resistance for a 2000 kg car traveling at a velocity of 100 km/h, if $f = 0.007$.

0-4-15 Determine the power required to overcome (a) the aerodynamic drag and (b) rolling resistance for a truck traveling at a velocity of 120 km/h, if $C_d = 0.8, A = 10\,m^2, \rho = 1.2\,kg/m^3, f = 0.01$, and $m = 20,000$ kg. Plot the power requirement—aerodynamic, rolling friction and total—against velocity within the range from 0 to 200 km/h.

0-4-16 Determine the power required to overcome (a) the aerodynamic drag and (b) rolling resistance for a bicyclist traveling at a velocity of 21 km/h, if $C_d = 0.8, A = 1.5\,m^2, \rho = 1.2\,kg/m^3, f = 0.01$ and $m = 100$ kg. Also determine (c) the metabolic energetic efficiency (work output/energy input) for the bicyclist if the rate at which calories are burned is measured at 650 calories/h.

0-4-17 Determine (a) the work transfer involved in compressing a spring with a spring constant of 150 kN/m from its rest position by 10 cm. (b) What is the work done in compressing the spring by an additional 10 cm?

0-4-18 An object with a mass of 200 kg is acted upon by two forces, 0.1 kN to the right and 0.101 kN to the left. Determine (a & b) the work done by the two faces and (c) the net work transfer as the system (the object) is moved a distance of 10 m.

0-4-19 A rigid chamber contains 100 kg of water at 500 kPa, 100°C. A paddle wheel stirs the water at 1000 rpm while an internal electrical resistance heater heats the water while consuming 10 amps of current at 110 Volts. At steady state, the chamber loses heat to the atmosphere at 27°C at a rate of 1.2 kW. Determine (a) $\dot{W}_{sh}$ in kW, (b) the torque in the shaft in N-m.

0-4-20 A piston-cylinder device containing a fluid is fitted with a paddle-wheel stirring device operated by the fall of an external weight of mass 50 kg. As the mass drops by a height of 5 m, the paddle wheel makes 10,000 revolutions, transferring shaft work into the system. Meanwhile the free-moving piston (frictionless and weightless) of 0.5 m diameter moves out by a distance of 0.7 m. Find the W_{net} for the system if the pressure outside is 101 kPa.

0-4-21 An insulated vertical piston-cylinder device contains steam at 300 kPa, 200°C, occupying a volume of 1 m³, and having a specific volume of 0.716 m³/kg. It is heated by an internal electrical heater until the volume of steam doubles due to an increase in temperature.

(a) Determine the final specific volume of steam.
(b) If the diameter of the piston is 20 cm and the outside pressure is 100 kPa, determine the mass of the weight placed on the piston to maintain a 300 kPa internal pressure.
(c) Calculate the W_B (magnitude only), which has been done by the steam in kJ.
(d) Calculate the amount of W_M (magnitude only) transferred into the weight in kJ.
(e) If you are asked to choose one of the three values—900 kJ, 300 kJ, 200 kJ—as the magnitude of W_{el}, which one will be your educated guess? Why?

100 kPa

50 kg

400 K

0-4-22 A gas in a vertical piston-cylinder device has a volume of 0.5 m³ and a temperature of 400 K. The piston has a mass of 50 kg and a cross-sectional area of 0.2 m². As the gas cools down to atmospheric temperature the volume decreases to 0.375 m³. Neglect friction and assume atmospheric pressure to be 100 kPa. Determine (a) the work transfer during the process. (b) *What-if scenario:* What would the work transfer be if the piston weight were considered negligible?

0-4-23 A man weighing 100 kg is standing on the piston head of a vertical piston-cylinder device containing nitrogen. The gas now is heated by an electrical heater until the man slowly is lifted by a height of 1 m. The piston is weightless and has an area of 1 m². The outside pressure is 100 kPa. Determine (a) the initial pressure inside the cylinder, (b) the final pressure inside, (c) the boundary work (magnitude only) performed by the piston-cylinder device (nitrogen is the system) assuming an average pressure of 101 kPa inside the cylinder during the heating process, and (d) the work (magnitude only) transferred to the man (man as the system). Assume the acceleration due to gravity to be 9.81 m/s².

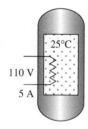

0-4-24 A 10 m³ insulated rigid tank contains 20 kg of air at 25°C. An electrical heater within the tank is turned on, which consumes a current of 5 Amps for 30 min from a 110 V source. Determine the work transfer in kJ.

25°C

110 V

5 A

0-4-25 A paddle wheel stirs a water tank at 500 rpm. The torque transmitted by the shaft is 20 N-m. At the same time, an internal electric resistance heater draws 2 Amps of current from a 110 V source as it heats the water. Determine (a) $\dot{w}$ in kW. (b) What is the total W in 1 hour?

0-4-26 Determine the power conducted by the crankshaft of a car, which is transmitting a torque of 0.25 kN-m at 3000 rpm.

0-4-27 An electric motor draws a current of 16 amp at 110 V. The output shaft delivers a torque of 10 N-m at a speed of 1500 RPM. Determine (a) the electric power transferred, (b) shaft power, and (c) the rate of heat transfer if the motor operates at steady state.

0-4-28 Determine the boundary work transfer in blowing up a balloon to a volume of $0.01\,\text{m}^3$. Assume that the pressure inside the balloon is equal to the surrounding atmospheric pressure, 100 kPa.

0-4-29 Air in a horizontal free-moving piston-cylinder assembly expands from an initial volume of $0.25\,\text{m}^3$ to a final volume of $0.5\,\text{m}^3$ as the gas is heated for 90 seconds by an electrical resistance heater consuming 1 kW of electric power. If the atmospheric pressure is 100 kPa, determine (a) W_B and (b) W_{net}. (c) **What-if scenario:** How would the answers change if the cylinder contained pure oxygen instead?

100 kPa

1 kW

0-4-30 A vertical piston-cylinder assembly (see figure) contains 10 L of air at 20°C. The cylinder has an internal diameter of 20 cm. The piston is 2 cm thick and is made of steel of density $7830\,\text{kg/m}^3$. If the atmospheric pressure outside is 101 kPa, (a) determine the pressure of air inside the cylinder. The air is now heated until its volume doubles. (b) Determine the boundary work transfer during the process. **What-if scenario:** What would the (c) pressure and (d) work be if the piston weight were neglected?

20°C
10 L

0-4-31 Air in the accompanying piston-cylinder device is initially in equilibrium at 200°C. The mass of the hanging weight is 10 kg and the piston diameter is 10 cm. As air cools due to heat transfer to the surroundings, at 100 kPa the piston moves to the left, pulling the weight up. Determine (a) the boundary work and (b) the work done in raising the weight for a piston displacement of 37 cm. (c) Explain why the two are different.

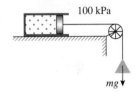

100 kPa

mg

0-4-32 A horizontal piston-cylinder device contains air at 90 kPa while the outside pressure is 100 kPa. This is made possible by pulling the piston with a hanging weight through a string-and-pulley arrangement (see accompanying figure). If the piston has a diameter of 20 cm:

(a) Determine the mass of the hanging weight in kg.
(b) The gas is now heated using an electrical heater and the piston moves out by a distance of 20 cm. Determine the boundary work (include sign) in kJ.
(c) What fraction of the boundary work performed by the gas goes into the hanging weight?
(d) How do you account for the loss of stored energy (in the form of PE) by the hanging weight?

0-4-33 An insulated, vertical piston-cylinder assembly (see figure) contains 50 L of steam at 105°C. The outside pressure is 101 kPa. The piston has a diameter of 20 cm and the combined mass of the piston and the load is 75 kg. The electrical heater and the paddle wheel are turned on and the piston rises slowly by 25 cm. Determine (a) the pressure of air inside the cylinder during the process (b) the boundary work performed by the gas and (c) the combined work transfer by the shaft and electricity if the net energy transfer into the cylinder is 3.109 kJ.

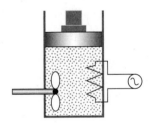

0-4-34 Steam is compressed from $p_1 = 100$ kPa, $V_1 = 1\,\text{m}^3$ to $p_2 = 200$ kPa, $V_2 = 0.6\,\text{m}^3$. The external force exerted on the piston is such that pressure increases linearly with a decrease in volume. Determine (a) the boundary work transfer and (b) show the work by shaded areas in a $p-V$ diagram.

0-4-35 A gas in a piston-cylinder assembly is compressed (through a combination of external force on the piston and cooling) in such a manner that the pressure and volume are related by $pV^n = \text{constant}$. Given an initial state of 100 kPa and $1\,\text{m}^3$ and a final volume of $0.5\,\text{m}^3$ evaluate the work transfer if (a) $n = 0$, (b) $n = 1$, (c) $n = 1.4$ and (d) plot a $p-v$ diagram for each processes and show the work by shaded areas.

1 m²
100 kPa

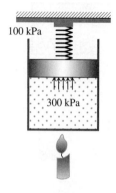

0-4-36   In the preceding problem the piston has a cross-sectional area of $0.05\,\text{m}^2\text{s}$. If the atmospheric pressure is 100 kPa and the weight of the piston and friction are negligible, plot how the external force applied by the connecting rod on the piston varies with the gas volume for (a) $n = 0$, (b) $n = 1$ and (c) $n = 1.4$.

0-4-37   A piston-cylinder device contains $0.03\,\text{m}^3$ of nitrogen at a pressure of 300 kPa. The atmospheric pressure is 100 kPa and the spring pressed against the piston has a spring constant of 256.7 kN/m. Heat now is transferred to the gas until the volume doubles. If the piston has a diameter of 0.5 m, determine (a) the final pressure of nitrogen, (b) the work transfer from nitrogen to the surroundings, and (c) the fraction of work that goes into the atmosphere.

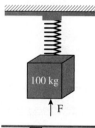

0-4-38   A 100 kg block of solid is moved upward by an external force F as shown in the accompanying figure. After a displacement of 10 cm, the upper surface of the block reaches a linear spring at its rest position. The external force is adjusted so that the displacement continues for another 10 cm. If the spring constant is 100 kN/m and acceleration due to gravity is $9.81\,\text{m/s}^2$, determine (a) the work done by the external force. (b) What fraction of the energy transferred is stored in the spring?

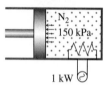

0-4-39   Nitrogen in a horizontal piston-cylinder assembly expands from an initial volume of $0.1\,\text{m}^3$ to a final volume of $0.5\,\text{m}^3$ as the gas is heated for 5 minutes by an electrical resistance heater consuming 1 kW of electric power. If the pressure remains constant at 150 kPa, and 70 kJ of heat is lost from the cylinder during the expansion process, determine (a) W_B (include sign), (b) W_{el} (include sign), and (c) W_{net} (absolute value) into the system through heat and work.

0-4-40   Water enters a horizontal system, operating at steady state, at 100 kPa, 25°C, 10 m/s at a mass flow rate of 200 kg/s. It leaves the system at 15 m/s, 1 MPa, 25°C. If the density of water is $1000\,\text{kg/m}^3$, determine (a) the rate of flow work ($\dot{W}$-F) at the inlet (magnitude only), (b) the diameter of the pipe at the inlet.

0-4-41   The rate of energy transfer due to flow work ($\dot{W}$-F) at a particular cross-section is 20 kW. If the volume flow rate ($\dot{V}$) is $0.2\,\text{m}^3/\text{min}$, determine the pressure at that location.

0-4-42   Water enters a pump at 100 kPa with a mass flow rate ($\dot{m}$) of 20 kg/s and exits at 500 kPa with the same mass flow rate. If the density of water is $1000\,\text{kg/m}^3$, determine the net flow work transfer in kW.

0-4-43   Water (density $1000\,\text{kg/m}^3$) flows steadily into a hydraulic turbine through an inlet with a mass flow rate ($\dot{m}$) of 500 kg/s. The conditions at the inlet are measured as 500 kPa, 25°C. (a) Determine the magnitude of the rate of flow work transfer in kW and (b) the sign of the work transfer treating the turbine as the system.

SECTION 0-5: HEAT TRANSFER

0-5-1   If a therm of heat costs $1.158 and a kW-h of electricity costs $0.106, then determine the prices of (a) heat and (b) electricity on the basis of GJ.

0-5-2   A power plant has an average load of 2000 MW (electrical). If the overall thermal efficiency is 38%, what is the annual cost of fuel for (a) natural gas ($1.26/Therm) and (b) No. 2 fuel oil ($9.40/MMBtu)?

0-5-3 A gas trapped inside a piston-cylinder device receives 20 kJ of heat as it expands, performing a boundary work (W_B) of 5 kJ. At the same time 10 kJ of electrical work (W_{el}) is transferred into the system. Evaluate (a) Q and (b) W with appropriate signs.

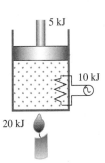

0-5-4 A gas station sells gasoline and diesel at $2.00/gallon and $1.75/gallon respectively. If the following data are known about the two fuels, compare the prices on the basis of MJ of energy: heating value: gasoline 47.3 MJ/kg, diesel 46.1 MJ/kg; density: gasoline 0.72 kg/L, diesel 0.78 kg/L.

0-5-5 A rigid cylindrical tank stores 100 kg of a substance at 500 kPa and 500 K while the outside temperature is 300 K. A paddle wheel stirs the system, transferring shaft work ($\dot{w}$-sh) at a rate of 0.5 kW. At the same time an internal electrical resistance heater transfers electricity ($\dot{w}$-el) at the rate of 1 kW. Determine $\dot{Q}$ necessary to ensure that no net energy enters or leaves the tank.

0-5-6 The nutrition label on a snack bar, which costs $1.00, reads—serving size 42 g; calories per serving 180. Determine (a) the heating value in MJ/kg, and (b) price in cents per MJ of heat release. (c) If gasoline with a heating value of 44 MJ/kg and a density of 750 kg/m^3 costs $2.50 a gallon, what is the gasoline price in cents/MJ?

0-5-7 A popular soda can contains 0.355 kg of soda, which can be considered to be composed of 0.039 kg of sugar and the rest water. If the calorific value is written as 140 calorie (bio), (a) calculate the heating value (maximum heat that will be released by 1 kg of the food) of sugar in MJ/kg. (b) Compare the heating value of sugar with different fuels listed in Table G-2.

0-5-8 Consider three options for heating a house. Electric resistance heating with electricity priced at $0.10/kWh, gas heating with gas priced at $1.10/Therm and oil (density 0.8 kg/L, heating value 46.5 MJ/kg) heating with oil priced at $1.50/gal. Energetic efficiencies are 100% for the electrical heating system, 85% for the gas-heating system and 80% for the oil-heating system. Determine the cost of delivering 1 GJ of energy by each system.

0-5-9 To determine which is a cheaper fuel, a student collects the following data for gasoline and diesel respectively. Price per gallon: $2.90 vs. $3.10; heating value: 44 MJ/kg vs. 43.2 MJ/kg; density: 740 kg/m^3 vs. 820 kg/m^3; 1 Gallon = 3.785 L; 1 $L = 10^{-3}$m^3. Determine (a) the price of gasoline in cents per MJ of heat release, (b) the price of diesel per MJ of heat release.

0-5-10 In 2003, the United States consumed (a) 20 MMbd (million barrels per day) of crude oil, (b) 21.9 tcf (trillion cubic feet) of natural gas, and (c) 1 billion tons (short) of coal. The Btu equivalents are as follows: 1 bbl crude oil: 5.8 million Btu; 1 Mcf gas: 1.03 million Btu; 1 ton coal: 21 million Btu. Compare the energy consumption in the consistent unit of Quad (1 Quad = 10^{15} Btu).

0-5-11 On Aug. 20, 2011, the prices for crude oil (heating value 43 MJ/kg) and natural gas were quoted as 82.26 USD/Bbl and 3.94 USD/MMBtu in the world market. Compare the prices on a comparable scale (USD/MJ). Use properties of heavy diesel to represent crude oil.

0-5-12 The United States consumed about 21 MMbd (million barrels per day) of crude oil (density 0.82 kg/L, heating value 47 MJ/kg), 67% of which is utilized in the transportation sector. Determine how many barrels of oil can be saved per year, if the fuel consumption in the transportation sector can be reduced by 20% through the use of hybrid technology.

SECTION 0-6: ENERGY CONVERSION

0-6-1 During charging, a battery pack loses heat at a rate of 0.2 kW. The electric current flowing into the battery from a 220 V source is measured as 10 amp. Determine (a) $\dot{Q}$, (b) W_{el}, and (c) W_{ext}. Include sign.

0-6-2 A car delivers its power to a winch, which is used to raise a load of 1000 kg at a vertical speed of 2 m/s. Determine (a) the work delivered by the engine to the winch in kW, and (b) the rate (g/s) at which fuel is consumed by the engine. Assume the engine to be 35% efficient with the heating value of fuel to be 40 MJ/kg.

0-6-3 A car delivers 96.24 kW to a winch, which is used to raise a load of 1000 kg. (a) Determine the maximum velocity in m/s with which the load can be raised. (b) If the heating value of the fuel used is 45 MJ/kg and the engine has an overall efficiency of 35%, determine the rate of fuel consumption in kg/h.

0-6-4 A semi-truck of mass 25,000 lb (1 kg = 2.2 lb) enters a highway ramp at 10 mph (1 m/h = 0.447 m/s). It accelerates to 75 mph while merging with the highway at the end of the ramp at an elevation of 15 m. (a) Determine the change in mechanical energy of the truck. (b) If the heating value of diesel is 40 MJ/kg and the truck engine is 30% efficient (in converting heat to mechanical energy), determine the mass of diesel (in kg) burned on the ramp.

0-6-5 We are interested in the amount of gasoline consumed to accelerate a car of mass 5000 kg from 5 to 30 m/s (about 67 mph) on a freeway ramp. The ramp has a height of 15 m. Assuming the internal energy of the car to remain constant, (a) determine the change in stored energy E of the car. (b) If the heating value of gasoline is 44 MJ/kg and the engine has a thermal efficiency of 30%, determine the amount of gasoline (in kg) that will be consumed as the car moves through the ramp. Assume all the engine power goes into the kinetic and potential energies of the car.

0-6-6 A jumbo jet with a mass of 5 million kg requires a speed of 175 mph for takeoff. Assuming an overall efficiency of 20% (from heat release to kinetic energy of the aircraft), determine the amount of jet fuel (heating value: 44 MJ/kg) consumed during the takeoff.

0-6-7 A 0.1 kg projectile travelling with a velocity of 200 m/s (represented by State-1) hits a stationary block of solid (represented by State-2) of mass 1 kg and becomes embedded (combined system is represented by State-3). Assuming momentum is conserved (there are no external forces, including gravity, on any system), (a) determine the velocity of the combined system (V_3) in m/s. (b) If the stored energies of the systems before and after the collision are same (that is, $E_1 + E_2 = E_3$), determine the change in internal energy after the collision ($U_3 - U_1 - U_2$) in kJ.

0-6-8 A semi-truck of mass 20,000 lb accelerates from 0 to 75 m/h (1 mi/h = 0.447 m/s) in 10 seconds. (a) What is the change in KE of the truck in 10 seconds? (b) If PE and U of the truck can be assumed constant, what is the average value of dE/dt of the truck during this period? (c) If 30% of the heat released from the combustion of diesel (heating value of diesel is 40 MJ/kg) is converted to KE, determine the average rate of fuel consumption in kg/s.

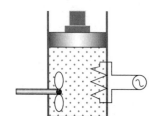

0-6-9 A gas trapped in a piston-cylinder device is subjected to the energy interactions shown in the accompanying figure for 30 seconds: the electric resistance draws 0.1 amp from a 100 V source, the paddle wheel turns at 60 rpm with the shaft transmitting a torque of 5 N-m and 1 kJ of heat is transferred into the gas from the candle. The volume of the gas increases by 6 L during the process. If the atmospheric pressure is 100 kPa and the piston can be considered weightless, determine (a) W_B, (b) W_{sh}, (c) W_{el}, (d) W_{net}.

0-6-10 A gas trapped inside a piston-cylinder device receives 20 kJ of heat while it expands, performing a boundary work of 5 kJ. At the same time 10 kJ of electrical work is transferred into the system. Evaluate (a) Q and (b) W with appropriate signs.

0-6-11 A gas at 300 kPa is trapped inside a piston-cylinder device. It receives 20 kJ of heat while it expands performing a boundary work of 5 kJ. At the same time 10 kJ of electrical work is transferred into the system. Evaluate (including sign) (a) Q, (b) W_{el}, (c) W_B, (d) W_{ext} with appropriate signs. (e) What is the net change in stored energy E of the system (magnitude only in kJ)? Neglect KE and PE.

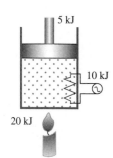

0-6-12 An iron block of 20 kg undergoes a process during which there is a heat loss from the block at 4 kJ/kg, an elevation increase of 50 m, and an increase in velocity from 10 m/s to 50 m/s. During the process, which also involves work transfer, the internal energy of the block decreases by 100 kJ. Determine the work transfer during the process in kJ if the total energy remains constant. Include sign..

0-6-13 A person turns on a 100-W fan before he/she leaves the warm room at 100 kPa, 30°C, hoping that the room will be cooler when he/she comes back after 5 hours. Heat transfer from the room to the surroundings occurs at a rate of 0.2t (t in minutes) watts. Determine the net energy transfer into the room in 5 hours.

0-6-14 A fully charged battery supplies power to an electric car of mass 3000 kg. Determine the amount of energy depleted (in kJ) from the battery as the car accelerates from 0 to 140 km/h.

0-6-15 A photovoltaic array produces an average electric power output of 20 kW. The power is used to charge a storage battery. Heat transfer from the battery to the surroundings occurs at 1.5 kW. Determine the total amount of energy stored in the battery (in MJ) in 5 hours of operation.

0-6-16 The heating value (maximum heat released as a fuel is burned with atmospheric air) of diesel is 43 MJ/kg. Determine the minimum fuel consumption necessary to accelerate a 20-ton (short ton) truck from 0 to 70 mph speed. Assume that all the work done by the engine is used to raise the KE of the truck and the efficiency of the engine is 35%.

0-6-17 In 2002, the United States produced 3.88 trillion kWh of electricity. If coal (heating value 24.4 MJ/kg) accounted for 51% of the electricity production at an average thermal efficiency (electrical work output/heat input) of 40%, determine the total amount of coal (in short tons) consumed by the power plants in 2002.

1 DESCRIPTION OF A SYSTEM: STATES AND PROPERTIES

In the Introduction, we defined a thermodynamic *system* as any entity with a boundary and discussed all possible *interactions* a system can have with its *surroundings*. In this chapter, we take a look inside the system and develop a methodology to describe it. The mathematical description of a system is called a *state*, which is a set of quantitative attributes or *properties*. The state of a uniform system or a flow, described by an *extended state,* is shown to build around the core concept of thermodynamic *equilibrium* represented by a *thermodynamic state*. A complex system, which may not be in equilibrium, can be described by an aggregate of local systems assumed to be in local thermodynamic equilibrium (LTE). Properties are classified in a systematic manner and *thermodynamic properties,* such as temperature, pressure, internal energy, enthalpy, and entropy that define a thermodynamic equilibrium are discussed with emphasis due to their importance in the evaluation of a state. The *zeroth law* of thermodynamics is introduced in connection with temperature and thermal equilibrium. Manual evaluation of a thermodynamic state will be discussed in Chapter 3. TESTcalcs are used as a numerical laboratory in this chapter to explore various properties in a quantitative manner and evaluate extended system and flow states.

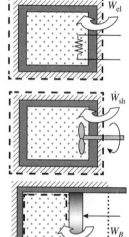

FIGURE 1.1 Different types of work transfer cause the same increase in the air temperature (see Anim. 1.A.*stateHistory*).

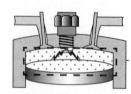

FIGURE 1.2 Exothermic combustion process in a spark ignition automobile can be analyzed by treating the system as an isolated system (see Anim. 7.B.*vConstHeating*).

1.1 CONSEQUENCES OF INTERACTIONS

Interactions have their consequences. If hot water is added to cold water in a bathtub, the addition of water not only increases the system's mass but also increases the system's stored energy. When heat is added to a system, we expect the system's temperature to rise, but this is not always so. A glass of ice-water does not warm up to room temperature despite heat transfer from the surroundings until the last chunk of ice melts. In this case, heat transfer is responsible for a phase change. Work transfer can also bring about the same changes that heat transfer can (see Anim. 1.A.*stateHistory*). Figure 1.1 shows three different ways of *heating* a gas using electric, shaft, and boundary work transfer. While the temperature rise in the air in Figure 1.1 can be readily attributed to Joule heating from electrical work and viscous friction driven by shaft work (through a paddle wheel), we are not accustomed to changes brought about by boundary work. For example, bicyclists are familiar with a hand pump becoming hot when inflating a tire. In fact, this rise in temperature through boundary work transfer is utilized to auto-ignite diesel as it is sprayed into the compressed air in the cylinders of a diesel engine, thus eliminating the need for spark plugs.

Sometimes, a system may spontaneously change without any interactions at all. In a spark ignition engine a mixture of gasoline vapor and air is trapped in the engine cylinder and an electrical spark initiates the combustion process (see Fig. 1.2 and Anim. 0.C.*isolatedSystem*). The electrical work transfer to the spark plug, however, is negligible compared to the heat released due to exothermic reaction, and the combustion takes place so rapidly that any heat or boundary work transfer during that process is insignificant. With no mass, heat, or work transfer, the system can be treated as an isolated system, simplifying a combustion analysis (Chapter 13).

A major goal of thermodynamic analysis is to predict how a system changes as a result of known interactions with its surroundings. Conversely, for a given set of changes that occur in a system, we may be interested in predicting the precise interactions in terms of mass, heat, and work transfer. Either way, thermodynamic analysis requires changes in a system to be quantitatively described.

1.2 STATES

A system is quantitatively described by its states and properties: A **state** is a mathematical description of the condition of a system or a flow at a given time expressed as a set of **properties**, which are measurable characteristics or attributes of the system or flow.

To explore the meaning of these thermodynamic terms, consider a system where a gas is trapped in a piston-cylinder device (Fig. 1.3). It is reasonable to assume that the system is *uniform*, that is, there is no significant variation of properties across locations within the system. The particular condition of the system at a given time can then be described by a single state, say, State-1, consisting of a set of properties such as mass m_1, volume V_1, pressure p_1, temperature T_1, elevation z_1, etc. Implicit in this description is the assumption that the system is in equilibrium. Thermodynamic equilibrium and its theoretical background will be thoroughly discussed in Chapters 3 and 11. Simply put, a system is in **thermodynamic equilibrium** when, blocked from all interactions, all its internal activities—internal motion, heat transfer, and chemical changes—spontaneously subside and eventually disappear as the properties assume fixed values (click Isolate in Anim. 1.A.*globalEquilibrium*). For example, if a system composed of a mixture of liquid water and ice is isolated (Fig. 1.4), some ice may melt and/or some water may freeze until the mixture reaches *equilibrium* where water and ice coexist without any further changes in properties such as temperature or density. If interactions are resumed, say, by heating the system, the system will depart from its original equilibrium state. If isolated again, it will assume a new equilibrium defined by a new set of properties. As another example, consider the gas in the piston-cylinder device of Figure 1.3 is in equilibrium and described by State-1. This means that if a system is kept isolated, properties such as temperature, pressure, density, etc., will not change over time. However, if the piston is moved to a new position, the system, given enough time, will arrive at a new equilibrium described by State-2. Properties that describe an equilibrium are called **thermodynamic properties** and a set of thermodynamic properties constitute a **thermodynamic state** (click Thermodynamic in Anim. 1.A.*systemState*).

With properties varying across locations and/or over time, a complex system may not always be in equilibrium. However, such a non-uniform, unsteady system, also referred to as a *global system*, can be divided into smaller sub-systems called *local systems*, each of which can be assumed to be in its LTE. Similarly, a global system evolving over time can be thought of as a system passing through a succession of local equilibrium states (see Anim. 1.A.*localEquilibrium*). A complex system, therefore, can be described thermodynamically by an aggregate of local systems in equilibrium, which is known as the **LTE hypothesis**. Without this implicit hypothesis, a mathematical description of a complex system becomes almost unmanageable, and thermodynamics would not be such a practical subject.

For engineering analysis, the description of a system often extends beyond the description of the underlying equilibrium state. An **extended state** (see Anim. 1.A.*extendedStates*) includes properties such as total mass, system volume, system velocity, etc., that do not necessarily characterize the core equilibrium state. For example, if the system of Figure 1.3 is placed at a higher elevation, the PE of the system will change without affecting the equilibrium state. Potential energy, therefore, is not a thermodynamic property. In a uniform system, every subsystem shares the same equilibrium with the global system. The mass and volume of the system, therefore, cannot be thermodynamic properties. An extended state that completely describes a system enclosed by a boundary is called a **system state** (click System State in Anim. 1.A.*propertyGroups*). If the extended system state of the gas in Figure 1.3 is known, we have a complete description of the system far beyond its underlying equilibrium. Two different system states do not necessarily represent two different equilibriums. If the gas in Figure 1.3 is compressed by the piston to a new state, State-2, the underlying equilibrium state also is changed. The change in a property, say, pressure, can be obtained from the two system states as $\Delta p = p_2 - p_1$. However, if a bigger cylinder (Fig. 1.5) contains a larger amount of the same gas at State-3 with $m_3 = 2m_1$, $p_3 = p_1$, and $T_3 = T_1$, many properties such as V, U, E, etc., will increase by a factor of 2, yet the core thermodynamic properties describing the equilibrium would be identical between states 1 and 3.

When describing the mass interactions that occur in an open system, it is necessary to describe a flow at the inlet and exit ports. Like the uniform velocity profile discussed in Sec. 0.5, other properties can be assumed to be uniform across the cross section of the flow. Moreover, if a small lump of fluid is suddenly isolated, say, at the inlet of a turbine (Fig. 1.6), the thermodynamic properties p_i, T_i, etc., would not change over time; that is, the flow can be assumed to be based on LTE at the inlet. A uniform flow, therefore, can be described by an extended state called **flow state** that builds upon the local equilibrium state by adding properties such as flow area, flow velocity, mass flow rate, etc (Anim. 1.A.*flowState*). The inlet and exit states, State-*i* and State-*e*, respectively, in Figure 1.6, are such flow states based on LTE. The change in a property, say, pressure, between the exit and inlet state can be obtained from the two flow states

FIGURE 1.3 The condition of the gas is represented by the system state State-1 at a given time (see Anim. 1.A.*systemStates*).

FIGURE 1.4 Ice and water can coexist in equilibrium when isolated from the surroundings. At equilibrium, all internal imbalances disappear and thermodynamic properties assume fixed values unique to that equilibrium (see Anim. 1.A.*thermodynamicState*).

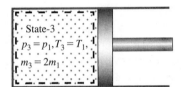

FIGURE 1.5 The system state of the trapped gas share the same equilibrium state as State-1 in Fig. 1.3.

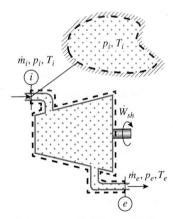

FIGURE 1.6 The inlet and exit conditions are defined by two flow states, which build upon the underlying equilibrium states at the inlet and exit ports (see Anim. 1.A.*localEquilibrium*).

as $\Delta p = p_e - p_i$. Note that the same sets of thermodynamic properties are at the core of both system and flow states (click Thermodynamic State in Anim. 1.A.*propertyGroups*).

Recall that the variables that constitute a state are called *properties* (see Anim. 1.A.*property*). We have already encountered several properties such as mass, pressure, temperature, and velocity. To be a property, first, a variable must describe a specific aspect or attribute of a state. Mass, velocity, and temperature are properties since they describe a system state. Similarly, mass flow rate, velocity, and temperature are properties of a flow state. However, work and heat transfer cannot be properties as they describe interactions, not states.

Second, a property is without memory and depends only on the current state of equilibrium (see Anim. 1.A.*stateHistory*). In mathematical terms, it is a **point function**, which means that its value is associated to the state so that the change in a property as a system goes from one equilibrium to another can be expressed as a difference. For example, when the temperature of a uniform system is measured as 100°C, it is not known whether the system was heated or cooled to achieve its current temperature. If a system migrates from State-1 to State-2, we can write $\Delta T = T_2 - T_1$, which ignores the specific path followed by the system. When the two states are infinitesimally close, $\Delta T \rightarrow dT$, which in mathematical terms is called an *exact differential*. Heat and work transfer, conversely, are **path functions** because their values depend on precisely how the system transitions from one state to another. To emphasize that it is not a property, a path function's differential, called an *inexact differential,* is expressed with a crossed d as in $đQ$ or $đW$. Unlike an exact differential, such as dT or dp, an inexact differential cannot be expressed as a difference.

Returning to our lake analogy introduced in Sec. 0.7, the amount of water in a lake (analogous to stored energy) behaves as a property—any change in the total amount of water in a given period can be determined from the water levels at the beginning and end of the period. However, the rainfall during that time, analogous to heat transfer during a process, is a *path function* as it requires continuous monitoring for a quantitative measurement.

Finally, different properties can be combined to create new properties for analytical convenience. Mass and volume are properties of a state (they describe a system, and they are point functions); therefore, we must accept their ratio, *density* ρ, and its inverse, *specific volume*: $v \equiv 1/\rho$, as legitimate properties. In summary, a variable is a property if it (a) represents an attribute of a state, (b) is a point function, or (c) is a combination of other properties.

Given that a uniform system or flow can be described by an extended system or flow state, a non-uniform system can be described by decomposing it into a number of uniform subsystems. For example, the non-uniform system of Figure 1.7—a solid block submerged in water—can be represented by two system states, State-1 and State-2, which describe the solid and liquid subsystems. The overall state of such a composite system is called the **global state** represented by the aggregate of the local states in LTE. When properties vary in a continuous manner throughout a system, the local subsystems have to be small to be considered uniform, but large enough for thermodynamic properties to be meaningful. To determine how small a local system can be, an understanding of the macroscopic nature of classical thermodynamics is necessary.

1.3 MACROSCOPIC VS. MICROSCOPIC THERMODYNAMICS

Consider a particular instant during the charging of an empty cylinder with propane (Fig. 1.8), which comes from a supply line that has constant properties. Suppose we are interested in the thermodynamic property density at any location in the tank (the system) at any instant. For the local system drawn around point A, the density can be expressed as

$$\rho_A = \frac{\Delta m}{\Delta V}; \quad \left[\frac{\text{kg}}{\text{m}^3}\right] \tag{1.1}$$

where ΔV is the volume and Δm is the mass of the subsystem. For a non-uniform system, ρ_A depends on the size of the subsystem, ΔV, and approaches the average value when ΔV approaches the tank's total volume, V (Fig. 1.9). However, as ΔV is made smaller around A, it approaches a constant value $\rho_{\text{local},A}$ that reflects the density at the local level. Similarly, ρ_B, the density calculated around B, approaches a different local value $\rho_{\text{local},B}$. In Figure 1.9, the volume axis uses a logarithmic scale and shows that local limits are approached long before ΔV approaches zero. When ΔV truly approaches zero, we reach the molecular scale, which explains the extreme density fluctuations near the origin. Regardless of a system's scale,

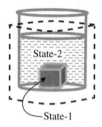

FIGURE 1.7 The global state of a non-uniform system can be described by the (local) states of the subsystems (see Anim. 5.B.*blockInWater*).

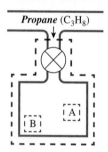

FIGURE 1.8 As propane enters the tank, the local state at *A* may be slightly different from that at *B* because the system is not truly *uniform*.

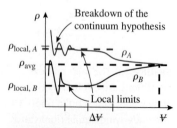

FIGURE 1.9 Density approaches the local limit as ΔV decreases.

classical thermodynamics is based on the hypothesis that a working substance can be treated as a **continuum**, a substance without any abrupt discontinuity. That is, a system, no matter how small, is assumed to consist of such a huge collection of constantly interacting molecules that the system's behavior can be predicted without any reference to individual molecular behavior. Under the continuum hypothesis, ρ_A and ρ_B will remain at their local values no matter how small is ΔV in Figure 1.9.

As long as ΔV is large enough to contain a huge number of molecules and the time of measurement is sufficiently large to allow a huge number of interactions among those molecules, the continuum hypothesis is applicable. To better understand what a sufficiently small volume and a sufficiently large duration are, consider a local state of air under room conditions having a tiny volume, $\Delta V = 1 \ \mu m^3$. Using kinetic theory, it can be shown that even such a small system will contain billions of molecules with trillions of collisions occurring every microsecond. Similarly, when a state is monitored over time, the concept of properties, such as temperature and pressure, breaks down as we approach a true instant. However, a microsecond can be shown to be long enough to have a sufficient number of collisions for local thermodynamic properties to be meaningful. A local thermodynamic equilibrium at a point at a given instant, therefore, is a valid statement as long as the 'point' contains a large number of molecules and the 'instant' allows a huge number of collisions among molecules. In the **macroscopic view** of classical thermodynamics, a continuum extends to the tiniest volume of a **macroscopic point** (a micrometer long) and time can be resolved down to a **macroscopic instant** (a microsecond duration). For the vast majority of engineering systems, including micro-mechanical systems, this hypothesis works well, and seldom is there any need for spatial resolution below a macroscopic point or time resolution smaller than a macroscopic instant. Notable exceptions, where the continuum assumption breaks down, include spacecraft re-entry through a rarefied atmosphere and shock waves that exhibit steep changes in properties within a distance comparable to molecular scales.

It should be mentioned that a parallel treatment of thermodynamics from a **microscopic view** is adopted in statistical thermodynamics, where every system, regardless of size, is treated as a collection of discrete particles (molecules or atoms). The laws of classical thermodynamics can be deduced by statistically averaging molecular phenomena. In this book, however, we will adopt the macroscopic framework, and only occasionally discuss the underlying microscopic structure when it helps us understand certain macroscopic behaviors of a working substance.

1.4 AN IMAGE ANALOGY

Just as we used the lake analogy to visualize energy, heat, and work, we will develop an image analogy to visualize the different types of states introduced earlier.

Consider a digital video of a flickering flame, which represents a non-uniform, time-dependent (*transient*) system. One of the frames from that video at a given instant $t = t_1$ is shown in Figure 1.10. An image such as this represents the *global state* of a non-uniform system at a given time. Each pixel, in this case, behaves as a *local system*, its state (composed of pixel color and brightness) being analogous to the *local state*, and its size representing a *macroscopic point*. The minimum exposure time necessary to record an image is analogous to the *macroscopic instant*—a truly instantaneous image is impossible, as zero photons will be captured by the camera.

Given that thermodynamic systems are three dimensional, it would be impossible for a simple camera to capture a system's global state. However, that cannot stop us from imagining a **state camera** that records the distribution of local states in a three-dimensional video. Sophisticated visualization software routinely uses false colors to represent field variables gathered through experiments or numerical analysis.

The image analogy can be quite effective in classifying systems into different categories (Fig. 1.11). A system whose global state changes with time is called an **unsteady system**; the flickering flame in that sense is an *unsteady system*. Conversely, a **steady system** does not change its global state over time, which means that the local states comprising the global state remain frozen in time. In our image analogy, the digital video of the system reduces to a single still image. Global *extensive* properties, properties such as a system's mass or total stored energy that can be obtained by analyzing the pixels (local systems), therefore, remain constant for a steady system. Another way of system classification is based on whether the local states vary spatially. In a **uniform system**, all local states are identical at a given time so that a single color will represent the global state in our image analogy (explore eight kinds of systems in Anim. 1.A.*systemsClassified*).

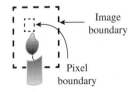

FIGURE 1.10 A frame from a digital video of a flickering flame represents a global system state.

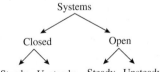

FIGURE 1.11 A simple classification of systems based on mass interaction and whether the global state changes over time (see Anim. 1.A.*systemsClassified*).

Note that a uniform system can be transient so that the color of the image may continually change over time. This simple classification scheme will help us simplify and customize (for specific applications) the governing equations, which will be formulated in Chapter 2 for a generic *open, unsteady, non-uniform* system, the most complex system imaginable.

1.5 PROPERTIES OF STATE

We have already introduced several properties; p, T, ρ, m, V, E, U, V, KE, PE, $\dot{m}$, $\dot{V}$, etc., while discussing equilibrium and extended states. This section formally introduces a number of important properties that constitute the bulk of an extended *system* or *flow* state. Although we will not be ready to manually evaluate properties until Chapter 3, we will use *state* TESTcalcs to develop a quantitative understanding of properties during our discussion.

Given the large number of properties that can be associated with a state, let us start by delineating a few important groups of properties. As already mentioned, the equilibrium state forms the core of an extended state and, therefore, it is important to differentiate *thermodynamic* properties from all other properties from the outset. An extended state that describes a system or flow beyond its core equilibrium contains many other convenient properties. Some of those properties, such as KE or PE, depend on the observer's reference frame. They are called **extrinsic properties**. Stored energy E must be extrinsic too since it depends on other extrinsic properties (KE and PE). Thermodynamic properties, such as pressure, p, and temperature, T, on the other hand, are **intrinsic** to a system, independent of whether an observer is stationed inside or outside a system (see Anim. 1.A.*intrinsicProperty*). **Material properties**, such as molar mass $\overline{M}$, also are intrinsic. However, they only identify the working substance, not its equilibrium state. Yet another pattern emerges if we look at the dependence of a property on the size or extent of a system. Properties, such as m, V, or E of a system scale, with the extent of the system are called **extensive properties** (see Anim. 1.A.*intensiveProperty*). If two identical systems are combined, their extensive properties double without disturbing the underlying equilibrium. Properties, such as p, T, or ρ, do not depend on the extent of the system. They are called **intensive properties**. Extensive properties per unit mass (or mole) are *intensive* and are called **specific properties**. While volume of a system is an extensive property, specific volume v, volume per unit mass, is intensive. With a few notable exceptions, the upper-case symbols are reserved for extensive properties and lower-case symbols for intensive properties. As we introduce additional system and flow properties in this chapter, we will classify them into categories as illustrated in Anim. 1.A.*propertyGroups*.

1.5.1 Property Evaluation by State TESTcalcs

State TESTcalcs in TEST are the building blocks of all other TESTcalcs. With the look and feel of a graphical spreadsheet, these TESTcalcs evaluate extended states of uniform systems and uniform flows for many working substances. State TESTcalcs are divided into two categories: system-state TESTcalcs and flow-state TESTcalcs, which are located in the states, system states and states, flow states branches, respectively. These TESTcalcs are sub-divided according to the material models used to classify the working substances (these models will be our topic of discussion in Chapter 3). Thus, to find a state involving a solid or a liquid we may use the solid/liquid (SL) model; for a gas the ideal gas (IG) model; and for a fluid that may undergo phase transition (e.g., steam) the phase-change (PC) model.

Beginning with the next section, we will use some of the state TESTcalcs as a numerical laboratory to explore the behavior of different properties of state. To illustrate some frequently used features of a state TESTcalc, let us launch the SL system-state TESTcalc, by following the *TESTcalcs > States > System* path from the TESTcalcs tab of the TEST task bar. The TESTcalc itself appears[1] in a rectangular box (Fig. 1.12) with a message panel in a separate box right below. Placing the cursor over any widget displays that widget's definition in the message panel; it also displays error messages and helpful tips during calculations. The global control panel at the top is used for updating a solution globally after changing a parameter or unit system. The tab panel allows you to switch between the state and I/O panels.

---

[1] TESTcalcs require the Java plug-in. If your browser does not display the TESTcalc, refer to the system requirements to make your browser Java compatible.

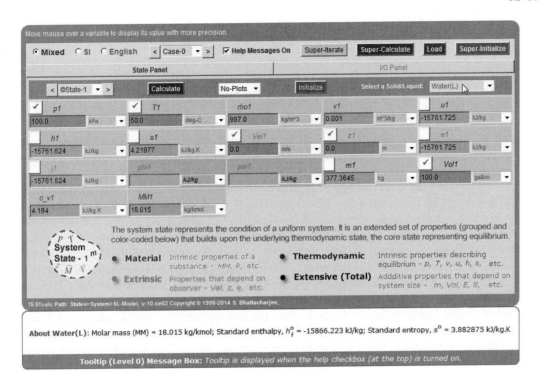

FIGURE 1.12 The SL system-state TESTcalc. Rolling the pointer over a property or the working substance brings up helpful information in the message panel at the bottom.

The first row of the state panel is the state control panel, which is used to pick a state number, select a working substance, initialize and calculate the state, and produce a variety of thermodynamic plots. A set of properties, 17 in this case, constitute the complete system state. Each property is encapsulated in a widget, and consists of the following: a check box that toggles between the input and display mode; a unique symbol with the state number as the suffix; a field for displaying the property's value, and a drop-down menu for selecting a unit. A property is input by clicking its check box (the background color of the value field changes to yellow in input mode), typing in a value, and selecting an appropriate unit (Fig. 1.13). When the Calculate button is clicked (or the Enter key is pressed), the property is read by the TESTcalc and the state is updated. To edit or change a property, the check box should be clicked twice to enter input mode. Properties are color coded: red for material; blue for thermodynamic; green for extrinsic; and black for extensive properties. Evaluating a state consists of four steps: identifying the state by selecting a state number; picking a working substance; entering the known (independent) properties; and clicking the Calculate button (or pressing Enter).

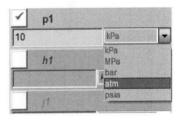

FIGURE 1.13 To enter a pressure of 10 atm, click on the pressure check box, type in the value in the yellow input field, select the unit, and press the Enter key. The yellow field turns green when the pressure is read by the TESTcalc. To modify the value, click the check box again.

As an exercise, select Water (L) as the working fluid, and evaluate State-1 for the following conditions: p1 = 10 atm, T1 = 50 deg-C, and Vol1 = 100 gallons. As soon as you select the working substance, the material properties (in this case, rho1, v1, c_v1, and MM1) in red are displayed. Also, Vel1 and z1 are initialized to zero value by default. Set those properties as unknown by clicking their check boxes twice. Now click on the check boxes of p1 and T1, two thermodynamic properties, to set those into input mode. If we try to click on a third thermodynamic (blue) property, say, u1, the TESTcalc issues a warning that u1 is not independent and can be calculated from p1 and T1. Input the values of p1 and T1 and as you click the Calculate button or press the Enter key, all the thermodynamic (blue) properties are calculated at once. Now input Vel1 and z1 as zeros to calculate all the extrinsic (green) properties with the exception of phi1 and psi1, which will be discussed in Sec. 1.5.10. Finally, enter the given volume, Vol1, and obtain the remaining extensive (black) properties. When evaluating additional states, use algebraic equations to relate properties whenever possible. To calculate a related state, State-2, select the state from the drop-down menu; enter T2, as '= T1 + 5', p2 as '= p1', and m2 as '= m1' to calculate the state provided State-1 has already been calculated. The I/O panel also serves as an engineering calculator. For instance, the stored energy in the system at State-1 can be calculated in the I/O panel by entering '= m1*e1'. Now suppose we have evaluated a number of states, each related to temperature at State-1, for a given working fluid. We can use T1 as a parameter in a "what-if scenario" by simply changing T1 to a new value and clicking Super-Calculate in the global control panel to update all calculated states. Likewise, to use the working substance as a parameter, simply select a new substance and click Super-Calculate.

Just as a state number stores the properties of a state, a **case** is used to preserve a set of states. The calculated states are automatically preserved as Case-0. To repeat a set of calculations with a different fluid, for instance, select a new case, say, Case-1, and then change the fluid type and click Super-Calculate. Selecting a calculated state or a calculated case (identified by an © sign in front of the state or case number), loads the state or the entire case. The Super-Calculate button also is used to convert the unit system of an entire solution, to produce a detailed output of the solution including spreadsheet-friendly tables in the I/O panel, and generating TEST-code (discussed later) for storing and reproducing the solution.

The flow-state TESTcalcs, which calculate extended flow states, are quite similar to the system-state TESTcalcs. As an exercise, launch the flow-state TESTcalc with the SL-model in a new browser window next to the system-state TESTcalc of Figure 1.12. The extensive total properties (marked in black) are replaced by the corresponding rate of transport: mass (m) by the mass flow rate (mdot); volume (Vol) by the volume flow rate (Voldot). However, all the intensive properties (marked in red, green, and blue colors) remain the same. Suppose we would like to evaluate the pipe diameter for a water flow identified by State-2 under the following conditions: p2 = 10 atm, T2 = 50 deg-C, Vel2 = 2 m/s, and Voldot2 = 100 gallons/hr. In the SL flow-state TESTcalc, select Water (L) from the working substance menu, select State-2, and enter the given properties. To change the default value of the velocity, you must de-select the Vel2 checkbox and then select it again to trigger the input mode. Click Calculate and the area is displayed as 0.5257 cm^2.

1.5.2 Properties Related to System Size (V, A, m, n, $\dot{m}$, $\dot{V}$, $\dot{n}$)

We begin the discussion of individual properties with extensive properties that are directly related to the extent or amount of matter in the system. While **volume** V and **mass** m are *extensive* properties of a *system state*, the *flow area* A, *volume flow rate* $\dot{V}$, and *mass flow rate* $\dot{m}$ are extensive properties of a *flow* state.

Although mass measures the quantity of matter in a system in kg (SI) or in lbm (English units), when we say a system is massive or heavy, we usually mean that the weight (the force with which the Earth pulls on the system) is large. Unlike weight, the mass of a closed system stays the same regardless of elevation. It is easier to intuitively understand force; hence, mass can be operationally defined by Newton's second law in terms of force, F and acceleration, a:

$$F = \frac{ma}{(1000 \text{ N/kN})}; \quad \left[\text{kN} = \text{kg} \cdot \frac{\text{m}}{\text{s}^2} \frac{\text{kN}}{\text{N}} \right] \tag{1.2}$$

The weight w of a mass, m, that is, the force with which the Earth pulls on an object of mass m, can be related to acceleration during free fall or the **acceleration due to gravity**, g, as:

$$w = \frac{mg}{(1000 \text{ N/kN})}; \quad \left[\text{kN} = \text{kg} \cdot \frac{\text{m}}{\text{s}^2} \frac{\text{kN}}{\text{N}} \right] \tag{1.3}$$

To prevent a free fall, an upward force, w, proportional to the mass of the body, must be exerted on the body. This gives us an appreciation for the mass of the body (Fig. 1.14). In space, mass can be viewed as attempting to alter the inertia of a body—the more massive the body, the greater the force that is necessary to affect the magnitude or direction of the body's velocity (think about how we can distinguish two identical looking baseballs that have different masses in the absence of gravity).

The acceleration due to gravity g can be regarded as a proportionality constant between the weight and mass of a system, which can be expressed by Newton's law of universal gravitation in the gravitational constant and the distance of the system from the Earth's center. The local value of g, therefore, is a function of the elevation, z, of a system. For example, at 45-degrees latitude and at sea level ($z = 0$), g has a value of 9.807 m/s². Most engineering problems involve only minor changes in elevation when compared with the Earth's radius, and a constant value of 9.81 m/s² for g, called the **standard gravity**, provides acceptable accuracy. In this book, we will assume standard gravity value for g unless mentioned otherwise.

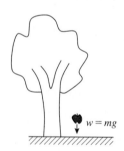

$w = mg$

FIGURE 1.14 Although equal and opposite forces act on the apple and Earth, the acceleration of Earth is negligible while that of the apple is g (see Anim. 0.A.*weight*).

DID YOU KNOW?

- The densest naturally occurring substance on Earth is Iridium, at about 22,650 kg/m³.
- Density of water is about 1000 kg/m³.
- Density of air at sea level is about 1.229 kg/m³.

Beside mass, the amount of matter in a system can be expressed by another extensive property called **mole**, which is the count of the smallest microscopic units that constitute a system (see Anim. 1.B.*massVsMole*). Often that smallest unit is a molecule for a pure substance (a substance with a uniform chemical composition), but sometimes it can be an atom (as in a pure metal) or even an electron or photon. To simplify the terminology, we will use the term *molecule* to describe the smallest discrete unit of a system. The symbol used to represent the mole of a system is *n*, and in SI units it is expressed in the unit of **kmol** (kilo-mol), that counts molecules just like the units dozen or century we use in our daily life. A mol is precisely 6.023×10^{23}, and is known as Avogadro's number, whereas a kmol is much larger: 1 kmol = 1000 mol (note the spelling difference—a *mole* is a property while *mol* or *kmol* is its unit). The unit of mole in English units is 2.737×10^{26}, which is called a *lbmol*.

Extensive properties per unit mass are called **specific properties** and they are usually expressed by a lowercase symbol. We have already encountered one (see Anim. 1.A. *specificProperty*) such property; the *specific volume*, which is the inverse of density, $v \equiv V/m$. When specific properties are expressed on a mole basis, volume-per-unit mole for example, they are called **molar specific properties** and are designated by a bar on top of their symbols. For example, in Table D-2, you will find several molar properties of hydrogen with units, such as kJ/kmol, kJ/(kmol·K), etc.

If the molecules comprising a system are identical, the ratio of the mass, *m*, to the mole *n*; that is, the mass-per-unit mole must be a constant for a given working substance. It is called the **molar mass** $\overline{M}$ (see Anim. 1.B.*massVsMole*):

$$\overline{M} \equiv \frac{m}{n} \left[\frac{\text{kg}}{\text{kmol}} \right]; \quad \text{therefore,} \quad n = \frac{m}{\overline{M}}; \quad \left[\text{kmol} = \text{kg} \frac{\text{kmol}}{\text{kg}} \right] \tag{1.4}$$

Oxygen has a molar mass of 32 kg/kmol, which means that the mass of 1 kmol of oxygen is 32 kg. Molar masses of several common substances are listed in Tables A-1, A-2, and C-1 (Fig. 1.15). Most state TESTcalcs display the molar mass once the working substance is selected. Although in chemistry molar mass is understood as a ratio—the relative mass of a molecule compared to the oxygen atom—in engineering thermodynamics, it is regarded as a material property of the working substance with a well-defined unit, kg/kmol. Even for a homogenous mixture, such as air, Eq. (1.4) can be used to define an apparent molar mass, which can be calculated as 29 kg/kmol for air (click Mixture in Anim. 1.B.*massVsMole*). If the molar mass of a substance is known, Eq. (1.4) can be used to convert a system's mass to mole and vice versa.

$\overline{M}_{H_2} = 2$ kg/kmol
$\overline{M}_{H_2O} = 18$ kg/kmol
$\overline{M}_{N_2} = 28$ kg/kmol
$\overline{M}_{O_2} = 32$ kg/kmol
$\overline{M}_{Air} = 28.97$ kg/kmol
$\overline{M}_{CO_2} = 44$ kg/kmol

FIGURE 1.15 Molar masses of a few common gases (see Anim. 1.B.*massVsMole*).

We have already encountered flow properties volume flow rate, $\dot{V}$, and mass flow rate, $\dot{m}$, (mdot and Voldot in flow-state TESTcalcs), expressed by Eqs. (0.1) and (0.2), while discussing mass interactions in Sec. 0.5. The amount of matter transported by a flow also can be expressed on a mole basis, which is called the **mole flow rate** represented by the symbol $\dot{n}$. An expression for $\dot{n}$ can be derived by converting the mass in the shaded region that passes through a given cross-section (Fig. 0.8) in a unit time into mole:

$$\dot{n} = \frac{\dot{m}}{\overline{M}} = \frac{\rho A V}{\overline{M}}; \quad \left[\frac{\text{kmol}}{\text{s}} = \frac{\text{kg}}{\text{s}} \frac{\text{kmol}}{\text{kg}} \right] \tag{1.5}$$

Although V, *m*, and *n* are relevant to a system state, $\dot{V}$, $\dot{m}$, and $\dot{n}$ are attributes of a flow. Merging two identical systems or flows would double such properties, hence, they are called extensive properties. Extensive properties of a system are also known as *total* properties, and represented by uppercase symbols, with the exception of total mass c and total mole (*n*). For a flow state, the corresponding properties are known as the *rate of transport*, symbolized by placing a dot on top of the property symbol. Although n or $\dot{n}$ do not explicitly appear as part of the TEST state panel, they can be calculated in the I/O panel using expressions such as '= m1/MM1' or '= mdot1/MM1'.

In Chapter 2, we will introduce the conservation of mass principle, a fundamental law of physics, in the form of a *mass balance equation* to track the inventory of mass for a general system. The mass equation will be used in almost all subsequent chapters as we analyze a variety of practical thermodynamic systems.

1.5.3 Density and Specific Volume (ρ, v)

Density, ρ, is a familiar property that expresses the concentration of matter, defined as the mass-per-unit volume of the working substance:

$$\rho \equiv \frac{m}{V}; \quad \left[\frac{\text{kg}}{\text{m}^3}\right] \tag{1.6}$$

For most solids and liquids, density does not vary much—for example, the density of water is about 1000 kg/m³ under atmospheric conditions and increases by only about 0.5% at 100 atm (at room temperature). A material is called **incompressible** if its density can be considered constant. While solids and liquids often are modeled as incompressible, gases and vapors, whose density can change significantly with changes in pressure and temperature, are considered *compressible* fluids.

A more convenient property for thermodynamic analysis is **specific volume**, v, which is defined as the volume of a substance per-unit mass, which is the reciprocal of density. The corresponding molar property is the **molar specific volume**, $\bar{v}$, defined as the volume-per-unit mole:

$$v \equiv \frac{V}{m} = \frac{1}{\rho}; \quad \left[\frac{\text{m}^3}{\text{kg}}\right]; \quad \bar{v} \equiv \frac{V}{n} = \frac{V\overline{M}}{m} = v\overline{M}; \quad \left[\frac{\text{m}^3}{\text{kmol}}\right] \tag{1.7}$$

Note the consistent use of the bar, even in the symbol for molar mass, to emphasize a mole-based property. While v and $\bar{v}$ are *specific properties* ρ is not. Both v and ρ, however, are independent of the system's size, and must be *intensive* properties (see Anim. 1.A.*intensiveProperty*). For a compressible substance (e.g., air), v and ρ also can be regarded as *thermodynamic* properties describing an equilibrium. For an incompressible substance (e.g., a copper block), however, they remain constant and can be regarded as *material* properties.

Another property that is related to density is **relative density**, ρ/ρ_{water} (water at some standard condition)—a dimensionless quantity that is also known as **specific gravity**. **Specific weight** is another related property that is defined as the weight of a unit volume [ρg from Eq. (1.3)] of material. The use of specific volume and density, however, are preferred in thermodynamics. The density of several solids and liquids are listed in Tables A-1 and A-2. For TESTcalcs based on the SL-model, the density (rho) and specific volume (v) fields are populated as soon as a working substance is selected.

EXAMPLE 1-1   Mass Versus Mole

Determine (a) the mass (in kg) and mole (in kmol) of a 1-m³ block of aluminum. *What-if scenario:* (b) What would the answers be if the block were made of iron? (See Fig. 1.16.)

SOLUTION

Obtain the density and molar mass of the working substance from Table A-1 or any SL-state TESTcalc and calculate the amounts of mass and mole.

Analysis

For aluminum (Al), Table A-1 lists ρ_{Al} = 2700 kg/m³ and $\overline{M}_{\text{Al}}$ = 27 kg/kmol. The block's mass is obtained from Eq. (1.6):

$$m_{\text{Al}} = \rho_{\text{Al}}V_{\text{Al}} = (2700)(1) = 2700 \text{ kg}$$

The mole of aluminum is found using Eq. (1.4):

$$n_{\text{Al}} = \frac{m_{\text{Al}}}{\overline{M}_{\text{Al}}} = \frac{2700}{27} = 100 \text{ kmol}$$

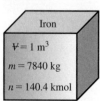

FIGURE 1.16 The iron block in Ex. 1-1 is heavier because its molecules (atoms) are heavier and more densely packed compared to aluminum.

TEST Analysis

Launch the SL system-state TESTcalc located in states, system states branch. Select Aluminum (Al) from the material selector, enter the volume with its appropriate units, and click Calculate (or press Enter). The mass m1 is calculated as 2700 kg. In the I/O panel, calculate the mole using the expression '= m1/MM1'; it is 100.07 kmol.

What-if scenario

Select Iron (Fe) from the working substance menu. When the working substance is changed, all properties are automatically updated. The new answers are 7840 kg and 140.38 kmol. Compared to aluminum, the same volume of iron has about 40.3% more molecules.

Discussion

The mass flow rate of Example 0-2 can, similarly, be converted to mole flow rate by applying Eq. (1.5). With $\overline{M}_{H_2O} = 18$ kg/kmol, the mole flow rate can be calculated as $\dot{n} = \dot{m}/\overline{M}_{H_2O} = 39.1/18 = 2.17$ kmol/s.

1.5.4 Velocity and Elevation (*V, z*)

The instantaneous **velocity**, *V*, of a uniform system, say, a projectile, describes the state of its motion and, therefore, is a state property. For non-uniform systems, the distribution of velocity among the local state is known as the *velocity field* in fluid mechanics. Even at a local level, the velocity is a *macroscopic* property of the continuum and should not be confused with the disorganized microscopic motion of molecules responsible for the temperature of the system.

A system's **elevation** or height, *z*, is the vertical distance of the system's center of gravity from an arbitrarily chosen horizontal level called the **datum**, ($z = 0$), which is standardized as zero at sea level. Velocity, *V*, elevation, *z*, and all other properties (such as KE or PE) that are directly based on *V* and *z* depend on the observer's frame of reference; hence, they are *extrinsic*. Note that changes in extrinsic properties do not necessarily reflect a change in equilibrium (see Anim. 1.A.*intrinsicProperty*).

1.5.5 Pressure (*p*)

Pressure, *p*, is the absolute compressive force exerted by a fluid per-unit-area on a surface in the perpendicular direction. For example, we feel pressure when we are underwater or when we are outside on a windy day. Pressure has the same unit as stress in mechanics—kN/m^2 or kPa (kilo-Pascal) in SI, and psi (lbf/in^2) in English units. Other common SI units are MPa (1 MPa = 1000 kPa), bar (1 bar = 100 kPa), and atm (1 atm = 101.325 kPa).

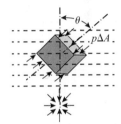

The compressive force due to pressure is not limited to physical walls. To appreciate the local pressure as a thermodynamic property, consider a small cubic element oriented at an angle θ to the vertical axis in a non-uniform system (Fig. 1.17). If ΔA is the area of each face, the compressive force ($\Delta F = p\Delta A$) on a face must be balanced by an equal and opposite force on the opposite face to keep the cube in mechanical equilibrium. Pressure at a point is the distribution of this compressive stress on the cube's surface as the cube's volume is tends to zero around that point. The force balance shown in Figure 1.17 is independent of the cube's orientation. As a result pressure at a point must be independent of direction.

FIGURE 1.17 Absolute pressure at a point is the thermodynamic pressure of the local state around that point (Anim. 1.B.*pressure*).

The pressure of atmospheric air varies around 101 kPa at sea level, but a **standard atmospheric pressure** is assumed to be 101.325 kPa, 14.696 psi, or 1 atm. To appreciate the magnitude of this pressure, consider that an unbalanced pressure of 100 kPa applied over an area of 1 m^2 can lift a large truck weighing more than 11.2 tons (10,194 kg) off the ground. Balanced by pressure inside our body, we cannot feel the surrounding pressure of atmosphere but we can detect slight changes when we climb a mountain or dive underwater. Similarly, it is easier for instruments to sense the difference between a system pressure, *p*, and the surrounding atmospheric pressure, p_0. Accordingly, the **gage** and **vacuum pressures** are defined (Fig. 1.18) as:

$$p_{\text{gage}} \equiv p - p_0 \quad \text{if} \quad p > p_0; \quad [\text{kPag}], \quad \text{and}$$

$$p_{\text{vac}} \equiv p_0 - p \quad \text{if} \quad p < p_0; \quad [\text{kPav}] \tag{1.8}$$

Gage and vacuum pressures, by definition, are always positive and have zero values at ambient condition. For clarity, the letter *g* or *v* is usually appended to their units (e.g., kPag or kPav). To distinguish from gage or vacuum pressures, pressure *p* is often called the **absolute pressure**, which has a zero value at absolute vacuum. When solving problems, it is a good practice to convert gage or vacuum pressure into absolute pressure.

In a stationary fluid, variation of pressure with depth is necessary to support the weight above each layer of fluid; this is commonly called **hydrostatic pressure** variation. A force

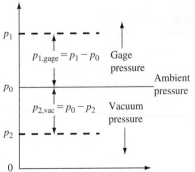

FIGURE 1.18 Gage and vacuum pressures depend on local ambient pressure.

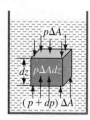

FIGURE 1.19 Vertical force balance on a local system produces the formula for the hydrostatic pressure variation, Eq. (1.10).

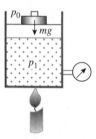

FIGURE 1.20 The pressure inside can be changed by changing the weight on the piston and/or by pinning the piston to the cylinder and heating the gas.

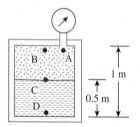

FIGURE 1.21 Schematic for Ex. 1-2.

DID YOU KNOW?

- For each 5.5 km rise in elevation, the atmospheric pressure halves.
- Lowest pressure ever created: 10^{-15} kPa.
- The highest and lowest atmospheric pressures ever recorded at sea level are 108.4 kPa and 92 kPa, respectively.
- Pressure inside a tire: 320 kPa.
- Pressure inside a soda can at 20°C: 250 kPa.
- Household water pressure: 350 kPa.
- Pressure of sunlight on Earth's surface: 3 μPa.
- Pressure at the Earth's center: 4 million atm.

balance on the local system of Figure 1.19 can be used to show that pressure cannot vary horizontally and that a differential change in the vertical direction can be expressed as:

$$dp = -\frac{\rho g \, dz}{(1000 \text{ N/kN})}; \quad \left[\text{kPa} = \frac{\text{kg}}{\text{m}^3} \frac{\text{m}}{\text{s}^2} \text{m} \frac{\text{kN}}{\text{N}} = \frac{\text{kN}}{\text{m}^2} \right] \qquad \textbf{(1.9)}$$

For incompressible fluids, Eq. (1.9) can be integrated to produce the pressure difference between any two points (local states) in a static medium:

$$\Delta p = p_2 - p_1 = \frac{\rho g (z_2 - z_1)}{(1000 \text{ N/kN})}; \quad [\text{kPa}] \qquad \textbf{(1.10)}$$

Hydrostatic pressure variation is a consequence of gravity, but pressure is a thermodynamic property that stems from momentum exchange between molecules and a wall (think of billiard balls bouncing off a wall) and does not rely on gravity for its existence. While the pressure in the piston-cylinder device of Figure 1.20 can be increased by putting additional weight on the piston, it also can be increased by simply heating the cylinder with the piston pinned to the wall, even in a gravity-free environment.

Often the hydrostatic pressure variation within a system can be considered negligible (in percentage terms) for systems with small variations in height or in systems in which the working fluid is a gas or a vapor (low-density fluids). The assumption of uniform pressure inside a system simplifies analysis as illustrated in Example 1-2.

EXAMPLE 1-2   Hydrostatic Pressure Variation

A tank of height 1 m holds equal volumes of liquid water and water vapor (Fig. 1.21). The gage pressure at the top of the tank is 200 kPag. (a) Determine the pressure at the vapor-liquid interface, and (b) evaluate the variation of pressure between the top and the bottom of the tank as a percentage of the measured pressure. Assume p_0 (ambient atmospheric pressure) to be 101 kPa, and the densities of the liquid and vapor phases to be 932 kg/m^3 and 1.655 kg/m^3, respectively.

SOLUTION

Label the points of interest as shown in Figure 1.21 and use the hydrostatic pressure variation formula, Eq. (1.10), to link the unknown pressure to ambient pressure, p_0.

Assumptions

Incompressible fluids. Pressure variation is hydrostatic.

Analysis

The Bourdon gage measures the gage pressure. Also A and B are at the same horizontal level. Therefore,

$$p_B = 200 + p_0 = 200 + 101 = 301 \text{ kPa} = p_A$$

We choose the bottom of the tank as the datum and apply Eq. (1.10) to obtain:

$$p_C = p_B + \frac{\rho_{\text{vap}} g (z_B - z_C)}{(1000 \text{ N/kN})}$$

$$= p_B + \frac{(1.655)(9.81)(1.0 - 0.5)}{(1000 \text{ N/kN})} = 301.01 \text{ kPa}$$

Similarly, $p_D = p_C + \dfrac{(932)(9.81)(0.5 - 0)}{(1000 \text{ N/kN})} = 305.58 \text{ kPa}$

The pressure variation, therefore, is calculated as:

$$\frac{p_D - p_A}{p_A} = \frac{305.578 - 301}{301} = 1.52\%$$

Discussion

Note that while properties, such as density, undergo discrete changes across an interface, pressure must be identical on the two sides of an interface to ensure that there is no net force (any net force will make the interface accelerate) on any surface element (a thin slice around the interface).

Let us now consider the pressure variation in a flow through a variable area duct (Fig. 1.22). Pressure may change along the flow to overcome wall friction or to adjust to a change in the flow cross section. Across a flow, fortunately, the pressure variation can be considered hydrostatic, and, therefore, negligible (Ex. 1-2) except for pipes of unusually large diameters.

Although hydrostatic pressure variation is either neglected or averaged out when assigning a value of pressure to a uniform system, it can be exploited to measure the pressure in a system or flow. For example, Figure 1.23 contains an apparatus, called an **open-tube manometer**, that utilizes hydrostatic pressure change in a column of water or mercury (Hg) to measure the gage pressure of a gas inside a tank. If the gas density is not too high, the hydrostatic pressure difference inside the tank can be neglected with the *thermodynamic* pressure of state 1 in the tank represented by a single pressure p_1. Through the convenient intermediate points (Fig. 1.23), p_1 can be related to p_0 using Eq. (1.10), as follows:

$$p_1 \cong p_A = p_B = p_C + \frac{\rho_{\text{liq}} g L}{(1000 \text{ N/kN})} = p_0 + \frac{\rho_{\text{liq}} g L}{(1000 \text{ N/kN})}; \quad [\text{kPa}] \qquad \textbf{(1.11)}$$

It is common to express the gage or vacuum pressure for the column length of the manometer's liquid. A pressure of *2 inches of water* or *10 mm vacuum of mercury* can be readily converted to absolute pressure in kPa using Eq. (1.11). Likewise, a blood pressure of 120/80 (the pressure is 120 mm Hg gage when the heart pushes blood out into the arteries, and 80 mm Hg gage when the heart relaxes between beats) can be expressed in kPa if the ambient pressure, p_0, is known:

A **barometer** (Fig. 1.24) is a manometer with a closed end that measures the absolute pressure by relating p_0 to the vapor pressure of the working fluid, usually mercury:

$$p_0 = p_A = p_B = p_C + \frac{\rho_{\text{Hg}} g L}{(1000 \text{ N/kN})} = p_{\text{vap}} + \frac{\rho_{\text{Hg}} g L}{(1000 \text{ N/kN})}; \quad [\text{kPa}]$$

The **Bourdon gage** (Fig. 1.25) is another commonly used pressure-measuring device. It is a bent hollow tube with an elliptical cross section that tends to straighten when subjected to a pressure. A gear-and-lever mechanism translates this bending into the movement of a pointer against a calibrated scale of gage pressure.

The manometer and the Bourdon gage are not suitable for measuring fluctuating pressures due to their slow response times. For transient measurements, a diaphragm-type pressure

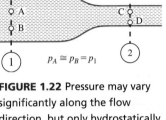

FIGURE 1.22 Pressure may vary significantly along the flow direction, but only hydrostatically across it. For small-diameter pipes, the hydrostatic variations can be neglected (see Anim. 1.B.*pressure*).

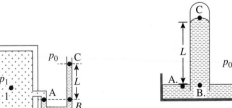

FIGURE 1.23 An open tube manometer. p_1 and p_0 are related by Eq. (1.11).

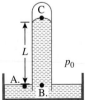

FIGURE 1.24 A barometer is a closed-top manometer. The vapor pressure of mercury can be obtained from a table if the temperature is known.

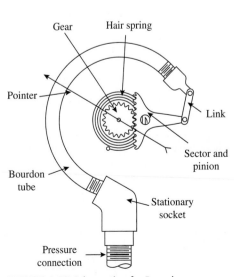

FIGURE 1.25 Schematic of a Bourdon gage.

transducer is used, which is sensitive and has a quick response time. Signals from such a transducer are digitized by an analog-to-digital (A/D) converter and read at high frequency by a computer.

EXAMPLE 1-3   Measuring Atmospheric Pressure

Determine the length of the water column in Figure 1.26 that is supported by a local atmospheric pressure of 100 kPa when the vapor pressure of water is 5 kPa. Assume the density of water to be 1000 kg/m^3 and standard gravity.

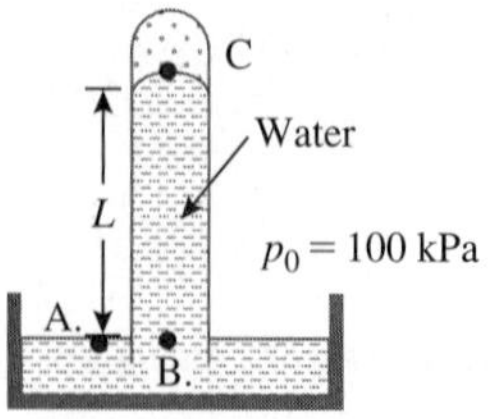

SOLUTION

Label the points of interest as shown in Figure 1.26, and use hydrostatic pressure variation to link the unknown pressure to the known pressure.

Assumptions

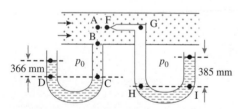

FIGURE 1.26 Schematic for Ex. 1-3

Density of water is constant.

Analysis

Pressures at A (atmospheric) and C (see Fig. 1.26) can be linked as follows:

$$p_0 = p_A = p_B = p_C + \frac{\rho_{liq}g(z_C - z_B)}{1000} = p_{vap} + \frac{\rho_{liq}gL}{1000}$$

Therefore,

$$L = \frac{(p_0 - p_{vap})(1000)}{\rho_{liq}g} = \frac{(100 - 5)(1000)}{(1000)(9.81)} \left[\frac{(kPa)(N/kN)}{(kg/m^3)(m/s^2)} \right] = 9.68 \text{ m}$$

Discussion

The vapor pressure of mercury is almost negligible at room temperature, and the standard atmospheric pressure can be shown to be equivalent to a column of 760 mm of mercury.

EXAMPLE 1-4   Measuring Pressure in a Flow

Carbon-dioxide (CO_2) gas at 30°C is flowing in a pipeline with diameter 0.1 m and at a velocity 50 m/s. Calculate the pressure read by the two mercury manometers (Fig. 1.27), when (a) one is connected to the wall and (b) the other is connected to what is called a *pitot tube* that points against the flow. The barometer reads 762 mm of Hg, g is 9.80 m/s^2, and $\rho_{Hg} = 13,640$ kg/m^3.

SOLUTION

Identify the principal points of interest as shown in Figure 1.27 and use hydrostatic pressure variation to link the unknown pressure to the known atmospheric pressure.

Assumptions

The density of CO_2 is assumed negligible when compared with the density of mercury.

Analysis

With reference to the principal points A–F shown in Figure 1.27:

$$p_A \cong^2 p_B \cong p_C = p_D = p_0 + \frac{\rho_{Hg}gL_1}{1000}$$

FIGURE 1.27 Schematic for Ex. 1-4.

$$= \frac{\rho_{Hg}gL_{atm}}{1000} + \frac{\rho_{Hg}gL_1}{1000} = \frac{(13640)(9.8)(0.762 + 0.366)}{1000} = 150.8 \text{ kPa}$$

---

[2]The symbol means almost equal.

At point F, the gas must be stationary as the manometer tube is effectively blocked by the stationary mercury column. The pressure at F, therefore, can be obtained using the hydrostatic formula, Eq. (1.10), as follows:

$$p_F \cong p_G \cong p_H = p_I = p_0 + \frac{\rho_{Hg}\, g\, L_2}{1000}$$

$$= \frac{(13640)(9.8)(0.762 + 0.385)}{1000} = 153.3 \text{ kPa}$$

Discussion

The local pressure at F is larger than the pressure anywhere else in the pipe because the gas is effectively brought to rest at F by the pitot tube. In fluid mechanics (discussed in Chapter 15), such a pressure is called the **stagnation pressure** to distinguish it from the pressure at the wall (point A), which is called the **static pressure**. The stagnation pressure is basically the thermodynamic pressure for the local state at F while the static pressure is the thermodynamic pressure of the local state at A. In the absence of the pitot tube, the flow is undisturbed and the static pressure represents the thermodynamic pressure of the flow state.

The ease with which pressure at a point can be measured makes it one of the two most commonly used independent properties in thermodynamics (the other being temperature). Absolute pressure is an essential property to describe thermodynamic equilibrium and is, therefore, a *thermodynamic* property.

1.5.6 Temperature (*T*)

Temperature, *T*, is a familiar property that conveys the degree of hotness or coldness of a system. Like pressure, temperature is an easily measurable thermodynamic property that helps define thermodynamic equilibrium. However, unlike pressure, which can be easily expressed in the fundamental quantities of force and area, temperature eludes a direct insight. For example, it is not easily understood what is meant by zero temperature in different scales, or, why the arbitrary marks on a mercury-in-glass thermometer can be an appropriate measure of temperature.

To establish temperature as a fundamental thermodynamic property, we have to first consider the concept of a thermal equilibrium. Consider Anim. 1.B.*thermalEquilibrium,* which shows two solid blocks, one warmer than the other, coming in thermal contact in an isolated enclosure. From our experience, we expect several properties—volume, electrical resistance, thermal conductivity, etc.—and, hence, the state of each subsystem to change due to *thermal interactions* between the two blocks. Eventually, the interactions subside and the states of the blocks show no further change; the subsystems are then said to be in **thermal equilibrium**. Without the benefit of daily life experience, however, we would not know that the two subsystems would be equally warm after thermal equilibrium is reached.

Now suppose system A is in thermal equilibrium with systems B and C simultaneously, as shown in Figure 1.28. Even though B and C are not in direct contact, we intuitively know from our experience that if we were to bring B and C together in thermal contact, there would be no change in their states. This fundamental truth cannot be proven. *If two systems are in thermal equilibrium with a third, then they are also in thermal equilibrium with each other*—this is called the **zeroth law** of thermodynamics because it logically precedes the first and second laws (to be introduced in Chapter 2). Anim. 1.B.*zerothLaw* contrasts thermal and chemical contact to underscore what is unique about the zeroth law of thermodynamics.

Starting with the zeroth law, mathematical arguments, which are outside the scope of this book, can be used to establish the existence of a new property that acts as an arbiter of thermal equilibrium—this new property is called *temperature*. If two systems have the same temperature, they must be in thermal equilibrium. The zeroth law can be summarized into the following postulates[3]:

i) Temperature is a thermodynamic property.
ii) If two systems are each in thermal equilibrium with a third body, all three must have the same temperature.

---

[3]A postulate is a statement that everyone accepts as true and does not require a proof.

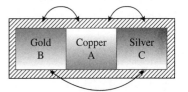

Thermal equilibrium

FIGURE 1.28 The zeroth law asserts that if system A is in thermal equilibrium with systems B and C, then B and C are also in thermal equilibrium with each other even though they are not in direct thermal contact (see Anim. 1.B.*zerothLaw*).

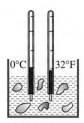

FIGURE 1.29 The length of mercury, a thermometric property, is an indirect measure of the temperature.

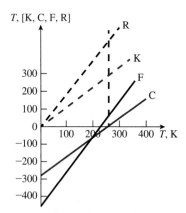

FIGURE 1.30 Relationship among different temperature scales in terms of absolute temperature in Kelvin.

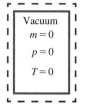

FIGURE 1.31 Absolute vacuum has zero absolute pressure, zero absolute temperature, and zero mass.

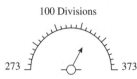

FIGURE 1.32 Schematic for Ex. 1-5.

The second postulate allows a way to compare temperatures of two systems at a distance. To determine if they have the same temperature, all we need to do is to construct a portable third body—a *thermometer*—and test if it is in thermal equilibrium with the other two systems.

Any property that responds to a thermal interaction is called a **thermometric property** and is a potential candidate for constructing a thermometer. For example, the expansion of mercury or alcohol is used in traditional *liquid-in-bulb* thermometers. *Gas thermometers* use the change of volume of a gas at constant pressure or the change of pressure of a gas at constant volume. Change in electric resistance is used in a *thermistor*, the flow of electricity due to the Seebeck effect in a *thermocouple*, and radiative emission in a radiation *pyrometer*.

Traditionally thermometers were calibrated using two readily reproducible temperatures at standard atmospheric pressure—the *ice point,* where ice and water can coexist (Fig. 1.29), and the *steam point,* where water and vapor can coexist in equilibrium. In the **Celsius scale**, for instance, the ice- and steam-point temperatures are arbitrarily marked as 0 and 100 degrees on a liquid-in-glass thermometer. The divisions in between depend on what fluid—mercury or alcohol—is used in the thermometer. Moreover, the marks between 0 and 100 may not necessarily be linear. Based on purely thermodynamic arguments (to be discussed in Chapter 2), Lord Kelvin developed an **absolute temperature** scale that does not depend on a specific thermometer. Known as the **Kelvin scale**, it has a minimum possible value of 0 K (note that the SI unit is K, not °K), is assumed linear, and, therefore, requires only one reference point for calibration. To be consistent with the Celsius scale in the "size" of the temperature unit (a temperature rise of 1°C is equivalent to a rise of 1 K), the triple point of water—a state where all three phases of water can coexist in equilibrium[4]—is assigned the value of 273.16 K. The Celsius scale assigns 0°C to the ice point (ice and water coexist in equilibrium at 273.15 K at standard atmospheric pressure). Hence the relationship between the two scales can be established as follows:

$$T_K = T_C + 273.15; \quad \text{and} \quad \Delta T_K = \Delta T_C \tag{1.12}$$

Rounding off the ice point and triple point of water to 273 K and the steam point to 373 K is an acceptable practice in engineering thermodynamics. Note that ΔT_K and ΔT_C are equivalent; therefore, Celsius and Kelvin are interchangeable for units that involve temperature differences. The specific heat (a property to be introduced in Chapter 3), for example, has the same value when expressed in kJ/(kg · °C) or kJ/(kg · K) because the temperature unit refers to a per-degree rise in temperature. Fundamental equilibrium relationships and thermodynamic laws (to be developed), conversely, involve absolute temperature. It is, therefore, a good practice to use absolute temperature (in K) in all calculations to avoid errors.

In English units, the Fahrenheit scale can be converted to an absolute scale called the Rankine scale ($T_R = T_F + 459.67 = (9/5)T_K$). The relationships among the four scales are graphically shown in Figure 1.30, where the ordinate assumes a variety of scales while the abscissa is marked in Kelvin. In this figure, the Celsius and Fahrenheit lines can be seen to intersect at −40 degrees (233.15 K). Although 0 K has never been achieved, its existence can be established by extrapolating measurements from a gas thermometer—in the limit of zero absolute pressure (pure vacuum), the absolute temperature approaches 0 K (Fig. 1.31).

EXAMPLE 1-5   **Non-Linear Behavior of a Thermometric Property**

The signal (e.m.f.) in mV produced by a thermocouple with its test junction at T K is given by $\varepsilon = -132 + 0.5T - 0.5 \times 10^{-4}T^2$. An engineer uses a millivoltmeter to measure the signal and calibrate it against the ice and steam points (273 K and 373 K, respectively). Unaware of the polynomial relationship, the engineer marks 100 equal intervals between the two reference points (Fig. 1.32). What will the thermometer read when the actual temperature is 323 K?

---

[4]Triple point will be discussed in Chapter 3.

SOLUTION

The signals produced at the two reference points, $T = 273$ K and $T = 373$ K are $\varepsilon = \varepsilon(273) = 0.774$ mV and $\varepsilon(373) = 47.54$ mV, respectively.

At $T = 323$ K, $\varepsilon(323) = -132 + 0.5 \times 323 - 0.5 \times 10^{-4} \times 323^2 = 24.28$ mV. The linear scale of the millivoltmeter will produce a reading of:

$$273 + \left[\frac{24.28 - 0.744}{47.54 - 0.774} \times 100 \right] = 323.26 \text{ K} \qquad \left[\frac{K}{mV}mV = K \right]$$

Discussion

A linear assumption—assuming that the change in output voltage is directly proportional to the change in temperature—produces only a slight error of 0.26 K in measuring an actual temperature of 323 K (less than 0.1).

1.5.7 Stored Energy (E, KE, PE, U, e, ke, pe, u, $\dot{E}$)

As part of our discussion of heat and work, we have already introduced stored energy E as the sum of internal energy, U, kinetic energy KE, and potential energy PE of a system:

$$E \equiv \underbrace{U}_{\text{Microscopic}} + \underbrace{\text{KE} + \text{PE}}_{\text{Macroscopic}}; \quad [\text{kJ}] \tag{1.13}$$

Whereas KE and PE are energy stored in the macroscopic organized motion and position of the system, U sums up the disorganized microscopic energies of the molecules. Energy is an extensive property stored in the mass of a system. Therefore several specific properties—**specific stored energy**, **specific internal energy**, **specific kinetic energy**, and **specific potential energy**—can be introduced, which are represented by the corresponding lowercase symbols:

$$e \equiv \frac{E}{m}; \quad u \equiv \frac{U}{m}; \quad \text{ke} \equiv \frac{\text{KE}}{m}; \quad \text{pe} \equiv \frac{\text{PE}}{m}; \quad \left[\frac{\text{kJ}}{\text{kg}} \right] \tag{1.14}$$

Let us start with kinetic energy. To obtain the familiar operational definition of KE, consider a net horizontal force, F, applied on a system at rest (Fig. 1.33). As mechanical work is transferred to the system, it accelerates in the x direction from 0 to V between the beginning and final locations, x_b and x_f, thereby converting work into kinetic energy. Application of Newton's law of motion produces:

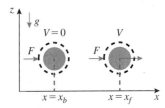

FIGURE 1.33 Work transferred in horizontal acceleration goes entirely into the KE (see Anim. 1.B.*ke*).

$$W_M = \int_{x_b}^{x_f} F dx = \int_{x_b}^{x_f} m \frac{dV}{dt} dx = \int_0^V m \frac{dx}{dt} dV = \int_0^V mV dV = \frac{mV^2}{2}; \quad \left[\frac{\text{kg.m}^2}{\text{s}^2} = \text{J} \right]$$

Therefore,

$$\text{KE} \equiv \frac{mV^2}{2(1000 \text{ J/kJ})}; \quad [\text{kJ}] \quad \text{and} \quad \text{ke} \equiv \frac{\text{KE}}{m} = \frac{V^2}{2(1000 \text{ J/kJ})}; \quad \left[\frac{\text{kJ}}{\text{kg}} \right] \tag{1.15}$$

Although potential energy has many different forms (depending on the force field), for most Earth-bound systems PE will symbolize the gravitational potential energy. To develop a formula for PE, consider a vertical external force that barely overcomes the system's weight (Fig. 1.34). With no net force to create acceleration, the work transferred into the system in raising it to an elevation z is completely stored in its potential energy:

$$W_M = \int_0^z F dz = \int_0^z w dz = \int_0^z mg dz = mgz \quad \left[\frac{\text{kg.m.m}}{\text{s}^2} = \text{J} \right];$$

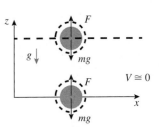

FIGURE 1.34 Work transferred in upward displacement is stored solely as PE (see Anim. 1.B.*pe*).

Therefore,

$$\text{PE} \equiv \frac{mgz}{(1000 \text{ J/kJ})}; \quad [\text{kJ}] \quad \text{and} \quad \text{pe} \equiv \frac{\text{PE}}{m} = \frac{gz}{(1000 \text{ J/kJ})}; \quad \left[\frac{\text{kJ}}{\text{kg}} \right] \tag{1.16}$$

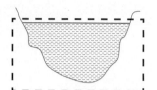

FIGURE 1.35 In the lake analogy (see Sec. 0.7 and Anim. 0.D.*lakeAnalogy*), change in water level is easier to measure than the absolute depth.

A constant value of g is an implicit assumption in these expressions.

As discussed in Sec. 0.7, absolute internal energy is almost impossible to quantify, just as the exact amount of water in the lake in our *lake analogy* is (Fig 1.35). Even though at 0 K the kinetic energy of molecules become zero, the lattice structure of a solidified working substance can store potential energy that is pressure dependent. Therefore, we cannot define an absolute zero for internal energy. For engineering purposes, however, it is sufficient to operationally define ΔU, the change in U between two states. Even this is made difficult due to the variability of the microscopic structure among materials, thus, denying a universal formula for ΔU for all working substances.

When a solid, liquid, or gas is heated, disorganized kinetic energy of the molecules, responsible for our sense of temperature (see Anim. 0.D.*uTranslationKE*), is the primary storage mechanism of internal energy. In this case, ΔU can be operationally related to ΔT—the relationship depends on the particular modes of molecular kinetic energy that are prominent. For most solids, liquids, and monatomic gases, U varies linearly with T.

The relationship between U and T becomes more complicated when a system undergoes a phase or chemical transformation. Attractive forces among molecules are larger in a solid, weaker in a liquid, and almost negligible in a gas because of the increasing separation among the molecules in these three phases. As a result, transformation from a denser phase is accompanied by a large increase in U due entirely to the increase in molecular PE while the molecular KE, and, hence, the temperature may remain unchanged. For instance, in a mixture of ice and water, U increases when ice melts due to heating even though the temperature may remain constant. When a change of state involves chemical reactions, molecular PE dominates as the molecular structures are rearranged. In Chapters 3, 11, and 13, we will develop formulas and, if necessary, charts and tables to relate ΔU with other measurable properties.

Combining contributions from its components, the specific stored energy (see Anim. 1.B.*storedEnergy*) can be expressed as follows:

$$e = u + \text{ke} + \text{pe}; \quad \text{and} \quad \Delta e = \Delta u + \Delta \text{ke} + \Delta \text{pe}; \quad \left[\frac{\text{kJ}}{\text{kg}} \right] \tag{1.17}$$

Although specific internal energy, u, is a thermodynamic property, ke, pe, and e are *extrinsic* because they depend on the observer's velocity and elevation.

Wherever mass flows, it transports e and all its components: ke, pe, and u. The rate of transport of different components of energy are expressed by the symbols $\dot{E}$, $\dot{\text{KE}}$, $\dot{\text{PE}}$, and $\dot{U}$. An expression for $\dot{E}$ can be obtained by revisiting the derivation of the $\dot{m}$ formula in Sec. 0.5 and using Figure 1.36:

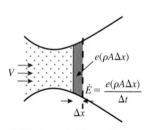

FIGURE 1.36 Schematic to derive Eq. (1.18). Also see Anim. 1.B.*transportEquation*.

$$\dot{E} = \lim_{\Delta t \to 0} \frac{e(\rho A \Delta x)}{\Delta t} = e \lim_{\Delta t \to 0} \frac{\rho A \Delta x}{\Delta t} = \dot{m} e; \quad \left[\frac{\text{kJ}}{\text{s}} = \text{kW} = \frac{\text{kg}}{\text{s}} \frac{\text{kJ}}{\text{kg}} \right] \tag{1.18}$$

The same template can be used to show that $\dot{\text{KE}} = \dot{m}(\text{ke})$ and $\dot{\text{PE}} = \dot{m}(\text{pe})$. Expressed in terms of a generic extensive property, $B = mb$, where b is the specific (per-unit mass) property, $\dot{B}$ is the rate of transport of that property and can be expressed as

$$\dot{B} = \dot{m} b; \quad \left[\frac{\text{Unit of } B}{\text{s}} \right] \tag{1.19}$$

Equation (1.19) is known as the **transport equation**, which will be revisited as part of a general discussion of the extended equations of state in Sec. 1.7.

State panels in all TESTcalcs initialize V (Vel) and z to zero so that ke and pe are neglected by default. In a flow-state TESTcalc, if a mass flow rate or volume flow rate is entered, the TESTcalc calculates a ridiculously large (its version of infinity) flow area using Eq. (0.1) or (0.2). If a non-zero value for V is input, a more realistic value of A is calculated. Also, ke can be calculated in the I/O panel with the expression '= e1 − u1' or '= Vel1^2/2000'. Similarly, $\dot{E}$ can be calculated by inputting '= mdot1*e1' in the I/O panel, once State-1 has been found.

EXAMPLE 1-6 Numerical Evaluation of Δe

Determine the change in specific stored energy in a block of copper due solely to (a) an increase in velocity from 0 to 10 m/s, (b) an increase in elevation by 10 m, and (c) an increase in temperature from 25°C to 35°C. For (c), use the SL system-state TESTcalc.

SOLUTION

Apply Eq. (1.17) to obtain the change in stored energy due to changes in the system through three different means.

Assumptions

The system is uniform so that two system states, State-1 and State-2, describe the initial and final states.

Analysis

The change in stored energy due to a change in velocity is given by:

$$\Delta e = \Delta u^0 + \Delta ke + \Delta pe^0 = ke_2 - ke_1$$

$$= \frac{V_2^2 - V_1^2}{2000} = \frac{10^2 - 0}{2000} = 0.05 \text{ kJ/kg}$$

Similarly, Δe due to a change in elevation is:

$$\Delta e = \Delta u^0 + \Delta ke^0 + \Delta pe = pe_2 - pe_1$$

$$= \frac{gz_2 - gz_1}{1000} = (9.81)\frac{10 - 0}{1000} = 0.098 \text{ kJ/kg}$$

TEST Analysis

Launch the SL system-state TESTcalc. Select Copper from the working substance menu, select State-1, and enter T1 = 25 deg-C. Properties Vel1 and z1 are set to zero by default. Calculate the state. Now evaluate State-2 with T2 as 35 deg-C. In the I/O panel, evaluate '= e2 − e1' (3.86 kJ/kg). To verify the results of the manual solution in parts a and b, calculate State-3 with u3 = u1 and Vel3 = 10 m/s. In the I/O panel, calculate '= e3 − e1'. Similarly, calculate State-4 with u4 = u1, z4 = 10 m, and evaluate '= e4 − e1'.

Discussion

The change in kinetic or potential energy does not depend on the nature of the working substance. This can be verified by changing the working substance to, say, aluminum and repeating the TEST solution process.

EXAMPLE 1-7 Heat Transfer and Stored Energy

Assuming that 20% of the heat released from gasoline goes into increasing the kinetic energy of a vehicle, determine the amount of fuel consumption for a 10,000 kg truck to accelerate from 0 to 70 mph. Assume the heating value of gasoline (heat released by 1 kg of fuel; see Anim. 0.D.heatingValue) is 40 MJ/kg.

SOLUTION

Equating 20% of the heat released to the change in kinetic energy of the truck, we obtain:

$$m_F(40,000)(0.2) = \frac{(10,000)(V_2^2 - V_1^2)}{2000} \text{ [kJ]}$$

$$\Rightarrow \quad m_F = \frac{(10,000)(70 \times 0.447)^2}{(40,000)(0.2)(2000)} = 0.612 \text{ kg}$$

Discussion

Generally, only about 40% of heat released in combustion is converted into shaft power; the rest is rejected into the atmosphere via the exhaust and radiator. The shaft power primarily goes into overcoming different types of frictional resistances, aerodynamic drag (see Example 0-4), and acceleration.

EXAMPLE 1-8 **Transport Rate of Kinetic Energy**

A pipe with a diameter of 10 cm transports water at a velocity of 5 m/s. Determine (a) the rate of transport of KE in kW. *What-if-scenario:* (b) What would the answer be if the velocity were 10 m/s (Fig. 1.37)?

SOLUTION

Apply the transport equation extensive properties, Eq. (1.19), to kinetic energy.

Assumptions

A single flow state, State-1, represents the uniform flow in LTE at a given cross-section.

Analysis

Using the $\dot{m}$ calculated in Example 0-2, we obtain:

$$\dot{\text{KE}}_1 = \dot{m}_1(ke_1) = \frac{\dot{m}_1 V_1^2}{2(1000 \text{ J/kJ})} = (39.15)\frac{5^2}{2000} = 0.489 \text{ kW} \quad \left[\frac{\text{kg}}{\text{s}}\frac{\text{kJ}}{\text{kg}} = \text{kW}\right]$$

TEST Analysis

Launch the SL flow-state TESTcalc. Follow the procedure described in Example 0-2 to calculate the state and evaluate the expression '= mdot1*Vel1^2/2000' in the I/O panel to verify the answer.

What-if scenario

Calculate State-2 with p2, Vel2, and A2 = A1. Obtain the transport rate of kinetic energy in the I/O panel (3.915 kW).

Discussion

Although kinetic energy is proportional to the square of velocity, its rate of transport is proportional to the cube of the flow velocity. This explains the cubic power law of a wind turbine (Fig. 1.38), which converts kinetic energy transported by wind into mechanical or electrical power. The conversion efficiency is called the *coefficient of performance,* which has a maximum theoretical limit of 59% (called the **Betz limit**).

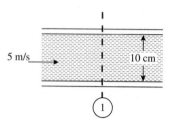

FIGURE 1.37 Illustration for Ex. 1-8.

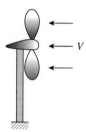

FIGURE 1.38 Power generated by a wind turbine is proportional to the wind speed's cube.

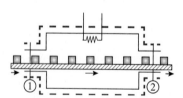

FIGURE 1.39 The blocks transport stored energy into and out of an electric oven.

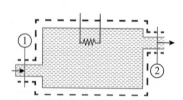

FIGURE 1.40 Water transports not only stored energy (such as the blocks in Fig. 1.39), but also transfers energy by performing considerable flow work in pushing the flow in and out at the ports (see Anim. 1.B.*transportEquation*).

1.5.8 Flow Energy and Enthalpy (j, $\dot{J}$, h, $\dot{H}$)

Consider the electric oven shown in Figure 1.39, where identical metal blocks transported at a rate $\dot{m}$ (equal to the number of blocks conveyed every second times the mass of each block) are heat-treated by exposing them to a high-temperature environment. Suppose we are interested in knowing how much energy is gained by the blocks as they pass through the oven. An indirect way to do so is to measure the electrical power consumption. A direct way is to use the transport equation. If the states of the blocks at the inlet and exit ports (State-1 and State-2) are known, the transport equation produces $\dot{E}_1 = \dot{m}e_1$ and $\dot{E}_2 = \dot{m}e_2$. The energy transfer to the blocks, therefore, must be $\dot{E}_2 - \dot{E}_1 = \dot{m}(e_2 - e_1)$.

A parallel situation that involves a fluid is shown in Figure 1.40, where a flow of water is electrically heated. Unlike the metal blocks, the energy transported by water must take into account not only the transport of stored energy, $\dot{E}$, but also a significant *flow work* transfer required to force the flow through the port. The contributions from these two components, expressed by Eqs. (1.18) and (0.18), are added to produce the total rate of energy transport (Fig. 1.41) by a flow at a given port as:

$$\dot{E} + \dot{W}_F = \dot{m}e + pAV = \dot{m}e + pv\frac{AV}{v} = \dot{m}(e + pv); \quad [\text{kW}] \qquad \textbf{(1.20)}$$

Equation (1.20) is applicable at either port of Figure 1.40 and $\dot{W}_F$ should be interpreted as the flow work transfer in the direction of the flow at a given port [not the net flow work transfer given by Eq. (0.20)]. What is interesting about this equation is that its right-hand side fits the format of the transport equation for the combination property $e + pv$. By treating it as a new property represented by the symbol j, the right side of Eq. (1.20) can be interpreted as the rate of transport of this new property j. The left hand side, therefore, must be represented by the symbol $\dot{J}$ according to the transport equation. Equation (1.20) therefore, can be rewritten as:

$$\dot{J} \equiv \dot{E} + \dot{W}_F = \dot{m}j; \quad [\text{kW}] \quad \text{where} \quad j \equiv e + pv = u + pv + \text{ke} + \text{pe}; \quad \left[\frac{\text{kJ}}{\text{kg}}\right] \qquad \textbf{(1.21)}$$

As evident from its unit, j is a specific property and is called the **specific flow energy**. $\dot{J}$ represents the rate of transport of flow energy (see Figure 1.41), bundling the stored energy transported by the flow with the flow work performed to sustain the flow. From Eq. (1.21), $\dot{W}_F$ can be expressed in local properties of the flow state:

$$\dot{W}_F = pAV = \dot{m}(j - e); \quad [\text{kW}] \qquad \textbf{(1.22)}$$

Using $\dot{J}$ to replace the summation of $\dot{E}$ and $\dot{W}_F$ offers some analytical advantages. As it already accounts for the flow work, only the external work transfer needs to be considered in an energy analysis of any open device. In a turbine analysis, for example, we can treat the invisible flow work as "internal work" at the ports and identify shaft work as the only relevant external work interaction as long as the energy transported by the flow is represented by $\dot{J}_i$ and $\dot{J}_e$ at the inlet and exit, respectively.

Moreover, in many practical flows, kinetic and potential energies are negligible and j, which is an extrinsic specific property, reduces to a more convenient combination property: $u + pv$. This combination property must be a thermodynamic property because p, v, and u are. It is called **specific enthalpy** and is represented by the symbol h. For h, the specific flow energy can be expressed as:

$$j = h + \text{ke} + \text{pe}; \quad \text{where} \quad h \equiv u + pv; \quad \left[\frac{\text{kJ}}{\text{kg}}\right] \qquad \textbf{(1.23)}$$

Specific enthalpy h, therefore, can be interpreted as the specific flow energy when contribution from ke and pe is negligible. Likewise, the energy transported by a flow $\dot{J}$ can be approximated by $\dot{H} = \dot{m}h$—the enthalpy transported by a flow.

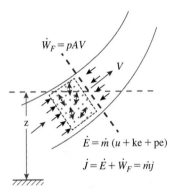

FIGURE 1.41 Flow energy combines stored energy and flow work into a single term (see Anim. 1.B.*flowEnergy*).

EXAMPLE 1-9 | Transport of Energy by Mass

Water flows through the system shown in Figure 1.42. The following data is given for the inlet and exit states. State-1: $\dot{m}_1 = 1000$ kg/min, $p_1 = 100$ kPa, $V_1 = 10$ m/s, and $z_1 = 5$ m; State-2: $\dot{m}_2 = 1000$ kg/min, $p_2 = 1.1$ MPa, $V_2 = 20$ m/s, and $z_2 = 15$ m. Assuming the internal energy u remains unchanged along the flow, determine the change in the flow energy transported by the flow between the exit and inlet.

SOLUTION

Apply the transport equation for flow energy, Eq. (1.21) at the inlet (State-1) and exit (State-2) to find $\dot{J}_1$ and $\dot{J}_2$, respectively, and calculate the difference between them.

Assumptions

Uniform flow states based on LTE at the inlet and exit. The density of water remains constant at 1000 kg/m^3.

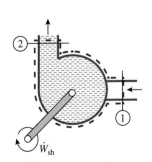

FIGURE 1.42 Schematic for Ex. 1-9 (see Anim. 4.A. *pump*).

Analysis

$$\dot{J}_2 - \dot{J}_1 = \dot{m}\Delta j = \dot{m}\Delta(u + pv + \text{ke} + \text{pe}) = \dot{m}[\Delta u^{0} + \Delta(pv) + \Delta\text{ke} + \Delta\text{pe}]$$

$$= \dot{m}\left[\left(\frac{p_2}{\rho_2} - \frac{p_1}{\rho_1}\right) + \frac{1}{2000}(V_2^2 - V_1^2) + \frac{g}{1000}(z_2 - z_1)\right]$$

$$= \left(\frac{1000}{60}\right)\left[\left(\frac{1100 - 100}{1000}\right) + \frac{(20^2 - 10^2)}{2000} + \frac{9.81 \times (15 - 5)}{1000}\right]$$

$$= 16.67 + 2.5 + 1.63 = 20.805 \text{ kW}$$

TEST Analysis

Launch the SL flow-state TESTcalc. Enter the known values of mdot1, vel1, p1 for State-1. (j1 is still an unknown.) Enter an arbitrary temperature, say, T1 = 25 deg-C. For State-2, enter mdot2 as '= mdot1', p2, Vel2, z2, and u2 as '= u1'. In the I/O panel, evaluate the expression '= mdot1*(j2 − j1)' to verify the answer obtained manually.

Discussion

To show that TEST results are independent of the temperature of State-1, enter a different value for T1 and click Super-Calculate. This updates all the calculated states. The same answer can be re-calculated in the I/O panel.

1.5.9 Entropy (*S*, *s*, *Ṡ*)

The concept of entropy as a property arises as a consequence of the second law of thermodynamics, just as temperature and internal energy owe their theoretical foundation to the zeroth and first law. Like temperature and internal energy that can be intuitively understood without referencing any fundamental laws, the entropy of a system can be introduced as a thermodynamic property without first discussing the second law, which will be introduced in the next chapter.

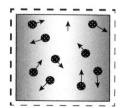

FIGURE 1.43 Entropy is a measure of molecular disorder associated with the distribution of the internal energy in a system (see Anim. 1.B.*entropy*).

Entropy, *S*, is a measure of molecular disorder (Fig. 1.43)—it is proportional to the logarithm of the number of ways in which the internal energy is distributed in the microscopic particles (molecules) of a system. We intuitively understand disorder at macroscopic scales—even a child can distinguish an orderly system from a chaotic one. When energy is stored in a system as kinetic or potential energy, all molecules participate in an organized manner, sharing the same velocity or elevation of the system. However, internal energy is stored by molecules in a disorganized manner with each molecule capable of storing a different amount of energy. Furthermore, because energy is quantized at microscopic scales, a molecule can store energy in a large number of discrete levels (Fig. 1.44) with contribution from translation, rotation, vibration, electric configuration, and many other modes (see Anim. 0.D.*uGasMoleculeKE*). Given the huge number of molecules in a system and the large number of discrete energy levels in each molecule (see Fig. 1.44), a system obviously has an extremely large, but finite, number of ways, say Ω, to distribute its total internal energy. Like the internal energy, entropy is an extensive property in its own right, which will be more thoroughly explored in Sec. 2.1.3. The **specific entropy** (entropy per unit mass) is represented by the lowercase symbol *s* and is a *thermodynamic* property.

An important differential relation known as the first *Tds* (pronounced T-d-s and discussed in detail in Chapters 3 and 10) relation, $Tds = du + pdv$, relates a differential increase in entropy (*ds*) to differential changes in *u* and *v* as a system moves from one equilibrium state to a neighboring equilibrium state (for any reason). An increase in the internal energy of a system increases the average share of energy stored by each molecule. The energized molecules also have more choices of energy levels (see Fig. 1.44). Therefore, a differential increase in the specific internal energy, *du*, causes a differential increase in specific entropy (by an amount *ds*). An increase in specific volume, *dv*, increases the average distance between the molecules, allowing more

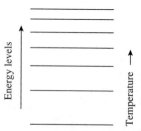

FIGURE 1.44 The ladder-like quantized energy levels in a molecule. The ladder grows taller and the steps (energy levels) become denser as temperature increases.

random distribution of molecular energy and hence entropy. The *Tds* relation also establishes the unit of specific entropy as kJ/(kg · K).

Like any other specific property, *s* can be related to the total entropy of a uniform system or the rate of entropy transport by a uniform flow (see Anim. 1.B.*entropyTransport*) as follows:

$$S = ms; \quad \left[kg \frac{kJ}{kg \cdot K} = \frac{kJ}{K} \right] \quad and \quad \dot{S} = \dot{m}s; \quad \left[\frac{kg}{s} \frac{kJ}{kg \cdot K} = \frac{kW}{K} \right] \quad \textbf{(1.24)}$$

Mass is not the only entity that can transport entropy; heat also can transport entropy as heat transfer involves chaotic transfer of energy through random molecular collisions (radiative heat transfer involves collisions of photons). Unlike energy, entropy cannot be transferred by work, which involves only organized motion of molecules.

To appreciate the significance of this abstract property, which after all, cannot be directly measured, consider several seemingly unrelated facts. First, there seems to be an asymmetry in nature regarding gradient-driven phenomena. Heat flows across a temperature drop, which reduces the temperature gradient. Electricity flows across a voltage drop, which reduces the potential difference. Viscous friction tends to destroy the velocity gradient, a gas expands to equalize a pressure difference; ink diffuses in clear water and diminishes any concentration gradient, chemical transformation occurs when there is a gradient of chemical potential—these events are directional and their opposites never occur naturally. The sole purpose behind these phenomena seems to be the destruction of the very gradients that drive these directed events. Second, a bouncing ball eventually comes to rest; the opposite never happens. Third, work can be completely converted into heat but not vice versa. Fourth, an *isolated system* moves toward a state of equilibrium. Fifth, fuel burns in air, but the combustion products do not spontaneously turn into fuel and air. And last, certain events are ordered which give us the sense of the passage of time. These apparently unrelated observations can all be explained and analyzed through an understanding of entropy and the second law of thermodynamics, which will be introduced in Chapter 2.

EXAMPLE 1-10   Numerical Exploration of *u* and *s*

A block of copper is heated from 25 °C to 1000 °C at constant pressure. Using the SL system-state TESTcalc, (a) determine Δu and Δs, and (b) plot how *u* and *s* vary with temperature when the pressure remains constant at 100 kPa. **What-if-scenario:** (c) How would the plots change if the block were kept in a pressurized chamber at 1 MPa?

TEST Analysis

Launch the SL system-state TESTcalc. Select Copper from the drop-down menu. Calculate the two states with p1 = 100 kPa, T1 = 25 deg-C, and p2 = p1, T2 = 1000 deg-C. Select *u-T* and *s-T* from the drop-down plot menu. In the plot window, click the p = c button to draw a constant-pressure line, such as the one shown in Figure 1.45. You can zoom in or zoom out, drag the plot around, and make other modifications to the plot to make it more useful.

What-if scenario

Change p1 to 1000 kPa, press Enter to register the change, and click Super-Calculate. The states are now updated. Select an appropriate plot option from the plot menu to visualize the states.

Discussion

Observe that for solids *u* and *s* are independent of pressure. While *u* increases linearly with temperature, entropy tapers off logarithmically (Fig. 1.45). Try any other solid or liquid (simply select a new substance and click Super-Calculate), and you will find that the trends are similar irrespective of the choice of liquid or solid.

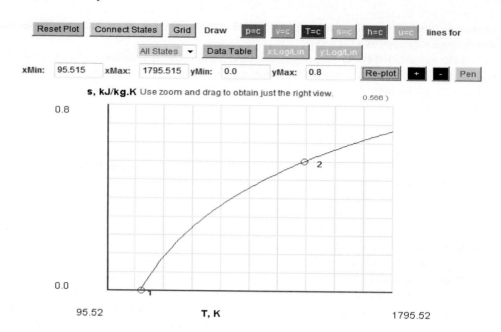

FIGURE 1.45 Logarithmic dependence of entropy on temperature for a solid copper block.

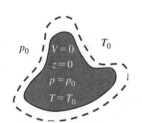

FIGURE 1.46 When a system is in equilibrium with its surroundings and has zero KE and PE, it is in its dead state (see Anim. 1.B.*storedExergy*).

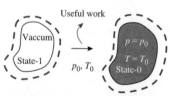

FIGURE 1.47 Whenever a system is not in its dead state, clever engineers can extract useful work. The upper limit of this useful work is the stored exergy Φ (see Anim. 1.B.*storedExergy*).

1.5.10 Exergy (ϕ, ψ)

Stagnant air and wind both have energy, yet it is much easier to extract useful work out of wind than stagnant air. Heat released from gasoline at high-temperature powers an automobile engine, yet the heat lost from the radiator is all but useless. All sources of energy, clearly, are not equally useful (see Anim. 1.B.*exergyAndKE,PE,U*). One of the major quests for engineers at all times has been delivery of useful work in the form of shaft or electrical power out of any source of available energy: wind, ocean waves, river streams, geothermal reserves, solar radiation, fossil fuels, and nuclear materials, to name a few. The amount of useful work that can be delivered from these sources is limited not only by practical difficulties, but also by the fundamental laws of thermodynamics. The maximum possible useful energy that can be theoretically delivered from a system is called its **stored exergy** (see Anim. 1.B.*storedExergy*). Likewise, the corresponding useful power that can be delivered from a flow is called **flow exergy** (see Anim. 1.B.*flowExergy*). Sometimes the term availability is used instead of exergy; however, in this book, we will use the term exergy in a consistent manner.

As expressed by Eq. (1.13), stored energy in a system has three components: KE, PE, and U. While the entire share of a system's kinetic and potential energies are readily convertible to exergy, the same is not true for internal energy only a fraction of which, if at all, can be converted to useful work. A calm ocean or an atmosphere without a wind may have a tremendous amount of internal energy by virtue of their huge mass alone, but disproportionately little exergy; that is why a ship or an airplane cannot extract any useful work out of these tremendous reservoirs of energy. For that matter, any stationary system at sea level, which is in equilibrium with atmospheric air, is said to be at its **dead state** with zero exergy (Fig. 1.46). When all the stored exergy of a system or the flow exergy transported by a flow is extracted, the system or the flow reaches its dead state.

Stored exergy, obviously, has the same unit as energy (kJ) and the rate of transport of exergy has the same unit as the rate of transport of energy (kW). Specific stored exergy is represented by the symbol ϕ (useful part of specific stored energy e) and the specific flow exergy by ψ (useful part of specific flow energy j). Like e and j, they are extrinsic properties as they depend on the system elevation and velocity. The total *stored exergy* and the *rate of transport of flow exergy* can be expressed in familiar formats:

$$\Phi = m\phi; \quad [\text{kJ}], \quad \text{and} \quad \dot{\Psi} = \dot{m}\psi; \quad [\text{kW}] \tag{1.25}$$

Exergy is easier to understand than energy. For example, if the stored exergy in a battery is known, we can calculate exactly how long it can power a given device, but the absolute value of stored energy E in the battery has no meaning since it depends on an arbitrarily selected datum for internal energy. The system in Figure 1.47 consists of a perfect vacuum and has zero energy, but it does not require much ingenuity to construct a device that can produce useful work when atmospheric air is allowed to rush in to fill the vacuum. The upper limit of that useful work is the stored exergy of the system. Similarly, exergy transported by a car exhaust, $\dot{\Psi}$, can tell us the upper limit of useful work that can be extracted from the exhaust stream while calculation of the rate of energy transport, $\dot{J}$, is arbitrary. In daily life, when we talk about energy, we often mean exergy—energy prices, energy crisis, and alternative energy are examples of phrases in which the word exergy would be scientifically proper.

Exergy is generally delivered through shaft or electrical work. Knowing the price of electricity tells us the price of any useful work. Another way to look at exergy is to treat it as the *quality of energy*. Exergy content of heat, for example, will be shown in Chapter 6 to be dependent on the source temperature—the higher the temperature, the higher the quality of heat. Stored exergy in a system tells us the maximum limit of how much energy can be delivered as useful work (electricity).

The stored exergy of a system depends on both its current state and its dead state. Hence the calculation of exergy requires evaluation of the dead state first. In TESTcalcs, State-0 is the designated dead state and must be evaluated at the pressure and temperature of the surrounding atmosphere. Only then can phi and psi of a given state be evaluated as extrinsic (green) properties. The following example numerically illustrates the difference between stored energy and stored exergy of a system.

EXAMPLE 1-11   Calculation of ΔE and $\Delta \Phi$ Using TEST

A granite rock of mass 1000 kg is sitting atop a hill at an elevation of 500 m. Solar radiation heats the rock from an initial temperature of 25°C to a final temperature of 80°C. Determine the change of (a) stored energy and (b) stored exergy, if the ambient conditions are 100 kPa and 25°C. Use the SL system-state TESTcalc.

TEST Analysis

Launch the SL system-state TESTcalc. Select Granite as the working substance. Calculate State-0 as the dead state with T0 = 25 deg-C and p0 = 100 kPa. Evaluate State-1 with T1 = T0, p1 = p0, m1 = 1000 kg, z1 = 500 m, and State-2 with T2 = 80 deg-C, p2 = p0, m2 = m1, z2 = z1. In the I/O panel, evaluate '= m1*(e2 − e1)' (43,450 kJ) and '= m1*(phi2 − phi1)' (3,574 kJ). (If you know the price of 1 kWh of electricity—as an engineer you ought to—you can assign a monetary value to the rock as a source of useful energy.)

Discussion

How much of the exergy gain comes from heating? You can eliminate the contribution of potential energy by simply setting z1 to zero and updating all states by clicking Super-Calculate.

1.6 PROPERTY CLASSIFICATION

We have already categorized most of the properties introduced in the last section. A tree diagram organizing different groups of properties is shown in Figure 1.48 and illustrated in Anim. 1.A.*propertiesClassified*.

To summarize, properties of an extended state (see Anim. 1.A.*propertyGroups*) can be divided into two broad categories: *extensive* and *intensive*. Those which depend on the extent of a system are *extensive* properties. When two identical systems or two identical flows are merged (Figs. 1.49 and 1.50), these properties double. *Total properties* such as $\dot{V}$, m, n, E, S, etc., of a system state, and *transport rates* $\dot{V}$, $\dot{m}$, $\dot{J}$, etc., of a flow state are the sub categories of extensive properties. With the exception of mass and mole (and their transport rates), extensive properties are represented by uppercase symbols. When total properties are divided by mass or transport rates are divided by mass flow rate, the resulting properties—v, e, u, j, h, ke, pe, etc., represented by lowercase symbols—are called *specific properties*. Properties that are independent of the extent of a system are called *intensive properties*. When two identical systems are merged, intensive properties remain unchanged. Properties T, p, ρ, V, z, $\overline{M}$, and all the *specific properties* are examples of intensive properties (see Anim. 1.A.*specificProperty* and *intensiveProperty*). Note that intensive properties are generally represented by lowercase symbols, although T, V, and $\overline{M}$ are exceptions.

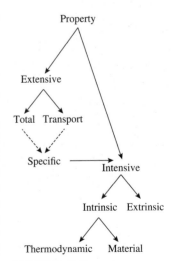

FIGURE 1.48 Property classification (click different nodes in Anim. 1.A.*propertiesClassified*).

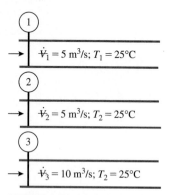

FIGURE 1.49 When two identical flows merge, the rate of transport of extensive properties doubles (click Flow State in Anim. 1.A.*intensiveProperty*).

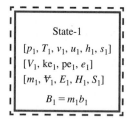

FIGURE 1.50 Thermodynamic, extrinsic, and extensive (total) properties of an extended system state (see Anim. 1.A.*propertyGroups*).

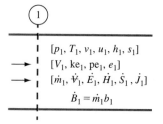

FIGURE 1.51 Thermodynamic, extrinsic, and system properties of a flow state (click Flow State in 1.A.*propertyGroups*).

Intensive properties such as V, z, ke, and pe are called *extrinsic properties* because their values depend on the observer (external factors) and are not intrinsic to the system. Any combination property that contains one or more extrinsic components—e, j, ϕ, ψ, etc.—is also extrinsic. *Intrinsic properties* (see Anim. 1.A.*intrinsicProperty*) such as $\overline{M}$, p, T, ρ, v, u, h, s, etc., are internal to a system and represent the equilibrium state of a system. An observer stationed inside a system can measure all its intrinsic properties while being oblivious of the surroundings. Some of the intrinsic properties, such as the molar mass $\overline{M}$, depend solely on the material composition of the system, and are constant for a given working substance. They are called *material properties*. The rest of the intrinsic properties—p, T, v, ρ, u, h, and s—are called *thermodynamic properties*, which constitute the *thermodynamic* state, the backbone of an extended state that defines its underlying equilibrium (click Thermodynamic State in Anim. 1.A.*propertyGroups*).

1.7 EVALUATION OF EXTENDED STATE

The relationship between the core thermodynamic state and an extended state is illustrated in Anim. 1.A.*propertyGroups* for a system as well as for a flow (Fig. 1.51). Once the thermodynamic state of a system or a flow is evaluated, extrinsic properties can be obtained from the knowledge of the velocity and elevation of the system or flow. Any extensive property can be obtained from the specific properties if the mass or the mass flow rate is known, completing a state evaluation. Evaluation of an extended state, therefore, reduces to evaluation of its core thermodynamic state.

An entire chapter (Chapter 3) will be devoted to the evaluation of the thermodynamic state of different groups of working substances. However, we have already introduced several equations relating thermodynamic properties with extrinsic or extensive properties, which are independent of the working substance. These equations, collectively known as **extended equations of state**, are summarized below.

Using the generic symbol B to represent any extensive property and b to represent its *specific* (per-unit mass) property and $\overline{b}$ the corresponding *molar specific* (per-unit mole) property, the total property B (e.g., m, KE, E, S, etc.) or its rate of transport $\dot{B}$ (e.g., $\dot{m}$, $\dot{KE}$, $\dot{J}$, $\dot{S}$, etc.) can be related to b and $\overline{b}$ as follows:

$$B = mb = n\overline{b}; \quad [\text{Unit of } B], \quad \text{and} \quad \dot{B} = \dot{m}b = \dot{n}\overline{b}; \quad \left[\frac{\text{Unit of } B}{\text{s}}\right] \tag{1.26}$$

where

$$b = \frac{B}{m} = \frac{\dot{B}}{\dot{m}}; \quad \left[\frac{\text{Unit of } B}{\text{kg}}\right], \quad \text{and} \quad \overline{b} = \frac{B}{n} = \frac{B\,m}{m\,n} = b\overline{M}; \quad \left[\frac{\text{Unit of } B}{\text{kmol}}\right] \tag{1.27}$$

As already mentioned in Sec. 1.5.7, $\dot{B} = \dot{m}b = \dot{n}\overline{b}$ is called the *transport equation*. Although these equations seem abstract, substituting b with familiar specific properties, such as v, u, e, ke, pe, s, etc., transforms them into familiar formulas for total (B) and transport ($\dot{B}$) properties. For example, $b = v$ produces formulas for total volume Ψ ($B = \Psi$) and volume flow rate $\dot{\Psi}$ ($\dot{B} = \dot{\Psi}$). Substitution of $b = s$ reproduces formulas for $B = S$ and $\dot{B} = \dot{S}$ as expressed by Eq. (1.24). For the special case of B representing mass, that is, for $B = m$ or $\dot{B} = \dot{m}$, Eqs. (1.4) and (1.5) are reproduced:

$$b = 1; \quad \left[\frac{\text{kg}}{\text{kg}}\right], \quad \overline{b} = \overline{M}; \quad \left[\frac{\text{kg}}{\text{kmol}}\right] \tag{1.28}$$

$$m = n\overline{M}; \quad [\text{kg}], \quad \text{and} \quad \dot{m} = \dot{n}\overline{M}; \quad \left[\frac{\text{kg}}{\text{s}}\right] \tag{1.29}$$

Note that the consistent use of the bar above all molar quantities, including molar mass.

Other frequently used extended state relations, introduced so far, include:

$$m = \rho\Psi; \quad [\text{kg}], \quad \dot{m} = \rho AV = \rho\dot{\Psi}; \quad \left[\frac{\text{kg}}{\text{s}}\right], \quad \rho = \frac{1}{v}; \quad \left[\frac{\text{kg}}{\text{m}^3}\right] \tag{1.30}$$

$$\text{ke} = \frac{V^2}{2\,(1000\ \text{J/kJ})}; \quad \text{pe} = \frac{gz}{(1000\ \text{J/kJ})}; \quad \left[\frac{\text{kJ}}{\text{kg}}\right] \tag{1.31}$$

Also, by definition

$$e \equiv u + ke + pe; \quad j \equiv h + ke + pe; \quad h \equiv u + pv; \quad \left[\frac{kJ}{kg}\right] \tag{1.32}$$

The extended equations of state can be used not only to evaluate extrinsic and system properties for a given equilibrium, but also to extract thermodynamic properties from extensive properties so that the underlying equilibrium of a system can be defined. However, these equations should not be confused with the **equations of state** that relate properties of the core thermodynamic state with one another, which will be developed in Chapter 3.

EXAMPLE 1-12 | Thermodynamic Properties

At the exit of a nozzle (Fig. 1.52) a gas has the following properties: $V_2 = 95$ m/s, $\dot{m}_2 = 0.5$ kg/min, and $\dot{J}_2 = 0.0635$ kW. If the nozzle exit diameter is 1 cm, determine the thermodynamic properties v_2 and h_2 at the exit. Neglect potential energy.

SOLUTION

Apply the extended equations of state to obtain thermodynamic properties from the given extrinsic and system properties.

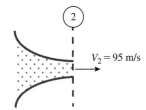

FIGURE 1.52 The flow state at the exit of the nozzle in Ex. 1-12.

Assumptions

The exit flow is described by a single flow state, State-2.

Analysis

From Eq. (1.30), the thermodynamic property v_2 is obtained as follows:

$$v_2 = \frac{1}{\rho_2} = \frac{A_2 V_2}{\rho_2 A_2 V_2} = \frac{A_2 V_2}{\dot{m}_2} = \frac{\pi (0.01^2)}{4}\left(\frac{95}{0.5/60}\right) = 0.895 \frac{m^3}{kg}$$

Equation (1.26) with $b = j$ yields:

$$j_2 = \frac{\dot{J}_2}{\dot{m}_2} = \frac{0.0635}{(0.5/60)} = 7.62 \frac{kJ}{kg}$$

Therefore, $h_2 = j_2 - ke_2 - pe_2^{\,0} = 7.62 - \dfrac{95^2}{2(1000 \text{ J/kJ})} = 3.11 \dfrac{kJ}{kg}$

TEST Analysis

The working substance is unknown; hence, we can use any flow-state TESTcalc, such as the IG-state TESTcalc (*TESTcalcs > States > Flow > IG-Model* page). Select any gas, such as nitrogen, and evaluate State-2 by entering Vel2, mdot2, A2 ('= PI*0.01^2/4'), and j2 (= .0635/mdot2). All thermodynamic properties, including v2 and h2, are calculated.

Discussion

In the TEST solution, all thermodynamic properties, not just v_2 and h_2, are calculated from the given input. By experimenting with this TESTcalc using different combinations of thermodynamic properties as input, we can verify that just two thermodynamic properties are sufficient to evaluate the complete thermodynamic state.

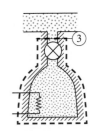

FIGURE 1.53 The electrical heating and charging process in Ex. 1-13 (see Anim. 0.C.*charging*).

EXAMPLE 1-13 | Energy Transfer and Extended State Calculation Using TEST

An insulated rigid tank (Fig. 1.53) contains 2 m³ of steam at State-1: 200 kPa and 400°C. Using the PC system-state TESTcalc, (a) determine the stored energy, E_1. A 2 kW electric heater is now turned on and allowed to heat the tank internally for 5 minutes. (b) Determine two thermodynamic properties of the final state, State-2. (c) Using the TESTcalc, calculate p_2 and T_2. (d) The valve is now opened and 0.2 kg of steam is allowed to enter the tank before the valve is closed. The state in the supply line remains fixed at State-3: 600 kPa, 600°C. Using the TESTcalc, determine (e) the flow energy transported by the mass, and (f) p_4 and T_4 of the final state, State-4.

SOLUTION

Evaluate the energy transfer through electrical work and mass transfer and obtain the desired properties from the extended equations of state and the PC system-state TESTcalc.

Assumptions

All the four states mentioned in this problem are in thermodynamic equilibrium. Negligible KE and PE. No change in volume of tank.

TEST Analysis

Launch the PC system-state TESTcalc and select H2O as the working substance. Calculate the four principal states by selecting the state number, entering the known properties, and clicking the Calculate button.

State-1 (given p_1, T_1, V_1):

$$u_1 = 2966.7 \frac{kJ}{kg}; \quad v_1 = 1.5493 \frac{m^3}{kg}; \quad m_1 = 1.291 \text{ kg}$$

Using Eqs. (1.26) and (1.32), we obtain:

$$E_1 = m_1 e_1 = m_1(u_1 + \cancel{ke_1}^0 + \cancel{pe_1}^0) = m_1 u_1$$
$$= (1.291)(2966.7) = 3830.0 \text{ kJ}$$

State-2: As the system transitions from State-1 to State-2, the mass of steam remains unchanged as mass cannot be created or destroyed. Also, the volume occupied by steam remains constant.

Therefore, $v_2 = \dfrac{V_2}{m_2} = \dfrac{V_1}{m_1} = v_1$, which is a thermodynamic property that can be obtained from the calculated state, State-1, as 1.5493 m³/kg.

During transition from State-1 to State-2, energy added by heat transfer can be calculated as:

$$Q = \dot{Q}\Delta t = (2)(5)(60) = 600 \text{ kJ}$$

Because energy cannot be created or destroyed, the specific internal energy at State-2, a second thermodynamic property, can be calculated as follows:

$$E_2 = E_1 + Q = 3830.0 + 600.0 = 4430.0 \text{ kJ}$$
$$\Rightarrow \quad U_2 + \cancel{KE}^0 + \cancel{PE}^0 = 4430 \text{ kJ}$$
$$\Rightarrow \quad u_2 = \frac{U_2}{m_2} = \frac{U_2}{m_1} = \frac{4430.0}{1.291} = 3431.4 \text{ kJ}$$

Substituting these properties into State-2 of the TESTcalc (we can input v2 as '= v1'), we obtain $p_2 = 281.7$ kPa, and $T_2 = 673.5°C$

State-3 (given p_3, T_3, m_3):

$$h_3 = 3700.9 \frac{kJ}{kg}; \quad j_3 = (h_3 + \cancel{ke_1}^0 + \cancel{pe_1}^0) = 3700.9 \frac{kJ}{kg}$$

Using Eq. (1.21), the flow energy transported by the flow can be calculated as:

$$J_3 = \int \dot{J}_3 dt = \int \dot{m}_3 j_3 dt = j_3 \int \dot{m}_3 dt = j_3 m_3 = (3700.9)(0.2) = 740.18 \text{ kJ}; \quad \textbf{(1.33)}$$

State-4: As we did for State-2, two thermodynamic properties can be evaluated from the known information.

$$v_4 = \frac{V_4}{m_4} = \frac{V_1}{(m_2 + m_3)} = \frac{V_1}{(m_1 + m_3)} = \frac{2}{(1.291 + 0.2)} = 1.341 \frac{m^3}{kg},$$

Also,

$$u_4 = \frac{U_4}{m_4} = \frac{E_4}{(m_2 + m_3)} = \frac{E_2 + J_3}{(m_1 + m_3)} = \frac{4430.0 + 740.18}{(1.291 + 0.2)} = 3467.6 \frac{kJ}{kg}$$

Substituting these properties into State-4 of the TESTcalc, we obtain $p_4 = 332.5$ kPa and $T_2 = 694.0°C$.

Discussion

Several aspects of this solution are noteworthy: (1). The final temperature is higher than the supply line temperature. This is because of the flow work done by the steam as it is introduced into the tank. (2). The net energy transported by steam can be broken up into two parts, E_3 and W_{F3}, which can be calculated from the given information at State-3. (3). If State-3 were located on the chamber side of the valve in Figure 1.53, the solution would be much more difficult as j_3 in the integral in Eq. (1.33) can no longer be assumed a constant and, therefore, cannot be taken outside the integral.

1.8 CLOSURE

In Chapter 1 we have developed a mathematical framework to describe a system or a flow of an extended system or flow state. An extended state was built upon the core equilibrium state, describing the underlying equilibrium. With the help of TEST animations, the concepts of thermodynamic equilibrium and local thermodynamic equilibrium (LTE) were established qualitatively. Properties of a state are discussed with an emphasis on a physics-based understanding. The zeroth law of thermodynamics was presented in connection with thermal equilibrium. TESTcalcs were used as a numerical laboratory to gain a quantitative understanding of properties, such as internal energy, entropy, and exergy. Properties were classified and relationships among different groups of properties, especially ones that relate an extended state with its core equilibrium state, were established. The manual evaluation of properties, delegated to Chapter 3, requires development of fundamental thermodynamic equations, which will be discussed in Chapter 2.

PROBLEMS

SECTION 1-1: UNDERSTANDING SYSTEM AND FLOW PROPERTIES

Mass, Mole, and Specific Volume

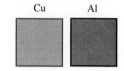

Cu   Al

1-1-1   Compare the mole of atoms in the unit of kmol in a 1 in^3 block of copper with that in an identical block of aluminum.

1-1-2   Determine the mass of one molecule of water (H_2O). Assume the density to be 1000 kg/m^3.

1-1-3   A chamber contains 14 kg of nitrogen (molar mass = 28 kg/kmol) at 100 kPa and 300 K. (a) How many kmol of N_2 is in there? (b) If 1 kmol of hydrogen (molar mass 2 kg/kmol) is added to the chamber, what is the total mole (in kmol) in the system now? (c) What is the average molar mass of the mixture? Assume universal gas constant to be 8.314 kJ/kmol · K.

1-1-4   1 kmol of nitrogen (N_2) is mixed with 2 kg of oxygen (O_2). Determine the total amount in (a) kg and (b) kmol.

1-1-5   10 kmol of a gas with a molar mass of 25 kg/kmol is mixed with 20 kg of another gas with a molar mass of 2 kg/kmol. If the pressure and temperature of the mixture are measured as 100 kPa and 400 K, respectively, determine the molar mass of the gas mixture.

1-1-6   A chamber contains a mixture of 2 kg of oxygen (O_2) and 2 kmol of hydrogen (H_2). (a) Determine the average molar mass of the mixture in kg/kmol. (b) If the specific volume of the mixture is 2 m^3/kg, determine the volume of the chamber in m^3.

1-1-7   2 kg of hydrogen (H_2) is mixed with 2 kg of oxygen (O_2). If the final mixture has a volume of 3 m^3: Determine (a) molar mass (b) specific volume and (c) the molar specific volume of the final mixture.

1-1-8   A 4 m × 5 m × 6 m room contains 120 kg of air. Determine (a) density, (b) specific volume, (c) mole, and (d) specific molar volume of air. Assume the molar mass of air to be 29 kg/kmol.

1-1-9   A 5 cm diameter upside-down piston-cylinder device contains 0.04 kg of an ideal gas at equilibrium at 100 kPa, 300 K, occupying a volume of 0.5 m^3. Determine (a) the gas density in kg/m^3, and (b) the specific volume in m^3/kg. (c) A mass is now hung from the piston so that the piston moves down and the pressure decreases. Determine the mass in kg necessary to reduce the pressure inside to 50 kPa after the piston comes to a new equilibrium. Assume the piston to be weightless.

(Data supplied: g = 9.81 m/s^2; Outside pressure: 100 kPa; Univ Gas Constant = 8.314 kJ/kmol · K).

1-1-10   10 kg of water (ρ = 1000 kg/m^3) and 5 kg of ice (ρ = 916 kg/m^3) are at equilibrium at 0°C. Determine the specific volume of the mixture.

1-1-11   If equal volumes of iron and copper (look up Table A-1 for densities) are melted together, determine (a) the specific volume and (b) density of the alloy. Assume no change in the final volume. (c) **What-if scenario:** What would the answers (specific volume and density) be if equal masses of iron and copper were melted together?

Thermodynamic Pressure

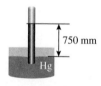

750 mm

Hg

1-1-12   What is the atmospheric pressure in kPa if a mercury barometer reads 750 mm? Assume ρ = 13,600 kg/m^3.

1-1-13   Determine the height of a mountain if the absolute pressures measured at the bottom and top are 760 mm and 720 mm of mercury, respectively. Assume ρ_{Hg} = 13,600 kg/m^3 and ρ_{air} = 1.1 kg/m^3.

1-1-14 A Bourdon gage measures the pressure of water vapor at the top of a cylindrical tank with a height of 6 m to be 300 kPa. Determine (a) the absolute pressure at the bottom of the tank if 50% of the tank volume is filled with water vapor. (b) What is the change in absolute pressure if the variation of pressure in the vapor phase was neglected? Assume $\rho_{vap} = 2.16$ kg/m^3.

1-1-15 An unknown mass is placed on the piston of a vertical piston-cylinder device containing nitrogen. The piston is weightless and has an area of 1 m^2. The outside pressure is 100 kPa and the pressure inside is measured at 101 kPa. Determine (a) the unknown mass. (b) Due to a leak, the piston gradually moves down until the leak is fixed, at which point half the gas (by mass) has escaped. What is the pressure inside in this new position of the piston? (c) *What-if scenario:* By what percentage will the pressure inside change if the mass placed on the piston is doubled?

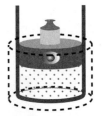

1-1-16 Determine the readings of the two Bourdon gages (outer first) if the inner tank (see figure) has a pressure of 500 kPa, the outer tank a pressure of 50 kPa and the atmosphere a pressure of 100 kPa.

1-1-17 (a) Determine the absolute pressure in kPa of the gas trapped in the piston cylinder device submerged in liquid water $\rho_{water} = 1000$ kg/m^3 if the piston is weightless and the depth of the piston from the free surface is 20 m. Assume the outside pressure to be 100 kPa. (b) *What-if scenario:* What would be the pressure if the piston (diameter: 1 m, thickness 10 cm, density 9000 kg/m^3) weight is included?

1-1-18 A piston separates two chambers in a horizontal cylinder as shown in the accompanying figure. Each chamber has a volume of 1 m^3 and the pressure of the gas is 5 kPa. The piston has a diameter of 50 cm and has a mass of 200 kg. If the cylinder is now set vertically, determine the pressure in each chamber. Assume pressure times volume to remain constant in each chamber.

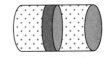

1-1-19 Two closed chambers A and B containing two gases are connected by a water manometer. If the gage pressure of chamber A is 10 mm of mercury vacuum and the height difference of the water columns is 10 cm: Determine the gage pressure in tank B if $\rho_{water} = 923$ kg/m^3.

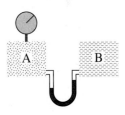

1-1-20 Air flows through a Venturi meter as shown in the accompanying figure. Determine the difference in pressure between points A and B if the height difference of the water columns is 3 cm. Assume $\rho_{air} = 1.2$ kg/m^3 and $\rho_{water} = 1000$ kg/m^3.

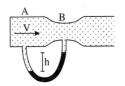

1-1-21 The maximum blood pressure of a patient requiring blood transfusion is found to be 110 mm Hg. What should be the minimum height of the IV to prevent a back flow? Assume $\rho_{blood} = 1050$ kg/m^3.

Temperature and the Zeroth Law

1-1-22 The temperatures assigned for the ice and steam points are 0 and 100 on the Celsius scale and 32 and 212 on the Fahrenheit scale. Both scales use a linear division. (a) Show that the two scales are related by $C/5 = (F - 32)/9$. (b) At what temperature do they show the same reading?

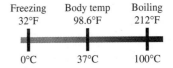

1-1-23 The temperature T on a thermometric scale is defined in property x by the function $T = a\ln(x) + b$, where a and b are constants. The values of x are found to be 1.8 and 6.8 at the ice point and the steam point, the temperatures of which are assigned numbers 0 and 100, respectively. Determine the temperature corresponding to a reading of $x = 2.4$ on this thermometer.

1-1-24 Two liquid-in-glass thermometers are made of identical materials and are accurately calibrated at 0°C and 100°C. While the first tube is cylindrical, the second tube has a conical bore, 15% greater in diameter at 0°C than at 100°C. Both tubes are subdivided uniformly between the two calibration points into 100 parts. If the conical bore thermometer reads 50°C: What will the cylindrical thermometer read? Assume the change in volume of the liquid is proportional to the change in temperature.

1-1-25 The signal (e.m.f.) produced by a thermocouple with its test junction at T °C is given by $\varepsilon = a + bT^2$ [mV], where $a = 0.2$ mV/°C and $b = 5.1 \times 10^{-4}$ mV/°C^2. Suppose you define a new scale after your name. The temperature in your scale is assumed to be linearly related to the signal through $T' = a' + b'\varepsilon$ with $T' = 25$ at ice point and $T' = 150$ at the steam point. Find the values of (a) a', (b) b', and (c) plot T' against T.

Energy and its Components

1-1-26 A projectile of mass 0.1 kg is fired upward with a velocity of 300 m/s. (a) Determine the specific kinetic energy (ke). If the sum of KE and PE is to remain constant, (b) determine the maximum vertical distance the projectile will travel.

1-1-27 A truck of mass 10 tonnes (metric) is traveling at a velocity of 65 miles per hour. Determine (a) kinetic energy (KE), (b) how high a ramp is necessary to bring it to a stop if the mechanical energy remains constant?

1-1-28 A projectile of mass 0.1 kg is moving with a velocity of 250 m/s. If the stored energy (E) in the solid is 1.5625 kJ, (a) determine the specific internal energy (u). Neglect PE and assume the solid to be uniform. (b) How do you explain a negative value for internal energy?

1-1-29 A rigid chamber contains 100 kg of water at 500 kPa, 100°C. A paddle wheel stirs the water at 1000 rpm with a torque of 100 N·m while an internal electrical resistance heater heats the water, consuming 10 amps of current at 110 Volts. Because of thin insulation, the chamber loses heat to the surroundings at 27°C at a rate of 1.0 kW. Determine (a) the rate of shaft work transfer ($\dot{W}_{sh}$) in kW, (b) the electrical work transfer (W_{el}) in kJ in 10 s, (c) the rate of heat transfer ($\dot{Q}$) in kW. Include the sign in all your answers. (d) Determine the change ($E_2 - E_1$) in stored energy E of the system in 10 s in kJ. Neglect KE and PE.

1-1-30 A truck with a mass of 20000 kg is traveling at 70 miles per hour.

(a) Determine its kinetic energy (KE) in MJ.
(b) The truck uses an electrical brake, which can convert the kinetic energy into electricity and charge a battery. If the efficiency of the system is 50%, what is the amount of energy in kWh that will be stored in the battery as the truck comes to a halt.

Transport Properties

1-1-31 Air flows through a pipe of diameter 50 cm with a velocity of 40 m/s. If the specific volume of air is 0.4 m^3/kg, determine (a) the specific kinetic energy (ke) of the flow, (b) the mass flow rate ($\dot{m}$), and (c) the rate of transport of kinetic energy ($\dot{KE}$) by the flow.

1-1-32 Steam flows through a pipe at 200 kPa, 379°C, with a velocity of 20 m/s and specific volume of 1.5 m^3/kg. If the diameter of the pipe is 1 m, determine (a) the mass flow rate ($\dot{m}$) in kg/s, (b) the mole flow rate ($\dot{n}$) in kmol/s, (c) $\dot{KE}$ (in kW), and (d) rate of flow work ($\dot{W}_F$) in kW.

1-1-33 Water flows down a 50 m long vertical constant-diameter pipe with a velocity of 10 m/s. If the mass flow rate ($\dot{m}$), is 200 kg/s, determine (a) the area of a cross-section of the pipe. Assuming the internal energy of water remains constant, determine the difference of rate of transport between the top and bottom for (b) kinetic energy ($\dot{KE}$), (c) potential energy ($\dot{PE}$), and (d) stored energy ($\dot{E}$).

1-1-34 Water enters a system operating at steady state, at 100 kPa, 25°C, 10 m/s at a mass flow rate ($\dot{m}$) of 200 kg/s. It leaves the system at 15 m/s, 1 MPa, 25°C. If the density of water is 1000 kg/m^3, determine

(a) the rate of flow work ($\dot{W}_F$) at the inlet (magnitude only)
(b) the rate of transport of kinetic energy ($\dot{KE}$) at the inlet, and
(c) the diameter of the pipe at the inlet.

1-1-35 Water ($\rho = 997$ kg/m^3) enters a system through an inlet port of diameter 10 cm with a velocity of 10 m/s. If the pressure at the inlet is measured as 500 kPa, determine (a) the mass flow rate ($\dot{m}$), (b) the rate of transport of kinetic energy ($\dot{KE}$) at the inlet, and (c) the rate of transport of flow energy ($\dot{J}$) assuming zero contribution from the potential and internal energy.

1-1-36 At the exit of a device, the following properties were measured—mass transport rate ($\dot{m}$): 51 kg/s; volume flow rate: 2 m³/s; specific flow energy (j): 230.8 kJ/kg; specific stored energy (e): 211.2 kJ/kg; flow area: 0.01 m². Determine (a) the specific volume (v) in m³/kg, (b) flow velocity in m/s, (c) the rate of flow work ($\dot{W}_F$) in kW, and (d) the pressure in kPa.

1-1-37 Betz's law states that only 53% of the KE transported by wind can be converted to shaft work by a perfect wind turbine. For a 50 m diameter turbine in a 20 mph wind, what is the maximum possible shaft power ($\dot{W}_{sh}$)? Assume density of air to be 1.1 kg/m³ (1 mph = 0.447 m/s).

1-1-38 For a 50 m diameter turbine in a 30 mph wind, determine: (a) the mass flow rate ($\dot{m}$) of air (in kg/s) intercepted by the turbine; (b) the specific kinetic energy (ke) of the flow (in kJ/kg); (c) the rate of transport of kinetic energy ($\dot{KE}$) kW; and (d) the maximum possible shaft power ($\dot{W}_{sh}$). Assume density of air as 1.1 kg/m³ and use Betz's law (see Problem 1-1-38). (e) *What-if scenario:* What will be the maximum shaft power if the wind velocity drops to 15 m/s?

1-1-39 A pipe of diameter 0.1 m carries a gas with the following properties at a certain cross-section: $p = 200$ kPa; $T = 30°C$; $v = 2$ m³/kg; $v = 15$ m/s. Determine (a) the mass flow rate ($\dot{m}$) in kg/s, (b) the volume flow rate (in m³/s), (c) the specific kinetic energy (ke) in kJ/kg, (d) the flow rate of kinetic energy ($\dot{KE}$) in W.

1-1-40 At the inlet of a steam turbine the flow state is as follows: $p_1 = 1$ MPa, $V_1 = 30$ m/s, $\dot{m}_1 = 9$ kg/s, and $A_1 = 0.1$ m². Determine (a) the rate of energy transfer due to flow work ($\dot{W}_F$) and (b) the rate of transport of kinetic energy ($\dot{KE}$) at the inlet port.

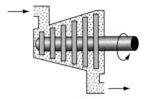

1-1-41 Steam flows into a steady adiabatic turbine at 10 MPa, 600°C and leaves at 58 kPa and 90% quality. The mass flow rate ($\dot{m}$) is 9 kg/s. Additional properties at the exit that are known are: $A = 1.143$ m², $v = 2.54$ m³/kg, $u = 2275.2$ kJ/kg, $e = 2275.4$ kJ/kg, and $h = 2422.4$ kJ/kg. If the turbine produces 1203 kW of shaft power, determine at the exit (a) the velocity in m/s. Also calculate the rate of transport of (b) $\dot{KE}$, (c) stored energy ($\dot{E}$), and (d) flow energy ($\dot{J}$). Neglect PE.

1-1-42 In an adiabatic nozzle, the specific flow energy j remains constant along the flow. The mass flow rate ($\dot{m}$) through the nozzle is 0.075 kg/s and the following properties are known at the inlet and exit ports. Inlet: $p = 200$ kPa, $u = 2820$ kJ/kg, $A = 100$ cm², $V = 10$ m/s; Exit: $h = 3013$ kJ/kg, $v = 1.67$ m³/kg. Determine (a) the exit velocity and (b) the exit area.

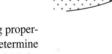

1-1-43 A superheated vapor enters a device with a mass flow rate ($\dot{m}$) of 5 kg/s with the following properties: $v = 30$ m/s, $p = 500$ kPa, $v = 0.711$ m³/kg, and $u = 3128$ kJ/kg. Neglecting PE, determine (a) the inlet area in m², (b) the rate of transport of KE in kW, (c) $\dot{E}$ in kW, (d) $\dot{J}$ in kW, and (e) the rate of flow work transfer ($\dot{W}_F$) in kW.

1-1-44 A mug contains 0.5 kg of coffee with specific stored energy (e) of 104.5 kJ/kg. If 0.1 kg of milk with specific stored energy (e) of 20.8 kJ/kg is mixed with the coffee, determine (a) the initial stored energy (E_i) of the system, (b) the final stored energy (E_f) of the system, and (c) the final specific energy (e_f) of the system.

Classification of Properties

1-1-45 What labels—extensive (0) or intensive (1) can be attached to the following properties: (a) m, (b) v, (c) p, (d) T, (e) ρ, (f) KE, (g) ke, (h) $\dot{m}$, and (i) V?

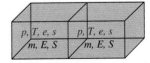

1-1-46 What labels—intensive, extensive, total, and flow—can be attached to the following properties: (a) m, (b) $\dot{m}$, (c) S, (d) $\dot{S}$, (e) h, (f) KE, (g) ke, and (h) $\dot{KE}$?

1-1-47 What labels—extrinsic (0) or intrinsic(1)—can be attached to the following properties: (a) u, (b) e, (c) j, (d) KE, (e) ke, (f) s, and (g) S?

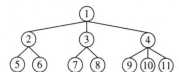

1-1-48   What labels—material, thermodynamic, intrinsic, and extrinsic—can be attached to the following properties: (a) m, (b) v, (c) p, (d) T, (e) ρ, (f) KE, (g) ke, (h) $\dot{m}$, and (i) V?

1-1-49   A rigid tank of volume 10 L contains 0.01 kg of a working substance (the system) in equilibrium at a gage pressure of 100 kPa. If the outside conditions are 25°C, 101 kPa, (a) how many independent *thermodynamic* properties of the system are supplied? (b) How many *extensive* properties of the system are supplied?

1-1-50   A rigid tank of volume 10 L contains 0.01 kg of a working substance in equilibrium at a gage pressure of 100 kPa. If the outside pressure is 101 kPa, determine two thermodynamic properties.

1-1-51   A vapor flows through a pipe with a mass flow rate ($\dot{m}$) of 30 kg/min. The following properties are given at a particular cross section, Area: 10 cm^2, Velocity: 60 m/s, Specific flow energy (j): 281.89 kJ/kg. If PE is negligible, determine two thermodynamic properties (v and h) for the flow state at the given cross section.

SECTION 1-2: NUMERICAL EXPLORATION OF PROPERTIES

1-2-1   A mug contains 0.5 kg of liquid water at 50°C. (a) Determine the stored energy (E) of the system, neglecting the KE and PE. If 0.1 kg of liquid water at 10°C mixed with the warm water, determine (b) the final stored energy (E_f) of the system assuming that no energy is lost during mixing. Use the SL state TESTcalc. (c) How do you explain a reduction of stored energy of the system?

1-2-2   A cup of coffee (system mass 1 kg) at 30°C rests on a table of height 1 m. An identical cup of coffee rests on the floor at a temperature of 25°C. Determine (a) the difference (cup 2 minus cup 1) in the stored energy ($E_2 - E_1$) in the two systems. (b) What fraction of the difference can be attributed to potential energy (PE)? Use the SL system-state TESTcalc. Assume properties of coffee to be similar to those of water.

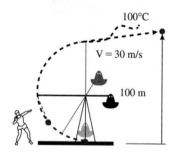

1-2-3   Using the SL system-state TESTcalc, determine the change in stored energy (E) in a block of copper of mass 1 kg due to (a) an increase in temperature from 25°C to 100°C, (b) an increase in velocity from 0 to 30 m/s (at a constant temperature) and (c) an increase in elevation by 100 m (at a constant temperature). (d) **What-if scenario:** What would the answer in (a) be if the working substance were granite instead?

1-2-4   For a 1 kg block of copper, determine the equivalent rise in stored energy (E) by 1 kJ for (a) increase in temperature (b) increase in velocity from rest and (c) an increase in height. (d) **What-if scenario:** What would the answer in (a) be if it were an aluminum block? Use the SL system-state TESTcalc. (Hint: enter *e2* as '=e1'.)

1-2-5   Use the SL system-state TESTcalc to determine specific entropy (s) of (a) aluminum, (b) iron, and (c) gold at 100 kPa, 298 K. Discuss why the entropy of gold is the lowest among these three metals.

1-2-6   Use the SL system-state TESTcalc to plot how the specific internal energy u and specific entropy s of aluminum changes with temperature at a constant pressure of 100 kPa. (Hint: Evaluate three states at, say, 300 K, 700 K, and 1200 K; draw the u-T and s-T diagrams using the pull-down menu. Use the p=c button to see the trend line.) Use the Log/Lin buttons to develop a functional relation for u and s for T.

1-2-7   Use the SL system-state TESTcalc to verify that u and s are independent of p for pure solids and liquids. (Hint: Select a solid or a liquid; Evaluate three states at, say, 100 kPa, 1 MPa, and 10 MPa at a constant temperature of, say, 300 K; draw the u-T and s-T diagrams using the pull-down menu. Use the p=c button. Repeat with different working substances and different temperatures.)

1-2-8   For copper, plot how the internal energy u, and entropy s, vary with T within the range 25°C–1000°C. Use the SL system-state TESTcalc.

1-2-9   For liquid water, plot how the internal energy u and entropy s vary with T within the range 25°C–100°C. Use the SL system-state TESTcalc.

1-2-10   Liquid water at 100 kPa, 30°C, enters a pump with a flow rate ($\dot{m}$) of 30 kg/s with a velocity of 2 m/s. At the exit the corresponding properties are 1000 kPa, 30.1°C, 30 kg/s, and 5 m/s. At the exit, determine (a) $\dot{KE}$, (b) $\dot{J}$ and (c) $\dot{S}$. Neglect PE.

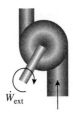

1-2-11   In Problem 1-2-10, determine the flow work ($\dot{W}_F$) in kW at the pump inlet and exit. What do you attribute the difference between the two quantities, if any, to?

1-2-12   A granite rock of mass 1000 kg is situated on a hill at an elevation of 1000 m. On a sunny day its temperature rises to 95°C. (a) Determine the maximum useful work that can be extracted from the rock if the atmospheric temperature is 30°C. (b) Compare the potential energy (PE) of the rock with its stored exergy (ϕ). Use the SL system-state TESTcalc.

1-2-13   A tank contains 2000 kg of water at 1000 kPa and 70°C. Using the SL model, determine (a) the stored exergy and (b) the stored energy in the water. Assume standard atmospheric conditions.

1-2-14   The cooling water in a power plant is discharged into a lake at a temperature of 35°C with a flow rate of 1000 kg/min. Determine the rate of discharge of exergy (ψ) using the SL flow-state TESTcalc.

1-2-15   A piston-cylinder device contains 1 kg of air at 100 kPa, 30°C. The gas is now compressed very slowly so that the temperature remains constant (isothermal). Calculate a series of states using the IG state TESTcalc as the volume decreases from the initial value to one tenth of the initial volume. Draw a constant-pressure line through the calculated states on a p-v diagram. What kind of conic section (parabola, hyperbola, etc) does the p-v diagram resemble? How does the stored energy (E) of air changes as the volume decreases?

1-2-16   A rigid tank contains 1 kg of air at 100 kPa, 30°C. Heat is now transferred to raise the temperature of the air. Calculate a series of states using the IG state TESTcalc as the temperature increases from the initial value to 1000°C. Plot how the pressure, internal energy (U), and entropy (S) change as a function of temperature.

1-2-17   A piston cylinder device contains 0.01 m³ of nitrogen at 500 kPa and 30°C. Determine the change in stored energy if (a) the pressure is doubled at constant temperature, (b) temperature is increased to 60°C at constant pressure. Use the IG system-state TESTcalc.

1-2-18   Determine the specific volume (v) of the gas in a 1 m³ chamber filled with (a) hydrogen, (b) carbon-dioxide. The pressure inside is 1 atm and the temperature is 25°C. Use the ideal gas (IG) state TESTcalc.

1-2-19   Use the IG system-state TESTcalc to determine entropy (s) of 1 kg of (a) hydrogen, (b) oxygen, and (c) carbon dioxide at 100 kPa, 298 K. Discuss why the entropy of hydrogen is the highest among these three gases.

1-2-20   A piston-cylinder device contains 1 kg of hydrogen at 100 kPa, 30°C (use the IG system-state TESTcalc). The gas is now compressed in such a manner that the entropy (s) remains constant. (a) Calculate the temperature when the volume becomes half the original volume. (b) Calculate a series of states using the as the volume decreases from the initial value to one tenth of the initial volume. Plot how the pressure of the gas changes as a function of volume. Compare the plot with the isothermal curve of Problem 1-2-15.

1-2-21 To explore how the internal energy (u) of a gas depends on temperature, volume, and pressure, evaluate the state of 1 kg of carbon dioxide at 100 kPa, 300 K using the IG state TESTcalc. Now holding mass and volume constant (hint: m2 = m1 and Vol2 = Vol1 in state-2), evaluate a series of states at different temperature. Plot how the internal energy u changes with temperature. In a similar manner, plot how u changes with temperature when pressure is held constant.

1-2-22 To explore how the entropy (s) of a gas depends on temperature, volume, and pressure, evaluate the state of 1 kg of nitrogen at 100 kPa, 300 K using the IG state TESTcalc. Now holding mass and volume constant (hint: Vol2 = Vol1 in state-2), evaluate a series of states at different temperatures. Plot how entropy (s) changes with temperature. In a similar manner, plot how entropy (s) changes with temperature when pressure is held constant.

1-2-23 A tank contains 5 kg of carbon dioxide at 2000 kPa and 25°C. Using the IG system-state TESTcalc, determine (a) the stored exergy (ϕ) and (b) the stored energy (E) in the gas. Assume the atmospheric conditions to be 100 kPa and 25°C. (c) How do you explain the negative sign of the stored energy calculated by the TESTcalc?

PC Model

1-2-24 A rigid tank contains 1 kg of H_2O at 1000 kPa, 1000°C. Heat is now transferred to the surroundings and the steam gradually cools down to a temperature of 30°C. Use the PC state TESTcalc to determine a series of states at intermediate temperatures, while holding the volume and mass constant. Plot how the pressure, internal energy (E), and entropy (S) change as a function of temperature. Compare these results with the results of Problem 1-2-16.

1-2-25 A piston cylinder device contains 0.01 m³ of steam at 500 kPa and 300°C. Determine the change in stored energy (E) if (a) the pressure is increased to 1 MPa at constant temperature (b) temperature is increased to 400°C at constant pressure. Use the PC (phase-change) system-state TESTcalc.

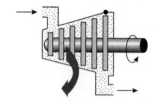

1-2-26 Steam flows into a turbine with a mass flow rate of 6 kg/s at a temperature of 500°C and a pressure of 1500 kPa. If the inlet area is 0.25 m², determine the transport properties (a) $\dot{J}$, (b) KE, (c) $\dot{U}$, (d) $\dot{E}$, (e) $\dot{H}$, and (f) $\dot{W}_F$ at the inlet. Use the PC (phase-change) flow-state TESTcalc. Neglect potential energy.

1-2-27 In Problem 1-2-26, determine the flow rate of (a) entropy ($\dot{S}$) and (b) exergy ($\dot{\psi}$) into the turbine if the atmospheric conditions are 100 kPa and 25°C.

2 DEVELOPMENT OF BALANCE EQUATIONS FOR MASS, ENERGY, AND ENTROPY: APPLICATION TO CLOSED-STEADY SYSTEMS

Classical thermodynamics is an axiomatic science. We start with a few basic axioms, or **laws** that are assumed to be self-evident and always true, from these laws, the behavior of thermodynamic systems can be predicted. A law is an abstraction of many observations distilled into concise statements that are self-evident and non-contradictory. For example, in Chapter 1, we discussed the zeroth law of thermodynamics, which establishes temperature as a thermodynamic property—an arbiter of thermal equilibrium between two objects.

In this chapter, we introduce the fundamental laws of thermodynamics: conservation of mass, conservation of energy (the *first law*), and the entropy principle (the *second law*). These laws are applied to an open-unsteady system (the most general form of a system) and translated into mass, energy, and entropy *balance equations*.

Closed systems operating at steady state are referred to as **closed-steady** systems and are analyzed in this chapter with the help of the balance equations. We will soon see that refrigerators, heat engines, and heat pumps are special cases of closed-steady systems. Performance-related analysis for these devices leads to the concept of reversibility and Carnot efficiency, and the establishment of the absolute temperature scale. Analysis of all other types of systems, which usually requires property evaluation, will be deferred until we develop systematic property evaluation methods in Chapter 3.

As in previous chapters, we will use the TEST animation module to illustrate important concepts. Use of TESTcalcs is not necessary for the analysis of closed-steady systems, which do not require property evaluation.

2.1 BALANCE EQUATIONS

Interactions between a system and its surroundings can lead to observable changes in a system. Consider a generic, unsteady, non-uniform, open system (Fig. 2.1) interacting in all possible ways with its surroundings. The global state, represented by the aggregate of local states, continuously changes (as indicated by the change of color with time in Anim. 2.A.*genericSystem*) as mass, heat, and work cross the system's boundary. Sometimes, even in the absence of any interaction, an isolated system may undergo a change of state spontaneously (see Anim. 0.C.*isolatedSystem*). Changes in a system, whether spontaneous or driven by interactions, are not arbitrary, but are governed by the fundamental laws of thermodynamics: conservation of mass, conservation of energy (the first law), and the entropy principle (the second law). Before we discuss these laws in detail, let's review some key points from the previous chapters relating to mass, heat, and work interactions.

Mass transfer through the inlet and exit ports of a system affect its total mass, which may increase, decrease, or, continuously remain constant, depending on the net rate of transfer. Recall from Sec. 0.5 that the mass flow rate is given by the familiar formula $\dot{m} = \rho AV$. But mass also transports all its attributes—properties such as energy, entropy, and all extensive properties—that can greatly affect the inventory of these properties in the system at a given instant. The transport equation, listed as part of the extended equations of state in Sec. 1.7, and reproduced as Eq. (2.1), can be used to evaluate the rate of transport for any extensive property (e.g., $\dot{KE}, \dot{J}, \dot{S}$) by a flow. As already established by Eq. (1.26), the rate of transport $\dot{B}$ (which symbolizes $\dot{KE}, \dot{J}, \dot{S}$, etc.), is related to the specific property b (which symbolizes ke, j, s, etc.) as follows:

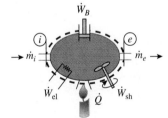

FIGURE 2.1 The *generic system*—an open-unsteady, non-uniform system— shown here with all possible interactions with its surroundings (see Anim. 2.A.*genericSystem*).

$$\dot{B} = \dot{m}b; \qquad \left[\frac{\text{unit of } B}{\text{s}} = \frac{\text{kg}}{\text{s}} \frac{\text{unit of } B}{\text{kg}} \right] \qquad (2.1)$$

Using this transport equation, energy or entropy transported by mass at a given inlet or exit port can be obtained from the flow state at that port.

Heat transfer, $\dot{Q}$, expressed by Eq. (0.5), not only transfers energy across a boundary, but also transfers entropy associated with the inherent disorganization at the microscopic level associated with the mechanism of heat transfer (collisions between molecules or absorption or emission of photons). Heat transfer through a system's inlet and exit ports can be neglected (Sec. 0.8) and only those portions of the boundary across which there is a temperature difference between the system and the surroundings needs to be considered.

Work transfer, $\dot{W}$, and its different components were introduced in Sec. 0.10 and described by Eq. (0.19) as the sum of two broad categories: external work, $\dot{W}_{ext}$, and flow work, $\dot{W}_F$. While the external work consists of boundary, electrical, and shaft work, flow work is absent in closed systems, and largely invisible in open systems. Energy transported by a flow involves two components: transport of stored energy and flow work. By adding them, we get the definition of the flow energy, j, where the rate of transport of the flow energy, $\dot{J} = \dot{m}j$, combines flow work with energy transport, thus, the need for evaluating $\dot{W}_F$ explicitly is eliminated. For any system, open or closed, only $\dot{W}_{ext}$ needs to be considered in place of $\dot{W}$, provided energy transport by mass is interpreted as *flow energy* transport $\dot{J}$.

In the sections that follow, each fundamental thermodynamic law will be expressed as a **balance equation**, a differential equation that keeps inventory of a certain extensive property of the system (mass for the conservation of mass, stored energy for the first law, and entropy for the second law). Each equation will be formulated for the most complex system possible: an unsteady, non-uniform, open system (Fig. 2.1)—hereafter called a **generic system**—that has all possible interactions with its surroundings. Once a governing equation is obtained in its most general form, this equation can be simplified and customized for any specific system or application. Before proceeding, we recommend reviewing Chapter 1's image analogy (Sec. 1.4) and exploring various types of systems shown in Anim. 1.A.*systemsClassified*.

2.1.1 Mass Balance Equation

The mass balance equation expresses an inventory of the mass of a generic system. It is based on the **conservation of mass principle**, stated as follows:

Mass can neither be created nor destroyed.

To translate this fundamental law into a balance equation, note that the mass of a system can change over time due to transport.

Consider the system, shown in Figure 2.2, at two different instants, t and $t + \Delta t$, where Δt is a short period of time. To simplify the derivation, only a single inlet and a single exit are considered. We also ignore the interactions that have no effect on the mass balance of the system (i.e., heat and various modes of work transfer). The open system of interest is enclosed within the fixed red external boundary through which the working substance enclosed in the black boundary passes through in time Δt.

Given that the system is unsteady, let the open system have a mass of $m(t)$ at time t and $m(t + \Delta t)$ at time $t + \Delta t$. Assuming the interval Δt to be small, the mass that enters and leaves the system within that period can be expressed as $\dot{m}_i \Delta t$ and $\dot{m}_e \Delta t$, where $\dot{m}_i$ and $\dot{m}_e$ are the mass transport rates at the inlet and exit at time t, respectively. There is no other mechanism through which the mass of the system can be affected (classical thermodynamics precludes conversions between mass and energy). As a result, $m(t + \Delta t)$ can be expressed as:

$$m(t + \Delta t) = m(t) + \dot{m}_i \Delta t - \dot{m}_e \Delta t; \quad [kg] \tag{2.2}$$

Re-arranging and taking the limit as Δt approaches zero, we obtain the following:

$$\lim_{\Delta t \to 0} \frac{m(t + \Delta t) - m(t)}{\Delta t} = \dot{m}_i - \dot{m}_e$$

$$\Rightarrow \quad \frac{dm}{dt} = \dot{m}_i - \dot{m}_e; \quad \left[\frac{kg}{s}\right]$$

For multiple inlets and exits, this equation can be generalized by summing all inlet and exit ports, yielding the **mass balance equation** in its most general form:

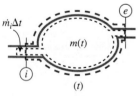

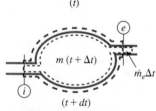

FIGURE 2.2 Schematic used for deriving the mass balance equation (see Anim. 2.B.*massBalanceEqn* and click each radio-button).

$$\underbrace{\frac{dm}{dt}}_{\substack{\text{Rate of change of}\\\text{mass in an open system.}}} = \underbrace{\sum_i \dot{m}_i}_{\substack{\text{Mass transport rate}\\\text{into the system.}}} - \underbrace{\sum_e \dot{m}_e}_{\substack{\text{Mass transport rate}\\\text{out of the system.}}} ; \quad \left[\frac{\text{kg}}{\text{s}}\right] \qquad \textbf{(2.3)}$$

The mass balance equation expresses the rate of change of mass in a *generic system* as the net rate of transport of mass into the system. A positive rate of change indicates accumulation, while a negative rate of change indicates depletion. Equation (2.3) can be visualized by the **flow diagram** shown in Figure 2.3, where the rate term, also called the **unsteady term**, is represented by the balloon, and the **transport terms** by the large arrows. Although trivial for the mass balance equation, a flow diagram can be a valuable visualization tool for the more complicated energy and entropy inventories of a system.

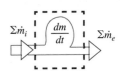

FIGURE 2.3 Flow diagram for the mass balance equation. The system is enclosed within the red boundary (see Anim. 2.B.*flowDiagram*).

A simple **checkbook analogy** (see Anim. 2.B.*checkbookAnalogy*) can be used to delineate different terms of a balance equation. In this analogy, the unsteady term can be interpreted as the rate of change of the net balance as a result of deposits (transport at the inlets) and withdrawals (transport at the exit). Because mass cannot be created or destroyed, there is no term in the mass balance equation that mimics interest accrued. The mass balance equation, therefore, resembles an interest-free checking account.

A balance equation derived for a generic system can be customized for simpler systems. First consider a *closed system*. With no possibility of mass transfer, the transport terms on the RHS (right-hand side) drop out, yielding:

$$\frac{dm}{dt} = 0, \quad \text{or} \quad m = \text{constant} \qquad \textbf{(2.4)}$$

The mass of a closed system remains constant, which is a trivial conclusion. For an *open-steady* system, the global state remains frozen in time (see Anim. 1.A.*systemsClassified*), and the system mass is constant irrespective of the transport of mass across the boundary. The reverse is not necessarily true: a system can be unsteady even if its mass does not change. By the definition of a steady state, the time derivative of all extensive properties, including mass, m, must be zero. Thus, the mass balance equation simplifies to:

$$\cancel{\frac{dm}{dt}}^{\,0} = \sum_i \dot{m}_i - \sum_e \dot{m}_e, \quad \text{or} \quad \underbrace{\sum_i \rho_i A_i V_i}_{\text{What goes in}} = \underbrace{\sum_e \rho_e A_e V_e}_{\text{What comes out}} ; \quad \left[\frac{\text{kg}}{\text{s}}\right] \qquad \textbf{(2.5)}$$

where $\dot{m} = \rho A V$ (Eq. 0.2) is substituted at all the inlet and exit ports. Thus, the equation reduces to *what goes in is what comes out*. For a system with only one inlet and one exit, a *single-flow* steady system, the summation symbols can be removed:

$$\dot{m}_i = \dot{m}_e; \quad \text{or,} \quad \rho_i A_i V_i = \rho_e A_e V_e; \quad \text{or,} \quad \frac{A_i V_i}{v_i} = \frac{A_e V_e}{v_e}; \quad \left[\frac{\text{kg}}{\text{s}}\right] \qquad \textbf{(2.6)}$$

For incompressible fluids (SL model in TEST), the density remains constant and the equation becomes $A_i V_i = A_e V_e$, which can be used to understand incompressible flow behavior in a variable area passage (Fig. 2.4). For example, a steady flow of a constant-density liquid through a converging duct must accelerate; conversely, the same liquid flowing through a diverging duct must decelerate. If the working fluid is a vapor or a gas, the possibility of a change in specific volume introduces complexities that will be addressed in later chapters (4 and 15).

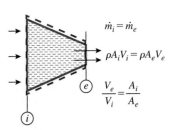

FIGURE 2.4 A constant-density fluid must accelerate in inverse proportion to flow area in a converging passage (see Anim. 4.A.*nozzleShapes*).

The mass balance equation involves several properties of a flow state including the *thermodynamic* property, v (or its inverse, density ρ). In the next chapter, we will evaluate the properties of a flow state; here, the properties are assumed known or are determined via state TESTcalcs.

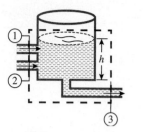

FIGURE 2.5 Schematic for Ex. 2-1.

EXAMPLE 2-1 Application of the Mass Balance Equation

Water enters a cylindrical tank through two inlets and leaves through a single exit as shown in Figure 2.5. The conditions at the three ports are as follows: State-1: $V_1 = 10$ m/s, $A_1 = 50$ cm^2; State-2: $V_2 = 300$ m/min, $A_2 = 0.011$ m^2; State-3: $p_3 = 500$ kPa, $T_3 = 20\,°C$, $V_3 = 7$ m/s, $D_3 = 14$ cm. Determine (a) the mass flow rate at each port and (b) the rate of change in height in mm/min if the tank has an inner diameter of 5 m. Assume the density of water to be constant at 997 kg/m^3.

SOLUTION

Analyze the mass balance equations for the open system enclosed by the red boundary of Figure 2.5.

Assumptions

The inlet and exit flows are uniform and in LTE (local thermodynamic equilibrium) so that three flow states: State-1, State-2, and State-3 can be used to describe the flow at the ports.

Analysis

The mass flow rates at the three ports are calculated from Eq. (0.2):

$$\dot{m}_1 = \rho_1 A_1 V_1 = (997)\left(\frac{50}{10,000}\right)(10) = 49.85\ \frac{\text{kg}}{\text{s}};$$

$$\dot{m}_2 = \rho_2 A_2 V_2 = (997)(0.011)\left(\frac{300}{60}\right) = 54.84\ \frac{\text{kg}}{\text{s}};$$

$$\dot{m}_3 = \rho_3 A_3 V_3 = (997)\frac{\pi}{4}\left(\frac{14}{100}\right)^2 (7) = 107.43\ \frac{\text{kg}}{\text{s}}$$

The mass balance equation, Eq. (2.3), produces:

$$\frac{dm}{dt} = \dot{m}_1 + \dot{m}_2 - \dot{m}_3 = 49.85 + 54.84 - 107.43 = -2.748\ \frac{\text{kg}}{\text{s}}$$

The mass of water in the tank can be expressed as a function of the water level (height, h, in Fig. 2.5); as a result, the unsteady term can be expressed as:

$$\frac{dm}{dt} = \frac{d(\rho \Psi)}{dt} = \frac{d(\rho A h)}{dt} = \rho A \frac{dh}{dt}$$

Therefore,

$$\frac{dh}{dt} = \frac{1}{\rho A}\frac{dm}{dt} = \frac{1}{997}\frac{4}{\pi}\left(\frac{1}{5^2}\right)(-2.748)\left(1000\ \frac{\text{mm}}{\text{m}}\right)\left(60\ \frac{\text{s}}{\text{min}}\right) = -8.4\ \frac{\text{mm}}{\text{min}}$$

Discussion

The negative sign for the rate of change indicates that the water level is declining. The mass of water contained in the sections of the pipes that fall within the system boundary can be ignored in this analysis since their amounts remain constant.

2.1.2 Energy Balance Equation

The energy balance equation keeps track of the energy inventory of a system. This is done by relating different ways in which energy can be transferred across a system boundary to the change in stored energy of a system through the application of the energy principle. The **conservation of energy principle**, also known as the **first law** of thermodynamics, is stated by the following postulates:

i) The specific internal energy u is a thermodynamic property.

ii) The stored energy $E = U + \mathrm{KE} + \mathrm{PE}$ of a system can neither be created nor destroyed, only transported by mass and transferred through heat and work across the system boundary. This indestructibility of energy is also known as the energy principle.

The first postulate establishes the specific internal energy, u, as one of the attributes of a thermodynamic equilibrium. As a corollary of the second postulate, the *energy of an isolated system must remain constant* because energy cannot be created or destroyed within the system. With no mass, heat, or work interaction, energy cannot be transported or transferred across the boundary. Therefore, energy can only be redistributed in its different modes within an isolated system, as shown in Anim. 0.C.*isolatedSystem2*.

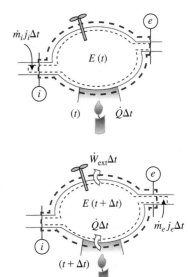

By providing a complete account of the interactions that lead to the change in E, the first law allows us to create an inventory of energy in the form of an *energy balance equation*. The *generic system* of Figure 2.1 is represented by simplified figures at two different instants, t and $t + \Delta t$, in Figure 2.6. As before, only a single inlet and a single exit are considered; the open system of interest is enclosed within the red boundary through which a working substance, enclosed within the black boundary, flows in time Δt. To further simplify the development of the energy balance equation, external work is represented by shaft work alone and heat transfer is assumed to take place from a single heating source—a formal name for an idealized source or sink of heat is a **thermal energy reservoir**, a **TER**, or simply a **reservoir**. A reservoir is considered so large that any amount of heat can be supplied or absorbed by it without affecting its state.

Within the short interval Δt, the amount of energy transported by mass at the inlet is $\dot{J}_i \Delta t = \dot{m}_i j_i \Delta t$, which includes both transport of stored energy and flow work (see Anim. 2.C.*energyTransport*). Similarly, the energy transported at the exit is given by $\dot{J}_e \Delta t = \dot{m}_e j_e \Delta t$. With the flow work already accounted for in the energy transport terms, energy transfer out of the system through work is given by $\dot{W}_{ext}\Delta t$, where $\dot{W}_{ext}$ is the net external work consisting of shaft, electrical, and boundary work (see Anim. 2.C.*externalWork*). The remaining mechanism, heat transfer, moves $\dot{Q}\Delta t$ energy into the system in time Δt, where $\dot{Q}$ is the net heat transfer rate, therefore, $E(t + \Delta t)$ can be related to $E(t)$ by taking into account the WinHip sign convention (Sec. 0.7.2) where heat added and work delivered are considered positive with respect to the system:

FIGURE 2.6 Schematic used for deriving the energy balance equation (see Anim. 2.C.*energyBalanceEqn*).

$$E(t + \Delta t) = E(t) + \dot{J}_i \Delta t - \dot{J}_e \Delta t + \dot{Q}\Delta t - \dot{W}_{ext}\Delta t; \quad [\mathrm{kJ}] \qquad (2.7)$$

Dividing by Δt, rearranging, and taking the limit as Δt tends to zero, we obtain:

$$\lim_{\Delta t \to 0} \frac{E(t + \Delta t) - E(t)}{\Delta t} = \dot{J}_i - \dot{J}_e + \dot{Q} - \dot{W}_{ext}$$

$$\Rightarrow \quad \frac{dE}{dt} = \dot{m}_i j_i - \dot{m}_e j_e + \dot{Q} - \dot{W}_{ext}; \quad \left[\frac{\mathrm{kJ}}{\mathrm{s}} = \mathrm{kW}\right]$$

For multiple inlets and exits, this equation can be generalized like the mass balance equation, yielding the **energy balance equation** in its most general form:

$$\underbrace{\frac{dE}{dt}}_{\substack{\text{Rate of change} \\ \text{of stored energy} \\ \text{of a generic system.}}} = \underbrace{\sum_i \dot{m}_i j_i}_{\substack{\text{Rate of flow energy} \\ \text{transport into the} \\ \text{system by mass at} \\ \text{the inlets.}}} - \underbrace{\sum_e \dot{m}_e j_e}_{\substack{\text{Rate of flow energy} \\ \text{transport out of the} \\ \text{system by mass at} \\ \text{the exits.}}} + \underbrace{\dot{Q}}_{\substack{\text{Rate of heat} \\ \text{transfer into} \\ \text{the system.}}} - \underbrace{\dot{W}_{ext}}_{\substack{\text{Rate of external} \\ \text{work transfer} \\ \text{out of the system.}}} ; \quad [\mathrm{kW}] \qquad (2.8)$$

where $j = h + \mathrm{ke} + \mathrm{pe} = h + \dfrac{V^2}{2(1000\ \mathrm{J/kJ})} + \dfrac{gz}{(1000\ \mathrm{J/kJ})}$, and $\dot{W}_{ext} = \dot{W}_B + \dot{W}_O$

The physical meaning of different terms of the energy equations can be explored in Anim. 2.C.*energyBalanceTerms* by clicking each term. As in the mass balance equation, the LHS, the time rate of stored energy, is the *unsteady* term. On the RHS, the first two terms are *transport* terms, flow energy transported by mass at the inlets and exits. The heat transfer $\dot{Q}$ is the net rate of heat addition, the algebraic sum, due to all possible mechanisms aggregated over the entire boundary. If a system loses heat, $\dot{Q}$ must have a negative value (following the WinHip convention). The external work consists mostly of shaft, electrical, and boundary work, $\dot{W}_{ext} = \dot{W}_{sh} + \dot{W}_{el} + \dot{W}_B$ (see Eq. 0.18). Each term is algebraic and assumed positive (following the WinHip convention) in the energy equation; therefore, if work is transferred into the system by a particular mechanism, it will contribute a negative value to the make up of $\dot{W}_{ext}$.

The energy equation can be interpreted as follows: *the rate of change of stored energy of an open system is equal to the net rate of transport of energy by the flows and the net rate of energy transfer across the boundary through heat and external work.* By representing the net rate of energy transport by $\dot{J}_{net}$, the energy equation can be expressed in an alternative form:

$$\underbrace{\frac{dE}{dt}}_{\substack{\text{Unsteady term.}}} = \underbrace{\dot{J}_{net}}_{\substack{\text{Net transport} \\ \text{of flow work.}}} + \underbrace{\dot{Q} - \dot{W}_{ext}}_{\substack{\text{Net transfer of heat} \\ \text{and external work.}}} ; \quad \text{where} \quad \dot{J}_{net} = \sum_i \dot{J}_i - \sum_i \dot{J}_e; \quad [\text{kW}] \qquad (2.9)$$

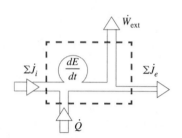

FIGURE 2.7 Flow diagram of the energy balance equation for a generic system (see Anim. 2.C. *flowDiagram*). Compare it with the mass flow diagram of Fig. 2.3.

An **energy flow diagram** can be constructed as shown in Figure 2.7, to visualize the different terms of the energy equation. In our *checkbook analogy* (click Energy in Anim. 2.B.*checkbookAnalogy*), the unsteady term is analogous to the rate of change of the account balance (the balloon in Fig. 2.7), and the transport terms are analogous to cash deposits and withdrawals. The new boundary transfer terms—heat and external work—can be represented by direct deposit of paychecks (for heat transfer) and wire transfer (for external work transfer). Like the mass balance equation, the account is interest free because creation or destruction of energy is forbidden by the first law.

Derived for a *generic system*, the energy equation, Eq. (2.8), can be readily adapted to a *closed system* by dropping the transport terms:

$$\frac{dE}{dt} = \dot{J}_{net}^{\;0} + \dot{Q} - \dot{W}_{ext} = \dot{Q} - \dot{W}_{ext}; \quad [\text{kW}] \qquad (2.10)$$

For a closed system, with no possibility of any flow work, $\dot{W} = \dot{W}_F^{\;0} + \dot{W}_{ext} = \dot{W}_{ext}$. As a result, the symbols $\dot{W}$ and $\dot{W}_{ext}$ are used interchangeably in a closed system's energy equation.

Many open devices operate at steady state and will be referred to as **open-steady** systems. The stored energy, like any other extensive property, remains constant as the global state of a steady system does not change with time. For *single-flow* devices the steady-state energy equation simplifies as follows:

$$\frac{dE}{dt}^{\;0} = \dot{m}_i j_i - \dot{m}_e j_e + \dot{Q} - \dot{W}_{ext}; \quad \Rightarrow \quad 0 = \dot{m}(j_i - j_e) + \dot{Q} - \dot{W}_{ext};$$

$$\Rightarrow \quad \underbrace{\dot{m}(h_i + ke_i + pe_i) + \dot{Q}}_{\text{Energy in}} = \underbrace{\dot{m}(h_e + ke_e + pe_e) + \dot{W}_{ext}}_{\text{Energy out}}$$

$$\Rightarrow \quad \underbrace{\dot{m}(h_e + ke_e + pe_e)}_{\substack{\text{Energy transported} \\ \text{out at the exit.}}} = \underbrace{\dot{m}(h_i + ke_i + pe_i)}_{\substack{\text{Energy transported in} \\ \text{at the inlet.}}} + \underbrace{\dot{Q}}_{\substack{\text{Heat} \\ \text{added to} \\ \text{the system.}}} - \underbrace{\dot{W}_{ext}}_{\substack{\text{External work} \\ \text{delivered by} \\ \text{the system.}}}; \quad [\text{kW}] \;\; (2.11)$$

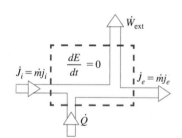

FIGURE 2.8 Flow diagram of the energy balance equation for a single-flow steady system (click Energy Flow Diagram in Anim. 4.A.*flowDiagram*).

In this derivation, $\dot{m}_i = \dot{m}_e = \dot{m}$ is substituted from the mass equation, Eq. (2.6). As shown in Figure 2.8, this simplified energy equation can be interpreted as a balance between incoming energy, through mass and heat, and outgoing energy, through mass and external work. A further simplification occurs when the kinetic and potential energy changes between the inlet and exit are insignificant. The flow energy can be replaced by *enthalpy* (Sec. 1.5.8), a thermodynamic property, which is relatively easier to evaluate.

The following examples are intended to highlight different terms of the energy balance equation; comprehensive analysis will be performed in later chapters.

EXAMPLE 2-2 **Energy Analysis: An Unsteady Closed System**

A cup of coffee is heated in a microwave oven (see Fig. 2.9). If heat (through radiation) is added at a constant rate of 200 W and the cup contains 0.3 kg of coffee, determine the rate of change of the specific internal energy of the coffee. Neglect evaporation.

SOLUTION

Simplify the energy equation for the closed, unsteady system and evaluate the unsteady term.

Assumptions

The entire amount of heat (in the form of microwave radiation) goes into the coffee, and the cup itself does not participate.

Analysis

With no mass transfer, no external work transfer, and no changes in KE and PE, the energy balance equation, Eq. (2.8), reduces to:

$$\frac{dE^U}{dt} = \dot{J}_{net}^{\ 0} + \dot{Q} - \dot{W}_{ext}^{\ 0}; \quad \Rightarrow \quad \frac{d(mu)}{dt} = \dot{Q}; \quad \Rightarrow \quad \frac{du}{dt} = \frac{\dot{Q}}{m};$$

$$\Rightarrow \quad \frac{du}{dt} = \frac{(200 \text{ W})}{(0.3 \text{ kg})(1000 \text{ W/kW})} = 0.667 \ \frac{\text{kW}}{\text{kg}}$$

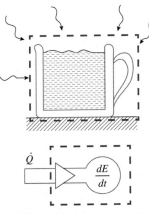

FIGURE 2.9 System schematic and the energy diagram for Ex. 2-2.

Discussion

The SL model (developed in the next chapter) expresses u as a function of T for an incompressible solid or liquid. Once such relationships are established, the rate of change of temperature can be calculated. Note that a positive value is used for heat transfer, consistent with the WinHip convention. If an electric heater with $\dot{W}_{el} = -200$ W was used to heat the coffee instead, a similar analysis would yield the same result.

EXAMPLE 2-3   Energy Analysis: Closed-Steady System

A thin wall separates two large chambers, the left chamber contains boiling water at 200°C and the right chamber contains a boiling refrigerant at 0°C. The heat transfer rate from the left chamber to the wall is measured at 2 kW. Assuming the wall to be at steady state, determine the rate of heat transfer (magnitude only) from the wall to the right chamber.

SOLUTION

Treating the wall as the system, heat transfer at the left and right boundaries is the only interaction between the system and the surroundings (the two chambers). Simplify the energy balance equation for the wall, a *closed-steady* system.

Assumptions

The wall maintains steady state so that all extensive properties, including the stored energy E, are constant.

Analysis

Heat transfer at the two surfaces are represented by the absolute values $\dot{Q}_1$ and $\dot{Q}_2$ whose directions are indicated by the arrows in Figure 2.10. The net rate of heat addition, therefore, is $\dot{Q} = \dot{Q}_1 - \dot{Q}_2$. Substituting this relationship into the energy balance equation, Eq. (2.8), and simplifying, we obtain:

$$\frac{dE}{dt}^{\,0,\text{ steady state}} = \dot{J}_{net}^{\ 0} + \dot{Q}_1 - \dot{Q}_2 - \dot{W}_{ext}^{\ 0};$$

$$\Rightarrow \quad \dot{Q}_2 = \dot{Q}_1 = 2 \text{ kW}$$

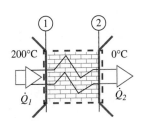

FIGURE 2.10 System schematic and energy flow diagram for Ex. 2-3. $\dot{Q}_1$ and $\dot{Q}_2$ are magnitudes of heat transfer with the directions indicated by arrows (see Anim. 2.E.*wall*).

Discussion

For the stored energy in the wall to remain constant, the energy balance equation dictates that heat must leave the wall at the same rate as it enters. The net heat transfer to the system is zero.

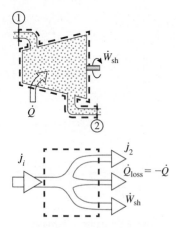

FIGURE 2.11 Schematic of the turbine and the energy flow diagram for Ex. 2-4 (see Anim. 4.A.*turbine*).

EXAMPLE 2-4   **Energy Analysis: An Open System**

A steam turbine with a single inlet and exit produces 1.132 MW of shaft power, while losing heat to the ambient atmosphere at a rate of 10 kW (see Fig. 2.11). The following data are supplied at the inlet and exit states: State-1: $v_1 = 0.1512$ m³/kg, $h_1 = 3247.6$ kJ/kg, $V_1 = 25$ m/s, $A_1 = 121$ cm², $z_1 = 6$ m; State-2: $v_2 = 1.694$ m³/kg, $h_2 = 2675.5$ kJ/kg, $V_2 = 50$ m/s, $A_2 = 680$ cm², $z_2 = 3$ m. Calculate each term of the (a) mass and (b) energy balance equations.

SOLUTION

Evaluate each term on the RHS of the mass and energy balance equations for this work-producing open system. Obtain the unsteady terms, from the balance equations.

Assumptions

The flow is uniform and in LTE at the inlet and exit. We cannot assume that the turbine is operating at steady state.

Analysis

Evaluate the mass flow rate and flow energy at the inlet and exit.

State-1:

$$ke_1 = \frac{V_1^2}{2(1000 \text{ J/kJ})} = 0.313 \, \frac{\text{kJ}}{\text{kg}}; \quad pe_1 = \frac{gz_1}{(1000 \text{ J/kJ})} = 0.06 \, \frac{\text{kJ}}{\text{kg}};$$

$$j_1 = h_1 + ke_1 + pe_1 = 3248.0 \text{ kJ/kg}; \quad \dot{m}_1 = \frac{A_1 V_1}{v_1} = 2.0 \text{ kg/s}$$

State-2:

$$ke_2 = \frac{V_2^2}{2(1000 \text{ J/kJ})} = 1.25 \, \frac{\text{kJ}}{\text{kg}}; \quad pe_2 = \frac{gz_2}{(1000 \text{ J/kJ})} = 0.03 \, \frac{\text{kJ}}{\text{kg}};$$

$$j_2 = h_2 + ke_2 + pe_2 = 2676.8 \text{ kJ/kg}; \quad \dot{m}_2 = \frac{A_2 V_2}{v_2} = 2.0 \text{ kg/s}$$

The unknown unsteady term of the mass equation can now be evaluated:

$$\frac{dm}{dt} = \underbrace{\dot{m}_1}_{2 \text{ kg/s}} - \underbrace{\dot{m}_2}_{2 \text{ kg/s}} = 0$$

Similarly, the unsteady terms of the energy balance equation are evaluated (to the nearest integer) as:

$$\frac{dE}{dt} = \underbrace{\dot{m}_1 j_1}_{j_1 = 6496 \text{ kW}} - \underbrace{\dot{m}_2 j_2}_{j_2 = 5354 \text{ kW}} + \underbrace{\dot{Q}}_{-10 \text{ kW}} - \underbrace{\dot{W}_{\text{ext}}}_{1132 \text{ kW}} = 0$$

Discussion

The net rate of transport of energy into the system is $\dot{J}_{\text{net}} = \dot{J}_i - \dot{J}_e = 6496 - 5354$ kW = 1142 kW, where 10 kW is lost as heat to the atmosphere (notice the negative sign), and the rest is delivered as shaft work. Even though mass and stored energy remain constant, we cannot declare the system to be steady. The reverse is always true.

In Ex. 1-9, $\dot{J}_{\text{net}}$ for the single-flow system was calculated as −20.80 kW. If the device operates at steady state in an adiabatic manner, the energy balance equation yields $\dot{W}_{\text{ext}} = \dot{J}_{\text{net}} = -20.80$ kW; the negative value (according to the WinHip convention) indicates that external work is transferred into the system—by a pump in this case—to increase the flow's pressure.

| **EXAMPLE 2-5** | **Energy Analysis of an Open Unsteady System** |

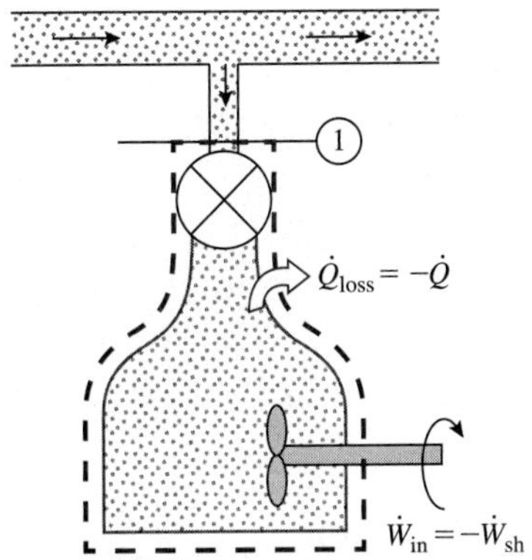

Steam from a supply line enters a rigid tank through a valve as shown in Figure 2.12. During the filling process, heat is lost to the surroundings at a rate of 1 kW. A paddle-wheel stirs the steam inside the tank at 500 rpm with a torque of 0.01 kN·m to overcome viscous friction. The inlet conditions on the supply side of the valve are: $\dot{m}_1 = 0.11425$ kg/s and $h_1 = 3069.26$ kJ/kg. If the mass and specific internal energy of the steam in the tank at a given instant are 2 kg and 2510 kJ/kg, respectively, determine the rate of change of (a) the total internal energy U and (b) specific internal energy, u, of the system. Neglect KE and PE at the inlet.

FIGURE 2.12 Schematic for Ex. 2-5.

SOLUTION

Perform a mass and energy balance on the tank and valve as a system (enclosed within the red boundary of Fig. 2.12).

Assumptions

Uniform flow based on LTE with negligible KE and PE at the port.

Analysis

Energy transport at the inlet and the work transfer by the shaft can be calculated as follows:

$$\dot{J}_1 = \dot{m}_1 j_1 \cong \dot{m}_1 h_1 = (0.11425)(3069.3) = 350.7 \text{ kW};$$

$$\dot{W}_{sh} = -2\pi \frac{N}{60} T = -2\pi \frac{500}{60}(0.01) = -0.524 \text{ kW}$$

The negative sign in the shaft work (recall the WinHip convention) indicates that work is transferred into the system. With the system losing heat, the heat transfer is also negative: $\dot{Q} = -1$ kW. The energy equation, Eq. (2.8), yields:

$$\frac{dE}{dt} \cong \frac{dU}{dt} = \dot{J}_i - \dot{J}_e^0 + \dot{Q} - \dot{W}_{ext} = 350.7 + (-1) - (-0.524) = 350.2 \text{ kW}$$

With the mass of the system changing, differentiating the LHS by parts, we get:

$$\frac{dU}{dt} = \frac{d(mu)}{dt} = m\frac{du}{dt} + u\frac{dm}{dt} = m\frac{du}{dt} + u\dot{m}_i$$

where the mass balance equation is used to substitute $\dot{m}_i$ for dm/dt in the last term. Simplifying, we get:

$$\frac{du}{dt} = \frac{1}{m}\left[\frac{dU}{dt} - u\dot{m}_i\right] = \frac{1}{2}[350.2 - (2510)(0.11425)] = 31.7 \frac{\text{kW}}{\text{kg}}$$

Discussion

An increase in u usually accompanies an increase in temperature for gases and vapors, in which case the temperature in the tank increases well above that in the supply line.

2.1.3 Entropy Balance Equation

Just as the mass and energy balance equations track the inventory of mass and stored energy of a system, the entropy balance equation keeps an inventory of entropy. Unlike mass and energy, entropy is not conserved and is governed by the **second law** of thermodynamics:

   i) Entropy, S, is an extensive property that quantifies a system's molecular disorder and can be related to the number of ways in which the system's internal energy can be distributed among its molecules; the specific entropy s is a thermodynamic property.

ii) Entropy is transferred across a boundary by heat at a rate of $\dot{Q}/T_B$, where $\dot{Q}$ is the rate of heat transfer and T_B is the boundary temperature. Work, however, does not transfer any entropy.

iii) Entropy cannot be destroyed. It is generated spontaneously when a system moves towards equilibrium by dissipating internal mechanical, thermal, and chemical imbalances. That is, $\dot{S}_{gen} \geq 0$, where $\dot{S}_{gen}$ represents the rate of entropy generation within a system. This is also known as the **entropy principle**.

In Chapter 1, we discussed entropy from the microscopic viewpoint as a measure of molecular disorder in a system. The first postulate establishes specific entropy as a thermodynamic property that helps define an equilibrium state. With molecules having discrete energy levels, a given amount of energy can be distributed among a system's molecules in numerous ways—more possibilities equals more disorder or entropy (see Anim. 1.B.*entropy*). Using statistical arguments, Boltzmann defined entropy as $S = k \ln \Omega$, where k is known as the **Boltzmann constant** and Ω is the number of **microstates**—the number of ways in which the internal energy of a system can be rearranged among the molecules without violating the first law. Because KE and PE involve organized motion or position of molecules, they have no effect on entropy. When two identical systems are merged (see Anim. 1.A.*intensiveProperty*), the total number of microstates is squared and the entropy doubles ($\ln \Omega^2 = 2 \ln \Omega$), therefore, entropy is an extensive property. The specific entropy, s (entropy per-unit mass) is a thermodynamic property just like u, and can be related to other properties of an equilibrium state without any reference to its microscopic origin.

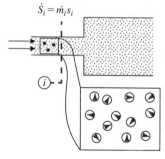

FIGURE 2.13 Transport of molecular disorder (or entropy) by mass (click Entropy in Anim. 2.D.*entropyTransport*).

Like energy, entropy can be transported by mass. The rate of transport of entropy can be deduced from the generic transport equation, Eq. (2.1), as $\dot{S} = \dot{m}s$ (see Fig. 2.13). Work does not transfer entropy; work transfer involves only organized, or coherent, motion as in the rotation of a shaft (shaft work), directed movement of electrons (electrical work), or displacement of a boundary (boundary work).

Energy transferred by work can be stored in an organized manner in a system (e.g., in system's KE or PE as shown in Fig. 2.14) without affecting the system's entropy. Energy transferred by heat transfers entropy (Fig. 2.15) across the boundary through random microscopic interactions. Heat, as we discussed in Sec. 0.8, is transferred when molecules of a warmer layer interact with adjacent molecules from a cooler layer (conduction and convection heat transfer) or through photons crossing the system boundary (radiation heat transfer). With the energy of the molecules (and photons) quantized, a huge number of possible interactions can result in the same amount of heat transfer across a plane. Energy transferred by heat inherently carries entropy in the direction of heat transfer. Furthermore, if the interface (boundary) temperature is lower, energy levels are more sparse (Fig. 1.44) and the same amount of heat can be transferred with more variation. This implies more entropy transfer. At high temperatures, the energy levels are so closely spaced that the transfer of heat takes place with less variation, that is, in a more organized manner with less entropy transfer. This explains why entropy transfer by heat is inversely proportional to the boundary temperature (second postulate).

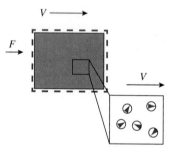

FIGURE 2.14 Mechanical work transfer to a rigid body affects all its molecules in an organized manner as the system's kinetic energy increases.

As an example, the electrical work transfer in Figure 2.16 does not carry any entropy into the system (in red boundary) containing the resistance heater; however, when work is converted to heat through Joule heating and the heat leaves the resistance heater, it must carry entropy with it. If the system is at steady state, how does the heater replenish its entropy to sustain a continuous entropy loss through heat? This is where the third postulate comes in—Joule heating is one of several distinctive mechanisms that generates entropy.

The third postulate states that a system has a natural tendency to generate entropy whenever it moves towards an equilibrium state. Even an *isolated* system without any interactions with its surroundings can spontaneously generate entropy until it reaches equilibrium and no further changes take place. Entropy generation can be looked upon as the permanent marker of spontaneous changes that lead a system to equilibrium. These spontaneous processes occur in a direction so as to dissipate or remove the internal imbalances—differences or gradients of certain properties (such as velocity, temperature, concentration, electric potential, chemical potential, etc.) within the system—and are always accompanied by relentless entropy generation. For example, coffee stirred in a cup spontaneously comes to a halt due to viscous friction. As another example, two different gases, initially separated as shown in Figure 2.17, spontaneously mix to form a uniform mixture when the valve is opened. Similarly, when the two bodies, initially at two different temperatures as shown in Figure 2.18, are placed in thermal contact, heat is transferred spontaneously so that the temperatures of the two bodies become equal. In all these

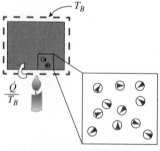

FIGURE 2.15 Entropy is transferred by heat because the energy transfer involves a huge number of random molecular interactions (see Anim. 2.D. *entropyTransferByHeat*).

cases, entropy is generated during the spontaneous processes leading to equilibrium. Friction is an easily identifiable mechanism of entropy generation. Any mechanism that tends to destroy an internal imbalance and thereby generate entropy will be referred to as some variation of mechanical friction:—**viscous friction** in the case of the coffee mug, **mixing friction** in the case of spontaneous mixing, **thermal friction** in the case of heat transfer over a finite temperature difference, **chemical friction** in the case of spontaneous phase change (ice melting to water) or chemical reaction (a fuel burning spontaneously), etc.

We will quantitatively explore different mechanisms of entropy generation (see animations starting with Anim. 2.D.*sGen-MechanicalFriction*) throughout this book, which will be collectively referred to as the **thermodynamic friction** with the rate of entropy generation $\dot{S}_{gen}$ (pronounced S-dot-gen) quantifying its intensity. Just as mechanical friction always opposes motion and at best is zero for a perfectly smooth interface, $\dot{S}_{gen}$ is never negative and at best is zero for idealized systems.

Because entropy cannot be destroyed, the generated entropy can be considered a permanent signature or a second-law footprint of a natural process. Once entropy is generated, there is no going back—entropy generation is irreversible and a fundamental marker of the passage of time. Systems or processes that involve entropy generation are called **irreversible**, and entropy generation is sometime referred to as the *arrow of time*. For a given system, $\dot{S}_{gen}$ can be viewed as the rate at which thermodynamic friction causes irreversible changes.

In Chapter 6, we will establish a one-to-one connection between entropy generation and destruction of *exergy* (Sec. 1.5.10), the useful part of the energy stored in a system. Whenever entropy is generated in a system, it is accompanied by destruction of a part of the stored exergy in the system. If electric energy can be stored in a battery in a reversible manner, without any entropy generation, the entire useful energy (electricity in this case) can be extracted back at a later time. However, due to nature's tendency to generate entropy, the stored exergy will decrease over time. The energy stored in a battery can be preserved by isolating it, but its exergy will spontaneously decay due to generation of entropy. Useful energy also can be stored in the kinetic energy of an isolated flywheel. Friction will ensure the stored exergy decreases over time even though the stored energy remains unchanged. Entropy generation, therefore, will be linked in Chapter 6 to the spontaneous degradation of the quality of stored energy in a system.

To express the second law of thermodynamics in the template of a balance equation, let's use the simplified generic system of Figure 2.6 that was used for the derivation of the energy equation. Within the interval Δt, the amount of entropy transported by mass at the inlet and exit are given by $\dot{S}_i \Delta t = \dot{m}_i s_i \Delta t$ and $\dot{S}_e \Delta t = \dot{m}_e s_e \Delta t$, respectively. By the second postulate, heat carries $\dot{Q}\Delta t / T_B$ amount of entropy across the boundary during the same interval, provided that the boundary temperature is uniform at T_B. By the third postulate, the entropy that is spontaneously generated is given by $\dot{S}_{gen} \Delta t$, where $\dot{S}_{gen} \geq 0$. Combining all contributions, $S(t + \Delta t)$ can be related to $S(t)$ by (see Anim. 2.D.*entropyBalanceEqn*):

$$S(t + \Delta t) = S(t) + \dot{S}_i \Delta t - \dot{S}_e \Delta t + \frac{\dot{Q}\Delta t}{T_B} + \dot{S}_{gen}\Delta t; \quad \left[\frac{kJ}{K}\right] \tag{2.12}$$

Rearranging and taking the limit yields:

$$\lim_{\Delta t \to 0} \frac{S(t + \Delta t) - S(t)}{\Delta t} = \dot{S}_i - \dot{S}_e + \frac{\dot{Q}}{T_B} + \dot{S}_{gen}$$

$$\Rightarrow \quad \frac{dS}{dt} = \dot{m}_i s_i - \dot{m}_e s_e + \frac{\dot{Q}}{T_B} + \dot{S}_{gen}; \quad \left[\frac{kJ}{s \cdot K} = \frac{kW}{K}\right]$$

For multiple inlets and exits, this equation leads to the **entropy balance equation** in its most general form:

$$\underbrace{\frac{dS}{dt}}_{\substack{\text{Rate of increase} \\ \text{of entropy in an} \\ \text{open system.}}} = \underbrace{\sum_i \dot{m}_i s_i}_{\substack{\text{Rate of entropy} \\ \text{transport into the} \\ \text{system by mass.}}} - \underbrace{\sum_e \dot{m}_e s_e}_{\substack{\text{Rate of entropy} \\ \text{transport out of the} \\ \text{system by mass.}}} + \underbrace{\frac{\dot{Q}}{T_B}}_{\substack{\text{Rate of entropy} \\ \text{transfer into the} \\ \text{system by heat.}}} + \underbrace{\dot{S}_{gen}}_{\substack{\text{Rate of entropy} \\ \text{generation within the} \\ \text{system boundary by} \\ \text{thermodynamic friction.}}} ; \quad \left[\frac{kW}{K}\right] \tag{2.13}$$

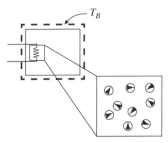

FIGURE 2.16 Entropy is generated in the electric heater through *electronic friction* and carried by heat (click Entropy Balance in Anim. 2.D. *sGen-ElectronicFriction*).

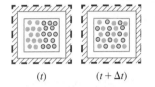

(t) $(t + \Delta t)$

FIGURE 2.17 Entropy is generated as the concentration gradient is destroyed spontaneously (through *mixing friction*) in this isolated system.

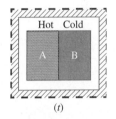

Hot   Cold

A   B

(t)

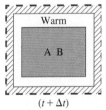

Warm

A B

$(t + \Delta t)$

FIGURE 2.18 Entropy is generated as the temperature gradient is destroyed by *thermal friction* spontaneously in this isolated system (see Anim.1.B. *thermalEquilibirium*).

As with the mass and energy equations, the LHS is the *unsteady* term—the rate of change of total entropy in an open unsteady system. The RHS consists of the contribution from *transport* of entropy by mass, transfer of entropy through heat, and spontaneous *generation* of entropy as dictated by the second law.

The entropy equation states that *the rate of change of entropy of a generic system is equal to the sum of the net rate of transport of entropy by mass, transfer of entropy by heat transfer, and spontaneous generation of entropy.* Representing the net rate of entropy transport by $\dot{S}_{net}$, Eq. (2.13) can be written in a concise form:

$$\underbrace{\frac{dS}{dt}}_{\substack{\text{Unsteady}\\\text{term}}} = \underbrace{\dot{S}_{net}}_{\substack{\text{Net}\\\text{transport}}} + \underbrace{\frac{\dot{Q}}{T_B}}_{\substack{\text{Boundary}\\\text{transfer}}} + \underbrace{\dot{S}_{gen}}_{\substack{\text{Generation}\\\text{term}}}; \quad \text{where} \quad \dot{S}_{net} = \sum_i \dot{S}_i - \sum_i \dot{S}_e; \quad \left[\frac{kW}{K}\right] \quad \textbf{(2.14)}$$

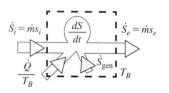

FIGURE 2.19 Flow diagram for the entropy balance equation (see Anim. 2.D.*flowDiagram*).

An **entropy flow diagram** can be useful in graphically representing the entropy balance equation. The only new type of term in this equation is the generation term $\dot{S}_{gen}$, which is represented by the dotted arrow in Figure 2.19. In our *checkbook analogy,* the generation term is analogous to the interest accrued in a savings account. Just as the interest rate cannot be negative, a negative $\dot{S}_{gen}$ is forbidden by the second law. We must emphasize that the entropy of a system can decrease, remain constant, or increase depending on the algebraic sum of the terms on the RHS of Eq. (2.14). For example, if a closed system ($\dot{S}_{net} = 0$) is losing heat to the surroundings ($\dot{Q}$ is negative), its entropy will most likely decrease despite the generation of entropy ($\dot{S}_{gen} \geq 0$).

The entropy equation can be simplified in the same manner as the energy and mass equations. In a *closed system*, for example, the transport terms drop out, resulting in:

$$\frac{dS}{dt} = \dot{S}_{net}^{\,0} + \frac{\dot{Q}}{T_B} + \dot{S}_{gen} = \frac{\dot{Q}}{T_B} + \dot{S}_{gen}; \quad \left[\frac{kW}{K}\right] \quad \textbf{(2.15)}$$

Furthermore, if the system is isolated, there is no heat transfer. While the energy of an isolated system remains constant, *the entropy of an isolated system increases* (or remains constant) because $\dot{S}_{gen} \geq 0$. Attributed to Clausius, this is one of several original proclamations of the second law.

Most open devices generally operate at steady state. For a single-flow open-steady device, the entropy equation can be further simplified as:

$$\frac{dS}{dt}^{\,0} = \dot{m}_i s_i - \dot{m}_e s_e + \frac{\dot{Q}}{T_B} + \dot{S}_{gen}; \quad \Rightarrow \quad 0 = \dot{m}(s_i - s_e) + \frac{\dot{Q}}{T_B} + \dot{S}_{gen};$$

$$\Rightarrow \quad \underbrace{\dot{m}s_e}_{\substack{\text{Entropy}\\\text{out}}} = \underbrace{\dot{m}s_i + \dot{Q}/T_B}_{\text{Entropy in}} + \underbrace{\dot{S}_{gen}}_{\substack{\text{Entropy}\\\text{generated}}}; \quad \left[\frac{kW}{K}\right] \quad \textbf{(2.16)}$$

From this simplified form, it can be easily established that the entropy at the inlet must be equal to that at the exit for an adiabatic, steady, single-flow device with no internal (thermodynamic) friction. This is an important conclusion that will be fully explored in Chapter 4.

2.1.4 Entropy and Reversibility

The transformation of a system from one equilibrium state to another is called a **process**. Restoring the system and its surroundings to their original conditions by reversing a process has been a long standing goal for engineers. Think how nice it would be if all polluting processes could be easily reversed or removing salt from saline water was as easy as dissolving salt in water. Most practical processes, however, are associated with one or more types of *thermodynamic friction*; each of which leaves an indelible footprint in the form of entropy generation within, and possibly, around the system. Such processes cannot be completely reversed as entropy once generated cannot be destroyed. That is why they are called **irreversible processes**. The degree of **irreversibility** of a process is directly related to the amount (or rate) of entropy generation—the more the entropy generation, the more the irreversibility of the process. A system with a non-zero $\dot{S}_{gen}$ is called an **irreversible system**, and the rate of entropy generation reflects the rate of irreversibility caused by thermodynamic friction at a given instant.

In most practical devices, entropy is generated not only inside the system but also in the immediate surroundings. This is typical for devices that exchange heat with the surroundings. For example, in the turbine of Figure 2.20, entropy is not only generated inside the turbine due to *mechanical* and *viscous* friction (different mechanisms of entropy generation are illustrated in animations starting with Anim. 2.D.*sGen-MechanicalFriction*) but also in the immediate vicinity due to **thermal friction**. Heat loss from the turbine into the atmosphere over a finite temperature difference is a source of entropy generation just like mechanical friction (this will be proven in Ex. 2-6). The entropy generation rate, $\dot{S}_{gen}$ depends on how the boundary is drawn—the external red boundary in Fig. 2.20 captures not only all the entropy generated inside the system but also that generated in the immediate surroundings. Entropy generated within a given system is called **internal irreversibility**. The generation of entropy outside the system is called **external irreversibility**. An extended enclosure, such as the red boundary of Figure 2.20, captures all the entropy generation, internal and external. The extended system is known as the **system's universe**. Entropy generation is additive and the sum of internal and external entropy generation defines the total entropy generation rate in the system's universe:

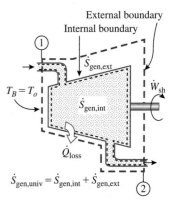

FIGURE 2.20 Entropy is generated internally and externally. The total entropy generated is called *entropy generation in the system's universe* (see Anim. 2.D. *reversibility*).

$$\dot{S}_{gen,univ} = \dot{S}_{gen,int} + \dot{S}_{gen,ext}; \quad \left[\frac{kW}{K}\right] \tag{2.17}$$

Because entropy cannot be destroyed, $\dot{S}_{gen,univ}$ cannot be less than $\dot{S}_{gen,int}$ or $\dot{S}_{gen,ext}$. It may appear that calculating $\dot{S}_{gen,univ}$ is a daunting task, given the subjective nature of drawing an external boundary to include the immediate surroundings. Actually, it is easier to calculate $\dot{S}_{gen,univ}$ than $\dot{S}_{gen,int}$ in most cases. For instance, the turbine's universe enclosed within the red external boundary of Figure 2.20. With the boundary passing through the surrounding air, T_B in the entropy balance equation can be replaced with T_0, the known temperature of the surrounding atmosphere. In contrast, calculation of $\dot{S}_{gen,int}$ would require a complete knowledge of the distribution of T_B over the internal boundary.

Recall that the second law does not mandate generation of entropy; it only stipulates that entropy cannot be destroyed. It is theoretically possible for a particular device (during a process or steady-state operation) to function without generating any entropy. A system is said to be **internally reversible** when $\dot{S}_{gen,int} = 0$, **externally reversible** when $\dot{S}_{gen,ext} = 0$, and **reversible** when $\dot{S}_{gen,univ} = 0$ (Fig. 2.21). Obviously, a reversible system must be both internally and externally reversible since $\dot{S}_{gen,univ} = 0$ implies $\dot{S}_{gen,int} = \dot{S}_{gen,ext} = 0$ from Eq. (2.17). For adiabatic systems (no heat transfer), there is no *thermal friction* in the immediate surroundings and $\dot{S}_{gen,univ} = \dot{S}_{gen,int}$. Although most practical systems are highly unlikely to be reversible, reversibility plays an important role in thermodynamic analysis by helping us define ideal limits of system performance.

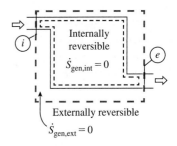

FIGURE 2.21 A *reversible* system must be internally and externally reversible (click Reversible in Anim. 2.D.*reversibility*).

For many practical systems with variable boundary temperatures, the boundary can be divided into k segments, each at a uniform temperature T_k, and the entropy equation can be modified as:

$$\frac{dS}{dt} = \sum_i \dot{m}_i s_i - \sum_e \dot{m}_e s_e + \sum_k \frac{\dot{Q}_k}{T_k} + \dot{S}_{gen}; \quad \left[\frac{kW}{K}\right] \tag{2.18}$$

Whether $\dot{S}_{gen}$ in this equation represents $\dot{S}_{gen,int}$ or $\dot{S}_{gen,univ}$ depends on what boundary, internal or external, is selected. Suppose we are interested in evaluating $\dot{S}_{gen,univ}$ for the closed-steady system in Figure 2.22 that exchanges heat with three different reservoirs, $\dot{Q}_1$ and $\dot{Q}_2$, coming from sources at T_1 and T_2, and $\dot{Q}_0$, which is rejected into the ambient atmosphere at T_0. In Figure 2.22, let the heat transfer symbols represent the magnitudes and the arrows indicate their directions. Equation (2.18) can now be simplified as follows:

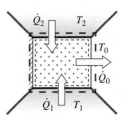

FIGURE 2.22 An entropy analysis is actually simpler for the system's universe (red boundary) than the internal system (black boundary) (see Anim. 2.D.*entropyTransfer ByHeat*).

$$\frac{dS^0}{dt} = \sum_i \dot{m}_i^0 s_i - \sum_e \dot{m}_e^0 s_e + \frac{\dot{Q}_1}{T_1} + \frac{\dot{Q}_2}{T_2} - \frac{\dot{Q}_0}{T_0} + \dot{S}_{gen,univ};$$

$$\Rightarrow \quad \dot{S}_{gen,univ} = \frac{\dot{Q}_0}{T_0} - \frac{\dot{Q}_1}{T_1} - \frac{\dot{Q}_2}{T_2}; \quad \left[\frac{kW}{K}\right]$$

The entire amount of entropy generation is due to *thermal friction*. Notice that a similar analysis to evaluate $\dot{S}_{gen,int}$ would be more difficult, since the temperature along the internal boundary most likely varies in a complicated manner.

In Example 2-6, we analyze closed and open-steady systems by exploring two important agents of entropy generation: *viscous* and *thermal* friction. Electrical resistance heating, free expansion, mixing, and chemical reactions are other prominent mechanisms of entropy illustrated in a series of animations starting with Anim. 2.D.*sGen-MechanicalFriction*.

EXAMPLE 2-6   Entropy Generation through Heat Transfer (Thermal Friction)

A thin wall separates two large chambers, the left one contains boiling water at 200°C and the right one contains a boiling refrigerant at 0°C. The heat transfer through the wall at steady state is measured at 20 kW. Assuming each wall surface to be at its respective chamber's temperature, determine the rate of entropy transfer (a) from the left chamber to the wall and (b) from the wall to the right chamber. (c) Explain the difference between the two answers.

SOLUTION

Analyze the entropy balance equation for the wall, a closed-steady system, where heat transfer at its two faces are the only interactions.

Assumptions

The wall remaining at steady state, all extensive properties, including the total entropy S of the system enclosed within the red boundary of Figure 2.23, remain constant.

Analysis

We have already established in Example 2-3 that the heat flow from the left chamber into the wall must be equal in magnitude to the heat flow from the wall into the right chamber. Using the same symbols as before, the absolute values of entropy transported by heat at the two wall faces can be obtained from the second law:

$$\text{Left face: } \frac{\dot{Q}_1}{T_1} = \frac{20}{(273 + 200)} = 0.0423 \ \frac{\text{kW}}{\text{K}}$$

$$\text{Right face: } \frac{\dot{Q}_2}{T_2} = \frac{20}{273} = 0.0733 \ \frac{\text{kW}}{\text{K}}$$

The direction of the entropy transfer coincides with the heat flow direction as shown in Figure 2.23.

At steady state, the total entropy of the wall remains constant, so the unsteady term of the entropy balance equation must be zero. Therefore, Eq. (2.18) simplifies to:

$$\frac{d\cancel{S}}{\cancel{dt}}^{\,0,\text{ steady state}} = \sum_i \cancel{\dot{m}_i}^{0} s_i - \sum_e \cancel{\dot{m}_e}^{0} s_e + \frac{\dot{Q}}{T_B} + \dot{S}_{gen}$$

$$\Rightarrow \quad 0 = \frac{\dot{Q}}{T_1} - \frac{\dot{Q}}{T_2} + \dot{S}_{gen}; \quad \Rightarrow \quad \dot{S}_{gen} = \frac{20}{273} - \frac{20}{473} = 0.031 \ \frac{\text{kW}}{\text{K}}$$

The generation of entropy in the wall explains the difference in the entropy transfer at the two wall faces.

Discussion

In this example, we have used the entropy balance equation to uncover one of the fundamental mechanisms of entropy generation: heat transfer across a finite temperature difference or *thermal friction* (see Anim. 2.D.*sGen-ThermalFriction*). For a given rate of heat transfer, $\dot{S}_{gen}$ depends on the magnitude of temperature difference and is independent of the thickness or any other properties of the wall.

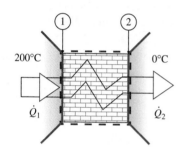

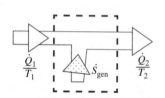

FIGURE 2.23 System schematic and entropy flow diagram for Ex. 2-6 (also see Anim. 2.D.*sGen-ThermalFriction*).

DID YOU KNOW?

• Clausius and Kelvin-Planck statements are the original pronouncements of the second law, which preceded the development of entropy as a property.

Imagine what would happen if the natural direction of heat transfer in Example 2-6 were from the right chamber to the left against the temperature gradient. An entropy analysis would then lead to a negative value of $\dot{S}_{\text{gen}}$, which is forbidden by the second law. The natural direction of heat transfer seems trivial, but without the second law, this simple fact is impossible to establish theoretically. One of the original pronouncements of the second law states:

It is impossible for any system to operate in such a way that the sole result would be an energy transfer by heat from a cooler body to a hotter body.

This is known as the **Clausius statement**, deduced as an exercise of the entropy balance equation.

If the temperature difference between the two chambers is reduced, $\dot{S}_{\text{gen}}$ decreases and approaches zero as $T_1 \rightarrow T_2$. In the limit of infinitesimal ΔT, heat transfer is said to be **reversible heat transfer** (click Reversible Heat Transfer in Anim. 2.D.*sGen-ThermalFriction*) as no permanent signature is generated in the form of $\dot{S}_{\text{gen}}$. However, temperature difference being the driving force of heat transfer, one may wonder if significant heat transfer is possible with an infinitesimal ΔT. Recall from Sec. 0.8, that Q not only depends on ΔT, but also on A, and the duration t. By providing a large area of contact and sufficient duration, a finite amount of reversible heat transfer is possible, however impractical that may be.

EXAMPLE 2-7   Entropy Generation through Friction and Heat Transfer

A fan recirculates air in a closed chamber while consuming 1 kW of electric power. The chamber temperature increases and reaches a steady value of 70°C at which point the electrical work is balanced by an equivalent amount of heat transfer out of the chamber into the surroundings at 25°C. Determine the entropy generation rate (a) inside the chamber and (b) in the chamber's universe. (c) Identify the mechanisms of entropy generation.

SOLUTION

Analyze the energy and entropy equation for the internal system and the system's universe.

Assumptions

The chamber wall is considered to be outside the system, part of the immediate surroundings.

Analysis

At steady state, all extensive properties, including the total stored energy E and entropy S, remain constant. The energy equation applied to the system or its universe (either boundary in Fig. 2.24) produces:

$$\frac{dE}{dt}^{0,\ \text{steady state}} = \sum_i \dot{m}_i^0 j_i - \sum_e \dot{m}_e^0 j_e + \dot{Q} - \dot{W}_{\text{ext}};$$

$$\Rightarrow \dot{Q} = \dot{W}_{\text{ext}} = \dot{W}_{\text{el}} = -1 \text{ kW}$$

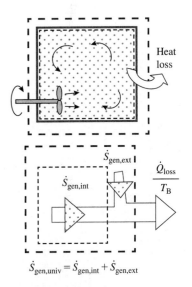

$$\dot{S}_{\text{gen,univ}} = \dot{S}_{\text{gen,int}} + \dot{S}_{\text{gen,ext}}$$

FIGURE 2.24 System schematic and entropy flow diagram for Ex. 2-7 (see Anim. 2D.*sGen-ViscousFriction*).

For the two boundaries shown in Fig. 2.24, the entropy balance equation produces:

$$\text{System: } \frac{dS}{dt}^{0,\ \text{steady state}} = \sum_i \dot{m}_i^0 s_i - \sum_e \dot{m}_e^0 s_e + \frac{\dot{Q}}{T_B} + \dot{S}_{\text{gen,int}};$$

$$\Rightarrow \dot{S}_{\text{gen,int}} = -\frac{\dot{Q}}{T_B} = -\frac{(-1)}{273 + 70} = 0.0029 \frac{\text{kW}}{\text{K}}$$

$$\text{Universe: } \frac{dS}{dt}^{0,\ \text{steady state}} = \sum_i \dot{m}_i^0 s_i - \sum_e \dot{m}_e^0 s_e + \frac{\dot{Q}}{T_B} + \dot{S}_{\text{gen,univ}};$$

$$\Rightarrow \dot{S}_{\text{gen,univ}} = -\frac{\dot{Q}}{T_B} = -\frac{(-1)}{273 + 25} = 0.0034 \frac{\text{kW}}{\text{K}}$$

Discussion

While $\dot{S}_{gen,int}$ can be attributed primarily to the dissipation of the air's KE into internal energy through *viscous friction*, the difference $\dot{S}_{gen,univ} - \dot{S}_{gen,int}$ is solely due to *thermal friction*. The solution would be identical if the fan was shaft-driven. Although work transfer, in this example, generates entropy, it is not always the case. For example, if work transfer is used to raise a weight, impart rotational KE into a flywheel, or charge a battery, entropy generation could be reduced or even eliminated. Unlike heat, work does not inherently transfer entropy.

EXAMPLE 2-8 Entropy Analysis of an Open-Steady System

The following additional data are supplied for the steam turbine of Ex. 2-4. State-1: $s_1 = 7.127 \text{ kJ/(kg} \cdot \text{K)}$; State-2: $s_2 = 7.359 \text{ kJ/(kg} \cdot \text{K)}$. Also, the temperature of the turbine surface remains constant at 150°C while the surrounding atmosphere is at 25°C. Assuming steady state operation, determine the entropy generation rate (a) inside the turbine and (b) in the turbine's universe.

SOLUTION

Analyze the entropy balance equation for the system and its universe.

Assumptions

The flow is uniform and in LTE at the inlet and exit.

Analysis

Simplifying the entropy balance equation, Eq. (2.13), for the internal system (Fig. 2.25) and substituting the given properties, we obtain:

$$\frac{d\cancel{S}^{0,\text{steady state}}}{dt} = \sum_i \dot{m}_i s_i - \sum_e \dot{m}_e s_e + \frac{\dot{Q}}{T_B} + \dot{S}_{gen,int}; \quad \left[\frac{\text{kW}}{\text{K}}\right]$$

$$\Rightarrow \quad \dot{S}_{gen,int} = -\dot{m}s_1 + \dot{m}s_2 - \frac{\dot{Q}}{T_B}$$

$$= -(2)(7.127) + (2)(7.359) - \frac{(-10)}{(273 + 150)}$$

$$= 0.488 \text{ kW/K}$$

For the system's universe, shown by the red boundary in Fig. 2.25, $T_B = T_0$. Therefore:

$$\frac{d\cancel{S}^{0}}{dt} = \sum_i \dot{m}_i s_i - \sum_e \dot{m}_e s_e + \frac{\dot{Q}}{T_B} + \dot{S}_{gen,univ}; \quad \left[\frac{\text{kW}}{\text{K}}\right]$$

$$\Rightarrow \quad \dot{S}_{gen,univ} = -\underbrace{\dot{m}s_1}_{\dot{S}_1 = 14.25 \text{ kW/K}} + \underbrace{\dot{m}s_2}_{\dot{S}_2 = 14.72 \text{ kW/K}} - \underbrace{\frac{\dot{Q}}{T_0}}_{-0.034 \text{ kW/K}} = 0.498 \text{ kW/K}$$

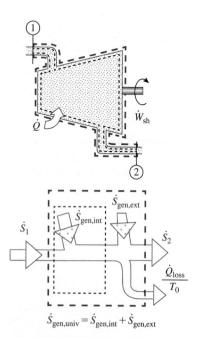

FIGURE 2.25 System schematic and entropy flow diagram for Ex. 2-8.

Discussion

Entropy transport rates $\dot{S}_1$ and $\dot{S}_2$ and the energy transfer $\dot{Q}$ in the entropy balance equation for the turbine are independent of the boundary—internal or external—that is selected for the analysis. Only the entropy generation rate is dependent on the selection of boundary. If a system is adiabatic, the internal and external boundaries can be interchangeably used without affecting the analysis.

2.2 CLOSED-STEADY SYSTEMS

Systems that are closed and steady can be analyzed without having to evaluate states, which involves understanding equilibrium, material models, and tables and charts (Chapter 3). A large class of important systems falls under this category. In this section, we analyze *closed-steady* systems with the help of the mass, energy, and entropy equations.

A closed system has no mass interactions with its surroundings, and a steady system does not change its global state over time. When both these conditions are satisfied, we have a closed-steady system. Notice how the generic open-unsteady system of Anim. 2.A.*genericSystem* differs from the generic closed-steady system illustrated in Anim. 2.E.*closedSteadyGeneric*. At a steady state, the thermodynamic picture of the system represented by the color composition does not change with time despite the presence of work and heat interactions. A steady system can be non-uniform as indicated by the variation of color across the system.

Although most well-known thermodynamic devices, such as turbines, pumps, compressors, etc., are open systems, we can identify plenty of closed-steady systems if we look around. This textbook is a trivial but legitimate closed (no mass transfer) steady (the global state does not change with time) system. Non-trivial examples include a light bulb, a gear box, an electric heater, an electrical adapter, or even a refrigerator (browse the series of animations depicting closed-steady systems in Sec. 2.E)—the *global states* of these closed systems remain unchanged over time during steady operation.

Extensive properties mass, stored energy, and entropy of a generic system can be obtained by aggregating the corresponding properties of local systems. Recall from our image analogy (Sec. 1.4) that each pixel of the global image represents a local system. At a steady state, by definition, the picture remains frozen so the local states (pixel colors) do not change with time. Total properties m, E, and S, which can be calculated by aggregating information from each local system, must also remain constant at a steady state. With the unsteady and transport terms set to zero, the balance equations simplify as follows:

$$\textbf{Mass} \quad m = \text{constant}; \quad [\text{kg}] \tag{2.19}$$

$$\textbf{Energy} \quad 0 = \dot{Q} - \dot{W}_{\text{ext}} = \dot{Q} - (\dot{W}_B + \dot{W}_{\text{sh}} + \dot{W}_{\text{el}}); \quad [\text{kW}] \tag{2.20}$$

$$\textbf{Entropy} \quad 0 = \frac{\dot{Q}}{T_B} + \dot{S}_{\text{gen}}; \quad \left[\frac{\text{kW}}{\text{K}}\right] \tag{2.21}$$

Even a cursory examination of the energy and entropy equations reveals that these equations do not involve any state properties, thereby, simplifying the analysis considerably. We have already analyzed a few closed-steady systems in Examples 2-3, 2-6, and 2-7.

Before we continue with any further analysis, let's define a commonly used performance parameter—the **energetic** or **first-law efficiency**. Efficiency is one of the frequently used terms in engineering and is loosely defined as *the ratio of desired output to required input*. We will come across different types of efficiencies, such as energetic or first-law efficiency, thermal efficiency, isentropic efficiency, exergetic or second-law efficiency, and combustion efficiency. Efficiency is used to compare the performance of a device with that of an idealized device, or to compare a desired output with the required input.

The energetic or first-law efficiency (Fig. 2.26), is defined as:

$$\eta_{\text{I}} \equiv \frac{\text{Desired Energy Output}}{\text{Required Energy Input}} \tag{2.22}$$

where the symbol η (eta) is reserved for efficiency of any kind and the subscript I symbolizes the first law. Device specific in nature, the numerator and denominator in this definition usually appear as individual terms of the energy balance equation for the system. Often, an energy flow diagram can be helpful in formulating the definition of the energetic efficiency.

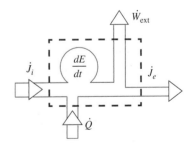

FIGURE 2.26 The energetic efficiency of a system is defined as the ratio of a desired energy term in the energy equation to the required energy transfer.

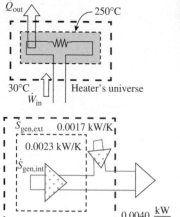

FIGURE 2.27 System schematics and entropy flow diagram for Ex. 2-9 (see Anim. 2.D. sGen-ElectronicFriction).

EXAMPLE 2-9 **Analysis of an Electric Heater**

An electric heater, operating steadily, maintains its chamber at 30°C while consuming a current of 10 A from a 120 V source. If the temperature of the heater surface is 250°C, determine (a) the rate of heat transfer, (b) the heater's energetic efficiency, (c) the rate of internal entropy generation, and (d) the rate of entropy generation in the heater's universe.

SOLUTION

Analyze the energy and entropy balance equations for the closed-steady system enclosed within the black (internal) and red (external) boundaries in Fig. 2.27.

Assumptions

The heater and its universe are at steady state.

Analysis

The electrical power consumption (absolute value) is:

$$\dot{W}_{in} = -\dot{W}_{el} = \frac{VI}{(1000 \text{ J/kJ})} = \frac{(120)(10)}{(1000 \text{ J/kJ})} = 1.2 \text{ kW}$$

The energy equation, Eq. (2.20), reduces to an identical form for either the system or its universe (red boundary), which is shown in Fig. 2.27:

$$\frac{dU^{\;0}}{dt} = \dot{Q} - \dot{W}_{ext} = (-\dot{Q}_{out}) - (\dot{W}_{in}); \quad \Rightarrow \quad \dot{Q}_{out} = \dot{W}_{in} = 1.2 \text{ kW}$$

The desired energy output is the heat produced by the heater while the required input is the electrical power consumed. Thus, the energetic efficiency is:

$$\eta_{I} = \frac{\dot{Q}_{out}}{\dot{W}_{in}} = \frac{1.2}{1.2} = 100\%$$

With the internal system boundary at 250°C, an entropy balance for the internal system (enclosed by the black boundary in Fig. 2.27) produces:

$$\frac{dS^{\;0}}{dt} = \frac{\dot{Q}}{T_B} + \dot{S}_{gen,int}; \quad \Rightarrow \quad \dot{S}_{gen,int} = -\frac{-1.2}{(273 + 250)} = 0.00229 \frac{\text{kW}}{\text{K}}$$

For the system's universe (enclosed within the red boundary of Fig. 2.27), which is also at steady state:

$$\frac{dS^{\;0}}{dt} = \frac{\dot{Q}}{T_B} + \dot{S}_{gen,univ}$$

$$\Rightarrow \quad \dot{S}_{gen,univ} = \frac{\dot{Q}_{out}}{T_B} = \frac{1.2}{(273 + 30)} = 0.00396 \frac{\text{kW}}{\text{K}}$$

Discussion

The difference between $\dot{S}_{gen,univ}$ and $\dot{S}_{gen,int}$ is $\dot{S}_{gen,ext}$, which is the external entropy generation in the heater's immediate surroundings. $\dot{S}_{gen,int}$ is caused primarily by the dissipation of electrical work into internal energy or what can be called *electrical friction* while $\dot{S}_{gen,ext}$ must be due to *thermal friction,* a mechanism investigated in Ex. 2-7. Thermal friction should not be confused with entropy transfer through heat.

| EXAMPLE 2-10 | Shaft Work Transfer in a Gear Box |

As shown in Fig. 2.28, a car engine delivers 150 kW of shaft power steadily at 4000 rpm into a gear box with a gear ratio of 2. The gear box's surface is 10°C warmer than the atmosphere, which is at 25°C. If the heat lost to the surroundings through convection is 3 kW, determine the (a) power and torque delivered by the output shaft, (b) energetic efficiency, (c) entropy generation rate in the gearbox, and (d) entropy generation rate in the system's universe.

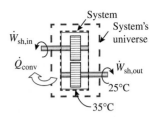

SOLUTION

Analyze the energy and entropy balance equations for the closed-steady systems enclosed within the black and red boundaries of Fig. 2.28.

Assumptions

The system and its universe are at steady state.

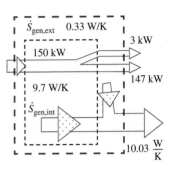

FIGURE 2.28 System schematic with energy and entropy flow diagrams for Ex.2-10 (see Anim. 2.D. *sGen-MechanicalFriction*).

Analysis

The energy equation, Eq. (2.20), for either boundary in Fig. 2.28 can be written as:

$$\cancel{\frac{dE}{dt}}^{0} = \dot{Q} - \dot{W}_{ext} = (-\dot{Q}_{conv}) - (\dot{W}_{sh,out} - \dot{W}_{sh,in}); \quad \text{or,} \quad \dot{W}_{sh,in} = \dot{W}_{sh,out} + \dot{Q}_{conv}$$

where $\dot{Q}_{conv}$, $\dot{W}_{sh,out}$, and $\dot{W}_{sh,in}$ are absolute values of the respective energy transfer rates. Solving for $\dot{W}_{sh,out}$ from the given quantities:

$$\dot{W}_{sh,out} = \dot{W}_{sh,in} - \dot{Q}_{conv} = 150 - 3 = 147 \text{ kW}$$

Given that the rotational speed of the output shaft is 4000/2 = 2000 rpm, the torque is calculated from Eq. (0.11) as:

$$T = \frac{60}{2\pi N} \dot{W}_{sh} = \frac{60}{2\pi (2000)} (147) = 0.702 \text{ kN} \cdot \text{m}$$

The energetic efficiency can be defined as the ratio of the shaft power delivered by the system to the required input; thus, the energetic efficiency is:

$$\eta_I = \frac{\dot{W}_{sh,out}}{\dot{W}_{sh,in}} = \frac{147}{150} = 98\%$$

An entropy balance for the internal system (black boundary) produces:

$$\cancel{\frac{dS}{dt}}^{0} = \frac{\dot{Q}}{T_B} + \dot{S}_{gen,int} = \frac{-\dot{Q}_{conv}}{T_B} + \dot{S}_{gen,int}$$

$$\Rightarrow \quad \dot{S}_{gen,int} = \frac{\dot{Q}_{conv}}{T_B} = \frac{3}{(273 + 35)} = 0.0097 \frac{\text{kW}}{\text{K}}$$

The entropy balance equation over the system's universe (enclosed by the red boundary in Fig. 2.28) yields:

$$\cancel{\frac{dS}{dt}}^{0} = \frac{\dot{Q}}{T_B} + \dot{S}_{gen,univ} = \frac{-\dot{Q}_{conv}}{T_0} + \dot{S}_{gen,univ}$$

$$\Rightarrow \quad \dot{S}_{gen,univ} = \frac{\dot{Q}_{conv}}{T_0} = \frac{3}{(273 + 25)} = 0.01007 \frac{\text{kW}}{\text{K}}$$

Discussion

In this example $\dot{S}_{gen,univ}$ marginally exceeds $\dot{S}_{gen,int}$, indicating that the majority of the entropy generation takes place due to the intense *mechanical friction* inside the gear box. The remainder is contributed by the moderate *thermal friction* in the immediate vicinity of the system due to heat transfer (see Anim. 2.D.*sGen-MechanicalFriction*). If the surface temperature were higher, the contribution from the thermal friction would also be higher.

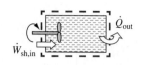

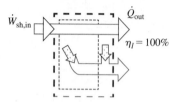

FIGURE 2.29 This device converts shaft power to heat with 100% efficiency (see Anim. 2.D.*sGen-ViscousFriction*).

2.3 CYCLES—A SPECIAL CASE OF CLOSED-STEADY SYSTEMS

It is easy to construct a device that converts work into heat. In fact, the electric heater of Ex. 2-9 converts electric work into heat with a perfect energetic efficiency of 100%. Another example is the paddle-wheel device of Fig. 2.29. As the paddle stirs the fluid, viscous friction dissipates the kinetic energy into internal energy, the temperature rises, and heat starts conducting across the wall into the outside atmosphere. Eventually, a steady state is reached. The simplified governing equations can be used to evaluate the energetic efficiency and the entropy generated in the system's universe:

$$\dot{W}_{sh,in} = \dot{Q}_{out}; \quad [kW];$$

therefore, $\quad \eta_I \equiv \dfrac{\dot{Q}_{out}}{\dot{W}_{sh,in}} = 100\%; \quad$ and, $\quad \dot{S}_{gen,univ} = \dfrac{\dot{Q}_{out}}{T_0}; \quad \left[\dfrac{kW}{K}\right]$

(2.23)

where $\dot{W}_{sh,in}$ and $\dot{Q}_{out}$ represent magnitudes of the energy transfers with the subscripts indicating the directions. As a consequence of the first law of thermodynamics, the energetic efficiency of the device must be 100%. Also, $\dot{Q}_{out}$ and T_0 are positive, so a positive $\dot{S}_{gen,univ}$ satisfies the second law.

While conversion of useful work into heat is a relatively simple to achieve (e.g., rubbing your hands against each other) at 100% efficiency, the reverse is not true. A device that can steadily convert heat to useful work is called a **heat engine** (see Anim. 2.F.*conceptHeatEngine*). Building heat engines has been such a historical obsession for engineers that the word *engin*eer owes its origin to the early pioneers of heat engines.

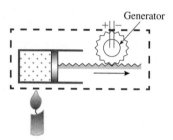

FIGURE 2.30 Although this device produces electricity from heat, it is not a heat engine because it is unsteady (click Unsteady Machine in Anim. 2.F.*perpetualMachines*).

At first glance, building a heat engine does not appear to be a difficult task. Useful work can be obtained by heating a gas in a piston-cylinder assembly as shown in Fig. 2.30; however, such a conversion is a one-time event and is not sustainable on a steady basis.

Many conceptual designs have been proposed for a heat engine (see Anim. 2.F.*perpetualMachines*). Let's consider one such device, shown in Fig. 2.31, where work is produced at a steady rate of $\dot{W}_{out}$ without any other interactions with the surroundings. With $\dot{Q} = 0$, the steady-state energy equation, Eq. (2.20), produces $\dot{W}_{ext} = \dot{W}_{out} = 0$. Any steady work output from this engine is a clear violation of the first law. Such a fictitious engine is called a **perpetual motion machine of the first kind** or **PMM1**. The second device, shown in Fig. 2.32, produces work at a steady rate of $\dot{W}_{out}$ as heat $\dot{Q}_H$ is added from a reservoir at T_H. An energy analysis yields $\dot{W}_{out} = \dot{Q}_H$, establishing an energetic efficiency of 100%. The first law is no longer violated by this improved design; however, the entropy balance equation, Eq. (2.21), when applied to the engine's universe produces, $\dot{S}_{gen,univ} = -\dot{Q}_H/T_H$, a negative quantity, violating the second law of thermodynamics (note that both Q_H and T_H are positive quantities). These fictitious engines violate the second law; hence, they are called **perpetual motion machines of the second kind** or **PMM2**.

The realization that PMM2 is forbidden by fundamental laws came before the development of the second law. The **Kelvin-Planck statement** of the second law asserts what we have just established from the entropy equation:

It is impossible to build a heat engine that exchanges heat with a single reservoir.

A reservoir or, more precisely, a thermal energy reservoir (TER) has been introduced in Sec. 2.1.2 as a large reservoir of energy, a heating source or sink, whose temperature does not change regardless of the amount of heat addition or rejection. The atmosphere is a reservoir at T_0.

FIGURE 2.31 Perpetual motion machine of the first kind violates the first law.

The Clausius statement, derived in Sec. 2.1.3, and the Kelvin-Planck statement are the pioneering pronouncements of the second-law of thermodynamics, which led to the discovery of entropy as a property. In our postulative approach, these statements are derived from the entropy balance equation. In Chapter 5, we will reconcile these two approaches by reconstructing the arguments that led to the concept of entropy.

2.3.1 Heat Engine

If the PPM2 discussed above is modified to reject heat $\dot{Q}_C$ (magnitude only with direction indicated by the arrow) to a second TER (see Fig. 2.33) at T_C, the energy and entropy balance equations, Eqs. (2.20) and (2.21), can both be satisfied:

$$\dot{Q}_{net} \equiv \dot{Q}_H - \dot{Q}_C = \dot{W}_{net}; \quad [kW]$$

Therefore, $\qquad\qquad \dot{W}_{net} > 0 \quad \text{if} \quad \dot{Q}_H > \dot{Q}_C;$

(2.24)

$$\dot{S}_{\text{gen,univ}} = \frac{\dot{Q}_C}{T_C} - \frac{\dot{Q}_H}{T_H}; \quad \left[\frac{\text{kW}}{\text{K}}\right]$$

Therefore, $$\dot{S}_{\text{gen,univ}} \geq 0 \quad \text{if} \quad \dot{Q}_C \geq \dot{Q}_H \frac{T_C}{T_H}; \qquad \textbf{(2.25)}$$

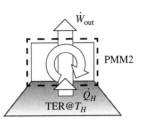

Here, $\dot{Q}_H$ and $\dot{Q}_C$ represent magnitudes only, and appropriate signs are added in accordance with the WinHip convention before they are substituted in the energy or entropy equation. To satisfy the energy equation, the analysis above establishes that the net work output must be equal to the net heat input. Moreover, the second law is satisfied. $\dot{S}_{\text{gen,univ}} \geq 0$, if $\dot{Q}_C \geq \dot{Q}_H (T_C/T_H) > 0$, that is, some of the heat received from the hot reservoir is rejected into the cold one. With the contradictions of the perpetual machines removed, the construct of Fig. 2.33 forms a fundamental framework for a practical heat engine (see Anim. 2.F.*conceptHeatEngine*). The cold reservoir is generally the atmosphere and the rejected heat $\dot{Q}_C$ is completely wasted as it dissipates into the internal energy of the atmosphere. Hence, the name **waste heat** is used in industries for $\dot{Q}_C$. Consequently, the net work output of such a conceptual **heat engine** has to be less than the required heat input, resulting in an energetic efficiency, $\dot{W}_{\text{net}}/\dot{Q}_H$, of less than 100%.

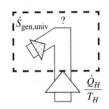

FIGURE 2.32 Perpetual motion machine of the second kind (top schematic) violates the second law. As shown by the entropy flow diagram (see Anim. 2.F.*perpetualMachines*), entropy will build up inside making the system unsteady.

Chapters 7 through 10 are devoted to analysis of heat engines and refrigeration cycles. For an overall analysis a system executing a cycle can be treated as a closed-steady system, which simplifies the analysis. Consider the schematic of a **Rankine cycle**, the analytical model of a steam power plant, shown in Fig. 2.34 and illustrated in Anim. 2.F.*openCycle*, where several open devices are connected in series to form a closed loop. External work is supplied to the **pump** at a rate $\dot{W}_{\text{in}}$ to raise the water pressure so as to force the water through the entire cycle. In the **boiler**, heat is added at a rate $\dot{Q}_H$ from an external high-temperature TER, such as a coal or gas-fired furnace, to transform water into vapor. The **turbine** extracts a significant fraction of the flow energy of the high-pressure, high-temperature vapor as useful shaft work $\dot{W}_{\text{out}}$. The low-pressure (sub-atmospheric) vapor, still hotter than atmospheric air, is condensed to liquid state by a heat exchanger called the **condenser**, where heat is rejected at a rate $\dot{Q}_C$ to a colder reservoir, usually the atmosphere or a water reservoir. The condensed water returns to the pump and completes the cycle. Power producing cycles of this type, where open-steady devices are connected back-to-back to form a loop, are called **open power cycles** and are covered in Chapters 8 (Gas Power) and 9 (Vapor Power).

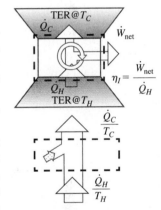

FIGURE 2.33 Energy and entropy flow diagrams for a heat engine (see Anim. 2.F.*conceptHeatEngine* and 2.F.*carnotEfficiency*).

There is only heat and work transfer no mass transfer, across the red boundary line of the system that is drawn around the Rankine cycle in Fig. 2.34 (click Closed-Steady Cycle in Anim. 2.F.*openCycle*). The resulting closed system internalizes all the open devices, leaving only heat and external work interactions with its surroundings. As complex as the interior of the heat engine may be, it is easy to determine if it is at steady state. The image of the global system, snapped with a *state camera,* will be quite colorful as the state varies from point-to-point within a device and from device-to-device. If the engine operates at steady state, any two snapshots taken at two different *macroscopic instants* will be identical. Temperature and pressure sensors connected at various locations along the steam's path will show constant readings, when fluctuations within a macroscopic instant are averaged out. The vapor power cycle qualifies as a closed-steady device as described by the abstract diagram of Fig. 2.33.

Heat engines commonly found in automobiles are called **reciprocating engines** due to the reciprocating motion of the pistons. There are two major types of reciprocating engines: the spark-ignition, or SI engines, used in gasoline-powered vehicles, which are modeled by the air-standard Otto cycle, and the compression-ignition, or CI engines, used in diesel powered vehicles, which are modeled by the air-standard Diesel cycle. To briefly describe a reciprocating cycle, for example, the Otto cycle, we consider the four processes executed by a fixed mass of air that is trapped in the piston-cylinder device of Fig. 2.35 (and in Anim. 0.B.*closedPowerCycles* and 2.F.*closedCycle*). We will revisit this topic in Chapter 7. In the compression process, the piston is pushed from the bottom location, called the **bottom dead center** or **BDC**, to the top location, called the **top dead center** or **TDC**, which requires W_{in} (magnitude only) amount of boundary work. At the end of the compression process, heat Q_H is transferred to the air almost instantly and at constant volume—this is to simulate the rapid combustion process that takes place in an actual spark-ignition engine. The high-pressure and high-temperature air expands, pushing the piston back to the BDC and transferring W_{out} in boundary work to the crankshaft. Finally, waste heat in the amount Q_C (magnitude only) is rejected at constant volume (to simulate the actual expulsion followed by fresh intake) until the air matches the original state. This process completes the cycle. The atmospheric work (work done by or against the atmospheric

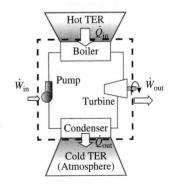

FIGURE 2.34 The Rankine cycle consists of open-steady devices connected in a loop (see Anim. 2.F.*openCycle*).

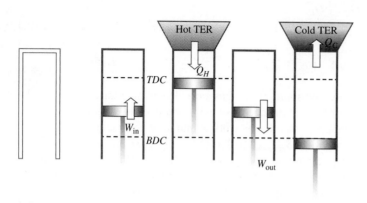

FIGURE 2.35 The Otto cycle consists of a series of processes forming a loop (see Anim. 7.B.*ottoCycle*).

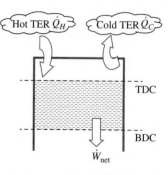

FIGURE 2.36 Averaged over many cycles, even a reciprocating system can be regarded as closed and steady (click Steady State in Anim. 2.F.*closedCycle*).

pressure) done during expansion and compression is equal and opposite in sign, which leaves the net work delivered to the shaft unaffected, i.e., $W_{net} = W_{out} - W_{in}$. Such cycles, which are executed by a closed system through a sequence of processes that form a loop, are known as **reciprocating** or **closed cycles**, and are analyzed in Chapter 7.

During a cycle, the closed system, which consists of the trapped gas, cannot be considered steady. Its state changes cyclically as the piston goes through the reciprocating motion. Suppose the shortest time scale we are interested in is long enough for a large number of cycles to complete. With the *state camera* exposed to multiple cycles in this *macroscopic instant*, the piston will appear as a blur between the TDC and BDC (Fig. 2.36); also click Closed-Steady Cycle in Anim 2.F.*closedCycle*), and the intermittent transfers of work and heat will appear continuous. If n is the fixed number of cycles executed every second, the cyclic quantities can be converted into rates through $\dot{W}_{net} = nW_{net}$, $\dot{Q}_H = nQ_H$, and $\dot{Q}_C = nQ_C$. Figure 2.33, once again, can be used to conceptually represent a heat engine, this time implemented by a closed cycle.

In terms of the absolute symbols used in Fig. 2.33 for the energy transfers, the energy balance equation for the closed-steady engine, regardless of how it is implemented (Rankine or Otto cycle), can be written as:

$$\text{Energy: } 0 = (\underbrace{\dot{Q}_H - \dot{Q}_C}_{\dot{Q}_{net}}) - \dot{W}_{net}; \quad \Rightarrow \quad \dot{Q}_{net} = \dot{W}_{net}; \quad [\text{kW}] \qquad \textbf{(2.26)}$$

The energetic efficiency of a heat engine is known as the **thermal efficiency** and is defined as follows (see Anim. 2.F.*efficiencyAndCOP*):

$$\eta_I = \eta_{th} \equiv \frac{\text{Desired Energy Output}}{\text{Required Energy Input}} = \frac{\dot{W}_{net}}{\dot{Q}_H} = \frac{\dot{Q}_H - \dot{Q}_C}{\dot{Q}_H} = 1 - \frac{\dot{Q}_C}{\dot{Q}_H} \qquad \textbf{(2.27)}$$

The operating expenses of a heat engine primarily depend on $\dot{Q}_H$, which is supplied from the burning of fossil fuels or other alternative sources. One of the goals in the design of an ideal heat engine is to maximize η_{th}, which also implies minimization of the ratio of waste heat to the heat supplied, $\dot{Q}_C/\dot{Q}_H$.

EXAMPLE 2-11   **Energy Analysis of Heat Engine**

The fuel efficiency of a compact car is rated at 40 mpg when the car has a constant speed of 70 mph and when its engine produces a power of 40 hp to overcome aerodynamic drag and other resistances. If the fuel has a heating value of 44 MJ/kg and a density of 700 kg/m³, determine (a) the thermal efficiency of the engine. ***What-if scenario:*** (b) What would the fuel mileage be if the thermal efficiency increased by 5 percent?

SOLUTION

Determine the rate of heat release $\dot{Q}_H$ and obtain the efficiency from Eq. (2.27).

Assumptions

The engine runs at steady state. The entire heating value of the fuel is converted to heat.

Analysis

Use the unit converter TESTcalc to express all quantities as SI units. The volume flow rate of fuel to the engine is:

$$\dot{V}_F = \frac{mph}{mpg} = \frac{70}{40} = 1.75 \frac{gal}{h} = 1.84 \times 10^{-6} \frac{m^3}{s}$$

The fuel consumption rate in kg/s, therefore, can now be obtained:

$$\dot{m}_F = \rho_F \dot{V}_F = \left(700 \frac{kg}{m^3}\right)\left(1.84 \times 10^{-6} \frac{m^3}{s}\right) = 1.288 \times 10^{-3} \frac{kg}{s}$$

The net power output of the engine is $\dot{W}_{net} = 40$ hp $= 29.83$ kW and the heat added due to fuel combustion is $\dot{Q}_H = \dot{m}_F(HV)$. Therefore, the thermal efficiency can be obtained from Eq. (2.27):

$$\eta_{th} = \frac{\dot{W}_{net}}{\dot{Q}_H} = \frac{\dot{W}_{net}}{\dot{m}_F(HV)} = \frac{29.83}{(1.288 \times 10^{-3})(44000)} = 52.64\%$$

What-if scenario

The mpg can be related to thermal efficiency as follows:

$$mpg \propto \frac{mph}{\dot{V}_F} = \rho_F \frac{mph}{\dot{m}_F} = \rho_F(HV)\frac{mph}{\dot{Q}_H} = \eta_{th}\rho_F(HV)\frac{mph}{\dot{W}_{net}} \qquad (2.28)$$

A 5% increase in the thermal efficiency will increase the fuel mileage by 5%. From Eq. (2.28), it might appear that an increase in the car's speed with increase its fuel mileage. But an increase in speed also increases the engine's power production $\dot{W}_{net}$ in a disproportionate manner (Ex. 0-4), and the fuel mileage actually decreases as a result, especially at a high velocity.

Discussion

The entire heating value is generally not released into heat due to incomplete combustion and other reasons that will be discussed in Chapter 13.

2.3.2 Refrigerator and Heat Pump

Being an idealized model of a steam power plant, the Rankine cycle is a reversible cycle. Conceptually, the Rankine cycle can be run backwards, causing heat to flow from the cold reservoir to the working fluid of the cycle and then from the working fluid to the hot reservoir. Such a cycle is emulated by a **refrigerator** to keep a refrigerated space colder than the environment, and by a **heat pump** to keep a heated space warmer than the environment.

A typical household refrigerator's vapor compression refrigeration cycle is shown in Fig. 2.37 and illustrated in Anim. 2.F.*refriCycle*. The working substance, which has a freezing temperature well below that of water, is called a **refrigerant**. In the **evaporator**, the temperature of the refrigerant is designed to be slightly below that of the refrigerated space, causing it to absorb $\dot{Q}_C$ (magnitude only), which is called the *cooling load* of a refrigerator. Change of phase (liquid to vapor) enables the refrigerant to remain at a constant temperature until it turns completely into vapor at the evaporator exit. An adiabatic **compressor**, consuming work at a rate of $\dot{W}_{net}$ (magnitude only), raises the pressure of the vapor to force it through the high-pressure condenser. An increase in pressure in the compressor accompanies an increase in temperature (as work transferred is converted to flow energy) above the ambient temperature. Consequently, as the vapor enters the **condenser**, heat is rejected into the atmosphere at a rate $\dot{Q}_H$. An **expansion valve** throttles the condensate into the evaporator, completing the cycle. While the pressure drop in the expansion valve (see Anim. 4.A.*expansionValve*) can be easily attributed to friction in the constricted passage of the valve, the accompanying drop in temperature is harder

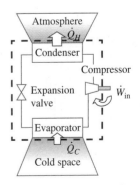

FIGURE 2.37 Vapor compression cycle used in a refrigerator or heat pump (see Anim. 2.F.*refriCycle*).

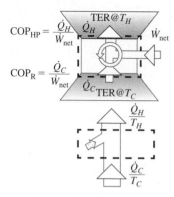

FIGURE 2.38 A refrigerator (or a heat pump) as a closed-steady system (see Anim. 2.F.*efficiencyAndCOP*).

to understand. Simply put, the equilibrium condition at a lower pressure requires some of the liquid to vaporize (the fluid to expand). With no external transfer of energy (the valve is insulated), the kinetic energy of the molecules (hence temperature) must drop to support the sudden increase of potential energy of the molecules due to increased separation in the vapor phase. We will analyze an expansion valve thoroughly in Chapter 4.

The refrigeration cycle also can be used for the opposite purpose—transferring heat from a cold space to a warm space. This device is called a **heat pump** and $\dot{Q}_H$ is called the desired *heating load*.

The closed-steady system of Fig. 2.38 can be used to model the refrigeration system shown in Fig. 2.37 for an overall analysis (click Closed-Steady Cycle in Anim. 2.F.*refriCycle*). The closed-steady system, operating in a cycle, extracts heat at a rate $\dot{Q}_C$ from the refrigerated space to keep it at a low temperature T_C against heat leakage or normal operational load. When viewed as a heat pump, the same closed-steady system supplies $\dot{Q}_H$ to the hot reservoir to keep it warm at T_H against any heat loss to the cold surroundings at T_C. The external work necessary to run the compressor is required in both configurations.

The energy equation, Eq. (2.26), derived for the heat engine, also applies to the refrigerator and heat pump described by the steady system of Fig. 2.38. The energetic efficiency for these devices can exceed 100%, and that is why it is called the **coefficient of performance** or **COP**:

$$\text{COP}_R \equiv \frac{\text{Desired Energy Transfer}}{\text{Required Input}} = \frac{\dot{Q}_C}{\dot{W}_{net}} = \frac{\dot{Q}_C}{\dot{Q}_H - \dot{Q}_C} = \frac{1}{\dot{Q}_H/\dot{Q}_C - 1} \quad \textbf{(2.29)}$$

$$\text{COP}_{HP} \equiv \frac{\text{Desired Energy Transfer}}{\text{Required Input}} = \frac{\dot{Q}_H}{\dot{W}_{net}} = \frac{\dot{Q}_H}{\dot{Q}_H - \dot{Q}_C} = \frac{1}{1 - \dot{Q}_C/\dot{Q}_H} \quad \textbf{(2.30)}$$

By manipulating these equations, it can be shown that $\text{COP}_{HP} = \text{COP}_R + 1$. From this relationship, we can see that the COP of a heat pump is always greater than 1. Although COP is a convenient measure of energetic performance for refrigerators and heat pumps, a different measure called the **Energy Efficiency Rating** or **EER** is often used in the U.S. An EER is the number of Btu's removed from a cooled space for every watt-hour of electricity consumed. Given 1 Wh $=$ 3.412 Btu, an EER can be related to the COP by EER $=$ 3.412 COP.

EXAMPLE 2-12   **Energy Analysis of a Refrigerator**

Due to ineffective sealing, heat leaks into a kitchen refrigerator at a rate of 5 kW. If the COP of the refrigerator is 3.5, determine (a) the rate of heat ejection by the refrigerator into the kitchen, and (b) the net rate of energy transfer into the kitchen. Treat the kitchen as a closed system.

SOLUTION

Perform an energy balance on the refrigerator as a closed-steady system, and the kitchen as a closed-unsteady system.

Assumptions

The kitchen can be considered an adiabatic closed system while the refrigerator runs a closed sub-system that is in a steady state.

Analysis

To keep the refrigerated space at a constant temperature, heat should be removed exactly at the same rate as the leakage rate. Therefore, $\dot{Q}_C = 5$ kW (see Anim. 2.F.*refrigerator*). The net work and the rejected heat can be obtained as follows:

$$\dot{W}_{net} = \frac{\dot{Q}_C}{\text{COP}_R} = \frac{5}{3.5} = 1.43 \text{ kW}; \quad \dot{Q}_H = \dot{Q}_C + \dot{W}_{net} = 6.43 \text{ kW}$$

As shown in the energy flow diagram of Fig. 2.39, the net rate of heat transfer from the refrigerator to the kitchen is $6.43 - 5 = 1.43$ kW.

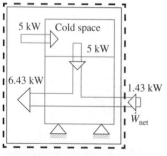

FIGURE 2.39 Schematic for Ex. 2-12 (see Anim. 2.F. *refrigerator*).

Discussion

Note that if we take the entire kitchen as a system, the only mode of energy transfer is electrical work. Regardless of how the electrical power is utilized inside the kitchen, $\dot{W}_{el} = \dot{W}_{net} = 1.43$ kW (absolute value).

2.3.3 The Carnot Cycle

Because a heat engine must reject some heat to satisfy the second law of thermodynamics, it can never achieve a thermal efficiency of 100%. What, then, is the maximum limit for η_{th}? Similar questions can be raised regarding the COP of a refrigerator or a heat pump.

A young engineer named Sadi Carnot answered those questions in 1811 using only deductive reasoning, about 30 years before the first law was formally established, and along the way gave a mathematical representation of the second law. The cycle he proposed, called the **Carnot cycle**, constitutes the most efficient heat engine possible. When its direction is reversed, it yields the highest possible COP for a refrigerator or heat pump.

A. CARNOT HEAT ENGINE The Carnot cycle is a completely reversible cycle, executed without any internal or external entropy generation. It operates between two reservoirs (TER), one at a constant high temperature T_H and another at a constant low temperature T_C (usually the temperature of the ambient atmosphere T_0). In a reversible cycle heat transfer $\dot{Q}_H$ and $\dot{Q}_C$ must take place reversibly (recall reversible heat transfer in Ex. 2-6).

Postponing specific implementations of the Carnot cycle to Chapter 7 (see Anim. 7.A.*carnotClosedCycle* for a preview), we can perform an energy and entropy analysis on the concept Carnot engine shown in Fig. 2.40. This differs from a generic heat engine (compare Fig. 2.40 to Fig. 2.33) in two respects: the temperatures of heat addition and rejection, T_H and T_C in a Carnot engine remain constant, and the engine's universe is completely devoid of any *thermodynamic friction*, that is, the engine is *reversible* or $\dot{S}_{gen,univ} = 0$. While the Rankine and Otto cycles are also internally reversible, heat transfers in those engines do not take place at constant temperatures.

The energy equation for the Carnot engine remains the same as Eq. (2.26), derived for the generic heat engine, however, when $\dot{S}_{gen,univ} = 0$ is substituted in the entropy equation, Eq. (2.25), a simple yet powerful result emerges:

$$0 = \frac{\dot{Q}_H}{T_H} - \frac{\dot{Q}_C}{T_C} + \cancelto{0}{\dot{S}_{gen,univ}}; \quad \left[\frac{kW}{K}\right] \quad \Rightarrow \quad \frac{\dot{Q}_C}{\dot{Q}_H} = \frac{T_C}{T_H} \tag{2.31}$$

To ensure reversibility, heat transfer in a Carnot cycle must be proportional to the absolute temperature of the reservoir. What is remarkable about this simple conclusion is that it is independent of how the actual cycle is implemented. Introducing this result into the definition of thermal efficiency, Eq. (2.27), results in the following expression for the **Carnot efficiency** (see Anim. 2.F.*carnotEfficiency*):

$$\eta_{Carnot} = \frac{\dot{W}_{net}}{\dot{Q}_H} = 1 - \frac{\dot{Q}_C}{\dot{Q}_H} = 1 - \frac{T_C}{T_H} \tag{2.32}$$

The Carnot efficiency is an elegantly simple formula that reduces a cycle's internal and external complexities into two easily measurable properties—the absolute temperatures of the two reservoirs. The cold reservoir is usually the atmosphere; therefore, only the highest temperature used in any heat engine needs to be known to estimate an upper limit of its thermal efficiency.

The following theorems, proved by Carnot, through deductive reasoning from the Kelvin-Planck statement, can be derived as corollaries of Eq. (2.32):

Theorem 1 *All Carnot engines operating between the same two temperatures, T_H and T_C, must have the same efficiency.*

Equation (2.32), which was derived without any reference to specific implementations of the cycle, shows that thermal efficiency of the Carnot engine is a function of the reservoir

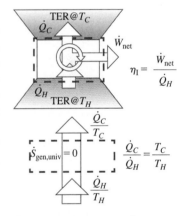

FIGURE 2.40 Energy and entropy flow in a Carnot heat engine (click Entropy Diagram in Anim. 2.F.*carnotEfficiency*).

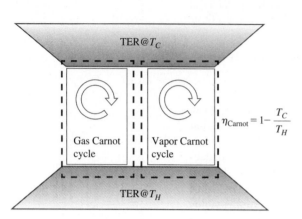

FIGURE 2.41 All Carnot cycles have the same efficiency regardless of how they are implemented.

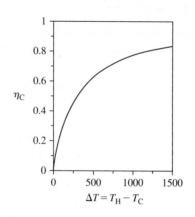

FIGURE 2.42 Carnot efficiency monotonically increases with the temperature difference between the two reservoirs.

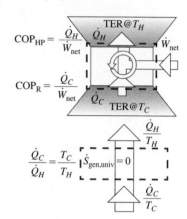

FIGURE 2.43 Energy and entropy flow in a Carnot refrigeration/heat pump cycle (see Anim. 2.F.*carnotCOP*).

temperatures alone (Fig. 2.41). Two Carnot engines, implemented differently (compare Anims. 7.A.*carnotClosedCycle* and 9.A.*carnotCycle*) but operating between the same pair of reservoirs must produce the same efficiency as dictated by Eq. (2.32).

Theorem 2 *The efficiency of a reversible engine is higher than any other engine operating between the same two temperatures.*

For an irreversible engine that operates between two fixed temperatures, the entropy generation term in the entropy equation, Eq. (2.25), cannot be neglected.

$$0 = \frac{\dot{Q}_H}{T_H} - \frac{\dot{Q}_C}{T_C} + \dot{S}_{gen,univ}; \quad \left[\frac{kW}{K}\right] \quad \Rightarrow \quad \frac{\dot{Q}_C}{\dot{Q}_H} = \frac{T_C}{T_H} + \dot{S}_{gen,univ}\frac{T_C}{\dot{Q}_H} \tag{2.33}$$

The thermal efficiency of an irreversible heat engine, therefore, is given by:

$$\eta_{th,irrev} = 1 - \frac{\dot{Q}_C}{\dot{Q}_H} = 1 - \frac{T_C}{T_H} - \dot{S}_{gen,univ}\frac{T_C}{\dot{Q}_H} = \eta_{Carnot} - \dot{S}_{gen,univ}\frac{T_C}{\dot{Q}_H} \tag{2.34}$$

with $\dot{S}_{gen,univ}$, $\dot{Q}_H$, and T_C all being positive, $\eta_{Carnot} \geq \eta_{th,irrev}$.

Theorem 3 *For the same cold reservoir temperature T_C, the Carnot engine that has the larger ΔT has the higher efficiency.*

This is obvious from the expression of η_{Carnot} derived in Eq. (2.32). A plot of η_{Carnot} against $\Delta T = T_H - T_C$, where T_C is assumed to be the standard atmospheric temperature T_0 is shown in Fig. 2.42. Notice how rapidly the Carnot efficiency increases with ΔT, especially when the efficiency is below 40%.

Although these theorems seem trivial to derive, we must remember that Carnot did not have the benefit of the entropy balance equation. In fact, it is Carnot's work that created the foundation for Clausius and Boltzmann to discover entropy.

B. CARNOT REFRIGERATOR AND HEAT PUMP   Being a completely reversible cycle, the Carnot heat engine can be perfectly reversed to serve as a refrigerator or a heat pump (compare Fig. 2.40 to Fig. 2.43). The entropy equation for this reversed configuration (Fig. 2.43) produces the same relationship between the heat transfers—proportionality of the magnitude of heat transfer rate with the absolute temperature of the reservoir—as in the case of the Carnot heat engine:

$$0 = \frac{\dot{Q}_C}{T_C} - \frac{\dot{Q}_H}{T_H} + \dot{S}_{gen,univ}^{\;\;0}; \quad \left[\frac{kW}{K}\right] \quad \Rightarrow \quad \frac{\dot{Q}_C}{\dot{Q}_H} = \frac{T_C}{T_H} \tag{2.35}$$

Substituting this result in Eqs. (2.29) and (2.30), the COP's of the Carnot refrigerator and heat pump can be expressed as follows (see the energy and entropy diagrams in Anim. 2.F.*carnotCOP*):

$$\text{COP}_{R,\text{Carnot}} = \frac{\dot{Q}_C}{\dot{W}_{\text{net}}} = \frac{\dot{Q}_C}{\dot{Q}_H - \dot{Q}_C} = \frac{1}{\dot{Q}_H/\dot{Q}_C - 1} = \frac{T_C}{T_H - T_C} \qquad (2.36)$$

$$\text{COP}_{HP,\text{Carnot}} = \frac{\dot{Q}_H}{\dot{W}_{\text{net}}} = \frac{\dot{Q}_H}{\dot{Q}_H - \dot{Q}_C} = \frac{T_H}{T_H - T_C} = 1 + \text{COP}_{R,\text{Carnot}} \qquad (2.37)$$

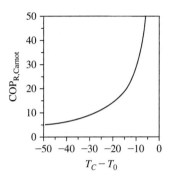

FIGURE 2.44 Carnot COP rapidly decreases when the temperature difference increases.

Again the simplicity of the formulas makes them some of the most elegant equations in engineering. The maximum possible COP of a refrigerator or heat pump can be calculated from the knowledge of the temperatures of reservoirs. Further simplification is possible when one of the reservoirs—the hot reservoir for a refrigerator or the cold reservoir for a heat pump—is the atmosphere. To study the sensitivity of the Carnot COP on the temperature difference between the reservoirs, the COP is calculated from Eq. (2.36) with T_H set to the standard atmospheric temperature T_0 and plotted in Fig. 2.44. The plot shows a dramatic increase in COP as T_C approaches T_0. Setting the temperature of the refrigerated space a few degrees closer to the outside temperature can result in considerable savings of network input for two reasons: improvement of COP and a reduction in $\dot{Q}_C$ (due to a reduction in the driving force, $T_0 - T_C$, for heat leakage).

Although actual refrigerators and heat pumps do not operate as Carnot cycles, Fig. 2.44 can be used as a qualitative guide for predicting actual COP. As an example, consider a Carnot heat pump designed for $T_H = 20°C$ (temperature inside a house), $T_C = 0°C$ (outside temperature), and $\dot{Q}_H = 5$ kW (heat pumped to the house). The compressor, therefore, must be rated at 0.34 kW as the COP can be calculated from Eq. (2.37) as 14.65. Now suppose the temperature one night drops to –20°C so that $T_C = -20°C$; consequently, the COP decreases to 7.325, almost half its original value. To pump heat at the same rate, the compressor must be able to operate at twice its original power rating. Compounding the problem, the heating load can be expected to increase due to increased heat loss to the colder surroundings. Heat pumps, therefore, are not suitable for extreme climates. The following examples highlight the power of overall cycle analysis in reaching conclusions that are independent of the specific implementation of the cycle.

EXAMPLE 2-13   Carnot Heat Engine

In the engine described in Ex. 2-11, the maximum temperature achieved during the cycle is 1500 K. If the atmospheric temperature is 298 K, determine if the fuel mileage of 40 mpg claimed by the manufacturer is reasonable.

SOLUTION

Evaluate the Carnot efficiency and compare it with the actual thermal efficiency of the engine.

Analysis

The most efficient engine for the car is a Carnot engine running between 1500 K and 298 K. The highest possible efficiency, therefore, is the Carnot efficiency:

$$\eta_{\text{th,Max}} = \eta_{\text{Carnot}} = 1 - \frac{T_C}{T_H} = 1 - \frac{298}{1500} = 80.1\%$$

Since actual thermal efficiency, calculated in Ex. 2-11, is only 52.64%, well below the Carnot efficiency. The fuel mileage claimed by the manufacturer is well within the theoretical limit and, therefore, reasonable.

Discussion

Approaching the Carnot efficiency is the ultimate goal of any energy efficient combustion engine. In Chapters 7 through 10, we will discuss various approaches to improving efficiencies of actual heat engine and refrigeration cycles. Fuel cells (Chapter 14), however, are not heat engines and many of the concepts developed in this section are not applicable to fuel cells.

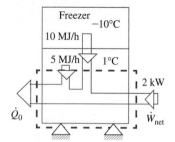

FIGURE 2.45 System schematic and energy flow diagram for Ex. 2-14.

EXAMPLE 2-14 Reversible Refrigerator

A kitchen refrigerator maintains the freezer compartment at a temperature $-10°C$ and the main compartment at $1°C$; the outside temperature is $30°C$. The refrigerator consumes 2 kW of power while removing heat from the two compartments at a rate of 10 MJ/h and 5 MJ/h, respectively (see Fig. 2.45). Determine (a) the COP of the refrigerator and (b) the rate of entropy generation in the refrigerator's universe. *What-if scenario:* (c) What would the power consumption be if the refrigerator operated in a reversible manner?

SOLUTION

Perform an energy and entropy balance on the refrigerator's universe—the closed system inside the red boundary of Fig. 2.45.

Assumptions

The refrigerator can be modeled as a closed-steady system.

Analysis

From Fig. 2.45 the known heat transfer rates are $\dot{Q}_F = 10$ MJ/h $= 2.778$ kW and $\dot{Q}_M = 5$ MJ/h $= 1.389$ kW. Also, $\dot{W}_{net} = 2$ kW so that $\dot{W}_{ext} = -2$ kW. By definition, the COP is:

$$\text{COP}_R = \frac{\text{Desired Energy Transfer}}{\text{Required Energy Input}} = \frac{\dot{Q}_F + \dot{Q}_M}{\dot{W}_{net}} = \frac{2.778 + 1.389}{2} = 2.08$$

An energy balance at steady state yields:

$$0 = (\dot{Q}_F + \dot{Q}_M - \dot{Q}_0) - \dot{W}_{ext};$$

$$\Rightarrow \quad \dot{Q}_0 = \dot{Q}_F + \dot{Q}_M - \dot{W}_{net} = 2.778 + 1.389 - (-2) = 6.167 \text{ kW}$$

An entropy balance over the system's universe (red boundary) produces:

$$0 = \frac{\dot{Q}_F}{T_F} + \frac{\dot{Q}_M}{T_M} - \frac{\dot{Q}_0}{T_0} + \dot{S}_{gen,univ}$$

$$\Rightarrow \quad \dot{S}_{gen,univ} = \frac{\dot{Q}_0}{T_0} - \frac{\dot{Q}_F}{T_F} - \frac{\dot{Q}_M}{T_M} = \frac{6.167}{303} - \frac{2.778}{263} - \frac{1.389}{274} = 0.0047 \frac{\text{kW}}{\text{K}}$$

When a reversible refrigerator replaces the actual one, $\dot{Q}_0$ and $\dot{W}_{net}$ would both change as $\dot{S}_{gen,univ}$ goes to zero. An entropy balance for the reversible system produces:

$$0 = \frac{\dot{Q}_F}{T_F} + \frac{\dot{Q}_M}{T_M} - \frac{\dot{Q}_{0,rev}}{T_0} + \cancel{\dot{S}_{gen,univ}}^{0}$$

$$\Rightarrow \quad \dot{Q}_{0,rev} = T_0\left(\frac{\dot{Q}_F}{T_F} + \frac{\dot{Q}_M}{T_M}\right) = 303\left(\frac{2.778}{263} + \frac{1.389}{274}\right) = 4.736 \text{ kW}$$

Substituting this value into the energy balance equation, Eq. (2.20), yields:

$$0 = (\dot{Q}_F + \dot{Q}_M - \dot{Q}_{0,rev}) - (-\dot{W}_{net,rev});$$

$$\Rightarrow \quad \dot{W}_{net,rev} = \dot{Q}_{0,rev} - \dot{Q}_F - \dot{Q}_M = 4.736 - 2.778 - 1.389 = 0.569 \text{ kW}$$

Next, the COP of the reversible refrigerator is calculated as:

$$\text{COP}_{R,rev} = \frac{\dot{Q}_F + \dot{Q}_M}{\dot{W}_{net,rev}} = \frac{2.778 + 1.389}{0.569} = 7.32$$

Discussion

Note that using an average cold-space temperature of $T_C = 268.5$ K, results in a Carnot COP of $268.5/(303 - 268.5) = 7.78$, which overestimates the answer.

EXAMPLE 2-15 **Heat Pump and its Alternatives**

When the outside temperature is 2°C, a house requires 600 MJ of heat per day to maintain its temperature at 20°C. The cost of electricity is 10 cents per kWh and the cost of natural gas is 75 cents per Therm. Compare the daily operational cost of the following alternative heating systems: (a) electrical heating, (b) a gas heating system with an energetic efficiency of 90%, and (c) a Carnot heat pump.

SOLUTION

Perform an energy analysis on each alternative, shown by the sub-systems in Fig. 2.46.

Assumptions

Each heating sub-systems behaves as a closed-steady system.

Analysis

Use the converter TESTcalc to convert different energy units into MJ: 1 kWh = 3.6 MJ, 1 Therm = 105.5 MJ. Therefore, the cost of electricity and natural gas per unit MJ can be calculated as 2.778 cents and 0.711 cents, respectively.

An energy balance for the electrical heater as a closed-steady system (see Fig. 2.46 where all symbols have absolute values) produces:

$$0 = (-\dot{Q}_H) - (-\dot{W}_{el,in}); \quad \Rightarrow \quad \dot{W}_{el,in} = \dot{Q}_H; \quad \Rightarrow \quad \frac{\dot{W}_{el,in}}{\Delta t} = \frac{\dot{Q}_H}{\Delta t};$$

$$\Rightarrow \quad \dot{W}_{el,in} = \dot{Q}_H = 600 \text{ MJ}$$

Electric Heating Cost = (600)(0.0278) = $16.68/day

For gas heating, the energetic efficiency relates the heat transferred to the house to the heat released by gas, $\dot{Q}_{gas}$. An energy balance for the furnace yields:

$$0 = (\eta_{I,heater}\dot{Q}_{gas} - \dot{Q}_H) - \dot{W}_{ext}^{\nearrow 0};$$

$$\Rightarrow \quad Q_{gas} = \dot{Q}_{gas}\Delta t = \frac{\dot{Q}_H}{\eta_{I,heater}}\Delta t = \frac{Q_H}{\eta_{I,heater}} = \frac{600}{0.9} = 666.7 \text{ MJ}$$

Gas Heating Cost = (666.7)(0.00711) = $4.74/day

A Carnot heat pump operating between the temperature inside and outside the house will have a COP of:

$$COP_{HP,Carnot} = \frac{T_H}{T_H - T_0} = \frac{293}{18} = 16.3$$

Therefore, the cost of the work input that is supplied by electrical power can be calculated as:

$$HP \text{ Electricity Cost} = \left(\frac{600}{16.3}\right)(0.0278) = \$1.02/day$$

Discussion

The best alternative is the reversible heat pump as far as the operating cost is concerned. There are a few caveats. A real heat pump has a much lower COP than its Carnot counterpart. Also, if the outside temperature decreases below 2°C, the COP of the heat pump deteriorates drastically (see Fig. 2.44).

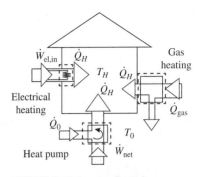

FIGURE 2.46 Energy flow diagrams for the three options (subsystems within the red boundaries) for heating a house. All symbols represent positive values.

2.3.4 The Kelvin Temperature Scale

While discussing temperature as a property in Sec. 1.5.6, we stated that the Kelvin scale for absolute temperature was a better alternative to the arbitrary Celsius or Fahrenheit markings on a thermometer. Having discussed the Carnot theorems, we are now in a position to understand the arguments that led Kelvin to establish the purely *thermodynamic* temperature scale that bears his name.

When Carnot proposed his theorem (see Sec. 2.3.3), he did not have the benefit of the balance equations. In fact, the concept of entropy or even heat as a form of energy in transit did not yet exist. Although by Carnot efficiency we routinely mean $\eta_{Carnot} = 1 - T_C/T_H$, what Carnot actually proposed as Theorem-1 is:

$$\eta_{th,Carnot} = \frac{\dot{W}_{net}}{\dot{Q}_H} = 1 - \frac{\dot{Q}_C}{\dot{Q}_H} = 1 - f(T_C, T_H); \quad \text{or,} \quad \frac{\dot{Q}_C}{\dot{Q}_H} = f(T_C, T_H) \tag{2.38}$$

The thermal efficiency of a reversible engine operating between two constant-temperature reservoirs is a function of the temperatures of the reservoirs alone. He arrived at this conclusion through deductive reasoning from the Kelvin-Planck statement of the second law that preceded his work.

After Carnot advanced his theorems, Kelvin realized that Carnot's conclusion was independent of the unit used for temperature and set out to explore the temperature dependent unknown function f of Eq. (2.38). To follow Kelvin's argument, consider the network of Carnot engines—two in series and one in parallel—shown in Fig. 2.47. Applying Eq. (2.38) to each engine, we obtain:

$$\frac{\dot{Q}_1}{\dot{Q}_2} = f(T_1, T_2); \quad \frac{\dot{Q}_2}{\dot{Q}_3} = f(T_2, T_3); \quad \text{and,} \quad \frac{\dot{Q}_1}{\dot{Q}_3} = f(T_1, T_3)$$

$$\Rightarrow \quad f(T_1, T_3) = \frac{\dot{Q}_1}{\dot{Q}_3} = \frac{\dot{Q}_1}{\dot{Q}_2}\frac{\dot{Q}_2}{\dot{Q}_3} = f(T_1, T_2)f(T_2, T_3) \tag{2.39}$$

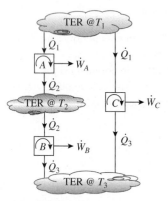

FIGURE 2.47 Illustration used to develop the absolute temperature scale.

Therefore, the nature of the function f must be such that T_2 disappears on the RHS of Eq. (2.39). This is possible only if f has the following form:

$$f(T_H, T_C) = \frac{g(T_H)}{g(T_C)}$$

so that
$$f(T_1, T_2)f(T_2, T_3) = \frac{g(T_1)\,g(T_2)}{g(T_2)\,g(T_3)} = f(T_1, T_3) \tag{2.40}$$

Kelvin proposed the simplest possible form, $g(T) = T$, for the function g so that Eq. (2.38) reduces to the familiar form:

$$\frac{T_H}{T_C} = \left(\frac{\dot{Q}_H}{\dot{Q}_C}\right)_{Carnot\ Cycle} \tag{2.41}$$

Equation (2.41) defines the absolute temperature scale known as the **Kelvin scale**. By taking the limit of $\dot{Q}_C \rightarrow 0$ or $\dot{Q}_H \rightarrow \infty$, we can see that the lower and upper limits of the Kelvin temperature are zero and infinity respectively—hence, the name **absolute temperature** for this thermodynamic scale of temperature.

2.4 CLOSURE

In this chapter we introduced the fundamental laws of thermodynamics, translated them into balance equations, and applied those to the study of closed-steady systems. The chapter began with the derivation of the mass, energy, and entropy balance equations from the fundamental laws for a generic open-unsteady system. Flow diagrams for energy and entropy were introduced to help visualize different terms of the balance equations. The first and second laws of thermodynamics were presented as postulates and the mechanisms of entropy generation, collectively termed thermodynamic friction, were discussed and established through examples and animations. Energetic efficiency and reversibility of different types of systems were defined in a general manner. The remainder of the chapter was devoted to comprehensive analysis of closed-steady systems, a unique class of systems that do not require property evaluation due to simplified forms of the balance equations. Heat engines, refrigerators, and heat pumps were studied as special cases of such systems. The thermal efficiency of heat engines and the COP of refrigerators and heat pumps were introduced as specialized energetic efficiencies. Carnot efficiency and Carnot COP's were derived from an energy and entropy analysis of the reversible Carnot cycles. Finally, the Kelvin temperature scale, introduced in Chapter 1, was established from a theoretical standpoint.

PROBLEMS

SECTION 2-1: MASS BALANCE EQUATION

2-1-1 Mass enters an open system, with one inlet and one exit, at a constant rate of 50 kg/min. At the exit, the mass flow rate is 60 kg/min. If the system initially contains 1000 kg of working fluid, determine (a) dm/dt, treating the tank as a system, and (b) the time when the system mass becomes 500 kg.

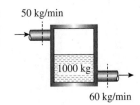

2-1-2 Steam enters an insulated tank through a valve. At a given instant, the mass of steam in the tank is found to be 10 kg and the conditions at the inlet are measured as follows: $A = 50$ cm^2, $V = 31$ m/s, and $\rho = 0.6454$ kg/m^3. Determine (a) dm/dt, treating the tank as a system. (b) Assuming the inlet conditions to remain unchanged, determine the mass of steam in the tank after 10 s.

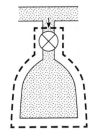

2-1-3 Air is introduced into a piston-cylinder device through a 10 cm-diameter flexible duct with a mass flow rate of 0.456 kg/s. The inlet's velocity and specific volume, at a given instant, are measured as 22 m/s and 0.1722 m^3/kg, respectively. At the same time air jets out through a 1 mm-diameter leak with a density of 1.742 kg/m^3, the piston rises with a velocity of 10 cm/s, and the mass of the air in the device increases at a rate of 0.415 kg/s. Determine (a) the mass flow rate ($\dot{m}$) of air at the inlet and (b) the jet velocity. (c) If the jet pressure is 100 kPa, use the IG flow-state TESTcalc to determine the temperature of the jet at the exit.

2-1-4 Air enters an open system with a velocity of 1 m/s and a density of 1 kg/m^3. The mass of air in the tank at a known instant is given by the expression $m = 5p/T$, where p is in kPa and T is in K. If the temperature in the tank remains constant at 500 K, due to heat transfer, (a) determine the rate of increase of pressure in the tank. (b) What is the sign of heat transfer, positive or negative?

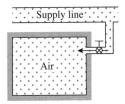

2-1-5 A propane tank is being filled at a charging station (see figure in Problem 2-1-2). At a given instant the mass of the tank is found to increase at a rate of 0.4 kg/s. Propane from the supply line at state-1 has the following conditions: $D = 5$ cm, $T = 25°C$, and $\rho = 490$ kg/m^3. Determine the velocity of propane in the supply line.

2-1-6 Mass leaves an open system with a mass flow rate of cm, where c is a constant and m is the system mass. If the mass of the system at $t = 0$ is m_0, derive an expression for the mass of the system at time t.

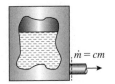

2-1-7 Water enters a vertical cylindrical tank with a cross-sectional area 0.01 m^2, at a constant mass flow rate of 5 kg/s. It leaves the tank through an exit near the base with a mass flow rate given by the formula $0.2h$ kg/s, where h is the instantaneous height in m. If the tank is initially empty, develop an expression for the liquid height h as a function of time t. Assume density of water to remain constant at 1000 kg/m^3.

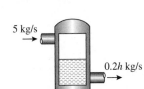

2-1-8 A conical tank, base diameter D and height H, is suspended in an inverted position to hold water. A leak at the apex of the cone causes water to leave with a mass flow rate of $c*sqrt(h)$, where c is a constant and h is the height of the water level from the leak at the bottom. (a) Determine the rate of change of height h. (b) Express h as a function of time t and other known constants, ρ (constant density of water), D, H, and c if the tank were completely full at $t = 0$.

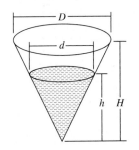

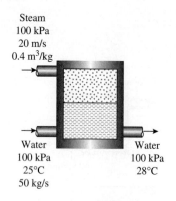

Steam
100 kPa
20 m/s
0.4 m³/kg

Water
100 kPa
25°C
50 kg/s

Water
100 kPa
28°C

2-1-9   Steam enters a mixing chamber at 100 kPa, 20 m/s with a specific volume of 0.4 m³/kg. Liquid water, at 100 kPa and 25°C, enters the chamber through a separate duct with a flow rate of 50 kg/s and a velocity of 5 m/s. If liquid water leaves the chamber at 100 kPa, 43°C, 5.58 m/s, and a volumetric flow rate of 3.357 m³/min, determine the port areas at (a) the inlets and (b) the exit. Assume liquid water density to be 1000 kg/m³ and steady state operation.

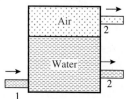

Air
2

Water

2

1

2-1-10   The diameter of the ports in the accompanying figures are 10 cm, 5 cm, and 1 cm at port 1, 2, and 3, respectively. At a given instant, water enters the tank at port 1 with a velocity of 2 m/s and leaves through port 2 with a velocity of 1 m/s. Assuming air to be insoluble in water and density of water and air to remain constant at 1000 kg/m³ and 1 kg/m³, determine:

(a)   the mass flow rate ($\dot{m}$) of water in kg/s at the inlet
(b)   the rate of change of mass of water $(dm/dt)_W$ in the tank
(c)   the rate of change of mass of air $(dm/dt)_A$ in the tank
(d)   the velocity of air port 3.

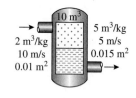

10 m³

2 m³/kg
10 m/s
0.01 m²

5 m³/kg
5 m/s
0.015 m²

2-1-11   Air is pumped into, and withdrawn from, a 10 m³ rigid tank as shown in the accompanying figure. The inlet and exit conditions are as follows. Inlet: $v_1 = 2$ m³/kg, $V_1 = 10$ m/s, $A_1 = 0.01$ m²; Exit: $v_2 = 5$ m³/kg, $V_2 = 5$ m/s, $A_2 = 0.015$ m². Assuming the tank to be uniform at all times with the specific volume and pressure related through $p*v = 9.0$ (kPa $-$ m³), determine the rate of change of pressure in the tank.

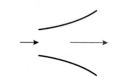

2-1-12   A gas flows steadily through a circular duct with a mass flow rate of 10 kg/s. The area of the duct's cross-sections is varied and not uniform. The inlet and exit conditions are as follows. Inlet: $V_1 = 400$ m/s, $A_1 = 179.36$ cm²; Exit: $V_2 = 582$ m/s, $v_2 = 1.1827$ m³/kg. (a) Determine the exit area. (b) Do you find the increase in velocity of the gas, accompanied by an increase in flow area, counter-intuitive? Why?

H₂
N₂
H₂

N₂ + H₂

2-1-13   A pipe, with a diameter of 10 cm, carries nitrogen at a velocity of 10 m/s and with a specific volume of 5 m³/kg, into a chamber. Surrounding the pipe, in an annulus of outer diameter 20 cm, is a flow of hydrogen entering the chamber at 20 m/s with a specific volume of 1 m³/kg. The mixing chamber operates at steady state with a single exit of diameter 5 cm. If the velocity at the exit is 62 m/s, determine: (a) the mass flow rate in kg/s at the exit, (b) the specific volume of the mixture at the exit in m³/kg, and (c) the apparent molar mass in kg/kmol of the mixture at the exit.

Cold air

Hot
air

Cold air

Warm air

2-1-14   A pipe, that has a diameter of 15 cm, carries hot air at a velocity of 200 m/s and a temperature of 1000 K, into a chamber. Surrounding the pipe, in an annulus that has an outer diameter of 20 cm, is a flow of cooler air entering the chamber at 10 m/s at a temperature of 300 K. The mixing chamber operates at steady state with a single exit with a diameter of 20 cm. If the air exits at 646 K and the specific volume of air is proportional to the temperature (in K), determine: (a) the exit velocity in m/s and (b) the dm/dt for the mixing chamber in kg/s.

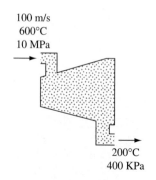

100 m/s
600°C
10 MPa

200°C
400 KPa

2-1-15   Steam enters a turbine through a duct with a diameter of 0.25 m at 10 MPa, 600°C, and 100 m/s. It exits the turbine through a duct with a diameter of 1 m at 400 kPa and 200°C. For steady state operation, determine (a) the exit velocity and (b) the mass flow rate of steam through the turbine. Use the PC flow-state TESTcalc to obtain the density of steam at the inlet and exit ports. (c) *What-if scenario:* What is the exit velocity if the exit area is equal to the inlet area?

2-1-16   Steam enters a turbine with a mass flow rate of 10 kg/s at 10 MPa, 600°C, and 30 m/s. It exits the turbine at 45 kPa, 30 m/s, and a quality of 0.9. Assuming steady-state operation, determine (a) the inlet area and (b) the exit area. Use the PC flow-state TESTcalc.

2-1-17 Refrigerant R-134 enters a device as saturated liquid at 500 kPa with a velocity of 10 m/s and a mass flow rate of 2 kg/s. At the exit the pressure is 150 kPa and the quality is 0.2. If the exit velocity is 65 m/s, determine the (a) inlet and (b) exit areas. Use the PC flow-state TESTcalc.

2-1-18 Air enters a 0.5 m diameter fan at 25°C, 100 kPa and is discharged at 28°C, 105 kPa with a volume flow rate of 0.8 m³/s. Determine, for steady-state operation, (a) the mass flow rate of air in kg/min, (b) the inlet, and (c) exit velocities. Use the PG flow-state TESTcalc.

2-1-19 Air enters a nozzle that has an inlet area of 0.1 m², at 200 kPa, 500°C, and 10 m/s. At the exit the conditions are 100 kPa and 443°C. If the exit area is 35 cm², determine the steady-state exit velocity. Use the IG flow-state TESTcalc.

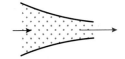

SECTION 2-2: ENERGY BALANCE EQUATION

2-2-1 A 20 kg block of a solid cools down by transferring heat at a rate of 1 kW to the surroundings. Determine the rate of change of (a) stored energy and (b) internal energy of the block.

2-2-2 A rigid chamber contains 100 kg of water at 500 kPa, 100°C. A paddle wheel stirs the water at 1000 rpm with a torque of 100 N-m. while an internal electrical resistance heater heats the water, consuming 10 amps of current at 110 Volts. Because of thin insulation, the chamber loses heat to the surroundings at 27°C at a rate of 1.2 kW. Determine the rate at which the stored energy of the system changes.

2-2-3 A closed system interacts with its surroundings and the following data are supplied: $\dot{W}_{sh} = -10$ kW. $\dot{W}_{el} = 5$ kW, $\dot{Q} = -5$ kW. (a) If there are no other interactions, determine dE/dt. (b) Is this system necessarily steady (yes: 1; no: 0)?

2-2-4 An electric bulb consumes 500 W of electricity. After it is turned on, the bulb becomes warmer and starts losing heat to the surroundings at a rate of $5t$ (t in seconds) watts until the heat loss equals the electric power input. (a) Plot the change in stored energy of the bulb with time. (b) How long does it take for the bulb to reach steady state?

2-2-5 Suppose the specific internal energy, in kJ/kg, of the solid in Problem 2-2-1 is related to its temperature through $u = 0.5T$, where T is the temperature of the solid in Kelvin, determine the rate of change of temperature of the solid. Assume the density of the solid is 2700 kg/m³.

2-2-6 A cup of coffee is heated in a microwave oven. If the mass of coffee (modeled as liquid water) is 0.2 kg and the rate of heat transfer is 0.1 kW, (a) determine the rate of change of internal energy (u). (b) Assuming the density of coffee is 1000 kg/m³ and the specific internal energy in kJ/kg is related to temperature through $u = 4.2T$, where T is in Kelvin, determine how long it takes for the temperature of the coffee to increase by 20°C.

2-2-7 At a given instant, a closed system is loosing 0.1 kW of heat to the outside atmosphere. A battery inside the system keeps it warm by powering a 0.1 kW internal heating lamp. A shaft transfers 0.1 kW of work into the system at the same time. Determine:

(a) the rate of external work transfer (include sign),
(b) the rate of heat transfer (include sign), and
(c) dE/dt of the system.

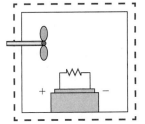

2-2-8 A semi-truck of mass 20,000 lb accelerates from 0 to 75 m/h (1 m/h = 0.447 m/s) in 10 seconds. (a) What is the change in kinetic energy of the truck in 10 seconds? (b) If PE and U of the truck can be assumed constant, what is the average value of dE/dt of the truck in kW during this period? (c) If 30% of the heat released from the combustion of diesel (heating value of diesel is 40 MJ/kg) is converted to kinetic energy, determine the average rate of fuel consumption in kg/s.

2-2-9 An insulated tank contains 50 kg of water, which is stirred by a paddle wheel at 300 rpm while transmitting a torque of 0.1 kN·m. At the same time, an electric resistance heater inside the tank operates at 110V, drawing a current of 2A. Determine the rate of heat transfer after the system achieves steady state.

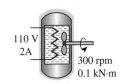

2-2-10 A drill rotates at 4000 rpm while transmitting a torque of 0.012 kN-m to a block of steel. Determine the rate of change of stored energy of the block initially.

2-2-11 A 20 kg slab of aluminum is raised by a rope and pulley, arranged vertically, at a constant speed of 10 m/min. At the same time the block absorbs solar radiation at a rate of 0.2 kW. Determine the rate of change of (a) potential energy ($\dot{PE}$), (b) internal energy ($\dot{U}$), and (c) stored energy ($\dot{E}$).

2-2-12 An external force F is applied to a rigid body with mass m. If its internal and potential energy remain unchanged, show that an energy balance on the body reproduces Newton's law of motion.

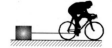

2-2-13 An insulated block, with a mass of 100 kg, is acted upon by a horizontal force of 0.02 kN. Balanced by frictional forces, the body moves at a constant velocity of 2 m/s. Determine (a) the rate of change of stored energy in the system and (b) power transferred by the external force. (c) How do you account for the work performed by the external force?

2-2-14 Do an energy analysis of a pendulum bob to show that the sum of its kinetic and potential energies remain constant. Assume internal energy to remain constant, neglect viscous friction and heat transfer. *What-if scenario:* Discuss how the energy equation would be affected if viscous friction is not negligible.

2-2-15 A rigid insulated tank contains 2 kg of a gas at 300 K and 100 kPa. A 1 kW internal heater is turned on. (a) Determine the rate of change of total stored energy (dE/dt). (b) If the internal energy of the gas is related to the temperature by $u = 1.1T$ (kJ/kg), where T is in Kelvin, determine the rate of temperature increase.

2-2-16 A 10 m³ rigid tank contains air at 200 kPa and 150°C. A 1 kW internal heater is turned on. Determine the rate of change of (a) stored energy, (b) temperature, and (c) pressure of air in the tank. Use the IG system-state TESTcalc. (Hint: Evaluate state-2 with stored energy incremented by the amount added during a small time interval, for example, 0.1 s.)

2-2-17 A 10 m³ rigid tank contains steam with a quality of 0.5 at 200 kPa. A 1 kW internal heater is turned on. Determine the rate of change of (a) stored energy, (b) temperature, and (c) pressure of the steam in the tank. Use the PC system-state TESTcalc. (Hint: Evaluate state-2 with stored energy incremented by the amount added over a small interval of time, for example, 0.1 s.)

2-2-18 A piston-cylinder device, containing air at 200 kPa, loses heat to the surrounding atmosphere at a rate of 0.5 kW. At a given instant, the piston, which has a cross-sectional area of 0.01 m², moves down with a velocity of 1 cm/s. Determine the rate of change of stored energy in the gas.

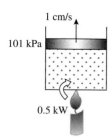

2-2-19 A piston-cylinder device contains a gas, which is heated at a rate of 0.5 kW from an external source. At a given instant the piston, which has an area of 10 cm² moves up with a velocity of 1 cm/s. (a) Determine the rate of change of stored energy (dE/dt) in the gas. Assume the atmospheric pressure to be 101 kPa, the piston to be weightless, and neglect friction. (b) *What-if scenario:* How would the answer change if the piston were locked in its original position with a pin.

2-2-20 A gas trapped in a piston-cylinder device is heated (as shown in figure of Problem 2-2-21) from an initial temperature of 300 K. The initial load on the massless piston with an area of 0.2 m², is such that the initial pressure of the gas is 200 kPa. When the temperature of the gas reaches 600 K, the piston velocity is measured as 0.5 m/s and dE/dt is measured as 30 kW. At that instant, determine

(a) the rate of external work transfer,
(b) the rate of heat transfer ($\dot{Q}$), and
(c) the load (in kg) on the piston. Assume the ambient atmospheric pressure is 100 kPa.

2-2-21 A piston-cylinder device is used to compress a gas by pushing the piston with an external force. During the compression process, heat is transferred out of the gas in such a manner that the stored energy in the gas remains unchanged. Also, the pressure is found to be inversely proportional to the volume of the gas. Determine an expression for the heat transfer rate in terms of the instantaneous volume and the rate of change of pressure of the gas.

2-2-22 A fluid flows steadily through a long insulated pipeline. Perform a mass and energy analysis to show that the flow energy j remains unchanged between the inlet and exit. *What-if scenario:* How would this conclusion differ if kinetic and potential energy changes were negligible?

2-2-23 Water enters a constant-diameter, insulated, horizontal pipe at 500 kPa. Due to the presence of viscous friction the pressure drops to 400 kPa at the exit. At steady state, determine the changes in (a) specific kinetic energy (ke) and (b) specific internal energy (u) between the inlet and the exit. Assume the water density to be 1000 kg/m³. Use the SL flow-state TESTcalc to verify the answer.

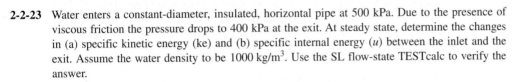

2-2-24 Water flows steadily through a variable diameter insulated pipe. At the inlet, the velocity is 20 m/s and at the exit, the flow area is half of the inlet area. If the internal energy of the water remains constant, determine the change in pressure between the inlet and exit. Assume the water density to be 1000 kg/m³.

2-2-25 Oil enters a long insulated pipe at 200 kPa and 20 m/s. It exits at 175 kPa. Assuming a steady flow, determine the changes in the following properties between the inlet and exit: (a) j, (b) ke and (c) h. Assume the oil density to be constant.

2-2-26 Nitrogen gas flows steadily through a pipe of diameter 10 cm. The inlet conditions are as follows: pressure 400 kPa, temperature 300 K, and velocity 20 m/s. At the exit, the pressure is 350 kPa (due to frictional losses). If the flow rate of mass and flow energy remain constant, determine (a) the exit temperature and (b) exit velocity. Use the IG (ideal gas) flow-state TESTcalc. (c) *What-if scenario:* What is the exit temperature if kinetic energy was neglected?

2-2-27 A 5 cm diameter pipe discharges water into the open atmosphere at a rate of 20 kg/s at an elevation of 20 m. The temperature of water is 25°C and the atmospheric pressure is 100 kPa. Determine (a) $\dot{J}$, (b) $\dot{E}$, (c) $\dot{KE}$, (d) $\dot{H}$, and (e) $\dot{W}_F$. (f) How important is the flow work transfer compared to kinetic and potential energy carried by the mass? Use the SL flow-state TESTcalc.

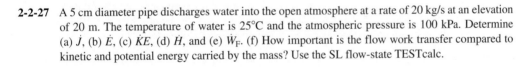
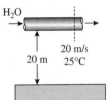

2-2-28 Water enters a pipe at 90 kPa, 25°C and a velocity of 10 m/s. At the exit the pressure is 500 kPa and velocity is 12 m/s while the temperature remains unchanged. If the volume flow rate is 10 m³/min, both at the inlet and exit, determine the flow rate of energy ($\dot{J}$) at (a) the inlet and (b) the exit. (c) *What-if scenario:* What would the answer in part (b) be if the exit velocity was 15 m/s?

2-2-29 Water at 1000 kPa, 25°C enters a 1-m-diameter horizontal pipe with a steady velocity of 10 m/s. At the exit the pressure drops to 950 kPa due to viscous resistance. Assuming steady-state flow, determine the rate of heat transfer ($\dot{Q}$) necessary to maintain a constant specific internal energy (u).

2-2-30 An incompressible fluid (constant density) flows steadily downward along a constant-diameter, insulated, vertical pipe. Assuming internal energy remains constant, show that the pressure variation is hydrostatic.

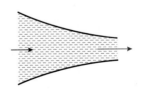

2-2-31 An incompressible fluid (constant density) flows steadily through a converging nozzle. (a) Show that the specific flow energy remains constant if the nozzle is adiabatic. (b) Assuming the internal energy remains constant and neglecting the inlet kinetic energy, obtain an expression for the exit velocity, in terms of pressures, at the inlet and exit and the fluid density.

2-2-32 Water flows steadily through an insulated nozzle. The following data is supplied: Inlet: $p = 200$ kPa, $v = 10$ m/s, $z = 2$ m; Exit: $p = 100$ kPa, $z = 0$. (a) Determine the exit velocity. Assume the density of water to be 1000 kg/m³. Also assume the internal energy remains constant. *What-if scenario:* What would the exit velocity be if (b) changes in potential energy or (c) inlet kinetic energy were neglected in the analysis?

2-2-33 Water, flowing steadily through a 2 cm diameter pipe at 30 m/s, goes through an expansion joint to flow into a 4 cm diameter pipe. Assuming the internal energy remains constant, determine (a) the change in pressure as the water goes through the transition. (b) Also determine the displacement of the mercury column in mm. Assume the water density to be 997 kg/m^3.

2-2-34 An adiabatic work producing device works at steady state with the working fluid entering through a single inlet and leaving through a single exit. Derive an expression for the work output in terms of the flow properties at the inlet and exit. *What-if scenario:* How would the expression for work simplify if changes in ke and pe were neglected?

2-2-35 A pump is a device that raises the pressure of a liquid at the expense of external work. (a) Determine the pumping power necessary to raise the pressure of liquid water from 10 kPa to 2000 kPa at a flow rate of 1000 L/min. Assume the density of water to be 1000 kg/m^3 and neglect changes in specific internal, kinetic and potential energies. (b) *What-if scenario:* What would the pumping power be if pe was not negligible and $z_1 = -10$, $z_2 = 0$?

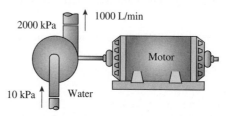

2-2-36 An adiabatic pump, working at steady state, raises the pressure of water from 100 kPa to 1 MPa, while the specific internal energy (u) remains constant. If the exit is 10 m above the inlet and the flow rate of the water is 100 kg/s, (a) determine the pumping power. Neglect any change in kinetic energy (ke). Assume the density of water to be 997 kg/m^3. (b) *What-if scenario:* What would the pumping power be if the changes in potential energy (pe) were also neglected?

2-2-37 Steam flows steadily through a single-flow device with a flow rate of 10 kg/s. It enters with an enthalpy of 3698 kJ/kg and a velocity of 30 m/s. At the exit, the corresponding values are 3368 kJ/kg and 20 m/s. If the rate of heat loss from the device is measured as 100 kW, (a) determine the rate of work transfer. Neglect any change in potential energy. (b) *What-if scenario:* What would the rate of work transfer be if the change in kinetic energy was also neglected?

2-2-38 Steam enters an adiabatic turbine with a mass flow rate of 5 kg/s at 3 MPa, 600°C, and 80 m/s. It exits the turbine at 40°C, 30 m/s, and a quality of 0.9. Assuming steady-state operation, determine the shaft power produced by the turbine. Use the PC flow-state TESTcalc to evaluate enthalpies at the inlet and exit.

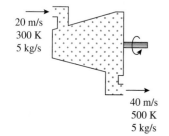

2-2-39 A gas enters an adiabatic work consuming device at 300 K, 20 m/s, and leaves at 500 K, 40 m/s. (a) If the mass flow rate is 5 kg/s, determine the rate of work transfer. Neglect changes in potential energy and assume the specific enthalpy of the gas to be related to its temperature in K through $h = 1.005T$. (b) *What-if scenario:* By what percent would the answer change if the change in kinetic energy was also neglected?

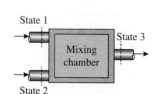

2-2-40 A refrigerant is compressed by an adiabatic compressor operating at steady state raising the pressure from 200 kPa to 750 kPa. The following data are supplied for the inlet and exit ports. Inlet: $v = 0.0835$ m^3/kg, $h = 182.1$ kJ/kg, $V = 30$ m/s; Exit: $v = 0.0244$ m^3/kg, $h = 205.4$ kJ/kg, $V = 40$ m/s. If the volume flow rate at the inlet is 3000 L/min, determine (a) the mass flow rate, (b) the volume flow rate at the exit, and (c) the compressor power. (d) *What-if scenario:* What would the power consumption be if the change in kinetic energy was neglected?

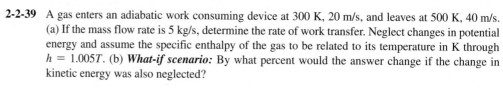

2-2-41 Two flows that have equal mass flow rates, one at state-1 and another at state-2 enter an adiabatic mixing chamber and leave through a single port at state-3. Obtain an expression for the velocity and specific enthalpy at the exit. Assume negligible changes in ke and pe.

2-2-42 Air at 500 kPa, 30°C from a supply line is used to fill an adiabatic tank. At a particular moment during the filling process, the tank contains 0.2 kg of air at 200 kPa and 50°C. If the mass flow rate is 0.1 kg/s, and the specific enthalpy of air at the inlet is 297.2 kJ/kg, determine (a) the rate of flow energy into the tank, (b) the rate of increase of internal energy in the tank and (c) the rate of increase of specific internal energy. Assume the tank to be uniform at all times and neglect kinetic and potential energies.

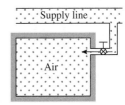

2-2-43 An insulated tank is being filled with a gas through a single inlet. At a given instant, the mass flow rate is measured as 0.5 kg/s and the enthalpy h as 400 kJ/kg. Neglecting ke and pe, determine the rate of increase of stored energy in the system at that instant.

2-2-44 Saturated steam at 200 kPa, which has a specific enthalpy (h) of 2707 kJ/kg is expelled from a pressure cooker at a rate of 0.1 kg/s. Determine the rate of heat transfer ($\dot{Q}$) necessary to maintain a constant stored energy E in the cooker. Assume that there is sufficient liquid water in the cooker at all time to generate the saturated steam. Neglect kinetic and potential energy of the steam.

SECTION 2-3: ENTROPY BALANCE EQUATION

2-3-1 Heat is transferred from a TER at 1500 K to a TER at 300 K at a rate of 10 kW. Determine the rates at which entropy (a) leaves the TER at higher temperature and (b) enters the TER at lower temperature. (c) How do you explain the discontinuity in the result?

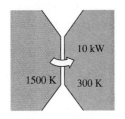

2-3-2 A wall separates a hot reservoir at 1000 K from a cold reservoir at 300 K. The temperature difference between the two reservoirs drives a heat transfer at the rate of 500 kW. If the wall maintains steady state, determine (a) $\dot{Q}$ (in kW), (b) $\dot{W}_{ext}$ (in kW), (c) dS/dt (in kW/K), and (d) $\dot{S}_{gen,univ}$ (kW/K).

2-3-3 A resistance heater operates inside a tank consuming 0.5 kW of electricity. Due to heat transfer to the ambient atmosphere at 300 K, the tank is maintained at a steady state. The surface temperature of the tank remains constant at 400 K. Determine the rate at which entropy (a) leaves the tank and (b) enters the tank's universe. (c) How does the system maintain steady state with regard to entropy?

2-3-4 Heat is conducted through a 2 cm thick slab. The temperature varies linearly from 500 K, on the left face, to 300 K, on the right face. If the rate of heat transfer is 2 kW, determine the rate of entropy transfer dS/dt (magnitude only) at the (a) left and (b) right faces. (c) Plot how the rate of entropy transfer varies from the left to the right face.

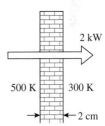

2-3-5 A 30 kg aluminum block cools down from its initial temperature of 500 K to the atmospheric temperature of 300 K. Determine the total amount of entropy transfer from the system's universe. Assume the specific internal energy of aluminum (in kJ/kg) is related to its absolute temperature (K) through $u = 0.9T$.

2-3-6 Water is heated in a boiler from a source at 1800 K. If the heat transfer rate is 20 kW, (a) determine the rate of entropy transfer into the boiler's universe. (b) Discuss the consequences of reducing the source temperature with respect to the boiler size and entropy transfer.

2-3-7 An insulated tank contains 50 kg of water at 30°C, which is stirred by a paddle wheel at 300 rpm while transmitting a torque of 0.2 kNm. Determine (a) the rate of change of temperature (dT/dt), (b) the rate of change of total entropy (dS/dt), and (c) the rate of generation of entropy ($\dot{S}_{gen}$) within the tank. Assume $s = 4.2 \ln T$ and $u = 4.2T$, where T is in Kelvin.

2-3-8 A rigid insulated tank contains 1 kg of air at 300 K and 100 kPa. A 1 kW internal heater is turned on. Determine the rate of (a) entropy transfer into the tank, (b) the rate of change of total entropy of the system, and (c) the rate of generation of entropy within the tank. Assume $s = \ln T$ and $u = T$, where T is in Kelvin.

2-3-9 A rigid tank contains 10 kg of air at 500 K and 100 kPa while the surroundings is at 300 K. A 2 kW internal heater keeps the gas hot by compensating the heat losses. At steady state, determine the rate of (a) heat transfer ($\dot{Q}$), (b) the rate of entropy generation ($\dot{S}_{gen}$) inside the tank, and (c) the rate of entropy generation ($\dot{S}_{gen,univ}$) in the system's universe.

2-3-10 A 10 m³ insulated rigid tank contains 30 kg of wet steam with a quality of 0.9. An internal electric heater is turned on, which consumes electric power at a rate of 10 kW. After the heater is on for one minute, determine (a) the change in temperature, (b) the change in pressure, and (c) the change in entropy of steam.

2-3-11 One kg of air is trapped in a rigid chamber of volume 0.2 m³ at 300 K. Because of electric work transfer, the temperature of air increases at a rate of 1°C/s. Using the IG system-state TESTcalc calculate: (a) the rate of change of stored energy, dE/dt, (b) the rate of external work transfer (include sign), (c) the rate of change of total entropy of the system, dS/dt, (d) the rate of entropy generation inside the system. (Hint: Evaluate two states separated by 1 s.)

2-3-12 A rigid cylindrical tank stores 100 kg of a substance at 500 kPa and 500 K while the outside temperature is 300 K. A paddle wheel stirs the system transferring shaft work ($\dot{W}_{sh}$) at a rate of 0.5 kW. At the same time an internal electrical resistance heater transfers electricity at the rate of 1 kW. (a) Do an energy analysis to determine the rate of heat transfer ($\dot{Q}$) in kW for the tank. (b) Determine the absolute value of the rate at which entropy leaves the internal system (at a uniform temperature of 500 K) in kW/K. (c) Determine the rate of entropy generation ($\dot{S}_{gen}$) for the system's universe.

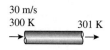

30 m/s
300 K   301 K

2-3-13 A 1 cm diameter insulated pipe carries a steady flow of water at a velocity of 30 m/s. The temperature increases from 300 K at the inlet to 301 K at the exit due to friction. If the specific entropy of water is related to its absolute temperature through $s = 4.2 \ln T$, determine the rate of generation of entropy within the pipe. Assume water density to be 1000 kg/m³.

2-3-14 Liquid water (density 997 kg/m³) flows steadily through a pipe with a volume flow rate of 30,000 L/min. Due to viscous friction, the pressure drops from 500 kPa at the inlet to 150 kPa at the exit. If the specific internal energy and specific entropy remain constant along the flow, determine (a) the rate of heat transfer ($\dot{Q}$) and (b) the rate of entropy generation in and around the pipe. Assume the temperature of the surroundings to be 300 K.

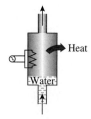

Heat
Water

2-3-15 An electric water heater works by passing electricity through an electrical resistance placed inside the flow of liquid water, as shown in the accompanying animation. The specific internal energy and entropy of water are correlated to its absolute temperature through $u = 4.2T$ and $s = 4.2\ln T$ (T in K). Water enters the heater at 300 K with a flow rate of 5 kg/s. At the exit the temperature is 370 K. Assuming steady state operation, negligible changes in ke and pe, and negligible heat transfer, determine (a) electrical power consumption rate and (b) the rate of entropy generation within the heater. **What-if scenario:** What would the answers be if the exit temperature was reduced by 20°C?

2-3-16 An open system with only one inlet and one exit operates at steady state. Mass enters the system with a flow rate of 5 kg/s and the following properties: $h = 3484$ kJ/kg, $s = 8.0871$ kJ/kg-K and $V = 20$ m/s. At the exit the properties are as follows: $h = 2611$ kJ/kg, $s = 8.146$ kJ/kg-K and $V = 25$ m/s. The device produces 4313 kW of shaft work while rejecting some heat to the atmosphere at 25°C. (a) Do a mass analysis to determine the mass flow rate at the exit. (b) Do an energy analysis to determine the rate of heat transfer (include sign). (c) Do an entropy analysis to evaluate the rate of entropy generation in the system's universe.

2-3-17 Steam flows steadily through a work-producing, adiabatic, single-flow device with a flow rate of 7 kg/s. At the inlet $h = 3589$ kJ/kg, $s = 7.945$ kJ/kg·K, and at the exit $h = 2610$ kJ/kg, $s = 8.042$ kJ/kg·K. If changes in ke and pe are negligible, determine (a) the work produced by the device, and (b) the rate of entropy generation within the device. *What-if scenario:* What would the answer in part (b) be if the device lost 5 kW of heat from its surface at 200°C?

2-3-18 The following information is supplied at the inlet and exit of an adiabatic nozzle operating at steady state: Inlet: $V = 30$ m/s, $h = 976.2$ kJ/kg, $s = 6.149$ kJ/kg·K; Exit: $h = 825.5$ kJ/kg. Determine (a) the exit velocity and (b) the minimum specific entropy (s) possible at the exit.

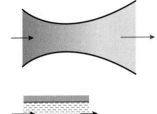

2-3-19 A refrigerant flows steadily through an insulated tube. Its entropy increases from 0.2718 kJ/kg·K at the inlet to 0.3075 kJ/kg·K at the exit. If the mass flow rate of the refrigerant is 0.2 kg/s and the pipe is insulated, determine (a) the rate of entropy generation within the pipe. (b) *What-if scenario:* What would the rate of entropy generation be if the mass flow rate doubled?

SECTION 2-4: CLOSED STEADY SYSTEMS ANALYSIS

2-4-1 A tank contains 50 kg of water, which is stirred by a paddle wheel at 300 rpm while transmitting a torque of 0.2 kNm. After the tank achieves steady state, determine (a) the rate of heat transfer, (b) the rate of entropy transfer, and (c) the rate of entropy generation in the tank's universe. Assume the atmospheric temperature to be 25°C.

2-4-2 A tank contains 1 kg of air at 500 K and 500 kPa. A 1 kW internal heater operates inside the tank at steady state to make up for the heat lost to the atmosphere which is at 300 K. Determine (a) the rate of entropy transfer into the atmosphere, (b) the rate of entropy generation in the system's universe, (c) the internal rate of entropy generation, and (d) the external rate of entropy generation.

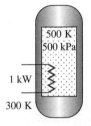

2-4-3 A rigid tank contains 1 kg of air, initially at 300 K and 100 kPa. A 1 kW internal heater is turned on. After the tank achieves steady state, determine (a) the rate of heat transfer ($\dot{Q}$), (b) the rate of entropy transfer ($\dot{S}$), and (c) the rate of entropy generation ($\dot{S}_{gen,univ}$) in the tank's universe. Assume the atmospheric temperature to be 0°C.

2-4-4 A closed chamber containing a gas is at steady state. The shaft transfers power at a rate of 2 kW to the paddle wheel and the electric lamp consumes electricity at a rate of 500 W. Using an energy balance determine (a) the rate of heat transfer. (b) If the surface temperature of the chamber is 400 K, determine the entropy generated within the chamber. (c) *What-if scenario:* What would the entropy generation within the chamber be if the surface temperature increased to 500 K?

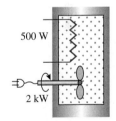

2-4-5 A copper block receives heat from two different sources: 5 kW from a source at 1500 K and 3 kW from a source at 1000 K. It looses heat to the 300 K atmosphere. Assuming the block to be at steady state, determine (a) the net rate of heat transfer in kW; (b) the rate of entropy generation in the system's universe. *What-if scenario:* Determine the entropy generation if the second source was also at 1500 K?

2-4-6 An electric bulb consumes 500 W of electricity at steady state. The outer surface of the bulb is warmer than the surrounding atmosphere by 75°C. If the atmospheric temperature is 300 K, determine (a) the rate of heat transfer between the bulb (the system) and the atmosphere. Also, determine the entropy generation rate (b) within the bulb, (c) in the system's universe, and (d) in the immediate surroundings outside the bulb.

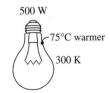

2-4-7 An electric heater consumes 2 kW of electricity, at steady state, to keep a house at 27°C. The outside temperature is −10°C. Using the heater inside the house as the system, determine (a) the maximum possible energetic efficiency of the heater and (b) the rate of entropy generation in the heater's universe in W/K. (c) If the surface of the heater is at 150°C, how much entropy is generated in the immediate surroundings of the the heater?

2-4-8 An electric adaptor for a notebook computer (converting 110 volts to 19 volts) operates 10°C warmer than the surroundings, which is 20°C. If the output current is 3 amps and heat is lost from the adapter at a rate of 10 W, determine (a) the energetic efficiency of the device, (b) the rate of internal entropy generation ($\dot{S}_{gen,int}$), and (c) the rate of external entropy generation ($\dot{S}_{gen,ext}$).

2-4-9 At steady state, the input shaft of a gearbox rotates at 2000 rpm while transmitting a torque of 0.2 kN-m. Due to friction, 1 kW of power is dissipated as heat and the rest is delivered to the output shaft. If the atmospheric temperature is 300 K and the surface of the gearbox maintains a constant temperature of 350 K, determine (a) the rate of entropy transfer into the atmosphere, (b) the rate of entropy generation in the system's universe, (c) the rate of entropy generation within the gearbox, and (d) the rate of entropy generation in the immediate surroundings.

2-4-10 A gearbox (a closed steady system that converts low-torque shaft power to high-torque shaft power) consumes 100 kW of shaft work ($\dot{W}_{sh}$). Due to lack of proper lubrication, the frictional losses amounts to 5 kW, resulting in an output power of 95 kW. The surface of the grearbox is measured to be 350 K while the surrounding temperature is 300 K. Determine (a) the first law efficiency (η) of the gearbox, (b) the heat transfer rate ($\dot{Q}$), (c) the rate of entropy generation ($\dot{S}_{gen,univ}$) in kW/K in the system's universe, and (d) the external rate of entropy generation (entropy that is generated in the immediate surroundings of the gear box).

2-4-11 A closed steady system receives 1000 kW of heat from a reservoir at 1000 K and 2000 kW of heat from a reservoir at 2000 K. Heat is rejected to the two reservoirs at 300 K and 3000 K, respectively. (a) Determine the maximum amount of heat that can be transferred to the reservoir at 3000 K. (b) The device clearly transfers heat to a high temperature TER without directly consuming external work. Is this a violation of the Clausius statement of the second law of thermodynamics? (1:Yes; 2:No)

SECTION 2-5: OVERALL ANALYSIS OF CYCLES

Energy Conversion

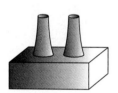

2-5-1 A steam power plant produces 500 MW of electricity with an overall thermal efficiency of 35%. Determine (a) the rate at which heat is supplied to the boiler and (b) the waste heat that is rejected by the plant. (c) If the heating value of coal (heat that is released when 1 kg of coal is burned) is 30 MJ/kg, determine the rate of consumption of coal in tons(US)/day. Assume that 100% of heat released goes to the cycle. (d) *What-if scenario:* What would the fuel consumption rate be if the thermal efficiency increased to 40%?

2-5-2 A utility company charges its residential customers 12 cents/kW·h for electricity and $1.20 per Therm for natural gas. Fed up with the high cost of electricity, a customer decides to generate his own electricity by using a natural gas fired engine that has a thermal efficiency of 35%. Determine the fuel cost per kWh of electricity produced by the customer. Do you think electricity and natural gas are fairly priced by your utility company?

2-5-3 A sport utility vehicle, with a thermal efficiency (η_{th}) of 20%, produces 250 hp of engine output while traveling at a velocity of 80 mph. (a) Determine the rate of fuel consumption in kg/s if the heating value of the fuel is 43 MJ/kg. (b) If the density of the fuel is 800 kg/m^3, determine the fuel mileage of the vehicle in the unit of miles/gallon.

2-5-4 A truck engine consumes diesel at a rate of 30 L/h and delivers 65 kW of power to the wheels. If the fuel has a heating value of 43.5 MJ/kg and a density of 800 kg/m^3, determine (a) the thermal efficiency of the engine and (b) the waste heat rejected by the engine. (c) How does the engine discard the waste heat?

2-5-5 Determine the rate of coal consumption by a thermal power plant with a power output of 350 MW in tons/hr. The thermal efficiency (η_{th}) of the plant is 35% and the heating value of the coal is 30 MJ/kg.

2-5-6 In 2003 the United States generated 3.88 trillion kWh of electricity, 51% of which came from coal-fired power plants. (a) Assuming an average thermal efficiency of 34% and the heating value of coal as 30 MJ/kg, determine the coal consumption in 2003 in tons. (b) *What-if scenario:* What would the coal consumption be if the average thermal efficiency had been 35%?

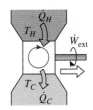

2-5-7 Determine the fuel cost per kWh of electricity produced by a heat engine with a thermal efficiency of 40% if it uses diesel as the source of heat. The following data is supplied for diesel: price = \$2.00 per gallon; heating value = 42.8 MJ/kg; density = 850 kg/m³.

2-5-8 A gas turbine with a thermal efficiency (η_{th}) of 21% develops a power output of 8 MW. Determine (a) the fuel consumption rate in kg/min if the heating value of the fuel is 50 MJ/kg. (b) If the maximum temperature achieved during the combustion of diesel is 1700 K, determine the maximum thermal efficiency possible. Assume the atmospheric temperature to be 300 K.

2-5-9 Two different fuels are being considered for a 1 MW (net output) heat engine which can operate between the highest temperature produced during the burning of the fuel and the atmospheric temperature of 300 K. Fuel A burns at 2500 K, delivering 50 MJ/kg (heating value) and costs \$2 per kilogram. Fuel B burns at 1500 K, delivering 40 MJ/kg and costs \$1.50 per kilogram. Determine the minimum fuel cost per hour for (a) fuel A and (b) fuel B.

2-5-10 A heat engine receives heat from a 2000 K source at a rate of 500 kW, and rejects the waste heat to a medium at 300 K. The net output from the engine is 300 kW. (a) Determine the maximum power that could be generated by the engine for the same heat input. (b) Determine the thermal efficiency of the engine, and its maximum possible limit.

2-5-11 A Carnot heat engine with a thermal efficiency of 60% receives heat from a source at a rate of 3000 kJ/min, and rejects the waste heat to a medium at 300 K. Determine (a) the power that is generated by the engine and (b) the source temperature.

2-5-12 A heat engine operates between a reservoir at 2000 K and an ambient temperature of 300 K. It produces 10 MW of shaft power ($\dot{W}_{sh}$). If it has a thermal efficiency (η_{th}) of 40%, (a) determine the rate of fuel consumption in kg/h if the heating value of the fuel is 40 MJ/kg. (b) If the heat engine is replaced by the most efficient engine possible, what is the minimum possible fuel consumption rate for the same power output?

2-5-13 A heat engine, operating between two reservoirs at 1500 K and 300 K, produces an output of 100 MW. If the thermal efficiency of the engine is measured at 50%, determine

    (a) The Carnot efficiency of the engine (in percent),
    (b) the rate of heat transfer ($\dot{Q}$) into the engine from the hot source in MW, and
    (c) the rate of fuel consumption (in kg/s) if the heating value of fuel is 40 MJ/kg.

2-5-14 A heat engine receives heat from two reservoirs: 50 MW from a reservoir at 500 K and 100 MW from a reservoir at 1000 K. If it rejects 90 MW to the atmosphere at 300 K, (a) determine the thermal efficiency of the engine. (b) Calculate the entropy generated in kW/K in the engine's universe.

2-5-15 A heat engine produces 1000 kW of power while receiving heat from two reservoirs: 1000 kW from a 1000 K source and 2000 kW from a 2000 K source. Heat is rejected to the atmosphere at 300 K. Determine (a) the waste heat (heat rejected) in kW and (b) the entropy generation rate ($\dot{S}_{gen}$) in kW/K in the system's universe.

2-5-16 A heat engine, operating between two reservoirs at 1500 K and 300 K, produces 150 kW of net power. If the rate of heat transfer ($\dot{Q}_H$) from the hot reservoir to the engine is measured at 350 kW, determine (a) the thermal efficiency of the engine, (b) the rate of fuel consumption in kg/h to maintain the hot reservoir at steady state (assume the heating value of the fuel to be 45 MJ/kg), and (c) the minimum possible fuel consumption rate (in kg/h) for the same output.

2-5-17 A solar-energy collector produces a maximum temperature of 100°C. The collected energy is used in a cyclic heat engine that operates in a 5°C environment. (a) What is the maximum thermal efficiency? (b) *What-if scenario:* What would the maximum efficiency be if the collector was redesigned to focus the incoming light to enhance the maximum temperature to 400°C?

2-5-18 The Ocean Thermal Energy Conversion (OTEC) system in Hawaii utilizes the surface water and deep water as thermal energy reservoirs. Assume the ocean temperature at the surface to be 20°C and at an unspecified depth to be 5°C; determine (a) the maximum possible thermal efficiency achievable by a heat engine. (b) *What-if scenario:* What would the maximum efficiency be if the surface water temperature increased to 25°C?

2-5-19 You have been hired by a venture capitalist to evaluate a concept engine proposed by an inventor, who claims that the engine consumes 100 MW at a temperature of 500 K, rejects 40 MW at a temperature of 300 K, and delivers 50 MW of mechanical work. What problems, if any, can you identify with this claim?

2-5-20 A heat engine produces 40 kW of power while consuming 40 kW of heat from a source at 1200 K, 50 kW of heat from a source at 1500 K, and rejecting waste heat to the atmosphere at 300 K. Determine (a) the thermal efficiency of the engine. (b) *What-if scenario:* What would the thermal efficiency be if all the irreversibilities could be magically eliminated? Assume no change in heat input from the two sources.

2-5-21 Two reversible engines, A and B, are arranged in series with the waste heat of engine A used to drive engine B. Engine A receives 200 MJ from a hot source at a temperature of 420°C. Engine B is in communication with a heat sink at a temperature of 4.4°C. If the work output of A is twice that of B, determine (a) the intermediate temperature between A and B and (b) the thermal efficiency of each engine.

2-5-22 A Carnot heat engine receives heat from a TER at T_{TER} through a heat exchanger where the heat transfer rate is proportional to the temperature difference as $(\dot{Q}_H) = A(T_{TER} - T_H)$. It rejects heat to a cold reservoir at T_C. If the heat engine is to maximize the work output, show why the high temperature in the cycle should be selected as $T_H = \sqrt{(T_{TER}T_C)}$.

2-5-23 Two Carnot engines operate in series. The first one receives heat from a TER at 2500 K and rejects the waste heat to another TER at a temperature T. The second engine receives the energy rejected by the first one, converts some of it to work, and rejects the rest to a TER at 300 K. If the thermal efficiency of both the engines are the same, (a) determine the temperature (T). *What-if scenario:* (b) What would the temperature be if the two engines produced the same output?

2-5-24 A reversible heat engine operates in outer space. The only way heat can be rejected is by radiation, which is proportional to the fourth power of the temperature and the area of the radiating surface. Show that for a given power output and a given source temperature (T_1), the area of the radiator is minimized when the radiating surface temperature is $T_2 = 0.75T_1$.

2-5-25 A heat engine receives heat at a rate of 3000 kJ/min from a reservoir at 1000 K and rejects the waste heat to the atmosphere at 300 K. If the engine produces 20 kW of power, determine (a) the thermal efficiency and (b) the entropy generated in the engine's universe.

Refrigeration Cycles (Sec. 5 Continued)

2-5-26 A household freezer operates in a 25°C kitchen. Heat must be transferred from the cold space at a rate of 2.5 kW to maintain its temperature at −25°C. What is the smallest (power) motor required to operate the freezer.

2-5-27 To keep a refrigerator in steady state at 2°C, heat has to be removed from it at a rate of 200 kJ/min. If the surrounding air is at 27°C, determine (a) the minimum power input to the refrigerator and (b) the maximum COP.

2-5-28 A Carnot refrigerator consumes 2 kW of power while operating in a 20°C room. If the food compartment of the refrigerator is to be maintained at 3°C, determine the rate of heat removal in kJ/min from the compartment.

2-5-29 An actual refrigerator operates with a COP that is half the Carnot COP. It removes 10 kW of heat from a cold reservoir at 250 K and dumps the waste heat into the atmosphere at 300 K. (a) Determine the net work consumed by the refrigerator. (b) *What-if scenario:* How would the answer change if the cold storage were to be maintained at 200 K without altering the rate of heat transfer?

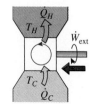

2-5-30 An inventor claims to have developed a refrigerator with a COP of 10 that maintains a cold space at −10°C, while operating in a 25°C kitchen. Is this claim plausible? (1:Yes; 0:No)

2-5-31 A refrigeration cycle removes heat at a rate of 250 kJ/min from a cold space maintained at −10°C while rejecting heat to the atmosphere at 25°C. If the power consumption rate is 0.75 kW, determine if the cycle is (1: reversible; 2: irreversible; 3: impossible).

2-5-32 A refrigeration cycle removes heat at a rate of 250 kJ/min from a cold space maintained at −10°C while rejecting heat to the atmosphere at 25°C. Assuming the power consumption rate is 1.5 kW, (a) do a first-law analysis to determine the rate of heat rejection to the atmosphere in kW. (b) Do a second-law analysis to determine the entropy generation rate in the refrigerator's universe.

2-5-33 In a cryogenic experiment a container is maintained at −120°C, although it gains 200 W due to heat transfer from the surroundings. What is the minimum power of a motor that is needed for a heat pump to absorb heat from the container and reject heat to the room at 25°C?

2-5-34 An air-conditioning system maintains a house at a temperature of 20°C while the outside temperature is 40°C. If the cooling load on this house is 10 tons, determine (a) the minimum power requirement. (b) *What-if scenario:* What would the minimum power requirement be if the interior was 5 degrees warmer?

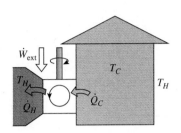

2-5-35 An air-conditioning system is required to transfer heat from a house at a rate of 800 kJ/min to maintain its temperature at 20°C. (a) If the COP of the system is 3.7, determine the power required for air conditioning the house. (b) If the outdoor temperature is 35°C, determine the minimum possible power required.

2-5-36 A solar-powered refrigeration system receives heat from a solar collector at T_H, rejects heat to the atmosphere at T_0 and extracts heat from a cold space at T_C. The three heat transfer rates are $\dot{Q}_H$, $\dot{Q}_O$ and $\dot{Q}_C$ respectively. Do an energy and entropy analysis of the system to derive an expression for the maximum possible COP, defined as the ratio $\dot{Q}_C/\dot{Q}_H$.

2-5-37 Assume $T_H = 425$ K, $T_0 = 298$ K, $T_C = 250$ K and $\dot{Q}_C = 20$ kW in the above system of problem 2-5-36. (a) Determine the maximum COP of the system. (b) If the collector captures 0.2 kW/m², determine the minimum collector area required.

2-5-38 A refrigerator with a COP of 2.0 extracts heat from a cold chamber at 0°C at a rate of 400 kJ/min. If the atmospheric temperature is 20°C, determine (a) the power drawn by the refrigerator and (b) the rate of entropy generation in the refrigerator's universe.

Heat Pump Cycles (Sec. 5 Continued)

2-5-39 On a cold night, a house is losing heat at a rate of 15 kW. A reversible heat pump maintains the house at 20°C while the outside temperature is 0°C. (a) Determine the heating cost for the night (8 hours). (b) Also determine the heating cost if resistance heating were used instead. Assume the price of electricity to be 15 cents/kWh.

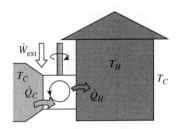

2-5-40 On a cold night, a house is losing heat at a rate of 80,000 Btu/h. A reversible heat pump maintains the house at 70°F, while the outside temperature is 30°F. Determine (a) the heating cost for the night (8 hours), assuming a price of 10 cents/kWh for electricity. Also determine (b) the heating cost if resistance heating were used instead.

2-5-41 A house is maintained at a temperature of 20°C by a heat pump pumping heat from the atmosphere. Heat transfer rate through the wall and roof is estimated at 0.6 kW per unit temperature difference between inside and outside. (a) If the atmospheric temperature is −10°C, what is the minimum power required to drive the pump? (b) It is proposed to use the same pump to cool the house in the summer. For the same room temperature, the same heat transfer rate, and the same power input to the pump, determine the maximum permissible atmospheric temperature. (c) *What-if scenario:* What would the answer in part (b) be if the heat transfer rate between the house and outside was estimated at 0.7 kW per unit temperature difference?

2-5-42 A house is maintained at a temperature T_H by a heat pump that is powered by an electric motor. The outside air at T_C is used as the low-temperature TER. Heat loss from the house to the surroundings is directly proportional to the temperature difference and is given by $\dot{Q}_{loss} = U(T_H - T_C)$. Determine the minimum electric power required to drive the heat pump as a function of the given variables.

2-5-43   A house is maintained at steady state (closed system) at 300 K while the outside temperature is 275 K. The heat loss ($\dot{Q}_C$) is measured at 2 kW. Two approaches are being considered: (A) electrical heating at 100% efficiency and (B) an ideal heat pump that operates with Carnot COP. The price of electricity is \$0.2/kWh. (a) Determine the cost of option A and (b) option B over a 10 hour operating period.

2-5-44   A house is maintained at a temperature of 25°C by a reversible heat pump powered by an electric motor. The outside air at 10°C is used as the low-temperature TER. Determine the percent savings in electrical power consumption if the house is kept at 20°C instead. Assume the heat loss from the house to the surroundings is directly proportional to the temperature difference.

2-5-45   A house is heated and maintained at 25°C by a heat pump. Determine the maximum possible COP if heat is extracted from the outside atmosphere at (a) 10°C, (b) 0°C, (c) −10°C and (d) −40°C. (e) Based on these results, would you recommend using heat pumps at locations with a severe climate?

Mixed Cycles (Sec. 5 Continued)

2-5-46   A Carnot heat engine receives heat at 800 K and rejects the waste heat to the surroundings at 300 K. The output from the heat engine is used to drive a Carnot refrigerator that removes heat from the −20°C cooled space at a rate of 400 kJ/min and rejects it to the same surroundings at 300 K. Determine (a) the rate of heat supplied to the heat engine and (b) the total rate of heat rejection to the surroundings. **What-if scenario:** (c) What would the rate of heat supplied be if the temperature of the cooled space was −30°C?

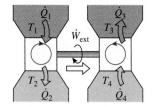

2-5-47   A reversible heat engine is used to drive a reversible heat pump. The power cycle takes in Q_1 heat units at T_1 and rejects Q_2 heat units at T_2. The heat pump extracts Q_4 from a heat sink at T_4 and discharges Q_3 at T_3. Develop an expression for Q_4/Q_1 in terms of the four given temperatures.

2-5-48   A heat engine with a thermal efficiency (η_{th}) of 35% is used to drive a refrigerator having a COP of 4. (a) What is the heat input $\dot{Q}_H$ to the engine for each MJ removed from the cold region by the refrigerator? (b) If the system is used as a heat pump, how many MJ of heat would be available for heating for each MJ of heat input into the engine?

2-5-49   A heat engine operates between two TERs at 1000°C and 20°C. Two-thirds of the work output is used to drive a heat pump that removes heat from the cold surroundings at 0°C and transfers it to a house kept at 20°C. If the house is losing heat at a rate of 60,000 kJ/h, determine (a) the minimum rate of heat supply to the heat engine. (b) **What-if scenario:** What would the minimum heat supply be if the outside temperature dropped to −10°C?

2-5-50   A heat engine is used to drive a heat pump. The waste heat from the heat engine and the heat transfer from the heat pump are used to heat the water circulating through the radiator of a building. The thermal efficiency of the heat engine is 30% and the COP of the heat pump is 4.2. Evaluate the COP of the combined system, defined as the ratio of the heat transfer to the circulating water to the heat transfer to the heat engine.

2-5-51   A furnace delivers heat at a rate of $\dot{Q}_{H1}$ at T_{H1}. Instead of directly using this for room heating, it is used to drive a heat engine that rejects the waste heat into atmosphere at T_0. The heat engine drives a heat pump that delivers $\dot{Q}_{H2}$ at T_{room}, using the atmosphere as the cold reservoir. Find the ratio $\dot{Q}_{H2}/\dot{Q}_{H1}$, the energetic efficiency of the system as a function of the given temperatures. Why is this a better set-up than direct room heating from the furnace?

2-5-52   A heat pump is used to heat a house in the winter and cool it in the summer by reversing the flow of the refrigerant. The interior temperature should be 20°C in the winter and 25°C in the summer. Heat transfer through the walls and ceilings is estimated to be 2500 kJ per hour per °C temperature difference between the inside and outside. (a) If the winter outside temperature is 0°C, what is the minimum power required to drive the heat pump? (b) For the same power input as in part (a), what is the maximum outside summer temperature at which the house can be maintained at 25°C?

3

EVALUATION OF PROPERTIES: MATERIAL MODELS

In Chapter 1, we discussed states and properties in a qualitative manner, and performed numerical exercises with system- and flow-state TESTcalcs to gain a quantitative understanding of properties describing an extended state, which builds upon the core *thermodynamic state*. Having introduced the governing equations in Chapter 2, we are now ready to develop a comprehensive methodology to evaluate thermodynamic states of a system for a wide range of working substances. To continue our discussion of equilibrium from Chapter 1, differential relations, which apply to all pure working substances for *thermodynamic properties* are introduced. The working substances are divided into groups or models based on the set of assumptions used to simplify the thermodynamic relations. These models include the solid/liquid (SL) model, phase-change (PC) model, ideal-gas (IG) model, perfect-gas (PG) model, and real-gas (RG) model. A set of equations known as the *equations of state*, that relate thermodynamic properties to each other are developed for some of these models. For others, tables and charts of thermodynamic and material properties are developed to facilitate manual evaluation of thermodynamic states. Wherever possible, manual calculations are verified through the use of state TESTcalcs. Two additional models, mixture models and moist air models, will be introduced in Chapters 11 and 12. Knowledge of all the models covered in this chapter is not required for subsequent chapters, and this chapter can be revisited whenever a new model is encountered during a system analysis.

3.1 THERMODYNAMIC EQUILIBRIUM AND STATES

The concept of thermodynamic equilibrium is central in evaluating states. In this section, we continue our qualitative discussion of equilibrium from Sec. 1.2, and define equilibrium in a more formal manner. Equilibrium will be presented in depth in Chapter 14.

3.1.1 Equilibrium and LTE (Local Thermodynamic Equilibrium)

Let's observe a freshly *isolated system*; one that contains a mixture of hydrogen, oxygen, and water vapor trapped in an enclosure (Fig. 3.1). Isolating a system requires cutting off all interactions—mass, heat, and work—between the system and its surroundings (see Anim. 3.A.*isolate*). Initially, assume the system to be absolutely non-uniform, that is, properties such as V, p, and T vary significantly from one local state to another. As time progresses, we can anticipate what will happen. With the system cut off from all external interactions, the internal motion will quickly abate. When the force imbalance between any two points within the system completely subsides, that is, the pressure becomes uniform, the *viscous friction* disappears, and all internal motions stop, the system achieves a state of **mechanical equilibrium**. Experience also suggests that the warmer regions of the system will become cooler and the cooler regions warmer, until the temperature becomes uniform everywhere. With temperature difference, which is the driving force for heat transfer (and the associated *thermal friction*) disappearing, the system reaches a state of **thermal equilibrium** other ways the system can still spontaneously change is through migration of different chemical components from one location to another (for example, water changing its phase from liquid to vapor or hydrogen diffusing from one side to another through an internal partition as shown in Anim. 0.C.*isolatedSystem*) or through chemical reactions (for example, hydrogen and oxygen reacting to produce water). Eventually, these chemical imbalances (and the associated *chemical friction*) disappear and **chemical equilibrium** is established with no further changes in the system's chemical or phase composition. There can be other types of imbalances in a system such as electrical and magnetic imbalances, but those are ruled out by the type of systems, called *simple systems* (to be defined in Sec. 3.1.2), we are going to study in this book. When the three major equilibriums—mechanical, thermal, and chemical—are established in a system, it is said

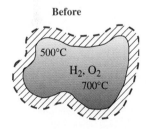

Before

After

FIGURE 3.1 An isolated system spontaneously seeks thermal, mechanical, and chemical equilibrium (see Anim. 0.C.*isolatedSystem*).

to be in **thermodynamic equilibrium** or simply **equilibrium**. To better understand equilibrium, study Anim. 3.A.*equilibrium*; in this animation, an isolated system's subsystems A and B are allowed to seek different types of equilibrium when different internal constraints are removed one at a time.

A system in thermodynamic equilibrium can be completely described by its **thermodynamic state** (see Anim. 3.A.*equilibriumState*), which is identified by a set of thermodynamic properties, such as p, T, v, u, h, and s. Once in equilibrium, an isolated system maintains its thermodynamic state. Any interactions in the form of mass, heat, or work, can change the equilibrium state, which is accompanied by a change in thermodynamic properties. Conversely, any change in thermodynamic properties signals a change in equilibrium. An important feature of an equilibrium state is that thermodynamic properties are related by the **equations of state**. By using these equations, we can drastically reduce the number of independent properties required to describe an equilibrium state.

We might be tempted to question the usefulness of evaluating a thermodynamic state, given that few of the practical systems encountered in thermodynamics—pumps, compressors, turbines, piston-cylinder devices, etc.—are isolated, and even fewer are in equilibrium. For thermodynamics to be a practical subject, states of non-uniform, unsteady systems have to be somehow related to the simplified state of equilibrium. This is possible because even the most complex thermodynamic system abides by the *LTE hypothesis* (local thermodynamic equilibrium) introduced in Sec. 1.2. Use of LTE allows us to represent a global, non-uniform state as a collection of local equilibrium states (Fig. 3.2) at a given *thermodynamic instant*. While the global system is in flux (for example, a gas accelerating in a nozzle), its constituent local systems are all in equilibrium. This apparent contradiction can be resolved by comparing the local relaxation time—the *macroscopic instant* over which equilibrium is reached in a local system—with the global relaxation time—the time over which the global system changes. Typically the local relaxation time, which depends on communication (through collisions) among molecules inside the tiny local system spanning only a *macroscopic point*, is much smaller than the time scale over which global changes take place in the system. While the global system is under severe imbalances, the local systems by the virtue of their small size can be considered to be in LTE.

The significance of the LTE hypothesis is far reaching. If a relationship can be established between a group of properties (e.g., p, ρ, and T) for a working substance in equilibrium, then the relationship can be applied locally in any type of systems—uniform, non-uniform, steady, unsteady, stationary, or flowing system. In other words; the state of any practical device at a given instant can be expressed as an aggregate of local equilibrium states for almost any type of system. A few notable exceptions include systems involving rarefied medium, shock waves, or slow chemical reactions, where the collision rate among molecules in a local system is not high enough to justify LTE within the observation period. For an overwhelming number of practical systems, describing local thermodynamic equilibrium, is at the core of describing an extended *system* or *flow* state (see Anim. 3.A.*LTE*).

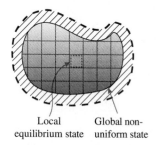

Local   Global non-
equilibrium state   uniform state

FIGURE 3.2 The global state can be expressed as a collection of local equilibrium states (see Anim. 3.A.*LTE*).

3.1.2 The State Postulate

In Sec. 1.7 we summarized the extended equations of state, Eqs. (1.26) through (1.32), which relate extrinsic and system properties of an extended state with its core equilibrium state. The task of evaluating a system, or flow state, reduces to evaluating the thermodynamic properties that define the underlying thermodynamic equilibrium. Beside p, T, v, u, h, and s, introduced in Chapter 1, there are many more thermodynamic properties yet to be introduced. Given their indefinite number, what is the minimum number of thermodynamic properties necessary to adequately describe an equilibrium state. The answer is remarkably simple and is provided by the **Gibbs state postulate**.

Before we can introduce this postulate we must expand our thermodynamic vocabulary to include the terms *simple system* and *pure substance*. Out of all different types of work transfer, a simple system allows only boundary work. Although this type of system may seem restrictive, it is not. For example, if we analyze how a rotating paddle-wheel transfers work (Fig. 3.3) into a working substance, we realize that energy is transferred through boundary work as the adjacent local systems are pushed by the blades. Similarly, when electric work is transferred to a heating element (resistance heater), what an adjacent local system actually experiences is heat transfer. Additional modes of work transfer, though rare, are still possible if the working

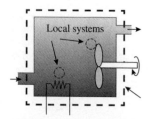

Local systems

FIGURE 3.3 The local systems experience only heat and boundary work transfer. Electrical and shaft work apply to the global system only.

substance contains magnetic dipoles or electric charge. In their absence, such a system is called a **simple compressible system** (or a **simple system**)—*simple* because there is only one mode of work transfer and *compressible* because only expansion or contraction of the working substance is allowed.

A working substance with a fixed chemical composition throughout a system is called a **pure substance**. Not only do pure solids, liquids, or gases qualify as pure substances, but so do mixtures with a fixed chemical composition. A mixture of ice and water constitutes a pure substance because the chemical composition does not change across phases. Similarly, atmospheric air, a mixture of approximately 77% N_2 and 23% O_2 by mass, can be considered a pure substance, unless of course, the temperature drops to such an extreme level that oxygen begins to condense (Fig. 3.4). By assuming the working substance to be pure, we can rule out any change in chemical composition across a system; therefore there is no driving force for a species transfer across the boundary of a local system. The time necessary for a tiny local system to break up is much larger than a macroscopic instant. Without any loss of generality, a local system can be treated as a closed system in equilibrium with the possibility of only heat and boundary work interactions with its surroundings at a given macroscopic instant. Such a simple local system can be scaled up to a large system for analytical purpose with the same set of thermodynamic properties describing the underlying equilibrium. The uniform system within the piston-cylinder assembly of Figure 3.5 is a typical simple system, basically a scaled-up version of the local systems in Figure 3.3. Any state relation (equation of state) derived for such a simplified system can be immediately applied to more general systems, open or closed, steady or unsteady, with or without different types of heat and work transfer.

Now consider a pure substance, such as, air, as the working fluid in the piston-cylinder assembly of Figure 3.6. For a given pressure, created by the amount of weight on the piston, and a given bath water temperature, set by the bath water heater, the gas achieves a particular equilibrium condition. If the weight and/or the bath temperature is changed, and the system is left undisturbed for a sufficiently long time, the trapped gas will seek a new equilibrium. Obviously we can control the pressure p and temperature T of the new equilibrium independently. Can a third property, for example, the specific volume v, be controlled independently without affecting p and T? If we try to squeeze the gas by placing some additional weights on the piston, any change in v will also be accompanied by a change in p. If air were magnetic, we could change the internal energy u by manipulating an external magnetic field without affecting the system's p or T. But this possibility is ruled out in a *simple system*. Hence it is impossible to independently control any third property once p and T are fixed. Experimental evidence of this kind is generalized by the **state postulate** (see Anim. 3.A.*statePostulate*):

> *The thermodynamic state of a simple compressible system composed of a pure substance is fixed by two independent thermodynamic properties.*

For the trapped air in Figure 3.6, knowing any independent pair of thermodynamic properties, such as (p, T), (p, h), (v, u), and (s, h), is sufficient to produce the system's complete thermodynamic state. A thermodynamic state can be uniquely located on two-dimensional

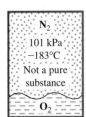

FIGURE 3.4 At −183°C air is not a pure substance, as oxygen begins to condense. Steam is always a pure substance regardless of condensation.

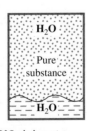

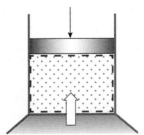

FIGURE 3.5 A large simple compressible system can be considered a magnified version of a local system shown in Fig. 3.3.

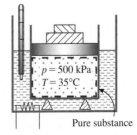

FIGURE 3.6 Up to two thermodynamic properties can be independently varied for a simple system containing a pure substance.

thermodynamic plots with coordinates such as *T-v*, *T-s*, *p-v*, *h-s*, etc. Identifying states on thermodynamic plots provides a visual check against inadvertent errors when evaluating states. Calculated states can be visualized in such thermodynamic plots (Figs. 3.7 and 1.45) by selecting a specific plot from the plot menu in the state panel of all TESTcalcs. In Chapter 14, which is exclusively devoted to the study of equilibrium, the theoretical basis of the state postulate will be developed and generalized to include multi-phase, multi-component working substances.

3.1.3 Differential Thermodynamic Relations

To exploit the state postulate for developing basic state relations, let the function $z(x,y)$ represent a thermodynamic property in terms of two independent thermodynamic properties x and y:

$$z = z(x,y) \tag{3.1}$$

Using Taylor's theorem from calculus (see Anim. 3.A.*taylorTheorem*), a differential change dz in $z(x,y)$ can be related to changes dx and dy of the independent variables (Fig. 3.8):

$$dz = \left(\frac{\partial z}{\partial x}\right)_y dx + \left(\frac{\partial z}{\partial y}\right)_x dy \tag{3.2}$$

where the partial derivatives with respect to x and y, which are slopes on the z surface along those coordinates, can be interpreted as new thermodynamic properties.

Application of Taylor's theorem to thermodynamic properties $u = u(T,v)$ and $h = h(T,p)$ yields:

$$du = \left(\frac{\partial u}{\partial T}\right)_v dT + \left(\frac{\partial u}{\partial v}\right)_T dv, \quad \text{and} \quad dh = \left(\frac{\partial h}{\partial T}\right)_p dT + \left(\frac{\partial h}{\partial p}\right)_T dp; \quad \left[\frac{\text{kJ}}{\text{kg}}\right] \tag{3.3}$$

Of the four new properties introduced by the partial derivatives in these relations, the slope of u with respect to T at constant volume (see Fig. 3.9) is called the **specific heat at constant volume** c_v, and the slope of h with respect to T at constant pressure is called the **specific heat at constant pressure**, c_p (see Anim. 3.A.*cv* and 3.A.*cp*). The names of these properties owe their origin to the extensive properties, *heat capacity* $C_v = mc_v$ at constant volume and heat capacity $C_p = mc_p$ at constant pressure, which are seldom used in engineering. A specific heat is a heat capacity per-unit mass. Specific heats have units of kJ/(kg · K), and they often behave as *material properties* for many substances. Although their definitions require v or p to be held constant in the partial derivative, c_v and c_p are thermodynamic properties in their own right, just like p or T, and can be used to describe an equilibrium state regardless of how the system arrived at that state. Neither volume nor pressure need be held constant when using c_v or c_p. Note the use of lowercase letters in the symbols for the specific heat, which is consistent with symbols used in other specific properties. (The symbol C_p is reserved for the pressure coefficient that is discussed in Chapter 15.)

The thermodynamic relations for du and dh can be expressed in terms of these newly defined properties as:

$$du = c_v dT + \left(\frac{\partial u}{\partial v}\right)_T dv; \quad \left[\frac{\text{kJ}}{\text{kg}}\right] \quad \text{where} \quad c_v \equiv \left(\frac{\partial u}{\partial T}\right)_v; \quad \left[\frac{\text{kJ}}{\text{kg} \cdot \text{K}}\right] \tag{3.4}$$

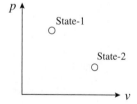

FIGURE 3.7 Two-dimensional thermodynamic plots, such as the *p-v* or the *T-s* diagram, are often used to graphically represent equilibrium states.

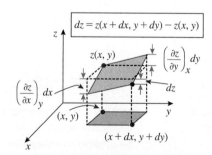

FIGURE 3.8 Geometric interpretation of Taylor's theorem (see Anim. 3.A.*taylorTheorem*).

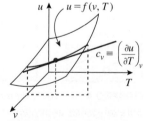

FIGURE 3.9 c_v can be interpreted as a slope on the constant-volume plane (see Anim. 3.A.*cv*).

$$dh = c_p dT + \left(\frac{\partial h}{\partial p}\right)_T dp; \quad \left[\frac{kJ}{kg}\right] \quad \text{where} \quad c_p \equiv \left(\frac{\partial h}{\partial T}\right)_p; \quad \left[\frac{kJ}{kg \cdot K}\right] \tag{3.5}$$

These relations serve as the most fundamental relations for evaluating u and h of any working substance. For evaluation of entropy, application of Taylor's theorem to $s = s(u,v)$ yields:

$$ds = \left(\frac{\partial s}{\partial u}\right)_v du + \left(\frac{\partial s}{\partial v}\right)_u dv; \quad \left[\frac{kJ}{kg \cdot K}\right] \tag{3.6}$$

In Chapter 5 we will analyze a differential process—a process where a system at equilibrium changes to a new equilibrium due to differential interactions of heat and work with the system's surroundings—and show that the slopes of entropy in Eq. (3.6) do not introduce new properties, but are $1/T$ and p/T respectively. With these as coefficients, Eq. (3.6) can be rewritten as a powerful thermodynamic relation called the first Tds relation: $Tds = du + pdv$. When $u = h - pv$ is substituted into this relation, use of the product rule of differentiation and some simplification leads to the second Tds equation. The two Tds equations are summarized as follows:

$$Tds = du + pdv; \quad \left[\frac{kJ}{kg}\right] \tag{3.7}$$

$$Tds = dh - vdp; \quad \left[\frac{kJ}{kg}\right] \tag{3.8}$$

Equations (3.4) through (3.8) are part of a larger set of fundamental thermodynamic relations that will be fully explored in Chapter 11. They relate two neighboring equilibrium states of any working substance separated in space (e.g., the two flow states of Fig. 3.10a) or time (e.g., the two system states of Fig. 3.10b) or both. The meaning of a differential relation is quantitatively explored in Examples 3-2 and 3-3.

The thermodynamic relations presented in this section will be used to obtain expressions for a difference in a thermodynamic property—Δu, Δh, etc.—in terms of other properties and differences. To define absolute values of such properties, we will introduce reference properties—properties that are assigned a certain value at a reference state. Although there are different standards in use today, we will use standard atmospheric conditions represented by $p_0 = 101.325$ kPa and $T_0 = 25°C$, which will be referred to as the **standard state**. Reference values will be further discussed in Sec. 3.7.

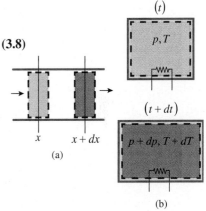

FIGURE 3.10 Neighboring states in various configurations: (a) Open-steady system; (b) unsteady-uniform system.

EXAMPLE 3-1 **Application of Taylor's Theorem**

The specific volume of a gas in m³/kg is expressed by: $v = 4.157\dfrac{T}{p}$ (see Fig. 3.11), where

T is in K and p is in kPa. Determine the change in v (a) exactly and (b) approximately (using Taylor's theorem) for the following change of state: State-1: $p = 101$ kPa, $T = 1000$ K; State-2: $p = 100$ kPa, $T = 1005$ K.

SOLUTION

The exact change in v can be calculated from the known functional relation:

$$\Delta v = v_2 - v_1 = 4.157\left(\frac{T_2}{p_2} - \frac{T_1}{p_1}\right) = 4.157\left(\frac{1005}{100} - \frac{1000}{101}\right) = 0.619 \frac{m^3}{kg}$$

Assuming all changes to be small, Taylor's theorem can be used to approximately relate a change in v with changes in T and p, since $v = v(T,p)$:

$$\Delta v \cong dv = \left(\frac{\partial v}{\partial T}\right)_p dT + \left(\frac{\partial v}{\partial p}\right)_T dp = 4.157\left(\frac{dT}{p} - \frac{Tdp}{p^2}\right)$$

$$= 4.157\left(\frac{5}{100} - \frac{(1000)(-1)}{101^2}\right) = 0.615 \frac{m^3}{kg}$$

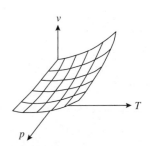

FIGURE 3.11 A surface plot showing v as a function of p and T (Ex. 3-1).

Discussion

The percent accuracy of the result from Taylor's theorem increases as the changes involved approach differential amounts. This can be verified by repeating the calculations with $p_2 = 100.5$ kPa and $T_2 = 1002$ K.

EXAMPLE 3-2   Use of Fundamental Relation for Testing Data

A piston-cylinder device contains steam at a pressure of 1 MPa and a temperature of 300°C. (a) Evaluate the difference between the left- and right-hand sides of the first Tds equation [Eq. (3.7)] when the system establishes a new equilibrium where its pressure is 1 kPa less and its temperature is 1°C warmer. (b) Estimate the steam's c_p value at this condition using a 0.5°C rise in temperature as a differential change in temperature. Use the PC system-state TESTcalc.

TEST Solution

After launching the TESTcalc, select H2O and evaluate the two states as State-1 and State-2 using the known properties. Calculate the property differences Δs, Δv, and Δu in the I/O panel using expressions such as '= s2 − s1', '= v2 − v1', etc. The property differences are:

$$\Delta s = 0.0041 \ \frac{\text{kJ}}{\text{kg}\cdot\text{K}}; \quad \Delta v = 7.561 \times 10^{-4} \ \frac{\text{m}^3}{\text{kg}}; \quad \Delta u = 1.659 \ \frac{\text{kJ}}{\text{kg}};$$

When evaluating the term Tds, either T_1, T_2 or their average can be used, since $(T \pm \Delta T)\Delta s = T\Delta s \pm \Delta T \Delta s \cong T\Delta s$ (the second term is an order of magnitude smaller). The two sides of the Tds equation and their difference are evaluated as follows:

$$\text{LHS} \cong T_1\Delta s = (300 + 273)0.0041 = 2.349 \text{ kJ/kg};$$

$$\text{RHS} \cong \Delta u + p_1\Delta v = 1.659 + (1000)(7.561 \times 10^{-4}) = 2.415 \text{ kJ/kg};$$

$$\text{Difference between LHS and RHS} = \frac{2.415 - 2.349}{2.349} \times 100 = 2.8\%$$

To estimate the value of c_p in Eq. (3.5), evaluate a neighboring state, State-3, at p3 = p1 and T3 = T1 + 0.5 deg-C. In the I/O panel, evaluate '= (h3 − h1)/(T3 − T1)'; the result, 2.13 kJ/kg·K, is an approximate value of steam's c_p at 1 MPa and 300°C.

Discussion

The discrepancy between the two sides of the Tds relation suggests that the two states may not be sufficiently close to each other to qualify as differential neighbors. If the states are gradually brought closer by reducing Δp and ΔT, the discrepancy continues to decrease down to about 1% and no further. TEST evaluates states by interpolating tables (just like a manual solution), and the error in interpolation can be as much as 2–3%. The exact value of any partial derivative cannot be numerically obtained from the TESTcalc (see Fig. 3.12).

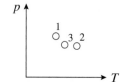

FIGURE 3.12 The three neighboring states of Ex. 3-2 on a p-T diagram.

3.2 MATERIAL MODELS

A *pure substance* can exist in any of the three principal phases (solids, liquids, and gases) or as a mixture of phases. A **phase** is a physically distinct homogenous (uniform) subsystem with a definite bounding surface. In our everyday experience, we find aluminum as a solid, tap water as a liquid, air as a gas, and boiling water as a mixture of liquid and water vapor. Working substances are so diverse and complex that different degrees of idealization are necessary to group them into a manageable set. Assumptions that go into creating such a group and the resulting state equations that lead to the evaluation of the complete thermodynamic state from any given pair of independent properties are collectively called the **material model** for the group. Thus, a block of copper can be represented by the solid/liquid (SL) model, boiling water by the phase-change (PC) model, and gaseous nitrogen by the ideal-gas (IG) model. In this chapter, we will introduce five such models for the evaluation of state. Although the choice of model is relatively straightforward for most working substances, sometimes several models are applicable to a given working substance, which requires consideration of the trade off between accuracy and complexity.

3.2.1 State TESTcalcs and TEST-Code

State TESTcalcs, introduced in Sec. 1.5.1, will be used throughout this chapter to verify manual results and perform "what-if" calculations. Once a TEST solution is obtained, clicking the Super-Calculate button generates a few lines of TEST-code in the I/O panel along with a detailed solution report. TEST-code describes the solution's algorithm by listing the known information and the analytical set up of the problem in a succint manner (see Fig. 3.13). Generated TEST-code can be saved by copying it from the I/O panel into a local text file. Later, the solution can be reproduced by copying the TEST-code into the appropriate TESTcalc's I/O panel, and clicking Load. TEST-code for all the TEST solutions presented in this textbook can be found in the *TEST > TEST-codes* module.

```
States {
    State-1: H2O;
    Given: { p1= 1.0 MPa; T1= 300.0
    deg-C; Vel1= 0.0 m/s; z1= 0.0 m;
    }

    State-2: H2O;
    Given: { p2= "p1-1" kPa; T2=
    "T1+1" deg-C; Vel2= 0.0 m/s; z2=
    0.0 m; }

    State-3: H2O;
    Given:     { p3= "p1" kPa;  T3=
    "T1+0.5" deg-C;   Vel3= 0.0 m/s;
    z3= 0.0 m;   }


}
```

FIGURE 3.13 TEST-code generated by the Super-Calculate button in Ex 3-2 can be used to regenerate a solution.

3.3 THE SL (SOLID/LIQUID) MODEL

Molecules in a solid are packed closely together resulting in large intermolecular attractive forces that are responsible for the natural resistance a solid has to deformation. Pinned by these forces into equilibrium positions, the mobility of the solid molecules are limited to vibration around their mean positions much like the structure of spring-mass systems shown in Figure 3.14. It is these molecules' vibrational kinetic energy that we feel with our tactile senses as temperature. Molecules of a *crystalline* solid, such as copper, are organized in a 3-D pattern called a *lattice* that is repeated throughout the solid. In a *non-crystalline* solid, such as glass or wood, local systems have fixed patterns, which are not necessarily replicated throughout. When the temperature of a solid is sufficiently raised, the increased vibrational momentum overcomes the attractive forces and chunks of molecules break away from the lattice, starting a phase-change phenomenon known as *melting*.

As a pure solid melts into a liquid, molecules retain their lattice structure in local neighborhoods. However, groups of molecules (Fig. 3.15) are free to move with respect to each other, offering no resistance to shear forces—a defining characteristic of a liquid phase (Click Liquid button in Anim. 3.B.*microStructure*). At the local level, there are remarkable similarities in molecular structures of a solid and a liquid. In both these phases, vibrational energy is the principle microscopic repository of internal energy, therefore, the relationship among properties must be similar between solids and liquids. Moreover, pressure has almost no affect on the molecular arrangement of an incompressible substance such as a solid or liquid. As a result we expect the SL model formulas for u and s (which depend on molecular energy and its distribution) to be independent of pressure.

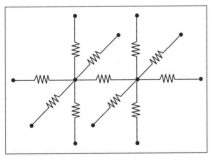

FIGURE 3.14 Solid molecules vibrate like a network of spring-mass systems (see Anim. 3.B.*microStructure*).

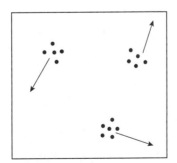

FIGURE 3.15 Liquid molecules move around in clusters that maintain the molecular structure of a solid locally (click Liquid in Anim. 3.B.*microStructure*).

3.3.1 SL Model Assumptions

All state equations for the SL model are derived from the following basic assumptions:

1. Specific volume v (and hence, density ρ) is a constant, that is, the working substance is incompressible.
2. Specific heat c_v is a constant.

A metal block expands when heated; however, the same block, when idealized by the SL model, must be considered incompressible. These model assumptions turn specific heat (we will soon establish that $c_p = c_v$ for this model) and density into material properties; Table A-1 lists several common solid and liquid properties. In SL state TESTcalcs, specific heats and specific volumes are populated as soon as a working substance is selected.

3.3.2 Equations of State

With specific volume v and specific heat c_v already tabulated for different working substances (Table A-1), our goal is to express the thermodynamic properties u, h, and s as functions of independent properties p and T. Such state functions are called *equations of state*.

Internal Energy (u) and Enthalpy (h):

By assumption, v and c_v are constant; therefore, Eq. (3.4) can be integrated rather easily, yielding:

$$du = c_v dT + \left(\frac{\partial u}{\partial v}\right)_T dv^0 = c_v dT; \quad \left[\frac{kJ}{kg}\right] \tag{3.9}$$

$$\Rightarrow \quad u = u(p_0, T_0) + \int_{p_0,T_0}^{p,T} c_v(T)\, dT = u_{ref}^o + c_v(T - T_0), \quad \text{where} \quad u_{ref}^o = u(T_0) \tag{3.10}$$

$$\Rightarrow \quad \Delta u = u_2 - u_1 = c_v(T_2 - T_1) = c_v \Delta T; \quad \left[\frac{kJ}{kg \cdot K}K = \frac{kJ}{kg}\right] \tag{3.11}$$

The superscript o in u_{ref}^o indicates that the reference values is assigned at the *standard condition*. As numerically predicted in Ex. 1-10, u is now established as a linear function of T by Eq. (3.10). While Δu can be easily obtained from these equations, the absolute value of u is difficult to determine; because $u_{ref}^o = u(T_0) = u(0) + c_v T_0$ cannot be theoretically or experimentally determined as the internal energy at absolute zero, $u(0)$, is an unknown quantity. Fortunately, in most analyses only Δu needs to be evaluated.

To derive a similar expression for $\Delta h = h_2 - h_1$, we start with the definition of enthalpy, use Eq. (3.9), and simplify using the fact that v is a constant:

$$dh = d(u + pv) = du + vdp + pdv^0 = c_v dT + vdp; \quad \left[\frac{kJ}{kg}\right] \tag{3.12}$$

$$\Rightarrow \quad h = h(p_0, T_0) + \int_{p_0,T_0}^{p,T} c_v dT + \int_{p_0,T_0}^{p,T} vdp = h_{ref}^o + c_v(T - T_0) + v(p - p_0) \tag{3.13}$$

$$\Rightarrow \quad \Delta h = h_2 - h_1 = c_v(T_2 - T_1) + v(p_2 - p_1) = c_v \Delta T + v\Delta p; \quad \left[\frac{kJ}{kg \cdot K}K = \frac{kJ}{kg}\right] \tag{3.14}$$

Enthalpy in the SL model is a bilinear function of temperature and pressure. The reference enthalpy $h_{ref}^o = h(p_0, T_0)$ is often expressed by the symbol $h^o(T_0)$ or simply h^o and is assigned a value of 0 in most property evaluation software including TEST. Obviously, u_{ref}^o cannot be independently chosen because $h_{ref}^o = u_{ref}^o + p_0 v$.

Differentiating $h = u + pv$ with respect to p while holding T constant, we obtain $(\partial h/\partial p)_T = v$ since $u = u(T)$ and v are constant for the SL model. Substituting this result into Eq. (3.5), we obtain a parallel expression for dh:

$$dh = c_p dT + \left(\frac{\partial h}{\partial p}\right)_T dp = c_p dT + vdp; \quad \left[\frac{kJ}{kg}\right] \tag{3.15}$$

A comparison with Eq. (3.12) establishes an important identity for the SL model:

$$c_p = c_v; \quad \left[\frac{kJ}{kg \cdot K}\right] \tag{3.16}$$

Because of this equality, which is graphically interpreted in Figure 3.16, the subscript is sometimes omitted when representing specific heats for a pure solid or liquid. With expressions for Δu and Δh evaluated, differences in extrinsic properties Δe and Δj can now be obtained from Eq. (1.32).

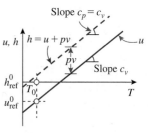

FIGURE 3.16 $u = u(T)$ while $h = h(p, T)$ for the SL model (see Anim. 3.B.*propertiesSL*).

Entropy (s):

By assumption, v and c_v are constants; therefore, the first Tds relation, Eq. (3.7), can be integrated:

$$ds = \frac{du}{T} + \frac{p\,d\!\!\!/v^{0}}{T}; \quad \left[\frac{kJ}{kg \cdot K}\right] \tag{3.17}$$

$$\Rightarrow \quad s = s(p_0, T_0) + \int_{p_0, T_0}^{p, T} \frac{du}{T} = s^{o}(T_0) + c_v \int_{T_0}^{T} \frac{dT}{T} = s^{o}(T_0) + c_v \ln \frac{T}{T_0} \tag{3.18}$$

$$\Rightarrow \quad \Delta s = s_2 - s_1 = \int_{T_1}^{T_2} \frac{c_v dT}{T} = c_v \int_{T_1}^{T_2} \frac{dT}{T} = c_v \ln \frac{T_2}{T_1}; \quad \left[\frac{kJ}{kg \cdot K}\right] \tag{3.19}$$

Like enthalpy, entropy is generally a function of temperature and pressure, and $s(p_0, T_0)$ is generally represented by the shorter symbol $s^{o}(T_0)$ or s^{o}. In the SL model, entropy can be seen to logarithmically increase (Fig. 3.17) with temperature, as was numerically predicted in Ex. 1-10. Unlike u or h, entropy does not require an arbitrary reference value because entropy is postulated to be zero at the absolute zero temperature. This postulate is known as the **third law of thermodynamics**, and will be further discussed in Chapter 13. With the help of this postulate, $s^{o}(T_0)$ is calculated and tabulated (Table G-1 where standard entropy of various solids, liquids, and gases are listed). Although entropy is independent of pressure in the SL model, the superscript in $s^{o}(T_0)$ is retained in Eq. (3.18) as a reminder that entropy at the standard temperature is evaluated (from measurement or theory) at the standard pressure. In most applications, we are interested in evaluating the difference in entropy, and Eq. (3.19) is most useful for that purpose. Note that the subscript ref is omitted for s^{o} because zero value of entropy is well defined unlike the arbitrary reference values for u and h.

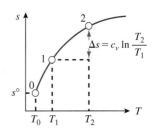

FIGURE 3.17 Graphical interpretation of Eq. (3.18) (see Anim. 3.B.*propertyDiffSL*).

3.3.3 Model Summary: SL Model

Frequently used formulas for manual evaluation of thermodynamic state of a pure solid or liquid can be summarized (see Anim. 3.B.*propertiesSL* and *propertyDiffSL*) as follows:

Obtain $v(=1/\rho)$ and $c_v(=c_p)$, which are material properties for this model, from Table A-1, A-2, or any SL state TESTcalc in TEST:

$$\Delta u = u_2 - u_1 = c_v(T_2 - T_1); \quad \Delta e = \Delta u + \Delta ke + \Delta pe; \quad \left[\frac{kJ}{kg}\right] \tag{3.20}$$

$$\Delta h = c_v(T_2 - T_1) + v(p_2 - p_1); \quad \Delta j = \Delta h + \Delta ke + \Delta pe; \quad \left[\frac{kJ}{kg}\right] \tag{3.21}$$

$$\Delta s = c_v \ln \frac{T_2}{T_1}; \quad \left[\frac{kJ}{kg \cdot K}\right] \tag{3.22}$$

Note that $u = u(T)$ and $s = s(T)$, while $h = h(p, T)$ in the SL model. Additionally, u and h vary linearly with T.

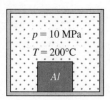

FIGURE 3.18 Illustration for Ex. 3-3.

EXAMPLE 3-3 Application of the SL Model

A 5-kg block of aluminum, initially at equilibrium with its surroundings at 30°C and 100 kPa, is placed in a pressurized chamber that maintains a pressure of 10 MPa and a temperature of 200°C (see Fig. 3.18). Determine the change in (a) the internal energy, and (b) the block's entropy after it reaches equilibrium with the chamber's pressure and temperature. ***What-if scenario[1]:*** What would the answers be if the block was made of (c) pure copper and (d) pure gold?

SOLUTION

Once the initial and final states are calculated, the answers can be obtained from simple differences since properties are point functions.

Assumptions

The block is at uniform temperature after equilibrium is reached; therefore, the SL model is applicable.

Analysis

The values for material properties v and c_v are retrieved from either Table A-1 or any SL state TESTcalc and are (1/2700) m³/kg and 0.9 kJ/kg · K, respectively. Taking advantage of the block's uniformity, the change in the system properties can be calculated from Eqs. (3.20) to (3.22):

$$\Delta U = U_2 - U_1 = m(u_2 - u_1) = mc_v(T_2 - T_1)$$

$$= (5)(0.9)(473 - 303)\left[(\text{kg})\left(\frac{\text{kJ}}{\text{kg} \cdot \text{K}}\right)(\text{K}) = \text{kJ}\right] = 765.0 \text{ kJ}$$

$$\Delta S = S_2 - S_1 = m(s_2 - s_1) = mc_v \ln\frac{T_2}{T_1}$$

$$= (5)(0.9)\ln\frac{473}{303}\left[(\text{kg})\left(\frac{\text{kJ}}{\text{kg} \cdot \text{K}}\right) = \frac{\text{kJ}}{\text{K}}\right] = 2.004\frac{\text{kJ}}{\text{K}}$$

TEST Analysis

To verify the solution with TEST, launch the SL system-state TESTcalc. Calculate State-1 and State-2 using the known temperatures and pressures. In the I/O panel, determine the answers by evaluating expressions '= m1 *(u2 − u1)' and '= m1 *(s2 − s1)'.

What-if scenario

Change the working substance and click Super-Calculate to update all calculations. The results are as follows:

	Aluminum	Copper	Gold
ΔU, (kJ)	765.0	328.1	109.7
ΔS, (kJ/K)	2.003	0.859	0.287

Discussion

Although the three blocks have the same mass, the aluminum block has the highest mole (evaluate '= m1 /MM1' in the I/O panel), which explains why its values are so much greater than the other blocks'.

EXAMPLE 3-4 Application of the SL Model

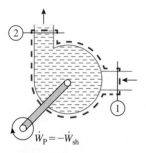

FIGURE 3.19 A pump raises flow energy of liquids while consuming shaft power (see Anim. 4.A.*pump*).

The pumping system shown in Figure 3.19 delivers high-pressure liquid water in an isentropic (constant-entropy) manner. At the inlet, $p_1 = 100$ kPa, $V_1 = 10$ m/s, and $z_1 = 0$ m; at the exit,

---

[1]Use of TEST is encouraged for verification and "what-if" studies, wherever possible.

$p_2 = 1000$ kPa, $V_2 = 15$ m/s, and $z_2 = 5$ m. Using the SL model, determine the changes in (a) specific internal energy, (b) specific stored energy, (c) specific enthalpy, and (d) specific flow energy between the exit and inlet states.

SOLUTION

Evaluate the states at the inlet and exit, State-1 and State-2, from the given information.

Assumptions

The inlet and exit states are uniform (based on LTE), and are described by two unique flow states (Fig. 3.19).

Analysis

From Table A-1 or any SL state TESTcalc, material properties v and c_v are obtained as $(1/997)$ m³/kg and 4.184 kJ/kg·K, respectively.

From Eq. (3.22), $\Delta s = c_v \ln(T_2/T_1) = 0$ implies $T_2 = T_1$. *Isentropic* states are also *isothermal* for a solid or liquid that is modeled by the SL model. Eq. (3.20) also yields:

$$\Delta u = c_v \cancel{(T_2 - T_1)}^0 = 0$$

Therefore, from Eq. (3.21):

$$\Delta h = \cancel{\Delta u}^0 + v(p_2 - p_1) = \frac{900}{997} = 0.90 \ \frac{\text{kJ}}{\text{kg}}$$

and

$$\Delta j = \Delta h + \Delta \text{ke} + \Delta \text{pe} = 0.9 + \frac{15^2 - 10^2}{2(1000)} + \frac{5}{1000}(9.81) = 1.01 \ \frac{\text{kJ}}{\text{kg}}$$

TEST Analysis

To verify the solution with TEST, launch the SL system-state TESTcalc and select Water(L) from the drop-down menu. Evaluate State-1 with the given properties and an arbitrary temperature of T1 = 25°C. To override a default property (i.e., the velocity), click its checkbox twice. Evaluate State-2 with p2 = 10*p1, s2 = s1, Vel2, and z2. In the I/O panel calculate the desired differences using expressions such as '= h2 − h1' and '= j2 − j1'. To verify that the answers are independent of T1, change T1's value, click Super-Calculate to update the states, and evaluate the differences again.

Discussion

The small difference between the calculated values of Δh and Δj is caused by the change in velocity and elevation between the ports.

3.4 THE PC (PHASE-CHANGE) MODEL

There are many engineering devices and processes—vapor power plants, refrigerators, liquefaction of gases, to name a few—where two different phases, mostly a liquid and a vapor, coexist in local thermodynamic equilibrium. Working substances such as steam, ammonia, and various refrigerants, whose phase composition can vary as a function of its thermodynamic state, are modeled by the phase-change or **PC model**.

To introduce this model from a microscopic point of view, as we did for the SL model, let's resume the discussion of molecular changes that occur as energy (heat or work) is added to a fixed mass of liquid kept at a constant pressure (Fig. 3.20). Recall that in the liquid phase, microscopic chunks of molecules freely move around the locally preserved lattice structure of a solid (see Anim. 3.B.*microStructure*). As more energy is added, the vibrational kinetic energy of molecules (and hence the temperature) keeps on increasing. Eventually a point is reached when the kinetic energy peaks and any additional transfer of energy must go into increasing the potential energy of molecules, which are now ready to break loose from the liquid phase's intermolecular bonds. At that point the liquid is said to be a **saturated liquid**. As even more energy

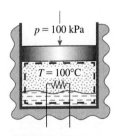

FIGURE 3.20 As energy (electrical work) is added, liquid turns into saturated mixture and then to superheated vapor (see Anim. 3.C.*propertiesPC*).

is added to the saturated liquid, the entire amount is absorbed (stored) by the increased potential energy associated with a change of phase from liquid to **vapor phase**. With the molecular kinetic energy remaining constant during this phase transformation, the temperature remains constant. Depending on the intensity of vapor formation, this process is called either **evaporation** (slow vapor formation) or **boiling** (vigorous formation of vapor).

Vapor molecules are far apart, have diminished intermolecular attractions, and are not only more energetic (due to more potential energy), but also more chaotic. This explains a step increase in v, u, and s as saturated liquid turns into vapor (Fig. 3.21). As the transition continues, eventually a point is reached when no more liquid is left. The phase composition is called **saturated vapor** because any removal of energy from the system causes **condensation**—the opposite of evaporation. Thermodynamic states at the beginning and end of any phase change process are called **saturated states**. If energy interaction is halted before vaporization is complete, saturated liquid and saturated vapor phases can coexist, forming an equilibrium mixture known as a **saturated mixture**. Molecules do not stop migrating from one phase to another in a saturated mixture, but the phase composition remains unchanged because of a dynamic balance—the migration into a phase is balanced by the migration out of a phase at equilibrium.

As energy is added to a saturated vapor, most of it goes into increasing the molecules' kinetic energy (mostly translational) and the temperature resumes rising. To distinguish it from saturated vapor, the vapor is called **superheated vapor**, whose u, v, and s increase monotonically with T. Conversely, a saturated liquid, when cooled, is described as a **subcooled** (or *compressed*) liquid. The PC model must be able to handle states involving *subcooled liquids*, *superheated vapors*, and *saturated mixtures* of liquid and vapor.

3.4.1 A New Pair of Properties—Qualities *x* and *y*

In a saturated mixture of liquid and vapor, the composition of the mixture is generally expressed by specifying the vapor fraction on a mass basis, and sometimes on a volume basis, resulting in two new properties: **quality *x*** and **volumetric quality *y*** (Fig. 3.22 and Anim. 3.C.*satMixture*) that are defined as follows:

$$x \equiv \frac{m_{\text{vapor}}}{m} \quad \text{and} \quad y \equiv \frac{V_{\text{vapor}}}{V} \tag{3.23}$$

where $m = m_{\text{vapor}} + m_{\text{liquid}}$ and $V = V_{\text{vapour}} + V_{\text{liquid}}$ are the total mass and volume of the saturated mixture. By definition, $0 \le x \le 1$ and $0 \le y \le 1$, with their limiting values describing saturated liquid ($x, y = 0$) and saturated vapor ($x, y = 1$). In industry the term **dry vapor**

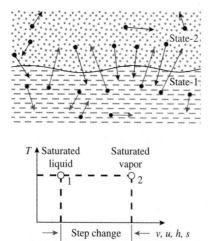

T ↑ Saturated liquid   Saturated vapor

→ | Step change | ← v, u, h, s

FIGURE 3.21 Temperature remains constant and specific properties display a step increase as saturated liquid (State-1) transforms into saturated vapor (State-2) at a constant pressure (click Liquid-Vapor in Anim. 3.B.*microStructure*).

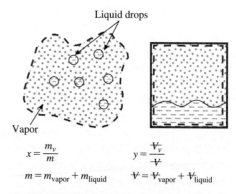

Liquid drops

Vapor

$$x = \frac{m_v}{m} \qquad y = \frac{V_v}{V}$$

$$m = m_{\text{vapor}} + m_{\text{liquid}} \qquad V = V_{\text{vapor}} + V_{\text{liquid}}$$

FIGURE 3.22 The composition of a saturated mixture (see Anim. 3.C.*satMixture*) is measured by properties *x* (quality) and *y* (volumetric quality).

is used to indicate saturated and superheated vapor, while a saturated mixture with a quality of $x < 1$ is called **wet vapor**.

Suppose a propane tank contains a saturated mixture of five gallons of saturated liquid and five gallons of saturated vapor at 30°C. The volumetric quality of the mixture can be easily calculated to be $y = 0.5$. The value of quality x must be much lower than y (see Fig. 3.23) seen from the nonlinear relation in Figure 3.23 between the two properties since liquid is more dense than vapor. (You can launch the PC system-state TESTcalc, and enter T and y to verify that $x = 0.046$.) Quality x is more useful for evaluating mixture properties through weighted averaging, as 1 kg of the mixture contains x kg of saturated vapor and $(1 - x)$ kg of saturated liquid.

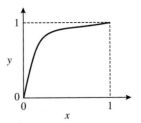

FIGURE 3.23 Although x and y have equal values for saturated fluids, $x < y$ for a mixture (see Anim. 3.C.*satMixture*).

3.4.2 Numerical Simulation

Consider the insulated piston-cylinder arrangement of Figure 3.24, in which the pressure can be controlled by adjusting the weights or the pull on the piston. Energy can also be added in desired increments through an electric heater. Initially, 0.1 kg of liquid water is kept at equilibrium at 1 atm and 25°C. We would like to visualize through thermodynamic plots (e.g., T-v and T-s) how the equilibrium state of the phase-change system of Figure 3.24 evolves as energy is added in small increments (see Anim. 3.C.*propertiesPC*). After each increment, sufficient time is allowed to for the system to arrive at a new equilibrium state. With the external forces on the piston remaining unchanged, the system pressure remains constant at its initial value, 1 atm. Suppose the energy increments are such that the system goes through the eight equilibrium states specified by p, T, and x in Table 3-1.

To numerically simulate the states, launch the PC system-state TESTcalc and select H2O from the fluid menu. For each state, select the state number, enter the known properties, and click Calculate. Plot T-v and T-s diagrams by selecting them from the diagram menu. To repeat the simulation for $p = 10$ atm, change p1 to its new value, click Super-Calculate to update all states, and then generate the plots again. The resulting plots from the two simulations are shown in Figure 3.25; (a) T-v and (b) T-s.

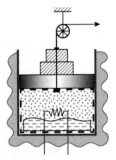

FIGURE 3.24 Pressure can be controlled by the push or the pull force on the piston (see Anims. 0.A.*vacuumPressure* and Anim. 3.C.*propertiesPC*).

Each of the plots show how the thermodynamic state changes as the system transitions from State-1 through State-8 (recall that, according to the state postulate, any pair of independent thermodynamic properties are sufficient to describe an equilibrium). For a given pressure, the calculated states congregate into three distinct zones with their locus forming a knee-shaped line, called the **constant-pressure line**. Between State-1 and State-3 the composition is a subcooled liquid. As T increases the specific volume v (Fig. 3.25a) shows a slight increase, capturing the effect of thermal expansion (recall that the SL model ignores this expansion). Also s (Fig. 3.25b) increases with temperature as expected from the SL model. The rapid increase in v and s after State-3 is caused by an increase in the proportion of vapor in the system. State-3, with $x = 0$, and State-5, with $x = 1$, represent saturated liquid and saturated vapor states, respectively. States beyond State-5 on the constant-pressure line are superheated and states in between the two saturated states (e.g., State-4) represent saturated mixtures of liquid and vapor. The constant temperature at which the two phases coexist in a saturated mixture for a given pressure is known as the **saturation temperature**, and is represented by the symbol $T_{sat@p}$ to emphasize its dependence on pressure. In Figure 3.25, for example, $T_{sat@1atm} \cong 100$°C and $T_{sat@10atm} \cong 180.5$°C. Along with the saturation temperature, the entire constant-pressure line shifts upward when the pressure is changed from 1 atm to 10 atm in Figure 3.25. Also, the distance between the two saturated states diminishes, creating the dome shape (as the locus) when the saturated states are joined for a number of different pressures. Note that the liquid states, State-1 and State-2, appear to be aligned with the saturated liquid lines in Figure 3.25. If we could zoom into those states, it would reveal that the subcooled states belong slightly to the left of the saturated liquid, a fact which is exaggerated in sketches (which are not to scale) of thermodynamic plots for clarity.

We could carry out a parallel set of simulations by holding temperature constant (by removing the insulation, locking the piston in its position, and letting the system of Figure 3.24 sit in a constant-temperature bath) and allowing the pressure to adjust as energy is added in increments at constant temperature. For a given

Table 3-1 Eight states are calculated using either pressure and temperature or pressure and quality as input.

State	P (atm)	T (deg-C)	x	Phase Composition
1	1	25		Subcooled Liquid
2	=p1	=T1+50		Subcooled Liquid
3	=p1		0	Saturated Liquid
4	=p1		0.1	Saturated Mixture
5	=p1		1	Saturated Vapor
6	=p1	=T5+50		Superheated Vapor
7	=p1	=T6+50		Superheated Vapor
8	=p1	=T7+50		Superheated Vapor

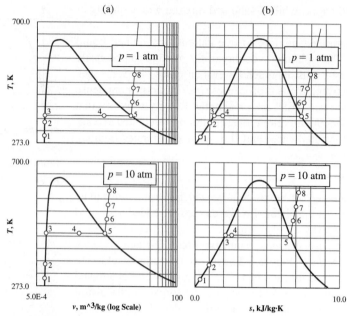

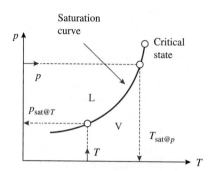

FIGURE 3.25 Thermodynamic plots generated by the PC state TESTcalc after the eight states of Table 3-1 are computed at two different pressures. (a) *T-v* diagrams, (b) *T-s* diagrams.

FIGURE 3.26 The saturation curve relates saturation pressure and saturation temperature (see Anim. 3.C.*pvT-H2O*) and separates states according to phase composition.

temperature, saturated mixtures can be in equilibrium only at a unique pressure called the **saturation pressure**, represented by the symbol $p_{\text{sat@}T}$. Because $p_{\text{sat@}T}$ and $T_{\text{sat@}p}$ are inverse functions of each other, their interdependence can be graphically depicted by what is known as a **phase diagram** (Fig. 3.26). Because every system has an inherent tendency to move towards equilibrium, knowing a fluid's saturation pressure at the temperature of its surroundings allows us to predict if the fluid, when exposed to the surroundings, will exist in a liquid form. If $p_{\text{sat@}T}$ is higher than the atmospheric pressure p_0, the exposed fluid will rapidly boil off (absorbing the required heat from the surroundings, if necessary) in its quest for equilibrium, and will raise the pressure of the immediate surroundings to $p_{\text{sat@}T}$. The intensity of boiling will depend on the difference between $p_{\text{sat@}T}$ and p_0. If $p_{\text{sat@}T} < p_0$, we would expect the fluid to be in liquid form under atmospheric pressure. At standard atmospheric temperature (25°C), water and propane have saturation pressures of 3.17 kPa and 945 kPa, respectively. This is why water exists in a liquid phase under atmospheric pressure (we will discuss evaporation in Chapter 12) while propane has to be in a pressurized tank to be in liquid phase.

3.4.3 Property Diagrams

States represented in a *T-v* or *T-s* diagram look remarkably similar; therefore, we will discuss only the *T-s* diagram, which turns out to be more useful for engineering analysis. Observations based on this diagram apply equally well to most other diagrams.

Consider the *T-s* diagram shown in Figure 3.27. Saturated states are usually represented by the letters *f* (saturated liquid from German word *Flüssigkeit*) and *g* (saturated vapor from German word *Gas*). The *phase dome*, is bounded by the *f*-line ($x = 0$) on the left and the *g*-line ($x = 1$) on the right, the loci of the *f*- and *g*-states, respectively. Even though the liquid branch of the constant pressure lines coincide with the *f*-line in the computed plot of Fig. 3.25, their invisible but minute separation from the *f*-line is exaggerated in Fig. 3.27 (and in all such diagrams in the rest of the book) to highlight the **compressibility effect**, dependence of liquid properties on pressure. The state, where the two lines meet at the peak of the dome, is called the **critical state**. Because there is only one critical state for a given fluid, the **critical properties**—p_{cr}, T_{cr}, v_{cr}, etc.—can also be regarded as *material properties*.

Any state in the zone $p > p_{\text{cr}}$ (Fig. 3.27) is called a **super critical** state. Super-critical states with $T \geq T_{\text{cr}}$ are called super-critical vapor (SCV) and super-critical states with $T < T_{\text{cr}}$ are called super-critical liquid (SCL). The transformation from super-critical liquid to super-critical vapor takes place over a range of temperatures and, as such, there is no boiling point for such transition. The sub-critical states ($p < p_{\text{cr}}$), which are more relevant in engineering analysis, are

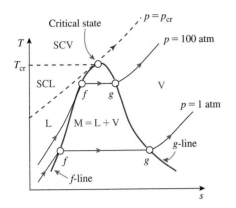

FIGURE 3.27 Different zones of phase composition on a *T-s* diagram.

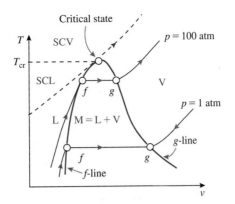

FIGURE 3.28 Different zones of phase composition on a *T-v* diagram.

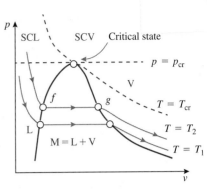

FIGURE 3.29 Different zones of phase composition on a *p-v* diagram.

divided into three distinct zones—**subcooled liquid** (L), **saturated mixture** (M = L + V), and **superheated vapor** (V)—with the mixture zone bounded by the distinctive phase dome. All the five regions—SCV, SCL, L, M, and V—are mapped in Figure 3.27 (and Anim. 3.C.*zones*). Once a state is located on this diagram, its phase composition can be read from the region to which it belongs.

The *T-v* and *p-v* diagrams (see Figs. 3.28 and 3.29) are also used to graphically represent states in the PC model, although not as often as the *T-s* diagram. While the *T-v* diagram is quite similar to the *T-s* diagram (compare the constant-pressure lines in Figs. 3.27 and 3.28), we leave it as an exercise to numerically establish the shape of the constant-temperature lines in a *p-v* diagram (Fig. 3.29). The *p-T* diagram (see Fig. 3.26) can be seen as the projection of a three-dimensional *p-v-T* surface on the *p-T* plane (see Anim. 3.C.*pvT-H2O*). In a saturated mixture, pressure and temperature are not independent properties, so all the saturated states bounded by the *f*- and *g*-states at a given pressure (or temperature) get projected into a single point in a *p-T* diagram. The saturation dome is projected into a line known as the **saturation curve** or the *phase diagram*, which graphically expresses the functional relationship between saturation temperature and pressure. Representing all the mixture states, the saturation curve separates the liquid (L) phase from the vapor (V) phase and extends up to the critical point where the mixture (M) region ends.

EXAMPLE 3-5 | **Using Thermodynamic Diagrams**

A pipe carrying refrigerant R-12 at 25°C as a saturated liquid develops a small leak (Fig. 3.30). If the atmospheric conditions are 100 kPa and 25°C, determine (a) the flow's direction at the leak, and (b) the refrigerant's temperature immediately after the refrigerant is exposed to the atmosphere.

SOLUTION

Compare the pressures inside and outside to determine the flow's direction. Also find the saturation temperature at the atmosphere's pressure.

Assumptions

Given the possibility of a phase change, the PC model is the appropriate model to use.

Analysis

Use Table B-11 to obtain $p_{sat@25°C} = 651.6$ kPa. Alternatively, launch the PC system-state TESTcalc, select R-12, enter T1 = 25 deg-C and x1 = 0, and click Calculate to obtain p1's value: 651.6 kPa. Given that the outside pressure is only 100 kPa, R-12 will be expelled from the pipe.

When exposed to the atmosphere, liquid R-12 quickly establishes mechanical equilibrium; its pressure drops from 651.6 kPa to 100 kPa almost instantly (near the vaporizing interface between R-12 and air). On the *T-s* diagram of Figure 3.30, the *p* = 100 kPa line

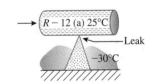

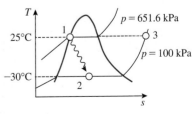

FIGURE 3.30 The leaking R-12 quickly vaporizes equilibriating to a temperature as low as −30°C.

can be seen to intersect the $T = 25°C$ line in the superheated vapor zone (State-3). R-12 must vaporize completely in its quest for equilibrium. During boiling, local equilibrium will ensure that the temperature remains constant at $T_{sat@100 kPa} = -30°C$ (enter p2 = 100 kPa, x2 = 0, and click Calculate to find T2) as liquid R-12 boils and absorbs heat from the warmer surroundings.

Discussion

Thermodynamic diagrams can be used to deduce many familiar and some unfamiliar consequences of thermodynamic equilibrium. For example, an increase in $T_{sat@p}$ with pressure is exploited by a pressure cooker to raise the cooking temperature. In *vacuum cooling*, just the opposite happens where supermarket produce is cooled to 0°C by reducing the pressure to 0.61 kPa ($p_{sat@0°C}$).

3.4.4 Extending the Diagrams: The Solid Phase

There are situations, although uncommon, where a solid phase may be present in an equilibrium mixture. The thermodynamic diagrams introduced in the previous section can be extended to include the solid phase.

Let's revisit the constant-pressure energy addition experiment described by Figure 3.24, this time starting with a block of ice whose state belongs to the S (solid) zone of the *T-s* diagram (Fig. 3.31). As energy is added, entropy accompanies the rise in temperature (see the constant-pressure lines, *i-l-f-g* in Fig. 3.31) until the solid becomes saturated at point *i* and **melting** begins. As more energy is added, melting continues at the constant **melting temperature** (microscopic kinetic energy having reached a maximum, any additional energy goes into supplying the sudden increase of potential energy of the molecules) until the saturated liquid state (transformation between liquid and solid), State-*l*, is reached. Thereafter, the constant-pressure line resumes the familiar pattern of going through the subcooled liquid region (L), becoming saturated liquid (transformation between liquid and vapor) at State-*f*, forming saturated mixtures of liquid and vapor (M = L + V), becoming saturated vapor at State-*g*, and finally transforming into superheated vapor (V).

Although the *i*-line (locus of *i* states) and the *l*-line (locus of *l*-states) do not form a dome shape, they bound all possible states involving a saturated mixture of solid and liquid (S + L) and separate the subcooled solid phase (S) from the subcooled liquid phase (L). Just as the projection of the *f* and *g* lines on the *p-T* plane form the saturation curve for liquid and vapor, the *i* and *l* lines project into a different saturation curve, separating a saturated solid from a liquid (see the expanded phase diagram of Fig. 3.32). The melting temperature also increases slightly with pressure for most substances (see the black line in Fig. 3.32), but decreases for a few special substances such as water, silicon, and PVC. For water, this anomalous behavior has some remarkable consequences. For example, the weight of an ice skater is supported by the small contact area of the skates, producing high enough pressure for the melting temperature to decrease below the temperature of the surrounding ice. As a result, the ice below the skates melts, providing a lubricating action. As the skater glides away, the pressure on the cold water

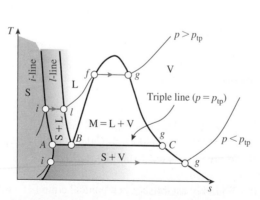

FIGURE 3.31 The *T-s* diagram extended to the solid phase (see Anim. 3.C.*triplePoint*).

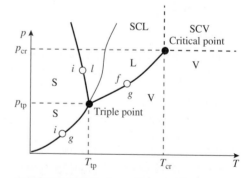

FIGURE 3.32 The phase diagram for water with the saturation curves separating different phases (see Anim. 3.C.*pvT-H2O*). The black curve, with positive slope is more common among most fluids than the red line separating liquid and solid phases for water.

returns to atmospheric pressure, and the water, still at the original temperature of the ice, is well below the melting (freezing) temperature, and refreezes. For most other fluids, the melting temperature increases with pressure.

The saturation curves in the p-T diagram meet at a single point called the **triple point** where all the three phases coexist in equilibrium (see Anim. 3.C.*triplePoint*). As can be seen from Figure 3.32, saturated states i, g, and f (l-state merges with f-state) meet at the triple point. As an example, ice, liquid water, and water vapor can stay in equilibrium at 0.61 kPa, 0.01°C. The corresponding range of states in the T-s diagram falls on the **triple line**, ABC in Figure 3.31. While the solid, liquid, and vapor phases are at equilibrium at States A, B, and C, an equilibrium saturated mixture can be anywhere between A and C based on the mixture composition. The unique values of temperature and pressure, T_{tp} and p_{tp}, at the triple point can be regarded as material properties.

For $p < p_{tp}$, a solid phase can directly transform into vapor phase (see the constant-pressure line passing through states i and g in Fig. 3.31). This phenomenon, known as **sublimation**, takes a solid from the i-state (saturated solid) directly into the g-state (saturated vapor) without having to go through an intermediate liquid phase, which is customary for $p > p_{tp}$. A cube of ice exposed to an ambient condition of 0.5 kPa and 25°C will not melt but *sublime*. Similarly, solidified CO_2 at atmospheric condition (with a triple-point pressure of 517 kPa) will sublime (Fig. 3.33) into superheated vapor without producing any liquid (hence the name dry ice). States between the i-line and g-line in Figure 3.31 constitute equilibrium mixtures of saturated solid and saturated vapor. In TEST, properties x and y can be interpreted as the quality of the vapor-solid mixture for $p < p_{tp}$. Try finding a state of H_2O using any PC state TESTcalc for T1 = −10 deg-C and x1 = 0.1 and plot it on a T-s diagram. In engineering applications transition between liquid and vapor is most common, which will be our focus in the following sections.

3.4.5 Thermodynamic Property Tables

Unlike the SL model, relations among thermodynamic properties for a PC fluid are too complex to be expressed by *equations of state*. The PC model requires use of specially formatted tables or charts. Taking water as a representative PC fluid, we will discuss how to look up water properties from the **steam tables**: Tables B-1 through B-4.

Saturation Table: The saturation table lists property values for the f- and g-states, sorted by convenient increments of the independent property p in the *pressure table* (Table B-1) and T in the *temperature table* (Table B-2). The first two columns list T and $p_{sat@T}$ in the temperature table, and p and $T_{sat@p}$ in the pressure table. Columns three through ten in each table list four pairs of specific properties: v_f and v_g, u_f and u_g, h_f and h_g, and s_f and s_g. The first row usually corresponds to the *triple line,* where solid, liquid, and vapor can coexist in equilibrium, and the last row represents the *critical state,* where the f and g lines converge and property values in the f and g columns become identical. Note that the enthalpy and entropy are assigned reference (zero) values not at the standard temperature and pressure, but at the triple point (Sec. 3.4.4) $p_{tp} = 0.6113$ kPa and $T_{tp} = 0.01$°C.

To illustrate how to use a saturation table, let's examine part of the temperature table reproduced in Figure 3.34. It is clear from the table that a saturated state—f- or g-state—can be determined from a single saturation property. For example, if the temperature of a saturated liquid is given, all other properties—p, v, u, h, and s —can be obtained from this saturation table as $p_{sat@T}$, v_f, u_f, h_f, and s_f. If the specific volume of a saturated vapor is known v_{given}, all other thermodynamic properties can be read from the row that corresponds to $v_g = v_{given}$. In the PC TESTcalcs, a saturated state is set by entering 0 or 1 for x or y. The critical state, the intersection of the f- and g-lines, can be obtained in TEST by specifying x as 0 (f-line) and y as 1 (g-line).

If a given property falls in between two rows, a linear interpolation, or for that matter, a visual estimate, produces sufficient accuracy for most engineering purposes. TEST can be used to gauge and improve the quality of such *guesstimate*.

Superheated Vapor and Compressed Liquid Tables: Pressure and temperature are the two independent properties that define an equilibrium state outside the mixture dome. The **superheated vapor table** and the **compressed liquid table** are organized into a series of pressure tables (see Tables B-3 and B-4) where thermodynamic properties are sorted according to temperature for a given pressure.

FIGURE 3.33 Dry ice sublimates because atmospheric pressure is below the triple point pressure of CO_2.

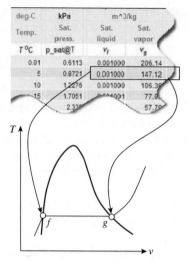

FIGURE 3.34 A partial listing of Table B-2. Besides the saturation pressure and temperature, each row contains saturation properties pairing *f*-state with *g*-state.

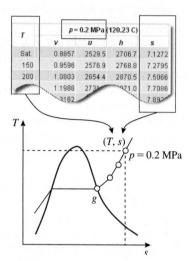

FIGURE 3.35 Each pressure table is a set of states on a single constant-pressure line in the superheated region.

Part of vapor Table B-3, for $p = 200$ kPa, is reproduced in Figure 3.35. To maintain continuity with the saturation table, $T_{\text{sat}@p}$ is displayed in the table header and the first row corresponds to the *g*-state at p. The last row of each liquid table corresponds to the *f*-state at p. Each column in the superheated or compressed liquid table represents a constant-pressure line as shown in Figure 3.35.

Superheated vapor states are often specified in terms of pressure p and a second property, such as T, v, u, h, or s. Suppose we need to find the value of u for a given pair of properties, p and v. If p is one of the pressures for which a pressure table is available, any property of the state, including u, can be interpolated from the given value of v. But if p falls between two consecutive pressure tables with pressures p^k and p^{k+1}, then two different values, u^k and u^{k+1}, of u can be interpolated from the two tables for the given value of v. A second round of interpolation for the given value of p results in:

$$u = u^k + (u^{k+1} - u^k)f; \quad \left[\frac{\text{kJ}}{\text{kg}}\right], \quad \text{where} \quad f = \frac{p - p^k}{p^{k+1} - p^k} \tag{3.24}$$

This interpolation scheme, known as *bilinear interpolation* and used in Ex. 3-6, can be applied to the compressed liquid table for subcooled liquids as well.

EXAMPLE 3-6 **Interpolating the Superheated Vapor Tables**

A rigid tank contains superheated steam at 120 kPa. If the specific enthalpy is 2750 kJ/kg, determine the steam's temperature.

SOLUTION

We use the superheated vapor table, Table B-3, for a manual solution and the PC system-state TESTcalc to verify the result.

Assumptions

The working fluid is in thermodynamic equilibrium, and a single state can be used to describe its condition within the tank.

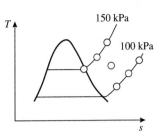

Analysis

The given pressure is bracketed between 100 kPa and 150 kPa for which pressure tables are available in Table B-3. For each table, we search the enthalpy column for the given enthalpy, locating its value between the second and third rows in the 100-kPa table and first and second rows in the 150-kPa table (see Fig. 3.36). Representing the two tables as the kth and $k + 1^{th}$ tables, temperature can be interpolated for the given enthalpy in each table as

$$T^k = 100 + (\overset{200}{150} - 100)\frac{(2750 - 2676.2)}{(2776.4 - 2676.2)} = 136.83°C$$

and

$$T^{k+1} = 120.23 + (\overset{200}{150} - 120.23)\frac{(2750 - 2706.7)}{(2768.8 - 2706.7)} = 140.99°C$$

FIGURE 3.36 States are listed at 100 and 150 kPa in Table B-3, requiring interpolation (Ex. 3-6).

Using Eq. (3.24), the desired temperature, now, can be found with another interpolation between the two pressures:

$$T = 136.83 + (140.99 - 136.83)\frac{(120 - 100)}{(200 - 100)} = 137.66°C$$

TEST Analysis

Launch the PC system-state TESTcalc. Select H_2O as the working fluid, enter p1 = 120 kPa and h1 = 2750 kJ/kg, and click Calculate. The phase composition is determined as *superheated vapor* and the temperature is calculated as 137.71°C, verifying the manual solution.

Discussion

Bilinear interpolations may be time consuming and may be replaced by good eye estimates if a slight sacrifice in accuracy can be tolerated.

3.4.6 Evaluation of Phase Composition

Evaluation of a thermodynamic state using the PC model begins with determination of the phase composition—locating the state on a thermodynamic diagram, such as the T-s diagram, and identifying which zone (L, V, M, SCL, or SCV in Fig. 3.37) that state belongs to. Depending on the pair of known properties, the strategy for determining the phase composition can vary. To simplify the discussion, we assume the phase compositions to be limited to subcritical liquids, vapors, or saturated mixtures bounded within the pressure range $p_{tp} < p \leq p_{cr}$.

a. **Given p and T:**

From the pressure saturation table, obtain $T_{sat@p}$ and compare it with the given temperature (see Figs. 3.37 and 3.38) to determine the phase composition from the following table:

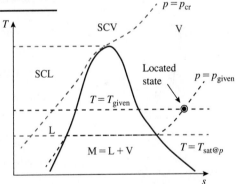

FIGURE 3.37 The phase composition (V, M, L, etc.) can be detected from the intersection of the constant-pressure and constant-temperature lines.

Table 3-4-6a   The state can now be graphically located at the intersection of the $p = p_{given}$ and $T = T_{given}$ lines on a T-s diagram (see Fig. 3.38). Note that while the phase composition of State-3 has been determined, its relative location between the f- and g-states is still unknown.

Condition		Phase Composition	Example (Fig. 3.38)
$p_{tp} < p \leq p_{cr}$	$T > T_{sat@p}$	Superheated Vapor (V)	State-1
	$T < T_{sat@p}$	Subcooled Liquid (L)	State-2
	$T = T_{sat@p}$	Saturated Mixture (M=L+V)	State-3

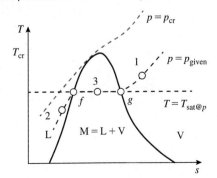

FIGURE 3.38 Locating states on a T-s diagram for a given pair of p and T.

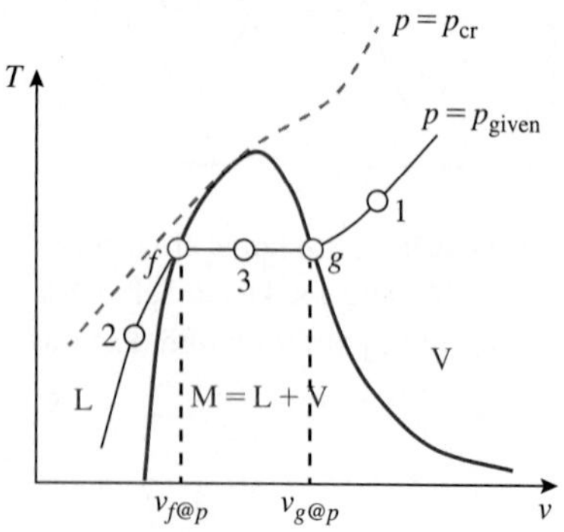

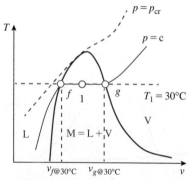

FIGURE 3.39 State spotting on the *T-s* diagram for a given pair of *p* and *s*.

b. Given *p* and *b* (*b* stands for *v, u, h,* or *s*):

From the pressure saturation table, obtain the saturation properties $b_{f@p}$ and $b_{g@p}$. Now use the following table to determine the phase composition:

Table 3-4-6b For a given *p* and *b*, their states are located in the three regions shown in Figure 3.39.

Condition		Phase Composition	Example (Fig. 3.39)
$p_{tp} < p \leq p_{cr}$	$b > b_{g@p}$	Superheated Vapor (V)	State-1
	$b < b_{f@p}$	Subcooled Liquid (L)	State-2
	$b_{f@p} \leq b \leq b_{g@p}$	Saturated Mixture (M)	State-3

c. Given *T* and *b* (*b* is *v, u, h,* or *s*):

This procedure is almost identical to the one described in (b) except that property *p* is replaced with *T* everywhere. As the temperature is known, the saturation temperature table is more convenient to use.

d. Given *x* or *y*:

If the quality *x* or the volumetric quality *y* is given, the phase composition has to be a *saturated mixture* (M = L + V) . Note that for subcooled liquid or superheated vapor these properties have no meaning.

e. Given b_1 and b_2 (*b* is *v, u, h,* or *s*):

Iterative procedures are necessary to determine the phase composition. Use of TEST is recommended.

EXAMPLE 3-7 Determining Phase Composition

A rigid tank with a volume of 0.02 m³ contains 0.5 kg of refrigerant ammonia (NH_3) at 30°C. Determine the pressure and the phase composition in the tank.

SOLUTION

From the given system properties, specific volume *v* can be calculated. With two independent thermodynamic properties (T and v) known, the complete state can be evaluated.

Assumptions

A single state can be used to describe the equilibrium within the tank.

Analysis

Suppose State-1 represents the desired state:

State-1: (given $m_1 = 0.5$ kg; $V_1 = 0.02$ m³; $T_1 = 30°C$)

$$v_1 = \frac{V_1}{m_1} = \frac{0.02}{0.5} = 0.04 \frac{m^3}{kg};$$

From Table B-12, the saturation temperature table for ammonia, we obtain:

$$p_{sat@30°C} = 1167 \text{ kPa}; \quad v_{f@30°C} = 0.00168 \frac{m^3}{kg}; \quad v_{g@30°C} = 0.1105 \frac{m^3}{kg};$$

The relationship $v_f \leq v_1 \leq v_g$ implies that State-1 must be a saturated mixture located somewhere between the *f*- and *g*-states (see Fig. 3.40).

TEST Analysis

To verify the solution with TEST, launch the PC system-state TESTcalc and select Ammonia (NH3). Enter T1, m1, and Vol1 and click Calculate. The phase composition is

FIGURE 3.40 The phase composition is a saturated mixture in Ex. 3-7 because $v_f \leq v_1 \leq v_g$.

displayed as part of the complete state. To obtain the saturated liquid and saturated vapor properties, evaluate State-2 with p2 = p1 and x2 = 0, and State-3 with p3 = p1 and x3 = 1.

Discussion

The high saturation pressure suggests that ammonia will rapidly evaporate when exposed to the atmosphere at 30°C. Although we have used the *T-v* diagram in this problem, the *T-s* diagram could be used just as effectively, since *v* and *s* behave similarly. You can select *T-v* from the TESTcalc's plot menu and verify visually that the state is located inside the dome region. By clicking the p = c button, you can draw the knee-shaped constant-pressure line.

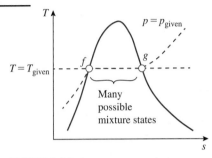

3.4.7 Properties of Saturated Mixture

Once the phase composition of a PC state is found (Sec. 3.4.6), the algorithm for evaluating any desired thermodynamic property is straightforward. In the following discussion, we will continue to use the generic symbol *b* to represent a specific property, such as *v*, *u*, *h* , or *s*, and use the *T-s* diagram as the representative thermodynamic diagram. For a saturated mixture, the state has been already located in the *M*-zone of the thermodynamic plot.

FIGURE 3.41 The pressure and temperature lines intersect along a line, revealing an infinite number of possible mixture states.

a. **Given *p* and *T*:** In a saturated mixture, pressure and temperature are not independent, and the state can be anywhere between the *f*- and *g*-states, as can be seen from the intersection of the constant-pressure and constant-temperature lines in Figure 3.41. A second independent property is necessary to determine the state.

b. **Given *T* and *x*:** The pressure of a saturated mixture is a function of its temperature. Obtain $p = p_{sat@T}$ from the temperature saturation table, using interpolation if necessary. For evaluating a specific property *b* (e.g., *v*, *u*, *s*, etc.) of the mixture, consider a lump of saturated mixture of mass *m* and quality *x* (see Fig. 3.42). The total extensive property *B* (e.g., Ψ, *U*, *S*, etc.) consists of the sum of the contributions from the constituent phases:

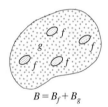

$$B = B_f + B_g; \quad [\text{Unit of } B] \quad \text{and} \quad m = m_f + m_g; \quad [\text{kg}] \tag{3.25}$$

$$\text{Therefore,} \quad b \equiv \frac{B}{m} = \frac{B_f + B_g}{m} = \frac{m_f b_f}{m} + \frac{m_g b_g}{m} = \frac{(m - m_g)b_f}{m} + \frac{m_g b_g}{m};$$

$$\Rightarrow \quad b = (1 - x)b_{f@T} + x b_{g@T}; \quad \left[\frac{\text{Unit of B}}{\text{kg}}\right] \tag{3.26}$$

FIGURE 3.42 System (extensive) property of a saturated mixture is the sum of contributions from its constituent phases (see Anim. 3.C.satMixture).

The specific property *b* can be interpreted as the mass-weighted average of saturation properties $b_{f@T}$ and $b_{g@T}$, which can be obtained from the temperature saturation table for a given *T*. Note that for *x* = 0.5, the formula reduces to a simple average.

A mixture's density ρ, which is not a specific property, cannot be directly obtained from Eq. (3.26); however, once *v* has been determined from Eq. (3.26), ρ = 1/*v* can be calculated. The volumetric quality *y* can be expressed in terms of *x* and *v* as:

$$y \equiv \frac{\Psi_g}{\Psi} = \frac{m_g v_g}{mv} = \frac{x v_g}{v} = \frac{x v_{g@T}}{(1 - x)v_{f@T} + x v_{g@T}} \tag{3.27}$$

c. **Given *p* and *x*:** Given the pressure, obtain the saturation temperature $T = T_{sat@p}$, from the temperature saturation table. Now that *T* and *x* are known, follow the procedure described in step **b**. A pressure saturation table is more convenient for looking up the values of $b_{f@p}$ and $b_{g@p}$, which can be used in Eq. (3.26) for calculating the mixture properties.

d. **Given *T* (or *p*) and *b*:** Obtain b_f and b_g at the known *T* (or *p*) from the temperature (or pressure) saturation table. Calculate the quality *x* from Eq. (3.26), and follow step **b** (or step **c**), now that *T* (or *p*) and *x* are known.

e. **Given *T* (or *p*) and *y*:** Obtain v_f and v_g from the known *T* (or *p*) and then *x* from Eq. (3.27). Next, follow step **b** (or step **c**).

f. **Given b_1 and b_2, b and y, or x and y:** If neither p nor T is known, the procedure has to involve iteration. Use of TEST is recommended.

g. **Critical State:** The critical state is a special case of a saturated mixture, a unique state, which appears as the last row of a saturation table. Recognizing that the f-line meets the g-line at the critical point, we can use the combination of $x = 0$ and $y = 1$ to obtain the critical state in any PC TESTcalc.

EXAMPLE 3-8   State Evaluation Using the PC Model

For Ex. 3-7, determine (a) x, (b) u, and (c) the volume Ψ_f occupied by the liquid phase.

SOLUTION

Continuing from Ex. 3-7, the quality of the mixture can be calculated as:

$$v_1 = (1 - x_1)v_{f@30°C} + x_1 v_{g@30°C};$$

$$\Rightarrow \quad 0.04 = (1 - x_1)(0.00168) + x_1(0.1105); \quad \text{or,} \quad x_1 = 0.352$$

Read $u_{f@30°C}$ and $u_{g@30°C}$ from the temperature saturation table (since T_1 is given) and substitute them into Eq.(3.26) to obtain:

$$u_1 = (1 - 0.352)(320.5) + 0.352(1337.4) = 678.6 \frac{kJ}{kg}$$

The volumetric quality can be obtained from Eq. (3.27):

$$y_1 = \frac{x_1 v_{g@30°C}}{v} = \frac{0.352 \times 0.1105}{0.04} = 0.972$$

Therefore, the volume occupied by the saturated liquid phase can be calculated as:

$$\Psi_{f1} = \Psi_1 - y_1\Psi_1 = 0.02 - 0.973 \times 0.02 = 0.00054 \text{ m}^3 = 540 \text{ mL}$$

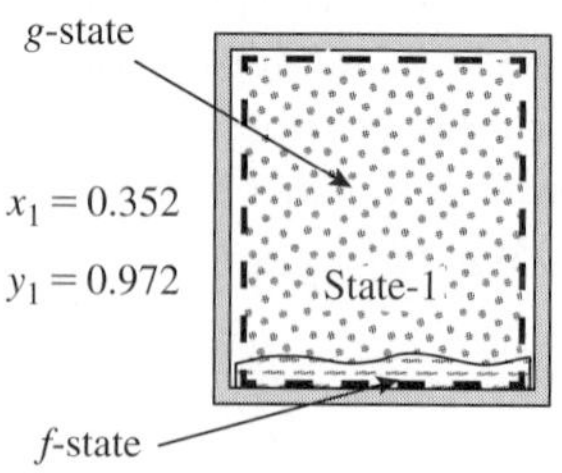

TEST Analysis

To verify the solution using TEST, launch the PC system-state TESTcalc. Select Ammonia, enter T1, m1, Vol1, and calculate the state. All the thermodynamic state properties including x1, u1, and y1 are calculated and displayed. The volume occupied by the saturated liquid can be obtained in the I/O panel from '= Vol1*(1 − y1)'.

Discussion

The quality of ammonia in the tank is 0.352; therefore, 64.8% of the mass is due to the liquid phase, which occupies only 2.8% of the total volume. Note that the two phases can be separated as shown in Figure 3.43 yet described by a single equilibrium state. Separate phases in equilibrium can be described by their respective f- and g-states (see Fig. 3.43).

g-state

$x_1 = 0.352$
$y_1 = 0.972$ State-1

f-state

FIGURE 3.43 While the saturated mixture in the tank in Ex. 3-8 is described by State-1, individual phases can be described by the f- and g-states at 30°C.

EXAMPLE 3-9   State Evaluation Using the PC Model

The tank in Ex. 3-7 is heated until all the liquid vaporizes. Determine (a) the pressure and (b) temperature in the tank. **What-if scenario:** (c) What would the answers be if the tank was only heated until the liquid volume diminished to 200 mL?

SOLUTION

Mass and volume remain constant in a closed tank; therefore v is known at the final state. The second property for the evaluation of state is provided by the quality x, which is 1 when the last drop of liquid vaporizes.

Analysis

State-1, evaluated in Ex. 3-7, and the saturated final state, State-2, are shown in the *T-v* diagram of Figure 3.44 for the constant-volume heating process:

$$\text{State-2: given}\left(x_2 = 1; \quad \text{and} \quad v_2 = v_{g@30°C} = v_1 = 0.04 \frac{m^3}{kg}\right)$$

From the saturation table of ammonia, Table B-12, the saturation pressure and temperature are interpolated for $v_{g@30°C} = 0.04\,m^3/kg$ to be:

$$p_2 = 3160 \text{ kPa}; \quad T_2 = 67.9°C$$

TEST Analysis

Having evaluated State-1, select State-2 and enter m2 = m1, Vol2 = Vol1, and x2 = 1. Click Calculate or press Enter to verify the answers.

What-if scenario

Designating the desired state as State-3, enter the known properties m3 = m2 and Vol3 = Vol1. Because the liquid volume is known, enter y3 as $y_3 = (V_3 - V_{f3})/V_3$, which evaluates to $(0.02 - 0.0002)/0.02 = 0.99$. Click Calculate or press Enter. The final pressure and temperature are calculated to be 2517 kPa and 58.5°C respectively. A *T-s* diagram generated by the TESTcalc with a constant-volume line passing through the calculated states is shown in Figure 3.45.

Discussion

The pressure of the saturation mixture in the tank varies with temperature as dictated by the *saturation curve* (Fig. 3.26). Constant-volume heating or cooling of a PC mixture can cause dramatic change in pressure, sometimes leading to explosions or implosions.

PROPERTIES OF SUPERHEATED VAPOR

When a state is located in the superheated region, the *V*-zone in a *T-s* diagram, the superheated tables are interpolated to find a desired property. In the following algorithm, the symbol ψ is used to represent any one of the following properties: *T*, *v*, *u*, *h*, or *s*.

a. **Given *p* and ψ:** Bracket the known *p* between two consecutive pressure tables and then bracket the given ψ between two consecutive rows in each table. Use the bilinear interpolation of Eq. (3.24) or visual estimates to calculate any third property.

b. **Given ψ_1 and ψ_2:** If the pressure is unknown, an iterative procedure must be used to determine the pressure first. Use of TEST is recommended.

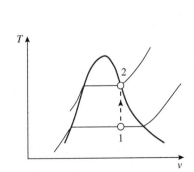

FIGURE 3.44 The *T-v* diagram indicates that the pressure must increase (Ex. 3-9).

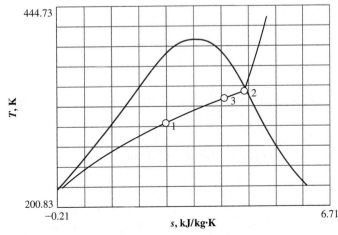

FIGURE 3.45 The *T-s* diagram with a constant volume line plotted by the PC state TESTcalc in Ex. 3-9.

EXAMPLE 3-10 **State Evaluation Using the PC Model**

Saturated vapor of refrigerant R-134a flows into an isentropic (constant entropy) compressor at a pressure of 100 kPa. If the exit pressure is 1 MPa, (a) determine the exit temperature. (b) Explain how entropy can remain unchanged despite a temperature rise at the exit.

SOLUTION

Pressure and entropy are the two independent properties supplied at the exit from which all other properties can be calculated.

Assumptions

The working fluid is uniform and at LTE at the inlet and exit, so that two flow states, State-1 and State-2, can be used to describe the flows.

Analysis

State-1:   (given $x_1 = 1$; $p_1 = 100$ kPa)

From Table B-10, $s_1 = s_{g@100 \text{ kPa}} = 0.9395 \dfrac{\text{kJ}}{\text{kg} \cdot \text{K}}$

State-2:   (given $p_2 = 1$ MPa) $s_2 = s_1 = 0.9395 \dfrac{\text{kJ}}{\text{kg} \cdot \text{K}}$

Given the higher pressure, State-2 must be superheated as evident from Figure 3.46. Interpolating from the superheated table at 1 MPa:

$$T_2 = 40 + (50 - 40)\frac{(0.9395 - 0.9066)}{(0.9428 - 0.9066)} = 49.1°C$$

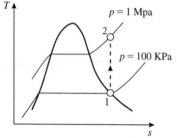

FIGURE 3.46 The *T-s* diagram indicates that the temperature must increase at the exit (Ex. 3-10).

TEST Analysis

To verify the solution with TEST, launch the PC system-state TESTcalc. Calculate State-1 with the given properties **p1** and **x1**. Select State-2 and enter **p2** as '=10*p1', **s2** as '= s1', and click Calculate. The temperature at the exit is calculated as 48.9°C. To visualize the calculated states, draw a *T-s* diagram by selecting it from the plot menu (see Fig. 3.46).

Discussion

The ratio of specific volumes, v_1/v_2, can be calculated in the I/O panel as 8.88. Such a high compression ratio means that the molecular spacing is greatly reduced at the exit, which tends to reduce molecular disorder and, hence, entropy. An increase in temperature tends to increase entropy at the exit. The two effects nullify each other in an isentropic process.

3.4.8 Subcooled or Compressed Liquid

When the phase composition is known to be subcooled (compressed liquid), the state is located in the L-zone in a *T-s* diagram, properties can be evaluated from the compressed liquid table. For most fluids, a compressed liquid table is not available. In such situations, a simplified model called the CL (compressed liquid) sub-model is commonly used that utilizes saturation tables to obtain subcooled state properties.

Compressed Liquid (CL) Sub-Model:

The simplest model for a liquid, the SL model, has a major drawback—it introduces significant error by assuming away the dependence of *v* on pressure and temperature. Also, the assumption of a constant c_v introduces error in the calculation of *u*, *h*, and *s*. An examination of the compressed liquid tables (e.g., Table B-4 for water), which lists accurate values of *v*, *u*, and *s* as functions of *p* and *T*, reveals that these properties have a weak dependence on pressure. The **compressed liquid (CL) sub-model** improves upon the SL model by approximately capturing the temperature effect, while neglecting the finer *compressibility effect*, that is, the effect of pressure on v, u, and s. It is called a *sub-model* because it can be used only under the framework of the PC model.

The model assumes that v, u, and s are functions of T only, so that $\phi(p,T) = \phi(T) = \phi_{f@T}$, where the generic symbol ϕ represents specific properties v, u, or s. The model can be summarized (see Fig. 3.47 and Anim. 3.C.*subcooledLiquid*) as:

$$v = v_{f@T}, \ u = u_{f@T}, \ s = s_{f@T}, \ h(p,T) = u_{f@T} + pv_{f@T} \qquad \textbf{(3.28)}$$

where the expression for enthalpy is obtained from its definition, $h \equiv u + pv$. Changes in v, u, and s can be evaluated via the temperature saturation table. If the temperature remains constant, it follows from Eq. (3.28) that v, u, and s of a subcooled liquid remain constant and the change in enthalpy depends on the change of pressure through $\Delta h = v_{f@T}\Delta p$. Depending on the known pair of independent properties, the complete state can be evaluated as follows:

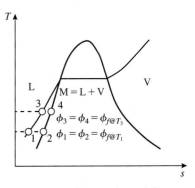

FIGURE 3.47 The CL sub-model relates States 1 and 3 to the corresponding saturated states, States 2 and 4, respectively (see Anim. 3.C.*subcooledLiquid*).

a. Given ψ_1 and ψ_2 ($\psi = T, v, u,$ or s): Since ψ is a function of T, only one independent property is actually given. From either ψ_1 or ψ_2, obtain the rest of the ψ's from the saturation table. To evaluate the state completely, p or h must be supplied.

b. Given p and ψ ($\psi = T, v, u,$ or s): From the known ψ, obtain all other ψ's using instructions given in step (a). The remaining property, h, can be evaluated from Eq. (3.28).

c. Given h and ψ ($\psi = T, v, u,$ or s): From the known ψ, obtain all other ψ's using (a). The remaining property, p, can be evaluated from $h = u + pv$ since h, u, and v are already known.

Compressed Liquid Table:

Available only for selected fluids, compressed liquid tables produce better accuracy than the CL sub-model because they account for change of density with temperature as well as pressure. Formatted the same way as the superheated vapor tables, compressed liquid tables, such as Table B-4 for compressed water, can be read and interpolated following the same procedure as described for superheated tables in the last section.

EXAMPLE 3-11 Use of CL Sub-Model for a PC Fluid

Water at 5 MPa and 200°C is pressurized at constant entropy to a pressure of 10 MPa. Determine the change in (a) temperature and (b) enthalpy. Use the CL sub-model for liquid water. **What-if scenario:** What would the answers be if (c) the compressed liquid table, or (d) the SL model, was used instead?

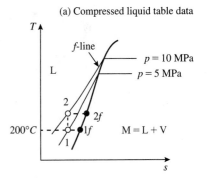

(a) Compressed liquid table data

SOLUTION

Using a T-s diagram, determine the phase composition of the two states, State-1 and State-2. Use tables and the CL sub-model to determine the desired properties.

Assumptions

Thermodynamic equilibrium at each state.

Analysis

State-1: (given $p_1 = 5$ MPa; $T_1 = 200°$C)

From Table B-1, $T_{sat@5MPa} = 264°$C $> T_1$

Therefore, water must be a subcooled liquid at this state. On a qualitative T-s diagram (Fig. 3.48a), draw the $p = p_1$ line, then locate State-1 and its corresponding saturated liquid state (State-1f) at T_1. Use the CL sub-model, Eq. (3.28), and the temperature saturation table for water to obtain $s_1 = s_{f@200°C} = 2.3309$ kJ/kg·K, $u_1 = u_{f@200°C} = 850.65$ kJ/kg, $v_1 = v_{f@200°C} = 0.001157$ m³/kg, and $h_1 = u_1 + p_1v_1 = 856.4$ kJ/kg.

State-2: (given $p_2 = 10$ MPa and $s_2 = s_1$)

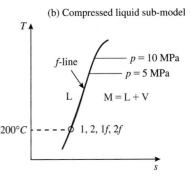

FIGURE 3.48 The subcooled states of Ex. 3-11: (a) Compressed liquid table data, (b) CL sub-model.

The $p = 10$ MPa line and the vertical $s = s_1$ line intersect at State-2, which belongs to the sub-cooled liquid zone. Using the CL sub-model:

$$s_2 = s_1; \quad \Rightarrow \quad s_{f@T_2} = s_{f@T_1}; \quad \Rightarrow \quad T_1 = T_2; \quad \Rightarrow \quad \Delta T = 0°C;$$

Therefore, $u_2 = u_1$ and $v_2 = v_1$.

Enthalpy also depends on pressure also and is given as:

$$h_2 = u_2 + p_2 v_2 = u_1 + p_2 v_1$$

$$= 850.65 + (10,000)(0.001157) = 862.22 \text{ kJ/kg}$$

Therefore, $\Delta h = h_2 - h_1 = 862.22 - 856.40 = 5.82$ kJ/kg

TEST Analysis

To verify the solution with TEST, launch the PC system-state TESTcalc. Select H2O as the working fluid, calculate State-1 from the known p1 and T1, and State-2 from p2 and s2 (= s1). To calculate the difference of temperature and enthalpy between the two calculated states, evaluate '= T2 − T1' and '= h2 − h1 ' in the I/O panel. The differences are found to be 0°C and 5.79 kJ/kg, respectively.

What-if scenario

To use the compressed liquid table in TEST, select H2O (Comp.Liq. Table) from the fluid selector. Clicking Super-Calculate produces the two updated states from which the differences are calculated as 0.83°C and 5.86 kJ/kg. Repeating the same procedure with the SL system-state TESTcalc, the differences can be shown to be 0°C and 5.02 kJ/kg.

Discussion

The compressed liquid table offers the highest accuracy. By comparison, the error in Δh is only about 1% for the CL sub-model and about 14% for the SL model. This result demonstrates once again that the separation among the constant-pressure lines in the liquid zone of Figure 3.48(a) due to *compressibility effect* is highly exaggerated. As the calculations suggest, T_2 is only slightly greater than T_1 (by less than one degree). The CL sub-model predicts them to be equal, collapsing all the constant-pressure lines in the liquid zone into the saturated liquid line (*f*-lines in Fig. 3.48b). Even when the CL sub-model is used in an analysis, it is a standard practice to plot the subcooled states as in Figure 3.48(a).

3.4.9 Supercritical Vapor or Liquid

The superheated vapor table and compressed liquid table usually extend to the super-critical region ($p > p_{cr}$). Note that pressure and temperature are independent of each other in the super-critical region, therefore, the procedure for the state evaluation is identical to those described for superheated vapor. Without a compressed liquid table, the CL sub-model can be used for a super-critical liquid, but with a significant loss in accuracy.

3.4.10 Sublimation States

The phase-change model developed for the saturated mixture of liquid and vapor phases is applicable to a saturated mixture of solid and vapor ($p < p_{tp}$). The *f*-point is replaced by the *i*-point as shown in Figure 3.31. For example, the saturation table for ice and water vapor, Table B-5, is similar to the corresponding saturation table, Table B-2, for saturated liquid and vapor. The procedure for evaluating liquid and vapor states can be leveraged to determine sublimation states. In TEST, the quality x also serves as the mass fraction of vapor in a saturated mixture of solid and vapor.

3.4.11 Model Summary—PC Model

State evaluation using the PC model consists of a two-step process: (i) determine the phase composition and, (ii) evaluate properties using the appropriate tables and the compressed liquid (CL) sub-model. For manual evaluation of a state, at least one saturation table (pressure or temperature table) and the superheated tables are required for a given working fluid.

3.5 GAS MODELS

The word **gas** is loosely used in thermodynamics to mean superheated vapor, supercritical vapor, saturated mixture, or sometimes liquid. In this section, we will introduce three gas models—ideal gas (IG), perfect gas (PG), and real gas (RG) note to editor: do we need to add the word 'model' at the end of this clause?—to capture the behavior of gases under a wide range of conditions. These models differ in the degree of simplicity and accuracy as there is always a trade-off between the two. The **IG model** is the most widely used model for gases, well suited for modeling most superheated vapors at low pressures and/or high temperatures (Fig. 3.49). The **PG model** is a simplified version of the IG model that has the additional assumption of constant specific heats. The **RG model** is a generalized model to handle vapor (and even saturated mixtures or subcooled liquid) of any working substance under all conditions. The generality of the RG model comes at the price of accuracy.

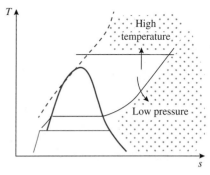

FIGURE 3.49 High T, low p, and small molecules turn superheated vapor into an ideal gas.

3.5.1 The IG (Ideal Gas) and PG (Perfect Gas) Models

Let's resume the discussion of microscopic behavior of superheated vapor from where we left off in Sec. 3.4. When a vapor expands due to an increase in temperature (and/or a reduction in pressure), the average spacing between its molecules (called the *mean free path*) increases. Consequently, the intermolecular force and the actual volume occupied by the molecules, themselves as a percentage of the system volume, decrease. Under such conditions (low pressure and/or high temperature), experiments by Boyle, Charles, and Gay-Lussac established that the specific volume v varies in direct proportion to T and in inverse proportion to p. Moreover, in the absence of intermolecular forces, the potential energy of molecules depends on the internal molecular structure only and remains unaffected, leaving molecular kinetic energy as the only storage mode of internal energy. In macroscopic terms, u becomes a function of T only, independent of pressure. These two observations—$v = v(T/p)$ and $u = u(T)$—constitute the basis for the IG model. A further simplification can be achieved if rotational and vibrational kinetic energy (see Anim. 0.D.*uGasMoleculeKE*) are also assumed to remain unaffected. In that case, u becomes directly proportional to the translational kinetic energy and temperature T of the gas. By definition [Eq. (3.4)], proportionality between u and T makes c_v a constant. This simplification forms the basis for the PG model. Monatomic gases such as Helium (He), Argon (Ar), etc., are truly perfect gases because vibration and rotation of single-atom molecules are impossible.

As in the case of a solid or a liquid, an increase in temperature of a gas is expected to increase its degree of disorder or entropy. From quantum physics, it can be established that an increase in the average distance among molecules increases the available molecular energy levels. Pressure has a large impact on entropy in the IG and PG models because the spacing among molecules can be significantly affected by pressure at a given temperature.

3.5.2 IG and PG Model Assumptions

The PG model is a simplified IG model; therefore, any assumption for the IG model also applies to the PG model (see Anim. 3.D.*PGModel* and *IGModel*).

i) Applicable to both IG and PG models, the molar specific volume $\bar{v}$ can be expressed as a function of p and T through the **IG (Ideal Gas) equation of state**:

$$\bar{v} = \bar{R}\frac{T}{p}; \quad \left[\frac{m^3}{kmol}\right], \quad \text{where} \quad \bar{R} = 8.314; \quad \left[\frac{kJ}{kmol \cdot K}\right] \tag{3.29}$$

or, on a mass basis,

$$pv = RT; \quad \left[kPa\frac{m^3}{kg} = \frac{kJ}{kg}\right] \quad \text{where} \quad v = \frac{\bar{v}}{M}; \quad \left[\frac{m^3}{kg}\right] \quad \text{and} \quad R \equiv \frac{\bar{R}}{M}; \quad \left[\frac{kJ}{kg \cdot K}\right] \tag{3.30}$$

The IG equation of state tells us that the volume occupied by 1 kmol of a perfect gas is independent of the identity of the gas, and depends only on the system's absolute temperature and pressure. In fact, at STP—used in chemistry; the standard temperature and pressure, $p = 100$ kPa, $T = 0°C$ is different from our standard atmospheric conditions—we can show from Eq. (3.29) that the volume occupied by 1 mol of any ideal gas is 22.4 liters. That is why $\bar{R}$ is called the **universal gas constant**; while R, known as the **gas constant**, is a material

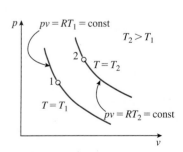

FIGURE 3.50 Two isotherms on a p-v diagram for a gas following the IG or PG model (click Isentropic Expansion in Anim. 3.D.*PGModel*).

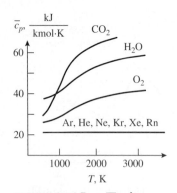

FIGURE 3.51 $\bar{c}_p = \overline{M}c_p$ for various gases as a function of temperature. Noble gases are true perfect gases.

property dependent on $\overline{M}$. We will refer to $pv = RT$ (see Fig. 3.50 and Anim. 3.D.*PGModel*) as the **IG equation of state**.

ii) Applicable to both IG and PG models, u is a function of temperature only, which will be shown in Chapter 11 to be a consequence of the IG equation of state.

iii) For the PG model, c_v is also assumed to be a material property and has a constant value for a given *perfect gas*.

The additional assumption of a constant c_v makes the PG model the simplest of all gas models. The error introduced by this assumption can be reduced if c_v is evaluated at an average temperature over the interval of interest. However, in this textbook we will use c_v evaluated at standard atmospheric temperature, 298 K, as the material property listed in Table C-1. For **monatomic gases**—helium, argon, neon, xenon, and krypton—c_v is truly a constant, given by $c_v = (3/2)R$ (see Fig. 3.51); therefore, monatomic gas is a *perfect gas*.

3.5.3 Equations of State

Using the model assumptions listed above, we can express v, u, h, and s in terms of the easily measurable independent properties p and T. The resulting equations of state can be used to evaluate a complete state from any pair of independent properties.

Specific Volume (v):

The IG equation of state, obeyed by the PG and IG models, provides the p-v-T relationship for determining the specific volume for a gas. Gas constant R and molar mass $\overline{M}$ for most common gases can be found in Table C-1. It is left as an exercise to manipulate the IG equation of state into various useful forms such as $p = \rho RT$, $p\forall = mRT$, $p\bar{v} = \overline{R}T$, and $p\forall = n\overline{R}T$. In this textbook, we will use $pv = RT$ as the preferred format from which other forms can be derived.

When a working substance can be modeled as an ideal or perfect gas, the IG equation of state not only allows evaluation of v, but also boundary work in isothermal (constant-temperature) processes, as the area under an **isotherm** (a constant-temperature line) can be evaluated through integration (click Isothermal Compression in Anim. 5.A.*pTsConstCompressionIG*) of $pd\forall = (mRT/\forall)d\forall$.

Internal Energy and Enthalpy (u, h):

Because $u = u(T)$ for the IG model, Eq. (3.4) simplifies to:

$$du = c_v dT + \underbrace{\left(\frac{\partial u}{\partial v}\right)_T}_{0} dv = c_v dT; \quad \left[\frac{kJ}{kg}\right] \tag{3.31}$$

With the left-hand side being a function of temperature only, so must be the right hand side, so that $c_v = du/dT = c_v(T)$. As with u, the specific heat at constant volume c_v of an ideal gas must be a function of temperature only. The differential relation for u, Eq. (3.31), can be integrated from a standard reference state at p_0, T_0 (Sec. 3.1.3) to obtain:

$$u^{IG} = u(p_0, T_0) + \int_{p_0, T_0}^{p, T} c_v(T)\,dT = u_{ref}^o + \int_{T_0}^{T} c_v(T)\,dT; \quad \left[\frac{kJ}{kg}\right] \tag{3.32}$$

$$\Rightarrow \quad \Delta u^{IG} = u_2^{IG} - u_1^{IG} = \int_{T_1}^{T_2} c_v(T)\,dT \quad \left[\frac{kJ}{kg} = \frac{kJ}{kg \cdot K}K\right] \tag{3.33}$$

With c_v acting as a material property for the PG model, these expressions can be further simplified:

$$u^{PG} = u(p_0, T_0) + \int_{p_0, T_0}^{p, T} c_v\,dT = u_{ref}^o + c_v(T - T_0); \quad \left[\frac{kJ}{kg}\right] \tag{3.34}$$

and $\quad \Delta u^{PG} = u_2^{PG} - u_1^{PG} = c_v(T_2 - T_1); \quad \left[\frac{kJ}{kg}\right] \tag{3.35}$

To obtain an expression for h, we substitute $pv = RT$ in the operational definition, $h \equiv u + pv$, of enthalpy:

$$dh = d(u + pv) = du + d(RT) = c_v dT + R dT = (c_v + R)dT \qquad \textbf{(3.36)}$$

The right-hand side of Eq. (3.36) is a function of temperature only; therefore, enthalpy of an ideal gas must be independent of pressure and is a function of temperature only, or, $h = h(T)$. Starting with Eq. (3.5), dh can be expressed in an alternative form in terms of c_p:

$$dh = c_p dT + \cancelto{0}{\left(\frac{\partial h}{\partial p}\right)_T} dp = c_p dT; \quad \left[\frac{kJ}{kg}\right] \qquad \textbf{(3.37)}$$

A comparison between these two expressions for dh leads to another important state equation valid for both IG and PG models:

$$\text{IG Model:} \quad c_p(T) = c_v(T) + R; \quad \text{PG Model:} \quad c_p = c_v + R; \quad \left[\frac{kJ}{kg \cdot K}\right] \qquad \textbf{(3.38)}$$

Because R cannot be negative, $c_p > c_v$. Integrating Eq. (3.37) from the standard reference state (Sec. 3.1.3) to the current state:

$$h^{IG} = h(p_0, T_0) + \int_{p_0, T_0}^{p, T} c_p(T) dT = h^{\circ}_{ref} + \int_{T_0}^{T} c_p(T) dT; \quad \left[\frac{kJ}{kg}\right] \qquad \textbf{(3.39)}$$

$$\text{and} \quad \Delta h^{IG} = h_2^{IG} - h_1^{IG} = \int_{T_1}^{T_2} c_p(T) dT; \quad \left[\frac{kJ}{kg}\right] \qquad \textbf{(3.40)}$$

Where the reference enthalpy and internal energy at *standard conditions* can be related as:

$$h^{\circ}_{ref} = h(p_0, T_0) = h(T_0) = u(T_0) + p_0 v = u^{\circ}_{ref} + RT_0; \quad \left[\frac{kJ}{kg}\right] \qquad \textbf{(3.41)}$$

Although the reference enthalpy and internal energy are functions of reference temperature only, the superscript 0 emphasizes the standard pressure for the sake of consistency. Obviously, only of the two reference values can be arbitrarily set. In the ideal gas tables and in TESTcalcs (see Tables D-3 through D-15) is usually set to zero at standard temperature with the exception of air.

For the PG model, c_p and c_v being constants, the expressions for enthalpy further simplify as:

$$h^{PG} = h(p_0, T_0) + \int_{p_0, T_0}^{p, T} c_p(T) dT = h^{\circ}_{ref} + c_p(T - T_0); \quad \left[\frac{kJ}{kg}\right] \qquad \textbf{(3.42)}$$

$$\text{and} \quad \Delta h^{PG} = h_2^{PG} - h_1^{PG} = c_p(T_2 - T_1); \quad \left[\frac{kJ}{kg}\right] \qquad \textbf{(3.43)}$$

Note that both u and h are linear functions of temperature T (Fig. 3.52) for the PG model. Their dependence on temperature for the IG model (Fig. 3.53) depends on the functional relationships $c_v(T)$ and $c_p(T)$, which are generally expressed in polynomials (Table D-1) or tabular form (Table D-2).

Substitution of the IG equation of state, Eq. (3.30), relates the enthalpy expression with the expression for internal energy for both an ideal gas or a perfect gas:

$$\Rightarrow \quad h^{IG,PG} = u^{IG,PG} + pv = u^{IG,PG} + RT; \quad \left[\frac{kJ}{kg}\right] \qquad \textbf{(3.44)}$$

This relation is graphically shown in Figure 3.52 for a perfect gas and Figure 3.53 for an ideal gas (see Anim. 3.D.*propertiesPG* and Anim. 3.D.*propertiesIG*). At a given temperature, the slope of the enthalpy curve (shifted vertically by an amount RT) must be steeper than the internal energy curve.

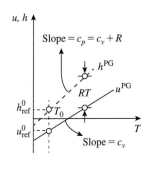

FIGURE 3.52 Linear change of u and h with T for a perfect gas (see Anim. 3.D.*propertyDiffsPG*).

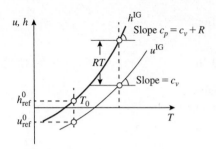

FIGURE 3.53 u and h are function of T for an ideal gas (see Anim. 3.D.*propertiesIG*).

Another important property is the **specific heat ratio**, defined as:

$$\text{IG Model:}\quad k(T) \equiv \frac{c_p(T)}{c_v(T)}; \quad \text{PG Model:}\quad k \equiv \frac{c_p}{c_v} \tag{3.45}$$

Being a ratio of two thermodynamic properties, the specific heat ratio k is also a thermodynamic property or an ideal gas, which varies mildly with temperature. For a perfect gas, it is a material property and can be obtained from the perfect gas table, Table C-1. Because $c_p > c_v$, k is always greater than one for all gases; for instance, $k = 1.4$ for air. The perfect gas table, Table C-1, lists five material properties—$\overline{M}$, c_p, c_v, k, and R; however, only two of those are independent. Equations (3.38) and (3.45) can be manipulated to express c_p and c_v in terms of k and R:

$$c_p = \frac{kR}{k-1} \quad \text{and} \quad c_v = \frac{R}{k-1}; \quad \text{where} \quad R = \frac{\overline{R}}{\overline{M}}; \quad \left[\frac{\text{kJ}}{\text{kg}\cdot\text{K}}\right] \tag{3.46}$$

Entropy (s):

Substituting $pv = RT$ and Eqs. (3.31) and (3.37) in the two Tds relations, we obtain:

$$ds = \frac{du}{T} + \frac{pdv}{T} = c_v\frac{dT}{T} + R\frac{dv}{v}; \quad \left[\frac{\text{kJ}}{\text{kg}\cdot\text{K}}\right] \tag{3.47}$$

$$ds = \frac{dh}{T} - \frac{vdp}{T} = c_p\frac{dT}{T} - R\frac{dp}{p}; \quad \left[\frac{\text{kJ}}{\text{kg}\cdot\text{K}}\right] \tag{3.48}$$

The first term depends on temperature only and the second term on specific volume or pressure. As a result, integration from (p_0,T_0) to (p,T) simplifies to integration from T_0 to T for the first term and either $v_0 = RT_0/p_0$ to $v = RT/p$ or p_0 to p for the second term. These terms can be independently integrated between two states. From Eq. (3.47), we obtain:

$$s^{\text{IG}} = s(p_0,T_0) + \int_{T_0}^{T}\frac{du}{T} + R\int_{v_0}^{v}\frac{dv}{v} = s^{\circ}(T_0) + \int_{T_0}^{T}\frac{c_vdT}{T} + R\ln\frac{v}{v_0} \tag{3.49}$$

$$\Rightarrow \quad \Delta s^{\text{IG}} = s_2^{\text{IG}} - s_1^{\text{IG}} = \int_{T_1}^{T_2}\frac{c_vdT}{T} + R\ln\frac{v_2}{v_1}; \quad \left[\frac{\text{kJ}}{\text{kg}\cdot\text{K}}\right] \tag{3.50}$$

Using the IG equation of state to replace v with p and T in Eq. (3.49), or directly integrating Eq. (3.48), we obtain the alternative expressions:

$$s^{\text{IG}} = s(p_0,T_0) + \int_{T_0}^{T}\frac{dh}{T} - R\int_{p_0}^{p}\frac{dp}{p} = s^{\circ}(T_0) + \int_{T_0}^{T}\frac{c_pdT}{T} - R\ln\frac{p}{p_0} \tag{3.51}$$

$$\Rightarrow \quad \Delta s^{\text{IG}} = s_2^{\text{IG}} - s_1^{\text{IG}} = \int_{T_1}^{T_2}\frac{c_pdT}{T} - R\ln\frac{p_2}{p_1}; \quad \left[\frac{\text{kJ}}{\text{kg}\cdot\text{K}}\right] \tag{3.52}$$

For the PG model (constant specific heats), the entropy expressions can be further simplified:

$$s^{\text{PG}} = s^{\circ}(T_0) + c_v\ln\frac{T_2}{T_0} + R\ln\frac{v}{v_0}; \quad \left[\frac{\text{kJ}}{\text{kg}\cdot\text{K}}\right] \tag{3.53}$$

$$\Delta s^{\text{PG}} = s_2^{\text{PG}} - s_1^{\text{PG}} = c_v\ln\frac{T_2}{T_1} + R\ln\frac{v_2}{v_1}; \quad \left[\frac{\text{kJ}}{\text{kg}\cdot\text{K}}\right] \tag{3.54}$$

$$s^{\text{PG}} = s^{\circ}(T_0) + c_p\ln\frac{T_2}{T_0} - R\ln\frac{p}{p_0}; \quad \left[\frac{\text{kJ}}{\text{kg}\cdot\text{K}}\right] \tag{3.55}$$

$$\Delta s^{\text{PG}} = s_2^{\text{PG}} - s_1^{\text{PG}} = c_p\ln\frac{T_2}{T_1} - R\ln\frac{p_2}{p_1}; \quad \left[\frac{\text{kJ}}{\text{kg}\cdot\text{K}}\right] \tag{3.56}$$

As we deduced from the microscopic perspective (Sec. 3.4), these relations confirm that entropy increases when temperature or volume increases, or when pressure decreases. Different terms in the entropy expressions are graphically explained in Figure 3.54. In most situations, we are interested in calculating a difference in entropy.

Thermodynamic Plots:

When modeling a working substance as an ideal or perfect gas, it is often necessary to present the calculated states, or processes, on thermodynamic diagrams. To explore how a constant-pressure line (**isobar**) differs from a constant-volume line (**isochor**) on a T-s diagram, compare their slopes at a given point. By substituting $dp = 0$ ($p = $ constant) in Eq. (3.48) and $dv = 0$ ($v = $ constant) in Eq. (3.47), the slope dT/ds can be shown to be T/c_p for an isobar and T/c_v for an isochor. Both constant-property lines must become steeper as T increases. Moreover, because $c_p > c_v$, an isochor must be steeper (see Fig. 3.55) than an isobar passing through a given T. The direction of increasing pressure in a family of isobars can be deduced if we consider a constant-pressure line (isobar) for a gas as an extension of the corresponding isobar in the superheated region (see Fig. 3.56). Alternatively, with the entropy expression, it can be seen that an increase in pressure (at a given temperature) reduces entropy. The isobar is shifted left as pressure increases. Likewise, an isochor shifts right as volume increases.

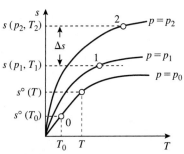

FIGURE 3.54 Unlike u or h, s is a function of p and T in the PG and IG models (see Anim. 3.D.*propertiesPG* and Anim. 3.D.*propertiesIG*).

Isentropic Relations for the PG Model:

States that have the same entropy are said to be **isentropic** to each other. For the PG model, simplified relations among properties of two isentropic states can be obtained by substituting $\Delta s^{PG} = 0$ into Eq. (3.54):

$$c_v \ln \frac{T_2}{T_1} = -R \ln \frac{v_2}{v_1}; \quad \left[\frac{kJ}{kg \cdot K} \right], \quad \Rightarrow \quad \frac{T_2}{T_1} = \left(\frac{v_1}{v_2} \right)^{k-1} \tag{3.57}$$

In a similar manner, two other algebraic equations can be developed—one between T and p by substituting $\Delta s^{PG} = 0$ in Eq. (3.56) and another between p and v by substituting the ideal gas equation of state in Eq. (3.57).

$$\frac{T_2}{T_1} = \left(\frac{p_2}{p_1} \right)^{(k-1)/k} = \left(\frac{v_1}{v_2} \right)^{k-1} = \left(\frac{\rho_2}{\rho_1} \right)^{k-1} \tag{3.58}$$

The isentropic relations should not be confused with the ideal gas equation of state. The latter can be applied to any pairs of state that involve a *perfect* or *ideal gas*; however, Eq. (3.58) can only be applied between isentropic states involving a perfect gas.

It is instructive to plot $pv^n = $ constant on a p-v diagram for various values of n, as shown in Figure 3.57 (see Anim. 5.A.*pTsConstExpansionIG* and *pTsConstCompressionIG*). For $n = k$, it reduces to the isentropic relation of Eq. (3.58) and for $n = 1$, it reproduces the isothermal relation of Eq. (3.30). Even the constant-pressure line can be

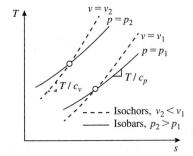

FIGURE 3.55 Isochors are steeper than isobars on a T-s diagram because $c_p > c_v$ for an ideal gas. To verify this, evaluate any state with the IG state TESTcalc and draw $p = $ c and $v = $ c lines on a T-s diagram.

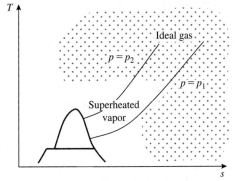

FIGURE 3.56 Relative locations of isobars in a thermodynamic diagram can be deduced by extending isobars in the superheated vapor region.

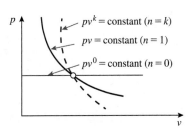

FIGURE 3.57 Isobaric ($n = 0$), isothermal ($n = 1$), and isentropic ($n = k$) lines on a p-v diagram for a perfect gas. Evaluate any state with the PG state TESTcalc and draw $s = $ c and $T = $ c lines on a p-v diagram (see Anim. 5.A.*pTsConstExpansionIG*).

obtained from this generic expression for $n = 0$. The relative shape of the isentropic line can be deduced from the fact that as n increases from 0 (constant pressure) to 1 (isothermal), the steepness increases (Fig. 3.57); therefore, the isentropic line must be steeper than the isothermal line since $k > 1$.

The isentropic relations derived for the PG model are sometimes extended to gas or vapor that goes through a quasi-isentropic process by substituting the coefficient n for the specific heat ratio k. Equation (3.58), where k has been replaced by an empirical n, is known as a **polytropic relation**. We will revisit polytropic relations in Chapter 5.

The IG (Ideal Gas) Tables:

Evaluating properties in the IG model using polynomial functions for c_p has a major drawback—finding T for a given u or h requires an iterative solution. The ideal gas reference tables, Tables D-3 through D-15, list values of molar specific properties $\bar{h}$, $\bar{u}$, and $\bar{s}^o$ (recall from Sec. 1 that $\bar{b} = bM$) as functions of T, where $\bar{s}^o$ is the molar specific entropy at the standard pressure (see Anim. 1.B.*molarProperties*). The expression for entropy, Eq. (3.51), is slightly modified by separating contribution from pressure and temperature:

$$s^{IG}(p,T) = [s^{IG}(p_0,T) - s^{IG}(p_0,0)^0] + [s^{IG}(p,T) - s^{IG}(p_0,T)] = s^o(T) - R\ln\frac{p}{p_0}$$

$$\Rightarrow \quad s^{IG}(p,T) = s^o(T) - R\ln\frac{p}{p_0} \quad \left[\frac{kJ}{kg \cdot K}\right] \tag{3.59}$$

where

$$s^o(T) = s^o(T_0) + \int_{T_0}^{T}\frac{c_p\,dT}{T}; \quad \left[\frac{kJ}{kg \cdot K}\right] \tag{3.60}$$

$s^o(T)$ can be interpreted (Fig. 3.54) as the specific entropy at p_0 and T, which is evaluated from specific heat data and tabulated for different gases (see Tables D-3 through D-15). This eliminates the need for integration. From Eq. (3.59), a change in entropy between two states, State-1 and State-2, can be expressed as:

$$\Delta s^{IG} = s^o(T_2) - s^o(T_1) - R\ln\frac{p_2}{p_1} \quad \left[\frac{kJ}{kg \cdot K}\right] \tag{3.61}$$

The corresponding molar expressions for entropy can be obtained by multiplying by $\bar{M}$:

$$\bar{s}^{IG}(p,T) = \bar{s}^o(T) - \bar{R}\ln\frac{p}{p_0}; \quad \Delta\bar{s}^{IG} = \bar{s}^o(T_2) - \bar{s}^o(T_1) - \bar{R}\ln\frac{p_2}{p_1} \quad \left[\frac{kJ}{kmol \cdot K}\right] \tag{3.62}$$

Given any one of the properties T, u, h, or s^o, the rest can be read from the ideal-gas table for a given species.

Isentropic Relations for the IG Model:

The isentropic relations for the PG model are based on the constant-specific heat assumption and cannot be used for an ideal gas. For the IG model, setting $\Delta s^{IG} = 0$, Eq. (3.61) can be rearranged for isentropic states, State-1 and State-2 as:

$$\frac{p_2}{p_1} = \frac{\exp[\bar{s}^o(T_2)/\bar{R}]}{\exp[\bar{s}^o(T_1)/\bar{R}]} = \frac{\exp[s^o(T_2)/R]}{\exp[s^o(T_1)/R]} \tag{3.63}$$

Note that the expression $\exp[s^o(T)/R]$ in this relation is solely a function of temperature and can be added to the ideal gas table in a new column as a new property. Represented by the symbol p_r, this dimensionless property relates the pressures of two isentropic states and is called the **relative pressure**:

$$\frac{p_2}{p_1} = \frac{p_{r2}(T_2)}{p_{r1}(T_1)}, \quad \text{where} \quad p_r(T) \equiv \frac{s^o(T)}{R} \tag{3.64}$$

This isentropic relation for an ideal gas [contrast it with the simpler relation for the PG model, Eq. (3.58)] can be used to find one unknown property from the set p_1, T_1, p_2, and T_2 as long as the ideal gas table that lists p_r as a function of T is available for the given gas. Table D-3 is a table for air.

The specific-volume ratio between two isentropic states can be similarly obtained by combining the pressure relation with the IG equation of state:

$$\frac{v_2}{v_1} = \left(\frac{RT_2}{p_2}\right)\left(\frac{p_1}{RT_1}\right) = \left(\frac{RT_2}{p_{r2}}\right)\left(\frac{p_{r1}}{RT_1}\right) = \frac{v_{r2}(T_2)}{v_{r1}(T_1)}, \quad \text{where} \quad v_r(T) \equiv \frac{RT}{p_r} \qquad (3.65)$$

Like the relative pressure, $v_r(T)$ is called the **relative volume** and is listed in the ideal gas tables. If any three properties out of the set v_1, T_1, v_2, and T_2 are known, the fourth can be found from this isentropic relation.

As an example, suppose air at $T_1 = 300$ K, $p_1 = 100$ kPa is compressed isentropically to $p_2 = 500$ kPa. If air is modeled as a perfect gas, the isentropic relation, Eq. (3.58), produces $T_2 = 300 \times 10^{(1.4-1)/1.4} = 475.4$ K. If we are to use the IG model, $p_{r1} = 1.386$ can be read from Table D-3 for air and p_{r2} calculated from Eq. (3.64) as $p_{r2} = 1.386(500/100) = 6.93$. Locate $p_{r2} = 6.93$ in Table D-3 between consecutive rows for 470 K and 480 K. Interpolating, $T_2 = 470 + 10(6.93 - 6.742)/(7.268 - 6.742) = 473.6$K. Contrast this calculation with the result of the PG model (see Figs. 3.58 and 3.59), which seem to be reasonably accurate given the formula's simplicity. Of course, if the temperature difference is large, the accuracy of the PG model would deteriorate. There is no need to evaluate relative pressure or volume while using the IG model in TESTcalcs. For both the PG and IG model (for that matter for all models), once a state, for example, State-1, is evaluated, an isentropic state, State-2, can be obtained by entering 's2 = s1' and supplying a second thermodynamic property, p2, T2, v2, u2, or h2.

 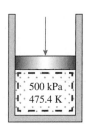

FIGURE 3.58 Isentropic-compression analysis for air with the PG model using Eq. (3.58) (see Anim. 5.A.*sConstCompression*IG).

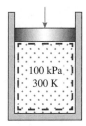 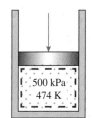

FIGURE 3.59 Isentropic-compression analysis for air with the IG model using Table D-3.

3.5.4 Model Summary: PG and IG Models

Frequently used formulas for manual evaluation of the thermodynamic state of a gas by using the IG or PG model can be summarized as follows (see Anim. *propertyDiffsPG* and *propertyDiffsIG*):

Obtain $\overline{M}$ and c_p, material properties for a perfect gas, from Table C-1 or any PG TESTcalc). Calculate other material properties from:

$$R = \frac{\overline{R}}{M}, c_v = c_p - R, \quad \text{and} \quad k = \frac{c_p}{c_v}, \quad \text{where} \quad \overline{R} = 8.314 \text{ kJ/kg} \cdot \text{K} \qquad (3.66)$$

Use the following equations of state for thermodynamic and important extrinsic properties:

IG/PG Models:

$$pv = RT; \quad \left[\frac{kJ}{kg}\right], \quad \Rightarrow \quad \Delta v = R\left(\frac{T_2}{p_2} - \frac{T_1}{p_1}\right); \quad \left[\frac{m^3}{kg}\right] \qquad (3.67)$$

PG Model:

$$\Delta u = u_2 - u_1 = c_v(T_2 - T_1); \quad \Delta e = \Delta u + \Delta \text{ke} + \Delta \text{pe}; \quad \left[\frac{kJ}{kg}\right] \qquad (3.68)$$

$$\Delta h = c_p(T_2 - T_1); \quad \Delta j = \Delta h + \Delta \text{ke} + \Delta \text{pe}; \quad \left[\frac{kJ}{kg}\right] \qquad (3.69)$$

$$\Delta s = s_2 - s_1 = c_p \ln\frac{T_2}{T_1} - R \ln\frac{p_2}{p_1}, \quad \text{and} \quad \Delta s = c_v \ln\frac{T_2}{T_1} + R \ln\frac{v_2}{v_1}; \quad \left[\frac{kJ}{kg \cdot K}\right] \qquad (3.70)$$

Note that $u = u(T)$, $h = h(T)$, and $s = s(p,T) = s(v,T)$. Additionally, u and h increase linearly with T.

Finally, between two isentropic states, for $s_2 = s_1$:

$$\frac{p_2}{p_1} = \left(\frac{T_2}{T_1}\right)^{\frac{k}{k-1}} = \left(\frac{\rho_2}{\rho_1}\right)^k = \left(\frac{v_1}{v_2}\right)^k \qquad (3.71)$$

IG Model:

Use the ideal gas look-up table (Tables D-3 through D-15) to obtain the set of T, u, h, and s^o if any one of these properties is given. Obtain $s = s(p,T)$ from $s^o(T)$ and using Eq. (3.59). Use relative pressure and volume as intermediate properties to relate isentropic states through Eqs. (3.64) and (3.65). Note that $u = u(T)$, $h = h(T)$, and $s = s(p,T)$ in both the IG and PG models.

EXAMPLE 3-12 State Evaluation Using the PG Model

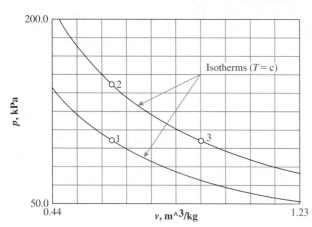

FIGURE 3.60 Computed states on a *p-v* diagram (Ex. 3-12) generated by the PG state TESTcalc.

A piston-cylinder device with a volume of 2 L contains argon at a pressure of 1 atm and has a temperature of 550°R. The gas is heated to 800°R in a constant-volume process. Determine (a) the mass of the gas, and the change in (b) pressure, and (c) entropy during the process. Assume the atmospheric conditions to be 100 kPa and 25°C. *What-if scenario:* (d) What would the answers in (c) be if the heating was carried out at a constant pressure?

SOLUTION

The problem involves state evaluation at the beginning and end of the process represented by the anchor states, State-1 and State-2.

Assumptions

The anchor states (Fig. 3.60) are in thermodynamic equilibrium. argon being a monatomic gas, the PG model is best suited for this problem.

Analysis

From Table C-1, or any PG state TESTcalc, obtain the material properties for argon: $\overline{M} = 39.95$ kg/kmol, $c_p = 0.5203$ kJ/kg·K, $R = 0.208$ kJ/kg·K, and $c_v = 0.3122$ kJ/kg·K. Properties in non-SI units are converted into SI units using the converter TESTcalc:

State-1: (given $V_1 = 2$ L $= 0.002$ m³; $p_1 = 101.3$ kPa; $T_1 = 305.6$ K)

$$\Rightarrow \quad v_1 = R\frac{T_1}{p_1} = (0.208)\frac{305.6}{101.3} = 0.627\ \frac{m^3}{kg}; \quad m_1 = \frac{V_1}{v_1} = \frac{0.002}{0.627} = 0.0032\ \text{kg}$$

State-2: (given $T_2 = 444.4$ K; $V_2 = V_1$; $m_2 = m_1$)

$$\Rightarrow \quad v_2 = v_1$$

Applying the IG equation of state to both states we obtain:

$$\Delta p = R\left(\frac{T_2}{v_2} - \frac{T_1}{v_1}\right) = \frac{R}{v_1}(T_2 - T_1) = \frac{0.208}{0.627}(444.4 - 305.6) = 46.0\ \text{kPa}$$

Note that for a constant-volume process, the second expression for entropy in Eq. (3.70) is more convenient:

$$\Delta s = c_v \ln\frac{T_2}{T_1} + R\ln\frac{v_2}{v_1}^{0} = c_v \ln\frac{T_2}{T_1} = 0.3122 \times \ln\frac{444.4}{305.6} = 0.1169\ \frac{\text{kJ}}{\text{kg}\cdot\text{K}}$$

Therefore, $\Delta S = m\Delta s = (0.0032)(0.1169) = 3.74 \times 10^{-4}\ \frac{\text{kJ}}{\text{K}}$

TEST Analysis

To verify the results, launch the PG system-state TESTcalc and select argon as the working fluid. Calculate State-1 from p1, T1, and Vol1 and State-2 from m2 = m1, Vol2 = Vol1, and T2. Use the I/O panel as a calculator to evaluate expressions such as '= p2 − p1' and '= m1*(s2 − s1)' and to verify answers.

What-if scenario

Represent the constant-pressure finish state as State-3. Enter p3 = p1, T3 = T2, and click Calculate. In the I/O panel, calculate ΔS by evaluating '= m1*(s3 − s2)' as 6.21×10^{-4} kJ/K.

Discussion

Observe how easy it is to visualize the calculated states on thermodynamic plots, such as the *p-v* diagram shown in Figure 3.60. Select a suitable diagram from the plot menu after the states have been calculated, then draw various constant-property lines on the thermodynamic plot.

EXAMPLE 3-13 | PG versus PC Model

Superheated steam at a pressure of 10 kPa and temperature of 200°C undergoes a process to a final condition of 50 kPa and 300°C. Assuming perfect gas behavior, determine (a) Δv, (b) Δu, (c) Δh, and (d) Δs. ***What-if scenario:*** (e) What would these results be if the PC model was used instead?

SOLUTION

The problem involves property differences between two states: State-1 and State-2.

Assumptions

The two states are in thermodynamic equilibrium, represented by system States-1 and 2.

Analysis

Material properties $\overline{M}$ and c_p are read from Table C-1 or any PG state TESTcalc as 18 kg/kmol and 1.8723 kJ/kg · K, respectively. Calculate $R = 8.314/18 = 0.462$ kJ/kg · K and $c_v = 1.8723 - 0.462 = 1.41$ kJ/kg · K. Using the PG model:

$$\Delta v = v_2 - v_1 = R\left(\frac{T_2}{p_2} - \frac{T_1}{p_1}\right) = 0.462\left(\frac{573}{50} - \frac{473}{10}\right)$$

$$= -16.56 \frac{m^3}{kg}\left[\left(\frac{kJ}{kg \cdot K}\right)\left(\frac{K}{kPa}\right) = \frac{kN \cdot m \cdot m^2}{kg \cdot kN} = \frac{m^3}{kg}\right]$$

$$\Delta u = u_2 - u_1 = c_v(T_2 - T_1)$$

$$= (1.41)(573 - 473)\left[\left(\frac{kJ}{kg \cdot K}\right)(K) = \frac{kJ}{kg}\right] = 141.0 \frac{kJ}{kg}$$

$$\Delta h = h_2 - h_1 = c_p(T_2 - T_1) = (1.872)(573 - 473) = 187.2 \frac{kJ}{kg}$$

$$\Delta s = s_2 - s_1 = c_p \ln\frac{T_2}{T_1} - R \ln\frac{p_2}{p_1}$$

$$= 1.872 \times \ln\frac{573}{473} - 0.462 \times \ln\frac{50}{10} = -0.385 \frac{kJ}{kg \cdot K}$$

TEST Analysis

To verify the results, launch the PG system-state TESTcalc and select H2O as the working fluid. Calculate the two states from the given properties. In the I/O panel, obtain the desired differences by evaluating expressions such as '= s2 − s1' (see Figs. 3.61 and 3.62).

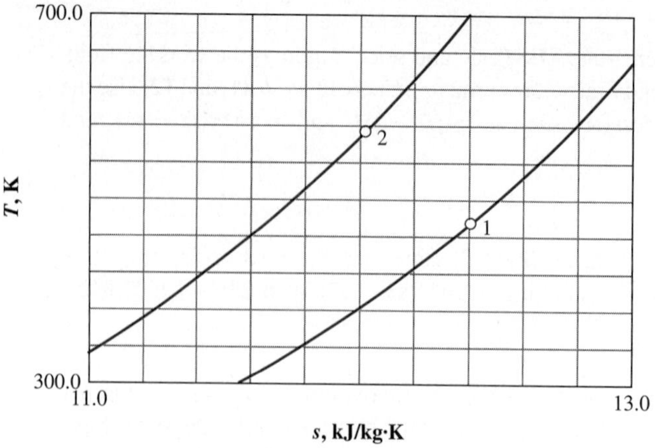

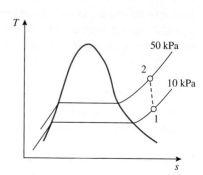

FIGURE 3.61 Computed states on a *T-s* diagram (Ex. 3-13) generated by the PG system-state TESTcalc.

FIGURE 3.62 The initial and final states on a *T-s* diagram (Ex. 3-13). The PC state TESTcalc can be used for a more accurate plot.

What-if scenario

Launch the PC system-state TESTcalc in a new browser tab. Create TEST-code in the I/O panel of the PG state TESTcalc by clicking Super-Calculate. Copy the code into the I/O panel of the PC TESTcalc, then click Load to obtain the PC states. Evaluate the property differences in the I/O panel. The results are compared in the table below:

	Δv (m³/kg)	Δu (kJ/kg)	Δh (kJ/kg)	Δs (kJ/kg · K)
PG Model	−16.56	141.0	187.2	−0.385
PC Model	−16.54	150.1	196.0	−0.366

Discussion

The comparison between the two models reveals that while the change in specific volume is predicted accurately by the PG model, other properties do not fare as well. In Sec. 3.8 we will seek the criterion that must be satisfied for a vapor to be reasonably modeled as a perfect or ideal gas.

EXAMPLE 3-14 State Evaluation Using the IG Model

A 10-m³ volume tank contains air at a pressure of 100 kPa and a temperature of 300 K. The air is heated (Fig. 3.63) to a temperature of 2000 K. Treat air as an ideal gas and (a) determine the final pressure and the change in the total internal energy of air. **What-if scenario:** (b) What would the answers be if the PG model had been used?

SOLUTION

Evaluate the two states using the IG model.

Assumptions

State-1 and State-2 represent the equilibrium states at the beginning and end of the heating process.

Analysis

The process is graphed on a *p-v* diagram in Figure 3.63. From Table C-1 or any PG state TESTcalc, obtain the material properties of air as $\overline{M} = 28.97$ kg/kmol, $c_p = 1.005$ kJ/kg · K, $R = 0.287$ kJ/kg · K, and $c_v = 0.718$ kJ/kg · K. From the air table, Table D-3, read u_1 and u_2:

State-1: (given $V_1 = 10$ m³; $p_1 = 100$ kPa; $T_1 = 300$ K); $\Rightarrow$ $u_1 = 214.07$ kJ/kg;

State-2: (given $V_2 = V_1$; $m_1 = m_2$; $T_2 = 2000$ K); $\Rightarrow$ $u_2 = 1678.7$ kJ/kg;

For the closed system, $v_2 = \dfrac{V_2}{m_2} = \dfrac{V_1}{m_1} = v_1$.

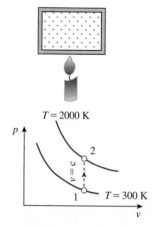

FIGURE 3.63 The constant-volume heating process is shown on a *p-v* diagram (Ex. 3-14) (see Anim. 5.A.*vConstHeatingIG*).

Using the IG equation of state:

$$v_1 = v_2, \quad \Rightarrow \quad p_2 = p_1 \frac{T_2}{T_1} = 100 \frac{2000}{300} = 667 \text{ kPa};$$

The mass of air can be obtained from either state. Using State-1, we obtain:

$$m = m_1 = \frac{V_1}{v_1} = V_1 \frac{p_1}{RT_1} = 10 \frac{100}{0.287 \times 300} = 11.61 \text{ kg}$$

Therefore, $\Delta U = m(u_2 - u_1) = 11.61(1678.7 - 214.07) = 17{,}004 \text{ kJ}.$

TEST Analysis

To verify the answers, launch the IG system-state TESTcalc. Select Air as the working fluid. For State-1, enter p1, T1, and Vol1, and click Calculate. Similarly, calculate State-2 from m2 = m1, Vol2 = Vol1, and T2. Evaluate the change in internal energy from the expression = m1*(u2 − u1) in the I/O panel. Note that Air and Air* used in the working fluid menu use different h_{ref}^o. If you select Air*, h and u from the TESTcalc will match those read from Table D-3. The differences, Δh and Δu do not depend on the reference values and will be the same whether you use Air or Air*.

What-if scenario

The final pressure does not change with the use of the PG model since both PG and IG models obey $pv = RT$. The change in internal energy from the perfect gas model can be calculated using Eq. (3.68) as $\Delta U = mc_v(T_2 - T_1) = 14{,}171 \text{ kJ}.$ The PG system-state TESTcalc can be used to verify this result.

Discussion

The significant difference (16.7%) in the prediction of ΔU by the two models reveals the weakness of the PG model, especially when the temperature variation is large. But the simplicity and the ability to produce algebraic expressions for results make the PG model attractive as an exploratory model for gases.

3.5.5 The RG (Real Gas) Model

By capturing the dependence of specific heats on temperature, the IG model produces better accuracy than the PG model, especially when the temperature change is significant. Both models rely heavily upon the ideal gas equation of state, $pv = RT$, whose validity becomes questionable when a state approaches the vapor dome. While the PC model is well suited for evaluating states in and around the vapor dome, it is useless without the saturation and superheated tables for the working fluid. In addition data in the supercritical region is generally hard to come by.

The real gas model attempts to overcome the limitations of the PC model by generalizing the gas models. While the simplicity of the IG or PG model is preserved, fluid-specific tables are discarded in the RG model; instead, a few generalized charts, independent of the working fluid, take their place. The generality comes at the expense of accuracy.

To understand the RG model, let us begin by comparing the IG model with the more accurate PC model for calculating superheated vapor states (see Anim. 3.D.RGModel). The failure of the ideal gas assumption as we approach the vapor dome can be examined by comparing the specific volume predicted by the IG equations of state, $v^{IG} = RT/p$, with the accurate value obtained from the superheated table. Their ratio should be close to 1 if a vapor can be modeled as an ideal gas:

$$Z \equiv \frac{v}{v^{IG}} = v / \left(\frac{RT}{p} \right) = \frac{pv}{RT} \tag{3.72}$$

For example, the superheated table for Nitrogen is used to plot Z against p with temperature as a parameter in Figure 3.64. Departure from the ideal gas behavior is displayed when Z deviates from its ideal value of one—$Z > 1$ signifies that the gas is more rarefied than the IG model prediction while $Z < 1$ indicates just the opposite. A ratio of thermodynamic

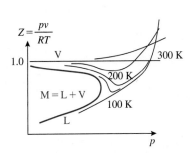

FIGURE 3.64 Compressibility factor Z of nitrogen as a function of p and T (see Anim. 3.D.RGModel).

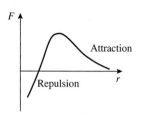

FIGURE 3.65 As r decreases, the attractive force between molecules turns into a repulsive force.

properties, Z is dimensionless and qualifies as a thermodynamic property itself. It is called the **compressibility factor**.

Observe in Figure 3.64 that the Z of Nitrogen vapor at 200 K is close to one at low pressure. It decreases as pressure increases and, finally, becomes greater than 1 at sufficiently high pressure. To understand this behavior, let's consider the microscopic structure once again. Recall from our earlier discussion that the ideal gas model assumes away all intermolecular forces. In reality, the intermolecular force F between two gas molecules, separated by a distance r (Fig. 3.65), is given by the Lennard-Jones equation:

$$F = \frac{a}{r^m} - \frac{b}{r^n}; \quad [\text{kN}], \quad \text{where} \quad a, b > 0 \quad \text{and} \quad n > m \tag{3.73}$$

The two terms, on the right-hand side with opposite signs, represent the attractive and repulsive components of F. With the intermolecular spacing r directly related to molar specific volume $\bar{v}$, a large value of $\bar{v} = \bar{R}T/p$ (caused by high temperature, low pressure, or a combination implies a large r and hence a small F according to Eq. (3.73). The IG assumption becomes justified at high temperatures and/or low pressures. This is confirmed by the behavior of the isotherms in Figure 3.64, which converge to $Z = 1$ for both these limits.

As the spacing r decreases due to an increase in pressure at a given temperature, the first term—it is the attractive component—becomes predominant since $m < n$. An increase in the attractive forces among molecules explains the compaction of molecules indicated by the downward slopes ($Z < 1$) of the isotherms in Figure 3.64. As r decreases further due to continued increase of pressure, a point is reached when the second term—the repulsive force—becomes significant. Thereafter, any further decrease in r causes F to decrease (Fig. 3.65), which causes the isotherms to turn around. When F reaches zero, the ideal gas limit ($Z = 1$) is recovered momentarily; however, any further increase in pressure makes the net force repulsive, and the departure from the ideal gas behavior resumes, this time in the opposite direction ($Z > 1$).

To explore if Z can be used as a universal correction coefficient for the IG equation of state, pressure and temperature are made dimensionless with the critical properties, p_{cr} and T_{cr}, respectively:

$$p_r \equiv \frac{p}{p_{cr}} \quad \text{and} \quad T_r \equiv \frac{T}{T_{cr}} \tag{3.74}$$

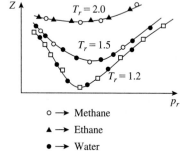

FIGURE 3.66 Universal behavior of Z in non-dimensional coordinates.

These new dimensionless properties, known as the **reduced pressure** and **reduced temperature** (not to be confused with the relative properties with the same symbols used in the IG model in Sec. 3.5.3), are used to re-plot Figure 3.64 along with data for several other fluids in Figure 3.66. The approximate alignment of the compressibility data of different working fluids into a single set of curves is known as the **principle of corresponding states** and the non-dimensional figure, as sketched in Figure 3.66, is called the **generalized compressibility chart**. In a similar manner, enthalpy and entropy of a real gas can be expressed through departure factors that can be correlated to reduced pressure and temperature through two universal charts. These generalized charts are the basis of the RG model.

3.5.6 RG Model Assumptions

The RG model is based upon the following assumptions:

i) p, v, and T are related by the **RG equation of state**:

$$pv = ZRT; \quad \left[\text{kPa}\frac{\text{m}^3}{\text{kg}} = \frac{\text{kJ}}{\text{kg}}\right], \quad \text{where} \quad R = \frac{\bar{R}}{M}; \quad \left[\frac{\text{kJ}}{\text{kg}\cdot\text{K}}\right]; \tag{3.75}$$

and Z, the compressibility factor, can be expressed as:

$$Z = Z(p_r, T_r), \quad \text{where} \quad p_r \equiv \frac{p}{p_{cr}} \quad \text{and} \quad T_r \equiv \frac{T}{T_{cr}} \tag{3.76}$$

ii) The enthalpy and entropy of a real gas can be related to the corresponding IG or PG model values through an **enthalpy departure factor** $Z_h(p_r, T_r)$ and **entropy departure factor** $Z_s(p_r, T_r)$, respectively, which are defined in the next section (see Anim. 3.D.*propertiesRG*). In Chapter 11 it will be established that these departure factors are consequences of the RG equation of state.

3.5.7 Compressibility Charts

An examination of a more detailed version of Figure 3.66, with more species, reveals that the *principle of corresponding states* is only approximately observed. Several approaches have been used to improve the accuracy of the compressibility charts. It is found that fluids whose molecules show spherical symmetry—argon, neon, xenon, and methane for example—collapse into a single plot in a phase diagram plotted on reduced coordinates (see Fig. 3.67). Such fluids are called **simple fluids** and the generalized charts based on this model, known as the **Lee-Kesler charts**, are reproduced in Tables E-2 through E-4 (see *TEST > Property Tables*). Asymmetric molecules such as H_2O deviate significantly from this universal plot. Another set of compressibility charts, derived by averaging data from 30 commonly used fluids, are called the **Nelson-Obert charts** (Tables E-5 through E-7 in *TEST > Property Tables*). Results from these two sets of charts do not always agree, underscoring the approximate nature of the RG model. The procedure for evaluating thermodynamic properties, described below, remains the same for the two sets of charts. Toggle buttons are used in the RG TESTcalcs in TEST to switch between the two types of charts.

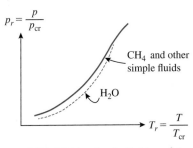

FIGURE 3.67 The simple fluid model produces significant error for fluids with asymmetric molecules such as water.

Specific Volume:

The compressibility charts, Table E-2 (Lee-Kesler ideal fluid model) and Table E-5 (Nelson-Obert averages), plot Z as a function of p_r and T_r at different magnification levels. For given values of p and T, p_r and T_r can be evaluated using critical property data from Figure E-1. Once Z is obtained from the compressibility chart, the real gas equation of state can be applied to evaluate v. In the Nelson-Obert chart lines of constant v_r, a **pseudo-reduced specific volume**, are defined as

$$v_r \equiv \frac{v}{v_{cr}^{IG}} = \frac{v}{(RT_{cr}/p_{cr})} \tag{3.77}$$

which is plotted to help evaluate the state if the specific volume is one of the known properties. It is called a *pseudo*-reduced property, because unlike the reduced properties p_r or T_r, a critical property is not used in the denominator of v_r.

Enthalpy:

In the development of the enthalpy equation for an ideal gas, the differential thermodynamic relation, Eq. (3.5), was integrated using the IG equation of state. A parallel development (to be carried out in Chapter 11) using the RG equation of state, Eq. (3.75), relates enthalpy h^{RG} and the corresponding ideal gas limit h^{IG} in terms of a dimensionless **enthalpy departure factor** Z_h as:

$$Z_h \equiv \frac{h^{IG} - h^{RG}}{RT_{cr}}, \quad \text{where} \quad Z_h = Z_h(p_r, T_r) \tag{3.78}$$

Therefore,

$$\Delta h^{RG} = h_2^{RG} - h_2^{RG} = (h_2 - h_1)^{IG} - RT_{cr}(Z_{h,2} - Z_{h,1})$$

$$\Rightarrow \Delta h^{RG} = \Delta h^{IG} - RT_{cr}\Delta Z_h; \quad \left[\frac{kJ}{kg}\right] \tag{3.79}$$

The ideal gas tables, Tables D-3 through D-15, are used to evaluate the first term, the IG contribution, while the generalized enthalpy departure chart, Table E-3 (Lee-Kesler) or E-6 (Nelson-Obert), is used to evaluate the real-gas correction. Notice how any error in the reading of Z_h is magnified by the factor T_{cr}. If the ideal gas table is unavailable, Δh^{IG} can be replaced with the corresponding PG expression $\Delta h^{PG} = c_p\Delta T$ without losing much accuracy—Redundant, which I believe you were going for. However, the end of the sentence does feel a bit odd to me, though. Almost as if you have forgotten what you just said. Perhaps add "once again" to the phrase after the dash. Something along the lines of: "provided, once again, that the temperature variation is not large." This shows recognition of the reiteration, and I believe that it places more emphasis on the point than it currently does.

Internal Energy:

To evaluate a change in u^{RG} for a real gas, the corresponding change in enthalpy is first evaluated. Then application of the relation $h = u + pv = u + ZRT$ produces:

$$\Delta u^{RG} = u_2^{RG} - u_1^{RG} = \Delta h^{RG} - (p_2 v_2 - p_1 v_1) = \Delta h^{RG} - R(Z_2 T_2 - Z_1 T_1); \quad \left[\frac{kJ}{kg}\right] \quad \textbf{(3.80)}$$

where Z is read from the compressibility chart.

Entropy:

Like Eq. (3.79) for enthalpy, Δs^{RG} for a real gas can be expressed (details to be worked out in Chapter 11) as the sum of contributions from the ideal gas model and a correction expressed in terms of an **entropy departure factor**, defined as:

$$Z_s \equiv \frac{s^{IG} - s^{RG}}{R}, \quad \text{where} \quad Z_s = Z_s(p_r, T_r) \quad \textbf{(3.81)}$$

Therefore,

$$\Delta s^{RG} = s_2^{RG} - s_1^{RG} = (s_2 - s_1)^{IG} - R(Z_{s,2} - Z_{s,1}) = \Delta s^{IG} - R\Delta Z_s; \quad \left[\frac{kJ}{kg \cdot K}\right] \quad \textbf{(3.82)}$$

where Δs^{IG} is obtained from the ideal gas model and ΔZ_s from the entropy departure chart, Figure E-4 (Lee-Kesler) or E-7 (Nelson-Obert). Often, Δs^{IG} is replaced with Δs^{PG} for simplicity.

3.5.8 Other Equations of State

The RG equation of state preserves the simplicity of the ideal gas equation by transferring the complexities into the compressibility factor Z, which cannot be expressed in a functional form. A complete functional form for the equation of state allows many analytical advantages, which will be explored in developing property relations in Chapter 11.

Historically, the **van der Waals equation** of state was the first successful attempt to improve the ideal gas equation:

$$\left(p + \frac{a}{v^2}\right)(v - b) = RT \left[\frac{kJ}{kg}\right], \quad \text{where} \quad a = \frac{27R^2 T_{cr}^2}{64 p_{cr}} \left[\frac{kPa \cdot m^6}{kg^2}\right]; \quad \text{and} \quad b = \frac{RT}{8 p_{cr}} [m^3] \quad \textbf{(3.83)}$$

The term a/v^2 compensates for the attractive intermolecular forces and b accounts for the volume occupied by the molecules themselves in a real gas. The two constants, expressed in terms of the critical properties in Eq. (3.83), are based on experimental information about a single state—the critical state—for a given gas; therefore, the accuracy of the equation is not sufficiently high. The Dieterici, Redlich-Kwong, Beatie-Bridgeman, and Benedict-Webb-Rubin (BWR) equations are a series of refinements to the van der Waals equation of state. Of these, the **BWR equation** is the most accurate, and is expressed in terms of eight empirical constants:

$$p = \frac{\overline{R}T}{\overline{v}} + \left(B_0 \overline{R}T - A_o - \frac{C_0}{T^2}\right)\frac{1}{\overline{v}^2} + \frac{b\overline{R}T - a}{\overline{v}^3} + \frac{a\alpha}{\overline{v}^6} + \frac{c}{\overline{v}^3 T^2}\left(1 + \frac{\gamma}{\overline{v}^2}\right)e^{-\gamma/\overline{v}^2}; \quad [kPa] \quad \textbf{(3.84)}$$

where the constants for different gases are listed in Table E-8. The BWR equation and its modified forms—the Strobridge equation with sixteen constants and Lee-Kesler equation with twelve constants—are frequently used.

An entirely different approach is to use the kinetic theory of microscopic thermodynamics to express Z as a power series known as a **virial expansion** [the word virial originates from the Latin word for force (in this case, intermolecular force)]:

$$Z = \frac{p\overline{v}}{\overline{R}T} = 1 + \frac{B(T)}{\overline{v}} + \frac{C(T)}{\overline{v}^2} + \frac{D(T)}{\overline{v}^3} + \cdots \quad \textbf{(3.85)}$$

where the temperature-dependent virial coefficients $B(T)$, $C(T)$, and $D(T)$ introduce corrections to the ideal gas equation.

3.5.9 Model Summary: RG Model

Frequently used formulas for manual evaluation of thermodynamic state using the RG model can be summarized (see Anim. 3.D.*propertiesRG*) as follows:

The RG model generalizes the IG model by using the compressibility factor Z, enthalpy departure factor Z_h, and entropy departure factor Z_s, which are correlated to reduced properties p_r and T_r (defined below) through generalized charts.

The real gas equation of state is written as:

$$pv = ZRT, \quad \text{where} \quad Z = Z(p_r, T_r), p_r \equiv \frac{p}{p_{cr}}, \quad \text{and} \quad T_r \equiv \frac{T}{T_{cr}} \tag{3.86}$$

Critical properties can be found in Table E-1, while charts for Z can be found (see *TEST > Property Tables*) in Figs. E-2 (Lee-Kesler chart) and E-5 (Nelson-Obert chart) as functions of p_r and T_r.

In the real gas model, $u = u(p,T)$, $h = h(p,T)$, and $s = s(p,T)$. Changes in enthalpy and entropy are calculated from:

$$\Delta h^{RG} = \Delta h^{IG} - RT_{cr}\Delta Z_h; \quad \left[\frac{kJ}{kg}\right] \quad \text{and} \quad \Delta s^{RG} = \Delta s^{IG} - R\Delta Z_s; \quad \left[\frac{kJ}{kg \cdot K}\right] \tag{3.87}$$

where charts for Z_h and Z_s can be found in Tables E-3 and E-4 (Lee-Kesler), and Tables E-6 and E-7 (Nelson-Obert) for given values of p_r and T_r (see *TEST > Property Tables*). Change in u can be related to Δh by:

$$\Delta u^{RG} = \Delta h^{RG} - (p_2 v_2 - p_1 v_1) = \Delta h^{RG} - R(Z_2 T_2 - Z_1 T_1); \quad \left[\frac{kJ}{kg}\right] \tag{3.88}$$

EXAMPLE 3-15 · State Evaluation Using the RG Model

A 10-m^3 volume tank contains nitrogen at a pressure of 3 MPa and a temperature of 125 K (Fig. 3.68). Determine the mass of nitrogen in the tank using the (a) ideal gas, (b) Lee-Kesler real gas, and (c) Nelson-Obert real gas models. *What-if scenario:* (d) What would the answers be if the PC model had been used instead?

FIGURE 3.68 Nitrogen can be treated as a perfect gas, an ideal gas, a real gas, or even a PC fluid (Ex. 3-15).

SOLUTION

Use the IG and RG equations of state to determine the system's mass.

Assumptions

Nitrogen is in thermodynamic equilibrium.

Analysis

From Table E-1, obtain the necessary material properties for nitrogen: $\overline{M} = 28$ kg/kmol, $p_{cr} = 3.39$ MPa, $T_{cr} = 126.2$ K, and $R = 8.314/28 = 0.297$ kJ/kg $\cdot$ K.

State-1: (given $V_1 = 10$ m^3; $p_1 = 3$ MPa; $T_1 = 125$ K)

From Eq. (3.86):

$$p_{r1} = 3/3.39 = 0.885 \quad \text{and} \quad T_{r1} = 125/126.2 = 0.99$$

From the Lee-Kesler compressibility chart, Figure E.2 (see TEST > *Property Tables*), $Z_1^{L-K} = 0.46$.
The IG equation of state produces:

$$v_1^{IG} = \frac{RT_1}{p_1} = \frac{0.297 \times 125}{3000} = 0.0124 \frac{m^3}{kg}$$

Therefore, $\quad m_1^{IG} = \dfrac{V_1}{v_1^{IG}} = \dfrac{10}{0.0124} = 806$ kg

The real gas equation, Eq. (3.86), produces:

$$v_1^{RG,\,L-K} = Z_1^{L-K} V_1^{IG} = 0.46 \times 0.0124 = 0.0057\,\frac{m^3}{kg}$$

Hence, $\quad m_1^{RG,\,L-K} = \dfrac{V_1}{v_1^{RG,\,L-K}} = \dfrac{10}{0.0057} = 1754\;kg$

From Nelson-Obert chart, Figure E-4, obtain $Z_1^{N-O} \cong 0.5$

Therefore, $\quad m_1^{RG,\,N-O} = \dfrac{V_1}{v_1^{RG,N-O}} = \dfrac{V_1}{Z_1^{N-O} v_1^{IG}} = \dfrac{10}{0.5 \times 0.0124} = 1613\;kg$

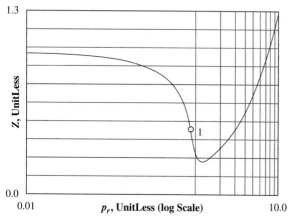

FIGURE 3.69 The calculated state is plotted on a Z-p_r diagram (Ex. 3-15) along with the isotherm passing through it in the plot panel of the RG system-state TESTcalc.

TEST Analysis

To verify the answers, launch the RG system-state TESTcalc. Select Nitrogen as the working fluid. Enter p1, T1, Vol1, and click Calculate. The mass in the tank is calculated as 1741 kg with the Lee-Kesler charts and 1704 kg with Nelson-Obert charts. To visualize the compressibility effect, plot the calculated state on a Z-p_r diagram by selecting it from the plot menu. Resize the plot by using the zoom feature or entering larger limits on the x axis. You can see how Z changes with p_r (Fig. 3.69) by tracing the graph with your mouse.

What-if scenario

Launch the PC system-state TESTcalc, and select nitrogen as the working fluid. Enter p1, T1, Vol1, and click Calculate. Alternatively, copy the TEST-code generated by the RG model into the I/O panel of the PC TESTcalc and click Load. The mass in the tank is calculated as 1636 kg.

Discussion

Being the most accurate model of all, the PC model serves as the benchmark in this problem. Given the high pressure and low temperature, the IG model is not applicable. Despite the discrepancy between the Lee-Kestler and Nelson-Obert charts, which underscores the approximate nature of these submodels, the RG model can be seen as an acceptable alternative to the PC model when the saturation and superheated tables are not available for a given working fluid.

3.6 MIXTURE MODELS

A mixture of gases can be treated as a pure substance as long as it has the same chemical composition everywhere (click Mixture in Anim. 1.B.*massVsMole*). We will introduce several mixture models in Chapter 11 to deal with general and binary gas mixtures. Air, being a frozen mixture of nitrogen and oxygen, is treated like a *pure gas*. However, moist air, is a special mixture of dry air and variable amounts of water vapor. An entire chapter (Chapter 12) is devoted to its analysis.

Mixtures of PC fluids are outside the scope of this book, although TEST offers a large selection of refrigerant mixtures (the suffix % after the refrigerant name indicates a mixture) under the PC-model category.

3.6.1 Vacuum

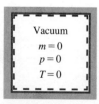

FIGURE 3.70 Absolute pressure, absolute temperature, and mass are all zero in a vacuum.

If the definition of a system is any entity with a well-defined boundary, then a vacuum is a legitimate system (see Anim. 5.B.*suddenExpansion*). Properties of such a system may seem irrelevant given that the system has no mass, however, a vacuum (Fig. 3.70) may have a finite volume and stored exergy (useful work can be extracted by allowing outside air to enter the system). A vacuum is specified in the TESTcalcs by setting the system mass to zero and by setting either the absolute pressure or absolute temperature to zero. Sometimes it may be necessary to use very small values of mass and pressure and to obtain limiting values of quantities, such as stored exergy $\Phi = m\phi$ (setting $m = 0$ would result in a 0/0 situation).

3.7 STANDARD REFERENCE STATE AND REFERENCE VALUES

As mentioned earlier, absolute values of internal energy and enthalpy depend on the choice of their reference values. Even entropy is sometimes given an arbitrary reference value to simplify the development of property tables. In the steam table, for example, zero values for internal energy and entropy are assigned to saturated liquid water at the triple-point temperature of 0.01°C. For refrigerants, R-134a, R-12, etc., enthalpy and entropy are assigned zero values at the saturated liquid state of −40°C, the temperature where the Celsius and Fahrenheit scales coincide.

For many substances, reference properties are set at the standard condition for temperature and pressure, which is referred to as standard state or abbreviated as STP (standard temperature and pressure). Several sets of such standards exist; however, in this book, standard reference state, or *standard state*, will mean state evaluated at $p = p_0 = 101.325$ kPa (1 atm) and $T = T_0 = 298.15$ K (25°C). In various TEST models, standard values of enthalpy are assigned to each substance at the standard state (more on this in Sec. 13.3.1); however, entropy is correctly assigned a zero value at a temperature of absolute zero (as dictated by the third law of thermodynamics).

Note that both h_{ref} and $u_{\text{ref}}^{\text{o}}$ cannot be arbitrarily assigned as they are related through $h = u + pv$. PG, IG, and SL models can be related as follows:

$$\text{PG/IG Models: } u_{\text{ref}}^{\text{o}} = h_{\text{ref}}^{\text{o}} - RT_{\text{ref}}; \quad \text{SL model: } \quad u_{\text{ref}}^{\text{o}} = h_{\text{ref}}^{\text{o}} - p_{\text{ref}}v; \quad \left[\frac{\text{kJ}}{\text{kg}}\right] \quad \textbf{(3.89)}$$

We must stress in most problems, it is the differences Δu, Δh, and Δs that appear in the analysis. If the same consistent set of reference values is used, the absolute values do not matter. But if two different models or tables from two different sources are used in conjunction, the difference in reference values, if any, must be accounted for. A failure to recognize differences in reference values (Fig. 3.71) can produce unexpected results which are difficult to detect.

3.8 SELECTION OF A MODEL

We have introduced five different models (Fig. 3.72) in this chapter for determining thermodynamic states of pure substances. Selection of a model for a particular substance is a trade-off between accuracy and simplicity, and is predicated by the availability of data. Consider the large selection of models that may be applicable to H_2O: the SL model for ice or liquid water; PC model for liquid, vapor, or saturated mixtures; PG or IG model for superheated vapor; and RG model for supercritical vapor. In such situations, the right choice depends on the analyst's experience and some general guidelines.

If there is no possibility of a phase transformation, the SL model can be applied to solids and pure liquids. For liquids, two other alternatives are available. If the temperature variation is substantial, the possibility of a phase change increases and it is safer to use the compressed liquid (CL) sub-model of the PC model. When available, the compressed liquid tables can be used for

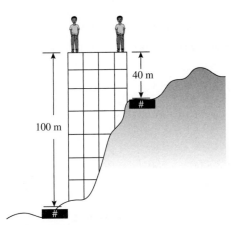

FIGURE 3.71 Elevations of the two observers depend on the reference elevation.

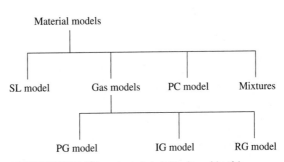

FIGURE 3.72 Different models introduced in this chapter. Mixture models will be discussed in Chapters 11 and 12.

better accuracy, especially in the super-critical liquid (SCL) region where the compressibility effect is relatively more severe.

With its tabulated data, the PC model is the most accurate for a saturated liquid, saturated mixture of liquid and vapor, saturated and superheated vapor, and even super-critical vapor. Error stemming from linear interpolation (or eye estimates) is generally well within acceptable limits. Superheated vapor under certain conditions can be treated by one of the simpler gas models with acceptable sacrifice in accuracy. Keeping in mind that this is only a guideline, the **ideal gas criteria** can be established in conjunction with the compressibility charts:

$$\text{a. } p_r < 0.05; \quad \text{or,} \quad \text{b. } T_r > 15; \quad \text{or,} \quad \text{c. } p_r < 10 \quad \text{and} \quad T_r > 2; \qquad \textbf{(3.90)}$$

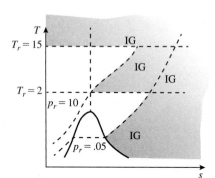

Any of these three conditions drives the value of Z close to one (Figs. 3.66 or E.2) so that the real gas equation of state reduces to the IG equation of state. On a T-s diagram, this can be translated into approximate regions—low-pressure high-temperature zones of the superheated region—as shown by the shaded area of Fig. 3.73.

The IG model can be further simplified into the PG model if the variation of c_p with temperature can be neglected. For small changes in temperature, acceptable results can be obtained if c_p is evaluated at an average temperature. For monatomic gases—helium, neon, argon, xenon, radon, and krypton—c_p is a material property and the IG model reduces to the PG model.

The RG model, which can be looked upon as a generalized IG model, can be used to model not only gases, but also super-critical vapor or liquid, and liquid-vapor mixtures. It is normally used for substances for which the PC tables are not available.

FIGURE 3.73 Different zones in the superheated vapor region where the IG approximation works well.

EXAMPLE 3-16   Comparison of Different Models

Superheated steam at a pressure of 500 kPa and a temperature of 450°C undergoes an isothermal process to a final pressure of 20 MPa. Employing the Lee-Kesler RG model, determine the change in (a) specific volume and (b) enthalpy. *What-if scenario:* What would the answers be if the (c) PC, (d) IG, or (e) PG models were used?

SOLUTION

The problem description is quite similar to that of Ex. 3-13. Here we add the RG model to the comparative study.

Assumptions

Equilibrium at each state.

Analysis

From Table E-1 (in an RG TESTcalc use p_r = 1 or T_r = 1), obtain $p_{cr} = 22.09$ MPa and $T_{cr} = 647.3$ K, Mbar $= 18.015$ kg/kmol, and R $= 8.314/18.015 = 0.462$ kJ/kg · K.

State-1:   (given $p_1 = 500$ kPa, $T_1 = 450°C = 723$ K, $p_{r1} = 0.5/22.09 = 0.023$, $T_{r1} = 723/647.3 = 1.117$)
From Figures E.2 and E.3 (see *TEST > Property Tables*); $Z_1 = 0.99$, $Z_{h1} = 0.0$.
State-2:   (given $p_1 = 20$ MPa, $T_2 = T_1$, $p_{r2} = 20/22.09 = 0.90$, $p_{r2} = 20/22.09 = 0.90$, $T_{r2} = T_{r1} = 1.117$)

From Figures E.2 and E.3; $Z_2 = 0.75$, $Z_{h2} = 0.9$.

The real gas equation, Eq. (3.86), produces:

$$\Delta v^{RG} = \frac{Z_2 R T_2}{p_2} - \frac{Z_1 R T_1}{p_1} = (0.462)(723)\left(\frac{0.75}{20000} - \frac{0.99}{500}\right) = -0.649 \frac{m^3}{kg}$$

From Eq. (3.87):

$$\Delta h^{RG} = \Delta h^{IG0} - R T_{cr} \Delta Z_h = -(0.462)(647.3)(0.9) = -269.1 \text{ kJ/kg}$$

Here the ideal gas contribution is zero because enthalpy is a function of temperature only according to the IG model.

TEST Analysis

Launch the RG system-state TESTcalc. Select H2O as the working fluid. Calculate State-1 for the given p1 and T1, and State-2 for the given p2 and T2. Use the I/O panel as a calculator to evaluate v2-v1 as -0.652 m^3/kg and h2-h1 as -274.5 kJ/kg.

What-if scenario

Click Super-Calculate to produce the TEST-code in the I/O panel. Launch the IG system-state TESTcalc in a separate tab and paste the TEST-code into its I/O panel, then click Load to generate the solution. Calculate the desired differences in the I/O panel. Repeat with the PC model. The results are summarized as follows:

	Δv (m^3/kg)	Δh (kJ/kg)
RG Model	−0.652	−274.5
IG Model	−0.651	0.0
PC Model	−0.651	−317.8

Discussion

All three models accurately agree in the prediction of volume change; however, enthalpy changes are quite different. First the IG model completely misses the effect of pressure. Second, the RG model shows a discrepancy of 14% when compared to the benchmark results from the PC model in predicting Δh. Finally, the RG model's advantage can be appreciated when we realize how little extra information—p_{cr}, T_{cr}, and $\overline{M}$—is necessary to model a new substance if the generalized charts and the IG table are available.

3.9 CLOSURE

In this chapter, we grouped different pure substances into a number of material models for systematic evaluation of properties. Having established the relation between an extended state with the core thermodynamic state in Chapter 1, we introduced the principle of *local thermodynamic equilibrium* (LTE) and the *state postulate* as the foundation for describing an extended system or flow state. Differential thermodynamic relations, known as *Gibbs equations*, were developed based on the state postulate. The rest of the chapter was devoted to evaluation of the core thermodynamic state composed of the property set p, T, v, u, h, and s for various working substances. Several material models—SL, PC, IG, PG, and RG models—were formulated based on the set of assumptions linked to the microscopic behavior of different groups of working substances. States evaluated through hand calculations were verified through the use of state TESTcalcs. The graphical capabilities of TESTcalcs to draw constant-property lines on a thermodynamic plot were frequently used to visualize relationships among calculated states.

PROBLEMS

SECTION 3-1: MATERIAL-INDEPENDENT PROPERTY RELATIONS

3-1-1   The enthalpy (H) of an ideal gas is a function of temperature (T) only, as can be seen from Table D-3 for air. (a) Using the data from the table, determine the specific heat at constant pressure (c_p) of air at 400 K. (b) **What-if scenario:** What would c_p be if the temperature was 800 K?

3-1-2   Use the PC system-state TESTcalc to evaluate (a) c_p and (b) c_v of steam at 100 kPa and 150°C. Assume a 1°C difference between the neighboring states. **What-if scenario:** What would the (c) c_p and (d) c_v be if the pressure was 300 kPa?

3-1-3   The c_p of a working substance is known to be 2.0 kJ/kg·K at 500 K. A software is used to simulate the heating of the substance at constant pressure (c_p). It produces the following behavior: At 500 K, it produces an enthalpy (H) of 2000 kJ/kg, while at 501 K, it produces a value of 2005 kJ/kg. Do you trust this software? (Yes:1; No:0) If not, why?

$$c_p \equiv \left(\frac{\partial h}{\partial T} \right)_p$$

3-1-4   The temperature of a gas is related to its absolute pressure and specific volume through $T = 3.488pv$, where T, p, and v are expressed in their standard SI units. Consider a gas at 100 kPa and 1 m³/kg. (a) Estimate the change in temperature (ΔT) using Taylor's theorem if the state of the gas changes to 101 kPa and 1.01 m³/kg. (b) Compare your estimate with the exact change in temperature.

3-1-5   The temperature of a gas is related to its absolute pressure and specific volume through $T = 3.488pv$, where T, p and v are expressed in their standard SI units. Consider a gas at 350 K and 1 m³/kg. (a) Estimate the change in pressure (Δp) using Taylor's theorem if the state of the gas changes to 355 K and 1.01 m³/kg. (b) Compare your estimate with the exact change in temperature.

$$dz = \left(\frac{\partial z}{\partial x} \right)_y dx + \left(\frac{\partial z}{\partial y} \right)_x dy$$

3-1-6   The second T-ds relation reduces to $Tds = dh$ at constant pressure (c_p). Use the superheated table of water (Table B-3) to evaluate the two sides of this equation at (a) 100 kPa, 200°C and (b) 50 MPa, 500°C. Calculate for $dT = 10$°C. (c) **What-if scenario:** What would the answer in part (b) be if the PC system-state TESTcalc was used with $dT = 10$°C?

3-1-7   The first T-ds relation reduces to $Tds = pdv$ if the internal energy remains constant. Use the PC system-state TESTcalc to evaluate the state of steam at 100 kPa and 200°C. Also evaluate two neighboring states at 99 kPa and 101 kPa, holding internal energy constant and compare the two sides of the equation. Why don't the two sides match exactly?

3-1-8   Using the PC state TESTcalc, calculate from first principle (a) c_p and (b) c_v of steam at 100 kPa and 320°C. Use a 1°C difference in setting up the neighboring states. (c) What is the ratio of $(c_p - c_v)/R$ for steam at this state? (d) **What-if scenario:** What would be the answer in part (c) if the steam is at 10 MPa, 320°C?

SECTION 3-2: SL (SOLID/LIQUID) MODEL

3-2-1   Determine the changes in (a) specific enthalpy (Δh) and (b) specific entropy (Δs) if the pressure of liquid water is increased from 100 kPa to 1 MPa at a constant temperature of 25°C. Use the SL model for water.

3-2-2   A cup of coffee of volume 0.3 L is heated from a temperature of 25°C to 60°C at a pressure of 100 kPa. Determine the change in the (a) internal energy (ΔU), (b) enthalpy (ΔH) and (c) entropy (ΔS). Assume the density (ρ) and specific heat of coffee to be 1100 kg/m³ and 4.1 kJ/kg · K. Employ the SL model. (d) **What-if scenario:** How would the answers change if the heating were done inside a chamber pressurized at 1 MPa? (1: increase; −1: decrease; 0: remain same)

3-2-3   A block of solid with a mass of 10 kg is heated from 25°C to 200°C. If the change in the specific internal energy (Δu) is found to be 67.55 kJ/kg, identify the material (aluminum: 1; copper: 2; iron: 3; wood: 4; sand: 5).

3-2-4   A system contains an unknown solid of unknown mass, which is heated from 300 K to 900 K at 100 kPa. The change in internal energy (ΔU) of the system is measured as 61.02 MJ. Using the SL model, determine the change in entropy (ΔS) of the system.

3-2-5   A block of aluminum with a mass of 10 kg is heated from 25°C to 200°C. Determine (a) the total change in internal energy (ΔU) and (b) entropy (ΔS) of the block. (c) *What-if scenario:* What would the change in entropy be if the block was made of copper?

3-2-6   A 2 kg block of aluminum at 600°C is dropped into a cooling tank. If the final temperature (T_2) at equilibrium is 25°C, determine (a) the change in internal energy (ΔU) and (b) the change in entropy (ΔS) of the block as the system. Use the SL model for aluminum ($c_v = 0.902$ kJ/kg-K).

3-2-7   A 3 kg block of iron at 800°C is dropped into 50 kg of water in an insulated cooling tank. If the final temperature (T_2) at equilibrium is 29.9°C, determine (a) the change in internal energy (ΔU) and (b) the change in entropy (ΔS) of the block and water as the system. Use the SL model for aluminum ($c_v = 0.45$ kJ/kg-K) and water ($c_v = 4.184$ kJ/kg-K).

3-2-8   The heat transfer necessary to raise the temperature of a constant-volume closed system is given by $Q = \Delta U$. Using the SL model, compare the heat necessary to raise the temperature by 10°C for such a system of 1 kg and composed of (a) liquid water, (b) liquid ethanol and (c) crude oil.

3-2-9   Repeat the above problem for the following solids as the working substance: (a) gold, (b) iron, (c) sand and (d) granite.

3-2-10   A block of iron, which has a volume of $1\,m^3$, undergoes the following change of state. State-1: $p_1 = 100$ kPa, $T_1 = 20$°C, $V_1 = 0$, $z_1 = 0$; State-2: $p_2 = 500$ kPa, $T_2 = 30$°C, $V_2 = 30$ m/s, $z_2 = 100$ m. Determine (a) ΔE, (b) ΔU, (c) ΔH and (d) ΔS.

3-2-11   A thermal storage, made of a granite rock bed of $10\,m^3$, which is heated to 425 K using solar energy. A heat engine receives heat from the bed and rejects the waste heat to the ambient surroundings at 290 K. During the process, the rock bed cools down. As it reaches 290 K, the engine stops working. The heat transfer from the rock is given as $Q_H = \Delta E$. Determine (a) the magnitude of the heat transfer (Q_H) and the maximum thermal efficiency (η_{th}) at (b) the beginning and (c) the end of the process.

3-2-12   A copper block, with a mass of 5 kg, is initially at equilibrium with the surroundings at 30°C and 100 kPa. It is placed in a pressurized chamber with a pressure of 20 MPa and a temperature of 200°C. Determine (a) the change in the internal energy (ΔU), (b) enthalpy (ΔH) and (c) entropy (ΔS) of the block after it comes to a new equilibrium. (d) *What-if scenario:* What would the change in internal energy be if the block was made of silver?

3-2-13   A 20 kg block of aluminum at 97.4°C is dropped into a tank containing 10 kg of water at 1°C. If the final temperature after equilibrium is 30°C, determine (a) ΔU and (b) ΔS for the combined system of aluminum and water after the process is complete.

3-2-14   A 20-kg block of iron (specific heat 0.45 kJ/kg·K) is heated by conduction at a rate of 1 kW. Assuming the block to be uniform at all time, determine the rate of change of temperature (dT/dt).

Fe Block

3-2-15   A copper bullet, with a mass of 0.1 kg, traveling at 400 m/s, hits a copper block, with a mass of 2 kg at rest, and becomes embedded. The combined system moves with a velocity of 19.05 m/s, in accordance with the conservation of momentum principle. Both the bullet and copper block are at 25°C initially. Assuming the stored energy of the combined system to remain constant during the collision, determine (a) the rise in temperature (ΔT) and (b) entropy (ΔS) change of the combined system. Assume the system to achieve a uniform temperature quickly after the collision and neglect any change in potential energy.

3-2-16   A cup of coffee cools down by transferring heat to the surroundings at a rate of 1 kW. If the mass of the coffee is 0.2 kg and coffee can be modeled as water, determine the rate of change of temperature (dT/dt) of coffee.

3-2-17   A pump raises the pressure of liquid water from 50 kPa to 5000 kPa in an isentropic manner. Determine (a) the change in temperature (ΔT) and (b) specific enthalpy (Δh) between the inlet and exit.

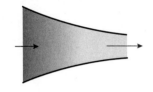

3-2-18 Water flows through an adiabatic pumping system at a steady flow rate ($\dot{m}$) of 5 kg/s. The conditions at the inlet are $p_1 = 90$ kPa, $T_1 = 15°C$, and $z_1 = 0$ m. The conditions at the exit are $p_2 = 500$ kPa, $T_2 = 17°C$ and $z_2 = 200$ m. (a) Simplify the energy equation to derive an expression for the pumping power. (b) Use the SL model to evaluate the pumping power ($\dot{W}_{net}$). Neglect any change in kinetic energy. *What-if scenario:* (c) What is the pumping power if the exit temperature is 18°C?

3-2-19 Oil ($c_v = 1.8$ kJ/kg-K, $\rho = 910$ kg/m^3) flows steadily through a long, insulated, constant-diameter pipe. The conditions at the inlet are $p = 3000$ kPa, $T = 20°C$, $V = 20$ m/s, and $z = 100$ m. The conditions at the exit are $p = 2000$ kPa and $z = 0$ m. (a) Evaluate the velocity (V) at the exit. (b) Determine the exit temperature (T_2).

3-2-20 Water flows steadily through a device at a flow rate of 20 kg/s. At the inlet, the conditions are 200 kPa and 10°C. At the exit, the conditions are 2000 kPa and 50°C. (a) Determine the difference between the entropy (S) transported by the flow at the exit and at the inlet. (b) What are the possible reasons behind the increase in entropy transport?

3-2-21 In an isentropic nozzle, operating at steady state, the specific flow energy (j) and specific entropy (s) remain constant along the flow. The following properties are known at the inlet and exit ports of an isentropic nozzle discharging water at a steady rate of 2 kg/s. Inlet: $p_1 = 300$ kPa, $A_1 = 4$ cm^2; Exit: $p_2 = 100$ kPa. Determine (a) the exit velocity (V_2) and (b) the exit area. Use the SL model for liquid water. (c) *What-if scenario:* What is the exit velocity if the inlet kinetic energy is neglected?

3-2-22 For copper, plot how the internal energy (U), and entropy (S), vary with T within the range 25°C–1000°C. Use the SL system-state TESTcalc.

3-2-23 For liquid water, plot how the internal energy (U), and entropy (S), vary with T within the range 25°C–100°C. Use the SL system-state TESTcalc.

SECTION 3-3: PC (PHASE CHANGE) MODEL

3-3-1 A cylinder contains H_2O only. The outside temperature is 298 K. To analytically determine the pressure inside, you assume that the temperature of H_2O must also be 298 K. You shake the cylinder and realize that it is only partially filled with liquid water. You consult a chart and come up with the answer. What is the pressure (in kPa)?

3-3-2 Saturated liquid water flows through a pipe at 10 MPa. Assuming thermodynamic equilibrium to exist at any given cross-section, (a) determine the temperature (T) and (b) the quality of water. A small leak develops and as water jets out, it quickly equilibrates with the outside pressure of 100 kPa. Some of the water evaporates and the composition of the jet is observed as a mixture of saturated liquid and vapor. Determine (c) the temperature of the jet in °C.

3-3-3 A 30-cm diameter pipe carries H_2O at a rate of 10 kg/s. At a certain cross-section, steam is found to be saturated vapor at 200°C. Determine (a) the pressure (kPa) and (b) velocity (m/s).

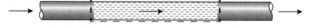

3-3-4 For H_2O, locate (qualitatively) the following states on a T-s and a p-v diagram. State-1: $p = 100$ kPa, $T = 50°C$; State-2: $p = 5$ kPa, $T = 50°C$; State-3: $p = 500$ kPa, $x = 50\%$.

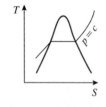

3-3-5 For H_2O, locate (qualitatively) the following states on a T-s and a p-v diagram. State-1: $p = 10$ kPa, saturated liquid; State-2: $p = 1$ MPa, $s = s_1$; State-3: $p = p_2$, $T = 500°C$ State-4: $p = p_1$, saturated vapor.

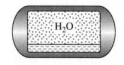

3-3-6 A sealed rigid tank contains saturated steam at 100 kPa. As the tank cools to the temperature of the surrounding atmosphere, the quality of the steam drops to 5%. Using a T-s diagram, explain why the pressure in the tank must decrease drastically to satisfy thermodynamic equilibrium.

3-3-7 An isentropic compressor is used to raise the pressure of a refrigerant entering the compressor as saturated vapor. Using a *T-s* diagram, explain why the temperature at the exit can be expected to be higher than that at the inlet.

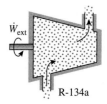

R-134a

3-3-8 A pipe carries saturated liquid water at a pressure of 500 kPa. Some water squirts out from the pipe through a small leak. As the water is expelled, it quickly achieves mechanical equilibrium with the atmosphere at 100 kPa. (a) Estimate the temperature of water inside and outside the pipe. *What-if scenario:* What would the answers be if the fluid were (b) R-134a or (c) R-12 instead?

3-3-9 An insulated piston inside an insulated rigid cylinder, closed at both ends, creates two chambers: one containing a two-phase mixture of H_2O and another containing a two-phase mixture of R-134a. If the temperature inside the chamber containing H_2O is 85.9°C, determine the temperature inside the other chamber. Assume mechanical equilibrium.

3-3-10 Determine the boiling temperature of water (a) at sea level and (b) on Mount Everest (elevation 8,848 m). Use Table H-3 to look up the pressure of atmosphere at different altitudes.

3-3-11 A vertical piston-cylinder assembly contains water. The piston has a mass of 2 kg and a diameter of 10 cm. Determine the vertical force necessary on the piston to ensure that water inside the cylinder boils at (a) 120°C or (b) 80°C. Assume atmospheric pressure to be 101 kPa. (c) *What-if scenario:* What would the answer, in part, (a) be if the piston mass was neglected?

3-3-12 A vertical piston-cylinder assembly contains a saturated mixture of water at 120°C and a gauge pressure of 108.5 kPa. The piston has a mass of 5 kg and a diameter of 12 cm. Determine (a) the atmospheric pressure outside and (b) the external force exerted on the piston to maintain a constant pressure.

3-3-13 A cooking pan, with an inner diameter of 20 cm, is filled with water. It is covered with a lid that has a mass of 5 kg. If the atmospheric pressure is 100 kPa, determine (a) the boiling temperature of water. (b) *What-if scenario:* What would the boiling temperature be if a 5 kg block were placed on top of the lid?

3-3-14 A heat engine cycle is executed with ammonia in the saturation dome. The pressure of ammonia is 1.5 MPa during heat addition and 0.6 MPa during heat rejection. What is the highest possible thermal efficiency? Based on the temperatures of heat addition and rejection, could you comment on possible application of such a low-efficiency cycle?

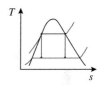

3-3-15 Plot how the saturation temperature of water increases with pressure. Use the full range from the triple point to critical point.

3-3-16 Plot the phase diagram (*p-T*) for the following refrigerants with a temperature range from −40°C to the critical temperature: (a) R-134a, (b) R-12 and (c) NH_3. Use a log scale for pressure and linear scale for temperature.

3-3-17 Complete the following property table for H_2O. Also locate the states on a *T-s* diagram (qualitatively).

State No	p, kPa	T, °C	x, %	v, m³/kg	u, kJ/kg	h, kJ/kg	s, kJ/kg·K
1	100	20	--	--	--	--	--
2	100	--	--	--	--	209.42	--
3	--	--	50	0.8475	--	--	--
4	--	100	--	--	2297.2	--	--
5	--	200	--	--	--	--	7.8342
6	--	--	--	3.5654	3131.5	--	--

3-3-18   Complete the following property table for Refrigerant-134a. Also locate the states on a *T-s* diagram (qualitatively).

State No	p, kPa	T, °C	x, %	v, m³/kg	u, kJ/kg	h, kJ/kg	s, kJ/kg·K
1	--	-20	100	--	--	--	--
2	1000	--	--	--	--	--	$= s_1$
3	$= p_2$	--	0	--	--	--	--
4	$= p_1$	--	--	--	$= h_3$	--	--

3-3-19   A 10 L rigid tank contains 0.01 kg of steam. Determine (a) the pressure (p), (b) stored energy (E), and (c) entropy (S), of steam if the quality is 50%. Neglect kinetic and potential energy. ***What-if scenario:*** What would the (d) pressure, (e) stored energy, (f) entropy of steam be if the steam quality were 100%?

3-3-20   A liquid-vapor mixture of water at 100 kPa has a quality of 1%. (a) Determine the volumetric quality of the mixture. (b) ***What-if scenario:*** What would the volumetric quality be if the working fluid was R-134a?

3-3-21   A 0.4 m³ vessel contains 10 kg of refrigerant-134a at 25°C. Determine the (a) phase composition (b) pressure (p), (c) total internal energy (U) and (d) total entropy (S) of the refrigerant.

3-3-22   A tank contains 1 L of saturated liquid and 99 L of saturated vapor of water at 200 kPa. Determine (a) the mass, (b) quality and (c) stored energy (E) of the steam.

3-3-23   A tank contains 20 kg of water at 85°C. If half of it (by mass) is in the liquid phase and the rest in vapor phase, determine (a) the volumetric quality, the stored energy (E) in (b) the liquid and (c) vapor phases.

3-3-24   A vessel having a volume of 0.5 m³ contains a 2 kg saturated mixture of H_2O at 500 kPa. Calculate the (a) mass of liquid, (b) mass of vapor and (c) volume of liquid (d) volume of vapor.

3-3-25   (a) What is the phase composition of H_2O at $p = 100$ kPa and $u = 1500$ kJ/kg? (b) How much volume does 2 kg of H_2O occupy in that state? (c) ***What-if scenario:*** What would the volume be if the mass of H_2O was 5 kg?

3-3-26   A rigid vessel contains 3 kg of refrigerant-12 at 890 kPa and 85°C. Determine (a) volume (V) of the vessel and (b) total internal energy (U).

3-3-27   For H_2O, plot how the volume fraction, y, changes with quality, x, over the entire possible range at (a) $p = 100$ kPa and (b) $p = 20$ MPa.

3-3-28   A rigid tank, volume 83 m³, contains 100 kg of H_2O at 100°C. The tank is heated until the temperature inside reaches 120°C. Determine the pressure (p) inside the tank at (a) the beginning and (b) the end of the heating process. (c) ***What-if scenario:*** What would the final pressure be if the tank's temperature increased to 125°C?

3-3-29   A tank contains 5 kg of saturated liquid and 5 kg of saturated vapor of H_2O at 500 kPa. Determine (a) its volume (V) in m³ an (b) its temperature (T) in Celsius. The tank is now heated to a temperature of 250°C. (c) Determine the pressure in MPa.

3-3-30   A tank contains 500 kg of saturated liquid and 5 kg of saturated vapor of H_2O at 500 kPa. Determine (a) the quality of the steam and (b) the volume of the tank.

3-3-31   In Problem 3-3-29, determine the change in (a) stored energy (ΔE) and (b) entropy (ΔS) of the system. (c) Using the entropy balance equation, explain why the entropy of the system increases.

3-3-32   A tank with a volume of 10 gallons contains a liquid-vapor mixture of propane at 30°C. Determine (a) the pressure and (b) the mass of propane inside. (c) If the tank is designed for a maximum pressure of 3000 kPa, determine the maximum temperature the tank will be able to withstand.

3-3-33   A piston cylinder device of volume 1 m³ contains 3 kg of water. The piston, which has an area of 100 cm², exerts a force of 1.7 kN on the pin to keep it stationary. Determine (a) the temperature and (b) quality of H_2O inside the cylinder. The water is now heated. (c) Determine the force on the pin when all the liquid in the tank vaporizes. Assume the atmospheric pressure to be 100 kPa and neglect the piston's mass.

3-3-34 A rigid tank, with a volume of $3.5\,m^3$, contains 5 kg of a saturated liquid-vapor mixture of H_2O at 80°C. The tank is slowly heated until all the liquid in the tank is completely vaporized. Determine the temperature at which this process occurred. Also show the process on a T-v diagram with respect to saturation lines.

3-3-35 A $10\,m^3$ rigid tank contains saturated vapor of H_2O at 200°C. The tank is cooled until the quality drops to 80%. Determine the (a) mass of H_2O in the tank, (b) drop in pressure (Δp), (c) drop in temperature (ΔT), and (d) drop in total entropy (ΔS). (e) What causes the entropy of this system to decrease?

3-3-36 A 1000 L rigid tank contains saturated liquid water at 40°C. (a) Determine the pressure (p) inside. The tank is now heated to 90°C. (b) Use the compressed liquid table to determine the pressure in the tank.

3-3-37 A lid with negligible weight is suddenly placed on a pan of boiling water and the heat is turned off. After about an hour, thermal equilibrium is reached between the water and the atmosphere, which is at 30°C and 101 kPa. If the inner diameter of the pan is 20 cm, determine (a) the force necessary to open the lid. (b) *What-if scenario:* What force would be needed for a 1 kg lid?

3-3-38 Superheated water vapor at 1.5 MPa and 280°C is allowed to cool at constant volume until the temperature drops to 130°C. At the final state, determine (a) the pressure (p_2), (b) the quality and (c) the specific enthalpy (h_2). Show the process on a T-s diagram.

3-3-39 A rigid tank contains steam at the critical state. Determine (a) the quality of the steam after the tank cools down to the atmospheric temperature, 25°C. (b) What percent of the volume is occupied by the vapor at the final state?

3-3-40 A rigid tank with a volume of $1\,m^3$ contains superheated steam at 500 kPa and 500°C. Determine (a) the mass and (b) total internal energy (U) of the steam. The tank is now cooled until the total internal energy decreases to 2076.2 kJ. Determine (c) the pressure (p_2) and (d) temperature (T_2) in the final state.

3-3-41 10 kg of ammonia is stored in a rigid tank with a volume of $0.64\,m^3$. Determine the pressure variation (Δp) in the tank as the ambient temperature swings between a nighttime temperature of (a) 5°C to (b) 40°C in the day time. (c and d) *What-if scenario:* What would the answer be if the volume of the tank were increased to $1\,m^3$?

3-3-42 A large industrial tank, which has a volume of $200\,m^3$, is filled with steam at 450°C and 150 kPa. Determine (a) the pressure (p) and (b) quality of steam when the temperature drops to 25°C. (c) Determine the heat transfer, using $Q = \Delta U$ (the formula for heat transfer at this constant volume).

3-3-43 A rigid tank, with a volume of $0.64\,m^3$, contains 10 kg of water at 5°C. (a) Plot how the state of water changes in a T-s diagram as the temperature is gradually increased to 500°C. (b) *What-if scenario:* How would the plot change if the tank held 0.1 kg of water?

3-3-44 Draw a line of constant specific volume that passes through the critical point on a T-s diagram for R-134a. (b) *What-if scenario:* How would the line shift if it passed through saturated vapor state at 50°C?

3-3-45 A $1\,m^3$ rigid tank contains 2.3 kg of a vapor-liquid mixture of water. The tank is heated to raise the quality of steam. Plot how the pressure and temperature in the tank vary as the quality of steam gradually increases to 100%.

3-3-46 Plot how (a) the pressure (p), (b) the stored energy (E), and (c) the entropy (S) of 4.5 kg of water contained in a $2\,m^3$ rigid vessel as the temperature is increased from 30°C to 300°C.

3-3-47 A piston-cylinder device contains 3 kg of saturated mixture of water with a quality of 0.8 at 180°C. Heat is added until all the liquid is vaporized. Determine (a) the pressure (p), (b) the initial volume (V_1), (c) the final volume (V_2) and (d) the work (W) performed by the vapor during the expansion process. (e) Show the process on a p-v diagram.

3-3-48 A piston-cylinder device initially contains $3\,ft^3$ of liquid water at 60 psia and 63°F. Heat is now transferred to the water at constant pressure until the entire liquid is vaporized. Determine (a) the mass of the water, (b) final temperature (T_2) and (c) total enthalpy change (ΔH). The T-s diagram is to be drawn.

3-3-49 A piston-cylinder device contains 0.6 kg of steam at 350°C and 1.5 MPa. The steam is cooled at constant pressure until half of the mass condenses. Determine (a) the final temperature (T_2) and (b) the boundary work transfer. (c) Show the process on a T-s diagram.

3-3-50 Water vapor (1 kg) at 0.2 kPa and 30°C is cooled in a constant pressure until condensation begins. Determine (a) the boundary work transfer and (b) change of enthalpy (ΔH) treating water as the system. *What-if scenario:* What would the (c) boundary work transfer and (d) change of enthalpy be if all the vapor condensed?

3-3-51 A piston cylinder device contains 10 L of liquid water at 100 kPa and 30°C. Heat is transferred at constant pressure until the temperature increases to 200°C. Determine the change in (a) the total volume (ΔV) and (b) total internal energy (ΔU) of steam. Show the process on a T-s and p-v diagram.

3-3-52 A piston-cylinder device contains a saturated mixture of water with a quality of 84.3% at 10 kPa. If the pressure is raised in an isentropic (constant entropy) manner to 5000 kPa, (a) determine the final temperature (T_2). (b) *What-if scenario:* What would the final temperature be if the water was at a saturated vapor state to start with?

3-3-53 A piston-cylinder device contains a saturated mixture of R-134a with a quality of 90% at 50 kPa. The quality of the mixture is raised to 100% by (i) compressing the mixture in an isentropic manner or, alternatively, (ii) by adding heat in a constant temperature process. (a) Draw the two processes on a T-s diagram. Determine the change in specific internal energy (Δu) (b) in (i) and (c) adding heat at constant temperature.

3-3-54 Draw the constant pressure line on a T-s diagram for H_2O at $p = 100$ kPa as the liquid H_2O is heated to superheated vapor. Redo the plot for a pressure of 500 kPa.

3-3-55 For H_2O, plot how the entropy (S) changes with T in the superheated vapor region for a pressure of (a) 10 kPa, (b) 100 kPa and (c) 10 MPa. Take at least 10 points from the saturation temperature to 800°C.

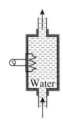

3-3-56 Water, at a pressure of 50 MPa, is heated in a constant pressure electrical heater from 50°C to 1000°C. Spot the states on a T-s diagram and determine (a) the change of specific enthalpy (Δh) and (b) specific entropy (Δs). Use compressed liquid model for liquid water.

3-3-57 Determine (a) the mass flow rate ($\dot{m}$) and (b) the volume flow rate ($\dot{V}$) of steam flowing through a pipe of diameter 0.1 m at 1000 kPa, 300°C and 50 m/s. (c) Also determine the rate of transport of energy by the steam. (d) *What-if scenario:* What would the answer in (c) be if the temperature were 400°C?

3-3-58 Refrigerant-134a flows through a pipe of diameter 5 cm with a mass flow rate of 0.13 kg/s at 100 kPa and 10 m/s. Determine (a) the temperature (T) and (b) quality of the refrigerant in the pipe. Also determine the rate of transport of (c) energy ($\dot{J}$) and (d) entropy ($\dot{S}$) by the flow.

3-3-59 Steam at a pressure of 2 MPa and 400°C flows through a pipe of diameter 10 cm with a velocity of 50 m/s. Determine the flow rates of (a) mass ($\dot{m}$), (b) energy ($\dot{J}$) and (c) entropy ($\dot{S}$).

3-3-60 Liquid water at 100 kPa, 30°C enters a boiler through a 2 cm-diameter pipe with a mass flow rate of 1 kg/s. It leaves the boiler as saturated vapor through a 20 cm-diameter pipe without any significant pressure loss. Determine (a) the exit velocity (V_2), the rate of transport of energy at (b) the inlet ($\dot{J}_1$) and (c) exit ($\dot{J}_2$). Neglect potential energy, but not kinetic energy. (d) *What-if scenario:* What would the rate of transport of energy at (d) inlet and (e) exit be if kinetic energy was neglected?

3-3-61 Repeat Problem 3-3-60 for a boiler pressure of 1 MPa.

3-3-62 Water is pumped in an isentropic (constant entropy) manner from 100 kPa, 25°C to 40 MPa. Determine the change in specific enthalpy (Δh) using (a) the compressed liquid table, (b) compressed liquid model and (c) solid/liquid model.

3-3-63   Water, at 30 MPa and 20°C, is heated at constant pressure until the temperature reaches 300°C. Determine (a) the change in specific volume (Δv) and (b) specific enthalpy (Δh). Use compressed liquid table for water. (c) *What-if scenario:* What would the specific enthalpy be if the water was only heated to 200°C?

3-3-64   Draw a constant pressure line ($p = 100$ kPa) on a T-v and a T-s diagram for H_2O. Repeat the problem with $p = 1000$ kPa.

3-3-65   In an isentropic nozzle, the specific flow energy (j) and entropy (s) remain constant along the flow. Superheated steam flows steadily through an isentropic nozzle for which the following properties are known at the inlet and exit ports: Inlet: $p = 200$ kPa, $T = 400°C$, $A = 100$ cm^2, $V = 5$ m/s; Exit: $p = 100$ kPa. Determine (a) the exit velocity (V_2), (b) the exit temperature (T_2) and (c) the exit area.

3-3-66   Steam enters a turbine, operating at steady state, at 5000 kPa and 500°C with a mass flow rate of 5 kg/s. It expands in an isentropic manner to an exit pressure of 10 kPa. Determine (a) the exit temperature (T_2), (b) exit quality and the volumetric flow rate ($\dot{V}$) at (c) the inlet and (d) exit .

Steam

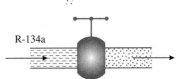

3-3-67   Refrigerant-134a enters a throttle valve as saturated liquid at 40°C and exits at 294 kPa. If enthalpy remains constant during the flow, determine (a) the drop in pressure (Δp) and (b) the drop in temperature (ΔT) in the valve.

R-134a

3-3-68   Saturated vapor of R-134a enters a compressor, operating at steady state, at 160 kPa with a volume flow rate of 10 L/min. The specific entropy remains constant along the flow. Determine (a) the exit temperature (T_2) if the compressor raises the pressure of the flow by a factor of 10. Also, determine the rate of transport of energy ($\dot{J}$) at (b) the inlet and (c) the exit. Neglect kinetic and potential energies.

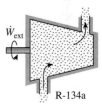

$\dot{W}_{ext}$

R-134a

3-3-69   Repeat Problem 3-3-68 with refrigerant-12 as the working fluid.

3-3-70   Water flows steadily through a 10 cm-diameter pipe with a mass flow rate of 1 kg/s. The flow enters the pipe at 200 kPa, 30°C and is gradually heated until it leaves the pipe at 300°C, without any significant drop in pressure. Plot the flow velocity against the temperature of the flow.

3-3-71   In Problem 3-3-70, plot how the rate of transport of energy increases as the flow temperature increases by (a) including and (b) neglecting kinetic energy.

SECTION 3-4: PG (PERFECT GAS) AND IG (IDEAL GAS) MODELS

3-4-1   Determine (a) the mass of air at 100 kPa, 25°C in a room with dimensions 5m × 5m × 5m. (b) How much air must leave the room if the pressure drops to 95 kPa at constant temperature? (c) How much air must leave the room if the temperature increased to 40°C at constant pressure?

3-4-2   Determine the specific enthalpy (h) of a gas (PG model: $k = 1.4$, $R = 4.12$ kJ/kg·K) given $u = 6001$ kJ/kg and $T = 1000$ K.

3-4-3   A tank of volume 1 m^3 contains 5 kg of an ideal gas with a molar mass of 44 kg/kmol. If the difference between the specific enthalpy (h) and the specific internal energy (u) of the gas is 200 kJ/kg, determine its temperature (T).

3-4-4   A tank contains helium (molar Mass = 4 kg/kmol) at 1 MPa and 20°C. If the volume of the tank is 1 m^3, determine (a) the mass and (b) the mole of helium in the tank. Use the PG or IG model.

3-4-5   Determine c_p of steam at 10 MPa, 350°C using (a) PG model if the specific heat ratio is 1.327, (b) the PC model (use the PC state TESTcalc to find a neighboring state, hotter by, 1°C at constant pressure, and numerically evaluate c_p from its definition). (c) *What-if scenario:* What is the answer in part b if the temperature separation between the states is reduced to 0.5°C?

$$c_p \equiv \left(\frac{\partial h}{\partial T} \right)_p$$

3-4-6 A cylinder, volume $2\,m^3$, contains 1 kg of hydrogen at 20°C. Determine the change in (a) pressure (Δp), (b) stored energy (ΔE) and (c) entropy (ΔS) of the gas as the chamber is heated to 200°C. Use the PG model for hydrogen. (d) *What-if scenario:* What would the (d) pressure, (e) stored energy, (f) entropy be if the chamber contained carbon-dioxide instead?

3-4-7 Air in an automobile tire, with a volume of 18 ft^3, is at 90°F and 25 psia. Determine (a) the amount of air to be added to bring the pressure up to 30 psig. Assume the atmospheric pressure to be 14.7 psia and the temperature and volume to remain constant. (b) *What-if scenario:* What would the answer in (a) be if the pressure went up to 40 psig?

3-4-8 The gauge pressure in an automobile tire is measured as 250 kPa when the outside pressure is 100 kPa and temperature is 25°C. If the volume of the tire is $0.025\,m^3$, (a) determine the amount of air that must be bled in order to reduce the pressure to the recommended value of 220 kPa gauge. Use the PG model for air. (b) *What-if scenario:* What would the answer be if the IG model was used instead?

3-4-9 A rigid tank, volume $10\,m^3$, contains steam at 200 kPa and 200°C. Determine the mass of steam inside the tank using (a) the PC model for steam, (b) PG model for steam and (c) IG model for steam. (d) Which answer is the most accurate?

3-4-10 To test the applicability of the ideal gas equation of state to calculate the density of saturated steam, compare the specific volume of saturated steam obtained from the steam table with the prediction from the IG model for the following conditions: (a) 50 kPa, (b) 500 kPa and (c) critical point. Express the comparisons as percentage errors, using the steam table results as benchmarks.

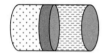 **3-4-11** A weightless piston separates an insulated horizontal cylindrical vessel into two closed chambers. The piston is in equilibrium with air on one side and H_2O on the other side, each occupying a volume of $1\,m^3$. If the temperature of both the chambers is 200°C, and the mass of H_2O is 15 kg, determine the mass of air. Treat air as a perfect gas and H_2O as a PC fluid.

3-4-12 Determine the mass of saturated steam stored in a rigid tank of volume $2\,m^3$ at 20 kPa using (a) the PC model and (b) the IG model. (c) *What-if scenario:* What would the answers using (c) PC model and (d) IG model be if the steam had a quality of 95% instead?

3-4-13 Calculate the change in specific internal energy (Δu) as air is heated from 300 K to 1000 K using (a) the PG model and (b) the IG model (for the IG model, use the IG system-state TESTcalc).

 3-4-14 A 1 L piston-cylinder device contains air at 500 kPa and 300 K. An electrical resistance heater is used to raise the temperature of the gas to 500 K at constant pressure. Determine (a) the boundary work transfer, the change in (b) stored energy (ΔE) and (c) entropy (ΔS) of the gas. (d) *What-if scenario:* Which part of the answers would not change if the IG model was used?

3-4-15 A piston-cylinder device contains 0.01 kg of nitrogen at 100 kPa and 300°C. Using (a) the PG model and (b) IG model, determine the boundary work transfer as the nitrogen cools down to 30°C. Show the process on a *T-s* and a *p-v* diagram.

 3-4-16 Oxygen at 100 kPa and 200°C is compressed to half its initial volume. Determine the final state, in terms of pressure (p_2) and temperature (T_2), if the compression is carried out in an (a) isobaric, (b) isothermal and (c) isentropic manner. Use the PG model for oxygen.

3-4-17 Repeat Problem 3-4-16 using the IG model for oxygen.

3-4-18 For nitrogen, plot how the internal energy (U) varies with T within the range 25°C–1000°C while the pressure is held constant at 100 kPa. Use (a) the PG model and (b) the IG model. (c) *What-if scenario:* Would any of the plots change if the pressure were 1 MPa instead?

3-4-19 For carbon dioxide, plot how the specific entropy (s) varies with T within the range 25°C–3000°C while the pressure is held constant at 100 kPa. Use (a) the PG model and (b) the IG model.

3-4-20 Superheated steam, at a pressure of 10 kPa and temperature 200°C, undergoes a process to reach a final pressure of 50 kPa and temperature of 300°C. Determine, magnitude only, (a) Δu, (b) Δh and (c) Δs. Assume superheated steam will behave as an ideal gas. *What-if scenario:* What would (d) Δu, (e) Δh, (f) Δs be if the phase-change and the perfect gas models had been used?

3-4-21   Air, at 300 K and 300 kPa, is heated at constant pressure to 1000 K. Determine the change in specific internal energy (Δu) using (a) perfect gas model with c_p evaluated at 300 K, (b) perfect gas model with c_p evaluated at the average temperature, (c) data from the ideal gas air table and (d) polynomial correlation between c_p and T.

3-4-22   Air, at 300 K and 300 kPa, is heated at constant pressure to 1000 K. Determine the change in specific entropy, Δs, using (a) perfect gas model with c_p evaluated at 298 K, (b) perfect gas model with c_p evaluated at the average temperature and (c) data from the ideal gas air table.

3-4-23   Air, at 15°C and 100 kPa, steadily enters the diffuser of a jet engine with a velocity of 100 m/s. The inlet area is $0.2\,\text{m}^2$. Determine (a) the mass flow rate of the air ($\dot{m}$). (b) *What-if scenario:* What is the mass flow rate if the entrance velocity is 150 m/s?

3-4-24   Air flows through a nozzle in an isentropic manner from $p = 400\,\text{kPa}$, $T = 25°C$ at the inlet to $p = 100\,\text{kPa}$ at the exit. Determine the temperature at the exit (T_2), modeling air as a perfect gas.

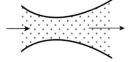

3-4-25 H_2O, at 500 kPa, 200°C, enters a long insulated pipe with a flow rate of 5 kg/s. If the pipe diameter is 30 cm, determine the flow velocity in m/s (a) using the PC model for H_2O, (b) using the PG model for H_2O (Molar Mass of $H_2O = 18\,\text{kg/kmol}$). (c) *What-if scenario:* What would be the answer if the IG model was used?

SECTION 3-5: RG (REAL GAS) MODEL

3-5-1   Determine the specific volume (v) of oxygen at 10 MPa and 175K based on (a) Lee-Kesler and (b) Nelson-Obert generalized compressibility chart.

3-5-2   Determine the compressibility factor of steam at 20 MPa, 400°C using (a) the LK chart, (b) the NO chart and (c) the PC model. (d) *What-if scenario:* What would the answers be if the steam was saturated at 1 MPa?

3-5-3   Compare the IG model and RG model (Lee Kesler chart) in evaluating the density of air at (a) 100 kPa, 30°C, (b) 10 MPa, −100°C and (c) 10 MPa, 500°C.

3-5-4   A $1\,\text{m}^3$ closed rigid tank contains nitrogen at 1 MPa and 200 K. Determine the total mass of nitrogen using (a) the RG model (LK chart), (b) the IG model and (c) the PC model. Discuss the discrepancy among the results.

3-5-6   Calculate the specific volume (v) of propane at a pressure of 8 MPa and a temperature of 40°C using (a) the IG model, (b) the RG model and (c) the PC model.

3-5-7   A $10\,\text{m}^3$ tank contains nitrogen at a pressure of 0.5 MPa and a temperature of 200 K. Determine the mass of nitrogen in the tank using (a) the ideal gas and (b) real gas model. (c) *What-if scenario:* What would the answer, in part, (a) be if the coditions in the tank were 3 MPa and 125 K?

3-5-8   Calculate the error (in percent) in evaluating the mass of nitrogen at 10 MPa, 200 K in a 100 L rigid tank while using (a) the IG model and (b) the RG model (LK chart). Use the PC model as the benchmark.

3-5-9   A 10 gallon tank contains 9 gallons of liquid propane and the rest is a vapor at 30°C. Calculate (a) the pressure (p) and (b) mass of the propane by using the Lee Kesler compressibility chart. *What-if scenario:* What would the (c) pressure and (d) mass be if the PC model was used?

3-5-10   A rigid tank, with a volume of $2\,\text{m}^3$, contains 180 kg of water at 500°C. Calculate the pressure (p) in the tank by using (a) the Lee-Kesler chart and (b) Nelson-Obert chart. (c) *What-if scenario:* What would the answer be if the PC model was used?

3-5-11   Determine the volume (Ψ) of 1 kg of water pressurized to 100 MPa at 1000°C. Use (a) the RG model with the Lee-Kesler chart, (b) the RG model with the Nelson-Obert chart and (c) the PC model.

3-5-12   A closed rigid tank contains carbon-dioxide at 10 MPa and 100°C. It is cooled until its temperature reaches 0°C. Determine the pressure at the final state (p_2). Use (a) the RG model with the Lee-Kesler chart, (b) the RG model with the Nelson-Obert chart and (c) the PC model.

3-5-13   A 15 L tank contains 1 kg of R-12 refrigerant at 100°C. It is heated until the temperature of the refrigerant reaches 150°C. Determine the change in (a) internal energy (ΔU), and (b) entropy, (ΔS). Use the RG model with Lee-Kesler charts.

3-5-14 A piston cylinder device contains 10 L of nitrogen at 10 MPa and 200 K. It is heated at a constant pressure to a temperature of 400 K. Determine (a) ΔH and (b) ΔS. Use the RG model with Lee-Kesler charts. (c) *What-if scenario:* What would the answers be if the PC model was used? If the PC model is always more accurate, why should anyone ever use the RG model?

3-5-15 For H_2O, plot how the specific volume (v), varies with T in the superheated vapor region. Assume pressures of (a) 10 kPa, (b) 50 kPa and (c) 10 MPa. Take at least 10 points from the saturation temperature to 600°C. To reduce the data into a simple correlation, plot pv against T. (d) Explain the behavior of the reduced data.

3-5-16 Use the PC model to generate v vs p data for (a) O_2 and (b) N_2 at 200 K over the reduced pressure range of 0.1 to 10. Evaluate Z and plot it against p_r. Compare your results with the prediction from the Lee-Kesler compressibility chart.

4 ∣ MASS, ENERGY, AND ENTROPY ANALYSIS OF OPEN-STEADY SYSTEMS

In Chapter 2, the fundamental laws of thermodynamics—the conservation of mass principle, the first law, and the second law—were expressed as *balance equations* of mass, energy, and entropy for a generic, open, unsteady system. Closed-steady systems were then analyzed as a special case by simplifying these balance equations. Analysis of other types of systems was postponed until the methodology for state evaluation was developed in Chapter 3. Now that material models for evaluating extended systems and flow states are in place, we are ready to analyze all types of thermodynamic systems.

This chapter is dedicated to the analysis of open-steady systems—systems that allow mass, heat, and work transfer, but whose local states remain invariant with time. Consequently, global properties such as total mass, stored energy, and entropy, which are aggregate of the corresponding local properties, do not change with time. Examples of open-steady systems can be found in flow through pipes, nozzles, diffusers, pumps, compressors, turbines, throttling valves, heat exchangers, mixing chambers, etc. The objective of this chapter is to gain insight into the steady-state operation of these devices through comprehensive mass, energy, and entropy analysis. The framework developed for the analysis—classify a system through suitable assumptions, customize the balance equations, select an appropriate material model for the working substance, make necessary approximations, obtain a manual solution, use an appropriate TESTcalc to verify manual solution, and conduct what-if studies when possible—will become a template for system analysis throughout this book.

4.1 GOVERNING EQUATIONS AND DEVICE EFFICIENCIES

Most engineering devices are designed to operate over long periods of time without much change in the operating conditions. Industrial systems such as gas turbines, steam power plants, refrigeration systems, and air-conditioning systems sometimes operate for months before scheduled maintenance or shut down occurs. Under the conditions they were designed for these systems and their components operate at *steady state*. This means the *global state* of the system, or any sub-system, that consists of the aggregate of local states, does not change with time. Although an overall system such as the Rankine cycle (see Sec. 2.3.1) may be *closed,* components such as pipes, turbines, pumps, compressors, nozzles, diffusers, heat exchangers, etc., allow mass transfer across their boundaries and are *open*. Open devices, operating at steady states are called **open-steady systems**.

Consider the steam turbine illustrated in Figure 4.1 and in Anim. 4.A.*turbine*, an open-steady system, which produces shaft work at the expense of flow energy as steam passes through an alternating series of stationary and rotating blades. The fixed blades attached to the turbine's stationary casing create nozzle-shaped passages, through which steam expands to a lower pressure and higher velocity (we discuss why a nozzle accelerates a flow shortly) before impinging on an array of blades attached to a central shaft. Transfer of momentum from steam to the blades creates a torque on the shaft that makes it turn. Steam then enters the next stage and the process is repeated until it leaves the turbine at a temperature and pressure much lower than those at the inlet. The large variation of property values across the device means the turbine is a non-uniform system; however, properties at a given point do not change with time at steady state—the global state, which is an aggregate of all the local states remains frozen in time.

This conclusion holds true for any open-steady system, a generic version of which is represented by Figure 4.2. The system is non-uniform, as indicated by the variation of color, but steady, which is indicated in Anim. 4.A.*singleFlowGeneric* by a color pattern that does not change with time. In a work-producing or work-consuming device, external work transfer

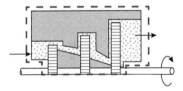

FIGURE 4.1 Steam passes through a series of nozzles attached to the casing and blades attached to a rotating shaft of the turbine (see Anims. 4.A.*turbine* and 8.A. *turbineBlades*).

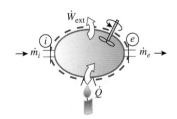

FIGURE 4.2 In a generic open-steady system, the global state is non-uniform and frozen in time (see Anim. 4.A. *singleFlowGeneric*).

generally consists of electrical and shaft work. Heat transfer occurs mostly with the atmospheric reservoir unless indicated otherwise.

Recall that the snapshot taken with a *state camera* (Sec. 1.4) does not change with time for a system at steady state. With the system image remaining frozen, all extensive properties of the open system, including m, E, and S, must remain constant at steady state. The unsteady terms in the balance equations—the time derivatives of mass, energy, and entropy—can be set to zero, which simplifies the general governing balance equations (Eqs. 2.3, 2.8, and 2.13). Furthermore, a large class of devices has only a single inlet and a single exit—a single flow through the system. For such **single-flow devices** the governing equations are further simplified, as the summation sign of the transport terms can be dropped. The resulting equations for a single-flow, open-steady device (Chapter 2) are shown below and illustrated in Anim. 4.A.*singleFlowGeneric*:

$$\text{Mass:} \quad \dot{m}_i = \dot{m}_e = \dot{m}; \quad \text{or,} \quad \rho_i A_i V_i = \rho_e A_e V_e; \quad \text{or,} \quad \frac{A_i V_i}{v_i} = \frac{A_e V_e}{v_e} \tag{4.1}$$

$$\text{Energy:} \quad \frac{dE^{0}}{dt} = \dot{m}(j_i - j_e) + \dot{Q} - \dot{W}_{\text{ext}}; \quad \text{where} \quad j \equiv h + \text{ke} + \text{pe} \tag{4.2}$$

$$\text{Entropy:} \quad \frac{dS^{0}}{dt} = \dot{m}(s_i - s_e) + \frac{\dot{Q}}{T_B} + \dot{S}_{\text{gen}} \tag{4.3}$$

Different terms of the energy and entropy equations are illustrated in the energy and entropy flow diagram of Anim. 4.A.*flowDiagram*. For many devices, changes in kinetic and potential energies are negligible, allowing the flow energy, j, to be replaced by the thermodynamic property enthalpy, h. For adiabatic systems, another simplification occurs when $\dot{Q}$ is set to zero. Note that by setting the transport terms to zero (no mass transfer), these equations reduce to those used in the analysis of closed-steady systems in Sec. 2.2.

4.1.1 TEST and the Open-Steady TESTcalcs

The **open-steady TESTcalcs** build upon the state TESTcalcs introduced in Section 1.5.1, and will be extensively used in this chapter to verify manual solutions and pursue *what-if* scenarios. They can be found at the Systems, Open, Steady, Generic, Single-Flow branch of the TESTcalcs module. As in the case of state TESTcalcs, selecting a material model that best suits the working substance launches the TESTcalc. All open-steady TESTcalcs have the same look and feel. They also work in a similar manner, so let us work with a particular TESTcalc, for example, the SL open-steady TESTcalc as an exercise. Launch it by clicking the SL-Model icon.

You will notice that the default view of the TESTcalc is similar to the SL flow-state TESTcalc located in the states, flow states branch. The control panels are almost identical and the state panel is displayed in both cases; however, the open-steady TESTcalc has two additional panels, device and exergy, that are accessible through the tabs. The inlet and exit states of the system are evaluated in the state panel, completely or partially, from the given information. As an exercise, select Water (L) as the working fluid and calculate two arbitrary states, State-1 and State-2.

Now switch to the device panel, where you will find a control panel, several device variables, and a sketch of a generic device with the simplified governing equations. In the control panel, you will see menus for inlet and exit states populated by the calculated states, from the state panel. To set up a particular device, select a device identification label (e.g., A, B, etc.), then select the inlet and exit states from the calculated state stacks. Enter the known device variables and click Calculate. The governing equations are solved to produce the unknown values. If an unknown state property (e.g., j, s, or mdot) is calculated, it is posted back to the appropriate state in the state panel. You can return to that state (select the state tab, then the state number) and calculate the state completely with the help of the newly evaluated property. Clicking the Super-Calculate button does the same by iterating between the device and state panels. It also generates a complete solution report and TEST-code in the I/O panel that can be stored and used at a later time to regenerate the solution. The Super-Calculate button also allows a simple way to pursue what-if scenarios. Just as the Calculate operation stores a state in the state stack (indicated by the © prefix), clicking the Super-Calculate button stores the entire solution as a calculated case (indicated by the © prefix) in the case menu. To study a new case, simply select a new case from this menu.

A discussion of the exergy panel will be presented in Chapter 6.

4.1.2 Energetic Efficiency

In Section 2.2, the energetic efficiency for closed systems was introduced as the ratio of desired energy outout to the required energy input for a closed-steady device. The same definition can be extended for an open-steady device. The **energetic efficiency**, also known as the **first-law efficiency**, is defined as:

$$\eta_{\text{I}} \equiv \frac{\text{Desired Energy Output}}{\text{Required Energy Input}} \tag{4.4}$$

Specific quantities that appear in the numerator and denominator depend on the purpose of a given device and how each term in its energy balance equation is interpreted physically. With the unsteady term dropping out due to the steady-state condition, the energy equation for a generic open-steady device, Eq. (2.8), can be interpreted as

$$0 = \underbrace{\sum_i \dot{m}_i j_i - \sum_e \dot{m}_e j_e}_{\substack{=\dot{J}_{\text{net}},\text{ Net rate of transport} \\ \text{of energy into the system.}}} + \underbrace{\dot{Q}}_{\substack{\text{Net rate of} \\ \text{heat transfer} \\ \text{into the} \\ \text{system.}}} - \underbrace{\dot{W}_{\text{ext}}}_{\substack{\text{Net rate of} \\ \text{external work} \\ \text{transfer out of} \\ \text{the system.}}}; \quad \text{or} \quad \dot{J}_{\text{net}} + \dot{Q} = \dot{W}_{\text{ext}}; \quad [\text{kW}] \tag{4.5}$$

That is, the net energy transported by the flow into the device, plus the energy transferred by heat, must equal the external work delivered by the system. This interpretation is also evident from the energy-flow diagram of Figure 4.3. Such diagrams are helpful when defining and visualizing the energetic efficiency for a given device. We have already calculated the efficiency of an electrical space heater—a closed-steady system—to be 100% in Example 2-9. Let's perform a similar analysis and define the energetic efficiency for an open-steady system, an electric water heater, in Example 4-1.

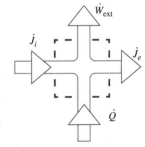

FIGURE 4.3 Energy flow diagram for the open-steady device shown in Figure 4.2 (click Energy in Anim. 4.A.*flowDiagram*).

EXAMPLE 4-1   Mass, Energy, and Entropy Analysis of a Water Heater

As shown in Figure 4.4, an electric water heater supplies hot water at 150 kPa and 70°C at a flow rate of 10 L/min. The water temperature at the inlet is 15°C. Due to poor insulation, heat is lost at a rate of 2 kW to the surrounding atmosphere at 25°C. Using the SL model for water and neglecting kinetic and potential energy transport, determine (a) the electrical power consumption, (b) the energetic efficiency, and (c) the rate of entropy generation in the heater's universe. Assume no pressure loss in the system.

SOLUTION

Analyze the heater's universe, enclosed within the red boundary of Figure 4.4, using the mass, energy, and entropy balance equations.

Assumptions

The heater is at steady state. Flow states at the inlet and exit, States-1 and 2, are uniform and at LTE (local thermodynamic equilibrium). Also ke and pe are negligible so that $j \equiv^1 h + \text{ke} + \text{pe} \cong^2 h$.

Analysis

The material properties for water, ρ and c_v, are obtained from Table A-1 or any SL state TESTcalc as 997 kg/m^3 and 4.184 kJ/kg · K, respectively.

The mass flow rate can be obtained from the volume flow rate via Eq. (0.2):

$$\dot{m} = \rho \dot{V} = 997 \, \frac{10 \times 10^{-3}}{60} = 0.166 \, \frac{\text{kg}}{\text{s}}$$

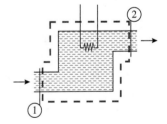
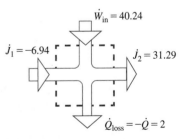

FIGURE 4.4 System schematic and energy flow diagram for Ex. 4-1 (see Anim. 4.A.*waterHeater* and click the Entropy Diagram button). All numbers are in kW.

---

[1]defined as.
[2]approximately equal.

The energy equation, Eq. (4.2), can be rearranged to produce the magnitude of the electricity consumption, $\dot{W}_{in} = -\dot{W}_{ext}$ (recall the WinHip sign convention from Sec. 0.7.2), as follows:

$$0 = \dot{m}(j_i - j_e) + \dot{Q} - \dot{W}_{ext}$$

$$\Rightarrow \quad \dot{W}_{in} = -\dot{W}_{ext} = -\dot{m}(j_1 - j_2) - \dot{Q} \cong \dot{m}(h_2 - h_1) - \dot{Q}$$

$$= \dot{m}c_v(T_2 - T_1) + \dot{m}v(p_2 - p_1)^0 - \dot{Q}$$

$$= (0.166)(4.184)(70 - 15) - (-2) = 40.2 \text{ kW}$$

In this derivation, the formula for enthalpy difference is substituted from Eq. (3.21). The signs for $\dot{W}_{ext}$ and $\dot{Q}$ must be negative since work is added and heat is lost.

The electrical work consumption is the required input while the energy supplied to the water, $\dot{m}(j_2 - j_1)$, is the desired output. The energetic efficiency; therefore, can be defined and evaluated as:

$$\eta_I = \frac{\text{Desired Energy Output}}{\text{Required Energy Input}} = \frac{\dot{m}(j_2 - j_1)}{\dot{W}_{in}} = \frac{38.2}{40.2} = 95.02\%$$

Note that in the accompanying energy-flow diagram (obtained from the TEST solution), we have used absolute values for the external work and heat transfer with the arrows indicating directions rather than using algebraic quantities. The negative value of $\dot{J}_1$ is due to the arbitrary reference temperature, 25°C, used for setting enthalpy to zero in TEST, and does not imply that energy is being transported in the opposite direction of the flow (which is impossible).

For the system's universe, enclosed within the red boundary of Figure 4.4, $T_B = T_0$ and the entropy equation, Eq. (4.3), produces:

$$\frac{ds^0}{dt} = \dot{m}(s_1 - s_2) + \frac{\dot{Q}}{T_0} + \dot{S}_{gen,univ}$$

$$\Rightarrow \quad \dot{S}_{gen,univ} = \dot{m}(s_2 - s_1) - \frac{\dot{Q}}{T_0} = \dot{m}c_v \ln \frac{T_2}{T_1} - \frac{\dot{Q}}{T_0}$$

$$= (0.166)(4.184) \ln \frac{(273 + 70)}{(273 + 15)} - \frac{(-2)}{(273 + 25)} = 0.128 \frac{\text{kW}}{\text{K}}$$

TEST Analysis

Launch the SL open-steady TESTcalc and select Water (L) from the working-fluid menu. Evaluate State-1 from p1, T1, and Voldot1, and State-2 from p2=p1, T2, and mdot2=mdot1. In the device panel, select State-1 and State-2 as the inlet and exit states, enter Qdot as –2 kW, and set T_B as 25 deg-C. Click Calculate. Wdot_ext and Sdot_gen are evaluated, verifying the manual results. By changing p1 to any other value, and super-calculating, the results are shown to be independent of the water heater's pressure.

Discussion

Entropy is generated inside the water heater. At the same time, electrical energy is dissipated into heat (through *electronic friction*) and heat is transferred across a finite temperature difference (*thermal friction*), both inside and in the immediate surroundings of the system. An energetic efficiency of 95% may seem satisfactory, however, as we will learn in Chapter 6, the accompanying entropy generation is a measure of degradation, or wastefulness, of useful energy. The same goal of raising the water temperature could be achieved with much less work if a *reversible* water heater could be invented.

4.1.3 Internally Reversible System

Although entropy generation through thermodynamic friction is inherent and pervasive in practical systems, idealized systems, where entropy generation can be neglected within the system boundary or even in the entire system's universe, serve as important benchmarks. Such systems were introduced in Sec. 2.1.4 as *internally reversible* and *reversible* systems. Devoid of any *thermodynamic friction,* an internally reversible system forms an ideal counterpart of an actual device. Analysis of such idealized systems (see Anim. 2.D.*reversibility*) can provide significant insight and help engineers determine the performance limit of actual devices.

To explore the implication of reversibility in an open-steady system, we can do an entropy analysis on the generic, internally reversible ($\dot{S}_{gen,int} = 0$) single-flow system shown in Figure 4.5. The suffix *int.rev.* indicate, that the system is internally reversible—there is no thermodynamic friction within the system's internal boundary. Consider a slice of the system bounded by flow State-1 and State-2 as a local internal system (Fig. 4.5 and click on Element in Anim. 4.A.*intReversible*). While the same mass flow passes through the local system, its share of heat and work transfer can be represented by differential amounts, $d\dot{Q}_{int.rev.}$ and $d\dot{W}_{ext}$, respectively, if State-1 and State-2 are separated infinitesimally. The entropy equation for the internal local system can be written as:

$$\frac{dS}{dt}^0 = \dot{m}(s_1 - s_2) + \frac{d\dot{Q}_{int.rev.}}{T} + d\dot{S}_{gen,int}^{\ 0}$$

$$\Rightarrow \quad d\dot{Q}_{int.rev.} = \dot{m}Tds; \quad [kW] \tag{4.6}$$

In deriving this equation, we have used $s_2 = s_1 + ds$ and substituted the local system temperature for the boundary temperature [$T_B = (T_1 + T_2)/2 = T + dT/2 = T$ as dT goes to zero]. For the single-flow system, Eq. (4.6) can be integrated from the inlet to the exit, yielding:

$$\dot{Q}_{int.rev.} = \dot{m} \int_i^e Tds; \quad [kW] \tag{4.7}$$

Although derived on purely theoretical grounds, this equation has an important implication for any *T-s* diagram. No matter what the working fluid is, the area under the *T-s* diagram (Fig. 4.6 and Anim. 4.A.*intReversible*) can be interpreted as the heat transfer per-unit mass of the working fluid, while the system is assumed to be internally reversible. The same interpretation also applies to reversible systems ($\dot{S}_{gen,univ} = 0$), which are internally and externally reversible. Heat transfer to an internally reversible system is called **reversible heat transfer** and is represented by Eq. (4.7). The significance of reversible heat transfer can be appreciated when we discuss reversible cycles in Chapters 7 through 10. For internally irreversible systems ($\dot{S}_{gen,int.} > 0$), it can be shown from Eq. (4.6) that the area under the *T-s* diagram can have contributions from heat transfer as well as entropy generated by thermodynamic friction inside the system. The area, if any, under the *T-s* diagram of an adiabatic (no heat transfer) system, therefore, must be attributed to internal irreversibilities.

The steady-flow energy equation for the local system in Figure 4.5 can be simplified for an *internally reversible* system, producing a parallel graphical interpretation for the external work transfer. For the internal system of Figure 4.5, the energy equation can be written as:

$$\frac{dE}{dt}^0 = \dot{m}(j_1 - j_2) + d\dot{Q}_{int.rev.} - d\dot{W}_{ext,\ int.rev.}$$

$$\Rightarrow \quad d\dot{W}_{ext,\ int.rev.} = -\dot{m}dj + d\dot{Q}_{int.rev.} = -\dot{m}dj + \dot{m}Tds; \quad [kW] \tag{4.8}$$

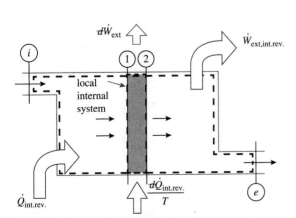

FIGURE 4.5 Locations of States 1 and 2 are so close to each other that the local system (dotted boundary) can be considered uniform (click Element in Anim. 4.A.*intReversible*).

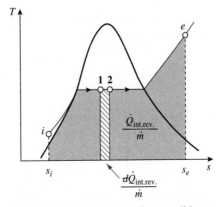

FIGURE 4.6 For an internally reversible system, the area under the *T-s* diagram can be associated with the heat transfer (click Element and Device buttons in Anim. 4.A.*intReversible*).

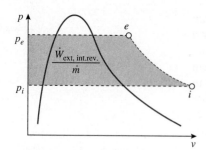

FIGURE 4.7 For an internally reversible system, the shaded area is proportional to the magnitude of external work transfer (see Anim. 4.A.*intReversible* and click on External Work Transfer).

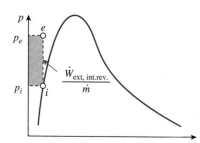

FIGURE 4.8 Compared to Figure 4.7, the work transfer is much lower here because the working fluid is a liquid, which has a much lower specific volume (see Anim. 4.A. *pumpVsCompressor*).

Note that we have substituted $d\dot{Q}_{int.rev.} = \dot{m}Tds$ from the entropy equation [Eq. (4.6)]; so an internally reversible device does not have to be *adiabatic*. Integrating from the inlet to the exit of the overall system, we obtain the external work transfer in an internally reversible open-steady device:

$$\dot{W}_{ext,\,int.rev.} = -\dot{m}\int_i^e dj + \dot{m}\int_i^e Tds = \dot{m}\int_i^e (Tds - dj) = \dot{m}\int_i^e \left[-vdp - d(\text{ke}) - d(\text{pe}) \right]$$

$$\Rightarrow \quad \dot{W}_{ext,\,int.rev.} = -\dot{m}\left[\int_i^e vdp + (\text{ke}_e - \text{ke}_i) + (\text{pe}_e - \text{pe}_i) \right]; \quad [\text{kW}] \qquad (4.9)$$

When changes in ke and pe between the inlet and exit are negligible, the equation simplifies to:

$$\dot{W}_{ext,\,int.rev.} = -\dot{m}\int_i^e vdp; \quad [\text{kW}] \qquad (4.10)$$

Notice this equation's misleading resemblance to the $pd\text{\textsterling}$ formula, Eq. (0.16), for boundary work involving a closed system. While the external work for an internally reversible system is proportional to the area projected on the p axis (Fig. 4.7), the $pd\text{\textsterling}$ work for a closed system is proportional to the projected area on a p-v diagram's v axis.

Equations (4.10) and (4.7) lend important physical meaning to thermodynamic plots. Even without knowing the details of how a device produces work or transfers heat, conclusions can be drawn from the graphical interpretations of these equations. For example, gas and vapor have a much larger v (specific volume) when compared to liquids. Assuming other parameters remain unchanged, the integral in Eq. (4.10) can be expected to be much larger in magnitude for a gas or a vapor when compared to a liquid. For the same mass flow rate, a compressor that raises the pressure of vapor, therefore, will require much more work than a pump that raises the pressure of a liquid by the same amount (compare Figs. 4.7 and 4.8). In the Rankine cycle (Fig. 2.34), the turbine and pump operate with the same pressure differences and same $\dot{m}$, yet the turbine produces much more power than what the pump consumes. Equation (4.10) provides mathematical insight into the operation of many work-consuming or work-producing devices or cycles. This graphical interpretation is repeated used (see 4.A.turbine for example) to explain the parametric behavior of open-steady devices.

4.1.4 Isentropic Efficiency

Systems can be idealized in many different ways. Because thermodynamic friction (Sec. 2.1.3), quantified by entropy generation, downgrades the performance of most devices, an ideal device can be assumed to be internally reversible, i.e., $\dot{S}_{gen,int} = 0$. Many single-flow devices such as pumps, compressors, turbines, nozzles, diffusers, etc., can be regarded as adiabatic and operate across fixed inlet and exit pressures under design conditions. Let's refer to the inlet state as State-1 and the actual exit state as State-2. At their ideal limit, these devices can be considered internally reversible. By representing the ideal exit state as State-3, the actual and ideal devices can be compared through an entropy analysis as follows:

$$\text{Actual device:} \quad \cancel{\frac{ds}{dt}}^{0,\text{ steady state}} = \dot{m}(s_1 - s_2) + \cancel{\frac{\dot{Q}}{T_B}}^{0,\text{ adiabatic}} + \dot{S}_{gen,int} \qquad (4.11)$$

$$\Rightarrow \quad s_2 \geq s_1 \text{ since } \dot{S}_{gen,int} \geq 0$$

$$\text{Ideal device:} \quad \cancel{\frac{ds}{dt}}^{0,\text{ steady state}} = \dot{m}(s_1 - s_3) + \cancel{\frac{\dot{Q}}{T_B}}^{0,\text{ adiabatic}} + \cancel{\dot{S}_{gen,int}}^{0,\text{ internally reversible}} \qquad (4.12)$$

$$\Rightarrow \quad s_1 = s_3$$

The ideal exit state has the same entropy as the actual inlet state. In addition, the pressure of the ideal exit state is assumed to be the same as the pressure at the actual exit state. The ideal single-flow device is called an **isentropic** device. It is mathematically possible for entropy to remain unchanged between the inlet and exit if entropy transfer through heat loss is exactly offset by internal entropy generation; however, such situations are not likely. An isentropic device implies that it is *steady, adiabatic,* and *internally reversible.* On a *T-s* diagram (Fig. 4.9) which shows actual and ideal states of a turbine, the ideal device is represented by the solid vertical line that joins the inlet and ideal exit states. For the corresponding actual device (represented by the dashed line), entropy generation being non-negative, $s_2 \geq s_1$ and the actual exit state, which is on the same constant-pressure line as State-2, is shifted to the right.

The **isentropic efficiency** compares an energetic term (work output, work input, exit ke, or inlet ke) from the energy equation of the actual device to the corresponding term of the isentropic device:

$$\eta_{\text{device}} = \frac{\text{An Energetic Term of the Actual (or Ideal) Device}}{\text{Corresponding Term of the Ideal (or Actual) Device}} \quad \textbf{(4.13)}$$

Whether the term involving the actual device appears in the numerator or denominator depends on the device and ensures that the maximum possible value of the isentropic efficiency does not exceed 100%. As we discuss individual devices in the sections that follow, we will introduce these device-specific definitions.

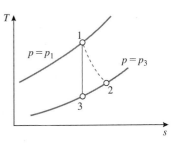

FIGURE 4.9 The isentropic and actual exit states share the same exit pressure and velocity (see Anim. 4.A.*turbineEfficiency*).

4.2 COMPREHENSIVE ANALYSIS

In the subsections that follow, we will discuss several open devices, and subject them to comprehensive analyses using the mass, energy, and entropy equations. Keeping the discussion of the inner workings to a minimum, we will introduce each device through its functionalities, with an emphasis on the device's interactions with the system's surroundings. Suitable assumptions and approximations will be made to simplify the governing equations to a set of equations customized for a particular device.

4.2.1 Pipes, Ducts, or Tubes

A **pipe**, **duct**, or a **tube**—to reduce confusion, we will simply use the term *pipe*—is a hollow cylinder used to conduct a liquid, gas, or finely divided solid particles. As simple as it may appear, a pipe can be turned into complex devices such as a hair dryer (Anim. 4.A.*hairDryer*), a water heater (Anim. 4.A.*waterHeater*), or a boiler tube (Anim. 4.A.*boilerTube*) depending on the heat and work interactions with the surroundings.

Consider a steady flow of a fluid through a constant-diameter pipe with work and heat interactions as shown in Figure 4.10. If wall friction in the pipe is not negligible, how will it impact the exit velocity? Before we jump to any conclusions, let's see where the balance equations lead us.

A steady-state mass balance between the inlet and exit yields $\dot{m} = A_i V_i / v_i = A_e V_e / v_e$, or $V_i / v_i = V_e / v_e$. Therefore, for an *incompressible fluid*— from Sec. 3.3 that the density of an *incompressible* fluid can be assumed to be constant and that the SL model can be used for such fluids—the mass equation yields $V_i = V_e$. The velocity must remain constant, regardless of friction, heat or work transfer, and orientation of the pipe. This is a simple yet powerful conclusion when we think about liquid water flowing downward through a vertical pipe.

For a *compressible* gas or vapor flow the situation becomes complicated, as v is a function of temperature and pressure. From the free-body diagram in Figure 4.11, we can deduce that the pressure decreases along the flow to overcome frictional resistance ($p_i A > p_e A$), regardless

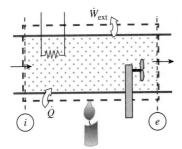

FIGURE 4.10 Many useful devices can be represented as a pipe flow with heat and/or external work transfer (see Anim. 4.A.*hairDryer*).

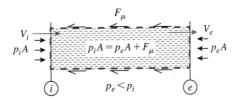

FIGURE 4.11 Free-body diagram of a fluid in a straight pipe. The force balance, independent of energy or entropy balance, establishes that pressure must decrease along a constant-diameter pipe.

of fluid density, or heat transfer. In the case of an isothermal (T = constant) or heated flow (T increases), v must increase along the flow according to the IG equation of state $v = RT/p$ (which is also approximately true for a vapor). An increase in specific volume results in an increase in velocity (recall from the mass equation that $V_i/v_i = V_e/v_e$). In fact, advanced gas dynamics analysis can be used to show that the gas velocity can increase up to the speed of sound despite the presence of friction.

EXAMPLE 4-2   Mass, Energy, and Entropy Analysis of a Duct Flow

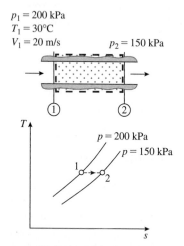

$p_1 = 200$ kPa
$T_1 = 30°C$
$V_1 = 20$ m/s
$p_2 = 150$ kPa

FIGURE 4.12 Schematic and T-s diagram for Ex. 4-2 (see Anim. 4.A.*longDuct*).

In Figure 4.12, helium flows steadily into a long and narrow adiabatic tube of diameter 1 cm with a velocity of 20 m/s at 200 kPa and 30°C. At the exit, the pressure drops to 150 kPa to overcome wall friction. Determine (a) the mass flow rate, (b) exit velocity, (c) exit temperature, and (d) the rate of entropy generation in the system's universe. Neglect changes in kinetic and potential energies. **What-if scenario:** (e) What would the exit temperature and velocity be if kinetic energy had not been neglected?

SOLUTION

Analyze the open-steady system (the pipe's universe enclosed within the red boundary of Fig. 4.12) using the mass, energy, and entropy balance equations.

Assumptions

Steady state, perfect-gas behavior for helium, uniform states based on LTE at the inlet and exit, and negligible changes in ke and pe.

Analysis

From Table C-1, or any PG TESTcalc, obtain the PG model constants R = 2.078 kJ/kg · K and c_p = 5.196 kJ/kg · K for helium.

The energy equation, Eq. (4.2), for the system shown in Figure 4.12 simplifies to:

$$\frac{dE^{0}}{dt} = \dot{m}(j_i - j_e) + \dot{Q} - \dot{W}_{ext} \cong \dot{m}(h_i - h_e) + \dot{Q}^0 - \dot{W}_{ext}^{0};$$

$$\Rightarrow \quad h_i - h_e = c_p(T_i - T_e) = 0; \quad 1 \quad T_e = T_i = 303 \text{ K}$$

For both the PG and IG models, h is a function of T only; therefore, the conclusion that T remains constant in an isenthalpic flow of a perfect gas must be valid for any ideal gas too.

The mass equation, $\dot{m}_i = \dot{m}_e = \dot{m}$, coupled with the application of the IG equation of state, $pv = RT$, produces:

$$\dot{m} = \frac{A_1 V_1}{v_1} = \frac{A_1 V_1 p_1}{RT_1} = \frac{(7.854 \times 10^{-5})(20)(200)}{(2.0785)(303)} = 4.99 \times 10^{-4} \frac{\text{kg}}{\text{s}};$$

$$\Rightarrow \quad V_2 = \frac{\dot{m} v_2}{A_2} = \frac{\dot{m} R T_2}{A_2 p_2} = \frac{(4.99 \times 10^{-4})(2.0785)(303)}{(7.854 \times 10^{-5})(150)} = 26.68 \frac{\text{m}}{\text{s}}$$

The entropy equation, Eq. (4.3), for the system's universe (red boundary in Fig. 4.12) yields:

$$\frac{dS^{0, \text{ steady state}}}{dt} = \dot{m}(s_1 - s_2) + \frac{\dot{Q}^0}{T_0} + \dot{S}_{gen,univ};$$

$$\Rightarrow \quad \dot{S}_{gen,univ} = \dot{m}(s_2 - s_1) = \dot{m}\left(c_p \ln \frac{T_2}{T_1} - R \ln \frac{p_2}{p_1}\right) = 3.0 \times 10^{-4} \frac{\text{kW}}{\text{K}} \qquad \textbf{(4.14)}$$

For an adiabatic system, there is no external entropy generation.

TEST Analysis

Launch the PG open-steady TESTcalc. Select He as the working fluid. Calculate the inlet state (State-1), from the known properties. For the exit state (State-2), make Vel2 an unknown. Enter p2, and set mdot2=mdot1, h2=h1, and A2=A1. Click Calculate to find the exit temperature and velocity. In the device panel, select the inlet and exit states, enter zero for Wdot_ext, then click Calculate to find Sdot_gen. The manual results are reproduced.

What-if scenario

To include the effect of the change in kinetic energy, use j2=j1 instead of h2=h1 in State-2. Click Super-Calculate to update all calculations. The exit temperature and velocity remain virtually unchanged at 29.97 deg-C and 26.66 m/s, respectively.

Discussion

Despite the presence of friction, it is the velocity, not temperature, which increases. This may appear counter-intuitive since we are accustomed to a steady flow of liquids. Although ke seems to have no impact in this example, $j \cong h$ cannot be automatically assumed. Use TEST to justify neglecting ke or pe on a case-by-case basis.

EXAMPLE 4-3 **Mass, Energy, and Entropy Analysis of a Hair Dryer**

A hair dryer can be modeled as a steady-flow duct, in which a small fan pulls in atmospheric air and forces it through electrical resistors where it is heated, (Fig. 4.13). Air enters at the ambient conditions of 100 kPa, 25°C, at a velocity 10 m/s, and leaves with negligible change of pressure at a temperature of 50°C. The cross-sectional area is 50 cm². Heat is lost from the dryer at a rate of 50 W. Determine (a) the flow's exit velocity, (b) rate of electrical power consumption, and (c) rate of entropy generation in the system's universe. Use the PG model for air. **What-if scenario:** (d) What would the answers be if the IG model had been used?

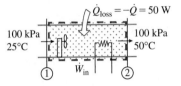

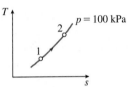

SOLUTION

Analyze the open-steady system (the hair dryer's universe enclosed within the red boundary in Fig. 4.13), using the mass, energy, and entropy balance equations.

Assumptions

Steady state, perfect-gas behavior of air, uniform states at the inlet and exit based on LTE, and negligible changes in ke and pe.

FIGURE 4.13 Schematic and *T-s* diagram for Ex. 4-3. A negative heat addition means that the system is losing heat (recall the WinHip convention).

Analysis

From Table C-1, or any PG TESTcalc, obtain $R = 0.287$ kJ/kg·K and $c_p = 1.005$ kJ/kg·K for air.

Like the previous example, the mass flow equation, Eq. (4.1), coupled with the ideal gas equation of state produces:

$$\dot{m} = \frac{A_2 V_2}{v_2} = \frac{A_1 V_1}{v_1} = \frac{A_1 V_1 p_1}{R T_1} = \frac{(0.005)(10)(100)}{(0.287)(298)} = 0.0585 \frac{\text{kg}}{\text{s}};$$

$$\Rightarrow \quad V_2 = \frac{\dot{m} v_2}{A_2} = \frac{\dot{m} R T_2}{A_2 p_1} = \frac{(0.0585)(0.287)(323)}{(0.005)(100)} = 10.85 \frac{\text{m}}{\text{s}}$$

With $\dot{W}_{in}$ and $\dot{Q}_{loss}$ representing absolute values, the energy equation, Eq. (4.2), yields:

$$\frac{d\cancel{E}}{dt}^{0,\text{ steady state}} = \dot{m}(j_i - j_e) + \dot{Q} - \dot{W}_{ext} \cong \dot{m}(h_i - h_e) + (-\dot{Q}_{loss}) - (-\dot{W}_{in});$$

$$\Rightarrow \quad \dot{W}_{in} = -\dot{W}_{ext} = -\dot{m} c_p (T_i - T_e) + \dot{Q}_{loss};$$

$$= -(0.0585)(1.005)(25 - 50) + 0.05 = 1.52 \text{ kW}$$

Notice how the absolute values are converted to algebraic values following the WinHip convention before substitution in the energy equation. The work calculated includes the power consumed by the fan.

The entropy equation, Eq. (4.3), produces the entropy generation rate in the system's universe:

$$\frac{d\cancel{S}}{dt}^{0,\text{ steady state}} = \dot{m}(s_1 - s_2) + \frac{\dot{Q}}{T_0} + \dot{S}_{gen,univ};$$

$$\Rightarrow \quad \dot{S}_{gen,univ} = \dot{m}\left(c_p \ln\frac{T_2}{T_1} - R \ln\frac{p_2}{p_1}\right) + \frac{\dot{Q}_{loss}}{T_0}$$

$$= (0.0585)(1.005)\ln\left(\frac{323}{298}\right) + \frac{0.05}{298} = 0.0049 \frac{\text{kW}}{\text{K}} \qquad \textbf{(4.15)}$$

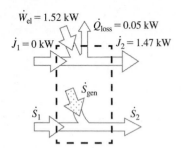

FIGURE 4.14 Energy and entropy flow diagram for Ex. 4-3.

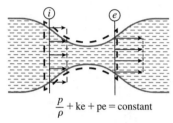

$$\frac{p}{\rho} + \text{ke} + \text{pe} = \text{constant}$$

FIGURE 4.15 Bernoulli's equation requires a flow to be steady, one-dimensional, incompressible, and internally reversible with no external work transfer (see Anim. 4.A.*bernoulliEquation*).

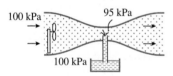

FIGURE 4.16 The difference in pressure creates the liquid spray, a phenomenon known as the Venturi effect.

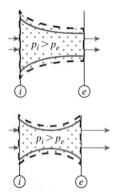

FIGURE 4.17 Nozzles are converging (subsonic) or converging/diverging (supersonic) variable area duct (see Anim. 4.A.*nozzleShapes*). Supersonic nozzles will be discussed in Chapter 15.

TEST Analysis

Launch the PG open-steady TESTcalc. Select air as the working fluid. Calculate State-1 and State-2 as described in the TEST-code for this problem (located in the *TEST > TEST-codes* page). The exit velocity is calculated as part of State-2. In the device panel, select the inlet and exit states, enter Qdot, and click Calculate to verify Wdot_ext and Sdot_gen.

What-if scenario

Launch the IG open-steady TESTcalc in a separate browser tab. Copy the TEST-code generated by the PG solution into the I/O panel and click Load. The results for the IG model are almost identical to the PG results since the temperature variation of the gas is not large enough to cause a significant change in the specific heats.

Discussion

An isentropic efficiency for a hair dryer makes little sense. Even in an ideal dryer, resistance heating will inherently generate entropy through electronic and thermal friction. However, an energetic efficiency (Fig. 4.14) can be defined and evaluated as follows:

$$\eta_I = \frac{\text{Energy Utilized (Desired)}}{\text{Required Energy Input}} = \frac{\dot{m}(j_2 - j_1)}{\dot{W}_{el}} = \frac{1.47}{1.52} = 96.7\%$$

Bernoulli Equation

For internally reversible (or reversible) flow, we have already seen how the expressions for external work and heat transfer simplify. The inlet and exit conditions can also be related by a simple formula, known in fluid mechanics as the **Bernoulli equation**.

Consider a steady *one-dimensional* flow through a variable-diameter pipe (see Fig. 4.15 and Anim. 4.A.*bernoulliEquation*) with no external work transfer. For an incompressible, internally reversible flow, Eq. (4.9) reduces to:

$$\dot{W}_{\text{ext, int.rev.}}^{\nearrow 0} = -\dot{m}\left[\int_i^e v\,dp + (\text{ke}_e - \text{ke}_i) + (\text{pe}_e - \text{pe}_i)\right]$$

$$\Rightarrow \quad 0 = \frac{p_e - p_i}{\rho} + (\text{ke}_e - \text{ke}_i) + (\text{pe}_e - \text{pe}_i) \quad \textbf{(4.16)}$$

$$\Rightarrow \quad \frac{p_i}{\rho} + \frac{V_i^2}{2(1000 \text{ J/kJ})} + \frac{gz_i}{(1000 \text{ J/kJ})} = \frac{p_e}{\rho} + \frac{V_e^2}{2(1000 \text{ J/kJ})} + \frac{gz_e}{(1000 \text{ J/kJ})}; \quad \left[\frac{\text{kJ}}{\text{kg}}\right]$$

The well-known Bernoulli equation relates any two flow states in a one-dimensional incompressible flow. Note that the flow does not have to be adiabatic as long as heat transfer, if any, takes place reversibly (without any thermal friction). If the working fluid is a gas or a vapor, the potential energy is often negligible and an increase in velocity (caused by a reduction in flow area $A_e = A_i V_i / V_e$) must accompany a decrease in pressure.

In the **Venturi effect**, the sub-atmospheric pressure created by a flow constriction (Fig. 4.16) is utilized to draw a liquid through a straw (a pipe); the atmosphere pushes the liquid through the straw. Many other applications of Bernoulli's equation can be found in any standard fluid mechanics textbook.

4.2.2 Nozzles and Diffusers

A **nozzle** is a specially designed variable-area duct that increases the flow velocity at the expense of a pressure drop (Fig. 4.17 and Anim. 4.A.nozzle). Generally, a nozzle operates adiabatically between known inlet conditions, p_i and T_i, and a known exit pressure, p_e. With no heat or external work transfer, the single-flow energy equation, Eq. (4.2), assumes the form:

$$\frac{d\cancel{U}^{\nearrow 0}}{dt} = \dot{m}j_i - \dot{m}j_e + \dot{Q}^{\nearrow 0} - \dot{W}_{\text{ext}}^{\nearrow 0}; \quad \Rightarrow \quad j_i = j_e \quad \textbf{(4.17)}$$

The specific flow energy remains constant along the flow in an adiabatic nozzle. Neglecting any change in pe, and realizing that ke_e is much larger than ke_i (recall the purpose of a nozzle), an expression for the exit velocity can be obtained:

$$j_i = j_e; \quad \Rightarrow \quad h_i \cong h_e + \text{ke}_e; \quad \Rightarrow \quad V_e \cong \sqrt{2(1000 \text{ J/kJ})(h_i - h_e)}; \quad \left[\frac{\text{m}}{\text{s}}\right] \quad \textbf{(4.18)}$$

It is the enthalpy difference that drives the exit kinetic energy; however, h_e cannot be independently controlled for a given nozzle.

Most nozzles are contoured in a smooth manner to minimize frictional losses which maximize, the exit kinetic energy. To be considered *isentropic,* an ideal nozzle must be *steady, adiabatic,* and *internally reversible.* Also recall from Sec. 4.1.4 that an isentropic device operates between the same inlet and exit pressures as the actual device; that is, $p_{es} = p_e$. By setting external work and change in potential energy to zero in Eq. (4.9), we obtain an expression for the exit kinetic energy of an isentropic nozzle, with the subscript *es* representing the isentropic exit state.

$$\text{Isentropic nozzle:} \quad \text{ke}_{es} = h_i - h_{es} = -\int_i^{es} v\,dp; \quad \left[\frac{\text{kJ}}{\text{kg}}\right] \tag{4.19}$$

From this equation, we see that dp must be negative, a pressure drop must occur for the exit velocity to be greater than the inlet velocity—the higher the pressure drop, the higher the exit kinetic energy. This will be established more rigorously from the momentum equation in Chapter 15, which exclusively deals with high-speed flow of gases.

We can obtain a simple closed-form expression for isentropic exit velocity if the working fluid can be modeled by the SL or the PG model. If the working fluid is a liquid, application of the SL model (constant v) in Eq. (4.19) results in:

$$\text{ke}_{es} = -\int_i^e v\,dp = \frac{p_i - p_e}{\rho}; \quad \Rightarrow \quad V_{es} = \sqrt{\frac{2(1000 \text{ J/kJ})(p_i - p_e)}{\rho}}; \quad \left[\frac{\text{m}}{\text{s}}\right] \tag{4.20}$$

The exit velocity is driven by the square root of the pressure difference between the inlet and exit. For a perfect gas, the isentropic relation, Eq. (3.71), can be applied, producing:

$$\text{ke}_{es} = h_i - h_{es} = c_p(T_i - T_{es}) = c_p T_i\left[1 - \left(\frac{p_e}{p_i}\right)^{(k-1)/k}\right]; \quad \left[\frac{\text{kJ}}{\text{kg}}\right] \tag{4.21}$$

When the working fluid is a gas, the exit velocity in a nozzle is a function of the pressure ratio rather than the pressure difference. A temperature drop must accompany the pressure drop along the flow, following the isentropic relation [Eq. (3.59)]. For a given pressure ratio, a nozzle with a higher inlet temperature will produce a greater exit velocity; this is also evident from Eq. (4.19), which predicts a higher exit kinetic energy for a working fluid that has a higher average specific volume. Even if the working fluid is an ideal gas or a vapor, these conclusions apply in a qualitative sense. For example, in a liquid-fueled rocket engine, hydrogen and oxygen must be ignited to create a high exit velocity (and thrust). A cold flow, operating under the same pressure difference, will produce a much smaller exit velocity (and thrust) due to a smaller average specific volume.

Using the isentropic nozzle as a benchmark, the performance of an actual nozzle can be measured by its isentropic efficiency (see Anim. 4.A.*nozzleEfficiency*):

$$\eta_{\text{nozzle}} \equiv \frac{\dot{KE}_e}{\dot{KE}_{es}} = \frac{\dot{m}(\text{ke}_e)}{\dot{m}(\text{ke}_{es})} = \frac{\text{ke}_e}{\text{ke}_{es}} = \frac{j_i - h_e}{j_i - h_{es}} \cong \frac{h_i - h_e}{h_i - h_{es}} \tag{4.22}$$

The *h-s* diagram of Figure 4.18, which is similar to a *T-s* diagram for the IG or PG model, is more convenient for nozzle flow because the difference in enthalpies can be directly interpreted as the kinetic energy. For an actual adiabatic nozzle $s_e \geq s_i$, as established by Eq. (4.11), and the exit kinetic energy is reduced as shown in Figure 4.18.

We will deduce the shapes of isentropic nozzles (see Anim. 4.A.*nozzleShapes*) through advanced gas dynamics analysis in Chapter 15. Subsonic (flow velocity less than speed of sound at the exit) nozzles will be shown to require a converging shape, while supersonic nozzles (flow velocity exceeding the speed of sound at the exit) require a converging-diverging shape. Well-contoured nozzles have typical efficiencies of 90% or more.

A **diffuser** works in the exact opposite manner as a nozzle (see Fig. 4.19 and Anim. 4.A.*diffuser*); its purpose is to *increase* a flow's pressure at the expense of its kinetic energy. As a result, the inlet kinetic energy is much greater than the exit kinetic energy, which can be considered negligible. Also, the exit pressure is not fixed and depends on the irreversibilities present in the diffuser. As in a nozzle, these irreversibilities are quantified through the **diffuser efficiency**, that is defined in terms of an isentropic ideal diffuser, which requires a lesser inlet

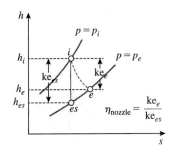

FIGURE 4.18 Graphical interpretation of nozzle efficiency (see Anim. 4.A.*nozzleEfficiency*).

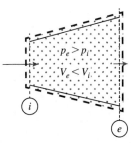

FIGURE 4.19 A diffuser is the exact opposite of a nozzle. It increases pressure at the expense of kinetic energy. A diverging diffuser is for subsonic flows only. Supersonic diffusers will be discussed in Chapter 15.

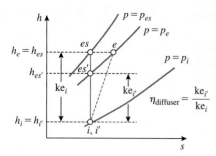

FIGURE 4.20 Isentropic efficiency for a diffuser (see Anim. 4.A.*diffuserEfficiency*).

kinetic energy to produce the same exit pressure as the actual diffuser. The ideal inlet flow state, State-i', has the same inlet temperature and pressure (and hence the same thermodynamic state as State-i in Fig. 4.20) but a lower velocity. While both State-es and State-es' in Figure 4.20 are isentropic to State-i, only State-es' has a pressure equal to the actual exit pressure, making the latter the ideal exit state. With no heat or external work transfer, the single-flow energy equation [Eq. (4.2)] for the actual, isentropic (but with a higher exit pressure than the design pressure), and ideal (isentropic and correct exit pressure) diffusers can be written as:

$$\text{Actual } (i - e): \quad j_i = j_e; \quad \Rightarrow \quad h_e \cong h_i + \text{ke}_i \tag{4.23}$$

$$\text{Isentropic } (i - es): \quad j_i = j_{es}; \quad \Rightarrow \quad h_{es} \cong h_i + \text{ke}_i = h_e \tag{4.24}$$

$$\text{Ideal } (i' - es'): \quad j_{i'} = j_{es'}; \quad \Rightarrow \quad h_{es'} \cong h_{i'} + \text{ke}_{i'} = h_i + \text{ke}_{i'} \tag{4.25}$$

For a given inlet state, the isentropic state, State-es, can be found from the independent properties $h_{es} \cong h_i + \text{ke}_i$ and $s_{es} = s_i$. If the exit pressure is given, State-es' can be evaluated from $p_{es'} = p_e$ and $s_{es'} = s_i$. The ideal inlet kinetic energy $\text{ke}_{i'}$ of State-i' can now be evaluated from the energy equation of the ideal device.

The isentropic efficiency is then defined (see Anim. 4.A.*diffuserEfficiency*) as the ratio of $\text{ke}_{i'}$ and ke_i, both of which produce the same exit pressure p_e:

$$\eta_{\text{diffuser}} \equiv \frac{\text{ke}_{i'}}{\text{ke}_i} = \frac{j_{es'} - h_i}{j_e - h_i} \cong \frac{h_{es'} - h_i}{h_e - h_i} \tag{4.26}$$

Notice how the efficiency is now expressed in terms of thermodynamic properties. If the working fluid is a perfect gas, the expression can be further simplified in terms of temperatures at State-i, State-e, and State-es'. An efficiency of 90% means that with 90% of the inlet kinetic energy, an ideal diffuser can produce the actual exit pressure. If the diffuser efficiency decreases, State-e shifts to the right along the $h = h_{es}$ (horizontal) line in Figure 4.20, reducing the exit pressure. State-es' moves down along the constant-entropy line so that $p_{es'}$ is equal to the reduced exit pressure. At 100% efficiency, States e, es, and es' merge to a single state as shown in Anim. 4.A.*diffuser*. A more rigorous analysis of the diffuser efficiency will be presented in Chapter 15.

EXAMPLE 4-4 **Mass, Energy, and Entropy Analysis of a Nozzle**

As shown in Figure 4.21, air at 0.15 MPa, 30°C enters an insulated nozzle at a velocity of 10 m/s, and leaves with a pressure of 0.1 MPa and a velocity of 200 m/s. (a) Determine the exit temperature of air. Assume air to behave as a perfect gas. (b) Is the nozzle isentropic?

SOLUTION

Analyze the open-steady system, the nozzle enclosed within the red boundary in Figure 4.21, using the mass and energy balance equations.

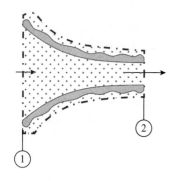

Assumptions

Steady state, PG model for air with $R = 0.287$ kJ/kg·K and $c_p = 1.005$ kJ/kg·K (Table C-1), uniform states at the inlet and exit based on LTE, and negligible pe's.

Analysis

Using the PG model to evaluate enthalpy difference, the energy equation for the nozzle, Eq. (4.18), can be simplified to produce the exit temperature:

$$j_i = j_e;$$

$$\Rightarrow \quad h_i + \text{ke}_i = h_e + \text{ke}_e; \quad \Rightarrow \quad h_i - h_e = \text{ke}_e - \text{ke}_i;$$

$$\Rightarrow \quad c_p(T_i - T_e) = \text{ke}_e + \text{ke}_i; \quad \Rightarrow \quad T_e = T_i - \frac{\text{ke}_e - \text{ke}_i}{c_p};$$

$$\Rightarrow \quad T_e = 30 - \frac{200^2 - 10^2}{(2000)(1.005)} = 10.1°C$$

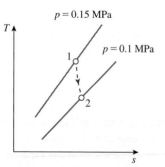

FIGURE 4.21 Schematic and *T-s* diagram for the nozzle in Ex. 4-4 (see Anim. 4.A.*nozzle*).

The change in entropy between the inlet and exit can be evaluated from Eq. (3.70):

$$\Delta s = s_2 - s_1 = c_p \ln \frac{T_2}{T_1} - R \ln \frac{p_2}{p_1};$$

$$= (1.005) \ln \frac{273 + 10.1}{273 + 30} - (0.287) \ln \frac{100}{150} = 0.048 \frac{kJ}{kg \cdot K}$$

The only way entropy can increase in an adiabatic, steady nozzle is through entropy generation; therefore, the nozzle is not isentropic.

TEST Analysis

Launch the PG open-steady TESTcalc, and select air as the working fluid. Calculate State-1 from p1, T1, and Vel1, and State-2 from p2, Vel2, and j2=j1. T2 and s2 are evaluated as part of the exit state. In the state panel, select T-s from the plot menu and draw p = c lines to visually verify the calculated states.

Discussion

Using the TESTcalc, you can determine the shape of the isentropic nozzle relative to a given inlet area (assume it to be 1 m²). As pressure continuously decreases along the nozzle [discussed in connection with Eq. (4.19)], an intermediate location between the inlet and exit can be identified by a pressure between 0.15 MPa and 0.1 MPa. To find an intermediate state, such as State-3, pick a suitable pressure for p3, and enter p3, mdot3=mdot1, j3=j1, and s3=s1. Click Calculate to find A3. Repeat with different values of p3 until you can deduce the shape of the nozzle (converging or converging-diverging).

EXAMPLE 4-5   Mass, Energy, and Entropy Analysis of a Nozzle

As shown in Figure 4.22, steam at 0.5 MPa, 250°C enters an adiabatic nozzle with a velocity of 30 m/s and leaves at 0.3 MPa, 200°C. If the mass flow rate is 5 kg/s, determine (a) the inlet and exit areas, (b) exit velocity, (c) isentropic exit velocity, and (d) isentropic efficiency.

SOLUTION

Analyze the open-steady system, the nozzle enclosed within the red boundary in Figure 4.22, using the mass, energy, and entropy balance equations.

Assumptions

Steady state, PC model for steam, uniform states based on LTE at the inlet and exit, negligible inlet ke, and negligible change in pe.

Analysis

Use TEST, or the manual approach, to determine the anchor states—State-1 for the inlet, State-2 for the actual exit, and State-3 for the isentropic exit (see Fig. 4.22).

State-1 (given p_1, T_1, V_1, $\dot{m}$):

$$\Rightarrow \quad v_1 = 0.4744 \frac{m^3}{kg}; \quad h_1 = 2960.6 \frac{kJ}{kg}; \quad s_1 = 7.271 \frac{kJ}{kg \cdot K};$$

$$j_1 = h_1 + ke_1 = 2960.6 + \frac{30^2}{2000} = 2961.1 \frac{kJ}{kg}; \quad A_1 = \frac{\dot{m}}{(V_1/v_1)} = 791 \text{ cm}^2$$

State-2 (given p_2, T_2, $\dot{m}$):

$$\Rightarrow \quad v_2 = 0.7163 \frac{m^3}{kg}; \quad h_2 = 2865 \frac{kJ}{kg}$$

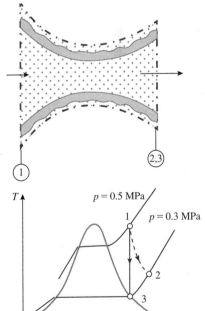

FIGURE 4.22 Schematic and T-s diagram for the nozzle in Ex. 4-5 (see Anim. 4.A.nozzleEfficiency).

The energy equation for the nozzle, Eq. (4.18), produces:

$$V_2 = \sqrt{2000(ke_2)} = \sqrt{2000(j_1 - h_2)} = \sqrt{2000(96.1)} = 438 \frac{m}{s}$$

Therefore, $A_2 = \dfrac{\dot{m}}{(V_2/v_2)} = 82 \text{ cm}^2$

For the isentropic exit state, State-3, the exit pressure and entropy are known. Also the energy equation for the adiabatic nozzle produces $j_3 = j_1$.

State-3 (given $p_3 = p_2$, $s_3 = s_1$, $j_3 = j_1$):

$$\Rightarrow \quad h_3 = 2847.3 \frac{kJ}{kg}$$

Once again, applying Eq. (4.18):

$$V_3 = \sqrt{2000(ke_3)} = \sqrt{2000(j_1 - h_3)} = \sqrt{2000(113.8)} = 477 \frac{m}{s}$$

The isentropic efficiency can now be calculated from Eq. (4.22):

$$\eta_{nozzle} = \frac{ke_e}{ke_{es}} = \frac{ke_2}{ke_3} = \frac{438^2}{477^2} = 84.3\%$$

TEST Analysis

Launch the PC open-steady TESTcalc and select H2O as the working fluid. Calculate State-1, State-2, and State-3 as described in the TEST-code (posted in *TEST > TEST-codes*). To calculate the isentropic efficiency, ke3 and ke2 must be manually calculated in the I/O panel by entering expressions '=Vel3²/2000' and '=Vel2²/2000'.

Discussion

From the TEST analysis, compare the exit areas of the actual and isentropic nozzles, A2=82 cm² and A3=74 cm². A nozzle contour that does not follow the isentropic shape can cause eddies and vortices (internal irreversibilities) in an otherwise streamlined flow, resulting in entropy generation associated with viscous friction.

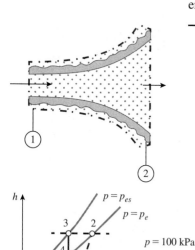

FIGURE 4.23 System schematic and the *h-s* diagram for Ex. 4-6 (see Anim. 4.A.*diffuserEfficiency*).

EXAMPLE 4-6   Mass, Energy, and Entropy Analysis of a Diffuser

In Figure 4.23, helium at 100 kPa, 300 K enters a 90% efficient adiabatic diffuser with a velocity of 200 m/s. If the exit velocity is negligible, determine the exit (a) temperature, and (b) pressure.

SOLUTION

Analyze the open-steady system, the diffuser enclosed within the red boundary in Figure 4.23, using the mass, energy, and entropy balance equations. State-1 and State-2 are the actual inlet and exit states, State-3 is isentropic to State-1, State-4 is the ideal exit state isentropic to State-1 and isobaric to State-2, and State-5 is the ideal inlet state with a reduced kinetic energy (at the same thermodynamic state as State-1) that would produce the same exit pressure as the actual exit when isentropically slowed down to State-4.

Assumptions

Steady state, PG model for helium, uniform states based on LTE at the inlet and exit, negligible exit ke, and negligible change in pe.

Analysis

From Table C-1 or any PG TESTcalc, obtain $R = 2.0785 \text{ kJ/kg} \cdot \text{K}$, $c_p = 5.1926 \text{ kJ/kg} \cdot \text{K}$, and $k = 1.667$ for helium.

The energy equation, Eq. (4.23), coupled with the PG model and applied between the actual inlet and exit states, State-1 and State-2 in Figure 4.23, yields the following exit temperature:

$$j_2 = j_1; \quad \Rightarrow \quad h_2 + \cancel{ke_2}^{\,0} \cong h_1 + ke_1; \quad \Rightarrow \quad c_p(T_2 - T_1) = ke_1;$$

$$\text{Therefore,} \quad T_2 = T_1 + \frac{V_1^2}{2000 c_p} = 300 + \frac{200^2}{(2000)(5.1926)} = 303.9 \text{ K};$$

The exit pressure p_2 is still unknown; however, the maximum possible exit pressure can be found by evaluating the isentropic exit state, State-3, with $V_3 = 0$. The energy equation with $j_3 = j_1$ and $V_3 = 0$ produces $T_3 = T_2$. The isentropic relation, Eq. (3.71), for the PG model, produces p_3:

$$\frac{T_3}{T_1} = \left[\frac{p_3}{p_1}\right]^{(k-1)/k} \quad \Rightarrow \quad p_3 = p_1\left[\frac{T_3}{T_1}\right]^{k/(k-1)} = 103.2 \text{ kPa}$$

Using the definition of the diffuser efficiency [Eq. (4.26)], and noting that enthalpy is proportional to temperature for a perfect gas, we can relate the actual exit temperature to the ideal exit temperature and the inlet temperature:

$$\eta_{\text{diffuser}} \cong \frac{h_4 - h_1}{h_3 - h_1}; \quad \Rightarrow \quad (T_4 - T_1) = (T_3 - T_1)\eta_{\text{diffuser}};$$

$$\Rightarrow \quad T_1\left[\left(\frac{p_4}{p_1}\right)^{(k-1)/k} - 1\right] = \eta_{\text{diffuser}} T_1\left[\left(\frac{p_3}{p_1}\right)^{(k-1)/k} - 1\right]$$

Except for p_4, all variables in this equation are known. Solving, we obtain $p_4 = p_2 = 102.9$ kPa.

TEST Analysis

Launch the PG open-steady TESTcalc and select He as the working fluid. Calculate State-1 through State-5 as described on the TEST-code (posted in *TEST > TEST-codes*). Obtain the actual exit pressure from State-2. State-5 is the ideal inlet state and Vel5 = 189.7 m/s is the minimum velocity that will create the same exit pressure as the actual diffuser. The ratio of Vel5^2/Vel1^2 must be same as the diffuser efficiency by definition.

Discussion

The temperature increases in a diffuser and decreases in a nozzle when the working fluid is a gas or a vapor. For a liquid entropy being a function of temperature only (SL model), isentropic flow also implies isothermal flow (see Anim. 4.A.*nozzleShapes*).

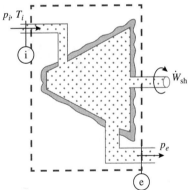

4.2.3 Turbines

A **turbine** is a rotary device which delivers shaft work $\dot{W}_T = \dot{W}_{\text{sh}}$ (positive according to WinHip sign convention) at the expense of the working fluid's flow energy (see Anim. 4.A.*turbine*). The inlet conditions, p_i and T_i, and the exit pressure p_e are generally known for a given turbine (Fig. 4.24 and Anim. 4.A.*turbine*). To simplify analysis, changes in ke and pe are often neglected.

Although an actual turbine is not reversible, Eq. (4.10) can be used as a guide to explore its idealized behavior. According to this equation, the work produced is proportional to the mass flow rate $\dot{m}$ and the integral of $v\,dp$—the shaded area in Figure 4.24. The output can be increased if the average specific volume of the working fluid can be increased. For a given mass flow rate and pressure difference, a vapor turbine will produce more work than a hydraulic turbine when liquid water is the working fluid. That is why hydraulic turbines handle much larger mass flow rates for comparable power output. For the same reason, an increase in the inlet temperature can be expected to increase the average specific volume and the output of a gas or vapor turbine.

The three main categories of turbines are based on the working fluid: steam (vapor) turbines, gas turbines, and hydraulic (water) turbines. In a steam turbine, steam exits at a sub-atmospheric pressure to a device called a condenser (Sec. 4.2.6). In a gas turbine, the gas is normally expelled to the atmosphere and the exhaust pressure is atmospheric. Heat transfer from a turbine is undesirable and is usually minimized with insulation.

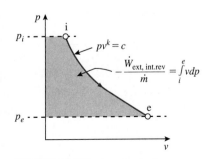

FIGURE 4.24 Work output of an internally reversible turbine is proportional to the shaded area (see Anim. 4.A.*turbine*).

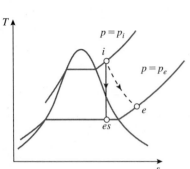

FIGURE 4.25 T-s diagram for an actual and isentropic steam turbine (see Anim. 4.A.*turbineEfficiency*).

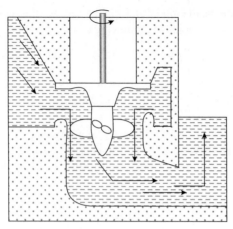

FIGURE 4.26 A hydraulic turbine.

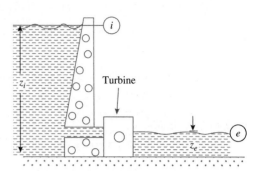

FIGURE 4.27 The driving force for a hydraulic turbine is the difference in the available head $z_i - z_e$.

Typical inlet, exit, and the isentropic exit states of a steam turbine are shown in the T-s diagram of Figure 4.25. The governing equations [Eqs. (4.2) and (4.3)] for an actual turbine and its isentropic counterpart can be simplified as follows:

$$\text{Actual:} \quad \dot{W}_T = \dot{W}_{ext} = \dot{m}(j_i - j_e) \cong \dot{m}(h_i - h_e); \quad s_e = s_i + \underbrace{\frac{\dot{S}_{gen}}{\dot{m}}}_{\geq 0} \tag{4.27}$$

$$\text{Isentropic:} \quad \dot{W}_{T,s} = \dot{m}(j_i - j_{es}) \cong \dot{m}(h_i - h_{es}); \quad s_{es} = s_i \tag{4.28}$$

Using these results, the isentropic **turbine efficiency**, which compares the performance of the actual turbine with the corresponding isentropic turbine, can be established as (see Anim. 4.A.*turbineEfficiency*):

$$\eta_T \equiv \frac{\dot{W}_T}{\dot{W}_{T,s}} = \frac{j_i - j_e}{j_i - j_{es}} \cong \frac{h_i - h_e}{h_i - h_{es}} \tag{4.29}$$

The isentropic exit state can be evaluated from a known $p_{es} = p_e$ and $s_{es} = s_i$. If the inlet and actual exit enthalpies are known, the turbine efficiency can be obtained from Eq. (4.29). Conversely, if the inlet state and the turbine efficiency are known, h_e can be obtained from Eq. (4.29) and the complete exit state can be calculated from p_e (which is usually known) and h_e. If the kinetic energy is not negligible, the exit velocity at the ideal exit state is assumed to be the same as the actual exit velocity.

For a gas or vapor turbine, a drop in pressure also accompanies a drop in temperature and an increase in specific volume (we can use the isentropic PG relation as a guide). With no significant change in velocity, the exit area must be greater than the inlet area for a vapor or gas turbine. In a steam turbine, the vapor quality is kept at 85% or higher at the turbine exit to avoid blade damage.

Equations (4.27) through (4.29) are applicable to both vapor and gas turbines. For a hydraulic turbine (Fig. 4.26), which is embedded between two reservoirs (Fig. 4.27), pressure and velocities at the inlet and exit are unknown. However, the analysis can be simplified by treating the surfaces on the two sides of a dam (Fig. 4.27) as the inlet and exit states. Equations (4.27) and (4.29) still apply if we realize that $j \neq h$ as pe can be quite different at the two surfaces. While $p_i = p_e$ (equals atmospheric pressure) and ke's can be neglected, due to large surface areas, pe must be included in the energy equation. Application of SL model produces:

$$j_i - j_e = (h_i - h_e) + (\text{pe}_i - \text{pe}_e) = (u_i - u_e) + \frac{\overbrace{(p_i - p_e)}^{0}}{\rho} + (\text{pe}_i - \text{pe}_e);$$

$$\Rightarrow \quad j_i - j_e = c_v(T_i - T_e) + g(z_i - z_e)/(1000 \text{ J/kJ}) \text{ [kJ/kg]} \tag{4.30}$$

The output of an actual hydraulic turbine, $\dot{W}_T = \dot{m}(j_i - j_e)$ depends on the difference in elevation $z_i - z_e$, which is known as the **available head**, and the temperature change brought about by friction. For an isentropic turbine, $s_i = s_e$ implies $T_i = T_e$ by application of the SL model; therefore, the power of an isentropic hydraulic turbine reduces to the simple expression: $\dot{W}_{T,s} = \dot{m}g(z_i - z_e)/1000$, which also can be deduced from Eq. (4.9). The power output depends mainly on the product of the mass flow rate and the available head. As a result, it is possible to produce considerable amounts of power with a small head if the available mass flow rate is large.

EXAMPLE 4-7   Mass, Energy, and Entropy Analysis of a Vapor Turbine

As shown in Figure 4.28, a steam turbine operates steadily with a mass flow rate of 5 kg/s. The steam enters the turbine at 500 kPa, 400°C and leaves at a pressure of 7.5 kPa with a quality of 0.95. Neglecting the changes in ke and pe, and any heat loss from the turbine to the surroundings, determine (a) the power generated by the turbine, (b) the isentropic efficiency, and (c) the entropy generation rate. *What-if scenario:* (d) What would the answers be if the steam left the turbine as saturated vapor?

SOLUTION

Analyze the open-steady system, the turbine enclosed within the red boundary in Figure 4.28, using the mass, energy, and entropy balance equations.

Assumptions

Steady state, PC model for steam, uniform states based on LTE at the inlet and exit, and negligible changes in ke and pe.

Analysis

Representing the inlet, exit, and the isentropic exit states by State-1, State-2 and State-3, respectively, use TEST, or the manual approach, to obtain the following state properties:

State-1 (given p_1, T_1):

$$j_1 \cong h_1 = 3271.8 \frac{kJ}{kg}; \quad s_1 = 7.794 \frac{kJ}{kg \cdot K};$$

State-2 (given p_2, x_2):

$$j_2 \cong h_2 = 2454.5 \frac{kJ}{kg}; \quad s_2 = 7.868 \frac{kJ}{kg \cdot K};$$

State-3 (given $p_3 = p_2$, $s_3 = s_1$):

$$j_3 \cong h_3 = 2431.1 \frac{kJ}{kg}$$

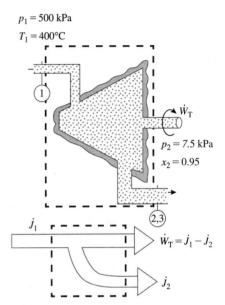

$p_1 = 500$ kPa
$T_1 = 400$°C

FIGURE 4.28 System schematic and the energy flow diagram for Ex. 4-7.

The energy equation, Eq. (4.27), for the actual and ideal turbines yields:

Actual (1-2): $\dot{W}_T = \dot{m}(j_1 - j_2) = (5)(3271.8 - 2454.5) = 4086.5$ kW;

Isentropic (1-3): $\dot{W}_{T,s} = \dot{m}(j_1 - j_3) = (5)(3271.8 - 2431.1) = 4203.5$ kW

The isentropic efficiency now can be evaluated as:

$$\eta_T \equiv \frac{\dot{W}_T}{\dot{W}_{T,s}} = \frac{4086.5}{4203.5} = 97.2\%$$

The entropy balance equation, Eq. (4.3), applied on the turbine's universe produces:

$$\frac{d\cancel{S}}{\cancel{dt}}^{\,0,\,\text{steady state}} = \dot{m}(s_1 - s_2) + \frac{\cancel{\dot{Q}}^{\,0}}{T_0} + \dot{S}_{\text{gen,univ}};$$

$$\Rightarrow \dot{S}_{\text{gen,univ}} = \dot{m}(s_2 - s_1) = (5)(7.868 - 7.794) = 0.370 \frac{kW}{K}$$

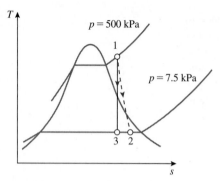

FIGURE 4.29 *T-s* diagram for Ex 4-7.

TEST Analysis

Launch the PC open-steady TESTcalc and select H2O as the working fluid. Evaluate the three states as described in the TEST-code (posted in *TEST* > *TEST-codes*). Draw the *T-s* diagram (Fig. 4.29) using the plot menu. In the device panel, set up Device-A as the actual turbine with State-1 and State-2 as the anchor states. Enter Qdot=0 and click Calculate to find the shaft work and entropy generation rate. Set up Device-B as the isentropic turbine between State-1 and State-3. Obtain the isentropic efficiency by finding the ratio of Wdot_ext from the two devices.

What-if scenario

Change x2 to 1 in State-2 and click Super-Calculate. Power output in the two device panels is updated to 3485.1 kW and 4203.4 kW, resulting in a reduced isentropic efficiency of 82.9%. The entropy generation rate increases significantly to 2.291 kW/K.

Discussion

An energetic efficiency for the turbine can be defined as $\dot{W}_{sh}/(\dot{J}_1 - \dot{J}_2)$, which must be 100% for an adiabatic turbine (see the energy flow diagram in Fig. 4.28). The isentropic efficiency is a better measure of the turbine's performance; from this, we can determine how much potential output is lost due to internal irreversibilities.

EXAMPLE 4-8 **Mass, Energy, and Entropy Analysis of a Hydraulic Turbine**

In Figure 4.30, water enters the intake pipe of a hydraulic turbine at 150 kPa, 15°C, 1 m/s, and exits at a point 100 m below the intake at 100 kPa, 10 m/s. If the turbine output is 700 kW for a flow rate of 1000 kg/s, determine (a) the exit temperature and (b) the isentropic efficiency.

SOLUTION

Analyze the open-steady system, the turbine enclosed within the red boundary in Figure 4.30, using the mass, energy, and entropy balance equations.

Assumptions

Steady state, SL model for water, uniform states based on LTE at the inlet and exit. Negligible heat transfer.

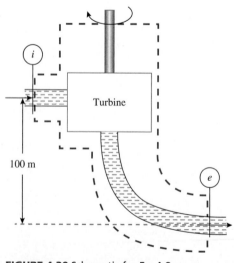

FIGURE 4.30 Schematic for Ex. 4-8.

Analysis

From Table A-1, or any SL TESTcalc, obtain $\rho = 997 \text{ kg/m}^3$ and $c_v = 4.187$ kJ/kg · K for liquid water.

The energy equation, Eq. (4.2), coupled with the SL model yields:

$$\dot{W}_T = \dot{m}(j_i - j_e) = \dot{m}(h_i - h_e) + \dot{m}(\Delta \text{pe} + \Delta \text{ke})$$

$$= \dot{m}(u_i - u_e) + \frac{\dot{m}(p_i - p_e)}{\rho} + \dot{m}(\text{pe}_i - \text{pe}_e) + \dot{m}(\text{ke}_i - \text{ke}_e)$$

$$= \dot{m}c_v(T_i - T_e) + \dot{m}\left[\frac{50}{997} + \frac{9.81\{0 - (-100)\}}{1000} + \frac{1 - 10^2}{2000}\right]$$

$$\Rightarrow T_e = T_i - \frac{\dot{W}_{sh}}{\dot{m}c_v} + \frac{0.9816}{c_v} = 15.08°C$$

The isentropic power can be obtained by substituting $T_i = T_e$ in the expression for turbine output:

$$\dot{W}_{\text{T,s}} = \frac{\dot{m}(p_i - p_e)}{\rho} + \dot{m}(\text{pe}_i - \text{pe}_e) + \dot{m}(\text{ke}_i - \text{ke}_e) = 981.6\ \text{kW}$$

The isentropic efficiency is:

$$\eta_{\text{T}} = \frac{\dot{W}_{\text{T}}}{\dot{W}_{\text{T,s}}} = \frac{700}{981.6} = 71.3\%$$

TEST Analysis

Launch the SL open-steady TESTcalc. Select Water (L) as the working fluid. Calculate State-1, State-2, and State-3, then analyze Device-A and Device-B as described in the TEST-code (see *TEST > TEST-codes*). The turbine efficiency is evaluated in the I/O panel by finding the ratio of Wdot_ext in the actual (A) device to that in the ideal (B) device.

Discussion

Although the temperature increase seems insignificant, it accounts for a significant attached to the rotating shaft (about 30%) loss of turbine power due to internal irreversibilities.

4.2.4 Compressors, Fans, and Pumps

Compressors, **fans**, **blowers**, and **pumps** raise the pressure of a fluid at the expense of useful work, $\dot{W}_{\text{C}} = \dot{W}_{\text{ext}}$ ($\dot{W}_{\text{ext}}$ is negative by WinHip sign convention), which is usually delivered through a shaft. The distinction among these devices (Fig. 4.31 and Anim. 4.A *.compressorTypes*) stems from the type of fluids they handle—a vapor or a gas for compressors, fans, and blowers, and a liquid for pumps. While a fan or blower increases the pressure of a gas just enough to create a desired mass flow, a compressor is capable of delivering a gas at a high pressure. From our discussion of reversible work when comparing Figures 4.7 and 4.8 in Sec. 4.1.3, we would expect a compressor to require more power than a pump for a given mass flow rate and a given pressure rise (see Anim. 4.A *.pumpVsCompressor*). As in a turbine, the inlet conditions p_i and T_i, and the exit pressure p_e, are generally given. Changes in ke and pe are quite small compared to the change in flow energy j, and are generally neglected.

Different types of compressors and pumps are illustrated and classified in Figures 4.31 and 4.32. The most common is the **dynamic** type, in which the rotating blades (called the impeller) impart kinetic energy to the incoming working fluid. The high-speed fluid passes through a diffuser section where it slows down and gains pressure. In an **axial** flow device, the direction of flow is parallel to the rotating shaft. In a **centrifugal** device the fluid enters axially through the central hub and exits in the radial direction. In a **reciprocating** device, the inlet valve opens during the intake stroke at the end of which a fixed volume of fluid is trapped. During the compression stroke, both valves are closed and the exit valve opens after the fluid reaches the desired pressure or volume. A reciprocating device is unsteady during a given cycle; however, when averaged over a specific period of time, the cyclic fluctuations disappear and the device can be treated as an open-steady system (this was discussed, with the aid of animations, in Sec. 2.3.1).

Given that the work requirement for reversible compression (Sec. 4.1.3) is proportional to the integral of vdp, it is desirable to have a small value for the specific volume to minimize the work required for compression. Heat rejection is the simplest way to cool the gas and reduce its specific volume. However, due to high-volume flow rates in axial compressors, sufficient time is not available for any significant heat rejection. One solution to this problem is to use multistage

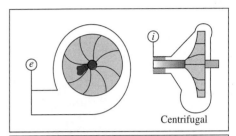

Centrifugal

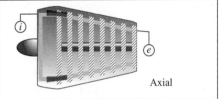

Axial

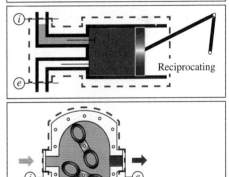

Reciprocating

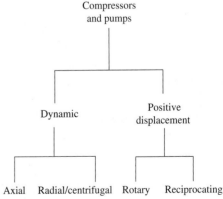

Rotary

FIGURE 4.31 Screen shots from Anim. 4.A.*compressorTypes*.

Compressors and pumps

Dynamic

Positive displacement

Axial   Radial/centrifugal   Rotary   Reciprocating

FIGURE 4.32 Categories of pumps and compressors (see Anim. 4.A.*compressorTypes*).

compression with **intercooling**, where the hot gas at the end of the first stage of compression is cooled, ideally, back to the original inlet temperature without any loss of pressure. We will revisit intercooling in Chapter 8 when we discuss large compressors for gas turbines. Unless mentioned otherwise, we will assume all compressors and pumps to be adiabatic.

The governing balance equation for an adiabatic compressor looks almost identical to those derived for a turbine, except $\dot{W}_C = -\dot{W}_{ext}$ is defined to convert the negative external work (recall the WinHip sign convention from Sec. 0.7.2) to absolute value of compressor power consumption.

$$\text{Actual:} \quad \dot{W}_C = -\dot{W}_{ext} = \dot{m}(j_e - j_i) \cong \dot{m}(h_e - h_i); \quad s_e = s_i + \underbrace{\frac{\dot{S}_{gen}}{\dot{m}}}_{\geq 0} \tag{4.31}$$

$$\text{Isentropic:} \quad \dot{W}_{C,s} = \dot{m}(j_{es} - j_i) \cong \dot{m}(h_{es} - h_i); \quad s_{es} = s_i \tag{4.32}$$

The isentropic compressor (or pump) efficiency can be defined as (see Anim. 4.A. *compressorEfficiency*):

$$\eta_{C/P} \equiv \frac{\dot{W}_{C,s}}{\dot{W}_C} \left(\text{or } \frac{\dot{W}_{P,s}}{\dot{W}_P} \right) = \frac{j_{es} - j_i}{j_e - j_i} \cong \frac{h_{es} - h_i}{h_e - h_i} \tag{4.33}$$

Compare this expression with a turbine's definition. The numerator and denominator are reversed for a work-consuming device to ensure that the efficiency is equal to or less than 100%. As in the case of a turbine, if the kinetic energy is non-negligible, the exit velocity at the ideal exit state is assumed to be the same as the actual exit velocity.

The expression for work required by a pump can be simplified by substituting Eq. (4.30) into Eq. (4.31). Furthermore, for an isentropic pump, $s_i = s_e$ implies $T_i = T_e$ for a liquid (using SL model). Neglecting the changes in ke and pe across the pump, the absolute value of the pumping power $\dot{W}_P = -\dot{W}_{ext}$ can be expressed as (see Anim. 4.A.*pumpVsCompressor*):

$$\text{Actual:} \quad \dot{W}_P = \dot{m}(j_e - j_i) \cong \dot{m}c_v(T_e - T_i) + \frac{(p_e - p_i)}{\rho} \tag{4.34}$$

$$\text{Isentropic:} \quad \dot{W}_{P,s} = \dot{m}(j_{e,s} - j_i) \cong \dot{m}\frac{(p_e - p_i)}{\rho}; \tag{4.35}$$

The temperature rise in an actual pump is a sign of an inefficient pump due to the internal irreversibilities caused by thermodynamic friction.

EXAMPLE 4-9 | Mass and Energy Analysis of a Pump

In the problem described in Example 3-4 (Fig. 4.33), determine the pump's power consumption per-unit mass of water by (a) including and (b) neglecting the effect of kinetic energy.

SOLUTION

Analyze the open-steady system, the pumping system enclosed within the red boundary in Figure 4.33, using the mass and energy balance equations.

Assumptions

Steady state, SL model for water, uniform states based on LTE at the inlet and exit. Isentropic condition implies that the system is adiabatic and internally reversible.

Analysis

Having evaluated the difference in enthalpy and flow energy in Example 3-4, we employ the energy equation for the pump, Eq. (4.34), to produce the pumping power per-unit mass:

$$\frac{\dot{W}_{P,s}}{\dot{m}} = \frac{-\dot{W}_{ext,s}}{\dot{m}} = (j_2 - j_1) = \Delta j = 1.01 \frac{kJ}{kg}$$

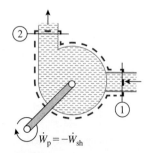

FIGURE 4.33 System schematic for Ex. 4-9 (see Anim. 4.A.*centrifugalCompressor*).

$\dot{W}_p = -\dot{W}_{sh}$

If the change in ke is neglected:

$$\frac{\dot{W}_{P,s}}{\dot{m}} = \Delta j = \Delta h + \Delta\text{pe} + \cancel{\Delta\text{ke}}^{0} = \Delta h + \Delta\text{pe}$$

$$= 0.9 + \frac{5}{1000}(9.81) = 0.95\ \frac{\text{kJ}}{\text{kg}}$$

TEST Analysis

Launch the SL open-steady TESTcalc and select Water. Evaluate the inlet and exit states, State-1 and State-2, respectively, (see TEST-code) using mdot1=1 kg/s as the basis. In the device panel, select the anchor states, enter Qdot=0, and click Calculate. The pumping power for the isentropic pump is verified. To determine the effect of neglecting kinetic energy, set Vel2=Vel1 and click Super-Calculate.

Discussion

As long as the inlet and exit conditions remain the same, the analysis is independent of the pump type. Also notice that while the external work transfer is negative (WinHip sign convention), its magnitude alone is sufficient to express the pumping power, whose direction of transfer relative to the system boundary is obvious.

EXAMPLE 4-10   **Mass, Energy, and Entropy Analysis of Compressors**

In Figure 4.34, air is compressed from an inlet condition of 100 kPa, 300 K to an exit pressure of 1000 kPa by an internally reversible compressor. Determine the compressor power per-unit mass flow rate if the device is (a) isentropic, (b) polytropic with $n = 1.3$, or (c) isothermal. Assuming compression to be polytropic with an exponent of 1.3 and an ideal intercooler, (d) determine the ideal intermediate pressure if two-stage compression intercooling is used.

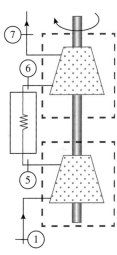

FIGURE 4.34 Compressors with intercooler in option (d) of Ex. 4-10.

SOLUTION

Analyze the open-steady systems—alternative compression devices—using the mass, energy, and entropy balance equations.

Assumptions

Steady state, PG model for air, uniform states based on LTE at the inlet and exit.

Analysis

Obtain $R = 0.287$ kJ/kg $\cdot$ K and $k = 1.4$ for air from Table C-1 or any PG TESTcalc.

(a) **Isentropic Compressor:** Let State-1 and State-2 be the inlet and exit states. Use the isentropic relation, Eq. (3.71), to simplify the energy equation, Eq. (4.31):

$$\dot{W}_{C,1-2} = \dot{m}(j_2 - j_1) \cong \dot{m}(h_2 - h_1) = \dot{m}c_p(T_2 - T_1)$$

$$= \dot{m}c_p T_1\left[\left(\frac{p_2}{p_1}\right)^{(k-1)/k} - 1\right] = \frac{\dot{m}kRT_1}{k-1}\left[\left(\frac{p_2}{p_1}\right)^{(k-1)/k} - 1\right] = 280.5\ \text{kW}$$

(b) Polytropic Compressor: Let State-3 ($p_3 = p_2$) represent the exit state. Recall from Sec. 3.5.3 that a polytropic relation has the same form as the corresponding isentropic relation where k has been replaced by the polytropic coefficient n. Therefore,

$$\dot{W}_{C,1-3} = \frac{\dot{m}nRT_1}{n-1}\left[\left(\frac{p_2}{p_1}\right)^{(n-1)/n} - 1\right] = 261.6\ \text{kW}$$

(c) Isothermal Compressor: Let State-4 ($p_4 = p_2$) represent the exit state. The energy and entropy equations produce:

$$\dot{W}_{C,1-4} = -\dot{W}_{ext,1-4} = \dot{m}(j_4 - j_1) - \dot{Q}_{1-4} \cong \dot{m}(h_4 - h_1) - \dot{Q}_{1-4}$$

$$= \dot{m}c_p(\cancel{T_4 - T_1})^0 - \dot{Q}_{1-4} = -\dot{Q}_{1-4}$$

$\dot{Q}_{1-4}$ can be evaluated from the entropy equation as follows:

$$0 = \dot{m}(s_1 - s_4) + \frac{\dot{Q}_{1-4}}{T_1} + \cancel{\dot{S}_{gen}}^0;$$

Therefore, $\quad \dot{Q}_{1-4} = \dot{m}T_1\left(c_p \ln\cancel{\frac{T_4}{T_1}}^1 - R\ln\frac{p_4}{p_1}\right) = -\dot{m}RT_1 \ln\frac{p_2}{p_1};$

$$\Rightarrow \quad \dot{W}_{C,1-4} = -\dot{Q}_{1-4} = 198.3 \text{ kW}$$

The same result can also be obtained by applying Eq. (4.10).

(d) Two-Stage Intercooled Polytropic Compressor: Let's represent the intermediate pressure by p_x (Fig. 4.35). Assuming that the inlet temperature is the same for each stage, the expression for the polytropic power can be applied to each stage, producing:

$$\dot{W}_{C,1-5-6-7} = \dot{W}_{C,1-5} + \dot{W}_{C,6-7} = \frac{\dot{m}nRT_1}{n-1}\left[\left(\frac{p_x}{p_1}\right)^{(n-1)/n} - 1 + \left(\frac{p_7}{p_x}\right)^{(n-1)/n} - 1\right]$$

To determine the value of p_x, that will minimize the total work, we differentiate the expression with respect to p_x and set it to zero. A little manipulation of the resulting equation produces $p_x = \sqrt{p_1 p_7} = 316.23$ kPa. Substituting this result into the work expression, we obtain $\dot{W}_{C,1-5-6-7} = 227.1$ kW.

TEST Analysis

Launch the PG open-steady TESTcalc and select air as the working fluid. Calculate States 1–7 as described in the TEST-code (TEST > TEST-codes). Notice how the polytropic relation is used to input v3, v5, and v7. Also note how both Qdot and Wdot_ext are calculated for an internally reversible device, set by assigning Sdot_gen to zero. TEST calculates the heat transfer from Eq. (4.7) for **internally reversible systems**. For polytropic compression (Device-B), the boundary temperature is entered as an average of the inlet and exit temperature. For two-stage compression with intercooling, p5 is initially assumed to be 500 kPa initially. Two devices—Device-D and Device-E—represent the two stages. The sum of Wdot_ext for the two stages is calculated as 225 kW. Now change p5 to a new value, then click Super-Calculate to find the net power. Repeat this iterative process until the optimum pressure is found.

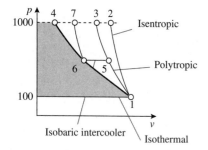

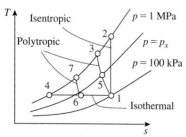

FIGURE 4.35 p-v and T-s diagrams for different compressors in Ex. 4-10. Work consumption is the least for an isothermal compressor (see Anim. 5.A.pTsconstCompressionIG).

Discussion

The isothermal compressor requires the least amount of work for a given compression ratio, which is evident from the p-v diagram in Figure 4.34 (the shaded area is minimized). One way to approach the isothermal limit is to indefinitely increase the number of stages of intercooling. In practice, the number of stages is decided by balancing the cost of an additional stage against the marginal power saving.

4.2.5 Throttling Valves

A **throttling device**, also known as an **expansion valve**, is a special restriction—an *orifice* in a plate, a *porous plug*, a *capillary tube*, or an *adjustable valve* (see Fig. 4.36 and Anim. 4.A.*expansionValve*)—that enhances frictional resistance to a flow and creates a large pressure drop.

There is no external work transfer and heat transfer is also negligible (due to insulation and high mass flow rate). For this adiabatic, single-flow, open-steady device Eqs. (4.2) and (4.3) simplify to:

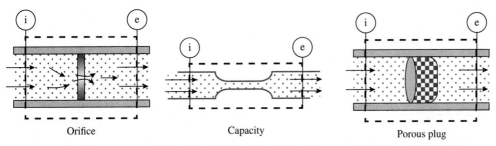

FIGURE 4.36 Different types of throttling devices. See Anim. 4.A.*expansionValve* or 4.A.*capillary.*

Energy: $j_i = j_e$; $\Rightarrow$ $h_i \cong h_e$;   **(4.36)**

Entropy: $s_e = s_i + \underbrace{\dot{S}_{gen}/\dot{m}}_{\geq 0}$; $\Rightarrow$ $s_e > s_i$ **(4.37)**

This simple form of the energy equation is often used to refer to throttling as an **isenthalpic process**. If the working fluid is an ideal (or perfect) gas, and enthalpy is a function of temperature only, we cannot expect any temperature change when a gas is throttled. But, for a PC fluid, the result can be drastically different. As can be seen from the *T-s* diagram of a PC fluid (Fig. 4.37), the constant-enthalpy lines inside the saturation dome have negative slopes. To verify this, draw a constant-enthalpy line by clicking h=constant button on a T-s plot after evaluating any saturated liquid state using a PC flow-state TESTcalc. This means that if a saturated liquid is throttled, the isenthalpic requirement produces a saturated mixture at the exit that must be at a much lower temperature.

Physically, the liquid is at a pressure much below the saturation pressure. It passes through the restriction and starts boiling as it seeks equilibrium at the lower pressure. Given the lack of heat transfer from the surroundings, the enthalpy of vaporization is supplied by the working fluid itself, which cools down rapidly as a result. The low temperature produced by throttling is exploited in applications such as refrigeration and air-conditioning. Frictional pressure drop is at the core of this phenomenon making a frictionless or isentropic valve is meaningless. Note that an increase in specific volume at lower pressure causes the flow velocity to increase to satisfy mass balance, but not to the extent to make exit ke significant as in a nozzle.

The rate of change of temperature, with respect to pressure, for an **isenthalpic process** is called the Joule-Thomson coefficient:

$$\mu_J = \left(\frac{\partial T}{\partial p}\right)_h \quad \left[\frac{K}{kPa}\right] \quad \textbf{(4.38)}$$

Obviously, μ_J is zero for an ideal, or perfect, gas. A positive μ_J means that temperature decreases during throttling, while a negative μ_J means that temperature increases. This newly defined *thermodynamic property* will be explored in Chapter 11.

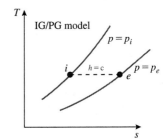

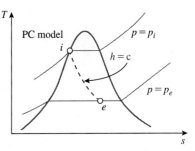

FIGURE 4.37 Constant-enthalpy lines on the *T-s* diagram reveal whether throttling is accompanied by a temperature drop (see Anim. 4.A.*expansionValve*).

EXAMPLE 4-11   **Mass, Energy, and Entropy Analysis of a Throttling Valve**

R-134a enters the throttling valve (Fig. 4.38) of a refrigeration system at 1.5 MPa and 50°C. The valve is set to produce an exit pressure of 150 kPa. Determine (a) the quality and (b) temperature at the exit, and (c) rate of entropy generation per-unit mass of refrigerant. Assume the atmospheric temperature to be 25°C. *What-if scenario:* (d) What would the answer in (c) be if the exit pressure had been reduced to 100 kPa?

SOLUTION

Analyze the open-steady system, the valve enclosed within the red boundary in Figure 4.38, using the mass, energy, and entropy balance equations.

Assumptions

Steady state, PC model for R-134a, uniform states based on LTE at the inlet and exit, and negligible changes in ke and pe.

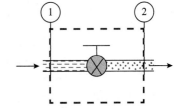

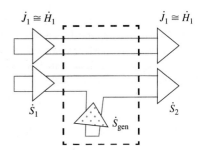

FIGURE 4.38 System schematic and energy and entropy flow diagrams for Ex 4-11.

Analysis

Use TEST, or the manual approach, described to evaluate the following states:

$$\text{State-1 (given } p_1, T_1\text{):} \quad h_1 = 123.0 \, \frac{\text{kJ}}{\text{kg}}; \quad s_1 = 0.439 \, \frac{\text{kJ}}{\text{kg}}$$

$$\text{State-2 (given } h_2 = h_1, p_2\text{):} \quad x_2 = 0.45; \quad T_2 = -17.3°\text{C}; \quad s_2 = 0.4856 \, \frac{\text{kJ}}{\text{kg}}$$

The entropy equation, Eq. (4.39), produces:

$$\dot{S}_{\text{gen,univ}}/\dot{m} = (s_e - s_i) = 0.0466 \, \frac{\text{kJ}}{\text{kg} \cdot \text{K}}$$

For the system to be adiabatic, there are no external irreversibilities, and all entropy generation takes place internally due to viscous friction.

TEST Analysis

Launch the PC open-steady TESTcalc and select R-134a. Calculate the inlet and exit states from the given conditions as described in the TEST-code (see *TEST > TEST-codes*). Analyze the device on the basis of a mass flow rate of 1 kg/s.

What-if scenario

Change p2 to **100 kPa**, press Enter, and click Super-Calculate to update all calculations. In the device panel, find the updated value of **Sdot_gen** as $0.0615 \, \text{kJ/kg} \cdot \text{K}$.

Discussion

The ratio of specific volume between the exit and inlet can be calculated as about 65.5. The velocity can increase significantly unless the exit area is made much larger. To evaluate the effect of ke, enter an arbitrary mass flow rate of 1 kg/s, an inlet velocity of 10 m/s, and use A2=A1, mdot2=mdot1, and j2=j1 for State-2. Despite a large change in velocity, the exit temperature is unchanged. Only the quality at the exit, and entropy generation rate, change slightly.

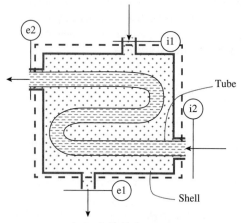

FIGURE 4.39 Schematic of a counter-flow tube-and-shell heat exchanger (see Anim. 4.B. *heatExchanger*). The multi-flow TESTcalcs in TEST allow two inlets and two exits.

4.2.6 Heat Exchangers

The purpose of a **heat exchanger** is to allow two streams of fluid to exchange heat without mixing and to minimize any heat transfer with the surroundings. The fluid that is heated may go through a phase transformation, as in the case of a steam power plant's boiler or a refrigerator's evaporator. Conversely, the fluid that is cooled may condense as in a condenser of a power plant or a refrigerator. The simplest form of heat exchanger is the **tube-and-shell** type shown in Figure 4.39 (see Anim. 4.B.*nonMixingGeneric*). In it, the inner stream flows through the tube and the outer stream flows through the shell, usually, in a counter-flow configuration. Loss of pressure in either stream is considered negligible, as are the changes in ke or pe. There is no possibility of any external work and heat transfer for the overall system is neglected, provided there is good insulation. If individual streams are selected as systems, the analysis reduces to paired pipe-flow analysis with heat rejected by on e stream being equal to the heat gained by the other.

4.2.7 TEST and the Multi-Flow, Non-Mixing TESTcalcs

In TEST, heat exchangers are classified as multi-flow, non-mixing devices and the relevant TESTcalcs are located in the open, steady, generic, multi-flow, non-mixing branch. The two streams, which are separated, may carry different fluids. A large number of composite models are offered by TEST in addition to the regular material models for identical fluids. For example, in the PC/IG model, there are two material selectors, one for a phase-change fluid (PC model) and another for a gas (IG model). In the device panel there is also a provision for two inlets and two exits. When the inlet and exit states are selected, and the non-mixing box is checked, the system schematic dynamically adjusts to the selected, configuration. The procedure for evaluating the anchor states and setting up a device remains the same as in the case of a single-flow device.

EXAMPLE 4-12 | Mass, Energy, and Entropy Analysis of a Heat Exchanger

Steam enters the condenser of a steam power plant at 15 kPa and a quality of 90% with a mass flow rate of 25,000 kg/h. It is cooled by circulating water from a nearby lake at 25°C as shown in Figure 4.40. If the water temperature does not rise above 35°C and the steam leave, the condenser as saturated liquid, determine (a) the mass flow rate of the cooling water, (b) the rate of heat removal from the steam, and (c) the rate of entropy generation in the system's universe. *What-if scenario:* (d) What would the mass flow rate be if the exit temperature of the cooling water was allowed to be 5°C warmer? Assume atmospheric conditions to be 100 kPa and 25°C.

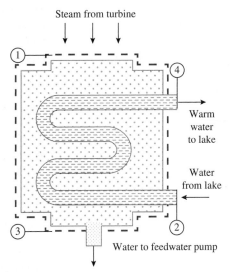

FIGURE 4.40 Condenser in Ex. 4-12 (see Anim. 4.B.*heatExchanger*).

SOLUTION

Analyze the open-steady system, the heat exchanger enclosed within the red boundary in Figure 4.40, using the mass, energy, and entropy balance equations.

Assumptions

Steady state, PC model for water, negligible changes in ke, pe, and pressure in each stream, uniform states based on LTE at the inlets and exits.

Analysis

Use TEST, or the manual approach, to determine the anchor states—State-1 and State-2 for the inlets, and State-3 and State-4 for the exits as shown in Figure 4.40:

State-1 (given p_1, x_1, $\dot{m}_1$):

$$j_1 \cong h_1 = 2361.7 \,\frac{kJ}{kg}; \quad s_1 = 7.284 \,\frac{kJ}{kg \cdot K};$$

State-2 (given p_2, T_2):

$$j_2 \cong h_2 = 105.0 \,\frac{kJ}{kg}; \quad s_2 = 0.3673 \,\frac{kJ}{kg \cdot K};$$

State-3 (given $p_3 = p_1$, $x_3 = 0$, $\dot{m}_3 = \dot{m}_1$):

$$j_3 \cong h_3 = 225.8 \,\frac{kJ}{kg}; \quad s_3 = 0.7543 \,\frac{kJ}{kg \cdot K};$$

State-4 (given $p_4 = p_2$, T_4):

$$j_4 \cong h_4 = 146.8 \,\frac{kJ}{kg}; \quad s_4 = 0.5052 \,\frac{kJ}{kg \cdot K}$$

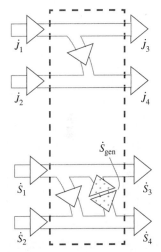

FIGURE 4.41 Energy and entropy flow diagrams in Ex. 4-12.

The mass equation for each stream produces $\dot{m}_1 = \dot{m}_3$ and $\dot{m}_2 = \dot{m}_4$. The energy equation for the entire heat exchanger, Eq. (2.8), yields:

$$\frac{dE^0}{dt} = \dot{m}_1 j_1 + \dot{m}_2 j_2 - \dot{m}_3 j_3 - \dot{m}_4 j_4; \quad \Rightarrow \quad \dot{m}_1(j_1 - j_3) = \dot{m}_2(j_4 - j_2);$$

$$\Rightarrow \quad \dot{m}_2 = \dot{m}_1(j_1 - j_3)/(j_4 - j_2) = 354.9 \text{ kg/s}$$

To obtain the internal heat exchange, analyze one of the streams, e.g., the circulating water, as a single-flow sub-system. By employing the energy equation of Eq. (4.2), we obtain:

$$0 = \dot{m}_2(j_2 - j_4) + \dot{Q}; \quad \Rightarrow \quad \dot{Q} = \dot{m}_2(j_4 - j_2) = 14.83 \text{ MW}$$

The entropy equation, Eq. (2.13), when applied to the overall adiabatic system produces:

$$\frac{ds^{0}}{dt} = \dot{m}_1 s_1 + \dot{m}_2 s_2 - \dot{m}_3 s_3 - \dot{m}_4 s_4 + \dot{S}_{\text{gen,univ}};$$

$$\Rightarrow \quad \dot{S}_{\text{gen,univ}} = \dot{m}_1(s_3 - s_1) + \dot{m}_2(s_4 - s_2) = 3.60 \, \frac{\text{kW}}{\text{K}}$$

The energy and entropy diagrams of Figure 4.41 visually illustrate how the two streams interact to generate entropy in the system's universe. While energy and entropy are transferred from the hotter stream to the cooler stream, entropy generated due to *thermal friction* boosts the entropy transported by the exit flows.

TEST Analysis

Launch the PC non-mixing, multi-flow TESTcalc (for identical fluids, there is no need for the PC/PC composite model) and select H2O as the working fluid. Calculate States 1–4 as described in the TEST-code (see *TEST > TEST-codes*). In the device panel, set up the overall heat exchanger as Device-A, select the appropriate inlet and exit states, and enter Qdot=Wdot_ext=0. Make sure that the non-mixing option is selected in the device panel and click Super-Calculate to evaluate mdot2 and Sdot_gen. To calculate the heat transfer between the streams, set up Device-B with State-2 and State-4 as the anchor states. Calculate Qdot after setting Wdot_ext=0.

What-if scenario

Change T4 to T2 + 15, press Enter, and click Super-Calculate. The new mass flow rate is calculated as mdot2 = 236.7 kg/s.

Discussion

The minimum pumping power necessary to circulate the cooling water can be shown [Eq. (4.34)] to be $\dot{W}_{\text{pump},s} = \dot{m}\Delta p/\rho$, where Δp is the pressure drop in the pipe. A 5°C increase in the exit temperature of the cooling water reduces the mass flow rate by 33.3% and the pumping power by as much. Generally, the warmer water is dumped into a water reservoir such as a lake or a river, which limits the allowable temperature rise to reduce adverse environmental consequences.

4.2.8 Mixing Chambers and Separators

A **mixing chamber** is a device where two streams of fluids merge to produce a single stream (Fig. 4.42 and Anim. 4.C.*mixingGeneric*). A simple T-elbow used to produce warm water in a shower head by mixing cold water with hot water can be considered a mixing chamber. In a mixing chamber, there is no external work transfer and heat transfer is usually negligible, as are changes in ke and pe between the inlets and exit. For the two streams to flow into the chamber, pressure at the two inlets must be equal; otherwise, fluid will back flow through the lower-pressure inlet. Frictional pressure loss in the chamber is often neglected, which allows us to assume a constant pressure throughout the system ($p_1 = p_2 = p_3$). When the two mixing streams have different temperatures, the mixing chamber works as a heat exchanger between the two. The term **direct-contact heat exchanger** is sometimes used for a mixing chamber. Mixing between two different fluids will be discussed in Chapter 11.

A **separator**, where a flow is bifurcated into two dissimilar streams, is the opposite of a mixing chamber. Separators are found in turbines with bleeding (Chapter 9) and flash chambers (Chapter 10) and their analysis is quite similar to mixing flow analysis.

4.2.9 TEST and the Multi-Flow, Mixing TESTcalcs

A mixing chamber in TEST is classified as a multi-flow, mixing, open-steady device. The corresponding TESTcalcs are located in the open, steady, generic, multi-flow mixing branch. The state panel of a mixing TESTcalc is identical to that of a single-flow open-steady TESTcalc. The device panel allows up to two inlets and two exits so that the same TESTcalc can be used as a mixing chamber or a separator. For a mixing chamber, only one

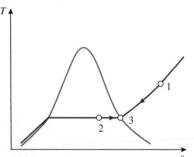

FIGURE 4.42 Schematic and *T-s* diagram for Ex. 4-13 (see Anim. 4.C.*mixingChamber*).

exit should be used—the other state should be left in its null-state (closed). Likewise, one of the inlet states should not be used for a separator. The procedure for evaluating the anchor states, and setting up a device, remains the same as in the case of a single-flow device (Sec. 4.1.1). Devices with more than two inlets can be creatively handled by splitting a multiple-inlet device into two or more devices, each with two inlets, connected in series.

EXAMPLE 4-13   **Mass, Energy, and Entropy Analysis of a Mixing System**

In Figure 4.42, superheated ammonia at 200 kPa, −10°C enters an adiabatic mixing chamber at a flow rate of 2 kg/s where it mixes with a flow of a saturated mixture of ammonia at a quality of 20%. The desuperheated ammonia exits as saturated vapor. Assuming that pressure remains constant, and the change in ke and pe can be neglected, determine (a) the mass flow rate of the saturated mixture and (b) the rate of entropy generation in the chamber's universe. *What-if scenario:* (c) What would the answer in (a) be if the superheated ammonia entered the chamber at 0°C?

SOLUTION

Analyze the open-steady system, the mixing chamber enclosed within the red boundary in Figure 4.42, using the mass, energy, and entropy balance equations.

Assumptions

Steady state, PC model for ammonia, no pressure drop in the chamber so that $p_1 = p_2 = p_3$, uniform states based on LTE at the inlet and exit of each device.

Analysis

Use TEST or a manual approach to determine the anchor states:

State-1 (given p_1, T_1, $\dot{m}_1$):

$$j_1 \cong h_1 = 1440.3 \,\frac{kJ}{kg}; \quad s_1 = 5.677 \,\frac{kJ}{kg \cdot K};$$

State-2 (given $p_2 = p_1$, x_2):

$$j_2 \cong h_2 = 359.2 \,\frac{kJ}{kg}; \quad s_2 = 1.428 \,\frac{kJ}{kg \cdot K};$$

State-3 (given $p_3 = p_2$, $x_3 = 1$):

$$j_3 \cong h_3 = 1419.5 \,\frac{kJ}{kg}; \quad s_3 = 5.599 \,\frac{kJ}{kg \cdot K}$$

The steady-state mass equation, Eq. (2.3), simplifies to $\dot{m}_3 = \dot{m}_1 + \dot{m}_2$. Substituting it in the energy balance equation of Eq. (2.8), we obtain:

$$0 = \dot{m}_1 j_1 + \dot{m}_2 j_2 - \dot{m}_3 j_3 = \dot{m}_1 j_1 + \dot{m}_2 j_2 - \dot{m}_1 j_3 - \dot{m}_2 j_3;$$

$$\Rightarrow \quad \dot{m}_2 = 0.039 \text{ kg/s}$$

The entropy equation, Eq. (2.13), when applied to the overall adiabatic system yields:

$$\frac{dS^0}{dt} = \dot{m}_1 s_1 + \dot{m}_2 s_2 - \dot{m}_3 s_3 + \dot{S}_{gen,univ};$$

$$\Rightarrow \quad \dot{S}_{gen,univ} = \dot{m}_1(s_3 - s_1) + \dot{m}_2(s_3 - s_2) = 0.0067 \,\frac{kW}{K}$$

TEST Analysis

Launch the PC multi-flow, mixing TESTcalc and select H2O as the working fluid. Set up States 1–3 and the device as described in the TEST-code (see *TEST > TEST-codes*). Leave mdot2 as an unknown. In the device panel, select the inlet and exit states, enter Qdot=Wdot_ext=0, and click Super-Calculate to update the solution. The mass flow rate of the saturated mixture is calculated and posted back in State-2.

What-if scenario

Change T1 to 0°C, press Enter, then click Super-Calculate. The new mass flow rate is mdot2 = 0.084 kg/s and the entropy generation rate is 0.0142 kW/K.

Discussion

As referred to in Sec. 2.1.3, *mixing friction* is now established as a fundamental entropy generating mechanism. The greater the differences between the two streams, in terms of temperature or phase composition, the greater are the mixing irreversibilities. This can be easily demonstrated by carrying out multiple what-if studies using the mixing TESTcalc.

EXAMPLE 4-14   **Mass, Energy, and Entropy Analysis of a Composite System**

As shown in Figure 4.43, saturated liquid water at 1.0 MPa is throttled in an expansion valve to a pressure of 500 kPa, which then flows into a flash chamber. The saturated liquid exits the chamber at the bottom while the saturated vapor, after exiting near the top, expands in an isentropic turbine to a pressure of 10 kPa. Assuming all components to be adiabatic, and a steady-flow rate of 10 kg/s at the inlet, determine (a) the quality at the turbine exit, (b) the power produced by the turbine in kW, and (c) the entropy generated in the system's universe. *What-if scenario:* (d) What would the turbine power and exit quality be if the saturated liquid entered the system at 1.5 MPa? Assume atmospheric conditions to be 100 kPa and 25°C.

SOLUTION

Analyze the open-steady system, the composite system enclosed within the red boundary in Figure 4.43, using the mass, energy, and entropy balance equations.

Assumptions

Steady state, PC model for water, negligible changes in ke and pe, uniform states based on LTE at the inlet and exit of each device.

Analysis

Use TEST, or a manual approach, to determine the five principal states shown in the schematic of Figure 4.43. To evaluate State-2, the energy equation for an adiabatic valve, $j_1 = j_2$ (Sec. 4.2.5), is used. The flash chamber separates the mixture into its component phases: saturated vapor and saturated liquid.

State-1 (given p_1, $x_1 = 0$, $\dot{m}_1$):

$$j_1 \cong h_1 = 762.8 \frac{kJ}{kg}; \quad s_1 = 2.139 \frac{kJ}{kg \cdot K};$$

State-2 (given p_2, $j_2 = j_1$):

$$\Rightarrow \quad j_2 \cong h_2 = 762.8 \frac{kJ}{kg}; \quad s_2 = 2.149 \frac{kJ}{kg \cdot K}; \quad x_2 = 5.823\%$$

State-3 (given $p_3 = p_2$, $x_3 = 0$):

$$j_3 \cong h_3 = 640.0 \frac{kJ}{kg}; \quad s_3 = 1.860 \frac{kJ}{kg \cdot K};$$

$$\dot{m}_3 = \dot{m}_2(1 - x_2) = 9.418 \frac{kg}{s}$$

State-4 (given $p_4 = p_2$, $x_4 = 1$):

$$j_4 \cong h_4 = 2748.6 \frac{kJ}{kg}; \quad s_4 = 6.822 \frac{kJ}{kg \cdot K}; \quad \dot{m}_4 = \dot{m}_2 x_2 = 0.582 \frac{kg}{s};$$

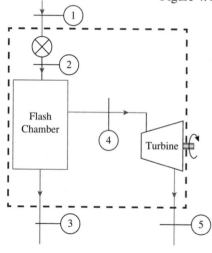

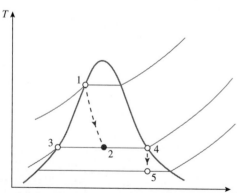

FIGURE 4.43 Schematic and *T-s* diagram for Ex. 4-14.

State-5 (given p_5, $s_5 = s_4$):

$$j_5 \cong h_5 = 2160.9 \frac{kJ}{kg}; \quad x_5 = 82.3\%$$

Energy and entropy balance on the combined system yields:

$$0 = \dot{m}_1 j_1 - \dot{m}_3 j_3 - \dot{m}_5 j_5 - \dot{W}_{sh} \quad \Rightarrow \quad \dot{W}_{sh} = 342 \text{ kW};$$

$$0 = \dot{m}_1 s_1 - \dot{m}_3 s_3 - \dot{m}_5 s_5 + \dot{S}_{gen,univ}; \quad \Rightarrow \quad \dot{S}_{gen,univ} = 0.103 \text{ kW/K}$$

TEST Analysis

Launch the PC mixing TESTcalc and select H2O as the working fluid. Set up the anchor states, individual (A–C), and the overall (D) devices as described in the TEST-code (see *TEST > TEST-codes*). Click Super-Calculate to obtain the net power and the entropy generation rate in the system's universe (Device-D).

What-if scenario

Change p1 to 1.5 MPa, press Enter, and click Super-Calculate. The turbine power is updated in Device-D as Wdot_ext = 570.8 kW.

Discussion

The solution illustrates the advantages of treating a network of multiple devices as a single composite system. The same solution could be obtained by analyzing each device separately, as illustrated in the TEST solution.

4.3 CLOSURE

This chapter presented a comprehensive mass, energy, and entropy analysis of open-steady systems based on the generalized governing equations developed in Chapter 2. Internally reversible systems were studied as a special class of systems and reversible heat and work transfer were discussed. Open-steady systems were classified into three groups: single-flow, non-mixing, multi-flow, and mixing, multi-flow systems. Analysis of single-flow devices—pipes, nozzles, diffusers, turbines, compressors, pumps and throttling valves—were emphasized. Isentropic device efficiencies for various single-flow devices were defined and explained with the help of animations. The single-flow TESTcalcs, multi-flow, mixing TESTcalcs, and the multi-flow, non-mixing TESTcalcs used for verification of manual solutions and pursuing what-if studies, will serve as the building blocks of more complicated cycle TESTcalcs in Chapters 8–10. Exergy analysis of open-steady systems will be discussed until Chapter 6.

PROBLEMS

SECTION 4-1: ANALYSIS OF SINGLE-FLOW OPEN-STEADY SYSTEMS

Flow Through Pipes and Ducts

4-1-1 An insulated electric water heater operates at a steady state with a constant pressure of 150 kPa, supplying hot water at a mass flow rate ($\dot{m}$) of 0.5 kg/s. If the inlet temperature is 20°C, and the exit temperature is maximized without creating any vapor, determine (a) the electrical power consumption ($\dot{W}_{el}$), (b) the temperature (T_2) of the water leaving the heater, (c) the volume flow rate ($\dot{V}$) in L/min at the inlet and (d) the volume flow rate in L/min at the exit. Use the PC model for water.

4-1-2 A steam heating system for a building 175 m high is supplied from a boiler located 20 m below ground level. Dry, saturated steam is supplied by the boiler at 300 kPa. It reaches the top of the building at 250 kPa. The heat loss from the supply line to the surroundings is 50 kJ/kg. Determine (a) the quality of steam at the 175 m elevation. Neglect any change in ke. (b) *What-if scenario:* What would be the change in quality if the PE was neglected in the energy balance?

4-1-3 Water flows steadily into a well-insulated electrical water heater with a mass flow rate ($\dot{m}$) of 1 kg/s at 100 kPa and 25°C. Determine (a) the electrical power consumption ($\dot{W}_{el}$) if the water becomes saturated (liquid) at the exit. Assume no pressure loss, neglect changes in ke and pe, and use the SL model (use $c_v = 4.184$ kJ/kg·K for water). (b) *What-if scenario:* What would the power consumption be if the PC model was used instead?

4-1-4 Water flows steadily down an insulated vertical pipe (of constant diameter). The following conditions are given at the inlet and exit. Inlet (state-1): $T_1 = 20°C$, $V_1 = 10$ m/s, $z_1 = 10$ m; Exit (state-2): $T_2 = 20°C$, $p_2 = 100$ kPa, $z_2 = 0$ m. Determine (a) the velocity (V_2) at the exit (m/s), (b) the pressure (p_1) at the inlet (in kPa). Use the SL model for water. Do not neglect KE and PE.

4-1-5 Water flows down steadily through a long, 10-cm diameter, vertical pipe. At the inlet: $p_1 = 300$ kPa, $T_1 = 50°C$, $V_1 = 5$ m/s, and $z_1 = 125$ m; At the exit: $p_2 = 1250$ kPa, $T_2 = 20°C$, and $z_2 = 10$ m. If the surrounding ambient temperature is 10°C, use SL model ($\rho = 997$ kg/m^3, $c_v = 4.184$ kJ/kg·K) to determine the (a) velocity (V_2) at the exit in m/s, (b) the mass flow rate ($\dot{m}$) in kg/s, (c) the rate of heat transfer ($\dot{Q}$) in kW, and (d) the rate of entropy generation ($\dot{S}_{gen}$) in the pipe's universe. (e) *What-if scenario:* What would be the answer in (c) if change in PE is neglected?

4-1-6 Water flows steadily into a well-insulated electrical water heater with a mass flow rate ($\dot{m}$) of 1 kg/s at 100 kPa, 25°C. Determine (a) the rate of entropy generation ($\dot{S}_{gen}$) in the water heater's universe if the water becomes saturated (liquid) at the exit. Assume no pressure loss, neglect changes in ke and pe, and use the SL model (use $c_v = 4.184$ kJ/kg·K for water). The ambient atmospheric conditions are 100 kPa and 20°C. (b) *What-if scenario:* What would the entropy generation rate be if the PC model was used instead?

4-1-7 Saturated liquid water flows steadily into a well-insulated electrical water heater with a mass flow rate ($\dot{m}$) of 1 kg/s at 100 kPa. Determine (a) the electrical power consumption ($\dot{W}_{el}$), and (b) the rate of entropy generation ($\dot{S}_{gen}$) in the water heater's universe if the heater turns water into saturated vapor at the exit. Assume no pressure loss, neglect changes in ke and pe, and use the PC model. The ambient atmospheric conditions are 100 kPa and 20°C.

4-1-8 Saturated water vapor flows steadily into a well-insulated electrical super heater with a mass flow rate ($\dot{m}$) of 1 kg/s at 100 kPa. Determine (a) the electrical power consumption ($\dot{W}_{el}$), and (b) the rate of entropy generation ($\dot{S}_{gen}$) in the water heater's universe if the vapor is superheated to 175°C at the exit. Assume no pressure loss, neglect changes in ke and pe, and use the PC model. The ambient atmospheric conditions are 100 kPa and 20°C.

4-1-9 An insulated high-pressure electric water heater operates at steady state at a constant pressure of 10 MPa, supplying hot water at a mass flow rate ($\dot{m}$) of 10 kg/s. If the inlet temperature is 20°C and the exit temperature is 200°C, determine (a) the electrical power consumption ($\dot{W}_{el}$), (b) the volume flow rate ($\dot{V}_1$) in L/min at the inlet, (c) the volume flow rate ($\dot{V}_2$) in L/min at the exit, and (d) the rate of entropy generation ($\dot{S}_{gen}$) in the heater. Use the PC model for water, neglect any pressure drop in the heater, and also neglect any change in KE or PE.

4-1-10 Water enters a boiler tube at 50°C, 10 MPa and at a rate of 10 kg/s. Heat is transferred from the hot surroundings which is maintained at 1000°C into the boiler tube (created by combustion of natural gas). Water exits the boiler as saturated vapor. Determine (a) the heating rate ($\dot{Q}$) in MW and (b) the rate of entropy generation ($\dot{S}_{gen}$) in the boiler tube's universe using kW/K. Assume a steady state. Neglect any pressure drop in the tube and changes in KE and PE.

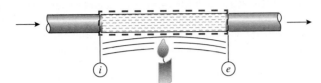

4-1-11 A coal-fired boiler produces superheated steam, steadily at 1 MPa, 500°C, from the feed water which enters the boiler at 1 MPa and 50°C. For a mass flow rate ($\dot{m}$) of 10 kg/s, determine (a) the rate of heat transfer ($\dot{Q}$) from the boiler to the water. (b) If the energetic efficiency of the boiler is 80% and the heating value of coal is 32.8 MJ/kg, determine the rate of fuel consumption in tons/hr. (c) *What-if scenario:* What would the rate of heat transfer be if the boiler pressure was 3 MPa?

4-1-12 Saturated steam at 40°C is to be cooled to saturated liquid in a condenser. If the mass flow rate ($\dot{m}$) of the steam is 20 kg/s, (a) determine the rate of heat transfer ($\dot{Q}$) in MW. Assume no pressure loss. (b) *What-if scenario:* What would the rate of heat transfer if the steam was at 90°C?

4-1-13 Water enters a boiler with a flow rate ($\dot{m}$) of 1 kg/s at 100 kPa, 20°C and leaves as saturated vapor. Assuming no pressure loss and neglecting changes in KE and PE, determine (a) the rate of heat transfer ($\dot{Q}$) in kW. (b) If the heating value of gasoline is 44 MJ/kg, what is the consumption rate ($\dot{W}$) of gasoline in kg/s? Use PC model for H$_2$O. *What-if scenario:* What would the fuel consumption rate if the water was preheated by solar radiation to 90°C?

4-1-14 Air at a pressure of 150 kPa, a velocity of 0.2 m/s, and a temperature of 30°C flows steadily in a 10 cm-diameter duct. After a transition, the duct is exhausted uniformly through a rectangular slot with a 3 cm × 6 cm in cross section. (a) Determine the exit velocity (V_2). Assume incompressible flow and use the PG model for air. (b) *What-if scenario:* What would the exit velocity be if the IG model was used?

4-1-15 Air is heated in a duct as it flows over resistance wires. Consider a 20 kW electric heating system. Air enters the heating section at 100 kPa and 15°C with a volumetric flow rate of 140 m^3/min. If heat is lost from the air in the duct to the surroundings at a rate of 150 W, determine (a) the exit temperature (T_2) of the air and (b) the energetic efficiency (η_1). Use the PG model for air and neglect the power consumed by the fan. (c) Draw an energy flow diagram for the system.

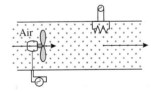

4-1-16 Air flows steadily through along insulated duct with a constant cross-sectional area of 100 cm^2. At the inlet, the conditions are 300 kPa, 300 K and 10 m/s. At the exit, the pressure drops to 100 kPa due to frictional losses in the duct. Determine (a) the exit temperature (T_2) and (b) the exit velocity (V_2). Use the PG model for air. (c) Explain why the temperature does not increase despite the presence of friction.

4-1-17 Air flows steadily through a long insulated duct that has a constant cross-sectional area of 100 cm^2. At the inlet, the conditions are 300 kPa, 300 K and 10 m/s. At the exit, the pressure drops to 50 kPa due to frictional losses in the duct. Determine (a) the mass flow rate ($\dot{m}$) of air, (b) the exit temperature (T_2) in K, and (c) the exit velocity (V_2) in m/s. Use the PG model for air. (d) Determine the entropy generated ($\dot{S}_{gen}$) due to thermodynamic friction in the duct.

4-1-18 Helium flows steadily through a long insulated duct with a constant cross-sectional area of 100 cm^2. At the inlet, the conditions are 300 kPa, 300 K and 10 m/s. A 5 kW internal electrical heater is used to raise the temperature of the gas. At the exit, the pressure drops to 100 kPa due to frictional losses in the duct. Determine (a) the exit velocity (V_2), and (b) the rate of entropy generation ($\dot{S}_{gen}$) in the system's universe. Use the PG model for air. (c) *What-if scenario:* How would the answer in part (b) change if the heater were turned off? (Assume the exit pressure to be 100 kPa.)

4-1-19 Steam enters a long horizontal pipe, with an inlet diameter of 14 cm, at 1 MPa, 250°C with a velocity of 1.2 m/s. Further downstream, the conditions are 800 kPa, 210°C, and the diameter is 12 cm. Determine (a) the rate of heat transfer ($\dot{Q}$) to the surroundings, which is 25°C, and (b) the rate of entropy generation ($\dot{S}_{gen}$) in the system's universe. (c) Draw an entropy flow diagram for the system.

4-1-20 Refrigerant R-134a flows through a long insulated pipe with a constant cross-sectional area. The inlet state is at 200 kPa and 50°C. At another section, further downstream, the state is 150 kPa, 34.3°C. Determine the velocity (V_1) of the refrigerant at the inlet.

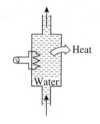

4-1-21 An electric water heater supplies hot water at 200 kPa, 80°C with a volumetric flow rate ($\dot{V}$) of 8 L/min while consuming 32 kW of electrical power, as shown in the accompanying figure. The water temperature at the inlet is 25°C, the same temperature as the surroundings. Using the SL (solid/liquid) model for water, and neglecting kinetic and potential energy changes, determine (a) the rate of heat loss ($\dot{Q}$) to the atmosphere, (b) the energetic efficiency (η_I) and (c) the rate of entropy generation ($\dot{S}_{gen,univ}$) in the heater's universe. Assume no pressure loss in the system.

4-1-22 Long steel rods of 5 cm diameter are heat-treated by drawing them at a velocity (V) of 5 m/s through an oven maintained at a constant temperature of 1000°C. The rods enter the oven at 25°C and leave at 800°C. Determine (a) the rate of heat transfer ($\dot{Q}$) to the rods from the oven and (b) the rate of entropy generation ($\dot{S}_{gen}$) in the system's universe. Use the SL model and assume steel has the same material properties as iron. (c) **What-if scenario:** What would be the rate of entropy generated in the system's universe be if the oven was maintained at 800°C?

Flow Through Nozzles

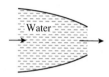

4-1-23 Liquid water flows steadily through an insulated nozzle. The following data is supplied. Inlet: $p = 500$ kPa, $T = 25$°C, $V = 10$ m/s; Exit: $p = 100$ kPa, $T = 25$°C. Determine the exit velocity by using (a) the SL model and (b) the PC model for water. (c) **What-if scenario:** What would the exit velocity be if the exit area was reduced without changing other inlet or exit conditions? (1: increase; −1: decrease; 0: remain same)

4-1-24 A water tower in a chemical plant holds 2000 L of liquid water at 25°C and 100 kPa in a tank on top of a 10 m tall tower. An insulated pipe leads to the ground level with a tap that can open a 1 cm diameter hole. Neglecting friction and pipe losses and assume temperature to remain constant. Estimate the time it will take to empty the tank. Use the SL model for water.

4-1-25 A large supply line carries water at a pressure of 1 MPa. A small leak, with an area of 1 cm^2 develops in the pipe, which is exposed to the atmosphere at 100 kPa. Assuming the resulting flow to be isentropic, determine (a) the velocity (V) of the jet and (b) the leakage rate in kg/min. **What-if scenario:** What would the (c) jet velocity and (d) leakage rate be if the leakage area increased to 2 cm^2?

4-1-26 Steam enters a nozzle operating steadily at 5 MPa, 350°C, 10 m/s and exits at 2 MPa, 267°C. The steam flows through the nozzle with negligible heat transfer and PE. The mass flow rate ($\dot{m}$) is 2 kg/s. Determine (a) the exit velocity (V_2) and (b) the exit area. (c) Draw an energy flow diagram for the nozzle.

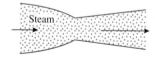

4-1-27 Steam enters an adiabatic nozzle steadily at 3 MPa, 670 K, 50 m/s and exits at 2 MPa, 200 m/s. If the nozzle has an inlet area of 7 cm^2, determine (a) the exit area. (b) **What-if scenario:** What must the exit area be for the exit velocity to be 400 m/s?

4-1-28 Superheated steam is stored in a large tank at 5 MPa and 800°C. The steam is exhausted isentropically through a converging-diverging nozzle. Determine (a) the exit pressure (p_2) and (b) the exit velocity (V_2) for condensation to begin at the exit. Use the PC model for steam. (c) **What-if scenario:** What would be the exit velocity necessary from part (b) be if steam was treated as a perfect gas?

4-1-29 Steam enters an insulated nozzle operating steadily at 5 MPa and 420°C with negligible velocity, and it exits at 3 MPa and 466 m/s. If the mass flow rate ($\dot{m}$) is 3 kg/s, determine (a) the exit area, (b) the exit temperature (T_2) and (c) the rate of entropy generation ($\dot{S}_{gen}$) in the nozzle. Neglect heat transfer and PE.

4-1-30 An adiabatic steam nozzle operates steadily under the following conditions. Inlet: superheated vapor, $p_1 = 1$ MPa, $T_1 = 300$°C, $A_1 = 78.54$ cm^2; Exit: saturated vapor, $p_2 = 100$ kPa. Determine (a) the exit velocity (V_2) in m/s, and (b) the rate of entropy generation ($\dot{S}_{gen}$) in kW/K. The mass flow rate ($\dot{m}$) is 1 kg/s.

4-1-31 Two options are available for reducing the pressure of superheated R-134a, flowing steadily at 500 kPa and 40°C, to a pressure of 100 kPa. In the first option, the vapor is allowed to expand through an isentropic nozzle. In the second option, a valve is used to throttle the flow down to the desired pressure. (a) Determine the temperature (T_2) of the flow at the exit for each option. (b) **What-if scenario:** What would the exit temperature be at the throttle if R-134a was treated as an ideal gas?

4-1-32 Steam enters an adiabatic nozzle steadily, at 3 MPa, 670 K, 50 m/s, and exits at 2 MPa. If the nozzle has an inlet area of 7 cm^2 and an adiabatic efficiency of 90%, determine (a) the exit velocity (V_2) and (b) the rate of entropy generation ($\dot{S}_{gen}$) in the nozzle's universe. Neglect PE. (c) *What-if scenario:* What would the exit velocity be if the adiabatic efficiency was 80%?

4-1-33 Steam enters an insulated nozzle steadily, at 0.7 MPa, 200°C, 50 m/s, and exits at 0.18 MPa, 650 m/s. If the nozzle has an inlet area of 0.5 m^2, determine (a) the exit temperature (T_2), (b) the mass flow rate ($\dot{m}$) and (c) the rate of entropy generation ($\dot{S}_{gen}$) in the nozzle's universe. Neglect PE. (d) Draw an entropy flow diagram for the device.

4-1-34 Steam flows steadily through an isentropic nozzle with a mass flow rate ($\dot{m}$) of 10 kg/s. At the inlet the conditions are: 1 MPa, 1000°C, and 10 m/s, and at the exit the pressure is 100 kPa. Determine the flow area at (a) the inlet and (b) the exit. (c) Determine the area of cross-section at an intermediate state where the flow velocity is 500 m/s. (d) Draw an approximate shape of the nozzle.

4-1-35 In the above problem, determine the minimum area of cross-section of the steam nozzle. (Hint: Use TEST and vary the flow velocity in the intermediate station until the flow area is minimized.)

4-1-36 CO$_2$ enters a nozzle at 35 psia, 1400°F, 250 ft/s and exits at 12 psia, 1200°F. The exit area of the nozzle is 2.8 in^2. Assuming the nozzle to be adiabatic and the surroundings to be at 14.7 psia, 65°F, determine (a) the exit velocity (V_2) and (b) the entropy generation rate ($\dot{S}_{gen,univ}$) in the nozzle's universe. *What-if scenario:* How would the answers change if the PG model were used instead?

4-1-37 Air (use the PG model) expands in an isentropic horizontal nozzle from inlet conditions of 1.0 MPa, 850 K, 100 m/s to an exhaust pressure of 100 kPa. (a) Determine the exit velocity (V_2). *What-if scenario:* What is the exit velocity (b) if the inlet kinetic energy is neglected in the energy equation? (c) What if the nozzle is vertical with the exhaust plane that is 4.0 m above the intake plane?

4-1-38 Air enters an adiabatic nozzle steadily at 400 kPa, 200°C, 35 m/s and leaves at 150 kPa, 180 m/s. The inlet area of the nozzle is 75 cm^2. Determine (a) the mass flow rate ($\dot{m}$), (b) the exit temperature (T_2) of the air, (c) the exit area of the nozzle and (d) the rate of entropy generation ($\dot{S}_{gen,univ}$) in the nozzle's universe. Use the IG model for air.

4-1-39 Steam enters an insulated nozzle operating steadily at 5 MPa and 700 K with negligible velocity, and it exits at 3 MPa. If the mass flow rate ($\dot{m}$) is 3 kg/s, using the nozzle simulation Interactive, determine (a) the exit velocity (V_2), (b) the exit temperature (T_2).

4-1-40 For the nozzle described in the above Problem 4-1-39, plot how exit velocity (V_2) varies with input temperature (T_1) varying from 400 K to 700 K, all other input parameters remaining unchanged.

4-1-41 For the nozzle described in Problem 4-1-39, plot how exit velocity (V_2) varies with input pressure (p_1) varying from 3 MPa to 5 MPa, all other input parameters remaining unchanged.

4-1-42 For the nozzle described in Problem 4-1-39, plot how exit velocity (V_2) varies with isentropic efficiency of nozzle varying from 3 MPa to 5 MPa, all other input parameters remaining unchanged.

4-1-43 Air expands in an isentropic horizontal nozzle from inlet conditions of 1.0 MPa, 850 K, 100 m/s to an exhaust pressure of 100 kPa. Using the nozzle simulation Interactive, (a) determine the exit velocity (V_2), (b) plot how exit velocity varies with inlet pressure varying from 500 kPa to 1 MPa.

Flow Through Diffusers

4-1-44 Air at 100 kPa, 15°C and 250 m/s steadily enters an insulated diffuser of a jet engine. The inlet area of the diffuser is 0.5 m^2. Air leaves the diffuser with a low velocity. Determine (a) the mass flow rate ($\dot{m}$) of air and (b) the temperature (T_2) of the air leaving the diffuser.

4-1-45 Air at 100 kPa, 12°C, and 300 m/s steadily enters the adiabatic diffuser of a jet engine. The inlet area of the diffuser is 0.4 m^2. Air leaves the diffuser at 150 kPa and 20 m/s. Determine (a) the mass flow rate ($\dot{m}$) of air, (b) the temperature (T_2) of air leaving the diffuser, and (c) the rate of entropy generation ($\dot{S}_{gen,univ}$) in the diffuser's universe. Neglect PE and use the PG model.

4-1-46 Nitrogen (use the PG model) enters an adiabatic diffuser at 75 kPa, −23°C and 240 m/s. The inlet diameter of the diffuser is 80 mm. Nitrogen leaves the diffuser at 100 kPa and 21 m/s. Determine (a) the mass flow rate ($\dot{m}$) of nitrogen, (b) the temperature (T_2) of nitrogen leaving the diffuser, and (c) the rate of entropy generation ($\dot{S}_{gen}$) in the diffuser's universe. Neglect PE.

4-1-47 Air enters an insulated diffuser operating at steady state at 100 kPa, −5°C and 250 m/s and exits with a velocity of 125 m/s. Neglecting any ΔPE and thermodynamic friction, determine (a) the temperature (T_2) of air at the exit, (b) the pressure (p_2) at the exit, and (c) the exit-to-inlet area ratio.

4-1-48 Repeat Problem 4-1-47 assuming thermodynamic friction in the diffuser causes its efficiency to go down to 85%.

4-1-49 Air at 100 kPa, 15°C, and 300 m/s enters the adiabatic diffuser of a jet engine steadily. The inlet area of the diffuser is 0.4 m², and exit area of the diffuser is 1 m². Using the diffuser simulation Interactive, calculate (a) exit pressure, (p_2) (b) exit temperature (T_2).

4-1-50 Using the diffuser described in Problem 4-1-49, plot how exit pressure (p_2) varies with inlet velocity (V_1) varying from 100 m/s to 600 m/s, all other input parameters remaining unchanged.

4-1-51 Refrigerant R-134a at 100 kPa, 300 K, and 100 m/s enters the adiabatic diffuser of a jet engine steadily. The inlet area of the diffuser is 0.1 m², and exit area of the diffuser is 1 m². Using the diffuser simulation Interactive, calculate (a) exit pressure (p_2), (b) exit temperature (T_2). Plot how exit pressure varies with inlet velocity varying from 20 m/s to 100 m/s.

Vapor, Gas, and Hydraulic Turbines

4-1-52 Steam enters a turbine operating at steady state with a mass flow rate ($\dot{m}$) of 1.5 kg/s. At the inlet, the pressure is 6 MPa, the temperature is 500°C, and the velocity is 20 m/s. At the exit, the pressure is 0.5 MPa, the quality is 0.95 (95%), and the velocity is 75 m/s. The turbine develops a power output of 2500 kW. Determine (a) the rate of heat transfer ($\dot{Q}$). (b) **What-if scenario:** What would the mass flow rate and exit velocity be if the turbine was redesigned with an exit area of 5000 cm²?

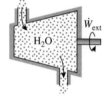

4-1-53 Steam enters an adiabatic turbine steadily at 6 MPa, 530°C and exits at 2.5 MPa, 420°C. The mass flow rate ($\dot{m}$) is 0.127 kg/s. Determine (a) the external power ($\dot{W}_{sh}$) and (b) the ratio of exit flow area to the inlet flow area to keep the exit velocity equal to the inlet velocity. Neglect KE and PE.

4-1-54 Steam enters a turbine steadily at 2.5 MPa, 350°C, 60 m/s and exits at 0.2 MPa, 100% quality, 230 m/s. The mass flow rate ($\dot{m}$) into the turbine is 1.7 kg/s and the heat transfer ($\dot{Q}$) from the turbine is 8 kW. Determine (a) the power output ($\dot{W}_{sh}$) of the turbine and (b) the energetic efficiency. (c) Draw an energy flow diagram for the turbine.

4-1-55 Steam enters a turbine steadily at 10 MPa, 550°C, 50 m/s and exits at 25 kPa, 95% quality. The inlet and exit areas are 150 cm² and 4000 cm². Heat loss of 50 kJ/kg occurs in the turbine. Determine (a) the mass flow rate ($\dot{m}$), (b) the exit velocity (V_2) and (c) the power output ($\dot{W}_{sh}$). (d) **What-if scenario:** What would the answers in parts (b) and (c) be if the turbine was redesigned with an exit area of 5000 cm²?

4-1-56 Steam enters an adiabatic turbine steadily at 6 MPa, 600°C, 50 m/s and exits at 50 kPa, 100°C, 150 m/s. The turbine produces 5 MW. Determine (a) the mass flow rate ($\dot{m}$). Neglect PE. (b) **What-if scenario:** What would the mass flow rate be if KE was also neglected?

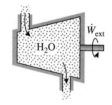

4-1-57 Steam enters an adiabatic turbine steadily at 2.5 MPa, 450°C and exits at 60 kPa, 100°C. If the power output of the turbine is 3 MW, determine (a) the mass flow rate ($\dot{m}$), (b) the isentropic efficiency and (c) the rate of internal entropy generation ($\dot{S}_{gen,univ}$) in the turbine.

4-1-58 Steam enters an adiabatic turbine, operating at steady state, with a flow rate ($\dot{m}$) of 10 kg/s at 1000 kPa, 400°C and leaves at 40°C with a quality of 0.9 (90%). Neglecting changes in KE and PE, determine (a) the pressure (p_2) (in kPa) at the turbine exit, (b) the turbine output ($\dot{W}_{sh}$) in MW, and (c) the rate of entropy generation ($\dot{S}_{gen}$), (d) **What-if scenario:** If the exit velocity was limited to 30 m/s, what would be the required exit area in m²?

4-1-59 Steam enters an adiabatic turbine steadily at 5 MPa, 540°C, 5 kg/s and leaves at 75 kPa. The isentropic efficiency of the turbine is 90%. Determine (a) the temperature (T_2) at the exit of the turbine and (b) the power output ($\dot{W}_{sh}$) of the turbine. (c) Draw an energy and an entropy flow diagram for the turbine.

4-1-60 A hydroelectric power plant operates at steady state. The difference of elevation between the upstream and downstream reservoirs is 600 m. For a discharge of 150 m³/s, determine the maximum power output ($\dot{W}_{sh}$). Use the SL model for water.

4-1-61 Hot gases enter a well-insulated jet engine turbine with a velocity of 50 m/s, a temperature of 1000°C, and a pressure of 600 kPa. The gases exit the turbine at a pressure of 250 kPa and a velocity of 75 m/s. Assume isentropic steady flow, and that each of the hot gases behaves as a perfect gas, with a mean molar mass of 25 and a specific heat ratio of 1.38. (a) Find the turbine power, per unit mass, flow rate of the working fluid. (b) *What-if scenario:* What would the specific power be if the hot gases were modeled as cold air?

4-1-62 A turbine, at steady state, receives air at a pressure of 5 bar and a temperature of 120°C. Air exits the turbine at a pressure of 1 bar. The work developed is measured as 100 kJ per kg of air flowing through the turbine. The turbine operates adiabatically and ΔKE and ΔPE can be neglected. Determine (a) the turbine efficiency and (b) the entropy generated (S_{gen}) per kg of air flow in the turbine's universe. Use the PG model for air. *What-if scenario:* What would be the (c) turbine efficiency and (d) rate of specific entropy generation if the IG model was used?

4-1-63 Steam at 4 MPa, 600°C enters an insulated turbine, operating at steady state, with a mass flow rate ($\dot{m}$) of 5 kg/s and exits at 200 kPa. Determine (a) the maximum theoretical power that can be developed by the turbine and (b) the corresponding exit temperature (T_2). Also, determine (c) the isentropic efficiency if steam exits the turbine at 220°C.

4-1-64 Combustion gases enter an adiabatic gas turbine steadily at 850 kPa, 850°C with a mass flow rate ($\dot{m}$) of 1 kg/s, and leave at 420 kPa. Treating the combustion gases as air with variable specific heat, and assuming an isentropic efficiency of 86%, determine the work output ($\dot{W}_{ext}$) of the turbine.

4-1-65 Steam (H_2O) enters a steady isentropic turbine with a mass flow rate ($\dot{m}$) of 10 kg/s at 3 MPa, 800°C and leaves at 100 kPa. Determine the power produced by the turbine using (a) the PC model for steam, (b) the PG model, and (c) the IG model.

4-1-66 Steam enters an adiabatic turbine steadily at 6 MPa, 600°C, 50 m/s and exits at 50 kPa, 100°C, 150 m/s. The turbine produces 5 kW. If the ambient conditions are 100 kPa and 25°C, determine (a) the entropy generation rate ($\dot{S}_{gen}$) by the device and the surroundings (turbine's universe). (b) Draw an entropy flow diagram for the turbine.

4-1-67 Steam enters an adiabatic turbine steadily at 2.5 MPa, 350°C, 10 m/s and exits at 1 MPa, 30 m/s. The mass flow rate ($\dot{m}$) is 5 kg/s. Using the turbine simulation Interactive, determine the shaft power ($\dot{W}_{sh}$).

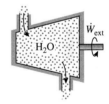

4-1-68 Using the Steam turbine described in Problem 4-1-67, plot how the shaft power ($\dot{W}_{sh}$) varies with input temperature (T_1) varying from 500 K to 1000 K, all other input parameters remaining unchanged.

4-1-69 Using the Steam turbine described in Problem 4-1-67, plot how the shaft power ($\dot{W}_{sh}$) varies with input pressure (p_1) varying from 1 MPa to 2.5 MPa, all other input parameters remaining unchanged.

4-1-70 Using the Steam turbine described in Problem 4-1-67, plot how the shaft power ($\dot{W}_{sh}$) varies with isentropic efficiency of turbine varying from 70% to 100%, all input parameters remaining unchanged.

4-1-71 Air enters an adiabatic turbine steadily at 6 MPa, 600°C, 50 m/s and exits at 50 kPa, 150 m/s with a mass flow rate ($\dot{m}$) of 6 kg/s. Assuming the turbine efficiency to be 90%, use the turbine simulation Interactive to (a) determine shaft power ($\dot{W}_{sh}$), (b) plot how shaft power ($\dot{W}_{sh}$) varies with turbine efficiency varying from 70% to 90%, all other input variables remaining unchanged.

Vapor and Gas Compressors

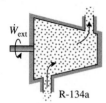

$\dot{W}_{ext}$

R-134a

4-1-72 Refrigerant R-134a enters a compressor at 175 kPa, −10°C and leaves at 1 MPa, 60°C. The mass flow rate ($\dot{m}$) is 0.02 kg/s and the power output to the compressor is 1.2 kW. Determine (a) the heat transfer rate ($\dot{Q}$) from the compressor. (b) Draw an energy diagram for the device. Assume steady-state operation.

4-1-73 Refrigerant-134a enters an adiabatic compressor as saturated vapor at 120 kPa, 1 m³/min and exits at 1 MPa. The compressor has an adiabatic efficiency of 85%. Assuming the surrounding conditions to be at 100 kPa and 25°C, determine (a) the actual power ($\dot{W}_{ext}$) and (b) the rate of entropy generation ($\dot{S}_{gen}$). (c) *What-if scenario:* What would the rate of entropy generation be if the compressor had an adiabatic efficiency of 70%?

4-1-74 Refrigerant-12 enters a compressor, operating at steady state, as saturated vapor at −7°C and exits at 1000 kPa. The compressor has an isentropic efficiency of 75%. Ignoring the heat transfer between the compressor and its surrounding, as well as KE and PE, determine (a) the exit temperature (T_2) and (b) the work input (w_{sh}) in kJ per kg of refrigerant flow.

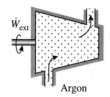

$\dot{W}_{ext}$

Argon

4-1-75 Argon gas enters an adiabatic compressor at 100 kPa, 25°C, 20 m/s and exits at 1 MPa, 550°C, 100 m/s. The inlet area of the compressor is 75 cm². Determine (a) the power of the compressor. (b) *What-if scenario:* What would the compressor power be if the inlet area were 100 cm² instead?

4-1-76 Argon gas enters an adiabatic compressor at 100 kPa, 25°C, 20 m/s and exits at 1 MPa, 550°C, 100 m/s. The inlet area of the compressor is 75 cm². Assuming the surroundings to be at 100 kPa and 25°C, determine (a) the internal entropy generation rate ($\dot{S}_{gen,int}$) by this device, (b) the external entropy generation ($\dot{S}_{gen,ext}$) in the immediate surroundings and (c) the entropy generation ($\dot{S}_{gen,univ}$) in the system's universe.

4-1-77 Air enters an adiabatic compressor at steady state at a pressure of 100 kPa, a temperature of 20°C, and a volumetric flow rate ($\dot{V}$) of 0.25 m³/s. Compressed air is discharged from the compressor at 800 kPa and 270°C. Given that the inlet and exit pipe diameters are 4 cm, determine (a) the exit velocity (V_2) of air at the compressor outlet and (b) the compressor power ($\dot{W}_{ext}$). Use PG model for air. (c) *What-if scenario:* What would be the compressor power if the pipe diameters at the inlet and exit were 5 cm?

4-1-78 Air, from the surrounding atmosphere at 100 kPa, 25°C, enters a compressor with a velocity of 7 m/s through an inlet of area 0.1 m². At the exit, the pressure is 600 kPa, the temperature is 250°C, and the velocity is 2 m/s. Heat transfer ($\dot{Q}$) from the compressor to its surrounding occurs at a rate of 3 kW. Determine (a) the power input to the compressor (magnitude only) and (b) the rate of entropy generation ($\dot{S}_{gen}$). Use the PG model for air. (c) *What-if scenario:* What would the power input need to be if the IG model was used instead?

4-1-79 A compressor, operating at steady state, receives air with a flow rate ($\dot{m}$) of 1 kg/s at 100 kPa and 25°C. The ratio of pressure at the exit to that at the inlet is 5. There is no significant heat transfer between the compressor and its surroundings. Also ΔKE and ΔPE are negligible. If the isentropic compressor efficiency is 75%, determine (a) the actual power (include sign) and (b) temperature at the compressor exit. Use the PG model for air. (c) *What-if scenario:* What would the compressor power be if the IG model had been used instead?

4-1-80 Air is compressed by an adiabatic compressor from 100 kPa, 25°C to 700 kPa, 300°C. Assuming variable specific heats and neglecting ΔKE and ΔPE, determine (a) the isentropic efficiency of the compressor and (b) the exit temperature (T_2) of air if the compressor were reversible.

4-1-81 Air from the surrounding atmosphere at 100 kPa, 25°C enters a compressor with a velocity of 7 m/s through an inlet of area 0.1 m². At the exit, the pressure is 600 kPa, and the velocity is 2 m/s. Heat transfer ($\dot{Q}$) from the compressor to its surrounding occurs at a rate of 3 kW. Using the the compressor simulation Interactive determine the shaft power ($\dot{W}_{sh}$) of the compressor.

4-1-82 Using the compressor described in Problem 4-1-81, plot how the shaft power ($\dot{W}_{sh}$) varies with input pressure (p_1) varying from 100 kPa to 600 kPa, all other input parameters remaining unchanged.

4-1-83 Using the compressor described in Problem 4-1-81, plot how the shaft power ($\dot{W}_{sh}$) varies with exit pressure (p_2) varying from 100 kPa to 600 kPa, all other input parameters remaining unchanged.

4-1-84 Refrigerant R-134a enters an adiabatic compressor, at 175 kPa, −10°C and leaves at 1 MPa. The mass flow rate ($\dot{m}$) is 0.02 kg/s. Using the compressor simulation Interactive, (a) determine the shaft power ($\dot{W}_{sh}$) of the compressor, (b) plot how shaft power varies with exit pressure varying from 175 kPa to 1 MPa.

Pumps

4-1-85 The free surface of the water in the well is 15 m below ground level. This water is to be pumped steadily to an elevation of 20 m above ground level. Assuming temperature to remain constant, and neglecting heat transfer and ΔKE, determine (a) power input to the pump required for steady flow of water at a rate of 2 m³/min. Use the SL model for water. (b) *What-if scenario:* What would the power input need to be if the flow rate was 1.5 m³/min?

4-1-86 Oil, with a density of 800 kg/m³, is pumped from a pressure of 0.6 bar to a pressure of 1.4 bar and the outlet is 3 m above the inlet. The flow rate ($\dot{m}$) is 0.2 m³/s and the inlet and exit areas are 0.06 m² and 0.03 m³, respectively. (a) Assuming the temperature to remain constant, and neglecting any heat transfer, determine the power input to the pump in kW. (b) *What-if scenario:* What would the necessary power input be if the change in KE had been neglected in the analysis?

4-1-87 A pump raises the pressure of water, flowing at a rate of 0.1 m³/s, from 70 kPa to 150 kPa. The inlet and exit areas are 0.05 m² and 0.02 m². Assuming the pump to be isentropic, and neglecting any ΔPE, determine (a) the power input ($\dot{W}_{in}$) to the pump in kW. (b) *What-if scenario:* What would the necessary power input be if the exit area was 0.01 m² instead?

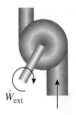

$\dot{W}_{ext}$

4-1-88 If the pump above has an adiabatic efficiency of 75%, determine (a) the power input ($\dot{W}_{in}$), (b) the exit temperature (T_2), and (c) the rate of entropy generation ($\dot{S}_{gen}$) in the pump. Assume the inlet and surroundings temperature to be 25°C.

4-1-89 Water, at 25°C, is being pumped at a rate of 1.5 kg/s from an open reservoir through a 10-cm pipe. The open end of the 5-cm discharge pipe is 15 m above the top of the reservoir's water surface. (a) Neglecting any losses, determine the power required in kW. Assume the temperature to remain unchanged and use the SL model for water. (b) *What-if scenario:* What would the required pump power be if the PC model had been used?

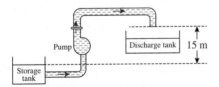

4-1-90 In the above problem, (a) determine the pumping power if the water temperature is 60°C throughout. (b) How high above the free surface of the storage tank can the pump be placed without vapor starting to form at the pump inlet? Use the PC model.

4-1-91 A 5 kW pump is raising water to an elevation of 25 m from the free surface of a lake. The water temperature increases by 0.1°C. Neglecting heat transfer and ΔKE, determine (a) the mass flow rate ($\dot{m}$) and (b) the entropy generated ($\dot{S}_{gen}$) in the system.

4-1-92 A small water pump is used in an irrigation system. The pump removes water was from a river at 10°C and 100 kPa at a rate of 4.5 kg/s. The exit line enters a pipe that goes up to 18 m above the pump and river, where water runs into an open channel. Assume the process is adiabatic and that the water stays at 10°C. Determine the required pump work rate ($\dot{W}_{in}$).

4-1-93 A 5 kW pump is raising water to a reservoir at 25 m above the free surface of a lake. The temperature of water increases by 0.1°C. Neglecting ΔKE and using the SL model, determine (a) the mass flow rate ($\dot{m}$). (b) *What-if scenario:* What would the mass flow rate be if the pumping power were 10 kW?

Throttling Devices

4-1-94 Saturated liquid water at 350°C is throttled to a pressure of 100 kPa at a mass flow rate ($\dot{m}$) of 10 kg/s. Neglecting ΔKE, determine (a) the exit temperature and (b) the amount of saturated vapor in kg/s produced by the throttling process.

4-1-95 In Problem 4-1-94, steam enters the throttling valve with a velocity of 10 m/s. If the exit area is 15 times as large as the area of the inlet, determine (a) the exit velocity (V_2) and (b) the vapor production rate.

4-1-96 Refrigerant-134a enters an insulated capillary tube of a refrigerator as saturated liquid at 0.8 MPa and is throttled to a pressure of 0.12 MPa. Determine (a) the quality of refrigerant at the final state and (b) the temperature drop (ΔT) during this process.

4-1-97 A pipe carries steam as a two phase liquid vapor mixture at 2.0 MPa. A small quantity is withdrawn through a throttling calorimeter, where it undergoes a throttling process to an exit pressure of 0.1 MPa. The temperature at the exit of the calorimeter is observed to be 120°C. Determine (a) the quality of the steam in the pipeline. (b) *What-if scenario:* What is the supply steam quality if the exit temperature is measured as 150°C?

R-12

4-1-98 Refrigerant-12 is throttled by a valve from the saturated liquid state at 800 kPa to a pressure of 150 kPa at a flow rate ($\dot{m}$) of 0.5 kg/s. Determine (a) the temperature (T_2) after throttling. (b) *What-if scenario:* What would the exit temperature be if Refrigerant-12 had been throttled down to 100 kPa?

4-1-99 Refrigerant-134a at 950 kPa is throttled to a temperature of −25°C and a quality of 0.5. If the velocity at the inlet and outlet remains constant at 10 m/s, determine (a) the quality at the inlet and (b) the ratio of exit-to-inlet area. (c) *What-if scenario:* What would the inlet quality and ratio of exit-to-inlet areas be if the inlet velocity was 20 m/s?

4-1-100 Refrigerant-12 is throttled by a valve from the saturated liquid state at 800 kPa to a pressure of 150 kPa at a mass flow rate ($\dot{m}$) of 0.5 kg/s. Assuming the surrounding conditions to be 100 kPa and 25°C, determine the rate of entropy generation ($\dot{S}_{gen}$).

4-1-101 Steam at 8 MPa and 500°C is throttled by a valve to a pressure of 4 MPa at a mass flow rate ($\dot{m}$) of 7 kg/s. Determine the rate of entropy generation ($\dot{S}_{gen}$).

4-1-102 Oxygen (model it as a perfect gas) is throttled by an insulated valve. At the inlet, the conditions are: 500 kPa, 300 K, 10 m/s, 1 kg/s. At the exit the conditions are: 200 kPa, 30 m/s. Determine (a) the exit area, and (b) the temperature (T_2) at the exit. Assume steady state, with no heat or external work transfer, and constant specific heats. Neglect ΔKE and ΔPE. (c) *What-if scenario:* What would the exit temperature be if the change in ke was not neglected. (d) Discuss if the use of IG model will affect your answers.

SECTION 4-2: ANALYSIS OF MIXING OPEN-STEADY SYSTEMS

4-2-1 Liquid water, at 100 kPa and 10°C, is heated by mixing it with an unknown amount of steam, at 100 kPa and 200°C. Liquid water enters the chamber at 1 kg/s and the chamber loses heat at a rate of 500 kJ/min with the ambient at 25°C. If the mixture leaves at 100 kPa and 50°C, determine (a) the mass flow rate ($\dot{m}$) of steam and (b) the rate of entropy generation ($\dot{S}_{gen}$) in the system and its immediate surroundings.

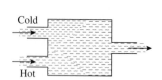
Cold
Hot

4-2-2 Consider an ordinary shower where hot water at 60°C is mixed with cold water at 10°C. A steady stream of warm water at 40°C is desired. The hot water enters at 1 kg/s. Assume heat losses from the mixing chamber to be negligible and the mixing to take place at a pressure of 140 kPa. Determine the mass flow rate ($\dot{m}$) of cold water using the SL model.

4-2-3 Superheated steam, with a state of 450°C, 1.8 MPa, flows into an adiabatic mixing chamber at a rate of 0.3 kg/s. A second stream of dry, saturated water vapor, at 1.8 MPa, enters the chamber at a rate of 0.1 kg/s. There is no pressure loss in the system and the exit pressure is also 1.8 MPa. Determine (a) the mass flow rate ($\dot{m}$), (b) temperature (T_2) of the exit flow, and (c) the entropy generation rate ($\dot{S}_{gen}$) during mixing.

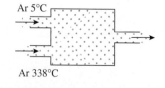
Ar 5°C
Ar 338°C

4-2-4 Argon gas flows steadily through a mixer nozzle device. At the first inlet, argon enters at 200 kPa, 5°C and 0.01 kg/s. At the second inlet, argon enters at 338°C, 200 kPa and 0.008 kg/s. At the exit, argon leaves at 94°C and 100 kPa. A stirrer transfers work into the device at a rate of 0.005 kW, and the heat transfer rate leaving the device is 0.007 kW. Determine (a) velocity (V_2) of argon at exit and (b) the entropy generation rate ($\dot{S}_{gen}$) during mixing.

4-2-5 A hot water stream at 75°C enters a mixing chamber, with a mass flow rate ($\dot{m}$) of 1 kg/s, where it is mixed with a stream of cold water at 15°C. If the mixture leaves the chamber at 40°C, determine (a) the mass flow rate ($\dot{m}$) of the cold water stream, and (b) the entropy generation rate ($\dot{S}_{gen}$) during mixing. Assume all streams are at a pressure of 300 kPa.

4-2-6 Liquid water, at 250 kPa and 20°C, is heated in a chamber by mixing with superheated steam, at 250 kPa and 350°C. The water enters the chamber at a rate of 2 kg/s. If the mixture leaves the chamber at 55°C, determine (a) the mass flow rate ($\dot{m}$) of the superheated steam and (b) the entropy generation rate ($\dot{S}_{gen}$) during mixing.

4-2-7 Water, at 350 kPa and 15°C, is heated in a chamber by mixing with saturated water vapor, at 350 kPa. Both streams enter the mixing chamber at a mass flow rate ($\dot{m}$) of 1 kg/s. Determine (a) the temperature (T_2) and (b) quality of the exiting stream. (c) Also find the entropy generation rate ($\dot{S}_{gen}$) during mixing.

4-2-8 Water, at 150 kPa and 12°C, is heated in a mixing chamber, at a rate of 3 kg/s, where it is mixed with steam entering, at 150 kPa and 120°C. The mixture leaves the chamber at 150 kPa and 55°C. Heat is lost to the surrounding air at a rate of 3 kW. (a) Determine the entropy generation rate ($\dot{S}_{gen}$) during mixing. (b) Draw an entropy flow diagram for the chamber.

SECTION 4-3: ANALYSIS OF NON-MIXING OPEN-STEADY SYSTEMS

4-3-1 Steam (use the PC model) enters a closed feedwater heater at 1.1 MPa, 200°C with a mass flow rate ($\dot{m}$) of 1 kg/s and leaves as saturated liquid at the same pressure. Feedwater enters the heater at 2.5 MPa, 50°C and leaves 12°C below the exit temperature of steam at the same pressure. Neglecting any heat losses, determine (a) the mass flow rate ratio ($\dot{m}_{water}$: $\dot{m}_{steam}$) and (b) the entropy generation rate ($\dot{S}_{gen}$) in the device and its surroundings. Assume surroundings to be at 20°C. (c) *What-if scenario:* What would the mass flow rate ratio and entropy generation rate be if the feedwater left 30°C below the exit temperature of steam?

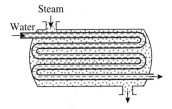

4-3-2 Refrigerant-134a, at 1.5 MPa, 90°C, is to be cooled by air to a state of 1 MPa, 27°C in a steady-flow heat exchanger. Air enters at 110 kPa, 25°C with a volume flow rate of 820 m³/min and leaves at 95 kPa, 62°C. Neglecting any heat losses, determine (a) the mass flow rate ($\dot{m}$) of refrigerant and (b) the entropy generation rate ($\dot{S}_{gen}$) in the system's universe.

4-3-3 Refrigerant-12 enters a counter-flow heat exchanger at −15°C, with a quality of 45% and leaves as saturated vapor at −15°C. Air at 100 kPa enters the heat exchanger in a separate stream with a mass flow rate ($\dot{m}$) of 5 kg/s and is cooled from 25°C to 10°C, with no significant change in pressure. The heat exchanger is at steady state and there is no appreciable heat transfer from its outer surface. Determine the rate of entropy generation ($\dot{S}_{gen}$).

4-3-4 Refrigerant-134a, at 900 kPa, 75°C and a mass flow rate ($\dot{m}$) of 9.5 kg/min, is to be cooled by water in an insulated condenser until it exits as a saturated liquid at the same pressure. The cooling water enters the condenser at 290 kPa, 11°C and leaves at 32°C with the same pressure. Neglecting any heat losses, determine (a) the mass flow rate ($\dot{m}$) of the cooling water and (b) the rate of entropy generation ($\dot{S}_{gen}$) in the system.

4-3-5 Steam enters the condenser of a steam power plant at 30 kPa, a quality of 90% and a mass flow rate ($\dot{m}$) of 300 kg/min. It leaves the condenser as saturated liquid at 30 kPa. It is to be cooled with water from a nearby river by circulating the water through the tubes within the condenser. To prevent thermal pollution, the river water is not allowed to be heated to a temperature above 5°C. Determine (a) the mass flow rate ($\dot{m}$) of the cooling water and (b) the entropy generation rate ($\dot{S}_{gen}$) in the heat exchanger.

SECTION 4-4: MULTIPLE INTERCONNECTED DEVICES

4-4-1 An irrigation pump takes water at 25°C from a lake and discharges it through a nozzle, located 20 m above the surface of the lake, with a velocity of 10 m/s. The exit area of the nozzle is 50 cm². Assuming adiabatic and reversible flow through the system, determine (a) the power input in kW. (b) *What-if scenario:* What would the required power input be for the exit velocity to double?

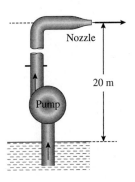

4-4-2 A water cannon sprays 50 L/min of liquid water, at a velocity of 100 m/s, horizontally out from a nozzle. It is driven by a pump that receives the water from a tank at 20°C and 100 kPa. The water in the tank and the nozzle exit are at the same elevation. Assuming adiabatic and reversible flow throughout the system, determine (a) the nozzle exit area, (b) the power input to the pump, and (c) the pressure (p_2) at the pump exit.

4-4-3 To operate a steam turbine in part-load power output, a throttling valve is used as shown in the figure below. The valve reduces the pressure of steam before it enters the turbine. The state of steam in the supply line remains fixed at 2 MPa, 500°C and the turbine exhaust pressure remains fixed at 10 kPa. Assuming the turbine to be adiabatic and reversible, determine (a) the full-load specific work output in kJ/kg, (b) the pressure (p) of the steam must be throttled to for 75% of full-load output, and (c) entropy generation ($\dot{S}_{gen}$) in the systems and their immediate surroundings per unit mass of steam.

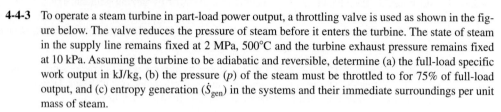

4-4-4 Repeat the above problem assuming the turbine to have an adiabatic efficiency of 90%.

4-4-5 An insulated mixing chamber receives 2 kg/s R-134a at 1 MPa, 100°C in a line (state-3). Another line brings 1 kg/s of R-134a as saturated liquid at 70°C (state-1), which is throttled to a pressure of 1 MPa (state-2) before it enters the mixing chamber. At the exit (state-4), the pressure is 1 MPa. Determine (a) the temperature (T_2) at the exit, (b) the entropy generated by the valve ($\dot{S}_{gen,valve}$), (c) the entropy generated by the mixing chamber, ($\dot{S}_{gen,mixing}$) and (d) the entropy generated in the system's universe ($\dot{S}_{gen,univ}$).

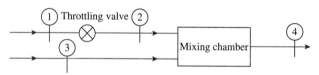

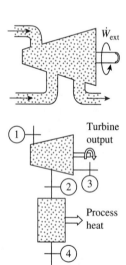

4-4-6 An adiabatic steam turbine receives steam from two boilers. One flow has a mass flow rate ($\dot{m}$) of 5 kg/s at 3 MPa and 600°C. The other flow has a mass flow rate ($\dot{m}$) of 5 kg/s at 0.5 MPa and 600°C. The exit flow is at 10 kPa with a quality of 100%. Neglecting ΔKE, determine (a) the total power output in MW and (b) the rate of entropy generation ($\dot{S}_{gen}$) in the turbine. *What-if scenario:* What would the (c) power output and (d) entropy generation rate be if the turbine worked in a reversible manner?

4-4-7 Steam is bled from a turbine to supply 2 MW of process heat in a chemical plant at 200°C, as shown in the schematic, so that state 4 is saturated liquid water at 200°C. At the turbine inlet, (state-1) steam is at 5 MPa, 500°C and at the turbine exit the pressure is 10 kPa. Determine (a) the quality of steam at the turbine exit, (b) the bleed pressure in MPa (assume no frictional losses), (c) the bleed rate, and (d) the mass flow rate ($\dot{m}$) at the turbine inlet if the power output is 2 MW. (e) *What-if scenario:* What would the mass flow rate at state-1 be if the process heating demand went down to 1 MW?

5 MASS, ENERGY, AND ENTROPY ANALYSIS OF UNSTEADY SYSTEMS

In Chapter 2, we derived the balance equations, in their most general forms, for an unsteady open system, called the *generic* system, and applied them to simplified analysis of closed-steady systems, which does not require property evaluation. After developing several material models for property evaluation in Chapter 3, we analyzed open-steady systems in Chapter 4. This chapter is dedicated to the analysis of the remaining class of systems—all types of unsteady systems: open or closed. Unsteady systems are divided into two categories—*transient* systems, where an instantaneous rate of change is important, and *processes,* where the overall changes in a system, transitioning from an initial or *beginning state* to a *final state*, are important. To analyze a process, the unsteady balance equations are integrated, over time, from the beginning to the end of the process, converting differential balance equations into algebraic forms. Given their importance, closed and open processes are studied separately; some of the closed processes analyzed in this chapter will find applications during subsequent chapters, especially during the analysis of reciprocating cycles in Chapter 7.

The framework of the analysis remains the same as established in Chapter 4—classify a system through suitable assumptions, customize the balance equations, select an appropriate material model for property evaluation of the working substance, make necessary approximations, obtain a manual solution, use TEST to verify results, and carry out "what-if" studies for further insight whenever possible.

5.1 UNSTEADY PROCESSES

Using the image analogy in Sec. 1.4, we defined an *unsteady system* as one in which the global image of the system changes during the time of observation. For a generic system, such as the one illustrated in Anim. 1.A.*systemsClassified,* the unsteady condition of a system is associated with the change of color over time. The color at a given location represents the local state in *local thermodynamic equilibrium* (LTE was introduced in Sec. 1.2). With the local state changing over time, global properties such as the total mass, m, total stored energy, E, or total entropy, S, of an unsteady system are expected to change with time. Mathematically, the *unsteady terms* in the mass, energy, and entropy balance equations—dm/dt, dE/dt, and dS/dt—cannot be set to zero as in the case of systems at steady state (Chapter 4).

Unsteady systems are best detected by monitoring the global image of the system. A change in a single property during the period of interest is sufficient to render a system unsteady. Several examples of unsteady systems are given in Figure 5.1. During the heating of a solid block [Fig. 5.1(a)], both mass and volume remain constant; however, the temperature increases. During the compression process in a piston-cylinder device [Fig. 5.1(b)], the volume decreases while mass remains constant. During the mixing between the two

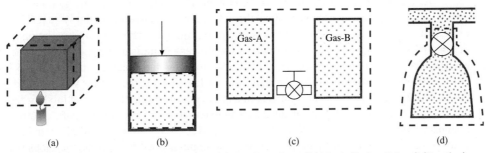

(a)   (b)   (c)   (d)

FIGURE 5.1 Examples of unsteady systems: (a) Heating of a solid block (Anim. 5.A.*solidHeating*); (b) Compressing a gas (Anim. 5.A.*pTsConstCompressionIG*); (c) Mixing two gases (Anim. 5.C.*vConstMixing*); (d) Charging a cylinder (Anim. 5.E.*charging*). In each case the *image* of the system, defined by the red boundary, changes with time.

gases in Figure 5.1(c), the total mass and total stored energy may not change; however, the composition of the system changes over time. During charging of an evacuated tank [Fig. 5.1(d)] the mass increases while the volume remains constant.

Sometimes, the time scale of observation determines whether the global image can be assumed to be frozen in time. The reciprocating power cycle, illustrated in Anim. 2.F.*closedCycle* (click the steady state radio-button), is clearly unsteady if we are interested in a single cycle; however, if the smallest period of interest is generally such that a large number of cycles are executed within that time, the blurry global image (time averaged over the period) does not change and the same system can be regarded as steady (click Closed-Steady Cycle in Anim. 2.F.*closedCycle*). For most open-steady systems, that were covered in the last chapter—a nozzle or a turbine for example—time-dependent fluctuations of properties occur due to a phenomenon called turbulence. Even then a steady analysis is possible because, in most cases, the period of fluctuation (turbulence time scale) is much shorter than the time of observation. Before doing an unsteady analysis, it is important to verify that changes in the global state are significant during the period of interest.

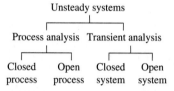

Unsteady systems

Process analysis   Transient analysis

Closed   Open   Closed   Open
process   process system   system

FIGURE 5.2 Of all unsteady systems (see 1.A.*systemsClassfied*), those executing closed processes are most commonly encountered.

There are two types of unsteady analysis (see Fig. 5.2). In **transient analysis**, we are interested in instantaneous rates of property changes, such as the rate of temperature increase as a cup of water is heated in a microwave oven. We will discuss transient analysis in Sec. 5.2. Most unsteady systems in engineering applications involve a **process**—a transition of a system over a certain period. In a process, the system starts from an initial or **beginning state** and ends in a **final state**. Represented by the symbols b and f (recall that symbols i and e are reserved for inlet and exit states), these states form the process's **anchor states**. Instead of instantaneous rates of changes, total changes in properties over the process duration and total transfer of mass, heat, and work (**process variables**), are of interest in a process analysis. A process analysis can be simplified, by integrating the balance equations between the anchor states b and f, eliminating unnecessary path-dependent details. We begin with **closed processes**—processes executed by closed systems (see Anim. 5.A.*closedProcessGeneric*).

5.1.1 Closed Processes

For a closed process, we are interested in properties of the **anchor states**—b and f states—of the system as well as the process variables total heat and external work transfer (there cannot be any mass transfer in a closed system). The mass equation is trivial—m must remain constant in the absence of mass transfer.

To integrate the energy equation for a closed system over the duration of the process, Eq. (2.10) is multiplied by dt and integrated from the b-state to the f-state:

$$dE = \dot{Q}dt - \dot{W}_{ext}dt = dQ - dW_{ext}; \quad \Rightarrow \quad \int_b^f dE = \int_b^f dQ - \int_b^f dW_{ext} \tag{5.1}$$

$$\Rightarrow \quad \underbrace{\Delta E = E_f - E_b}_{\substack{\text{Change in } E \text{ during} \\ \text{the process.}}} = \underbrace{Q}_{\substack{\text{Heat added to} \\ \text{the system.}}} - \underbrace{W_{ext}}_{\substack{\text{Work transferred} \\ \text{out of the system.}}}; \quad [\text{kJ}] \tag{5.2}$$

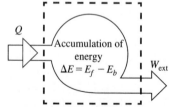

FIGURE 5.3 An energy flow diagram explaining Eq. (5.2) (see Anim. 5.A.*flowDiagram*).

The change in the stored energy of the system ΔE is due to the net energy transfer into the system, consisting of heat received minus the external work delivered by the system. This is also explained with the help of the energy flow diagram (Fig. 5.3), where the accumulation (positive or negative) of energy in the system during the process is indicated by a balloon. The stored energy E is a property; therefore, we do not need to know the precise path the system follows during the process to evaluate the left hand side of Eq. (5.2). If the anchor states can be broken down into local states of the sub-systems, that are in equilibrium, the material models developed in Chapter 3, can be used to evaluate ΔE. The external work consists of shaft, electricity, and boundary work, as described by Eq. (0.20); that is, $W_{ext} = W_{sh} + W_{el} + W_B$. For stationary systems, in the absence of mechanical work W_M, the boundary work W_B is synonymous with $pd\forall$ work. Unless stated otherwise, W_M is assumed to be zero for all systems. Note that the system does not have to be in quasi-equilibrium during the process for Eq. (5.2) to apply. However, evaluation of the boundary work $W_{pd\forall}$, if any, requires the quasi-equilibrium assumption to establish p as a function of $\forall$ for evaluating the integral of Eq. (0.16).

Flow work is absent (by definition) in a closed system, $W_{\text{ext}} = W$, and Eq. (5.2) is alternatively expressed as $\Delta E = Q - W$ in many textbooks. We will continue the use of W_{ext} to emphasize that the same energy balance equation is used in balance different forms for open-steady systems and closed processes.

The entropy equation for a closed system, Eq. (2.15), can be customized for a closed process to produce the following result:

$$dS = \frac{\dot{Q}}{T_B}dt + \dot{S}_{\text{gen}}dt; \quad \Rightarrow \quad \int_b^f dS = \int_b^f \frac{dQ}{T_B} - \int_b^f dS_{\text{gen}} \tag{5.3}$$

$$\Rightarrow \quad \underbrace{\Delta S = S_f - S_b}_{\substack{\text{Change in } S \text{ during} \\ \text{the process.}}} \quad = \underbrace{\frac{Q}{T_B}}_{\substack{\text{Entropy added} \\ \text{to the system} \\ \text{by heat.}}} + \underbrace{S_{\text{gen}}}_{\substack{\text{Entropy generated} \\ \text{in the system} \\ \text{during the process}}}; \quad \left[\frac{\text{kJ}}{\text{K}}\right] \tag{5.4}$$

For a closed process, the change in system entropy is the sum of entropy transfer through heat and entropy generation during the process. See the entropy-flow diagram in Figure 5.4 to visualize the different terms of the customized entropy equation for a closed process. In integrating the entropy transfer term, we have assumed that the boundary temperature is uniform and does not change during the process. If the process is adiabatic ($Q = 0$), the entropy generated can be indirectly obtained from the entropy change of the working substance using Eq. (5.4). Also note that if a closed system is *adiabatic* and *reversible* ($S_{\text{gen}} = 0$), there cannot be any accumulation of entropy in the system, making the process *isentropic* ($S_f = S_b$).

It is convenient to divide closed processes into two categories—**uniform closed processes** in which two unique states, each based on LTE, can describe the process's anchor states (*b*-state and *f*-state) and **non-uniform closed processes** where a composite state based on more than one LTE must be used to describe at least one of the anchor states. The processes depicted in Figs. 5.1(a) and 5.1(b) can be treated as uniform processes. In each case a pair of states, based on LTE, can describe the initial and final states. The mixing between the two gases in Figure 5.1(c) must be regarded as a non-uniform process as the *b*-state involves two different LTE's; one for Gas-A and one for Gas-B.

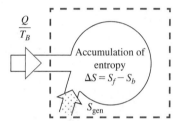

FIGURE 5.4 An entropy flow diagram explaining Eq. (5.4). Also see Anim. 5.A.*flowDiagram*.

5.1.2 TEST and the Closed-Process TESTcalcs

The TESTcalcs offered by TEST to analyze closed systems undergoing uniform processes are called **closed-process TESTcalcs**. These are located in the systems, closed, process, generic, uniform systems branch. Like the *open-steady TESTcalcs* introduced in Sec. 4.1.1, these TESTcalcs build upon the state TESTcalcs. The state panel is visible by default, where the anchor states with a given material model are evaluated. In the *process analysis* panel, a process is identified by a letter (A–Z) and it is set up by importing the anchor states from a stack of fully or partially calculated states. Known process variables (Q, W_B, W_O, T_B, and S_gen) are entered, and energy and entropy equations are solved. Given the importance of boundary work in a closed process, external work is separated into W_B and W_O, the latter being the sum of electrical and shaft work (Sec. 0.10.6). If any unknown property of an anchor state is calculated, it is posted in the appropriate state panel. Clicking Super-Calculate iterates between the state and process panels, completing the analysis. Non-uniform process TESTcalcs and open-process TESTcalcs will be introduced in latter sections. (The exergy panel will be discussed in the next chapter.)

5.1.3 Energetic Efficiency and Reversibility

Depending on what can be considered the desired output and required input, an **energetic efficiency** can be defined for a process like that of an open-steady device (Sec. 4.1.2). An energy flow diagram can be helpful in defining such an efficiency, which is illustrated in Example 5-1 (which parallels Example 4-1).

EXAMPLE 5-1   Energy and Entropy Analysis of a Water Heater

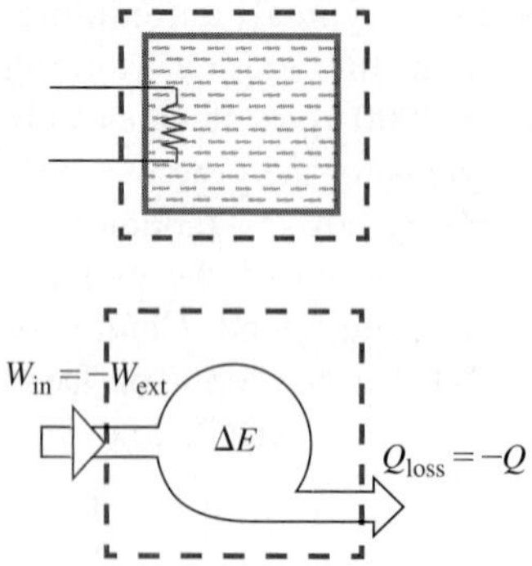

FIGURE 5.5 System schematic and energy flow diagram for Ex. 5-1 (click Energy Diagram in Anim. 5.A.*waterHeater*).

As shown in Figure 5.5, an electric heater is used to raise the temperature of 1000 L of water stored in a tank from 15°C to 70°C. Due to poor insulation, 18 MJ of heat is lost to the atmosphere, which is at 15°C. Using the SL model for water, determine (a) the electrical work consumption in kWh, (b) the energetic efficiency, and (c) the entropy generation in the heater's universe.

SOLUTION

Analyze the heater's universe, enclosed by the red boundary in Figure 5.5, using the energy and entropy balance equations customized for a closed process.

Assumptions

The beginning and final states are based on LTE and represented by State-1 and State-2, respectively.

Analysis

The material properties for water, ρ and c_v, are retrieved from Table A-1, or any SL system-state TESTcalc, as 997 kg/m^3 and 4.184 kJ/kg·K.

The mass of the water in the tank is:

$$m = \rho \forall = 997(1000 \times 10^{-3}) = 997 \text{ kg}$$

With no change in volume, there cannot be any boundary work transfer. There is no change in KE or PE; therefore, the energy equation, Eq. (5.2) simplifies as:

$$W_{in} = -W_{ext} = \Delta E - Q = \Delta U - (-Q_{loss}) = mc_v\Delta T + Q_{loss}$$

$$= (997)(4.184)(70 - 15) + 18,000 = 247,430 \text{ kJ} = 68.73 \text{ kWh}$$

While the external work and heat transfer are negative (recall the WinHip sign convention from Sec. 0.7.2) W_{in} and Q_{loss} are absolute values. Also note that the SL model (Sec. 3.3.3) is used to relate ΔU to ΔT.

The external work is the required energy input while the desired output is the energy supplied to the water, $\Delta E = \Delta U$. The energetic efficiency can be defined and evaluated as:

$$\eta_I = \frac{\Delta E}{W_{in}} = \frac{W_{in} - Q_{loss}}{W_{in}} = \frac{247,430 - 18,000}{247,430} = 92.7\%$$

For the system's universe, enclosed within the red boundary of Figure 5.5, the entropy equation, Eq. (5.4), coupled with the SL model, produces:

$$S_{gen,univ} = \Delta S - \frac{Q}{T_0} = \dot{m}c_v \ln\frac{T_2}{T_1} + \frac{Q_{loss}}{T_0} \tag{5.5}$$

$$= (997)(4.184)\ln\frac{(273 + 70)}{(273 + 15)} + \frac{18,000}{(273 + 15)} = 792 \frac{\text{kJ}}{\text{K}}$$

TEST Analysis

Launch the SL closed-process TESTcalc and select **Water(L)** from the working substance menu. Evaluate State-1 and State-2 from the given properties as described in TEST-code (see *TEST > TEST-codes*). In the process panel, select State-1 and State-2 as the *b*-state and *f*-state, enter W_B=0, Q, and T_B. Click Calculate to verify the manual solution.

Discussion

Entropy is generated internally, as electrical energy is dissipated through *electrical friction* into internal energy, and externally, through *thermal friction* in the tank's immediate surroundings. An energetic efficiency of 93% seems quite satisfactory, but as we will see in Chapter 6, the accompanying entropy generation is an indicator that the objectives can be met with a smaller consumption of useful work. In this solution, we could have used the more accurate PC model; however, the simplicity of the SL model allows a closed-form solution.

Had there been any phase change involved, the SL model must be rejected in favor of the PC model.

A process can be described graphically by a **process curve** created on a thermodynamic plot, such as a *T-s* diagram (Fig. 5.6), by joining the anchor states with a dotted line—dotted to emphasize the fact that the intermediate states the system passes through may not be in equilibrium and cannot be uniquely identified on the plot. A process that is executed as a series of small steps, allowing sufficient time at each step for the system to relax to equilibrium, is called a **quasi-equilibrium** or a **quasi-static** process. The process diagram in that case reduces to a succession of equilibrium states and is represented by a solid line in the limit of infinite intermediate steps. For the process in Example 5-1, the quasi-equilibrium line in Figure 5.6 will follow the logarithmic relation, $s = s_{ref} + c_v \ln(T/T_{ref})$, as dictated by the SL model. Although practical processes are not extremely slow, a quasi-equilibrium assumption is justifiable on the ground that the time to achieve equilibrium is often quite small. This assumption is routinely used for modeling processes in reciprocating cycles, which will be introduced in Chapter 7.

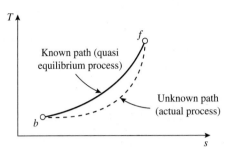

FIGURE 5.6 The process curves on a *T-s* diagram for the process described in Ex. 5-1.

An **internally reversible process** is another idealized process, where there is no internal entropy generation in the system during the process, that is, $S_{gen,int} = 0$. A departure from equilibrium is generally caused by some kind of internal imbalance, which triggers one or more of the entropy generating mechanisms (e.g., viscous friction, thermal friction, etc.). To remain internally reversible, a process must follow a quasi-equilibrium path, a path that is a succession of many equilibrium states. The reverse is true only for *simple compressible systems* (Sec. 3.1.2), where boundary work is the only mode of work transfer. For non-simple systems—the water heater in Example 5-1 for instance—entropy is generated even if the process is carried out in a quasi-equilibrium manner. Finally, a **reversible process**, like a reversible system (Sec. 4.1.3), requires all internal and external entropy generation to be eliminated during the process, $S_{gen,univ} = 0$. Therefore, a reversible process is the most restrictive of all idealized processes (Fig. 5.7).

Actual process

↓

Quasi-equilibrium process

↓

Internally reversible process

↓

Reversible process

FIGURE 5.7 Different types of processes with increasing degree of restrictions (from top to bottom).

Revisiting the derivation of the $pd\mathbb{V}$ formula for boundary work in Sec. 0.8.4, we now can say that the formula is accurate only for quasi-equilibrium processes. Without this condition, p cannot be routinely expressed as a function of $\mathbb{V}$. The formula was derived for a simple compressible system, which allows only boundary work transfer; hence, the process must be internally reversible. The $pd\mathbb{V}$ work formula of Eq. (0.16) should only be applied to internally reversible (or reversible) processes:

$$W_{pd\mathbb{V}} = W_{B,\,int.rev.} = \int_b^f pd\mathbb{V} = m\int_b^f pdv; \quad dW_{B,\,int.rev.} = pd\mathbb{V}; \quad [kJ] \qquad (5.6)$$

The magnitude of the work is related to the area under the *p-v* process curve, as shown in Figure 5.8 [you should compare this formula with the corresponding expression of Eq. (4.10) for $\dot{W}_{ext,\,int.rev.}$]. While the work required to compress a gas in a piston-cylinder device is proportional to the area under the *p-v* diagram, the power required to compress the gas by an internally reversible open-steady device is proportional to the area to the left of the curve (Fig. 5.8).

To develop a heat transfer formula for an internally reversible process (as we did for open-steady systems in Sec. 4.1.3), the entropy equation can be simplified by recognizing that the temperature must be uniform inside the system at any instant to prevent entropy generation through thermal friction. Substituting $T_B = T$ for a uniform system undergoing an internally reversible process, the entropy equation, Eq. (2.15), simplifies to:

$$\frac{dS}{dt} = \frac{\dot{Q}}{T_B} + \dot{S}_{gen,int}^{\;0} = \frac{\dot{Q}_{int.rev.}}{T}$$

$$\Rightarrow \quad TdS = \dot{Q}_{int.rev.}dt = dQ_{int.rev.}; \quad \text{or}, \quad Q_{int.rev.} = \int_b^f TdS = m\int_b^f Tds; \quad [kJ] \qquad (5.7)$$

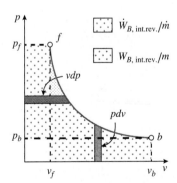

FIGURE 5.8 The shaded areas are proportional to the magnitude of the work transfer: black for an internally reversible closed process and red for an internally reversible open-steady device (compare Anim. 4.A.*intReversible* and 5.A.*intReversible*).

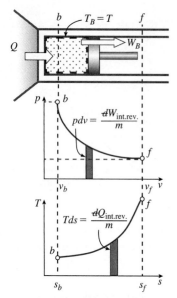

FIGURE 5.9 An internally reversible process lends itself to insightful graphical interpretation of its heat and boundary work interactions (see Anim. 5.A.*intReversible*).

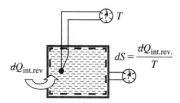

FIGURE 5.10 A concept meter for entropy measurement based on the relation $dQ_{int.rev.} = TdS$.

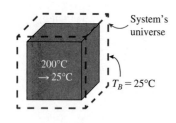

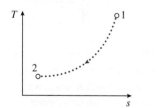

FIGURE 5.11 Cooling of a block of a solid in Ex. 5-2 (see Anim. 5.A.*solidCooling*).

The total heat transfer for an internally reversible process, $Q_{int.rev.}$, can be related to the area under the T-s diagram like that of $\dot{Q}_{int.rev.}$ for an internally reversible device (compare Fig. 5.9 with Fig. 4.6).

The differential relations $dQ_{int.rev.} = TdS$ and $dW_{B,int.rev.} = pd\forall$ are significant because they relate the path-dependent differentials dQ and dW_B to properties which are *point functions* and path independent. Historically, $dS = dQ_{int.rev.}/T$ provided the definition of entropy and established its unit as kJ/K. Although there is no real device that can directly measure entropy, that does not prevent us from imagining a conceptual entropy meter as shown in Figure 5.10. By monitoring the boundary temperature, and the heat added in an internally reversible manner, with an embedded chip that integrates the quantity $dQ_{int.rev.}/T$, the change in entropy can be calculated provided the system is internally reversible. There are practical difficulties in measuring small amounts of heat transfer and ensuring internal reversibility by eliminating thermal friction. An easier approach is to theoretically exploit the Tds relations that relates entropy to more easily measurable quantities such as temperature, pressure, and specific volume.

5.1.4 Uniform Closed Processes

Often a process involving a closed system is anchored in two unique, but *uniform*, beginning and final states, based upon thermodynamic equilibrium. Only two system states—b and f—are sufficient to describe the end states of a system that is uniform (homogenous). Such a process is called a **uniform closed process** or, simply, a **uniform process**. While the governing energy and entropy equations, Eqs. (5.2) and (5.4), relate ΔE and ΔS to process variables Q, W_{ext}, and S_{gen}, a suitable material model (Chapter 3) must be used to connect these variables to easily measurable properties of the anchor states. The resulting set of equations, when solved, produces the desired unknowns, usually a process variable or an anchor-state property. At this point, you should browse the series of animations depicting the processes involving uniform closed systems starting with Anim. 5.A.*waterHeater*.

Let's go over a few examples, then we will classify uniform processes into some commonly occurring groups.

EXAMPLE 5-2 **Energy and Entropy Analysis in Heat Transfer**

As shown in Figure 5.11, an aluminum block with a mass of 2000 kg, cools down from an initial temperature of 200°C to the atmospheric temperature of 25°C. Determine (a) the heat transfer and (b) entropy generated in the system's universe. **What-if scenario:** (c) What would the answer, in part, (a) be if the initial temperature had been 800°C?

SOLUTION

Analyze the block's universe, enclosed by the red boundary in Figure 5.11, using the energy and entropy balance equations customized for a closed uniform process.

Assumptions

The anchor states of the process are described by State-1 and State-2 based on LTE. No work transfer and no changes in KE or PE.

Analysis

From Table A-1, or any SL TESTcalc, we obtain $c_v = 0.9$ kg/kg·K (the exact value is 0.902 kJ/kg·K). The energy equation, Eq. (5.2), for the closed process, coupled with the SL model, produces:

$$\Delta E = \Delta U + \Delta KE^0 + \Delta PE^0 = Q - W_{ext}^{0};$$

$$Q = U_f - U_b = mc_v(T_f - T_b) = (2000)(0.9)(25 - 200)$$

$$= -315{,}000 \text{ kJ} = -315 \text{ MJ}$$

The entropy equation, Eq. (5.4), applied over the system's universe produces:

$$\Delta S = \frac{Q}{T_B} + S_{gen,univ}; \quad \Rightarrow \quad S_{gen,univ} = mc_v \ln\frac{T_2}{T_1} - \frac{Q}{T_0};$$

$$\Rightarrow \quad S_{gen,univ} = (2000)(0.9)\ln\left(\frac{298}{473}\right) - \frac{(-315,000)}{298} = 225.44\,\frac{kJ}{K}$$

TEST Solution

Launch the SL uniform closed-process TESTcalc and select aluminum from the working substance menu. Calculate State-1 (b-state) with T1=200 deg-C, m1=2000 kg, and State-2 (f-state) with T2=25 deg-C and m2=m1. In the process panel, load State-1 as the b-state and State-2 as the f-state, enter W_O=0, then click Calculate to verify the manual solution.

What-if scenario

Change T1 to 800 deg-C, press Enter, and click Super-Calculate. Heat transfer is updated in the process panel as −1395 MJ.

Discussion

The solid is assumed to be in LTE only at the beginning and the end of the process. Notice that the answers are independent of how long the process takes, what modes of heat transfer (conduction or radiation) are involved, and if the process is internally reversible. The exact path the process follows is unknown, so a dotted line is used to depict the process in the thermodynamic plot of Figure 5.11.

EXAMPLE 5-3 Energy and Entropy Analysis of Heat Addition from a TER

As shown in Figure 5.12, a 5-L piston-cylinder device contains steam at a pressure of 200 kPa and a quality of 25%. Heat is transferred, at constant pressure, from a reservoir (TER) at 1000 K until the system volume doubles. No heat transfer occurs between the steam and the atmosphere, which is at 25°C and 100 kPa. Determine (a) the heat transfer, (b) work transfer, and (c) entropy generation in the system's universe. Assume the process is internally reversible.

SOLUTION

Analyze the system's universe, enclosed by the red boundary in Figure 5.12, using the energy and entropy balance equations customized for a closed uniform process.

Assumptions

The system is always at LTE. KE and PE are constant.

Analysis

Using the steam tables, or any PC TESTcalc, evaluate the anchor states—State-1 as the b-state and State-2 as the f-state.

State-1: (given p_1, x_1, V_1):

$$v_1 = 0.2224\,\frac{m^3}{kg}; \quad u_1 = 1010.7\,\frac{kJ}{kg}; \quad s_1 = 2.9293\,\frac{kJ}{kg \cdot K}; \quad m_1 = 0.0225\,kg;$$

State-2: (given $p_2 = p_1, V_2 = V_1, m_2 = m_1$):

$$v_2 = 0.4447\,\frac{m^3}{kg}; \quad u_2 = 1519.4\,\frac{kJ}{kg}; \quad s_2 = 4.3353\,\frac{kJ}{kg \cdot K}$$

As the pressure remains constant during the internally reversible process, the boundary work can be evaluated from the rectangular area under the p-V diagram of Figure 5.12:

$$W_B = p_1(V_2 - V_1) = p_1 m_1(v_2 - v_1) = 1\,kJ$$

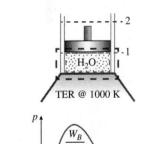

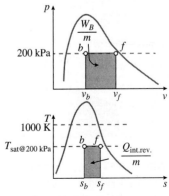

FIGURE 5.12 Constant-pressure expansion process of Ex. 5-3 (see Anim. 5.A.*pConstHeatingIG*).

The energy equation, Eq. (5.2), yields:

$$\Delta U + \Delta \cancel{KE}^0 + \Delta \cancel{PE}^0 = Q - (W_B + \cancel{W_O}^0)$$

$$\Rightarrow \quad Q = \Delta U + W_B = m_1(u_2 - u_1) + W_B = 12.45 \text{ kJ}$$

The entropy equation, Eq. (5.4), applied over the system's universe produces:

$$\Delta S = \frac{Q}{T_B} + S_{gen,univ}; \quad \Rightarrow \quad S_{gen,univ} = m(s_2 - s_1) - \frac{Q}{T_{TER}}$$

$$\Rightarrow \quad S_{gen,univ} = (0.0225)(4.335 - 2.929) - \frac{12.45}{1000} = 0.0192 \frac{\text{kJ}}{\text{K}}$$

TEST Solution

Launch the PC uniform closed-process TESTcalc and select H2O as the working fluid. Calculate State-1 and State-2 as described in the TEST-Code (see *TEST > TEST-codes*). Import the calculated states as the anchor states into the process panel, enter W_O=0, T_B=1000 K, and click Calculate to find Q and S_gen.

Discussion

For a constant-pressure process, the energy equation is simplified in Eq. (5.9) to produce $Q = \Delta H = m_1(h_2 - h_1)$, which leads to the answer without having to calculate the boundary work. Another way to calculate the heat transfer is to realize that it is the area under the process curve of a *T-s* diagram for an internally reversible process. From Figure 5.12, $Q = Q_{int.rev.} = m_1 T_1(s_2 - s_1) = 12.44 \text{ kJ}$.

EXAMPLE 5-4 Energy Analysis of an Expansion Process

A piston-cylinder device (Fig. 5.13), contains 0.2 m³ of R-134a at −20°C and a quality of 20%. The spring is linear and has a spring constant of 100 kN/m. The refrigerant is heated until its temperature reaches 25°C. If the cross-sectional area of the piston is 0.2 m², determine (a) the mass of the refrigerant, (b) the pressure when the piston reaches the pins at a volume of 0.4 m³, (c) the final pressure, and (d) the work and heat transfer during the entire process.

SOLUTION

Calculate the anchor states—State-1 at the beginning, State-2 when the piston hits the pins, and State-3 at the end of the process. The boundary work can be directly calculated from the *p-V* diagram. Perform mass and energy analyses of the two back-to-back processes.

Assumptions

The piston is weightless. R-134a can be modeled by the PC model. No changes in KE or PE. The process is internally reversible.

Analysis

Evaluate the states, fully or partially, from the given data.

State-1 (given T_1, x_1, V_1):

$$v_1 = 0.0299 \frac{\text{m}^3}{\text{kg}}; \quad m_1 = \frac{V_1}{v_1} = 6.6916 \text{ kg}; \quad p_1 = 133.7 \text{ kPa}; \quad u_1 = 63.23 \text{ kJ/kg};$$

State-2 (given $m_2 = m_1$, V_2): $\quad v_2 = 0.0598 \frac{\text{m}^3}{\text{kg}};$

State-3 (given T_3, $m_3 = m_1$, $V_3 = V_2$): $\quad p_3 = 374.4 \text{ kPa}; \quad u_3 = 247.72 \text{ kJ/kg}$

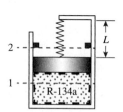

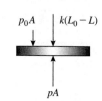

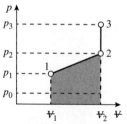

FIGURE 5.13 A linear spring ensures that the p versus V graph is linear during compression of the spring (see Anim. 5.A.pConst-pLinearIG).

For the entire process, from State-1 to State-3, the energy equation, Eq. (5.2), can be simplified as:

$$\Delta U + \Delta KE^0 + \Delta PE^0 = Q - (W_B + W_O^{\,0}) \quad \Rightarrow \quad \Delta U = Q - W_B; \quad [\text{kJ}] \qquad (5.8)$$

Although ΔU can be evaluated from the state properties, there are still two unknowns: Q and W_B. To evaluate W_B, observe that the functional relation between p and V must be linear for a linear spring (Fig. 5.13). Instead of integrating, the boundary work can be obtained from the area of the trapezoid under the process curve. Before we can calculate the area, we need one additional property, p_2.

A force balance on the piston at State-1 and State-2 produces:

$$p_1 A = p_0 A + k(L_0 - L_1); \quad \text{and} \quad p_2 A = p_0 A + k(L_0 - L_2)$$

where L_0 is the length of the spring at rest. Subtracting the first equation from the second yields:

$$p_2 - p_1 = \frac{k(L_1 - L_2)}{A} = k\frac{(V_2 - V_1)}{A^2}$$

$$\Rightarrow \quad p_2 = 133.7 + (100)\frac{(0.4 - 0.2)}{(0.2)^2} = 633.7 \text{ kPa}$$

The boundary work now can be calculated from the shaded area of Fig. 5.13 as:

$$W_B = \frac{1}{2}(V_2 - V_1)(p_2 + p_1) = \frac{(0.2)(633.7 + 133.7)}{2} = 76.74 \text{ kJ}$$

The heat transfer in the entire process is now obtained from Eq. (5.8):

$$Q = m(u_3 - u_1) + W_B = (6.692)(247.7 - 63.23) + 76.74 = 1311.3 \text{ kJ}$$

TEST Solution

Launch the PC uniform closed-process TESTcalc and select R-134a as the working fluid. Calculate the anchor states as described in the TEST-Code (see *TEST > TEST-codes*). Import the anchor states, State-1 and State-3, as the *b*-state and *f*-state of the overall process, then enter W_O=0. The TESTcalc fails to automatically calculate the boundary work because it has no knowledge of the linear nature of the process curve. We must enter the manually calculated value of W_B and press Calculate to produce $Q = 1311$ kJ.

Discussion

If the atmospheric pressure is given, $W_{atm} = p_0(V_2 - V_1)$ can be calculated and the distribution of the boundary work among the recipients, spring and atmosphere, must satisfy $W_B = W_{sp} + W_{atm}$. We can calculate the spring work from Eq. (0.15) and verify this equation.

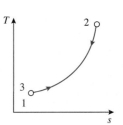

EXAMPLE 5-5   **Energy and Entropy Analysis of Paddle Wheel Work**

As shown in Figure 5.14, a 2-m³, insulated, rigid tank contains hydrogen at the atmospheric conditions of 100 kPa and 20°C. A paddle wheel transfers 100 kJ of shaft work into the system. Determine (a) the final temperature of the gas and (b) the entropy generated in the system during the process. Use the PG model. *What-if scenario:* (c) What would the entropy generation in the system's universe be if the tank were not insulated and sufficient time were allowed for heat transfer?

SOLUTION

Analyze the system, enclosed by the red boundary in Figure 5.14, using the energy and entropy balance equations customized for a closed uniform process.

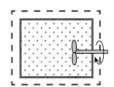

FIGURE 5.14 Paddle-wheel work followed by heat loss brings the system back to its original state in Ex. 5-5 (see Anim. 5.A. *vConstPaddleWheel*).

Assumptions

The system is in equilibrium at the anchor states—State-1 in the beginning and either State-2 (insulated) or State-3 (non-insulated) at the end. Hydrogen can be treated as a perfect gas. No changes in KE or PE.

Analysis

From Table C-1, or any PG TESTcalc, we obtain $R = 4.124$ kJ/kg $\cdot$ K and $c_v = 10.18$ kJ/kg $\cdot$ K.

$$\text{State-1 (given } p_1, T_1, V_1\text{):}\quad m_1 = \frac{V_1}{v_1} = \frac{p_1 V_1}{R T_1} = \frac{(100)(2)}{(4.124)(20 + 273)} = 0.166 \text{ kg}$$

The energy equation, Eq. (5.2), with the application of PG model yields:

$$\Delta U + \Delta KE^0 + \Delta PE^0 = Q^0 - (W_B^0 + W_{el}^0 + W_{sh}) \quad \Rightarrow \quad \Delta U = -W_{sh};$$

$$\Rightarrow \quad mc_v(T_2 - T_1) = -W_{sh}; \quad \Rightarrow \quad T_2 = T_1 - \frac{W_{sh}}{mc_v} = 293 - \frac{(-100)}{(0.166)(10.18)} = 352.3 \text{ K}$$

The entropy equation, Eq. (5.4), similarly, produces:

$$\Delta S = \frac{Q^0}{T_B} + S_{gen,univ} \quad \Rightarrow \quad S_{gen,univ} = mc_v \ln\frac{T_2}{T_1} - mR \ln\frac{v_2}{v_1}^0$$

$$= (0.166)(10.18) \ln\left(\frac{352.3}{293}\right) = 0.311 \frac{\text{kJ}}{\text{kg} \cdot \text{K}}$$

There is no heat transfer in this process; therefore, the result is independent of how the boundary is drawn—internally or externally—for this analysis.

TEST Solution

Launch the PG uniform closed-process TESTcalc. Select hydrogen as the working fluid. Calculate the anchor states as described in the TEST-code (posted in *TEST > TEST-codes*). Set up the process (process-A) in the process panel by importing the anchor states. Then enter Q=0 and W_el=W_O=−100 kJ. Click Calculate, then Super-Calculate. The manual results are reproduced.

What-if scenario

If there is no insulation, the final temperature will reach the initial temperature after sufficient time. State-1 and State-3, therefore, must be identical since two thermodynamic properties— temperature and specific volume—remain unchanged between the two states. The energy and entropy equations for the modified process 1−3 yield:

$$\Delta U^0 = Q - (W_B^0 + W_O); \quad \Rightarrow \quad Q = W_O = -100 \text{ kJ}$$

$$\Delta S^0 = \frac{Q}{T_B} + S_{gen,univ}; \quad \Rightarrow \quad S_{gen,univ} = -\frac{Q}{T_0} = 0.3413 \frac{\text{kJ}}{\text{kg} \cdot \text{K}}$$

In the process panel of the TEST solution, set up processes B (2−3) and C (1−3), where State-3 is evaluated with identical conditions as State-1. For process-C, enter W_O=−100 kJ and click Calculate to reproduce the above result.

Discussion

Process-C (1−3) can be broken into two steps—adiabatic process-A (1−2) for the insulated tank followed by process-B (2−3) with heat transfer. In the first process, entropy is generated as work is dissipated into the internal energy through *viscous friction*. In the second process, entropy is generated by *thermal friction* as heat dissipates into the internal energy of the surroundings through heat transfer over a finite temperature difference. The sum of entropy generated in these constituent processes is equal to the entropy generated in process-C, the overall process.

| EXAMPLE 5-6 | **Energy and Entropy Analysis of an Isentropic Process** |

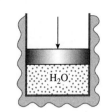

As shown in Figure 5.15, an insulated piston-cylinder device contains 0.01 kg of steam at 500 kPa, 400°C, which expands in an internally reversible manner until it is saturated. Determine (a) the pressure and (b) temperature of the steam at the end of the expansion process. Also, determine (c) the boundary work transfer.

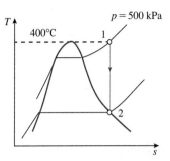

FIGURE 5.15 The external force on the piston must adjust (decrease) as steam expands (see Anim. 5.A.*sConstExpansionIG*) in this quasi-equilibrium process.

Solution

Analyze the system, enclosed by the red boundary in Figure 5.15, using the energy and entropy balance equations customized for this closed uniform process.

Assumptions

The system is uniform and in quasi-equilibrium throughout the process. There is no heat transfer, no internal entropy generation, and no change in KE or PE.

Analysis

Using the steam tables, or any PC TESTcalc, evaluate the anchor states—State-1 as the *b*-state and State-2 as the *f*-state.

State-1 (given p_1, T_1, m_1): $u_1 = 2963.2 \, \dfrac{\text{kJ}}{\text{kg}}; \quad s_1 = 7.7937 \, \dfrac{\text{kJ}}{\text{kg} \cdot \text{K}};$

State-2 (given $x_2, m_2 = m_1$): A second thermodynamic property is required before this state can be evaluated.

The energy equation, Eq. (5.2), yields:

$$\Delta U + \Delta KE^0 + \Delta PE^0 = \cancel{Q}^0 - \left(W_B + \cancel{W_O}^{\,0} \right);$$
$$\Rightarrow \quad W_B = -\Delta U = m_1(u_1 - u_2)$$

With State-2 still an unknown, the boundary work cannot be obtained from this equation or from the $pd\Psi$ integral. However, the entropy equation, Eq. (5.4), provides the second property at State-2:

$$\Delta S = \frac{\cancel{Q}^0}{T_B} + \cancel{S_{\text{gen}}}^{\,0}; \quad \Rightarrow \quad m(s_2 - s_1) = 0; \quad \Rightarrow \quad s_2 = s_1$$

The process is isentropic because it is adiabatic and internally reversible. Evaluation of State-2 can now proceed as x_2 and s_2 are known:

$$p_2 = 27.92 \text{ kPa}; \quad T_2 = 67.46°C; \quad u_2 = 2466.3 \, \frac{\text{kJ}}{\text{kg}}$$

Returning to the energy equation, the boundary work can now be evaluated:

$$W_B = m_1(u_1 - u_2) = (0.01)(2963.2 - 2466.3) = 4.97 \text{ kJ}.$$

TEST Solution

Launch the PC uniform closed-process TESTcalc and select H2O as the working fluid. Calculate State-1 and State-2 as described in the TEST-code (see *TEST > TEST-codes*). Import the calculated states as the anchor states in the process panel, enter W_O=0, Q=0, S_gen=0, and click Calculate. Manual results are verified.

Discussion

An internally reversible process executed by an adiabatic system is *isentropic*. This conclusion parallels the isentropic behavior of internally reversible, adiabatic, open-steady devices.

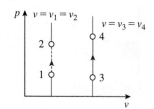

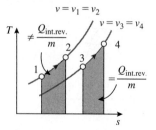

FIGURE 5.16 p-v and T-s diagrams for two constant-volume processes—one irreversible and another internally reversible (see Anim. 5.A.vConstJouleHeatingIG).

A. Constant-Volume Process

If the working substance can be considered incompressible, or is contained in a rigid chamber, the volume remains constant during any process performed by a closed system. With no change in volume, there is no possibility of any boundary work. Because mass cannot change for a closed system, specific volume, $v = V/m$, also remains constant. Such a closed process is called a **constant-volume** or **isochoric process**. Several isochoric processes are illustrated in TEST starting with Anim. 5.A.solidHeating.

Representing a **constant-volume process** on a p-v diagram is trivial—the process curves are parallel to the p axis. However, to sketch the process curve on a T-s diagram, constant-volume lines, or **isochors**, passing through the anchor states are helpful. Two isochoric processes with a single-phase working substance (solid, liquid, vapor, or a gas) are shown in Figure 5.16. Process curve 1−2 represents an irreversible isochoric process and 3−4 represents an internally reversible process. As explained in Sec. 5.1.3, the area under the process curves in the p-v and T-s diagrams can be interpreted as the boundary work and heat transfer per-unit mass of the working substance for an internally reversible process. For the irreversible process 1−2, the p-v diagram correctly predicts the zero-boundary work. The area under the curve 1−2 in the T-s diagram does not necessarily represent the heat transfer in the process. In fact, the same process 1−2 could be carried out adiabatically by transferring shaft or electrical work, producing an identical final state with zero heat transfer. The area under curve 3−4, conversely, is proportional to the heat transfer. We have already encountered constant-volume processes in Exs. 5-1, 5-2, and 5-5. Example 5-7 is another analysis that involves such a process.

EXAMPLE 5-7 | **Energy Analysis of a Constant-Volume Process**

As shown in Figure 5.17, a rigid tank with a volume of 0.1 m³ contains steam of quality 2% at a temperature of 25°C. A resistance heater within the tank is turned on and operates for 30 minutes drawing a current of 2 A from a 120-V source. At the same time, 50 kJ of heat is lost from the tank to the atmosphere at 25°C. Determine (a) the final pressure and (b) final temperature. **What-if scenario:** (c) What would the answers be if the initial quality was 1%?

SOLUTION

Analyze the system, enclosed by the red boundary in Figure 5.17, using the energy balance equation customized for an isochoric process.

Assumptions

Equilibrium at the anchor states, PC model for H_2O, and no changes in KE or PE.

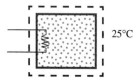

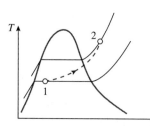

FIGURE 5.17 A schematic and the process curve in Ex. 5-7 (see Anim. 5.A.vConstHeatingPC).

Analysis

The energy equation, Eq. (5.2), yields:

$$\Delta U + \Delta KE^0 + \Delta PE^0 = Q - (W_B^0 + W_O) \implies \Delta U = Q - W_{el};$$

$$\implies m(u_2 - u_1) = -50 - \left(-\frac{120 \times 2 \times 30 \times 60}{(1000 \text{ W/kW})}\right); \implies u_2 = u_1 + \frac{382}{m}$$

The anchor states—State-1 (b-state) and State-2 (f-state)—can be evaluated from any PC TESTcalc (or manually) to produce:

State-1 (given T_1, x_1, V_1):

$$v_1 = 0.8713 \frac{\text{m}^3}{\text{kg}}; \quad u_1 = 150.98 \frac{\text{kJ}}{\text{kg}}; \quad m_1 = 0.1148 \text{ kg};$$

$$\text{State-2}\left(\text{given } V_2 = V_1; m_2 = m_1; u_2 = u_1 + \frac{382}{m} = 3479 \frac{\text{kJ}}{\text{kg}}\right):$$

$$v_2 = 0.872 \frac{\text{m}^3}{\text{kg}}; \quad p_2 = 515.2 \text{ kPa}; \quad T_2 = 700°C$$

TEST Solution

Launch the PC closed-process TESTcalc and select H2O as the working fluid. Calculate the anchor states, then set up the process panel as described in the TEST-Code (see *TEST > TEST-codes*). Click Calculate, then click Super-Calculate to update states and verify the manual results.

What-if scenario

Change x1 to 1% and click Super-Calculate. The new answers are 259 kPa and 128 deg-C. A reduction in the initial quality means the system has more liquid at the start. Much of the energy transfer is expended in vaporizing the additional liquid, resulting in a much lower final temperature.

Discussion

If the process is reversed, and the tank is allowed to cool down to the atmospheric temperature of 25°C, the final pressure can be shown to be only 3.2 kPa (in TEST set T2, m3, and Vol3 to calculate p3). The difference in pressure between the outside and inside can crush a tank like a soda can unless its wall is designed to withstand the predictable compressive load.

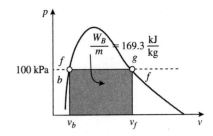

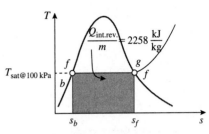

FIGURE 5.18 The boundary work done during the phase transformation of saturated water at 100 kPa is 7.5% of the heat added (the PC state TESTcalc is used to obtain the *b*- and *f*-states).

B. Constant-Pressure Process

A **constant-pressure** process, also known as an **isobaric** process, is generally associated with a piston-cylinder device when the piston is free to move (see animations starting with Anim. 5.A.*pConstHeatingIG*). A free-body diagram of the piston (Sec. 0.1) reveals whether the pressure inside the cylinder remains constant during the process. A quasi-equilibrium assumption is often reasonable in a constant-pressure process and the rectangular area under the *p-v* diagram makes boundary work easy to evaluate (Fig. 5.18 and Anim. 5.A.*pTsConstExpansion WorkIG*):

$$W_B = p(v_f - v_b)m = p(V_f - V_b); \quad [\text{kJ}]$$

Substituting this expression in the energy equation, Eq. (5.2), produces a special form of the energy equation for an isobaric process:

$$\Delta E = \Delta U + \Delta KE^0 + \Delta PE^0 = Q - (W_B + W_{sh} + W_{el})$$
$$\Rightarrow \quad (U_f - U_b) = Q - p(V_f - V_b) - (W_{sh} + W_{el})$$
$$\Rightarrow \quad (U_f + p_f V_f) - (U_b + p_b V_b) = Q - W_O$$
$$\Rightarrow \quad H_f - H_b = \Delta H = m\Delta h = Q - W_O; \quad [\text{kJ}] \tag{5.9}$$

Calculation of boundary work can be avoided by equating enthalpy change with net energy transfer by heat and $W_O = W_{sh} + W_{el}$. Observe that this equation is not limited to a piston-cylinder device, or any particular working substance, as the expression used for the boundary work is quite general. It can be applied to any fixed mass going through a constant-pressure process, even for complex processes such as vaporization or boiling of a working substance at constant pressure.

To illustrate the usefulness of Eq. (5.9), let's determine the amount of heat necessary to boil off some saturated liquid water from an open pan. With transfer of mass from one phase to another, an open-system analysis seems unavoidable. The analysis can be greatly simplified by visualizing an invisible, weightless piston at the interface between the vaporized steam and air. With the atmospheric pressure remaining constant, the boiling process can be treated as a constant-pressure process. Representing the beginning and final states by saturated states *f* and *g*, Eq. (5.9) can be applied (Fig. 5.18), producing $Q = m\Delta h = m(h_g - h_f)$, where the values for h_g and h_f at atmospheric pressure can be read from the saturation steam table. A significant portion of the heat goes into supplying the boundary work (Fig. 5.18) necessary to displace the surrounding air to accommodate the water vapor. In the following example, we analyze a constant-pressure boiling process. Similar analysis can be carried out for other practical phase-change phenomena such as droplet vaporization or sublimation of dry ice.

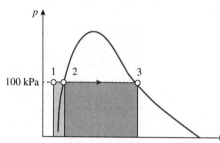

FIGURE 5.19 A system schematic and the process curve for a sequential process in Ex. 5-8.

EXAMPLE 5-8 Mass and Energy Analysis of Boiling Process

As shown in Figure 5.19, an insulated, electrically heated teapot, with a power consumption rate of 1 kW, contains 1 L of water at 25°C. (a) Determine the time it takes for the water to reach its boiling point after the heater is turned on. (b) How much longer should the heat be applied to completely boil off the water? Assume the atmospheric pressure to be 100 kPa. **What-if scenario:** (c) What would the answers be if the water had been heated on top of a mountain, where the atmospheric pressure is 90 kPa?

SOLUTION

Analyze the process using the energy balance equation customized for an **isobaric process**. The problem can also be analyzed as an open process, which will be discussed in Sec. 5.1.7.

Assumptions

Quasi-equilibrium constant-pressure process, PC model for H_2O, and no changes in KE or PE.

Analysis

The two consecutive processes, sensible heating (process A) and boiling (process B), are shown on the p-v diagram (Fig. 5.19). Using the steam tables, or any PC TESTcalc, evaluate the three states—State-1 for the initial state, State-2 for the saturated liquid, and State-3 for the saturated vapor.

State-1 (given p_1, T_1, V_1):

$$v_1 = 0.001 \frac{m^3}{kg}; \quad m_1 = 1 \text{ kg}; \quad h_1 = 104.98 \frac{kJ}{kg}; \quad s_1 = 0.3673 \frac{kJ}{kg \cdot K};$$

State-2 (given $p_2 = p_1, x_2, m_2 = m_1$): $h_2 = 417.44 \frac{kJ}{kg};$

State-3 (given $p_3 = p_1, x_3, m_3 = m_1$): $h_3 = 2675.5 \frac{kJ}{kg}$

Equation (5.9) for the first process, process-A, produces the time of heating Δt_A:

$$\Delta H = m\Delta h = \cancel{Q}^0 - W_O; \quad \Rightarrow \quad -W_O = -\dot{W}_{el}\Delta t_A = m(h_2 - h_1);$$

$$\Rightarrow \quad \Delta t_A = \frac{m(h_2 - h_1)}{-\dot{W}_{el}} = \frac{(1)(417.44 - 104.98)}{-(-1)} = 312 \text{ s} = 5 \text{ min } 12 \text{ s}$$

For the boiling process, process-B, the same form of the energy equation yields:

$$-\dot{W}_{el}\Delta t_B = m(h_3 - h_2);$$

$$\Rightarrow \quad \Delta t_A = \frac{m(h_3 - h_2)}{-\dot{W}_{el}} = \frac{(1)(2675.5 - 417.44)}{-(-1)} = 2258 \text{ s} = 37 \text{ min } 38 \text{ s}$$

TEST Solution

Launch the PC closed-process TESTcalc and select H2O as the working fluid. Calculate the anchor states as described in the TEST-Code (see *TEST > TEST-codes*). In the process panel, set up process-A by importing the beginning and final states (1 and 2). The boundary work is automatically calculated. Enter Q=0 and calculate W_O as −311.5 kJ. Select process-B, then load State-2 and State-3 as the *b*-state and *f*-state, respectively. Enter Q=0 and click Calculate. W_O is found as −2251 kJ. The heating time can now be manually calculated from these results.

What-if scenario

Change p1 to 90 kPa, press Enter, then click Super-Calculate. The new results can be evaluated from the process panel as 300 s and 2258 s.

Discussion

We generally associate heat transfer with a temperature change, which is sometime called **sensible heat**. As opposed to that, heat absorbed or released during a phase change at constant-pressure is called **latent heat** because the temperature remains constant. The **latent heat of vaporization**, which is also known as heat of evaporation, **heat of vaporization**, or **enthalpy of vaporization**, is defined as the energy necessary to transform the unit mass of a saturated liquid into saturated vapor at constant pressure, $h_{fg@p} \equiv h_{g@p} - h_{f@p}$. In a parallel manner, **latent heat of fusion** can be defined for transformation from liquid to solid.

C. Constant-Temperature Process

The process curves for a constant-temperature or **isothermal** process, are called **isotherms**. Familiar examples of an **isothermal process** are boiling and condensation, which also happen to be *isobaric* (assuming quasi-equilibrium condition). Without phase transformation, temperature can be held constant only if a transfer of heat is offset by an equal amount of work.

In isothermal processes involving a gas there is no distinction between the IG or PG model. For an ideal (or perfect) gas, an isotherm on a *T-s* diagram is trivial, while that on a *p-v* diagram (Fig. 5.20) can be obtained from the ideal gas equation of state: $pv = RT = $ constant. The corresponding boundary work can be evaluated from:

$$W_{B,isothermal}^{IG/PG} = m \int_b^f p\,dv = m \int_b^f \frac{RT}{v}\,dv = mRT \ln \frac{v_f}{v_b} = mRT \ln \frac{p_b}{p_f}; \quad \text{[kJ]} \quad \textbf{(5.10)}$$

The heat transfer in the process, similarly, can be obtained from Eq. (5.7) after substituting entropy change formula from Eq (3.50) or (3.54):

$$Q_{int.rev.}^{IG/PG} = m \int_b^f T\,ds = mT(s_f - s_b) = mRT \ln \frac{v_f}{v_b} = mRT \ln \frac{p_b}{p_f}; \quad \text{[kJ]} \quad \textbf{(5.11)}$$

It is not a coincidence that the two expressions for work and heat transfer are identical. With $u = u(T)$ remaining constant in an isothermal process, the energy equation yields:

$$\Delta U^0 + \Delta KE^0 + \Delta PE^0 = Q - (W_B + W_O^0); \quad \Rightarrow \quad Q = W_B; \quad \text{[kJ]} \quad \textbf{(5.12)}$$

The equality between heat and boundary work transfer also means that the shaded areas in Figure 5.20 must be equal.

D. Constant-Entropy Process

In a **constant-entropy** or **isentropic process**, entropy remains constant. As we have done for an isentropic device in Chapter 4, the necessary conditions for a process to maintain constant entropy can be established from the entropy balance equation. If there is no heat transfer and no internal entropy generation in a closed system (Fig. 5.21) during the process, the entropy equation, Eq. (5.4), produces the isentropic equality:

$$\Delta S = S_f - S_b = \cancelto{0, \text{ adiabatic}}{\frac{Q}{T_B}} + \cancelto{0, \text{ reversible}}{S_{gen}}; \quad \Rightarrow \quad S_b = S_f; \quad \left[\frac{kJ}{kg}\right]$$

$$\Rightarrow \quad s = \text{const.}; \quad \left[\frac{kJ}{kg \cdot K}\right] \quad \textbf{(5.13)}$$

In an adiabatic system, there cannot be any external entropy generation either; therefore $S_{gen,univ} = S_{gen} = 0$, making the process *reversible*. An isentropic closed process, therefore, is *reversible* and *adiabatic*. The reversibility condition also means that the process must be in quasi-equilibrium, as explained in Sec. 5.1.1. The isentropic assumption is routinely used to model practical processes involving rapid compression or expansion in reciprocating cycles (Chapter 7) even though these processes are not quasi-static. While the rapidity of a process may contribute to internal entropy generation, the time available for heat transfer is reduced making the adiabatic assumption more reasonable. An isentropic efficiency, similar to device efficiencies discussed in Sec. 4.1.4, can be defined for a process, relating the work transfer in an

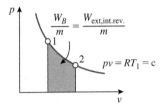

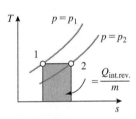

FIGURE 5.20 For an internally reversible isothermal process, the two shaded areas (in both are shaded in gray) must be equal (see Anim. 5.A.*pTsConstExpansionIG*).

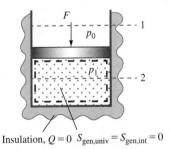

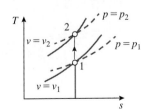

FIGURE 5.21 Entropy remains constant in a compression process if the system is (a) adiabatic and (b) internally reversible.

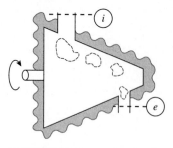

FIGURE 5.22 In an open-steady isentropic device, closed systems such as the one identified by the red dye, undergo isentropic processes.

actual process with that in the corresponding isentropic process. Notice the remarkable similarity between an isentropic open device, discussed in Sec. 4.1.4, and an isentropic process. This is not a coincidence. If we follow a lump of fluid in an isentropic device, for example, the closed system shown by the red boundary in Figure 5.22, the system is subjected to an isentropic process with the i and e states of the device serving as the b and f states of the closed process.

The isentropic relation for a perfect gas, Eq. (3.71), produces a special relation between pressure and volume:

$$p_f v_f^k = p_b v_b^k = p v^k = C; \quad \left[\text{kPa} \left(\frac{\text{m}^3}{\text{kg}} \right)^k \right] \tag{5.14}$$

The value of C, a constant for a given isentropic process, can be obtained from the known values of p and v at the b-state or f-state. To maintain this special condition, the external force F on the piston in Figure 5.21 must vary with the system volume in a specific manner as dictated by a force balance on the piston:

$$\frac{F}{A} = p - p_0 = p_b \left(\frac{v_b}{v} \right)^k - p_0 = p_b \left[\left(\frac{v_b}{v} \right)^k - \frac{p_0}{p_b} \right]; \quad \left[\frac{\text{kN}}{\text{m}^2} \right] \tag{5.15}$$

where p_0 is the pressure of the atmosphere outside and A is the piston's area.

The isentropic boundary work (click Isentropic Expansion radio-button of Anim. 5.A. *pTsConstExpansionWorkIG*) can then be directly evaluated from Eq. (5.6):

$$W_{B,\text{isentropic}}^{\text{PG}} = m \int_b^f p\, dv = m \int_b^f \frac{C}{v^k} dv = m \left[\frac{C v_f^{1-k} - C v_b^{1-k}}{1 - k} \right]$$

$$= m \left[\frac{p_f v_f^k v_f^{1-k} - p_b v_b^k v_b^{1-k}}{1 - k} \right] = m \left[\frac{p_f v_f - p_b v_b}{1 - k} \right] = \frac{mR(T_f - T_b)}{1 - k}; \quad [\text{kJ}] \tag{5.16}$$

The sign of the work is consistent with the WinHip sign convention—positive for expansion and negative for compression. The superscript PG in this equation emphasizes that this formula applies only to gases that obey the PG model.

E. Polytropic Processes

In an actual expansion or compression process of a gas or vapor, heat transfer may not be negligible, and true internal reversibility never exists. In addition, the PG model cannot be applied if accuracy is a concern, and the IG or PC model may be needed to model realistic fluids. Given the simplicity and elegance of the perfect-gas isentropic relations, attempts are made to correlate pressure versus volume data from actual processes using the perfect gas formulas of Eq. (5.14). It turns out that such correlations work quite well for many situations if the specific heat ratio k is replaced by an experimentally determined exponent, n, known as the **polytropic coefficient**. The resulting model is called a **polytropic process**. All the isentropic relations for the PG model can then be extended to a polytropic process with k replaced by n in the corresponding formulas:

$$\frac{T_f}{T_b} = \left(\frac{p_f}{p_b} \right)^{(n-1)/n} = \left(\frac{v_b}{v_f} \right)^{n-1} \tag{5.17}$$

$$W_{B,\text{polytropic}}^{\text{PG, IG, PC}} = m \int_{v_b}^{v_f} p\, dv = \frac{mR(T_f - T_b)}{1 - n} \quad (n \neq 1); \quad [\text{kJ}] \tag{5.18}$$

These formulas are general as they can be applied to PG, IG, or even PC models (superheated vapor) undergoing realistic processes with heat transfer and internal irreversibilities.

When drawing a polytropic process curve on a thermodynamic diagram, it is helpful to draw a family of curves for different values of n spanning known limits. Table 5-1, and the accompanying p-v and T-s diagrams of Figure 5.23, describe the various processes discussed so far through the appropriate choice of the polytropic coefficient n. The processes curves rotate clockwise as n increases from 0 to ∞, which makes their relative orientations easy to deduce.

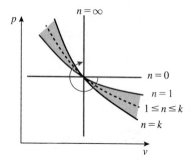

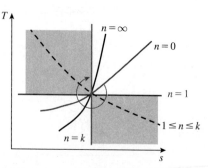

FIGURE 5.23 Polytropic and other standard processes for a perfect gas on p-v and T-s diagrams (see Table 5-1, Anim. 5.A.*pTsConstExpansionIG*).

Table 5-1 Polytropic coefficients used for various processes.

$n = 0$	Isobaric process	$p = $ constant
$n = 1$	Isothermal process	$T = $ constant
$n = k$	Isentropic process	$s = $ constant
$n = \infty$	Isochoric process	$v = $ constant
$1 < n < k$	Polytropic process	

EXAMPLE 5-9 Energy and Entropy Analysis of Isentropic, Isothermal, and Polytropic Processes

A handheld bicycle pump contains 50 cc of air at 25°C and 100 kPa. The exit hole is blocked and the piston is pressed until the pressure increases to 300 kPa. Determine the work and heat transfer during the process if it is carried out (a) very rapidly, (b) very slowly, and (c) at an intermediate speed. A polytropic coefficient of $n = 1.3$ is applicable. Assume perfect gas behavior for the air and make suitable assumptions for the quick and slow processes.

SOLUTION

Obtain the boundary work directly from the integral of $p d\mathcal{V}$ or the energy equation for the processes depicted in Figure 5.24.

Assumptions

The quick process (State-1 to State-2) is adiabatic and reversible, i.e., isentropic, and the slow process (State-1 to State-3) is isothermal. The system is in quasi-equilibrium during all processes. Use the PG model for air and neglect any changes in KE or PE.

Analysis

From Table C-1 or any PG TESTcalc, we obtain $R = 0.287$ kJ/kg·K, $c_v = 0.718$ kJ/kg·K, and k as 1.4. The anchor states are now evaluated as accurately as possible.

State-1 (given p_1, T_1, $\mathcal{V}_1$):

$$m_1 = \frac{\mathcal{V}_1}{v_1} = \frac{p_1 \mathcal{V}_1}{RT_1} = \frac{(100)(50 \times 10^{-6})}{(0.287)(25 + 273)} = 58.5 \times 10^{-6} \text{ kg}$$

State-2 (given $p_2 = 3p_1$; $s_2 = s_1$; $m_2 = m_1$):

$$T_2 = T_1\left(\frac{p_2}{p_1}\right)^{(k-1)/k} = 298(3^{0.2857}) = 408 \text{ K};$$

State-3 (given $p_3 = 3p_1$; $T_3 = T_1$; $m_3 = m_1$):

$$v_3 = \frac{RT_3}{p_3} = 0.2851 \frac{\text{m}^3}{\text{kg}};$$

State-4 (given $p_4 = 3p_1$; $m_4 = m_1$):

$$T_4 = T_1\left(\frac{p_2}{p_1}\right)^{(n-1)/n} = 298(3^{0.231}) = 384 \text{ K}$$

For the isentropic process, Eq. (5.16) produces:

$$W_B = W_{B,\text{isentropic}}^{PG} = \frac{mR(T_f - T_b)}{1 - k} = \frac{mR(T_2 - T_1)}{1 - k} = -0.0046 \text{ kJ}$$

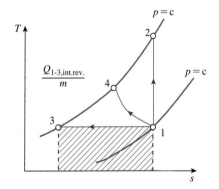

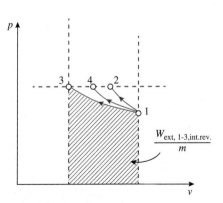

FIGURE 5.24 The three processes in Ex. 5-9 in T-s and p-v diagrams. The areas under the process curves are proportional to the heat and work transfer if the processes are internally reversible (click the polytroic radio-button in Anim. 5.A.*pTsConstCompressionIG*).

Alternatively, the energy equation for the process, Eq. (5.2), yields:

$$\Delta U = \cancel{Q}^0 - (W_B + \cancel{W_O}^0); \quad \Rightarrow \quad W_B = -mc_v\Delta T = -0.0046 \text{ kJ}$$

For the isothermal process, we use Eq. (5.10):

$$W_B = W_{B,\text{isothermal}}^{\text{PG}} = mRT_1 \ln\frac{p_1}{p_3} = -0.0055 \text{ kJ}$$

The heat transfer can be evaluated from the energy equation as:

$$\cancel{\Delta U}^0 = Q - (W_B + \cancel{W_O}^0); \quad \Rightarrow \quad Q = W_B = -0.0055 \text{ kJ}$$

For the polytropic process, Eq. (5.18) yields:

$$W_B = W_{B,\text{polytropic}} = \frac{mR(T_4 - T_1)}{1 - n} = -0.0048 \text{ kJ}$$

Finally, the heat transfer can be evaluated from the energy equation:

$$\Delta U = Q - (W_B + \cancel{W_O}^0); \quad \Rightarrow \quad Q = W_B + mc_v(T_4 - T_1) = -0.0012 \text{ kJ}$$

TEST Solution

Launch the PG closed-process TESTcalc and select air as the working fluid. Calculate the anchor states and set up the processes as described in the TEST-code (see *TEST > TEST-codes*). Click Super-Calculate to verify the manually obtained results.

Discussion

A polytropic process is assumed to be a quasi-equilibrium process, which is not necessarily internally reversible. Therefore, the interpretation of the areas under the process curves (Fig. 5.8) is only approximately applicable to a polytropic process. Note that the signs of the work and heat transfers are consistent with the WinHip convention.

5.1.5 Non-Uniform Systems

When more than one state is necessary to describe one or more of the anchor states (State-*b* and State-*f*) of a process, the system is said to be **non-uniform**. The procedure developed for the closed-process analysis can be extended to a closed, non-uniform system, which can then be decomposed into uniform subsystems.

In a **non-mixing, non-uniform system**, the subsystems maintain their separate identities, and there is no mass transfer between them. Examples of such non-mixing processes include two solid blocks exchanging heat with each other (Anim. 5.B.*blocksInContact*), a hot solid block dropped into a liquid that raises the liquid's temperature (Anim. 5.B.*blockInWater*), etc.

Conversely, in a **mixing, non-uniform system**, a phenomenon where mass transfer is driven by concentration difference is allowed among the subsystems. Fluids at two different states, stored in two separate tanks undergo a mixing process if a connecting valve is opened and the two fluids are allowed to meet (see animations beginning with Anim. 5.C.*mixingGeneric*). Even if the two fluids are at the same pressure, mixing occurs spontaneously through **diffusion**. As we will see shortly, mixing is encouraged by the second law because entropy generation is accompanied by a destruction of concentration imbalance. The underlying entropy generation mechanism is called **mixing friction**. The irreversibility associated with entropy generation is evident from the fact that a mixture does not spontaneously separate. Mixing is said to be complete if the subsystems are exposed to one another long enough for the new composite system to reach equilibrium. In that case, the final state can be represented by a single uniform system state. If the valve connecting the two tanks is closed prematurely, before mixing is complete, the mixing process remains unfinished and the final state has to be represented by a composite state consisting of the final states of the two subsystems.

5.1.6 TEST and the Non-Uniform Closed-Process TESTcalcs

The **non-mixing, closed-process TESTcalcs** in TEST are located in the closed, process, generic, non-uniform branch and allow up to two states to define a composite beginning or final state. Depending on the extent of mixing, you can pick mixing, semi-mixing, and non-mixing

TESTcalcs. These TESTcalcs are quite similar to the uniform-process TESTcalcs (Sec. 5.1.2) with a few notable differences.

When two dissimilar working substances are involved, two working substance menus are provided, one for each model. You must select the working substance from an appropriate model for each anchor state. For example, to analyze the heat exchange between a gas and a solid, the IG/SL composite model may be suitable. With the system assumed to be made up of two subsystems, A and B, four possible anchor states—bA, bB, fA, and fB—appear in the process panel. The mixing TESTcalcs only provide a single *f*-state since the system is uniform at the end of the process. As in the case of uniform process TESTcalcs, setting up the process panel means importing calculated states from the state panel as anchor states and entering the known process variables. Clicking Super-Calculate iterates between the state and process panels, producing the complete solution for a well-posed problem.

EXAMPLE 5-10   **Energy and Entropy Analysis of a Non-uniform, Non-mixing, Closed Process**

A 40-kg aluminum block at 100°C is dropped into an insulated tank that contains an unknown mass of liquid water at 20°C. After thermal equilibrium is established, the temperature is measured at 22°C. Determine (a) the mass of water and (b) entropy generated in the system's universe. *What-if scenario:* (c) What would the answer in part (b) be if the temperature of the block was 200°C?

SOLUTION

Analyze the system, enclosed by the red boundary in Figure 5.25, using the energy and entropy balance equations customized for a non-mixing, non-uniform, closed process.

Assumptions

Subsystems A and B, in Figure 5.25, are uniform and in equilibrium at the beginning and final states. The SL model is applicable to both working substances. No changes in KE and PE. No heat transfer between the tank and the surroundings.

Analysis

From any SL TESTcalc, or Table A-1, we obtain $c_{vA} = 0.9$ kJ/kg·K for aluminum and $c_{vB} = 4.184$ kJ/kg·K for liquid water. The energy equation, Eq. (5.2), is applied to the composite system by treating it as an aggregate of its subsystems, yielding:

$$\Delta U = \cancel{Q}^0 - \cancel{W}^0; \quad \Rightarrow \quad (U_{fA} + U_{fB}) - (U_{bA} + U_{bB}) = 0;$$

$$\Rightarrow \quad m_A c_{vA}(T_f - T_{bA}) + m_B c_{vB}(T_f - T_{bB}) = 0;$$

$$\Rightarrow \quad (40)(0.9)(22 - 100) + m_B(4.184)(22 - 20) = 0;$$

$$\Rightarrow \quad m_B = 335.6 \text{ kg}$$

The entropy balance equation, Eq. (5.4), similarly, can be simplified to produce the entropy generation in the system's universe:

$$\Delta S = S_f - S_b = \frac{\cancel{Q}^0}{T_B} + S_{gen,univ};$$

$$\Rightarrow \quad S_{gen,univ} = (S_{fA} + S_{fB}) - (S_{bA} + S_{bB}) = (S_{fA} - S_{bA}) + (S_{fB} - S_{bB})$$

$$\Rightarrow \quad S_{gen,univ} = m_A c_{vA} \ln \frac{T_f}{T_{bA}} + m_B c_{vB} \ln \frac{T_f}{T_{bB}} = 1.1053 \frac{\text{kJ}}{\text{K}}$$

TEST Solution

Launch the SL/SL non-mixing, closed-process TESTcalc. Evaluate the four anchor states as described in the TEST-Code (see *TEST > TEST-codes*), making sure that the appropriate working substance is chosen for each state. Note that m2, mass of water at State-2, is an unknown, while m4=m2 is entered for State-4. In the process panel, import State-1 and State-2 as the bA and bB

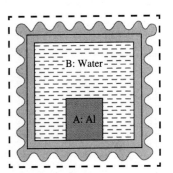

FIGURE 5.25 In the non-uniform non-mixing system of Ex. 5-10 subsystems A and B are the aluminum block and liquid water, respectively (see Anim. 5.B.*blockInWater*).

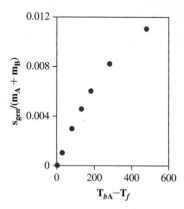

FIGURE 5.26 Entropy generation perunit mass of the system (*thermal friction*) increases as the initial temperature difference between the subsystems (the driving force) is increased.

states and State-3 and State-4 as the fA and fB states. Enter W=0, Q=0, and T_B=20 deg-C. Click Calculate, then Super-Calculate. The water mass m2 and S_gen are calculated, confirming the manual results. If the final temperature is an unknown, the TESTcalc has to be run iteratively: Set T4=T3, guess T3, and calculate Q; repeat until Q is close to zero.

What-if scenario

Change T1 to 200 deg-C, press Enter, then click Super-Calculate. The new values of S_gen increase to 4.104 kJ/K. The "what-if" study can be repeated for a number of different values of T1 and the entropy generation can be plotted against the temperature difference T1 − T2 as shown in Figure 5.26.

Discussion

Notice how the entropy generation increases with an increase in the temperature difference between the two subsystems. Once again, the mechanism of entropy generation is *thermal friction*—irreversibilities associated with heat transfer across a finite temperature difference.

EXAMPLE 5-11 Energy and Entropy Analysis of a Mixing Process

Two rigid tanks, A and B, each with a volume of 1 m^3, contain R-134a at a pressure of 300 kPa and 100 kPa, respectively, and are connected by a valve. The tanks are in thermal equilibrium with the atmosphere, which is at 25°C. The valve is opened and an equilibrium state at 25°C is established after sufficient time. Determine (a) the final pressure in the tank, (b) heat transfer, and (c) entropy generated during the process. *What-if scenario:* (d) What would the entropy generation be if the pressure in tank A was 500 kPa initially?

SOLUTION

Analyze the system, enclosed by the red boundary in Figure 5.27, using the energy and entropy balance equations customized for a mixing, non-uniform, closed process.

Assumptions

The composite beginning state, State-1 and State-2, and the uniform final state, State-3, are assumed to be in equilibrium. The PC model is applicable to R-134a. The work transfer in opening the valve is negligible. No changes in KE or PE.

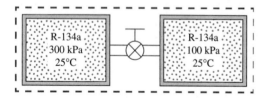

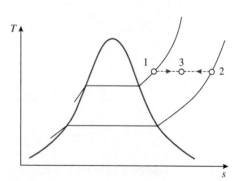

FIGURE 5.27 Non-uniform mixing system of Ex. 5-11 (see Anim. 5.C.*vConstMixing*).

Analysis

Evaluate the anchor states manually or use any PC TESTcalc.

State-1 (given p_1, T_1, V_1):

$$m_1 = \frac{V_1}{v_1} = 13.16 \text{ kg}; \quad u_1 = 248.8 \frac{\text{kJ}}{\text{kg}}; \quad s_1 = 1.003 \frac{\text{kJ}}{\text{kg} \cdot \text{K}};$$

State-2 (given p_2, $T_2 = T_1$, $V_2 = V_1$):

$$m_2 = \frac{V_2}{v_2} = 4.198 \text{ kg}; \quad u_2 = 251.5 \frac{\text{kJ}}{\text{kg}}; \quad s_2 = 1.102 \frac{\text{kJ}}{\text{kg} \cdot \text{K}};$$

State-3 (given $m_3 = m_1 + m_2$; $T_3 = T_1$; $V_3 = V_1 + V_2$):

$$v_3 = 0.1153 \frac{\text{m}^3}{\text{kg}};$$

$$p_3 = 202.3 \text{ kPa}; \quad u_3 = 250.2 \frac{\text{kJ}}{\text{kg}}; \quad s_3 = 1.040 \frac{\text{kJ}}{\text{kg} \cdot \text{K}}$$

Observe the mild variation in u among the three states, despite the temperature remaining constant. The energy equation, Eq. (5.2), for the composite system yields:

$$\Delta U + \Delta \cancel{KE}^0 + \Delta \cancel{PE}^0 = U_f - U_b = Q - (\cancel{W_B}^0 + \cancel{W_O}^0)$$

$$\Rightarrow \quad Q = m_3 u_3 - (m_1 u_1 + m_2 u_2) = 12.97 \text{ kJ}$$

To keep the temperature constant, the atmosphere supplies the heat.

An entropy balance using Eq. (5.4) over the system's universe yields:

$$\Delta S = S_f - S_b = \frac{Q}{T_B} + S_{\text{gen,univ}}$$

$$\Rightarrow \quad S_{\text{gen,univ}} = m_3 s_3 - (m_1 s_1 + m_2 s_2) - \frac{Q}{T_0} = 0.1831 \frac{\text{kJ}}{\text{K}}$$

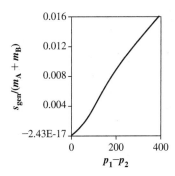

FIGURE 5.28 Entropy generation per unit mass (*mixing friction*) of the system increases with pressure difference (the driving force).

TEST Solution

Launch the PC non-uniform, mixing, closed-process TESTcalc and select R-134a as the working fluid. Evaluate the three anchor states as described in the TEST-Code (see *TEST > TEST-codes*). In the process panel, import State-1 and State-2 as the bA and bB states, and State-3 as the *f*-state. Enter W_ext=0 and T_B=25 deg-C. Click Calculate, then Super-Calculate. The manual results are verified.

What-if scenario

Change p1 to 500 kPa, press Enter, then click Super-Calculate. S_gen is updated to 0.519 kJ/K. Repeating this process for a number of different values of p1, the entropy generation perunit system mass is plotted against the pressure difference p1−p2 in Figure 5.28. Observe how entropy generation increases with an increase in the initial pressure difference, the driving force for *mixing friction* in this problem.

Discussion

If the tanks were insulated, the final temperature could be evaluated by assigning Q = 0 and leaving T3 as an unknown. The final temperature in that case can be shown to be T3 = 22.44 deg-C—slightly cooler than the surrounding temperature. This is not a violation of the second law because, on balance, entropy is still generated by the adiabatic mixing process.

EXAMPLE 5-12	**Energy and Entropy Analysis of a Sudden (Free) Expansion Process**

A well-insulated, rigid tank is partitioned into two chambers, with volumes 0.1 m³ and 0.8 m³. The smaller chamber contains steam at 100 kPa and 50% quality, while the larger chamber is completely evacuated. The stop is removed and steam is allowed to expand to the entire volume of the tank. Determine (a) the change in temperature of the steam and (b) entropy generated in the process. *What-if scenario:* (c) What would the temperature change be if the steam was modeled as an ideal gas?

SOLUTION

Analyze the composite system, enclosed by the red boundary in Figure 5.29, using the energy and entropy balance equations customized for a mixing, non-uniform, closed process.

Assumptions

Equilibrium at the anchor states—State-1 and State-2, the composite *b*-state, and State-3, the *f*-state. The mass of the partition is negligible. No changes in KE or PE.

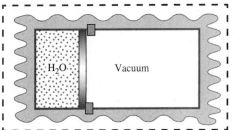

FIGURE 5.29 Free expansion as a non-uniform mixing process (see Anim. 5.B.*suddenExpansion*).

Analysis

The energy equation, Eq. (5.2), for the overall system can be simplified as:

$$\Delta U + \Delta \cancel{KE}^0 + \Delta \cancel{PE}^0 = \cancel{Q}^0 - \cancel{W}^0;$$

$$\Rightarrow \quad U_f = U_b; \quad \Rightarrow \quad mu_3 = m_1u_1 + \cancel{m_2}^0 u_2; \quad \Rightarrow \quad u_3 = u_1$$

With the help of this relation, all the states can now be completely evaluated.

State-1 (given p_1, x_1, V_1):

$$T_1 = T_{sat@p_1} = 99.6°C; \quad u_1 = 1461.7 \frac{kJ}{kg}; \quad s_1 = 4.331 \frac{kJ}{kg \cdot K};$$

$$v_1 = 0.8475 \frac{m^3}{kg}; \quad m_1 = \frac{V_1}{v_1} = 0.118 \text{ kg};$$

State-2 (given m_2, V_2): $\quad p_2 = 0$ kPa;

State-3 (given $m_3 = m_1, V_2, u_3 = u_1$):

$$v_3 = \frac{V_3}{m_3} = 7.628 \frac{m^3}{kg}; \quad T_3 = 47.48°C; \quad s_3 = 4.862 \frac{kJ}{kg \cdot K};$$

Therefore $\Delta T = 47.48 - 99.62 = -52.14°C$.

The entropy equation, Eq. (5.4), applied over the system's universe, produces:

$$\Delta S = \frac{\cancel{Q}^0}{T_B} + S_{gen,univ}; \quad \Rightarrow \quad S_{gen,univ} = S_f - S_b = m_3s_3 - m_1s_1 - \cancel{m_2}^0 s_2$$

$$S_{gen,univ} = m_1(s_3 - s_1) = (0.118)(4.862 - 4.331) = 0.0627 \text{ kJ/K}$$

TEST Solution

Launch the PC mixing, closed-process, TESTcalc and select H2O. Evaluate the anchor states as described in the TEST-Code (see *TEST > TEST-codes*). Switch to the process panel and import State-1 and State-2 as the bA and bB states, and State-3 as the *f*-state. Click Super-Calculate to reproduce the manual results.

What-if scenario

The specific internal energy u of an ideal gas is a function of T only; therefore, the temperature will not change during a sudden expansion process as $u_3 = u_1$. To verify this using TEST, copy the TEST-code generated by the PC TESTcalc into the I/O panel of the corresponding IG TESTcalc, launch in a separate browser tab, and click Load to update the results.

Discussion

This problem could also be considered a closed process executed by a uniform system in the smaller chamber (State-1 to State-3). The boundary work, in that case, would be zero as the unresisted piston can move without doing any work. Expansion against the vacuum is known as **sudden or free expansion**. As evident from this solution, free expansion is yet another fundamental mechanism for entropy generation, hence, another component of *thermodynamic friction*.

5.1.7 Open Processes

As in closed systems, most unsteady problems encountered in open thermodynamic devices generally involve a transition from a *beginning state* to a *final state*. Such a process is known as an **open process**. Charging and discharging a gas cylinder, opening a pressure cooker's pressure-relief valve, inflating a tire, etc., are examples of open processes (see animations starting with Anim. 5.E.*openProcessGeneric*).

To derive a custom set of governing equations, we repeat the procedure followed for the closed process in Sec. 5.1.1. The mass equation for an open system is non-trivial. Multiplying

Eq. (2.3), by dt and integrating across the duration of the process—from the b-state to the f-state—produces the algebraic form of the mass equation for an open process:

$$dm = \sum_i \dot{m}_i dt - \sum_e \dot{m}_e dt;$$

$$\Rightarrow \int_b^f dm = \int_b^f \sum_i \dot{m}_i dt - \int_b^f \sum_e \dot{m}_e dt = \sum_i \int_b^f \dot{m}_i dt - \sum_e \int_b^f \dot{m}_e dt; \quad [\text{kg}]$$

Therefore,

$$\underbrace{\Delta m = m_f - m_b}_{\substack{\text{Change in } m \text{ during} \\ \text{the process.}}} = \underbrace{\sum_i m_i - \sum_e m_e}_{\substack{\text{Mass transported into} \\ \text{the system.}}}; \quad [\text{kg}] \tag{5.19}$$

Although this equation may look complex, it states the obvious; the mass gained by the system must be equal to the net transport of mass into the system during the process. The equation simplifies further for processes that involve a single inlet or a single exit, which is usually the case. For example, in the charging process of an evacuated cylinder, Eq. (5.19) reduces to $m_f = m_i$.

The energy equation can be similarly customized. Multiplying Eq. (2.8) by dt and integrating across the duration of the process, we obtain:

$$dE = \sum_i \dot{m}_i j_i dt - \sum_e \dot{m}_e j_e dt + \dot{Q} dt - \dot{W}_{\text{ext}} dt;$$

$$\Rightarrow \int_b^f dE = \int_b^f \sum_i \dot{m}_i j_i dt - \int_b^f \sum_e \dot{m}_e j_e dt + \int_b^f \dot{Q} dt - \int_b^f \dot{W}_{\text{ext}} dt; \tag{5.20}$$

$$\Rightarrow \Delta E = E_f - E_b = \sum_i \int_b^f \dot{m}_i j_i dt - \sum_e \int_b^f \dot{m}_e j_e dt + Q - W_{\text{ext}}; \quad [\text{kJ}]$$

As in a closed process, Q represents the total heat transfer during the process and W_{ext} represents the total external work transfer, usually the sum of shaft and electrical work (boundary work tends to be absent in most situations).

To evaluate the integrals of the transport terms a further assumption is necessary. In many applications, the thermodynamic states at the inlets and exits do not vary significantly for the duration of the process. For example, the supply-line pressure does not change when an evacuated tank is filled (Fig. 5.30). During the discharge from a pressure cooker, the exit state remains saturated as long as there is liquid in the chamber (see Anim. 5.E.*dischargePC*).The assumption of an invariant exit or inlet state, known as the **uniform-flow assumption**, can be quite convenient in simplifying the energy equation; the specific flow energy, j, can be taken outside the integrals, and Eq. (5.20) can be simplified to:

$$\underbrace{\Delta E = E_f - E_b}_{\substack{\text{Change in } E \text{ during} \\ \text{the process.}}} = \underbrace{\sum_i m_i j_i - \sum_e m_e j_e}_{\substack{J_{\text{net}}: \text{ Net energy} \\ \text{transported into the} \\ \text{system.}}} + \underbrace{Q}_{\substack{\text{Energy} \\ \text{added} \\ \text{through} \\ \text{heat.}}} - \underbrace{W_{\text{ext}}}_{\substack{\text{Energy lost} \\ \text{through external} \\ \text{work transfer.}}}; \quad [\text{kJ}] \tag{5.21}$$

where m_i and m_e are the total amount of mass transferred at the ports during the process with j_i and j_e remaining invariant during the process. The energy flow diagram (Fig 5.31) graphically explains the meaning of each term in Eq. (5.21).

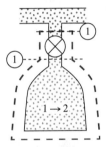

FIGURE 5.30 State-1 is the same as that in the supply line while State-2 changes with time (see Anim. 5.E.*charging*.)

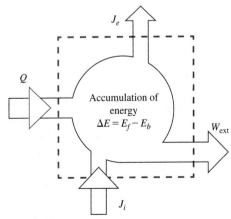

FIGURE 5.31 An energy flow diagram explaining Eq. (5.21).

The entropy equation, Eq. (2.13), can be customized for an open process using the same procedure—integrating across the process duration and applying the uniform-flow assumption:

$$dS = \sum_i \dot{m}_i s_i dt - \sum_e \dot{m}_e s_e dt + \frac{\dot{Q}}{T_B} dt + \dot{S}_{gen} dt;$$

$$\Rightarrow \int_b^f dS = \sum_i \int_b^f \dot{m}_i s_i dt - \sum_e \int_b^f \dot{m}_e s_e dt + \int_b^f \frac{\dot{Q}}{T_B} dt + \int_b^f \dot{S}_{gen} dt;$$

$$\Rightarrow \Delta S = S_f - S_b = \sum_i m_i s_i - \sum_e m_e s_e + \frac{Q}{T_B} + S_{gen}; \qquad \left[\frac{kJ}{kg \cdot K}\right] \qquad \textbf{(5.22)}$$

The change in entropy is not only due to entropy transfer by heat and internal entropy generation, as in the case of closed processes, but also to the transport of entropy by mass flow.

Evaluation of the left-hand sides of Eqs. (5.21) and (5.22) can be further simplified if the system can be assumed to be uniform (based on LTE) at the beginning and end of the process. With only a pair of system states representing the b and f states, ΔE can be evaluated from $\Delta E = E_f - E_b = m_f e_f - m_b e_b$. Likewise, $\Delta S = m_f s_f - m_b s_b$. The inlet and exit ports are described by uniform flow states based on LTE. The resulting simplified open process is known as a **uniform-state**, **uniform-flow** process. When selecting the system boundary, care should be taken to avoid violating the *uniform-flow* assumption. For example, in the charging process shown in Figure 5.30, if the valve is not included within the system boundary, the new inlet state, State-3, will have time-dependent properties and the uniform-state, uniform-flow assumption will break down.

5.1.8 TEST and Open-Process TESTcalcs

In TEST, the **open-process TESTcalcs** are located in the systems, open, unsteady processes branch of the TESTcalc module. As in a closed-process TESTcalc (Sec. 5.1.1), the anchor states—in this case b, f, i, and/or e states—are evaluated in the state panel before being imported to the process panel as anchor states. Known process variables are entered and the Super-Calculate button is used to iterate between the state and analysis panels until the desired unknowns are calculated. Note that in the state panel, system states are used for all anchor states, including inlet and exit states. Flow states are not used because the process equations involve cumulative amounts of mass transfer rather than mass flow rates.

EXAMPLE 5-13 **Mass, Energy, and Entropy Analysis of a Charging Process**

A 1-m³ evacuated, insulated, rigid tank is being filled from a supply line carrying steam at 100 kPa and 400°C (Fig. 5.32). Determine (a) the final temperature and (b) entropy generated during the process. Assume the final pressure in the tank is 100 kPa. Use the PG model for steam. ***What-if scenario:*** (c) What would the temperature be if the PC model was used instead?

SOLUTION

Analyze the system, enclosed by the red boundary in Figure 5.32, using the energy and entropy balance equations customized for an open process.

Assumptions

The process is assumed to be a uniform-state, uniform-flow process with the b-state and f-state represented by State-1 and State-2, and the inlet state by State-3, chosen at the valve's supply side. Neglect KE and PE.

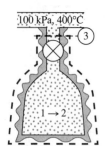

FIGURE 5.32 The charging process in Ex. 5-13 (see Anim. 5.E.*charging*).

Analysis

From Table C-1, or any PG TESTcalc, we obtain $R = 0.461$ kJ/kg · K, $c_p = 1.868$ kJ/kg · K, and $k = 1.328$.

The energy equations, Eq. (5.21), with $m_1 = 0$ reduces to:

$$\Delta E = m_3 j_3; \quad \Rightarrow \quad \Delta U + \Delta\cancel{KE}^0 + \Delta\cancel{PE}^0 = m_3 h_3 + m_3 \cancel{ke_3}^0 + m_3 \cancel{pe_3}^0;$$

$$\Rightarrow \quad m_2 u_2 - \cancel{m_1}^0 u_1 = m_3 h_3; \quad \Rightarrow \quad u_2 = h_3;$$

State-3 can be evaluated from p_3 and T_3 and State-2 from $p_2 = p_3$ and $u_2 = h_3$. For the PG model, the task becomes simpler since $u_2 = h_3$ can be manipulated as follows:

$$u_2 = h_3 = u_3 + p_3 v_3 = u_3 + RT_3;$$

$$\Rightarrow \quad u_2 - u_3 = RT_3; \quad \Rightarrow \quad c_v(T_2 - T_3) = RT_3;$$

$$\Rightarrow \quad T_2 = \left(1 + \frac{R}{c_v}\right)T_3 = \left(\frac{c_p}{c_v}\right)T_3 = kT_3 = (1.328)(673) = 894 \text{ K}$$

The ideal gas equation of state yields:

$$m_2 = \frac{p_2 V_2}{RT_2} = \frac{(100)(1)}{(0.461)(894)} = 0.243 \text{ kg}$$

The entropy balance equation of Eq. (5.22) yields:

$$S_f - S_b = \sum_i m_i s_i - \sum_i \cancel{m_e}^0 s_e + \frac{\cancel{Q}^0}{T_B} + S_{gen,univ};$$

$$\Rightarrow \quad S_{gen,univ} = m_2 s_2 - \cancel{m_1}^0 s_1 - m_3 s_3 = m_2(s_2 - s_3) = m_2 c_v \ln\frac{T_2}{T_1};$$

$$\Rightarrow \quad S_{gen,univ} = (0.243)(1.868)\ln\frac{894}{673} = 0.129 \frac{\text{kJ}}{\text{K}}$$

TEST Solution

Launch the PG open-process TESTcalc and select H2O. Evaluate the states as described in the TEST-code (see *TEST > TEST-codes*). Specify the evacuated tank's state by setting p1 and m1 to zero. In the process panel, import the *b, f,* and *i* states, enter Q=W_ext=0, click Calculate, then Super-Calculate to obtain T2 and S_gen.

What-if scenario

Launch the PC open-process TESTcalc in a separate browser tab. Copy the TEST-code generated in the PG solution into the I/O panel of the PC TESTcalc and click Load. The PC solution produces a final temperature of 859 K, which is more accurate than the PG solution. To explore if the difference can be explained by temperature dependence of specific heats (neglected by the PG model), repeat the solution with the IG model. The final temperature produced by the IG model is 859 K, identical to that of the PC model. The inaccuracy in the PG value is entirely due to the assumption that specific heats are constant.

Discussion

The rise in temperature can be explained by the transfer of flow work into the insulated tank by the incoming charge. The mechanism for entropy generation can be understood when you compare the charging process with the unresisted expansion of Example 5-12. In this case, steam is partially resisted as the pressure inside gradually increases.

EXAMPLE 5-14 **Mass, Energy, and Entropy Analysis of a Discharge Process**

A 5-L pressure cooker has an operating temperature of 120°C. It contains a saturated mixture of liquid water and vapor; one-half of the volume is liquid water. Heat is supplied at a rate of 1 kW. Determine (a) the operating pressure, (b) the mass of the pressure regulator that can be supported by a 10-mm² flow area, and (c) the time required for all the liquid to completely vaporize. *What-if scenario:* (d) What would the answer in part (c) be if the operating pressure had been the same as the ambient pressure of 100 kPa?

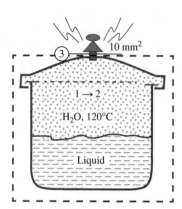

FIGURE 5.33 Schematic for Ex. 5-14 (see Anim. 5.E.*dischargePC*).

SOLUTION

Analyze the system, enclosed by the red boundary in Figure 5.33, using the energy and entropy balance equations customized for an open process.

Assumptions

The discharge process is assumed to be a uniform-state, uniform-flow process so that the *b*-state and *f*-state are represented by State-1 and State-2 and the exit state by State-3, which is chosen right below the regulator mass. Neglect KE and PE.

Analysis

The exit and final states are both composed of saturated vapor at the same temperature. Even though the former is a flow state and the latter is a system state, the underlying equilibrium in each state must be identical. From the known volumetric vapor fraction y_1 at State-1, the quality x_1 can be calculated by following the procedure developed in Sec. 3.4.7:

State-1 (given T_1, $y_1 = 50\%$, V_1):

$$p_1 = 198.5 \text{ kPa}; \quad m_1 = 2.3612 \text{ kg}; \quad u_1 = 505.9 \frac{\text{kJ}}{\text{kg}};$$

State-2 (given $T_2 = T_1$, $x_2 = 100\%$, $V_2 = V_1$): $\quad m_2 = 0.0056 \text{ kg}; \quad u_2 = 2529.2 \frac{\text{kJ}}{\text{kg}};$

State-3 (given $T_3 = T_1$, $x_3 = 100\%$): $\quad j_3 \cong h_3 = 2706.3 \frac{\text{kJ}}{\text{kg}};$

A force balance on the weight produces: $\quad \dfrac{mg}{(1000 \text{ N/kN})} = A(p_1 - p_0); \quad \Rightarrow \quad m = 0.1 \text{ kg};$

The mass balance equation for the open process, Eq. (5.19), produces:

$$m_2 - m_1 = -m_3; \quad \Rightarrow \quad m_3 = 2.3612 - 0.0056 = 2.3556 \text{ kg};$$

The energy equation, Eq. (5.21), yields:

$$-Q = U_b - U_f - m_3 j_3 - W_{\text{ext}}^{\,0} = m_1 u_1 - m_2 u_2 - m_3 h_3;$$

$$\Rightarrow \quad Q = -(2.3612)(505.9) + (0.0056)(2529.2) + (2.3556)(2706.3) = 5194.6 \text{ kJ}$$

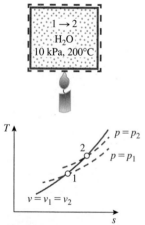

FIGURE 5.34 In the TEST solution, two neighboring states, 1 and 2, on the constant volume line are evaluated.

Therefore, the time of vaporization is:

$$t = \frac{Q}{\dot{Q}} = \frac{5194.6}{1} \text{ s} = 1.44 \text{ h}$$

TEST Solution

Launch the PC open-process TESTcalc and select H2O as the working fluid. Evaluate the states as described in the TEST-code (see *TEST > TEST-codes*). In the process panel, import the calculated states as b, f, and e states, enter W_ext=0, and click Super-Calculate. Q is evaluated as 5194.7 kJ from which *t* can be manually calculated.

What-if scenario

Make T1 an unknown and enter p1 as 100 kPa. Click Super-Calculate to produce Q = 5412.3 kJ. The new time of vaporization can be calculated as 1.5 h.

Discussion

In the TEST solution, the value of S_gen calculated is negative, which is a gross violation of the second law. The reason behind this contradiction is the default value for T_B, 25 deg-C—a boundary temperature that is cooler than the steam inside the system (120°C). This makes it impossible to transfer heat into the system. By setting T_B to any realistic temperature, greater than 120°C, and clicking Super-Calculate, S_gen can be shown to be positive.

5.2 TRANSIENT ANALYSIS

The time derivatives in the balance equations are set to zero for steady systems and integrated away for unsteady systems undergoing a process. In a transient analysis, instantaneous behavior of the system is of interest and these derivates play an important role. We have already performed a few transient analyses in Examples 2-1, 2-2, 2-4, and 2-5 with the goal of understanding various terms of the balance equations. We will revisit the analysis, more formally in this section, for both closed- and open-transient systems.

5.2.1 Closed-Transient Systems

The mass equation for a closed system, steady or unsteady, is trivial—mass remains constant for a closed system. Without the transport terms, the energy and entropy balance equations— Eqs. (2.10) and (2.15)—can be interpreted as follows:

$$\underbrace{\frac{dE}{dt}}_{\substack{\text{Rate of change of} \\ \text{stored energy of a} \\ \text{closed system.}}} = \underbrace{\dot{Q}}_{\substack{\text{Rate of energy} \\ \text{transfer into the} \\ \text{system by heat.}}} - \underbrace{\dot{W}_{\text{ext}}}_{\substack{\text{Rate of energy} \\ \text{transfer out of the} \\ \text{system by exemal} \\ \text{work.}}} ; \quad [\text{kW}] \tag{5.23}$$

$$\underbrace{\frac{dS}{dt}}_{\substack{\text{Rate of change of} \\ \text{entropy for a closed} \\ \text{system.}}} = \underbrace{\frac{\dot{Q}}{T_B}}_{\substack{\text{Rate of entropy} \\ \text{transfer by heat.}}} + \underbrace{\dot{S}_{\text{gen}}}_{\substack{\text{Rate of generation} \\ \text{of entropy inside the} \\ \text{system boundary.}}} ; \quad \left[\frac{\text{kW}}{\text{K}}\right] \tag{5.24}$$

The instantaneous rate of change of energy is equal the net rate of energy transfer through heat and work across the system boundary. The rate of change of entropy is determined by the rate at which entropy is transferred by heat across the boundary. And by the rate of spontaneous generation due to *thermodynamic friction* within the system. Assuming quasi-equilibrium at any given instant during the evolution of the system, allows us to relate U and S to other properties of the system. This is illustrated in Example 5-15.

EXAMPLE 5-15 **Energy Analysis of a Transient System**

Heat is added at a rate of 1 kW to a 1-m^3 rigid tank that contains steam. Determine the rate of increase of (a) temperature and (b) pressure at the instant when the pressure and temperature inside are 10 kPa and 200°C, respectively. Use the PG model for steam.

SOLUTION

Analyze the system, steam within the rigid tank, using the energy and entropy balance equations customized for a closed-transient system.

Assumptions

The system is considered uniform and in quasi-equilibrium at any given time. The volume remains constant.

Analysis

The energy equation, Eq. (5.23), for the constant mass of the system reduces to:

$$\frac{dU}{dt} + \cancelto{0}{\frac{dKE}{dt}} + \cancelto{0}{\frac{dPE}{dt}} = \dot{Q} - \cancelto{0}{\dot{W}_{ext}} \quad \Rightarrow \quad m\frac{du}{dt} = \dot{Q}$$

Substituting $du = c_v dT$ [Eq. (3.31) of the IG/PG model], we obtain:

$$mc_v\frac{dT}{dt} = \dot{Q} \quad \text{where,} \quad m = \frac{\forall}{v} = \frac{p\forall}{RT}$$

From Table C-1, or any PG TESTcalc, we obtain $R = 0.462$ kJ/kg·K and $c_v = 1.4104$ kJ/kg·K. Therefore,

$$m = \frac{p\forall}{RT} = \frac{(10)(1)}{(0.462)(273 + 200)} = 0.0458 \text{ kg}$$

$$\text{and} \quad \frac{dT}{dt} = \frac{\dot{Q}}{mc_v} = \frac{1}{(0.0458)(1.41)} = 15.5 \frac{K}{s}$$

The rate of change of pressure is obtained as follows:

$$p = \frac{mRT}{\forall}; \quad \frac{dp}{dt} = \frac{mR}{\forall}\frac{dT}{dt} = \frac{(0.0458)(0.462)}{(1)}(15.5) = 0.328 \frac{kPa}{s}$$

TEST Solution

Launch the PG closed-process TESTcalc and select H2O as the working fluid. Calculate State-1 from the known values of p1, T1, and Vol1. For a neighboring state, select State-2, enter m2=m1, Vol2=Vol1, and calculate the state partially. In the process panel, import State-1 as the *b*-state and State-2 as the *f*-state, enter W_O=W_B=0 and Q=1 kJ (heat transfer over 1 s), and click Calculate. Now click Super-Calculate to iterate between the process and state panels to produce p2 as 10.375 kPa and T2 as 215.5 deg-C. The rates can be calculated in the I/O panel as (T2−T1)/1=15.49 K/s, and (p2−p1)/1=0.327 kPa/s. To refine the results, reduce the duration of the process (and, consequently, Q) and repeat the solution.

Discussion

The particular form of energy equation $mc_v\frac{dT}{dt} = \dot{Q}$ can be shown to apply to the SL model as well. In many transient heat transfer analyses, this equation is the starting point for what is known in heat transfer textbooks as the **lumped capacity** solution.

5.2.2 Isolated Systems

An **isolated system**—a system that has absolutely no interactions with the surroundings (Fig. 5.35)—is the simplest closed system that is amenable to a transient analysis. With no mass, heat, or work transfer, the energy and entropy equations, Eqs. (5.23) and (5.24), simplify to:

$$\frac{dE}{dt} = 0; \quad [\text{kW}], \quad \text{and} \quad \frac{dS}{dt} = \dot{S}_{gen} \geq 0; \quad \left[\frac{\text{kW}}{\text{K}}\right] \tag{5.25}$$

Like mass, the stored energy of an isolated system remains constant. However, because entropy is indestructible and can be spontaneously generated inside a system, the entropy of the isolated system must increase until all spontaneous changes subside and the system achieves thermodynamic equilibrium (see Anim. 1.A.*globalEquilibrium*). This is known as the **increase of entropy principle**, which applies only to isolated systems. Thus, if heat is lost from a closed system, its entropy can decrease despite internal entropy production. We will revisit isolated systems in Chapter 14.

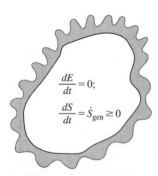

$$\frac{dE}{dt} = 0;$$

$$\frac{dS}{dt} = \dot{S}_{gen} \geq 0$$

FIGURE 5.35 The increase of entropy principle applies to isolated systems only and can be derived from the entropy balance equation (see Anim. 3.A.*equilibrium*).

5.2.3 Mechanical Systems

Although most thermal systems are stationary, it is instructive to analyze a transient mechanical system, motion of a point mass, from a thermodynamic standpoint. The energy equation that we derive here for transient **mechanical systems** will be used for advanced state analysis in Chapter 11.

Consider a simple mechanical system, a rigid body that can be represented by a **point mass** of mass m moving along a curved path as shown in Figure 5.36. An external force vector, **F**, besides the weight mg, acts on the body whose instantaneous velocity is represented by the vector **V**. At a given instant, the tangential velocity is V and the tangential external force is F_s. Now suppose that the body is heated by external radiation while in motion. We would like to know if the heat transfer has any effect on the system's mechanical energy.

Newton's second law of motion along the tangential direction (Fig. 5.36) at a given instant can be written as:

$$F_s - \frac{mg}{(1000 \text{ N/kN})} \sin\theta = \frac{1}{(1000 \text{ N/kN})} m\frac{dV}{dt}; \quad \left[kN = \frac{\text{kg} \cdot \text{m}}{\text{s}^2} \frac{kN}{N} \right] \qquad (5.26)$$

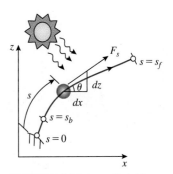

FIGURE 5.36 Decoupling of mechanical energy from internal energy in rigid-body dynamics.

Multiplying both sides by $V = \dfrac{ds}{dt}$ and expressing $\sin\theta$ as the ratio of dz to ds (Fig. 5.36), we obtain:

$$F_s\frac{ds}{dt} = \frac{1}{(1000 \text{ N/kN})}\left[m\frac{dV}{dt}\frac{ds}{dt} + mg\frac{dz}{ds}\frac{ds}{dt} \right]; \quad [kW]$$

$$\Rightarrow \quad F_s V = \frac{1}{(1000 \text{ N/kN})}\left[m\frac{VdV}{dt} + mg\frac{dz}{dt} \right] = \frac{1}{(1000 \text{ N/kN})}\left[\frac{d}{dt}\left(\frac{mV^2}{2}\right) + \frac{d}{dt}(mgz) \right]; \quad [kW]$$

Treating the rigid body as a thermodynamic system, the left-hand side can be interpreted as the negative of the rate of mechanical work transfer (Eq. 0.10), and the two terms inside the parentheses on the right-hand side are the system's KE and PE. After rearranging, we obtain what is called the **mechanical energy equation**:

$$\underbrace{\frac{d}{dt}(\text{KE}) + \frac{d}{dt}(\text{PE})}_{\substack{\text{Rate of change of mechanical} \\ \text{energy of a rigid body.}}} = \underbrace{-\dot{W}_M}_{\substack{\text{Rate of mechanical work} \\ \text{transfer into the system.}}} ; \quad [kW] \qquad (5.27)$$

Mechanical work, transferred into a rigid closed system, goes into increasing the system's kinetic and potential energies. In our definition for mechanical work, the gravitational force is not considered an external force; instead, the effect of gravity is incorporated into the system's potential energy.

What is remarkable about the mechanical energy equation is that it is not affected by heat transfer or any other form of work transfer such as electrical work. Subtracting the mechanical energy equation [Eq. (5.27)] from the general energy equation [Eq. (5.23)] for a closed system, then substituting the expression for external work from Eq. (0.20), we obtain:

$$\frac{dE}{dt} - \left[\frac{d}{dt}(\text{KE}) + \frac{d}{dt}(\text{PE}) \right] = \dot{Q} - \dot{W}_{\text{ext}} + \dot{W}_M$$

$$= \dot{Q} - [\dot{W}_{\text{sh}} + \dot{W}_{\text{el}} + \dot{W}_B] + \dot{W}_M = \dot{Q} - [\dot{W}_{\text{sh}} + \dot{W}_{\text{el}} + \dot{W}_{pdV} + \dot{W}_M] + \dot{W}_M;$$

$$\Rightarrow \quad \frac{dU}{dt} = \dot{Q} - [\dot{W}_{\text{sh}} + \dot{W}_{\text{el}} + \dot{W}_{pdV}]; \quad [kW] \qquad (5.28)$$

In deriving Eq. (5.28), we have not assumed KE and PE to be negligible; yet this equation has the same form as the energy equation derived from Eq. (5.23) for a stationary system. In a closed system analysis, neglecting changes in KE and PE, is not an approximation if mechanical work is also ignored. Most thermodynamic systems, however, are stationary and decoupling the energy equation into mechanical and thermal components is seldom necessary.

For the mechanical system of Figure 5.36, the energy transfer by radiation, according to Eq. (5.28), goes into the internal energy while the mechanical work transfer goes directly into the system's mechanical energy.

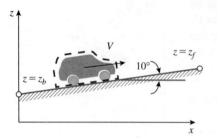

FIGURE 5.37 The power provided by the engine helps the car climb a slope, accelerate, and overcome drag and rolling resistances.

EXAMPLE 5-16 Energy Analysis of a Mechanical Process

As shown in Figure 5.37, a subcompact car with a mass of 1000 kg accelerates from 0 to 100 km/h in 30 s on a road with an upward slope of 10°. Determine the additional power requirement due to climbing and acceleration.

SOLUTION

Solve the unsteady energy equation treating the car, shown in Figure 5.37, as a closed mechanical system.

Assumptions

The car is a rigid mechanical system with a constant mass. The sole purpose of the engine is to provide the necessary "push" to this mechanical system. The additional engine power, $\dot{W}_E$, must be equal to the negative of the mechanical work transfer, calculated from Eq. (5.27) for the car.

Analysis

The mechanical energy equation, Eq. (5.27), for this closed mechanical system reduces to:

$$\frac{d(\text{KE})}{dt} + \frac{d(\text{PE})}{dt} = -\dot{W}_M;$$

$$\Rightarrow \quad \dot{W}_E = \frac{\text{KE}_f - \text{KE}_b}{\Delta t} + \frac{\text{PE}_f - \text{PE}_b}{\Delta t} = \frac{m}{(1000 \text{ J/kJ})}\left[\frac{V_f^2 - V_b^2}{2\Delta t} + \frac{g(z_f - z_b)}{\Delta t}\right]$$

The subscripts b and f mark the beginning and final locations of the car over the period of acceleration. The change in elevation can be calculated from the average velocity of the time interval, given that the acceleration remains constant. From Figure 5.37:

$$z_f - z_b = s \sin 10° = \sin 10°\left(\frac{V_f + V_b}{2}\right)\Delta t = \frac{\sin 10°}{2}V_f\Delta t;$$

$$= \frac{\sin 10°}{2}\left(\frac{100 \times 1000}{3600}\right)30 = 72.35 \text{ m}$$

The additional power requirement now can be calculated from the expression for $\dot{W}_E$:

$$\frac{1000}{1000}\left[\frac{(27.78^2)}{2(30)} + \frac{(9.81)(72.36)}{30}\right] = 12.86 + 23.66 = 36.5 \text{ kW}$$

Discussion

Although the power delivered by the engine is treated as external mechanical work, the mass of the engine must be included in the system mass. In general, the engine power $\dot{W}_E$ of a car is expended in acceleration and climbing $(-\dot{W}_M)$, and to overcome viscous drag $(\dot{W}_d)$ and rolling resistance $(\dot{W}_{RR})$:

$$\dot{W}_E = -\dot{W}_M + \dot{W}_d + \dot{W}_{RR} = \frac{d\text{KE}}{dt} + \frac{d\text{PE}}{dt} + \dot{W}_d + \dot{W}_{RR}$$

We have already evaluated the power required to overcome aerodynamic drag in Example 0-4.

5.2.4 Open-Transient Systems

Transient analysis of open systems is sometimes necessary when analyzing the filling or emptying (charging or discharging) of vessels; and the starting or stopping behavior of devices such as turbines, automobile intake and exhaust manifolds, etc. The governing equations are the balance equations (developed in Chapter 2) in their most general forms. With all terms of the balance equations present, analysis of open-transient systems often requires a computer solution. When a simple material model, such as the PG model, can be applied, an analytical solution is possible as shown in Example 5-17.

EXAMPLE 5-17 | Mass and Energy Analysis of an Open-Transient System

As shown in Figure 5.38, a 1-m³ insulated, rigid tank is being filled at the rate of 0.01 kg/s from a supply line that carries steam at 100 kPa and 400°C. Determine (a) the rate of change of temperature and (b) pressure at the instant the temperature and pressure inside the tank are measured at 400°C and 50 kPa. Assume that the steam behaves as a perfect gas.

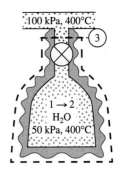

Solution

Analyze the unsteady mass and energy equations for the open system enclosed within the red boundary of Figure 5.38.

Assumptions

At any given time, the system is considered uniform and at thermodynamic equilibrium.

Analysis

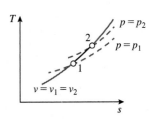

The mass equation, Eq. (2.3), with the application of the IG equation of state, $pv = RT$, can be manipulated to produce:

$$\frac{dm}{dt} = \dot{m}_i; \quad \Rightarrow \quad \frac{d}{dt}\left(\frac{p\forall}{RT}\right) = \dot{m}_i; \quad \Rightarrow \quad \frac{\forall}{RT}\frac{dp}{dt} - \frac{p\forall}{RT^2}\frac{dT}{dt} = \dot{m}_i;$$

$$\Rightarrow \quad \frac{dp}{dt} = \frac{p}{T}\frac{dT}{dt} + \frac{RT}{\forall}\dot{m}_i; \quad \Rightarrow \quad \frac{dp}{dt} = 0.0743\frac{dT}{dt} + 3.109$$

Notice $R = 8.314/\overline{M}_{H_2O} = 0.462$ kJ/kg·K has been substituted. The energy equation, Eq. (2.8), can be simplified:

$$\frac{dU}{dt} + \frac{dKE^0}{dt} + \frac{dPE^0}{dt} = \dot{m}_i j_i + \dot{Q}^0 - \dot{W}_{ext}^0;$$

$$\Rightarrow \quad \frac{d(mu)}{dt} = \dot{m}_i(h_i + ke_i^0 + pe_i^0); \quad \Rightarrow \quad m\frac{du}{dt} + u\frac{dm}{dt} = \dot{m}_i h_i;$$

$$\Rightarrow \quad mc_v\frac{dT}{dt} + \dot{m}_i u = \dot{m}_i(u_i + p_i v_i); \quad \Rightarrow \quad mc_v\frac{dT}{dt} + \dot{m}_i u = \dot{m}_i(u_i + RT_i);$$

$$\Rightarrow \quad mc_v\frac{dT}{dt} = \dot{m}_i(u_i - u + RT_i); \quad \Rightarrow \quad mc_v\frac{dT}{dt} = \dot{m}_i c_v(T_i - T) + \dot{m}_i RT_i;$$

$$\Rightarrow \quad \frac{dT}{dt} = \frac{\dot{m}_i}{m}(T_i - T) + \frac{\dot{m}_i RT_i}{mc_v} = 13.7\frac{K}{s}$$

where $m = p\forall/(RT) = 0.161$ kg have been used. Substituting this result into the mass equation yields:

$$\frac{dp}{dt} = 4.12\frac{kPa}{s}$$

TEST Solution

Launch the PG open-process TESTcalc and select H2O as the working fluid. To set up a differential process lasting 0.1 s calculate the b-state, State-1, from the known values of p1, T1, and Vol1. Only Vol2 = Vol1 is known about the f-state, State-2, at this point. Calculate the inlet state, State-3, from the known properties p3, T3, and m3 = 0.001 kg (mass transfer in 0.1 s). In the process panel, import the anchor states (i, b, and f states), set Q=W_ext=0, and click Super-Calculate. In State-2, p2 and T2 are calculated as 50.41 kPa and 401.36 deg-C. The desired rates can be calculated as (T2−T1)/(0.1)=13.6 K/s and (p2−p1)/(0.1)=4.1 kPa/s. These

FIGURE 5.38 Transfer of mass brings about changes in temperature and pressure inside the tank of Ex. 5-17.

answers can be further refined by restricting the differential process to an shorter period. A similar solution for the PC model is quite difficult to obtain manually.

Discussion

A simple manual solution is not possible without the simplification offered by the PG model. For more accurate results, the PC model can be used with the TEST solution to yield the rates as 11.71 K/s and 3.98 kPa/s. Complex transient processes can be accurately modeled by assuming the system passes through a sequence of quasi-equilibrium states and by, essentially, repeating the calculations outlined in this example.

5.3 DIFFERENTIAL PROCESSES

Complex thermodynamic systems can be described by an aggregate of local systems in LTE (Sec. 3.1.1). Two neighboring local systems, separated by space or time, can be described by two equilibrium states which are differentially apart. With equilibrium states being "memoryless" (a system at equilibrium has no knowledge of how it arrived at a particular state), we can set up a hypothetical closed process that connects any two neighboring equilibrium states of a pure substance. Such a process is known as a **differential process**.

The energy and entropy equation for a differential-closed process can be customized from Eqs. (5.1) and (5.3). Recall that local systems can only experience heat and boundary work transfer, even if the global system involves electrical and shaft work transfer. The boundary work may consist of both mechanical work, which is responsible for the acceleration or deceleration of the local system, and pdV work, which is responsible for its deformation. Additionally, a local system can be considered closed as diffusive mixing is not possible in a system with uniform chemical composition (pure substance). Without neglecting KE or PE of the local system, the energy equation for a differential closed process, Eq. (5.1), can be written:

$$\underbrace{dE}_{\text{Differential change in stored energy.}} = \underbrace{dQ}_{\substack{\text{Heat added to} \\ \text{the system.}}} - \underbrace{dW_B}_{\substack{\text{Boundary work transferred out} \\ \text{of the system.}}} \quad ; \quad [\text{kJ}]$$

$$\Rightarrow \quad dU + d(\text{KE}) + d(\text{PE}) = dQ - dW_{pdV} - dW_M; \quad [\text{kJ}]$$

Substituting $d(\text{KE}) + d(\text{PE}) = -dW_M$ from Eq. (5.27), the energy equation for a differential process reduces to:

$$dU = dQ - dW_{pdV}; \quad [\text{kJ}] \tag{5.29}$$

The change in internal energy is entirely due to heat and pdV work transfer, regardless of changes in KE or PE of the local system.

An equilibrium state is described by thermodynamic properties, which are intensive, independent of the size of the system. Therefore, a differential process involving two local states can be scaled up to a large uniform system, undergoing differential changes due to heat and pdV work transfer as shown in Figure 5.39. Out of myriad of processes that can cause the observed differential changes, dp, dV, dT, dU, dS, etc., in the system, let's consider an internally reversible path—a path that has no entropy generation within the system (red) boundary. It can be shown that for any pair of neighboring states, an infinite number of internally reversible paths exist. To avoid entropy generation, due to friction through eddies or vortices in the trapped gas, the piston in Figure 5.39 must be kept at equilibrium at all times, displaced in increments only at extremely slow speed. The pdV work transfer along such an internally reversible path can be written:

$$dW_{pdV,\text{int.rev.}} = pdV; \quad [\text{kJ}] \tag{5.30}$$

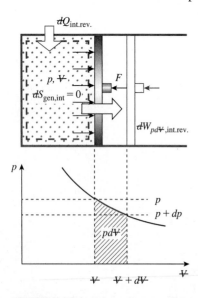

FIGURE 5.39 An internally reversible differential process (see Anim. 5.G.*TdsEquation*).

The entropy equation, Eq. (5.3), for a differential process can be simplified by realizing, for a local system, the internal boundary temperature is same as the system temperature (which must be uniform as in Fig. 5.39).

$$\underbrace{dS}_{\substack{\text{Change in } S \text{ during} \\ \text{the differential process.}}} = \underbrace{dQ/T}_{\substack{\text{Entropy added to} \\ \text{the system by heat.}}} + \underbrace{dS_{\text{gen,int}}}_{\substack{\text{Entropy generated in the system} \\ \text{during the process.}}} ; \quad \left[\frac{kJ}{K}\right] \qquad \textbf{(5.31)}$$

Valid for any kind of differential process, Eq. (5.31) can be applied along an internally reversible path with $dS_{\text{gen,int.rev.}} = 0$ to express the differential heat transfer in terms of differential change in entropy:

$$dQ_{\text{int.rev.}} = TdS; \quad [kJ] \qquad \textbf{(5.32)}$$

Substituting Eqs. (5.30) and (5.32) into the energy equation, Eq. (5.29), of an internally reversible process, we obtain the TdS equation, which was first introduced (without a proof) in Sec. 3.1.3:

$$dU = dQ_{\text{int.rev.}} - dW_{pdV,\text{int.rev.}} = TdS - pdV; \quad [kJ]$$

$$\Rightarrow \quad TdS = dU + pdV; \quad [kJ]$$

$$\underbrace{TdS}_{\substack{\text{Reversible} \\ \text{heat transfer.}}} = \underbrace{dU}_{\substack{\text{Change in the} \\ \text{internal energy.}}} + \underbrace{pdV}_{\substack{\text{Reversible work} \\ \text{transfer.}}} ; \quad [kJ] \qquad \textbf{(5.33)}$$

Although each term of this equation can be interpreted as a differential amount of energy—increase in internal energy caused by reversible heat and work transfer—the TdS equation relates change in entropy to other differential changes in properties. The relation involves only properties of equilibrium states without any identifying details of the actual process that connects the two states. This makes the relation very useful, as it can be applied to any two neighboring equilibrium states, connected by any type of differential process, internally reversible or irreversible. In Chapter 11, we will further generalize this proof by directly establishing it for an irreversible differential process.

5.4 THERMODYNAMIC CYCLE AS A CLOSED PROCESS

In Chapter 2, we treated heat engines, refrigerators, and heat pumps—devices that operate in cycles—as special cases of closed-steady systems. Cyclic systems were divided into two categories: open-device based cycles, or *open cycles*, where open-steady systems are connected back-to-back to form a loop and closed-process cycles, or *closed cycles*, where a fixed mass executes a series of closed processes ending up in the same state. If we follow a closed parcel of fluid in an open cycle, a sequence of closed-processes forming a cyclic process can be identified. An open cycle is fundamentally no different from a closed cycle. A **thermodynamic cycle**, therefore, can be interpreted as a sequence of quasi-equilibrium, cyclic, closed processes executed by a working fluid where the beginning and final states are identical (see the generic T-s diagram of a heat engine cycle in Fig. 5.40) in each cycle. Every conclusion drawn in Sec. 2.3, that was based on a closed-steady analysis of cycles can also be reached by treating a cycle as a cyclic closed process. Our purpose here, is not to reproduce the same results using a different approach, but to discuss the origin of two important properties—u and s—which are at the core of the postulative approach adopted in this textbook.

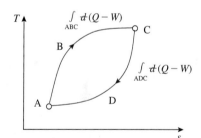

FIGURE 5.40 Integral of $(Q - W)$ between any two states is path independent and must be a point function or *property* of the state.

Even without a detailed knowledge of the processes that constitute a cycle, the governing equations for a differential closed process, Eqs. (5.1) and (5.3), can be integrated over a cycle. Noting that the cyclic integral of any property is zero by definition and $W = W_{\text{ext}}$ for a closed system, the integrated energy and entropy equations for a cyclic process can be simplified as:

$$\text{Energy:} \quad \oint dE = \oint dQ - \oint dW; \Rightarrow \oint dQ = \oint dW; \quad [kJ] \qquad \textbf{(5.34)}$$

$$\text{Entropy:} \quad \oint dS = \oint \frac{dQ}{T_B} + \oint dS_{\text{gen}}; \Rightarrow \oint \frac{dQ}{T_B} = -\oint dS_{\text{gen}}; \quad \left[\frac{kJ}{K}\right] \qquad \textbf{(5.35)}$$

The limits of the integral from b to f in a process are replaced by the cyclic integral symbols to indicate that the b-state and f-state are identical in a cycle. The origin of internal energy and entropy as thermodynamic properties is intricately connected to the energy and entropy equations for a cyclic process as expressed by Eqs. (5.34) and (5.35).

5.4.1 Origin of Internal Energy

The origin of internal energy goes back to Joules' classic paddle-wheel experiment in 1845. He demonstrated that work transfer (using a falling weight to drive the paddle wheel) is equivalent to heat transfer in raising the temperature of a system. The first law of thermodynamics was originally proposed in the form of Eq. (5.34):

$$Q_{cycle} = W_{cycle}; \quad \text{or,} \quad \oint dQ = \oint dW; \quad [\text{kJ}] \tag{5.36}$$

To retrace the origin of internal energy as a property, observe any cyclic process depicted on a thermodynamic diagram such as the one shown on the T-s diagram of Figure 5.40. Equation (5.36) can be rewritten for this cycle A-B-C-D-A:

$$\int_{\text{ABC}} d(Q - W) + \int_{\text{CDA}} d(Q - W) = 0; \quad [\text{kJ}]$$

$$\text{Therefore,} \quad \int_{\text{ABC}} d(Q - W) = -\int_{\text{CDA}} d(Q - W) = \int_{\text{ADC}} d(Q - W); \quad [\text{kJ}] \tag{5.37}$$

Locations of B and D could be anywhere, making the integral of the combination variable $Q - W$ in Eq. (5.37) path independent. Only a point function exhibits such a behavior; therefore, $d(Q - W)$ must be an *exact differential*. That is, $d(Q - W_{ext}) = dE$, where E must be a property. It represents the difference between the heat added and the work rejected along the process curve, which must be the energy stored in the system. In the absence of KE and PE—the well known repositories of mechanical energy—another component of E must exist to account for the difference $dQ - dW$. Thus, the concept of **internal energy** U as a property was born to satisfy $dU \equiv dQ - dW$. In our postulative approach u was asserted to be a thermodynamic property from the very beginning, which led to the energy balance equation.

5.4.2 Clausius Inequality and Entropy

Like internal energy, entropy s was postulated to be a thermodynamic property in formulating the second law of thermodynamics (Sec. 2.1.3). The concept of entropy was developed from a theorem known as the Clausius inequality.

To retrace the origin of entropy as a property, observe a closed uniform system undergoing a cycle. The entropy balance equation for a closed cycle, Eq. (5.35), can be rewritten for the cycle as:

$$\oint \frac{dQ}{T} \leq 0; \quad \left[\frac{\text{kJ}}{\text{K}}\right] \tag{5.38}$$

Arriving at Eq. (5.38), we assume the system to be uniform at a given time, so that $T_B = T$, and utilize the fact $S_{\text{gen,cycle}}$, the entropy generated in the system during the cyclic process, must be non-negative as postulated by the second law. This inequality is known as the **Clausius inequality**. Although this appears to be a rather simple application of the entropy balance equation to a cyclic process, in 1865 Clausius did not have the benefit of the entropy balance equation—in fact, the concept of entropy did not exist at the time. He derived this inequality using deductive reasoning from the Carnot theorems (Sec. 2.3.3).

To retrace Clausius' arguments, consider a differential process (Fig. 5.41) in which a reservoir (TER) at T_H supplies dQ_H to a Carnot heat engine, A, which rejects dQ as waste heat to a control system, B, at T. The energy equation, Eq. (5.1), applied to this differential process for the combined system (within the red boundary) can be expressed using the Carnot relation, Eq. (2.31):

$$dE_C = dQ_H - dW_C = \frac{dQ_H}{dQ}dQ - dW_C = \frac{T_H}{T}dQ - dW_C; \quad [\text{kJ}] \tag{5.39}$$

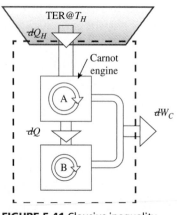

FIGURE 5.41 Clausius inequality is proved from this differential process.

Suppose the control system B also executes an integral number of cycles in sync with a single cycle of the Carnot heat engine. Integrating Eq. (5.39) over the larger of the two cycle periods, we obtain:

$$\oint dE_C^{\;0} = T_H \oint \frac{dQ}{T} - \oint dW_C; \quad [kJ] \quad \Rightarrow \quad \oint \frac{dQ}{T} = \frac{W_{C,Cycle}}{T_H}; \quad \left[\frac{kJ}{K}\right] \qquad \textbf{(5.40)}$$

However, the combined system is forbidden by the Kelvin-Planck statement (Sec. 2.3), the original pronouncement of the second law, to produce work in cycles because it exchanges heat with only a single reservoir (the TER at T_H). It can consume net work without violating the second law; therefore, $W_{C,Cycle}$ must be negative or, at best, zero. This establishes the Clausius inequality, Eq. (5.38), without the use of the entropy balance equation.

Clausius went a step further and reasoned that if the combined system is internally reversible, all the directions of energy transfer can be reversed without affecting any of the conclusions drawn above. This means that $W_{C,Cycle}$, which must be non-positive according to Eq. (5.40), must be non-negative for the reversed cycle. Only $W_{C,int.rev.Cycle} = 0$ can satisfy these contradictory conditions. For an internally reversible system, Eq. (5.40) reduces to:

$$\oint \frac{dQ_{int.rev.}}{T} = 0; \quad \left[\frac{kJ}{K}\right] \qquad \textbf{(5.41)}$$

Using the same logic as used in the previous section for internal energy, we can show that the integrand in this equation must be a point function (Fig. 5.42) so $\dfrac{dQ_{int.rev.}}{T}$ must be the differential of a new property. Clausius proposed the term **entropy** for this property for the first time.

Instead of starting from the Kelvin-Planck statement and uncovering the concept of entropy by tracing its historical route, in this book we have followed the postulative approach to get to the core of the fundamental laws of thermodynamics. Use of internal energy before a discussion of thermodynamic cycles, and entropy before the derivation of Clausius inequality, is no different than the use of temperature as a property before a discussion of the zeroth law. In the postulative approach, we began with the physical meaning of the properties T, u, and s, and through them introduced the zeroth, first, and second laws of thermodynamics.

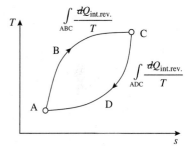

FIGURE 5.42 Integral of $\dfrac{dQ_{int.rev.}}{T}$ between any two states is path independent; therefore, must be an exact differential of a point function or property of the state (see Anim. 1.A.*property*).

5.5 CLOSURE

In this chapter we presented a comprehensive analysis of closed and open unsteady systems. Unsteady systems that involve a closed process, a closed system goes from a beginning state to a final state, were discussed first. A number of examples were used to illustrate closed processes involving uniform systems. Different types of closed processes such as isochoric, isobaric, isothermal, isentropic, and polytropic processes were analyzed. Processes involving non-uniform systems were divided into three categories depending upon whether the subsystems undergo complete mixing, partial mixing, or no mixing. Open processes were discussed with an emphasis on the charging and discharging processes. Transient systems, where the interest lies in instantaneous behavior of unsteady systems, were also studied with examples from both transient closed and open systems. A differential process was analyzed to develop the TdS equation (introduced in Chapter 3 without a proof). Finally, thermodynamic cycles were revisited as a special case of closed processes to explore the origin of internal energy and entropy as thermodynamic properties.

PROBLEMS

SECTION 5-1: SIMPLE PROCESSES INVOLVING CLOSED UNIFORM SYSTEMS

Constant Volume Processes

Al Block

5-1-1   Determine (a) the amount of heat (Q) necessary to raise the temperature of 1 kg of aluminum from 30°C to 100°C. (b) *What-if scenario:* What would the answer in part (a) be if the working substance were wood instead?

5-1-2   Suppose the aluminum block in the above problem was heated by a reservoir (TER) at 200°C. Determine (a) the change in entropy (ΔS) of the block and (b) the entropy transferred from the reservoir to the block. (c) Explain the discrepancy between the two results. (d) *What-if scenario:* What would the answers in part (a) and (b) be if the reservoir was at 500°C?

5-1-3   A mug contains 0.5 kg of coffee (properties: $\rho = 1000 \ \text{kg/m}^3$, $c_v = 1 \ \text{kJ/kg} \cdot \text{K}$) at 20°C. Determine: (a) The amount of heat (Q) in kJ necessary (there is no work transfer) to raise the temperature to 70°C, (b) the amount of work (W) in kJ necessary to move it up by a height of 10 m, (c) and the amount of work (W) in kJ necessary to accelerate the mug from 0 to 30 m/s. Assume the mug has no mass. For the last two parts assume the mug to be adiabatic.

5-1-4   A block of iron (specific heat: 0.45 kJ/kg · K) with a mass of 20 kg is heated from its initial temperature of 10°C to a final temperature of 200°C by keeping it in thermal contact with a thermal energy reservoir (TER) at 300°C. All other faces of the block are insulated. Determine the amount of (a) heat (Q) and (b) entropy transfer (S) from the reservoir. (c) Also calculate the entropy generated ($S_{\text{gen,univ}}$) in the system's universe due to this heat transfer.

5-1-5   A block of solid with a mass of 2 kg is heated from 300 K to a final temperature of 550 K by transferring 50 kJ of heat from a reservoir at 1000 K. Determine (a) the specific heat (c) of the solid and (b) the entropy generation ($S_{\text{gen,univ}}$) in the universe due to this heating process.

5-1-6   A block of aluminum with $m = 0.5 \ \text{kg}$, $T = 20°C$ is dropped into a reservoir at a temperature of 90°C. Calculate (a) the change in stored energy (ΔE), (b) the amount of heat transfer (Q), (c) the change in entropy (ΔS), (d) the amount of entropy transfer by heat, and (e) the entropy generation ($S_{\text{gen,univ}}$) in the system's universe during the heat transfer process.

5-1-7   A 0.8 m³ rigid tank initially contains refrigerant-134a in saturated vapor form at 0.9 MPa. As a result of heat transfer from the refrigerant, the pressure drops to 250 kPa. Determine (a) the final temperature (T_2), (b) the amount of refrigerant that condenses, and (c) the heat transfer (Q).

5-1-8   A rigid tank with a volume of 50 m³ contains superheated steam at 100 kPa and 200°C. The tank is allowed to cool down to the atmospheric temperature of 10°C. Determine (a) the final pressure (p_2), (b) the heat transfer (Q), (c) the change in entropy (ΔS) of steam and (d) the entropy generation ($S_{\text{gen,univ}}$) in the tank's universe. (e) *What-if scenario:* How would the final pressure change if the steam were initially at 500 kPa? (1: increase; −1: decrease; 0: remain same)

5-1-9   A rigid tank, volume 10 m³, contains superheated steam at 1 MPa and 400°C. Due to heat loss to the outside atmosphere, the tank gradually cools down to the atmospheric temperature of 25°C. Determine (a) the heat transfer (Q) and (b) the entropy generated ($S_{\text{gen,univ}}$) in the system's universe during this cooling process. (c) Plot how the pressure changes as temperature decreases during the process (pressure vs. temperature plot). (d) Can you explain the discontinuity in the plot?

5-1-10   A rigid tank contains 3.2 kg of refrigerant-134a at 26°C and 140 kPa. The refrigerant is cooled until its pressure drops to 100 kPa. Determine (a) the entropy change (ΔS) of the refrigerant, (b) the entropy transfer to a reservoir at −50°C, and (c) the entropy generation ($S_{\text{gen,univ}}$) in the universe due to the process.

5-1-11   A steam radiator (used for space heating) has a volume of 20 L and is filled with steam at 200 kPa and 250°C. As the inlet and exit ports are closed, the radiator cools to a room temperature of 20°C. Determine (a) the heat transfer (Q) and show the process on a p-v diagram. (b) *What-if scenario:* What would the heat transfer be if the steam pressure in the radiator was 400 kPa instead?

5-1-12   A steam radiator has a volume of 25 L and is filled with steam at 350 kPa and 280°C. When the inlet and exit ports are closed, the radiator pressure drops to 180 kPa. Determine (a) the heat transfer (Q) and show the process on a p-v diagram. (b) *What-if scenario:* What would the heat transfer be if the steam pressure in the radiator was 500 kPa?

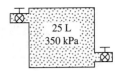

5-1-13   The radiator of a steam heating system has a volume of 15 L and is filled with superheated water vapor at 225 kPa and 230°C. The inlet and exit ports are then closed. After a while the temperature of steam drops to 88°C as a result of heat transfer to the room air. Determine (a) the heat transfer (Q) and (b) the entropy change (ΔS) of the steam during this process.

5-1-14   A steam radiator has a volume of 20 L and is filled with steam at 200 kPa and 250°C. The inlet and exit ports are then closed. As the radiator cools down to a room temperature of 20°C, determine (a) the final pressure (p_2) and (b) the entropy generated ($S_{gen,univ}$) in the universe.

5-1-15   A well insulated, rigid tank contains 6 kg of saturated liquid vapor mixture of water at 150 kPa. Initially, half of the mass is in liquid phase. An electric resistance heater placed in the tank is turned on and kept on until all the liquid is vaporized. Determine (a) the electrical work (W_{el}), (b) the entropy change (ΔS) of the steam during this process, and (c) the entropy generated ($S_{gen,univ}$) in the system's universe.

5-1-16   A 0.4 m^3 rigid tank contains refrigerant-134a at 250 kPa and 45% quality. Heat is then transferred to the refrigerant from a source at 37°C until the pressure rises to 420 kPa. Determine (a) the entropy change (ΔS) of the refrigerant, (b) the entropy change (ΔS) of the heat source, and (c) the total entropy generation ($S_{gen,univ}$) in the universe due to the process.

5-1-17   A rigid tank filled with 3 kg of H_2O at 150 kPa, $x = 0.2$ is heated with 1000 kJ. Determine (a) the final pressure (p_2) and (b) phase composition of H_2O. (c) *What-if scenario:* What would the final pressure be if the tank was heated with 500 kJ?

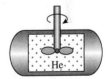

5-1-18   A rigid chamber with a volume of 1 m^3 contains steam at 100 kPa, 200°C. (a) Determine the mass of steam. (b) Determine the amount of heat loss (magnitude only) necessary for the steam, at constant volume, to cool down, to 100°C. Use the PC model for H_2O.

5-1-19   Repeat Problem 5-1-18 using the PG model for H_2O. (c) *What-if scenario:* How would the answer in part (b) change if the IG model was used?

5-1-20   A rigid tank contains 1 kg of H_2O at 100 kPa and $x = 0.1$. The tank can withstand a maximum internal pressure of 5 MPa. Determine (a) the maximum temperature to which the steam in the tank can be heated and (b) the amount of heat transfer (Q) necessary to reach the maximum internal pressure.

5-1-21   An insulated, rigid tank contains 1.5 kg of helium at 30°C and 500 kPa. A paddle wheel with a power rating of 0.1 kW is operated within the tank for 30 minutes. Determine (a) the final temperature (T_2) and (b) pressure (p_2). (c) Also determine the entropy generated (S_{gen}) in the tank.

5-1-22   A 2 m^3 insulated, rigid tank contains 3 kg of carbon dioxide at 110 kPa. Then paddle wheel work is done on the system until the pressure in the tank rises to 127 kPa. Determine (a) the entropy change (ΔS) of the carbon dioxide, (b) work (W) done by paddle wheel, and (c) the entropy generated (S_{gen}) in the tank and its immediate surroundings. Use the PG model. (d) *What-if scenario:* What would the work done be if the IG model was used instead?

5-1-23   Air is contained in an insulated, rigid tank at 25°C and 180 kPa. A paddle wheel, inserted in the tank, does 800 kJ of work to the air. If the volume is 2 m^3, determine (a) the entropy increase (ΔS), (b) the final pressure (p_2), and (c) final temperature (T_2). Use the IG model for air.

5-1-24   A person living in a 4m × 5m × 5m room turns on a 100 W fan before he leaves the warm room at 100 kPa and 30°C, hoping that the room will be cooler when he comes back after 5 hours. Disregarding any heat transfer and using the PG model for air, determine (a) the temperature (T_2) he discovers when he comes back. (b) *What-if scenario:* What would the temperature be if the fan power was 50-W instead?

5-1-25   A piston-cylinder device contains 0.01 kg of steam at a pressure of 100 kPa and a quality of 10%. Determine the heat transfer necessary to increase the quality to 100% when heating is carried out (a) in a constant pressure manner by allowing the piston to move freely, or (b) in a constant volume manner by locking the piston in its initial position.

5-1-26 Saturated water vapor, with a mass of 10 kg, at 300 kPa is heated at constant pressure until the temperature reaches 500°C. Calculate (a) the work (W) done by the steam during the process and (b) the amount of heat transfer (Q). (c) *What-if scenario:* What would the heat transfer be if the pressure remained constant at 600 kPa?

5-1-27 In Problem 5-1-26, determine the minimum average value of the boundary temperature for which the second law is not violated.

5-1-28 A vertical piston-cylinder assembly contains 10 L of air at 20°C. The cylinder has an internal diameter of 20 cm. The piston is 2 cm thick and is made of steel with density of 7830 kg/m³. If the atmospheric pressure is 101 kPa and the volume of air doubles, determine (a) the heat (Q) and (b) work transfer (W) during the process. (c) *What-if scenario:* What are the answers if the piston weight is neglected?

5-1-29 A frictionless piston-cylinder device contains 0.1 kg of refrigerant-12 as a saturated liquid. The piston is free to move, and its mass maintains a pressure of 200 kPa on the refrigerant. Due to heat transfer from the atmosphere, the temperature of the refrigerant gradually rises to the atmospheric temperature of 25°C. Calculate (a) the work transfer (W), (b) the heat transfer (Q), (c) the change of entropy (ΔS) of the refrigerant, and (d) the entropy generated (S_{gen}) in the system's universe.

5-1-30 A mass of 2 kg of liquid water is completely vaporized at a constant pressure of 1 atm. Determine (a) the heat added. (b) *What-if scenario:* What is the heat added if the pressure is 2 atm?

5-1-31 A frictionless piston is used to provide a constant pressure of 500 kPa in a cylinder containing steam originally at 250°C with a volume of 3 m³. Determine (a) the final temperature (T_2) if 3000 kJ of heat is added and (b) the work (W) done by piston (magnitude only).

5-1-32 A piston-cylinder device initially contains 2 kg of liquid water at 140 kPa and 25°C. The water is then heated at a constant pressure by the addition of 3600 kJ of heat. Determine (a) the final temperature (T_2), (b) the entropy change (ΔS) of water during this process, and (c) the boundary work.

5-1-33 A frictionless piston-cylinder device beginning at 1 m³ contains saturated steam at 100°C. During a constant pressure process, 700 kJ of heat is transferred to the surrounding air at 25°C. As a result, part of the water vapor contained in the cylinder condenses. Determine (a) the entropy change (ΔS) of the water and (b) the total entropy generated (S_{gen}) during this heat transfer process.

5-1-34 A frictionless piston-cylinder device contains 10 kg of superheated vapor at 550 kPa and 340°C. Steam is then cooled at constant pressure until 60 percent of it, by mass, condenses. Determine (a) the work (W) done during the process. (b) *What-if scenario:* What is the work done if the steam is cooled at constant pressure until 80 percent of it, by mass, condenses?

5-1-35 A piston-cylinder device contains 8 kg of refrigerant-134a at 850 kPa and 70°C. The refrigerant is then cooled at constant pressure until it comes to thermal equilibrium with the atmosphere, which is at 20°C. Determine the amount of (a) heat transfer (Q), (b) entropy transfer (S) into the atmosphere, (c) the change of entropy (ΔS) of the refrigerant, and (d) the entropy generated ($S_{gen,univ}$) in the system's universe.

5-1-36 An insulated piston-cylinder device contains 3 L of saturated liquid water at a constant pressure of 180 kPa. An electric resistance heater inside the cylinder is turned on and 2000 kJ of energy is transferred to the water. Determine (a) the final temperature (T_2), (b) the boundary work transfer, and (c) entropy change (ΔS) of water in this process.

5-1-37 A piston-cylinder device initially contains 20 g of saturated water vapor at 300 kPa. A resistance heater is operated within the cylinder with a current of 0.4 A from a 240 V source until the volume doubles. At the same time a heat loss of 4 kJ occurs. Determine (a) the final temperature (T_2) and (b) the duration of the process. (c) *What-if scenario:* What is the final temperature if the piston-cylinder device initially contains saturated water?

5-1-38 An insulated container contains a 1 ton (US) block of ice at 0°C. The insulation is removed, and the ice gradually melts and comes to thermal equilibrium with the surroundings at 25°C. Assuming the pressure to remain constant at 100 kPa, determine (a) the boundary work and (b) the heat transfer (Q) during the process. (c) *What-if scenario:* By what percentage will the heat transfer increase if the initial temperature of the ice block is −25°C ?

5-1-39 In Problem 5-1-38, determine (a) the change of entropy (ΔS) of the system, (b) the entropy transfer (S) from the surroundings, and (c) the entropy generation ($S_{gen,univ}$) in the system's universe during the process.

Temperature Regulated Processes

5-1-40   1.5 kg of air at 160 kPa and 15°C is contained in a gas-tight, frictionless piston-cylinder device. The air is then compressed to a final pressure (p_2) of 650 kPa. During the process heat is transferred from the air so that the temperature inside the cylinder remains constant. (a) Calculate the work (W) done during this process. Use the PG model for air. (b) *What-if scenario:* What is the work done if the IG model is used?

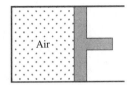

5-1-41   In Problem 5-1-40 determine (a) the change of entropy (ΔS) of the air, (b) the entropy transferred (S) into the atmosphere at 10°C, and (c) the entropy generation ($S_{gen,univ}$) in the system's universe during the compression process.

5-1-42   A piston-cylinder device contains 0.1 kg of a gas (PG model: $c_v = 10.18$, kJ/kg·K, k = 1.4, R = 4.12 kJ/kg·K) at 1000 kPa and 300 K. The gas undergoes an isothermal (constant temperature) expansion process to a final pressure (p_2) of 500 kPa. (a) Determine the boundary work and (b) the heat transfer (Q). (c) If heat is transferred from the surroundings at 300 K, determine the entropy generation (S_{gen}) during the expansion process.

5-1-43   Nitrogen, at an initial state of 80°F, 25 psia and 6 ft^3, is pressed slowly in an isothermal process to a final pressure of 110 psia. Determine (a) the work (W) done during the process. (b) *What-if scenario:* What is the work done if the final pressure is 100 psia?

5-1-44   An insulated piston-cylinder device contains 0.04 m^3 of steam at 300 kPa and 200°C. The steam is then compressed in a reversible manner to a pressure of 1 MPa. Calculate (a) the work done (W). (b) *What-if scenario:* What is the work done if the initial temperature is 800°C?

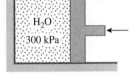

5-1-45   Air at 20°C, 95 kPa is compressed in a piston-cylinder device of volume 1 L in a frictionless and adiabatic manner. If the volumetric compression ratio is 10, determine (a) the final temperature (T_2) and (b) the boundary work transfer (W_B). Use the PG model for air. *What-if scenario:* What would the (c) final temperature and (d) boundary work transfer be if the initial temperature was 45°?

5-1-46   Nitrogen at 300 K, 100 kPa is compressed in a piston-cylinder device of volume 1 m^3 in a frictionless and adiabatic manner. If the volumetric compression ratio is 15, determine (a) the final pressure (p_2) and (b) the boundary work transfer (W_B). Use the PG model.

5-1-47   Solve the Problem 5-1-56, using the IG model.

5-1-48   A piston-cylinder device initially contains 0.1 m^3 of N$_2$ at 100 kPa and 300 K. Determine the work transfer involved in compressing the gas to one-fifth of its original volume in an (a) isothermal, (b) isentropic and (c) isobaric manner. Show the processes on p-v diagrams. (d) *What-if scenario:* What would the answers be if the IG model was used instead?

5-1-49   A piston-cylinder device initially contains 10 ft^3 of argon gas at 25 psia and 70°F. Argon is then compressed in a polytropic process (pv^n = constant) to 70 psia and 300°F. Determine (a) if the process is 1: reversible; 2: impossible; or 3: irreversible. (b) Also determine the change of entropy for argon.

5-1-50   Air is compressed in a frictionless manner in an adiabatic piston-cylinder device from an initial volume of 1000 cc to the final volume of 100 cc. If the initial conditions are 100 kPa, 300 K, (a) what is the temperature (T_2) (in K) right after the compression? (b) Determine the change in internal energy (ΔE) of the gas, and (c) the work transfer (W) during the process.

5-1-51   0.5 kg of air is compressed to 1/10th its original volume in a piston-cylinder device in an isentropic manner. If the original volume of the piston is 0.5 m^3, molar mass of air is 29 kg/kmol, and k = 1.4, determine the final density. Use the PG model for air.

5-1-52   Steam at 100 kPa and 200°C is compressed to a pressure of 1 MPa in an isentropic manner. Determine the final temperature (T_2) in Celcius using (a) the PC model and (b) the PG model with k = 1.327.

5-1-53   A piston-cylinder device initially contains 0.8 kg of O$_2$ at 100 kPa and 27°C. It is then compressed in a polytropic process ($pv^{1.3}$ = constant) to half the original volume. Determine (a) the change of entropy (ΔS) for the system, (b) entropy transfer (S) to the surroundings at 27°C and (c) the entropy generated ($S_{gen,univ}$) in the system's universe due to this process. (d) *What-if scenario:* What would the answer in part (c) be if the surrounding temperature was 0°C?

5-1-54 A piston-cylinder device contains 0.1 kg of hydrogen gas (PG model: $c_v = 10.18$ kJ/Kg·K, $k = 1.4$, $R = 4.12$ kJ/kg·K) at 1000 kPa and 300 K. The gas undergoes an expansion process and the final conditions are 500 kPa, 270 K. If 10 kJ of heat is transferred into the gas from the surroundings at 300 K, determine (a) the boundary work (W_b), and (b) the entropy generated (S_{gen}) during the process.

5-1-55 Nitrogen at 300 K, 100 kPa is compressed in a piston-cylinder device of volume 1 m³ in an adiabatic manner. If the volumetric compression ratio is 15 and the final temperature after compression is measured as 910 kPa, determine (a) the boundary work transfer (W_B) and (b) the entropy generated during the process (S_{gen}). Use the PG model.

5-1-56 A cylinder fitted with a piston has an initial volume of 0.1 m³ and contains nitrogen at 100 kPa and 25°C. The piston is pushed, compressing the nitrogen until the pressure is 1.5 MPa and the temperature is 200°C. During this compression process heat is transferred from nitrogen to the atmosphere at 25°C, and the work done on the nitrogen is 30 kJ. Determine (a) the amount of heat transfer (Q) (include sign) and (b) the entropy generation (S_{gen}) during the process. Use the PG model for nitrogen.

5-1-57 A rubber ball of mass m is dropped from a height h onto a rigid floor. It bounces back and forth, finally coming to rest on the floor. What is the entropy generation ($S_{gen,univ}$) in the universe due to this irreversible phenomenon? Assume the atmospheric temperature to be T_0.

Sequential Processes (Sec-1 Continued)

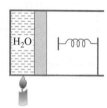

5-1-58 A piston-cylinder device contains 40 kg of water at 150 kPa and 30°C. The cross-sectional area of the piston is 0.1 m². Heat is then added causing some of the water to evaporate. When the volume reaches 0.2 m³, the piston reaches a linear spring with a spring constant of 120 kN/m. More heat is added until the piston moves another 25 cm. Determine (a) the final pressure (p_2), (b) final temperature (T_2), and (c) the total heat transfer (Q).

5-1-59 A piston-cylinder device contains 40 kg of water at 150 kPa and 30°C. The cross-sectional area of the piston is 0.1 m². Heat is added causing some of the water to evaporate. When the volume reaches 0.2 m³, the piston reaches a spring with a constant spring constant. More heat is added until the volume increases to 0.3 m³ and the pressure increases to 1.35 MPa. Determine (a) the spring constant, (b) the final temperature (T_2), and (c) the total heat transfer (Q).

5-1-60 If the heat addition takes place from a source at 500°C during the entire process in Problem 5-1-59 determine (a) the change of entropy (ΔS) of the steam, (b) entropy (S) transfer from the source, and (c) entropy generated ($S_{gen,univ}$) in the universe during the process.

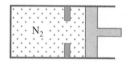

5-1-61 A piston-cylinder device contains 1 m³ of N_2 at 100 kPa and 300 K (atmospheric conditions). (a) Determine the mass of N_2 (in kg). The trapped N_2 is now rapidly compressed (adiabatically) with negligible friction (reversibly) to one eighth its original volume. Determine (b) the temperature and (c) pressure (kPa) right after the compression process is over. The volume is now held constant while the air cools down to the atmospheric temperature of 300 K. Determine (d) the pressure at the end of this cooling process. Finally, the gas is allowed to expand back to its original pressure of 100 kPa in an isentropic manner. (e) Determine the final temperature right after the expansion (K). Also determine (f) the work required for compression and (g) work produced by expansion. Use the PG model for N_2.

5-1-62 10 kg of saturated liquid-vapor mixture of R-12 is contained in a piston-cylinder device at 0°C. Initially half of the mixture is in the liquid phase. Heat is then transferred, the piston, which is resting on a stop-ring, starts moving when the pressure reaches 500 kPa. Heat transfer continues until the total volume doubles. Determine (a) the final pressure (p_2), (b) work (W_2) and (c) heat transfer (Q_2) in the entire process. Also, show the process on a p-v diagram.

5-1-63 An insulated cylinder with a frictionless piston contains 10 L of CO_2 at ambient conditions, 100 kPa and 20°C. A force is then applied on the piston, compressing the gas until it reaches a set of stops, at which point the cylinder volume is 1 L. The insulation is then removed from the walls, and the gas cools down to the ambient temperature of 20°C. Calculate (a) the work (W_2) and (b) heat transfer (Q_2) for the overall process. Also, show the process on a T-s diagram. Use the PG model.

5-1-64 A piston-cylinder device with a set of stops contains 12 kg of refrigerant-134a. Initially 9 kg of refrigerant is in the liquid form and the temperature is $-12°C$. Heat is then transferred from the atmosphere at 25°C to the refrigerant until the piston hits the stop, at which point the volume is 380 L. Heat continues to transfer until the temperature of the refrigerant reaches the atmospheric value. Determine (a) the final pressure (p_2), (b) the work transfer (W), (c) the heat transfer, (Q) and (d) the entropy generation (S_{gen}) during the entire process.

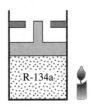

5-1-65 A piston-cylinder device has a ring to limit the expansion stroke. Initially, the mass of air is 2 kg at 500 kPa and 30°C. Heat is then transferred until the piston touches the stop, at which point the volume is twice the original volume. More heat is transferred until the pressure inside doubles. Using the IG model, determine (a) the amount of heat transfer (Q) and (b) the final temperature (T_2).

5-1-66 An insulated container contains a 1 ton (US) (1 US ton is 907.2 kg) block of ice at $-20°C$. The insulation is removed. The ice gradually melts and comes to thermal equilibrium with the surroundings at 20°C. Assuming the pressure to remain constant at 100 kPa, determine the heat transfer (Q) during (a) the sensible heating of ice, (b) the melting of ice and (c) the sensible heating of water. (d) *What-if scenario:* If the melting of ice in part (b) takes 24 hours, what is the rate of heat transfer in kW?

5-1-67 Three alternative processes, shown in the accompanying figure, are suggested to change the state of one kg of air from 0.8 MPa and 25°C (state-1) to 0.3 MPa and 60°C (state-5). (a) Process 1-2-5 consists of a constant pressure expansion followed by a constant volume cooling, (b) process 1-3-5 an isothermal expansion followed by a constant pressure expansion, and (c) process 1-4-5 an adiabatic expansion followed by a constant volume heating. Derive expressions for heat and work transfer for each alternative.

5-1-68 An aluminum (Al) block of mass 2000 kg is heated from a temperature of 25°C. Using the SL model Interactive (Interactives, unsteady, generic branch) (a) determine the final temperature (T_2) of the block if the heat input is 300 MJ. (b) *What-if scenario:* How would the answer change if the block were made of silver (Ag)?

5-1-69 A container containing 5 kg of CO_2 is heated from a temperature of 300 K. Using the SL model Interactive (Interactives, unsteady, generic branch), (a) determine the final temperature (T_2) of the block if the heat input is 1 MJ. (b) *What-if scenario:* What would be the final temperature if heat input was 5 MJ?

SECTION 5-2: CLOSED PROCESSES INVOLVING NON-UNIFORM MIXING SYSTEMS

5-2-1 An insulated, rigid tank is divided into two equal parts by a membrane. At the beginning, one part contains 3 kg of nitrogen at 500 kPa and 50°C, while the other part is completely evacuated. The membrane is punctured and the gas expands into the entire tank. Determine the final (a) temperature (T_2) and (b) pressure (p_2). Use the PG model.

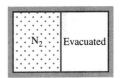

5-2-2 For Problem 5-2-1, (a) determine the entropy generated (S_{gen}) during the expansion process. (b) *What-if scenario:* What would the answer be if the temperature of nitrogen were 100°C?

5-2-3 An insulated, rigid tank of volume 1 m^3 is separated into two chambers by a membrane. One chamber contains 1 kg of saturated liquid water at 100 kPa while the other chamber is completely evacuated. The membrane is punctured and water expands to occupy the entire tank. Determine the final (a) temperature (T_2) and (b) pressure (p_2). (c) Determine the entropy generation (S_{gen}) in the tank's universe. Use the PC model.

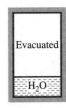

5-2-4 A rigid, well insulated tank consists of two compartments, each having a volume of 1.5 m^3, separated by a valve. Initially, one of the compartments is evacuated and the other contains nitrogen gas at 700 kPa and 100°C. The valve is opened and nitrogen expands to fill the total volume, eventually achieving an equilibrium state. Determine the final (a) temperature (T_2) and (b) pressure (p_2). Also, determine (c) the entropy generated (S_{gen}). Use the IG model.

5-2-5 An insulated cylinder is divided into two parts of 1 m³ each by a membrane. Side A has air at 200 kPa, 25°C and side B has air at 1 MPa, 1000°C. The membrane is punctured so the air comes to a uniform temperature. Determine (a) the final pressure (p_2), (b) temperature (T_2), (c) the entropy generated (S_{gen}) in this process. Use the PG model (d) *What-if scenario:* What would the (d) final pressure and (e) temperature, (f) entropy generated be if the IG model were used?

5-2-6 A rigid tank has two compartments, one 500 times larger than the other. The smaller part contains 2 kg of compressed liquid water at 1 MPa and 25°C, while the other part is completely evacuated. The partition is removed and the water expands to fill the entire tank. Heat transfer with the atmosphere at 25°C is allowed, as a result the final temperature after mixing is 25°C. Determine the final (a) pressure (p_2) and (b) quality of the mixture. Find (c) the entropy change (ΔS_w) of water and (d) the entropy generation (S_{gen}) during the process.

5-2-7 An insulated, rigid tank has two compartments, one 100 times larger than the other. The smaller part contains 2 kg of compressed liquid water at 400 kPa and 50°C, while the other part is completely evacuated. The partition is then removed and water expands to fill the entire tank. Determine (a) the final pressure (p_2), (b) temperature (T_2), and (c) the entropy generation (S_{gen}) during the process.

5-2-8 A tank, with an unknown volume, is divided into two parts by a partition. One side contains 0.02 m³ of saturated liquid R-12 at 0.7 MPa, while the other side is evacuated. The partition is then removed and R-12 fills up the entire volume. If the final state is 200 kPa and 30°C, determine (a) the volume (V) of the tank and (b) the heat transfer (Q). (c) *What-if scenario:* What would the volume of the tank be if the final state was 300 kPa and 30°C?

5-2-9 Four ice cubes (3 cm × 2 cm × 1 cm) at −15°C are added to an insulated glass of cola (water) at 15°C. The volume of cola is 1.5 L. Determine (a) the equilibrium temperature (T_2) and (b) the total entropy change (ΔS) for this process.

5-2-10 Two tanks are connected by a valve and line. The volumes are both 1 m³ with R-134a at 21°C, quality 25% in tank A and tank B is evacuated. The valve is opened and saturated vapor flows from A to B until the pressure becomes equal. The process occurs slow enough that all temperatures remain at 21°C during the process. Determine (a) the entropy generated (S_{gen}) and (b) the total heat transfer (Q).

5-2-11 A tank, with an unknown volume, is divided into two parts by a partition. One side contains 0.02 m³ of saturated liquid R-12 at 0.7 MPa, while the other side is evacuated. The partition is then removed and R-12 fills up the entire volume. If the final state is 200 kPa, quality 90%; determine (a) the volume (V) of the tank, (b) the heat transfer (Q) and (c) the entropy generated (S_{gen}). The atmospheric temperature is 30°C. (d) *What-if scenario:* What would the heat transfer be if the final pressure was 300 kPa?

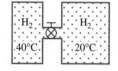

5-2-12 Two tanks are connected; tank A of 0.5 m³ containing hydrogen at 40°C, 200 kPa is connected to tank B of 1 m³ rigid tank containing hydrogen at 20°C, 600 kPa. The valve is opened and the system is allowed to reach thermal equilibrium with the surroundings at 15°C. Determine (a) the final pressure (p_2), (b) heat transfer (Q) and (c) the entropy generated (S_{gen}) during the mixing process. Use the PG model.

5-2-13 Two rigid tanks are connected by a valve. Tank A contains 0.4 m³ of water at 330 kPa and 90 percent quality. Tank B contains 0.5 m³ of water at 250 kPa and 250°C. The valve is then opened, and the two tanks eventually come to equilibrium while exchanging heat with the surroundings at 25°C. Determine (a) the final pressure (p_2), (b) heat transfer (Q) and (c) the entropy generation (S_{gen}) during the process.

5-2-14 Two insulated tanks are connected, both contain H₂O. Tank-A is at 200 kPa, $v = 0.4\,\text{m}^3/\text{kg}$, $V = 1\,\text{m}^3$ and tank B contains 3.5 kg at 0.5 MPa, 400°C. The valve is then opened and the two come to a uniform state. Find (a) the final pressure (p_2), (b) temperature (T_2) and (c) the entropy generated (S_{gen}) by the mixing process.

5-2-15 Two rigid tanks are connected by a valve as shown in the accompanying figure. Tank A is insulated and contains 0.1 m³ of steam at 500 kPa and 90% quality. Tank B is uninsulated and contains 2 kg of steam at 100 kPa and 300°C. The valve is then opened, and steam flows from tank A to tank B. As the pressure in tank A drops to 300 kPa, the valve is closed. During the process 500 kJ of heat is transferred from tank B to the surroundings at 25°C. Assuming the steam in tank A to have undergone a reversible adiabatic process (isentropic), determine (a) the final temperature (T_2) in (a) tank A and (b) tank B, (c) the final pressure (p_2) in tank B and (d) the entropy generated (S_{gen}) in the system's universe.

5-2-16   Two rigid tanks are connected by a valve. Tank A contains 1 m³ of air at 1 MPa and 200°C. Tank B contains 3 m³ of air at 100 kPa and 25°C. The valve is then opened and air flows from tank A to tank B. Before the two gases come to mechanical equilibrium, the valve is closed. After sufficient time, air in both tanks comes to thermal equilibrium with the surroundings at 10°C. The pressure in tank A is measured as 500 kPa. Using the PG model for air, determine (a) the final pressure (p_2) in tank B, (b) the heat transfer (Q) and (c) the entropy generated (S_{gen}) in the system's universe.

5-2-17   An insulated tank containing 0.5 m³ of R-134a at 500 kPa and 90% quality is connected to an initially evacuated insulated piston-cylinder device, as shown in the accompanying figure. The force balance on the piston is such that a pressure of 120 kPa is required to lift the piston. The valve is then opened slightly and part of the refrigerant flows to the cylinder, pushing the piston up. The process ends when the pressure in the tank drops to 120 kPa. Assuming the refrigerant in the tank undergoes an isentropic (reversible and adiabatic) process, determine the final quality in the (a) tank and (b) cylinder. Also, calculate (c) the boundary work (W_B) and (d) the entropy generated (S_{gen}) during the process.

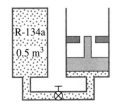

SECTION 5-3: CLOSED PROCESS INVOLVING NON-UNIFORM NON-MIXING SYSTEMS

5-3-1   A 40 kg aluminum block at 90°C is dropped into an insulated tank that contains 0.5 m³ of liquid water at 20°C. Determine the equilibrium temperature (T_2).

5-3-2   In Problem 5-3-1, determine the entropy generated (S_{gen}) the process.

5-3-3   A 25 kg aluminum block, initially at 225°C, is brought into contact with a 25 kg block of iron at 150°C in an insulating enclosure. Determine (a) the equilibrium temperature (T_2) and (b) the total entropy change (ΔS) for this process.

5-3-4   A half kg bar of iron, initially at 782°C, is removed from an oven and quenched by immersing it in a closed tank containing 10 kg of water at 21°C. Heat transfer from the tank can be neglected. Determine (a) the equilibrium temperature (T_2) and (b) the total entropy change (ΔS) for this process.

5-3-5   A 15 kg block of copper at 100°C is dropped into an insulated tank that contains 1 m³ of liquid water at 20°C. Determine (a) the equilibrium temperature (T_2) and (b) the entropy generated (S_{gen}) in this process.

5-3-6   An unknown mass of iron at 80°C is dropped into an insulated tank that contains 0.1 m³ of liquid water at 20°C. Meanwhile, a paddle wheel driven by a 200 W motor is used to stir the water. When equilibrium is reached after 20 min, the final temperature (T_2) is 25°C. Determine (a) the mass of the iron block. (b) *What-if scenario:* What is the mass if the iron block is at 150°C at the time of dropping?

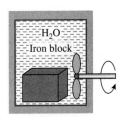

SECTION 5-4: PROCESSES INVOLVING OPEN SYSTEMS

5-4-1   An insulated rigid tank is evacuated. A valve is opened and air, at 100 kPa and 25°C, enters the tank until the pressure in the tank reaches 100 kPa and the valve is closed. (a) Determine the final temperature (T_2) of the air in the tank. Use the PG model. (b) *What-if scenario:* What is the final temperature if the IG model is used?

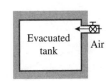

5-4-2   In Problem 5-4-1, determine the entropy generated (S_{gen}) during the process if the volume of the tank is 2 m³.

5-4-3   A 3 m³ tank initially contains air at 100 kPa and 25°C. The tank is connected to a supply line at 550 kPa and 25°C. The valve is opened and air is allowed to enter the tank until the pressure in the tank reaches the line pressure, at which point the valve is closed. A thermometer placed in the tank indicates that the air temperature at the final state is 65°C. Treating air as a perfect gas, determine (a) the mass of air that has entered the tank, (b) the heat transfer (Q) and (c) the entropy generated (S_{gen}).

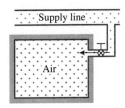

5-4-4   An insulated rigid tank is evacuated. It is then connected through a valve to a supply line that carries steam at 2 MPa and 350°C. The valve is then opened and steam is allowed to flow slowly into the tank until the pressure reaches 2 MPa, at which point the valve is closed. Determine the final temperature (T_2) of the steam in the tank.

5-4-5   A completely evacuated, insulated, rigid tank, with a volume of 8 m³, is filled from a steam line transporting steam at 450°C and 3.5 MPa. Determine (a) the temperature (T_2) of steam in the tank when its pressure reaches 3.5 MPa. Find (b) the mass of the steam that flows into the tank.

5-4-6   A 0.2 m³ tank contains R-12 at 1 MPa and $x = 1$. The tank is charged to 1.2 MPa, $x = 0$ from a supply line that carries R-12 at 1.5 MPa and 30°C. Determine (a) the final temperature (T_2) and (b) the heat transfer (Q).

5-4-7   In the charging process, described in Problem 5-4-6, determine (a) the change of entropy (ΔS) of refrigerant in the tank and (b) the entropy generation (S_{gen}) by the device and its surroundings. Assume the surrounding temperature to be 50°C.

5-4-8   A 0.5 m³ rigid tank contains refrigerant-134a at 0.8 MPa and 100 percent quality. The tank is connected by a valve to a supply line that carries refrigerant-134a at 1.5 MPa and 30°C. The valve is opened and the refrigerant is allowed to enter the tank. The valve is closed when it is observed that the tank contains saturated liquid at 1.5 MPa. Determine (a) the heat transfer (Q) and (b) mass of refrigerant that has entered the tank.

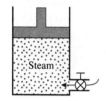

5-4-9   A piston-cylinder device contains 0.1 m³ of steam at 200°C. The force on the piston is such that it maintains a constant pressure of 400 kPa inside. Due to heat rejection, into the ambient air, the temperature of the steam drops down to 25°C. To restore the steam temperature to its original value, superheated steam from a supply line at 1 MPa, 600°C is introduced through a valve into the cylinder, as shown in the accompanying figure. Neglecting any heat transfer during this charging process, determine (a) the heat transfer (Q) during the cooling process and (b) the mass of steam introduced.

5-4-10   An insulated piston-cylinder device contains 0.01 m³ of steam at 200°C. The force on the piston is such that it maintains a constant pressure of 400 kPa inside. A valve is then opened and steam, at 1 MPa, 200°C, is allowed to enter the cylinder until the volume inside increases to 0.04 m³. Determine (a) the final temperature (T_2) of the steam and (b) the amount of mass transfer.

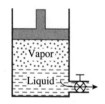

5-4-11   An insulated piston-cylinder device contains 0.2 m³ of R-134a, half (by volume) of which is in the vapor phase. The mass of the piston maintains a constant pressure of 200 kPa inside. A valve is opened and all the liquid refrigerant is allowed to escape. Determine (a) the mass of liquid refrigerant in the beginning, (b) the mass withdrawn and (c) the entropy generated (S_{gen}) during the process.

5-4-12   A 0.5 m³ tank contains saturated liquid water at 200°C. A valve in the bottom of the tank is opened and half the liquid is drained. Heat is transferred from a 300°C source to maintain constant temperature inside the tank. Determine (a) the heat transfer (Q). (b) *What-if scenario:* What would the heat transfer be if the 0.5 m³ tank initially contained saturated liquid water at 100°C?

5-4-13   In Problem 5-4-12, determine the entropy generated ($S_{gen,univ}$) in the system's universe during the discharge.

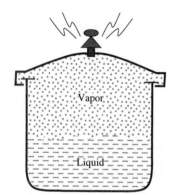

5-4-14   A 0.2 ft³ pressure cooker has an operating pressure of 40 psia. Initially, 50% of its volume is taken up by vapor and the rest is liquid water. Determine (a) the heat transfer (Q) necessary to vaporize all the water in the cooker. (b) *What-if scenario:* What would the heat transfer be if, initially, 20% of the volume was vapor?

6 | EXERGY BALANCE EQUATION: APPLICATION TO STEADY AND UNSTEADY SYSTEMS

We developed the governing balance equations for mass, energy, and entropy in Chapter 2, introduced material models for evaluation of system and flow states in Chapter 3, and carried out comprehensive mass, energy, and entropy analysis of steady and unsteady systems in Chapters 2, 4, and 5. The comprehensive analysis revealed the fundamental tendency of all types of systems to generate entropy spontaneously due to different types of *thermodynamic friction*.

In this chapter, we will manipulate the mass, energy, and entropy equations to develop expressions for a new pair of properties—*stored exergy* and *flow exergy* introduced in Chapter 1—complete with an exergy balance equation. Stored exergy, an extensive property of a system state, is the maximum useful work that can be extracted as the system is brought to its *dead state*, a state that is in equilibrium with the standard atmosphere at sea level. The exergy balance equation establishes an inventory of a system's exergy and links entropy generation due to thermodynamic friction to spontaneous degradation of energy, or exergy destruction, in the system. Although not as fundamental a property, as energy or entropy, exergy is the most sought after value in energy. When we say energy crisis, we really mean exergy crisis. For example, ocean water has endless energy but not enough exergy, that can be exploited by ships.

After developing the balance equation for exergy, we explore the meaning of different terms and establish that exergy is transported and transferred like energy. Unlike energy, exergy is spontaneously destroyed. Analysis of a system using the exergy balance equation gives rise to new quantities: *reversible work*, *irreversibility*, and *exergetic efficiency*. The different types of systems analyzed in Chapters 2, 4, and 5—open, closed, steady, and unsteady systems—are revisited for an exergy analyses. Building upon the problem solving framework from Chapters 4 and 5, the open-steady and closed-process TESTcalcs in TEST are employed for verification of manual results and pursuing "what-if" studies.

6.1 EXERGY BALANCE EQUATION

In Chapter 1, exergy was introduced as an extrinsic property that describes an extended system or flow state (see animations beginning at Anim. 1.B.*exergyAndKE*). **Stored exergy** of a system is a measure of the maximum useful work the system is capable of delivering before it reaches its *dead state,* a state in equilibrium with the atmosphere at sea level. **Flow exergy** is the measure of useful work a flow is capable of delivering as it comes to its dead state. By definition, exergy of a system, or a flow, is zero at the dead state. Due to the fact that exergy represents the part of energy that can be potentially converted into shaft or electrical work, it can be looked upon as the *quality* of stored or flow energy. Unlike energy or entropy, the concept of exergy is not rooted in any fundamental law; instead, it is a convenient property, like enthalpy, obtained by manipulating the mass, energy, and entropy equations. Although the derivation is somewhat lengthy, it is a worthwhile analytical exercise that produces mathematical definitions of the dead state, stored exergy, and flow exergy. It also leads to the exergy balance equation in the same format as the energy and entropy equations.

We begin the derivation by revisiting the concept of a thermal energy reservoir (**TER**), introduced in Sec. 2.1.2. A TER or, a *reservoir,* is a very large system whose temperature is not affected by heat interactions with its surroundings. The atmosphere, identified as TER-0, is a special reservoir with a constant known temperature T_0. Another example of a TER is a furnace that maintains a constant temperature by burning fossil fuels. In Figure 6.1, the heat and entropy transfer through the internal boundary of TER-k is represented by $\dot{Q}_k$ and $\dot{Q}_k/T_k$, respectively. The algebraic sign (using the WinHip rule) associated with $\dot{Q}_k$ is with respect to the system, not the reservoir k. Another concept that will be used in this derivation is that of a dead state, introduced in Sec. 1.5.10. When a system reaches equilibrium with the quiescent atmosphere

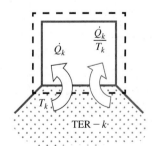

FIGURE 6.1 The temperature of a TER does not change due to energy transfer. It is an infinite reservoir of energy (see Anim. 2.D.*entropyTransferByHeat*).

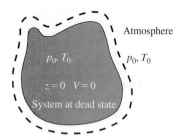

FIGURE 6.2 A stationary system at equilibrium, with its atmosphere at sea level, is at its dead state.

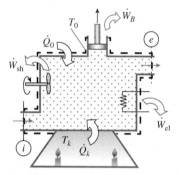

FIGURE 6.3 Schematic of an unsteady, open system used in the derivation of the exergy balance equation (see Anim 6.A.*genericSystem*).

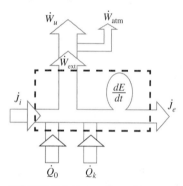

FIGURE 6.4 Energy-flow diagram and external work classification for the system in Fig. 6.3.

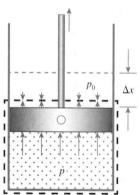

FIGURE 6.5 Schematic for evaluation of atmospheric work transfer in Eq. (6.5) (see Anim. 0.D.*atmosphericWork*).

(TER-0) at sea level (Fig. 6.2), it is said to be at its **dead state** because there is no mechanical, thermal, or chemical imbalance left between the system and the environment that can be exploited to produce useful work. The further a system is from its dead state, the more potential there is for extracting useful work.

Consider the generic, unsteady, open system shown in Figure 6.3 (and Anim. 6.A.*genericSystem*). Like the systems used for the derivation of energy and entropy equations in Chapter 2, this system has only one inlet and one exit—a restriction that will be removed shortly. The surroundings are divided into two reservoirs—TER-0, which represents the atmosphere at T_0, and TER-k that maintains a fixed temperature T_k, and represents an external heat source such as a furnace. The number of TER's can be extended just like the number of inlet or exit ports.

The red boundary of Figure 6.3 encloses the system's universe—the system and its immediate surroundings that are affected by the system. The net heat transfer $\dot{Q}$ is split into contributions $\dot{Q}_0$ (from the atmospheric reservoir: TER-0) and $\dot{Q}_k$ (from the reservoir-k: TER-k). The energy and entropy equations, Eqs. (2.8) and (2.13), for the single-flow, unsteady, open system of Figure 6.3 can be written as:

$$\frac{dE}{dt} = \dot{m}_i j_i - \dot{m}_e j_e + \dot{Q}_0 + \dot{Q}_k - \dot{W}_{\text{ext}}; \quad [\text{kW}] \tag{6.1}$$

$$\frac{dS}{dt} = \dot{m}_i s_i - \dot{m}_e s_e + \frac{\dot{Q}_0}{T_0} + \frac{\dot{Q}_k}{T_k} + \dot{S}_{\text{gen,univ}}; \quad \left[\frac{\text{kW}}{\text{K}}\right] \tag{6.2}$$

With the boundary of the system's universe passing through two reservoirs, the entropy transfer by heat is split into two terms. Entropy generated in the system's universe, $\dot{S}_{\text{gen,univ}}$, includes the contributions of all sources of *irreversibilities* (due to *thermodynamic friction*), internal or external to the system.

Multiplying Eq. (6.2) by the constant T_0 and subtracting the resulting equation from Eq. (6.1) yields:

$$\frac{d(E - T_0 S)}{dt} = \dot{m}_i(j_i - T_0 s_i) - \dot{m}_e(j_e - T_0 s_e) + \dot{Q}_k\left(1 - \frac{T_0}{T_k}\right) - \dot{W}_{\text{ext}} - T_0 \dot{S}_{\text{gen,univ}}; \quad [\text{kW}] \tag{6.3}$$

Different components of external work were discussed in Sec. 0.8 and summarized in Anim. 6.A.*externalWork*. While most types of external work can be readily converted into useful work, a portion of the boundary work that involves displacement of the atmosphere cannot be considered useful. Thus, external work can be divided into a **useful** and an **atmospheric** component (see Anim. 6.A.*usefulWork*):

$$\dot{W}_{\text{ext}} = \dot{W}_u + \dot{W}_{\text{atm}}; \quad [\text{kW}] \tag{6.4}$$

This is graphically shown in the energy-flow diagram of Figure 6.4 for the open, unsteady system shown in Figure 6.3. Although $\dot{W}_{\text{ext}}$ is the work that leaves the system, a part of it, $\dot{W}_{\text{atm}}$, is spent in lifting the atmosphere as the piston in Figure 6.3 rises, leaving only a reduced amount, $\dot{W}_u$, for any useful purpose (for example, to generate electric work).

To develop an expression for $\dot{W}_{\text{atm}}$, study the displacement Δx of the piston against the atmospheric pressure p_0 in time Δt (Fig. 6.5). Note that the external boundary includes a thin layer of atmosphere. The $pd\forall$ formula with $p = p_0$ can be applied to determine the work done by the moving boundary against the atmosphere. Referring to Figure 6.5, the rate of work transfer to the atmosphere can be expressed as:

$$\dot{W}_{\text{atm}} = \lim_{\Delta t \to 0} \frac{(p_0 A \Delta x)}{\Delta t} = \lim_{\Delta t \to 0} \frac{p_0 \Delta \forall}{\Delta t} = p_0 \frac{d\forall}{dt}; \quad [\text{kW}] \tag{6.5}$$

Observe that the system pressure p can be quite different from the atmospheric pressure p_0.

Although derived for a piston-cylinder device, Eq. (6.5) can be generalized for systems with an irregular boundary by imagining it to be composed of a large number of discrete piston-cylinder devices. For any generic system, summing up the contribution from the individual elements leads to Eq. (6.5). The same expression is applicable to a contracting system, where Eq. (6.5) produces a negative value for $\dot{W}_{\text{atm}}$, which is consistent with the sign convention.

Using Eqs. (6.4) and (6.5), Eq. (6.3) can be rearranged:

$$\underbrace{\frac{d(E - T_0 S + p_0 V\!\!\!/)}{dt}}_{A} = \underbrace{\dot{m}_i(j_i - T_0 s_i) - \dot{m}_e(j_e - T_0 s_e)}_{B} + \underbrace{\dot{Q}_k\left(1 - \frac{T_0}{T_k}\right)}_{C}$$

$$- \underbrace{\dot{W}_u}_{D} - \underbrace{T_0 \dot{S}_{gen,univ}}_{E}; \quad [\text{kW}] \tag{6.6}$$

Letters A through E identify the five terms. The transport term B is the net rate of transports of a combination property $j - T_0 s$ into the system. At the dead state, the combination property assumes the following form:

$$(j - T_0 s)_0 = j_0 - T_0 s_0 = h_0 + \cancel{ke_0}^{0} + \cancel{pe_0}^{0} - T_0 s_0 = h_0 - T_0 s_0; \quad \left[\frac{\text{kJ}}{\text{kg}}\right] \tag{6.7}$$

Note that properties h_0 and s_0 in Eq. (6.7) are properties of the working substance, not surrounding air, at the ambient conditions, p_0 and T_0.

Given Eq. (6.7), term B in Eq. (6.6) can be modified by introducing the dead state, State-0, as a reference state for the combination property $(j - T_0 s)$. As a result, term B in Eq. (6.6) gives rise to three new terms F, G, and H as follows:

$$A = \underbrace{\dot{m}_i[(j_i - T_0 s_i) - (j_0 - T_0 s_0)]}_{F} - \underbrace{\dot{m}_e[(j_e - T_0 s_e) - (j_0 - T_0 s_0)]}_{G}$$

$$+ \underbrace{(j_0 - T_0 s_0)(\dot{m}_i - \dot{m}_e)}_{H} + C - D - E; \quad [\text{kW}] \tag{6.8}$$

The new combination property, transported by $\dot{m}_i$ and $\dot{m}_e$, is represented by the symbol $\psi \equiv (j - T_0 s) - (j_0 - T_0 s_0)$. Like the specific flow energy j, ψ is an extrinsic specific property of the flow with units of kJ/kg and is called the **specific flow exergy** (see Anim. 6.A.*flowExergy*). We will explore the physical meaning of ψ after the final balance equation has been derived.

Recognizing that at the dead state $j_0 = h_0$, term H in Eq. (6.8) can be further simplified by applying Eq. (6.7) and substituting the mass-balance equation, Eq. (2.3):

$$H = (j_0 - T_0 s_0)(\dot{m}_i - \dot{m}_e) = (h_0 - T_0 s_0)\frac{dm}{dt} = \frac{d}{dt}(mh_0) - T_0\frac{d}{dt}(ms_0) = \frac{dH_0}{dt} - T_0\frac{dS_0}{dt}$$

$$= \frac{d}{dt}(H_0 - T_0 S_0) = \frac{d}{dt}(U_0 + p_0 V\!\!\!/_0 - T_0 S_0) = \frac{d}{dt}(E_0 - T_0 S_0 + p_0 V\!\!\!/); \quad [\text{kW}]$$

Therefore,

$$A - H = \frac{d}{dt}[(E - T_0 S + p_0 V\!\!\!/) - (E_0 - T_0 S_0 + p_0 V\!\!\!/_0)]; \quad [\text{kW}] \tag{6.9}$$

The combination system property that appears inside the derivative is represented by the uppercase symbol $\Phi \equiv (E - T_0 S + p_0 V\!\!\!/) - (E_0 - T_0 S_0 + p_0 V\!\!\!/_0)$. Known as the **stored exergy**, this system property has units of kJ like the stored energy E. We will postpone a discussion of the physical meaning of Φ, and the corresponding **specific stored exergy** (see Anim. 6.A.*storedExergy*) $\phi = \Phi/m$, until later in the chapter.

Substituting Eq. (6.9) into Eq. (6.8), and generalizing the number of inlets, exits, and reservoirs (TER's), we obtain the **exergy balance equation** for an unsteady, open system (see Anim. 6.A.*exergyBalanceTerms*):

$$\underbrace{\frac{d\Phi}{dt}}_{\substack{\text{Rate of change} \\ \text{of stored exergy } \Phi \\ \text{of an open system.}}} = \underbrace{\sum_i \dot{m}_i \psi_i}_{\substack{\text{Flow exergy} \\ \text{transported} \\ \text{in by mass flow.}}} - \underbrace{\sum_e \dot{m}_e \psi_e}_{\substack{\text{Flow exergy} \\ \text{transported out} \\ \text{by mass flow.}}} + \underbrace{\sum_k \dot{Q}_k\left(1 - \frac{T_0}{T_k}\right)}_{\substack{\text{Exergy transferred in by} \\ \text{heat from all TER's.}}}$$

$$- \underbrace{\dot{W}_u}_{\substack{\text{Exergy transferred} \\ \text{out by useful} \\ \text{external work.}}} - \underbrace{\dot{I}}_{\substack{\text{Exergy destroyed} \\ \text{by thermodynamic} \\ \text{friction.}}} \quad [\text{kW}] \tag{6.10}$$

DID YOU KNOW?

- Saturated steam at 1 atm in a pipe, with a flow rate of 1 kg/s, can produce useful work at a maximum possible rate of 488 kW.

where $\quad \Phi = \oint\limits_{sys} \phi \rho d\forall = (E - T_0 S + p_0 \forall) - (E_0 - T_0 S_0 + p_0 \forall_0)$

$$= (U - U_0) - T_0(S - S_0) + p_0(\forall - \forall_0) + KE + PE; \quad [\text{kJ}]$$

$$\phi = (u - u_0) - T_0(s - s_0) + p_0(v - v_0) + ke + pe$$

$$= (e - e_0) - T_0(s - s_0) + p_0(v - v_0); \quad \left[\frac{\text{kJ}}{\text{kg}}\right]$$

$$\psi = (j - T_0 s) - (j_0 - T_0 s_0) = (h - h_0) - T_0(s - s_0) + ke + pe$$

$$= (j - j_0) - T_0(s - s_0); \quad \left[\frac{\text{kJ}}{\text{kg}}\right]$$

and $\quad \dot{I} \equiv T_0 \dot{S}_{\text{gen,univ}}; \quad [\text{kW}]$

Although the equation looks formidable, it lends itself to interpretation like all other balance equations (see Anim. 6.A.*exergyBalanceTerms*). As we explore the terms of this equation in the next section, the physical meaning of exergy will be clarified and several new exergy-related definitions will emerge.

6.1.1 Exergy, Reversible Work, and Irreversibility

The exergy balance equation, Eq. (6.10), has been derived through manipulation of the mass, energy, and entropy balance equations. The striking similarity of this equation to the mass, energy, and entropy equations (compare Anim. 6.A.*flowDiagram* with 2.C. *flowDiagram* or 2.D. *flowDiagram*) suggests that the inventory of exergy in a system is no different from that of mass, energy, or entropy.

The left side of Eq. (6.10), as usual, represents the unsteady term involving a total extensive property, which in this case is the system's stored exergy Φ. To explore the physical meaning of Φ, consider a closed system (Fig. 6.6 and Anim. 6.A.*storedExergy*) attached to a concept heat engine that exploits the system's kinetic, potential, and internal energy to deliver useful work. The engine works until the system loses all its KE and PE and achieves equilibrium with its surroundings at p_0 and T_0. At that point, the system arrives at its dead state. Work extracted by this engine is maximized if the engine operates reversibly, if $\dot{I} = T_0 \dot{S}_{\text{gen,univ}} = 0$ throughout the process. Accounting for heat ejection to the surroundings, a necessity for any practical heat engine, Eq. (6.10) reduces to:

$$\frac{d\Phi}{dt} = \dot{Q}_0\left(1 - \frac{T_0}{T_0}\right) - \dot{W}_{u,\max} = -\dot{W}_{u,\max}; \quad [\text{kW}] \quad \Rightarrow \quad d\Phi = -\dot{W}_{u,\max} dt; \quad [\text{kJ}] \quad \textbf{(6.11)}$$

Integrating Eq. (6.11) from the current state to the final equilibrium state (the dead state) and recognizing that, by definition, $\Phi_0 = 0$, we obtain:

$$\int d\Phi = -\int \dot{W}_{u,\max} dt; \quad [\text{kJ}] \quad \Rightarrow \quad \Phi_0 - \Phi = -W_{u,\max}; \quad [\text{kJ}]$$

$$W_{u,\max} = \Phi; \quad [\text{kJ}] \quad \text{and} \quad \phi = \frac{\Phi}{m} = \frac{W_{u,\max}}{m}; \quad \left[\frac{\text{kJ}}{\text{kg}}\right] \quad \textbf{(6.12)}$$

Therefore, a system's stored exergy Φ is the maximum possible useful work that can be extracted as the system is brought to its dead state in a reversible manner, while allowing heat and work interactions with the surrounding atmosphere.

To explore the physical meaning of the transport terms, let's look at a steady device, which extracts useful work out of a flow as shown in Figure 6.7 (and Anim. 6.A.*flowExergy*). If the maximum possible useful work is to be extracted from the flow, then the exit state must be the dead state and the device must be reversible so that $\dot{I} = T_0 \dot{S}_{\text{gen,univ}} = 0$. In addition, $\psi_0 = 0$; therefore, Eq. (6.10) simplifies to:

$$\frac{d\Phi^0}{dt} = \dot{m}\psi_i - \dot{m}\psi_0^0 + \dot{Q}_0\left(1 - \frac{T_0}{T_0}\right) - \dot{W}_{u,\max} - \dot{I}^0; \quad [\text{kW}]$$

$$\Rightarrow \quad \dot{W}_{u,\max} = \dot{\Psi}_i = \dot{m}\psi_i; \quad [\text{kW}]; \quad \text{and} \quad \psi_i = \frac{\dot{W}_{u,\max}}{\dot{m}}; \quad \left[\frac{\text{kJ}}{\text{kg}}\right] \quad \textbf{(6.13)}$$

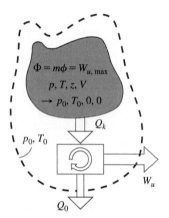

FIGURE 6.6 The stored exergy Φ of a system is the maximum possible useful work it can potentially deliver as it approaches the dead state (see Anim. 6.A.*storedExergy*).

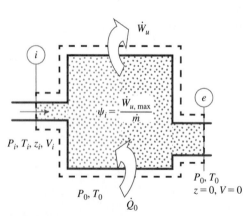

FIGURE 6.7 The specific flow exergy is the maximum possible useful work a flow can potentially deliver (per-unit mass) as it approaches its dead state (see Anim. 6.A.*flowExergy*).

Thus, the specific flow exergy ψ is the maximum useful work that can be extracted from the flow (per-unit mass) in a reversible steady manner, while still allowing heat and work interactions with the surrounding atmosphere.

Therefore the transport terms $\dot{\Psi}_i \equiv \dot{m}_i \psi_i$ and $\dot{\Psi}_e \equiv \dot{m}_e \psi_e$ in Eq. (6.10) are the rate of flow exergy transported into and out of the system at the inlet i and exit e, just like any other transport terms (see Anim. 6.A.*exergyTransport*). In this context, note the parallel (Fig. 6.8) between stored energy and stored exergy (E and Φ), and between flow energy and flow exergy (j and ψ).

From now on, we will interchangeably use the terms *maximum useful work* and *exergy,* as their equivalence is evident from Eqs. (6.12) and (6.13). With no atmospheric work involved, electrical and shaft work are 100% useful. Transfer of such work, therefore, can be treated as direct transfer of exergy. Also, the price of electricity, which is usually well publicized (on a kWh basis), can be used as a benchmark for pricing exergy.

To explore the exergy equation's remaining terms, consider a reversible closed system that extracts useful work steadily as heat $\dot{Q}_k$ from a source maintained at T_k is transferred into the system (Fig. 6.9). To satisfy the second law, some heat must be rejected into the surroundings. Equation (6.10) simplifies to produce the maximum work output:

$$\frac{d\Phi^{0}}{dt} = \dot{Q}_k\left(1 - \frac{T_0}{T_k}\right) - \dot{Q}_k\left(1 - \frac{T^{1}_{0}}{T_0}\right) - \dot{W}_{u,\,max} - \dot{I}^{0}$$

$$\Rightarrow \quad \dot{W}_{u,\,max} = \dot{Q}_k\left(1 - \frac{T_0}{T_k}\right); \quad [\text{kW}] \qquad \qquad \textbf{(6.14)}$$

In this expression, we can identify the thermal efficiency to be the *Carnot efficiency* (Sec. 2.3.3). To obtain the maximum useful work from heat, the exergy extracting device of Figure 6.9 must be a reversible Carnot heat engine. The above derivation helps us interpret heat transfer in terms of exergy. As can be seen from Eq. (6.14), the maximum useful work or exergy that can be extracted from heat depends on its source temperature. More precisely, only a fraction, $1 - T_0/T_k$, of $\dot{Q}_k$ can be converted into exergy, which can be interpreted as the **quality of heat**. All heat transfers are not of the same quality, even if each transfer involves the same amount of energy. In the limiting case of $T_k \rightarrow \infty$, the quality of heat approaches that of pure exergy— useful work such as electrical or shaft work. At the other extreme, heat transferred to or from the atmospheric reservoir has a quality of $1 - T_0/T_0 = 0$. An interesting situation arises for cold reservoirs where $T_k < T_0$. The negative quantity $1 - T_0/T_k$ signifies that the direction of exergy flow is opposite of heat transfer's. Exergy is deposited, not withdrawn, as heat is extracted from a cold space.

It is instructive to compare (Fig. 6.10) the different means of transferring 1 kW of exergy into a closed system through heat and external work. From Eq. (6.14), we can verify that a 2-kW heat flow from a source at 600 K, a 1.5-kW heat flow from a source at 900 K, a shaft-work

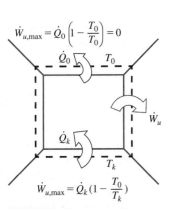

FIGURE 6.8 Exergy and energy stored in a system vs. exergy and energy transported by a flow (see Anim. 6.A.*exergyTransport*).

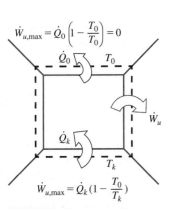

FIGURE 6.9 Exergy transfer by heat can be interpreted as the quality of heat (see Anim. 6.A.*exergyTransfer*).

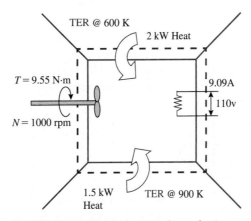

FIGURE 6.10 Different means of transferring 1 kW of exergy with the atmosphere at standard conditions—heat transfer from two reservoirs, shaft work, and electrical work (see Anim. 6.A.*exergyTransfer*).

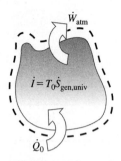

FIGURE 6.11 Exergy is destroyed spontaneously with or without interactions between the system and its surroundings (see Anim. 6.A.*viscousAndElectronicFriction*).

transfer of 1 kW, and an electrical power of 1 kW are all equivalent as far as transfer of useful work or exergy is concerned.

In deriving expressions for the maximum useful work output in Eqs. (6.12) through (6.14), we set $\dot{I} = T_0 \dot{S}_{gen,univ}$ to zero. In actual systems $\dot{S}_{gen,univ} \geq 0$ and $\dot{I} \geq 0$. To understand the physical meaning of the term $\dot{I}$, look at a closed system (Fig. 6.11), which has no useful work interactions and that does not communicate with a thermal reservoir other than the atmosphere. With only heat and work transfer with the atmosphere, Eq. (6.10) reduces to:

$$\frac{d\Phi}{dt} = -\dot{I}; \quad [\text{kW}] \tag{6.15}$$

A positive value for $\dot{I}$ means that the system's stored exergy depletes spontaneously. This depletion is illustrated by the everyday experience of an unused battery losing its power over time. Exergy stored in a hot block of solid depletes as heat is lost to the atmosphere— exergy must be destroyed in this process, as it cannot be stored in the atmosphere (which is perpetually at dead state). Even an isolated system can lose its exergy since $\dot{I} = T_0 \dot{S}_{gen,univ}$ and $\dot{S}_{gen,univ} \geq 0$. $\dot{I}$ is called the **rate of exergy destruction** or the **rate of irreversibility**. The proportionality between entropy generation and exergy destruction means that the thermodynamic friction discussed in Sec. 2.1.3 is behind the destruction of exergy.

In a work-consuming or work-producing steady device, inlet and exit conditions and the heat transfer $\dot{Q}_k$ remain fixed, and the sum of $\dot{W}_u$ and $\dot{I}$ [from Eq. (6.10)] remain constant as well. This sum is called the **reversible work** and is represented by the symbol $\dot{W}_{rev}$:

$$\underbrace{\dot{W}_{rev}}_{\substack{\text{Maximum useful} \\ \text{work transfer.}}} \equiv \underbrace{\dot{W}_u}_{\substack{\text{Actual useful} \\ \text{work transfer.}}} + \underbrace{\dot{I}}_{\substack{\text{Exergy destruction} \\ \text{in the system's} \\ \text{universe.}}} ; \quad [\text{kW}] \tag{6.16}$$

The reversible work (see Anim. 6.A.*reversibleWork*) can be interpreted as the maximum work (in an algebraic sense) $\dot{W}_{u,\,max}$ if all the irreversibilities can be eliminated ($\dot{I} = 0$). For example, for a turbine, a work-producing device, it is the maximum power that can be delivered by the turbine without any change in the inlet or exit conditions. If the exit state happens to be the dead state, the reversible power approaches the flow exergy transported into the turbine in the absence of heat exchange with non-atmospheric reservoirs.

Just as we developed flow diagrams to understand the energy and entropy balance equations, we can construct similar flow diagrams for the exergy balance equation. To do so, we rewrite the exergy equation, Eq. (6.10), in a shorter form:

$$\frac{d\Phi}{dt} = \dot{\Psi}_{net} + \sum_k \dot{Q}_k \left(1 - \frac{T_0}{T_k}\right) - \dot{W}_u - \dot{I}; \quad \text{where,} \quad \dot{\Psi}_{net} = \sum_i \dot{\Psi}_i - \sum_e \dot{\Psi}_e; \quad [\text{kW}] \tag{6.17}$$

$$\text{or,} \quad \frac{d\Phi}{dt} = \dot{\Psi}_{net} + \sum_k \dot{Q}_k \left(1 - \frac{T_0}{T_k}\right) - \dot{W}_{rev}; \quad [\text{kW}] \tag{6.18}$$

The exergy flow diagram in Figure 6.12 is keyed to this equation and can lead to device-specific definitions of *exergetic* or *second-law efficiency*. As with *energetic efficiency*, **exergetic efficiency** is defined as the ratio of two specific terms of the exergy balance equation:

$$\eta_{II} = \frac{\text{Desired Exergy Output}}{\text{Required Exergy Input}} \tag{6.19}$$

For a steam turbine, operating at steady state, the exergy diagram can be simplified (Fig. 6.13), where the desired output and the required input are identified. Consequently, exergetic efficiency can be defined as the ratio of useful work, $\dot{W}_{sh}$, to the reversible work, $\dot{W}_{rev}$, where Eq. (6.18) can be used to show that $\dot{W}_{rev} = \dot{\Psi}_i - \dot{\Psi}_e$. Although an isentropic turbine is reversible, exergetic efficiency is quite different from the isentropic device efficiency discussed in Sec. 4.1.4. In defining the isentropic

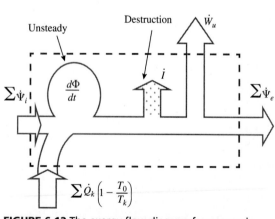

FIGURE 6.12 The exergy flow diagram for a generic, open, unsteady system (see Anim. 6.A.*flowDiagram*).

efficiency, the ideal device was assumed to be internally reversible, adiabatic, and to operate with the actual device's inlet conditions and exit pressure. For the exergetic efficiency, however, the ideal device must operate between the same inlet and exit conditions and be reversible, which allows reversible heat transfer with the surroundings. The difference between these two types of efficiency is highlighted in Example 6-5.

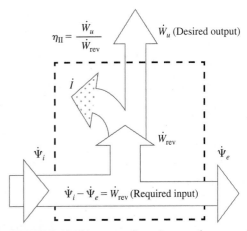

FIGURE 6.13 The exergy flow diagram for a turbine at steady state helps explain the meaning of the exergetic efficiency.

6.1.2 TESTcalcs for Exergy Analysis

Specific stored exergy and specific flow exergy are integrated into state panels of all material models as part of an extended system or flow state. Dependent on kinetic and potential energies, specific exergies are extrinsic properties and are color coded accordingly in the TESTcalcs. In all state and system TESTcalcs, State-0 is assigned to be the designated dead state. To calculate the dead state, we select the working fluid, set pressure and temperature to that of the surroundings, and initialize velocity and elevation to zero. With the dead-state set, specific exergies phi (ϕ) and psi (ψ) are automatically calculated as part of any new state. The single-flow open-steady TESTcalcs and uniform-process TESTcalcs contain a separate exergy panel that displays the appropriate form of the exergy equation and all the calculated exergy terms. To analyze a particular process or device, evaluate the dead state in addition to the anchor states, perform energy, and entropy analysis in the process or device panel, then switch to the exergy panel. There, if necessary, the net heat transfer (Qdot in the device or Q in the process panel) can be distributed between two reservoirs, TER-0 (designated for the surroundings) and TER-1 (at a temperature different from atmospheric temperature). By default, all heat transfer is assumed to take place with TER-0.

EXAMPLE 6-1   **Calculation of Stored and Flow Exergy**

A tank contains 1 kg of superheated steam at 10 MPa and 500°C. Calculate (a) the energy and (b) exergy stored in the tank. If steam, at the same state, flows through a pipe with a flow rate of 1 kg/s, evaluate the rate of transport of (c) energy and (d) exergy. Assume the atmospheric conditions to be 100 kPa and 25°C.

SOLUTION

Evaluate the system and flow states of the steam.

Assumptions

The steam is in equilibrium and the PC model is the appropriate model. Negligible ke and pe.

Analysis

Use TEST, or the manual approach, to determine the dead state, State-0, the system state, State-1, and the flow state, State-2.

State-0 (given p_0, T_0):

$$v_0 = 0.001 \, \frac{m^3}{kg}; \quad u_0 = 104.9 \, \frac{kJ}{kg}; \quad h_0 = 105.0 \, \frac{kJ}{kg}; \quad s_0 = 0.3673 \, \frac{kJ}{kg \cdot K};$$

State-1 (given p_1, T_1):

$$v_1 = 0.0328 \, \frac{m^3}{kg}; \quad u_1 = 3046 \, \frac{kJ}{kg}; \quad s_1 = 6.596 \, \frac{kJ}{kg \cdot K};$$

$$\phi_1 = (u_1 - u_0) - T_0(s_1 - s_0) + p_0(v_1 - v_0) + \cancel{ke_1} + \cancel{pe_1}$$

$$= (3046 - 104.9) - (298)(6.596 - 0.3673) + (100)(0.0328 - 0.001)$$

$$= 1088 \, kJ/kg$$

Therefore, the stored energy and exergy in the tank are:

$$E_1 \cong U_1 = m_1 u_1 = (1)(3046) = 3046 \text{ kJ};$$

$$\text{and} \quad \Phi_1 = m_1 \phi_1 = 1088 \text{ kJ};$$

State-2 (given $p_2 = p_1$, $T_2 = T_1$): $h_2 = 3374 \dfrac{\text{kJ}}{\text{kg}}$; $s_2 = s_1$;

$$\psi_2 = (h_2 - h_0) - T_0(s_2 - s_0) + \cancel{ke_2} + \cancel{pe_2}$$

$$= (3374 - 105) - (298)(6.596 - 0.3673) = 1413 \text{ kJ/kg}$$

The rates of energy and exergy transports are:

$$\dot{J}_2 = \dot{m}_2 j_2 \cong \dot{m}_2 h_2 = (1)(3374) = 3374 \text{ kW}; \quad \text{and} \quad \dot{\Psi}_2 = \dot{m}_2 \psi_2 = 1413 \text{ kW}$$

TEST Solution

Launch the PC system-state TESTcalc. Select H2O as the working fluid. Calculate State-0 and State-1 from the given properties (see TEST-code in *TEST* > *TEST-codes*). In the I/O panel, calculate E1 and Phi1 by evaluating expressions '= m1*e1' and '= m1*phi1'. Using the PC flow-state TESTcalc, calculate '= mdot2*j2' and '= mdot2*psi2' after evaluating State-0 and State-2.

Discussion

The calculation of exergy may be cumbersome, but its physical meaning is easier to understand than energy. The steam in the tank has the potential of producing 1088 kJ of useful work worth $0.06 at 20 cents/kWh. The flow of steam carries 1.413 MW of power, capable of producing 1.413 MW of electric power. Calculated energies, however, depend on the reference value used for the internal energy, and cannot be interpreted in absolute terms. Property tables from different sources can produce different answers for E or $\dot{J}$, but not for Φ or $\dot{\Psi}$.

6.2 CLOSED-STEADY SYSTEMS

In Chapter 2, we carried out energy and entropy analysis of systems that can be regarded as closed and steady. Examples included an electrical heater, a gearbox, heat transfer through a wall, a heat engine, and a refrigerator. We revisit this class of systems with the added tool of the exergy balance equation. As with the energy and entropy equation, we simplify the exergy equation, Eq. (6.10), by dropping the transport terms and setting the unsteady term to zero:

$$0 = \sum_k \dot{Q}_k \left(1 - \frac{T_0}{T_k} \right) - \dot{W}_u - \dot{I}; \quad \text{[kW]} \quad \text{whereas} \quad \dot{W}_{\text{rev}} = \dot{W}_u + \dot{I}; \quad \text{[kW]} \quad \textbf{(6.20)}$$

The closed system is at steady state, so the net exergy that enters the system with heat must be equal to the reversible work output, which is the sum of exergy leaving the system through useful work, and the exergy destroyed. To explore the exergy inventory of a closed-steady system, review Example 2-6, which deals with steady-state heat transfer through a wall.

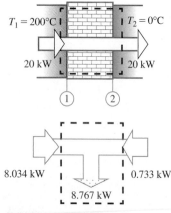

$T_1 = 200°C$ $T_2 = 0°C$

20 kW   20 kW

① ②

8.034 kW   0.733 kW

8.767 kW

FIGURE 6.14 Energy and exergy flow diagrams for Ex. 6-2 (see Anim. 6.A.*exergyDest-thermalFriction*).

EXAMPLE 6-2 **Exergy Transfer and Destruction through Heat Transfer (Thermal Friction)**

A thin wall separates two large chambers. The left chamber contains boiling water at 200°C and the right chamber contains a boiling refrigerant at 0°C. Heat transfer through the wall, at steady state, is measured at 20 kW. Assuming each wall surface is at its respective chamber's temperature, and the atmospheric temperature is 10°C, determine (a) the rate at which exergy enters the wall through the left face, (b) the rate at which exergy leaves the wall through the right face, (c) the rate of exergy destruction in the wall and its immediate vicinity, and (d) the reversible work.

SOLUTION

Analyze the exergy equation for the wall's universe that is enclosed within the red boundary of Figure 6.14.

Assumptions

The wall is at steady state; therefore, all extensive properties, including the total stored exergy Φ of the system, enclosed within the red boundary of Figure 6.14, remain constant.

Analysis

Exergy transferred with the heat transfer $\dot{Q}_{wall}$ (direction of heat transfer is from left-to-right at both faces; so $\dot{Q}_{wall}$ represents absolute value) is calculated as:

$$\text{Left face:}\quad \dot{Q}_{wall}\left(1 - \frac{T_0}{T_1}\right) = (20)\left(1 - \frac{283}{473}\right) = 8.034 \text{ kW}$$

$$\text{Right face:}\quad \dot{Q}_{wall}\left(1 - \frac{T_0}{T_2}\right) = (20)\left(1 - \frac{283}{273}\right) = -0.733 \text{ kW}$$

The negative value indicates that exergy is being transferred from right-to-left in the opposite direction of heat.

Equation (6.20) yields:

$$\dot{I} = \sum_k \dot{Q}_k\left(1 - \frac{T_0}{T_k}\right) - \dot{W}_u^0 = \dot{Q}_{wall}\left(1 - \frac{T_0}{T_1}\right) - \dot{Q}_{wall}\left(1 - \frac{T_0}{T_2}\right)$$

$$= 8.034 - (-0.733) = 8.767 \text{ kW}$$

Alternatively, $\dot{I}$ can be obtained by multiplying $\dot{S}_{gen,univ}$ (Ex. 2-6) by T_0, resulting in the same answer.

The reversible work in this heat transfer is:

$$\dot{W}_{rev} = \dot{W}_u + \dot{I} = \dot{I} = 8.767 \text{ kW}$$

Obviously, there is no useful work output from a passive wall. However, if the same heat transfer could be carried out in a reversible manner, useful work or exergy at the rate of 8.767 kW could be extracted as a by-product. The passive wall, in this case, simply destroys the exergy that enters the wall from both faces.

Discussion

It is easy to understand that heat flowing from a high-temperature chamber carries exergy with it. But how can exergy flow in against the direction of heat transfer at the right face? The answer lies in the fact that useful work can be extracted from a cold system as easily as from a hot system. This is accomplished by exploiting the temperature difference between a system and its surroundings through a heat engine. Heat transfer at the right face in this example transfers exergy to the wall by cooling it below atmospheric temperature and creating a temperature difference.

6.2.1 Exergy Analysis of Cycles

An inventory of exergy for heat engines, refrigerators, and heat pumps, which are special cases of closed-steady systems, can be useful when comparing actual cycles with their ideal counterpart—the reversible Carnot cycle. For a heat engine (Fig. 6.15), the atmospheric TER is the cold reservoir, so $T_C = T_0$. The entire amount of net work is useful, that is, $\dot{W}_u = \dot{W}_{net}$. Treating the engine as a closed-steady system (Sec. 2.3), the exergy balance equation, Eq. (6.20), and the reversible work can be expressed in terms of the Carnot efficiency derived in Sec. 2.3.3:

$$0 = \dot{Q}_H\left(1 - \frac{T_0}{T_H}\right) - \dot{W}_{net} - \dot{I}; \quad \Rightarrow \quad \dot{W}_{net} = \dot{Q}_H\left(1 - \frac{T_0}{T_H}\right) - \dot{I}; \quad [\text{kW}] \qquad \textbf{(6.21)}$$

$$\dot{W}_{rev} \equiv \dot{W}_u + \dot{I} = \dot{W}_{net} + \dot{I} = \dot{Q}_H\left(1 - \frac{T_0}{T_H}\right) = \dot{Q}_H\eta_{Carnot}; \quad [\text{kW}] \qquad \textbf{(6.22)}$$

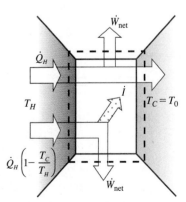

FIGURE 6.15 Energy and exergy flow diagrams for a heat engine where $T_C = T_0$ (click Exergy Diagram in Anim. 6.B.*heatEngine*).

As shown in Figure 6.15, exergy is transferred to the engine from the TER at T_H while the heat rejected to the TER at $T_C = T_0$ carries no exergy. Part of the exergy that enters the engine is destroyed and the remainder is transferred out as useful net work. The reversible power is equal to the Carnot power. Using Eq. (6.19), the **exergetic efficiency** can be related to the thermal efficiency (Sec. 2.3) and expressed as:

$$\eta_{II} = \frac{\text{Desired Exergy Output}}{\text{Required Exergy Input}} = \frac{\dot{W}_{net}}{\dot{Q}_H\left(1 - \dfrac{T_0}{T_H}\right)} = \frac{\dot{W}_{net}}{\dot{W}_{rev}} = \frac{\eta_{th}\dot{Q}_H}{\eta_{th,Carnot}\dot{Q}_H} = \frac{\eta_{th}}{\eta_{th,Carnot}} \quad \textbf{(6.23)}$$

As expected from this expression, the Carnot engine has an exergetic efficiency of 100%.

For a refrigerator, recognizing that the net work $\dot{W}_{net}$ is the magnitude of the useful work input to the device, a similar exergy balance on the system shown in Figure 6.16 yields:

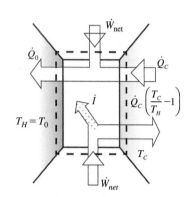

FIGURE 6.16 Energy and exergy flow for a refrigerator cycle with $T_H = T_0$ (see Anim. 6.B.*refrigeration AndHP*).

$$0 = \dot{Q}_C\left(1 - \frac{T_0}{T_C}\right) - (-\dot{W}_{net}) - \dot{I}; \quad \Rightarrow \quad \dot{W}_{net} = \dot{Q}_C\left(\frac{T_0}{T_C} - 1\right) + \dot{I}; \quad [\text{kW}] \quad \textbf{(6.24)}$$

$$\dot{W}_{rev} \equiv \dot{W}_u + \dot{I} = -\dot{W}_{net} + \dot{I} = \dot{Q}_C\left(1 - \frac{T_0}{T_C}\right) = \frac{-\dot{Q}_C}{\text{COP}_{R,Carnot}}; \quad [\text{kW}] \quad \textbf{(6.25)}$$

The negative reversible work represents the minimum amount of work that must be supplied to the cycle. Note that in the exergy flow diagram, exergy travels in the opposite direction of $\dot{Q}_C$ since $T_0 > T_C$ (Ex. 6-2). In defining the exergetic efficiency for a refrigeration cycle, we must realize that the exergy delivered to the cold space is the desired quantity while the net work input is the required exergy input. The **exergetic efficiency** for a refrigerator can be written:

$$\eta_{II,R} = \frac{\dot{Q}_C\left(\dfrac{T_0}{T_C} - 1\right)}{\dot{W}_{net}} = \frac{\dot{Q}_C}{\dot{W}_{net}}\left(\frac{T_0 - T_C}{T_C}\right) = \frac{|\dot{W}_{rev}|}{\dot{W}_{net}} = \frac{\text{COP}_R}{\text{COP}_{R,Carnot}} \quad \textbf{(6.26)}$$

The upper limit of η_{II} is 100% when the actual cycle performs like the Carnot refrigeration cycle.

For a heat pump, the cold reservoir is the atmosphere, so $T_C = T_0$. Part of the exergy supplied in the form of useful work is destroyed and the rest enters the warm reservoir (see Anim. 6.B.*refrigerationAndHP*). The **exergetic efficiency** can be expressed from an exergy balance similar to Eq. (6.24) as:

$$\eta_{II,HP} = \frac{\dot{Q}_H\left(1 - \dfrac{T_0}{T_H}\right)}{\dot{W}_{net}} = \frac{|\dot{W}_{rev}|}{\dot{W}_{net}} = \frac{\text{COP}_{HP}}{\text{COP}_{HP,Carnot}} \quad \textbf{(6.27)}$$

EXAMPLE 6-3   Exergy Balance of a Heat Engine

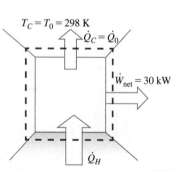

FIGURE 6.17 The heat engine analyzed in Ex. 6-3.

As shown in Figure 6.17, a heat engine operates between a reservoir at 1500 K and the atmosphere at 298 K, while producing a net power of 30 kW with a thermal efficiency of 40%. Determine (a) the rate of exergy destruction, (b) exergetic efficiency, and (c) reversible power.

SOLUTION

Conduct an exergy analysis of the closed-steady system enclosed within the red boundary of Figure 6.17.

Assumptions

The heat engine can be treated as a closed-steady system.

Analysis

The exergy equation, Eq. (6.21), can be simplified to produce the rate of exergy destruction:

$$0 = \dot{Q}_H\left(1 - \frac{T_0}{T_H}\right) - \dot{W}_{net} - \dot{i}$$

$$\Rightarrow \quad \dot{i} = \frac{\dot{W}_{net}}{\eta_{th}}\left(1 - \frac{T_0}{T_H}\right) - \dot{W}_{net} = \frac{30}{0.4}\left(1 - \frac{298}{1500}\right) - 30 = 60.1 - 30 = 30.1 \text{ kW}$$

Recognizing that the first term on the right side of the above equation is the exergy delivered to the engine (required exergy), the exergetic efficiency for the engine is:

$$\eta_{II} = \frac{\text{Desired Exergy Output}}{\text{Required Exergy Input}} = \frac{30}{60.1} = 49.92\%$$

We could also arrive at this result by dividing the thermal efficiency with the Carnot efficiency:

$$\eta_{Carnot} = 1 - T_C/T_H = 1 - 298/1500 = 80.1\%.$$

The reversible power, the power produced by the engine if all the irreversibilities could be eliminated, is:

$$\dot{W}_{rev} = \dot{W}_{net} + \dot{i} = 30 + 30.1 = 60.1 \text{ kW}$$

TEST Solution

Launch the closed-steady TESTcalc located in the systems, closed, steady branch. Enter Wdot_net, eta_th, T_C, and T_H. Press Enter or click Calculate to evaluate all relevant quantities.

Discussion

The TEST solution also produces $\dot{S}_{gen,univ}$ for the engine to be 0.101 kW/K from an entropy analysis. As expected, the rate of exergy destruction obtained from the exergy analysis is consistent with $\dot{i} = T_0\dot{S}_{gen,univ} = 298(0.101) = 30.1$ kW.

6.3 OPEN-STEADY SYSTEMS

Chapter 4 was devoted to the mass, energy, and entropy analysis of open-steady systems, a category of devices that includes nozzles, turbines, compressors, pumps, expansion valves, heat exchangers, and mixing chambers. In this section, we complete that analysis by including a thorough exergy analysis.

For a generic single-flow open system, operating at steady state, the exergy balance equation, Eq. (6.10), can be simplified:

$$\cancel{\frac{d\phi^0}{dt}} = \underbrace{\dot{m}(\psi_i - \psi_e)}_{\substack{\psi_{net}\text{: Exergy transported} \\ \text{into the system.}}} + \underbrace{\sum_k \dot{Q}_k\left(1 - \frac{T_0}{T_k}\right)}_{\substack{\text{Exergy transferred into} \\ \text{the system by heat from} \\ \text{all reservoirs.}}} - \underbrace{\dot{W}_u}_{\substack{\text{Exergy transferred} \\ \text{out of the system} \\ \text{through useful} \\ \text{external work.}}} - \underbrace{\dot{i}}_{\substack{\text{Exergy} \\ \text{destroyed.}}} ; \quad [\text{kW}] \qquad \textbf{(6.28)}$$

Introducing the definition of reversible work from Eq. (6.16), the above equation can be rearranged:

$$\dot{W}_{rev} = \dot{m}(\psi_i - \psi_e) + \sum_k \dot{Q}_k\left(1 - \frac{T_0}{T_k}\right); \quad \text{where} \quad \dot{W}_{rev} = \dot{W}_u + \dot{i}; \quad [\text{kW}] \qquad \textbf{(6.29)}$$

If the system is adiabatic ($\dot{Q}_k = 0$), Eq. (6.29) establishes equality between the reversible work and the net exergy transported into the system.

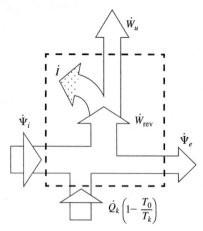

FIGURE 6.18 Exergy flow diagram for a generic, open-steady system with a single flow and heat transfer with the atmosphere and a reservoir at T_k.

The exergy flow diagram (Fig. 6.18) visually illustrates the exergy inventory of an open-steady system. Exergy is transported by mass and transferred by heat into the system. It is then transported out by mass, transferred out by useful work, and spontaneously destroyed due to generalized friction in the system's universe. Depending on what is considered the desired exergy output and the required exergy input, a device-specific exergetic efficiency can be defined from Eq. (6.28) as illustrated in Anim. 6.B.*turbineExergetics*, for example. In the following examples, we will examine the exergy inventory of several open-steady systems.

EXAMPLE 6-4 **Exergy Analysis of a Water Heater**

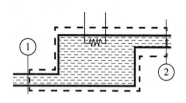

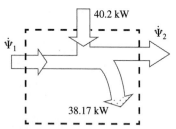

FIGURE 6.19 System schematic and exergy flow diagram for the water heater of Ex. 6-4 (click Exergy in Anim. 6.B.*openWaterHeater*).

An electric water heater supplies hot water at 150 kPa, 70°C with a flow rate of 10 L/min, as shown in Figure 6.19. The water temperature at the inlet is 15°C. Due to poor insulation, heat is lost at the rate of 2 kW to the surrounding atmosphere at 100 kPa, 25°C. Using the SL model for water, and neglecting ke and pe changes, determine (a) the electrical power consumption, (b) energetic efficiency, (c) exergetic efficiency, (d) rate of exergy destruction, and (e) reversible power. Assume no pressure loss in the system.

SOLUTION

Continue the analysis of Example 4-1 to include an exergy inventory over the heater's universe, enclosed within the red boundary of Figure 6.19.

Assumptions

The heater is at steady state. Flow states at the inlet and exit, States-1 and 2, are uniform and at LTE.

Analysis

In Example 4-1, we have already solved parts (a) and (b) and determined:

$$\dot{m} = 0.166\,\frac{\text{kg}}{\text{s}}; \quad \dot{W}_{\text{ext}} = -40.2\,\text{kW}; \quad \eta_{\text{I}} = 95.02\%; \quad \dot{S}_{\text{gen,univ}} = 0.128\,\frac{\text{kW}}{\text{K}}$$

The exergy gained by the stream can be evaluated using the SL model as follows:

$$\psi_2 - \psi_1 = [(h_2 - h_0) - T_0(s_2 - s_0) + \text{ke}_2 + \text{pe}_2] - [(h_1 - h_0) - T_0(s_1 - s_0) + \text{ke}_1 + \text{pe}_1]$$

$$= (h_2 - h_1) - T_0(s_2 - s_1) = (u_2 - u_1) + p(v_2 - v_1) - T_0(s_2 - s_1)$$

$$= c_v(T_2 - T_1) - T_0 c_v \ln\frac{T_2}{T_1} = 4.184\left[(70 - 15) - 298\ln\frac{343}{288}\right] = 12.21\,\text{kJ/kg}$$

$$\Rightarrow \quad \dot{m}(\psi_2 - \psi_1) = (0.166)(12.21) = 2.027\,\text{kW}$$

The exergetic efficiency, therefore, is calculated as

$$\eta_{\text{II}} = \frac{\text{Desired Exergy Output}}{\text{Required Exergy Input}} = \frac{2.027}{40.2} = 5.04\%$$

The exergy equation for the single-flow device, Eq. (6.27), can be used to obtain the rate of exergy destruction:

$$\dot{I} = \dot{m}(\psi_1 - \psi_2) + \dot{Q}_0\left(1 - \frac{T_0}{T_0}\right)^0 - \dot{W}_u = -2.027 - (-40.2) = 38.17\,\text{kW}$$

Therefore, the reversible work is:

$$\dot{W}_{\text{rev}} = \dot{W}_u + \dot{I} = \dot{m}(\psi_1 - \psi_2) = -40.2 + 38.17 = -2.027\,\text{kW}$$

TEST Analysis

Continuing from the TEST analysis of Example 4-1, evaluate State-0 as the dead state from the known atmospheric conditions, 25 deg-C and 100 kPa. Click Super-Calculate to update exergy values at each calculated state. In the exergy panel, obtain Idot, Wdot_rev, and Psidot_net to verify the manual solution.

Discussion

The exergetic efficiency for the water heater works out to be the ratio of the reversible work to the actual useful work, meaning that only 5.04% of the actual work (which is equal to the reversible work) would have been required to raise the water temperature by the same amount—if a reversible device could be used instead of electrical heating. The analysis does not tell us how to build such a device, but establishes its feasibility from a thermodynamic standpoint.

EXAMPLE 6-5   Exergy Analysis of a Vapor Turbine

A steam turbine operates steadily with a mass flow rate of 5 kg/s. The steam enters the turbine at 500 kPa, 400°C and exits at 7.5 kPa with a quality of 0.95. Neglecting the changes in ke and pe, as well as any heat loss from the turbine to the surroundings, which is at 100 kPa and 25°C, determine (a) the reversible work output and (b) exergetic efficiency. If the cost of exergy is valued at $0.10 per kWh, (c) determine the daily cost of exergy destruction in the turbine's universe. **What-if scenario:** (d) What would the exergetic efficiency be if the turbine were isentropic?

SOLUTION

Continue the analysis of Example 4-7 to include an exergy inventory over the turbine's universe enclosed within the red boundary in Figure 6.20.

Assumptions

Steady state, PC model for steam, uniform states based on LTE at the inlet and exit, and negligible changes in ke and pe.

Analysis

Evaluate the dead state (State-0) and the specific flow exergy at the inlet and exit states, State-1 and State-2:

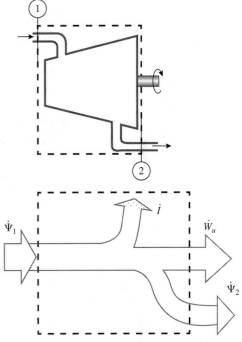

FIGURE 6.20 System schematic and the exergy flow diagram for Ex. 6-5 (see Anim. 4.A.*turbine*).

State-0 (given p_0, T_0):

$$j_1 \cong h_0 = 104.98 \,\frac{kJ}{kg}; \quad s_0 = 0.3673 \,\frac{kJ}{kg \cdot K};$$

$$\psi_1 = (h_1 - h_0) - T_0(s_1 - s_0) + ke_1^{0} + pe_1^{0}$$

$$= (3271.8 - 104.98) - (298)(7.7937 - 0.3673) = 953.8 \text{ kJ/kg}$$

$$\psi_2 = (h_2 - h_0) - T_0(s_2 - s_0) + ke_2^{0} + pe_2^{0}$$

$$= (2454.5 - 104.98) - (298)(7.8682 - 0.3673) = 114.3 \text{ kJ/kg}$$

Using the exergy balance equation, Eq. (6.28), we obtain

$$\dot{W}_{rev} = \dot{W}_u + \dot{I} = \dot{m}(\psi_1 - \psi_2) = 5(953.8 - 114.3) = 4197.5 \text{ kW}$$

The actual work output was evaluated in Example 4-7 as 4086.5 kW; therefore, the exergetic efficiency of the turbine is:

$$\eta_{II} = \frac{\text{Desired Exergy Output}}{\text{Required Exergy Input}} = \frac{\dot{W}_u}{\dot{m}(\psi_1 - \psi_2)} = \frac{\dot{W}_u}{\dot{W}_{rev}} = \frac{4086.5}{4197.5} = 97.35\%$$

The rate of exergy destruction in the turbine's universe is:

$$\dot{I} = \dot{m}(\psi_1 - \psi_2) - \dot{W}_u = 4197.5 - 4086.5 = 111.0 \text{ kW}$$

The daily cost of exergy destruction can now be calculated:

$$(111.0 \text{ kW})\left(\frac{24 \text{ h}}{\text{day}}\right)\left(\frac{\$0.1}{\text{kW} \cdot \text{h}}\right) = \$266.4$$

TEST Solution

Continuing the TEST analysis of Example 4-7, evaluate the dead state, State-0, and click Super-Calculate to update all calculated states. In the exergy panel, different terms of the exergy balance equations are evaluated. The exergetic efficiency can be manually calculated from these results.

What-if scenario

Make x2 an unknown (click its checkbox twice), enter s2 = s1, and click Super-Calculate. Wdot_rev and Wdot_u are calculated to be equal for the isentropic turbine. The exergetic efficiency is 100%.

Discussion

The exergetic efficiency (see Anim. 6.B.*turbineExergetics*) is not required to be equal to the isentropic efficiency, which was calculated as 97.2% in Example 4-7. Note that the isentropic work, 4203.5 kW in Example 4-7, is slightly higher than the reversible work; this is because the reversible device operates with the actual exit conditions and has a higher entropy than the isenentropic exit state. For an isentropic turbine, which must also be reversible, the two efficiencies are both 100%.

EXAMPLE 6-6 Exergy Analysis of a Throttling Valve

As shown in Figure 6.21, R-134a enters the throttling valve of a refrigeration system at 1.5 MPa and 50°C. The valve is set to create an exit pressure of 150 kPa. Determine the exergy destruction per-unit-mass of the refrigerant. Assume the atmospheric conditions to be 25°C, 100 kPa.

Solution

Continue the analysis of Example 4-11 to include an exergy inventory over the expansion valve's universe, enclosed within the red boundary in Figure 6.21.

Assumptions

Steady state, PC model for R-134a, uniform states based on LTE at the inlet and exit, and negligible changes in ke and pe.

Analysis

With no heat transfer and external work transfer, the energy equation yields $j_1 = j_2$, and the exergy equation, Eq. (6.28), can be simplified to express the desired exergy destruction rate:

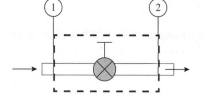
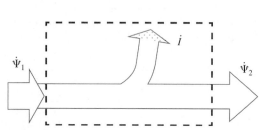

$$\frac{\dot{I}}{\dot{m}} = (\psi_1 - \psi_2) = (j_1 - j_2) - T_0(s_1 - s_2) = -T_0(s_1 - s_2)$$

$$= -(298)(0.439 - 0.4856) = 13.9 \text{ kJ/kg}$$

FIGURE 6.21 System schematic and exergy diagram for Ex. 6-6 (see Anim. 4.A.*expansionValve*).

With the entropy generation rate already calculated in Example 4-11, we could have obtained the same result from:

$$\frac{\dot{I}}{\dot{m}} = \frac{T_0 \dot{S}_{\text{gen,univ}}}{\dot{m}} = 298(0.0466) = 13.9 \text{ kJ/kg}$$

TEST Solution

Extend the TEST analysis of Example 4-11 by evaluating the dead state (State-0). Click Super-Calculate to update all calculated states for exergy. In the exergy panel, different terms of the exergy equation are evaluated from which the manual results can be verified.

Discussion

Exergy destruction occurs in the throttling device even though the flow energy remains constant. From Eq. (6.28), it can be seen that the reversible work per-unit-mass of the refrigerant must also be 13.9 kJ/kg. This means that the same exit state could be created while producing useful work at the rate of 13.9 kJ for every kg of R-134a if the throttling valve could be replaced with a reversible device.

EXAMPLE 6-7   Exergy Analysis of a Mixing System

Superheated ammonia at 200 kPa, $-10\,°C$ enters an adiabatic mixing chamber with a flow rate of 2 kg/s where it mixes with a flow of saturated mixture of ammonia at a quality of 20%. The mixture leaves as a saturated vapor without any loss of pressure. Neglecting any change in ke or pe, determine (a) the rate of exergy destruction and (b) reversible power. Assume the atmosphere to be at 25°C.

SOLUTION

Continue the analysis of Example 4-13 to include the exergy inventory of the mixing device, which is enclosed within the red boundary in Figure 6.22.

Assumptions

Steady state, PC model for ammonia, no pressure drop in the chamber so that $p_1 = p_2 = p_3$, uniform states based on LTE at the inlet and exit of each device.

Analysis

The exergy equation, Eq. (6.10), can be simplified for the adiabatic, steady, mixing chamber:

$$\frac{d\cancel{\Phi}^0}{dt} = \sum_i \dot{m}_i \psi_i - \sum_e \dot{m}_e \psi_e + \sum_k \dot{\cancel{Q}}_k^0 \left(1 - \frac{T_0}{T_k}\right) - \dot{\cancel{W}}_u^0 - \dot{I}$$

$$\Rightarrow \quad \dot{I} = \dot{m}_1 \psi_1 + \dot{m}_2 \psi_2 - (\dot{m}_1 + \dot{m}_2)\psi_3 = \dot{m}_1(\psi_1 - \psi_3) + \dot{m}_2(\psi_2 - \psi_3)$$

From the properties already evaluated in Example 4-13, we obtain:

$$\psi_1 - \psi_3 = (j_1 - j_3) - T_0(s_1 - s_3) = (1440.3 - 1419.5) - (298)(5.677 - 5.599)$$

$$= -2.444 \text{ kJ/kg} \quad \text{and} \quad \psi_2 - \psi_3 = (j_2 - j_3) - T_0(s_2 - s_3)$$

$$= (359.16 - 1419.5) - (298)(1.428 - 5.599) = 182.62 \text{ kJ/kg}$$

Therefore,

$$\dot{I} = \dot{m}_1(\psi_1 - \psi_3) + \dot{m}_2(\psi_2 - \psi_3) = (2)(-2.444) + (0.039)(182.62) = 2.23 \text{ kW}$$

and $\quad \dot{W}_{rev} = \dot{\cancel{W}}_u^0 + \dot{I} = 2.23 \text{ kW}$

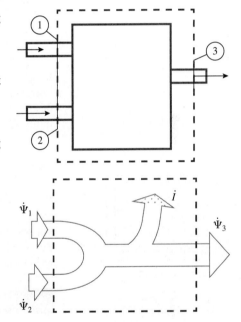

FIGURE 6.22 System schematic and exergy diagram for Ex. 6-7 (see Anim. 4.C.mixingChamber).

TEST Solution

Extend the TEST analysis of Example 4-13 by evaluating the dead state, State-0. Click Super-Calculate to update all calculated states for exergy. In the I/O panel, evaluate the expression '= mdot1*psi1 + mdot2*psi2 − mdot3*psi3' as 2 kW. The discrepancy between the manual and TEST results suggests that more significant digits should be used in an exergy analysis, especially for entropy values.

Discussion

The exergy analysis reveals that 2.23 kW of reversible power, which is entirely wasted through exergy destruction, could be tapped without changing the inlet or exit conditions or violating any fundamental law. Although the analysis does not tell us how to accomplish this, it does guide an inventor by establishing what is thermodynamically possible. Note that mixing or non-mixing open-steady TESTcalcs do not have an exergy panel.

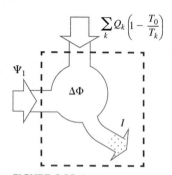

FIGURE 6.23 An energy flow diagram explaining Eq. (6.30). The balloon represents accumulation (or depletion) of stored exergy during the process.

6.4 CLOSED PROCESSES

In an unsteady system, we are often interested in a process where the system passes from a unique beginning state (b-state) to a unique final state (f-state). Various unsteady processes were analyzed in Chapter 5 with the help of mass, energy, and entropy balance equations. We now extend this analysis to include exergy accounting for a closed process (Fig. 6.23).

Following the development of energy and entropy equations for a closed process in Chapter 5, the exergy equation, Eq. (6.10), is multiplied by dt and integrated across the duration of the process, from the b-state to f-state, yielding:

$$d\Phi = \sum_k \dot{Q}_k dt \left(1 - \frac{T_0}{T_k}\right) - \dot{W}_u dt - \dot{I} dt;$$

$$\Rightarrow \quad \int_b^f d\Phi = \int_b^f \sum_k \dot{Q}_k dt \left(1 - \frac{T_0}{T_k}\right) - \int_b^f \dot{W}_u dt - \int_b^f \dot{I} dt$$

$$\Rightarrow \quad \underbrace{\Delta\Phi = \Phi_f - \Phi_b}_{\substack{\Phi_{net}: \text{ Change in} \\ \text{stored exergy.}}} = \underbrace{\sum_k Q_k \left(1 - \frac{T_0}{T_k}\right)}_{\substack{\text{Exergy transferred in} \\ \text{by heat.}}} - \underbrace{W_u}_{\substack{\text{Exergy} \\ \text{transferred out} \\ \text{through useful} \\ \text{external work.}}} - \underbrace{I}_{\substack{\text{Exergy} \\ \text{destroyed.}}} ; \quad \text{[kJ]} \quad \textbf{(6.30)}$$

In this customized exergy balance equation for a closed process, the left side represents the net exergy stored in the system (due to exergy transfer by heat), exergy transfer by work, and exergy destruction through thermodynamic friction (irreversibilities). Extending the concept of reversibility from Eq. (6.16) to a process, Eq. (6.30) can be written as:

$$W_{rev} = -\Delta\Phi + \sum_k Q_k \left(1 - \frac{T_0}{T_k}\right); \quad \text{[kJ]}, \quad \text{where} \quad W_{rev} = W_u + I; \quad \text{[kJ]} \quad \textbf{(6.31)}$$

It is evident from Eq. (6.31) that the reversible work output equals the net decrease in stored exergy for an adiabatic process. Depending on what can be considered the desired exergy output and the required exergy input, process-specific exergetic efficiency can be defined using Eq. (6.30) just as in the case of steady devices. Let's consider a few examples to illustrate exergy accounting in a closed process.

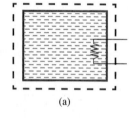

(a)

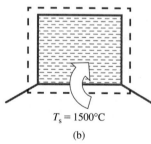

$T_s = 1500°C$

(b)

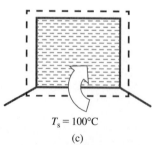

$T_s = 100°C$

(c)

FIGURE 6.24 The three heating options of Ex. 6-8 (also see Anim. 5.A.*waterHeater*).

EXAMPLE 6-8   **Exergy Analysis of a Closed Water Heater**

Three different options are being considered to raise the temperature of 1000 L of water stored in an insulated tank from 15°C to 70°C. (a) Use of electrical resistance heating, (b) heat transfer from a source at 1500°C, and (c) heat transfer from a source at 100°C. Based on the exergetic efficiency, evaluate the best option. Assume atmospheric conditions to be 100 kPa, 25°C (Fig. 6.24).

SOLUTION

Perform exergy analyses for the three configurations and compare their exergetic efficiencies to determine the best alternative.

Assumptions

The water in the tank is in equilibrium at the beginning and final states. The SL model for liquid water is applicable. Changes in KE or PE, if any, can be neglected.

Analysis

The material properties for water, ρ and c_v, are obtained from Table A-1, or any SL TESTcalc, as 997 kg/m^3 and 4.184 kJ/kg · K, respectively.
    The mass of water in the tank is:

$$m = \rho V = 997(1000 \times 10^{-3}) = 997 \text{ kg}$$

The energy equation, Eq. (5.2), for the closed process produces the electrical work necessary for the heating process:

$$\Delta E = Q - W_{ext}; \quad \Rightarrow \quad \Delta U = \cancel{Q}^0 - W_{el};$$

$$\Rightarrow \quad W_{el} = -\Delta U = -mc_v\Delta T = -(997)(4.184)(70 - 15) = -229{,}430 \text{ kJ}$$

Similarly, for (b) and (c), the energy equation yields:

$$\Delta E = Q - \cancel{W_{ext}}^0; \quad \Rightarrow \quad \Delta U = Q;$$

$$\Rightarrow \quad Q = mc_v\Delta T = (997)(4.184)(70 - 15) = 229{,}430 \text{ kJ}$$

The same amount of energy transfer is required, regardless of the source temperature in (b) and (c).

The anchor states (*b* and *f*) are identical for each process, so the gain in the stored exergy, $\Delta\Phi$, of the system must be the same for all processes. Using the definition of Φ from Eq. (6.10) and the SL-model formulas (Sec. 3.3.3), we obtain:

$$\Delta\Phi = \Delta U + \cancel{\Delta KE}^0 + \cancel{\Delta PE}^0 - T_0\Delta S + p_0\cancel{\Delta V}^0$$

$$\Rightarrow \quad \Delta\Phi = mc_v(T_2 - T_1) - T_0 mc_v \ln\frac{T_2}{T_1} = 12{,}169 \text{ kJ}$$

The exergy supplied in (a) is the electrical work, 229,430 kJ. In (b), the exergy supplied by heat is given by:

$$Q\left(1 - \frac{T_0}{T_s}\right) = (229{,}430)\left(1 - \frac{25 + 273}{1500 + 273}\right) = 190{,}868 \text{ kJ}$$

And in (c) it is:

$$Q\left(1 - \frac{T_0}{T_s}\right) = (229{,}430)\left(1 - \frac{25 + 273}{100 + 273}\right) = 46{,}132 \text{ kJ}$$

The exergetic efficiencies of the three options are $12{,}169/229{,}430 = 5.3\%$, $12{,}169/190{,}868 = 6.4\%$, and $12{,}169/46{,}132 = 26.4\%$, respectively. The last option, utilizing the highest fraction of the supplied exergy (which translates to cost), is the best choice.

TEST Analysis

Launch the SL closed-process TESTcalc and select Water (L) from the working substance menu. Evaluate State-0, State-1, and State-2 from the given properties as described in TEST-code (see *TEST > TEST-codes*). In the process panel, set up Process-A with State-1 and State-2 as the *b*- and *f*-states, enter Q = 0, and click Calculate. The exergy destroyed is found in the exergy panel. Now set up Process-B with the same anchor states and enter W_O = 0 and T_B = 1500 deg-C. In the exergy panel, set T1 = 1500 deg-C and Q1 = 229,430 (calculated in the process panel). Click Calculate to find the exergy destroyed. Follow the same procedure for process-C. Notice that the exergy analysis displayed on the exergy panel applies to the selected process in the process panel.

Discussion

For the two heating sources evaluated in this example, the one with a lower $\Delta T = T_s - T_u$ (T_u is the utilization or desired temperature while T_s is the source temperature) has the higher exergetic efficiency, hence, lower fuel costs. This is shown in the qualitative-cost estimation plot of Figure 6.25. A smaller value of ΔT requires a larger contact area for the same amount of heat transfer in a given time. Therefore, the size of the heat exchanger, and the capital cost, will increase as ΔT is reduced. The total cost, which is the sum of the capital and fuel costs, assumes a typical U-shaped (rather than V-shaped) behavior on the cost curve. As can be seen in Figure 6.25, there is an optimum zone instead of an optimal point. A designer can choose a point slightly towards the left or right, changing the distribution of cost between capital and fuel costs without significantly affecting the overall cost. An exergy analysis, therefore, is desirable in optimizing a thermal system.

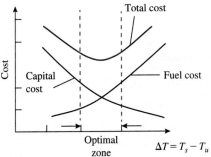

FIGURE 6.25 Fuel cost increases and the capital cost decreases as the difference between the source and utilization temperature increases.

EXAMPLE 6-9 Indirect Evaluation of Entropy Generation in an Inelastic Collision

A rubber ball of mass m is dropped from a height h onto a rigid floor. It bounces up and down several times and finally comes to rest on the floor. Derive an expression for the entropy generation in the universe due to this irreversible phenomenon. Assume the atmospheric temperature to be T_0.

SOLUTION

Exergy destruction being directly related to the entropy generation in a system's universe, the exergy balance equation for the inelastic process can be solved to obtain $S_{gen,univ}$.

Assumptions

The heat generated during this closed process, the ball going from its initial state to the resting state, is dissipated into the atmosphere.

Analysis

Observing that the only difference between the beginning and final state of the ball is the elevation, Eq. (6.30) can be simplified to obtain:

$$S_{gen,univ} = \frac{\Phi_b - \Phi_f - \cancel{W_u^0}}{T_0} = -\frac{\Delta\Phi}{T_0}$$

$$= -\frac{1}{T_0}\left[\cancel{\Delta U^0} - T_0\cancel{\Delta S^0} + p_0\cancel{\Delta V^0} + \cancel{\Delta KE^0} + \Delta PE\right]$$

$$= -\frac{1}{T_0}[PE_f - PE_b] = -\frac{1}{T_0}\frac{mg(z_f - z_b)}{1000}$$

$$= \frac{1}{T_0}\frac{mgh}{1000}\frac{kJ}{K}$$

Discussion

For many mechanical systems, the thermodynamic equilibrium remains undisturbed so that none of the thermodynamic properties change during mechanical processes. With thermodynamic properties u, v, and s remaining constant, Eq. (6.10) produces $\Delta\phi = \Delta ke + \Delta pe$. If there are no irreversibilities, ϕ must remain constant, resulting in the familiar mechanical energy equation $\Delta ke + \Delta pe = 0$.

EXAMPLE 6-10 Exergy Analysis of a Non-uniform, Non-mixing Closed Process

A 40-kg aluminum block at 100°C is dropped into an insulated tank containing an unknown mass of 20°C liquid water. After thermal equilibrium is established, the temperature is measured as 22°C. Determine (a) the change of stored exergy in the combined system, and (b) exergy destroyed during this process. Assume atmospheric temperature to be 25°C.

SOLUTION

Continue the analysis of Example 5-10 to include an exergy accounting of the combined system enclosed within the red boundary in Figure 6.26.

Assumptions

Subsystems A and B, in Figure 6.26, are uniform and in equilibrium at the beginning and final states. The SL model is applicable to both working substances. No changes in KE and PE. No heat transfer between the tank and the surroundings.

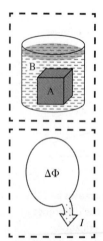

FIGURE 6.26 System schematic and exergy diagram for Ex. 6-10 (see Anim. 5.B.*blockInWater*).

Analysis

The mass of the liquid water, subsystem-B, has been evaluated in Example 5-10 as $m_B = 335.6$ kg.

The change of exergy of the combined system can be evaluated using the SL model as:

$$\Delta\Phi = \Delta U^0 + \Delta KE^0 + \Delta PE^0 - T_0\Delta S + p_0\Delta V^0$$

$$= -T_0[\Delta S_A + \Delta S_B] = -T_0\left[m_A c_{v,A}\ln\frac{T_f}{T_{bA}} + m_B c_{v,B}\ln\frac{T_f}{T_{bB}}\right]$$

$$= -(298)\left[(40)(0.9)\ln\frac{295}{373} + (335.6)(4.184)\ln\frac{295}{293}\right]$$

$$= -329.7 \text{ kJ}$$

The exergy balance equation, Eq. (6.29), applied over the composite system yields:

$$\Delta\Phi = \sum_k Q_k^0\left(1 - \frac{T_0}{T_k}\right) - W_u^0 - I$$

$$\Rightarrow \quad I = 329.7 \text{ kJ}$$

TEST Solution

Note that the exergy destroyed can also be obtained from $I = T_0 S_{\text{gen,univ}}$ with $S_{\text{gen,univ}}$ already evaluated in Example 5-10. Obtain Sdot_gen by using the TEST solution of Example 5-10 and multiply it with T_0 to verify the manual result.

Discussion

As the composite system comes to thermal equilibrium, the combined exergy of the system decreases. This is due to the destruction of exergy by thermal friction.

6.5 OPEN PROCESSES

The exergy equation can be simplified for an open process by using the *uniform-flow, uniform-state* assumption (Sec. 5.1.7). Multiplying the exergy equation, Eq. (6.10), by dt and integrating from the *b*-state to the *f*-state, we obtain an equation similar to Eq. (6.30) with the following additional transport terms:

$$\int_b^f \sum_i \dot{m}_i\psi_i\,dt - \int_b^f \sum_e \dot{m}_e\psi_e\,dt = \sum_i \psi_i\int_b^f \dot{m}_i\,dt - \sum_e \psi_e\int_b^f \dot{m}_e\,dt = \sum_i \psi_i m_i - \sum_e \psi_e m_e; \quad \text{[kJ]}$$

The open-process exergy equation can, therefore, be expressed as:

$$\underbrace{\Delta\Phi = \Phi_f - \Phi_b}_{\substack{\Phi_{\text{net}}:\text{ Change in stored}\\ \text{exergy.}}} = \underbrace{\sum_i m_i\psi_i - \sum_e m_e\psi_e}_{\text{Exergy transported into the system.}} + \underbrace{\sum_k Q_k\left(1 - \frac{T_0}{T_k}\right)}_{\text{Exergy added by heat.}}$$

$$- \underbrace{W_u}_{\substack{\text{Exergy transferred out}\\ \text{of the system through}\\ \text{useful external work.}}} - \underbrace{I}_{\substack{\text{Exergy}\\ \text{destroyed.}}} ; \quad \text{[kJ]} \qquad \textbf{(6.32)}$$

Comparing this equation with Eq. (6.20), we see that the first two terms on the right-hand side are the only new terms; they represent the net exergy transported into the system by the flows. The similarity of this equation with the corresponding energy and entropy equations, [Eqs. (5.21) and (5.22)], makes exergy analysis similar to the energy and entropy analysis carried out in Chapter 5 for several open processes. In fact, if the entropy generation is already known from an entropy analysis, it may be simpler to obtain the exergy destruction term directly from $I = T_0 S_{\text{gen,univ}}$.

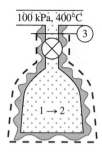

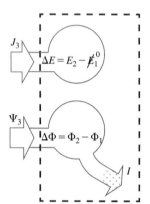

FIGURE 6.27 System schematic and exergy diagram for Ex. 6-11 (see Anim. 5.E.*charging*).

EXAMPLE 6-11 **Exergy Analysis of a Charging Process**

A 1-m^3 evacuated-and-insulated, rigid tank is being filled from a supply line carrying steam at 100 kPa and 400°C to a pressure of 100 kPa. Determine (a) the initial exergy and (b) the final exergy stored in the tank. Also, determine (c) the exergy destroyed during the process. Use the PG model for steam and assume the atmospheric conditions to be 100 kPa, 25°C.

SOLUTION

Continue the analysis of Example 5-13 to include an exergy accounting of the tank shown in Figure 6.27 as it is charged.

Assumptions

The process is assumed to be a uniform state, uniform flow process with the b- and f-states represented by State-1 and State-2, and the inlet state by State-3, chosen at the valve's supply side. Neglect KE and PE.

Analysis

In Example 5-13, we already evaluated the mass of steam in the tank as $m_2 = 0.243$ kg and the final temperature as $T_2 = 894$ K.

The stored exergy of the vacuum in State-1 may appear to be zero since vacuum has no mass ($m_1 = 0$) and $\Phi_1 = m_1\phi_1$. But we know that there is useful energy in an evacuated chamber. The surrounding air will rush in when the valve is opened and it can be exploited to produce useful work. From the expression of ϕ in Eq. (6.10), it can be seen that, as $m_1 \to 0$, $v_1(= V_1/m_1)$ and, consequently, ϕ_1 approach infinity; therefore, we must re-evaluate Φ in the limit of $m_1 \to 0$:

$$\Phi_1 = m_1\phi_1 = m_1[(u_1 - u_0) - T_0(s_1 - s_0) + p_0(v_1 - v_0) + \mathrm{ke}_1 + \mathrm{pe}_1]$$

$$= m_1^{0}[(u_1 - u_0) - T_0(s_1 - s_0) - p_0v_0 + \mathrm{ke}_1 + \mathrm{pe}_1] + \lim_{m_1 \to 0} m_1 p_0\left(\frac{V_1}{m_1}\right)$$

$$= p_0 V_1 = (100)(1) = 100 \text{ kJ}$$

Exergy stored in steam at the f-state can be calculated in a straightforward manner using the definition of ϕ and applying the PG model formulas (Sec. 3.5.4):

$$\Phi_2 = (u_2 - u_0) - T_0(s_2 - s_0) + p_0(v_2 - v_0) + \mathrm{ke}_2^{0} + \mathrm{pe}_2^{0}$$

$$= c_v(T_2 - T_0) - T_0\left[c_p \ln\frac{T_2}{T_0} - R\ln\frac{p_2}{p_0}\right] + p_0R\left(\frac{T_2}{p_2} - \frac{T_0}{p_0}\right)$$

$$= c_v(T_2 - T_0) - T_0c_p \ln\frac{T_2}{T_0} + R(T_2 - T_0)$$

$$= (1.406)(894 - 298) - (298)(1.868)\ln\frac{894}{298} + (0.461)(894 - 298)$$

$$= 501.2 \text{ kJ}$$

Therefore,

$$\Phi_2 = m_2\phi_2 = (0.243)(501.2) = 121.8 \text{ kJ}$$

With the entropy generated by the process already calculated in Example 5-13, the exergy destruction can be obtained as:

$$I = T_0 S_{\mathrm{gen,univ}} = (298)(0.129) = 38.4 \text{ kJ}$$

TEST Solution

Recreate the TEST solution of Example 5-13 from the TEST-code. Evaluate the dead state, State-0, and click Super-Calculate. To correctly predict phi1 use p1 = 0.0001 kPa and m1 = 0.0001 kg (small numbers instead of zeros to obtain the limiting value). Calculate Phi1 and Phi2 by evaluating the expressions '= m1*phi1' and '= m2*phi2' in the I/O panel.

Discussion

The exergy destroyed during the process can also be evaluated from the exergy balance equation if the exergy transported is calculated as $m_i\psi_i = m_2\psi_3 = 121.6$ kJ and substituted in Eq. (6.32).

6.6 CLOSURE

In this chapter we further developed the concept of flow and stored exergy, introduced in Chapter 1, for non-reacting systems. The exergy balance equation was developed by manipulating the mass, energy, and entropy equations. Each term of the equation was explored for its physical meaning, leading to the definition of reversible work, exergy destruction, and exergetic efficiency. All four types of systems covered in the previous chapters—closed-steady and open-steady systems, and unsteady systems executing closed and open processes—were analyzed for their exergy inventory. The similarity of the exergy balance equation with the mass, energy, and entropy equations was leveraged by simplifying the exergy equation for a given system or process. Chemical contributions to exergy will be discussed in Chapter 13.

PROBLEMS

SECTION 6-1: EXERGY BALANCE EQUATION

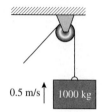

6-1-1 A crane lifts a 1000-kg load vertically at a speed of 0.5 m/s. Assuming no destruction of exergy, determine (a) the rate at which exergy is transferred into the system (load) and (b) the rate of change of exergy ($d\Phi/dt$) of the system.

6-1-2 A closed system operating at steady state receives 5000 kW of heat from a source at 1500 K, produces 2000 kW of useful power, and rejects the remaining heat into the atmosphere. Determine the rate at which exergy (a) enters and (b) leaves the system. Assume the atmospheric temperature to be 300 K.

6-1-3 Heat is transferred from a TER at 1500 K to a TER (thermal energy reservoir) at 400 K. The rate of transfer is 10 kW. If the atmospheric temperature is 300 K, determine the rate at which exergy (a) leaves the TER at higher temperature and (b) enters the TER at lower temperature. (c) How do you explain the discontinuity in the result?

6-1-4 Heat is transferred at a rate of 5 kW from a TER at 1500 K to a TER at 500 K. Determine the rate at which exergy (a) leaves the high-temperature TER, (b) enters the low-temperature TER and (c) is destroyed in the thermodynamic universe. Assume the atmospheric temperature to be 300 K.

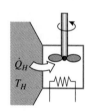

6-1-5 A closed system receives heat at a rate of 2 kW from a source at 1000 K, electrical power at a rate of 2 kW, and shaft power at a rate of 2 kW. Determine the net rate of (a) energy transfer and (b) exergy transfer into the system. (c) What is the maximum possible rate of exergy rise in the system?

6-1-6 Three kilograms of water undergo a process from an initial state where the water is saturated vapor at 140°C, the velocity is 40 m/s, and the elevation is 5 m to a final state where the water is saturated liquid at 15°C, the velocity is 25 m/s, and the elevation is 2 m. Determine the exergy (Φ) at (a) the initial state, (b) the final state and (c) the change in exergy ($\Delta\Phi$). Take $T_0 = 25°C$, $p_0 = 1$ atm.

6-1-7 A balloon filled with helium at 25°C, 100 kPa and a volume of 0.1 m^3 is moving at 10 m/s. The balloon is at an elevation of 1 km relative to an exergy reference environment of $T_0 = 20°$, $p_0 = 100$ kPa. Determine the specific exergy (ϕ) of helium.

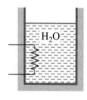

6-1-8 An insulated tank contains 20 kg of liquid water at the ambient condition of 100 kPa and 25°C. An internal electric heater is turned on. It consumes 20 kW of electric power. Determine the rate of change of (a) temperature (dT/dt) and (b) rate of change of exergy ($d\Phi/dt$) of the water in the tank. (c) How do you explain the fact that the rate of exergy increase is much less than the rate of exergy transfer into the system?

6-1-9 A rigid tank contains 20 kg of liquid water at 100 kPa and 25°C. Heat at a rate of 10 kW is transferred into the tank from a source at 1000 K. Determine (a) the rate of exergy transfer from the heating source, (b) the rate of increase of temperature, and (c) the rate of exergy change ($d\Phi/dt$) of the water in the tank. Assume the atmospheric temperature to be 300 K.

6-1-10 A rigid tank contains 1 kg carbon dioxide at 500 kPa and 200°C. Heat at a rate of 5 kW is transferred into the tank from a source at 200°C. Determine (a) the rate of exergy transfer from the heating source, (b) the rate of increase of temperature and (c) the rate of exergy of the carbon dioxide in the tank. Assume the atmospheric temperature to be 300 K.

6-1-11 A cylinder of an internal combustion engine contains 3000 cm^3 of gaseous products at a pressure of 10 bars and a temperature of 800°C just before the exhaust valve opens. Determine (a) the specific exergy (ϕ) of the gas and (b) the exergy stored (Φ) in the cylinder. Model the combustion products as air, a perfect gas. Take $T_0 = 25°C, p_0 = 1$ atm. (c) *What-if scenario:* What would the exergy stored in the cylinder be if the ideal gas model had been used for air?

6-1-12 A 1 m^3 tank contains air (use the IG model). (a) Plot how the total stored energy (E) and total stored exergy (Φ) of air in the tank change as the pressure is increased from 10 kPa to 10 MPa with temperature held constant at 25°C. (Assume atmospheric conditions to be 100 kPa, 25°C.) (b) Repeat the plot for the specific stored energy and specific stored exergy. (c) Plot how the stored energy and stored exergy of air in the tank change as the temperature is increased from −100°C to 1000°C with pressure held constant at 100 kPa. (d) Repeat the plot for the specific stored energy and specific stored exergy.

6-1-13 Determine (a) the exergy (Φ) of air at 150°C, 50 kPa in a 2 m^3 storage tank. (b) *What-if scenario:* What would the exergy be if the tank were filled with steam instead?

6-1-14 A granite rock ($\rho = 2700$ kg/m^3, specific heat $= 1.017$ kJ/kg·K), with $m = 5000$ kg, heats up to $T = 45°C$ during daytime due to solar heating. Assuming the surroundings to be at 20°C, determine (a) the maximum amount of useful work (W_u) that could be extracted from the rock. *What-if scenario:* What would the answer be if (b) the rock temperature was increased by 5°C or (c) the ambient temperature fell by 5°C?

6-1-15 Steam at 1 MPa, 500°C flows through a 10-cm-diameter pipe with a velocity of 25 m/s. Determine the rate of (a) exergy ($\dot{\Psi}$) transported by the flow. Use the PC model and assume atmospheric conditions to be 100 kPa and 25°C. (b) *What-if scenario:* What would the rate of exergy be if the IG model were used for steam?

6-1-16 A 20-cm-diameter pipe carries water at 500 kPa, 25°C and 30 m/s. (a) Determine the flow rate of exergy ($\dot{\Psi}$) through the pipe. Assume atmospheric conditions to be 100 kPa and 25°C. Use the SL model for water. (b) *What-if scenario:* What would the flow rate of exergy be if the kinetic energy was neglected?

SECTION 6-2: EXERGY ANALYSIS OF CLOSED SYSTEMS

Closed Steady Systems

6-2-1 A house is electrically heated by a resistance heater that draws 15 kW of electric power. The house is kept at a temperature of 20°C while the outside is 5°C. Assuming steady state, determine (a) the reversible power and (b) the rate of irreversibility.

6-2-2 Heat is conducted steadily through a house's 5 m × 10 m × 10 cm brick wall. When the temperature outside is −5°C, the temperature inside is maintained at 25°C. The temperature of the inner and outer surfaces of the wall are measured at 20°C and 0°C respectively. If the rate of heat transfer ($\dot{Q}$) is 1 kW, determine the rate of exergy destruction ($\dot{I}$) (a) in the wall and (b) in its universe.

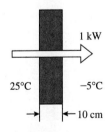

6-2-3 A refrigerator has a second-law efficiency of 45%, and heat is removed from it at a rate of 200 kJ/min. If the refrigerator is maintained at 2°C, while the surrounding air is at 27°C, determine the power input to the refrigerator.

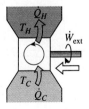

6-2-4 An air-conditioning system must transfer heat from a house at a rate of 800 kJ/min to maintain its temperature at 20°C while the outside temperature is 40°C. If the COP of the system is 3.7, determine (a) the power required for air conditioning the house and (b) the rate of exergy destruction ($\dot{I}$) in the universe.

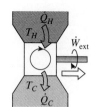

6-2-5 A heat engine receives heat from a source at 2000 K at a rate of 500 kW and rejects the waste heat to the atmosphere at 300 K. The net output from the engine is 300 kW. Determine (a) the reversible power output, (b) the rate of exergy input into the engine, (c) the rate of exergy destruction ($\dot{I}$) in the universe and (d) the exergetic (second-law) efficiency.

6-2-6 A heat engine produces 40 kW of power while consuming 40 kW of heat from a source at 1200 K, 50 kW of heat from a source at 1500 K, and rejecting the waste heat to the atmosphere at 300 K. Determine (a) the reversible power and (b) the rate of exergy destruction ($\dot{I}$) in the engine's universe.

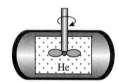

6-2-7 An insulated, rigid tank contains 1.5 kg of helium at 30°C and 500 kPa. A paddle wheel with a power rating of 0.1 kW, is operated within the tank for 30 minutes. Determine (a) the minimum work in which this process could be accomplished and (b) the exergy destroyed (I) in the universe during the process. Assume the surroundings to be at 100 kPa and 25°C.

6-2-8 An insulated, rigid tank contains 1.0 kg of air at 130 kPa and 20°C. A paddle wheel inside the tank is rotated by an external power source until the temperature in the tank rises to 54°C. If the surrounding air is at 20°C, determine (a) the exergy destroyed (I) and (b) the reversible work (W_{rev}). Use the IG model for air. (c) *What-if scenario:* What would the exergy destroyed be if the initial pressure had been 180 kPa?

6-2-9 A steam radiator (used for space heating) has a volume of 20 L and is filled with steam at 200 kPa and 250°C. The inlet and exit ports are then closed. As the radiator cools down to a room temperature of 20°C, determine (a) the heat transfer (Q) and (b) reversible work (W_{rev}). (c) *What-if scenario:* What would the reversible work be if the steam pressure in the radiator had been 400 kPa?

6-2-10 A piston cylinder device initially contains 10 ft^3 of helium gas at 25 psia and 40°F. The gas is then compressed in a polytropic process ($pv^{1.3}$ = constant) to 70 psia. Determine (a) the minimum work with which this process could be accomplished and (b) 2nd law efficiency. Assume the surroundings to be at 14.7 psia and 70°F.

6-2-11 A piston-cylinder device contains 0.1 kg of steam at 1.4 MPa and 290°C. The steam expands to a final state of 220 kPa and 150°C, doing boundary work. Heat losses from the system to the surroundings are estimated to be 4 kJ during this process. Assume the surroundings to be at 25°C and 100 kPa. Determine (a) the exergy change ($\Delta\Phi$) of steam and (b) the exergy destroyed (I) during the process.

6-2-12 Water, initially a saturated liquid at 95°C, is contained in a piston-cylinder assembly. The water undergoes a process to the corresponding saturated vapor state, during which the piston moves freely in the cylinder. The change in state is brought about adiabatically by the stirring action of paddle wheel. Determine, on a unit of mass basis, the (a) change in stored exergy ($\Delta\Phi$), (b) the exergy transfer accompanying work, (c) the exergy transfer accompanying heat and (d) the exergy destruction (I). Let $T_0 = 20$°C and $p_0 = 1$ bar.

6-2-13 An insulated piston cylinder device contains 20 L of O$_2$ (use the IG model) at 300 kPa and 100°C. It is then heated for 1 min by a 200 W resistance heater placed inside the cylinder. The pressure of O$_2$ remains constant during the process. Determine (a) the change in stored exergy ($\Delta\Phi$) and (b) the irreversibility during the process. Assume the surroundings to be at 100 kPa and 25°C. (c) *What-if scenario:* What was the answer in part (b) be if the volume was 50 L?

6-2-14 A piston-cylinder device initially contains 20 g of saturated water vapor at 300 kPa. A resistance heater is operated within the cylinder with a current of 0.4 A from a 240 V source until the volume doubles. At the same time a heat loss of 4 kJ occurs. Determine (a) the final temperature (T_2), (b) duration of the process and (c) second-law efficiency. Assume the surrounding temperature and pressure to be 20°C and 100 kPa. (d) *What-if scenario:* What would the efficiency be if the heat loss was negligible during the process?

6-2-15 A piston-cylinder device contains 0.1 kg of steam at 900 kPa and 320°C. The steam expands to a final state of 180 kPa and 135°C, doing work. Heat losses from the system to the surroundings are estimated to be 4 kJ during this process. Assuming the surroundings to be at 25°C and 100 kPa, determine (a) the exergy of the steam at the initial and final states, (b) the exergy change ($\Delta\Phi$) of steam, (c) the exergy destroyed (I) and (d) the exergetic (second-law) efficiency.

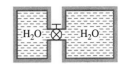

6-2-16 Two insulated tanks are connected, both containing H$_2$O. Tank-A is at 200 kPa, $v = 0.4$ m^3/kg, $V = 1$ m^3 and tank B contains 3.5 kg at 0.5 MPa, 400°C. The valve between the tanks is opened and the two tanks come to a uniform state. Determine (a) the final pressure (p_2), (b) temperature (T_2) and (c) irreversibility of the process. Assume the surroundings to be at 100 kPa and 25°C.

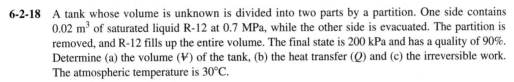

6-2-17 A 0.5 m^3 rigid tank, containing hydrogen at 40°C and 200 kPa, is connected to another 1 m^3 rigid tank containing hydrogen at 20°C and 600 kPa. The valve between the tanks is opened and the system is allowed to reach thermal equilibrium with the surroundings at 15°C. Determine the irreversibility in this process. Assume variable c_p.

6-2-18 A tank whose volume is unknown is divided into two parts by a partition. One side contains 0.02 m^3 of saturated liquid R-12 at 0.7 MPa, while the other side is evacuated. The partition is removed, and R-12 fills up the entire volume. The final state is 200 kPa and has a quality of 90%. Determine (a) the volume (V) of the tank, (b) the heat transfer (Q) and (c) the irreversible work. The atmospheric temperature is 30°C.

6-2-19 An insulated, rigid tank has two compartments, one 10 times as large as than the other, divided by a partition. At the beginning, the smaller side contains 4 kg of H$_2$O at 200 kPa, 90°C and the other side is evacuated. The partition is removed and the water expands to a new equilibrium condition. (a) Determine the irreversibility during the process. Assume the surrounding conditions to be 100 kPa and 25°C. (b) **What-if scenario:** What would the irreversibility be if the larger chamber was 100 times larger?

6-2-20 A 40 kg aluminum block at 90°C is dropped into an insulated tank that contains 0.5 m^3 of liquid water at 20°C. Determine the irreversibility in the resulting process if the surrounding temperature is 27°C.

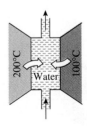

6-2-21 A 4 kg iron block, initially at 300°C, is dropped into an insulated tank that contains 80 kg of water at 25°C. Assume the water that vaporizes during this process will condense back into the tank. The surroundings are at 20°C and 100 kPa. Determine (a) the final equilibrium temperature (T_f), (b) the exergy (Φ) of the combined system at initial and final states and (c) the wasted work potential during this process.

SECTION 6-3: EXERGY ANALYSIS OF OPEN SYSTEMS

6-3-1 A 5 kW pump is raising water to an elevation of 25 m from the free surface of a lake. The temperature of the water increases by 0.1°C. Neglecting the KE, determine (a) the mass flow rate ($\dot{m}$), (b) the minimum power (magnitude only) required and (c) the exergetic (second-law) efficiency of the system. Assume the ambient temperature to be 20°C.

6-3-2 Measurements during steady state operation indicate that warm air exits a hand held hair dryer at a temperature of 90°C with a velocity of 10 m/s through an area of 20 cm^3. Air enters the dryer at 25°C, 100 kPa with a velocity of 3 m/s. No significant change in pressure is observed and no significant heat transfer between the dryer and its surroundings occurs. Determine (a) the external power and (b) the exergetic efficiency. Let $T_0 = 25°$, $p_0 = 1$ atm. Use the PG model.

6-3-3 A feedwater heater has water with a mass flow rate of 5 kg/s at 5 MPa, 40°C flowing through it. The water is being heated from two sources. One source adds 900 kW from a 100°C reservoir and the other source adds heat from a 200°C reservoir so that the water exit conditions are 5 MPa and 180°C. Determine the exergetic efficiency of the device. Let $T_0 = 25°$, $p_0 = 1$ atm

6-3-4 Argon gas enters an adiabatic compressor at 100 kPa, 25°C, 20 m/s and exits at 1 MPa, 550°C, 100 m/s. The inlet area of the compressor is 75 cm^2. Assuming the surroundings to be at 100 kPa and 25°C, determine (a) the reversible power ($\dot{W}_{rev}$), (b) irreversibility for this device and (c) the exergetic efficiency. (d) **What-if scenario:** What would the answers be if the compressor lost 5 kW of heat to the atmosphere due to poor insulation?

6-3-5 Refrigerant-134a is to be steadily compressed from 0.2 MPa and −5°C to 1 MPa and 50°C by an adiabatic compressor. If the environment's conditions are 20°C and 95 kPa, determine (a) the specific exergy change ($\Delta\phi$) of the refrigerant and (b) the minimum work input (magnitude only) that needs to be supplied to the compressor per unit mass of the refrigerant.

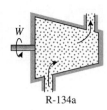

R-134a

6-3-6 Refrigerant-134a enters an adiabatic compressor as saturated vapor at 120 kPa at a rate of 1 m³/min and exits at 1 MPa. The compressor has an adiabatic efficiency of 85%. Assume the surrounding conditions to be 100 kPa and 25°C. Determine (a) the actual power ($\dot{W}_{rev}$) and (b) the second-law efficiency of the compressor. (c) *What-if scenario:* What would the actual power be if the compressor had an adiabatic efficiency of 70%?

6-3-7 Consider an air compressor that receives ambient air at 100 kPa and 25°C. It compresses the air to a pressure of 2 MPa and it exits at a temperature of 800 K. Since the air and compressor housing are both hotter than the ambient temperature, the compressor loses 80 kJ per kilogram air flowing through the compressor. Determine (a) the reversible work (W_{rev}) and (b) the irreversibility in the process. Use the IG model for air.

6-3-8 Carbon dioxide (CO_2) enters a nozzle at 35 psia, 1400°F, 250 ft/s and exits at 12 psia, 1200°F. Assume the nozzle to be adiabatic and the surroundings to be at 14.7 psia and 65°F. Determine (a) the exit velocity (V_2) and (b) the availability drop between the inlet and the exit. (c) *What-if scenario:* What would the exit velocity be if carbon dioxide entered the nozzle at 500 ft/s?

6-3-9 Steam enters a turbine with a pressure of 3 MPa, a temperature of 400°C, and a velocity of 140 m/s. Steam exits as saturated vapor at 100°C with a velocity of 105 m/s. At steady state, the turbine develops work at a rate of 500 kJ per kg of steam flowing through the turbine. Heat transfer between the turbine and its surroundings occurs at an average outer surface temperature of 450 K. Determine the irreversibility per unit mass of steam flowing through the turbine in kJ/kg. Neglect PE.

6-3-10 An insulated steam turbine, receives 25 kg of steam per second at 4 MPa and 400°C. At the point in the turbine where the pressure is 0.5 MPa, steam is bled off for processing equipment at a rate of 10 kg/s. The temperature of this steam is 230°C. The balance of steam leaves the turbine at 30 kPa, 95% quality. Determine (a) the specific flow exergy (ψ) at each port and (b) the exergetic efficiency of the turbine.

Steam

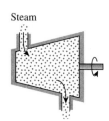

6-3-11 Steam enters an adiabatic turbine steadily at 6 MPa, 600°C, 50 m/s and exits at 50 kPa, 100°C, 150 m/s. The turbine produces 5 MW. If the ambient conditions are 100 kPa and 20°C, determine (a) the maximum possible power output, (b) the second-law (exergetic) efficiency and (c) the irreversibility. (d) *What-if scenario:* What would the irreversibility be at an ambient temperature of at 40°C?

6-3-12 Steam enters a turbine steadily at 2 MPa, 400°C, 6 kg/s and exits at 0.3 MPa, 150°C. The steam is losing heat to the surrounding air at 100 kPa and 25°C at a rate of 200 kW. Determine (a) the actual power output ($\dot{W}_{ext}$), (b) the maximum possible power output ($\dot{W}_{ext}$), (c) the second law efficiency and (d) the exergy destroyed ($\dot{I}$).

6-3-13 Steam enters an adiabatic turbine steadily at 8 MPa, 500°C, 50 m/s and exits at 30 kPa, 150 m/s. The mass flow rate is 1 kg/s, the adiabatic efficiency is 90%, and the ambient temperature is 300 K. Determine (a) the second law efficiency of the turbine. (b) *What-if scenario:* What is the second law efficiency if the adiabatic efficiency is 85%?

6-3-14 A steam turbine has inlet conditions of 5 MPa, 500°C and an exit pressure of 12 kPa. Assuming the atmospheric conditions to be 100 kPa and 25°C, plot how the exergetic efficiency of the turbine changes as the isentropic efficiency decreases from 100% to 75%.

R-12

6-3-15 Refrigerant-12 is throttled by a valve from the saturated liquid state at 800 kPa to a pressure of 150 kPa with a mass flow rate of 0.5 kg/s. Assuming the surrounding conditions to be 100 kPa and 25°C, determine (a) the rate of exergy destruction ($\dot{I}$) and (b) the reversible power.

6-3-16 Superheated water vapor enters a valve at 3445 kPa, 260°C and exits at a pressure of 551 kPa. Determine (a) the specific flow exergy (ψ) at the inlet and exit and (b) the rate of exergy destruction in the valve per unit of mass. Let $T_0 = 25°$, $p_0 = 1$ atm.

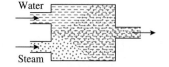

Water

Steam

6-3-17 Water at 140 kPa and 280 K enters a mixing chamber at a rate of 2 kg/s, where it is mixed steadily with steam entering at 140 kPa and 400 K. The mixture leaves the chamber at 140 kPa and 320 K. Heat is lost to the surrounding air at 20°C at a rate of 3 kW. Determine (a) the reversible work and (b) the rate of exergy destruction ($\dot{I}$).

6-3-18   Liquid water at 100 kPa, 10°C, 1 kg/s is heated by mixing it with an unknown amount of steam at 100 kPa and 200°C in an adiabatic mixing chamber. The surrounding atmosphere is at 100 kPa and 25°C. If the mixture leaves at 100 kPa and 50°C, determine (a) the mass flow rate ($\dot{m}$) of steam. Also determine the rate of transport exergy ($\dot{\Psi}$) (b) into and (c) out of the chamber.

6-3-19   Steam enters a closed feedwater heater at 1.1 MPa, 200°C and leaves as saturated liquid at the same pressure. Feedwater enters the heater at 2.5 MPa, 50°C and in an isobaric manner leaves 12°C below the exit temperature of the steam. Neglecting any heat losses, determine (a) the mass flow rate ($\dot{m}$) ratio and (b) the exergetic efficiency of the heat exchanger. Assume the surroundings to be at 100 kPa and 25°C.

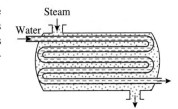

6-3-20   A 0.5 m³ tank initially contains saturated liquid water at 200°C. A valve on the bottom of the tank is opened and half the liquid is drained. Heat is transferred from a source at 300°C to maintain constant temperature inside the tank. Determine (a) the heat transfer (Q) and (b) the reversible work. Assume the surroundings to be at 25°C and 100 kPa.

6-3-21   A 100 m³ rigid tank, initially containing atmospheric air at 100 kPa and 300 K, is to be used as a storage vessel for compressed air at 2 MPa and 300 K. Compressed air is to be supplied by a compressor that takes in atmospheric air at 100 kPa and 300 K. Determine the reversible work (W_{rev}). Use the PG model for air.

6-3-22   A 0.2 m³ tank initially contains R-12 at 1 MPa and $x = 1$. The tank is charged to 1.2 MPa, $x = 0$ from a supply line that carries R-12 at 1.5 MPa and 30°C. Determine (a) the heat transfer (Q) and (b) the wasted work potential associated with the process. Assume the surrounding atmospheric temperature to be 50°C.

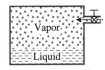

7 | RECIPROCATING CLOSED POWER CYCLES

Having analyzed cycles as a special case of closed-steady systems in Chapter 2, open-steady systems in Chapter 4, and closed processes in Chapter 5, we are now ready for detailed analyses of specific thermodynamic cycles. Thermodynamic cycles, idealized models for *heat engines*, *refrigerators*, and *heat pumps*, are generally divided into two classes: **power cycles**, executed by heat engines, are discussed in Chapters 7–9, and **refrigeration cycles**, executed by refrigerators and heat pumps, are covered in Chapter 10. Conversion of heat to work at a steady rate is the goal of any *heat engine*. In combustion engines, heat is supplied by burning fossil fuels. Depending on how the combustion takes place—externally as in a boiler, or internally as in a car engine—a combustion heat engine is called either an **external combustion** or **internal combustion** engine. Internal combustion engines are usually **reciprocating** in nature. The working fluid, generally a gas, trapped inside a piston-cylinder device executes a series of processes before being replaced by a fresh charge. Idealized power cycles used to model such reciprocating engines through a series of closed processes are called **closed power cycles**. In this chapter, several closed power cycles are constructed to model different types of reciprocating engines encountered in automobiles, locomotives, and ships. Because the working fluid is a gas, considerable simplification is achieved by using the perfect gas (PG) model in manual calculations. TEST solutions are used for verifying manual results, switching the gas model from the PG to the IG model for improved accuracy, and carrying out "what-if" studies.

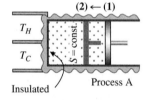

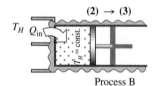

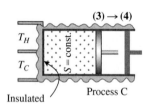

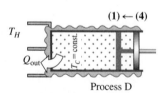

FIGURE 7.1 A gas going through a closed Carnot cycle (see Anim. 7.A.*CarnotClosed Cycle*).

7.1 THE CLOSED CARNOT HEAT ENGINE

In Sec. 2.3.3, we established that the reversible *Carnot heat engine*—a heat engine operating between two reservoirs at constant temperatures without any thermodynamic friction in its universe (no entropy generation)—is the most efficient of all heat engines. If an automobile engine could be built on the Carnot cycle, the gas mileage would be the highest possible. Even if such an engine is yet to be a reality, a concept Carnot engine can serve as a guide for a more fuel-efficient design of any practical heat engine cycle.

A conceptual implementation of the *Carnot cycle* as a closed cycle is shown in Figure 7.1 (also see Anim. 7.A.*CarnotClosedCycle*). As a brief explanation, consider a gas trapped in a piston-cylinder device with removable insulation so that the cylinder can be rendered adiabatic at a moment's notice. Also suppose that by selectively exposing the cylinder, through a highly conductive wall, to a low or high temperature reservoir (or **TER**: thermal energy reservoir), the working fluid can be maintained in an isothermal condition at the temperature of the reservoir, T_C or T_H. Four consecutive reversible processes make up the cycle as follows:

Reversible Adiabatic Compression (Process A: State-1 to State-2): The cylinder is completely insulated, and the gas is compressed adiabatically so that $Q_{12} = 0$. The process is reversible ($S_{\text{gen,univ}} = 0$); therefore, it must be isentropic (Sec. 5.1.3D). Summarizing:

$$s_1 = s_2; \quad \left[\frac{\text{kJ}}{\text{kg} \cdot \text{K}}\right]; \quad Q_{12} = 0; \quad [\text{kJ}] \tag{7.1}$$

External boundary work is necessary for the compression process and the temperature is expected to increase during this isentropic compression.

Reversible Isothermal Heat Addition (Process B: State-2 to State-3): The compressed gas is brought in contact with the reservoir at T_H and is allowed to expand in an isothermal manner. As derived in Sec. 5.1.3C, the heat and work transfer in this reversible process are given as

$$Q_{23} = W_{B,23} = mT_H(s_3 - s_2); \quad [\text{kJ}] \tag{7.2}$$

Note that for reversible heat transfer to occur from the reservoir to the working gas, T_H must be only differentially greater than T_2. To approach the reversible limit, the exposed surface area must be very large for any significant amount of heat transfer to occur in a reasonable amount of time (Sec. 2.1.3).

Reversible Adiabatic Expansion (Process C: State-3 to State-4): The cylinder is completely insulated and the gas is allowed to expand in a reversible manner. As in Process-A, the adiabatic and reversible conditions imply an isentropic process.

$$s_3 = s_4; \quad \left[\frac{\text{kJ}}{\text{kg} \cdot \text{K}}\right]; \quad Q_{34} = 0; \quad [\text{kJ}] \tag{7.3}$$

External boundary work is delivered during this expansion process and the temperature is expected to decrease according to the isentropic relations obeyed by the working fluid (Sec. 5.1.3D).

Reversible Isothermal Heat Rejection (Process D: State-4 to State-1): The rarefied gas is brought in contact with the TER at T_C. Heat is rejected in an isothermal manner until the gas returns to State-1, completing the cycle. As with Process-B, the heat and boundary work transfer can be expressed as

$$Q_{41} = W_{B,41} = mT_C(s_1 - s_4); \quad [\text{kJ}] \tag{7.4}$$

Consistent with the WinHip sign convention, Q_{41} and $W_{B,41}$ are negative since $s_1 < s_4$. Note that the expressions for heat or work transfer developed so far in Eqs. (7.1)–(7.4) are independent of the model used for the working fluid.

Cycle Analysis (Cycle 1-2-3-4-1): Now that each process of the closed Carnot cycle has been analyzed for heat transfer, the overall cycle parameters can be found by summing up contributions from the constituent processes. In absolute terms, heat addition Q_{in} and heat rejection Q_{out} (the subscripts indicating the direction of heat transfer) can be expressed as

$$Q_{in} = Q_{23} = mT_H(s_3 - s_2); \quad Q_{out} = -Q_{41} = mT_C(s_4 - s_1); \quad [\text{kJ}] \tag{7.5}$$

Note that the symbols Q_H and Q_C for Carnot cycle in Chapter 2 are being replaced with Q_{in} and Q_{out} to generalize these symbols so they can be used in all types of cycles, not just Carnot cycle. By treating a cyclic device as a closed-steady system, we have shown in Sec. 2.3.1 that the net heat transferred into the system in a cycle must be equal to the net work output (see Eq. 2.24). This can also be established by considering the cycle as a process where the beginning state and the final state are identical. Applying the energy equation for a closed process, Eq. (5.2), we obtain

$$\Delta E^\theta = Q_{in} - Q_{out} - W_{ext}; \quad \Rightarrow \quad \underbrace{Q_{net}}_{\substack{\text{Net heat transferred} \\ \text{into the working fluid} \\ \text{during the cycle.}}} = \underbrace{W_{net}}_{\substack{\text{Net external work} \\ \text{during the cycle.}}}; \quad [\text{kJ}] \tag{7.6}$$

The net heat transferred to the working fluid during a cycle equals the net work delivered by the cycle as energy cannot be stored in a cyclic process. Equation (7.6) is applicable not only to the Carnot cycle under consideration, but to any cycle executed by a fixed mass.

The **thermal efficiency** [see Eq. (2.27)], which is a widely used *energetic efficiency* for a heat engine, for the Carnot cycle can be expressed as

$$\eta_{th,C} = \frac{W_{net}}{Q_{in}} = \frac{Q_{net}}{Q_{in}} = \frac{Q_{in} - Q_{out}}{Q_{in}} = 1 - \frac{Q_{out}}{Q_{in}} = 1 - \frac{mT_C(s_4 - s_1)}{mT_H(s_3 - s_2)} = 1 - \frac{T_C}{T_H} \tag{7.7}$$

In simplifying the expression, we recognize that $s_4 - s_1 = s_3 - s_1$ as evident from the *T-s* diagram of Fig. 7.2. Not surprisingly, it is identical to the expression, Eq. (2.32), derived in Chapter 2 by treating the Carnot engine as a closed-steady system.

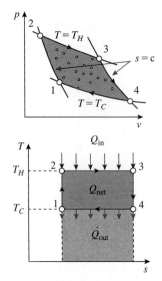

FIGURE 7.2 Thermodynamic diagrams for the Carnot cycle executed by an ideal gas (see Anim. 7.A.*CarnotClosedCycle*).

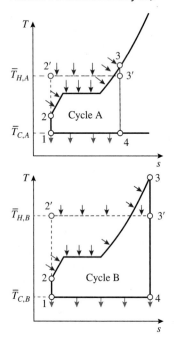

FIGURE 7.3 Cycle B is more efficient because $\overline{T}_{H,B} > \overline{T}_{H,A}$ and $\overline{T}_{C,B} = \overline{T}_{C,A}$. The equivalent Carnot cycles are represented by the red dashed lines.

The net power of the Carnot cycle, and for that matter any reciprocating cycle, consists of the sum of positive and negative boundary ($pd\text{\textbrokenbar}$) work during the constituent processes. In Chapter 6, we learned that not all of the boundary work is useful, as some of it goes into the useless task of displacing atmospheric air. However, in the context of a cyclic process, the net atmospheric work turns out to be zero, as any boundary work transferred to the atmosphere during expansion of the system is recovered during the contraction processes. Atmospheric work (Anim. 0.D.*atmosphericWork*), therefore, can be routinely ignored in calculating the net work transfer from a cyclic process.

7.1.1 Significance of the Carnot Engine

The Carnot cycle is sketched on both a *p-v* and a *T-s* diagram in Figure 7.2 with an ideal gas as the working fluid. Being a reversible cycle, each of its constituent processes must also be reversible and, hence, internally reversible ($S_{gen,univ} = 0$ implies $S_{gen,int} = 0$: see Sec. 5.1.3). For each internally reversible process, the area projected on the abscissa under each process curve in Figure 7.2 can be interpreted as the work transfer on the *p-v* diagram and heat transfer on the *T-s* diagram per unit mass of the working fluid (revisit the derivation of Eqs. (5.6) and (5.7) and browse Anim. 5.A.*intReversible*). The area enclosed by the cycles, therefore, must be proportional to the net work and net heat transfer, which are equal according to Eq. (7.6). Due to its rectangular shape, the *T-s* diagram is easier to interpret. Heat added, rejected, and the net work can be quickly deduced from Figure 7.2, as

$$Q_{in} = mT_H(s_4 - s_1); \quad Q_{out} = mT_C(s_4 - s_1); \quad [\text{kJ}] \tag{7.8}$$

$$\text{and} \quad W_{net} = Q_{net} = m(T_H - T_C)(s_4 - s_1); \quad [\text{kJ}] \tag{7.9}$$

The Carnot efficiency, $1 - T_C/T_H$, can be easily derived for this gas cycle by using Eqs. (7.9) and (7.8).

The algebraic formula can also be visually associated with the separation between the heat addition and rejection temperatures—the bigger the separation, the smaller is T_C/T_H, and the higher is the thermal efficiency. This interpretation can be generalized into a rule of thumb for comparing efficiencies of two reversible cycles, which are not necessarily Carnot cycles. For instance, consider the two reversible cycles, each marked as 1-2-3-4-1, shown on the *T-s* diagrams of Figure 7.3. Although reversible, these are not Carnot cycles because the heat addition in both these cycles does not take place at a constant temperature. However, an **effective temperature** of heat addition can be loosely defined as some sort of an equivalent temperature of a reservoir (TER), which provides the same amount of heat as the original process (area under 2-3 is equal to area under the equivalent process 2'-3'). Any reversible cycle, thus, can be converted to an **equivalent Carnot cycle** with two effective temperatures, $\overline{T}_H$ and $\overline{T}_C$ replacing the variable temperature of heat addition and heat rejection of a given cycle. Comparing two equivalent Carnot cycles is a much easier task than comparing the original cycles for the purpose of estimating their thermal efficiencies.

Going back to Figure 7.3, the main difference between the two cycles is that the maximum temperature, T_3, of cycle B is much higher. How does that affect the thermal efficiency? The net output W_{net} from cycle B must be higher, as its total enclosed area is greater than that of cycle A. However, Q_{in} is also higher for cycle B; therefore, it is not clear how the efficiency, the ratio of W_{net} to Q_{in}, would differ between the two cycles. But if we take advantage of the Carnot analogy and compare the effective temperatures, we can quickly deduce that cycle B must be more efficient since $\overline{T}_{H,B} > \overline{T}_{H,A}$ while $\overline{T}_{C,B} = \overline{T}_{C,A}$.

7.2 IC ENGINE TERMINOLOGY

Some of the frequently used terms in a reciprocating cycle are explained through the operation of a four-stroke spark ignition engine in Anim. 7.B.*SIEngine*. Although most engines have more than one cylinder, analysis can be based on a single cylinder, taking advantage of the fact that they are identical. Therefore, at its core, an IC engine can be represented by a piston-cylinder device as depicted in Figure 7.4. During various stages of the cycle, the gas trapped inside can be the oxidizer (pure air), the fuel and oxidizer mixture that burns internally (hence, the name **internal combustion**), or the products of combustion. The reciprocating motion (hence the name **reciprocating engines**) of the piston is converted to rotary motion through the crank mechanism, delivering the power to the **crank shaft**. The end of the cylinder with the valves is called the top of the cylinder regardless of its orientation, and the cylinder diameter is known as the **bore**. The position of the piston where the system volume reaches a maximum is called the bottom dead center or **BDC**, and the position where the system volume reaches the minimum or the **clearance volume** V_c is called the top dead center or **TDC**. The volume between BDC and TDC, the volume swept by the piston, is called the displacement volume V_d or simply the **displacement**, and the length of the sweep is called the **stroke**. The **compression ratio** r is defined as the ratio of the maximum to minimum volume of the gas that can be trapped in the device:

$$r = \frac{V_{\text{BDC}}}{V_{\text{TDC}}} = \frac{V_c + V_d}{V_c} = 1 + \frac{V_d}{V_c} \qquad (7.10)$$

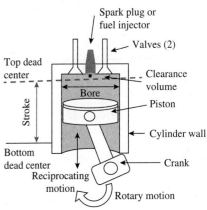

FIGURE 7.4 The piston-cylinder device is at the core of a reciprocating internal combustion engine (see Anim. 7.B.*SIEngine*).

Depending on how the ignition occurs, IC engines are divided into two categories. In the **spark ignition** or **SI engines**, a mixture of fuel and air, known as the charge, is ignited with the help of the spark plug. In a **compression ignition** or **CI engine**, the charge—in this case ambient air—is compressed to a relatively high compression ratio raising the temperature to such a point (recall Ex. 5-9) that spontaneous ignition occurs when fuel is sprayed into the hot and compressed air by the fuel injector. Yet another way to classify the IC engines is through the number of strokes it takes to complete the mechanical cycle. Accordingly, the cycles are called **four-stroke** or **two-stroke** cycles (compare Anim. 7.B.*SIEngine* with 7.D.*twoStrokeEngine*). In a four-stroke cycle, the net work W_{net} is delivered in every two revolutions of the crank shaft, while in a two-stroke cycle the same is accomplished in every revolution. The net power, $\dot{W}_{\text{net,total}}$, developed by an engine can be expressed in terms of the number of cylinders n_C, the number of stroke per cycle, and the rotational speed N of the crank shaft expressed in **rps** (revolution per second) or **Hz** (hertz) as

$$\dot{W}_{\text{net,total}} = n_C \dot{W}_{\text{net}} = n_C \frac{N}{2} W_{\text{net}}; \quad \text{(4stroke)} \quad [\text{kW}] \qquad (7.11)$$

$$\dot{W}_{\text{net,total}} = n_C \dot{W}_{\text{net}} = n_C N W_{\text{net}}; \quad \text{(2stroke)} \quad [\text{kW}] \qquad (7.12)$$

The power delivered by the crankshaft is called the **brake power** as measured by a **dynamometer**, a device that absorbs the power of the engine by providing an external load through frictional, hydraulic, or electric dissipation. Due to frictional losses in the engine, the brake power can be expected to be smaller than the total net power developed in the cylinders. However, the difference is not very significant, and in this text we will make no distinction between the brake power and the net power. Therefore, the torque T developed in the crankshaft can be related to the net power as

$$\dot{W}_{\text{net,total}} = \dot{W}_{\text{brake}} = 2\pi N T; \quad [\text{kW}] \qquad (7.13)$$

While $\dot{W}_{\text{net}}$, the net work output per cylinder per cycle, is independent of engine speed or number of cylinders in an engine, it still depends on the cylinder size. To compare two engines with different displacements, the **mean effective pressure** (MEP) is defined as the ratio of net work to cylinder displacement.

$$\text{MEP} = \frac{W_{\text{net}}}{V_d} = \frac{W_{\text{net}}}{m(v_{\text{BDC}} - v_{\text{TDC}})}; \quad \left[\text{kPa} = \frac{\text{kJ}}{\text{m}^3} = \frac{\text{kN} \cdot \text{m}}{\text{m}^2 \cdot \text{m}} \right] \qquad (7.14)$$

It has the unit of pressure and can be interpreted as the average pressure that produces the same network output in a single expansion stroke regardless of the cylinder size. The specific fuel consumption, **sfc**, another important engineering parameter, is defined as the fuel flow rate per-unit total power.

$$\text{sfc} = \frac{\dot{m}_F}{\dot{W}_{\text{net,total}}} = \frac{\dot{m}_F}{n_C \dot{W}_{\text{net}}} = \frac{\dot{m}_F}{2\pi N T} \left[\frac{\text{kg}}{\text{kJ}} = \frac{\text{kg}}{\text{s} \cdot \text{kW}} \right] \qquad (7.15)$$

Representing the **heating value**, heat released by combustion of 1 kg of fuel, by the symbol q_{Comb} (see Anim. 0.D.*heatingValue*), the thermal efficiency can be related to sfc as follows:

$$\eta_{\text{th}} = \frac{\dot{W}_{\text{net}}}{\dot{Q}_{\text{in}}} = \frac{n_C \dot{W}_{\text{net}}}{n_C \dot{Q}_{\text{in}}} = \frac{\dot{W}_{\text{net,total}}}{\dot{m}_F q_{\text{comb}}} = \frac{1}{\text{sfc} \cdot q_{\text{comb}}}; \quad \left[\frac{\text{kW} \cdot \text{kg} \cdot \text{s}}{\text{kg} \cdot \text{kJ}} = 1 \right] \qquad (7.16)$$

DID YOU KNOW?

- The United States consumed 20.68 million barrels of fuel per day in 2007, 43% of which was gasoline.
- 1 barrel is 42 gallons or 159 liters. A barrel of oil has the energy equivalence of 6119 MJ.

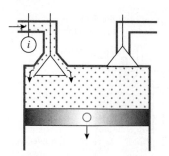

FIGURE 7.5 Schematic used for the definition of the volumetric efficiency.

For a given fuel (given q_{comb}), therefore, the specific fuel consumption is directly related to the thermal efficiency. Typical thermal efficiency of an Otto cycle is in the 25% to 30% range while that of a Diesel cycle is generally higher, in the range of 33% to 38%.

Yet another useful performance parameter is the **volumetric efficiency** defined as the ratio of the actual mass of air inducted into the cylinder to the maximum possible amount, which is the mass that would occupy the displaced volume if it were at the same conditions as the charger inside the intake manifold (State-i in Fig. 7.5). For a four-stroke engine it can be expressed as

$$\eta_v = \frac{2\dot{m}_a}{\rho_i n_C \Psi_d N}; \quad \left[\frac{kg \cdot m^3 \cdot s}{s \cdot kg \cdot m^3} = 1 \right] \qquad (7.17)$$

The volumetric efficiency is a mass ratio: the actual mass flow rate $\dot{m}_a$ divided by the maximum possible flow rate the cylinder can accommodate. The factor 2 is needed to account for the fact that the charge is inducted once every two revolutions of the crankshaft. Due to frictional pressure drop around the valve, the density of the charge is less inside the cylinder than at the intake manifold, causing η_v to be less than one. The more the volumetric efficiency, the more is the amount of charge that can be inducted in every cycle, resulting in greater power for a given engine displacement. Because engine size and weight scale with displacement, a high volumetric efficiency can produce a lighter engine with the same net power output. The thermal efficiency of the cycle, however, is not affected by η_v.

EXAMPLE 7-1   Operational Parameters

A six-cylinder four-stroke engine produces a torque of 1200 N • m at a speed of 2000 rpm. It has a bore of 150 mm and a stroke of 155 mm. Determine (a) the power developed by the engine, and (b) the MEP. *What-if scenario*: (c) What would the MEP be if the engine speed were doubled without any change in the torque?

SOLUTION

Use the definitions of the basic operational parameters to obtain the answers.

Assumptions

No distinction is made between the shaft power and the power developed by the engine through boundary work.

Analysis

Using Eq. (7.13):

$$\dot{W}_{net,total} = 2\pi NT = 2\pi \left(\frac{2000}{60 \text{ s/min}} \right) \left(\frac{1200}{1000 \text{ N/kN}} \right) = 251.3 \text{ kW}$$

Also, from Eq. (7.11) the net work output from a single cylinder can be obtained as

$$W_{net} = \frac{2}{N} \frac{\dot{W}_{net,total}}{n_C} = \frac{2 \times 60}{2000} \frac{251.3}{6} = 2.513 \text{ kJ}$$

The MEP can now be obtained from Eq. (7.14) as

$$\text{MEP} = \frac{W_{net}}{\Psi_d} = \frac{4W_{net}}{\pi d^2 L} = \frac{4 \times 2.513}{\pi (150 \times 10^{-3})^2 (155 \times 10^{-3})} = 917.6 \text{ kPa}$$

What-if scenario

If the engine speed N doubles, so does $\dot{W}_{net,total}$. However, W_{net} is not affected by the engine speed. Likewise, MEP is also independent of the engine speed.

Discussion

The torque produced by an engine usually varies with the engine speed. Typical torque-speed and power-speed characteristic curves of a four-stroke engine is shown in Figure 7.6. The torque can be seen to increase with the engine speed. However, at very high speed viscous friction

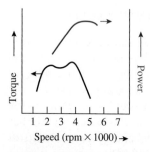

FIGURE 7.6 Typical performance curves of a four-stroke engine.

rapidly rises, causing the torque to fall. Being the product of torque and speed, the power curve also eventually falls.

EXAMPLE 7-2 Operational Parameters

A 4 L four-stroke engine with a thermal efficiency of 25% and a volumetric efficiency of 85% has a power output of 100 kW at 3500 rpm. If the heating value of the fuel is 40 MJ/kg, determine (a) the sfc, (b) the MEP, and (c) the air-fuel ratio on a mass basis. Assume the intake manifold conditions to be the same as the atmospheric conditions: 100 kPa, 298 K (see Fig. 7.7).

SOLUTION

Use the definitions of the basic operational parameters to obtain the answers.

Assumptions

Air can be modeled as an ideal gas with $R = 0.287$ kJ/kg·K.

Analysis

Using Eq. (7.16)

$$\text{sfc} = \frac{1}{\eta_{th} \cdot q_{comb}} = \frac{1}{(0.25)(40,000)} = 1 \times 10^{-4} \frac{\text{kg}}{\text{kJ}} = 0.36 \frac{\text{kg}}{\text{kW} \cdot \text{hr}}$$

The MEP can be obtained from Eq. (7.14) as

$$\text{MEP} = \frac{W_{net}}{V_d} = \frac{2 \, \dot{W}_{net,total}}{N \quad n_C V_d} = \frac{2(100)(60)}{(3500)(4 \times 10^{-3})} = 857.1 \text{ kPa}$$

The flow rate of air now can be calculated from Eq. (7.17) using the IG equation of state.

$$\dot{m}_a = \eta_v \frac{N\rho_i n_C V_d}{2} = \eta_v \frac{N p_i n_C V_d}{2RT_i} = (0.85)\frac{(3500)(100)(4 \times 10^{-3})}{(60)(2)(0.287)(298)} = 0.116 \frac{\text{kg}}{\text{s}}$$

The fuel consumption rate can be obtained from Eq. (7.15)

$$\dot{m}_F = \text{sfc} \cdot \dot{W}_{net,total} = (1 \times 10^{-4})(100) = 0.01 \frac{\text{kg}}{\text{s}}$$

Therefore, the air-fuel ratio is 11.6.

Discussoin

The specific fuel consumption relates fuel consumption in kg with the work delivered in kWh. Comparing the local electricity price per kWh with the price of gasoline per kg establishes the advantage electric vehicles offer in fuel cost. The limited range of electric vehicles, however, is a major drawback.

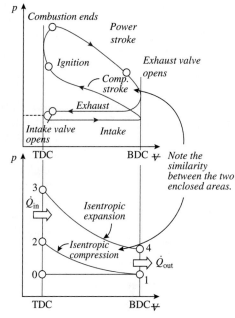

FIGURE 7.7 p-V diagram for the 4-stroke SI engine and the idealized Otto cycle (see Anim. 7.B.*SIEngine* and 7.B.*ottoCycle*).

7.3 AIR-STANDARD CYCLES

In an IC engine, the working fluid actually does not execute a complete cycle; if it did there would be no such thing as car exhaust. The charge undergoes permanent changes through chemical reaction during the course of the cycle before being ejected into the surroundings. There are many other complexities that arise in a real engine due to water-cooling of the cylinders, different types of thermodynamic friction, heat transfer and chemical reactions, non-uniform global state, complex flow patterns, etc.

The **air-standard cycle**, however, is conceived as a simplified model that incorporates the essential physics, allowing reasonably sophisticated thermodynamic analysis without the need for complex computer simulations. The framework for the air-standard cycle consists of the following elements: 1. A fixed mass of air, assumed to be the working fluid at all times, executes a

DID YOU KNOW?

• The heat of combustion of gasoline and diesel are 44.5 MJ/kg and 42.5 MJ/kg respectively. However, diesel being denser on a volume basis has a heating value of 35 MJ/L, as opposed to 32.9 MJ/L for gasoline.

complete cycle, eliminating the need for the intake and exhaust strokes; 2. Air can be modeled as an ideal gas; 3. Heat release through internal combustion is replaced with heat addition from an external reservoir; 4. Similarly, heat is rejected to an external reservoir eliminating the exhaust stroke; 5. All processes are assumed internally reversible. Further simplification is possible if the PG model replaces the IG model with specific heats evaluated at the ambient temperature. Not only does property evaluation become easier, but the use of the isentropic relations (Sec. 3.5.4) can provide direct insight into the compression and expansion processes. The resulting simplified model is called the **cold air-standard cycle**, as properties of fuel air mixture and combustion generated hot gases are modeled as cold air (PG model with properties of air evaluated at STP). Before we discuss the specifics of any cold air-standard cycle, let us introduce the reciprocating cycle TESTcalc, which will be used to analyze different types of IC engines.

7.3.1 TEST and the Reciprocating Cycle TESTcalcs

The **reciprocating closed-cycle TESTcalcs**, the TESTcalcs used for modeling reciprocating cycles, are located in the systems, closed, process, specific, reciprocating cycle branch. While the IG and PG models are mostly used to simulate the air-standard cycles, several mixture models including the PC model are also available for more specialized cycles (a steam engine, for example). All reciprocating cycle TESTcalcs have similar interfaces. They build upon the closed-process TESTcalcs introduced in Sec. 5.1.2. After the anchor states are calculated in the state panel, the constituent processes are analyzed in the process panel. The only new addition is the cycle panel, which sums up the energy transfers in the sequential processes and displays the overall cycle parameters once the cycle is completely constructed. The same reciprocating cycle TESTcalc can be used to model different types of cycles such as the Otto, Diesel, and Stirling cycles.

Simulation Interactives of the Otto and Diesel cycles are located in the Interactives, unsteady processes, specific branch. State properties and process variables can be directly entered in the schematic of the cycle. Cycle properties are calculated once all the processes have been defined. Parametric studies can be performed using the default setting through a few clicks to evaluate the effect of any parameter (say, the compression ratio) on the performance of the cycle. Although not as flexible as a TESTcalc, an Interactive can be used for exploring the behavior of an IC engine even before acquiring a detailed knowledge of the cycle.

7.4 OTTO CYCLE

In a **spark ignition (SI)** engine, combustion of fuel-air mixture is initiated by a spark plug. The **Otto cycle**, named after the inventor of the four-stroke spark ignition (SI) engine Nicolaus Otto (1832–1891), models both the 4-stroke and 2-stroke (to be discussed later) SI engines.

Process	SI Engine (Anim. 7.B.*SIEngine*)	Otto Cycle (Anim. 7.B.*OttoCycle*)
0-1	**Intake Stroke**: With the intake valve open, the motion of the piston towards the BDC draws in a fresh charge of fuel air mixture.	This open process is not part of the Otto cycle, which begins with a fixed mass of air trapped at State-1 with the piston already at the BDC.
1-2	**Compression Stroke**: With the valves closed, the charge is compressed as the piston moves from the BDC to the TDC.	**Isentropic Compression**: The compression process has a known volumetric compression ratio and is assumed internally reversible and adiabatic, i.e., isentropic. $v_2 = v_1/r$; $s_2 = s_1$.
2-3	**Spark Ignition**: A spark ignites the mixture towards the end of the compression stroke and heat is released almost instantaneously raising the pressure to its maximum.	**Constant-Volume Heat Addition:** With the piston at the TDC, heat is added instantaneously (volume remains constant) from an external source until the system reaches State-3. $v_3 = v_2$.
3-4	**Power Stroke**: The high pressure, high temperature gas mixture expands and pushes the piston towards the BDC while delivering work.	**Isentropic Expansion**: Like the compression process, expansion is also assumed isentropic, which continues until the piston reaches the BDC. $v_4 = v_1$, $s_4 = s_3$.

Process	SI Engine (Anim. 7.B.*SIEngine*)	Otto Cycle (Anim. 7.B.*OttoCycle*)
4-1	**Blow Down:** With the piston at the BDC, the exhaust valve opens and the pressure drops almost instantly due to sudden communication with outside.	**Constant-Volume Heat Rejection:** With the piston at the BDC, heat is rejected, instantly bringing the system back to State-1, completing the cycle. $v_4 = v_1$.
1-0	**Exhaust Stroke:** With the exhaust valve open, the piston sweeps out the remaining burned gases completing the mechanical cycle.	This open process, like the intake stroke, is not part of the Otto cycle. As shown in Fig. 7.7, the intake and exhaust strokes tend to cancel each other, having little effect on the cycle.

Mechanical operations of a 4-stroke SI engine and the corresponding idealized processes that constitute the air-standard Otto cycle are described in the table below with the aid of Figure 7.7 and Anim. 7.B.*SIEngine* and 7.B.*OttoCycle*.

7.4.1 Cycle Analysis

We model the working fluid air as a perfect gas (PG) in this manual analysis, a restriction that can be easily lifted in a TEST solution. Changes in KE and PE are routinely neglected in cycle analysis, resulting in the following energy balances for the two processes involving heat transfer.

$$\text{Process } 2-3: \quad Q_{\text{in}} = Q_{23} = m(u_3 - u_2) = mc_v(T_3 - T_2); \quad [\text{kJ}] \tag{7.18}$$

$$\text{Process } 4-1: \quad \dot{Q}_{\text{out}} = -Q_{41} = -m(u_1 - u_4) = mc_v(T_4 - T_1); \quad [\text{kJ}] \tag{7.19}$$

The net work transfer being equal to the net heat transfer [Eq. (7.6)], the thermal efficiency for the Otto cycle can be expressed as

$$\eta_{\text{th,Otto}} = \frac{W_{\text{net}}}{Q_{\text{in}}} = \frac{Q_{\text{net}}}{Q_{\text{in}}} = \frac{Q_{\text{in}} - Q_{\text{out}}}{Q_{\text{in}}} = 1 - \frac{Q_{\text{out}}}{Q_{\text{in}}} = 1 - \frac{(T_4 - T_1)}{(T_3 - T_2)} \tag{7.20}$$

Note that unlike the Carnot cycle, heat is not added or rejected at constant temperatures. That is why, despite being reversible, the Otto cycle is not a Carnot cycle. This is also evident from the fact that unlike the Carnot cycle, the thermodynamic diagram of the Otto cycle in Figure 7.8 is not a rectangle. However, because each constituent process is assumed to be internally reversible, the areas under the T-s diagram can be interpreted as the reversible heat transfer. Using the isentropic relations for a perfect gas (Eq. 3.71),

The thermal efficiency, $\eta_{\text{th,Otto}}$, can be expressed in terms of the compression ratio r (see Eq. (7.10)) as follows:

$$r = \frac{V_{\text{BDC}}}{V_{\text{TDC}}}; \quad \frac{V_1}{V_2} = \frac{V_4}{V_3}; \quad \Rightarrow \quad \frac{T_2}{T_1} = \frac{T_3}{T_4} = r^{k-1}; \quad \Rightarrow \quad \frac{T_4}{T_1} = \frac{T_3}{T_2} \tag{7.21}$$

$$\text{Therefore,} \quad \eta_{\text{th,Otto}} = 1 - \frac{T_1(T_4/T_1 - 1)}{T_2(T_3/T_2 - 1)} = 1 - \frac{T_1}{T_2} = 1 - \frac{1}{r^{k-1}} \tag{7.22}$$

The volumetric compression ratio r is the only controllable parameter that appears in the thermal efficiency expression of in Eq. (7.22). The higher the compression ratio, the higher is the thermal efficiency of an Otto cycle and, by extension, the fuel efficiency of an SI engine. Although k, the specific heat ratio, is considered a material property in the PG model, it generally decreases with an increase in temperature—$k = 1.4$ at 298 K and about 1.3 at 1500 K for air—as shown in Figure 7.9. The efficiency, calculated from Eq. (7.22), is plotted against the compression ratio in Figure 7.10 for two different values of k. An increase in r clearly leads to a monotonic increase in efficiency for both values of k. For a given r, however, the gas mixture with a lower k can be seen to produce a slightly lower efficiency. More accurate calculation with the IG model that accounts for variation of specific heats with temperature, therefore, is expected to produce lower thermal efficiency (verified in Ex. 7-3) than predicted by the PG model or from the cold air-standard expression of Eq. (7.22). Moreover, the presence of irreversibilities due to different types of thermodynamic friction, incomplete combustion, leakage past the piston and

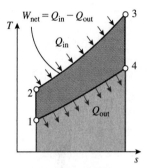

FIGURE 7.8 Geometric interpretation of the T-s diagram of the Otto cycle (see Anim. 5.A.*pTsConstCompressionPG*).

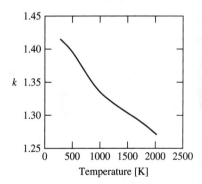

FIGURE 7.9 Variation of k with temperature for air.

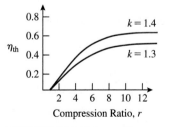

FIGURE 7.10 Efficiency of an Otto cycle depends on r and k. Use the Otto cycle simulator (TEST.*Interactives*) to perform a parametric study (select compression ratio and click Analyze).

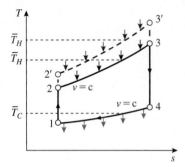

FIGURE 7.11 As the compression ratio is increased, $\overline{T}_H$ increases for an Otto cycle. Use the Otto cycle simulator (TEST.*Interactives*) to verify numerically.

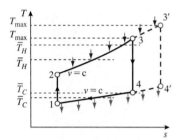

FIGURE 7.12 At higher T_{max} both $\overline{T}_H$ and $\overline{T}_C$ increase. Therefore, the effect on the thermal efficiency is uncertain. Use the Otto cycle simulator (TEST.*Interactives*) to verify numerically.

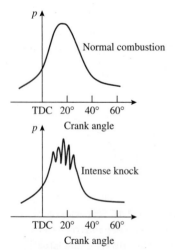

FIGURE 7.13 Pressure profiles for normal combustion and an intense knock. BDC corresponds to 180° (see Anim. 7.B.*engineKnock*).

valves, etc., are expected to lower the actual efficiency of a spark-ignition engine even further and Eq. (7.22) should be treated as only an upper bound for a given cycle.

7.4.2 Qualitative Performance Predictions

As explained in Sec. 7.1.1, identifying the heat transfer processes and estimating the net enclosed area in the T-s diagram of a cycle can give us insight into the **engine performance**—behavior of thermal efficiency and net output of the cycle. In this discussion, any change in the T-s diagram due to a modification will be indicated by red lines and red symbols, as opposed to black lines and black symbols for the base cycle.

Consider first the effect of an increase in the compression ratio r for an Otto cycle. As r is increased, the heat addition curve in Figure 7.11 is shifted upward from 2-3 to 2'-3', raising the effective temperature $\overline{T}_H$ of heat addition (red vs. black symbols). The effective heat rejection temperature, $\overline{T}_C$, on the other hand, remains unaffected. The equivalent Carnot efficiency, $1 - \overline{T}_C/\overline{T}_H$, and, hence, the Otto cycle efficiency must increase with r. The net work also increases with an increase in r as evident from the enlargement of the enclosed area in Figure 7.11. At higher compression ratio, an Otto cycle will deliver greater power at the same cycle frequency or rpm of the engine. Note that while Eq. (7.22) is true only for the PG model, the qualitative conclusions we draw from the T-s diagram are quite general and independent of the particular gas model (PG or IG) used in an analysis.

How does the maximum temperature of the cycle affect its performance? To answer that, the Otto cycle is drawn for two different values of T_{max} in the T-s diagram of Figure 7.12, keeping the inlet conditions (State-1) and compression ratio (State-2) unaltered. As evident from the figure, $\overline{T}_H$, $\overline{T}_C$, and the enclosed area all increase with an increase in T_{max}. The net work, proportional to the enclosed area, clearly increases with T_{max}; however, no definitive conclusion regarding efficiency can be drawn due to the difficulty in predicting how the ratio of $\overline{T}_H$ and $\overline{T}_C$ would vary with T_{max}.

Notice that in Figure 7.11 the compression ratio as well as T_{max} are changed simultaneously. However, if T_{max} is held constant, which is a more practical scenario, and r is increased, it is left as an exercise to show that $\overline{T}_H$ would increase and $\overline{T}_C$ would decrease in the modified cycle, thereby, increasing the thermal efficiency.

A change in the ambient temperature or pressure would simply change the location of State-1 on the T-s diagram, relocating the entire cycle without affecting $\overline{T}_H$, $\overline{T}_C$, or the enclosed area. On the other hand, the net work, which is proportional to m, the mass of the system [see Eq. (7.20)], is affected by the ambient conditions. A decrease in the ambient pressure, for instance, will reduce m and hence the net power output. This is why automobiles tend to lose power at high altitudes where air is rarefied due to low ambient pressure. This phenomenon is exploited by the turbocharger found in some automobiles, which raises m through pre-compression of the incoming charge to extract greater power without increasing engine displacement. The volumetric efficiency also plays an important role in this regard.

7.4.3 Fuel Consideration

Equation (7.22) firmly establishes the compression ratio r as the single most important parameter that controls the thermal efficiency of a SI engine. Mechanically, an increase in r does not pose any significant problem—the engine simply has to be built stronger and heavier. Unfortunately an increase in r, according to Eq. (7.21), also raises the temperature after the compression stroke (T_2 in Fig. 7.11). At high enough values of r it is possible for the fuel-air mixture to **auto-ignite**, preempting the carefully timed spark ignition process. The fluctuating pressure that results—this is shown qualitatively in Figure 7.13 with the piston position represented by the crank angle—makes a pinging noise, known as a **knock**. Sustained knocking can lead to serious engine damage through erosions at the piston surface caused by the high temperature ignition spots.

The knocking phenomenon limits the maximum value of r that can be safely used; therefore a fuel with higher anti-knock properties is desirable for a SI engine. Ever since Nicolaus Otto used gasoline in 1876, it has been the fuel of choice for SI engines due to its relatively low boiling temperature, between 40°C and 200°C, that ensures quick vaporization. Different blends have been tried since then to increase the anti-knock characteristics. The **octane number** is a measure of the resistance to knock. To provide a standard measure of knocking characteristics, an octane number is assigned to a fuel based on its resistance to knocking. As a reference, isooctane (2,2,4-trimethylpentane) is assigned an octane number of 100 and heptane (C_7H_{16}) a zero. Unrefined gasoline from the distillation column typically has an octane number of 70 (as if it contained 70% isooctane and 30% heptane). To improve the octane rating to around 90, different refinement processes such as *cracking* and *isomerization* are carried out. The number displayed on the pump is actually an anti-knock index (AKI), which is the average of the octane numbers obtained under two different standards named *research* and *motored*. The octane number of gasoline reached a peak value of 103 in the 1960s with the additive **tetraethyl lead**, which has been found to pose serious health hazards. Leaded gasoline was phased out in the 1970s; consequently, the compression ratios had to be lowered because the alternative additives were not quite as good. The typical octane number of unleaded fuel in use today is about 90, the compression ratio is about 8–10, and the thermal efficiency of an actual engine ranges between 25–30 percent.

DID YOU KNOW?

- Use of high-octane gasoline does not improve the gas mileage or performance unless an engine is specifically designed for high-octane gasoline. For most automobiles, use of the lowest grade of 87 octane is recommended.

EXAMPLE 7-3 | Air-Standard Otto Cycle

A 4-cylinder SI engine with a cylinder displacement of 0.5 L and a clearance volume of 62.5 mL is running at 3000 rpm. At the beginning of the compression process, air is at 100 kPa, 20°C. The maximum temperature during the cycle is 1800 K. Employing the cold air-standard Otto cycle, determine (a) the power developed by the engine, (b) the thermal efficiency, and (c) the MEP. *What-if scenario:* (d) What would the answer in part (b) be if the maximum temperature were raised to 2200 K? (e) What would the answer in part (b) be if the IG model were used?

SOLUTION

Perform energy analysis for the four closed processes constituting the Otto cycle and compile the results to determine the desired cycle variables.

Assumptions

Ideal Otto cycle assumptions. Adopt the PG model for manual solution and the more accurate IG model for TEST solution.

Analysis

From any PG TESTcalc or Table C-1, we obtain material properties for cold air: $\overline{M} = 29$ kg/kmol, $c_v = 0.717$ kJ/kg·K, and $k = 1.4$. From the isentropic relations (Eq. 3.64), evaluate the pressure and temperature of the anchor states shown in the accompanying T-s diagram (see Fig. 7.14).

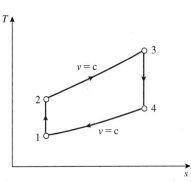

FIGURE 7.14 *T-s diagram for the Otto cycle in Ex. 7-3.*

State-1 (given p_1, T_1, V_1):

$$V_1 = V_{clearance} + V_{displ} = 562.5 \text{ mL}$$

$$m_1 = \frac{\overline{M}p_1 V_1}{\overline{R}T_1} = \frac{(29)(100)(0.563 \times 10^{-3})}{(8.314)(293)} = 0.670 \times 10^{-3} \text{ kg}$$

State-2 (given r, T_1): Use of isentropic relation leads to:

$$r = \frac{V_c + V_d}{V_c} = \frac{562.5}{62.5} = 9; \quad T_2 = (293)9^{1.4-1} = 705.6 \text{ K}$$

State-4 (given r, T_3):

$$T_4 = T_3/r^{k-1} = 1800/9^{1.4-1} = 747.4 \text{ K}$$

An energy analysis for the heat addition and rejection processes yields:

Process 2–3: $Q_{in} = Q_{23} = m(u_3 - u_2) = mc_v(T_3 - T_2)$

$$= (0.670 \times 10^{-3})(0.717)(1800 - 705.6) = 0.526 \text{ kJ}$$

Process 4–1: $Q_{out} = -Q_{41} = -m(u_1 - u_4) = mc_v(T_4 - T_1)$

$$= (0.670 \times 10^{-3})(0.717)(747.4 - 293) = 0.218 \text{ kJ}$$

The net work, efficiency and MEP can now be calculated as

$$W_{net} = Q_{in} - Q_{out} = 0.526 - 0.218 = 0.308;$$

$$\eta_{th,Otto} = \frac{W_{net}}{Q_{in}} = \frac{0.308}{0.526} = 58.6\%; \quad MEP = \frac{W_{net}}{V_{disp}} = \frac{0.308}{0.5 \times 10^{-3}} = 616 \text{ kPa}$$

The engine power developed by the 4 cylinders is

$$\dot{W}_{net,total} = n_{cyl}\frac{N}{2}W_{net} = 4\left(\frac{3000}{2 \times 60}\right)(0.308) = 30.8 \text{ kW}$$

TEST Analysis

Launch the PG reciprocating-cycle TESTcalc. Calculate the four states and processes as described in the TEST-code (see *TEST > TEST-codes*). In the cycle panel, press Calculate and compare the results with the manual solution.

What-if scenario

Change T3 to 2200 K and click Super-Calculate. The efficiency is calculated as 58.5%—not much of a change from the earlier result. Change T3 back to 1800 K. Generate the TEST-code using Super-Calculate button, launch the reciprocating-cycle IG TESTcalc on a separate browser tab, copy the TEST-codes into the I/O panel, and click Load. The new value for the efficiency is found to be 53.55%.

You can use the Otto Cycle Interactive, located in the Interactives, closed processes, specific branch, simple simulations and parametric studies. Launch the Otto Cycle Interactive, enter the ambient air conditions p1 and T1, engine displacement Vol1, maximum temperature T3, and set the compression ratio. Click Calculate to produce the thermal efficiency as 53.55% along with all other cycle variables. Note how easy it is to perform a parametric study by selecting a control variable (say, the compression ratio), setting its range (say, from 2 to 15), and selecting an output parameter, say, the overall efficiency. Click the Analyze button to display the trend graphically.

Discussion

A rise in the maximum temperature does not affect the thermal efficiency significantly because the effective temperatures of heat addition and rejection rise simultaneously. From the second part of the "what-if" study, it can be seen that the PG model over-predicts the efficiency compared to the IG model, as anticipated from the discussion of Figure 7.10 in Sec. 7.4.1.

EXAMPLE 7-4 **Exergy Analysis of an Otto Cycle**

In Example 7-3, do a TEST analysis and determine (a) the process that carries the biggest penalty in terms of exergy destruction. Assume heat is added from a source at 1900 K and rejected to the surroundings at 100 kPa and 20°C. (b) What is the rate of exergy destruction for the entire engine? (c) Determine the maximum possible power that can be extracted from the exhaust gases. (d) Develop a balance sheet for exergy for the entire cycle.

TEST Analysis

Continuing from the analysis described in Example 7-3, evaluate the dead state, State-0, from the given conditions. Click Super-Calculate to update exergy ϕ for all calculated states. For manual evaluation of ϕ at each state, follow the procedure described in Chapter 6.

Exergy destroyed in each process for the closed system shown in Figure 7.15 can be obtained from either a second law analysis or an exergy analysis. Here we obtain $S_{gen,univ}$ for each process from Eq. (5.4).

$$\text{Process-A } (1-2): \quad S_{gen,univ} = \Delta S - \frac{\cancel{Q}^0}{T_B} = m(s_2 - \cancel{s_1^0}) = 0$$

$$\text{Process-B } (2-3): \quad S_{gen,univ} = m(s_3 - s_2) - \frac{Q}{T_B} = 1.721 \times 10^{-4} \frac{\text{kJ}}{\text{K}}$$

$$\Rightarrow \quad I = T_0 S_{gen,univ} = (20 + 273)(1.721 \times 10^{-4}) = 0.050 \text{ kJ}$$

$$\text{Process-C } (3-4): \quad S_{gen,univ} = \Delta S - \frac{\cancel{Q}^0}{T_B} = m(s_4 - \cancel{s_3^0}) = 0$$

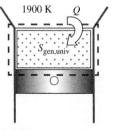

FIGURE 7.15 System boundary for the entropy analysis of process-B in Ex. 7-4.

Process-D (4–1): $S_{gen,univ} = m(s_4 - s_1) - \dfrac{Q}{T_B} = 2.931 \times 10^{-4} \dfrac{kJ}{K}$

$$\Rightarrow \quad I = T_0 S_{gen,univ} = (293)(2.931 \times 10^{-4}) = 0.086 \text{ kJ}$$

Clearly, the heat rejection process, Process-D, carries the maximum penalty in terms of exergy destruction. To obtain these results directly from the TESTcalc, make sure to change the default value of T_B to the appropriate value for each process.

For the entire engine the rate of exergy destruction can be calculated as:

$$\dot{I} = n_{cyl}\frac{N}{2} I_{cycle} = 4\left(\frac{3000}{2 \times 60}\right)(0.050 + 0.086) = 13.6 \text{ kW}$$

The exergy at State-4 can be calculated (or directly obtained from the TESTcalc) as

$$\phi_4 = (u_4 - u_0) - T_0(s_4 - s_0) + p_0(v_4 - v_0) = 128.5 \frac{kJ}{kg};$$

The maximum power that could be extracted from this waste is, by definition, the purge rate of stored exergy and can be calculated as

$$\dot{W}_{rev,\,max} = n_{cyl}\frac{N}{2} m\phi_4 = 4\left(\frac{3000}{2 \times 60}\right)(6.70 \times 10^{-4})(128.5) = 8.60 \text{ kW}$$

The exergy balance sheet can be expressed on a rate basis by using the constant factor $n_{cyl}\dfrac{N}{2}m$ for the four-stroke cycle.

Exergy supplied to the gas:

$$\dot{Q}_{in}\left(1 - \frac{T_0}{T_B}\right) = \left(n_{cyl}\frac{N}{2}\right)Q_{in}\left(1 - \frac{T_0}{T_B}\right) = (100)(0.524)\left(1 - \frac{293}{1900}\right) = 44.32 \text{ kW}$$

Exergy gained by the gas: $n_{cyl}\dfrac{N}{2}m(\phi_3 - \phi_2) = 39.24$ kW

Exergy delivered by power stroke: $n_{cyl}\dfrac{N}{2}m(\phi_3 - \phi_4) = 45.47$ kW

Exergy lost by heat rejection: $n_{cyl}\dfrac{N}{2}m(\phi_4 - \phi_1) = 8.59$ kW

Exergy gained during compression: $n_{cyl}\dfrac{N}{2}m(\phi_2 - \phi_1) = 14.82$ kW

Discussion

The exergy flow diagram of Figure 7.16 shows the exergy inventory for the cycle. Note how part of the exergy of the power stroke is recycled back into the compression stroke. The exergetic efficiency for the cycle can be evaluated as the net exergy delivered, $45.47 - 14.82 = 30.65$ kW, which is same (within computational accuracy) as the net power calculated in Example 7-3, divided by the exergy supplied, 44.324 kW, or 69.2%. In reality, fuel is burned internally and the fuel exergy, to be discussed in Chapter 13, is used in the denominator of the exergetic efficiency for a closed cycle.

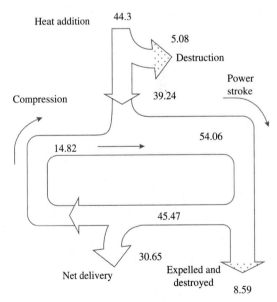

FIGURE 7.16 Exergy flow diagram (in kW) in Ex. 7-4.

7.5 DIESEL CYCLE

About twenty years after the SI engine was developed, Rudolph Diesel developed an engine where ignition is initiated through compression of the charge, eliminating the need for spark plugs used in SI engines, as liquid fuel is directly injected into the cylinder. The engine is known

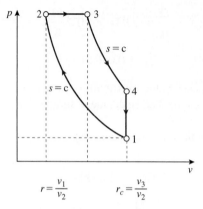

FIGURE 7.17 *p-v* diagram for the Diesel cycle (see Anim. 7.C.*CIEngine*).

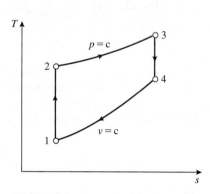

FIGURE 7.18 *T-s* diagram for the Diesel cycle (see Anim. 7.C.*DieselCycle*).

as a **Diesel** or a **compression ignition (CI)** engine, and the air-standard cycle that models a CI engine, is called the **Diesel cycle**. Mechanical operations of a 4-stroke CI engine and the corresponding idealized processes in the Diesel cycle are described below with the aid of Figures 7.17 and 7.18, and Anim. 7.C.*CIEngine* and 7.C.*DieselCycle*.

Process	CI Engine (Anim. 7.C.*CIEngine*)	Diesel Cycle (Anim. 7.C.*dieselCycle*)
0-1	**Intake Stroke**: Same as in a SI Engine.	Not a part of the closed cycle.
1-2	**Compression Stroke**: Same as in SI engine, except a nozzle sprays diesel fuel into the compressed air near the end of the stroke.	**Isentropic Compression**: Same as in the Otto cycle, except r is large, typically 18-20 as opposed to 8-10 in a typical Otto cycle. $v_2 = v_1/r$; $s_2 = s_1$.
2-3	**Combustion/Power Stroke**: Evaporation, mixing, and heat release due to combustion of fuel during the first part of the stroke resulting in expansion and delivery of power to the shaft.	**Constant-Pressure Heat Addition**: Combustion over a finite period is modeled by constant-pressure heat addition until the volume expands to $v_3 = v_2 r_c = v_1(v_2/v_1)r_c = v_1 r_c/r$, where, $r_c (>1)$ is called the **cut-off ratio** (see Fig. 7.17). $p_3 = p_2$.
3-4	**Power Stroke**: The high-pressure, high-temperature gas mixture continues to expand and delivers power until the piston reaches the BDC.	**Isentropic Expansion**: With combustion over, the expansion continues in an adiabatic and internally reversible, i.e., isentropic manner. $v_4 = v_1$, $s_4 = s_3$.
4-1	**Blow Down**: Same as in SI Engine.	**Constant-Volume Heat Rejection**: Same as in the Otto cycle.
1-0	**Exhaust Stroke**: Same as in a SI Engine.	As in the Otto cycle, this process is not part of the air-standard cycle.

7.5.1 Cycle Analysis

The analysis of the Diesel cycle differs slightly from the Otto cycle in that the power stroke involves two different processes: a constant-pressure heat addition process followed by isentropic expansion. An energy analysis for the heat addition (at constant pressure) and heat rejection (at constant volume) processes (refer to cycle 1-2-3-4-1 in Fig. 7.18) leads to:

$$\text{Process } 2-3: \quad Q_{23} - W_{B,23} = \Delta U$$
$$\Rightarrow \quad Q_{in} = Q_{23} = m(u_3 - u_2) + p_2(v_3 - v_2) \tag{7.23}$$
$$\Rightarrow \quad Q_{in} = m(h_3 - h_2) = mc_p(T_3 - T_2); \quad [\text{kJ}]$$

$$\text{Process } 4-1: \quad Q_{out} = -Q_{41} = -m(u_1 - u_4) = mc_v(T_4 - T_1); \quad [\text{kJ}] \tag{7.24}$$

Note that Q_{23}, $W_{B,23}$, and Q_{41} are algebraic quantities following the WinHip sign convention (Sec. 0.7.2) while Q_{in} and Q_{out} are absolute values. Using these expressions for heat transfer in $W_{net} = Q_{net} = Q_{in} - Q_{out}$ [see Eq. (7.6)], we can express the thermal efficiency as:

$$\eta_{th,Diesel} = \frac{W_{net}}{Q_{in}} \frac{Q_{net}}{W_{net}} = 1 - \frac{Q_{out}}{Q_{in}} = 1 - \frac{c_v(T_4 - T_1)}{c_p(T_3 - T_2)} = 1 - \frac{T_1(T_4/T_1 - 1)}{kT_2(T_3/T_2 - 1)} \quad \text{(7.25)}$$

Although this expression is similar to the corresponding expression for $\eta_{th,Otto}$, the compression ratios are different for the two isentropic processes in a Diesel cycle and, therefore, $T_4/T_1 \neq T_3/T_2$. In terms of the cut-off ratio $r_c = v_3/v_2$, and compression ratio $r = v_1/v_2$, Eq. (7.25) can be simplified using the isentropic relations, Eq. (3.71), for the PG model.

$$\frac{T_2}{T_1} = r^{k-1}; \quad \frac{T_3}{T_4} = \left(\frac{v_4}{v_3}\right)^{k-1} = \left(\frac{v_1}{v_3}\right)^{k-1} = \left(\frac{v_1}{v_2}\frac{v_2}{v_3}\right)^{k-1} = \left(\frac{r}{r_c}\right)^{k-1}; \quad \frac{T_3}{T_2} = \frac{v_3}{v_2} = r_c$$

$$\Rightarrow \quad \eta_{th,Diesel} = 1 - \frac{1}{r^{k-1}}\left[\frac{r_c^k - 1}{k(r_c - 1)}\right] \quad \text{(7.26)}$$

It is left as an exercise to show that the quantity inside the bracket of Eq. (7.26) is greater than unity since $k > 1$ ($k \approx 1.4$). With this information, a comparison of Eq. (7.26) with the expression for $\eta_{th,Otto}$ derived in (7.22) reveals that $\eta_{th,Otto} > \eta_{th,Diesel}$ for a given compression ratio.

Using the qualitative approach developed in Sec. 7.4.2, the same conclusion can be reached by comparing the T-s diagrams of the Otto and Diesel cycles for a given compression ratio (see Fig. 7.19). The efficiency of the Otto cycle (1-2-3-4-1 in Fig. 7.19) is greater since $\bar{T}_{H,Otto} > \bar{T}_{H,Diesel}$ while $\bar{T}_C$ is identical for the two cycles. The power output is also expected to be greater for the Otto cycle as the enclosed area is clearly larger.

However, this is not a fair comparison owing to the fact that CI engines typically operate at twice the compression ratio of that used in SI engines. Given the constraint of the same maximum temperature—a more realistic limitation faced by designers of both types of engines— it is left as an exercise to qualitatively establish that the Diesel cycle is the more efficient and more powerful compared to the Otto cycle.

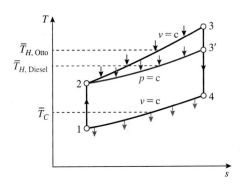

FIGURE 7.19 T-s diagram comparing a Diesel cycle (1-2-3'-4-1) with Otto cycle (1-2-3-4-1) for the same compression ratio.

7.5.2 Fuel Consideration

In the compression ignition engine, the fuel is not premixed with air. Instead, it is directly sprayed into the cylinder towards the end of the compression stroke to be auto-ignited as the temperature rises through isentropic compression. The situation, clearly, is the reverse of the Otto cycle where self-ignition leads to premature combustion; an undesirable phenomenon. The fuel used for Diesel engine is diesel, which consists of a blend of hydrocarbons with boiling temperature in the range between 180°C and 360°C. The ignition quality of diesel fuel is given by the **cetane number**—the higher the number the easier it is for the fuel to ignite. Typical values of cetane number of diesel fuel in use today range from approximately 40 to 55. Additives such as nitrate esters are used to increase the number.

EXAMPLE 7-5 | Air-Standard Diesel Cycle

An air-standard Diesel cycle has a compression ratio of 18. The heat transferred to the working fluid per cycle is 2000 kJ/kg. At the beginning of the compression process the pressure is 100 kPa and the temperature is 25°C. Employing the PG model for the working gas, determine (a) the pressure at each point in the cycle, (b) the cut-off ratio, (c) the thermal efficiency, (d) the net work per unit mass, and (e) the MEP. **What-if scenario:** (f) What would the thermal efficiency be if the heat transfer were 1500 kJ/kg?

SOLUTION

Perform four closed-process energy analyses and compile the results to determine the desired cycle variables.

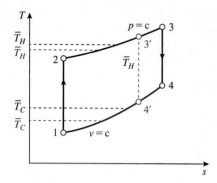

FIGURE 7.20 *T-s* diagram for Ex. 7-5. Cycle 1-2-3'-4'-1 is the modified cycle of the "what-if" study. Use the Diesel cycle simulator (TEST.*Interactives*) to perform different parametric studies (select a parameter and click Analyze).

Assumptions

Ideal Diesel cycle assumptions.

Analysis

From Table C-1 or any PG TESTcalc, we obtain the material properties for air: $\overline{M} =$ 29 kg/kmol; $c_v = 0.718$ kJ/kg·K; $c_p = 1.005$ kJ/kg·K; $R = 0.287$ kJ/kg·K; and $k = 1.4$. Evaluate the pressure and temperature of the anchor states shown in the accompanying *T-s* diagram (see Fig. 7.20).

State-2 (given r, p_1, T_1):   Use of isentropic relations leads to

$$T_2 = T_1 r^{k-1} = (298)18^{1.4-1} = 947 \text{ K}; \quad p_2 = p_1 r^k = (100)18^{1.4} = 5720 \text{ kPa};$$

State-3 (given p_2, T_2, $p_3 = p_2$):   The energy equation can be used to find T_3 as the heat transfer per unit mass is given. From Eq. (7.23),

$$Q_{in} = mc_p(T_3 - T_2); \quad \Rightarrow \quad T_3 = T_2 + \frac{Q_{in}}{mc_p} = 947 + \frac{2000}{1.005} = 2937 \text{ K};$$

$$r_c = \frac{v_3}{v_2} = \frac{T_3 \not{p_2}}{T_2 \not{p_3}} = \frac{T_3}{T_2} = \frac{2937}{947} = 3.1;$$

State-4 (given r, T_3):

$$\frac{v_4}{v_3} = \frac{v_4}{v_2}\frac{v_2}{v_3} = \frac{r}{r_c} = \frac{18}{3.1} = 5.806;$$

Therefore $T_4 = T_3/5.806^{1.4-1} = 1453.3 \text{ K}; \quad p_4 = p_3/5.806^{1.4} = 487.5 \text{ kPa};$

An energy analysis for the heat addition and rejection processes yields:

Process 2−3: $Q_{in} = Q_{23} = 2000m$;

Process 4−1: $Q_{out} = -Q_{41} = -m(u_1 - u_4) = mc_v(T_4 - T_1)$
$$= (0.718)(1453.3 - 298)m = 829.5m \text{ kJ};$$

The net work, efficiency, and MEP can now be calculated as

$$W_{net} = Q_{in} - Q_{out} = 2000m - 829.5m = 1170.5m;$$

$$W_{net}/m = 1170.5 \frac{\text{kJ}}{\text{kg}}; \quad \eta_{th,Diesel} = \frac{W_{net}}{Q_H} = \frac{1170.5m}{2000m} = 58.5\%;$$

$$\text{MEP} = \frac{W_{net}}{V_{disp}} = \frac{1170.5m}{(v_1 - v_2)m} = \frac{1170.5}{v_2(r-1)} = \frac{(1170.5)p_2}{(17)RT_2} = 1449 \text{ kPa};$$

TEST Analysis

Launch the PG reciprocating-cycle TESTcalc. Calculate the four states and four processes as described in the TEST-code (see *TEST > TEST-codes*) based on 1 kg of air. In the cycle panel, click Calculate to obtain all the cycle related variables. The cut-off ratio can be obtained by evaluating v3/v2 in the I/O panel. The results are quite close to those found in the manual solution.

You can also use the Diesel cycle Interactive, located in the Interactives, closed processes, specific branch, to simulate ideal cycles. The simulator is not as versatile as a TESTcalc, but is very simple to use for parametric studies. Enter the ambient air conditions p1 and T1, an arbitrary engine displacement Vol1, maximum temperature T3 or the cut-off ratio, and set the compression ratio. Click Calculate to produce the thermal efficiency as 49.88% (the Interactive uses the IG model) along with all other cycle variables. You can perform a parametric study by simply selecting a control parameter and clicking the Analyze button.

What-if scenario

Reduce Q for Process-B to 1500 kJ, press the Enter key, and click Super-Calculate. The net work decreases to 910 kJ/kg while the efficiency slightly increases to 60.7%.

Discussion

A decrease in heat addition reduces the net enclosed area and, therefore, the net work output of the cycle as shown in Figure 7.20. The maximum temperature also decreases, causing a decrease in both $\overline{T}_H$ and $\overline{T}_C$; hence, it is not so obvious how the thermal efficiency would change in this case. However, due to the fact that the $v = $ constant line is steeper than the $p = $ constant line (see Fig. 3.55), the drop in $\overline{T}_C$ must be more severe, resulting in a slight increase in the equivalent Carnot efficiency $\eta_C = 1 - \overline{T}_C/\overline{T}_H$. This is numerically confirmed by the TEST solution.

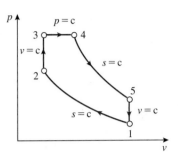

7.6 DUAL CYCLE

One of the main drawbacks of the air-standard cycles is the simplified manner in which combustion (heat addition) is modeled through a constant-volume process in the Otto cycle and a constant-pressure process in the Diesel cycle. In reality, heat release during combustion is too fast for the constant-pressure assumption, but not fast enough for a constant-volume assumption. To more accurately reflect the complex process of combustion, the **dual cycle** distributes the heat addition between two consecutive processes—a constant-volume internally reversible process followed by a constant-pressure internally reversible process as shown in Figure 7.21. The rest of the cycle is basically identical to the Otto or Diesel cycle. The example below illustrates the dual nature of the cycle which bridges Otto and Diesel cycles.

FIGURE 7.21 *p-v* diagram of a dual cycle. Compare it with the realistic diagram of an actual SI engine shown in Fig. 7.7.

EXAMPLE 7-6 Air-Standard Dual Cycle

The Diesel cycle described in Example 7-5 is modified into a dual cycle by breaking the heat addition process into two equal halves so that 1000 kJ/kg of heat is added at constant volume followed by 1000 kJ/kg of heat addition at constant pressure. Determine (a) the temperature at each anchor point in the cycle, (b) the thermal efficiency, (c) the net work per unit mass, and (d) the MEP. Use the PG model. **What-if scenario:** (e) What would the thermal efficiency be if 75% of the total heat transfer took place at constant volume?

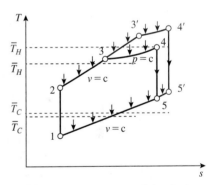

SOLUTION

Perform five closed-process energy analyses and combine them to determine the desired cycle variables.

Assumptions

Ideal dual cycle assumptions. Adopt the PG model for the cold air-standard cycle. Use IG model for TEST solution to obtain more accurate answers.

FIGURE 7.22 *T-s* diagram of the dual cycle in Ex. 7-6. The modified cycle (in red) has higher thermal efficiency because $\overline{T}_H$ is clearly higher while $\overline{T}_C$ is barely changed.

Analysis

From Table C-1 or any PG TESTcalc, we obtain the material properties for air: $\overline{M} = 29$ kg/kmol; $c_v = 0.718$ kJ/kg·K; $c_p = 1.005$ kJ/kg·K; $R = 0.287$ kJ/kg·K; and $k = 1.4$. Evaluate the states of the dual cycle as sketched on the *T-s* diagram of Figure 7.22.

State-2 (given r, p_1, T_1):   Use isentropic relations for the PG model to obtain

$$T_2 = T_1 r^{k-1} = (298)18^{1.4-1} = 947 \text{ K}; \quad p_2 = p_1 r^k = (100)18^{1.4} = 5720 \text{ kPa};$$

$$v_2 = \frac{v_1}{18} = \frac{RT_1}{(18)p_1} = 0.0475 \frac{\text{m}^3}{\text{kg}};$$

State-3 (given $p_2, T_2, v_3 = v_2$):   The energy equation, Eq. (7.18), for the constant-volume process can be used to find T_3 as the heat transfer per unit mass is given.

$$Q_{\text{in}} = mc_v(T_3 - T_2); \quad \Rightarrow \quad T_3 = T_2 + \frac{Q_{\text{in}}}{mc_v} = 947 + \frac{1000}{0.718} = 2340 \text{ K};$$

State-4 (given $v_3 = v_2$, T_3, $p_4 = p_3$):   The energy equation, Eq. (7.23), yields T_4.

$$Q_{in} = mc_p(T_4 - T_3); \implies T_4 = T_3 + \frac{Q_{in}}{mc_p} = 2340 + \frac{1000}{1.005} = 3335 \text{ K};$$

$$v_4 = v_3 \frac{T_4}{T_3} \frac{p_3}{p_4} = (0.0475) \frac{3335}{2340} = 0.0677 \frac{\text{m}^3}{\text{kg}};$$

State-5 (given r, T_4):

$$\frac{v_5}{v_4} = \frac{v_1}{v_4} = \frac{(18)v_2}{v_4} = \frac{(18)(0.0475)}{(0.0677)} = 12.63;$$

Therefore, $T_5 = T_4/12.63^{1.4-1} = 1209$ K;

An energy analysis for the heat addition and rejection processes yields:

Process 2−4: $Q_{in} = Q_{23} + Q_{34} = m(2000) = 2000m;$

Process 5−1: $Q_{out} = -Q_{51} = -m(u_1 - u_5) = mc_v(T_5 - T_1)$

$$= (0.718)(1209 - 298)m = 654m \text{ kJ};$$

Therefore, the net work, efficiency and MEP can be calculated as

$$W_{net} = Q_{in} - Q_{out} = 2000m - 654m = 1346m;$$

$$W_{net}/m = 1346 \frac{\text{kJ}}{\text{kg}}; \quad \eta_{th,Dual} = \frac{W_{net}}{Q_H} = \frac{1346m}{2000m} = 67.3\%;$$

$$\text{MEP} = \frac{W_{net}}{V_{disp}} = \frac{1346m}{(v_1 - v_2)m} = \frac{1346}{v_2(r - 1)} = \frac{(1346)}{(0.0475)(17)} = 1667 \text{ kPa};$$

TEST Analysis

Launch the PG reciprocating cycle TESTcalc. Calculate the states and processes as described in the TEST-code (see *TEST > TEST-codes*). To continue iteration between the process and state panels, click Super-Calculate followed by Super-Iterate. The net work, thermal efficiency, and MEP are calculated as 1347 kJ, 67%, and 1666 kPa respectively.

What-if scenario

Change Q for Process-B to 1500 kJ, and Q for Process-C to 500 kJ. Press the Enter key, click Super-Calculate and then Super-Iterate. The net work increases to 1366 kJ/kg while the efficiency slightly increases to 68.3%.

Discussion

The reason for the slight increase in the efficiency can be understood from the *T-s* diagram of Figure 7.22. With the redistribution of heat addition, the modified cycle is represented by 1-2-3'-4'-5'. Comparing the two cycles of Figure 7.22, it can be seen that $\overline{T}_H$ increases substantially for the modified cycle while $\overline{T}_C$ increases only slightly. As a result, the thermal efficiency is expected to increase, confirming the numerical study. Note that the dual cycle approaches the Otto or Diesel cycle in the limiting cases of zero heat addition in the constant-pressure branch and constant-volume branch respectively.

7.7 ATKINSON AND MILLER CYCLES

The geometry of the standard crank-shaft mechanism ensures that that the compression ratio and expansion ratio are equal in an IC engine. However, if the expansion ratio of an IC engine could be increased without changing the compression ratio, as shown in Figure 7.23, the effective temperature of heat rejection $\overline{T}_C$ could be lowered without changing the effective heat addition temperature, $\overline{T}_H$. As a result, the thermal efficiency (estimated as $1 - \overline{T}_C/\overline{T}_H$) as well as the net

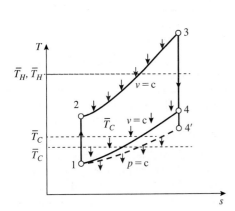

FIGURE 7.23 The Atkinson cycle (in red) has a lower effective temperature of heat rejection compared to an Otto cycle (see Anim. 7.D.*AtkinsonCycle*).

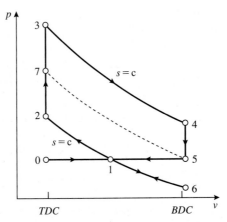

FIGURE 7.24 The Miller cycle (1-2-3-4-5-1) has a higher efficiency compared to the corresponding Otto cycle (5-7-3-4-5) as the compression work is reduced (compare the area under 1-2 with that in 5-7) while the expansion work remains unchanged (see Anim. 7.D.*MillerEngine*).

power output (proportional to the enclosed area in the *T-s* or *p-v* diagram) of the cycle would be increased. Such an approach is used in the Atkinson and Miller cycles. In the **Atkinson cycle**, which is a modified Otto cycle, the expansion stroke continues as shown in Figure 7.23 (and Anim. 7.D.*AtkinsonCycle*) until the pressure at point 4 becomes equal to the atmospheric pressure so that $p_4 = p_1$. As a result, $\overline{T}_C$ is clearly lowered, resulting in improved thermal efficiency.

Another variation of the Otto cycle is the **Miller cycle**, shown in the *p-v* diagram of Figure 7.24 and illustrated in Anim. 7.D.*MillerEngine*. During the intake stroke the cylinder pressure follows the line from State-0 to State-1 at which point the intake valve is prematurely closed. As the piston continues toward the BDC the pressure decreases through isentropic expansion to sub-atmospheric pressure of State-6. During the isentropic compression process 6-2, the path 1-6 is retraced with no net work—work required in creating the vacuum is balanced by the work produced during compression from state 6 to 1. This means the effective compression ratio v_1/v_2 (compression from v_6 to v_1 is free) is less than the expansion ratio v_4/v_3, which is not affected by the modification. At the end of the expansion stroke (State-4) at BDC, when the exhaust valve opens, the cylinder pressure cannot go down below the atmospheric pressure. The constant-pressure process therefore follows the constant-volume blow-down process, 4-5, 5-1, to complete the cycle. An equivalent Otto cycle for the same compression ratio is enclosed by 5-7-3-4-5 in Figure 7.24. Note that the boundary work produced during the power stroke, 3-4, is identical for the two cycles while the work of compression for the Miller cycle, proportional to the area under 5-1-2, is clearly less than the work of compression for the Otto cycle, proportional to the area under the curve 5-7 in Figure 7.24. The net work and efficiency, therefore, are both higher in the Miller cycle.

7.8 STIRLING CYCLE

The Stirling cycle increases efficiency in a gas power cycle by using **regeneration**, a process in which heat from one part of the cycle, which otherwise would be rejected to the surroundings, is transferred to another part of the cycle. A thermal energy storage device, called the **regenerator**, is used for this purpose. It is generally composed of a ceramic mesh or any other material with a high value of mc_v, known as the **thermal mass** of a system. Compared to the regenerator, the mass of the working fluid can be considered negligible. An implementation of the Stirling engine is shown in Figure 7.25 (also see Anim. 7.D.*StirlingEngine*), where two pistons reciprocate in the same cylinder. To start the cycle, let us assume that the left chamber contains all

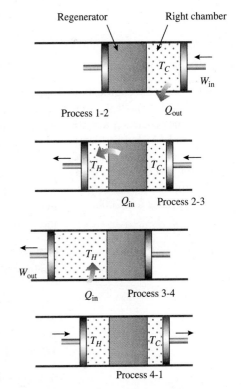

FIGURE 7.25 Implementation of the Stirling cycle. (see Anim. 7.D.*StirlingEngine*).

the working fluid, generally a gas, at the highest pressure and temperature of the cycle. With the aid of the accompanying *p-v* and *T-s* diagrams the actual and idealized processes of the cycle are described in the following table.

Process	Stirling Engine	Stirling Cycle (Anim. 7.D.*StirlingEngine*)
1-2	**Heat Rejection/Compression**: The gas, completely on the right chamber, is isothermally compressed, cooled by transferring heat to a reservoir at $T_C - \Delta T_C$. Work is transferred to the gas during this contraction process.	**Isothermal Heat Rejection**: The pressure increases and volume decreases during the isothermal compression as shown in the *p-v* diagram (Fig. 7.26). For heat transfer to be reversible, the sink temperature must be $T_C - \Delta T_C$.
2-3	**Internal Heat Transfer**: The pistons move to the left forcing the cold gas through the regenerator, which heats the gas from T_C at the right to T_H at the left. The regenerator has just the right temperature distribution T_C at the right and T_H at the left for reversible heat transfer.	**Constant-Volume Isothermal Heating**: The volume remains constant as the pistons move to the left in unison. Heat transfer from the regenerator to the gas is reversible because at any point, the temperature of the regenerator is only differentially greater than the gas temperature, eliminating thermal friction.
3-4	**Power/Heat Addition**: The left piston moves outward delivering external work as heat is added from a source at $T_H + \Delta T_H$. The right piston remains stationary.	**Isothermal Heat Addition**: The pressure decreases and volume increases during the isothermal expansion as shown in the *p-v* diagram. Internal heat transfer is assumed reversible.
4-1	**Internal Heat Transfer**: Both pistons move to the right simultaneously and the hot gas is cooled to a temperature T_C as it transfers heat to the regenerator. No work is transferred during the process.	**Constant-Volume Cooling**: The volume remains constant as the pistons move together. The regenerator temperature varies from T_H at the left face to T_C at the right to ensure reversible heat transfer between the gas and the regenerator.

The thermal mass (mc_v) of the regenerator is much greater than that of air so that the temperature distribution of the regenerator remains unchanged over time. The heat rejected by the gas to the regenerator during the process 4-1 is recovered by the gas during process 2-3. Due to the fact that each process in the cycle is reversible and heat transfer between the reservoirs and the engine takes place at constant temperatures, the Stirling cycle can be considered a variation of the Carnot cycle with an efficiency of $1 - T_C/T_H$. Also, unlike the Otto or Diesel cycles, there is no intermittent explosion as heat is added in a continuous manner. This makes the Stirling engine quieter and suitable for use in submarines.

The efficiency of a practical Stirling cycle, however, is significantly less than that of a Carnot cycle (but considerably greater than the Otto or Diesel cycle) for several reasons: heat transfer does not take place reversibly; the regenerator is not ideal; and pistons are not frictionless. The biggest shortcoming of the cycle is a long warm up time as the heat transfer from an external source takes time to work its way through the cycle. The long response time also makes the engine sluggish, unsuitable for applications in automobiles.

EXAMPLE 7-7   Air-Standard Stirling Cycle

An ideal Stirling cycle, running on a closed system, has air at 100 kPa, 300 K at the beginning of the isothermal compression process. Heat supplied from a source at 1700 K is 800 kJ/kg. Determine (a) the thermal efficiency, and (b) the net work output per kg of air. Use the IG model. *What-if scenario:* (c) What would the efficiency be if the working fluid were argon?

SOLUTION

Perform energy analysis of the four constituent closed processes and combine them to determine the desired cycle variables. Use TEST (system state TESTcalc) for obtaining properties.

Assumptions

Ideal Stirling cycle assumptions.

Analysis

Evaluate the anchor states of the cycle as sketched on the p-v and T-s diagram of Figure 7.26 using the ideal gas table for air or any IG TESTcalc. Assume a system mass of 1 kg.

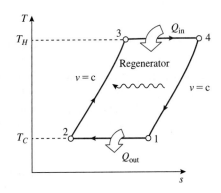

State-1 (given p_1, T_1):

$$u_1 = -84.22 \text{ kJ/kg}; \quad h_1 = 1.873 \text{ kJ/kg}; \quad v_1 = 0.861 \text{ m}^3\text{/kg};$$

State-4 (given $v_4 = v_1, T_4$): $\quad u_4 = 1096.0 \text{ kJ/kg};$

State-3 (given $T_3 = T_4$): $\quad u_3 = u_4 = 1096.0 \text{ kJ/kg};$

The energy equation for process 3-4 yields

$$W_{B,34} = Q_{34} - m(e_4 - e_3)^0 = Q_{34} = 800 \text{ kJ};$$

For the constant-temperature process involving an ideal gas, Eq. (5.10) can be used to determine v_3.

$$W_{B,34} = mRT_4 \ln\frac{v_4}{v_3}; \quad \Rightarrow \quad v_3 = v_4 \exp\left(\frac{-W_{B,34}}{mRT_4}\right) = 0.167 \text{ } \frac{\text{m}^3}{\text{kg}};$$

State-2 (given $T_2 = T_1$, $v_2 = v_3$): $\quad p_2 = \dfrac{RT_2}{v_2} = \dfrac{RT_2}{v_3} = 515.4 \text{ kPa};$

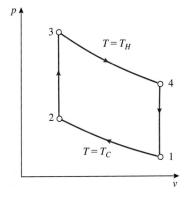

FIGURE 7.26 T-s and p-v diagrams of the Stirling cycle.

Using Eq. (5.10), W_{12} in process 1-2 can be calculated.

$$W_{12} = Q_{12} - m(e_2 - e_1)^0 = W_{B,12} = mRT\ln\frac{p_1}{p_2} = -141 \text{ kJ};$$

The net work, thermal efficiency, and MEP now can be calculated.

$$W_{\text{net}} = W_{34} + W_{12} = 800.0 - 141.2 = 658.8 \text{ kJ};$$

$$Q_{\text{in}} = Q_{34} = 800 \text{ kJ}; \quad \eta_{\text{th,Stirling}} = \frac{W_{\text{net}}}{Q_{\text{in}}} = \frac{658.8}{800} = 82.3\%;$$

TEST Analysis

Launch the IG reciprocating-cycle TESTcalc. Calculate the states and processes as described in the TEST-code (see *TEST > TEST-codes*). Set up the processes with manually calculated boundary work for Process-A (1-2). In the cycle panel, set Process-D (4-1) as the heat donor and Process-B (2-3) as the regenerator receiver, then click Super-Calculate to iterate between the state and process panels. The net work and thermal efficiency are reproduced as in the manual solution.

What-if scenario

In the state panel, select argon as the working fluid. Click Super-Calculate and then Super-Iterate to further iterate between the state and process panels. The thermal efficiency remains unchanged at 82.3%.

Discussion

During processes 2-3 and 4-1, heat transfer takes place internally between the gas and the regenerator without affecting the external heat addition or rejection. The external reservoir temperatures being 1700 K and 300 K respectively, the Carnot efficiency can be calculated as $1 - 300/1700 = 0.824$, or 82.4%, which is identical to the Stirling efficiency. Although a change of the working substance is shown to have no effect on the efficiency, a high-conductivity gas such as hydrogen is preferred in a Stirling engine. Higher conductivity means that for the same heat transfer, the contact surface of the regenerator can be compact.

7.9 TWO-STROKE CYCLE

The two-stroke cycle, developed by Dugland Clerk in 1878, accomplishes a complete spark ignition or compression ignition cycle in one revolution of the crankshaft. With one power stroke in every mechanical cycle, the two-stroke engine can produce more power compared to a four-stroke engine for the same engine mass. Because of its favorable power-to-weight ratio and mechanical simplicity, two-stroke engines have found usage in motorcycles, chain saws, lawn mowers, boats, and model airplane engines. The principle of operation of a crank case scavenged two-stroke engine is shown in Anim. 7.D.*twoStrokeEngine*.

During part of the cycle known as **scavenging**, both the exhaust and intake ports open simultaneously (observe the downward stroke in the animation) and the compressed charge from the crankcase flows into the cylinder, pushing out the remaining exhaust gas. Clearly, scavenging cannot be a perfect operation as it short-circuits some of the fresh charge through the exhaust port. This means a loss of efficiency, and, more importantly, harmful hydrocarbon emissions.

7.10 FUELS

Gasoline and diesel, two of the most widely used fuels, are quite complex chemically. The crude oil contains about 25,000 different compounds, and a refinery uses fractional distillation (see Anim. 13.A.*fuelsClassified* and 13.A.*distillation*) and other chemical processes to separate crude oil into engine fuels (gasoline, diesel, jet fuel), heating fuels (burner, coke, kerosene, residual), chemical feedstock (aromatics, propylene), and asphalt. A typical refinery may distill 40% of the crude oil into gasoline, 20% into diesel, 15% into residual fuel oil, 5% into jet fuel, and the rest into other hydrocarbons. Even after distillation, gasoline or diesel fuel is still made of 100 or more hydrocarbons. The energy density by volume (MJ/L) of diesel is about 8% greater than that of gasoline. Thermodynamic properties of fuel will be discussed in Chapter 13.

The oxygenated fuels program and the reformulated gasoline program, set up by the U.S. Clean Air Act of 1990, attempt to reduce the carbon monoxide and hydrocarbon level in the emissions by increasing the oxygen content of the fuel. The most common additives used for this purpose are methyl tertiary-butyl ether (MTBE) and ethanol (C_2H_5OH). However, the contamination of water supply by MTBE puts its use in question. Gasohol is a gasoline-ethanol blend with about 10% ethanol by volume. The energy density of ethanol is about two-thirds that of gasoline, lowering the heating value of gasohol. The octane rating, on the other hand, is 111, allowing use of higher compression ratio.

Propane (C_3H_8) has been in use as a vehicular fuel since the 1930s. When mixed with butane (C_4H_{10}) and ethane (C_2H_6), the blend is known as liquefied petroleum gas (LPG). Liquid propane (the saturation pressure at 27°C is 1 MPa) has three-fourths the energy density of gasoline; however, the CO_2 emissions from propane is 90% that of gasoline on an equivalent energy basis. In vehicles, propane is stored as a compressed liquid at about 0.9 to 1.4 MPa (use the PC state TESTcalc to verify that propane is a liquid at room temperature in that pressure range). With an octane rating of 112, relatively high compression ratios can be used. Another advantage of using propane or LPG in vehicles is the elimination of sulfur-based compounds present in gasoline or diesel, making the catalytic converter cheaper to build.

Natural gas is generally found in oil fields and is primarily composed of methane (CH_4). It is a **greenhouse gas** with a global warming potential about 10 times as high as carbon dioxide, despite the fact that its CO_2 emissions are about 25% lower than that of gasoline. Compressed natural gas (CNG) is stored at about 20 MPa for vehicular applications and has about one-third the volumetric energy density as that of gasoline. It has an octane rating of 127 allowing relatively high compression ratio.

Hydrogen is produced from many different sources, including natural gas, coal, biomass, and water. Currently used mostly as a rocket fuel, hydrogen combustion does not produce any greenhouse gases. Hydrogen is also used

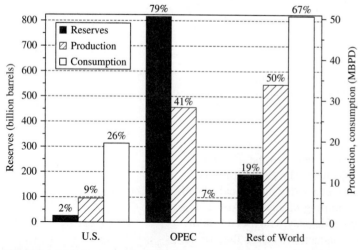

FIGURE 7.27 Reserve and production of oil in 2000 (more fuel statistics can be found in Anim. 0.D.*energyStats*).

as a fuel in fuel cells (discussed in Chapter 14). Fuel cell powered automobiles are still in the demonstration stage. However, the high cost of hydrogen fuel and its low energy density (compressed hydrogen requires 11 times more volume to store the same energy as gasoline) remain major challenges for hydrogen powered vehicles to become a reality.

A snapshot of oil production, consumption (in the unit of million barrel per day) and reserve (billion barrel) is given in Fig. 7.27 with data from 2000. The major changes in this picture until now come from a significant rise in consumption in the developing countries and an increase in production due to the advent of a new technology called **hydraulic fracturing** (commonly known as fracking). Previously inaccessible oil and gas can now be extracted by injecting high pressure fluid (water mixed with chemicals) into the well, creating small-diameter permanent fractures. Environmental concerns including the possibility of ground water contamination has brought international scrutiny of this rapidly spreading technology.

7.11 CLOSURE

Various implementations of the reciprocating heat engine are studied in this chapter as an extension of the overall analysis of cycles that began in Chapter 2. After introducing basic reciprocating engine terminology, the Carnot cycle, implemented by a reciprocating engine, is analyzed for its significance in serving as the benchmark for all other engines. Closed-cycles such as Otto, Diesel, Atkinson, Miller, and Stirling cycles are developed as idealized (cold air-standard) models for the various reciprocating engines. Particular attention is paid to the calculation of thermal efficiency and net power. We have introduced the concept of an equivalent Carnot cycle to qualitatively predict how a modification of a cycle parameter affects the engine performance. Beside verification of manual results, the reciprocating TESTcalcs are used for quantitative "what-if" studies involving various cycle parameters to confirm qualitative theoretical conclusions.

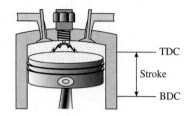

PROBLEMS

SECTION 7-1: ENGINE TERMINOLOGY

7-1-1 A four-cylinder four-stroke engine operates at 4000 rpm. The bore and stroke are 100 mm each, the MEP is measured as 0.6 MPa, and the thermal efficiency is 35%. Determine (a) the power produced ($\dot{W}_{net}$) by the engine in kW, (b) the waste heat in kW, (c) and the volumetric air intake per cylinder in L/s.

7-1-2 A six-cylinder four-stroke engine operating at 3000 rpm produces 200 kW of total brake power. If the cylinder displacement is 1 L, determine (a) the net work output in kJ per cylinder per cycle, (b) the MEP and (c) the fuel consumption rate in kg/h. Assume the heat release per kg of fuel to be 30 MJ and the thermal efficiency to be 40%.

7-1-3 A four-cylinder two-stroke engine operating at 2000 rpm produces 50 kW of total brake power. If the cyclinder displacement is 1 L, determine (a) the net work output in kJ per cylinder per cycle, (b) the MEP and (c) the fuel consumption rate in kg/h. Assume the heat release per kg of fuel to be 35 MJ and the thermal efficiency to be 30%.

7-1-4 A six-cylinder engine with a volumetric efficiency of 90% and a thermal efficiency of 38% produces 200 kW of power at 3000 rpm. The cylinder bore and stroke are 100 mm and 200 mm respectively. If the condition of air in the intake manifold is 95 kPa and 300 K, determine (a) the mass flow rate ($\dot{m}$) of air in kg/s, (b) the fuel consumption rate in kg/s and (c) the specific fuel consumption in kg/kWh. Assume the heating value of the fuel to be 35 MJ/kg of fuel.

SECTION 7-2: CARNOT CYCLES

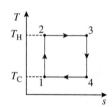

7-2-1 A Carnot cycle running on a closed system has 1.5 kg of air. The temperature limits are 300 K and 1000 K, and the pressure limits are 20 kPa and 1900 kPa. Determine (a) the efficiency and (b) the net work output. Use the PG model. (c) *What-if scenario:* How would the answer in part (b) change if the IG model were used intead?

7-2-2 Consider a Carnot cycle executed in a closed system with 0.003 kg of air. The temperature limits are 25°C and 730°C, and the pressure limits are 15 kPa and 1700 kPa. Determine (a) the efficiency and (b) the net work output per cycle. Use the PG model for air.

7-2-3 An air standard Carnot cycle is executed in a closed system between the temperature limits of 300 K and 1000 K. The pressure before and after the isothermal compression are 100 kPa and 300 kPa, respectively. If the net work output per cycle is 0.22 kJ, determine (a) the maximum pressure in the cycle, (b) the heat transfer to air, and (c) the mass of air. Use the PG model. (d) *What-if scenario:* What would the mass of air be if the IG model were used?

7-2-4 An air standard Carnot cycle is executed in a closed system between the temperature limits of 350 K and 1200 K. The pressure before and after the isothermal compression are 150 kPa and 300 kPa respectively. If the net work output per cycle is 0.5 kJ, determine (a) the maximum pressure in the cycle, (b) the heat transfer to air and (c) the mass of air. Use the IG model for air.

SECTION 7-3: OTTO CYCLES

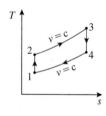

7-3-1 An ideal Otto cycle has a compression ratio of 9. At the beginning of compression, air is at 14.4 psia and 80°F. During constant-volume heat addition 450 Btu/lbm of heat is transferred. Calculate (a) the maximum temperature, (b) efficiency and (c) the net work output. Use the IG model. (d) *What-if scenario:* What would the efficiency be if the air were at 100°F at the beginning of compression?

7-3-2 An ideal Otto cycle with air as the working fluid has a compression ratio of 8. The minimum and maximum temperatures in the cycle are 25°C and 1000°C respectively. Using the IG model, determine (a) the amount of heat transferred per unit mass of air during the heat addition process, (b) the thermal efficiency and (c) the mean effective pressure.

7-3-3   An ideal Otto cycle has a compression ratio of 7. At the beginning of the compression process, air is at 98 kPa, 30°C and 766 kJ/kg of heat is transferred to air during the constant-volume heat addition process. Determine (a) the pressure (p_2) and temperature (T_2) at the end of the heat addition process, (b) the net work output, (c) the thermal efficiency and (d) the mean effective pressure for the cycle. Use the IG model.

7-3-4   A two-stroke engine equipped with a single cylinder having a bore of 12 cm and a stroke of 50 cm operates on an Otto cycle. At the beginning of the compression stroke air is at 100 kPa, 25°C. The maximum temperature in the cycle is 1100°C. (a) If the clearance volume is 1500 cc, determine the air standard efficiency. (b) At 300 rpm, determine the engine output in kW. Use the PG model. (c) *What-if scenario:* What would the efficiency and engine output be if the clearance volume were reduced to 1200 cc?

7-3-5   The temperature at the beginning of the compression process of an air standard Otto cycle with a compression ratio of 8 is 27°C, the pressure is 101 kPa, and the cylinder volume is 566 cm³. The maximum temperature during the cycle is 1726°C. Determine (a) the thermal efficiency and (b) the mean effective pressure. Use the PG model for air. (c) *What-if scenario:* What would the thermal efficiency be if the IG model were used?

7-3-6   The compression ratio of an air standard Otto cycle is 8. Prior to isentropic compression, the air is at 100 kPa, 20°C and 500 cm³. The temperature at the end of combustion process is 900 K. Determine (a) the highest pressure in the cycle, (b) the amount of heat input in kJ, (c) thermal efficiency and (d) MEP. Use the PG model. (e) *What-if scenario:* What would the efficiency be if the compression ratio were increased to 10? Explain the change with the help of a *T-s* diagram.

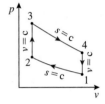

7-3-7   At the beginning of the compression process of an air standard Otto cycle, pressure is 100 kPa, temperature is 16°C, and volume is 300 cm³. The maximum temperature in the cycle is 2000°C and the compression ratio is 9. Determine (a) the heat addition in kJ, (b) the net work (W_{net}) in kJ, (c) thermal efficiency and (d) MEP. Use the PG model.

7-3-8   The compression ratio in an air standard Otto cycle is 8. At the beginning of the compression stroke, the pressure is 101 kPa and the temperature is 289 K. The heat transfer to the air per cycle is 1860 kJ/kg. Determine (a) the thermal efficiency and (b) the mean effective pressure. Use the PG model.

7-3-9   An air standard Otto cycle has a compression ratio of 9. At the beginning of the compression, pressure is 95 kPa and temperature is 30°C. Heat addition to the air is 1 kJ, and the maximum temperature in the cycle is 750°C. Using the IG model for air, determine (a) the net work (W_{net}) in kJ, (b) thermal efficiency and (c) MEP. Assume mass of air as 0.005 kg.

7-3-10   The compression ratio of an air standard Otto cycle is 8.7. Prior to the isentropic compression process, air is at 120 kPa, 19°C, and 660 cm³. The temperature at the end of the isentropic expansion process is 810 K. Using the PG model, determine (a) the highest temperature and pressure in the cycle, (b) the amount of heat transfer in kJ, (c) the thermal efficiency and (d) MEP.

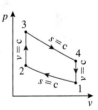

7-3-11   The compression ratio in an air standard Otto cycle is 8. At the beginning of the compression stroke the pressure is 0.1 MPa and the temperature is 21°C. The heat transfer to the air per cycle is 2000 kJ/kg. Determine (a) the thermal efficiency and (b) the mean effective pressure. Use the PG model for air.

7-3-12   An ideal Otto cycle with argon as the working fluid has a compression ratio of 8.5. The minimum and maximum temperatures in the cycle are 350 K and 1630 K, and the minimum pressure is 100 kPa. Accounting for variation of specific heats with temperature (that is, using the IG model for air), determine (a) the amount of heat transferred to 1 kg of air during the heat addition process, (b) the thermal efficiency and (c) the thermal efficiency of a Carnot cycle operating between the same temperature limits.

7-3-13   An ideal Otto cycle has a compression ratio of 8.3. At the beginning of the compression process, air is at 100 kPa and 25°C, and 1000 kJ/kg of heat is transferred to air during the constant volume heat addition process. Using the IG model for air, determine (a) the maximum temperature and pressure that occur during the cycle, (b) the thermal efficiency and (c) the mean effective pressure for the cycle.

7-3-14 In Problem 7-3-13, assume the heat addition can be modeled as heat transfer from a source at 1700°C. Using the PG model for air, determine (a) the exergy (Φ) transferred from the reservoir and (b) the exergy (Φ) rejected to the atmosphere from the engine per unit mass of the gas. Assume the atmospheric conditions to be 100 kPa and 25°C.

7-3-15 A four-stroke engine equipped with a single cylinder having a bore of 12 cm and a stroke of 50 cm operates on an Otto cycle. At the beginning of the compression stroke air is at the atmospheric conditions of 100 kPa, 25°C. The maximum temperature in the cycle is 1100°C and the heat addition can be assumed to take place from a reservoir at 1500°C. If the clearance volume is 1500 cm^3 and the engine runs at 300 rpm, determine (a) the engine output in kW, (b) the exergy destruction (I) over an entire cycle and (c) the rate of exergy destruction ($\dot{I}$) in kW. Use the PG model.

SECTION 7-4: DIESEL CYCLES

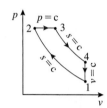

7-4-1 An ideal cold air standard Diesel cycle has a compression ratio of 20. At the beginning of compression, air is at 95 kPa and 20°C. If the maximum temperature during the cycle is 2000°C, determine (a) the thermal efficiency and (b) the mean effective pressure. Use the PG model. (c) *What-if scenario:* What would the MEP be if the compression ratio were reduced to 10? Explain the change with the help of a *T-s* diagram.

7-4-2 The displacement volume of an internal combustion engine is 3 L. The processes within each cylinder of the engine are modeled as an air standard Diesel cycle with a cut off ratio of 2. The state of air at the beginning of the compression is fixed by $p_1 = 100$ kPa, $T_1 = 25$°C, and $V_1 = 3.5$ L. Determine (a) the net work (W_{net}) per cycle, (b) efficiency and (c) the power developed by the engine, if the cycle is executed 1500 times per min. (d) *What-if scenario:* What would the efficiency and power developed be if the cut-off ratio were 2.5? Explain the changes with the help of a *T-s* diagram.

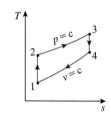

7-4-3 An air standard Diesel cycle has a compression ratio of 15 and cutoff ratio of 3. At the beginning of the compression process, air is at 97 kPa and 30°C. Using the PG model for air, determine (a) the temperature (T_3) after the heat addition process, (b) the thermal efficiency and (c) the mean effective pressure. (d) *What-if scenario:* What would the thermal efficiency be if the IG model were used?

7-4-4 An air standard Diesel cycle has a compression ratio of 16 and cutoff ratio of 2. At the beginning of the compression process, air is at 100 kPa, 15°C and has a volume of 0.014 m^3. Determine (a) the temperature (T_3) after the heat addition process, (b) the thermal efficiency and (c) the mean effective pressure. Use the PG model. (d) *What-if scenario:* What would the efficiency be if the IG model were used?

7-4-5 At the beginning of the compression process of an air standard Diesel cycle operating with a compression ratio of 10, the temperature is 25°C and the pressure is 100 kPa. The cutoff ratio of the cycle is 2. Determine (a) the thermal efficiency and (b) the mean effective pressure. Use the PG model. (c) *What-if scenario:* What would the efficiency be if the compression ratio were increased to 15?

7-4-6 The conditions at the beginning of the compression process of an air standard Diesel cycle are 150 kPa and 100°C. The compression ratio is 15 and the heat addition per unit mass is 750 kJ/kg. Determine (a) the maximum temperature, (b) the maximum pressure, (c) the cutoff ratio, (d) the net work (W_{net}) per unit mass of air and (e) the thermal efficiency.

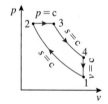

7-4-7 An air standard Diesel cycle has a compression ratio of 20 and cutoff ratio of 3. At the beginning of the compression process, air is at 90 kPa and 20°C. Using the PG model for air, determine (a) the temperature (T_3) after the heat addition process, (b) the thermal efficiency and (c) the mean effective pressure. *What-if scenario:* What would the answers be if the cutoff ratio were (d) 2 and (e) 4?

7-4-8 An air standard Diesel cycle has a compression ratio of 17.9. Air is at 85°F and 15.8 psia at the beginning of the compression process and at 3100°R at the end of the heat addition process. Accounting for the variation of specific heats with temperature, determine (a) the cutoff ratio, (b) the heat rejection per unit mass and (c) the thermal efficiency. Use the IG model for air.

7-4-9   An air standard Diesel cycle has a compression ratio of 18 and cutoff ratio of 3. At the beginning of the compression process, air is at 100 kPa and 20°C. Using the PG model for air, determine (a) the net work (W_{net}) per cycle and (b) the thermal efficiency. (c) *What-if scenario:* What would the answers be if the compression ratio were 21?

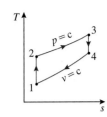

7-4-10   An ideal diesel engine has a compression ratio of 20 and uses nitrogen gas as working fluid. The state of nitrogen gas at the beginning of the compression process is 95 kPa and 20°C. If the maximum temperature in the cycle is not to exceed 2200 K, determine (a) the thermal efficiency and (b) the mean effective pressure. Use the PG model for nitrogen. (c) *What-if scenario:* What would the efficiency and MEP be if carbon-dioxide were used as the working substance?

7-4-11   An ideal diesel engine has a compression ratio of 22 and uses air as working fluid. The state of air at the beginning of the compression process is 95 kPa and 22°C. If the maximum temperature in the cycle does not exceed 1900°C, determine (a) the thermal efficiency and (b) the mean effective pressure. Use the PG model.

7-4-12   At the beginning of the compression process of the standard Diesel cycle, air is at 100 kPa and 298 K. If the maximum pressure and temperature during the cycle are 7 MPa and 2100 K, determine (a) the compression ratio, (b) the cutoff ratio, (c) the thermal efficiency, and (d) the mean effective pressure.

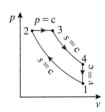

7-4-13   A four cylinder 3-L (maximum volume per cylinder) diesel engine that operates on an ideal Diesel cycle has a compression ratio of 18 and a cutoff ratio of 3. Air is at 25°C and 95 kPa at the beginning of the compression process. Using the cold-air standard assumptions, determine (a) how much power the engine will deliver at 1700 rpm. (b) *What-if scenario:* What would the power be if the engine speed decreased to 1500 rpm?

7-4-14   A four-stroke diesel engine operates at 3000 rpm on a standard Diesel cycle has a compression ratio of 14. The state of air at the beginning of the compression process is 98 kPa and 24°C. If the maximum temperature in the cycle does not exceed 1700°C, determine (a) the thermal efficiency and (b) the specific fuel consumption. Assume diesel fuel has a heating value of 45 MJ/kg. Use the PG model. (c) *What-if scenario:* What would the answers be if the engine operates at 2500 rpm instead?

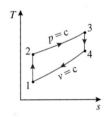

7-4-15   An air standard Diesel cycle has a compression ratio of 19, and heat transfer to the working fluid per cycle is 2000 kJ/kg. At the beginning of the compression process the pressure is 105 kPa and the temperature is 20°C. Determine (a) the net work (W_{net}), (b) the thermal efficiency and (c) the mean effective pressure.

SECTION 7-5: OTHER RECIPROCATING POWER CYCLES

7-5-1   An ideal dual cycle has a compression ratio of 14 and uses air as working fluid. The state of air at the beginning of the compression process is 100 kPa and 300 K. The pressure ratio is 1.5 during the constant volume heat addition process. If the maximum temperature in the cycle is 2200 K, determine (a) the thermal efficiency and (b) the mean effective pressure. Use the PG model.

7-5-2   At the beginning of the compression process of an air standard dual cycle with a compression ratio of 18, $p = 100$ kPa and $T = 300$ K. The pressure ratio for the constant volume part of the heating process is 1.5 and the volume ratio of the constant pressure part is 1.2. Determine (a) the thermal efficiency and (b) the MEP. Use the PG model.

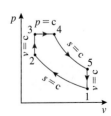

7-5-3   An air standard dual cycle has a compression ratio of 17. At the beginning of compression, $p_1 = 100$ kPa and $T_1 = 15$°C and volume is 0.5 ft³. The pressure doubles during the constant volume heat addition process. For maximum cycle temperature of 1400°C, determine (a) the thermal efficiency (η_{th}) and (b) the MEP. Assume variable c_p (IG model).

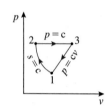

7-5-4 An air standard dual cycle has a compression ratio of 15 and a cutoff ratio of 1.5. At the beginning of compression, $p_1 = 1$ bar and $T_1 = 290$ K. The pressure doubles during the constant volume heat addition process. If the mass of air is 0.5 kg, determine (a) the net work (W_{net}) of the cycle, (b) the thermal efficiency (η_{th}) and (c) the MEP. Use the PG model. (d) *What-if scenario:* What would the net work and efficiency be if the compression ratio were increased to 18? Explain the changes with the help of a *T-s* diagram.

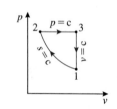

7-5-5 A 3-stroke cycle is executed in a closed system with 1 kg of air, and it consists of the following three processes: (1) Isentropic compression from 100 kPa, 300 K to 800 kPa, (2) $p =$ constant during heat addition in amount of 2000 kJ, (3) $p = cv$ during heat rejection to initial state. Calculate (a) the maximum temperature and (b) efficiency. Show the cycle on *T-s* and *p-v* diagrams. Use the PG model for air. (c) *What-if scenario:* What would the efficiency be if the constant pressure heat addition amounted to 1000 kJ?

7-5-6 An air standard cycle is executed in a closed system with 0.005 kg of air, and it consists of the following three processes: (1) Isentropic compression from 200 kPa, 30°C to 2 MPa, (2) $p =$ constant during heat addition in the amount of 2 kJ, (3) $p = c_1v + c_2$ heat rejection to initial state. Calculate (a) the heat rejected, and (b) the thermal efficiency (η_{th}). Assume constant specific heats at room temperature.

7-5-7 An air standard cycle is executed in a closed system with 0.001 kg of air, and it consists of the following three processes: (1) $v =$ constant during heat addition from 95 kPa 20°C to 450 kPa, (2) isentropic expansion to 95 kPa, (3) $p =$ constant heat rejection to initial state. Using PG model calculate (a) the net work (W_{net}) per cycle in kJ, and (b) the thermal efficiency (η_{th}).

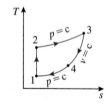

7-5-8 An air standard cycle is executed in a closed system with 1 kg of air, and it consists of the following three processes: (1) Isentropic compression from 100 kPa, 27°C to 700 kPa, (2) $p =$ constant during heat addition to initial specific volume, (3) $v =$ constant during heat rejection to initial state. Calculate (a) the maximum temperature and (b) efficiency. Show the cycle on *T-s* and *p-v* diagrams. Use the PG model. (c) *What-if scenario:* What would the maximum temperature be if isentropic compression took place from 100 kPa, 27°C to 500 kPa?

7-5-9 An air standard cycle is executed in a closed system with 1 kg of air, and it consists of the following three processes: (1) Isentropic compression from 100 kPa, 27°C to 700 kPa, (2) $p =$ constant during heat addition to initial specific volume, (3) $v =$ constant during heat rejection to initial state. Calculate (a) the maximum temperature and (b) efficiency. Show the cycle on *T-s* and *p-v* diagrams. Use the IG model.

7-5-10 An air standard cycle is executed in a closed system and is composed of the following four processes: (1) 1-2: Isentropic compression from 110 kPa and 30°C to 900 kpa, (2) 2-3: $p =$ constant during heat addition in the amount of 3000 kJ/kg, (3) 3-4: $v =$ constant during heat rejection to 110 kPa, (4) 4-1: $p =$ constant during heat rejection to initial state. (a) Calculate the maximum temperature in the cycle and (b) determine the thermal efficiency (η_{th}). Use the PG model.

7-5-11 An air standard cycle is executed in a closed system and is composed of the following four processes: (1) 1-2: $v =$ constant during heat addition from 15 psia and 85°F in the amount of 320 Btu/lbm, (2) 2-3: $p =$ constant during heat addition to 3500°R, (3) 3-4: Isentropic expansion to 15 psia, (4) 4-1: $p =$ constant during heat rejection to initial state. (a) Calculate the amount of heat addition in the cycle and (b) determine the thermal efficiency (η_{th}). Use the IG model.

7-5-12 An air standard cycle with a variable specific heats is executed in a closed system and is composed of the following four processes: (1) 1-2: Isentropic compression from 95 kPa and 25°C to 900 kPa, (2) 2-3: $v =$ constant during heat addition to 1200°C, (3) 3-4: Isentropic expansions to 95 kPa, (4) 4-1: $p =$ constant during heat rejection to initial state. (a) Calculate the net work (W_{net}) output per unit mass and (b) determine the thermal efficiency (η_{th}). Use the IG model for air.

7-5-13 An air standard cycle is executed in a closed system with 0.001 kg of air and is composed of the following three processes: (1) 1-2: Isentropic compression from 110 kPa and 30°C to 1.1 MPa, (2) 2-3: $p =$ constant during heat addition in the amount of 1.73 kJ, (3) 3-1: $p = c_1v + c_2$ during heat rejection to initial state (c_1 and c_2 are constant). (a) Calculate the heat rejected and (b) determine the thermal efficiency (η_{th}). Use the PG model.

7-5-14 An ideal Stirling cycle running on a closed system has air at 200 kPa, 300 K at the beginning of the isothermal compression process. Heat supplied from a source of 1700 K is 800 kJ/kg. Determine (a) the efficiency and (b) the net work (W_{net}) output per kg of air. Use the PG model.

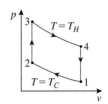

7-5-15 Consider an ideal Stirling cycle engine in which the pressure and temperature at the beginning of the isothermal compression process are 95 kPa, 20°C, the compression ratio is 5, and the maximum temperature in the cycle is 1000°C. Determine (a) maximum pressure and (b) the thermal efficiency (η_{th}) of the cycle. Use the PG model.

7-5-16 An ideal Stirling engine using helium as the working fluid operates between the temperature limits of 38°C and 850°C and pressure limits of 102 kPa and 1020 kPa. Assuming the mass used in the cycle is 1 kg, determine (a) the thermal efficiency (η_{th}) of the cycle and (b) the net work (W_{net}). (c) **What-if scenario:** What would the efficiency and net work be if argon were used as the working fluid?

7-5-17 Consider an ideal Stirling cycle engine in which the pressure, temperature and volume at the beginning of the isothermal compression process are 100 kPa, 15°C and 0.03 m³, the compression ratio is 8, and the maximum temperature in the cycle is 650°C. Determine (a) the net work (W_{net}), (b) the thermal efficiency (η_{th}) and (c) the mean effective pressure. Use the PG model.

7-5-18 Fifty grams of air undergoes a Stirling cycle with a compression ratio of 4. At the beginning of the isothermal process, the pressure and volume are 100 kPa and 0.05 m³, respectively. The temperature during the isothermal expansion is 990 K. Determine (a) the net work (W_{net}) output per kg and (b) the mean effective pressure. Use the PG model.

7-5-19 An ideal Stirling engine using helium as the working fluid operates between the temperature limits of 300 K and 1800 K and pressure limits of 150 kPa and 1200 kPa. Assuming the mass used in the cycle is 1.5 kg, determine (a) the thermal efficiency (η_{th}) of the cycle, (b) the amount of heat transfer in the regenerator, and (c) the work output per cycle.

7-5-20 At the beginning of the compression process of a Miller cycle with a compression ratio of 8 air is at 25°C, 101 kPa. The maximum temperature during the cycle is 1600°C. The minimum pressure during the cycle is 80 kPa. Determine (a) the net work (W_{net}) output, (b) the thermal efficiency (η_{th}), and (c) the mean effective pressure. Use the PG model for air. (d) **What-if scenario:** What would the answers be for an Otto cycle operating under the same conditions?

SECTION 7-6: EXERGY ANALYSIS OF RECIPROCATING POWER CYCLES

7-6-1 A Carnot cycle running on a closed system has 1 kg of air and executes 20 cycles every second. The temperature limits are 300 K and 1000 K, and the pressure limits are 20 kPa and 1900 kPa. Atmospheric conditions are 100 kPa and 300 K. Using the PG model for air, perform a complete exergy inventory and draw an exergy flow diagram for the cycle on a rate (kW) basis. What is the exergetic efficiency of the Carnot engine?

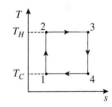

7-6-2 Consider a Carnot cycle executed in a closed system with 0.5 kg of air. The temperature limits are 50°C and 750°C, and the pressure limits are 15 kPa and 1700 kPa. Heat addition takes place from a reservoir at 775°C and heat rejection takes place to the atmosphere at 100 kPa, 25°C. (a) What is the exergetic efficiency of the Carnot engine? (b) Perform a complete exergy inventory and draw an exergy flow diagram for the cycle on unit mass basis (kJ/kg). Use the PG model for air.

7-6-3 In Problem 7-3-2, (a) perform a complete exergy inventory and draw an exergy flow diagram for the cycle on unit mass basis (kJ/kg). Assume the heat addition to take place from a reservoir at 1500°C and heat rejection to the atmosphere at 100 kPa, 25°C. Use the PG model for air. (b) What is the overall exergetic efficiency of the engine?

7-6-4 A four-stroke IC engine with 4 cylinders operates at 3000 RPM in an air-standard Otto cycle. Data for a single cylinder are given as follows. The compression ratio is 8.7. Prior to the isentropic compression process, air is at the atmospheric conditions of 100 kPa, 20°C and 660 cm³. The temperature at the end of the isentropic expansion process is 810 K. Assume the heat addition to take place from a reservoir at 1827°C and heat rejection to the atmosphere. Using the PG model, (a) perform a complete exergy inventory and draw the exergy flow diagram on a rate (kW) basis. Determine (b) the overall exergetic efficiency and (c) the thermal (energetic) efficiency (η_{th}) of the cycle.

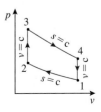

7-6-5 For each process in Problem 7-3-15, (a) develop an exergy inventory on a rate basis (in kW) and draw an exergy flow diagram for the cycle, and (b) determine the exergetic efficiency of the engine. Assume the heat addition and heat rejection to take place with reservoirs at the maximum and minimum temperature of the cycle respectively.

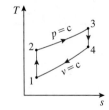

7-6-6 A four cylinder, four-stroke 3-L (maximum volume per cylinder) diesel engine that operates at 1500 rpm on an ideal Diesel cycle has a compression ratio of 18 and a cutoff ratio of 3. Air is at 25°C and 100 kPa (atmospheric conditions) at the beginning of the compression process. Assume heat addition takes place from a reservoir at 2600°C and heat rejection to the atmosphere. Using the PG model, (a) perform a complete exergy inventory and draw the exergy flow diagram on a rate (kW) basis. Determine (b) the overall exergetic efficiency and (c) the thermal (energetic) efficiency (η_{th}) of the cycle.

7-6-7 In Problem 7-4-15 assume that heat is added from a reservoir at 1800°C and the atmospheric conditions are 100 kPa and 20°C. (a) Determine the process that carries the biggest penalty in terms of exergy destruction. (b) Also develop a balance sheet for exergy for the entire cycle on a rate (kW) basis, including an exergy flow diagram.

7-6-8 In Problem 7-5-8 assume that heat is added from a reservoir at 2000°C and the atmospheric conditions are 100 kPa and 27°C. Determine (a) the thermal efficiency (η_{th}) and (b) exergetic efficiency of the cycle. (c) Also develop a balance sheet for exergy for the entire cycle on unit mass (kJ/kg) basis complete with an exergy flow diagram.

8 OPEN GAS POWER CYCLE

In Chapter 7, we studied reciprocating internal combustion engines based on closed power cycles. A large class of heat engines operates with open-steady devices connected back to back through which a working fluid circulates in a loop. The resulting cycles are called **open cycles**. This chapter is devoted to **gas turbines**—power systems running on open cycles with a gas mixture as the working fluid. Major gas turbine applications include modern jet propulsion, helicopters, tanks, oil and gas pipe pumping, offshore platforms, combined cycle power plants, utility peak load power generators, and private electricity generators for hospitals, industry, etc. Depending on the application, gas turbines are divided into two categories: industrial turbines and aircraft. The output of the industrial turbine is shaft power, while the output of the aircraft engine is high velocity jets producing thrust. The biggest advantage of a gas turbine over a reciprocating engine is its favorable power-to-weight and power-to-size ratios, making it lightweight and compact, which is ideal for aircraft. An industrial engine based on a gas turbine is built more durably and can be heavy, but the rapid-start capability of a gas turbine makes it ideal for peak load power generation. However, the poor fuel efficiency, particularly at off-design conditions, makes gas turbines unsuitable as an automobile engine.

In this chapter, we cover the basic and advanced gas turbine cycles. Emphasis is given to quantitative and qualitative prediction of work output and thermal efficiency. As in Chapter 7, TEST solutions are used for verifying manual results, studying the effect of using different gas models, and carrying out "what-if" studies. A dedicated gas turbine simulation Interactive, located in the Interactives, open-steady, specific branch, is used for some of the parametric studies for its sophistication and ease of use. Aircraft propulsion is briefly discussed after introducing the momentum balance equation in the same format as all the other balance equations developed in Chapter 2. Analysis of high-speed flows is delegated to Chapter 15.

8.1 THE GAS TURBINE

The basic operation of a simple gas turbine is shown in Figure 8.1 and illustrated in Anim. 8.A. *simpleGasTurbine*. It consists of three open devices—a **compressor**, a **combustor**, and a **turbine**—connected in series. The intake air from the surrounding atmosphere is delivered to the combustor, at a significantly higher pressure (by about a factor of 10), by the compressor. There, fuel such as natural gas, diesel, jet fuel, etc., is burned, raising the temperature to a high value (about 1000°C) without any significant loss of pressure. The high-temperature and high-pressure gas expands through the turbine, transferring part of its flow energy into shaft power. Connected to the compressor through a single shaft, the turbine supplies necessary power to the compressor while delivering the excess power as net shaft power or *propulsive power*. Although the pressure drop across the turbine is the same as the pressure rise across the compressor, the turbine generates excess power because the working fluid is rarefied (higher specific volume) at the higher average temperature of the turbine compared to that in the compressor (Anim. 8.A.*turbineVsCompressor* and Sec. 4.1.3). Let's briefly discuss the three essential components of a gas turbine before constructing a suitable thermodynamic cycle to model a gas turbine based power plant.

Compressor. There are two types of compressors used in gas turbines: **centrifugal** and **axial flow**. The centrifugal compressor (see Anim. 8.A.*centrifugalCompressor*) is more rugged, but the axial flow compressor (see Anim. 8.A.*compressor*) is smaller in diameter, longer, and has a better efficiency.

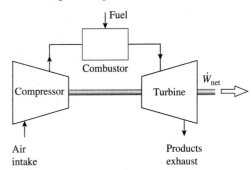

FIGURE 8.1 A simple three-component industrial gas turbine (see Anim. 8.A.*simpleGasTurbine*).

FIGURE 8.2 Impeller of a centrifugal compressor (see Anim. 8.A.*centrifugal Compressor*).

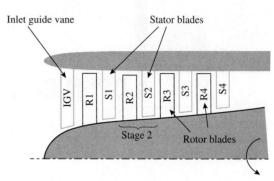

FIGURE 8.3 A multistage axial compressor (see Anim. 8.A.*axialCompressor*).

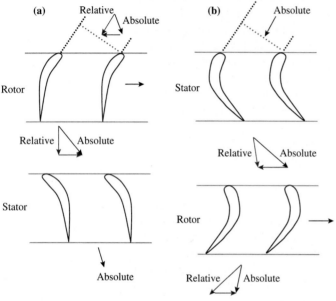

FIGURE 8.4 Typical shapes of (a) rotor and stator blades of an axial compressor (see Anim. 8.A.*compressorBlades*) and (b) stator and rotor blades of an axial turbine (see Anim. 8.A.*turbineBlades*).

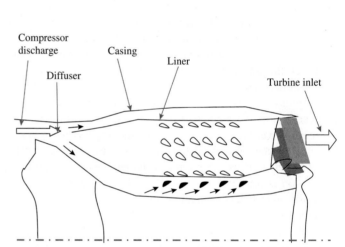

FIGURE 8.5 An annular combustor (see Anim. 8.A.*combustor*).

In a centrifugal compressor, air is drawn into the center or hub of the rapidly rotating impeller (Fig. 8.2) that imparts high kinetic energy into the air and also raises its pressure through centrifugal action. It is then radially discharged into a diffuser where it decelerates due to an increase in flow area, causing the pressure to rise even further.

In an axial compressor, sketched in Figure 8.3, the air flows parallel to the rotating shaft through alternating rows of rotating and stationary blades, called **rotors** and **stators**, each pair forms what is known as a *stage*. The first row is a special stationary row called the inlet guide vanes (IGV). As shown by the velocity vector diagram of Figure 8.4, the blades are shaped such that the air enters the passage between the moving blades with a zero angle of attack to minimize losses. The kinetic energy imparted by the rotor blades to the air is recovered as the pressure rise [recall the Bernoulli equation, Eq. (4.16)] in the stationary blades.

Unlike a centrifugal compressor, which requires only one or two stages, an axial compressor may require as many as 15 stages to achieve the operating pressure of a gas turbine. With density of air increasing along with pressure [recall the isentropic equation for the PG model, Eq. (3.71)], the air passage is made progressively narrower (Fig. 8.3) to maintain a constant average velocity.

Combustor. Typical air velocity entering the combustor (see Fig. 8.5) can be as high as 60 m/s (134 miles/hr), and the space available for combustion is relatively small. A well-designed combustor ensures continuous, stable, and efficient combustion while providing a uniform temperature distribution at the exit to prevent localized overheating. If the combustion process is not efficient, unburned or incompletely burned, hydrocarbons will increase hazardous emissions and reduce engine efficiency due to a reduction in heat release. Another major problem of incomplete combustion is the formation of carbon particles. These particles, through their relentless bombardment, erode the turbine blades and decrease the useful life of a turbine.

Air is generally supplied to a combustor in stages (see Anim. 8.B.*combustor*). In the primary zone of the burner, about 20% of the air is introduced with a swirl around the fuel nozzle which injects a fine spray of fuel (diesel, jet fuel, etc.) droplets. The swirling creates a good mix necessary for rapid combustion at high temperatures. About 30% of the air is introduced in the secondary zone where the combustion becomes complete. The remaining air is introduced in the third zone to cool down the combustion temperature to the turbine inlet temperature. Such cooling is necessary in a gas turbine due to the continuous nature in which the turbine blades are subjected to a high temperature flow. In a reciprocating engine, by contrast, exposure to high temperature is intermittent in a cylinder, allowing a much higher peak temperature.

For most hydrocarbon fuels, about 15 kg of air is required to burn 1 kg of fuel in a theoretical reaction (more about this in Chapter 13). The *air-fuel ratio* on a mass basis is around 60 in a gas turbine; about four times the amount of air needed for theoretical combustion. Treating the combustion gases as air is a very reasonable assumption in the cycle analysis, which dispels the common misconception that heat addition increases the pressure (as in the Otto cycle). Pressure does not increase in a combustor despite the heat release. Turbulence in the combustion chamber, necessary for mixing fuel with oxidizer, actually increases viscous friction and hence creates a slight pressure drop.

Turbine.   As with compressors, there are two basic types of turbines: **radial** and **axial**. The radial turbine, which is exactly the opposite of a centrifugal compressor, is not suitable for handling the high-temperature gas mixture. The axial turbine is similar to an axial compressor except the flow is in the reverse direction (see Anim. 8.A.*turbine* and Fig. 8.6). As the combustion gases encounter the stator blades connected to the casing, shaped so that the passage between two blades forms the contours of a converging-diverging nozzle (Fig. 8.4), they expand converting flow energy into kinetic energy. As explained in Sec. 4.2.2 (and in Anim. 4.A.*nozzle*), it is not only the pressure drop, but also the high inlet temperature [Eq. (4.21)] that causes a high-exit, even supersonic, velocity, for a compressible fluid such as air or steam. High velocity gases hitting rotor blades turns them along with the turbine shaft to which they are connected.

Depending on how the rotor blades are shaped, axial turbines can be divided into two categories: **impulse** and **reaction** turbines. In an impulse turbine the flow passage between the blades is of constant cross-sectional area resulting in no significant changes in pressure, temperature, or kinetic energy. The turning of the flow imparts a net force on the rotor blade creating the driving torque that rotates the turbine shaft. In a reaction turbine the flow passage is contoured like a nozzle so that the flow accelerates as it turns and expands. The impulse turbine is more economic to manufacture, but the reaction turbine is more efficient. In general the reaction type is used in gas turbines and the impulse type for steam turbines (Chapter 9). With density of the compressible fluid decreasing along the turbine pressure, the fluid passage is made progressively larger in an axial turbine to maintain a constant velocity, giving it its signature diverging shape.

On land based industrial turbines, where compactness or power-to-weight ratio is not important, thermal efficiency can be improved through the use of components such as intercoolers, reheat chambers, and regenerators. These devices will be introduced after the basic gas turbine is analyzed.

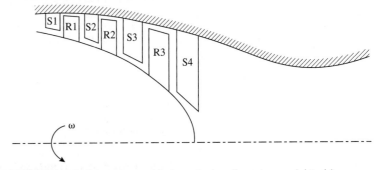

FIGURE 8.6 Stator and rotor blades pairs in a four-stage axial turbine.

8.2 THE AIR-STANDARD BRAYTON CYCLE

The *air-standard assumption*—treating the working fluid as air and using the PG (cold air) or IG (variable specific heats) model—used for analyzing reciprocating engines (Chapter 7) can also be applied to gas turbines. The theoretical model used to analyze a simple gas turbine is called the **Brayton cycle**. In the Brayton cycle, the working gas mixture is assumed to be represented by air throughout the cycle. Combustion is replaced by constant-pressure heat addition, and to complete the loop, a constant-pressure heat exchanger is added to cool the turbine exhaust to the

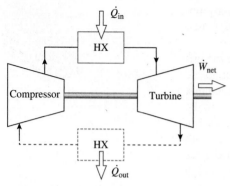

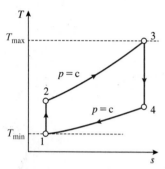

FIGURE 8.7 The Brayton cycle to model the simple gas turbine of Fig. 8.1.

FIGURE 8.8 A thermodynamic diagram of the Brayton cycle (see Anim. 8.A.*idealBraytonCycle*).

ambient temperature (compare Fig. 8.7 to Fig. 8.1). Each component device is considered steady and internally reversible, and changes in kinetic and potential energies are neglected. Shown in the *T-s* diagram of Figure 8.8 and Anim. 8.A.*idealBraytonCycle*, the Brayton cycle is composed of isentropic compression in the compressor (1-2), constant-pressure heat addition in the heat exchanger (2-3), isentropic expansion in the turbine (3-4), and constant-pressure heat rejection in the heat exchanger (4-1).

To derive expressions for the thermal efficiency and other performance parameters of the cycle, let us analyze each open-steady component with the *cold air-standard* assumption, i.e., using the PG model for air. In the following analysis, the symbols $\dot{W}_T$, $\dot{W}_C$, $\dot{Q}_{in}$, and $\dot{Q}_{out}$ represent absolute values of respective energy transfers and appropriate signs are added following the WinHip sign convention (Sec. 0.7.2) before they are used in the energy balance equations.

Device A (1-2: Isentropic compression):

$$\frac{d\cancel{E}^0}{dt} = \dot{m}j_i - \dot{m}j_e + \dot{\cancel{Q}}^0 - \dot{W}_{ext} = \dot{m}j_i - \dot{m}j_e + \dot{\cancel{Q}}^0 - (-\dot{W}_C);$$

$$\Rightarrow \quad \dot{W}_C = \dot{m}(j_2 - j_1) \cong \dot{m}(h_2 - h_1) = \dot{m}c_p(T_2 - T_1); \quad [\text{kW}] \quad (8.1)$$

Device B (2-3: Constant-pressure heat addition):

$$\frac{d\cancel{E}^0}{dt} = \dot{m}j_i - \dot{m}j_e + \dot{Q} - \dot{\cancel{W}}_{ext}^0 = \dot{m}j_i - \dot{m}j_e + \dot{Q}_{in};$$

$$\Rightarrow \quad \dot{Q}_{in} = \dot{m}(j_3 - j_2) \cong \dot{m}(h_3 - h_2) = \dot{m}c_p(T_3 - T_2); \quad [\text{kW}] \quad (8.2)$$

Device C (3-4: Isentropic expansion):

$$\frac{d\cancel{E}^0}{dt} = \dot{m}j_i - \dot{m}j_e + \dot{\cancel{Q}}^0 - \dot{W}_{ext} = \dot{m}j_i - \dot{m}j_e + \dot{\cancel{Q}}^0 - (\dot{W}_T);$$

$$\Rightarrow \quad \dot{W}_T = \dot{m}(j_3 - j_4) \cong \dot{m}(h_3 - h_4) = \dot{m}c_p(T_3 - T_4); \quad [\text{kW}] \quad (8.3)$$

Device-D (4-1: Constant-pressure heat rejection):

$$\frac{d\cancel{E}^0}{dt} = \dot{m}j_i - \dot{m}j_e + \dot{Q} - \dot{\cancel{W}}_{ext}^0 = \dot{m}j_i - \dot{m}j_e + (-\dot{Q}_{out});$$

$$\Rightarrow \quad \dot{Q}_{out} = \dot{m}(j_4 - j_1) \cong \dot{m}(h_4 - h_1) = \dot{m}c_p(T_4 - T_1); \quad [\text{kW}] \quad (8.4)$$

The **back work ratio (BWR)** is defined as the fraction of power output from the turbine that is consumed by the compressor. The net power $\dot{W}_{net}$, BWR, and thermal efficiency $\eta_{th,Brayton}$ can now be obtained as follows:

$$\dot{W}_{net} = \dot{W}_T - \dot{W}_C = \dot{Q}_{in} - \dot{Q}_{out} = \dot{m}c_p(T_3 - T_2 - T_4 + T_1); \quad [\text{kW}] \quad (8.5)$$

$$\text{BWR} \equiv \frac{\dot{W}_C}{\dot{W}_T} = \frac{h_2 - h_1}{h_3 - h_4} = \frac{T_2 - T_1}{T_3 - T_4} \quad (8.6)$$

$$\eta_{th,Brayton} \equiv \frac{\dot{W}_{net}}{\dot{Q}_{in}} = \frac{\dot{Q}_{in} - \dot{Q}_{out}}{\dot{Q}_{in}} = 1 - \frac{h_4 - h_1}{h_3 - h_2} = 1 - \frac{T_1(T_4/T_1 - 1)}{T_2(T_3/T_2 - 1)} \quad (8.7)$$

Typical BWR of a gas turbine is generally quite high, more than 50%. The isentropic relation for the PG model, Eq. (3.71), can be used to further simplify the expression for $\eta_{th,Brayton}$.

$$\frac{T_2}{T_1} = \left(\frac{p_2}{p_1}\right)^{(k-1)/k} = r_p^{(k-1)/k}; \quad \frac{T_3}{T_4} = \left(\frac{p_3}{p_4}\right)^{(k-1)/k} = \left(\frac{p_2}{p_1}\right)^{(k-1)/k} = \frac{T_2}{T_1};$$

$$\Rightarrow \quad \eta_{th,Brayton} = 1 - \frac{1}{r_p^{(k-1)/k}}; \quad \text{where} \quad r_p = \frac{p_2}{p_1} \quad (8.8)$$

The thermal efficiency of a simple Brayton cycle, thus, is a function of the pressure ratio r_p and the specific heat ratio k. Note that the degree of compression in an open compressor is expressed through the **pressure ratio** r_p, as opposed to the volumetric compression ratio r used for the closed compression processes in reciprocating devices (Chapter 7). For the same value of r_p and r, a Brayton cycle is less efficient (see Fig. 8.9) than an Otto cycle. For a given compression ratio $\eta_{th,Brayton}$ is independent of the maximum or minimum temperature. This surprising result will be discussed when we explore qualitatively the performance of the Brayton cycle with the help of an equivalent Carnot cycle in Sec. 8.2.3.

In the United States, the energy conversion efficiency is often expressed in dimensional **heat rate**, defined as the Btu's of heat supplied for generating 1 kWh of electricity. Disregarding any losses in the generator, the heat rate can be related to η_{th} as

$$\text{Heat Rate} = \frac{\dot{Q}_{in}(0.9478 \text{ Btu/kJ})}{\dot{W}_{net}/(3600 \text{ kJ/kWh})};$$

$$\Rightarrow \quad \text{Heat Rate} = \frac{3412}{\eta_{th}} \frac{\text{Btu}}{\text{kWh}} \quad (8.9)$$

Due to the fact that the size of a gas turbine scales with the mass flow rate of air, another useful quantity is the **specific power**, defined as

$$w_{net} \equiv \frac{\dot{W}_{net}}{\dot{m}}; \quad \left[\frac{\text{kJ}}{\text{kg}}\right] \quad (8.10)$$

The specific power becomes an important performance parameter for gas turbines used in propulsion.

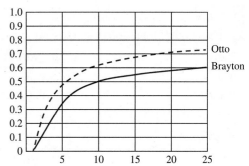

FIGURE 8.9 Thermal efficiency as a function of r_p and r for the cold air-standard Brayton and Otto cycles. Use the gas turbine simulator. (linked from TEST.*Interactives*) to reproduce this result.

8.2.1 TEST and the Open Gas Power Cycle TESTcalcs

The **open power cycle TESTcalcs**, located in the systems, open, steady, specific, power cycles branch, can be used to analyze gas as well as vapor power cycles. For modeling gas turbines, the PG, IG, and several gas mixture models can be used. The PC model is used for vapor power plants, and the binary models (PC/PC and PC/IG) are used for combined gas and vapor cycles. The open power cycle TESTcalcs build upon the *open-steady TESTcalcs* (Chapter 4) by adding a cycle panel. Once the component devices are analyzed in the device panel and the loop is complete, overall cycle variables are automatically updated in the cycle panel. If the mass flow rate is not given, a unit flow (1 kg/s) must be assumed for the device and cycle variables to be calculated on the basis of 1 kg of working fluid. The power output in kW in that case can be interpreted as the *specific power* output in kJ/kg.

The gas turbine simulator linked from the Interactive module is a visual simulator of the Brayton cycle; it covers the basic cycle and its several modifications. State properties and device variables can be directly entered, cycle properties are calculated once a system is completely defined, and parametric studies can be performed visually to evaluate the effect of any parameter (say, the compressor efficiency) on the performance of the cycle. Although not as flexible as a TESTcalc, the Interactive can be used for exploring the behavior of a gas turbine without a detailed knowledge of the cycle and by using the default values of all input parameters.

8.2.2 Fuel Consideration

As in reciprocating engines, the **specific fuel consumption (sfc)**, is defined as

$$\text{sfc} = \frac{\dot{m}_F}{\dot{W}_{net}}; \quad \left[\frac{\text{kg/s}}{\text{kW}} = \frac{\text{kg}}{\text{kJ}}\right] \quad (8.11)$$

Like an overall efficiency, it relates the amount of fuel consumed to the net work. The sfc is often used in place of thermal efficiency as it directly involves the fuel consumption rate, the major source of the operating cost of a gas turbine plant. The thermal efficiency is a more fundamental parameter independent of the type of fuel. If the **heating value** (defined in Chapter 13

DID YOU KNOW?

• A giant gas turbine can produce an output of 340 MW, which equals that of 13 jumbo jet engines.

Table 8-1   Lower heating values of various fuels used in gas turbines (see Table G-2).

Fuel	q_{comb} (MJ/kg)
JP-4	39.4
Natural Gas	45.0
Propane	46.4
Heavy Diesel	42.8

as the energy released by combustion of 1 kg of fuel) is represented by the symbol q_{comb}, the thermal efficiency and sfc can be related as:

$$\eta_{th} = \frac{\dot{W}_{net}}{\dot{Q}_{in}} = \frac{\dot{W}_{net}}{\dot{m}_F\, q_{comb}} = \frac{1}{sfc \cdot q_{comb}}; \quad \left[\frac{1}{(kg/kJ) \cdot (kJ/kg)} = 1\right] \quad (8.12)$$

Typical values of q_{comb} for different fuels used in gas turbines are listed in Table 8-1. For a more comprehensive listing, see Table G-2.

EXAMPLE 8-1   **Thermal Efficiency Related Parameters**

A gas turbine engine with a thermal efficiency of 30% produces 50 MW of power. If diesel, which has a heating value of 42.8 MJ/kg and costs $0.45 per kg, is used as the fuel, determine (a) the heat rate, (b) the fuel consumption rate, (c) the sfc, and (d) the fuel cost per kWh of power produced.

SOLUTION

Use the definitions of the basic operational parameters to obtain the answers.

Assumptions

The shaft power produced by the gas turbine is assumed to be completely converted to electrical power.

Analysis

From the basic definitions, the answers can be evaluated:

$$\dot{Q}_{in} = \frac{\dot{W}_{net}}{\eta_{th}} = \frac{50}{0.3} = 166.7 \text{ MW};$$

$$\dot{m}_F = \frac{\dot{Q}_{in}}{q_{comb}} = \frac{(166.7)(3600)}{42.8} = 14,019 \frac{kg}{h}$$

$$sfc = \frac{1}{\eta_{th} \cdot q_{comb}} = \frac{1}{(0.3)(42.8)} = 0.078 \frac{kg}{MJ} = \frac{0.078}{(1/3.6)} \frac{kg}{kWh} = 0.2808 \frac{kg}{kWh}$$

Therefore, the fuel cost is:

$$\frac{cost}{kWh} = \frac{cost}{kg \text{ of Fuel}} \cdot \frac{kg \text{ of Fuel}}{kWh} = (\$0.45)(0.2808) = \$0.13 \frac{1}{kWh}$$

Discussion

The cost to operate a gas turbine depends primarily on fuel cost, which can be seen to be directly proportional to sfc and inversely proportional to the thermal efficiency.

8.2.3 Qualitative Performance Predictions

Understanding how the performance of a gas turbine depends on different design and operational parameters is an important goal of gas turbine cycle analysis. We will use the approach established in Chapter 7, for closed power cycle analysis, to qualitatively predict the behavior of thermal efficiency and net power of various gas turbine cycles with respect to variation in different controlling parameters. The results obtained in this section will be used in later sections for modifications of the basic cycle in order to improve performance of gas turbines.

As with the reciprocating cycle, reversible heat transfers in the Brayton cycle can be interpreted as projected areas shown in the T-s diagram of Figure 8.10 (Sec. 4.1.3). As each device

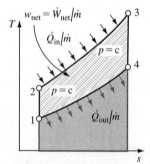

FIGURE 8.10 Geometric interpretation of the diagram of the Brayton cycle (see Anim. 8.A.*idealBraytonCycle*).

in the ideal Brayton cycle is internally reversible, Eq. (4.7) can be applied to simplify the expressions for the net power and specific power output from the cycle.

$$\dot{W}_{net} = \dot{Q}_{in} - \dot{Q}_{out} = \dot{m}\left[\int_2^3 Tds - \int_1^4 Tds\right]; \quad [\text{kW}] \qquad \textbf{(8.13)}$$

$$w_{net} = \frac{\dot{W}_{net}}{\dot{m}} = \frac{\dot{Q}_{in}}{\dot{m}} - \frac{\dot{Q}_{out}}{\dot{m}} = \int_2^3 Tds - \int_1^4 Tds; \quad \left[\frac{\text{kJ}}{\text{kg}}\right] \qquad \textbf{(8.14)}$$

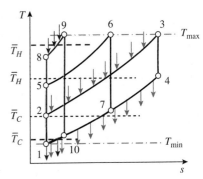

FIGURE 8.11 *T-s* diagrams comparing Brayton cycles with different pressure ratios but the same maximum and minimum temperatures. Use the gas turbine simulator (TEST.*Interactives*) to plot how the compression ratio affects the thermal efficiency or net power.

The area under the heat addition process, 2-3, represents the reversible heat transfer per unit mass of the flow $\dot{Q}_{in}/\dot{m}$; similarly, the shaded area under the heat rejection process represents $\dot{Q}_{out}/\dot{m}$. The enclosed area, which is the difference between the two areas, represents the specific power $w_{net} = \dot{W}_{net}/\dot{m}$. A modification that increases the enclosed area will indicate an increase in the specific power output and, hence, the net power for a given mass flow rate.

Now consider the effect of the pressure ratio r_p on the thermal efficiency and the net power. With the cold air-standard assumption, Eq. (8.8) tells us precisely how η_{th} increases with an increase in r_p, which is plotted in Figure 8.9. Given the same maximum and minimum temperatures, Brayton cycles with three different values of r_p are superimposed on a single *T-s* diagram in Fig 8.11. For cycles 1-2-3-4-1 and 1-8-9-10-1, corresponding to two extreme values of r_p, the heat addition and rejection processes are identified by a series of arrows. As the pressure ratio r_p is increased, the cycle becomes more slender, and the effective temperature of heat addition $\overline{T}_H$ can be seen to increase while the effective temperature of heat rejection $\overline{T}_C$ decreases. The equivalent Carnot efficiency being $1 - \overline{T}_C/\overline{T}_H$, the Brayton cycle efficiency must increase monotonically with an increase in r_p. Additionally, it can be deduced from Figure 8.11 that the enclosed area increases and then decreases as r_p is increased. In fact, the enclosed area approaches zero as r_p approaches its minimum or maximum limit of 1 and $(T_{max}/T_{min})^{\frac{k}{k-1}}$ respectively. For a given mass flow rate the net power is expected to have an inverted U-shaped profile as r_p is increased, reaching a peak at a certain pressure. Clearly, the mass flow rate, which has no effect on the efficiency, can be increased to boost the output; however, it should be kept in mind that the size and weight of the components generally scale with the mass flow rate. Increasing the power by boosting the mass flow rate carries the penalty of a bulkier power plant at much higher cost.

To see how T_{max} affects the performance we hold the compression ratio constant and construct two cycles with two different value of T_{max} in Figure 8.12. Both $\overline{T}_H$ and $\overline{T}_C$ can be seen to increase as T_{max} is increased, leaving the effect on efficiency inconclusive. This is consistent with the analytical expression for the thermal efficiency, Eq. (8.8), which shows that $\eta_{th,Brayton}$ is independent of T_{max} (in the PG model, k is assumed to be a constant). The net power, however, clearly increases as indicated by the enlargement of the enclosed area.

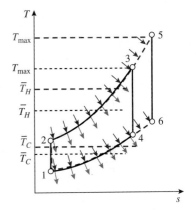

FIGURE 8.12 As T_{max} is raised, both $\overline{T}_H$ and $\overline{T}_C$ increase.

EXAMPLE 8-2 **Analyze Ideal Brayton Cycle**

Air enters the compressor of a 50 MW simple gas turbine at 100 kPa, 298 K. The pressure ratio of the compressor is 11 and the turbine inlet temperature is 1500 K. Using the cold air-standard Brayton cycle to model (PG model) the power plant, determine (a) the mass flow rate, (b) thermal efficiency, and (c) back-work ratio. **What-if scenario:** (d) What would the efficiency and mass flow rate be if the compression ratio varied over the range 4–50?

SOLUTION

Use the PG model to evaluate the four principal states and perform an energy balance for each open-steady device.

Assumptions

Cold air-standard assumption as discussed in Sec. 8.2.

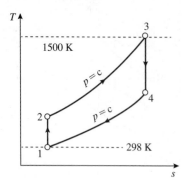

FIGURE 8.13 *T-s* diagram used in Ex. 8-2.

Analysis

From the perfect gas isentropic relations (Eq. 3.71), evaluate the pressure and temperature of the principal states shown in the accompanying *T-s* diagram (Fig. 8.13).

State-1 (given p_1, T_1): $\quad v_1 = \dfrac{\overline{R}T_1}{Mp_1} = \dfrac{(8.314)(298)}{(29)(100)} = 0.8543 \dfrac{m^3}{kg}$

State-2 (given r_p, $s_2 = s_1$): Use of isentropic relation leads to:

$$T_2 = T_1 r_p^{(k-1)/k} = (298)11^{(1.4-1)/1.4} = 591 \text{ K}$$

State-4 (given r_p, T_3): $T_4 = T_3/r^{(k-1)/k} = 1500/11^{(1.4-1)/1.4} = 756 \text{ K}$

An energy analysis is carried out for each device as follows:

$$\text{Device-A } (1\text{--}2): \quad \dot{W}_C = \dot{m}c_p(T_2 - T_1) = 294.6\dot{m} \text{ kW}$$

$$\text{Device-B } (2\text{--}3): \quad \dot{Q}_{in} = \dot{m}c_p(T_3 - T_2) = 913.3\dot{m} \text{ kW}$$

$$\text{Device-C } (3\text{--}4): \quad \dot{W}_T = \dot{m}c_p(T_3 - T_4) = 747.0\dot{m} \text{ kW}$$

$$\text{Device-D } (4\text{--}1): \quad \dot{Q}_{out} = \dot{m}c_p(T_4 - T_1) = 460.3\dot{m} \text{ kW}$$

The mass flow can now be obtained from the net power:

$$\dot{W}_{net} = \dot{W}_T - \dot{W}_C = 452.4\dot{m} = 50,000 \text{ kW}; \quad \Rightarrow \quad \dot{m} = 110.5 \dfrac{kg}{s}$$

The thermal efficiency and the back work ratio can now be calculated as

$$\eta_{th,Brayton} = \dfrac{\dot{W}_{net}}{\dot{Q}_{in}} = \dfrac{452.4\dot{m}}{911.6\dot{m}} = 49.6\% \quad \text{and} \quad BWR = \dfrac{\dot{W}_C}{\dot{W}_T} = 39.4\%$$

TEST Analysis

Launch the PG power-cycle TESTcalc. Calculate the four principal states based on a mass flow rate of 1 kg/s and analyze the devices as described in the TEST-code (see *TEST > TEST-codes*). In the cycle panel the net power is calculated as 452.4 kW. Evaluate $\dot{m}$ manually as 50,000/452.4 = 110.5 kg/s and use this value to replace mdot1. Click Super-Calculate to update the solution.

What-if scenario

Set the mass flow rate back to 1 kg/s. Change p2 to different values over a wide range—from '= 5*p1' to '= 30*p1'—updating the solution every time using the Super-Calculate button and calculating the mass flow rate manually. The results are listed in Table 8-2. The gas turbine simulation Interactive can be used for visual parametric studies. Simply set the inlet conditions, compression ratio, and turbine inlet temperature and click Calculate to obtain the cycle parameters, which are slightly different from the manual results since the Interactive uses the IG model. Click on the Parametric Study button and select a control parameter, a range, and click Analyze to develop graphical display of how any cycle variable depends on the control parameter.

Table 8-2 Parametric study in Ex. 8-2. Use the gas turbine simulator (TEST.*Interactives*) to reproduce this study visually (select Configuration tab, click Parametric Study, and click Analyze after selecting a parameter).

r_p	$\eta_{th,Brayton}$	$\dot{m}$ (kg/s)
5	36.88	131.4
10	48.23	111.7
15	53.90	108.3
20	57.54	108.5
30	62.19	112.6

Discussion

As predicted in the qualitative analysis, the thermal efficiency monotonically increases with an increase in the pressure ratio. The mass flow rate, however, decreases and reaches a minimum around $r_p = 15$ and then reverses its trend. Given a constant $\dot{W}_{net}$, we would expect this behavior as $\dot{m} = \dot{W}_{net}/w_{net}$ and w_{net}, the enclosed area inside the cycle, exhibits just the opposite behavior (the enclosed area can be seen to increase and then decrease as r_p is increased in Fig. 8.11).

8.2.4 Irreversibilities in an Actual Cycle

An actual gas turbine differs from the ideal Brayton cycle in several ways. Entropy generation in the compressor and turbine and pressure drop in the heat exchangers make the cycle distorted on a T-s diagram, as shown in Figure 8.14. The frictional losses in the heat exchangers, however, contribute relatively little towards this distortion and are ignored in the simplified T-s diagram of Figure 8.15, which is the basis for the subsequent analysis. The degree of irreversibilities in the compressor and turbine are represented by their respective isentropic efficiencies (Sec. 4.1.4), which range between 80–90% in modern turbo machineries. If the kinetic and potential energy changes are neglected, the enthalpies at the inlet, exit, and the isentropic exit state can be related through the compressor and turbine efficiencies as defined in Eqs. (4.29) and (4.33). In terms of the T-s diagram of Figure 8.15, they can be expressed as follows:

$$\eta_T = \frac{\dot{W}_{T,\text{actual}}}{\dot{W}_{T,\text{isentropic}}} = \frac{j_i - j_e}{j_i - j_{es}} \cong \frac{h_i - h_e}{h_i - h_{es}} = \frac{h_4 - h_6}{h_4 - h_5} \tag{8.15}$$

$$\eta_C = \frac{\dot{W}_{C,\text{isentropic}}}{\dot{W}_{C,\text{actual}}} = \frac{j_{es} - j_i}{j_e - j_i} \cong \frac{h_{es} - h_i}{h_e - h_i} = \frac{h_2 - h_1}{h_3 - h_1} \tag{8.16}$$

The following example shows the sensitivity of the cycle performance on these component efficiencies. Note that the combustion irreversibilities cannot be addressed in an air-standard analysis as heat is assumed to be added externally through a heat exchanger in the Brayton cycle.

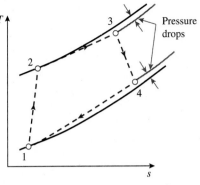

FIGURE 8.14 An actual Brayton cycle with irreversibilities in every component (see Anim. 8.A.*actualBraytonCycle*).

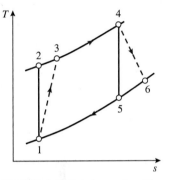

FIGURE 8.15 Brayton cycle with irreversibilities in the turbine and compressor (see Anim. 8.A.*actual BraytonCycle*).

EXAMPLE 8-3 | **Brayton Cycle Irreversibilities**

A stationary power plant, operating on a simple Brayton cycle, has a pressure ratio of 10. The compressor efficiency is 80% and its inlet conditions are 100 kPa, 300 K with a flow rate of 10 m³/s. The turbine efficiency is 85% and the turbine inlet temperature is 1400 K. Using the air-standard (with variable specific heat) Brayton cycle to model the power plant, determine (a) the power output, (b) thermal efficiency, and (c) back-work ratio. ***What-if scenario:*** (d) What would the thermal efficiency and net power be if the compressor efficiency dropped to 70%?

SOLUTION

Use the IG model to evaluate the six states (Fig. 8.15) and perform an energy balance for each open-steady device.

Assumptions

Air-standard assumptions outlined in Sec. 8.2.

Analysis

Use the IG power cycle TESTcalc (or Table D-3) to evaluate the following states.

State-1 (given p_1, T_1, $\dot{V}_1$):

$$\dot{m}_1 = \frac{\dot{V}_1}{v_1} = \frac{\overline{M}p_1\dot{V}}{\overline{R}T_1} = 11.62 \, \frac{\text{kg}}{\text{s}}; \quad h_1 = 1.873 \, \frac{\text{kJ}}{\text{kg}}$$

State-2 (given $p_2 = r_p p_1$, $s_2 = s_1$):

$$T_2 = 572 \text{ K}; \quad h_2 = 281.1 \, \frac{\text{kJ}}{\text{kg}}$$

State-3 (given η_C, $p_3 = p_2$): Use Eq. (8.16) to derive an algebraic expression for h_3:

$$h_3 = h_1 + \frac{h_2 - h_1}{\eta_C} = 350.9 \, \frac{\text{kJ}}{\text{kg}}; \quad T_3 = 638.1 \text{ K}$$

State-4 (given $p_4 = p_3$, T_4):

$$h_4 = 1217.8 \frac{kJ}{kg}$$

State-5 (given $p_5 = p_4/r_p$, $s_5 = s_4$):

$$T_5 = 786.8 \text{ K}; \quad h_5 = 511.3 \frac{kJ}{kg}$$

State-6 (given η_T, $p_6 = p_5$): Use Eq. (8.15) to derive an algebraic expression for h_6:

$$h_6 = h_4 - \eta_T(h_4 - h_5) = 617.3 \frac{kJ}{kg}; \quad T_6 = 882.9 \text{ K}$$

An energy analysis is carried out for each open-steady device:

Device-A (1–3): $\dot{W}_C = -\dot{W}_{1-3} = \dot{m}(j_3 - j_1) \cong \dot{m}(h_3 - h_1) = 4054$ kW

Device-B (3-4): $\dot{Q}_{in} = \dot{Q}_{3-4} = \dot{m}(j_4 - j_3) \cong \dot{m}(h_4 - h_3) = 10{,}069$ kW

Device-C (4–6): $\dot{W}_T = \dot{W}_{4-6} = \dot{m}(j_4 - j_6) \cong \dot{m}(h_4 - h_6) = 6975$ kW

Device-D (6–1): $\dot{Q}_{out} = -\dot{Q}_{6-1} = \dot{m}(j_6 - j_1) \cong \dot{m}(h_6 - h_1) = 7147$ kW

The thermal efficiency, back-work ratio, and the net output are:

$$\dot{W}_{net} = 6975 - 4054 = 2921 \text{ kW}; \quad \eta_{th,Brayton} = \frac{\dot{W}_{net}}{\dot{Q}_{in}} = \frac{2921}{10{,}069} = 29.0\%;$$

$$\text{BWR} = \frac{\dot{W}_C}{\dot{W}_T} = 58.1\%$$

TEST Analysis

Launch the IG power-cycle TESTcalc. Calculate the states and analyze the devices as described in the TEST-code (see *TEST > TEST-codes*). The manual answers are reproduced in the cycle panel.

What-if scenario

Change the compressor efficiency to 70% by changing the expression for h3 to 'h3 = h1 + (h2 − h1)*0.7'. Press the Enter key and click Super-Calculate. The new answers are calculated as 2342 kW, 66.4%, and 24.68% respectively.

Discussion

To explore the importance of the compressor efficiency, the thermal efficiency and the net output are calculated for different values of η_C (by modifying the expression for h3 with different values of the compressor efficiency) and compiled in Table 8-4. The table shows that the gas turbine ceases to produce any net power when the compressor efficiency dips below 47%. The gas turbine simulation Interactive can also be used for such parametric studies.

Table 8-3 Parametric study in Ex. 8-3, exploring the effect of compressor efficiency on the cycle performance. Use the gas turbine simulator (TEST.*Interactives*) to reproduce this study.

η_C	$\eta_{th,Brayton}$	$\dot{W}_{net}$ (MW)
0.8	0.290	2.92
0.7	0.247	2.34
0.6	0.180	1.570
0.5	0.064	0.489
0.47	0.010	0.075

Table 8-4 Parametric study in Ex. 8-4 with the turbine inlet temperature as a parameter.

T_{max} (K)	$\eta_{th,Brayton}$	$\dot{W}_{net}$ (MW)
1800	0.318	5.00
1400	0.290	2.92
1200	0.258	1.88
1000	0.184	0.854
850	0.034	0.091
832	0.001	0.002

EXAMPLE 8-4   **Brayton Cycle Parametric Studies**

In Example 8-3 determine how (a) the power output and (b) the thermal efficiency vary with the turbine inlet temperature.

TEST Analysis

Launch the IG power-cycle TESTcalc, copy the TEST-code generated in Example 8-3 into the I/O panel, and click the Load button. The solution of Example 8-3 is reproduced. Now change T4 to a new value, press the Enter key, and click Super-Calculate. Obtain the thermal efficiency and net output from the cycle panel then repeat the procedure with different values of T4. The results are listed in Table 8-4.

Discussion

As can be seen from Table 8-4, the thermal efficiency and the net output clearly decrease with a decrease in the maximum cycle temperature. The net power and the efficiency go down to zero as the turbine inlet temperature drops below 832 K. According to the cold air-standard (PG model) solution for the ideal Brayton cycle, the thermal efficiency is independent of the turbine inlet temperature as can be seen from Eq. (8.8). The current cycles, however, differ in two aspects; use of the IG model and irreversibilities of the compressor and turbine. To further isolate the reason behind the observed behavior, repeat the parametric study with the IG model after setting the turbine and compressor efficiencies to 100% by using h3 = h2 and h6 = h5. With the irreversibilities removed, the thermal efficiency can be found to increase slightly as the turbine inlet temperature is reduced. Therefore, it is the irreversibilities in the turbine and compressor that must be responsible for the increase of thermal efficiency with an increase in the turbine inlet temperature.

Note that the equivalent Carnot cycle analogy cannot be used for an irreversible cycle. At the turbine inlet temperature of 832 K (Table 8-4), the net work output of the cycle is almost zero even though the T-s diagram of Figure 8.16 still indicates a finite area enclosed within the loop. This is not a contradiction, as the area under an irreversible T-s curve can no longer be interpreted as the heat transfer (Sec. 4.1.3).

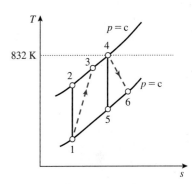

FIGURE 8.16 For a cycle that is irreversible, the net area is not equal to the specific net work.

8.2.5 Exergy Accounting of Brayton Cycle

The previous example illustrates how detrimental irreversibilities can be to the performance of a gas turbine. To evaluate the relative severity of irreversibilities in different components, there is no substitute for a complete exergy inventory. An exergy balance sheet is analogous to a financial balance sheet. It tells us exactly where in the cycle exergy is destroyed. In the Brayton cycle, exergy is not only destroyed through irreversibilities in the turbine and compressor, but is also discarded with the exhaust gas. Combustion is another major source of irreversibilities (*chemical friction*, to be addressed in Chapter 13) and exergy destruction.

To avoid an exergy analysis of the combustion process, to be discussed in Chapter 13, we will replace the combustion process with a heat transfer (equal to the heat released by the fuel) from an external source at an effective temperature. The source temperature is selected such that the exergy destruction due to *thermal friction* is equivalent to the exergy destruction due to thermodynamic friction (*thermal, chemical* and *mixing friction*) in an actual combustion chamber. A combustion analysis (Chapter 13) can later be used to obtain this equivalent source temperature, but for now it will be assumed known.

For a single-flow steady device that exchanges heat with a single non-atmospheric reservoir at T_k, the exergy balance equation, Eq. (6.10), simplifies to:

$$0 = \underbrace{\dot{m}(\psi_i - \psi_e)}_{\substack{\text{Net flow exergy} \\ \text{transported into} \\ \text{the system}}} + \underbrace{\dot{Q}_k\left(1 - \frac{T_0}{T_k}\right)}_{\substack{\text{Net exergy transfer} \\ \text{by heat from the} \\ \text{external source}}} - \underbrace{\dot{W}_{sh}}_{\substack{\text{Exergy transfer} \\ \text{through shaft} \\ \text{work}}} - \underbrace{\dot{i}}_{\substack{\text{Exergy} \\ \text{destruction}}} \quad ; \quad [\text{kW}] \qquad \textbf{(8.17)}$$

This equation (see Anim. 6.A.*exergyBalanceTerms*) can be applied to each component of the gas turbine cycle. The flow exergy ψ is evaluated as a state property once the dead state is specified as State-0 in the TESTcalcs. Once the principal states are evaluated, the exergy destruction rate $\dot{i}$ can be calculated from Eq. (8.17). This is illustrated in Example 8-5.

EXAMPLE 8-5 **Brayton Cycle Exergy Accounting**

For the gas turbine described in Example 8-3, develop a balance sheet for exergy accounting as the air passes through various components. Assume the heat addition to take place from a source at 1800 K and the inlet conditions are same as the ambient atmospheric conditions.

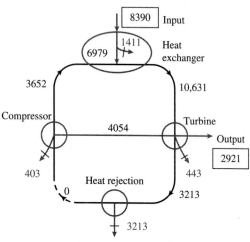

FIGURE 8.17 Exergy flow diagram (all numbers in kW) for Ex. 8-5. Crossed lines indicate exergy destruction.

TEST Analysis

In the TEST solution of Example 8-3, evaluate State-0 as the dead state for the given atmospheric conditions, 100 kPa and 300 K. Click Super-Calculate to evaluate the specific flow exergy of all principal states. Coupled with the heat and work transfer calculated for each device in Example 8-3, the exergy balance equation, Eq. (8.17), can now be used to construct the following balance sheet (Fig. 8.17).

Exergy supplied to the gas by the heat exchanger:

$$\dot{Q}_{in}\left(1 - \frac{T_0}{T_B}\right) = (10,069)\left(1 - \frac{300}{1800}\right) = 8390 \text{ kW}$$

Exergy gained by the gas from heat addition:

$$\dot{m}(\psi_4 - \psi_3) = (11.615)(915.3 - 314.4) = 6979 \text{ kW}$$

Exergy destroyed during heat addition:

$$8390 - 6979 = 1411 \text{ kW}$$

Exergy delivered to the turbine by the gas:

$$\dot{m}(\psi_4 - \psi_6) = (11.615)(915.3 - 276.7) = 7418 \text{ kW}$$

Exergy delivered by the turbine to the shaft:

$$\dot{W}_T = 6975 \text{ kW}$$

Exergy destroyed in the turbine:

$$\dot{I} = 7418 - 6975 = 443 \text{ kW}$$

Exergy lost by the gas through heat rejection:

$$\dot{m}(\psi_6 - \psi_1) = (11.615)(276.6 - 0) = 3213 \text{ kW}$$

Exergy supplied to the atmosphere:

$$\dot{Q}_{out}\left(1 - \frac{T_0}{T_0}\right) = 0 \text{ kW}$$

Exergy destroyed during heat rejection:

$$\dot{I} = 3213 - 0 = 3213 \text{ kW}$$

Exergy gained by the gas during compression:

$$\dot{m}(\psi_3 - \psi_1) = (11.615)(314.4 - 0) = 3651 \text{ kW}$$

Exergy input into the compressor as shaft work:

$$\dot{W}_C = 4054 \text{ kW}$$

Exergy destroyed in the compressor:

$$\dot{I} = 4054 - 3651 = 403 \text{ kW}$$

Discussion

The relative significance of exergy destruction in various components is clear from this balance sheet. While a considerable amount of exergy is destroyed in the turbine and compressor, the bulk of the destruction takes place during the heat rejection (exhaust) and heat addition (combustion) processes. The exergy flow diagram of Figure 8.17 shows different pathways of exergy destructions. From the exergy flow diagram, the exergetic efficiency for the cycle, defined as the ratio of the desired exergy output to required input, can be calculated as $2921/8390 = 34.8\%$.

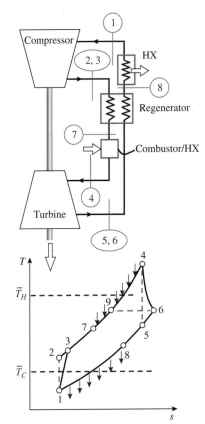

FIGURE 8.18 A gas turbine with a regenerator (see Anim. 8.A.*regenCycle*). The red arrows represent external heat addition and rejection in the modified cycle.

8.3 GAS TURBINE WITH REGENERATION

It is clear from the exergy analysis of Example 8-5 that the turbine exhaust is the single biggest source of exergy destruction. One elaborate way to capture some of the wasted exergy is to use a secondary cycle, usually a vapor power plant, with the gas turbine exhaust as the heating source. The resulting system known as the *combined cycle* will be discussed in Chapter 9. A simpler and more common approach is the use of a **regenerator**.

Regeneration means taking heat from one part of the cycle and putting it into another to increase the effective temperature of heat addition, and/or reduce the effective temperature of heat rejection, so as to increase the thermal efficiency. The regenerator incorporated in the Brayton cycle of Figure 8.18 is simply a counterflow heat exchanger (see Anim. 4.B.*heatExchanger*) that utilizes the hot exhaust gas to preheat the air entering the combustor. The turbine exhaust gas entering the regenerator at State-6 is cooled to State-8 as the air is heated from State-3 to State-7 without burning any additional fuel. With external heat addition eliminated between State-3 and State-7, the effective temperature of heat addition $\overline{T}_H$ is raised by regeneration. At the same time the effective temperature of heat rejection $\overline{T}_C$ is lowered as the highest temperature at which heat is rejected decreases from T_6 to T_8. Even for an irreversible cycle like this one, the thermal efficiency can be expected to increase with regeneration.

Obviously, T_7, the highest temperature of the pre-heated flow, cannot be greater than $T_9 = T_6$, the temperature of the turbine exhaust (Fig. 8.18). Enthalpy being a function of temperature only for an ideal gas, $T_9 = T_6$ implies $h_9 = h_6$ and the maximum possible enthalpy increase for air in the regenerator must be $h_9 - h_3 = h_6 - h_3$. The **regenerator effectiveness** is defined as the ratio of the actual enthalpy rise to the maximum possible:

$$\varepsilon_{\text{reg}} \equiv \frac{h_8 - h_3}{h_6 - h_3}; \quad \text{For the PG model,} \quad \varepsilon_{\text{reg}} = \frac{T_8 - T_3}{T_6 - T_3} \qquad \textbf{(8.18)}$$

The expression for the regenerator simplifies to a ratio of temperature differences only if the PG model is used for air (cold air-standard model).

The compactness of a regenerator depends on the area of contact between the two streams. For the same amount of heat transfer, the size of the regenerator can be reduced by increasing the temperature difference ΔT maintained between the hot and cold stream (Fig. 8.19). However, for a given exhaust temperature T_6, raising ΔT means lowering T_8, which results in a lower h_8, and hence, lower η_{reg}. As discussed in Chapter 6, a greater ΔT also makes the heat exchanger exergetically less efficient. A large heat exchanger is clearly more desirable from an efficiency standpoint. However, the large size has an undesirable side effect—the pressure drop due to friction becomes significantly large, reducing the expansion ratio and the turbine output. Typical values of ε_{reg} in industrial gas turbines range from 60% to 80%.

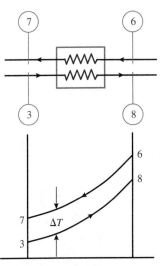

FIGURE 8.19 Temperature distribution in a counter flow heat exchanger (see Anim. 4.B.*heatExchanger*).

EXAMPLE 8-6	**Brayton Cycle Modified with Regeneration**

In the gas turbine of Example 8-3, a regenerator with an effectiveness of 80% is incorporated as shown in Figure 8.18. Determine (a) the exergetic efficiency of the regenerator and (b) the thermal efficiency of the cycle. *What-if scenario:* (c) How would the thermal efficiency vary if the regenerator effectiveness varied from 0 through 1?

Assumptions

To complete the air-standard cycle, a heat exchanger is incorporated between the regenerator and compressor to cool the air from State-7 to State-1 (Fig. 8.18).

TEST Analysis

Reproduce the TEST solution of Example 8-3 as was done in Example 8-4. Evaluate State-0 as the dead state for the given atmospheric conditions, 100 kPa and 300 K. Click Super-Calculate. The introduction of the regenerator gives rise to two new states, State-8 and State-7, while all other states remain unchanged.

State-8 (given $p_8 = p_3$): Use the definition of regenerator effectiveness to derive $h_8 = h_3 + \varepsilon_{reg}(h_6 - h_3)$. Enter p8 = p3 and h8 as '= h3 + 0.8*(h6 − h3)' and calculate State-8.

State-7 (given $p_7 = p_6$): Use an energy balance on the adiabatic regenerator, a multi-flow, non-mixing, open-steady system, to relate h_7 with other known properties.

$$\frac{dE^0}{dt} = \sum_i \dot{m}_i j_i - \sum_e \dot{m}_e j_e + \dot{Q}^0 - \dot{W}_{ext}^0;$$

$$\Rightarrow \quad \dot{m}(j_6 - j_7) = \dot{m}(j_8 - j_3); \quad \Rightarrow \quad h_6 - h_7 \cong h_8 - h_3 \tag{8.19}$$

Enter p7 = p6 and h7 as '= h6 − (h8 − h3)' and calculate State-7. Now that all the principal states are completely evaluated, the exergetic efficiency of the regenerator can be calculated in the I/O panel by simply typing in the expression '= (psi8 − psi3)/(psi6 − psi7)'.

$$\eta_{II,reg} = \frac{\text{Desired Exergy Output}}{\text{Required Exergy Input}} = \frac{\dot{m}(\psi_8 - \psi_3)}{\dot{m}(\psi_6 - \psi_7)} = 95.8\%$$

Before the cycle efficiency can be evaluated, each device must be analyzed separately. In the device panel, analysis of the compressor and turbine remain the same as in Example 8-3. However, the inlet state of the combustor (Device-B) and heat exchanger (Device-D) must be adjusted to reflect the presence of the regenerator. To set up the regenerator, choose a new device, Device-E, select the Non-Mixing button, select the appropriate inlet and exit states, enter Wdot_ext = 0, and click Calculate. The heat transfer is calculated as zero as expected. Super-Calculate and obtain the thermal efficiency from the cycle panel as 38.47%.

Table 8-5 A parametric study with the regenerator efficiency as a parameter in Ex. 8-6.

ε_{reg}	$\eta_{th,Brayton}$	$\dot{W}_{net}$ (MW)
0	0.290	2.92
0.25	0.314	2.92
0.50	0.343	2.92
0.75	0.377	2.92
1.0	0.419	2.92

What-if scenario

Change the regenerator effectiveness that appears in the expression of h8 to zero. Press the Enter key and click Super-Calculate. As expected, the significantly reduced thermal efficiency, 29.0%, is the same as the efficiency of the Brayton cycle of Example 8-3 without a regenerator. Table 8-5 is created by repeating this calculation procedure with different values of ε_{reg}. The thermal efficiency can be seen to increase by 5.3% for a regenerator with $\varepsilon_{reg} = 0.5$ and by 12.9% for an ideal regenerator with $\varepsilon_{reg} = 1.0$. The gas turbine simulation Interactive can also be used to produce this result.

Discussion

The highest thermal efficiency, achieved with a perfect regenerator, is 41.9% as listed in Table 8-5. Moreover, if the compressor and turbine are made ideal (modify h3 and h6 with an isentropic efficiency of 100% and click Super-Calculate), the thermal efficiency of an ideal regenerative cycle can be shown to reach a respectable 60.48%, even though it is well short of the Carnot efficiency of 78.57% for a reversible engine running between the maximum and minimum temperatures of the cycle. Also note from Table 8-5 that the net power developed by the cycle is not affected by regeneration as the enclosed area in the T-s diagram in Figure 8.18 is not altered by the introduction of the regenerator.

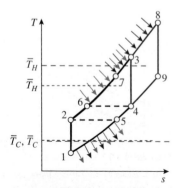

FIGURE 8.20 If T_{max} is increased, $\overline{T}_H$ increases but $\overline{T}_C$ remains constant in a regenerative cycle.

8.4 GAS TURBINE WITH REHEAT

The maximum temperature T_{max} of a gas turbine cycle, alternatively known as the combustor exit temperature or the turbine inlet temperature, is controlled by adjusting the air-fuel ratio, more specifically, by controlling the fuel injection rate for a given mass flow of air. A higher T_{max} increases both $\overline{T}_H$ and $\overline{T}_C$ (see Fig. 8.12) without having much effect on the thermal efficiency. If a regenerator is present as shown in the T-s diagram of Figure 8.20,

$\overline{T}_H$ still increases with an increase in T_{max}, but $\overline{T}_C$ becomes independent of T_{max} as heat rejection takes place between State-1 and State-5, regardless of the value of T_{max}. In a regenerative cycle, therefore, the thermal efficiency increases with an increase in T_{max}. The net power also increases as the net area inside the T-s diagram can be seen to increase in Figure 8.20. For an irreversible cycle, we have already seen in Example 8-4 that the thermal efficiency and the net power both increase with T_{max} even in the absence of a regenerator. It is, therefore, quite desirable to have as high a turbine inlet temperature as possible in a gas turbine regardless of whether a regenerator is used or not.

If hydrocarbon fuels are burned with the right amount of air (in Chapter 13 the theoretical air-fuel ratio will be shown to be about 15) in a constant-pressure combustor, it is possible to reach a maximum temperature greater than 2000 K at the turbine inlet. For metallurgical reasons, the maximum turbine inlet temperature is limited to 1500 K or less. To bring the temperature down to the desired turbine inlet temperature, excess air is used in the combustor, sometimes three to four times as much as the theoretical amount (see Anim. 8.A.*Combustor*).

The idea of reheat is to utilize the excess air at the turbine exit to burn additional fuel to improve the performance of a gas turbine. In the arrangement shown in Figure 8.21 (and Anim. 8.A.*reheatCycle*), the turbine is split into two stages with a **reheat combustor** placed in between to raise the temperature of the gas mixture back to T_{max}. In the corresponding air-standard cycle, the air is partially expanded in the first turbine to an intermediate pressure p_4, at which it is reheated in a heat exchanger at constant pressure from State-4 to State-5 while the expansion to $p_6 = p_1$ is completed in the second turbine. It is clear from the T-s diagram of Figure 8.21 that the net area and, hence, the net power, are increased by reheat. The efficiency may not be affected much as both $\overline{T}_H$ and $\overline{T}_C$ increase simultaneously. However, the turbine exhaust temperature is significantly raised by reheat—T_6 with reheat as opposed to T_7 without reheat as in Figure 8.21—increasing the regeneration potential. This is illustrated in the following example and in Anim. 8.A.*reheatRegenCycle*.

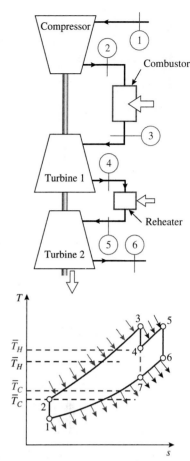

FIGURE 8.21 $\overline{T}_H$ and $\overline{T}_C$ both increase in a reheat cycle without regeneration.

EXAMPLE 8-7 Brayton Cycle Modified with Regeneration and Reheat

The following data are supplied for an ideal air-standard (variable c_p) Brayton cycle modified with a reheater and regenerator. Compressor inlet conditions: 100 kPa, 298 K; compressor pressure ratio: 12; first turbine inlet temperature: 1200 K; second turbine inlet conditions: 350 kPa, 1200 K. Assume the regenerator to have an effectiveness of 100%. Determine (a) the thermal efficiency of the cycle. *What-if scenario*: (b) How would the thermal efficiency vary if the intermediate pressure between the turbine stages varied through the entire possible range?

SOLUTION

Use the IG model to evaluate the eight principal states (Fig. 8.22), perform energy analysis on each open-steady device, and obtain the cycle parameters.

Assumptions

For an ideal cycle, the compression and expansion devices are isentropic and there is no pressure drop in the combustors.

Analysis

Use the IG power-cycle TESTcalc or the IG table for air, Table D-3, to evaluate the enthalpy at each principal state.

State-1 (given p_1, T_1): $h_1 = -0.13$ kJ/kg

State-2 (given $p_2 = r_p p_1$, $s_2 = s_1$): $h_2 = 307.7$ kJ/kg

State-3 (given T_3, $p_3 = p_2$): $h_3 = 979.9$ kJ/kg

State-4 (given p_4, $s_4 = s_3$): $h_4 = 614.5$ kJ/kg

State-5 (given T_5, $p_5 = p_4$): $h_5 = h_3 = 979.9$ kJ/kg

State-6 (given $p_6 = p_1$, $s_6 = s_5$): $h_6 = 609.3$ kJ/kg

State-7 (given $p_7 = p_2$, $\varepsilon_{reg} = 1.0$): $h_7 = h_2 + (h_6 - h_2)\varepsilon_{reg} = h_6 = 609.3$ kJ/kg

State-8 (given $p_8 = p_1$): An energy balance on the regenerator (Eq. (8.19)) produces

$$h_8 = h_2 = 307.7 \text{ kJ/kg}$$

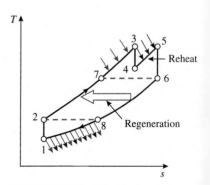

FIGURE 8.22 Ideal Brayton cycle with reheat and regeneration analyzed in Ex. 8-7 (also see Anim. 8.A.*reheatRegenCycle*).

Table 8-6 A parametric study with the reheat pressure as a parameter in Ex. 8-7. Use the gas turbine simulator (TEST.*Interactives*) to reproduce this study.

p_4 (kPa)	η_{th}	$\dot{W}_{net}$ (kW)
1200	0.515	327
800	0.554	383
500	0.577	419
350	0.582	428
200	0.570	408
100	0.515	327

Based on unit mass flow rate the specific net power and external heat addition can be calculated as:

$$\dot{W}_{net}/\dot{m} = (h_3 - h_4) + (h_5 - h_6) - (h_2 - h_1) = 428.2 \text{ kJ/kg};$$

$$\dot{Q}_{in}/\dot{m} = (h_3 - h_7) + (h_5 - h_4) = 736.0 \text{ kJ/kg}$$

Therefore, $\quad \eta_{th} = \dfrac{\dot{W}_{net}}{\dot{Q}_{in}} = \dfrac{428.2}{736.0} = 58.2\%$

TEST Analysis

Use a mass flow rate of 1 kg/s as the basis. After evaluating the eight principal states, analyze all the seven devices as described in the TEST-code (see *TEST > TEST-codes*). For Device-B, the regenerator, the 'Non-Mixing' button must be checked. Use the Super-Calculate button to obtain the thermal efficiency in the cycle panel.

What-if scenario

Change the reheat pressure, p4, to different values spanning the entire possible range: from 1200 kPa to 100 kPa. For each new pressure, click Super-Calculate to produce the cycle efficiency and net output (based on a flow rate of 1 kg/s). The results are listed in Table 8-6. The gas turbine simulation Interactive can also be used to produce this table. Both the efficiency and net work can be seen to reach a peak when the intermediate pressure is close to 346.4 kPa, which is equal to $\sqrt{p_3 p_6}$.

Discussion

The reheat cycle reduces to the same regenerative Brayton cycle in the two limits, as one of the turbines is eliminated in each limit. This is confirmed from the identical results at the two extreme limits—$p_4 = 1200$ kPa and $p_4 = 100$ kPa (Table 8-6). Although we have numerically verified that $p_4 \approx \sqrt{p_3 p_6}$ for the optimal performance of a reheat cycle, this important result can be theoretically established by deriving an expression for efficiency using the PG model, differentiating it with respect to p_4, and equating the result to zero. For the IG model, however, there is no alternative to a numerical parametric study.

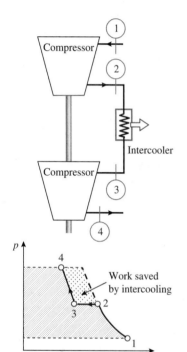

FIGURE 8.23 A two-stage compressor with an intercooler.

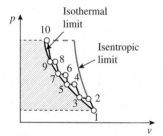

FIGURE 8.24 If the number of stages is increased, the isothermal limit is reached, minimizing the compressor work.

8.5 GAS TURBINE WITH INTERCOOLING AND REHEAT

The benefits of multistage turbines, with reheat and regeneration in terms of increased output and efficiency, can be duplicated by replacing the compressor with **intercooled multistage compressors**. In Example 4-10 we demonstrated that the compressor work is minimized when the intermediate pressure of a two-stage compressor is chosen as the square root of the product of the end pressures of the original compressor, $\sqrt{p_1 p_4}$ in Figure 8.23. The saving can be increased if the compression is broken into even more stages, as shown in Figure 8.24. In the limit of infinite number of stages, the total compressor power approaches the minimum power required by an internally reversible isothermal compressor operating between the same end pressures (see Anim. 4.A.*isothermalCompressor*).

An ideal Brayton cycle with intercooled two-stage compression, two-stage expansion with reheat and regeneration is depicted on the *T-s* diagram of Figure 8.25 (and Anim. 8.A.*reheatRegenIntercooled*). The turbine and compressors are assumed isentropic, the regenerator effectiveness is assumed to be one, pressure loss due to friction is neglected, and the reheat and intercooling are assumed to be ideal so that $T_8 = T_6$ and $T_3 = T_1$. Obviously, the reduction of compressor work through multi-staging does not affect the turbine power as the inlet and exit conditions of the turbines remain unaffected. Intercooling also does not affect external heat addition and $\overline{T}_H$ remains unaltered. It does reduce the compressor work and increases the net output, as is evident from a comparison of enclosed areas 1-2-3-4-6-7-8-9-1 (intercooled) and 1-4′-6-7-8-9-1 (original) in Figure 8.25. To understand the effect of intercooling on the thermal efficiency, let us see how $\overline{T}_C$ is affected. Without intercooling, regeneration would have eliminated external heating from 4′-5 and external heat rejection from 9-10′. Intercooling lowers the compressor exit temperature from $T_{4'}$ to T_4 and the enhanced regeneration eliminates external heat rejection from State-10' to State-10, thereby reducing $\overline{T}_C$ significantly (shown in red in Fig. 8.25), to an increase in the cycle efficiency when compared the same arrangement without intercooling. The thermal efficiency of the cycle, as estimated by the equivalent Carnot cycle, $1 - \overline{T}_C/\overline{T}_H$, therefore, increases due to intercooling.

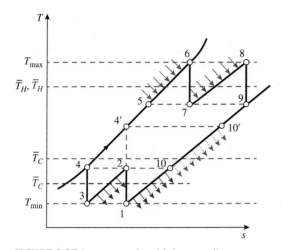

FIGURE 8.25 Brayton cycle with intercooling, reheat, and regeneration (see Anim. 8.A.*reheat RegenIntercooled*).

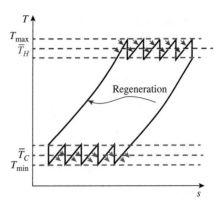

FIGURE 8.26 As the number of compressor and turbine stages is increased, $\overline{T}_C \rightarrow T_{min}$ and $\overline{T}_H \rightarrow T_{max}$.

The Carnot efficiency based on the maximum and minimum temperatures of the cycle, $1 - T_{max}/T_{min}$, is still higher than the two-stage cycle of Figure 8.25. However, if the number of stages are increased indefinitely $\overline{T}_C \rightarrow T_{min}$ and $\overline{T}_H \rightarrow T_{max}$, as can be deduced from Figure 8.26. In the limit of an infinite number of stages, the multistage intercooled compressors can be replaced with a single isothermal compressor and the multistage turbine with a single isothermal turbine.

To maintain constant temperature, heat must be added to the turbine and rejected from the compressor, which then constitute the only external heat transfers for the cycle, eliminating the need for a separate combustor. The necessary preheating of the gas before it enters the turbine is accomplished through a regenerator as shown in Figure 8.27. The resulting cycle, known as the **Ericsson cycle**, has the same efficiency as the Carnot cycle as $1 - \overline{T}_C/\overline{T}_H = 1 - T_{min}/T_{max}$. To show this more rigorously, the external heat transfer to the compressor and turbine can be calculated using the PG model (cold-air standard) as follows:

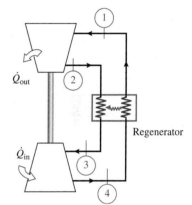

FIGURE 8.27 Implementation of the Ericsson cycle (see Anim. 8.A.*EricssonCycle*).

$$\dot{Q}_{in} = \dot{m} \int_i^e T ds = \dot{m} T_{max}(s_4 - s_3) = \dot{m} T_{max} c_p \ln \frac{p_3}{p_4}; \quad [\text{kW}] \tag{8.20}$$

$$\dot{Q}_{out} = \dot{m} T_{min}(s_1 - s_2) = \dot{m} T_{min} c_p \ln \frac{p_2}{p_1} = \dot{m} T_{min} c_p \ln \frac{p_3}{p_4}; \quad [\text{kW}] \tag{8.21}$$

Therefore,

$$\eta_{th,Ericsson} = \frac{\dot{W}_{net}}{\dot{Q}_{in}} = \frac{\dot{Q}_{in} - \dot{Q}_{out}}{\dot{Q}_{in}} = 1 - \frac{\dot{Q}_{out}}{\dot{Q}_{in}} = 1 - \frac{T_{min}}{T_{max}} \tag{8.22}$$

Although the Ericsson cycle is reversible and has the same efficiency as the Carnot cycle, it is not a Carnot cycle, because external heat transfers do not take place at constant temperatures (the *T-s* diagram is not a rectangle).

8.6 REGENERATIVE GAS TURBINE WITH REHEAT AND INTERCOOLING

Reheat and intercooling have each been found to increase the net output and, coupled with regeneration, each have been shown to boost the thermal efficiency of the gas turbine. A combination of reheat and intercooling along with regeneration leads to increase in both thermal efficiency and net output. In this section we study a regenerative gas turbine with reheat and intercooling with the help of an example (see Anim. 8.A.*reheatRegenMultiStage*).

EXAMPLE 8-8 **Brayton Cycle with Regeneration, Reheat, and Intercooling**

Air steadily enters the first compressor of the gas turbine shown in Figure 8.28 at 100 kPa and 300 K with a mass flow rate of 50 kg/s. The pressure ratio across the two-stage compressor and turbine is 15. The intercooler and reheater each operate at an intermediate pressure given by the

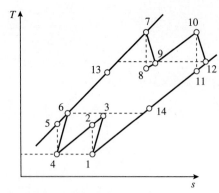

FIGURE 8.28 *T-s* diagram used in Ex. 8-8 (see Anim. 8.A.*reheatRegenMultiStage*).

square root of the product of the first compressor and turbine inlet pressures. The inlet temperature of each turbine is 1400 K and that of the second compressor is 300 K. The isentropic efficiency of each compressor and turbine and the regenerator effectiveness are both 80%. Determine (a) the thermal efficiency. **What-if scenario:** (b) What would the thermal efficiency of the cycle be if the turbine and compressor efficiency increased to 90%? Use the IG model for air.

SOLUTION

Use the IG model to evaluate the fourteen principal states (see Fig. 8.28) and perform an energy balance for each open-steady device to obtain the cycle parameters.

Assumptions

No pressure drop in heat exchangers, combustor, and reheater.

Analysis

Use the IG power-cycle TESTcalc or the IG table for air, Table D-3, to evaluate the enthalpy of all principal states as shown below.

State	Given	h (kJ/kg)	State	Given	h (kJ/kg)
1	p_1, T_1	1.873	8	$p_8 = p_2, s_8 = s_7$	754.1
2	$p_2, s_2 = s_1$	144.0	9	$p_9 = p_2,$	846.8
				$h_9 = h_7 -$	
				$(h_7 - h_8)\eta_T$	
3	$p_3 = p_2,$	179.6	10	$p_{10} = p_2,$	1217.8
	$h_3 = h_1 +$			$T_{10} = T_7$	
	$(h_2 - h_1)/\eta_C$				
4	$p_4 = p_2, T_4 = T_1$	1.873	11	$p_{11} = p_1,$	754.1
				$s_{11} = s_{10}$	
5	$p_5, s_5 = s_4$	144.0	12	$p_{12} = p_1,$	846.8
				$h_{12} = h_{10} -$	
				$(h_{10} - h_{11})\eta_T$	
6	$p_6 = p_5,$	179.6	13	$p_{13} = p_5,$	713.4
	$h_6 = h_4 +$			$h_{13} = h_6 +$	
	$(h_5 - h_4)/\eta_C$			$(h_{12} - h_6)\varepsilon_{reg}$	
7	$p_7 = p_5, T_7$	1218	14	$p_{14} = p_1,$	313.0
				$h_{14} = h_{12} -$	
				$(h_{13} - h_6)$	

Steady-state analysis of each device (see Ex. 8-2) produces the following table of external energy transfers (algebraic signs according to WinHip sign convention).

Device	$\dot{Q}$ (MW)	$\dot{W}_{ext}$ (MW)	Device	$\dot{Q}$ (MW)	$\dot{W}_{ext}$ (MW)
A (1-3): Comp.-I	0	−8.885	F (7-9): Turb.-I	0	18.55
B (3-4): Intercooler	−8.885	0	G (9-10): Reheater	18.55	
C (4-6): Comp.-II	0	−8.885	H (10-12): Turb.-II	0	18.55
D (6,12-13,14): Regenerator	0	0	I (14-1): Heat Exchanger	−15.56	0
E (13-7): Combustor	25.22	0			

The thermal efficiency now can be calculated as:

$$\eta_{th} = \frac{\dot{W}_{net}}{\dot{Q}_{in}} = \frac{2(18.55 - 8.885)}{25.22 + 18.55} = \frac{19.33}{43.77} = 44.2\%$$

Table 8-7 A parametric study with the reheat pressure as a parameter in Ex. 8-8. Use the gas turbine simulator (TEST.*Interactives*) to reproduce this study.

p_5/p_1	η_{th}	$\dot{W}_{net}$ (kW)
5	0.426	13.78
10	0.444	17.64
15	0.442	19.33
20	0.436	20.30
25	0.430	20.80
30	0.423	21.23

TEST Analysis

Evaluate the fourteen states and analyze the nine devices with the IG power-cycle TESTcalc. Use the TEST-code posted in the *TEST > TEST-codes* page.

What-if scenario

Change the compressor output pressure, p_5, to a new value, press the Enter key and click Super-Calculate. Repeat the procedure to produce the results compiled in Table 8-7. The gas turbine simulation Interactive can also be used for this purpose.

Discussion

The parametric study shows that while the net output monotonically increases, the overall cycle efficiency only slightly decreases with the overall pressure ratio. If the pressure ratio is increased indefinitely, it is possible to get a compressor inlet temperature higher than the turbine exhaust temperature, making it impossible for the regenerator to work. In such a situation, the heat transfer for the regenerator (Device-D) will be non-zero.

8.7 GAS TURBINES FOR JET PROPULSION

Except for very low speed aircraft, most aircraft in use today are powered by gas turbine driven **jet propulsion**. Gas turbine engines are generally of three types—**turbojet, turbofan** and **turboprop**. While in a land-based gas turbine the entire useful power is delivered to the shaft, in gas turbine propulsion, the useful work output is produced wholly or in part through expansion of the already partially expanded turbine exhaust in a propelling nozzle.

In the **turbojet engine** (see Anim. 8.B.*turboJet*), the turbine produces just enough power to run the compressor. The rest of the expansion takes place in the nozzle creating a very high speed jet with velocity about twice that of the speed of sound (to be discussed in Chapter 15). The next section demonstrates that it is the momentum of the jet that really matters as far as the propulsive thrust is concerned. High momentum flow can be created not only by high jet speed, but also by increasing the mass flow rate of air. The **turbofan engine** accomplishes this by using a fan (see Anim. 8.B.*turboFan*), also driven by the turbine, in front of the compressor. The flow created by the fan bypasses the engine through a duct, called a *cowl*, surrounding the engine. The relatively cooler and slower fan exhaust combines with the high speed turbine exhaust to deliver the necessary propelling force, or *thrust*, at a lower average jet velocity. Beside a considerable reduction in noise, a turbofan engine also offers better efficiency. The **bypass ratio**, the ratio of the mass flow rate through the cowl to that through the turbine, is about 5 or 6 for a turbofan engine. Increasing the bypass ratio generally increases the thrust. In a **turboprop engine** (see Anim. 8.B.*turboProp*), the fan is made even larger as *propellers* and the cowling cover is removed, increasing the bypass ratio to a value as high as 100. To understand and define the thrust developed by an engine, we must first develop the momentum balance equation, Newton's second law of motion for an open system.

8.7.1 The Momentum Balance Equation

Just as the mass, energy, and entropy balance equations are expressions of fundamental laws of thermodynamics applied to an open unsteady system, the **momentum balance equation** is an expression of Newton's second law of motion applied to an open unsteady system. To derive that, we begin with Newton's second law for a closed system:

The rate of change of momentum of a closed system is equal to the net external force applied on the system.

Momentum and force are vectors, and the momentum equation can be split into three independent equations along the x, y, and z directions in the Cartesian coordinates. In a representative x-direction, Newton's second law can be expressed as

$$\frac{dM_x}{dt} = \sum F_x \quad [\text{kN}];$$

where $\quad M_x = \dfrac{mV_x}{(1000 \text{ N/kN})} \quad \left[\text{kg}\dfrac{\text{m}}{\text{s}}\dfrac{\text{kN}}{\text{N}} = \left(\text{kg}\dfrac{\text{m}}{\text{s}^2}\right)\text{s}\dfrac{\text{kN}}{\text{N}} = \text{N} \cdot \text{s} \dfrac{\text{kN}}{\text{N}} = \text{kN} \cdot \text{s} \right]$ **(8.23)**

In this equation for a closed system, the net force $\sum F_x$ is expressed in kN, and M_x, the x-momentum, is divided by a conversion factor of 1000 N/kN to make the equation dimensionally consistent. Only for constant mass, that is, a rigid closed system, the equation reduces to the familiar form—force equals mass times acceleration.

To obtain the momentum equation for an open system, as we did for the energy, entropy, and exergy equations, consider a simplified sketch of an open, non-uniform, unsteady system at two different instants; t and $t + \Delta t$ (Fig. 8.29). As before, only a single inlet and a single exit are considered in the beginning. The net external force on the closed system, enclosed in the black boundary, passing through the open system, enclosed within the red boundary, is shown by a single red arrow and includes all surface (pressure and shear force) and body (gravitational force) forces.

The momentum M_x of the open system can be treated as an extensive system property just as stored energy E or entropy S. Now consider different reasons for the momentum to change from $M_x(t)$ to $M_x(t + \Delta t)$ within an infinitesimally small interval Δt, in which the fixed mass of the closed system passes through the open system. First, the net force in the x-direction can increase the momentum of the closed system by an amount $\sum F_x \Delta t$ according to Eq. (8.23). The net force can be directly evaluated by considering the open system alone, as there is no distinction between it and the closed system passing through at time t. Second, the momentum within the open system can also be affected by the transport of momentum with mass flow at the inlet and exit. The rate of transport, of any extensive property by a flow, can be calculated from the *transport equation*, Eq. (2.1), if the mass flow rate and the specific property are known at a given location. The specific x-momentum, the momentum per unit mass in the x-direction, can be written as $M_x/m = V_x/(1000 \text{ N/kN})$, and the momentum transported at the inlet and exit during the time interval Δt are given by $\dot{M}_{x,i}\Delta t = \dot{m}_i V_{x,i}\Delta t/(1000 \text{ N/kN})$ and $\dot{M}_{x,e}\Delta t = \dot{m}_e V_{x,e}\Delta t/(1000 \text{ N/kN})$ respectively. $M_x(t + \Delta t)$ can be related to $M_x(t)$ as follows:

$$M_x(t + \Delta t) = M_x(t) + (\dot{m}_i V_{x,i}\Delta t - \dot{m}_e V_{x,e}\Delta t)/1000 + \sum F_x \Delta t \quad [\text{kN}] \quad \textbf{(8.24)}$$

Dividing this equation by Δt, rearranging, and taking the limit as Δt tends to zero, we obtain:

$$\lim_{\Delta t \to 0} \frac{M_x(t + \Delta t) - M_x(t)}{\Delta t} = \frac{\dot{m}_i V_{x,i} - \dot{m}_e V_{x,e}}{(1000 \text{ N/kN})} + \sum F_x;$$

$$\Rightarrow \quad \frac{dM_x}{dt} = \frac{1}{(1000 \text{ N/kN})}(\dot{m}_i V_{x,i} - \dot{m}_e V_{x,e}) + \sum F_x \quad \left[\frac{\text{kJ}}{\text{s}} = \text{kN}\right]$$

For multiple inlets and exits, this equation can be generalized yielding the **momentum balance equation** in its most general form for an unsteady, open, generic system:

$$\underbrace{\frac{dM_x}{dt}}_{\substack{\text{Rate of change of} \\ x\text{-momentum of} \\ \text{an open system.}}} = \underbrace{\sum_i \frac{\dot{m}_i V_{x,i}}{(1000 \text{ N/kN})}}_{\substack{\text{Rate of transport of} \\ x\text{-momentum into} \\ \text{the system.}}} - \underbrace{\sum_e \frac{\dot{m}_e V_{x,e}}{(1000 \text{ N/kN})}}_{\substack{\text{Rate of transport of} \\ x\text{-momentum} \\ \text{out of the system.}}} + \underbrace{\sum F_x}_{\substack{\text{Net rate of} \\ \text{generation of} \\ x\text{-momentum.}}} ; \quad [\text{kN}] \quad \textbf{(8.25)}$$

The momentum equation in the y or z directions, can be written by changing the subscript x into y or z. Comparing the momentum balance equation with the entropy balance equation, Eq. (2.13), the net external force can be interpreted (see Anim. 8.B.*momBalanceTerms*) as a generation term for momentum. In a system at rest, application of net force will generate momentum. Momentum can also be transported just like energy and entropy. Unlike mass and entropy the specific momentum reduces to another fundamental property, the velocity of the flow.

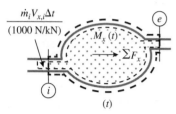

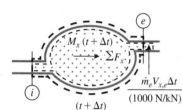

FIGURE 8.29 Schematic used for deriving the momentum balance equation (see Anim. 8.B.*momBalanceEqn*).

If the working fluid is a gas, the gravitational force can be ignored, and the net force is entirely due to pressure and shear forces (due to friction) exerted on the system boundary. The net rate of change of course depends on generation as well as transport of momentum at the ports. At steady state, with M_x remaining unchanged, like all other global properties of the open system (see Anim. 4.A.*singleFlowGeneric*), the unsteady term drops out and the net force on the system can be related to the net transport of momentum.

EXAMPLE 8-9 | Application of Momentum Balance Equation

A water jet impinges upon a perpendicular plate with a mass flow rate of 50 kg/s and a velocity of 20 m/s as shown in Figure 8.30. Determine the force R necessary to hold the plate in place.

SOLUTION

Perform a momentum analysis in the x-direction on the system shown within the red boundary of Figure 8.30.

Assumptions

The system is assumed to be at steady state so that all system properties, including momentum M_x, remain constant. Neglect viscous friction between the jet and the surroundings. Atmospheric pressure produces no net force and can be completely ignored.

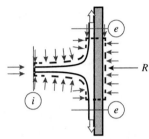

FIGURE 8.30 Schematic for Ex. 8-9. An external force is necessary to hold the plate stationary (see Anim. 8.B.*thrust*).

Analysis

Realizing that the flows at the two exits transport only vertical momentum, the horizontal momentum equation can be simplified:

$$\frac{dM_x^{\;0}}{dt} = \sum_i \dot{m}_i \frac{V_{x,i}}{(1000 \text{ N/kN})} - \sum_e \dot{m}_e \frac{V_{x,e}^{\;0}}{(1000 \text{ N/kN})} + \sum F_x;$$

$$\Rightarrow \quad \sum F_x = -\sum_i \dot{m}_i \frac{V_{x,i}}{(1000 \text{ N/kN})} = \frac{-\dot{m}_i V_{x,i}}{(1000 \text{ N/kN})}$$

The force in the x-direction consists of the external force R and the force due to atmospheric pressure, as shown in Figure 8.30. Across the jet, which is exposed to the atmosphere, the pressure must be atmospheric. The atmospheric pressure, however, does not contribute to the net force since it is uniform and the projected areas from the left and right are equal. Therefore, $\sum F_x = -R$ so that

$$R = -\sum F_x = \frac{\dot{m}_i V_{x,i}}{(1000 \text{ N/kN})} = \frac{(50)(20)}{1000} = 1 \text{ kN}$$

Discussion

As the jet impinges on the plate, a pressure field is created on top of the plate which pushes the plate in the x-direction. Evaluating the pressure distribution over the plate is not a simple task, as it often involves computer solution of partial differential equations (called the Navier-Stokes equations). By including the plate within the system, we evaluate the net effect of this pressure field on the wall. We will use the same approach in evaluating the *thrust* of an engine in the next section.

8.7.2 Jet Engine Performance

While thermal efficiency and net power are of principal interest in analyzing the performance of a land-based gas turbine, the thrust generated is an additional consideration for a jet engine. It is not the net power but the propulsive power—power used to propel the aircraft—that is useful for a jet engine.

To appreciate the concept of thrust as a propelling force, consider a closed tube of cross-sectional area A_t (Fig. 8.31) containing air at an internal pressure of p_t, which is higher than the atmospheric pressure p_0. With forces in the x or y direction completely balanced, there is no net force on the tube. Suppose the right half of the tube is instantly removed (see Anim. 8.B.*thrust*). At $t = 0$, the instant after the end is removed, the imbalance between the internal and external pressures will propel the

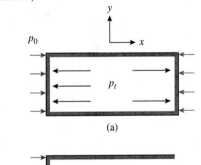

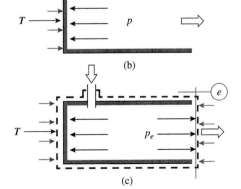

FIGURE 8.31 (a) Closed tube, no thrust. (b) Thrust after the end is removed. (c) Steady-state thrust (see Anim. 8.B.*thrust*).

tube in the negative x-direction with the unbalanced force $T = A_t(p - p_0)$ where $p_t \geq p \geq p_0$ is the internal pressure. T, the net propulsive force, is called the **thrust** for this device. Another way to interpret the magnitude of the thrust is to regard it as the external force is necessary to keep the tube stationary. In the unsteady system of Fig. 8.31, the thrust cannot last long as the pressure imbalance drives a flow through the open end, causing the internal pressure to rapidly drop, to the ambient value. However, if fresh air is supplied through a flexible tube to replenish the lost fluid and maintain a high internal pressure, the thrust can be maintained indefinitely. Knowing the exact distribution of the internal pressure, especially for a jet engine with complex geometry, is almost impossible. That is where a momentum analysis that internalizes the pressure distribution becomes invaluable.

The application of the momentum equation over the external (red) boundary in the x-direction in Figure 8.31(c) (see Anim. 8.B.*rocketThrust*) produces:

$$0 = -\frac{\dot{m}}{(1000 \text{ N/kN})}V_e + T - A_t(p_e - p_0); \quad [\text{kN}]$$

$$\Rightarrow \quad T = \frac{\dot{m}V_e}{(1000 \text{ N/kN})} + A_t(p_e - p_0); \quad [\text{kN}] \tag{8.26}$$

where $\dot{m}$ is the mass flow rate, V_e is the average velocity, p_e is the pressure at the exit, and T is the magnitude of the thrust, the external force necessary to hold the system stationary. When the system is free to move, the thrust acts as a propelling force, accelerating the system and overcoming the viscous drag.

Contribution of the first term to the thrust is the **momentum thrust** and that from the second term is called the **pressure thrust**. This result is directly applicable to chemical rockets. Although the exit pressure is different from atmospheric (such situations may arise in a non-isentropic choked flow to be discussed in Chapter 15), in most situations $p_e = p_0$. The thrust is then entirely due to the momentum thrust, which can be obtained without any knowledge of the internal pressure distribution.

To obtain the formula for thrust generated by a jet engine, see the control volume around the simplified schematic of a jet engine shown in Figure 8.32. Air enters the engine intake at a velocity V_a, which is usually equal and opposite to the cruising velocity of the aircraft. The power unit accelerates the flow to an exit velocity V_j, relative to the engine. The net forward force, produced by the pressure distribution over the internal engine surfaces, is the thrust T, which is transmitted to the aircraft fuselage and propels the aircraft forward.

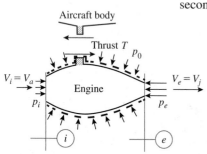

FIGURE 8.32 A control volume around a turbojet engine to calculate thrust (see Anim. 8.B.*jetThrust*).

As in the case of the propelled tube of Figure 8.31, we can avoid the complexity of internal pressure distribution by drawing the system around the entire engine where T is shown as a reaction force from the rest of the aircraft on the engine, or the force necessary to keep the engine from accelerating. In addition to T, force p_iA_i at the inlet, p_eA_e at the exit, and force due to atmospheric pressure p_a (which at the cruising altitudes can be much smaller than the standard atmospheric pressure p_0) over the rest of the boundary contribute to the net force in the x-direction. Although in most situations involving low speed flights, $p_i = p_e = p_0$, we will not make this assumption. In Chapter 15 we will discuss how p_e and, sometimes, even p_i can be substantially different from p_0 in a high-speed flight. To find the net force in the x-direction due to the pressure distribution around the engine, we realize that if p_0 is subtracted from pressure all around the engine, it will have no effect on the net force, leaving $A_i(p_i - p_0) - A_e(p_e - p_0)$ as the net force in the x-direction beside the thrust T (see Anim. 8.B.*jetThrust*). Assuming the mass flow rate to remain constant (neglecting the fuel mass flow rate), the steady-state momentum equation in the x-direction, Eq. (8.25), produces:

$$\sum F_x = T + A_i(p_i - p_0) - A_e(p_e - p_0) = \frac{\dot{m}}{(1000 \text{ N/kN})}(V_j - V_a); \quad [\text{kN}]$$

Therefore, the magnitude of the thrust in kN can be expressed as:

$$T = \underbrace{\frac{\dot{m}}{(1000 \text{ N/kN})}(V_j - V_a)}_{\text{Momentum thrust}} + \underbrace{A_e(p_e - p_0) - A_i(p_i - p_0)}_{\text{Pressure thrust}}; \quad [\text{kN}] \tag{8.27}$$

Note that the momentum equation is applied to an open system (control volume) fixed to the engine, and V_a and V_j are velocities relative to the engine. Pressure, on the other hand, is a thermodynamic property and is independent of the engine speed. In the rest of this analysis, we will assume that the gases are fully expanded to p_0, that is, $p_e = p_0$ and the slight drop in pressure at the inlet is negligible so that $p_i = p_0$. With only the momentum component left, Eq. (8.27) simplifies to:

$$T \cong \underbrace{\frac{\dot{m}}{(1000 \text{ N/kN})}(V_j - V_a)}_{\text{Momentum thrust}} ; \quad [\text{kN}] \qquad (8.28)$$

To achieve a desired thrust, Eq. (8.28) suggests that the mass flow rate and the jet velocity can be independently selected in designing an engine. While a turbojet depends on a high jet velocity for developing the thrust, turboprop and turbofan engines augment the thrust by creating airflow that bypasses the core engine. For a given thrust and flight speed, it can be shown that a turboprop engine uses significantly less fuel than a turbojet engine.

The **propulsive power**, the useful work output of the engine, is the rate of mechanical work (see Anim. 0.D.*mechanicalWork*) done by the thrust (Fig. 8.33) in propelling the aircraft with a velocity V_a with respect to ground, and is given by:

$$\dot{W}_P = F\frac{\Delta x}{\Delta t} = T\frac{V_a \Delta t}{\Delta t} = TV_a = \frac{\dot{m}V_a}{(1000 \text{ N/kN})}(V_j - V_a) \quad \left[\text{kW} = \text{kN}\,\frac{\text{m}}{\text{s}}\right] \qquad (8.29)$$

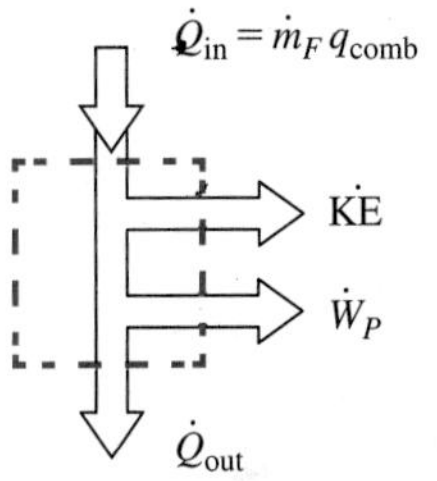

That is not the only mechanical power delivered by the jet engine. A significant portion goes into the kinetic energy $\dot{\text{KE}} = \dot{m}(V_j - V_a)^2/2000$, of the exhaust stream, which has a ground velocity of $V_j - V_a$.

The **propulsive efficiency** is defined as the ratio of the useful propulsive power to the net mechanical power developed by the engine:

$$\eta_P = \frac{\dot{W}_P}{\dot{W}_P + \dot{\text{KE}}} = \frac{1}{1 + \dot{\text{KE}}/\dot{W}_P} = \frac{2}{1 + (V_j/V_a)} \qquad (8.30)$$

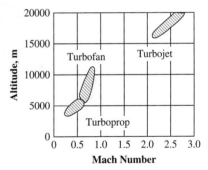

FIGURE 8.33 Energy flow diagram for a jet engine.

The propulsive efficiency measures the effectiveness with which the engine provides the useful propulsive power. Compare the thrust equation, Eq. (8.27), with the above relation for η_P. Under static condition, i.e., $V_a = 0$, the thrust T is a maximum. However, the propulsive efficiency η_P is then zero. On the other hand, η_P reaches a maximum of 1 for $V_j = V_a$ — when the exit flow becomes stationary with respect to an observer on the ground—which is accompanied by a zero thrust. Therefore, in choosing an engine for an aircraft, care should be taken so that the difference between V_j and V_a is not too high. If V_j is much higher than V_a, the increased thrust comes at the expense of the propulsive efficiency, which means that much of the mechanical energy is wasted as the kinetic energy of the exhaust gas. The requirement of matching V_j with the aircraft cruise speed V_a explains why a turbojet engine with high V_j is suitable for fighter airplanes designed for high V_a, while a turboprop or a piston engine is preferred for a relatively low velocity commuter airplane. This is indicated in the flight regime diagram of Figure 8.34, which employs the **Mach number**, the non-dimensional aircraft speed with respect to the speed of sound (to be discussed in Chapter 15) and the cruising altitude as the selection criteria.

FIGURE 8.34 Flight regime diagram for different types of jet engines.

The propulsive efficiency is not a measure of energy conversion from heat to work. With the aid of the energy flow diagram of Figure 8.33, the **overall efficiency** η_o is defined as the ratio of the propulsive power to the heat released by the fuel with a heating value q_{comb}:

$$\eta_o = \frac{\dot{W}_P}{\dot{Q}_{\text{in}}} = \frac{TV_a}{\dot{m}_F\, q_{\text{comb}}} \qquad (8.31)$$

The overall efficiency is intricately related to the velocity of the aircraft. Comparing two aircraft engines, therefore, poses a problem as different aircraft are designed for different cruise speeds. The **specific fuel consumption** or **sfc** for jet engines is defined as the fuel consumption per unit thrust, which can be related to the overall efficiency.

$$\text{sfc} = \frac{\dot{m}_F}{T} = \frac{\dot{m}_F\, q_{\text{comb}}}{TV_a}\frac{V_a}{q_{\text{comb}}} = \frac{V_a}{\eta_o\, q_{\text{comb}}} \quad \left[\frac{\text{kg/s}}{\text{kN}} = \frac{\text{m}}{\text{s}}\frac{\text{kg}}{\text{kJ}}\right] \qquad (8.32)$$

For a given fuel, q_{comb} is constant and the overall efficiency is proportional to V_a/sfc rather than $1/sfc$ as in the case of land based gas turbine. Another important performance parameter is the **specific thrust**, defined as the thrust per unit mass flow of air.

$$T_s = \frac{T}{\dot{m}} = \frac{T}{\dot{m}_F}\frac{\dot{m}_F}{\dot{m}} = \frac{f}{sfc} \quad \left[\frac{kN}{kg/s} = \frac{kN \cdot s}{kg}\right] \tag{8.33}$$

In this equation, f is the fuel-air ratio on a mass basis. Engine dimensions usually scale with the mass flow rate of air. An engine with a larger specific thrust will produce more thrust per-unit mass of the engine. To produce the same thrust, an engine with a smaller specific thrust has to be larger, which not only increases the engine weight but also creates larger frontal area and, consequently, larger drag.

8.7.3 Air-Standard Cycle for Turbojet Analysis

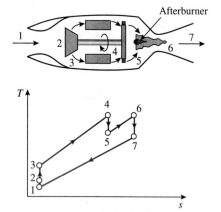

FIGURE 8.35 Air-standard cycle for a turbojet engine (see Anim. 8.B.*turboJet*).

In this section we analyze the turbojet engine as a representative aircraft engine. A turbojet engine with an afterburner along with the corresponding ideal cycle is shown on the T-s diagram of Figure 8.35. The diffuser section decelerates the flow, raising the pressure (Ex. 4-6) through what is known as the **ram effect**. The compressor, combustor, and turbine function in the same manner as the ideal Brayton cycle, except the turbine produces just enough power to run the compressor. The gases leaving the turbine have a significantly higher pressure than atmospheric pressure and complete the expansion in a nozzle creating the desired jet velocity. Some turbojets are fitted with an **afterburner**, which is a reheat combustor to increase the jet velocity by increasing the inlet temperature to the nozzle (Eq. 4.21).

In accordance with the air-standard assumption, the mass flow rate is assumed constant, the gas mixture is modeled as an ideal or perfect gas, the diffuser, compressor, turbine, and nozzle are assumed isentropic, pressure drop in the combustors is neglected, and an imaginary constant-pressure heat exchanger is added at the end of the nozzle to complete the cycle.

EXAMPLE 8-10 **Analysis of an Ideal Turbojet Cycle**

A turbojet engine has a cruising speed of 300 m/s at an altitude of 5000 m where the ambient conditions are 54 kPa, 256 K. The diffuser inlet area is 0.2 m^2, the compressor pressure ratio is 8, and the turbine inlet temperature is 1200 K. Employ the cold air-standard assumptions to determine (a) the nozzle exit velocity, (b) thrust, (c) propulsive efficiency, and (d) the overall efficiency. *What-if scenario:* (e) How would the thrust and the propulsive efficiency vary if the cruising speed varied from 100 m/s through 900 m/s?

SOLUTION

Use the PG model and an energy balance for each open-steady device to evaluate the principal states from which the thrust and all other engine parameters can be calculated.

Assumptions

Cold air-standard assumptions for the ideal turbojet cycle. The gases expand in the nozzle isentropically to the ambient pressure.

Analysis

With reference to the accompanying T-s diagram (Fig. 8.36), the principal states are evaluated as follows:

State-1 (given p_1, T_1, V_1, A_1):

$$\dot{m} = \frac{AV_1}{v_1} = \frac{AV_1\overline{M}p_1}{\overline{R}T_1} = \frac{(0.2)(300)(29)(54)}{(8.314)(256)} = 44.1 \frac{kg}{s}$$

State-2 (given $V_2 = 0$): The energy and entropy equation for the steady diffuser produces (Ex. 4-6):

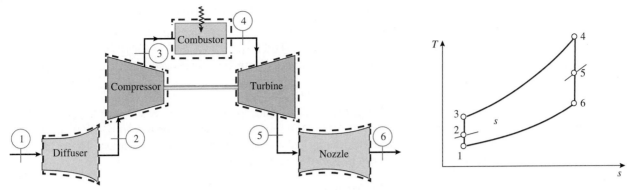

FIGURE 8.36 The ideal turbojet cycle in Ex. 8-10 (also see Anim. 8.B.*turboJet*).

$$j_2 = j_1; \quad \Rightarrow \quad h_2 + \cancel{ke_2}^0 = h_1 + ke_1;$$

$$\Rightarrow \quad c_p(T_2 - T_1) = \frac{V_1^2}{2(1000 \text{ N/kN})}; \quad \Rightarrow \quad T_2 = T_1 + \frac{V_1^2}{2(1000 \text{ N/kN})c_p} = 300.8 \text{ K};$$

$$s_2 = s_1; \quad \Rightarrow \quad \frac{T_2}{T_1} = \left(\frac{p_2}{p_1}\right)^{(k-1)/k}; \quad \Rightarrow \quad p_2 = p_1\left(\frac{T_2}{T_1}\right)^{k/(k-1)} = 94.95 \text{ kPa}$$

State-3 (given $p_3 = r_p p_2$, $s_3 = s_2$):

$$p_3 = 8p_2 = 759.6 \text{ kPa}; \quad T_3 = T_2 r_p^{(k-1)/k} = 545.3 \text{ K}$$

State-4: (given $p_4 = p_3$, $T_4 = 1200$ K)

State-5 (given $s_5 = s_4$): The power output from the turbine can be equated to the power input to the compressor. Alternatively, the two systems can be treated as a combined system with two inlets and two exits with no heat or external work transfer.

$$\dot{m}(j_4 - j_5) = \dot{m}(j_3 - j_2); \quad \Rightarrow \quad h_4 - h_5 = h_3 - h_2;$$

$$\Rightarrow \quad c_p(T_4 - T_5) = c_p(T_3 - T_2); \quad \Rightarrow \quad T_5 = 955.6 \text{ K};$$

From the isentropic relation, $T_4 = T_5\left(\dfrac{p_4}{p_5}\right)^{\frac{k-1}{k}}; \quad \Rightarrow \quad p_5 = 342.5$ kPa

State-6 (given p_6): The entropy and energy equation for the steady nozzle produces (Ex. 4-4).

$$s_6 = s_5; \quad \Rightarrow \quad \frac{T_6}{T_5} = \left(\frac{p_6}{p_5}\right)^{(k-1)/k}; \quad \Rightarrow \quad T_6 = 563.4 \text{ K};$$

$$j_6 = j_5; \quad \Rightarrow \quad h_6 + ke_6 = h_5 + \cancel{ke_5}^0;$$

$$\Rightarrow \quad \frac{V_6^2}{2(1000 \text{ N/kN})} = c_p(T_5 - T_6); \quad \Rightarrow \quad V_6 = 887.2 \text{ m/s}$$

The thrust generated can be calculated from Eq. (8.27)

$$T \cong \frac{\dot{m}}{(1000 \text{ N/kN})} (V_j - V_a) = \frac{44.1}{1000} (887 - 300) = 25.89 \text{ kN}$$

The propulsive and overall efficiency can be obtained as

$$\dot{W}_P = TV_a = TV_1 = 7767 \text{ kW}; \quad \dot{\text{KE}}_{\text{out}} = \frac{\dot{m}(V_6 - V_1)^2}{2000} = 7602 \text{ kW};$$

$$\dot{Q}_{\text{in}} = \dot{m}(h_4 - h_3) = \dot{m}c_p(T_4 - T_3) = 28,974 \text{ kW};$$

Therefore, $\quad \eta_P = \dfrac{\dot{W}_P}{\dot{W}_P + \dot{\text{KE}}_{\text{out}}} = 50.5\%; \quad \eta_o = \dfrac{\dot{W}_P}{\dot{Q}_{\text{in}}} = 26.8\%$

Table 8-8 A parametric study of a turbojet engine in Ex. 8-10.		
V_a (m/s)	η_P (%)	F_s (kN·s/kg)
100	21.6	0.724
200	38.0	0.652
300	50.5	0.587
400	60.5	0.520
600	77.0	0.357
900	99.8	0.002

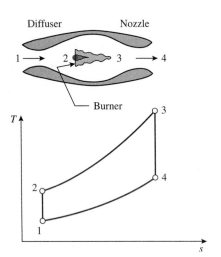

FIGURE 8.37 A ramjet engine where compression is accomplished by the ram effect (see Anim. 8.B.*ramJet*).

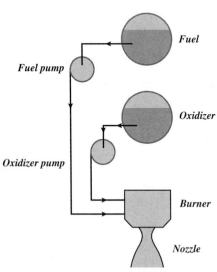

FIGURE 8.38 A rocket engine. The burner raises the temperature, not pressure (see Anim. 8.B.rocketThrust).

TEST Analysis

Launch the PG power-cycle TESTcalc. Calculate the six principal states as described in the TEST-code (see *TEST > TEST-codes*). Use the I/O panel as a calculator to obtain the answers and verify the manual results.

What-if scenario

Change Vel1 to a new value, press the enter key and click Super-Calculate. Evaluate '= 2/(1 + Vel6/Vel1)' and '= (Vel6 − Vel1)/2000' in the I/O panel as the propulsive efficiency and specific thrust. The results are shown in Table 8-8 as a function of V_a.

Discussion

Although we have assumed that the gases expand down to the ambient pressure in the nozzle, this is not always the case, especially when phenomenon such as choking, underexpansion, or overexpansion occurs. Nozzle flows will be further analyzed in Chapter 15.

8.8 OTHER FORMS OF JET PROPULSION

In a very fast moving jet engine, the **ram compression**—compression achieved by using a *diffuser* (see Anim. 4.A.*diffuser*) alone—may be sufficient for generating high enough pressure for the nozzle to generate the necessary thrust. The compressor and turbine can be eliminated resulting in what is known as a **ramjet**, a jet engine with no moving parts. However, the engine, sketched in Figure 8.37, must be already in flight for the ram effect to kick in. That is why a ramjet engine is suitable for missiles launched from high-speed aircraft. Although the cruising speed of a ramjet may be Mach 2 or 3, the flow slows down to subsonic speed in the combustor. At higher cruising speed, at Mach 6 or above, the flow is supersonic even in the combustion chamber and the engine is called a **scramjet**.

All the engines discussed in this chapter require air for the combustion of the fuel and are known as **air-breathing engines**. For space travel or very high altitude flights, rockets are used for propulsion. A **rocket** engine is simply a combustion chamber connected with a nozzle as shown in Figure 8.38. Pressurized fuel and oxidizer (liquid hydrogen and oxygen, for example) are ignited in the combustion chamber (burner) to maximize the nozzle inlet temperature and, hence, the exit velocity (Eq. 4.21) for a given pressure difference. From the same equation it can be seen that a high value of c_p also leads to high exit velocity. Hydrogen, therefore, is preferred as rocket fuel and is also used in nuclear-powered rockets.

8.9 CLOSURE

In this chapter, we have studied gas turbines for land based power plants and for jet propulsion. After a brief explanation of the practical aspects of the main components—compressor, combustor, and turbine—the air-standard Brayton cycle was introduced to model a gas turbine. Modifications to the gas turbine through regeneration, reheating, and intercooling were evaluated qualitatively through the use of the equivalent Carnot cycle concept introduced in the last chapter. TEST was used for verification of manual solutions and a simulation Interactive was used for some of the parametric studies. Application of gas turbines to jet propulsion was discussed in the last section of the chapter with special emphasis on an ideal turbojet cycle. The momentum balance equation was derived from Newton's second law, complementing the open system formulation of all other fundamental laws presented in Chapter 2. The concept of thrust was established, and thrust developed by a jet was studied with the help of the momentum equation.

PROBLEMS

SECTION 8-1: BASIC AND MODIFIED BRAYTON CYCLES

8-1-1 Air enters the compressor of an ideal air standard Brayton cycle at 100 kPa, 25°C, with a volumetric flow rate ($\dot{V}$) of 8 m³/s. The compressor pressure ratio is 12. The turbine inlet temperature is 1100°C. Determine (a) the thermal efficiency (η_{th}), (b) net power output ($\dot{W}_{net}$) and (c) back work ratio. Use the PG model for air. (d) *What-if scenario:* What would the answers be if we used the IG model?

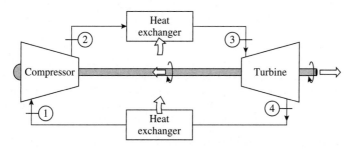

8-1-2 A stationary power plant, operating on an ideal Brayton cycle, has a pressure ratio of 7. The gas temperature is 25°C at the compressor inlet and 1000°C at the turbine inlet. Utilizing the air standard assumptions, determine (a) the gas temperature (T) at the exits of the compressor, (b) back work ratio and (c) the thermal efficiency ($\eta_{th,Brayton}$). Use the PG model.

8-1-3 In an air standard Brayton cycle the air enters the compressor at 0.1 MPa and 20°C. The pressure leaving the compressor is 1 MPa and the maximum temperature in the cycle is 1225°C. Determine (a) the compressor work (w_C), (b) turbine work (w_T), and (c) the cycle efficiency ($\eta_{th,Brayton}$) per unit mass of air. Use the PG model. (d) *What-if scenario:* What would the answers be if the compressor exit pressure was 1.5 Mpa?

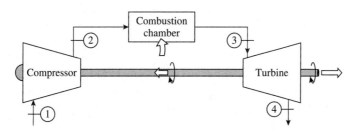

8-1-4 Air (use the PG model) enters the compressor of an ideal air standard Brayton cycle at 100 kPa and 305 K with a volumetric flow rate of 5 m³/s. The compressor pressure ratio is 10. The turbine inlet temperature is 1000 K. Determine (a) the thermal efficiency ($\eta_{th,Brayton}$), (b) net power output ($\dot{W}_{net}$) and (c) back work ratio. Use the PG model for air.

8-1-5 A gas turbine power plant operates on a simple Brayton cycle with air (use the PG model) as the working fluid. The air enters the turbine at 1 MPa and 1000 K and leaves at 125 kPa, 610 K. Heat is rejected to the surroundings at a rate of 8000 kW and air flow rate ($\dot{m}$) is 25 kg/s. Assuming a compressor efficiency of 80%, determine (a) the net power output ($\dot{W}_{net}$). Use the PG model for air. (b) *What-if scenario:* What would the net power output be if the compressor efficiency dropped to 75%?

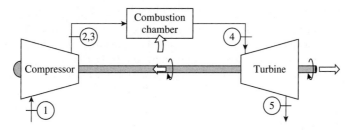

8-1-6   Repeat Problem 8-1-4 assuming a compressor efficiency of 80% and a turbine efficiency of 90%. Determine (a) back work ratio, (b) the thermal efficiency ($\eta_{th,Brayton}$) and (c) the turbine exit temperature. Use the PG model.

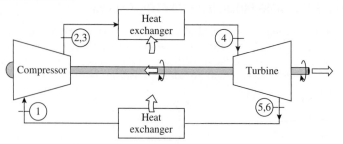

8-1-7   In a air standard Brayton cycle the air enters the compressor at 0.1 MPa, 20°C. The pressure leaving the compressor is 1 MPa. and the maximum temperature in the cycle is 1225°C. Assume a compressor efficiency of 80%, a turbine efficiency of 85%, and a pressure drop between the compressor and turbine of 25 kPa. Determine (a) the compressor work (w_C), (b) the turbine work (w_T) and (c) the cycle efficiency ($\eta_{th,Brayton}$). Use the PG model.

8-1-8   Air enters the compressor of an ideal air standard Brayton cycle at 195 kPa, 290 K, with a volumetric flow rate of $6\,m^3/s$. The compressor pressure ratio is 9. The turbine inlet temperature is 1400 K. The compressor has an efficiency of 90% and the turbine has an efficiency of 75%. Determine (a) the thermal efficiency ($\eta_{th,Brayton}$), (b) net power output ($\dot{W}_{net}$) and (c) back work ratio. Use the PG model for air.

8-1-9   A gas turbine power plant operates on a simple Brayton cycle with air as the working fluid. Air enters the turbine at 800 kPa and 1200 K, and it leaves the turbine at 100 kPa and 750 K. Heat is rejected to the surroundings at a rate of 6800 kW and air flows through the cycle at a rate of 20 kg/s. Assuming a compressor efficiency of 80%, determine (a) net power output ($\dot{W}_{net}$) of the plant. Use the PG model.

8-1-10 Air is used as the working fluid in a simple ideal Brayton cycle that has a pressure ratio of 12, a compressor inlet temperature of 310 K, and a turbine inlet temperature of 900 K. Determine (a) the required mass flow rate ($\dot{m}$) of air for a net power of 25 MW, assuming both the compressor and the turbine have an isentropic efficiency of 90%. Use the PG model. (b) *What-if scenario:* What would the mass flow rate be if both compressor and the turbine had 100% efficiency?

8-1-11 A gas turbine power plant operates on the simple Brayton cycle with air as the working fluid and delivers 10 MW of power. The minimum and maximum temperatures in the cycle are 300 K and 1100 K, and the pressure of air at the compressor exit is 9 times the value at the compressor inlet. Assuming an adiabatic efficiency of 80% for the compressor and 90% for the turbine, determine (a) the mass flow rate ($\dot{m}$) of air through the cycle. Use the PG model. (b) *What-if scenario:* What would the mass flow rate be using the IG model?

8-1-12 Air enters the compressor of a simple gas turbine at 100 kPa, 25°C, with a volumetric flow rate ($\dot{V}$) of $6\,m^3/s$. The compressor pressure ratio is 10 and its isentropic efficiency is 80%. At the turbine inlet's pressure is 100 kPa temperature is 1000°C. The turbine has an isentropic efficiency of 88% and the exit pressure is 100 kPa. On the basis of air standard analysis using the PG model, determine (a) the thermal efficiency (η_{th}) of the cycle, (b) net power developed ($\dot{W}_{net}$), in kW. (c) *What-if scenario:* What would the answers be if the inlet temperature increased to 50°C?

8-1-13 Air enters the compressor of a simple gas turbine at 95 kPa, 310 K, where it is compressed to 800 kPa and 600 K. Heat is transferred to air in the amount of 1000 kJ/kg before it enters the turbine. For a turbine efficiency of 90%, determine (a) the thermal efficiency (η_{th}) of the cycle, and (b) the fraction of turbine work output used to drive the compressor. Use the PG model for air.

8-1-14 Air enters the compressor of a simple gas turbine at 0.1 MPa, 300 K. The pressure ratio is 9 and the maximum temperature is 1000 K. The turbine process is divided into two stages, each with a pressure ratio of 3, with intermediate reheating to 1000 K. Determine (a) the cycle efficiency (η_{th}) and (b) the net work output per unit mass (w_{net}). Use the PG model. (c) *What-if scenario:* What would the answers be if the reheating was eliminated?

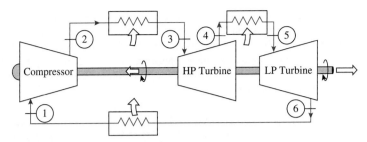

8-1-15 Repeat Problem 8-1-14 for the net output per kg of air, assuming the pressure ratio of the first stage turbine before reheat to be (a) 7, (b) 5, (c) 3, (d) 2. (e) Use a T-s diagram to explain why the output increases and then decreases.

8-1-16 Air enters the compressor of an ideal air standard Brayton cycle at 100 kPa and 290 K with a mass flow rate ($\dot{m}$) of 6 kg/s. The compressor pressure ratio is 10. The turbine inlet temperature is 1500 K. If a regenerator with an effectiveness of 70% is incorporated in the cycle, determine (a) the thermal efficiency ($\eta_{th,Brayton}$) of the cycle. Use the PG model for air. (b) *What-if scenario:* What would the thermal efficiency be if the regenerator effectiveness increased to 90%?

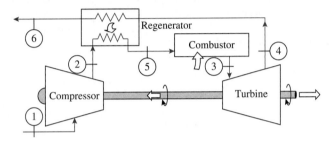

8-1-17 Repeat Problem 8-1-16 using the IG model for air.

8-1-18 A Brayton cycle with regeneration, and air at 100 kPa as the working fluid, operates on a pressure ratio of 8. The minimum and maximum temperatures of the cycle are 300 and 1200 K. The isentropic efficiencies of the turbine and the compressor are 80% and 82% respectively. The regenerator effectiveness is 65%. Determine (a) the thermal efficiency ($\eta_{th,Brayton}$) and (b) net work output per unit mass (w_{net}). Use the PG model. (c) *What-if scenario:* What would the thermal efficiency be if the regenerator effectiveness increased to 75%?

8-1-19 A 100-hp, regenerative, Brayton-cycle gas turbine operates between a source at 840°C and the reference atmosphere at 21°C. Air enters the compressor at 21°C, 101 kPa. The air is then compressed to 345 kPa and then heated to 840°C. Part of this heating is accomplished in a regenerator whose effectiveness is 90%. Determine (a) the thermal efficiency ($\eta_{th,Brayton}$) of the cycle, (b) work done by compressor (w_C) and (c) work done by turbine (w_T). Use the PG model.

8-1-20 Repeat Problem 8-1-19 assuming the regenerator has an effectiveness of 85%, determine the thermal efficiency ($\eta_{th,Brayton}$). Use the IG model.

8-1-21 Air enters the compressor of a regenerative gas turbine engine at 100 kPa and 290 K, where it is compressed to 750 kPa and 550 K. The regenerator has an effectiveness of 70% and the air enters the turbine at 1200 K. For a turbine efficiency of 80%, determine (a) the amount of heat transfer (q) in the regenerator and (b) the thermal efficiency (η_{th}) of the cycle. Assume variable specific heats of the air.

8-1-22 Air is compressed from 100 kPa and 310 K to 1000 kPa in a two stage compressor with intercooling between stages. The intercooler pressure is 350 kPa. The air is cooled back to 310 K in the intercooler before entering the second compressor stage. Each compressor stage is isentropic. Determine (a) the temperature (T) at the exit of the second compressor stage and (b) the total compressor work in kJ/kg. Use the PG model.

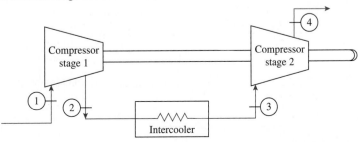

8-1-23 Air enters the compressor of an ideal air standard Brayton cycle at 100 kPa, 25°C with a volumetric flow rate of $8\,m^3/s$ and is compressed to 1000 kPa. The temperature at the inlet to the first turbine stage is 1000°C. The expansion takes place isentropically in two stages, with reheat to 1000°C between the stages at a constant pressure of 300 kPa. If a regenerator having an effectiveness of 100% is incorporated in the cycle, determine (a) the thermal efficiency of the cycle ($\eta_{th,Brayton}$). Use the PG model.

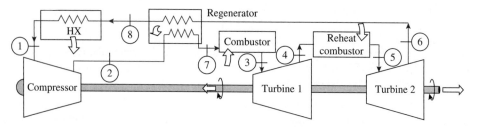

8-1-24 Consider an ideal gas turbine cycle with two stages of compression and two stages of expansion. The pressure ratio across each stage of the compressor and the turbine is 2. Air (use the IG model) enters each stage of the compressor at 310 K and each stage of the turbine at 1100 K. Determine (a) the thermal efficiency (η_{th}) of the cycle and (b) back work ratio. Use the IG model. (c) *What-if scenario:* What would the thermal efficiency and BWR be if the pressure ratio across each stage of the compressor and the turbine was 4?

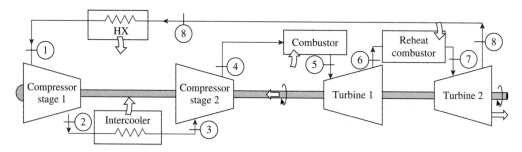

8-1-25 Repeat Problem 8-1-24 assuming a regenerator with 80% effectiveness is added at the end of the last compressor. Determine (a) the thermal efficiency (η_{th}) of the cycle and (b) back work ratio.

8-1-26 Consider a regenerative gas turbine power plant with two stages of compression and two stages of expansion. The overall pressure ratio of the cycle is 9. Air enters each stage of compressor at 290 K and each stage of turbine at 1400 K. The regenerator has an effectiveness of 100%. Determine (a) the minimum mass flow rate ($\dot{m}$) of air needed to develop a net power output of 50 MW. Use the IG model. (b) *What-if scenario:* What would the mass flow rate be if argon was used as the working fluid ?

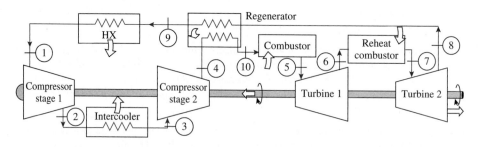

8-1-27 Repeat 8-1-24 assuming an efficiency of 80% for each compressor stage and an efficiency of 85% for each turbine stage. Determine (a) the thermal efficiency (η_{th}) of the cycle and (b) back work ratio.

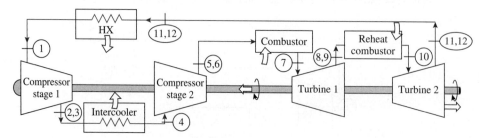

8-1-28 Repeat Problem 8-1-24 assuming an efficiency of 80% for each compressor stage and an efficiency of 85% for each turbine stage, and a regenerator with 80% effectiveness (see modified diagram below). Determine (a) the thermal efficiency (η_{th}) of the cycle and (b) the back work ratio.

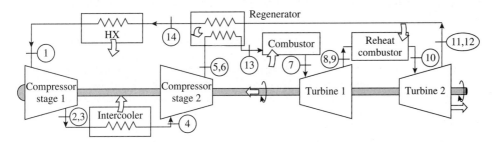

8-1-29 A regenerative gas turbine with intercooling and reheat operates at steady state. Air enters the compressor at 100 kPa, 300 K, with a mass flow rate ($\dot{m}$) of 8 kg/s. The pressure ratio across the two stage compressor and two stage turbine is 12. The intercooler and reheater each operates at 300 kPa. At the inlets to the turbine stages the temperature is 1500 K. The temperature at the inlet to the second compressor stage is 300 K. The efficiency of each compressor is 85% and turbine stages is 80%. The regenerator effectiveness is 75%. Determine (a) the thermal efficiency (η_{th}), (b) the back work ratio and (c) the net power developed ($\dot{W}_{net}$). Use the IG model.

8-1-30 Air enters steadily the first compressor of the gas turbine at 100 kPa and 300 K with a mass flow rate ($\dot{m}$) of 50 kg/s. The pressure ratio across the two-stage compressor and turbine is 15. The intercooler and reheater each operate at an intermediate pressure given by the square root of the product of the first compressor and turbine inlet pressures. The inlet temperature of each turbine is 1400 K and that of the second compressor is 300 K. The isentropic efficiency of each compressor and turbine is 80% and the regenerator effectiveness is also 80%. Determine (a) the thermal efficiency (η_{th}). Use the IG model for air. (b) *What-if scenario:* What would the thermal efficiency of the cycle be if the turbine and compressor efficiency increased to 90%?

8-1-31 Consider an ideal Ericsson cycle, with air as working fluid, executed in a steady-flow system. Air is at 30°C and 115 kPa at the beginning of the isothermal compression process during which 155 kJ/kg of heat is rejected. Heat transfer to air occurs at 1250 K. Determine (a) the maximum pressure in the cycle, (b) the net work output per unit mass (w_{net}) of air and (c) the thermal efficiency (η_{th}) of the cycle. Use the PG model.

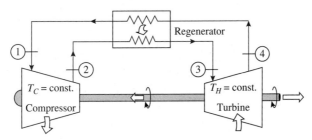

8-1-32 Air enters the turbine of an Ericsson cycle at 1000 kPa and 1200 K, with a mass flow rate of 1 kg/s. The temperature and pressure at the inlet to the compressor are 250 K and 100 kPa. Determine (a) the thermal efficiency (η_{th}) of the cycle, (b) the net work output (w_{net}) per unit mass of air. Use the PG model.

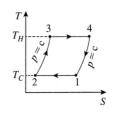

8-1-33 An ideal Ericsson cycle, with air as the working fluid, has a compression ratio of 10. Isothermal expansion takes place at 1000 K. Heat transfer from the compressor occurs at 350 K. Determine (a) the net work (w_{net}) per kg of air in the cycle, and (b) the thermal efficiency (η_{th}).

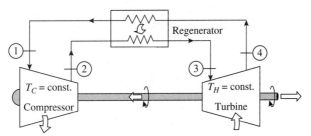

8-1-34   Hydrogen enters the turbine of an Ericsson cycle at 1500 kPa and 900 K, with a mass flow rate of 1 kg/s. The temperature and pressure at the inlet to the compressor are 320 K and 150 kPa. Determine (a) the thermal efficiency (η_{th}) of the cycle, (b) the net work (w_{net}) output per unit mass of air. Use the PG model. (b) **What-if scenario:** How would the answers change using the IG model?

8-1-35   A gas turbine power plant operates on a simple Brayton cycle with air as the working fluid having a pressure ratio of 8. The compressor efficiency is 80% and its inlet conditions are 100 kPa and 300 K with a mass flow rate ($\dot{m}$) of 25 kg/s. The turbine efficiency is 85% and its inlet temperature is 1400 K. Heat is rejected to the surroundings at a rate of 8000 kW. Using Gas Turbine Simulator Interactive (linked from left margin) determine (a) the net power output ($\dot{W}_{net}$), (b) the thermal efficiency (η_{th}), and (c) the back work ratio.

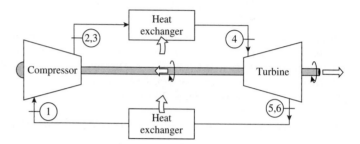

8-1-36   Using the gas turbine power plant described in Problem 8-1-35, plot how thermal efficiency ($\eta_{th,Brayton}$) varies with pressure ratio varying from 5 to 15, all other input parameters remaining unchanged.

8-1-37   Using the gas turbine power plant described in Problem 8-1-35, plot how thermal efficiency ($\eta_{th,Brayton}$) varies with compressor efficiency varying from 70% to 90%, all other input parameters remaining unchanged.

8-1-38   Air enters the compressor of an ideal air standard Brayton cycle at 100 kPa, 290 K, with a mass flow rate ($\dot{m}$) of 6 kg/s. The compressor pressure ratio is 10. The turbine inlet temperature is 1500 K. If a regenerator with an effectiveness of 70% is incorporated in the cycle, using Gas Turbine Simulator Interactive determine (a) the net power output ($\dot{W}_{net}$), (b) the thermal efficiency ($\eta_{th,Brayton}$) of the cycle.

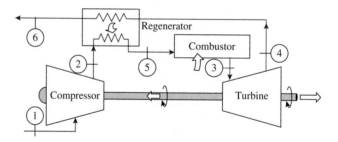

8-1-39   Using the air standard brayton cycle described in Problem 8-1-38, plot how thermal efficiency varies with pressure ratio varying from 5 to 15, all other input parameters remaining unchanged.

8-1-40   Using the air standard brayton cycle described in Problem 8-1-38, plot how thermal efficiency ($\eta_{th,Brayton}$) varies with regenerative efectiveness varying from 70% to 90%, all other input parameters remaining unchanged.

8-1-41   The gas turbine power plant described in Problem 8-1-35, is converted into a multi-stage cycle with two stages of compression and two stages of expansion. The pressure ratio across each stage of the compressor and the turbine is 2. Using the Gas Turbine Simulator IA, plot how thermal efficiency ($\eta_{th,Brayton}$) varies with first compressor efficiency varying from 70% to 90%.

8-1-42   Repeat Problem 8-1-41, assuming a regenerator with 80% effectiveness is added at the end of second compressor. Plot how thermal efficiency ($\eta_{th,Brayton}$) varies with first compressor efficiency varying from 70% to 90%.

SECTION 8-2: MOMENTUM EQUATION AND JET ENGINES

8-2-1   Liquid water flows through a pipe at a mass flow rate of 100 kg/s. If the cross-sectional area of the pipe is 0.01 m², determine the flow rate of momentum through the pipe.

8-2-2   A firefighter is trying to hold a fire hose steady while spraying water. If the jet of water ($\rho = 997\,\text{kg/m}^3$) is coming from the 6.5-cm diameter fire hose at 400 GPM (0.025 m³/s), what is the force required by the firefighter to hold the hose steady?

8-2-3   A rocket motor is fired on a test stand. Hot exhaust gases leave the exit with a velocity of 700 m/s at a mass flow rate ($\dot{m}$) of 10 kg/s. The exit area is 0.01 m² and the exit pressure is 50 kPa. For an ambient pressure of 100 kPa, determine the rocket motor thrust that is transmitted to the stand. Assume steady state and one-dimensional flow.

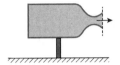

8-2-4   A jet engine is traveling through the air with a velocity of 150 m/s. The exhaust gas (treat as air using the PG model) leaves the nozzle with an exit velocity of 450 m/s with respect to the nozzle. Pressure at the inlet and exit are 30 kPa and 100 kPa, respectively, and the ambient pressure is 30 kPa. If the mass flow rate ($\dot{m}$) is 10 kg/s and the exit area is 0.2 m², determine the jet thrust.

8-2-5   A turbojet aircraft is flying with a velocity of 300 m/s at an altitude of 6000 m, the ambient conditions are 45 kPa and −15°C. The compressor pressure ratio is 14 and the turbine inlet temperature is 1500 K. Assuming ideal operation of all components and constant specific heats, determine (a) the pressure at the turbine exit, (b) the velocity of the exhaust gases and (c) the propulsive efficiency (η_P). (d) **What-if scenario:** What would the velocity of the exhaust gases be with an aircraft velocity of 200 m/s?

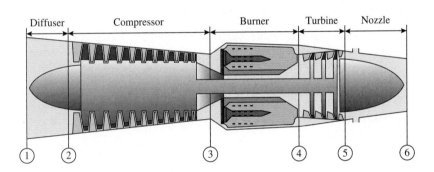

8-2-6   A jet engine is being tested on a test stand. The inlet area to the compressor is 0.2 m² and air enters the compressor at 90 kPa, 100 m/s. The pressure of the atmosphere is 100 kPa. The exit area of the engine is 0.1 m², and the products of combustion leave the exit at a pressure of 130 kPa and a temperature of 1000 K. The air fuel ratio is 45 kg air/kg fuel, and the fuel enters with a low velocity. The mass flow rate of air entering the engine is 10 kg/s. Determine (a) the inlet temperature, (b) the exit velocity of combustion products and (c) the thrust on the engine.

8-2-7   Consider an ideal jet propulsion cycle in which air enters the compressor at 100 kPa and 20°C. The pressure leaving the compressor is 1100 kPa, and the maximum temperature in the cycle is 1200°C. Air expands in the turbine; the turbine work is equal to the compressor work. On leaving the turbine, air expands in a nozzle to 100 kPa. The process is reversible and adiabatic. Determine (a) the velocity of the air leaving the nozzle. Use the PG model.

8-2-8   Consider an ideal jet propulsion cycle in which air enters the compressor at 100 kPa and 25°C. The exit pressure is 1 MPa and the maximum temperature in the cycle is 1000°C. Air expands in the turbine; the turbine work is just equal to the compressor work. On leaving the turbine, air expands in a nozzle to 100 kPa. The process is reversible and adiabatic. Determine (a) the velocity of the air leaving the nozzle. Use the PG model. (b) **What-if scenario:** What would the exit velocity be if the maximum temperature achieved in the cycle was 800°C?

8-2-9   A turbojet aircraft is flying with a velocity of 350 m/s at an altitude of 9150 m. The ambient conditions are 30 kPa and −30°C. The pressure ratio across the compressor is 10 and the temperature at the turbine inlet is 1200 K. Air enters the compressor at a rate ($\dot{m}$) of 30 kg/s and the jet fuel has a heating value of 42000 kJ/kg. Using the PG model, determine (a) the velocity of the exhaust gases, (b) the propulsive power developed and (c) the rate of fuel consumption.

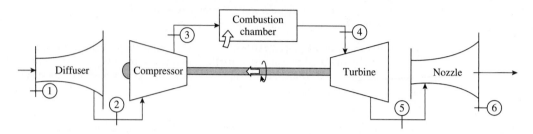

8-2-10   A turbojet aircraft is flying with a velocity of 250 m/s at an altitude where the ambient conditions are 20 kPa and −25°C. The pressure ratio across the compressor is 12 and the temperature at the turbine inlet is 1000°C. Air enters the compressor at a rate of 50 kg/s. Determine (a) the temperature (T) and pressure (p) of the gases at the turbine exit, (b) the velocity (V) of the gases at the nozzle exit and (c) the propulsive efficiency of the cycle. Use the PG model. (d) **What-if scenario:** What would the exit velocity be if the aircraft velocity was 300 m/s?

8-2-11   Consider an aircraft powered by a turbojet engine that has a pressure ratio of 10. The aircraft is on the ground, held stationary by its brakes. The ambient air is 25°C, 100 kPa, and enters the engine at a rate of 20 kg/s. The jet fuel has a heating value of 42700 kJ/kg and it is burned completely at a rate of 0.35 kg/s. Neglecting the effect of the diffuser, and disregarding the slight change in the mass at the engine exit as well as the inefficiencies of the engine components, determine (a) the force that must be applied on the brakes to hold the plane stationary. Use the PG model.

8-2-12   10°C air enters a turbojet engine at a rate ($\dot{m}$) of 15 kg/s with a velocity of 320 m/s (relative to the engine). Air is heated in the combustion chamber at a rate of 25000 kJ/s and exits the engine at 420°C. Determine the momentum thrust produced by this turbojet engine. Use the IG model.

8-2-13   Repeat Problem 8-2-9 using a compressor efficiency of 80% and turbine efficiency of 85%. Determine (a) the velocity (V_2) of the exhaust gases, (b) the propulsive power ($\dot{W}_P$) developed and (c) the rate of fuel consumption.

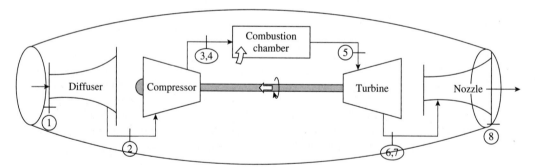

8-2-14   Air enters the diffuser of a turbojet engine with a mass flow rate ($\dot{m}$) of 65 kg/s at 80 kPa, −40°C and a velocity of 245 m/s. The pressure ratio for the compressor is 10 and its isentropic efficiency is 87%. Air enters the turbine at 1000°C with the same pressure as the exit of the compressor. Air exits the nozzle at 80 kPa. The turbine has an isentropic efficiency of 90%. Determine (a) the rate of heat addition ($\dot{Q}$), (b) the compressor power input ($\dot{W}_{in}$) and (c) the velocity of the air leaving the nozzle. Use the IG model.

8-2-15   Air at 30 kPa, 250 K, and 250 m/s enters a turbojet engine while in flight at an altitude of 10,000 m. The pressure ratio across the compressor is 12. The turbine inlet temperature is 1400 K and the pressure at the nozzle exit is 30 kPa. The diffuser and nozzle processes are isentropic, the compressor and turbine have isentropic efficiencies of 85% and 80%, respectively, and there is no pressure drop for flow through the combustor. Using the IG model, determine (a) the velocity (V) of the exhaust gases.

8-2-16   Consider the addition of an afterburner to the turbojet engine in Problem 8-2-15 that raises the temperature at the inlet of the nozzle to 1300 K. Using the IG model, determine (a) the velocity (V) at the nozzle exit. (b) **What-if scenario:** What would the exit velocity be if the temperature at the inlet of the nozzle was 1200 K?

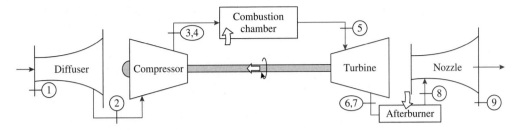

8-2-17   Air enters the diffuser of a ramjet engine at 50 kPa, 230 K, with a velocity of 480 m/s, and is decelerated to a velocity of zero. After combustion, the gases reach a temperature of 1000 K before being discharged through the nozzle at 50 kPa. Using the PG model, determine (a) the pressure (p) at the diffuser exit and (b) the velocity (V) at the nozzle exit.

8-2-18   Air enters the diffuser of a ramjet engine at 25 kPa, 200 K, with a velocity of 3000 km/h, and is decelerated to a negligible velocity. On the basis of an air standard analysis, the heat addition is 870 kJ/kg of air passing through the engine. Air exits the nozzle at 25 kPa. Using the IG model, determine (a) the pressure (p) at the diffuser exit, and (b) the velocity (V) at the nozzle exit. (c) **What-if scenario:** What would the exit velocity be with a heat addition as 1200 kJ/kg?

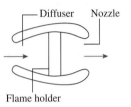

SECTION 8-3: EXERGY ANALYSIS OF GAS TURBINES

8-3-1   In an air (use the PG model) standard Brayton cycle, air enters the compressor at 0.1 MPa, 20°C (atmospheric conditions), and a mass flow rate ($\dot{m}$) of 10 kg/s. The pressure leaving the compressor is 1 MPa and the maximum temperature in the cycle is 1225°C. If the heat addition is assumed to take place from a reservoir at 1500°C, (a) perform an exergy inventory and draw an exergy flow diagram for the cycle. Determine (b) the thermal efficiency ($\eta_{th,Brayton}$) and (c) the exergetic efficiency.

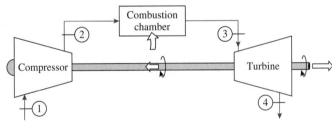

8-3-2   Repeat Problem 8-3-1 assuming a compressor efficiency of 80% and a turbine efficiency of 80%.

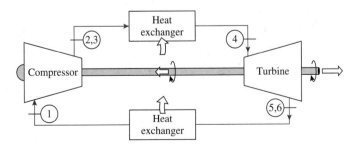

8-3-3   Air (use the IG model) enters the compressor of an ideal air standard Brayton cycle at 100 kPa, 290 K and a mass flow rate ($\dot{m}$) of 6 kg/s. The compressor pressure ratio is 10. The turbine inlet temperature is 1500 K. A regenerator with an effectiveness of 70% is incorporated in the cycle. If the heat addition can be assumed to take place from a reservoir at 1800 K, (a) perform an exergy inventory and draw an exergy flow diagram for the cycle. Determine (b) the thermal efficiency ($\eta_{th,Brayton}$) and (c) the exergetic efficiency. (d) Calculate the rate of exergy destruction (I) in the regenerator.

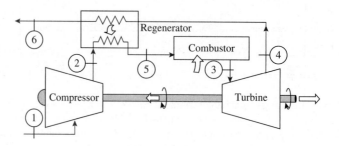

9 | OPEN VAPOR POWER CYCLES

Reciprocating engines, gas turbines, hydroelectric plants, and vapor power plants deliver practically all of the mechanical and electrical power used worldwide. The hydroelectric power plant, briefly discussed in Chapter 4, converts potential energy into electricity, therefore, is not a heat engine. In Chapter 7 we studied reciprocating engines based on closed power cycles and in Chapter 8 we studied rotary gas turbines that run on open power cycles. In this chapter, we will introduce the remaining class of heat engines, the **vapor power cycles**, which operate with a phase-change substance (PC fluid) as the working fluid. If the working fluid of choice is steam, the heat engine is called a **steam power plant**. Like gas turbines, the vapor power cycles run on an **open cycle**, where open-steady devices are connected in series to form a loop. Unlike reciprocating engines or gas turbines, which are *internal combustion* engines, vapor power cycles are powered by the *external combustion* of fossil fuels or external heat addition through alternative means, such as nuclear reaction or solar radiation.

The objective of this chapter is to study the basic and more advanced vapor power cycles. Emphasis will be given to quantitative and **qualitative predictions** of work output and thermal efficiency. Modifications to the basic cycles will be introduced in steps as means of improving the thermal efficiency. **Cogeneration** and **combined cycles** will also be studied as modifications to the vapor power cycle that help to achieve better system efficiency. TEST solutions will be used for verification of manual results and studying "what-if" scenarios.

9.1 THE STEAM POWER PLANT

A simplified sketch of a fossil fuel fired steam power plant is broken into four logical subsystems as shown in Figure 9.1 and Anim. 9.A.*steamPowerPlant*.

Subsystem I in this figure (see the dashed boundaries) is the *heat engine*, introduced in Sec. 2.3.1 in connection with the overall analysis of a heat engine as a closed-steady system. This is where the energy conversion from heat to work takes place, and it is the focus of our study in this chapter. Water is pumped

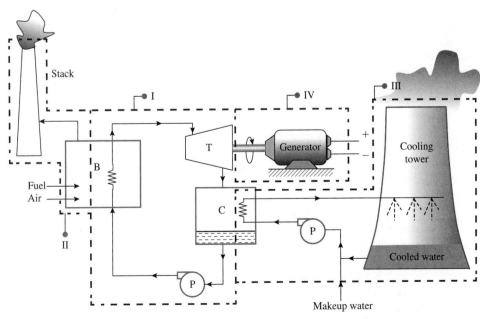

FIGURE 9.1 A steam power plant and its subsystems (see Anim. 9.A.*steamPowerPlant*).

at high pressure—around 10 MPa—by the **feedwater pump** into the **boiler**. The boiler is basically a heat exchanger where external heat addition takes place at a constant pressure to transform liquid water into superheated vapor at high temperature, around 500°C. The high temperature, high pressure vapor expands through the **turbine** producing shaft work, a small part of which is used to power the feedwater pump. To maximize power output from the turbine, the water vapor leaves the turbine at the lowest temperature possible. The vapor is then condensed by circulating water at atmospheric temperature, from a water reservoir in a heat exchanger called the **condenser**. The turbine exit temperature is generally above the temperature of the circulating water by about 5–10°C for effective heat transfer. The pressure of the steam in the condenser must be sub-atmospheric—$p_{sat@T}$ at 40°C is only about 7.4 kPa (Table B-2). The saturated liquid water produced by the condenser is fed back to the feedwater pump, completing the cycle.

The power required for pumping is much smaller than the power produced by the turbine. This can be deduced from the fact that the external work transfer in an isentropic device with negligible changes in ke and pe is given by [see Eq. (4.10) and Anim. 4.A.*pumpVsCompressor*]:

$$\dot{W}_{ext,\ int.rev.} = -\dot{m} \int_i^e v\, dp; \quad [\text{kW}] \tag{9.1}$$

Table 9-1 Different sources of heat used for producing electricity in the U.S. in 2006 and 2009.

	2006	2009	2012
Coal	48.9%	44.9%	37%
Natural Gas	20.0%	23.4%	30%
Nuclear	19.3%	20.3%	19%
Hydro	7.1%	6.9%	7%
Renewables	2.4%	3.5%	5%
Petroleum	2.3%	1.0%	1%
Total:	100.0%	100.0%	100%

Source: U.S. Energy Information Administration.

The average specific volume of steam in the turbine is much greater than that of liquid water in the pump; therefore, given the same mass flow rate and pressure difference, the pumping power can be expected to be only a fraction of the turbine output.

Subsystem II in Figure 9.1 supplies the external heat to the heat engine typically through burning of fossil fuel. Heat is transferred from the hot gases of combustion to the water as it passes through the boiler tubes. The exhaust that leaves the stack drains exergy and is a major source of **air pollution** (emissions from fossil fuel burning will be discussed in Chapters 13 and 14). In nuclear power plants, the external heat is provided by controlled nuclear reaction, while solar plants may have reflectors that concentrate solar energy into a receiver.

The heat rejection system, subsystem III, circulates cooling water through the condenser. The large flow rate of water necessary to absorb the heat from the condensing steam often comes from a nearby water reservoir such as a lake or a river. The resulting rise in water temperature must be within an acceptable limit to minimize **thermal pollution**. A more environmentally friendly alternative is shown in Figure 9.1, where the warm water leaving the condenser is sprayed inside a **cooling tower** (discussed in Chapter 12) through which ambient air circulates. Heat is rejected to the ambient air through evaporative cooling, and the cooled water is pumped back to the condenser after new water is added to make up for the loss through evaporation. The remaining subsystem is the electrical generator, subsystem IV in Figure 9.1, which converts shaft power to electricity.

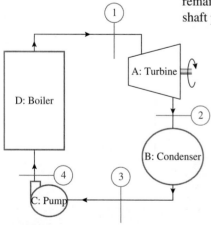

FIGURE 9.2 Ideal Rankine cycle (see Anim. 9.A.*idealRankineCycle*).

9.2 THE RANKINE CYCLE

The ideal cycle for modeling a simple vapor power cycle, the subsystem I of Figure 9.1, is the **Rankine cycle** shown in Figure 9.2 and Anim. 9.A.*idealRankineCycle*. As in the case of the ideal Brayton cycle, each device is assumed steady and internally reversible. Pressure drops in the boiler, condenser, and the connecting pipes are neglected, as are changes in ke or pe across any device. Isentropic compression in the pump is followed by constant-pressure heat addition in the boiler, isentropic expansion in the turbine, and constant-pressure heat rejection in the condenser, as shown in the *T-s* diagram of Figure 9.3. The working fluid, which changes phase during the cycle, must be modeled as a PC fluid (Sec. 3.4). While the superheated vapor and saturation tables must be used to determine State-1, 2, and 3, the compressed liquid sub-model (Sec. 3.4.8) is employed for its simplicity in evaluating the subcooled state, State-4.

To derive expressions for thermal efficiency, and other performance parameters of the Rankine cycle, an energy analysis is carried out for each open-steady device of Figure 9.2. In this analysis, $\dot{W}_T$, $\dot{W}_P$, $\dot{Q}_{in}$, and $\dot{Q}_{out}$ represent absolute values, with the direction of energy transfer evident from the nature of the device (T for turbine and P for pump) or from the subscript. They can be easily converted to $\dot{Q}$ or $\dot{W}_{ext}$, terms in the energy balance equation, using the WinHip sign convention (Sec. 0.7.2).

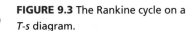

FIGURE 9.3 The Rankine cycle on a T-s diagram.

Device-A (1-2: Isentropic expansion in the turbine):

$$\frac{dE^0}{dt} = \dot{m}j_i - \dot{m}j_e + \dot{Q}^0 - \dot{W}_{ext} = \dot{m}j_1 - \dot{m}j_2 - \dot{W}_T$$

$$\Rightarrow \quad \dot{W}_T = \dot{m}(j_1 - j_2) \cong \dot{m}(h_1 - h_2); \quad [\text{kW}] \tag{9.2}$$

Device-B (2-3: Heat rejection in the condenser at constant pressure):

$$\frac{dE^0}{dt} = \dot{m}j_i - \dot{m}j_e + \dot{Q} - \dot{W}_{ext}^0 = \dot{m}j_2 - \dot{m}j_3 + (-\dot{Q}_{out})$$

$$\Rightarrow \quad \dot{Q}_{out} = \dot{m}(j_2 - j_3) \cong \dot{m}(h_2 - h_3); \quad [\text{kW}] \tag{9.3}$$

Device-C (3-4: Isentropic compression in the pump):

$$\frac{dE^0}{dt} = \dot{m}j_i - \dot{m}j_e + \dot{Q}^0 - \dot{W}_{ext} = \dot{m}j_3 - \dot{m}j_4 - (-\dot{W}_P)$$

$$\Rightarrow \quad \dot{W}_P = \dot{m}(j_4 - j_3) \cong \dot{m}(h_4 - h_3); \quad [\text{kW}] \tag{9.4}$$

Although in the T-s diagram T_4 is shown to be significantly greater than T_3, the actual rise in temperature during isentropic compression of a liquid is almost negligible. For example, the temperature increase across an isentropic feedwater pump raising the pressure from 5 kPa to 5 MPa is less than 0.5°C. The compressed liquid model (Sec. 3.4.10) stipulates that for a liquid, the properties v, u, and s are functions of temperature only, implying $T_4 \cong T_3$ and $u_4 \cong v_{f@T_3}$ for the isentropic process. The expression for the pumping power, therefore, can be simplified by using Eq. (3.28) for the enthalpy difference.

$$\dot{W}_P \cong \dot{m}(h_4 - h_3) = \dot{m}(u_4 - u_3)^0 + \dot{m}v_{f@T_3}(p_4 - p_3) \quad [\text{kW}]$$

$$\Rightarrow \quad \dot{W}_P \cong \dot{m}v_{f@T_3}(p_4 - p_3); \quad [\text{kW}] \tag{9.5}$$

Device-D (4-1: Constant-pressure heat addition in the boiler):

$$\frac{dE^0}{dt} = \dot{m}j_i - \dot{m}j_e + \dot{Q} - \dot{W}_{ext}^0 = \dot{m}j_4 - \dot{m}j_1 + \dot{Q}_{in}; \quad [\text{kW}]$$

$$\Rightarrow \quad \dot{Q}_{in} = \dot{m}(j_1 - j_4) \cong \dot{m}(h_1 - h_4); \quad [\text{kW}] \tag{9.6}$$

The net power, the specific power, the **back-work ratio** BWR (the fraction of the turbine power that is consumed by the pump), and the thermal efficiency can now be obtained as follows:

$$\dot{W}_{net} = \dot{W}_T - \dot{W}_P = \dot{Q}_{in} - \dot{Q}_{out} = \dot{m}(h_1 - h_2 + h_3 - h_4); \quad [\text{kW}] \tag{9.7}$$

$$w_{net} \equiv \frac{\dot{W}_{net}}{\dot{m}} = w_T - w_P = \frac{\dot{Q}_{in}}{\dot{m}} - \frac{\dot{Q}_{out}}{\dot{m}} = \int_2^3 Tds - \int_1^4 Tds; \quad \left[\frac{\text{kJ}}{\text{kg}}\right] \tag{9.8}$$

$$\text{BWR} \equiv \frac{\dot{W}_P}{\dot{W}_T} = \frac{h_4 - h_3}{h_1 - h_2};$$

$$\eta_{th,\text{Rankine}} \equiv \frac{\dot{W}_{net}}{\dot{Q}_{in}} = \frac{\dot{Q}_{in} - \dot{Q}_{out}}{\dot{Q}_{in}} = 1 - \frac{h_2 - h_3}{h_1 - h_4} \tag{9.9}$$

The heat rate defined by Eq. (8.9) and sfc (specific fuel consumption) defined by Eq. (8.11) for gas turbines are also applicable to vapor power cycles.

Table 9-2 Approximate heating values of various fossil fuels used in steam power plants.

Fuel	q_{comb} (MJ/kg)
Coal	27
Natural Gas	45.0
Petroleum	42.8
Heavy Diesel	42.8

9.2.1 Carbon Footprint

Greenhouse gases (Sec. 7.10) have been directly linked to global warming. Because carbon dioxide is the primary greenhouse gas, the emission of greenhouse gases is measured in the unit of equivalent mass of carbon dioxide. A **carbon footprint** or **carbon emission** is the total amount of CO_2 and other greenhouse gases emitted over the full life cycle of a product or process. It is expressed in the unit of mass (g, kg, or metric ton) of CO_2 equivalent per kWh of electric power generation, taking into account the different global warming potential of different gases. The life cycle of a product means all causes of carbon emission must be included in the analysis. For example, energy used in the delivery of the fuel to the power plant should be included. However, calculation of direct emissions due to the burning of fossil fuels can provide a lower limit of the carbon footprint of a power plant.

Hydrocarbon fuels will be discussed in more detail in Chapter 13. In combustion analysis, natural gas is represented by methane (CH_4), gasoline by octane (C_8H_{18}), and diesel by dodecane ($C_{12}H_{26}$). Symbolizing this class of fossil fuels (known as *paraffins*) by the generic formula C_nH_{2n+2}, it will be shown in Chapter 13 that for complete combustion of a unit mole of the fuel, n times as much carbon dioxide is generated as part of the combustion products. Using Eq. (1.5) to convert mass flow rate to mole flow rate, the carbon dioxide production rate can be expressed in terms of n as follows:

$$\dot{m}_{CO_2} = \dot{n}_{CO_2}\overline{M}_{CO_2} = \dot{n}_F\overline{M}_{CO_2}\frac{\dot{n}_{CO_2}}{\dot{n}_F} = \dot{n}_F\overline{M}_{CO_2}n = \frac{\dot{m}_F}{\overline{M}_F}\overline{M}_{CO_2}n; \quad \left[\frac{\text{kg of } CO_2}{\text{s}}\right]$$

$$\Rightarrow \quad \frac{\dot{m}_{CO_2}}{\dot{m}_F} = n\frac{\overline{M}_{CO_2}}{\overline{M}_F}; \tag{9.10}$$

Given an overall thermal efficiency η_{th} and a heating value of q_{comb}:

$$\dot{m}_F = \frac{\dot{Q}_{in}}{q_{comb}} = \frac{\dot{W}_{net}}{\eta_{th}q_{comb}}; \quad \left[\frac{\text{kg of Fuel}}{\text{s}} = \frac{\text{kW}}{\text{kJ/(kg of Fuel)}}\right] \tag{9.11}$$

The CO_2 produced per kJ of electrical energy can be obtained by combining Eqs. (9.10) and (9.11):

$$\frac{m_{CO_2}}{W_{net}} = \frac{\dot{m}_{CO_2}}{\dot{W}_{net}} = \frac{\dot{m}_{CO_2}}{\dot{m}_F\eta_{th}q_{comb}} = \frac{1}{\eta_{th}q_{comb}}\frac{n\overline{M}_{CO_2}}{\overline{M}_F}; \quad \left[\frac{\text{kg of } CO_2}{\text{kJ}}\right] \tag{9.12}$$

In the unit of g/kWh, the carbon footprint CF can now be expressed as:

$$CF = \frac{m_{CO_2}(1000 \text{ g/kg})}{W_{net}(1/3600 \text{ kWh/kJ})} = (3.6 \times 10^6)\frac{1}{\eta_{th}q_{comb}}\frac{n\overline{M}_{CO_2}}{\overline{M}_F} \quad \left[\frac{\text{g of } CO_2}{\text{kWh}}\right] \tag{9.13}$$

It should be stressed that Eq. (9.13) is treated as a lower estimate of the actual carbon footprint. Chemical formula, molar mass, and heating value of different fuels can be found in Table G-2.

9.2.2 TEST and the Open Vapor Power Cycle TESTcalcs

All open power cycles—gas turbine cycles, steam power cycles, and combined cycles—are treated by the same class of TESTcalcs—the open power cycle TESTcalcs discussed in Sec. 8.2.1. The only difference between the various power cycle TESTcalcs stems from the model used for the working fluid: PC model for vapor power, IG or PG model for gas power, and PC/IG model for combined cycles.

EXAMPLE 9-1   **Overall Performance Parameters**

A coal fired steam power plant produces 150 MW of electric power with a thermal efficiency of 35%. If the energetic efficiency of the boiler is 75%, the heating value of coal is 30 MJ/kg, and the temperature rise of the cooling water in the condenser is 10°C, determine (a) the fuel consumption rate in kg/h, (b) carbon footprint (CF) in kg/kWh, and (c) mass flow rate of cooling water.

SOLUTION

Use the definitions of the basic operational parameters to obtain the answers.

Assumptions

The net power output is converted to electricity without any losses. The boiler efficiency is the ratio of the desired heat transfer to the heat release rate.

Analysis

The burning rate of coal can be obtained as:

$$\dot{m}_F = \frac{(\dot{Q}_{in}/\eta_{boiler})}{q_{comb}} = \frac{(\dot{W}_{net}/\eta_{th})}{\eta_{boiler}\,q_{comb}} = \frac{(150)(10^3)(3600)}{(0.75)(0.35)(30000)} = 68{,}571 \frac{kg}{h} \qquad \textbf{(9.14)}$$

Using Eq. (9.10), we obtain:

$$\dot{m}_{CO_2} = \dot{m}_F n \frac{\overline{M}_{CO_2}}{\overline{M}_F} = (68{,}571)(1)\frac{44}{12} = 251{,}420 \frac{kg}{h}$$

In one hour the plant produces 150,000 kWh of electricity, while producing 251,420 kg of CO_2. The CF can, therefore, be calculated as:

$$CF \equiv \frac{m_{CO_2}\text{in g}}{W_{net}\text{ in kWh}} = \frac{251{,}420(1000 \text{ g/kg})}{150{,}000 \text{ kWh}} = 1680 \frac{\text{g of }CO_2}{\text{kWh}}$$

A steady-state energy balance for the circulating water in the condenser (Fig. 9.4) yields:

$$\dot{Q}_{out} = \dot{Q} = \dot{m}(j_e - j_i) \cong \dot{m}(h_6 - h_5)$$

Application of the SL model (Sec. 3.3) for the cooling water with the assumption of negligible pressure drop in the condenser results in:

$$\dot{Q}_{out} = \dot{Q} = \dot{m}(j_e - j_i) \cong \dot{m}(h_6 - h_5) = \dot{m}c_v(T_6 - T_5) = \dot{m}c_v\Delta T$$

$$\Rightarrow \quad \dot{m} = \frac{\dot{Q}_{out}}{c_v \Delta T} = \frac{\dot{Q}_{in} - \dot{W}_{net}}{c_v \Delta T} = \frac{\dot{W}_{net}}{c_v \Delta T}\left(\frac{1}{\eta_{th}} - 1\right) = 23.9 \times 10^6 \frac{kg}{h}$$

Discussion

The carbon footprint, calculated based on direct emission, is only a lower estimate. In fact, a **life cycle analysis**, also known as *cradle-to-grave analysis,* has to take into account many other factors, as sketched in Figure 9.5.

The pumping power for the cooling water can be calculated from Eq. (9.5) as $\dot{m}v\Delta p$ for a pressure drop Δp in the entire circulation system. Note that the efficiency of the boiler, power

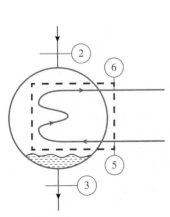

FIGURE 9.4 The circulating water constitutes the system in Example 9-1.

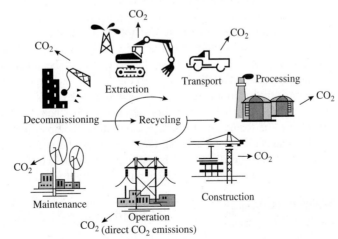

FIGURE 9.5 Sources of CO_2 that must be considered in a life-cycle analysis.

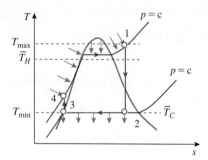

FIGURE 9.6 Heat is rejected at a constant temperature and is added at an effective temperature of $\overline{T}_H$.

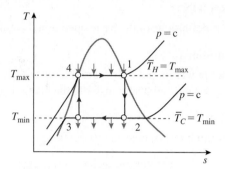

FIGURE 9.7 Open-cycle implementation of Carnot cycle (see Anim. 9.A.*CarnotCycle*).

requirement of the heating or cooling subsystems, and losses in the electrical subsystem, are outside the heat engine cycle, and therefore do not enter into the definition of the thermal efficiency of the Rankine cycle.

9.2.3 Qualitative Performance Predictions

As with any reversible heat transfer (Sec. 4.1.3), the area under the heat addition line 4-1 in the T-s diagram of Figure 9.6 represents the heat input per unit mass, $\dot{Q}_{in}/\dot{m}$, and the area under the heat rejection line 2-3 represents the magnitude of heat rejection per unit mass, $\dot{Q}_{out}/\dot{m}$ (see Anim. 4.A.*intReversible*). Therefore, as established by Eq. (9.8), the net area enclosed can be interpreted as the specific power $w_{net} = \dot{W}_{net}/\dot{m}$, the net work per unit mass.

Like Otto, Diesel, and Brayton cycles, the Rankine cycle is not a Carnot cycle, despite being reversible. The concept of the *equivalent Carnot cycle*, introduced in Sec. 7.4.2 for a reciprocating cycles and in Sec. 8.2.3 for a Brayton cycle, can also be applied to the Rankine cycle to predict how its efficiency and net work depend on various parameters. Heat rejection in the Rankine cycle (Figure 9.6) is taking place at a constant temperature; the effective temperature of heat rejection $\overline{T}_C$ is also the actual minimum temperature T_{min} of the cycle. Heat addition takes place over a wide range of temperatures—from T_4 to T_{max}. The effective heat addition temperature, $\overline{T}_H$, is defined so that the thermal efficiency of the Rankine cycle is given by the Carnot efficiency, $1 - \overline{T}_C/\overline{T}_H$. For our qualitative purpose, it suffices to understand that $\overline{T}_H$ must be closer to T_{max} than to T_4 since more heat addition takes place at the higher end of the heat addition curve—examine how the area under the heat addition curve 4-1 in Figure 9.6 is distributed. By observing how a modification to the basic cycle affects $\overline{T}_H$, $\overline{T}_C$, and the enclosed area in a T-s diagram, we will be able to deduce how a particular cycle performance is affected by the modification.

A special case of the Rankine cycle is shown in Figure 9.7, where the turbine inlet state is not superheated and the pump inlet state is such that the pump exit temperature is the same as the turbine inlet temperature. The resulting cycle, a rectangle on the T-s diagram, is obviously an open cycle implementation of the **Carnot cycle** (the closed-cycle implementation was discussed in Sec. 7.1). Unfortunately, there are several drawbacks that make such an implementation of the Carnot cycle impractical. First, the vapor cannot be superheated to keep the temperature of heat addition a constant, limiting T_H to the critical temperature at best. Second, an increase in T_H will reduce the enclosed area on the T-s diagram (by making it slender) and, hence, the specific power of the cycle. Third, at quality well below 90% near the turbine exit, liquid droplets can be expected to be formed in the steam. Continuous bombardments of the turbine blades by these droplets can cause severe erosion. Fourth, it is quite difficult, if not impossible, to design a compressor that must handle a two-phase mixture. All these difficulties are avoidable in the Rankine cycle, making it much more practical although slightly less efficient than a Carnot cycle.

EXAMPLE 9-2 **Analysis of an Ideal Rankine Cycle**

A steam power plant operates on an ideal Rankine cycle. Superheated steam flows into the turbine at 2 MPa, 500°C with a flow rate of 100 kg/s and exits the condenser at 50°C as saturated water. Determine (a) the net power output, (b) the thermal efficiency, and (c) the quality of steam at the turbine exit. ***What-if scenario:*** (d) What would the efficiency be if the condenser temperature dropped to 30°C?

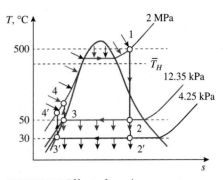

SOLUTION

Evaluate the four principal states of the ideal Rankine cycle from the given information. The condenser temperature cannot be assumed constant as steam at the turbine exit may be superheated. Once the states are obtained, perform an energy analysis on each open-steady device to determine the cycle's parameters (Fig. 9.8).

FIGURE 9.8 Effect of condenser temperature studied in Ex. 9-2.

Assumptions

Ideal Rankine cycle assumptions as outlined in Sec. 9.2.

Analysis

Evaluate the four principal states using the PC model manually or the PC flow-state TESTcalc. The condenser inlet temperature is an unknown and can be obtained only after State-3 is determined.

State-1 (given $\dot{m}$, p_1, T_1): $h_1 = 3467.5$ kJ/kg;

State-3 (given $x_3 = 0$, T_3): $h_3 = 209.33$ kJ/kg; $v_3 = 0.00101$ m³/kg; $p_3 = 12.35$ kPa;

State-2 (given $p_2 = p_3$, $s_2 = s_1$): $h_2 = 2383.7$ kJ/kg; $x_2 = 91.3\%$;

State-4 (given $p_4 = p_1$, $s_4 = s_3$): Using the compressed liquid sub-model (Eq. 3.28)

$$h_4 = h_3 + v_{f@T_3}(p_4 - p_3) = 211.34 \text{ kJ/kg}; \quad v_3 = 0.00101 \text{ m}^3/\text{kg}$$

A steady-state energy analysis is carried out for each device:

Device-A (1−2): $\dot{W}_T = \dot{m}(h_1 - h_2) = 108{,}380$ kW

Device-B (2−3): $\dot{Q}_{out} = \dot{m}(h_2 - h_3) = 217{,}439$ kW

Device-C (3−4): $\dot{W}_P = \dot{m}(h_4 - h_3) = 201$ kW

Device-D (4−1): $\dot{Q}_{in} = \dot{m}(h_1 - h_4) = 325{,}619$ kW

The net power and the thermal efficiency are:

$$\dot{W}_{net} = \dot{W}_T - \dot{W}_P = 108{,}179 \text{ kW}; \quad \eta_{th,Rankine} = \frac{\dot{W}_{net}}{\dot{Q}_{in}} = 33.2\%$$

TEST Analysis

Launch the PC power cycle TESTcalc. Evaluate the four states and analyze the devices as described in the TEST-code (see *TEST > TEST-codes*). Click Super-Calculate to obtain the cycle related variables in the cycle panel.

What-if scenario

Change T3 to 30 deg-C. Click Super-Calculate to obtain the new thermal efficiency as 36.5% and the net output as 121,901 kW.

Discussion

Lowering the condenser temperature by 20°C results in a significant increase in the thermal efficiency. This could be predicted as $\overline{T}_C$, which is the same as the, is substantially lowered, while $\overline{T}_H$ is barely affected (State-4 and 4' are almost identical in Fig. 9.8). The net area of the cycle also increases, explaining the increase in the net output. There is a lower limit of the condenser temperature set by the condensing steam, which must be hotter than the circulating cooling water (by about 5–10°C) in the condenser. A detrimental side effect of lowering the condenser pressure is that the quality of steam at the turbine exit decreases, in this example to 87.2%. A remedy for this problem will be provided in Sec. 9.3 where we discuss various modifications to the basic cycle.

9.2.4 Parametric Study of the Rankine Cycle

In a **parametric study** only one controllable parameter is varied at a time over a practical range while holding all other independent variables constant. In the last example, the condenser temperature was the parameter, and we established how it affects the thermal efficiency and the net power output of the cycle. The working fluid is a saturated mixture during condensation; therefore, the condenser pressure and temperature are not independent. For instance, if the cooling water is available at 30°C in a steam power plant, the condensation temperature must be above 40°C to provide a meaningful temperature difference (about 10°C) for effective heat transfer. The condenser pressure in that case must be $p_{sat@40°C} = 7.39$ kPa or higher, but still well below the atmospheric pressure. A study of condenser temperature as a parameter is equivalent to a parametric study of condenser pressure.

Let's now look at the turbine inlet temperature. Increasing the turbine inlet temperature increases the effective temperature of heat addition $\overline{T}_H$ and the net area of the cycle, as shown in Figure 9.9. $\overline{T}_C$ remains unchanged as long as the turbine exit state is not superheated. Both thermal efficiency and net specific power are expected to increase with T_{max}. This is numerically demonstrated in Table 9-3 with results obtained by carrying out a parametric study using the TEST-code of Example 9-2. Another beneficial effect of increasing the turbine inlet temperature is the increase in steam quality at the turbine inlet. Metallurgical considerations, however, limit the maximum turbine inlet temperature, which is presently about 625°C. In nuclear power plants, on the other hand, the turbine inlet temperature is set by safety considerations and is usually significantly lower than that in fossil-fuel fired plants. This explains why the thermal efficiency of a nuclear power plant is about 5% lower than the efficiency of typical external combustion plants, which can be as high as 40 percent.

The effect of boiler pressure, similarly, can be qualitatively deduced from Figures 9.10 and 9.11. The condenser pressure and T_{max} remaining unchanged, an increase in the boiler pressure clearly raises $\overline{T}_H$ without affecting $\overline{T}_C$. If the boiler pressure is raised beyond the critical pressure, $\overline{T}_H$ will reverse its trend and start to decrease as the share of heat addition, at relatively lower temperature, increases in the absence of constant-temperature boiling. The trend for the enclosed area of the cycle is somewhat unclear from these T-s diagrams, but the area, and net specific power, decreases as the boiler pressure is increased beyond the critical pressure.

Using the TEST-code from Example 9-2, a parametric study is performed by changing p_1, clicking Super-Calculate, then repeating the process. The results, tabulated in Table 9-4, confirm the predicted trend. Although the performance seems to deteriorate at super-critical pressure, with modifications such as reheat, regeneration, etc., (Sec. 9.3), $\overline{T}_H$ can be made to increase with an increase in the boiler pressure, even in the super-critical region. Modern steam power plants may be designed for a boiler pressure as high as 30 MPa, much in excess of the critical pressure, 22.09 MPa, of water.

Table 9-3 Parametric study of the Rankine cycle with turbine inlet temperature as the parameter.

T_{max} (°C)	$\eta_{th,Rankine}$	$\dot{W}_{net}$ (MW)
400	0.332	108.2
500	0.350	121.7
600	0.368	136.5
700	0.386	152.1

Table 9-4 Parametric study of the Rankine cycle with boiler pressure as the parameter.

p_{max} (MPa)	$\eta_{th,Rankine}$	$\dot{W}_{net}$ (MW)
5	0.356	106
10	0.381	110
15	0.392	108
20	0.394	102
30	0.362	69.3

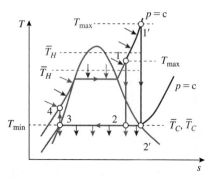

FIGURE 9.9 As T_{max} is increased $\overline{T}_H$ increases, while $\overline{T}_C$ is unaffected or barely affected.

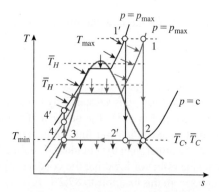

FIGURE 9.10 As p_{max} is increased $\overline{T}_H$ also increases, but $\overline{T}_C$ remains unchanged.

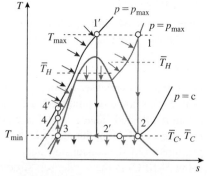

FIGURE 9.11 As p_{max} is increased beyond the critical pressure $\overline{T}_H$ begins to decrease.

9.2.5 Irreversibilities in an Actual Cycle

An actual vapor power cycle is compared with the ideal Rankine cycle in Figure 9.12 (and Anim. 9.A.*actualRankineCycle*). Even though a turbine can be reasonably assumed to be adiabatic, entropy generation due to friction, and other irreversibilities, result in s_3 being greater than s_2. The actual exit state (State-3) is related to the ideal isentropic state (State-2) through the isentropic turbine efficiency (Sec. 4.2.3 and Anim. 4.A.*turbineEfficiency*):

$$\eta_T = \frac{\dot{W}_{T,\text{actual}}}{\dot{W}_{T,\text{isentropic}}} = \frac{j_i - j_e}{j_i - j_{es}} \cong \frac{h_i - h_e}{h_i - h_{es}} = \frac{h_1 - h_3}{h_1 - h_2} \qquad (9.15)$$

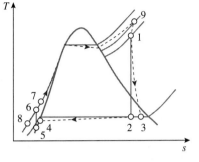

FIGURE 9.12 Actual vs. ideal Rankine cycle (see Anim. 9.A.*actualRankine Cycle*).

Viscous friction in the condenser passages causes a slight pressure drop in the condenser so that $p_4 < p_3$. Notice that the saturated liquid at State-4 is not directly fed to the feedwater pump to prevent **cavitation**—an undesirable phenomenon of vapor bubble formation at the low pressure side of the impeller where the pressure can dip below the saturation pressure at T_4. Instead, the saturated liquid formed in the condenser is subcooled to State-5 so that $p_{\text{sat}@T_5}$ is well above the lowest pressure in the pump.

The isentropic efficiency of the pump (Anim. 4.A.*pumpEfficiency*) relates the actual pump exit state (State-7) with the corresponding isentropic state (State-6):

$$\eta_P = \frac{\dot{W}_{P,\text{isentropic}}}{\dot{W}_{P,\text{actual}}} = \frac{j_{es} - j_i}{j_e - j_i} \cong \frac{h_{es} - h_i}{h_e - h_i} = \frac{h_6 - h_5}{h_7 - h_5} \qquad (9.16)$$

Poor isentropic efficiencies reduce the turbine output and increase the pump input reducing the thermal efficiency.

Pressure and temperature decrease slightly from the pump exit condition, State-7, to the boiler inlet, State-8, as the compressed liquid is piped into the boiler. The most significant pressure loss happens in the boiler due to friction in the narrow boiler tubes. A further reduction in pressure and temperature occurs from State-9 to State-1 as the superheated steam is routed into the turbine. The available pressure difference for the turbine and its output is clearly diminished by the cumulative pressure drop. Heat loss from the steam to the surroundings, which must be compensated through additional heat transfer in the boiler, also has an adverse effect on the thermal efficiency. Other factors that contribute to the loss of efficiency include steam leaks in the boiler, air leaks into the condenser, friction in bearings, and power consumed by auxiliary equipment.

DID YOU KNOW?

- The biggest power plant in the world is China's Three Gorges Dam, with a capacity of 18.2 GW. This is closely followed by the Itaipu hydroelectric plant of Brazil/Paraguay with a capacity of 14.75 GW.

EXAMPLE 9-3 Rankine Cycle with Irreversibilities

A power plant operates on a simple Rankine cycle producing a net power of 100 MW. The turbine inlet conditions are 15 MPa and 600°C, and the condenser pressure is 10 kPa. If the turbine and pump each have an isentropic efficiency of 85%, and there is a 5% pressure drop in the boiler, determine (a) the thermal efficiency, (b) the mass flow rate of steam in kg/h, and (c) the back-work ratio. **What-if scenario:** Perform a parametric study on how the thermal efficiency depends on (d) the boiler pressure drop, (e) pump efficiency, and (f) turbine efficiency.

SOLUTION

Use the PC model to evaluate the six principal states of the working fluid (Fig. 9.13) and perform an energy balance for each ideal and actual open-steady device to obtain the cycle variables.

Assumptions

The cycle is ideal in all respects except for the irreversibilities mentioned.

Analysis

Use the manual approach (Sec. 3.4), or the PC flow-state TESTcalc, to evaluate the enthalpy of each principal state, as tabulated below.

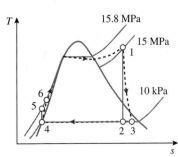

FIGURE 9.13 Actual cycle analyzed in Ex. 9-3.

State	Given	h (kJ/kg)	State	Given	h (kJ/kg)
1	p_1, T_1	3582	4	$p_4 = p_2, x_4 = 0$	191.8
2	$p_2, s_2 = s_1$	2115	5	$p_5 = p_1/0.95,$	207.8
3	$p_3 = p_2,$ $h_3 = h_1 - (h_1 - h_2)/\eta_T$	2408		$s_5 = s_4,$ or, $h_5 = h_4 + v_{f@T_4}(p_5 - p_4)$	
			6	$p_6 = p_5,$ $h_6 = h_4 + (h_5 - h_4)/\eta_P$	211.8

Steady-state energy analysis of each device (Sec. 9.2) produces the following table of external energy transfers (WinHip sign convention applied).

Device	$\dot{Q}/\dot{m}$ (MJ/kg)	$\dot{W}_{ext}/\dot{m}$ (MJ/kg)	Device	$\dot{Q}/\dot{m}$ (MJ/kg)	$\dot{W}_{ext}/\dot{m}$ (MJ/kg)
A: Turbine (1-3)	0	1.174	C: Pump (4-6)	0	−0.020
B: Condenser (3-4)	−2.217	0	D: Boiler (6-1)	3.371	0

The mass flow rate of steam, now, can be obtained as:

$$\dot{W}_{net} = \dot{W}_T - \dot{W}_P = \dot{m}\left(\frac{\dot{W}_T}{\dot{m}} - \frac{\dot{W}_P}{\dot{m}}\right); \quad \Rightarrow \quad \dot{m} = \frac{100}{1.174 - 0.02} = 86.65 \frac{kg}{s};$$

$$\eta_{th} = \frac{\dot{W}_{net}}{\dot{Q}_{in}} = \frac{100}{(3.371)(86.65)} = 34.2\%; \quad BWR = \frac{\dot{W}_T}{\dot{W}_P} = 1.7\%$$

TEST Analysis

Using a mass flow rate of 1 kg/s, evaluate the six principal states, and analyze the four devices with the PC power cycle TESTcalc as outlined in the TEST-code (see *TEST > TEST-codes*). In the cycle panel, Wdot_net is calculated as **1.154 MW**. Therefore, for a net output of 100 MW, the flow rate must be 100/1.154 = 86.65 kg/s. Now change mdot1 to the new value and click Super-Calculate to verify the manual results.

What-if scenario

For the sensitivity study, change p5 to '= p1' to eliminate boiler pressure drop and click Super-Calculate. No noticeable change is observed in the thermal efficiency in the cycle panel. Now change the pump efficiency to 100% by substituting '= h5' for h6. Click Super-Calculate to show that the thermal efficiency does not change much. Change the turbine efficiency to 100% by substituting '= h2' for h3 and click Super-Calculate. This time the thermal efficiency increases significantly, to 43.0%.

Discussion

The pressure drop in the boiler, and the isentropic efficiency of the pump, do not seem to have any noticeable impact on the net specific power output or the thermal efficiency of the cycle. The extra pumping power required to compensate for these losses is insignificant compared to the turbine output, given the typically small value of the back-work ratio of a Rankine cycle. Irreversibility in the turbine has a much bigger impact on the cycle performance.

9.2.6 Exergy Accounting of Rankine Cycle

So far, the performance of a Rankine cycle has been analyzed with the help of mass and energy analysis. However, an energy analysis completely ignores the quality of heat—heat lost from a high-pressure turbine is considered completely equivalent to the same amount of heat lost from a low-pressure turbine. Only an exergy accounting can compare the quality of various energy transfers and determine the relative significance of irreversibilities present in various components of a vapor power plant.

Exergy that enters the plant with fuel is the required input to a power plant, a significant part of which is delivered as the desired electrical power output. Their ratio is the overall **exergetic efficiency**. Exergy that is not converted to electrical power is transported out by the stack gases, transferred to the cooling water in the condenser, and destroyed through thermodynamic friction (irreversibilities). Combustion irreversibilities, or chemical friction, and the exergy transported by the stack flow will be studied in Chapter 13. In this section we will model the boiler with a simple heat addition process from a reservoir (Fig. 9.14) at a known effective temperature. The irreversibilities associated with this heat addition (thermal friction) are equivalent to the combustion generated irreversibilities (mixing and chemical friction). The exergy that accompanies the heat from the reservoir will be the starting point of our exergy analysis.

For a single-flow, device exchanging heat with a single non-atmospheric reservoir, the exergy balance equation for a steady device, Eq. (6.10) simplifies to:

$$0 = \underbrace{\dot{m}(\psi_i - \psi_e)}_{\substack{\text{Net exergy} \\ \text{transport into the} \\ \text{system by mass.}}} + \underbrace{\dot{Q}_k\left(1 - \frac{T_0}{T_k}\right)}_{\substack{\text{Net exergy} \\ \text{transfer into the} \\ \text{system by heat.}}} - \underbrace{\dot{W}_{sh}}_{\substack{\text{Exergy delivered} \\ \text{by the system} \\ \text{through shaft work.}}} - \underbrace{\dot{I}}_{\substack{\text{Exergy destruction} \\ \text{rate in the system's} \\ \text{universe.}}} \quad ; \quad \text{[kW]} \quad \textbf{(9.17)}$$

In the TESTcalcs, the flow exergy ψ is evaluated as a state property once the dead state, State-0, is calculated at the pressure and temperature of the surroundings with zero ke and pe. The exergy destruction rate, $\dot{I}$, can be calculated from Eq. (9.17), after the principal states are completely evaluated. This is illustrated in the following example.

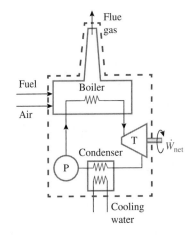

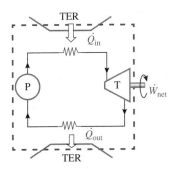

FIGURE 9.14 Simplification of a power plant for the purpose of exergy analysis.

EXAMPLE 9-4   Rankine Cycle Exergy Accounting

For the cycle described in Example 9-3, develop a balance sheet for exergy inventory. Assume the heat addition to take place from a reservoir (TER) at 1500 K and heat rejection to a reservoir (TER) at 310 K. The atmospheric conditions are 100 kPa and 300 K.

TEST Analysis

In the TEST solution of Example 9-3, evaluate State-0 as the dead state with the atmospheric conditions 100 kPa and 300 K. Also use a mass flow rate of 86.655 kg/s in State-1. Click Super-Calculate to evaluate the specific flow exergy of all principal states. The exergy balance equation, Eq. (9.17), can now be used to construct the following balance sheet for exergy.

Exergy supplied to the working fluid in the boiler:

$$\dot{Q}_{in}\left(1 - \frac{T_0}{T_H}\right) = (292,072)\left(1 - \frac{300}{1500}\right) = 233,658 \text{ kW} = 233.66 \text{ MW}$$

Exergy gained by the working fluid from heat addition:

$$\dot{m}(\psi_1 - \psi_6) = (86.655)(1584.2 - 18.449)/1000 = 135.68 \text{ MW}$$

Therefore, exergy destroyed during heat addition: $\dot{I} = 233.66 - 135.68 = 97.98 \text{ MW}$

Exergy delivered to the turbine by the working fluid:

$$\dot{m}(\psi_1 - \psi_3) = (86.655)(1584.2 - 134.30)/1000 = 125.64 \text{ MW}$$

Exergy delivered by the turbine to the shaft: $\dot{W}_T = 101.73 \text{ MW}$

Therefore, exergy destroyed in the turbine: $\dot{I} = 125.64 - 101.73 = 23.91$ MW

Exergy lost by the steam to the condenser:

$$\dot{m}(\psi_3 - \psi_4) = (86.655)(134.30 - 2.244)/1000 = 11.443 \text{ MW}$$

Exergy transferred by the heat lost in the condenser:

$$\dot{Q}_{\text{out}}\left(1 - \frac{T_0}{T_C}\right) = (192.07)\left(1 - \frac{300}{310}\right) = 6.20 \text{ MW}$$

Therefore, exergy destroyed during heat rejection: $\dot{I} = 11.443 - 6.20 = 5.243$ MW

Note that the exergy transferred to the circulating coolant in the condenser is usually destroyed through thermal friction.

Exergy gained by water in the pump:

$$\dot{m}(\psi_6 - \psi_4) = (86.655)(18.450 - 2.244)/1000 = 1.404 \text{ MW}$$

Exergy input into the pump as shaft work: $\dot{W}_P = 1.726$ MW

Therefore, exergy destroyed in the pump: $\dot{I} = 1.726 - 1.404 = 0.322$ MW

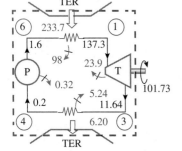

FIGURE 9.15 Detailed exergy flow diagram (in MW) for Ex. 9-4. Crossed arrows indicate exergy destroyed.

Discussion

Different exergy values are shown in the exergy flow diagram for the cycle in Figure 9.15. The overall exergetic efficiency of the plant is $100/233.66 = 42.8\%$. That means 57.2% of the available exergy is wasted due to irreversibilities in various components and through losses into the surroundings. The bulk of the exergy destruction takes place in the boiler, followed by the turbine, while the exergy destruction in the pump is almost insignificant. The huge amount of heat that is lost to the cooling water, 192 MW has an exergy content of only 6.2 MW, or just 3.2%. That is why it is called **waste,** or **low grade, heat** in industry.

9.3 MODIFICATION OF RANKINE CYCLE

The parametric study of the ideal Rankine cycle, conducted in Sec. 9.2.4, suggests that the thermal efficiency of the cycle can be improved by manipulating some of the principal parameters, such as the turbine inlet temperature and condenser pressure. Exergy accounting of an actual cycle, on the other hand, pinpoints the sources of exergy destruction in an actual cycle and tells us how to improve the performance of a given cycle. We will systematically introduce a series of modifications to the basic cycle to improve the thermal efficiency and reduce the fuel cost, which is the primary operating cost of a power plant. Due to the fact that power plants mostly run on fossil fuels, improving the efficiency has a direct impact on their carbon footprint.

The specific work output, $w_{\text{net}} = \dot{W}_{\text{net}}/\dot{m}$, is not as much of a concern for land based units as it is in applications where the size and weight of the plant are a concern. Used mostly for land based power plants, the Rankine cycle can be scaled up by increasing $\dot{m}$ and, hence, $\dot{W}_{\text{net}}$.

9.3.1 Reheat Rankine Cycle

We have already established (Sec. 9.2.3) that an increase in the turbine inlet temperature leads to a higher effective temperature of heat addition without significantly affecting the effective temperature of heat rejection (unlike the Brayton cycle), resulting in improved thermal efficiency. Saturated steam produced by the boiler is superheated in a separate heat exchanger, known as the **superheater**, to increase the steam temperature as much as possible. Unfortunately, material consideration limits that temperature to only about 600°C. The boiler and superheater together are called the **steam generator**.

An increase in the boiler pressure also raises the thermal efficiency until the pressure exceeds the critical limit (Table 9-4). A detrimental side effect of the boiler pressure rise is the lowering of the quality of the steam at the turbine exit (compare States 2 and 2' in Figs. 9.10 or 9.11).

The modification that allows higher boiler pressure and yet avoids low-quality steam at the turbine exit, is called **reheat**. In the ideal reheat cycle, shown in Figure 9.16 (and

Anim. 9.B.*reheatCycle*), the turbine is split into two stages with the steam expanding through the first-stage to an intermediate pressure between the boiler and condenser pressures. It is then reheated in the steam generator, usually to the inlet temperature of the first turbine stage. Pressure drop in the reheater, if any, is neglected in an ideal reheat cycle. The reheated steam completes its expansion to the condenser pressure in the second stage and leaves the turbine with an improved quality at State-4 as opposed to State-4' (no reheat) in Figure 9.16.

As a result of reheat, more heat is added at the relatively high temperature of the reheater, while the heat rejection temperature $\overline{T}_C$ remains unaffected; therefore, the thermal efficiency is improved by as much as 5%. By increasing the number of stages of reheat it seems that $\overline{T}_H$ can be indefinitely increased. However, this may result in superheated exhaust, raising $\overline{T}_C$. An increase in $\overline{T}_C$ nullifies any improvement resulting from an elevated $\overline{T}_H$. Another drawback of multiple stages of reheat is the pressure drop in the reheater may become significant. As a result, two reheat stages are generally used in modern power plants that employ very high, even super-critical, boiler pressure.

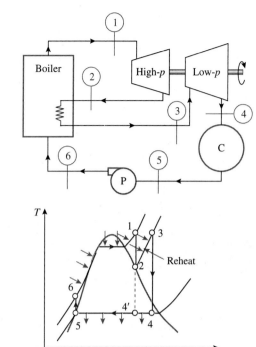

FIGURE 9.16 The reheat Rankine cycle (see Anim. 9.A.*reheatCycle*).

EXAMPLE 9-5   **Rankine Cycle Modified with Reheat.**

A steam power plant operates on an ideal reheat Rankine cycle. Steam enters the high pressure turbine at 15 MPa, 525°C with a flow rate of 100 kg/s and exits as saturated vapor. The quality of steam at the exit of the second-stage turbine is 90% and the condenser pressure is 8 kPa. Determine (a) the reheat pressure, (b) the reheat temperature, and (c) the thermal efficiency. *What-if scenario:* (d) How would the thermal efficiency vary if the boiler pressure varied from 5 MPa through 30 MPa?

SOLUTION

Use the PC model to evaluate the six principal states (Fig. 9.17) and perform a steady state energy balance for each device to obtain the cycle parameters.

Assumptions

Ideal reheat cycle.

Analysis

Use the manual approach (Sec. 3.4), or the PC flow-state TESTcalc, to evaluate the enthalpy of all principal states.

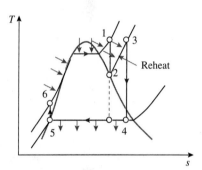

FIGURE 9.17 *T-s* diagram for the reheat Rankine cycle of Ex 9-5.

State	Given	h (kJ/kg)	State	Given	h (kJ/kg)
1	$p_1, T_1, \dot{m}_1$	3379	4	$p_4, x_4 = 0.9$	2337
2	$x_2, s_2 = s_1$	2793	5	$p_5 = p_4, x_5 = 0$	174
		$p = 1.56$ MPa	6	$p_6 = p_1,$	189.0
3	$p_3 = p_2,$	$h = 3409$		$s_6 = s_5$ or	
	$s_3 = s_4$	$T = 471°C$		$h_6 = h_5$	
				$+ v_{f@T_5}(p_6 - p_5)$	

Steady-state energy analysis of each device (Sec. 9.2) produces the following results.

Device	$\dot{Q}$, (MW)	$\dot{W}_{ext}$, (MW)	Device	$\dot{Q}$, (MW)	$\dot{W}_{ext}$, (MW)
A: Turbine-I (1-2)	0	58.53	D: Condenser (4-5)	−216.3	0
B: Reheater (2-3)	61.62	0	E: Pump (5-6)	0	−1.51
C: Turbine-II (3-4)	0	107.3	F: Boiler (6-1)	319.0	0

Keeping in mind that the heat addition takes place in two separate processes, the thermal efficiency is evaluated as:

$$\eta_{th} = \frac{\dot{W}_{net}}{\dot{Q}_{in}} = 1 - \frac{\dot{Q}_{out}}{\dot{Q}_{in}} = 1 - \frac{216.3}{319 + 61.62} = 43.17\%$$

TEST Analysis

Evaluate the six principal states and analyze the six devices with the PC power cycle TESTcalc, as described in the TEST-code (see *TEST > TEST-codes*). Thermal efficiency and all other cycle parameters are automatically evaluated in the cycle panel once the analyzed devices complete a loop.

What-if scenario

Change p1 to a new value and click Super-Calculate to update the solution. Repeat with different values of p1. The results are listed in Table 9-5.

Discussion

Both the thermal efficiency and net power output are found to increase monotonically with the boiler pressure. Contrast this to the results of a similar parametric study, Table 9-4, for the ideal Rankine cycle. With the addition of reheat, the performance no longer deteriorates for a boiler pressure above the critical pressure. In this study, we allowed T_3 to vary while preserving the quality of steam at State-2 and State-4. It is left as an exercise to show that the quality at the exhaust of the second turbine would deteriorate as the boiler pressure is increased if T_3 is held constant.

Table 9-5   Parametric study of the Rankine cycle with reheat while varying boiler pressure as the parameter.

p_{max} (MPa)	$\eta_{th,Rankine}$	$\dot{W}_{net}$ (MW)
5	0.382	134
10	0.411	151
15	0.432	164
20	0.448	175
30	0.473	194

9.3.2 Regenerative Rankine Cycle

A cursory look at any *T-s* diagram of a simple Rankine cycle, say Figure 9.6, immediately reveals that the heat addition takes place over a wide range of temperatures, from T_4, the feedwater temperature, to T_1, the maximum temperature of the cycle. The separation between State-4 and State-3 is insignificant and generally exaggerated in most diagrams. In reality, $T_4 \approx T_3$ even for an inefficient pump. Clearly a large part of heat addition occurs at a relatively low temperature in raising the temperature to the boiling point, depressing the effective temperature of heat addition $\overline{T}_H$ and, hence, the thermal efficiency.

While introducing different modifications to the Brayton cycle (Chapter 8), we discussed **regeneration**, in which heat is transferred from one part of the cycle to another to avoid low-temperature external heating and elevate $\overline{T}_H$. For the Rankine cycle, one possibility is to extract heat from the combustion gases at the stack by using a heat exchanger. But this is impractical, as the stack is generally far removed from the feedwater, and the piping necessary to have the paths of the two fluids cross in a heat exchanger would cause significant pressure drop. Besides, the exiting combustion gases are used in a different regenerator, called the **economizer**, to heat the ambient air before it is used in combustion. The economizer raises the energetic efficiency of the boiler without affecting the performance of the heat engine. To use regeneration in the heat engine, another approach is to circulate the feedwater in a counter-flow heat exchanger around the turbine before it enters the boiler. The problem with this approach is that heat loss from the turbine may cause the quality of steam at the turbine exhaust to fall below acceptable levels.

A more practical solution for regeneration is to extract, or **bleed**, steam from an intermediate point in the turbine and use it to heat up the feedwater in an open or closed heat exchanger called the **feedwater heater** (FWH). Extraction of steam from the turbine obviously reduces the turbine output. The external heat input to the cycle is also reduced as water enters the boiler at a much higher temperature due to regenerative heating. Despite a loss of turbine output, the thermal efficiency increases as regeneration raises $\overline{T}_H$ without affecting $\overline{T}_C$. Two configurations that accomplish regeneration, one with an open heater and another with a closed heater, are illustrated in Figs. 9.18 (Anim. 9.B.*openFWH*) and 9.19 (Anim. 9.B.*closedFWH*).

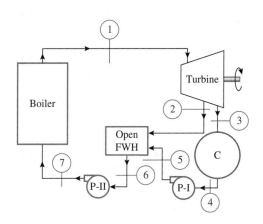

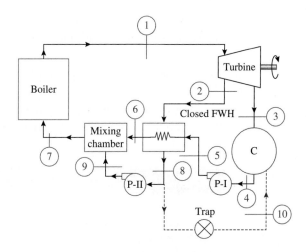

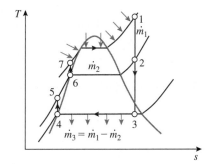

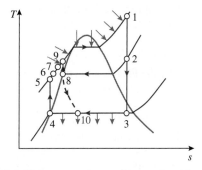

FIGURE 9.18 Regenerative Rankine cycle with one open feedwater heater (see Anim. 9.A.*openFWH*).

FIGURE 9.19 Regenerative Rankine cycle with one closed feedwater heater (see Anim. 9.A.*closedFWH*).

OPEN FEEDWATER HEATER   An **open feedwater heater**, shown in Figure 9.18 (see Anim. 9.B.*openFWH*), is simply a mixing chamber where the feedwater is heated by mixing with the steam bled from the turbine. As with any mixing chamber (Sec. 4.2.8), the pressure of the extracted steam must be the same as that of the feedwater to prevent any back flow, that is, $p_2 = p_5 = p_6$. Pumping is done in two stages, with the first pump raising the pressure of the feedwater from p_4 to p_5 and the second pump raising the pressure to the boiler pressure, from p_6 to p_7. The amount of steam extracted is controlled so that the water leaving the heater is subcooled or, at best, saturated. Presence of any vapor may cause cavitations, damaging the pump.

In an ideal regenerative cycle with an open heater, the steam expands in the turbine isentropically before and after bleeding; however, the turbine is no longer a single-flow device. As in the case of simple Rankine cycle, water leaves the condenser as saturated water at State-4 without any loss of pressure. The first isentropic pump raises the water pressure to the intermediate bleed pressure. The ideal bleeding flow rate ensures that after mixing with steam, the feedwater exiting the heater is saturated at State-6. The second isentropic pump raises the pressure of the feedwater to that in the boiler at State-7, after which external heat is added at constant pressure in the boiler from State-7 through State-1. Neglecting any changes in ke or pe, the steady state energy balance for each device produces the following expressions for heat and work transfers:

$$\dot{Q}_{in} = \dot{Q} \cong \dot{m}_1(h_1 - h_7); \quad [kW] \tag{9.18}$$

$$\dot{W}_T \cong \dot{m}_1(h_1 - h_2) + \dot{m}_3(h_2 - h_3); \quad [kW]$$
$$= \dot{m}_1[(h_1 - h_2) + (1 - r)(h_2 - h_3)], \quad \text{where,} \quad r \equiv \dot{m}_2/\dot{m}_1 \text{ and } \dot{m}_3 = \dot{m}_1(1 - r) \tag{9.19}$$

$$\dot{Q}_{out} = -\dot{Q} \cong \dot{m}_3(h_3 - h_4) = \dot{m}_1(1 - r)(h_3 - h_4); \quad [kW] \tag{9.20}$$

$$\dot{W}_{P,I} \cong \dot{m}_3(h_5 - h_4) = \dot{m}_1(1 - r)v_{f@T_4}(p_5 - p_4); \quad [kW] \tag{9.21}$$

$$\dot{W}_{P,II} \cong \dot{m}_1(h_7 - h_6) = \dot{m}_1 v_{f@T_6}(p_7 - p_6); \quad [kW] \tag{9.22}$$

Finally, an energy balance on the adiabatic heater yields an equation to obtain $r = \dot{m}_2/\dot{m}_1$, the fraction of steam extracted from the turbine:

$$\dot{m}_1 h_6 \cong \dot{m}_2 h_2 + (\dot{m}_1 - \dot{m}_2)h_5; \quad [\text{kW}] \quad \text{or}, \quad h_6 \cong rh_2 + (1 - r)h_5; \quad \left[\frac{\text{kJ}}{\text{kg}}\right] \quad \textbf{(9.23)}$$

The following example explores the effect of bleed pressure on an ideal regenerative Rankine cycle with a single open feedwater heater.

EXAMPLE 9-6 **Rankine Cycle Modified with Open Feedwater Heater**

A steam power plant operates on an ideal regenerative Rankine cycle with a single open feed-water heater (Fig. 9.18). The turbine inlet conditions are 10 MPa, 650°C, and the condenser pressure is 10 kPa. If bleeding takes place at 2 MPa, determine (a) the fraction of total flow through the turbine that is bled and (b) the thermal efficiency. *What-if scenario:* (c) How would the thermal efficiency vary if the bleeding pressure varied over the entire possible range?

SOLUTION

Use the PC model to evaluate the seven principal states of Figure 9.18 and perform a steady state energy balance for each device to obtain the thermal efficiency.

Assumptions

Ideal regenerative Rankine cycle.

Analysis

Use the manual approach (Sec. 3.4), or the PC flow-state TESTcalc, to evaluate the enthalpy of each principal state.

State	Given	h (kJ/kg)	State	Given	h (kJ/kg)
1	p_1, T_1	3748	6	$p_6 = p_2, x_6 = 0$	908.8
2	$p_2, s_2 = s_1$	3191	7	$p_7 = p_1$	918.2
3	$p_3, s_3 = s_1$	2230		$s_7 = s_6$ or	
4	$p_4 = p_3, x_4 = 0$	191.8		$h_7 = h_6$	
5	$p_5 = p_2,$	193.8		$\quad + v_{f@T_6}(p_7 - p_6)$	
	$s_5 = s_4$ or				
	$h_5 = h_4$				
	$\quad + v_{f@T_4}(p_5 - p_4)$				

An energy balance on the FWH, Eq. (9.23), yields:

$$r = \frac{h_6 - h_5}{h_2 - h_5} = \frac{908.9 - 193.8}{3191 - 193.8} = 0.239$$

Using Eqs. (9.18) and (9.20), the thermal efficiency is obtained:

$$\eta_{\text{th}} = 1 - \frac{\dot{Q}_{\text{out}}}{\dot{Q}_{\text{in}}} = 1 - \frac{\dot{m}_1(1 - r)(h_3 - h_4)}{\dot{m}_1(h_1 - h_7)} = 1 - \frac{(0.761)(2038)}{3748 - 918.2} = 45.19\%$$

TEST Analysis

Evaluate the principal states and analyze the six devices with the PC power cycle TESTcalc as described in the TEST-code (see *TEST > TEST-codes*). Note that while mdot1 is assumed to be1 kg/s, mdot2 is entered as '= mdot1*(h6 − h5)/(h2 − h5)' from the energy balance on the heater. In the device panel, the turbine and FWH are treated as mixing devices by selecting the appropriate radio-button. The efficiency and all other cycle parameters are automatically calculated and displayed in the cycle panel once all the devices are analyzed.

What-if scenario

Change p2 to a new value, and click Super-Calculate, to obtain updated results. Repeat this process for the entire possible range of p2—from condenser pressure to boiler pressure. The new values of efficiency and net power are listed in Table 9-6.

Discussion

The regenerative cycle can be converted to a simple Rankine cycle by setting mdot2 to zero and using the Super-Calculate button. The resulting thermal efficiency of 42.49% is exactly the same as the result we obtained for a FWH pressure of 10 kPa in Table 9-6 (for an explanation, redraw Figure 9.18 for a FWH pressure of 10 kPa). With an increase in the FWH pressure the efficiency increases reaching a peak at around 1 MPa, then begins to decrease, although the effective temperature of heat addition continues to increase monotonically. In this cycle, the mass flow rates are not equal through the boiler and condenser. Therefore, the thermal efficiency is no longer a function of $\overline{T}_C$ and $\overline{T}_H$ alone. For the same reason, the net work output is not necessarily proportional to the enclosed area on the T-s diagram.

Table 9-6 Parametric study of regenerative Rankine cycle with turbine bleed pressure as the parameter.

p_{FWH} (MPa)	$\eta_{th,Rankine}$	$\dot{W}_{net}/\dot{m}_1$ (MW)
0.01	0.4249	1508
0.5	0.453	1403
1	0.454	1350
1.5	0.453	1310
2	0.452	1278
5	0.442	1143
10	0.4252	995.2

CLOSED FEEDWATER HEATER   The second kind of feedwater heater used in power plants is basically a closed type heat exchanger, known as a **closed feedwater heater**, where heat is transferred by the steam extracted from the turbine to the feedwater coming from the condenser, without any direct contact between the two streams. A regenerative cycle, with a single closed FWH, is shown in Figure 9.19. The extracted steam leaves the heater as saturated water ($x = 0$) while the feedwater is heated from State-5 to State-6. From practical consideration T_8 must be slightly higher (by about 5–10°C) than T_6, but are assumed to be equal in an ideal configuration. The amount of steam extracted is calculated from an energy balance on the FWH with the condition that State-8 must be saturated.

There are two ways in which the condensate can be circulated back to the main loop. In the first arrangement, path 8-9-7 in Figure 9.19 (Anim. 9.B.*closedFWH*), the feedwater and the condensate leaving the heater are mixed in a mixing chamber (open FWH) after a pump raises the condensate pressure to the boiler pressure. States 5, 6, 7, 9, and 1 are all on the same constant-pressure line. A simpler alternative is to throttle the condensate back to the condenser pressure by using a valve called a vapor **trap** (see path 8–10 in Fig. 9.19 and Anim. 9.A.*useOfTrap*), which permits only liquid to pass through to the low pressure side. The trap eliminates the second pump and the mixing chamber altogether. The irreversible throttling process (Sec. 4.2.5) taking place in the trap is indicated by the dotted line 8–10 in Figure 9.19. There is a slight drop in the boiler inlet temperature (from State-7 to State-6) when the trap is used. However, this difference is too little to make any appreciable difference in $\overline{T}_H$ to affect the thermal efficiency of the cycle and the second alternative is often used in power plants.

With extra tubing required to separate the flows, a closed FWH is more complex. Its main advantage over an open FWH is that a pump is not mandatory to handle the condensate. On the other hand, an open FWH is simpler in design and relatively inexpensive. It also has better heat transfer characteristics, resulting from direct contact between the two streams. By adjusting the heater pressure to atmospheric level, an open FWH can be exposed to the atmosphere and used as a **deaerator**, a device where any air that leaks into the cycle in the condenser or other contaminants in the working fluid can be removed. Without a deaerator, air bubbles would accumulate, affecting the cycle performance, and contaminants would corrode the boiler tubes.

Regeneration is used in all modern power plants. Although the thermal efficiency monotonically increases with the number of heaters used, the improvement diminishes with the addition of each new heater. This is known as the *law of diminishing return* in economics and is illustrated in the next example. Many large power plants use as many as eight FWH's with a combination of the closed and open types, with at least one open FWH for deaeration.

EXAMPLE 9-7   **Rankine Cycle Modified with Reheat and Two Feedwater Heaters**

A steam power plant operates on an ideal reheat-regenerative Rankine cycle with two feedwater heaters, one open and one closed, as shown in Figure 9.20. Steam enters the first turbine at 15 MPa and 500°C, expands to 1 MPa, and is reheated to 450°C before it enters the second

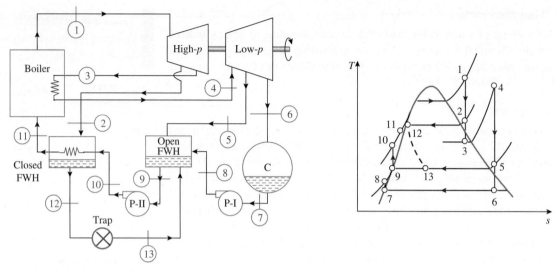

FIGURE 9.20 Reheat-regenerative Rankine cycle with two FWH's for Ex. 9-7.

turbine, where it expands to 10 kPa. Some steam is extracted from the first turbine at 3 MPa and sent to the closed FWH, where the feedwater leaves at 5°C below the temperature at which the saturated condensate leaves. The condensate is fed through the trap to the open FWH, which operates at 0.5 MPa. Steam is also extracted from the second turbine at 0.5 MPa and fed to the open FWH. The flow out of the open FWH is saturated liquid at 0.3 MPa. If the power output of the cycle is 150 MW, determine (a) the thermal efficiency, (b) the mass flow rate through the boiler, (c) the bleeding rate from the first turbine, and (d) the bleeding rate from the second turbine. *What-if scenario:* What would the thermal efficiency be if (e) the closed FWH was eliminated, (f) both the FWH's were eliminated?

SOLUTION

Use the PC model to evaluate the thirteen principal states of Figure 9.20 and perform an energy balance for each open-steady device to obtain the cycle parameters.

Assumptions

Ideal reheat and regenerative cycle.

Analysis

Use the PC flow-state TESTcalc to evaluate the enthalpy of each principal state.

State	Given	h (kJ/kg)	State	Given	h (kJ/kg)
1	p_1, T_1	3309	9	$p_9 = p_5, x_9 = 0$	640.1
2	$p_2, s_2 = s_1$	2887	10	$p_{10} = p_1,$	655.8
3	$p_3, s_3 = s_1$	2668		$s_{10} = s_9$ or	
4	p_4, T_4	3371		$h_{10} = h_9$	
5	$p_5, s_5 = s_4$	3156		$+ v_{f@T_9}(p_{10} - p_9)$	999.8
6	$p_6, s_6 = s_5$	2413	11	$p_{11} = p_1, T_{11} = T_{12} - 5$	1008.4
7	$p_7 = p_6, x_7 = 0$	192.0	12	$p_{12} = p_2, x_{12} = 0$	
8	$p_8 = p_5,$	192.3	13	$p_{13} = p_5, h_{13} \cong h_{12}$	1008.4
	$s_8 = s_7$ or				
	$h_8 = h_7$				
	$+ v_{f@T_7}(p_8 - p_7)$				

An energy balance on the closed FWH produces:

$$\dot{m}_2(h_2 - h_{12}) \cong \dot{m}_1(h_{11} - h_{10}); \implies \dot{m}_2 = 0.183\dot{m}_1;$$

The open FWH produces:

$$\dot{m}_5 h_5 + (\dot{m}_1 - \dot{m}_2 - \dot{m}_5)h_8 + \dot{m}_2 h_{13} \cong \dot{m}_1 h_9;$$

$$\Rightarrow \dot{m}_5 = \dot{m}_1 \frac{h_9 - h_8}{h_5 - h_8} + \dot{m}_2 \frac{h_8 - h_{13}}{h_5 - h_8}; \quad \Rightarrow \dot{m}_5 = 0.101\dot{m}_1$$

Therefore, $\dot{m}_3 = \dot{m}_4 = \dot{m}_1 - \dot{m}_2 = 0.817\dot{m}_1$, and $\dot{m}_6 = \dot{m}_4 - \dot{m}_5 = 0.716\dot{m}_1$

Using these mass flow relations, the net output can be expressed:

$$\begin{aligned}
\dot{W}_{net} &= \dot{W}_{T,I} + \dot{W}_{T,II} - \dot{W}_{P,I} - \dot{W}_{P,II} \\
&= (\dot{m}_1 h_1 - \dot{m}_2 h_2 - \dot{m}_3 h_3) + (\dot{m}_4 h_4 - \dot{m}_5 h_5 - \dot{m}_6 h_6) \\
&\quad - \dot{m}_7(h_8 - h_7) - \dot{m}_9(h_{10} - h_9) \\
&= 1291.6\dot{m}_1
\end{aligned}$$

The mass flow rate can now be obtained:

$$\dot{m}_1 = \frac{\dot{W}_{net}}{1291.6} = \frac{150,000}{1291.6} = 116.13 \frac{kg}{s};$$

The external heat addition can be determined: $\dot{Q}_{in} = \dot{m}_1(h_1 - h_{11}) + \dot{m}_3(h_4 - h_3) = 2883\dot{m}_1$

Therefore, the thermal efficiency is: $\eta_{th} = \dfrac{\dot{W}_{net}}{\dot{Q}_{in}} = \dfrac{1291.6\dot{m}_1}{2883\dot{m}_1} = 44.8\%$

TEST Analysis

Evaluate the principal states, and analyze the devices with the PC power cycle TESTcalc, as described in the TEST-code (see *TEST > TEST-codes*). While mdot1 is assumed to be 1 kg/s, mdot2 and mdot5 are entered as expressions. To evaluate the thermal efficiency, only the devices that participate in external heat or work transfer need to be analyzed.

What-if scenario

To eliminate the closed FWH, overwrite mdot2 as zero. To make State-11 identical to State-10, make T11 an unknown and specify any independent property, for example, h11 as '= h10'. Click Super-Calculate and then Super-Iterate, if necessary. The new value of the thermal efficiency is calculated as 43.8%. To eliminate both the heaters, set mdot2=mdot5=0, set T11 as unknown, then enter h11 as '= h8'. Click Super-Calculate to obtain the efficiency as 41.4%.

Discussion

The example clearly illustrates the law of diminishing return (Fig. 9.21): the efficiency jumps by more than 2 points with the first FWH, while the second heater only adds about 1 point. In a closed FWH, exergy is destroyed through thermal friction. The trap is also an irreversible device, its operation depending on the frictional losses. Nevertheless, their presence makes the overall cycle more efficient—the rise in the effective temperature of heat addition (which reduces thermal friction in the boiler) outweighs the thermal and viscous friction introduced by these irreversible devices.

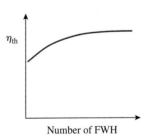

FIGURE 9.21 The cost eventually outweighs the benefit as the number of FWH is indefinitely increased.

9.4 COGENERATION

Many processes or systems require energy input in the form of heat, also known as **process heat**. Process heating is vital to nearly all manufacturing processes which supply heat to produce basic materials and commodities (see Anim. 0.D.*energyStats*). It accounts for about 17% of all energy used by industries. In many industries—for example oil refining, textile, pulp and paper, and food processing industry—the process heat is supplied by superheated steam, usually in a temperature range of 150°C to 200°C and a pressure range of 500 kPa to 700 kPa.

One straightforward set-up for process heating is shown in Figure 9.22, where heat is added to the boiler from a source maintained at a high temperature, typically at about 1500 to

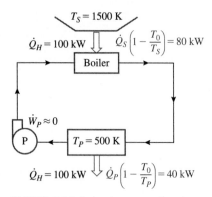

FIGURE 9.22 A simple process heating system. Numbers in red are exergy transfer rates.

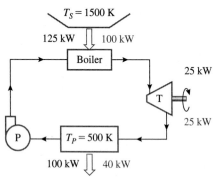

FIGURE 9.23 An ideal cogeneration plant for process heating (see Anim. 9.B.*coGeneration*). Numbers in red are exergy transfer rates. The system has a utilization factor of 1.0 and an exergetic efficiency of 65%.

1700 K, by burning fossil fuels. The process heater is basically a condenser operating at the saturation pressure at the desired temperature, i.e., $P_{sat@T_p}$. The pump simply supplies enough pressure difference to overcome the frictional losses to maintain the flow. If the heat and friction losses in the pipes are not significant, the *energetic efficiency*, defined as $\dot{Q}_p/\dot{Q}_s$, can nearly approach 100%.

Energetic efficiency completely misses the fact that the heat added at the source is of much higher quality than that of the process heat supplied at a much lower temperature. A better measure of how well the useful energy is being utilized is the *exergetic efficiency* of the system. In the arrangement of Figure 9.22, the exergy that is transferred by heat (Sec. 6.1.1) to the boiler is given by $(100)(1 - 300/1500) = 80$ kW, if the ambient temperature is assumed to be 300 K. Exergy delivered by the steam for process heating is only $(100)(1 - 300/500) = 40$ kW; therefore, the exergetic efficiency of the process heating system is $\eta_{II} = 50\%$. Exergetic efficiency increases as the source temperature T_s is matched closer to the utilization temperature T_p (see Example 6-8). Unfortunately the flame temperature for fossil fuel burning in air is thermodynamically determined (Chapter 13) and cannot be adjusted based on specific applications.

Industries that are heavy on process heating also consume large amounts of electric power. The Rankine cycle can be modified to meet the power demand, while also meeting the process heating need, through raising the condenser pressure to match the desired process temperature at high exergetic efficiencies. This is called a **cogeneration** plant; a plant that meets two energy conversion goals—produce electric power and deliver process heat—from the same source of energy. Comparing the arrangements of Figures 9.22 and 9.23, we observe that, in addition to supplying the required process heating at the desired temperature, the cycle produces 25 MW of electric power for an additional 25 MW of heat input. The exergetic efficiency can be shown to increase to 65%. The energetic efficiency for process heating is redefined as the **utilization factor**:

$$\varepsilon_u \equiv \frac{\text{Desired Heat and Work Output}}{\text{Required Energy Input}} = \frac{\dot{W}_{net} + \dot{Q}_p}{\dot{Q}_{in}} = 1 - \frac{\dot{Q}_{loss}}{\dot{Q}_{in}} \qquad \textbf{(9.24)}$$

For the ideal cogeneration system of Figure 9.23, this definition produces a utilization factor of 1, if the pump work is neglected. Such a system is not suitable in a practical environment where the demand for process heat may vary. In a more practical configuration, drawn in Figure 9.24 (and Anim. 9.B.*coGeneration*), a variable demand for process heat can be accommodated.

Under normal load, the expansion valve is closed and part of the steam flowing into the turbine is extracted at the desired intermediate pressure p_2 for supplying the process heat. The rest is expanded further in the turbine, for additional output, and routed through a condenser and pump before mixing with the condensate returning from the process heater. Heat rejected in the condenser, obviously, should be treated as *waste heat*, as it cannot be utilized because of its low exergy content. As the process heating load increases, more steam is diverted to the process heater, leading to a complete cutoff ($\dot{m}_5 \approx 0$) of the condenser in the limit. If the load increases even further, some steam from the boiler is throttled through the expansion valve to the desired pressure p_2, without any loss of enthalpy (process 2–9 in Fig. 9.24), thereby augmenting the energy supply to the process heater. Obviously, the turbine output goes down as the bypass ratio increases. At the peak load, supply to the turbine is completely cut off ($\dot{m}_3 = 0$) and the system becomes a pure heating system. At the other extreme, if there is no demand for any process heat, the system operates purely as a simple Rankine cycle.

EXAMPLE 9-8 **Analysis of a Cogeneration Plant**

The cogeneration plant, shown in Figure 9.24, has a peak capacity of 25 MW for process heating. Steam leaves the boiler at 8 MPa and 600°C. The process heater and the condenser operate at a pressure of 400 kPa and 8 kPa, respectively. Assuming ideal behavior for each component, determine (a) the mass flow rate of steam through the boiler and (b) the maximum turbine output. **What-if scenario:** To meet a certain load condition, 10% of the steam is diverted to the

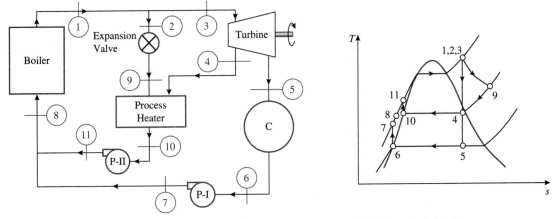

FIGURE 9.24 A practical cogeneration plant used in Example 9-8 (see Anim. 9.B.*coGeneration*).

expansion valve, and half the steam flowing through the turbine is extracted for process heating. Determine (c) the process heat load, and (d) the utilization factor.

SOLUTION

Use the PC model to evaluate the principal states and perform an energy analysis of each open-steady device to obtain the desired answers.

Assumptions

The turbine and the pumps are isentropic, no pressure drop in the boiler, condenser, heaters and piping. No stray heat losses from any device. Condensate leaves the condenser and the process heater as saturated liquid. Negligible changes in ke and pe across any device, so that $j \cong h$.

Analysis

Use the PC flow-state TESTcalc to evaluate the enthalpy of each principal state. Perform steady-state energy balance for the adiabatic mixing chamber and expansion valve.

State	Given	h (kJ/kg)	St.	Given	h (kJ/kg)
1-3	p_1, T_1	3642	8	$p_8 = p_1,$	419.0
4	$p_4, s_4 = s_3$	2793		$h_8 = (\dot{m}_7 h_7 + \dot{m}_{11} h_{11})/\dot{m}_1$	
5	$p_5, s_5 = s_3$	2197	9	$p_9 = p_4, h_9 = h_2$	3642
6	$p_6 = p_5, x_6 = 0$	173.9	10	$p_{10} = p_4, x_{10} = 0$	604.7
7	$p_7 = p_1,$	181.9	11	$p_{11} = p_1,$	612.9
	$s_7 = s_6$ or			$s_{11} = s_{10}$ or	
	$h_7 = h_6 + v_{f@T_6}(p_7 - p_6)$			$h_{11} = h_{10} + v_{f@T_{10}}(p_{11} - p_{10})$	

At the peak process load, the entire flow rate passes through the expansion valve. Therefore, an energy balance on the process heater yields the mass flow rate through the boiler:

$$\dot{m}_9 h_9 + \dot{m}_4^{\;0} h_4 \cong \dot{m}_{10} h_{10} + \dot{Q}_p;$$

$$\Rightarrow \quad \dot{m}_1 = \frac{\dot{Q}_p}{h_9 - h_{10}} = \frac{25000}{3642 - 604.7} = 8.231 \, \frac{\text{kg}}{\text{s}}$$

The entire flow is sent through the turbine for peak power output:

$$\dot{W}_{T,\text{max}} \cong \dot{m}_1(h_3 - h_5) = (8.231)(3642 - 2197) = 11.9 \text{ MW}$$

TEST Analysis

Evaluate the principal states and analyze the devices with the PC power cycle TESTcalc as described in the TEST-code (see *TEST > TEST-codes*). Set mdot2 and mdot4 to zero to obtain the maximum turbine output.

What-if scenario

Adjust mdot2 as '= 0.1*mdot1' and mdot4 as '= 0.5*mdot1'. Click Super-Calculate. In the device panel, obtain Qdot for the boiler as 26.53 MW, Qdot for the condenser as –7.49 MW, and Qdot for the process heater as –10.61 MW. The heating load, therefore, is 10.61 MW and the utilization factor can be obtained from Eq. (9.24) as $1 - 7.49/26.53 = 71.8\%$.

Discussion

The heat input remaining unchanged, the utilization factor obviously changes with the process heating load. When the process heating is completely stopped ($\dot{m}_9 = \dot{m}_4 = 0$), the utilization factor becomes equal to the thermal efficiency of the Rankine cycle.

9.5 BINARY VAPOR CYCLE

Water is used as the working fluid of choice in most vapor power cycles, as it offers several advantages over other fluids. It is non-corrosive, non-toxic, chemically stable, inexpensive, and has a relatively large enthalpy of vaporization (h_{fg}) at the pressure typically used in steam generators. A large h_{fg} means a lower mass flow rate can be used for the same power output, limiting the component size. The difference in specific volume between saturated liquid and saturated vapor makes the **back work ratio** quite low.

The main drawback of water is that its critical temperature is only 374.14°C, and if super-critical pressure is to be avoided, the bulk of the heat addition takes place below the critical temperature. Even raising the boiler pressure to above the critical pressure, the effective temperature of heat addition cannot be increased significantly after a certain point. The **binary vapor cycle** attempts to overcome this drawback by coupling two Rankine cycles, one with a working fluid that has good characteristics at high temperatures and another that is suitable for the low temperature region. The combination results in a much higher effective temperature of external heat addition, thereby, improving the thermal efficiency.

A mercury-water binary vapor cycle is depicted in Figure 9.25. Heat rejected by the high temperature mercury cycle, also called the *topping cycle*, is transferred to the low-temperature, or *bottom cycle*, through a heat exchanger. Although water is superheated through external heat addition, the majority of heat addition takes place at the elevated temperature of the mercury cycle, as shown in the *T-s* diagram. A thermal efficiency of 50% or higher is possible with binary cycles; therefore, the high initial cost can be justified given the increasing cost of fuel.

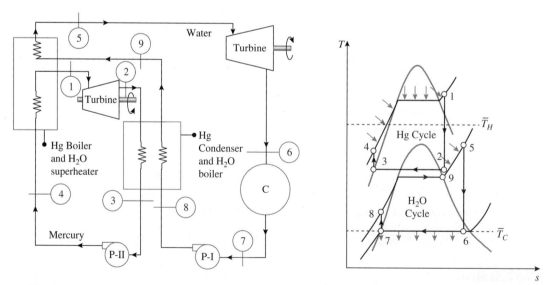

FIGURE 9.25 A Hg-H₂O binary vapor cycle. Gray arrows in the *T-s* diagram show external heat addition and rejection.

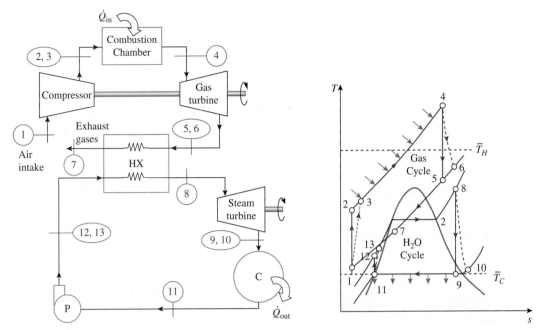

FIGURE 9.26 A combined gas-vapor power cycle analyzed in Ex. 9-9 (see Anim. 9.B.*combinedCycle*). Gray arrows on the *T-s* diagram show external heat addition and rejection.

9.6 COMBINED CYCLE

Just as the binary vapor cycle couples two Rankine cycles to increase the effective temperature of external heat addition, a **combined gas-vapor power cycle**, or **combined cycle**, accomplishes the same goal by having a Brayton cycle as the **topping cycle** over a Rankine cycle at the bottom (see Anim. 9.B.*combinedCycle*).

Sophisticated gas turbine blades can be internally cooled by blowing air through internal passages, resulting in turbines that can withstand a maximum inlet temperature of over 1200°C. However, a higher inlet temperature also causes a higher exhaust temperature, which is fixed by atmospheric pressure and the isentropic expansion condition between the inlet and exit. For instance, isentropic expansion of air from 900 kPa, 1200°C to 100 kPa produces an exhaust temperature of 582°C. The effective temperature of heat rejection, $\overline{T}_C$, also increases; therefore, despite an increase in the effective temperature $\overline{T}_H$ of heat addition, the thermal efficiency of a basic gas turbine cycle does not necessarily increase.

In a Rankine cycle, steam expands through the turbine to sub-atmospheric pressure (about 5 kPa); therefore, the effective temperature of heat rejection is quite low, only a few degrees above the atmospheric temperature. The turbine inlet temperature of a steam power plant is typically only about 650°C. The strength of the steam power cycle, therefore, is a relatively low $\overline{T}_C$, while that of a gas turbine is a relatively high $\overline{T}_H$. The **combined cycle**, shown in Figure 9.26, harnesses the strengths of both cycles by eliminating external heat addition to the steam cycle, and significantly reducing the external heat rejection of the gas cycle, by coupling the two cycles through a heat exchanger. The thermal efficiency of the combined cycle far exceeds that of its constituent cycles.

DID YOU KNOW?

- A combined cycle power plant can produce an overall thermal efficiency of more than 60%.

EXAMPLE 9-9   **Combined Gas-Vapor Power-Cycle**

A combined gas turbine-steam power plant produces a net power output of 500 MW. Air enters the compressor of the gas turbine at 100 kPa, 300 K. The compressor has a compression ratio of 12 and an isentropic efficiency of 85%. The turbine has an isentropic efficiency of 90%, inlet conditions of 1200 kPa, 1400 K, and an exit pressure of 100 kPa. Air from the turbine exhaust passes through a heat exchanger and exits at 400 K. On the steam turbine side, steam at 8 MPa, 400°C enters the turbine, which has an isentropic efficiency of 85%, and expands to the condenser pressure of 8 kPa. Saturated water at 8 kPa is circulated back to the heat exchanger by a pump with an isentropic efficiency of 80%. Determine (a) the ratio of mass flow rates in the two cycles, (b) the mass flow rate of air, and (c) the thermal efficiency. *What-if scenario:* (d) What would the thermal efficiency be if the turbine inlet temperature increased to 1600 K?

SOLUTION

Use the IG and the PC models to evaluate the principal states of the combined cycle, depicted in Figure 9.26, and perform an energy analysis of each open-steady device for a complete solution of the problem.

Assumptions

No frictional pressure drop in any component or piping. No stray heat losses from any device. Negligible changes in ke and pe across any device so that $j \cong h$. Air-standard Brayton cycle and Rankine cycle to model the two constituent cycles. Use variable specific heats for air, i.e., the IG model for the Brayton cycle.

Analysis

Use the manual approach, or the PC/IG power cycle TESTcalc, to evaluate the enthalpies of each principal state. Steady-state energy balance for individual devices is employed to develop enthalpy relations as listed in the table below.

State	Given	h (kJ/kg)	State	Given	h (kJ/kg)
1	p_1, T_1	1.9	8	p_8, T_8	3138.2
2	$p_2, s_2 = s_1$	311.7	9	$p_9, s_9 = s_8$	1990.0
3	$p_3 = p_2,$	366.5	10	$p_{10} = p_9,$	2162.2
	$h_3 = h_1 +$			$h_{10} = h_8 -$	
	$(h_2 - h_1)/\eta_C$			$(h_8 - h_9)\eta_T$	
4	$p_4 = p_2, T_4$	1217.8	11	$p_{11} = p_9, x_{11} = 0$	173.9
5	$p_5 = p_1, s_5 = s_4$	471.1	12	$p_{12} = p_8, s_{12} = s_{11}$	181.9
6	$p_6 = p_5,$	545.8	13	$p_{13} = p_{12},$	183.9
	$h_6 = h_4 -$			$h_{13} = h_{11} +$	
	$(h_4 - h_5)\eta_T$			$(h_{12} - h_{11})/\eta_P$	
7	$p_7 = p_6, T_7$	103.0			

An energy balance on the adiabatic heat exchanger produces:

$$\dot{m}_1(h_6 - h_7) \cong \dot{m}_8(h_8 - h_{13});$$

$$\Rightarrow \quad \dot{m}_1 = \frac{h_8 - h_{13}}{h_6 - h_7}\dot{m}_8 = \frac{3138.2 - 183.9}{545.8 - 103.0}\dot{m}_8 = 6.673\dot{m}_8$$

With this relation, the net power output can be written:

$$\dot{W}_{net} = \dot{W}_{T,I} + \dot{W}_{T,II} - \dot{W}_C - \dot{W}_P$$
$$\cong \dot{m}_1(h_4 - h_6) + \dot{m}_8(h_8 - h_{10})$$
$$- \dot{m}_1(h_3 - h_1) - \dot{m}_8(h_{13} - h_{11})$$
$$= 452.2\dot{m}_1$$

Given the net output as 500 MW, the mass flow rate of air can be calculated as:

$$\dot{m}_1 = 500{,}000/452.2 = 1107 \text{ kg/s}$$

To obtain the thermal efficiency the external heat addition $\dot{Q}_{in}$ is evaluated first:

$$\dot{Q}_{in} \cong \dot{m}_1(h_4 - h_3) = 1107(1217.8 - 366.5) = 942.39 \text{ MW}$$

Therefore,

$$\eta_{th} = \frac{\dot{W}_{net}}{\dot{Q}_{in}} = \frac{500}{942.4} = 53.1\%$$

TEST Analysis

Evaluate the principal states and analyze the devices with the PC/IG power cycle TESTcalc, as described in the TEST-code (see *TEST > TEST-codes*). Once the net power is found on the basis of mdot1 = 1 kg/s, calculate mdot1 manually and enter its new value. Click Super-Calculate to update the efficiency and all other cycle variables.

What-if scenario

Increase T4 to 1600 K and click Super-Calculate. The thermal efficiency is recalculated as 54.8%. An increase in efficiency is expected as the effective temperature of heat addition is increased by an increase in T_{max}.

Discussion

The mass flow ratio of the two streams in a combined cycle is generally much greater than one, 6.673 in this problem. Instead of having one huge gas turbine, several gas turbine plants are generally used in parallel to supply the necessary flow to the heat exchanger.

DID YOU KNOW?

- In 1973 oil-fired power plants accounted for 18% of total electric power in the U.S. Today it is less than 1%.

9.7 CLOSURE

In this chapter we analyzed the vapor power cycle, the primary cycle used for electric power production in the world. Components of a steam power plant were briefly discussed, and the heat engine at its center, the Rankine cycle was examined thoroughly by extending the overall analysis of a heat engine that began in Chapter 2. TEST solutions was used to verify lengthy manual solutions and to perform numerical studies that explored parametric behavior. For qualitative prediction of cycle performance, the use of an equivalent Carnot cycle, introduced in Chapter 7 was continued. Influence of different parameters, such as the turbine inlet temperature, condenser pressure, boiler pressure, etc., were explored to optimize the cycle performance. Various modifications to the basic cycle, including reheat and regeneration using open and closed feedwater heaters, were discussed in order to improve the cycle efficiency, reduce the production of greenhouse gases, and increase the net power output. Cogeneration, binary cycles, and combined cycles were discussed and analyzed as major contenders for high-efficiency alternatives.

TABLE 9-7 World consumption of average power in TW (10^6MW) in 1980 and 2006.

	1980	2006
Petroleum	4.38	5.74
Coal	2.34	4.27
Natural Gas	1.80	3.61
Hydro	0.60	1.00
Nuclear	0.25	0.93
Renewables	0.02	0.16
Total:	9.48	15.8

Source: U.S. Energy Information Administration.

PROBLEMS

SECTION 9-1: BASIC AND MODIFIED RANKINE CYCLES

9-1-1 Water is the working fluid in a Carnot vapor power cycle. Saturated liquid enters the boiler at a pressure of 10 MPa and enters the turbine as saturated vapor. The condenser pressure is 9 kPa. Determine (a) the thermal efficiency (η_{th}), (b) the back-work ratio, (c) the heat transfer to the working fluid per unit mass (q_{in}) passing through the boiler in kJ/kg.

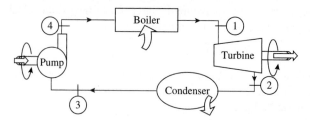

9-1-2 Water enters the boiler of a steady flow Carnot engine as a saturated liquid at 800 kPa, and leaves with a quality of 0.95. Steam leaves the turbine at a pressure of 100 kPa. Determine (a) the thermal efficiency (η_{th}) and (b) the net work (w_{net}) output.

9-1-3 Water is the working fluid in a Carnot vapor power cycle. Saturated liquid enters the boiler at a pressure of 10 MPa and saturated vapor enters the turbine. The condenser pressure is 8 kPa. The effects of irreversibilities in the adiabatic expansion and the compression processes are taken into consideration. The turbine and pump efficiencies are 80% and 75%, respectively. Determine (a) the thermal efficiency (η_{th}), (b) the back-work ratio, (c) the heat transfer (q) to the working fluid per unit mass passing through the boiler in kJ/kg and (d) the heat transfer (q) from the working fluid per unit mass passing through the condenser in kJ/kg.

9-1-4 Consider a steam power plant operating on the simple ideal Rankine cycle. The steam enters the turbine at 4 MPa, 400°C, and is condensed in the condenser at a pressure of 100 kPa. Determine (a) the thermal efficiency ($\eta_{th,Rankine}$) of the cycle. (b) *What-if scenario:* What would the thermal efficiency be if steam entered the turbine at 5 MPa and the condenser pressure was 90 kPa?

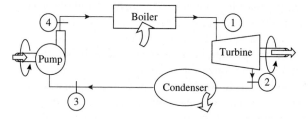

9-1-5 A steam power plant operates on the simple ideal Rankine cycle. Steam enters the turbine at 4 MPa, 500°C, and is condensed in the condenser at a temperature of 40°C. (a) Show the cycle on a T-s diagram. If the mass flow rate ($\dot{m}$) is 10 kg/s, determine (b) the thermal efficiency (η_{th}) of the cycle and (c) the net power output ($\dot{W}_{net}$) in MW.

9-1-6 Water is the working fluid in an ideal Rankine cycle. Saturated vapor enters the turbine at 6.9 MPa. The condenser pressure is 6.9 kPa. Determine (a) the net work per unit mass (w_{net}) of steam flow in kJ/kg, (b) the heat transfer (q) to the steam passing through the boiler in kJ/kg, (c) the thermal efficiency (η_{th}) and (d) the back-work ratio.

9-1-7 Consider a steam power plant operating on the ideal Rankine cycle. The steam enters the turbine at 5 MPa, 350°C, and is condensed in the condenser at a pressure of 15 kPa. Determine (a) the thermal efficiency ($\eta_{th,Rankine}$) of the cycle. (b) *What-if scenario:* What would the thermal efficiency be if the steam was superheated to 750°C instead of 350°C ?

9-1-8 Steam is the working fluid in an ideal Rankine cycle. Saturated vapor enters the turbine at 9 MPa and saturated liquid exits the condenser at 0.009 MPa. The net power output ($\dot{W}_{net}$) of the cycle is 100 MW. Determine (a) the thermal efficiency (η_{th}) of the cycle, (b) the back-work ratio, (c) the mass flow rate ($\dot{m}$) of steam, (d) the heat transfer ($\dot{Q}$) into the working fluid as it passes through the boiler and (e) the heat transfer ($\dot{Q}$) from the condenser to the steam as it passes through the condenser.

9-1-9   Steam is the working fluid in an ideal Rankine cycle. Saturated vapor enters the turbine at 10 MPa and saturated liquid exits the condenser at 0.01 MPa. The net power output ($\dot{W}_{net}$) of the cycle is 150 MW. The turbine and the pump each have an isentropic efficiency of 85%. Determine (a) the thermal efficiency (η_{th}) of the cycle, (b) the mass flow rate ($\dot{m}$) of steam, (c) the heat transfer ($\dot{Q}$) into the working fluid as it passes through the boiler and (d) the heat transfer ($\dot{Q}$) from the condenser to the steam as it passes through the condenser.

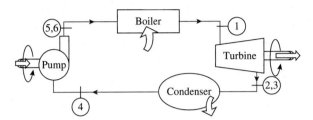

9-1-10   Propane is the working fluid in a supercritical power plant. The turbine inlet pressure is 10 MPa, the temperature is 150°C, and it exits at −30°C. The net power output ($\dot{W}_{net}$) of the cycle is 2 kW. The turbine and the pump have isentropic efficiencies of 90% and 80%, respectively. Determine (a) the thermal efficiency (η_{th}) of the cycle and (b) the mass flow rate ($\dot{m}$) of propane.

9-1-11   Water is the working fluid in an ideal Rankine cycle. Superheated vapor enters the turbine at 12 MPa and 500°C. The condenser pressure is 8 kPa. The turbine and the pump have isentropic efficiencies of 85% and 75%, respectively. Determine (a) the thermal efficiency (η_{th}) of the cycle, (b) the net power output ($\dot{W}_{net}$), (c) the heat transfer ($\dot{Q}$) into the working fluid as it passes through the boiler and (d) the heat transfer ($\dot{Q}$) from the condenser to the steam as it passes through the condenser. (e) **What-if scenario:** What would the net power output be if the mass flow rate of the working fluid was 100 kg/s?

9-1-12   In a steam power plant, operating on a Rankine cycle, steam enters the turbine at 3 MPa, 350°C and is condensed in the condenser at a pressure of 75 kPa. If the adiabatic efficiencies of the pump and turbine are 80% each, determine (a) the thermal efficiency ($\eta_{th,Rankine}$) of the cycle. (b) **What-if scenario:** What would the thermal efficiency be if the boiler pressure increased to 5 MPa?

9-1-13   Water is the working fluid in a vapor power plant. Superheated steam leaves the steam generator at 8.2 MPa, 540°C, and enters the turbine at 7.5 MPa, 500°C. The steam expands through the turbine, exiting at 8 kPa with a quality of 94%. Condensate leaves the condenser at 5 kPa, 30°C and is pumped to 9 MPa before entering the steam generator. The pump efficiency is 80%. Determine (a) the thermal efficiency (η_{th}) of the cycle and (b) the net power ($\dot{W}_{net}$) developed. (c) **What-if scenario:** What would the net power developed be if the mass flow rate of steam was 15 kg/s?

9-1-14   Water is the working fluid in a vapor power plant. Superheated steam enters the turbine at 18 MPa and 580°C. Steam expands through the turbine, exits at 6 kPa, and the turbine efficiency is 82%. Condensate leaves the condenser at 4.5 kPa, 25°C, and is pumped to 18.5 MPa before entering the steam generator. The pump efficiency is 77%. Determine (a) the net work per unit mass (w_{net}) of steam flow, (b) the heat transfer per unit mass (q) of steam passing through the boiler, (c) the thermal efficiency (η_{th}) and (d) the heat transfer per unit mass (q) of steam passing through the condenser. (e) **What-if scenario:** What would the thermal efficiency be if efficiencies of both the turbine and pump were 99%?

9-1-15   A steam power plant operates on the following cycle, producing a net power ($\dot{W}_{net}$) of 25 MW. Steam enters the turbine at 16 MPa, 550°C, and enters the condenser as saturated mixture at 10 kPa. Subcooled liquid enters the pump at 9 kPa, 35°C, and leaves at 17 MPa, which then enters the boiler at 16.8 MPa, 33°C, and exits at 16.2 MPa, 575°C. If the isentropic efficiency of the turbine is 90%, and that of the pump is 83%, determine (a) the mass flow rate ($\dot{m}$) of steam and (b) the mass flow rate ($\dot{m}$) of cooling water in the condenser in which temperature rises from 20°C to 30°C.

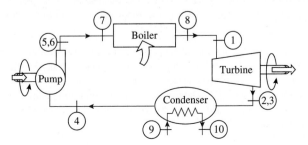

9-1-16 Water is the working fluid in a vapor power plant. Steam enters the turbine at 4 MPa, 540°C, and exits the turbine as a two-phase, liquid vapor mixture at 27°C. The condensate exits the condenser at 25°C. The turbine efficiency is 90% and the pump efficiency is 80%. If the power developed is 1 MW, determine (a) the steam quality at the turbine exit, (b) the mass flow rate ($\dot{m}$) and (c) the thermal efficiency (η_{th}).

9-1-17 Water is the working fluid in an ideal Rankine cycle. The pressure and temperature at the exit of the steam generator are 9 MPa and 480°C. A throttle valve placed between the steam generator and the turbine reduces the turbine inlet pressure to 7 MPa. The condenser pressure is 7 kPa, and the mass flow rate ($\dot{m}$) of the steam is 170 kg/s. The turbine and the pump each have an isentropic efficiency of 90%. Determine (a) the net power developed ($\dot{W}_{net}$), (b) the heat transfer ($\dot{Q}$) to the steam passing through the boiler and (c) the thermal efficiency (η_{th}). (d) *What-if scenario:* What would the net power developed be if the mass flow rate of steam was 100 kg/s?

9-1-18 Consider a steam power plant operating on a reheat Rankine cycle. Steam enters the high pressure turbine at 16 MPa, 550°C, and is condensed in the condenser at 10 kPa. If the moisture content of the steam at the exit of the low pressure turbine is not to exceed 5%, determine (a) the pressure at which the steam should be reheated and (b) the thermal efficiency ($\eta_{th,Rankine}$) of the cycle. Assume the steam is reheated to the inlet temperature of the high pressure turbine. (c) *What-if scenario:* What would the thermal efficiency be if the moisture tolerance of the turbine were increased to 10%?

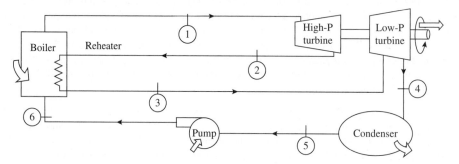

9-1-19 Consider a steam power plant that operates on a reheat Rankine cycle. Steam enters the high pressure turbine at 9 MPa, 600°C, and leaves as a saturated vapor. The steam is then reheated to 500°C, before entering the low pressure turbine, and is condensed in a condenser at 7 kPa. The mass flow rate ($\dot{m}$) is 150 kg/s. Determine (a) the net power developed ($\dot{W}_{net}$), (b) the rate of heat transfer ($\dot{Q}$) to the working fluid in the reheat process and (c) the thermal efficiency (η_{th}). (d) *What-if scenario:* What would the rate of heat transfer be if the steam was reheated to 550°C?

9-1-20 Consider a steam power plant, operating on an ideal Rankine cycle, that has reheat at a pressure of one-fifth the pressure entering the high pressure turbine. Steam enters the high pressure turbine at 17 MPa and 500°C. The steam is reheated to 500°C before entering the low pressure turbine, and is condensed in a condenser at 10 kPa. Determine (a) the thermal efficiency (η_{th}) and (b) the steam quality at the exit of the second turbine stage.

9-1-21 An ideal reheat cycle operates with steam as the working fluid. The reheat pressure is 2 MPa. Steam enters the high pressure turbine at 13 MPa and 600°C. The steam is reheated to 600°C before entering the low pressure turbine and is condensed in a condenser at 6 kPa. Determine (a) the thermal efficiency (η_{th}) and (b) the steam quality (x) at the exit of the second turbine stage. (c) *What-if scenario:* How would the answer in (b) change if the reheat pressure was 7 MPa?

9-1-22 In a steam power plant operating on a reheat Rankine cycle, steam enters the high pressure turbine at 15 MPa, 620°C, and is condensed in the condenser at a pressure of 15 kPa. If the moisture content in the turbine is not to exceed 10%, determine (a) the reheat pressure and (b) the thermal efficiency (η_{th}) of the cycle. (c) *What-if scenario:* What would the thermal efficiency be if the moisture tolerance of the turbine increased to 15%?

9-1-23 Steam is the working fluid in an ideal Rankine cycle with superheat and reheat. Steam enters the first stage turbine at 10 MPa, 500°C, and expands to 700 kPa. It is then reheated to 450°C before entering the second stage turbine, where it expands to the condenser pressure of 8 kPa. The net power output ($\dot{W}_{net}$) is 100 MW. Determine (a) the thermal efficiency (η_{th}) of the cycle, (b) the mass flow rate ($\dot{m}$) of steam, and (c) the rate of heat transfer ($\dot{Q}$) from the condensing steam as it passes through the condenser.

9-1-24 In a steam power plant, operating on the ideal regenerative Rankine cycle, with one open feedwater heater, steam enters the turbine at 9 MPa, 480°C, and is condensed in the condenser at a pressure of 7 kPa. Bleeding from the turbine to the FWH occurs at 0.7 MPa. The net power output of the cycle is 100 MW. Determine (a) the thermal efficiency (η_{th}) of the cycle, (b) the mass flow rate ($\dot{m}$) entering the turbine and (c) the rate of heat transfer ($\dot{Q}$) to the working fluid passing through the steam generator. (d) **What-if scenario:** What would the net power developed be if the bleeding pressure increased to 1.2 MPa?

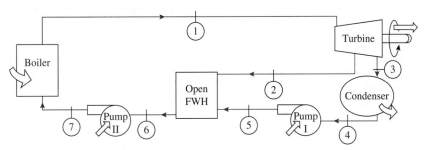

9-1-25 In a steam power plant, operating on the ideal regenerative Rankine cycle, with one open feedwater heater, steam enters the turbine at 15 MPa, 620°C, and is condensed in the condenser at a pressure of 15 kPa. Bleeding from the turbine to the FWH occurs at 1 MPa. Determine (a) the fraction of steam extracted and (b) the thermal efficiency (η_{th}) of the cycle. (c) **What-if scenario:** What would the thermal efficiency be if the bleeding pressure increased to 1.5 MPa?

9-1-26 A power plant operates on a regenerative vapor power cycle with one open feedwater heater. Steam enters the first turbine stage at 11 MPa, 600°C, and expands to 1 MPa, where some of the steam is extracted and diverted to the open feedwater heater, operating at 1 MPa. The remaining steam expands through the second turbine stage to a condenser pressure of 6 kPa. Saturated liquid exits the open feedwater heater at 1 MPa. The net power output ($\dot{W}_{net}$) is 264 MW. Determine (a) the thermal efficiency (η_{th}) of the cycle, (b) the mass flow rate ($\dot{m}$) into the first turbine stage and (c) the fraction of flow extracted where bleeding occurs. (d) **What-if scenario:** What would the net power developed be if the bleeding pressure increased to 1.2 MPa?

9-1-27 Consider a steam power plant, operating on the ideal regenerative Rankine cycle, with one open feedwater heater. Steam enters the turbine at 14 MPa, 610°C, and is condensed in the condenser at a pressure of 12 kPa. Some steam leaves the turbine at a pressure of 1.2 MPa and enters the open feedwater heater. Determine (a) the thermal efficiency (η_{th}) of the cycle and (b) the fraction of flow extracted where bleeding occurs. (c) **What-if scenario:** What would the answer in (b) be if the bleeding pressure increased to 1.5 MPa?

9-1-28 A steam power plant operates on an ideal regenerative Rankine cycle. Steam enters the turbine at 5 MPa, 450°C, and is condensed in the condenser at 15 kPa. Steam is extracted from the turbine at a pressure of 0.4 MPa and enters the open feedwater heater. Water leaves the feedwater heater as a saturated liquid. Determine (a) the thermal efficiency (η_{th}) of the cycle and (b) the net work output per kilogram of steam (w_{net}) flowing through the boiler.

9-1-29 A regenerative vapor power cycle has two turbine stages with steam entering the first turbine stage at 8 MPa, 550°C and expanding to 700 kPa, where some of the steam is extracted and diverted to the open feedwater heater operating at 700 kPa. The remaining steam expands through the second turbine stage to the condenser at a pressure of 7 kPa. Saturated liquid exits the open feedwater heater at 700 kPa. Each turbine stage has an isentropic efficiency of 88% and each pump has an isentropic efficiency of 80%. Determine (a) the thermal efficiency (η_{th}) of the cycle, (b) the net work developed per kilogram of steam (w_{net}) and (c) the fraction of flow extracted where bleeding occurs. (d) **What-if scenario:** What would the net power developed be if the mass flow rate of steam entering the first stage of turbine was 170 kg/s?

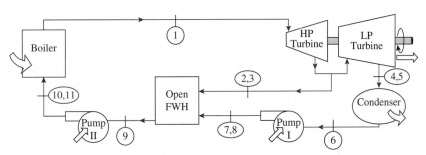

9-1-30 A regenerative vapor power cycle has two turbine stages, with steam entering the first turbine stage at 12 MPa, 600°C, and expands to 1 MPa, where some of the steam is extracted and diverted to the open feedwater heater operating at 1 MPa. The remaining steam expands through the second turbine stage to a condenser at pressure of 6 kPa. Saturated liquid exits the open feedwater heater at 6 kPa. Each turbine stage and pump has an isentropic efficiency of 80%. The mass flow rate ($\dot{m}$) into the first turbine stage is 100 kg/s. Determine (a) the thermal efficiency (η_{th}) of the cycle, (b) the net power developed ($\dot{W}_{net}$) and (c) the heat transfer ($\dot{Q}$) to the steam in the steam generator. (d) *What-if scenario:* What would the net power developed be if the feedwater pressure was 1.4 MPa?

9-1-31 A power plant operates on a regenerative vapor power cycle with one closed feedwater heater. Steam enters the first turbine stage at 10 MPa, 500°C, and expands to 1 MPa, where some of the steam is extracted and diverted to a closed feedwater heater. Condensate, exiting the feedwater heater as saturated liquid at 1 MPa, passes through a trap into the condenser. The feedwater exits the heater at 10 MPa with a temperature of 175°C. The condenser pressure is 6 kPa. The mass flow rate ($\dot{m}$) into the first stage turbine is 270 kg/s. For isentropic processes in each turbine stage and the pump, determine (a) the mass flow rate ($\dot{m}$) of steam extracted from the turbine, (b) the thermal efficiency (η_{th}) of the cycle, and (c) the net power developed ($\dot{W}_{net}$).

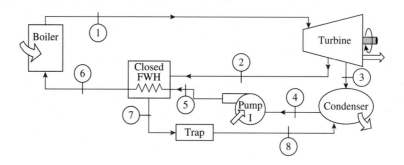

9-1-32 Repeat Problem 9-1-31, replacing the trap with a pump and a mixing chamber as shown in the schematic below.

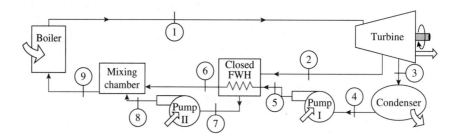

9-1-33 A power plant operates on a regenerative vapor power cycle with one closed feedwater heater. Steam enters the first turbine stage at 7 MPa, 550°C, and expands to 700 kPa, where some of the steam is extracted and diverted to a closed feedwater heater. Condensate exiting the feedwater heater as saturated liquid at 700 kPa passes through a trap into the condenser. The feedwater exits the heater at 7 MPa with a temperature of 175°C. The condenser pressure is 8 kPa. If the power developed is 100 MW, determine (a) the thermal efficiency (η_{th}) of the cycle and (b) the mass flow rate ($\dot{m}$) into the first stage turbine. (c) *What-if scenario:* What would the thermal efficiency be if the extraction pressure was 600 kPa?

9-1-34 A power plant operates on a regenerative vapor power cycle with one closed feedwater heater. Steam enters the first turbine stage at 12 MPa, 520°C, and expands to 1200 kPa, where some of the steam is extracted and diverted to a closed feedwater heater. Condensate, exiting the feedwater heater as saturated liquid at 120 kPa, passes through a trap into the condenser. The feedwater exits the heater at 7 MPa with a temperature of 170°C. The condenser pressure is 12 kPa. If the mass flow rate into first stage of turbine is 300 kg/s and each turbine stage has an isentropic efficiency of 80%, determine (a) the thermal efficiency (η_{th}) of the cycle and (b) the net power developed ($\dot{W}_{net}$). (c) *What-if scenario:* What would the net power developed be if the condenser pressure was 9 kPa?

9-1-35 A power plant operates on an ideal reheat-regenerative Rankine cycle and has a net power output ($\dot{W}_{net}$) of 100 MW. Steam enters the high pressure turbine stage at 12 MPa, 550°C, and leaves at 0.9 MPa. Some steam is extracted at 0.9 MPa to heat the feedwater in an open feedwater heater with the water leaving the FWH as saturated liquid. The rest of the steam is reheated to 500°C and is expanded in the low pressure turbine to the condenser at a pressure of 8 kPa. Determine (a) the thermal efficiency (η_{th}) of the cycle and (b) the mass flow rate ($\dot{m}$) of steam through the boiler. (c) *What-if scenario:* What would the thermal efficiency be if the steam entered the turbine at 15 MPa?

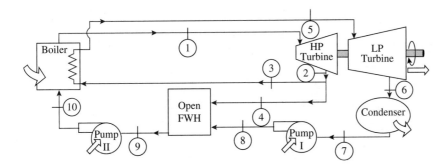

9-1-36 Repeat Problem 9-1-35, but replace the open feedwater heater with closed feedwater heater. Assume that the feedwater leaves the heater at the condensation temperature of the extracted steam and that the extracted steam leaves the heater at state-10 as a saturated liquid before it is pumped to the line carrying the feedwater. Determine (a) the thermal efficiency (η_{th}) of the cycle and (b) the mass flow rate ($\dot{m}$) of steam through the boiler.

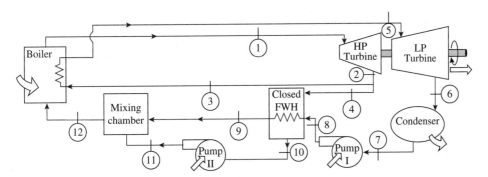

9-1-37 A steam power plant operates on a reheat-regenerative Rankine cycle with a closed feedwater heater. Steam enters the turbine at 12 MPa, 500°C at a rate ($\dot{m}$) of 25 kg/s and is condensed in the condenser at a pressure of 20 kPa. Steam is reheated at 5 MPa to 500°C. Steam at a rate ($\dot{m}$) of 5 kg/s is extracted from the high pressure turbine at 1.2 MPa, is completely condensed in the closed feedwater heater, then pumped to 12 MPa before it mixes with the feedwater at the same pressure. Assuming an isentropic efficiency of 88% for both the turbine and the pump, determine (a) the temperature at the inlet of the closed feedwater heater, (b) the thermal efficiency ($\eta_{th,Rankine}$) of the cycle and (c) the net power output ($\dot{W}_{net}$).

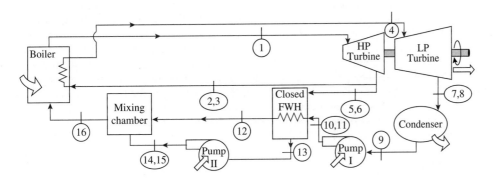

9-1-38 A steam power plant operates on an ideal reheat-regenerative Rankine with one reheat and two open feedwater heaters. Steam enters the high pressure turbine at 10 MPa, 600°C and leaves the low pressure turbine at 7 kPa. Steam is extracted from the turbine at 2 MPa and 275 kPa, and it is reheated to 540°C at a pressure of 1 MPa. Water leaves both feedwater heaters as a saturated liquid. Heat is transferred to the steam in the boiler at a rate of 6 MW. Determine (a) the mass flow rate ($\dot{m}$) of steam through the boiler, (b) the net power ($\dot{W}_{net}$) output and (c) the thermal efficiency (η_{th}) of the cycle.

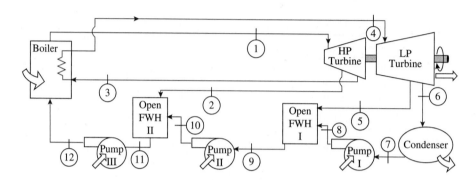

9-1-39 Consider a reheat-regenerative vapor power cycle with two feedwater heaters, a closed feedwater heater, and an open feedwater heater. Steam enters the first turbine at 10 MPa, 500°C, and expands to 0.8 MPa. The steam is reheated to 440°C before entering the second turbine, where it expands to the condenser at a pressure of 0.007 MPa. Steam is extracted from the first turbine at 2 MPa and fed to the closed feedwater heater. Feedwater leaves the closed heater at 205°C, 10 MPa, and condensate exits as saturated liquid at 2 MPa. The condensate is trapped in the open feedwater heater. Steam extracted from the second turbine at 0.3 MPa is also fed into the open feedwater heater, which operates at 0.3 MPa. The steam exiting the open feedwater heater is saturated liquid at 0.3 MPa. The net power output ($\dot{W}_{net}$) of the cycle is 100 MW. There is no stray heat transfer from any component to its surroundings. If the working fluid experiences no irreversibilities as it passes through the turbines, pumps, steam generator, reheater and condenser, determine (a) the thermal efficiency (η_{th}) of the cycle and (b) the mass flow rate ($\dot{m}$) of steam entering the first turbine.

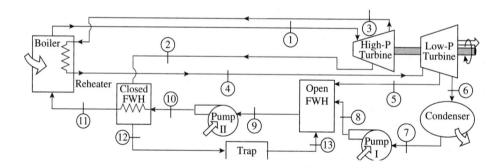

SECTION 9-2: COGENERATION, COMBINED AND BINARY CYCLES

9-2-1 Water is the working fluid in a cogeneration cycle that generates electricity and provides heat for campus buildings. Steam at 2.5 MPa, 320°C, with a mass flow rate ($\dot{m}$) of 1 kg/s, expands through a two-stage turbine. Steam at 0.2 MPa is extracted with a mass flow rate ($\dot{m}$) of 0.3 kg/s between the two stages and provided for heating. The remaining steam expands through the second stage to the condenser at a pressure of 6 kPa. The condensate returns from the campus buildings at 0.1 MPa, 60°C, and passes through a trap into the condenser. Each turbine stage has an isentropic efficiency of 80%. Determine (a) the net heat transfer rate ($\dot{Q}$) to the working fluid passing through the steam generator, (b) the net power developed ($\dot{W}_{net}$) and (c) the rate of heat transfer ($\dot{Q}$) for building heating. (d) *What-if scenario:* What would the rate of heat transfer be if the inlet conditions at the turbine were 3 MPa and 400°C?

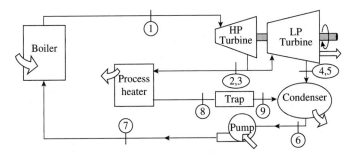

9-2-2 Water is the working fluid in a cogeneration cycle. Steam generator provides 280 kg/s of steam at 9 MPa, 500°C, of which 110 kg/s is extracted between the first and second stages at 1.5 MPa and diverted to a process heating load. Condensate returns from the process heating load at 1 MPa, 120°C, and is mixed with the liquid exiting the lower pressure pump at 1 MPa. The entire flow is then pumped to the steam generator pressure. Saturated liquid at 8 kPa leaves the condenser. The turbine stages and pumps operate with isentropic efficiencies of 90% and 80%, respectively. Determine (a) the net heat transfer rate ($\dot{Q}$) to the working fluid passing through the steam generator, (b) the net power ($\dot{W}_{net}$) developed and (c) the heating load.

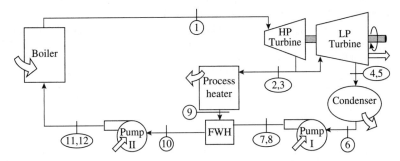

9-2-3 A large food processing plant requires 3.5 kg/s of saturated or slightly superheated steam at 550 kPa, which is extracted from the turbine of a cogeneration plant. The boiler generates steam at 7 MPa, 540°C, at a rate of 9 kg/s and the condenser pressure is 14 kPa. Steam leaves the process heater as saturated liquid. It is then mixed with the feedwater at the same pressure and this mixture is pumped to the boiler pressure. Assuming both the pumps and the turbine have adiabatic efficiencies of 86%, determine (a) the net heat transfer rate ($\dot{Q}$) to the working fluid passing through the steam generator and (b) the power output ($\dot{W}_{net}$) of the cogeneration plant. (c) **What-if scenario:** What would the power output be if the efficiencies of both the pumps and the turbine were 100%?

9-2-4 Consider a cogeneration plant. Steam enters the turbine at 8 MPa and 600°C. 20% of the steam is extracted before it enters the turbine and 60% of the steam is extracted from the turbine at 500 kPa for process heating. The remaining steam continues to expand to 6 kPa. Steam is then condensed at constant pressure and pumped to the boiler pressure of 8 MPa. Steam leaves the process heater as a saturated liquid at 500 kPa. The mass flow rate ($\dot{m}$) of steam through the boiler is 20 kg/s. Determine (a) the rate of process heat supply ($\dot{Q}$) and (b) the net power ($\dot{W}_{net}$) developed. (c) **What-if scenario:** What would the net power developed be if no process heat was supplied?

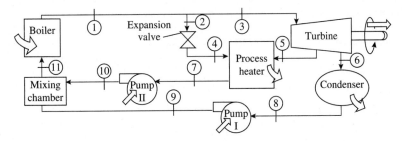

9-2-5 Repeat Problem 9-2-4 to determine (a) the maximum rate at which process heat can be supplied.

9-2-6 Steam is generated in the boiler of a cogeneration plant at 4 MPa and 480°C with a rate of 7 kg/s. The plant is to produce power while meeting the process steam requirements for a certain industrial application. One-third of the steam leaving the boiler is throttled to a pressure of 820 kPa then routed to the process heater. The rest of the steam is expanded in an isentropic turbine to a pressure of 820 kPa and is also routed to the process heater. Steam leaves the process heater as saturated liquid. Determine (a) the net power ($\dot{W}_{net}$) produced and (b) the rate of process heat supply. (c) *What-if scenario:* What would the net power produced be if one-half instead of one-third of the steam leaving the boiler was throttled?

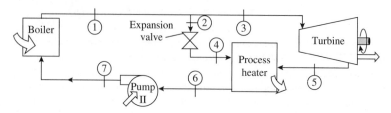

9-2-7 Consider a cogeneration power plant modified with regeneration. Steam enters the turbine at 5 MPa, 450°C, and expands to a pressure of 0.6 MPa. At this pressure, 65% of the steam is extracted from the turbine, and the remainder expands to 10 kPa. Part of the extracted steam is used to heat the feedwater in an open feedwater heater. The rest of the extracted steam is used for process heating and leaves the process heater as saturated liquid at 0.6 MPa. It is subsequently mixed with the feedwater leaving the feedwater heater and the mixture is pumped to the boiler pressure. Assuming the turbines and the pumps to be isentropic, (a) determine the mass flow rate ($\dot{m}$) of the steam through the boiler for a net power output ($\dot{W}_{net}$) of 15 MW. (b) *What-if scenario:* What would the mass flow rate of steam be if only 50% of the steam was extracted from the turbine?

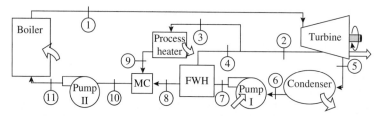

9-2-8 Consider a cogeneration power plant modified with regeneration. Steam enters the turbine at 7 MPa, 440°C, at a rate of 20 kg/s and expands to a pressure of 0.4 MPa. At this pressure 60% of the steam is extracted from the turbine and the remainder expands to 10 kPa. Part of the extracted steam is used to heat the feedwater in an open feedwater heater. The rest of the extracted steam is used for process heating and leaves the process heater as a saturated liquid at 0.4 MPa. It is subsequently mixed with the feedwater leaving the feedwater heater, and the mixture is pumped to the boiler pressure. Assuming the turbines and the pumps to be isentropic, determine (a) the total power output ($\dot{W}_{net}$) of the turbine, (b) the mass flow rate ($\dot{m}$) of the steam through the process heater, (c) the rate of heat supply ($\dot{Q}$) from the process heater per unit mass of steam passing through it, and (d) the rate of heat transfer ($\dot{Q}$) to the steam boiler.

9-2-9 The gas-turbine portion of a combined gas-steam power plant has a pressure ratio of 15. Air enters the compressor at 300 K and 1 atm at a rate of 13 kg/s and is heated to 1500 K in the combustion chamber. The combustion gases leaving the gas turbine are used to heat the steam to 400°C at 10 MPa in a heat exchanger. The combustion gases leave the heat exchanger at 420 K. The steam leaving the turbine is condensed at 15 kPa. Assuming all the compression and expansion processes to be isentropic, determine (a) the mass flow rate ($\dot{m}$) of steam, (b) the net power output ($\dot{W}_{net}$) and (c) the thermal efficiency (η_{th}) of the combined cycle. (d) *What-if scenario:* What would the thermal efficiency be if the compression ratio increased to 17?

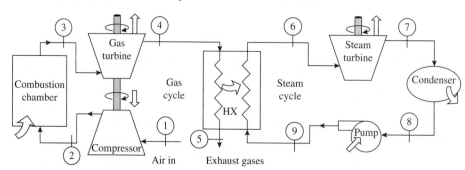

9-2-10 Consider a combined gas-steam power plant that has a net power output ($\dot{W}_{net}$) of 600 MW. The pressure ratio of the gas turbine cycle is 16. Air enters the compressor at 300 K and the turbine at 1600 K. The combustion gases leaving the gas turbine are used to heat the steam to 400°C at 10 MPa in a heat exchanger. The combustion gases leave the heat exchanger at 400 K. An open feedwater heater incorporated with the steam cycle operates at a pressure of 0.6 MPa. The condenser pressure is 15 kPa. Assuming all the compression and expansion processes to be isentropic, determine (a) the mass flow rate ($\dot{m}$) of steam and (b) the thermal efficiency (η_{th}) of the combined cycle. (c) *What-if scenario:* What would the thermal efficiency be if the compression ratio increased to 17?

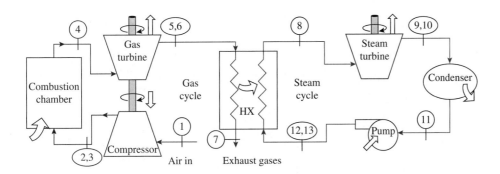

9-2-11 Repeat Problem 9-2-10 assuming isentropic efficiencies of 100% for the pump, 82% for the compressor, 86% for the gas and steam turbines. Determine (a) the mass flow rate of steam and (b) the thermal efficiency of the combined cycle.

9-2-12 A combined gas turbine-vapor power plant has a net power output ($\dot{W}_{net}$) of 15 MW. Air enters the compressor of the gas turbine at 100 kPa, 290 K, and is compressed to 1100 kPa. The isentropic efficiency of the compressor is 80%. The conditions at the inlet to the turbine are 1100 kPa and 1400 K. Air expands through the turbine, that has an isentropic efficiency of 88%, to a pressure of 100 kPa. Air then passes through the interconnecting heat exchanger, and is finally discharged at 420 K. Steam enters the turbine of the vapor power cycle at 8 MPa, 390°C, and expands to the condenser at a pressure of 8 kPa. Water enters the pump as saturated liquid at 8 kPa. The turbine and pump have isentropic efficiencies of 90% and 80%, respectively. Determine (a) the thermal efficiency (η_{th}) of the combined cycle, the mass flow rates ($\dot{m}$) of (b) air and (c) water, and (d) the rate of heat transfer ($\dot{Q}$) to the combined cycle.

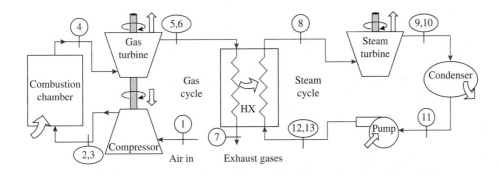

9-2-13 A simple gas turbine is the topping cycle for a simple vapor power cycle. Air enters the compressor of the gas turbine at 101 kPa, 15°C, and mass flow rate ($\dot{m}$) of 23 kg/s. The compressor pressure ratio is 10 and the turbine inlet temperature is 1100°C. The compressor and turbine have an isentropic efficiency of 85%. Air leaves the interconnecting heat exchanger at 200°C and 101 kPa. Steam enters the turbine of the vapor power cycle at 7 MPa, 480°C, and expands to the condenser pressure of 7 kPa. Water enters the pump as saturated liquid at 7 kPa. The turbine and pump have isentropic efficiencies of 90% and 80%, respectively. Determine (a) the thermal efficiency (η_{th}) of the combined cycle, (b) the mass flow rate ($\dot{m}$) of water and (c) the net power output ($\dot{W}_{net}$).

9-2-14 Consider a combined cycle power plant using helium and water as the working fluids. Helium enters the compressor of the gas turbine at 1.4 MPa, 350 K, and is compressed to 5.5 MPa. The isentropic efficiency of the compressor is 80%. The conditions at the inlet to the turbine are 5.5 MPa and 760°C. Helium expands through the turbine to a pressure of 1.4 MPa. The turbine has an isentropic efficiency of 80%. The mass flow rate ($\dot{m}$) of the gas is 100 kg/s. Saturated vapor at 8 MPa exits the heat exchanger, which is superheated to 425°C, before it enters the turbine of the vapor power cycle, and expands to the condenser at a pressure of 7 kPa. The steam exits the turbine at a quality of 0.9. Water enters the pump as saturated liquid at 7 kPa. Determine (a) the thermal efficiency (η_{th}) of the combined cycle, (b) the mass flow rate ($\dot{m}$) of steam and (c) the net power ($\dot{W}_{net}$) developed. (d) *What-if scenario:* What would the thermal efficiency be if air was used as working fluid for the gas phase?

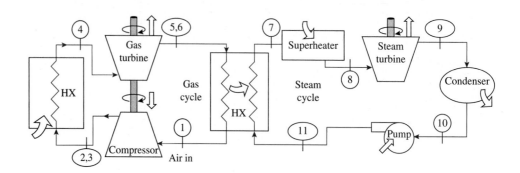

9-2-15 Steam and ammonia are the working fluids in a binary vapor power cycle consisting of two ideal Rankine cycles. The heat rejected from the steam cycle is provided to the ammonia cycle. In the steam cycle, steam at 6 MPa, 650°C, enters the turbine and exits at 60°C. Saturated liquid at 60°C enters the pump and is pumped to the steam generator pressure. Saturated vapor of ammonia enters the turbine at 50°C and exits at 1 MPa. It then enters the condenser and condenses to saturated liquid. The saturated liquid is pumped through the heat exchanger. The power output of the binary cycle is 25 MW. Determine (a) the mass flow rates ($\dot{m}$) of steam and ammonia, (b) the power outputs ($\dot{W}$) of the steam and ammonia turbines, (c) the rate of heat addition ($\dot{Q}_{in}$) to the cycle and (d) the thermal efficiency (η_{th}).

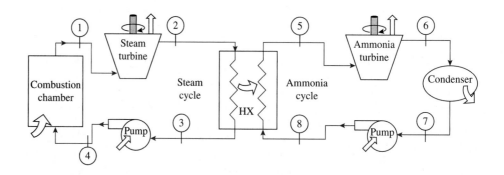

9-2-16 Water and the refrigerant R-134a are the working fluids in a binary cycle used for cogeneration of power and process steam. In the steam cycle, superheated vapor enters the turbine with a mass flow rate ($\dot{m}$) of 5 kg/s at 4 MPa, 470°C, and expands isentropically to 150 kPa. Half of the flow is extracted at 150 kPa and is used for industrial process heating. The rest of the stream passes through a heat exchanger, which serves as the boiler for the refrigerant cycle and the condenser of the steam cycle. The condensate leaves the heat exchanger as saturated liquid at 100 kPa, which is combined with the return flow from the process, at 100 kPa and 65°C, before being pumped isentropically to the steam generator pressure. Refrigerant 134a is in an ideal Rankine cycle with refrigerant entering the turbine at 1.5 MPa, 101°C, and saturated liquid leaving the condenser at 800 kPa. Determine (a) the rate of heat transfer ($\dot{Q}$) to the working fluid passing through the steam generator of the steam cycle, (b) the net power output ($\dot{W}_{net}$) of the binary cycle, (c) the mass flow rate ($\dot{m}$) of the refrigerant and (d) the rate of heat transfer ($\dot{Q}$) to the industrial process. (e) *What-if scenario:* What would the mass flow rate be using the refrigerant R-12 instead of R-134a?

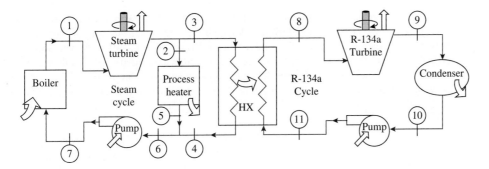

9-2-17 A geothermal resource exists as saturated liquid at 200°C. The geothermal liquid is withdrawn from a production well, at a rate of 200 kg/s, and is flashed to a pressure of 500 kPa by an essentially isenthalpic flashing process, where the resulting vapor is separated from the liquid in a separator and is directed to the turbine. Steam leaves the turbine at 12 kPa, with a moisture content of 14%, and enters the condenser, where it is condensed and routed to a reinjection well, along with the liquid coming from the separator. Determine (a) the mass flow rate ($\dot{m}$) of steam through the turbine and (b) the power output ($\dot{W}_{ext}$) of the turbine.

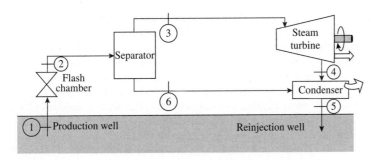

SECTION 9-3: EXERGY ANALYSIS OF VAPOR POWER, COGENERATION, AND BINARY CYCLES

9-3-1 Consider a steam power plant operating on the simple ideal Rankine cycle. Steam enters the turbine at 4 MPa, 400°C, and is condensed in the condenser at a pressure of 100 kPa. The mass flow rate is 10 kg/s. If the boiler receives heat from a source at 1200°C and the condenser rejects heat to a reservoir at 25°C, determine (a) the thermal efficiency of the cycle, (b) the exergetic efficiency of the cycle and (c) draw an exergy flow diagram for the cycle. Assume the atmospheric conditions to be 100 kPa and 25°C.

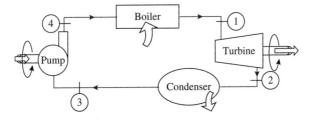

9-3-2 Steam is the working fluid in an ideal Rankine cycle. Saturated vapor enters the turbine at 10 MPa and saturated liquid exits the condenser at a pressure of 0.01 MPa. The net power output ($\dot{W}_{net}$) of the cycle is 150 MW. The turbine and the pump both have an isentropic efficiency of 85%. If the boiler receives heat from a source at 1200°C, and the condenser rejects heat to a reservoir at 25°C, (a) identify the device of maximum exergy destruction (boiler: 0; turbine: 1; condenser: 3; pump: 4), (b) determine the exergetic efficiency of the cycle and (c) draw an exergy flow diagram for the cycle. Assume the atmospheric conditions to be 100 kPa and 25°C.

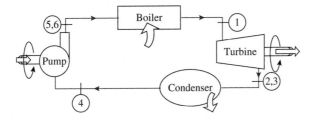

9-3-3 In a steam power plant, operating on the ideal regenerative Rankine cycle, with one open feed-water heater, steam enters the turbine at 9 MPa, 480°C, and is condensed in the condenser at a pressure of 7 kPa. Bleeding from the turbine to the FWH occurs at 0.7 MPa. The net power output ($\dot{W}_{\text{net}}$) of the cycle is 100 MW. The boiler receives heat from a source at 1200°C and the condenser rejects heat to atmosphere at 100 kPa, 25°C. (a) Perform an exergy inventory and draw an exergy flow diagram for the cycle. Determine (b) the thermal efficiency (η_{th}) and (c) exergetic efficiency of the cycle.

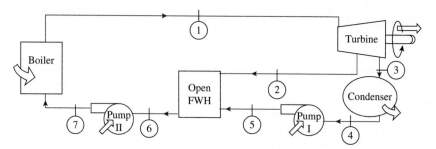

9-3-4 Repeat Problem 9-3-3, assuming no steam is bled from the turbine for regeneration.

9-3-5 Water is the working fluid in a cogeneration cycle that generates electricity and provides heat for campus buildings. Steam at 2.5 MPa, 320°C, and a mass flow rate ($\dot{m}$) of 1 kg/s, expands through a two-stage turbine. Steam at 0.2 MPa, with a mass flow rate ($\dot{m}$) of 0.3 kg/s, is extracted between the two stages and provided for heating. The remaining steam expands through the second stage to the condenser at a pressure of 6 kPa. The condensate returns from the campus buildings at 0.1 MPa, 60°C, and passes through a trap into the condenser. Each turbine stage has an isentropic efficiency of 80%. Heat addition to the boiler takes place from a source at 1200°C, the building is maintained at 50°C, and the atmospheric conditions are 100 kPa, 20°C. (a) Perform an exergy analysis and draw an exergy flow diagram for the system. (b) Define and evaluate the exergetic efficiency of the system.

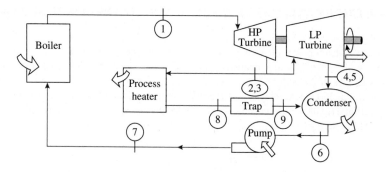

10 REFRIGERATION CYCLES

In Chapter 2 we studied power and refrigeration cycles as closed-steady systems for the purpose of an overall analysis. In Chapters 6 through 8 power cycles were implemented as closed and open cycles. This chapter is devoted to the study of refrigeration and heat pump cycles implemented by open cycles. **Refrigeration** is the cooling of a system below the temperature of its surroundings. A **heat pump**, on the other hand, keeps a system warmer than its surroundings by pumping heat from the surroundings into the system. Both the refrigeration and the heat pump systems operate on the same basic **refrigeration cycle**, with only the desired output being different for the two systems. Three major kinds of refrigeration cycles—**vapor compression**, **vapor absorption**, and **gas refrigeration**—are presented in this chapter. In a vapor compression cycle, which resembles a reversed Rankine cycle, a compressor is used to raise the pressure of the refrigerant vapor. In the vapor absorption cycle, the refrigerant is absorbed in a liquid, and a pump is used to raise the pressure, thereby eliminating the compressor. The gas refrigeration cycle is simply a reversed Brayton cycle and the working fluid remains as a gas throughout the cycle.

The objective of this chapter is to study both the basic and more advanced vapor and gas refrigeration cycles with applications to practical refrigeration and heat pump systems. Throughout the chapter, animations are used to illustrate complex cycles, and TEST solutions are used for verifying manual results and carrying out what-if studies.

10.1 REFRIGERATORS AND HEAT PUMP

Long before the advent of refrigeration cycles, melting of ice or snow was used for refrigeration purposes. In that traditional method, still practiced in places where electricity is not readily available, a block of ice is placed within the system (which is warmer than 0°C) to be cooled. As heat flows into the ice, it starts melting at a constant temperature of 0°C (at atmospheric pressure). The latent heat (enthalpy of fusion) of ice is 334 kJ/kg (Fig. 10.1); so a ton (2000 lbm) of ice melting over a day from 0°C ice to 0°C liquid water at atmospheric pressure has the **cooling capacity** of 303.6 MJ/day, 212 kJ/min, or 3.51 (more precise value is 3.517) kW. In practice 1 **ton of refrigeration** is assumed equivalent to 3.517 kW. A 3 ton refrigeration system, thus, has a cooling capacity of 10.55 kW and is equivalent to having 3 tons of ice melting over a 24 hour period. Cooling capacity is frequently expressed in this unit, but keep in mind that outside the United States a tonne of refrigeration refers to melting of 1 tonne (1 metric ton or 1000 kg) of ice, which is equivalent to 234 kJ/min. Another traditional medium of refrigeration is solid carbon dioxide, or **dry ice**. As shown by the *T-s* diagram of Figure 10.2, CO_2 cannot exist in liquid state below 5.11 atm. At atmospheric pressure dry ice must sublimate, absorbing the latent heat of sublimation of 620 kJ/kg at a constant temperature of –78.5°C.

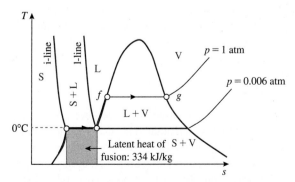

FIGURE 10.1 *T-s* diagram for H_2O extended to the solid phase (see Anim. 3.C.*zones*).

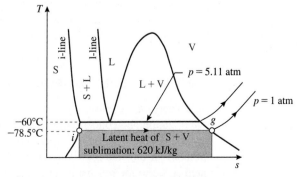

FIGURE 10.2 *T-s* diagram for CO_2 highlighting sublimation—transformation from solid to vapor phase.

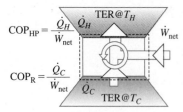

$$COP_{HP} = \frac{\dot{Q}_H}{\dot{W}_{net}}$$

$$COP_R = \frac{\dot{Q}_C}{\dot{W}_{net}}$$

FIGURE 10.3 The same refrigeration cycle can be used for cooling the cold region or heating the warm region (see Anim. 10.A.*COPvsEfficiency*).

A modern refrigerator is a cyclic device that removes heat from the refrigerated space and pumps it into the surroundings as shown in Figure 10.3 (see Sec. 2.3.2 for an overall analysis of the refrigeration cycle). The cold region is the refrigerated space and the warm region is usually the surroundings for a refrigerator. In the case of heat pump, the surroundings serve as the hot reservoir. Instead of using subscripts 'in' and 'out' as in the case of power cycles, we will use H or C to associate heat transfer with a specific reservoir, hot or cold. The desired output of a refrigeration cycle, thus, is the rate of heat removal, $\dot{Q}_C$, shown in Figure 10.3, which is also called the **cooling load**. It is initially due to the transient cooling of the refrigerated space from a certain initial temperature to a desired final temperature, and at steady state is entirely due to leakage of heat from the surroundings to the refrigerated space (see Anim. 10.A.*refrigerator*). A steady-state energy balance on the closed system produces the heat rejected to the surroundings as $\dot{Q}_H = \dot{Q}_C + \dot{W}_{net}$. If the same cycle is used for a heat pump, $\dot{Q}_H$ becomes the **heating load**, supplying heat to the warm region with the cold region representing the cooler surroundings. In both these configurations, the net work supplied to the cycle is the required input. Therefore, as derived in Sec. 2.3.2 and illustrated in Anim. 10.A.*COPvsEfficiency,* we can express COP's of the refrigeration and heat pump cycles as follows:

$$\dot{W}_{net} = \dot{Q}_H - \dot{Q}_C; \quad [kW]$$

$$COP_R = \frac{\text{Desired Energy Output}}{\text{Required Energy Input}} = \frac{\dot{Q}_C}{\dot{W}_{net}} = \frac{\dot{Q}_C}{\dot{Q}_H - \dot{Q}_C} = \frac{1}{\dot{Q}_H/\dot{Q}_C - 1} \qquad \textbf{(10.1)}$$

$$COP_{HP} = \frac{\text{Desired Energy Output}}{\text{Required Energy Input}} = \frac{\dot{Q}_H}{\dot{W}_{net}} = \frac{\dot{Q}_H}{\dot{Q}_H - \dot{Q}_C} = \frac{1}{1 - \dot{Q}_C/\dot{Q}_H} \qquad \textbf{(10.2)}$$

$$\text{From Eqs. (10.1) and (10.2), } COP_{HP} = 1 + COP_R \qquad \textbf{(10.3)}$$

In these expressions, $\dot{Q}_H$, $\dot{Q}_C$, and $\dot{W}_{net}$ are the magnitude of the respective energy transfer with the subscripts indicating the direction. The COP of a heat pump is always greater than one. Therefore, the heat pump will work, at worst, like an electric heater even if $COP_R \rightarrow 0$. However, the heat pump is generally housed outside the residence and heat loss in the piping and other components make it possible for COP_{HP} to drop below unity. It is important not to confuse Eqs. (10.1) and (10.2) with Carnot COP given by Eqs. (2.36) and (2.37) and summarized in Anim. 10.A.*CarnotCOP*.

10.2 TEST AND THE REFRIGERATION CYCLE TESTCALCS

The refrigeration cycle TESTcalcs located in the systems, open, steady, specific, refrigeration cycles branch of the TESTcalcs module are meant for detailed analysis of both vapor and gas compression cycles. Specifically, the PC TESTcalc, which contains about 40 refrigerants, should be used for the vapor compression cycles; the IG, PG, and n-IG TESTcalcs for the gas refrigeration cycle, and the PC/PC TESTcalc for two stage cascade refrigeration systems involving two different refrigerants.

The refrigeration TESTcalc builds upon the single-flow open-steady TESTcalcs (introduced in Chapter 4) by adding a cycle panel just like the power cycle TESTcalcs. Once the various steady devices are analyzed in the device panel and the loop is complete, overall cycle variables are automatically evaluated in the cycle panel. Note that the mass flow rate must be supplied or set to 1 kg/s for the cycle panel to work.

10.3 VAPOR-REFRIGERATION CYCLES

In a vapor-refrigeration cycle the working fluid, called a **refrigerant**, undergoes alternating vaporization and condensation. The vaporization pressure is maintained at such a value that the corresponding saturation temperature is slightly—usually about 10°C—below the desired temperature of the cold region T_C. As a result, heat can be effectively transferred from the refrigerated space to the refrigerant at a rate $\dot{Q}_C$, the *cooling load*. Similarly, condensation takes place at such a pressure that the saturation temperature at that pressure is a few degrees above the temperature of the warm region T_H (usually the surrounding atmosphere), making it practical for the heat transfer $\dot{Q}_H$ to occur from the refrigerant to the warm region.

10.3.1 Carnot Refrigeration Cycle

In Sec. 2.3.3B (see Anim. 10.A.*CarnotCOP*), we analyzed the Carnot refrigerator and heat pump by treating the overall system as a closed-steady system. Here we revisit the Carnot refrigeration cycle with a vapor cycle implementation as depicted in Figure 10.4 and Anim. 10.A.*CarnotCycle*. It is essentially the same as the Carnot power cycle, discussed in Sec. 9.2.3, except the direction of the flow is reversed. The working fluid, in this case a refrigerant, enters the **condenser** as a saturated vapor at $T_H + dT_H$, differentially higher than the temperature T_H of the surroundings. Heat is rejected at constant temperature (and pressure) as the vapor condenses into saturated liquid at State-2. Driven by an infinitesimal temperature difference, heat transfer takes place reversibly, that is, without any generation of entropy, and is given by the area under the T-s curve, $\dot{Q}_H = \dot{m}T_H(s_1 - s_2)$, where $\dot{m}$ is the mass flow of the refrigerant [see Eq. (4.7) and Anim. 4.A.*intReversible*]. The saturated liquid leaving the condenser expands in an **isentropic turbine** and leaves the turbine at State-3 as a saturated mixture at $T_C - dT_C$, differentially below the desired temperature T_C of the refrigerated space. The temperature remains constant in the **evaporator** as heat is removed in a reversible manner from the refrigerated space at a rate of $\dot{Q}_C = \dot{m}T_C(s_4 - s_3) = \dot{m}T_C(s_1 - s_2)$, the cooling capacity or the cooling load of the cycle. The vapor-liquid mixture that leaves the evaporator at State-4 can now be isentropically compressed back to State-1 by the compressor, completing the cycle. The COP's for the Carnot refrigerator and heat pump can be obtained from Eq. (10.1) as

$$\text{COP}_{R,\text{Carnot}} = \frac{\dot{Q}_C}{\dot{Q}_H - \dot{Q}_C} = \frac{\dot{m}T_C(s_4 - s_3)}{\dot{m}T_H(s_1 - s_2) - \dot{m}T_C(s_4 - s_3)} = \frac{T_C}{T_H - T_C} \quad (10.4)$$

$$\text{COP}_{HP,\text{Carnot}} = \frac{\dot{Q}_H}{\dot{Q}_H - \dot{Q}_C} = \frac{\dot{m}T_H(s_1 - s_2)}{\dot{m}T_H(s_1 - s_2) - \dot{m}T_C(s_4 - s_3)} = \frac{T_H}{T_H - T_C} \quad (10.5)$$

As expected, these expressions are identical to those derived in Sec. 2.3.3B, purely from an overall closed-steady analysis.

A real refrigeration cycle differs from the Carnot cycle in a number of ways. Heat transfer over an infinitesimally small temperature difference requires an infinitely large surface area of contact, which is impractical. For effective heat transfer, the pressure in the evaporator is maintained at such a level that the evaporation temperature is a few degrees cooler than the temperature of the cold region so that $T_C' = T_C - \Delta T$. Similarly, the condensation temperature is maintained at a few degrees warmer than T_H so that $T_H' = T_H + \Delta T$ (Fig. 10.5). A careful examination of the derivation of the Carnot COP shows that it is the refrigerant temperatures that appear in Eqs. (10.4) and (10.5); therefore, the COP clearly deteriorates as T_C' and T_H' are substituted for T_C and T_H in these equations. Another practical problem with this implementation is **wet compression**, which is the compression of a two-phase mixture that often results in blade damage due to impingement of liquid droplets. The turbine and compressor power being proportional to $\int_i^e vdp$ [see Eq. (4.10) and browse Anim. 4.A.*pumpVsCompressor*], the turbine output is relatively small compared to the compressor work input. Also, irreversibilities in the turbine reduce this small output even further. The vapor compression cycle overcomes these limitations by completely eliminating the turbine.

10.3.2 Vapor Compression Cycle

By replacing the turbine of the Carnot cycle with an **expansion valve**, the problems associated with the turbine can be eliminated. Also, by letting evaporation continue until the refrigerant leaves the evaporator as saturated vapor, **dry compression**, compression of superheated vapor, can be achieved (see the T-s diagram of Fig. 10.6). The resulting cycle, shown in Figure 10.6 (and Anim. 10.A.*idealVapCompCycle*), is called the **vapor-compression cycle**. It is the cycle of choice in most refrigeration and heat pump systems in use today.

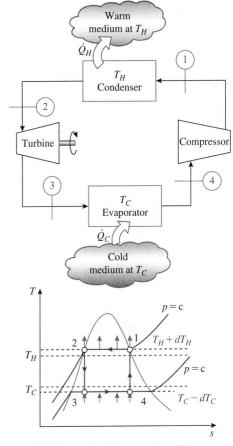

FIGURE 10.4 An implementation of the Carnot refrigeration cycle (see Anim. 10.A.*CarnotCycle*).

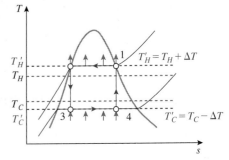

FIGURE 10.5 Effective heat transfer requires $T_H' > T_H$ in the condenser and $T_C' < T_C$ in the evaporator.

10.3.3 Analysis of an Ideal Vapor-Compression Refrigeration Cycle

In an ideal vapor-compression refrigeration cycle, shown in Figure 10.6, the internal irreversibilities are neglected in all devices except for the expansion valve, whose operation depends on the pressure drop caused by frictional losses. Pressure drop in the condenser and evaporator are assumed negligible, resulting in isothermal heat transfer in both these devices. Phase composition at the condenser and evaporator exits is assumed known, saturated liquid at the condenser exit and saturated vapor at the evaporator exit. The compressor is assumed internally reversible and adiabatic; that is, isentropic. Frictional pressure losses in the piping are assumed negligible. Also, any change in ke or pe across all devices is neglected.

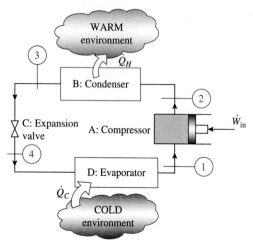

With these simplifications, an energy balance for each open-steady device in Figure 10.6 can now be carried out. In the following derivations $\dot{W}_{net}(=\dot{W}_{in})$, $\dot{Q}_H$, and $\dot{Q}_C$ are absolute values, and appropriate signs according to the WinHip sign convention (Sec. 0.7.2) are added before they are substituted in the energy equations.

Compressor (Device-A, 1-2: Isentropic compression):

$$\frac{d\cancel{E}^0}{dt} = \dot{m}j_i - \dot{m}j_e + \dot{\cancel{Q}}^0 - \dot{W}_{ext} = \dot{m}(j_1 - j_2) - (-\dot{W}_{net})$$

$$\Rightarrow \quad \dot{W}_{net} = \dot{m}(j_2 - j_1) \cong \dot{m}(h_2 - h_1); \quad [kW] \tag{10.6}$$

Condenser (Device-B, 2-3: Constant-pressure heat rejection):

$$\frac{d\cancel{E}^0}{dt} = \dot{m}j_i - \dot{m}j_e + \dot{Q} - \dot{\cancel{W}}_{ext}^0 = \dot{m}(j_2 - j_3) + (-\dot{Q}_H);$$

$$\Rightarrow \quad \dot{Q}_H = \dot{m}(j_2 - j_3) \cong \dot{m}(h_2 - h_3); \quad [kW] \tag{10.7}$$

Valve (Device-C, 3-4: Isenthalpic pressure reduction):

$$\frac{d\cancel{E}^0}{dt} = \dot{m}j_i - \dot{m}j_e + \dot{\cancel{Q}}^0 - \dot{\cancel{W}}_{ext}^0;$$

$$\Rightarrow \quad j_4 = j_3 \Rightarrow h_4 \cong h_3; \quad \left[\frac{kJ}{kg}\right] \tag{10.8}$$

Evaporator (Device D, 4-1: Constant-pressure heat addition):

$$\frac{d\cancel{E}^0}{dt} = \dot{m}j_i - \dot{m}j_e + \dot{Q} - \dot{\cancel{W}}_{ext}^0 = \dot{m}(j_4 - j_1) + \dot{Q}_C;$$

$$\Rightarrow \quad \dot{Q}_C = \dot{m}(j_1 - j_4) \cong \dot{m}(h_1 - h_4); \quad [kW] \tag{10.9}$$

The **COP** can now be expressed in terms of enthalpies as

$$COP_R = \frac{\dot{Q}_C}{\dot{W}_{net}} = \frac{h_1 - h_4}{h_2 - h_1} \tag{10.10}$$

The following example illustrates the analysis of an ideal vapor compression cycle.

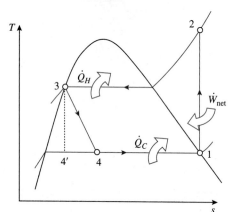

FIGURE 10.6 The vapor-compression refrigeration cycle (see Anim. 10.A.*idealVapCompCycle*).

EXAMPLE 10-1 Analysis of an Ideal Vapor Compression Cycle

An ideal vapor-compression refrigeration cycle uses R-12 as the working fluid with a mass flow rate of 0.1 kg/s. The temperature of the atmosphere, the warm region where heat rejection from the condenser takes place, is 25°C and that of the refrigerated space is −10°C. If a temperature difference of 5°C is maintained for effective heat transfer in the evaporator and condenser, determine (a) the cooling power in tons, (b) the net power input, and (c) the COP. *What-if scenario:* (d) What would the answers be if R-12 were replaced with more environmentally friendly R-134a?

SOLUTION

Evaluate the four principal states as shown on the *T-s* diagram of Figure 10.6, and perform an energy analysis of each open-steady device before calculating the cycle parameters.

Assumptions

Ideal vapor compression cycle.

Analysis

Use the manual approach or the PC flow-state TESTcalc to evaluate the enthalpy for each state as tabulated below.

State	Given	h (kJ/kg)	State	Given	h (kJ/kg)
1	$T_1 = -15°C$, $x_1 = 100\%$	180.97	2	$p_2 = p_3$, $s_2 = s_1$	205.72
3	$T_1 = 30°C$, $x_1 = 0$	64.59	4	$p_4 = p_1$, $h_4 = h_3$	64.59

The refrigeration capacity is given by

$$\dot{Q}_C = \dot{m}(h_1 - h_4) = (0.1)(180.97 - 64.59) = 11.64 \text{ kW} = 3.3 \text{ tons}$$

The net power consumption is due to the compressor and is given by

$$\dot{W}_{net} = \dot{m}(h_2 - h_1) = (0.1)(205.72 - 180.97) = 2.48 \text{ kW}$$

Therefore, the COP can be obtained as

$$\text{COP}_R = \frac{\dot{Q}_C}{\dot{W}_{net}} = \frac{11.64}{2.48} = 4.70$$

Test Analysis

Evaluate the four principal states and analyze the four devices with the PC refrigeration cycle TESTcalc as outlined in the TEST-code (see *TEST > TEST-codes*). All the cycle related variables are displayed in the cycle panel after the constituent devices are analyzed.

What-if scenario

In the state panel, change the working fluid to R-134a and click Super-Calculate. The cooling load, compressor power, and COP are recalculated as 4.19 tons, 3.2 kW, and 4.60 respectively.

Discussion

Refrigerant-12 has been found to be detrimental to the protective ozone layer of the atmosphere. Replacing R-12 with the more environmentally friendly R-134a does not seem to carry any significant penalty in terms of the COP or the cooling capacity of the refrigerator.

10.3.4 Qualitative Performance Predictions

The ideal vapor-compression refrigeration cycle is not a reversible cycle because one of its essential components, the expansion valve, is not internally reversible (see Anim. 10.A.*expansionValve*). The rest of the components, however, are internally reversible. Therefore, as shown in the *T-s* diagram of Figure 10.7, the area under the evaporation and condensation processes can still be interpreted as the heat transfer per unit mass of the refrigerant. However, the difference between these two areas, which is proportional to the net work, is no longer equal to the net area inside the loop due to entropy generation in the expansion valve. Heat transfer processes being internally reversible in this cycle, we can define effective temperatures of heat addition and heat rejection, $\bar{T}_H$ and $\bar{T}_C$, as we did in the case of the Rankine or Brayton cycle. The area under the irreversible expansion process 3-4 in Figure 10.7, which represents the degree of irreversibilities (entropy generated in the valve), is generally small compared to the cooling load, represented by the area under the evaporative process 4-1.

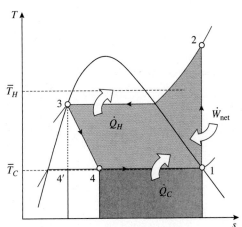

FIGURE 10.7 $\dot{W}_{net}/\dot{m} = (\dot{Q}_H - \dot{Q}_C)/\dot{m}$ is greater than the net area enclosed by the cycle due to inherent irreversibilities in the throttling process.

Hence the Carnot cycle analogy can be approximately applied to a vapor-compression cycle also. In terms of the effective temperatures, the equivalent Carnot expression $\overline{T}_C/(\overline{T}_H - \overline{T}_C)$ can be used to represent the COP of a refrigeration cycle and $\overline{T}_H/(\overline{T}_H - \overline{T}_C)$ for the COP of a heat pump. Another useful formula that helps predict the required power input, is $\dot{m}\int_i^e vdp$, the power consumption formula for an internally reversible device (Sec. 4.1.3). By realizing how a change in a parameter would affect the average specific volume in the compressor, the change in $\dot{W}_{net}$ can be predicted.

As an example of using this qualitative approach, consider how lowering the temperature of a refrigerated space would affect the net power requirement of a refrigeration cycle. As $\overline{T}_C$ decreases, $\overline{T}_H - \overline{T}_C$ increases, and the COP can be expected to decrease. Moreover, the temperature difference between the refrigerated space and the surroundings being higher, the cooling load $\dot{Q}_C$ probably goes up as the heat leakage increases. The power consumption, therefore, significantly increases if the temperature setting of a refrigerator is lowered.

EXAMPLE 10-2 **Analysis of an Ideal Vapor Compression Cycle**

Table 10-1 Parametric study with the temperature of the refrigerated space as a parameter.

T_C (°C)	COP$_R$
0	7.91
−5	6.53
−10	5.51
−15	4.70
−20	4.06
−25	3.54
−30	3.11

In Example 10-1 determine how the COP is affected if the temperature of the refrigerated space is varied within the range 0°C to –30°C.

TEST Analysis

Launch the refrigeration cycle PC TESTcalc. Copy the TEST-code of Example 10-1 (see *TEST > TEST-codes*) into the I/O panel and click the Load button to reproduce the previous solution. Change **T1** to **0 deg-C** and click Super-Calculate. The new COP can be found in the Cycle Panel as 7.91. Repeat the procedure for different values of **T1**. The resulting COP's are listed in Table 10-1.

Discussion

A reduction in the temperature of the refrigerated space clearly lowers the COP of the cycle as anticipated from the equivalent Carnot efficiency introduced earlier. For the same cooling load, the compressor power must increase in inverse proportion to the COP. Also, the mass flow rate of the refrigerant must increase to meet the cooling load $\dot{Q}_C$ as $\dot{Q}_C/\dot{m}$, the area under the evaporation process in the *T-s* diagram, decreases (Fig. 10.7). The components of a given refrigerator may not be designed to handle the higher mass flow rate. That is why the temperature of the refrigerated space cannot be arbitrarily reduced significantly below the design temperature.

10.3.5 Actual Vapor-Compression Cycle

An actual vapor-compression cycle differs from the ideal cycle in several ways (see Anim. 10.A.*actualVapCompCycle*). In an ideal cycle, the refrigerant is assumed to leave the evaporator as saturated vapor, and the condenser as saturated liquid. In practice, however, it is difficult to precisely control the quality of the working fluid at a device exit. Therefore, the refrigerant is allowed to be slightly superheated—State-1 in Figure 10.8—at the exit of the evaporator to ensure *dry compression*. The cost of ensuring dry compression is the increased compressor power, $\dot{W}_{int.rev} = -\dot{m}\int_i^e vdp$ [see Eq. (4.10)], which goes up due to an increase in the average specific volume during compression at higher temperature. At the condenser end, the refrigerant is slightly subcooled to State-4 to prevent excessive frictional pressure drop and, hence, a premature temperature drop in the piping when a two-phase mixture is allowed to leave the condenser. Subcooling also increases the cooling power as the enthalpy at the evaporator inlet $h_5 = h_4$, is slightly reduced. The downside of subcooling is a loss of COP due to the fact that the refrigerant temperature at the end of the condenser must be a few degrees higher than the temperature of the warm region T_H for effective heat transfer, therefore raising $\overline{T}_H$.

Subcooling at the end of the condenser and superheating at the end of the evaporator can be more meaningfully depicted in a *p-h* diagram of the cycle, shown

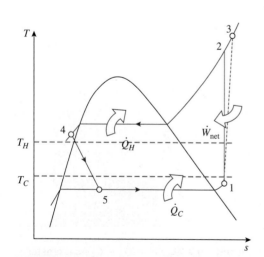

FIGURE 10.8 Actual vapor-compression cycle (see Anim. 10.A.*actualVapCompCycle*).

in Figure 10.9, because the isenthalpic expansion process becomes a simple vertical line. That is why handbooks on refrigerants often adopt the *p-h* diagram over *T-s* as a standard practice. Use the PC state TESTcalc to explore constant-entropy and constant-temperature lines on a *p-h* diagram by evaluating arbitrary states in the vapor, mixture, and liquid regions and then using the constant-property buttons.

One simple way to accomplish the superheating and subcooling simultaneously is to use the flow leaving the evaporator to exchange heat with the flow leaving the condenser in a heat exchanger, as shown in Figure 10.10 (see Anim. 10.A.*heatExchanger*).

In an actual cycle (Fig. 10.11 and Anim. 10.A.*actualVapCompCycle*), compression is not be isentropic. Effect of irreversibilities in an adiabatic compression process can be incorporated in the analysis through the isentropic compression efficiency (Sec. 4.1.4).

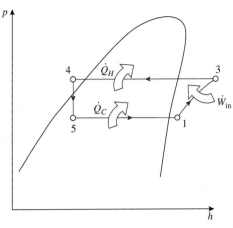

FIGURE 10.9 Actual vapor-compression cycle of Fig. 10.8 on a *p-h* diagram.

$$\eta_C = \frac{\dot{W}_{C,\text{isentropic}}}{\dot{W}_{C,\text{actual}}} = \frac{j_{es} - j_i}{j_e - j_i} \cong \frac{h_{es} - h_i}{h_e - h_i} = \frac{h_2 - h_1}{h_3 - h_1} \qquad \textbf{(10.11)}$$

Note that heat loss from a compressor is not entirely undesirable. A lower temperature of the vapor accompanies a reduced average specific volume during compression, reducing its power consumption (Sec. 4.1.3).

Additional departure from an ideal cycle occurs due to frictional pressure drop in the condenser, evaporator and piping, and stray heat loss (or gain) in the expansion valve and piping. These effects are not included in Figure 10.11 and are neglected in the subsequent analysis.

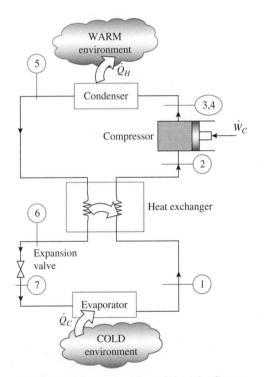

FIGURE 10.10 Superheating of the inlet flow to the compressor and subcooling of the inlet flow to the expansion valve is accomplished by a heat exchanger (see Anim. 10.A.*heatExchanger*).

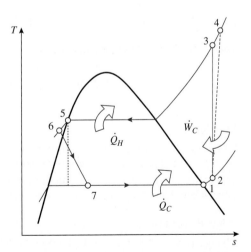

FIGURE 10.11 Actual cycle analyzed in Ex. 10-3.

EXAMPLE 10-3   Analysis of an Actual Vapor Compression Cycle

A refrigerator with a cooling capacity of 5 ton uses ammonia as the refrigerant. The condenser and evaporator maintain a pressure of 1500 kPa and 200 kPa respectively. The compressor has an isentropic efficiency of 80%. If the vapor leaving the evaporator is superheated by 5°C and the liquid leaving the condenser is supercooled by 5°C, determine (a) the compressor power, (b) the mass flow rate of ammonia, and (c) the COP. *What-if scenario:* (d) What would the COP be if the vapor were superheated 10°C above the saturation temperature at the evaporator exit?

SOLUTION

Use the PC model to evaluate the seven principal states (Fig. 10.11) and perform an energy balance for the open-steady evaporator and compressor.

Assumptions

The cycle is ideal in all respect excepts, for the irreversibilities mentioned.

Analysis

Use the manual approach or the PC flow-state TESTcalc to evaluate the enthalpy for each state as tabulated below.

State	Given	h (kJ/kg)	State	Given	h (kJ/kg)
1	$p_1, x_1 = 100\%$	$T_1 = -18.9°C$	5	$p_5 = p_3,$ $x_5 = 0$	$T_5 = 38.7°C$
2	$p_2 = p_1,$ $T_2 = T_1 + 5$	$h_2 = 1431.2$	6	$p_6 = p_3,$ $T_6 = T_5 - 5$	$h_6 = h_{f@6} + v_{f@T_6}$ $(p_6 - p_{sat@T_6})$
3	$p_3, s_3 = s_2$	$h_3 = 1741.8$			$\approx h_{f@T_6} = 340.8$
4	$p_4 = p_3,$ $h_4 = h_2 +$ $(h_3 - h_2)/\eta_C$	$h_4 = 1819.5$	7	$p_7 = p_1,$ $h_7 = h_6$	$h_7 = 340.8$

The cooling capacity supplied is 5 ton, or 5(3.516) = 17.58 kW. Steady-state energy balance on the evaporator produces

$$\dot{Q}_C = \dot{m}(h_2 - h_7); \quad \Rightarrow \quad \dot{m} = \frac{\dot{Q}_C}{(h_2 - h_7)} = \frac{17.58}{(1431.2 - 340.8)} = 0.016\,\frac{kg}{s}$$

The compressor power now can be obtained from an energy balance.

$$\dot{W}_C = \dot{m}(h_4 - h_2) = (0.016)(1819.5 - 1431.2) = 6.21 \text{ kW}$$

Therefore, the COP can be calculated as

$$\text{COP}_R = \frac{\dot{Q}_C}{\dot{W}_{net}} = \frac{\dot{Q}_C}{\dot{W}_{comp}} = \frac{17.58}{6.21} = 2.83$$

Test Analysis

Using mdot1 = 1 kg/s and expressing all other mass flow rates in terms of mdot1, evaluate the six principal states, and set up the devices (see TEST-code in *TEST > TEST-codes*). Use the evaporator energy balance to obtain the correct value of mdot1, substitute it in State-1, and click Super-Calculate to reproduce the manual solution.

What-if scenario

Change T2 to = 'T1 + 10' and click Super-Calculate. The new value of COP can be found in the cycle panel as 2.78.

Discussion

An increase in the compressor inlet temperature increases the average specific volume of vapor inside the compressor and, hence, the compressor power. The cooling capacity is also slightly increased as can be deduced from the *T-s* diagram. From the point of view of the equivalent Carnot cycle, it can be argued that the rise in $\overline{T}_H$ is more significant than the rise in $\overline{T}_C$, causing the equivalent COP, $\overline{T}_C/(\overline{T}_H - \overline{T}_C)$, to decrease.

10.3.6 Components of a Vapor-Compression Plant

Let us briefly discuss the four major components of a vapor-compression refrigeration plant.

Condenser: The condenser desuperheats and then condenses the compressed refrigerant at constant pressure by rejecting heat to the surrounding air or circulating water. Air cooled condensers are used in small units while water-cooled condensers, which operate on a smaller temperature difference, are preferred in larger units. As will be shown in Example 10-4, heat transfer across a smaller temperature difference improves the system performance significantly.

Expansion Valve: An expansion valve is a pressure reducing device that also controls the flow rate of the refrigerant in the cycle. In small refrigeration units, a simple capillary tube of a fixed size and length causes the desired pressure drop. No modification of the operating condition is possible in that case. In larger units, **throttle valves** (or thermostatic expansion valve, see Anim. 10.A.*expansionValve*) that can regulate the flow according to the load on the evaporator are used.

Compressor: When the volume flow rate at the compressor inlet is large, a **centrifugal compressor** (Sec. 8.1) is used. The rotation of its impeller exerts centrifugal force on the refrigerant raising its pressure as the refrigerant is forced against the side of the volute. A **rotary compressor**, in which a roller rotates eccentrically trapping and squeezing the vapor, is used in smaller units (see Anim. 10.A.*compressors*).

Reciprocating compressors: These use the reciprocating action of a piston to raise the pressure of the refrigerant, and are the most widely used compressors in plants up to 100 ton capacity (Fig. 10.12). Several factors—presence of clearance volume, leakage past piston and valves, and throttling effects in the intake and exhaust valves—contribute to a reduction of the actual volume of vapor that enters the compressor as compared to the volume displaced by the piston. These imperfections are accounted for by the *volumetric efficiency* of the compressor, which is defined as

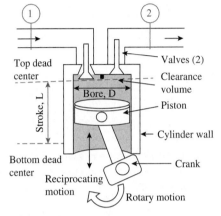

FIGURE 10.12 A reciprocating compressor (see Anim. 10.A.*compressors*).

$$\eta_v = \frac{\text{Actual volume drawn at evaporator exit state}}{\text{Piston displacement}} = \frac{\dot{m}v_1}{n_C \Psi_d N} \quad \textbf{(10.12)}$$

where $\dot{m}$ is the mass flow rate of the refrigerant, v_1 is the specific volume at the exit of the evaporator, n_C is the number of cylinders, $\Psi_d = \pi D^2 L/4$ is the piston displacement, D and L are the piston diameter and stroke of the compressor (Fig. 10.12), and N is the rotational speed in revolutions per second.

Evaporator: The most common type of evaporator is a **plate evaporator**, which is a coil brazed on a plate. In an indirect expansion coil, water or brine—for temperatures down to 0°C and –21°C respectively—are chilled in a central evaporator, and the chilled water or brine is piped to various regions meeting the distributed cooling load through heat exchangers.

10.3.7 Exergy Accounting of Vapor Compression Cycle

A complete exergy accounting of a vapor-compression cycle helps us rank different sources of irreversibilities according to their relative severity. For instance, exergy is destroyed in the expansion valve as well as in the compressor. With an exergy inventory, it is easy to compare the benefit of improving the compressor efficiency against replacing the expansion valve with a turbine.

For a single-flow device exchanging heat with a single non-atmospheric reservoir, the exergy balance equation for an open-steady device, Eq. (6.10), can be written as

$$0 = \underbrace{\dot{m}(\psi_i - \psi_e)}_{\substack{\text{Net exergy flow} \\ \text{through mass into} \\ \text{the system.}}} + \underbrace{\dot{Q}_k\left(1 - \frac{T_0}{T_k}\right)}_{\substack{\text{Net exergy transfer} \\ \text{by heat into the} \\ \text{system.}}} - \underbrace{\dot{W}_{sh}}_{\substack{\text{Exergy transfer} \\ \text{by shaft work} \\ \text{out of the system.}}} - \underbrace{\dot{I}}_{\substack{\text{Exergy} \\ \text{destruction in} \\ \text{the system's} \\ \text{universe.}}} \quad ; \quad [\text{kW}] \quad \textbf{(10.13)}$$

System boundaries that include the immediate surroundings of each system are shown in Figure 10.13. Also shown are the energy and exergy flow diagrams for the overall system. The heat transfer terms are relevant only for the condenser and evaporator. For the condenser, $T_k = T_H = T_0$ is the temperature of the warm region; therefore, heat loss into the surroundings

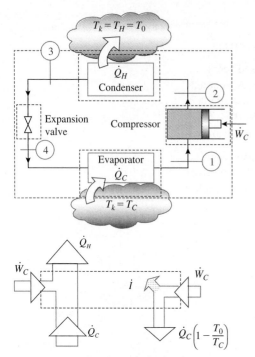

FIGURE 10.13 Energy and exergy diagrams for the vapor-compression refrigeration cycle (see Anim. 6.B.*dualSystem*).

does not transfer any exergy out of the condenser's universe. For the evaporator, $T_k = T_C$ is the temperature of the cold region, which is obviously less than T_0. Therefore, the exergy transfer by heat, the second term in Eq. (10.13), is negative; meaning the direction of exergy transfer is opposite to that of $\dot{Q}_C$ (compare the energy and exergy flow diagrams in Fig. 10.13). In other words, useful energy is delivered to the cold region by the evaporator while energy is flowing in the opposite direction as heat.

In TEST, the flow exergy ψ is evaluated as a state property once the designated dead state, State-0, is evaluated. The exergy destruction rate, $\dot{I}$, can be calculated from Eq. (10.13) after the principal states are completely evaluated. This is illustrated in the following example.

EXAMPLE 10-4 **Exergy Analysis of a Vapor-Compression Cycle**

Develop a balance sheet for exergy accounting for the cycle described in Example 10-3. Assume the atmospheric conditions to be 100 kPa, 25°C and the temperature of the refrigerated space to be –10°C.

TEST Analysis

Launch the PC refrigeration cycle TESTcalc. Copy the TEST-code from Example 10-3 into the I/O panel and click the Load button to reproduce the previous solution. Evaluate State-0 as the dead state with the given atmospheric conditions, 100 kPa and 25°C. Click Super-Calculate to evaluate the specific flow exergy of all principal states. The exergy balance equation, Eq. (10.13), can now be used to construct the following balance sheet for exergy.

Exergy supplied to the compressor: $\dot{W}_C = 6.251$ kW

Exergy gained by the refrigerant in the compressor:

$$\dot{m}(\psi_4 - \psi_2) = (0.0161)(439.6 - 105.9) = 5.374 \text{ kW}$$

Therefore, exergy destroyed in the compressor: $\dot{I} = 6.251 - 5.374 = 0.877$ kW

Exergy lost by the refrigerant in the condenser:

$$\dot{m}(\psi_4 - \psi_5) = (0.0161)(439.7 - 323.4) = 1.871 \text{ kW}$$

Exergy delivered by the condenser to the warm region, the surrounding atmosphere:

$$\dot{Q}_H\left(1 - \frac{T_0}{T_H}\right) = \dot{Q}_H\left(1 - \frac{T_0}{T_0}\right) = 0$$

Therefore, exergy destroyed in the condenser: $\dot{I} = 1.871$ kW

Exergy destroyed in the expansion valve:

$$\dot{I} = \dot{m}(\psi_6 - \psi_7) = (0.0161)(322.6 - 293.7) = 0.465 \text{ kW}$$

Exergy lost by the refrigerant in the evaporator:

$$\dot{m}(\psi_7 - \psi_2) = (0.0161)(293.7 - 105.8) = 3.02 \text{ kW}$$

Exergy delivered to the evaporator:

$$\dot{Q}_C\left(1 - \frac{T_0}{T_C}\right) = (17.58)\left(1 - \frac{298}{263}\right) = -2.34 \text{ kW}$$

The negative sign signifies that exergy flows opposite to the direction of $\dot{Q}_C$, that is, from the evaporator to the cold TER. Therefore, out of the 3.02 kW of exergy lost by the refrigerant, 2.34 kW is delivered to the cold region. The remaining amount must be the exergy that is destroyed during transfer of heat through thermal friction.

$$\dot{I} = 3.02 - 2.34 = 0.68 \text{ kW}$$

The desired exergy output of this cycle is, obviously, the exergy that flows into the cold region, which is 2.34 kW. The required exergy input being the compressor power, 6.251 kW, the exergetic efficiency can be evaluated as

$$\eta_{II} = \frac{\text{Desired Exergy Ouput}}{\text{Required Exergy Input}} = \frac{2.34}{6.251} = 36.8\%$$

Discussion

From these results, an exergy flow diagram is constructed in Figure 10.14. Notice that the highest rate of exergy destruction takes place in the condenser. Surprisingly, the expansion valve, whose operation depends on *viscous friction,* ranks at the bottom in this regard. Improving the performance of the overall system should begin at reducing the penalty associated with *thermal friction* in the condenser—temperature difference maintained for heat transfer from the condenser to the surroundings. Reducing this temperature difference through better design is therefore more prudent than replacing the valve with a turbine or even improving the compressor efficiency.

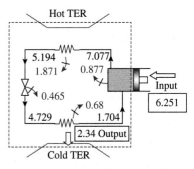

FIGURE 10.14 Detailed exergy flow diagram (all numbers in kW) for Ex.10-4.

10.3.8 Refrigerant Selection

Halogenated hydrocarbons (chlorine and fluorine, for example, are halogens), marketed under various trade names such as freon, genetron, suva, etc., are the most common refrigerants in use today. These are either methane (CH_4) based or ethane (C_2H_6) based with the hydrogen atoms replaced with chlorine and/or fluorine atoms. Methane-based refrigerants are represented by the trade symbol R$-mn$, where $m - 1$ represents the number of hydrogen atoms, n the number of fluorine atoms with the remaining (5-m-n) atoms representing the number of chlorine atoms. The chemical formula for R-12, therefore, is CCl_2F_2 (Fig. 10.15) and that of R-22 is $CHClF_2$. Ethane-based compounds are represented by the symbol R$-1mn$, where m and n carry the same meaning as before and the first number, 1 in this case, represents the number of carbon atoms minus one. R-142, thus, represents $C_2H_3ClF_2$.

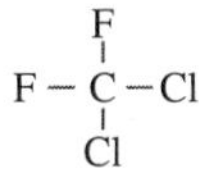

FIGURE 10.15 The chlorine atoms in CFC's such as R-12 have been linked to ozone depletion.

Choosing a refrigerant for a particular application depends on several factors. Sub-atmospheric pressure anywhere in the cycle is undesirable as any leakage becomes more difficult to locate. The lowest pressure in the cycle occurs in the evaporator. If the evaporator pressure is to be maintained above 101 kPa, the temperature in the evaporator cannot drop below $-26°C$, if the refrigerant is R-134a. Refrigerant-14, on the other hand, has a saturation temperature of $-128°C$ at atmospheric pressure and is capable of creating quite a low temperature in the cold region. Excessive pressure in the condenser is also undesirable and the selection of refrigerant must take into account if the saturation pressure at the temperature of the warm region is within an acceptable range.

Chemical stability, toxicity and corrosiveness can also be important factors in the choice of refrigerant. For instance, use of ammonia was discontinued except in *absorption chillers* (described later) due to its toxicity and flammability. Refrigerant-12 (CCl_2F_2), one of the most popular refrigerants in use until the end of the last century, has been linked to the depletion of stratospheric ozone layer that protects Earth by absorbing ultra violet rays from the sun. It is the chlorine atom in R-12 that actually attacks the ozone molecule. Under the Clean Air Act of 1990, heavily chlorinated **CFC** (chlorofluorocarbon) refrigerants have been phased out in the U.S. and replaced with **HCFC's** (hydrochlorofluorocarbon), refrigerants containing lower level of chlorine, and **HFC's** (hydrofluorocarbon), refrigerants that do not contain harmful chlorine or bromine at all. The Montreal Protocol, a landmark international agreement signed in 1987, phased out all CFC productions by 2005 and stipulated that all HCFC's be phased out by 2030. The chemical industry has responded by rapidly developing alternatives. An acceptable substitute for R-12 in use today is ethane based R-134a (CF_3CH_2F), a HCFC, which has comparable properties.

Cost is obviously a major consideration in selecting a refrigerant. One cost-effective way to develop safe refrigerants that match CFC performance and properties is by blending two or more pure refrigerants. Blends are not new; R-502 is a blend of 22/115 developed in the 1950s to improve on R-22's low-temperature performance. Blends have 400 or 500 series ASHRAE (American Society of Heating, Refrigeration, and Air-Conditioning Engineers) numbers, e.g., 401A, 404A, 409A, 507. In the TESTcalcs all blends have a % suffix, and moving the pointer over the selected working fluid displays the composition on the message panel.

10.3.9 Cascade Refrigeration Systems

The simple vapor-compression cycle discussed so far is adequate for most applications. However, if the temperature difference between the warm and cold region becomes too large, significant improvement in the system performance can be achieved by performing the refrigeration in stages in what is called a **cascading refrigeration cycle**.

A two-stage cascading cycle is shown in Figure 10.16 (and Anim. 10.A.*cascadeCycle*), which couples two simple vapor-compression cycles through a heat exchanger. In drawing the thermodynamic diagram of Figure 10.17, the working fluids of the two cycles are assumed to be identical, although that is not necessary. The evaporator of the *topping cycle* (cycle A) and the condenser of the *bottoming cycle* (cycle B) are housed in a closed type heat exchanger. With no possibility of mixing, the two fluids need not be identical and can be chosen to match the temperature requirement of the cold and warm regions.

Assuming no heat loss and neglecting any change in ke or pe, an energy balance on this device relates the mass flow rates in the two component cycles as

$$\dot{m}_A(h_2 - h_3) = \dot{m}_B(h_5 - h_8); \quad [\text{kW}]; \quad \Rightarrow \quad \dot{m}_B = \dot{m}_A \frac{h_2 - h_3}{h_5 - h_8}; \quad \left[\frac{\text{kg}}{\text{s}}\right] \qquad (10.14)$$

The COP of the cascading cycle can be calculated as

$$\text{COP}_{\text{R,cascade}} = \frac{\dot{Q}_C}{\dot{W}_{\text{net}}} = \frac{\dot{m}_A(h_1 - h_4)}{\dot{m}_A(h_2 - h_1) + \dot{m}_A(h_6 - h_5)} \qquad (10.15)$$

The *T-s* diagram shows that cooling capacity is increased due to cascading. Also, the compressor power consumption being proportional to $\int_i^e v\,dp$ (Sec. 4.1.3), the second compressor is expected to consume less power as the average specific volume is lower between states 5–6 than between states 2–6′. The COP, therefore, is expected to increase with cascading. If the thermodynamic plots are redrawn with more numbers of stages, it can be shown that the compressor work keeps decreasing as the compressor curves approach the saturated vapor line. The cooling power also keeps on increasing as the throttling curves approaches the saturated liquid line. The improvement in COP, however, follows the *law of diminishing returns* with each additional stage. Refrigeration systems with three or four stages of cascading are not uncommon.

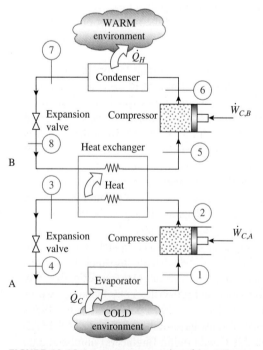

FIGURE 10.16 A two-stage cascade refrigeration cycle (see Anim. 10.A.*cascadeCycle*).

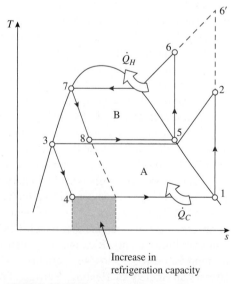

FIGURE 10.17 The two-stage cascade refrigeration cycle described in Ex. 10-5.

| **EXAMPLE 10-5** | **Analysis of a Cascade Refrigeration Cycle** |

A two-stage cascade refrigeration plant uses R-22 as the working fluid in both the stages. The lower cycle operates between the pressure limits of 120 kPa and 380 kPa and the topping cycle has a condenser pressure of 1200 kPa. The heat exchanger that couples the two cycles requires a minimum temperature difference of 5°C between the heating and the heated streams. If the mass flow rate in the lower cycle is 0.08 kg/s, determine (a) the mass flow rate in the upper cycle, (b) the cooling capacity, and (c) the COP. *What-if scenario:* (d) What would the COP be if a single cycle operated between 1200 kPa and 120 kPa?

SOLUTION

Use the PC model to evaluate the eight principal states of Figure 10.17 and perform an energy balance on the open-steady evaporator and the compressors to obtain the COP.

Assumptions

Each stage follows an ideal vapor-compression cycle.

Analysis

Use the manual approach or the PC flow-state TESTcalc to evaluate the enthalpy of each principal state.

State	Given	h (kJ/kg)	State	Given	h (kJ/kg)
1	$p_1, x_1 = 100\%$	234.5	5	$T_5 = T_3 - 5, x_1 = 100\%$	244.9
2	$p_2, s_2 = s_1$	262.0	6	$p_6, s_6 = s_5$	278.2
3	$p_3 = p_2, x_3 = 0$	35.3	7	$p_7 = p_6, x_3 = 0$	81.6
4	$p_4 = p_1, h_4 = h_3$	35.3	8	$p_8 = p_5, h_8 = h_7$	81.6

An energy balance on the heat exchanger yields

$$\dot{m}_B = \dot{m}_A \frac{h_2 - h_3}{h_5 - h_8} = (0.08)\frac{262 - 35.3}{244.9 - 81.6} = 0.11 \text{ kg/s}$$

The cooling capacity and the compressor power can be obtained as

$$\dot{Q}_C = \dot{m}_A(h_1 - h_4) = (0.08)(234.5 - 35.3) = 15.94 \text{ kW} = 4.53 \text{ ton};$$

$$\dot{W}_{net} = \dot{W}_{C,A} + \dot{W}_{C,B} = \dot{m}_A(h_2 - h_1) + \dot{m}_B(h_6 - h_5)$$

$$= (0.08)(262 - 234.5) + (0.11)(278.2 - 244.9) = 5.86 \text{ kW}$$

Therefore, $\text{COP}_{R,cascade} = \dfrac{\dot{Q}_C}{\dot{W}_{net}} = \dfrac{15.94}{5.86} = 2.72$

Test Analysis

Launch the PC refrigeration cycle TESTcalc. Evaluate the principal states and analyze each device as described in the TEST-code (see *TEST > TEST-codes*). While mdot1 through mdot4 are given, mdot5 is entered as an expression representing the energy balance on the heat exchanger. Note that mdot6 through mdot8 are equated to mdot5, a single unknown. Once all the states are calculated, click Super-Calculate button to update all mass flow rates. The manual solutions are reproduced in the cycle panel. The heat exchanger is analyzed to produce Qdot = 0, which serves as a consistency check of the overall results.

What-if scenario

To remove the topping cycle, set mdot5 to zero. Also change p2 to 1200 kPa, and click Super-Calculate. The new COP is calculated as 2.58.

Discussion

Even with the new source of exergy destruction brought about by thermal friction in the additional heat exchanger (that couples the two cycles), cascading is shown to improve the COP of the overall cycle. The electrical power consumption being directly proportional to the COP, a 4.65% improvement (from a COP of 2.58 to 2.70) may economically justify the extra equipment cost.

10.3.10 Multistage Refrigeration with Flash Chamber

The COP of a cascading refrigeration system can be improved further if the temperature difference necessary for effective heat transfer in the heat exchanger can be reduced. When the refrigerants in the component cycles are the same, the heat exchanger can be replaced by a combination of a **flash chamber** and a **mixing chamber**, also known as a **direct-contact heat exchanger** as shown in Figure 10.18 (and Anim. 10.A.*flashChamber*). The flash chamber separates the saturated mixture at State-8 into its components, saturated vapor at State-9, and saturated liquid at State-3 so that $\dot{m}_9 = x_8\dot{m}_8$ and $\dot{m}_3 = (1 - x_8)\dot{m}_8$. The direct contact between the working fluids of the two cycles eliminates the temperature difference that has to be maintained in the heat exchanger of a cascading system. State-5, however, is slightly superheated as dictated by the energy balance on the mixing chamber, which can be written as:

$$\dot{m}_5 h_5 = \dot{m}_2 h_2 + \dot{m}_9 h_9 = \dot{m}_5(1 - x_8)h_2 + \dot{m}_5 x_8 h_9; \quad [\text{kW}]$$

$$\Rightarrow \quad h_5 = (1 - x_8)h_2 + x_8 h_9 = h_2 - x_8(h_2 - h_9); \quad \left[\frac{\text{kJ}}{\text{kg}}\right] \qquad \textbf{(10.16)}$$

The two-stage compression in this cycle is analogous to multistage compression with intercooling discussed in connection with gas turbines in Sec. 8.5. Instead of circulating water, the intercooling here is accomplished by the introduction of the cooler saturated vapor into the mixing chamber, resulting in direct-contact heat exchange between the superheated vapor at State-2 and saturated vapor at State-9.

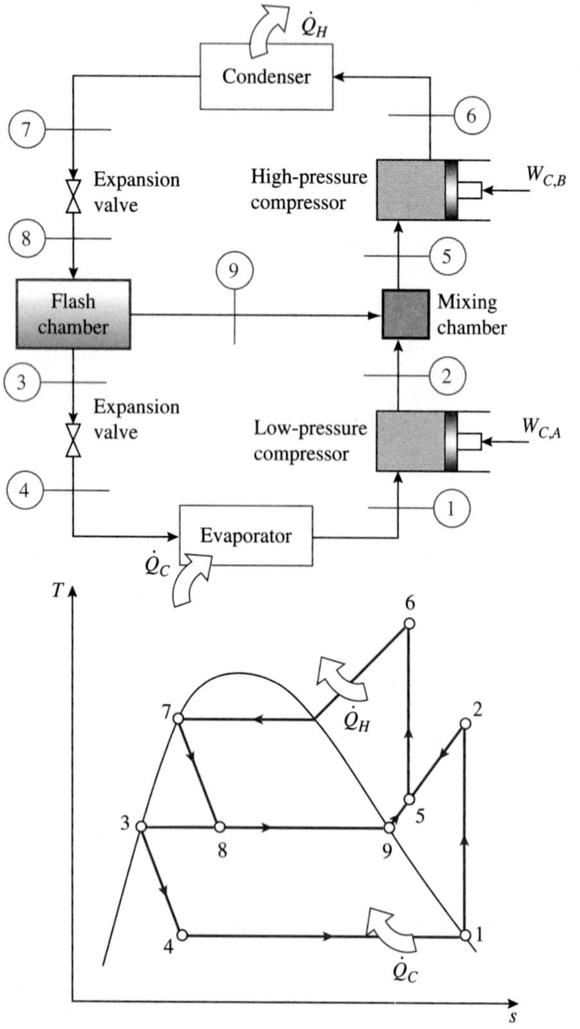

FIGURE 10.18 A two-stage refrigeration system with a flash chamber (see Anim. 10.A.*flashChamber*).

EXAMPLE 10-6 Analysis of a Refrigeration Cycle Modified with a Flash Chamber

A two-stage compression refrigeration plant uses R-22 as the working fluid and operates between the pressure limits of 120 kPa 1200 kPa, with the intermediate pressure being 380 kPa. Assuming that the cycle operates ideally as shown in Figure 10.18, determine (a) the fraction of refrigerant that flows through the evaporator, and (b) the COP. *What-if scenario:* (c) How would the COP vary if the intermediate pressure were varied through the entire possible range, from 120 kPa through 1200 kPa?

SOLUTION

Use the PC model to evaluate the nine principal states of Figure 10.18, and perform an energy balance on the open-steady evaporator and the compressors to obtain the COP.

Assumptions

Isentropic compressors, adiabatic flash chamber and mixing chamber, no pressure loss and stray heat transfer in pipes. Refrigerant leaves the evaporator as saturated vapor and the condenser as saturated liquid.

Analysis

Use the manual approach or the PC flow-state TESTcalc to evaluate the enthalpy of each principal state. Note that State-1 through State-4 and State-7 are same as in Example 10-5. The remaining states are calculated using an energy balance on the mixing chamber and assuming a mass flow rate of 1 kg/s through the condenser. However, State-8 and State-9 must be evaluated first in order to evaluate State-5 and State-6.

State	Given	h(kJ/kg)	State	Given	h(kJ/kg)
8	$p_8 = p_5, h_8 = h_7$	81.6	5	$p_5 = p_2,$ $h_5 = (1 - x_8)h_2 + x_8 h_9$	258.3
9	$p_9 = p_8, x_1 = 100\%,$ $\dot{m}_9 = x_8\dot{m}_5$	246.9	6	$p_6, s_6 = s_5$	289.3

The fraction of mass that flows through the lower stage is given as

$$\frac{\dot{m}_7}{\dot{m}_8} = 1 - x_8 = 0.78$$

The COP can be obtained as

$$\dot{Q}_C = \dot{m}_A(h_1 - h_4) = (0.08)(234.5 - 35.3) = 15.94 \text{ kW} = 4.53 \text{ ton};$$

$$\dot{W}_{net} = \dot{W}_{C,A} + \dot{W}_{C,B} = \dot{m}_A(h_2 - h_1) + \dot{m}_B(h_6 - h_5)$$

$$= (0.08)(262 - 234.5) + (0.11)(278.2 - 244.9) = 5.86 \text{ kW}$$

Therefore,

$$COP_R = \frac{\dot{Q}_C}{\dot{W}_{net}} = \frac{\dot{m}_4(h_1 - h_4)}{\dot{m}_4(h_2 - h_1) + \dot{m}_5(h_6 - h_5)} = \frac{(h_1 - h_4)}{(h_2 - h_1) + (\dot{m}_5/\dot{m}_4)(h_6 - h_5)}$$

$$= \frac{(234.5 - 35.3)}{(262 - 234.5) + (1/0.78)(289.8 - 258.7)} = 2.96$$

Test Analysis

Evaluate the principal states and analyze the evaporator and the two compressors with the PC refrigeration cycle TESTcalc, as described in the TEST-code (see *TEST > TEST-codes*). Note how the mass flow rates in the bottom cycle are expressed in terms of mdot3, and h5 is expressed through an energy balance on the mixing chamber as '= (mdot2*h2 + mdot9*h9)', given mdot5 = 1 kg/s. With many interrelated states, it is safer to use the Super-Iterate button, more than once if necessary, after clicking Super-Calculate to continue iterations, until the calculated value of COP (in the cycle panel) converges.

What-if scenario

Change p2 to a new value, click Super-Calculate and then Super-Iterate. Obtain the new COP from the cycle panel. Repeat with new value of p2. The results are tabulated in Table 10-2.

Discussion

When the intermediate pressure is equal to the maximum or minimum pressure of the cycle, the cycle reduces to a single-stage simple vapor compression cycle. This explains why the COP's at the two extremes are equal. It reaches a peak when the flash chamber pressure is about 500 kPa, which is considerably greater than $\sqrt{p_{max}p_{min}} = 379.5$ kPa; the ideal intercooling pressure for a perfect gas (see Ex. 8-7).

Table 10-2 Variation of COP with the pressure in the flash chamber.

p_2 (MPa)	COP_R
1.2	2.583
1.0	2.719
0.8	2.797
0.7	2.897
0.6	2.940
0.5	2.974
0.4	2.965
0.3	2.923
0.2	2.801
0.12	2.583

10.4 ABSORPTION REFRIGERATION CYCLE

In a vapor-compression cycle, the required input involves compression of vapor (Sec. 4.1.3), which has a relatively high specific volume. In an **absorption refrigeration cycle**, the compressor is replaced by an **absorber-generator** assembly that requires much less work.

An absorption cycle with ammonia as the refrigerant and water as the absorbent is shown in Figure 10.19 (and Anim. 10.A.*ammoniaAbsorption*). In this **aqua-ammonia** absorption system, the vapor leaving the evaporator is readily absorbed by a weak solution of ammonia in water coming from the generator. Just as condensation of vapor releases latent heat, so does absorption of vapor. Absorption capacity, however, decreases with an increase in temperature—just try opening a soda can when it is warm. To keep the temperature of the solution from rising, the heat released as ammonia dissolves in water, $\dot{Q}_A$ (proportional to *enthalpy of solution*), is rejected to

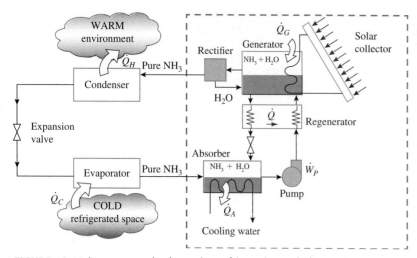

FIGURE 10.19 Aqua-ammonia absorption refrigeration cycle (see Anim. 10.A.*ammoniaAbsorption*).

the circulating cooling water. The ammonia rich solution is pumped to the generator at the condenser pressure. The pump, which handles a high-density liquid solution, requires much less work than a compressor, which has to handle low-density vapor (Sec. 4.1.3). In the generator, heat, $\dot{Q}_G$, is supplied from an external source. The boiling temperature of ammonia is much less than that of water; as a result, mostly ammonia vapor is given off by the solution. The remaining weak solution returns to the absorber after its pressure is reduced back to the absorber pressure by passing it through an expansion valve. The heat exchanger works as a regenerator by transferring heat from the weak solution to the strong solution, thereby, reducing both $\dot{Q}_G$ and $\dot{Q}_A$.

Although the generator operates at a temperature that boils ammonia off the solution, it is impossible to completely prevent water from evaporating. Water vapor, in significant amount, can form ice in the expansion valve blocking the refrigerant flow. To minimize the water vapor content, the vapor flowing out of the generator is sent through a **rectifier**, which is essentially a water-cooled heat exchanger that condenses the water vapor and returns it to the generator through a drip line.

The rest of the cycle is the same as a simple vapor-compression cycle. Ammonia vapor condenses in the condenser, rejecting the heat of condensation $\dot{Q}_H$ to the warm region. The saturated liquid is throttled by the expansion valve to the evaporator pressure, where it absorbs the cooling load $\dot{Q}_C$ from the cold region. The saturated vapor leaving the evaporator enters the absorber, thus, completing the cycle. The pumping work is often neglected in the analysis of an absorption refrigeration cycle and the COP is expressed as

$$\text{COP}_R = \frac{\text{Desired Energy Output}}{\text{Required Energy Input}} = \frac{\dot{Q}_C}{\dot{Q}_G + \dot{W}_{\text{net}}} \cong \frac{\dot{Q}_C}{\dot{Q}_G} \qquad \textbf{(10.17)}$$

The COP of an absorption refrigeration cycle is typically less than one, significantly lower than that of a vapor compression cycle. Moreover, the absorption system is much more complex, expensive, and requires more space and higher maintenance. However, the advantage of the system becomes apparent if one takes into account the quality of energy that is required as input. Instead of shaft work, much cheaper process steam, solar energy, or waste heat from other systems (such as a power plant) can be used as input. If the temperature of the external heating source can be matched (see Ex. 6-8) to the generator temperature, high exergetic efficiency can result despite a low value of the COP. Another advantage is that the vibrations and noise associated with a compressor is completely eliminated. Large air-conditioning units often employ absorption refrigeration, known as an **absorption chiller**. Absorption chillers also use water as the refrigerant with lithium bromide as the absorbent. Obviously, the evaporator temperature cannot go below 0°C in such units.

The maximum possible COP of an absorption refrigeration cycle can be obtained from an exergy analysis of an idealized system. A steady-state exergy balance for the entire system produces

$$\underbrace{\dot{Q}_G\left(1 - \frac{T_0}{T_S}\right)}_{\substack{\text{Exergy transfer by heat} \\ \text{into the generator.}}} = \underbrace{\dot{Q}_C\left(\frac{T_0 - T_C}{T_C}\right)}_{\substack{\text{Exergy transfer by heat} \\ \text{into the evaporator.}}} + \underbrace{\dot{W}_P}_{\substack{\text{Exergy transfer by} \\ \text{the pump.}}} + \underbrace{\dot{I}}_{\substack{\text{Cumulative exergy} \\ \text{destruction.}}} ; \quad [\text{kW}] \qquad \textbf{(10.18)}$$

where T_S, T_C, and T_0 are the absolute temperatures of the external heating source, cold region, and ambient atmosphere respectively. Neglecting the pump work and the cumulative exergy destruction $\dot{I}$ in the entire cycle, a maximum limit of the COP can be obtained from the above equation as

$$\text{COP}_{R,\text{max}} \cong \frac{\dot{Q}_C}{\dot{Q}_G} = \left(1 - \frac{T_0}{T_S}\right)\left(\frac{T_C}{T_0 - T_C}\right) \qquad \textbf{(10.19)}$$

It is not a coincidence that the two factors in parentheses in the above expression are the thermal efficiency of a Carnot heat engine operating between T_0 and T_S, and the COP of a Carnot refrigeration cycle operating between T_0 and T_C. The absorption-generator assembly can be looked upon as a heat engine that supplies the necessary work to run the refrigeration cycle.

10.5 GAS REFRIGERATION CYCLES

In gas refrigeration cycles, a gas such as air is the working fluid. Beside specialized applications as in **aircraft cabin cooling**, a gas refrigeration system is used for creating extremely low temperatures necessary for **liquefaction of gases**.

10.5.1 Reversed Brayton Cycle

The implementation of Carnot refrigeration cycle, discussed in Sec. 10.3.1, consists of a reversed-Rankine cycle in which the working fluid flows in the opposite direction to that in the power cycle. The vapor-compression refrigeration cycle is also, in essence, a modified reversed-Rankine cycle.

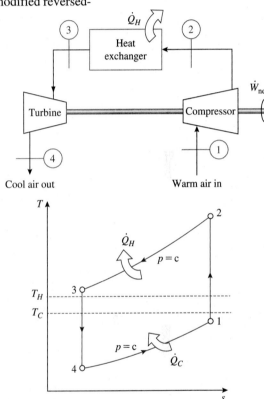

Similarly, the simple **gas refrigeration cycle** is the reverse of a **Brayton cycle** (Sec. 8.2) used for modeling gas turbines. In an ideal reversed-Brayton cycle, shown in Figure 10.20 (and Anim. 10.B.*reversedBrayton*), the gas at State-1 returning from the heat exchanger in the refrigerated space is isentropically compressed to State-2 in the compressor. The high-temperature compressed gas rejects heat at constant pressure to the warm region (usually the atmosphere) at temperature T_H. For effective heat transfer, the temperature at State-3 must be a few degrees warmer than T_H. The gas then isentropically expands in a turbine to State-4, delivering shaft work while cooling down as a result (see Eq. 3.71). The average specific volume of the gas in the turbine being smaller than that in the compressor, the turbine output must be less than the compressor input. However, unlike the vapor compression cycle, the turbine cannot be replaced with an expansion valve. Enthalpy being a function of temperature alone for an ideal gas, isenthalpic expansion (Sec. 4.2.5) in an expansion valve cannot cause any temperature change. The cold gas created at the turbine exit (State-4 in Fig. 10.20) through isentropic expansion flows through a heat exchanger in the cold region absorbing the cooling load. At the exit the gas temperature increases back to T_1, completing the cycle. Obviously, T_1 must be a few degrees cooler than T_C for effective heat removal from the cold region (see the *T-s* diagram of Fig. 10.20). If a counter-flow heat exchanger (see Anim. 4.B.*heatExchanger*) is used, T_C can be significantly lowered to a few degrees above T_4.

In the ideal cycle, the compressor and turbine are assumed isentropic, and there is no stray heat loss or pressure drop due to friction. Changes in ke and pe are also neglected across each device. Under these assumptions the cooling capacity, net power input, and COP can be derived from steady-state energy balance on the cold heat exchanger, compressor, and turbine. Referring to Figure 10.20

FIGURE 10.20 A gas refrigeration cycle (see Anim. 10.B.*reversedBrayton*) is identical to that of the Brayton cycle except for the direction of flow.

$$\dot{Q}_C = \dot{Q} = \dot{m}(j_1 - j_4) \cong \dot{m}(h_1 - h_4); \quad [\text{kW}] \qquad (10.20)$$

$$\dot{W}_{net} = \dot{W}_C - \dot{W}_T \cong \dot{m}(h_2 - h_1) - \dot{m}(h_3 - h_4); \quad [\text{kW}] \qquad (10.21)$$

$$\text{COP}_R = \frac{\dot{Q}_C}{\dot{W}_{net}} = \frac{h_1 - h_4}{(h_2 - h_1) - (h_3 - h_4)} \qquad (10.22)$$

It is left as an exercise to simplify the above expression using the PG model to obtain

$$\text{COP}_R^{PG} = \frac{T_4}{T_3 - T_4} = \frac{1}{(p_1/p_2)^{(k-1)/k} - 1} \qquad (10.23)$$

In the next example we evaluate the COP of a gas refrigeration cycle by treating the gas as an ideal gas and compare that with the prediction of Eq. (10.23).

EXAMPLE 10-7 Analysis of an Ideal Reversed-Brayton Cycle

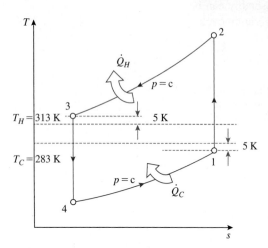

An ideal gas refrigeration, based on the reversed-Brayton cycle, is used to maintain a cold region at 10°C while rejecting heat to a warm region at 40°C. The minimum temperature difference between the working fluid, air, and the cold or warm region is 5°C. Air enters the compressor, which has a compression ratio of 3, at 101 kPa with a volumetric flow rate of 100 m³/min. Using the PG model, determine (a) the cooling capacity in ton, (b) the net power input, and (c) the COP. *What-if scenario:* (d) What would the COP be if the IG model were used?

FIGURE 10.21 *T-s* diagram for the refrigeration cycle of Ex. 10-7.

SOLUTION

Use the PG model to evaluate the temperature of each principal state. Use an energy balance on the open-steady components to evaluate the desired answers.

Assumptions

Ideal reversed-Brayton cycle assumptions.

Analysis

Use the PG model to evaluate the following states. It can be deduced from Figure 10.21 that $T_1 = 278$ K and $T_3 = 318$ K.

State-1 (given $p_1, T_1, \dot{V}_1$): $\dot{m}_1 = \dfrac{\dot{V}_1}{v_1} = \dfrac{\overline{M}p_1\dot{V}_1}{\overline{R}T_1} = \dfrac{(29)(101)(100/60)}{(8.314)(273+5)} = 2.11$ kg/s;

State-2 (given $p_2 = 3p_1, s_2 = s_1$):   Use of isentropic relation gives us $T_2 = (278)3^{(1.4-1)/1.4} = 380.8$ K;

State-3 (given $p_3 = p_2, T_3 = 318$ K);

State-4 (given $p_4 = p_3/3, s_4 = s_3$): $T_4 = (318)/3^{(1.4-1)/1.4} = 232.4$ K

An energy analysis is carried out for the following devices:

Heat Exchanger (4−1): $\dot{Q}_C = \dot{Q} = \dot{m}(h_1 - h_4) = \dot{m}c_p(T_1 - T_4)$
$$= (2.11)(1.00)(278.0 - 232.4) = 96.9 \text{ kW} = 27.5 \text{ ton}$$

Compressor (1−2): $\dot{W}_C = -\dot{W}_{\text{ext}} = \dot{m}(h_2 - h_1) = \dot{m}c_p(T_2 - T_1)$
$$= (2.11)(1.00)(380.8 - 278.0) = 217.3 \text{ kW}$$

Turbine (3−4): $\dot{W}_T = \dot{W}_{\text{ext}} = \dot{m}(h_3 - h_4) = \dot{m}c_p(T_3 - T_4)$
$$= (2.11)(1.00)(318.0 - 232.4) = 181.5 \text{ kW}$$

Therefore, the COP is obtained as:

$$\text{COP}_R = \frac{\dot{Q}_C}{\dot{W}_{\text{net}}} = \frac{96.9}{217.3 - 181.5} = 2.71$$

Also, using Eq. (10.23), we can directly obtain $\text{COP}_R^{\text{PG}} = \dfrac{1}{3^{(1.4-1)/1.4} - 1} = 2.71$

Test Analysis

Evaluate the principal states and analyze all the four devices as described in the TEST-code (*TEST > TEST-codes*) using the PG refrigeration cycle TESTcalc. Click Super-Calculate and obtain the overall results in the cycle panel.

What-if scenario

Click Super-Calculate to produce the TEST codes in the I/O panel. Launch the IG refrigeration cycle TESTcalc in a separate browser tab. Paste the TEST codes on the I/O panel of the new TESTcalc. Use the Load button and the new COP is calculated as 2.72.

Discussion

The temperature variation in this problem is not severe enough to have any significant impact on the specific heat. As a result, the PG model agrees quite well with the more accurate IG model. The large variation of temperature in the heat exchangers (as opposed to constant-temperature heat addition and rejection in Carnot cycle) explains the significantly lower COP of 2.7, compared to the Carnot COP of 9.43 for a refrigerator operating between 10°C and 40°C. Need a separation line after this discussion paragraph (as in any other example).

Unlike the vapor-compression cycle, each device in the gas refrigeration cycle is completely reversible. As a result, the net area enclosed by the loop is equal to the net specific work input, $\dot{W}_{net}/\dot{m}$. Also, the equivalent Carnot cycle analogy can be applied to predict the performance of the gas refrigeration cycle qualitatively. With $\overline{T}_H$ and $\overline{T}_C$ representing the effective temperatures of heat transfer in the warm and cold regions, the equivalent Carnot COP is given by $\overline{T}_C/(\overline{T}_H - \overline{T}_C)$. From Figure 10.22, it can be seen that $\overline{T}_C$ is much lower than the cold region temperature T_C, while $\overline{T}_H$ is much higher than the warm region temperature T_H. This explains why the COP of the gas refrigeration cycle is generally low compared to a vapor-compression cycle.

Although vapor compression cycles can deliver much better COP, with suitable modifications gas refrigeration systems can achieve a cold region temperature down to approximately −150°C, well below the capability of any vapor-compression cycle. One way to achieve low T_C is simply to raise the pressure ratio of the compressor and turbine, but that requires larger and more expensive equipment. Moreover, irreversibilities in the compressor and turbine (expressed through their isentropic efficiencies) would adversely affect the cycle COP.

A much better approach is to use a regenerative heat exchanger as shown in Figure 10.23. Instead of expanding the gas coming out of the heat exchanger at State-3, it is cooled to State-4 by the gas exiting the heat exchanger in the cold region. With a lower inlet temperature, expansion in the turbine produces much cooler exit state at State-5. Cooling takes place at much lower average temperature, from State-5 to State-6. Without the heat exchanger, the basic cycle 1-2-3-7, shown in the T-s diagram of Fig 10.23, would produce cooling from State-7 to State-1, at a much higher average temperature. Therefore, it can be concluded that the regenerative cycle has a lower COP; but this is not an appropriate comparison, because the regenerative cycle in this case produces a much colder temperature. For the same minimum temperature, it is left as an exercise to show that the basic cycle will have a much lower COP than the corresponding regenerative cycle.

An application of the gas refrigeration system for the cabin cooling of an aircraft is shown in Figure 10.24. The ambient air, at State-1, may be at a much lower pressure than the desired cabin pressure and, therefore, cannot be directly introduced into the cabin. Instead, a small amount of compressed air is extracted from the main jet engine compressor. As shown in the T-s diagram, the temperature at State-2 is much higher than the desired temperature due to isentropic compression. The extracted air is cooled by heat transfer to the ambient at constant pressure. Part of the air is then isentropically expanded to the cabin pressure through an auxiliary turbine, which supplies the necessary power to the fan. The rest of the air is bypassed through an expansion valve. Enthalpy remains constant across a valve (Sec. 4.2.5); therefore, temperature does

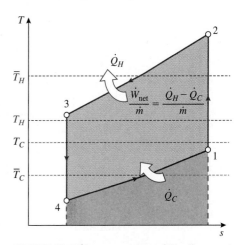

FIGURE 10.22 Interpretation of T-s diagram for an ideal gas refrigeration cycle.

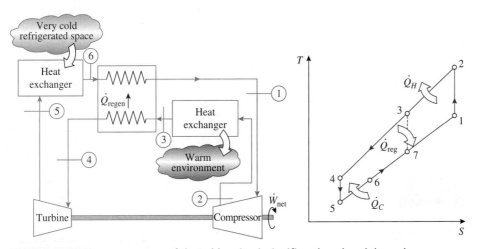

FIGURE 10.23 The temperature of the cold region is significantly reduced through regenerative heat exchange.

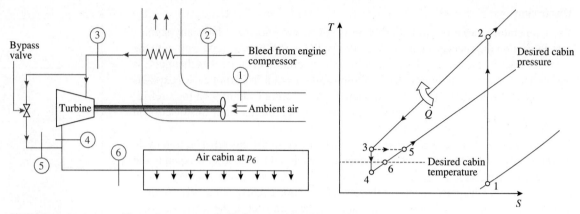

FIGURE 10.24 A gas refrigeration system applied to aircraft cabin cooling.

not change between State-3 and State-5 as $h = h(T)$ for an ideal gas. The bypass ratio is controlled so as to produce the desired temperature after mixing at State-6.

10.5.2 Linde-Hampson Cycle

In **cryogenic processes**—processes that involve temperature below $-100°C$—liquefied gases play an important role. Application of cryogenic processes includes separation of oxygen and nitrogen from air, production of liquid hydrogen as a liquid propellant, and study of superconductivity.

How can a gas be liquefied? Compressing a vapor isentropically raises its temperature and moves its state further from the dome region of the phase diagram. On the other hand, when compressed isothermally, the vapor turns into a liquid as long as its temperature is below the critical temperature. The critical temperatures of water, propane, and nitrogen, for instance, are $373°C$, $97°C$, and $-147°C$, respectively. Water exists in the liquid form at atmospheric pressure. Propane can be liquefied at $25°C$ by raising its pressure to the corresponding saturation pressure, 956 kPa. Unfortunately ordinary refrigeration techniques discussed so far cannot attain a low enough temperature to liquefy nitrogen or, for that matter, most gases.

One of the cycles used for gas liquefaction is the **Linde–Hampson Cycle** shown in Figure 10.25 (and Anim. 10.B.*gasLiquefaction*). Fresh supply of gas at State-1 is mixed with the

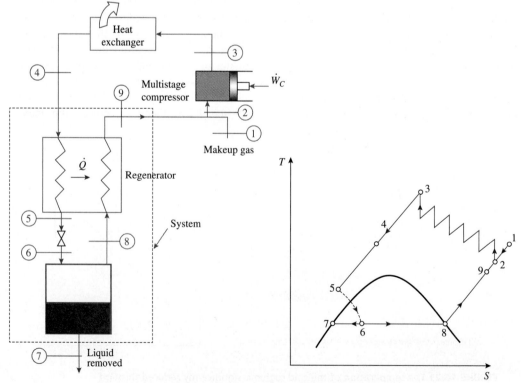

FIGURE 10.25 Linde–Hampson cycle for gas liquefaction. (see Anim. 10.B.*gasLiquefaction*)

unused gas from the cycle at State-9. The resulting mixture, at State-2, is compressed in stages with intercooling, approaching isothermal compression (see Anim. 4.A.*isothermalCompressor*) in the ideal limit, to State-3. The high pressure gas is then cooled in a regenerative counterflow heat exchanger to State-5 by the saturated vapor leaving the flash chamber. It is now throttled through an expansion valve into a two-phase mixture in a **flash chamber**, also known as a **separator**. Saturated liquefied gas at State-5 is collected as the desired product and the remaining vapor at State-8 is routed to the regenerator. Notice that, at steady state, the flow rate of makeup gas becomes equal to the rate of production of the liquefied gas.

The yield, Y, of a gas liquefaction system is defined as the ratio of the mass of liquid produced to the mass of gas compressed. The theoretical yield can be determined from an energy analysis of the control system of Figure 10.25, also called the **Joule–Kelvin refrigeration system**. At steady state,

$$\dot{m}h_4 = \dot{m}_7 h_7 + \dot{m}_9 h_9 = \dot{m}_f h_7 + (\dot{m} - \dot{m}_f)h_9; \quad [\text{kW}] \tag{10.24}$$

$$\Rightarrow \quad Y = \frac{\dot{m}_f}{\dot{m}} = \frac{h_4 - h_9}{h_7 - h_9} = \frac{h_9 - h_4}{h_9 - h_7}$$

Equation (10.24) provides the criterion for liquefaction by the Linde–Hampson cycle. State-9 and State-7 are on the same constant-pressure line (see the accompanying *T-s* diagram) with $h_9 > h_7$; therefore, for Y to be positive h_9 must be greater than h_4.

The same cycle can be used for solidification of gases, particularly for production of dry ice, by throttling State-6 down to a pressure below the triple-line pressure.

10.6 HEAT PUMP SYSTEMS

The objective of a heat pump system is to keep the warm region warm by transferring or *pumping* heat from the cold region into the warm region. As established in Section 10.3.1, the COP of the Carnot heat pump cycle is inversely proportional to the temperature difference between the warm and cold region. Using the equivalent Carnot cycle analogy, it can be said that the COP of any practical heat pump cycle must decrease with an increase in the temperature difference between the two regions. As a result, heat pump systems work best for moderate temperature difference between the warm and cold regions.

Vapor compression refrigeration cycles, as well as vapor absorption cycles, are used in heat pump systems. A typical vapor compression heat pump system for residential heating is shown in Figure 10.26. It has the same basic components and underlying cycle as the corresponding refrigeration system depicted in Figure 10.6. The main difference stems from the fact that the desired heat transfer now occurs at the condenser. Many possible sources of energy can be used to transfer heat to the refrigerant in the evaporator—the higher the source temperature, the

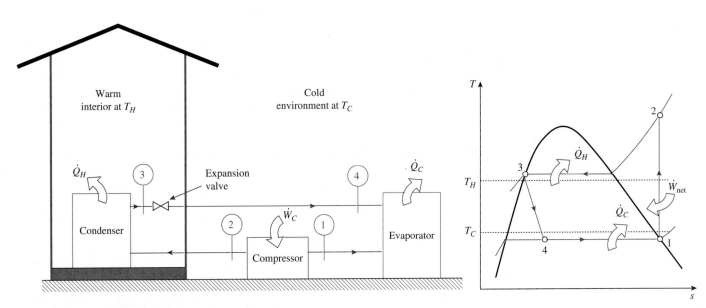

FIGURE 10.26 Air source heat pump system for residential application.

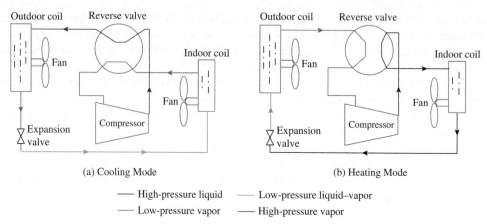

FIGURE 10.27 Dual use of a heat pump system for heating and cooling purposes (see Anim. 10.B.*dualSystem*)

better the performance. Therefore, while the outside air is the readily available thermal energy reservoir, other possibilities include the ground and water from lakes, rivers, or wells. Liquid heated by a solar collector and stored in an insulated tank can provide even better performance; however, the cost increase must be justified against the improvement in performance. Industrial heat pumps often use the available waste heat and are capable of achieving high condenser temperatures or high COP.

Heat pumps are called **air-source heat pumps** when outside air is used as the source of heat. When the outside temperature becomes too low, the COP and the heating capacity go down significantly ($\dot{Q}_H = \text{COP}_{HP}\dot{W}_C$). Supplementary heating systems such as electric resistance heaters and oil or gas furnaces are, therefore, necessary in very cold climate.

During the summer, the refrigerant flow direction can be reversed using a reversing valve, and the same hardware can be used to provide cooling, as illustrated in Figure 10.27. The condenser and the evaporator switch their roles as shown in the figure. Air-source heat pumps are cost effective in areas that have a large cooling load in the summer and a moderate heating load in the winter.

10.7 CLOSURE

Building upon the overall analysis introduced in Chapter 2, we have analyzed refrigeration and heat pump cycles in a comprehensive manner. Two types of refrigeration cycles are discussed—the vapor compression cycles where the refrigerant is alternately vaporized and condensed, and the gas cycles where the refrigerant remains as a gas throughout the cycle. The vapor cycle implementation of the Carnot refrigeration cycle is presented first, as it serves as a benchmark for qualitative prediction of performance for the subsequent cycles. Both vapor compression and vapor absorption cycles are discussed, although detailed analysis is limited to the widely used vapor compression cycles only. Two major modifications to the basic cycle—the cascade refrigeration cycle and the multistage cycle—are analyzed with examples and parametric studies using the TESTcalcs. The gas refrigeration cycle is introduced as a reversed-Brayton cycle. Closed form expression for the COP is derived using the PG model for the working gas. Application of the gas refrigeration cycle to aircraft cabin cooling and gas liquefactions is also presented. The chapter ends with a discussion of the vapor compression heat pump system. It is shown how the same hardware can be used for both cooling and heating purposes.

PROBLEMS

SECTION 10-1: BASIC AND MODIFIED VAPOR COMPRESSION CYCLES

10-1-1 A Carnot vapor refrigeration cycle uses R-134a as the working fluid. The refrigerant enters the condenser as saturated vapor at 30°C and leaves as saturated liquid. The evaporator operates at a temperature of −5°C. Determine, in kJ per kg of refrigerant flow, (a) the work input to the compressor, (b) the work developed by the turbine, (c) the heat transfer to the refrigerant passing through the evaporator and (d) the coefficient of performance of the cycle. (e) *What-if scenario:* What would the answer in part (d) and (c) be if R-134a were replaced with R-12?

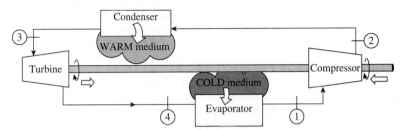

10-1-2 Refrigerant R-22 is the working fluid in a Carnot vapor refrigeration cycle for which the evapora- tor temperature is 0°C. Saturated vapor enters the condenser at 40°C, and saturated liquid exits at the same temperature. The mass flow rate of refrigerant is 4 kg/min. Determine (a) the rate of heat transfer ($\dot{Q}_C$) to the refrigerant passing through the evaporator, (b) the net power input ($\dot{W}_{net}$) (magnitude only) to the cycle in kW and (c) the coefficient of performance (COP_R). (d) *What-if scenario:* What would the answer in part (a) be if the mass flow rate were 1 kg/min?

10-1-3 A Carnot vapor refrigeration cycle operates between thermal reservoirs at 40°F and 100°F. For (a) R-12, (b) R-134a, (c) water, (d) R-22 and (e) ammonia as the working fluid, deter- mine the operating pressures in the condenser and evaporator, in lbf/in², and the coefficient of performance.

10-1-4 A Carnot vapor refrigeration cycle is used to maintain a cold region at 0°F where the ambient temperature is 75°F. Refrigerant R-134a enters the condenser as saturated vapor at 100 lbf/in² and leaves as saturated liquid at the same pressure. The evaporator pressure is 20 lbf/in². The mass flow rate of refrigerant is 12 lbm/s. Calculate (a) the compressor and turbine power, in Btu/min and (b) the coefficient of performance.

10-1-5 A steady-flow Carnot refrigeration cycle uses refrigerant-134a as the working fluid. The refriger- ant changes from saturated vapor to saturated liquid at 30°C in the condenser as it rejects heat. The evaporator pressure is 120 kPa. (a) Show the cycle on a *T-s* diagram relative to saturation lines, determine (b) the coefficient of performance (COP_R), (c) the amount of heat absorbed from the refrigerated space per kg of flow ($\dot{Q}_c$) and (d) the net work input per kg of flow ($\dot{W}_{net}$).

10-1-6 Refrigerant R-134a enters the condenser of a steady-flow Carnot refrigerator as a saturated vapor at 100 psia, and it leaves as saturated liquid. The heat absorption from the refrigerated space takes place at a pressure of 30 psia and the mass flow rate is 1 kg/s. (a) Show the cycle on a *T-s* diagram relative to saturation lines, determine (b) the coefficient of performance (COP_R), (c) the quality at the beginning of the heat-absorption process and (d) the net work input ($\dot{W}_{net}$).

10-1-7 A refrigerator uses R-12 as the working fluid operates on an ideal vapor compression refrigera- tion cycle between 0.15 MPa and 1 MPa. If the mass flow rate ($\dot{m}$) is 0.04 kg/s, determine (a) the tonnage of the system, (b) compressor power ($\dot{W}_{net}$) and (c) the COP_R. (d) *What-if scenario:* What would the COP be if R-12 were replaced with R-134a, a more environmentally benign refrigerant?

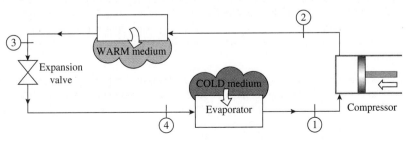

10-1-8 A refrigerator uses R-134a as the working fluid and operates on an ideal vapor-compression refrigeration cycle between 0.15 MPa and 1 MPa. For a cooling load of 10 kW, determine the mass flow rate ($\dot{m}$) of the refrigerant through the evaporator.

10-1-9 Refrigerant R-134a enters the compressor of an ideal vapor-compression refrigeration system as saturated vapor at $-10°C$ and leaves the condenser as saturated liquid at $35°C$. For a cooling capacity of 20 kW, determine (a) the mass flow rate ($\dot{m}$), (b) the compressor power ($\dot{W}_C$) in kW and (c) the coefficient of performance (COP_R).

10-1-10 A refrigerator uses R-12 as the working fluid and operates on an ideal vapor-compression refrigeration cycle between 0.15 MPa and 0.8 MPa. The mass flow rate of the refrigerant is 0.04 kg/s. (a) Show the cycle on a T-s diagram with respect to saturation lines. Determine (b) the rate of heat removal ($\dot{Q}_c$) from the refrigerated space, (c) the power input to the compressor ($\dot{W}_{net}$), (d) the rate of heat rejection ($\dot{Q}_n$) to the environment and (e) the coefficient of performance (COP_R). (f) *What-if scenario:* What would the answer in part (b) and (c) be if the mass flow rate were doubled?

10-1-11 An ideal vapor-compression refrigeration cycle operates at steady state with refrigerant R-134a as the working fluid. Saturated vapor enters the compressor at $-5°C$, and saturated liquid leaves the condenser at $35°C$. The mass flow rate ($\dot{m}$) of refrigerant is 5 kg/min. Determine (a) the compressor power ($\dot{W}_{net}$) in kW, (b) the refrigerating capacity in tons and (c) the coefficient of performance (COP_R). (d) *What-if scenario:* What would the COP be if the condenser operated at $50°C$?

10-1-12 Repeat Problem 10-1-11 by varying the condenser exit temperature from $0°C$ through $60°C$. Plot how the COP and the Carnot COP (based on maximum and minimum temperature of the cycle) vary with the condenser exit temperature.

10-1-13 A large refrigeration plant is to be maintained at $-18°C$, and it requires refrigeration at a rate of 200 kW. The condenser of the plant is to be cooled by liquid water, which experiences a temperature rise of $8°C$ as it flows over the coils of the condenser. Assuming the plant operates on the ideal vapor-compression cycle using R-134a as the working fluid between the pressure limits of 120 kPa and 700 kPa, determine (a) the mass flow rate ($\dot{m}$) of the refrigerant, (b) the power input ($\dot{W}_{net}$) to the compressor and (c) the mass flow rate ($\dot{m}$) of cooling water.

10-1-14 An ideal vapor-compression refrigeration system operates at steady state with refrigerant R-12 as the working fluid. Superheated vapor enters the compressor at 25 lbf/in², $10°F$ and saturated liquid leaves the condenser at 200 lbf/in². The refrigeration capacity is 5 tons. Determine (a) the compressor power ($\dot{W}_C$) in horsepower, (b) the rate of heat transfer ($\dot{Q}_H$) from the working fluid passing through the condenser, in Btu/min, and (c) the coefficient of performance (COP_R). (d) *What-if scenario:* What would the compressor power be if the refrigeration capacity were 10 tons?

10-1-15 Refrigerant R-12 enters the compressor of an ideal vapor-compression refrigeration system as saturated vapor at $-10°C$ with a volumetric flow rate ($\dot{V}$) of 1 m³/min. The refrigerant leaves the condenser at $35°C$ and 10 bar. Determine (a) the compressor power ($\dot{W}_{net}$), in kW, (b) the refrigerating capacity in tons and (c) the coefficient of performance (COP_R).

10-1-16 A refrigerator using R-134a as the working fluid operates on an ideal vapor compression refrigeration cycle between 0.15 MPa and 1 MPa. If the mass flow rate ($\dot{m}$) is 1 kg/s, determine (a) the net power ($\dot{W}_{net}$) necessary to run the system and (b) the COP_R. (c) *What-if scenario:* What would the COP be if the expansion valve were replaced with an isentropic turbine?

10-1-17 A vapor-compression refrigeration system, using ammonia as the working fluid, has evaporator and condenser pressures of 1 bar and 14 bar, respectively. The refrigerant passes through each heat exchanger with a negligible pressure drop. At the inlet and exit of the compressor, the temperatures are $-12°C$ and $210°C$, respectively. The heat transfer rate from the working fluid passing through the condenser is 15 kW, and liquid exits the condenser at 12 bar, $28°C$. If the compressor operates adiabatically, determine (a) the compressor power input in kW and (b) the coefficient of performance. (c) *What-if scenario:* What would the compressor power be if the condenser temperature rose to $250°C$?

10-1-18 A vapor-compression refrigeration system, with a capacity of 15 tons, has superheated refrigerant R-134a vapor entering the compressor at $15°C$, 4 bar and exiting at 12 bar. The compression process can be taken as polytropic, with $n = 1.01$. At the condenser exit, the pressure is 11.6 bar and the temperature is 44°C. The condenser is water-cooled, with water entering at 20°C and leaving at 30°C with a negligible change in pressure. Heat transfer from the outside of the condenser can be neglected. Determine (a) the compressor power input ($\dot{W}_{net}$) in kW, (b) the coefficient of performance (COP$_R$) and (c) the irreversibility rate of the condenser, in kW, for $T_0 = 20$°C.

10-1-19 An ideal vapor-compression refrigeration cycle, with ammonia as the working fluid, has an evaporator temperature of −25°C and a condenser pressure of 20 bar. Saturated vapor enters the compressor, and saturated liquid exits the condenser. The mass flow rate ($\dot{m}$) of the refrigerant is 3 kg/min. Determine (a) the coefficient of performance (COP$_R$) and (b) the refrigerating capacity, in tons. (c) **What-if scenario:** What would the COP be if the evaporator temperature were −40°C?

10-1-20 Consider a 500 kJ/min refrigeration system that operates on an ideal vapor-compression refrigeration cycle with R-134a as the working fluid. The refrigerant enters the compressor as saturated vapor at 150 kPa and is compressed to 800 kPa. (a) Show the cycle on a T-s diagram with respect to saturation lines, and determine (b) the quality of the refrigerant at the end of the throttling process, (c) the coefficient of performance (COP$_R$) and (d) the work input per unit mass of flow ($\dot{W}_{net}$) to the compressor. (e) **What-if scenario:** What would the answers be if R-12 were the working fluid?

10-1-21 Refrigerant R-12 enters the compressor of a refrigerator as superheated vapor at 0.14 MPa, −20°C at a rate of 0.04 kg/s, and leaves at 0.7 MPa, 50°C. The refrigerant is cooled in the condenser to 24°C, 0.65 MPa and is throttled to 0.15 MPa. Disregarding any heat transfer and pressure drops in the connecting lines between the components, (a) show the cycle on a T-s diagram with respect to saturation lines, determine (b) the rate of heat removal ($\dot{Q}_C$) from the refrigerated space, (c) the power input to the compressor ($\dot{W}_{net}$), (d) the isentropic efficiency of the compressor and (e) the COP of the refrigerator.

10-1-22 Refrigerant R-12 enters the compressor of a refrigerator at 140 kPa, −10°C at a rate of 0.3 m³/min and leaves at 1 MPa. The compression process is isentropic. The refrigerant enters the throttling valve at 0.95 MPa, 30°C and leaves the evaporator as saturated vapor at −18.5°C. (a) Show the cycle on a T-s diagram with respect to saturation lines, determine (b) the power input ($\dot{W}_{net}$) to the compressor, (c) the rate of heat removal from the refrigerated space, (d) the pressure drop and rate of heat gain in the line between the evaporator and compressor.

10-1-23 A vapor-compression refrigeration system for a household refrigerator has a refrigerating capacity of 1500 Btu/h and uses R-12 as the refrigerant. The refrigerant enters the evaporator at 21.422 lbf/in² and exits at 5°F. The isentropic compressor efficiency is 70%. The refrigerant condenses at 122.95 lbf/in² and exits the condenser as subcooled at 90°F. There are no significant pressure drops in the flows through the evaporator and condenser. Determine (a) the mass flow rate ($\dot{m}$) of refrigerant in lb/min, (b) the compressor power input in horsepower and (c) the coefficient of performance.

10-1-24 The refrigerator-freezer unit, shown in the schematic below, uses R-134a as the working fluid and operates on an ideal vapor-compression cycle. The temperatures in the condenser, refrigerator, and freezer are 25°C, 2°C, and −20°C, respectively. The mass flow rate of the refrigerant is 0.1 kg/s. If the refrigerant quality at the refrigerator exit is 0.4, determine the rate of heat removal from (a) the refrigerator and (b) freezer. Also, determine (c) the compressor power input and (d) the COP of the unit.

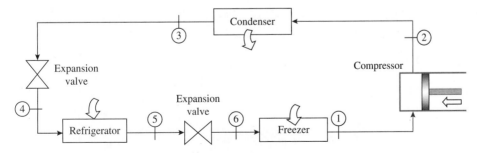

10-1-25 Refrigerant-134a is the working fluid in a vapor-compression refrigeration system with two evaporators. The system uses only one compressor. Saturated liquid leaves the condenser at 11 bar, one part of the liquid is throttled to 3 bar, the second part is throttled to the second evaporator at a temperature of $-15°C$. Vapor leaves the first evaporator as saturated vapor and is throttled to the pressure of the second evaporator. The refrigerating capacity in the first evaporator is 1 ton, in the second is 2 tons. All processes of the working fluid are internally reversible, except for the expansion through each valve. The compressor and valves operate adiabatically. Kinetic and potential energy effects are negligible. Determine (a) the mass flow rates through each evaporator, (b) the compressor power input, (c) the heat transfer from the refrigerant passing through the condenser. (d) *What-if scenario:* What would the compressor power input be if the entire cooling capacity of the first evaporator were shifted to the second?

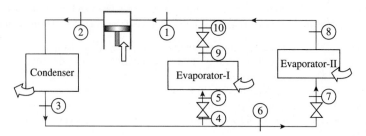

10-1-26 An ideal vapor-compression cycle uses R-134a as a working fluid and operates between the pressures of 0.1 MPa and 1.5 MPa. The refrigerant leaves the condenser at $30°C$ and the heat exchanger at $10°C$. The refrigerant is then throttled to the evaporator pressure. Refrigerant leaves the evaporator as saturated vapor and goes to the heat exchanger. The mass flow rate ($\dot{m}$) is $1\,kg/s$. Determine (a) the rate of heat removal ($\dot{Q}_C$) from the refrigerated space per unit of the mass flow and (b) the COP. (c) *What-if scenario:* What would the answers be if the heat exchanger were removed?

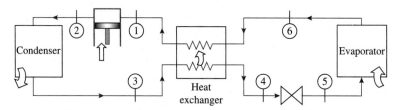

10-1-27 Repeat Problem 10-1-26 with R-12 as the working fluid.

10-1-28 Consider a two-stage R-12 refrigeration system operating between 0.15 MPa and 1 MPa. The refrigerant leaves the condenser as saturated liquid and is throttled to a flash chamber operating at 0.4 MPa. The vapor from the flash chamber is mixed with the refrigerant leaving the low-pressure compressor and the mixture is compressed by the high-pressure compressor to the condenser pressure. The liquid in the flash chamber is throttled to the evaporator pressure where the cooling load is handled through evaporation. Assuming the refrigerant leaves the evaporator as saturated vapor and both compressors are isentropic, determine (a) the fraction of refrigerant that evaporates in the flash chamber, (b) the cooling load and (c) the COP. (d) *What-if scenario:* What would the COP be if the intermediate pressure were changed to 0.8 MPa?

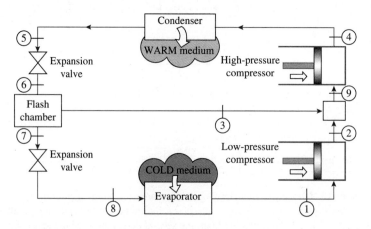

10-1-29 Repeat Problem 10-1-28 with R-134a as the working fluid.

10-1-30 Consider an ideal two-stage refrigeration system (Fig. 10-1-32) that uses R-12 as the working fluid. Saturated liquid leaves the condenser at $40°C$ and is throttled to $-20°C$. The liquid and vapor at this temperature are separated, and the liquid is throttled to the evaporator temperature at −70°C. Vapor leaving the evaporator is compressed to the saturation pressure corresponding to −20°C, after which it is mixed with the vapor leaving the flash chamber. Determine (a) the coefficient of performance (COP$_R$) of the system. (b) *What-if scenario:* What would the COP be if the flash chamber were removed with the entire flow directed to the second expansion valve?

10-1-31 Consider a two-stage compression refrigeration system operating between the pressure limits of 1.2 MPa and 0.08 MPa. The working fluid is R-12. The refrigerant leaves the condenser as saturated liquid with a mass flow rate of 1 kg/s and is throttled to a flash chamber operating at 0.4 MPa. Part of the refrigerant evaporates during this flashing process, and this vapor is mixed with the refrigerant leaving the low-pressure compressor. The mixture is then compressed to the condenser pressure by the high-pressure compressor. The liquid in the flash chamber is throttled to the evaporator pressure, and it cools the refrigerated space as it vaporizes in the evaporator. Assuming the refrigerant leaves the evaporator as saturated vapor and both compressors are isentropic, determine (a) the fraction of the refrigerant that evaporates as it is throttled to the flash chamber, (b) the amount of heat removed from the refrigerated space, (c) the compressor power and (d) the coefficient of performance. (e) *What-if scenario:* What would the answers be if R-134a were used instead?

10-1-32 A two-stage compression refrigeration system operates between the pressure limits of 1 MPa and 0.12 MPa. Refrigerant R-134a leaves the condenser as saturated liquid and is throttled to a flash chamber operating at 0.7 MPa. The refrigerant leaving the low-pressure compressor at 0.7 MPa is also routed to the flash chamber. The vapor in the flash chamber is then compressed to the condenser pressure by the high-pressure compressor, and the liquid is throttled to the evaporator pressure. Assuming the refrigerant leaves the evaporator as saturated vapor and both the compressors are isentropic, determine (a) the fraction of the refrigerant that evaporates as it is throttled to the flash chamber, (b) the rate of heat removed from the refrigerated space for a mass flow rate of 1 kg/s through the condenser and (c) the coefficient of performance. (d) *What-if scenario:* Do a parametric study of how the COP changes with the flash chamber pressure as it increases from 0.5 MPa to 0.9 MPa.

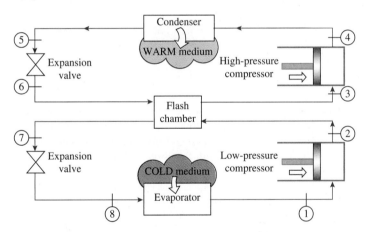

10-1-33 Consider a two-stage cascade refrigeration system operating between the pressure limits of 2 MPa and 0.05 MPa. Each stage operates on an ideal vapor-compression refrigeration cycle with R-134a as the working fluid. Heat rejection from the lower cycle to the upper cycle takes place in an adiabatic counterflow heat exchanger where both streams enter at 0.5 MPa. If the mass flow rate ($\dot{m}$) of the refrigerant through the upper cycle is 0.25 kg/s, determine (a) the mass flow rate ($\dot{m}$) of the refrigerant through the lower cycle, (b) the rate of heat removal ($\dot{Q}$) from the refrigerated space, (c) the power input ($\dot{W}_{in}$) to the compressor in the lower cycle and (d) the coefficient of performance (COP$_R$) of this cascade refrigerator. *What-if scenario:* What would the COP be if the heat exchanger pressure were (e) 0.4 MPa or (f) 0.7 MPa?

10-1-34 Consider a two-stage cascade refrigeration system operating between −80°C and 80°C. Each stage operates on an ideal vapor-compression refrigeration cycle. Upper cycle use R-12 as working fluid, lower cycle use R-13. In the lower cycle refrigerant condenses at 0°C, in the upper cycle refrigerant evaporates at −5°C. If the mass flow rate in the lower cycle is 1 kg/s, determine (a) the mass flow rate ($\dot{m}$) through the upper cycle, (b) the amount of heat removed from the refrigerated space and (c) COP. (d) *What-if scenario:* What would the COP be if we consider a one-stage ideal vapor-compression system between −80°C and 80°C with R-13 as the working fluid?

10-1-35 Consider a two-stage cascade refrigeration system operating between 0.1 MPa and 1 MPa. Each stage operates on the ideal cycle with R-134a as the working fluid. Heat rejection from the lower to the upper cycle occurs at 0.4 MPa. If the mass flow rate ($\dot{m}$) in the upper cycle is 0.1 kg/s, determine (a) the mass flow rate ($\dot{m}$) through the lower cycle and (b) the COP. (c) **What-if scenario:** What would the COP be if the intermediate pressure were changed to 0.6 MPa?

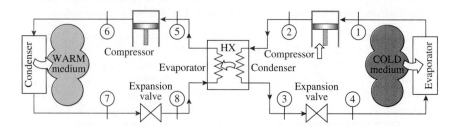

10-1-36 Consider a two-stage cascade refrigeration system operating between −60°C and 50°C. Each stage operates on an ideal vapor-compression refrigeration cycle. The upper cycle uses R-134a as working fluid, lower cycle uses R-22. In the lower cycle refrigerant condenses at 10°C, in the upper cycle refrigerant evaporates at 0°C. If the mass flow rate in the upper cycle is 0.5 kg/s, determine (a) the mass flow rate ($\dot{m}$) through the lower cycle, (b) the rate of cooling in tons, (c) the coefficient of performance (COP_R) and (d) the compressor power input ($\dot{W}_{net}$) in kW.

SECTION 10-2: GAS REFRIGERATION CYCLES

10-2-1 In a gas refrigeration system air enters the compressor at 10°C, 50 kPa and the turbine at 50°C, 250 kPa. The mass flow rate ($\dot{m}$) is 0.08 kg/s. Assuming variable specific heat, determine (a) the rate of cooling, (b) the net power input ($\dot{W}_{net}$) and (c) the COP. (d) **What-if scenario:** What would the COP be if the compressor inlet temperature were 15°C?

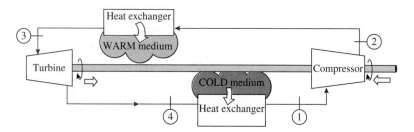

10-2-2 In an ideal Brayton refrigeration cycle air enters the compressor at 100 kPa and 300 K. The compression ratio is 4, and air enters the turbine inlet at 350 K. The mass flow rate ($\dot{m}$) of air is 0.05 kg/s. Determine (a) the rate of cooling, (b) the net power input ($\dot{W}_{net}$), (c) the COP, and (d) the Carnot COP. (e) **What-if scenario:** What would the answers be if air enters the turbine inlet at 370 K instead?

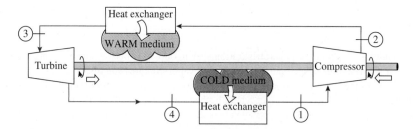

10-2-3 Air enters the compressor of a perfect-gas refrigeration cycle at 45°F, 10 psia and the turbine at 120°F, 30 psia. The mass flow rate ($\dot{m}$) of air through the cycle is 0.5 lbm/s. Determine (a) the rate of refrigeration, (b) the net power input ($\dot{W}_{ext}$) and (c) the coefficient of performance (COP_R).

10-2-4 Air enters the compressor of a perfect-gas refrigeration cycle at 15°C, 50 kPa and the turbine at 50°C, 300 kPa. The mass flow rate ($\dot{m}$) through the cycle is 0.25 kg/s. Assuming constant specific heats for air (PG model), determine (a) the rate of refrigeration, (b) the net power input ($\dot{W}_{net}$) and (c) the coefficient of performance (COP_R). (d) **What-if scenario:** What would the COP be if the IG model were used for air?

10-2-5 Air enters the compressor of an ideal Brayton refrigeration cycle at 200 kPa, 270 K, with a volumetric flow rate ($\dot{V}$) of 1 m³/s, and is compressed to 600 kPa. The temperature at the turbine inlet is 330 K. Treating air as a perfect gas, determine (a) the net power input ($\dot{W}_{net}$) in kW, (b) the refrigeration capacity in kW and tons and (c) the coefficient of performance (COP_R). (d) *What-if scenario:* What would the COP be if a reversible cycle could be operated between the highest and lowest temperatures of the cycle?

10-2-6 Repeat Problem 10-2-5 if the compressor and turbine have an isentropic efficiency of 80%.

10-2-7 An ideal-gas refrigeration cycle uses air as the working fluid to maintain a refrigerated space at −30°C while rejecting heat to the surrounding medium at 30°C. If the pressure ratio of the compressor is 4, determine (a) the maximum and minimum temperatures in the cycle, (b) the coefficient of performance (COP_R) and (c) the rate of refrigeration for a mass flow rate of 0.05 kg/s. (d) *What-if scenario:* What would the COP be if the PG model were used for air?

10-2-8 In an ideal Brayton refrigeration cycle air enters the compressor at 18 lbf/in² and 400°R. The compression ratio is 5, and air enters the turbine inlet at 600°R. The mass flow rate of air is 2 lb/min. Determine (a) the refrigeration capacity in tons, (b) the net power input in Btu/min, and (c) the COP. (d) *What-if scenario:* What would the refrigeration capacity be if air enters the turbine inlet at 700°R instead?

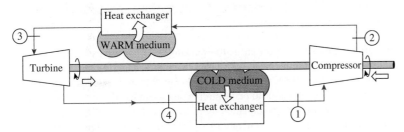

10-2-9 An ideal Brayton refrigeration cycle has a compressor pressure ratio of 6. At the compressor inlet, the pressure and temperature of the entering air are 55 lbf/in² and 600°R. The temperature at the exit of the turbine is 370°R. For a refrigerating capacity of 15 tons, determine (a) the net power input ($\dot{W}_{net}$), in Btu/min, (b) the coefficient of performance (COP_R) and (c) the specific volumes of the air at the compressor and turbine inlets, each in ft³/lb.

10-2-10 Air enters the compressor of an ideal Brayton refrigeration cycle at 120 kPa and 275 K. The compressor pressure ratio is 3, and the temperature at the turbine inlet is 325 K. Treating air as a perfect gas, determine (a) the net work input per unit mass of air flow (w_{net}), in kJ/kg, (b) the refrigeration capacity per unit mass of air flow, in kJ/kg and (c) the coefficient of performance (COP_R). (d) *What-if scenario:* What would the COP be if the IG model were used for air?

10-2-11 A gas refrigeration system uses helium as the working fluid operates with a pressure ratio of 3.5. The temperature of the helium is −10°C at the compressor inlet and 50°C at the turbine inlet. Assuming an adiabatic efficiency of 80% for both the compressor and the turbine, determine (a) the minimum temperature of the cycle, (b) mass flow rate ($\dot{m}$) for a refrigeration rate of 1 ton and (c) the COP. (d) *What-if scenario:* What would the COP be if the adiabatic efficiency increased to 85%?

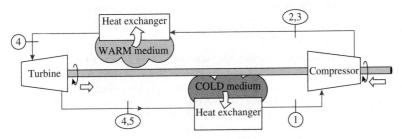

10-2-12 In Problem 10-2-9 consider that the compressor and turbine each has an isentropic efficiency of 85%. Determine for the modified cycle, (a) the mass flow rate ($\dot{m}$) of air, in lb/s and (b) the coefficient of performance (COP_R). (c) *What-if scenario:* Do a parametric study of how the COP would change if the isentropic efficiency varied from 50% to 100%.

10-2-13 In Problem 10-2-10 consider that the compressor and turbine have isentropic efficiencies of 80% and 90%, respectively. Determine for the modified cycle, (a) the coefficient of performance and (b) the irreversibility rates, per unit mass of air flow, in the compressor and turbine, each in kJ/kg, for $T_0 = 300$ K.

10-2-14 A gas refrigeration cycle with a pressure ratio of 3 uses helium as the working fluid. The temperature of the helium is $-15°C$ at the compressor inlet at $50°C$ at the turbine inlet. Assuming adiabatic efficiencies of 85% for both the turbine and the compressor, determine (a) the minimum temperature in the cycle, (b) the coefficient of performance (COP$_R$) and (c) the mass flow rate ($\dot{m}$) of the helium for a refrigeration rate of 10 kW. (d) **What-if scenario:** What would the COP be if the turbine inlet temperature were increased to $60°C$?

10-2-15 An ideal gas refrigeration system with a regenerative HX uses air as the working fluid. Air enters the compressor at 270 K, 150 kPa at a flow rate of 0.1 m³/min and exits at 400 kPa. Compressed air enters the regenerative HX at 300 K and is cooled down to 270 K at the turbine inlet. Determine (a) the cooling capacity, (b) the net power input ($\dot{W}_{net}$), and (c) the COP.

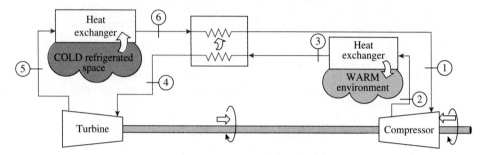

10-2-16 A gas refrigeration system uses air as the working fluid has a pressure ratio of 4. Air enters the compressor at $-7°C$. The high-pressure air is cooled to $30°C$ by rejecting heat to the surroundings. It is further cooled to $-15°C$ by regenerative cooling before it enters the turbine. Assuming both the turbine and the compressor to be isentropic and using the PG model for air, determine (a) the lowest temperature that can be obtained by this cycle, (b) the coefficient of performance (COP$_R$) of the cycle and (c) the mass flow rate ($\dot{m}$) of air for a refrigeration rate of 12 kW. (d) **What-if scenario:** What would the COP be if the pressure ratio were 5?

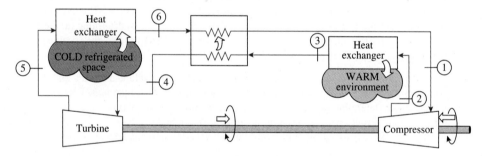

10-2-17 In Problem 10-2-16 evaluate the effect of regeneration on the COP by changing the turbine inlet temperature to (a) $-10°C$, (b) $0°C$, (c) $10°C$ and (d) $20°C$.

10-2-18 Helium undergoes a Stirling refrigeration cycle, which is a reverse Stirling power cycle. At the beginning of isothermal compression helium is at 100 kPa, 275 K. The compression ratio is 4 and during isothermal expansion the temperature is 150 K. Determine per kg of helium, (a) the net work per cycle, (b) the heat transfer during isothermal expansion, and (c) the COP.

SECTION 10-3: HEAT PUMP SYSTEMS

10-3-1 A heat pump which operates on an ideal vapor-compression cycle with R-12 is used to heat water from $5°C$ to $30°C$ at a mass flow rate ($\dot{m}$) of 0.2 kg/s. The condenser and evaporator pressures are 0.8 MPa and 0.2 MPa, respectively. Determine (a) the power input ($\dot{W}_{net}$) to the heat pump and (b) the COP of the heat pump. (c) **What-if scenario:** Could we heat the water if R-12 were replaced with R-22 (Answer 1 if yes and 2 if no.)?

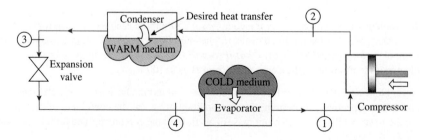

10-3-2 Ammonia is the working fluid in a vapor-compression heat pump system with a heating capacity of 25,000 Btu/h. The condenser operates at 250 lbf/in^2, and the evaporator temperature is $-10°F$. The refrigerant is a saturated vapor at the evaporator exit and a liquid is 100°F at the condenser exit. Pressure drops in the flows through the evaporator and condenser are negligible. The compression process is adiabatic, and the temperature at the compressor exit is 400°F. Determine (a) the mass flow rate ($\dot{m}$) of refrigerant, in lb/min, (b) the compressor power input ($\dot{W}_{net}$) in horsepower, (c) the isentropic compressor efficiency and (d) the coefficient of performance (COP$_{HP}$).

10-3-3 An ideal vapor-compression heat pump cycle with refrigerant R-134a as the working fluid provides 10 kW to maintain a building at 22°C when the outside temperature is 5°C. The refrigerant leaves the condenser as saturated liquid. Calculate (a) the power input ($\dot{W}_{net}$) to the compressor in kW, (b) the coefficient of performance and (c) the coefficient of performance of a reversible heat pump cycle operating between thermal reservoirs at 22°C and 5°C. (d) *What-if scenario:* What would the answers in part (b) and (c) be if the outside temperature were at 0°C?

10-3-4 A heat pump which operates on the ideal vapor-compression cycle with R-12 is used to heat a house and maintain it at 20°C using underground water at 25°C as the heat source. The house is losing heat at a rate of 90,000 kJ/h. The evaporator and condenser pressures are 0.35 MPa and 0.8 MPa, respectively. Determine (a) the power input ($\dot{W}_{net}$) to the heat pump. (b) *What-if scenario:* If an electric resistance heater is used instead of a heat pump, calculate the increase in electric power input.

10-3-5 A heat pump with a heating capacity of 8 kW uses R-134a as the working fluid in an ideal vapor-compression cycle. The refrigerant leaves the condenser at 900 kPa, 26°C and the evaporator at 200 kPa, $-12°C$. Determine (a) the mass flow rate of refrigerant, (b) the power input to the compressor, (c) the COP, and (d) the Carnot COP.

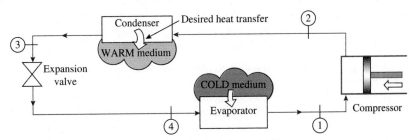

10-3-6 A heat pump operates on a vapor-compression cycle with R-134a as the working fluid has a heating capacity of 40,000 Btu/h. The refrigerant leaves the compressor at 160 lbf/in^2, 160°F. Saturated liquid leaves the condenser at 90°F. The evaporator operates at 4°F and saturated vapor leaves the evaporator. Assume adiabatic compression. Determine the (a) mass flow rate of R-134a in lb/min, (b) power input to the compressor in hp, (c) compressor isentropic efficiency, and (d) COP of the system.

10-3-7 A vapor-compression heat pump with a heating capacity of 500 kJ/min is driven by a power cycle with a thermal efficiency of 25%. For the heat pump, R-134a is compressed from saturated vapor at $-10°C$ to the condenser pressure of 10 bar. The isentropic compressor efficiency of 85%. Liquid enters the expansion valve at 9.6 bar and 34°C. For the power cycle, 90% of the heat rejected is transferred to the heated space. Determine (a) the power input ($\dot{W}_{net}$) to the heat pump compressor in kW and (b) evaluate the ratio of the total rate at which heat is delivered to the heated space to the rate of heat input to the power cycle.

10-3-8 The refrigerant R-22 is used as the working fluid in a conventional heat pump cycle. Saturated vapor enters the compressor of this unit at 15°C and its exit temperature from the compressor is 90°C. If the isentropic efficiency of the compressor is 75%, (a) what is the coefficient of performance (COP$_{HP}$) of the heating pump? (b) *What-if scenario:* What would the coefficient of performance be if the the isentropic compressor efficiency changed to 65%?

10-3-9 A heat pump uses R-12 to heat a house by using underground water at 8°C as the heat source. The house is losing heat at a rate of 30,000 kJ/h. The refrigerant enters the compressor at 250 kPa, $-2°C$ and it leaves at 1.2 MPa, 70°C. The refrigerant leaves the condenser at 30°C. Determine (a) the power input ($\dot{W}_{net}$) to the heat pump and (b) the rate of heat absorption from the water. (c) *What-if scenario:* If an electric resistance heater were used instead of heat pump, calculate the increase in electric power input.

10-3-10 A heat pump which operates on the ideal vapor-compression cycle with R-134a is used to heat the house and maintain it at 25°C while the outside temperature is 5°C. The house is losing heat at a rate of 10 kW. The evaporator and condenser pressures are 300 kPa and 1 MPa, respectively. Determine (a) the mass flow rate of refrigerant, (b) the power input to the compressor, and (c) the COP. (d) *What-if scenario:* What would the answers be if the house were losing heat at a rate of 20 kW instead?

10-3-11 In an actual refrigeration cycle, a heat pump uses R-12 as the working fluid at mass flow rate ($\dot{m}$) of 0.1 kg/s. Vapor enters the compressor at 200 kPa, −5°C, and leaves at 1.5 MPa, 90°C. The power input ($\dot{W}_{net}$) to the compressor is 2.6 kW. The refrigerant enters the expansion valve at 1.2 MPa, 40°C and leaves the evaporator at 200 kPa, −12°C. Determine (a) the irreversibility during the compression process, (b) the heating capacity and (c) COP of the heating pump. Assume atmospheric temperature to be 298 K.

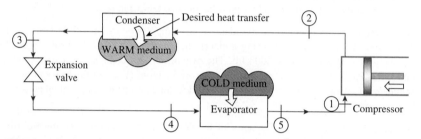

10-3-12 A vapor-compression heat pump system uses refrigerant R-134a as the working fluid. The refrigerant enters the compressor at 2.4 bar, 0°C and a volumetric flow rate of 0.8 m³/min. Compression is adiabatic to 10 bar, 50°C and saturated liquid exits the condenser at 9 bar. Determine (a) the power input ($\dot{W}_{net}$) to the compressor in kW, (b) the heating capacity of the system in kW and tons, (c) the coefficient of performance (COP_{HP}) and (d) the isentropic compressor efficiency. (e) *What-if scenario:* What would the COP be if the refrigerant temperature at the compressor exit were 70°C?

10-3-13 A vapor-compression heat pump uses R-134a as the working fluid has a heating capacity of 12 kW. Refrigerant enters the compressor at −15°C, 1.4 bar and exits at 8 bar. The isentropic compressor efficiency is 72%. There are no significant pressure drops as the refrigerant flows through the condenser and evaporator. The refrigerant leaves the condenser at 8 bar and 22°C. Ignoring the heat transfer between the compressor and its surroundings, determine (a) the mass flow rate of R-134a, (b) the COP and (c) the irreversibility rates of the compressor and expansion valve in kW for an ambient temperature of 20°C.

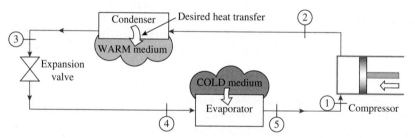

SECTION 10-4: EXERGY ANALYSIS OF REFRIGERATION AND HEAT PUMP CYCLES

10-4-1 A refrigerator uses R-134a as the working fluid operates on an ideal vapor compression refrigeration cycle between 0.15 MPa and 1 MPa. A temperature difference of 5°C is maintained for effective heat exchange between the refrigerant and its surroundings at the evaporator and condenser. The atmospheric conditions are 100 kPa and 25°C. If the mass flow rate ($\dot{m}$) is 0.04 kg/s, (a) perform an exergy inventory on a rate (kW) basis for the entire cycle complete with an exergy flow diagram. Determine (b) the exergetic efficiency and (c) COP of the system. (d) Identify the device with the highest rate of exergy destruction.

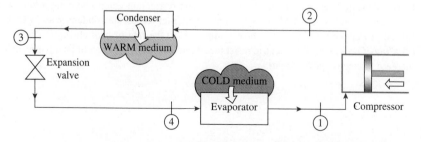

10-4-2 A vapor-compression refrigeration system circulates R-134a at a rate ($\dot{m}$) of 10 kg/min. The refrigerant enters the compressor at $-10°C$, 1.2 bar and exits at 7 bar. The isentropic compressor efficiency is 68%. There are no significant pressure drops as the refrigerant flows through the condenser and evaporator. The refrigerant leaves the condenser at 7 bar and 24°C. Ignoring the heat transfer between the compressor and its surroundings, determine (a) the coefficient of performance (COP$_R$), (b) the refrigerating capacity in tons and (c) the irreversibility rates of the compressor and expansion valve each in kW for an ambient temperature of 20°C.

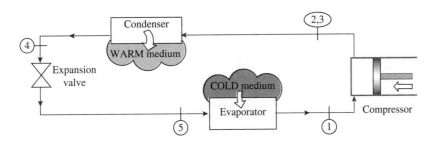

10-4-3 A vapor-compression refrigeration system circulates R-134a at a rate ($\dot{m}$) of 10 kg/min. The refrigerant enters the compressor at $-10°C$, 1.2 bar and exits at 7 bar. The isentropic efficiency of the adiabatic compressor is 68%. There are no significant pressure drops as the refrigerant flows through the condenser and evaporator. The refrigerant leaves the condenser at 7 bar and 24°C. A temperature difference of 5°C is maintained for effective heat exchange between the refrigerant and its surroundings at the evaporator and condenser. The atmospheric conditions are 100 kPa and 25°C. (a) Perform an exergy inventory on a rate (kW) basis for the entire cycle complete with an exergy flow diagram. Determine (b) the exergetic efficiency and (c) COP of the system. (d) Identify the device with the highest rate of exergy destruction.

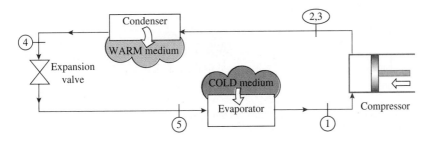

10-4-4 A heat pump which operates on the vapor-compression cycle uses R-134a as the working fluid. Refrigerant enters the compressor at 20 lbf/in^2, 10°F and is compressed adiabatically to 200 lbf/in^2, 180°F. Saturated liquid enters the expansion valve at 200 lbf/in^2, 100°F and exits at 20 lbf/in^2. The atmosphere temperature is 20°F. (a) Perform an exergy inventory on a rate (Btu/min) basis for the entire cycle complete with an exergy flow diagram. Determine (b) the exergetic efficiency and (c) COP of the system. (d) Identify the device with the highest rate of exergy destruction.

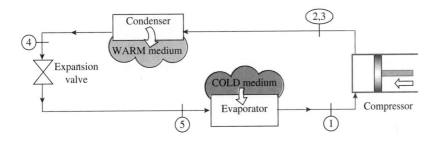

10-4-5   An ideal vapor-compression cycle uses R-134a as a working fluid and operates between 0.1 MPa and 1.5 MPa. The refrigerant leaves the condenser at 30°C and the heat exchanger at 10°C. The refrigerant is then throttled to the evaporator pressure. Refrigerant leaves the evaporator as a saturated vapor and goes to the heat exchanger. The mass flow rate ($\dot{m}$) is 1 kg/s. A temperature difference of 5°C is maintained for effective heat exchange between the refrigerant and its surroundings at the evaporator and condenser. The atmospheric conditions are 100 kPa and 25°C. (a) Perform an exergy inventory on a rate (kW) basis for the entire cycle complete with an exergy flow diagram. Determine (b) the cooling capacity, (c) exergetic efficiency and (d) COP of the system.

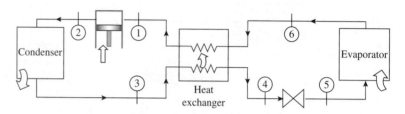

10-4-6   Repeat Problem 10-4-5 with the heat exchanger removed.

10-4-7   A heat pump which operates on the ideal vapor-compression cycle with R-134a is used to transfer heat at a rate of 20 kW to a space maintained at 50°C from outside atmosphere at 0°C. A temperature difference of 5°C is maintained for effective heat exchange between the refrigerant and its surroundings at the evaporator and condenser. The atmospheric conditions are 100 kPa and 0°C. (a) Perform an exergy inventory on a rate (kW) basis for the entire cycle complete with an exergy flow diagram. Determine (b) the power consumption rate ($\dot{W}_{net}$), (c) the exergetic efficiency and (d) the COP of the system.

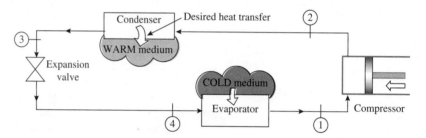

10-4-8   Repeat Problem 10-4-7 with the outside atmosphere at 100 kPa and −5°C.

11 EVALUATION OF PROPERTIES: THERMODYNAMIC RELATIONS

This chapter can be regarded as a continuation of Chapter 3, where we developed a number of material models for evaluation of thermodynamic states of various working substances. Central to this development were *thermodynamic relations* such as the *Tds* equations and state postulate which were stated without proof. We begin this chapter by establishing these fundamental relations from the first principle. Application of a few well-known theorems of differential calculus leads to the development of Maxwell relations, which are frequently used in manipulating thermodynamic relations and expressing them in terms of readily measurable properties, such as pressure, temperature, and specific volume. The relations developed in this chapter will be used in completing the real gas model, briefly introduced in Chapter 3. We also introduce mixture models—PG mixture, IG mixture, and RG mixture—including a new set of properties called partial properties. These models will be used in the subsequent chapters in connection with psychrometry, combustion, and chemical equilibrium.

11.1 THERMODYNAMIC RELATIONS

In Chapter 3, we introduced several thermodynamic relations, including the *Tds* equations and differential relations for *u* and *h*. These relations were at the heart of development of different material models such as the SL, PC, PG, IG, and RG models. For any given model, a procedure was developed to calculate a complete thermodynamic state from any pair of independent thermodynamic properties. In this section, we will explore the root of those relations and develop a framework for thermodynamic relations, which can be used for state evaluation and developing property tables.

11.1.1 The *Tds* Relations

The importance of the *Tds* relations in thermodynamics cannot be overstated. Introduced in Sec. 3.1.3, and derived in Sec. 5.3 with the help of a reversible differential process (see Anim. 5.G.*TdsEquation*), the two *Tds* relations provided the fundamental relations behind developing entropy expressions for different material models. Although derived by analyzing a reversible differential process, these relations hold true for irreversible processes too, because the end states are in equilibrium and have no knowledge of the processes bridging them. Given their importance, let us examine these relations one more time, this time in a more general setting.

When a large system such as the gas trapped in the piston-cylinder device of Figure 11.1 goes from one equilibrium state to a neighboring one due to differential interactions, the end states are related by the *Tds* equations. Now consider the irreversible expansion of Figure 11.2 where a large pressure difference drives the piston after the pin keeping it in place is suddenly removed. Obviously, the overall system cannot be in equilibrium during the unrestrained expansion. However, any pair of adjacent local systems, for instance local systems A and B in Figure 11.2, can be considered in LTE (local thermodynamic equilibrium) at a given time. These states are neighbors not only at a given instant, but also between time frames, as long as the difference in properties is infinitesimal between any two states. By the LTE hypothesis (Sec. 3.1.1), *Tds* equations can be applied between neighboring local states in open or closed systems, steady or unsteady systems, uniform or non-uniform systems, reversible, internally reversible or irreversible systems, accelerating or decelerating systems, in the presence or absence of gravity, even states separated by time, space, or both. The only restrictions are that the local system must be *simply compressible,* that is, changes

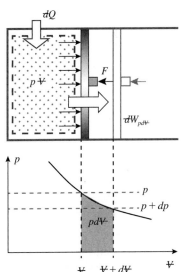

FIGURE 11.1 A simple system which allows heat and boundary work transfer. For an irreversible process, $dW_{pdV} < pdV$ (see Anim. 5.G.*TdsEquation*).

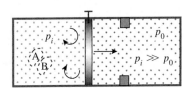

FIGURE 11.2 The two local systems, A and B, can be in LTE while the system expands irreversibly after the pin holding the piston (internal constraint) is removed.

between the two neighboring states can be brought about only through heat and boundary work and cannot involve any change in chemical composition.

A general proof of such versatile relations, which encompass so many diverse situations, may seem to be a daunting task. However, the hypothesis of local thermodynamic equilibrium (LTE) can be effectively used to simplify our task. According to this hypothesis, a *local system* in thermodynamic equilibrium at a given *thermodynamic instant* can be treated like a closed system within the time of interest, a macroscopic instant (which is much smaller than typical mixing time). A local system, by definition, is uniform—so, its thermodynamic properties are independent of its size. The system, therefore, can be scaled up to any size. Another simplification stems from the assumption that the system is *simply compressible*—that is, the system interacts with the surroundings only through heat and boundary work transfer, the latter consisting of $pd\kern-0.5ex\forall$ work only (Sec. 5.3). But what about shaft and electrical work transfers? If a large system undergoes differential change in equilibrium due to electrical or shaft work transfer, the global system is not *simply compressible*, and the *Tds* equations cannot be applied between the two equilibrium states. However this is not a big restriction as far as local systems are concerned, where boundary work is the only mode of work transfer (Sec. 3.1.2). Finally, the working fluid is assumed to be a *pure substance*, that is, the chemical composition is assumed to be uniform and invariant. When two different gases mix with each other or chemically react, the composition will vary with location and time between two neighboring states. Therefore, even if the states are considered to be in LTE at all times, the *Tds* equations in their current form cannot be applied across such states.

The simplifications described above lead us to construct a simple system in Figure 11.1—we select the familiar piston-cylinder device as it allows only heat and boundary work transfer with the surroundings—composed of a pure substance, say, a gas or a vapor. The system, identified by the red dashed boundary in Figure 11.1, is assumed to be in a certain equilibrium state, State-1, described by the properties T, p, $\forall$, U, H, and S.

Now consider a differential process in which dQ, amount of heat, and dW_B, amount of boundary work, are transferred across the boundary and the system arrives at a neighboring equilibrium state, State-2, with only differential changes in properties as sketched in the *T-s* diagram of Figure 11.3. For the sake of generality, the process is assumed to be irreversible, i.e., entropy is allowed to be generated inside and in the immediate surroundings of the system through *thermodynamic friction*. In Figure 11.4 different mechanisms contributing to thermodynamic friction responsible for the internal entropy generation, $dS_{gen,int}$, are represented by a single mechanism: mechanical friction between the sliding blocks, one connected to the piston and one to the cylinder. Regardless of which direction the piston moves, some frictional work, $dW_{B,\,lost}$, is performed, which affects the boundary work transfer dW_B (see Anim. 11.A.*lostWork*).

The energy and entropy equations for any process, reversible or irreversible, were derived in Eqs. (5.29) and (5.31) as

$$dU = dQ - dW_B; \quad \text{[kJ]} \tag{11.1}$$

$$dS = \frac{dQ}{T} + dS_{gen,int}; \quad \left[\frac{\text{kJ}}{\text{K}}\right] \tag{11.2}$$

Note that the kinetic and potential energies of the local system were not neglected in deriving the energy equation for a differential process in Sec. 5.3.

For a given differential process 1–2, the anchor states 1 and 2, which are the beginning and final states of the process (Fig. 11.3), are fixed. However, the choices for the path followed by the system to traverse from State-1 to State-2 are many, each with a unique combination of dQ and dW_B. Out of the myriad possible paths the differential process may take to reach the same destination, it is entirely possible to construct reversible ones. After all, generation of entropy, though pervasive in any real process, is not required by the second law in a successful transition. If the transition takes place through such a reversible path, internal thermodynamic friction and entropy generation would be completely absent to give

$$dS_{gen,int.rev.} = 0; \quad \left[\frac{\text{kJ}}{\text{K}}\right], \quad dQ_{int.rev.} = TdS, \quad \text{and} \quad dW_{B,int.rev.} = pd\kern-0.5ex\forall; \quad \text{[kJ]} \tag{11.3}$$

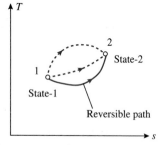

FIGURE 11.3 Many combinations of dQ and $dW_{pd\forall}$ result in the same finish state.

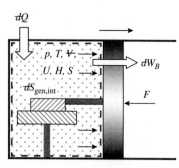

FIGURE 11.4 Internal irreversibilities are represented by friction between the blocks as they slide over each other (see Anim. 11.A.*lostWork*).

In a realistic process, however, we would expect $dS_{gen,int}$ to be positive and the presence of internal friction as depicted in Figure 11.4 to reduce (algebraically) the $pd\Psi$ work transfer by an amount $dW_{B,\,lost}$

$$dW_{B} = dW_{B,int.rev.} - dW_{B,lost} = pd\Psi - dW_{B,lost}; \quad [\text{kJ}] \tag{11.4}$$

where the lost work $dW_{B,lost}$ is always positive, regardless of the direction of the piston movement. During compression, $dW_{B,lost}$ can be seen to increase the magnitude of the work transfer so that, algebraically $dW_B \leq dW_{B,int.rev.}$. Likewise, the heat transfer in a realistic process can be related to its reversible limit by using Eqs. (11.2) and (11.3) since $dQ \leq dQ_{int.rev.}$.

Eliminating dQ between Eqs. (11.1) and (11.2) and substituting the expression for dW_B from Eq. (11.4), we obtain

$$
\begin{aligned}
TdS &= dU + dW_B + TdS_{gen,int} \\
&= dU + dW_{B,int.rev.} - dW_{B,lost} + TdS_{gen,int} \\
\Rightarrow \quad TdS &= dU + pd\Psi + \underbrace{(TdS_{gen,int} - dW_{B,\,lost})}_{d\epsilon}; \quad [\text{kJ}]
\end{aligned} \tag{11.5}
$$

Although the term inside the parentheses, $d\epsilon$, is dependent on the irreversibilites associated with the path, its value must be a constant for a given pair of neighboring equilibrium states. This is because the rest of the terms in the equation, which do not have any path dependency, are fixed for the given pair of states: State-1 and State-2 in this case. If we could determine the value of $d\epsilon$, which is a constant, along a particular path, that must be its only value. With that goal, let us recall that we have already established $dS_{gen,int}$ and $dW_{B,lost}$ to be zero along a reversible path. Therefore, $d\epsilon = 0$ is true not only for an internally reversible transition, but also for all possible paths connecting the two equilibrium states. Using $d\epsilon = 0$ in Eq. (11.5) results into two important relations, the first one is called the **Guy–Stodola theorem** and the second one is the familiar first *Tds* relation.

$$dW_{B,lost} = TdS_{gen,int}; \quad [\text{kJ}] \quad \text{(Guy–Stodola Theorem)} \tag{11.6}$$

$$TdS = dU + pd\Psi; \quad [\text{kJ}] \quad \text{or}, \quad Tds = du + pdv; \quad \left[\frac{\text{kJ}}{\text{kg}}\right] \quad \text{(First } Tds \text{ Relation)} \tag{11.7}$$

The Guy–Stodola theorem tells us that the lost work is proportional to the severity of *thermodynamic friction* as quantified by the internal entropy generation. The exergy destruction in a process, $I = T_0 S_{gen,univ}$, derived in Chapter 6, can be interpreted as an extension of this theorem.

Substitution of $u = h - pv$ in Eq. (11.7) and a little manipulation leads to the second *Tds* relation.

$$TdS = dH - \Psi dp; \quad [\text{kJ}] \quad \text{or}, \quad Tds = dh - vdp; \quad \left[\frac{\text{kJ}}{\text{kg}}\right] \quad \text{(Second } Tds \text{ Relation)} \tag{11.8}$$

As already mentioned, these *Tds* relations are applicable between any two differentially separated equilibrium states of a *simple compressible system* with a pure substance as the working fluid. For example, in steam flowing through a nozzle, a local system may contain a mixture of saturated mixture of liquid and vapor, accelerate, and undergo irreversible changes. The powerful assumption of LTE ensures that we can apply the *Tds* relations to relate two neighboring local systems. For non-simple systems that involve mass transfer or other modes of work transfer such as magnetic work, the *Tds* relations can be extended by including additional terms.

Beside the *Tds* relations, we have encountered other differential thermodynamic relations for u and h in Chapter 3 (Sec. 3.1.3). Before we can derive these relations, let us go over the mathematical framework necessary to develop further functional relationships among thermodynamic properties.

11.1.2 Partial Differential Relations

Suppose z is a continuous point function of independent variables x and y, i.e., $z = z(x, y)$, where x, y, and z represent any three thermodynamic properties. Taylor's theorem, Eq. (3.2) discussed in Sec. 3.1.3 (and Anim. 11.A.*TaylorTheorem*), can be used to relate differential changes in x and y with those in z:

$$dz = M_{(x,y)}dx + N_{(x,y)}dy; \quad \text{where,} \quad M = \left(\frac{\partial z}{\partial x}\right)_y \quad \text{and} \quad N = \left(\frac{\partial z}{\partial y}\right)_x \tag{11.9}$$

The subscript used in a partial derivative explicitly indicates the variable that is held constant. In calculus, dz is called an **exact differential**, when a differentiable function z must exist for dz to be expressed in this particular way. For example, du and dh can be expressed in this format because u and h are properties, but dQ and dW are path dependent and, therefore, not exact differentials. With both M and N being functions of x and y, we can take the partial derivatives of these functions with respect to y (while holding x constant) and x (while holding y constant).

$$\left(\frac{\partial M}{\partial y}\right)_x = \frac{\partial}{\partial y}\left[\left(\frac{\partial z}{\partial x}\right)_y\right]_x = \frac{\partial^2 z}{\partial y \partial x}; \quad \left(\frac{\partial N}{\partial x}\right)_y = \frac{\partial}{\partial x}\left[\left(\frac{\partial z}{\partial y}\right)_x\right]_y = \frac{\partial^2 z}{\partial x \partial y};$$

The order of differentiation being immaterial, we arrive at our first useful mathematical identity.

$$\left(\frac{\partial M}{\partial y}\right)_x = \left(\frac{\partial N}{\partial x}\right)_y \tag{11.10}$$

This is also known as a **test of exactness** because, starting with Eq. (11.10), dz can be established as an exact differential.

By changing the independent variables, the function $z = z(x, y)$ can also be expressed as $x = x(y, z)$. Applying Taylor's theorem, Eq. (11.9), dx can be written as

$$dx = \left(\frac{\partial x}{\partial y}\right)_z dy + \left(\frac{\partial x}{\partial z}\right)_y dz \tag{11.11}$$

Substituting Eq. (11.11) in Eq. (11.9) dx can be eliminated. Rearranging, we obtain

$$\left[1 - \left(\frac{\partial z}{\partial x}\right)_y\left(\frac{\partial x}{\partial z}\right)_y\right]dz = \left[\left(\frac{\partial z}{\partial x}\right)_y\left(\frac{\partial x}{\partial y}\right)_z + \left(\frac{\partial z}{\partial y}\right)_x\right]dy$$

Noting that any two instances of x, y, and z can be independently varied, we select y and z to be independent. Hence, values of dy and dz can be arbitrary chosen. Under this scenario, the identity above can be ensured only if each of the terms within the brackets is identically zero. The argument in the first parentheses when set to zero yields what is known as the **reciprocatory relation**.

$$\left(\frac{\partial z}{\partial x}\right)_y = \frac{1}{(\partial x/\partial z)_y} \tag{11.12}$$

Equating the term inside the second parentheses to zero and then applying the reciprocatory relation to the second term, we obtain the **cyclic relation**.

$$\left(\frac{\partial z}{\partial x}\right)_y\left(\frac{\partial x}{\partial y}\right)_z\left(\frac{\partial y}{\partial z}\right)_x = -1 \tag{11.13}$$

Thermodynamic properties being point functions (Sec. 1.2), their differentials are exact as opposed to differentials of Q or W, which are path functions and **inexact**. While $\Delta S = S_2 - S_1$, the equation $\Delta Q = Q_2 - Q_1$ does not make any sense. To emphasize that, an infinitesimal amount of heat or work transfer is expressed with a crossed d as in dQ or dW_B. Use of the Taylor's theorem with thermodynamic properties has been illustrated in Example 3-1. The following example demonstrates the validity of the reciprocatory and cyclic relations.

EXAMPLE 11-1   Reciprocatory and Cyclic Relations

The IG (ideal gas) equation of state is used to express the specific volume as a function of the independent variables p and T. Verify the cyclic and reciprocatory relations with the help of this equation.

Assumptions

A one percent variation in a property can be regarded as a differential change.

SOLUTION

The functional relationship $v = v(p, T)$ is given by the IG equation of state $v = RT/p$, where R is the gas constant. Obtain the relevant partial derivatives of this function to verify the reciprocatory and cyclic relations.

Analysis

Treating $v = v(p, T)$ as $z = v(x, y)$, we would like to establish that

$$\left(\frac{\partial v}{\partial p}\right)_T = \frac{1}{(\partial p/\partial v)_T} \quad \text{and} \quad \left(\frac{\partial v}{\partial p}\right)_T \left(\frac{\partial p}{\partial T}\right)_v \left(\frac{\partial T}{\partial v}\right)_p = -1$$

The four partial derivatives that appear in these equations can be directly obtained by differentiating the ideal gas equation.

$$v = v(p, T) = \frac{RT}{p}; \quad \Rightarrow \quad \left(\frac{\partial v}{\partial p}\right)_T = -\frac{RT}{p^2}$$

$$p = p(v, T) = \frac{RT}{v}; \quad \Rightarrow \quad \left(\frac{\partial p}{\partial v}\right)_T = -\frac{RT}{v^2} = -\frac{RT}{(RT/p)^2} = -\frac{p^2}{RT}$$

Therefore, $\left(\dfrac{\partial v}{\partial p}\right)_T = \dfrac{1}{(\partial p/\partial v)_T}$

Now, $p = p(v, T) = \dfrac{RT}{v}; \quad \Rightarrow \quad \left(\dfrac{\partial p}{\partial T}\right)_v = \dfrac{R}{v} \quad \text{and} \quad T = T(v, p) = \dfrac{pv}{R}; \quad \Rightarrow \quad \left(\dfrac{\partial T}{\partial v}\right)_p = \dfrac{p}{R}$

Therefore, $\left(\dfrac{\partial v}{\partial p}\right)_T \left(\dfrac{\partial p}{\partial T}\right)_v \left(\dfrac{\partial T}{\partial v}\right)_p = -\dfrac{RT}{p^2}\dfrac{R}{v}\dfrac{p}{R} = -\dfrac{RT}{pv} = -1$

TEST Analysis

Launch the IG system-state TESTcalc. Select any gas, say, air. Evaluate an anchor state, State-1, with p1 = 100 kPa and T1 = 300 K. To evaluate the partial derivatives of v while holding T constant, evaluate State-2 with p2=1.01*p1 and T2=T1. Evaluate (v2−v1)/(p2−p1)=−.0085 unit in the I/O panel. By definition, it is the reciprocal of $(p2 − p1)/(v2 − v1)$. To determine the rest of the partial derivatives, evaluate State-3 with p3=1.01*p1 and v3=v1, and State-4 with T4=1.01*T1 and p4=p1. In the I/O panel, calculate (p3−p1)/(T3−T1) as 0.333 unit and (T4−T1)/(v4−v1) as 348.448 unit. The product of these three values is found to be −0.99. The cyclic rule is thus verified within a reasonable accuracy.

Discussion

Expanding the case study to different gases and different anchor states, the validity of the reciprocatory and cyclic rule can be extensively verified. Conversely, thermodynamic data from an unknown source can be tested against the cyclic relation to verify its consistency.

11.1.3 The Maxwell Relations

The Tds relations are actually part of four fundamental relations known as **Gibbs equations**. The other two involve two new properties; the specific **Helmholtz function** f, and the specific **Gibbs function** g, defined as

$$f \equiv u - Ts; \quad \left[\frac{kJ}{kg}\right]; \quad \text{and} \quad g \equiv h - Ts; \quad \left[\frac{kJ}{kg}\right] \tag{11.14}$$

Defined as combinations of thermodynamic properties, f and g are also thermodynamic specific properties, with the corresponding extensive properties represented by the uppercase symbols F and G. Using the Tds relations, the differentials of Helmholtz and Gibbs functions can be simplified, leading to the four *Gibbs equations:*

$$du = Tds - pdv; \quad \left[\frac{kJ}{kg}\right] \tag{11.15}$$

$$dh = Tds + vdp; \quad \left[\frac{kJ}{kg} \right] \qquad \qquad (11.16)$$

$$df = -sdT - pdv; \quad \left[\frac{kJ}{kg} \right] \qquad \qquad (11.17)$$

$$dg = -sdT + vdp; \quad \left[\frac{kJ}{kg} \right] \qquad \qquad (11.18)$$

The properties that appear on the left-hand side of the Gibbs equations can be regarded as a particular type of specific energy with the unit kJ/kg. While u and h can be easily identified as the specific stored energy and specific flow energy respectively (for negligible ke and pe), f and g are known as Helmholtz and Gibbs *free energy* respectively, which will be shown to reduce to special types of specific energy under certain restrictions.

Maxwell's relations relate partial derivatives of p, v, T, and s to each other for a simple compressible system and can be derived from Gibbs equations. Each differential on the left-hand side of a Gibbs equation is *exact* (recall from Sec. 1.2 that thermodynamic properties are point functions) and can be expressed in the generic format

$$z \equiv z(x, y); \quad dz = Mdx + Ndy \quad \text{with} \quad \left(\frac{\partial M}{\partial y} \right)_x = \left(\frac{\partial N}{\partial x} \right)_y \qquad (11.19)$$

Comparing Eq. (11.19) with Gibbs equations, Eqs. (11.15)–(11.18), produce

$$u = u(s, v); \quad \Rightarrow \quad M = T = \left(\frac{\partial u}{\partial s} \right)_v \quad \text{and} \quad N = -p = \left(\frac{\partial u}{\partial v} \right)_s \qquad (11.20)$$

$$h = h(s, p); \quad \Rightarrow \quad M = T = \left(\frac{\partial h}{\partial s} \right)_p \quad \text{and} \quad N = v = \left(\frac{\partial h}{\partial p} \right)_s \qquad (11.21)$$

$$f = f(T, v); \quad \Rightarrow \quad M = -s = \left(\frac{\partial f}{\partial T} \right)_v \quad \text{and} \quad N = -p = \left(\frac{\partial f}{\partial v} \right)_T \qquad (11.22)$$

$$g = g(T, p); \quad \Rightarrow \quad M = -s = \left(\frac{\partial g}{\partial T} \right)_p \quad \text{and} \quad N = v = \left(\frac{\partial g}{\partial p} \right)_T \qquad (11.23)$$

Subjecting Eqs. (11.20) through (11.23) to the equality of partial derivatives dictated by Eq. (11.19), we obtain the **Maxwell relations**:

$$\left(\frac{\partial T}{\partial v} \right)_s = -\left(\frac{\partial p}{\partial s} \right)_v \qquad \qquad (11.24)$$

$$\left(\frac{\partial T}{\partial p} \right)_s = \left(\frac{\partial v}{\partial s} \right)_p \qquad \qquad (11.25)$$

$$\left(\frac{\partial s}{\partial v} \right)_T = \left(\frac{\partial p}{\partial T} \right)_v \qquad \qquad (11.26)$$

$$\left(\frac{\partial s}{\partial p} \right)_T = -\left(\frac{\partial v}{\partial T} \right)_p \qquad \qquad (11.27)$$

Despite their abstract form, Gibbs equations and Maxwell relations offer some of the most useful equations in thermodynamics. For instance, entropy, which cannot be directly measured, is related to changes in p, v, and T—properties which can be easily measured.

The property diagram of Figure 11.5 is called the **thermodynamic square** and is an excellent mnemonic to reconstruct Eqs. (11.15)–(11.27). The specific energies—u, f, g, and h—appear on the sides of the square, each flanked by their respective independent variables as expressed in the Gibbs equations. With the properties arranged in this manner, it can be readily deduced from the square that $h = h(s, p)$, $f = f(T, v)$, etc. The Gibbs equations themselves can be reconstructed by using the directed diagonals, which point from independent variables to their respective coefficients, with the reverse direction indicating a negative coefficient. For instance, in the Gibbs equation $dh = vdp + Tds$, v is the coefficient of dp and T is the coefficient of ds, as indicated by the arrows pointing from the independent variables p and s towards the coefficients

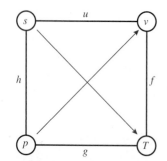

FIGURE 11.5 The thermodynamic square (see Anim. 11.A.*thermoSquare*) from which all thermodynamic relations can be readily obtained.

v and T. Likewise, we can reconstruct $dg = vdp - sdT$ as another Gibbs equation, where the negative sign before s appears as the diagonal arrow points away from the coefficient s.

Relations (11.20)–(11.23) can be obtained by remembering that a property appearing at a corner (take s for example) can be expressed as a partial derivative of either specific energy appearing on the far sides of the square (f or g in this example) with respect to the diagonally opposite property (with respect to T), while holding the remaining independent variable (v in the case of f and p in the case of g) constant. A negative sign is added if the diagonal points away from the property. Thus, we can express s as

$$s = -\left(\frac{\partial f}{\partial T}\right)_v \quad \text{or} \quad s = -\left(\frac{\partial g}{\partial T}\right)_p$$

Similarly, all other **property relations** expressed in Eqs. (11.20)–(11.23) can be recovered from the thermodynamic square.

Maxwell relations involve partial derivatives of the independent variables (s, v, T, p) of the four energies (u, f, g, h). For a given energy (take u for example), the relation is constructed by equating the partial of each independent property (s and v) with respect to the property at the neighboring corner (p and T) while holding the diagonally opposite property (T and p) constant. A negative sign is added to one side of the equation if the arrows are asymmetric. Internal energy u, thus, leads to the fourth relation, Eq. (11.27).

EXAMPLE 11-2 Property Relations and Maxwell Equations

Using TEST, verify the first relation of Eq. (11.20) and the first Maxwell equation, Eq. (11.24), for R-12 at 200 kPa and 50% quality.

SOLUTION

Using the PC system-state TESTcalc, evaluate the anchor state from the given conditions. Evaluate a neighboring state while holding the relevant property constant as dictated by the partial derivative. Evaluate the partial derivative in the I/O panel.

Assumptions

A one percent variation in a property can be regarded as a differential change.

TEST Analysis

The two relations (Fig. 11.6) to be verified are:

$$T = \left(\frac{\partial u}{\partial s}\right)_v \quad \text{and} \quad \left(\frac{\partial T}{\partial v}\right)_s = -\left(\frac{\partial p}{\partial s}\right)_v$$

Launch the PC system-state TESTcalc and select R-12 as the working fluid. Evaluate the anchor state, State-1, with p1=200 kPa and x1=50%. Evaluate a neighboring state, State-2, with p2=1.01*p1 and v2=v1. The partial derivative can be calculated in the I/O panel as (u2 − u1)/(s2 − s1) = 261.5 K, which is approximately equal to T1 = 260.6 K.

To verify the Maxwell relation, Eq. (11.24), evaluate the right hand side as (p2 − p1)/ (s2 − s1) = 651 unit. To create a constant entropy process, evaluate State-3 with T3=1.01*T1 and s3=s1. Now evaluate the left-hand side as (T3 − T1)/(v3 − v1) = −649.4 unit.

Discussion

The inability of the TEST solution to completely confirm the Maxwell relation reveals the limitation of the TEST database, which relies on linear interpolation among discrete data points with an inherent 1–2% tolerance. Moreover, errors in the numerator and denominator may get while evaluating a ratio. If $z = x/y$, then the error in z can be expressed (by taking logarithms and then evaluating differentials on both side of the equation) as

$$\frac{dz}{z} = \frac{dx}{x} - \frac{dy}{y}; \quad \text{therefore,} \quad \left|\frac{dz}{z}\right| = \left|\frac{dx}{x}\right| + \left|\frac{dy}{y}\right| \tag{11.28}$$

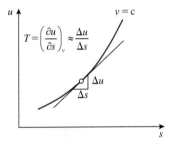

FIGURE 11.6 Temperature as the slope in Ex. 11-2 according to Eq. (11.20).

If the two neighboring states are brought closer together by reducing the difference between p2 and p1, the round-off error in v will eventually approach $v2 - v1$, at which point the ratio between ΔT and Δv will become meaningless. This is a problem inherent in any discretized data set.

11.1.4 The Clapeyron Equation

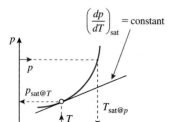

FIGURE 11.7 The saturation curve relates saturation pressure and saturation temperature.

In building the saturation tables of the PC model (Sec. 3.4) for a given substance, not all properties need to be directly measured. The Clapeyron equation is an excellent application of thermodynamic relations that allows determination of the enthalpy change $h_{fg} \equiv h_g - h_f$ during phase transformation of a PC fluid from the behavior of measurable properties p, v, and T.

To derive this equation, consider the third Maxwell equation, Eq. (11.26). During a phase-change process, a saturated mixture of liquid and vapor can be in equilibrium. The pressure and temperature are called saturation pressure p_{sat} and saturation temperature T_{sat}, respectively, which are functions of each other (Sec. 3.4.7), that is, $p_{sat} = p_{sat@T}$ and $T_{sat} = T_{sat@p}$. Therefore, for a saturated mixture the partial derivative $(\partial p / \partial T)_v$ reduces to a simple derivative, $(dp/dT)_{sat}$, which can be interpreted as the slope at a given temperature on the saturation curve in the phase diagram (Sec. 3.4.4) of Figure 11.7. During the phase change process as the state migrates from a saturated liquid state to a saturated vapor state (States f and g in Fig. 11.8), this slope remains unchanged. Therefore, Eq. (11.26) can be integrated between the saturated states to yield:

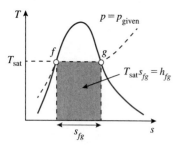

FIGURE 11.8 Graphical interpretation of Eq. (11.30) (see Anim. 11.A.*ClapeyronEqn*).

$$\int_f^g ds = \int_f^g \left(\frac{dp}{dT}\right)_{sat} dv; \quad \Rightarrow \quad s_g - s_f = \left(\frac{dp}{dT}\right)_{sat}(v_g - v_f); \quad \left[\frac{kJ}{kg \cdot K}\right] \quad \textbf{(11.29)}$$

$$\Rightarrow \quad \left(\frac{dp}{dT}\right)_{sat} = \frac{s_{fg}}{v_{fg}}; \quad \left[\frac{kPa}{K}\right]$$

$$\text{where} \quad s_{fg} \equiv s_g - s_f; \quad \left[\frac{kJ}{kg \cdot K}\right], \quad \text{and} \quad v_{fg} \equiv v_g - v_f; \quad \left[\frac{m^3}{kg}\right]$$

As the pressure also remains constant during the phase transition, the Tds relation yields

$$Tds = dh - v\,dp^{0}; \quad \Rightarrow \quad \int_f^g Tds = \int_f^g dh; \quad \Rightarrow \quad T_{sat}s_{fg} = h_{fg}; \quad \left[\frac{kJ}{kg}\right] \quad \textbf{(11.30)}$$

Substituting this result in Eq. (11.29), we obtain the **Clapeyron equation**:

$$\Rightarrow \quad \left(\frac{dp}{dT}\right)_{sat} = \frac{h_{fg}}{T_{sat}v_{fg}}; \quad \left[\frac{kPa}{K}\right] \quad \textbf{(11.31)}$$

By simply measuring the slope of the saturation curve on a p-T diagram and the change in specific volume v_{fg} during phase transformation, the enthalpy of vaporization at a given temperature can be calculated from this equation. Although the equation is written for liquid-vapor phase transformation, the derivation is q uite general and Eq. (11.31) can be applied for any other type of constant-temperature phase transformation, such as melting. If the two phases are represented by 1 and 2, Eq. (11.31) can be generalized by replacing h_{fg} with h_{12} and v_{fg} with v_{12}.

EXAMPLE 11-3 Application of Clapeyron Equation

Estimate h_{fg}, u_{fg}, and s_{fg} for ammonia, using the Clapeyron equation at 100 kPa. Use TEST.

SOLUTION

Use the PC system-state TESTcalc to evaluate various terms of the Clapeyron equation to obtain h_{fg}. From the definition of h, evaluate u_{fg} and then use Eq. (11.30) to determine s_{fg}.

Assumptions

A one percent variation in a property can be regarded as a differential change.

TEST Analysis

Launch the PC system state TESTcalc and select NH3 as the working fluid. Evaluate the saturated states: *f*-state, State-1, with p1=100 kPa and x1=0, and and *g*-state, State-2, with p2=p1 and x2=100%. To obtain the slope of the saturation curve, evaluate a neighboring state at a different pressure—State-3 with p3=1.01*p1 and x3=100%. In the I/O panel, evaluate the slope as (p3−p2)/(T3−T2)=4.988 kPa/K. Estimate h$_{fg}$ as T2*(v2−v1)*4.988=1363 kJ/kg, u$_{fg}$ as h$_{fg}$-p2*(v2−v1)=1249 kJ/kg, and s$_{fg}$ as h$_{fg}$/T2=5.692 kJ/kg·K. These results compare reasonably well with the values obtained directly from the TESTcalc, h2 − h1 = 1370 kJ/kg, u2 − u1 = 1257 kJ/kg, and s2 − s1 = 5.723 kJ/kg·K.

Discussion

The approach used in this problem is often applied in developing saturation tables for phase-change substances. Clearly, the accuracy of such tables depends heavily on the precision of the saturation curve in the *p-T* diagram.

11.1.5 The Clapeyron–Clausius Equation

The Clapeyron equation can be further simplified for a liquid-vapor or solid-vapor transition by utilizing the fact that at low pressures $v_g \gg v_f$ (or, for sublimation, $v_g \gg v_i$), and the ideal gas equation (Sec. 3.5.3) can be used to evaluate v_g approximately. That is,

$$v_{fg} = v_g - v_f^{\text{negligible}} \cong v_g = \frac{RT}{p}; \quad \left[\frac{\text{m}^3}{\text{kg}}\right]$$

Substituting this approximate value for v_{fg} in Eq. (11.31) results in

$$\left(\frac{dp}{dT}\right)_{\text{sat}} = \frac{p h_{fg}}{RT^2}; \quad \left[\frac{\text{kPa}}{\text{K}}\right] \quad \Rightarrow \quad \left(\frac{dp}{p}\right)_{\text{sat}} = \frac{h_{fg}}{R}\left(\frac{dT}{T^2}\right)_{\text{sat}}; \qquad \textbf{(11.32)}$$

Replacing h_{fg} with an average value $h_{fg,\text{avg}}$ (note that the symbol $\bar{h}_{fg}$ is reserved for molar specific enthalpy by our convention) between two saturation pressures p_1 and p_2, Eq. (11.32) can be integrated to functionally relate saturation pressure with temperature.

$$\ln\left(\frac{p_2}{p_1}\right)_{\text{sat}} = \frac{h_{fg,\text{avg}}}{R}\left(\frac{1}{T_1} - \frac{1}{T_2}\right)_{\text{sat}} \qquad \textbf{(11.33)}$$

This is known as **Clapeyron–Clausius equation**, which predicts the nature of the saturation curve in the *p-T* diagram at relatively low pressures. For a given fluid, if the saturation pressure $p_{\text{sat}@T_1}$ and $h_{fg@T_1}$ are known at a particular state (State-1), then p_{sat} can be expressed as a function of temperature T as $\ln p_{\text{sat}} = c_1 - c_2/T$, where c_1 and c_2 can be regarded as known constants (Fig. 11.9). For solid-vapor transition, this relation can be modified by replacing $h_{fg,\text{avg}}$ with $h_{ig,\text{avg}}$, the enthalpy of sublimation.

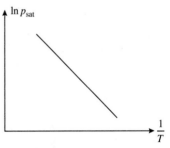

FIGURE 11.9 The Clapeyron–Clausius equation establishes an approximate logarithmic relation between saturation pressure and reciprocal of temperature.

EXAMPLE 11-4   **Application of Clapeyron-Clausius Equation**

In the pressure saturation table for water (Table B-1) suppose there is no entry for 1.5 kPa. Using the entry for 1 and 2 kPa, determine the saturation temperature (a) by linear interpolation, and (b) by application of the Clapeyron–Clausius equation.

SOLUTION

Using h_{fg} from the 2 kPa row, use Eq. (11.33) to determine the saturation temperature and compare it with the linearly interpolated result.

Assumptions

At 1.5 kPa, which is very low compared to the critical pressure of water, saturated vapor can be modeled as an ideal gas, allowing the use of Clapeyron–Clausius equation.

Analysis

A linear interpolation produces

$$T_{sat@1.5\ kPa} = \frac{T_{sat@1\ kPa} + T_{sat@2\ kPa}}{2} = \frac{6.98 + 17.50}{2} = 12.24°C$$

From the saturation table of water or the PC system-state TESTcalc, obtain $h_{fg} = h_{fg@2\ kPa} = 2460.1\ kJ/kg$ and $T_1 = T_{sat@2\ kPa} = 290.63\ K$, then calculate the gas constant as $R = \overline{R}/\overline{M} = 8.314/18 = 0.462\ kJ/kg \cdot K$. Substituting these values and $p_2 = 1.5\ kPa$ in the Clapeyron–Clausius equation, we obtain

$$\ln\left(\frac{1.5}{2}\right) = \frac{2460.1}{0.462}\left(\frac{1}{290.65} - \frac{1}{T_2}\right); \quad \Rightarrow \quad T_2 = 286.1\ K = 13.0°C$$

Discussion

The saturation table (Table B-1) lists $T_{sat@1.5\ kPa}$ as 13.03°C, which is quite accurately predicted by the Calpyeron–Clausius equation. Verify that picking h_{fg} as $(h_{fg@2\ kPa} + h_{fg@1\ kPa})/2$ would not change the answer significantly.

11.2 EVALUATION OF PROPERTIES

Beside the *Tds* relations, the differential relations involving *u* and *h* also have played a crucial role in the development of material models in Chapter 3. In fact, the *state postulate* allowed us to express any thermodynamic property as a function of two others for a pure, simply compressible substance in the format of Eq. (11.9). We will use some of the concepts developed in this chapter to refine the expressions for Δu, Δv, Δs and c_v as functions of *p*, *v*, *T*, and c_p in this section. For any working substance, the goal is to reduce the necessary experimental information. It will be shown that finding an accurate $p-v-T$ relation and a correlation for c_p are sufficient to obtain all other thermodynamic properties.

11.2.1 Internal Energy

Expressing the internal energy as a function of *T* and *v*, i.e., $u = u(T, v)$, and recalling the definition of c_v (see Sec. 3.1.3 and Fig. 11.10), Eq. (3.4) can be rewritten as

$$du = c_v dT + \left(\frac{\partial u}{\partial v}\right)_T dv; \quad \left[\frac{kJ}{kg}\right], \quad \text{where} \quad c_v \equiv \left(\frac{\partial u}{\partial T}\right)_v; \quad \left[\frac{kJ}{kg \cdot K}\right] \tag{11.34}$$

From the *Tds* relation we can express *du* as

$$du = Tds - pdv; \quad \left[\frac{kJ}{kg}\right] \tag{11.35}$$

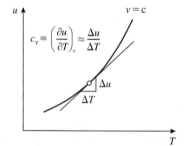

Also, expressing *s* as $s = s(T, v)$, *ds* can be expanded using Taylor's theorem.

$$ds = \left(\frac{\partial s}{\partial T}\right)_v dT + \left(\frac{\partial s}{\partial v}\right)_T dv; \quad \left[\frac{kJ}{kg \cdot K}\right] \tag{11.36}$$

Substituting this into Eq. (11.35) leads to

$$du = T\left(\frac{\partial s}{\partial T}\right)_v dT + \left[T\left(\frac{\partial s}{\partial v}\right)_T - p\right]dv; \quad \left[\frac{kJ}{kg}\right] \tag{11.37}$$

FIGURE 11.10 Specific heat at constant volume is the slope of the constant volume line in the *u-T* diagram at a given temperature (see Anim. 11.A.cv).

The coefficients of Eqs. (11.34) and (11.37) can be equated to produce the following two relations:

$$\frac{c_v}{T} = \left(\frac{\partial s}{\partial T}\right)_v; \quad \text{and} \quad \left(\frac{\partial u}{\partial v}\right)_T = T\left(\frac{\partial s}{\partial v}\right)_T - p \tag{11.38}$$

The second of these relations can be further simplified by using the third Maxwell relation, Eq. (11.26):

$$\left(\frac{\partial s}{\partial v}\right)_T = \left(\frac{\partial p}{\partial T}\right)_v; \quad \text{therefore,} \quad \left(\frac{\partial u}{\partial v}\right)_T = T\left(\frac{\partial p}{\partial T}\right)_v - p \qquad \textbf{(11.39)}$$

Substituting this result into Eq. (11.34), we obtain the final expression for du in terms of p, v, T, and c_v (in Sec. 11.2.5 we will express c_v in terms of p, v, T, and c_p):

$$du = c_v dT + \left[T\left(\frac{\partial p}{\partial T}\right)_v - p\right]dv; \quad \left[\frac{kJ}{kg}\right] \qquad \textbf{(11.40)}$$

It should be noted that expressing u in terms of four properties, p, v, T, and c_v, is not a violation of the *state postulate*. The p–v–T relation and a correlation for c_v in terms of any pair of independent properties from p, v, and T can be used to reduce the number of independent variables in Eq. (11.40) to two.

The change in u between two states—State-1 and State-2—can be obtained by integrating Eq. (11.40):

$$\Delta u = u_2 - u_1 = \int_1^2 c_v dT + \int_1^2 \left[T\left(\frac{\partial p}{\partial T}\right)_v - p\right]dv; \quad \left[\frac{kJ}{kg}\right] \qquad \textbf{(11.41)}$$

It should be stressed that this is a general relation that applies to all *pure substances*—solids, liquids, vapors, gases, and mixtures of saturated phases—as long as the systems involved are *simply compressible* (Sec. 3.1.2). The power of this general relation can be appreciated if we apply Eq. (11.41) to any specific substance, such as an ideal gas. Application of $pv = RT$ eliminates the second term and Eq. (11.41) reduces to Eq. (3.33), that is,

---

EXAMPLE 11-5 Behavior of u for an Ideal Gas

A gas that follows the IG equation of state $pv = RT$ is defined as an ideal gas. Show that if the temperature is held constant, the internal energy cannot be a function of (a) v or (b) p.

SOLUTION

Beginning with $u = u(v, T)$ and $u = u(p, T)$, use the ideal gas equation $pv = RT$ to evaluate the partial derivative of u with respect to v, and u with respect to p, holding T constant.

Assumptions

Thermodynamic equilibrium at all states.

Analysis

R being a constant, use of $pv = RT$ yields:

$$\left(\frac{\partial p}{\partial T}\right)_v = \left(\frac{\partial(RT/v)}{\partial T}\right)_v = \frac{R}{v}$$

Substituting this in the second relation of Eq. (11.39) produces:

$$\left(\frac{\partial u}{\partial v}\right)_T = T\left(\frac{\partial p}{\partial T}\right)_v - p = \frac{TR}{v} - p = \frac{pv}{v} - p = 0$$

Therefore, u cannot be a function of v when T is held constant.

Using the chain rule of calculus and the above conclusion, we obtain:

$$\left(\frac{\partial u}{\partial p}\right)_T \left(\frac{\partial p}{\partial v}\right)_T \left(\frac{\partial v}{\partial u}\right)_T = 1;$$

$$\Rightarrow \quad \left(\frac{\partial u}{\partial p}\right)_T \left(\frac{\partial p}{\partial v}\right)_T = \frac{1}{(\partial v/\partial u)_T} = \left(\frac{\partial u}{\partial v}\right)_T = 0$$

The second factor in the left-hand side $\left(\dfrac{\partial p}{\partial v}\right)_T = \left(\dfrac{\partial (RT/v)}{\partial v}\right)_T = -\dfrac{RT}{v^2} \neq 0$

Therefore, first factor, $\left(\dfrac{\partial u}{\partial p}\right)_T$, must be zero.

This implies that u cannot be a function of p when T is held constant.

Discussion

Although we have proved that u cannot be a function of v or p when T is held constant, we have yet to prove that u is a function of T alone, which will be done in Example 11-7.

11.2.2 Enthalpy

The derivation of an expression for Δh parallels the procedure used for Δu in the previous section. This time, we select p and T as the independent variables so that $h = h(p, T)$ and begin with Eq. (3.5), which introduces the specific heat at constant pressure c_p (Fig. 11.11).

$$dh = c_p dT + \left(\frac{\partial h}{\partial p}\right)_T dp; \quad \left[\frac{\text{kJ}}{\text{kg}}\right], \quad \text{where} \quad c_p \equiv \left(\frac{\partial h}{\partial T}\right)_p; \quad \left[\frac{\text{kJ}}{\text{kg} \cdot \text{K}}\right] \tag{11.42}$$

From the second Tds relation we express dh as

$$dh = Tds + vdp; \quad \left[\frac{\text{kJ}}{\text{kg}}\right] \tag{11.43}$$

Expressing s as $s = s(T, p)$, ds can be expanded as

$$ds = \left(\frac{\partial s}{\partial T}\right)_p dT + \left(\frac{\partial s}{\partial p}\right)_T dp; \quad \left[\frac{\text{kJ}}{\text{kg} \cdot \text{K}}\right] \tag{11.44}$$

Substituting this into Eq. (11.43) and rearranging, we obtain

$$dh = T\left(\frac{\partial s}{\partial T}\right)_p dT + \left[T\left(\frac{\partial s}{\partial p}\right)_T + v\right]dp \tag{11.45}$$

Equating the coefficients of Eqs. (11.42) and (11.45) results in

$$\frac{c_p}{T} = \left(\frac{\partial s}{\partial T}\right)_p; \quad \text{and} \quad \left(\frac{\partial h}{\partial p}\right)_T = T\left(\frac{\partial s}{\partial p}\right)_T + v \tag{11.46}$$

The second of these relations can be further simplified by using the fourth Maxwell relation, Eq. (11.27).

$$\left(\frac{\partial s}{\partial p}\right)_T = -\left(\frac{\partial v}{\partial T}\right)_p; \quad \text{therefore,} \quad \left(\frac{\partial h}{\partial p}\right)_T = v - T\left(\frac{\partial v}{\partial T}\right)_p \tag{11.47}$$

Substituting this result in Eq. (11.42) yields

$$dh = c_p dT + \left[v - T\left(\frac{\partial v}{\partial T}\right)_p\right]dp; \quad \left[\frac{\text{kJ}}{\text{kg}}\right] \tag{11.48}$$

This equation can be integrated between two states—State-1 and State-2—to produce an expression for Δh. Of course, we can also use $h \equiv u + pv$ and get a different expression for Δh in terms of Δu

$$\Delta h = h_2 - h_1 = \int_1^2 c_p dT + \int_1^2 \left[v - T\left(\frac{\partial v}{\partial T}\right)_p\right]dp; \quad \left[\frac{\text{kJ}}{\text{kg}}\right] \tag{11.49}$$

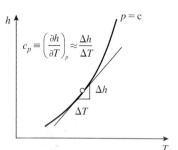

FIGURE 11.11 Specific heat at constant pressure is the slope of the constant pressure line in the h-T diagram at a given temperature (see Anim. 11.A.cp).

$$\Delta h = h_2 - h_1 = u_2 - u_1 + (p_2 v_2 - p_1 v_1); \quad \left[\frac{kJ}{kg}\right] \tag{11.50}$$

In the second relation $u_2 - u_1$ can be substituted from Eq. (11.41). Thus, the change in enthalpy is expressed in terms of the measurable properties p, v, T, and c_p or c_v by these two relations.

11.2.3 Entropy

To get an expression for ds, the two partial derivatives of Eq. (11.36) can be expressed in terms of p, v, T, and c_v by using the first equality of Eq. (11.38) and the third Maxwell relation, Eq. (11.26), resulting in

$$ds = \frac{c_v}{T}dT + \left(\frac{\partial p}{\partial T}\right)_v dv; \quad \left[\frac{kJ}{kg \cdot K}\right] \tag{11.51}$$

An alternative expression can be obtained by starting with Eq. (11.44), using the first equality of Eq. (11.46), and the fourth Maxwell relation, Eq. (11.27), yielding

$$ds = \frac{c_p}{T}dT - \left(\frac{\partial v}{\partial T}\right)_p dp; \quad \left[\frac{kJ}{kg \cdot K}\right] \tag{11.52}$$

Integrating Eqs. (11.51) and (11.52) from State-1 to State-2, we obtain the following general expressions for Δs

$$\Delta s = s_2 - s_1 = \int_1^2 \frac{c_v}{T}dT + \int_1^2 \left(\frac{\partial p}{\partial T}\right)_v dv; \quad \left[\frac{kJ}{kg \cdot K}\right] \tag{11.53}$$

$$\Delta s = s_2 - s_1 = \int_1^2 \frac{c_p}{T}dT - \int_1^2 \left(\frac{\partial v}{\partial T}\right)_p dp; \quad \left[\frac{kJ}{kg \cdot K}\right] \tag{11.54}$$

Like changes in u or h, the change in entropy is expressed in terms of the measurable properties p, v, T, and c_p or c_v by these relations.

EXAMPLE 11-6 **Enthalpy Change of a Non-Ideal Gas**

The $p-v-T$ behavior of a gas under certain conditions is found to follow the following equation of state: $(p + a/v^2)(v - b) = RT$, where a and b are constants. Also, the specific heat at constant volume is found to vary linearly with temperature according to $c_v = c_0 + c_1 T$. Derive an expression for the change in (a) internal energy, and (b) enthalpy.

SOLUTION
Use the relations developed in this section to obtain the desired expressions.

Analysis
The equation of state can be differentiated as follows:

$$\left(\frac{\partial p}{\partial T}\right)_v = \left[\frac{\partial}{\partial T}\left(\frac{RT}{v - b} - \frac{a}{v^2}\right)\right]_v = \frac{R}{v - b}$$

Therefore,

$$\left[T\left(\frac{\partial p}{\partial T}\right)_v - p\right] = \frac{RT}{v - b} - p = \frac{RT}{v - b} - \frac{RT}{v - b} + \frac{a}{v^2} = \frac{a}{v^2}$$

Substituting this in Eq. (11.41), we obtain:

$$\Delta u = \int_1^2 c_v dT + \int_1^2 \left[T \left(\frac{\partial p}{\partial T} \right)_v - p \right] dv = \int_{T_1}^{T_2} (c_0 + c_1 T) dT + \int_{v_1}^{v_2} \frac{a}{v^2} dv$$

$$= c_0 (T_2 - T_1) + \frac{c_1}{2} (T_2^2 - T_1^2) + a \left(\frac{1}{v_1} - \frac{1}{v_2} \right)$$

The change in enthalpy can be obtained from Eq. (11.50).

$$\Delta h = u_2 - u_1 + (p_2 v_2 - p_1 v_1)$$

$$= c_0 (T_2 - T_1) + \frac{c_1}{2} (T_2^2 - T_1^2) + a \left(\frac{1}{v_1} - \frac{1}{v_2} \right) + (p_2 v_2 - p_1 v_1)$$

Discussion

The change in enthalpy is expressed in terms of three properties, one of which can be eliminated by the application of the equation of state. The changes in u and h are not functions of temperature only, as in the case of the IG model.

11.2.4 Volume Expansivity and Compressibility

We define three new properties in this section that will be helpful in expressing thermodynamic relations in a concise format. For a single-phase simply compressible substance, p and T are independent properties and v can be expressed as $v = v(p, T)$. The differential of v, therefore, can be written using Taylor's theorem (see Anim. 11.A.*TaylorTheorem*) as:

$$dv = \left(\frac{\partial v}{\partial p} \right)_T dp + \left(\frac{\partial v}{\partial T} \right)_p dT; \quad \left[\frac{\mathrm{m}^3}{\mathrm{kg}} \right] \tag{11.55}$$

Like specific heats, the two coefficients of this equation can be used to define two new thermodynamic properties: **isothermal compressibility** κ_T (pronounced kappa) and **volume expansivity** β (pronounced beta). Additionally, the partial derivative of v with respect to p at constant entropy is used to define a third property, the **isentropic compressibility** κ_s.

$$\beta \equiv \frac{1}{v} \left(\frac{\partial v}{\partial T} \right)_p; \quad \left[\frac{1}{K} \right] \quad \kappa_T \equiv -\frac{1}{v} \left(\frac{\partial v}{\partial p} \right)_T; \quad \kappa_s \equiv -\frac{1}{v} \left(\frac{\partial v}{\partial p} \right)_s; \quad \left[\frac{1}{kPa} \right] \tag{11.56}$$

The volume expansivity is a measure of the change in volume as temperature is changed, holding pressure constant. The isothermal and isentropic compressibility indicate the change in volume when pressure is changed at constant temperature and entropy respectively. The negative signs in the definition ensure that the compressibility is a positive quantity. It can be seen from these equations that both compressibilities have the same unit of the reciprocal of pressure, and the volume expansivity has the unit of the reciprocal of temperature.

11.2.5 Specific Heats

By state postulate, specific heats, like any other thermodynamic properties, can be expressed as functions of two independent properties (see Anim. 11.A.*cp* and 11.A.*cv*). In this section we will express the difference between the specific heats $c_p - c_v$ and their ratio $k = c_p/c_v$ in terms of directly measurable properties. We will also derive a few more useful relations involving specific heats.

The two differential relations for ds, Eqs. (11.51) and (11.52), can be equated, resulting in

$$(c_p - c_v) dT = T \left(\frac{\partial p}{\partial T} \right)_v dv + T \left(\frac{\partial v}{\partial T} \right)_p dp; \quad \left[\frac{\mathrm{kJ}}{\mathrm{kg}} \right] \tag{11.57}$$

Expressing the equation of state as $p = p(T, v)$, the differential of p can be written as

$$dp = \left(\frac{\partial p}{\partial T} \right)_v dT + \left(\frac{\partial p}{\partial v} \right)_T dv$$

Substituting this expression for dp in Eq. (11.57) and rearranging produces

$$\left[(c_p - c_v) - T\left(\frac{\partial v}{\partial T}\right)_p\left(\frac{\partial p}{\partial T}\right)_v\right]dT = T\left[\left(\frac{\partial p}{\partial T}\right)_v + \left(\frac{\partial v}{\partial T}\right)_p\left(\frac{\partial p}{\partial v}\right)_T\right]dv$$

Because T and v are independent properties, dT and dv can assume arbitrarily different values. The only way this identity can be satisfied is for each of the coefficients to be zero. Therefore,

$$(c_p - c_v) = T\left(\frac{\partial v}{\partial T}\right)_p\left(\frac{\partial p}{\partial T}\right)_v; \quad \text{and} \quad \left(\frac{\partial p}{\partial T}\right)_v = -\left(\frac{\partial v}{\partial T}\right)_p\left(\frac{\partial p}{\partial v}\right)_T \tag{11.58}$$

Inserting the second relation into the first, we obtain a second form. The resulting expressions are:

$$c_p - c_v = T\left(\frac{\partial v}{\partial T}\right)_p\left(\frac{\partial p}{\partial T}\right)_v; \quad \text{and} \quad c_p - c_v = -T\left(\frac{\partial v}{\partial T}\right)_p^2\left(\frac{\partial p}{\partial v}\right)_T \tag{11.59}$$

From observed $p-v-T$ data, the difference of specific heats can be obtained from either relation above. For an ideal gas it can be quickly verified that the use of $pv = RT$ reduces the right-hand side of the first equation to R, as was established in Sec. 3.5.3, with the additional assumption that u is a function of T alone (which will be proved in Ex. 11-7).

Using the definitions of β and κ_T given in Eq. (11.56), the right-hand side of the second relation can be expressed in a compact form

$$c_p - c_v = -Tv\left[\frac{1}{v^2}\left(\frac{\partial v}{\partial T}\right)_p^2\right]\Big/\left[\frac{1}{v}\left(\frac{\partial v}{\partial p}\right)_T\right] = vT\frac{\beta^2}{\kappa_T}; \quad \left[\frac{kJ}{kg \cdot K}\right] \tag{11.60}$$

A number of material independent conclusions can be drawn from this equation regarding the nature of specific heats.

1. The isothermal compressibility κ_T is a positive quantity for all materials, as are v and absolute temperature T. Although the volume expansivity can be negative for some special cases—for instance, liquid water at a temperature below 4°C—β^2 must be always positive. Therefore, $c_p - c_v \geq 0$ or $c_p \geq c_v$ for all working substances, not just ideal or perfect gases as was established in Chapter 3.
2. As the absolute temperature approaches zero, c_p approaches c_v.
3. When specific volume reaches a minimum, as it happens for water at 4°C, $\beta = 0$ and the specific heats become equal.
4. For truly incompressible substances, both κ_T and β are zero, because v is a constant. However, as β^2 appears in the numerator, the limiting value of β^2/κ_T goes to zero, making the specific heats equal. For an incompressible solid or liquid $\beta = 0$, justifying the conclusion $c_p = c_v$ we arrived at during the development of the SL model in Chapter 3.

To derive a relation for $k \equiv c_p/c_v$, we rewrite Eqs. (11.38) and (11.46) using the *cyclic relation*

$$\frac{c_v}{T} = \left(\frac{\partial s}{\partial T}\right)_v = \frac{-1}{(\partial v/\partial s)_T\,(\partial T/\partial v)_s};$$

$$\text{and} \quad \frac{c_p}{T} = \left(\frac{\partial s}{\partial T}\right)_p = \frac{-1}{(\partial p/\partial s)_T\,(\partial T/\partial p)_s}$$

The ratio of the two expressions can be simplified by using the *reciprocatory relation*, yielding

$$k \equiv \frac{c_p}{c_v} = \left[\left(\frac{\partial v}{\partial s}\right)_T\left(\frac{\partial s}{\partial p}\right)_T\right]\left[\left(\frac{\partial p}{\partial T}\right)_s\left(\frac{\partial T}{\partial v}\right)_s\right] \tag{11.61}$$

Finally applying the chain rule, we can substitute $(\partial v/\partial p)_T$ for $(\partial v/\partial s)_T(\partial s/\partial p)_T$ and $(\partial p/\partial v)_s$ for $(\partial p/\partial T)_s(\partial T/\partial v)_s$, and obtain

$$k \equiv \frac{c_p}{c_v} = \left(\frac{\partial v}{\partial p}\right)_T\left(\frac{\partial p}{\partial v}\right)_s = \frac{\kappa_T}{\kappa_s} \tag{11.62}$$

where isothermal and isentropic compressibility are introduced from Eq. (11.56).

Two important relations about the behavior of specific heats can be obtained by using the *test of exactness* on the differential relations of Eqs. (11.51) and (11.52).

$$\left(\frac{\partial c_v}{\partial v}\right)_T = T\left(\frac{\partial^2 p}{\partial T^2}\right)_v; \quad \text{and} \quad \left(\frac{\partial c_p}{\partial p}\right)_T = -T\left(\frac{\partial^2 v}{\partial T^2}\right)_p \tag{11.63}$$

The second of these relations can be used to isolate the effect of pressure on c_p at a given temperature. Representing the low pressure limit of c_p by c_{p0} we can integrate the second relation with respect to p while holding T constant, obtaining

$$(c_p - c_{p0})_T = -T\int_0^p \left(\frac{\partial^2 v}{\partial T^2}\right)_p dp; \quad \left[\frac{\text{kJ}}{\text{kg}\cdot\text{K}}\right] \tag{11.64}$$

As for c_v, a separate relation is unnecessary, as we have already established a relation between c_p and c_v in terms of p, v, and T in Eq. (11.59). In the following example, we will show that for an ideal gas $u = u(T)$, eliminating one of the major assumptions used in constructing the IG and PG models in Chapter 3.

EXAMPLE 11-7   **Proof that $u = u(T)$ and $c_p = c_p(T)$ for an Ideal Gas**

Show that the (a) internal energy, (b) enthalpy, and (c) specific heats of an ideal gas are functions of temperature only.

SOLUTION

Use the ideal gas equation $pv = RT$ in the general relation, Eq. (11.63). Also use findings from Example 11-5.

Assumptions

Thermodynamic equilibrium prevails during any differential change of state.

Analysis

Using $p = \dfrac{RT}{v}$, the differential relation for u, Eq. (11.40), can be simplified as

$$du = c_v dT + \left[T\left(\frac{\partial p}{\partial T}\right)_v^{\,p} - p\right]dv = c_v dT$$

This relation suggests that u is a function of c_v and T. Expressing c_v as a function of T and V, $c_v = c_v(T, v)$ we apply $p = RT/v$ to the first relation in Eq. (11.63), yielding

$$\left(\frac{\partial c_v}{\partial v}\right)_T = T\left(\frac{\partial^2 p}{\partial T^2}\right)_v = T\left[\frac{\partial}{\partial T}\left(\frac{\partial p}{\partial T}\right)_v\right]_v = \frac{TR}{v}\left[\frac{\partial}{\partial T}\left(\frac{\partial T}{\partial T}\right)_v^{\,1}\right]_v = 0$$

Therefore, c_v must be a function of T only. Consequently u must also be a function of T alone.

For an ideal gas, $h = u + pv = u + RT = u(T) + RT = h(T)$. Hence, like internal energy, enthalpy is a function of temperature only for an ideal gas.

To show that c_p is also a function of T alone, we could start with the second relation of Eq. (11.63). Alternatively, we apply the IG equation of state to Eq. (11.59).

$$\left(\frac{\partial v}{\partial T}\right)_p = \left(\frac{\partial(RT/p)}{\partial T}\right)_p = \frac{R}{p}; \quad \text{and} \quad \left(\frac{\partial p}{\partial T}\right)_v = \left(\frac{\partial(RT/v)}{\partial T}\right)_v = \frac{R}{v}$$

Inserting these results in Eq. (11.59), we obtain

$$c_p - c_v = T\left(\frac{\partial v}{\partial T}\right)_p\left(\frac{\partial p}{\partial T}\right)_v = T\frac{RR}{pv} = \frac{RT}{pv}R = R$$

Having already proved that c_v is a function of T alone, $c_p = c_v + R$ must also be a function of T alone as R is a constant for a given gas.

Discussion

In developing the PG and IG model in Chapter 3, we used the assumption that u is a function of T alone. Obviously, this is no longer necessary and the ideal gas equation of state alone is sufficient to develop the IG model.

11.2.6 Joule–Thomson Coefficient

In Chapter 4, we analyzed throttling devices (Sec. 4.2.5) whose purpose is to create a pressure drop by passing a fluid through a restriction such as a valve, porous plug, or a capillary tube. It was shown that adiabatic, steady-state throttling can be approximated as an isenthalpic (constant enthalpy) process. As we saw in Example 4-11, a large drop in temperature may accompany the pressure drop, especially when the fluid experiences a change of phase as it comes to equilibrium at the exit pressure at constant enthalpy. In throttling involving an ideal gas, h being a function of T alone, we do not expect any change in temperature. Under certain circumstances, however, the temperature of a working fluid may even increase during throttling.

The **Joule–Thomson coefficient** measures the throttling behavior of a working fluid at a given state by the change in temperature caused by unit change in pressure when the enthalpy is held constant.

$$\mu_J \equiv \left(\frac{\partial T}{\partial p}\right)_h; \quad \left[\frac{K}{kPa}\right] \tag{11.65}$$

The Joule–Thomson coefficient μ_J at a given p and T can be geometrically interpreted as the slope on the constant-enthalpy line at the coordinates (p, T) on a $p\text{-}T$ diagram (Fig. 11.12). A positive slope indicates a drop in temperature and a negative slope indicates just the opposite, since pressure always decreases in throttling. At the maximum of the curve μ_J is zero, which is called the **inversion state** for a given enthalpy.

A family of constant enthalpy lines is also shown in Figure 11.12 for a particular working fluid. The locus of the inversion points, shown by the dotted line, is called the **inversion line**. The intersection of the inversion line and the ordinate is the **maximum inversion temperature** for a given working fluid. Above this temperature, the slope of the constant-enthalpy line is negative at all states. Therefore, a fluid must warm up during throttling if its temperature is above the maximum inversion temperature.

With h remaining constant, the differential relation for enthalpy, Eq. (11.48), can be simplified

$$dh^0 = c_p dT + \left[v - T\left(\frac{\partial v}{\partial T}\right)_p\right] dp; \quad \left[\frac{kJ}{kg}\right] \tag{11.66}$$

Dividing by dp and keeping in mind that h must be held constant, μ_J can be expressed in terms of T, v, and c_p. The same relation can also be rearranged to express c_p in terms of T, v, and μ_J.

$$\mu_J = -\frac{1}{c_p}\left[v - T\left(\frac{\partial v}{\partial T}\right)_p\right]; \quad \left[\frac{K}{kPa}\right], \quad \text{and}$$

$$c_p = -\frac{1}{\mu_J}\left[v - T\left(\frac{\partial v}{\partial T}\right)_p\right]; \quad \left[\frac{kJ}{kg \cdot K}\right] \tag{11.67}$$

The Joule-Thomson coefficient can be hypothetically determined through a simple experiment. For this, the fluid under investigation at a given state is forced through a porous plug (Fig. 11.12) and the exit temperature and pressure are measured. By repeating these measurements with several plugs of varying size, each producing a different pressure drop, the constant-enthalpy line around the inlet state can be constructed. The value of μ_J at the inlet state can now be obtained from Eq. (11.65). Conversely, if μ_J and the $p\text{-}v\text{-}T$ behavior are known, c_p can be evaluated from Eq. (11.67).

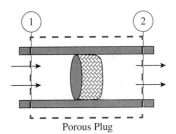

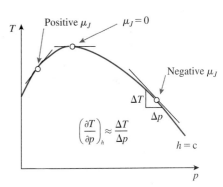

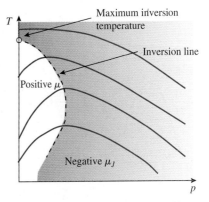

FIGURE 11.12 Joule–Thomson coefficient and the inversion line (see Anim. 11.A.*JouleThomsonCoeff*).

EXAMPLE 11-8 **Numerical Determination of μ_J**

Using a suitable TESTcalc, evaluate the Joule–Thomson coefficient for nitrogen at the following states: (a) 100 kPa and 0% quality, (b) 100 kPa and 100% quality, and (c) 60 MPa and $-100°$C.

SOLUTION

Numerically evaluate the derivative in Eq. (11.65) at the given states.

Assumptions

The PC model is the most accurate model for nitrogen under the given conditions. A one percent drop in pressure can be regarded as a differential change.

TEST Analysis

Launch the PC system-state TESTcalc and select nitrogen as the working fluid. Evaluate State-1 with p1=100 kPa and x1=0%. For the neighboring state, State-2, enter p2=0.99*p1 and h2=h1. In the I/O panel, calculate the Joule–Thomson coefficient by evaluating the expression '=(T2−T1)/(p2−p1)' as 0.0834 K/kPa. Change State-1 to p1=100 kPa and x1=100%. Click Super-Calculate to update State-2, and calculate the coefficient in the I/O panel as 0.0270 K/kPa. Now change p1 and T1 to 60 MPa and −100 deg-C. Update State-2. Evaluate the new coefficient as −0.00045 K/kPa.

Discussion

TEST can be used to construct a constant-enthalpy line passing through a fixed state, State-1. Note that application of the ideal gas model will produce a zero value for the coefficient as a constant enthalpy ensures a constant temperature during throttling.

11.3 THE REAL GAS (RG) MODEL

In Chapter 3, we introduced the RG (real gas) model as a generalization of the IG (ideal gas) model. The IG equation of state was modified with the help of the compressibility factor Z, which was correlated to the reduced temperature and pressure through the use of the generalized compressibility chart (see Anim. 11.B.*RGmodel*). The RG (real gas) equation of state, Eq. (3.75), was used to evaluate the specific volume of a real gas (Anim. 11.B.*IGvsRG*). Two new factors, the enthalpy and entropy departure factors, were introduced by Eqs. (3.78) and (3.81) to provide corrections to the enthalpy and entropy change formulas of the IG model. In this section, we will use the newly developed general relations to explore the theoretical foundation of compressibility and the departure factors to gain new insight into the RG model.

Enthalpy: A close look at the general relation for enthalpy, Eq. (11.49), suggests that the evaluation of enthalpy change can be greatly simplified for a constant-pressure ($dp = 0$) or a constant-temperature ($dT = 0$) process; only one of the integrals survives and the resulting equation can be easily integrated for the restricted process. For a non-restricted process from any given state, State-1, to any other, State-2, as shown in the *T-s* diagram of Figure 11.13, a convenient path can be selected to carry out the integration, since enthalpy, like any other property, is a point function. One such suitable path is shown in Figure 11.13, which consists of a constant temperature process with $T = T_1$ to a third pressure p^*, followed by a constant-pressure process with $p = p^*$, ending with a constant-temperature process with $T = T_2$. The change in enthalpy between State-1 and State-2, therefore, can be split into three separate groups, each corresponding to one path segment.

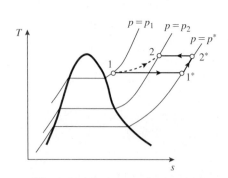

FIGURE 11.13 Regardless of the path (1–2 vs. 1-1*-2*-2) taken, the change in a property between state-1 and state-2 is the same.

$$h_2 - h_1 = (h_2 - h_2^*) - (h_1 - h_1^*) + (h_2^* - h_1^*); \quad \left[\frac{kJ}{kg}\right] \tag{11.68}$$

Clearly, this equality is independent of the choice of p^*. Selecting a very small or zero value (in the limit, of course) for p^* allows an important simplification. Given that at very low pressure a real gas can be accurately modeled as an ideal gas (Sec. 3.8),

c_p can be replaced by its ideal gas limit, c_p^*. Using the general expression, Eq. (11.49), for Δh, each term in Eq. (11.68) can be simplified as follows:

$$h_2 - h_2^* = \int_{2^*}^{2} c_p dT^0 + \int_{2^*}^{2}\left[v - T\left(\frac{\partial v}{\partial T}\right)_p\right]dp$$

$$h_1 - h_1^* = \int_{1^*}^{1} c_p dT^0 + \int_{1^*}^{1}\left[v - T\left(\frac{\partial v}{\partial T}\right)_p\right]dp$$

and

$$h_2^* - h_1^* = \int_{1^*}^{2^*} c_p dT + \int_{1^*}^{2^*}\left[v - T\left(\frac{\partial v}{\partial T}\right)_p\right]dp^0 = \int_{1}^{2} c_{p0} dT = \int_{T_1}^{T_2} c_{p0} dT$$

The last term can be evaluated based purely on temperature integral as c_{p0} for the IG model is a function of T alone. the IG model. Each of the first two terms in Eq. (11.68) displays the same pattern in terms of $(h^* - h)_T$, the difference between the ideal gas and real gas enthalpies at a given state, which is called the **enthalpy departure**. If the enthalpy departure can be calculated, enthalpy difference for a real gas can be obtained from Eq. (11.68). Using the real gas equation of state to model the p–v–T behavior, $v = ZRT/p$ and using the limit $p^* - > 0$, the enthalpy departure from the ideal gas limit at any p and T can be expressed in terms of the compressibility factor Z as follows:

$$(h^* - h)_T = (h^{IG} - h)_T = -\int_{0}^{p}\left[v - T\left(\frac{\partial v}{\partial T}\right)_p\right]dp = RT^2\int_{0}^{p}\left(\frac{\partial Z}{\partial T}\right)_p\frac{dp}{p}; \quad \left[\frac{kJ}{kg}\right] \quad \textbf{(11.69)}$$

Substituting $T_r \equiv T/T_{cr}$ and $p_r \equiv p/p_{cr}$, this equation can be generalized in a non-dimensional form, along the way, defining the **enthalpy departure factor** as

$$Z_h = \frac{(h^{IG} - h)_T}{RT_{cr}} = \frac{(\bar{h}^{IG} - \bar{h})_T}{\bar{R}T_{cr}} = T_r^2\int_{0}^{p_r}\left(\frac{\partial Z}{\partial T_r}\right)_{p_r}\frac{dp_r}{p_r} \quad \textbf{(11.70)}$$

The enthalpy departure factor Z_h, expressed both on mass and molar basis in this equation, was introduced originally in Chapter 3. The integration is carried out numerically with Z data from the compressibility chart, and Z_h is plotted as a function of p_r and T_r in the generalized **enthalpy departure chart** (compare Table E-3 with Anim. 11.B.*propertiesRG*). Using the superscript IG instead of the superscript * to represent an ideal gas behavior, the enthalpy departure factor can be used to simplify Eq. (11.68). In mass and molar terms, the enthalpy difference for a real gas can now be expressed as

$$h_2 - h_1 = (h_2 - h_1)^{IG} - RT_{cr}(Z_{h,2} - Z_{h,1}); \quad \left[\frac{kJ}{kg}\right] \quad \textbf{(11.71)}$$

$$\bar{h}_2 - \bar{h}_1 = \bar{M}(h_2 - h_1) = (\bar{h}_2 - \bar{h}_1)^{IG} - \bar{R}T_{cr}(Z_{h,2} - Z_{h,1}); \quad \left[\frac{kJ}{kmol}\right] \quad \textbf{(11.72)}$$

The term inside the first parenthesis must be obtained from the ideal gas table of the working gas and the term in the second parenthesis evaluated from the enthalpy departure chart.

Internal Energy: Using the definition $h \equiv u + pv$ or, in molar terms $\bar{h} \equiv \bar{u} + p\bar{v}$, the change in internal energy can be expressed as

$$\Delta u = u_2 - u_1 = \Delta h - (p_2 v_2 - p_1 v_1) = \Delta h - R(Z_2 T_2 - Z_1 T_1); \quad \left[\frac{kJ}{kg}\right] \quad \textbf{(11.73)}$$

$$\Delta \bar{u} = \bar{u}_2 - \bar{u}_1 = \Delta \bar{h} - (p_2 \bar{v}_2 - p_1 \bar{v}_1) = \Delta \bar{h} - \bar{R}(Z_2 T_2 - Z_1 T_1); \quad \left[\frac{kJ}{kmol}\right] \quad \textbf{(11.74)}$$

Entropy: Following the same procedure we adopted for enthalpy poses a problem for entropy; the entropy at zero pressure for an ideal or perfect gas is infinity (see Eq. 3.48).

Realizing that our goal is to evaluate the departure of the real gas results from the IG model, let State-1^{IG} and State-2^{IG} in Figure 11.14 represent the states calculated by application of the IG model. The entropy difference for the RG model, then, can be expressed as

$$s_2 - s_1 = (s_2 - s_2^{IG}) - (s_1 - s_1^{IG}) + (s_2^{IG} - s_1^{IG}); \quad \left[\frac{kJ}{kg \cdot K}\right] \tag{11.75}$$

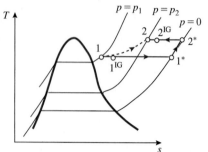

FIGURE 11.14 Entropy difference between states 1 and 2 for a real gas can be related to the difference predicted by the IG model.

The last term on the right hand side poses no problem and can be evaluated using the IG model developed in Chapter 3. To evaluate the first two terms in parentheses, we recognize that states 1 and 1^{IG} may have different values of entropy, but are based on the same pressure and temperature—p_1 and T_1. Likewise states 2 and 2^{IG} are based on p_2 and T_2. So the task reduces to finding the difference $s - s^{IG}$ between entropies of the same state, evaluated by two different models at pressure p and temperature T. We construct fictitious isothermal paths as shown in Figure 11.14, connecting State-1 and State-2 with corresponding zero-pressure states 1* and 2* in Figure 11.14. Representing properties at the zero-pressure states with the subscript *, we express $s - s^{IG}$ (which is applicable to both State 1 and 2) as

$$(s - s^{IG})_{p,T} = [s(p, T) - s_*^{IG}(0, T)] - [s^{IG}(p, T) - s_*^{IG}(0, T)] \tag{11.76}$$

The equation can be simplified by applying Eq. (11.54) at constant temperature ($dT = 0$).

$$(s - s^{IG})_{p,T} = -\int_0^p \left(\frac{\partial v}{\partial T}\right)_p dp + \int_0^p \left(\frac{\partial v^{IG}}{\partial T}\right)_p dp$$

where, according to the RG and IG models, $v = ZRT/p$ and $v^{IG} = RT/p$ respectively. After differentiation of the second term, the right hand side can be expressed as a single integral.

$$(s - s^{IG})_{p,T} = \int_0^p \left[\frac{R}{p}(1 - Z) - \frac{RT}{p}\left(\frac{\partial Z}{\partial T}\right)_p\right] dp; \quad \left[\frac{kJ}{kg \cdot K}\right] \tag{11.77}$$

The problem of finding entropy at zero pressure is now avoided, replaced by finding an integral instead. By substituting $T = T_r T_{cr}$ and $p = p_r p_{cr}$, Eq. (11.77) can be generalized in a non-dimensional form as in the case of enthalpy.

$$Z_s = \frac{(s^{IG} - s)_{p,T}}{R} = \frac{(\bar{s}^{IG} - \bar{s})_{p,T}}{\bar{R}} = \int_0^{p_r} \left[Z - 1 + T_r\left(\frac{\partial Z}{\partial T_r}\right)_{p_r}\right] \frac{dp_r}{p_r} \tag{11.78}$$

The **entropy departure factor** Z_s was defined originally in Chapter 3. The integration is carried out numerically with Z data from the compressibility chart and Z_s is plotted as a function of p_r and T_r in the generalized **entropy departure chart** (compare Table E-4 with Anim. 11.B.*propertiesRG*). Switching to superscript IG to indicate ideal gas, the entropy departure factor can be used to simplify Eq. (11.75), producing the following mass and mole based entropy difference formulas for a real gas.

$$s_2 - s_1 = (s_2 - s_1)^{IG} - R(Z_{s,2} - Z_{s,1}); \quad \left[\frac{kJ}{kg \cdot K}\right] \tag{11.79}$$

$$\bar{s}_2 - \bar{s}_1 = \bar{M}(s_2 - s_1) = (\bar{s}_2 - \bar{s}_1)^{IG} - \bar{R}(Z_{s,2} - Z_{s,1}); \quad \left[\frac{kJ}{kmol \cdot K}\right] \tag{11.80}$$

The term in the first parenthesis must be obtained from the ideal gas table of the working gas, and the term in the second parenthesis evaluated from the entropy departure chart.

It should be noted that the PG model can be used as a substitute for the IG model if the temperature change between the states is moderate. As has been discussed in Sec. 3.5.7, there are two sets of compressibility charts; the Lee-Kesler and Nelson-Obert charts, which can be used to evaluate the compressibility factor along with enthalpy and entropy departures.

EXAMPLE 11-9 Real Gas (RG) Model

Nitrogen at 10 MPa and 150 K flows steadily through a tube with a mass flow rate of 1 kg/s, receiving heat from the surroundings at 300 K. At the end of the tube, it enters an expansion valve and leaves at 1 MPa and 125 K. Using the RG model, determine (a) the heat transfer, and (b) entropy generation rate in the thermodynamic universe. *What-if scenario:* What would the answers be if (c) the IG or (d) the PC model were used instead? Use Lee–Kesler charts.

SOLUTION

Perform a steady-state energy and entropy analysis of the complete system. Use the enthalpy and entropy departure charts to evaluate properties with the RG model.

Assumptions

LTE at the inlet and exit states, State-1 and State-2. The ideal gas model, necessary as a sub-model in evaluating enthalpy and entropy differences, can be replaced with the PG model when the IG tables (Tables D-4 - D-15) do not extend to the temperatures of interest.

Analysis

The energy and entropy equations, Eqs. (4.2) and (4.3), for the system's universe, enclosed within the red boundary of Figure 11.15, can be simplfied as follows:

$$\frac{d\cancel{U}^0}{dt} = \dot{m}(j_i - j_e) + \dot{Q} - \cancel{\dot{W}_{ext}}^0 \cong \dot{m}(h_i - h_e) + \dot{Q}$$

$$\Rightarrow \quad \dot{Q} = \dot{m}(h_e - h_i) = \dot{m}(h_2 - h_1) \quad \textbf{(11.81)}$$

$$\frac{d\cancel{S}^0}{dt} = \dot{m}(s_i - s_e) + \frac{\dot{Q}}{T_B} + \dot{S}_{gen,univ}; \quad \Rightarrow \quad \dot{S}_{gen,univ} = \dot{m}(s_2 - s_1) - \frac{\dot{Q}}{T_0} \quad \textbf{(11.82)}$$

To evaluate the states, we obtain the necessary material properties of nitrogen: $p_{cr} = 3.39$ MPa, $T_{cr} = 126.2$ K, $c_p = 1.04$ kJ/kg·K, and $R = 8.314/28 = 0.297$ kJ/kg·K from Tables C-1 and E-1. To obtain critical properties from TEST, launch the flow state RG TESTcalc, select nitrogen, and calculate a state with $p_r = 1$ and $T_r = 1$. Using the departure charts (Tables E-3 and E-4) the states can now be evaluated.

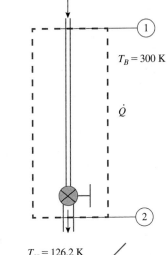

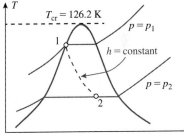

FIGURE 11.15 Schematic and the *T-s* diagram for Ex. 11-9.

State-1: (given p_1, T_1):

$$\Rightarrow \quad p_{r,1} = 10/3.39 = 2.95; \quad T_{r,1} = 150/126.2 = 1.19; \quad Z_{h,1} = 2.85; \quad Z_{s,1} = 1.77$$

State-2: (given p_2, T_2):

$$\Rightarrow \quad p_{r,2} = 1/3.39 = 0.295; \quad T_{r,2} = 125/126.2 = 0.99; \quad Z_{h,2} = 0.3; \quad Z_{s,2} = 0.2$$

The enthalpy difference now can be calculated from Eq. (11.71) as

$$h_2 - h_1 = (h_2 - h_1)^{IG} - RT_{cr}(Z_{h,2} - Z_{h,1})$$

$$\cong (h_2 - h_1)^{PG} - RT_{cr}(Z_{h,2} - Z_{h,1}) = c_p(T_2 - T_1) - RT_{cr}(Z_{h,2} - Z_{h,1})$$

$$= (1.04)(125 - 150) - (0.297)(126.2)(0.3 - 2.8) = 67.7 \frac{kJ}{kg}$$

The entropy difference can be similarly obtained from Eq. (11.79)

$$s_2 - s_1 = (s_2 - s_1)^{IG} - R(Z_{s,2} - Z_{s,1})$$

$$\cong (s_2 - s_1)^{PG} - RT_{cr}(Z_{s,2} - Z_{s,1}) = c_p \ln\frac{T_2}{T_1} - R \ln\frac{p_2}{p_1} - R(Z_{s,2} - Z_{s,1})$$

$$= (1.04) \ln\frac{125}{150} - (0.297) \ln\frac{1}{10} - (0.297)(0.2 - 1.7)$$

$$= 0.94 \frac{kJ}{kg \cdot K}$$

The heat transfer and entropy generation rates can now be obtained as

$$\dot{Q} = \dot{m}(h_2 - h_1) = 67.7 \text{ kW};$$

$$\dot{S}_{\text{gen,univ}} = \dot{m}(s_2 - s_1) - \frac{\dot{Q}}{T_0} = (1)(0.94) - \frac{67.7}{300} = 0.71\frac{\text{kW}}{\text{K}}$$

TEST Analysis

Launch the RG single-flow open-steady TESTcalc and select nitrogen as the working fluid. Select the Lee–Kesler radio-button. Evaluate State-1 with p1=10 MPa, T1=150 K, and mdot1=1 kg/s, and State-2 with p2=1 MPa, T2=125 K, and mdot2=mdot1. In the device panel, set the two states as the inlet and exit states, correct T_B to 300 K, enter Wdot_ext=0, and click Calculate. Qdot and Sdot_gen are calculated as 68.0 kW and 0.719 kW/K respectively. These results agree closely with the manual solution. Click Super-Calculate to generate TEST codes.

What-if scenario

Launch the corresponding IG TESTcalc on a separate browser tab and paste the TEST codes generated by the RG TESTcalc on the I/O panel. Use the Load button to update the solution and obtain the following results in the device panel: Qdot = −25.7 kW and Sdot_gen = 0.58 kW/K. Similarly, use of the PC TESTcalc produces Qdot = 69.6 kW and Sdot_gen = 0.727 kW/K.

Discussion

The PC model being the most accurate of all, its results can be used as benchmarks. Use of IG model produces totally erroneous results; it predicts heat loss when heat is actually gained. The IG model assumes away any dependence of enthalpy on pressure, while the RG model handles the pressure dependence through the enthalpy departure term.

11.4 MIXTURE MODELS

In Chapter 3, we developed several models—the SL, PC, PG, IG, and RG models—to evaluate thermodynamic properties of pure substances. However, a large number of devices and processes involve some type of **mixture**. For instance, air is treated as a mixture of dry air and water vapor in air-conditioning (Chapter 12), and exhaust gas is treated as a mixture of its component species in reacting flows (Chapter 13 and 14).

In this section our focus will be on the mixture of gases, where the constituents can be modeled by the PG, IG, or RG model. As with all other models, our goal is to develop expressions for change in volume, internal energy, enthalpy, and entropy for a mixture. It should be recalled that a mixture can be treated as a *pure substance* as long as its chemical composition remains unchanged. In the following sub-sections, we first discuss a few general features of a mixture followed by development of specific mixture models.

11.4.1 Mixture Composition

A mixture of a given number of gases can be formed in three different manners: by mixing known masses, known moles (Sec. 1.5.2), or known volumes of the constituents together. Sufficient time should be allowed so that the mixture formed is chemically uniform. It is assumed that there is no chemical reaction among the components of the mixture. To describe a mixture, a few new properties are necessary, which are introduced next.

Mass Fraction: If the mass of the kth component of a mixture is represented by the symbol m_k, the mixture can be described by specifying m_k for each component. Consequently, the total mass is given by the following summation taken over all components.

$$m = \sum_k m_k; \quad [\text{kg}] \tag{11.83}$$

Alternatively, a mixture composition can be expressed in terms of the *specific* component mass, called the **mass fraction** (Fig. 11.16).

$$x_k \equiv \frac{m_k}{m}; \quad \sum_k x_k = \sum_k \frac{m_k}{m} = \frac{1}{m}\sum_k m_k = 1 \quad \text{(11.84)}$$

The sum of mass fractions of all species must be one in a mixture. Notice the similarity in the definition of mass fraction with that of the quality (Sec. 3.4.1) of a vapor-liquid mixture; quality x therefore, can be interpreted as the mass fraction of vapor in a saturated mixture of liquid and vapor. A **gravimetric analysis** of a mixture is the experimental determination of mass fractions of the components.

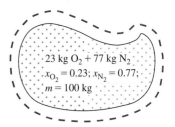

FIGURE 11.16 A mixture of O_2 and N_2 is described by the mass fractions of its constituents and the total mass (see Anim. 11.C.*massFraction*).

Mole Fraction: The amount of a component in a mixture can be expressed in terms of its mass or mole. If no chemical reaction is allowed among the constituents, the total mole n in a mixture must be equal to the sum of its components' moles. Representing the mole of the species k in a mixture by the symbol n_k, the mixture can be described by specifying n_k or the **mole fraction** y_k, which is defined in a parallel manner (Fig. 11.17) as the mass fraction.

$$n = \sum_k n_k; \quad [\text{kmol}],$$

$$y_k \equiv \frac{n_k}{n}, \quad \sum_k y_k = \sum_k \frac{n_k}{n} = \frac{1}{n}\sum_k n_k = 1 \quad \text{(11.85)}$$

The sum of mole fractions of all components must be in unity, just like the sum of mass fractions in a mixture.

The mass of a pure species per unit mole (kg/kmol) depends on its chemical formula and is a material property known as the molar mass (Fig. 11.18). Without any chemical reaction, the composition of a mixture remains frozen. An **apparent molar mass** $\overline{M}$ can be attributed to the mixture as the average mass of a unit mole of the mixture. Similarly, an **apparent gas constant** can be defined in terms of $\overline{M}$ as follows:

$$\overline{M} \equiv \frac{m}{n} = \frac{\sum_k m_k}{n} = \sum_k \frac{n_k \overline{M}_k}{n} = \sum_k y_k \overline{M}_k; \quad \left[\frac{\text{kg}}{\text{kmol}}\right] \quad \text{(11.86)}$$

$$\text{or} \quad \overline{M} = \frac{m}{\sum_k n_k} = \frac{1}{\sum_k n_k/m} = \frac{1}{\sum_k m_k/(\overline{M}_k m)} = \frac{1}{\sum_k x_k/\overline{M}_k}; \quad \left[\frac{\text{kg}}{\text{kmol}}\right] \quad \text{(11.87)}$$

$$\text{where} \quad R \equiv \frac{\overline{R}}{\overline{M}}; \quad \left[\frac{\text{kJ}}{\text{kg} \cdot \text{K}}\right] \quad \text{(11.88)}$$

The mixture molar mass, $\overline{M}$, from Eq. (11.86) can be interpreted as the mole-fraction weighted average of the component molar masses. It will be shown shortly that any molar property of a

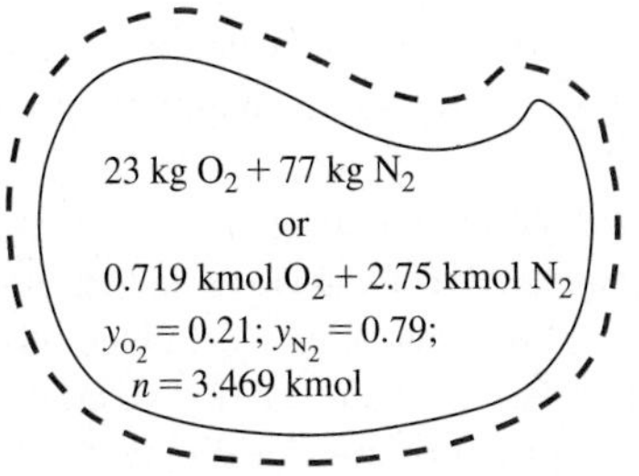

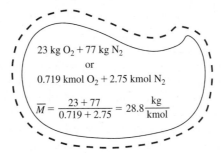

FIGURE 11.17 A mixture of O_2 and N_2 is described by the mole fractions of its constituents and the total mole (see Anim. 11.C.*moleFraction*).

FIGURE 11.18 Molar mass of a mixture is the ratio of its mass to its mole (see Anim. 11.C.*molarMass*).

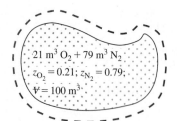

FIGURE 11.19 A mixture of O_2 and N_2 is described by the volume fractions of its constituents and the total volume.

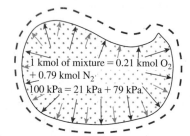

FIGURE 11.20 The total pressure of 100 kPa in a mixture is the sum of the partial pressures of its components (see Anim. 11.C.*DaltonLawIG*).

mixture can be similarly expressed as the mole-fraction weighted average of the molar specific properties of the constituents. The second formula, Eq. (11.87) can be interpreted simply as the reciprocal of total mole (kmol of mixture) in a unit mass (1 kg) of the mixture.

The mass fraction of species k can be obtained from the molar composition of a mixture as

$$x_k = \frac{m_k}{m} = \frac{m_k}{\sum_k m_k} = \frac{n_k \overline{M}_k}{\sum_k n_k \overline{M}_k} = \frac{(n_k/n)\overline{M}_k}{\sum_k (n_k/n)\overline{M}_k} = \frac{y_k \overline{M}_k}{\sum_k y_k \overline{M}_k} = \frac{y_k \overline{M}_k}{\overline{M}} \qquad \textbf{(11.89)}$$

The final expression can be interpreted as the ratio of the species mass (kg of k) to the total mass (kg of mixture) in a unit mole (1 kmol) of the mixture (see Anim. 11.C.*moleFraction*). Similarly, the mole fraction of species k can be obtained when the composition is given in terms of mass fractions.

$$y_k = \frac{n_k}{n} = \frac{(m_k/\overline{M}_k)}{\sum_k (m_k/\overline{M}_k)} = \frac{m_k/(m\overline{M}_k)}{\sum_k m_k/(m\overline{M}_k)} = \frac{(x_k/\overline{M}_k)}{\sum_k (x_k/\overline{M}_k)} = \frac{(x_k/\overline{M}_k)}{(1/\overline{M})} \qquad \textbf{(11.90)}$$

The final expression can be interpreted as the ratio of the mole of component k (kmol of k) to the total mole (kmol of mixture) in a unit mass (1 kg) of the mixture (see Anim. 11.C.*massFraction*).

Volume Fraction: A gas mixture can also be formed by mixing known volumes (Fig. 11.19) of species at the same temperature and pressure of the mixture. Representing the volume of component k, called the **partial volume**, by the symbol V_k, the **volume fraction** z_k is defined as

$$z_k \equiv \frac{V_k}{V} \qquad \textbf{(11.91)}$$

In Sec. 11.4.3, we will prove that the total volume is the sum of the partial volumes.

Partial Pressure: Momentum transfer by molecules being responsible for the thermodynamic property pressure, each species in a gas mixture contributes according to its molar presence—the higher the mole fraction of a particular species, the higher is its contribution to the total pressure p (Fig. 11.20). The **partial pressure** p_k of a constituent component k is defined as the pressure of the species when it alone occupies the total volume of the system at the temperature of the mixture. We will revisit the concepts of partial pressure and partial volume and relate these two properties in the next section.

11.4.2 Mixture TESTcalcs

TEST offers a suite of calculators for analyzing mixtures of gases. Mixture TESTcalcs can be found in almost all branches of the TESTcalc map. The binary mixture TESTcalcs allow only two components while the general mixture TESTcalcs allow any number of components. The moist air TESTcalc handles a binary mixture of dry air and water vapor and will be discussed in Chapter 12. The IGE state TESTcalcs involve chemical equilibrium of a gas mixture and will be discussed in Chapter 14. The remainder of the mixture TESTcalcs are identified by the underlying material model used. For example, a PG/PG TESTcalc indicates a mixture of two gases, each modeled as a perfect gas. The n-PG or n-IG TESTcalcs, similarly, can handle any number of perfect or ideal gases in a mixture.

The mixture TESTcalcs differ from their pure-substance counterparts in the state panel only. The binary mixture TESTcalcs (PG/PG, IG/IG, and RG/RG) have two selectors for components labeled as A and B, and two additional properties—mass and mole fractions of component A. Composition of a mixture is specified by entering the mass or mole fraction of component A. To create a state with purely A or B, use a mass (or mole) fraction of 1 or 0 respectively. In the n-PG or n-IG TESTcalcs, the mixture is composed by selecting each component and specifying its amount (in mass, mole, or volume) in a composition box in the state panel. In the binary mixture TESTcalcs, the composition can be changed between states, while in the n-IG or n-PG TESTcalcs, once a mixture is composed, it is frozen for all subsequent state calculations. When a state is evaluated by the n-IG or n-PG TESTcalc, the partial specific properties and partial molar properties of each component are listed in a table in the composition panel.

EXAMPLE 11-10 **Mass and Mole Fractions of a Gas Mixture**

A gas mixture has the following composition: N_2: 2 kg; H_2: 1 kg; and CO_2: 4 kg. Determine (a) the mass fraction and (b) mole fraction of H_2, and the (c) apparent molar mass of the mixture.

SOLUTION

From the given gravimetric analysis obtain the total mole of the mixture, from which the desired answers can be found using the definitions of mole fraction and molar mass.

Assumptions

The mixture is uniform so that a single equilibrium state describes the mixture.

Analysis

The mass fraction of hydrogen can be obtained from its definition.

$$x_{H_2} = \frac{m_{H_2}}{m} = \frac{m_{H_2}}{m_{N_2} + m_{H_2} + m_{CO_2}} = \frac{1}{2 + 1 + 4} = 0.143$$

To find the mole fraction, the total mole of the mixture, which is simply the sum of the component moles, is determined first.

$$n = \sum_k n_k = \frac{m_{N_2}}{\overline{M}_{N_2}} + \frac{m_{H_2}}{\overline{M}_{H_2}} + \frac{m_{CO_2}}{\overline{M}_{CO_2}} = \frac{2}{28} + \frac{1}{2.02} + \frac{4}{44} = 0.657 \text{ kmol}$$

Therefore,

$$y_{H_2} = \frac{n_{H_2}}{n} = \frac{0.495}{0.657} = 0.753 \quad \text{and} \quad \overline{M} = \frac{m}{n} = \frac{7}{0.657} = 10.65 \frac{\text{kg}}{\text{kmol}}$$

TEST Analysis

Launch the n-PG system-state TESTcalc and compose the mixture of State-1 by entering the mass of each component. To remove a component from the mixture, add zero amount of that component. As each additional component is added, the composition and partial properties are updated in the composition panel. Once the mixture is composed, enter m1=7 kg, and click Calculate to obtain the molar mass and other material properties of the mixture, verifying the manual solution.

Discussion

If the mass fractions were given instead of species masses, the same analysis could be carried out on the basis of 1 kg of the mixture. Observe that even a small amount of hydrogen on a mass basis has a strong influence on the molar mass of the mixture.

EXAMPLE 11-11 **Mass and Mole Fractions of a Gas Mixture**

A gas mixture contains H_2O, CO_2, and N_2. If the mole fractions of CO_2 and N_2 are 0.2 and 0.3 respectively, determine (a) the mass fractions of H_2O and (b) apparent molar mass of the mixture.

SOLUTION

From a molar analysis, obtain the mass of one kmol of the mixture, and find the desired answers by using the definitions of mass fraction and molar mass.

Assumptions

The mixture is uniform so that a single equilibrium state describes the mixture.

Analysis

Using $\sum_k y_k = 1$, we obtain $y_{H_2O} = 1 - y_{CO_2} - y_{N_2} = 1 - 0.3 - 0.2 = 0.5$.

The mass of 1 kmol of mixture can be found as

$$m = m_{H_2O} + m_{CO_2} + m_{N_2} = y_{H_2O}\overline{M}_{H_2O} + y_{CO_2}\overline{M}_{CO_2} + y_{N_2}\overline{M}_{N_2}$$

$$= (0.5)(18) + (0.2)(44) + (0.3)(28) = 26.2 \text{ kg};$$

This must be the apparent molar mass of the mixture. Therefore, $\overline{M} = 26.2$ kg/kmol.

The mass fraction of H_2O now can be calculated as

$$x_{H_2O} = \frac{m_{H_2O}}{m} = \frac{m_{H_2O}\overline{M}_{H_2O}}{m} = \frac{(0.5)(18)}{26.2} = 0.344$$

TEST Analysis

The procedure for TEST solution is identical to the one described in the previous example except the amount of each component must be entered as '% by Mole'.

Discussion

Either the n-PG or n-IG TESTcalc can be used in the TEST Analysis. Note that the solution does not require for the mixture to be a gas mixture.

11.4.3 PG and IG Mixture Models

The conditions under which a pure gas behaves like an ideal gas were discussed in Sec. 3.8. Basically, a gas is called an ideal gas when it obeys the IG equation of state $pv = RT$. When different gases, each of which can be treated as an ideal gas, are mixed, the presence of dissimilar types of molecules does not affect the molecular interactions in any special way as long as there are no chemical reactions. Therefore, such a mixture satisfies the ideal gas criteria, $pv = RT$, and is called an **ideal gas mixture**. If the constituent gases have constant specific heats, it will be shown that the mixture also has constant specific heats, and is called a **perfect gas mixture**. Obviously, formulas derived for the IG mixture model will also apply to the PG mixture model, except further simplification may be possible in the latter due to the assumption of constant specific heats.

Two sub-models are generally used to analyze a PG or IG mixture. In the **Dalton model**, a mixture at a given state is looked upon (Fig. 11.21) as an aggregate of pure constituent gases, each sharing the same volume and temperature of the mixture but having its own partial pressure. Microscopically speaking, the temperature of a gas depends on the kinetic energy of its molecules while the pressure depends on the collision frequency. Therefore, if all the constituents except the kth species of a gas mixture occupying a system are instantly removed, the remaining component will still have the same volume and temperature of the original system, but its pressure will decrease to p_k, the **partial pressure** of the component k, which can be predicted by applying the IG equation of state.

The IG equation, $pv = RT$ applied to the mixture and the kth component alone yields:

$$p\Psi = n\overline{R}T; \quad p_k\Psi = n_k\overline{R}T$$

$$\text{Therefore,} \quad \frac{p_k}{p} = \frac{n_k}{n} = y_k \tag{11.92}$$

The partial pressure of an ideal gas (hence, perfect gas), therefore, can be obtained from the total pressure if the mole fraction of the species in the mixture is known. Summing the partial pressure over all the components, we establish the **Dalton's law** of partial pressure, which states that the sum of partial pressures add up to the total pressure.

$$\sum_k p_k = \sum_k \frac{n_k\overline{R}T}{\Psi} = \frac{\overline{R}T}{\Psi}\sum_k n_k = \frac{n\overline{R}T}{\Psi} = p; \quad [\text{kPa}] \tag{11.93}$$

The **Amagat model** treats a mixture in a completely different way. Suppose all components of a mixture suddenly separate into subsystems with imaginary walls (Fig. 11.22) so that the temperature and pressure in each subsystem are still the same as that of the mixture. Then the volume occupied by the kth species is called the **partial volume** Ψ_k. The ideal gas equation applied to the mixture and the kth species yields

$$p\Psi = n\overline{R}T; \quad p\Psi_k = n_k\overline{R}T; \quad \Rightarrow \quad \frac{\Psi_k}{\Psi} = \frac{n_k}{n} = y_k \tag{11.94}$$

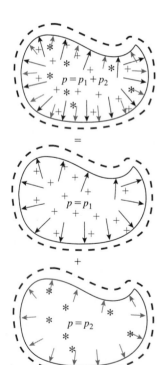

FIGURE 11.21 Dalton model: the subsystems share the temperature and volume of the original system, but have different partial pressures (see Anim. 11.C.models DaltonAmagat).

The relation between the partial volumes and total volume can be obtained by summing V_k over all the components.

$$\sum_k V_k = \sum_k \frac{n_k \overline{R} T}{p} = \frac{\overline{R} T}{p} \sum_k n_k = \frac{n \overline{R} T}{p} = V; \quad [\text{m}^3] \qquad (11.95)$$

The sum of the partial volumes is equal to the total volume for an ideal gas (hence, perfect gas) mixture. This is known as **Amagat's law** of partial volume.

Dalton's law and Amagat's law can be combined to produce a simple yet powerful relation.

$$\frac{p_k}{p} = \frac{V_k}{V} = \frac{n_k}{n} = y_k \qquad (11.96)$$

The mole fraction of a component is equal to its partial pressure fraction and its volume fraction. We will use this equation frequently to determine composition of a mixture if either the partial pressures or partial volumes of the components are known. Note that Dalton's and Amagat's laws cannot be applied to RG mixtures.

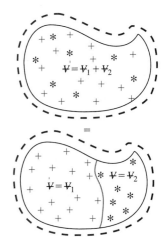

FIGURE 11.22 Amagat model: the subsystems share the temperature and pressure of the original system, but have different partial volumes (see Anim. 11.C. modelsDaltonAmagat).

EXAMPLE 11-12   Partial Properties of a Gas Mixture

The gas mixture descried in Example 11-10 is kept in a tank of volume 2 m^3 at 500 kPa and 400 K. Determine (a) the total mass, (b) partial pressure, and (c) partial volume of each constituent. Use the PG mixture model.

SOLUTION

Using results from Example 11-10, evaluate the mixture gas constant and species mole fractions. Evaluate partial pressures and partial volumes using Eq. (11.96).

Assumptions

The mixture is uniform so that a single equilibrium state describes the mixture. Dalton's and Amagat's laws are applicable.

Analysis

The molar mass of the mixture was evaluated as 10.65 kg/kmol in Example 11-10. Therefore, the gas constant can be calculated as

$$R = \overline{R}/\overline{M} = 8.314/10.65 = 0.781 \text{ kJ/kg} \cdot \text{K}$$

Using IG equation of state, the mass of the mixture can be calculated as

$$m = \frac{V}{v} = \frac{pV}{RT} = \frac{(500)(2)}{(0.781)(400)} = 3.201 \text{ kg}$$

The partial pressures and volumes are calculated by applying Eq. (11.96).

	N$_2$	H$_2$	CO$_2$
$n_k = \dfrac{m_k}{\overline{M}_k}$ [kmol]	2/28 = 0.071	1/2 = 0.5	4/44 = 0.091
$y_k = \dfrac{n_k}{n}$	0.071/0.657 = 0.108	0.5/0.657 = 0.761	0.091/0.657 = 0.139
$p_k = p y_k$ [kPa]	(0.108)(500) = 54.0	(0.761)(500) = 380.5	(0.139)(500) = 69.5
$V_k = V y_k$ [m^3]	(0.108)(2) = 0.216	(0.761)(2) = 1.522	(0.139)(2) = 0.278

TEST Analysis

Compose the mixture using the n-PG system-state TESTcalc as described in Example 11-10. Enter the known volume, pressure, and temperature and click Calculate. Some of the partial properties must be manually calculated from the mole fractions displayed in the composition panel.

Discussion

The partial volumes provide a simple way to produce a mixture from its components and to experimentally determine the composition of a mixture. For the latter, which is known as a **volumetric analysis**, the mixture is passed through different chemicals that selectively absorb different components. The reduction in volume as a fraction of the original mixture volume yields the partial volume of the species absorbed, and the molar composition of the mixture then can be deduced using Eq. (11.96).

Internal Energy:

A system with an IG mixture as the working fluid can be divided into as many number of subsystems as the number of components of the mixture in accordance with either the Dalton or Amagat model (Figs. 11.21 and 11.22). Any extensive property, therefore, can be expressed by summing the corresponding property of all sub-systems. Therefore, the total internal energy U of the system can be expressed in terms of mass- or mole- based specific internal energy of each component.

$$U = \sum_k U_k = \sum_k m_k u_k(p_k, T) = m \sum_k x_k u_k(T) = \sum_k n_k \bar{u}_k = n \sum_k y_k \bar{u}_k(T); \quad [\text{kJ}] \quad \textbf{(11.97)}$$

$$\Rightarrow \quad u = \frac{U}{m} = \sum_k x_k u_k(T); \quad \left[\frac{\text{kJ}}{\text{kg}}\right] \quad \text{and} \quad \bar{u} = \frac{U}{n} = \sum_k y_k \bar{u}_k(T); \quad \left[\frac{\text{kJ}}{\text{kmol}}\right] \quad \textbf{(11.98)}$$

The contribution from unit mass of component k in a mixture $u_k(p_k, T)$ towards the total internal energy U is called the **partial specific internal energy**, which is equal to the specific internal energy, $u_k(T)$, of the pure component k, as given by the IG or PG model (Sec. 3.5.3). Likewise, $\bar{u}_k(p_k, T) = \bar{u}_k(T)$ is the **partial molar specific internal energy**. We can use the n-IG state TESTcalcs to explore how $u_k(p_k, T)$ or $\bar{u}_k(p_k, T)$ depends only on the temperature of the mixture.

For a PG mixture, the change in internal energy can be simplified by using the perfect gas model for the components and defining an average c_v for the mixture.

PG Mixture:

$$\Delta u^{\text{PG}} = \sum_k x_k \Delta u_k = \left(\sum_k x_k c_{v,k}\right)\Delta T = c_v \Delta T \quad \text{where} \quad c_v = \sum_k x_k c_{v,k}; \quad \left[\frac{\text{kJ}}{\text{kg}}\right] \quad \textbf{(11.99)}$$

$$\Delta \bar{u}^{\text{PG}} = \sum_k y_k \Delta \bar{u}_k = \left(\sum_k y_k \bar{c}_{v,k}\right)\Delta T = \bar{c}_v \Delta T \quad \text{where} \quad \bar{c}_v = \sum_k y_k \bar{c}_{v,k}; \quad \left[\frac{\text{kJ}}{\text{kmol}}\right]$$
$$\textbf{(11.100)}$$

For an IG mixture, the change in internal energy can be expressed in terms of component specific heat or the mixture specific heat at constant volume as defined above.

IG Mixture:

$$\Delta u^{\text{IG}} = \sum_k x_k[u_k(T_2) - u_k(T_1)] = \sum_k x_k \int_{T_1}^{T_2} c_{v,k} dT = \int_{T_1}^{T_2}\sum_k x_k c_{v,k} dT = \int_{T_1}^{T_2} c_v(T) dT; \quad \left[\frac{\text{kJ}}{\text{kg}}\right]$$
$$\textbf{(11.101)}$$

$$\Delta \bar{u}^{\text{IG}} = \sum_k y_k[\bar{u}_k(T_2) - \bar{u}_k(T_1)] = \sum_k y_k \int_{T_1}^{T_2} \bar{c}_{v,k} dT = \int_{T_1}^{T_2}\sum_k y_k \bar{c}_{v,k} dT = \int_{T_1}^{T_2} \bar{c}_v(T) dT; \quad \left[\frac{\text{kJ}}{\text{kmol}}\right]$$
$$\textbf{(11.102)}$$

where $u_k(T)$ or $\bar{u}_k(T)$ can be obtained from the ideal gas tables (Tables D-4 through D-15) for species k, and $c_v(T) = c_p(T) - R$ from the polynomial relations for c_p with temperature (Table D-1). Both PG and IG gas mixture formulas can be seen to simplify into the same form as the pure gas model, the only difference being the specific heat is now evaluated by suitable averaging as prescribed by Eq. (11.99).

Enthalpy:

The development of formulas for enthalpy difference parallels that of internal energy. The resulting formulas are summarized below.

$$H = \sum_k H_k = \sum_k m_k h_k(p_k, T) = m \sum_k x_k h_k(T) = \sum_k n_k \bar{h}_k = n \sum_k y_k \bar{h}_k(T); \quad [\text{kJ}] \quad \textbf{(11.103)}$$

$$\Rightarrow \quad h = \frac{H}{m} = \sum_k x_k h_k(T); \quad \left[\frac{\text{kJ}}{\text{kg}}\right] \quad \text{and} \quad \bar{h} = \frac{H}{n} = \sum_k y_k \bar{h}_k(T); \quad \left[\frac{\text{kJ}}{\text{kmol}}\right] \quad \textbf{(11.104)}$$

where $h_k(p_k, T) = h_k(T)$ and $\bar{h}_k(p_k, T) = \bar{h}_k(T)$ are called the **partial specific enthalpy** and **partial molar specific enthalpy**, respectively, and are equal to the corresponding specific enthalpies of a pure component since enthalpy, like internal energy, is a function of temperature alone for an ideal gas. We can use the n-PG or n-IG state TESTcalcs to explore how $h_k(p_k, T)$ or $\bar{h}_k(p_k, T)$ depends only on the temperature of the mixture.

For a mixture with constant composition, the enthalpy difference for the PG and IG mixture models can be expressed as follows:

PG Mixture:

$$\Delta h^{\text{PG}} = \sum_k x_k \Delta h_k = \left(\sum_k x_k c_{p,k}\right) \Delta T = c_p \Delta T; \quad \left[\frac{\text{kJ}}{\text{kg}}\right] \quad \text{where} \quad c_p = \sum_k x_k c_{p,k} \quad \textbf{(11.105)}$$

$$\Delta \bar{h}^{\text{PG}} = \sum_k y_k \Delta \bar{h}_k = \left(\sum_k y_k \bar{c}_{p,k}\right) \Delta T = \bar{c}_p \Delta T; \quad \left[\frac{\text{kJ}}{\text{kmol}}\right] \quad \text{where} \quad \bar{c}_p = \sum_k y_k \bar{c}_{p,k} \quad \textbf{(11.106)}$$

IG Mixture:

$$\Delta h^{\text{IG}} = \sum_k x_k [h_k(T_2) - h_k(T_1)] = \sum_k x_k \int_{T_1}^{T_2} c_{p,k} dT = \int_{T_1}^{T_2} \sum_k x_k c_{p,k} dT = \int_{T_1}^{T_2} c_p(T) dT; \quad \left[\frac{\text{kJ}}{\text{kg}}\right] \quad \textbf{(11.107)}$$

$$\Delta \bar{h}^{\text{IG}} = \sum_k y_k [\bar{h}_k(T_2) - \bar{h}_k(T_1)] = \sum_k y_k \int_{T_1}^{T_2} \bar{c}_{p,k} dT = \int_{T_1}^{T_2} \sum_k y_k \bar{c}_{p,k} dT = \int_{T_1}^{T_2} \bar{c}_p(T) dT; \quad \left[\frac{\text{kJ}}{\text{kmol}}\right] \quad \textbf{(11.108)}$$

As with $u_k(T)$, specific enthalpy $h_k(T)$ can be obtained from the ideal gas tables. As in the case of internal energy formulas, a perfect gas mixture can be regarded as a pseudo-pure gas with suitably averaged specific heats for the evaluation of enthalpy.

Entropy:

Entropy of an ideal or perfect gas is a function of temperature and pressure. For the kth subsystem of a mixture, analyzed by the Dalton model, the pressure used in the entropy expression of Eqs. (3.51) or (3.55) for a pure gas, must be the partial pressure $p_k = y_k p$. The contribution of component k towards the total entropy, the **partial specific entropy**, $s_k(p_k, T)$, therefore, is different (higher) than the specific entropy of the pure component, $s_k(p, T)$, at the total pressure p and temperature T of the mixture.

$$S = \sum_k S_k = \sum_k m_k s_k(p_k, T) = m \sum_k x_k s_k(p_k, T) = \sum_k n_k \bar{s}_k = n \sum_k y_k \bar{s}_k(p_k, T) \quad \textbf{(11.109)}$$

$$\Rightarrow \quad s = \frac{S}{m} = \sum_k x_k s_k(p_k, T); \quad \left[\frac{\text{kJ}}{\text{kg} \cdot \text{K}}\right] \quad \text{and} \quad \bar{s} = \frac{S}{n} = \sum_k y_k \bar{s}_k(p_k, T); \quad \left[\frac{\text{kJ}}{\text{kmol} \cdot \text{K}}\right] \quad \textbf{(11.110)}$$

The **partial molar specific entropy**, $\bar{s}_k(p_k, T)$, is higher than the molar specific entropy $\bar{s}_k(p, T)$ of the pure component k. We can use the n-PG or n-IG state TESTcalcs to numerically verify this.

For a PG or IG mixture with constant composition, x_k and y_k remaining constant, the entropy difference formulas also reduce to their pure-gas counterparts.

PG Mixture:

$$\Delta s^{PG} = \sum_k x_k \Delta s_k = \sum_k x_k \left(c_{p,k} \ln \frac{T_2}{T_1} - R_k \ln \frac{p_{k,2}}{p_{k,1}} \right) = c_p \ln \frac{T_2}{T_1} - R \ln \frac{p_2}{p_1}; \quad \left[\frac{kJ}{kg \cdot K} \right]$$
$$(11.111)$$

$$\Delta \bar{s}^{PG} = \sum_k y_k \Delta \bar{s}_k = \sum_k y_k \left(\bar{c}_{p,k} \ln \frac{T_2}{T_1} - \bar{R} \ln \frac{p_{k,2}}{p_{k,1}} \right) = \bar{c}_p \ln \frac{T_2}{T_1} - \bar{R} \ln \frac{p_2}{p_1}; \quad \left[\frac{kJ}{kmol \cdot K} \right]$$
$$(11.112)$$

IG Mixture:

$$\Delta s^{IG} = \sum_k x_k \Delta s_k = \sum_k x_k [(s_k^o(T_2) - s_k^o(T_1))] - R \ln \frac{p_2}{p_1}; \quad \left[\frac{kJ}{kg \cdot K} \right] \qquad (11.113)$$

$$\Delta \bar{s}^{IG} = \sum_k y_k \Delta \bar{s}_k = \sum_k y_k [(\bar{s}_k^o(T_2) - \bar{s}_k^o(T_1))] - \bar{R} \ln \frac{p_2}{p_1}; \quad \left[\frac{kJ}{kmol \cdot K} \right] \qquad (11.114)$$

In the IG mixture model, $s_k^o(T)$, the temperature dependent part of entropy, is obtained from the ideal gas tables, Tables D-4 through D-15.

Although we have developed formulas for differences in properties, expressions for absolute values can be obtained if State-1 is regarded as a reference state with properties $h_{k,\text{ref}}$ and $s_{k,\text{ref}}$ specified (or evaluated) at a reference temperature T_{ref} and pressure p_{ref}. As with a pure gas model, $u_{k,\text{ref}}$ can be obtained from the relation $h_{k,\text{ref}} = u_{k,\text{ref}} + p_{\text{ref}} v_{k,\text{ref}} = u_{k,\text{ref}} + R_k T_{\text{ref}}$. Note that the mixture composition may not remain frozen during a process. In such situations, the pseudo-pure gas model and the difference formulas derived above cannot be applied and mixture properties have to be evaluated by summing contribution from each species at a given state. This approach will be illustrated in Example 11-13. In Chapter 13, we will develop expressions for absolute values of u_k, h_k, and s_k taking into account the possibility of chemical transformation (see Anim. 11.C.*partialPropertiesIG*).

Partial Molar Properties:

The partial molar properties $\bar{u}_k(p_k, T)$, $\bar{h}_k(p_k, T)$, and $\bar{s}_k(p_k, T)$ introduced in this section through the use of the Dalton model can be generalized for a mixture of all types of pure substances, not just ideal gases. Such a generalized partial molar property, represented by the symbol $\bar{b}_k$, describes how an extensive property B (such as U, H, S, etc.) of a mixture changes as unit mole of component k is added to the mixture while holding p, T, and mole fractions of all other components, $y_i (i \neq k)$, constant. Mathematically, $\bar{b}_k$ is defined as

$$\bar{b}_k \equiv \left(\frac{\partial B}{\partial n_k} \right)_{p,T,n_{i \neq k}}; \quad \left[\frac{\text{Unit of B}}{\text{kmol of } k} \right] \qquad (11.115)$$

Use of Euler's theorem, a discussion of which is outside the scope of this book, leads to a general expression of the total extensive property B of a mixture as the sum of the contribution, $\bar{b}_k n_k$, from each component (see Anim. 11.C.*partialMolarProperty*).

$$B = \sum_k \bar{b}_k n_k; \quad [\text{Unit of B}] \qquad (11.116)$$

B being a function of p, T, and n_i, a general functional relation for $\bar{b}$ must be $\bar{b}_k = \bar{b}_k(p, T, n_i)$. Only for an ideal gas mixture does the application of the Dalton model produce the simplified functional relation, $\bar{b}_k = \bar{b}_k(p_k, T) = \bar{b}_k(y_k, p, T)$. For a pure substance, which can be regarded as a single-component mixture, $\bar{b}_k$ reduces to molar specific property $\bar{b}$ as Eq. (11.116) reduces to $B = n\bar{b}$ [see Eq. (1.26)].

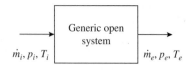

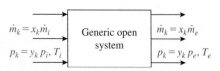

FIGURE 11.23 A mixture flowing at an inlet or exit can be split into its components using the Dalton model.

11.4.4 Mass, Energy, and Entropy Equations for IG-Mixtures

The governing equations developed in Chapter 2 can be easily customized when the working fluid is an ideal gas or perfect gas mixture. Realizing that mixture properties can be expressed in terms of partial properties, the flow of a mixture can be split into separate flows of pure components at their respective partial pressures and temperature of the mixture (Fig. 11.23). The summation over inlets or exits then can be alternatively carried out by summation over the mixture components.

The mass balance equation, Eq. (2.3), thus can be written as

$$\frac{dm}{dt} = \sum_{k(\text{inlet})} \dot{m}_k - \sum_{k(\text{exit})} \dot{m}_k = \sum_{k(\text{inlet})} \overline{M}_k \dot{n}_k - \sum_{k(\text{exit})} \overline{M}_k \dot{n}_k; \quad \left[\frac{\text{kg}}{\text{s}}\right] \qquad \textbf{(11.117)}$$

Here, $\dot{n}_k$ represents the mole flow rate (in kmol/s) of component k.

The transport terms in the energy equation can, similarly, be expressed as summation over the components at the inlets and exits at their local partial pressures. Simplifying Eq. (2.8) by neglecting ke and pe, and recalling (from Ex. 11-7) that $h_k(p_k, T) = h_k(T)$ or $\overline{h}_k(p_k, T) = \overline{h}_k(T)$ for an ideal gas, we obtain

$$\frac{dE}{dt} = \sum_{k(\text{inlet})} \dot{m}_k h_k - \sum_{k(\text{exit})} \dot{m}_k h_k + \dot{Q} - \dot{W}_{\text{ext}} = \sum_{k(\text{inlet})} \dot{n}_k \overline{h}_k - \sum_{k(\text{exit})} \dot{n}_k \overline{h}_k + \dot{Q} - \dot{W}_{\text{ext}}; \quad [\text{kW}]$$

$$\textbf{(11.118)}$$

In this equation, the partial specific enthalpy can be interpreted as the specific enthalpy of the pure component at the temperature of the port.

The entropy equation, Eq. (2.13) can be written as

$$\frac{dS}{dt} = \sum_{k(\text{inlet})} \dot{m}_k s_k(p_k, T) - \sum_{k(\text{exit})} \dot{m}_k s_k(p_k, T) + \frac{\dot{Q}}{T_B} + \dot{S}_{\text{gen}}$$

$$= \sum_{k(\text{inlet})} \dot{n}_k \overline{s}_k(p_k, T) - \sum_{k(\text{exit})} \dot{n}_k \overline{s}_k(p_k, T) + \frac{\dot{Q}}{T_B} + \dot{S}_{\text{gen}}; \quad \left[\frac{\text{kW}}{\text{K}}\right] \qquad \textbf{(11.119)}$$

Note that unlike enthalpy $h_k(T)$, $\overline{s}_k(p_k, T)$ depends on the partial pressure of component k at the ports.

The simplification of these balance equations for specific systems follows the same methodologies as developed in Chapter 4 for open-steady systems and Chapter 5 for unsteady systems.

EXAMPLE 11-13   **Mixing Friction**

A 1 m³ insulated rigid tank is divided by a partition into two chambers. The left chamber, which has a volume of 0.6 m³, contains N_2 at 300 kPa and 300°C. The right chamber contains CO_2 at the same pressure and temperature. The partition is removed and the two gases are allowed to mix. After the mixture comes to equilibrium, determine (a) the final temperature, (b) final pressure, and (c) entropy generated during the mixing process. Use the PG mixture model. *What-if scenario:* (d) What would the answers be if the right chamber contained the same amount of CO_2 at 400°C instead?

SOLUTION

Treating the entire tank as a system, perform an energy and entropy balance for the closed, non-uniform process. Use the PG mixture model to evaluate mixture properties.

Assumptions

The composite beginning state, State-1 for the left chamber and State-2 for the right, and the final state, State-3, are in thermodynamic equilibrium (Fig. 11.24). Neglect KE and PE. No chemical reactions.

Analysis

From Table C-1 or any PG TESTcalc we obtain the following material properties for the two gases: $c_{p,N_2} = 1.03$, $R_{N_2} = 0.297$, $c_{p,CO_2} = 0.844$, and $R_{CO_2} = 0.189$, each in the unit of kJ/kg·K. Evaluate the three anchor states partially as follows

State-1 (given p_1, T_1, and V_1):

$$\Rightarrow m_1 = \frac{V_1}{v_1} = \frac{p_1 V_1}{R_{N_2} T_1} = 1.058 \text{ kg}; \quad n_1 = \frac{m_1}{\overline{M}_{N_2}} = 0.038 \text{ kmol}$$

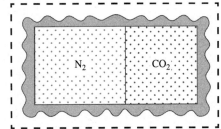

FIGURE 11.24 Schematic for Ex. 11-13.

State-2 (given p_2, T_2, and V_2):

$$\Rightarrow \quad m_2 = \frac{V_2}{v_2} = \frac{p_2 V_2}{R_{CO_2} T_2} = 1.108 \text{ kg}; \quad n_2 = \frac{m_2}{M_{CO_2}} = 0.025 \text{ kmol}$$

State-3 (given $m_3 = m_1 + m_2$, and $V_3 = V_1 + V_2$):

$$\Rightarrow \quad n = n_3 = n_1 + n_2 = 0.063 \text{ kmol}; \quad y_{N_2,3} = \frac{n_{N_2,3}}{n} = \frac{n_1}{n_3} = 0.6;$$

$$y_{CO_2,3} = \frac{n_{CO_2,3}}{n} = \frac{n_2}{n_3} = 0.4; \quad \overline{M}_3 = \sum_k y_{k,3}\overline{M}_k = 34.40 \frac{\text{kg}}{\text{kmol}};$$

$$R_3 = \frac{\overline{R}}{\overline{M}_3} = 0.242 \frac{\text{kJ}}{\text{kg} \cdot \text{K}};$$

Energy Analysis

With no heat or work transfer the energy equation, Eq. (5.2), for the closed process executed by the non-uniform system yields

$$\Delta U = \cancel{Q}^0 - \cancel{W_{ext}}^0; \quad \Rightarrow \quad U_3 - (U_1 + U_2) = 0;$$
$$\Rightarrow \quad (m_{N_2,3}u_{N_2,3} + m_{CO_2,3}u_{CO_2,3}) - (m_1 u_1 + m_2 u_2) = 0$$
$$\Rightarrow \quad (m_1 u_{N_2,3} + m_2 u_{CO_2,3}) - (m_1 u_{N_2,1} + m_2 u_{CO_2,2}) = 0$$
$$\Rightarrow \quad m_1(u_{N_2,3} - u_{N_2,1}) + m_2(u_{CO_2,3} - u_{CO_2,2}) = 0$$
$$\Rightarrow \quad m_1 c_{v,N_2}(T_3 - T_1) + m_2 c_{v,CO_2}(T_3 - T_2) = 0$$
$$\Rightarrow \quad (m_1 c_{v,N_2} + m_2 c_{v,CO_2})(T_3 - T_1) = 0; \quad \Rightarrow \quad T_3 = T_1 = 300°C$$

Applying the ideal gas equation of state to the mixture

$$p_3 = \frac{R_3 T_3}{v_3} = \frac{R_3 T_3 m_3}{V_3} = 300 \text{ kPa}$$

The entropy balance equation, Eq. (5.4), similarly can be simplified to produce the entropy generated during mixing.

$$\Delta S = S_f - S_b = \frac{\cancel{Q}^0}{T_B} + S_{gen};$$
$$\Rightarrow \quad S_{gen} = S_3 - (S_1 + S_2) = (m_{N_2,3}s_{N_2,3} + m_{CO_2,3}s_{CO_2,3}) - (m_1 s_1 + m_2 s_2)$$
$$= (m_1 s_{N_2,3} + m_2 s_{CO_2,3}) - (m_1 s_{N_2,1} + m_2 s_{CO_2,2})$$
$$= m_1(s_{N_2,3} - s_{N_2,1}) + m_2(s_{CO_2,3} - s_{CO_2,2})$$
$$= m_1\left(c_{p,N_2}\ln\cancel{\frac{T}{T_1}}^1 - R_{N_2}\ln\frac{p_{N_2,3}}{p_1}\right) + m_2\left(c_{p,CO_2}\ln\cancel{\frac{T}{T_2}}^1 - R_{CO_2}\ln\frac{p_{CO_2,3}}{p_2}\right)$$
$$= -m_1 R_{N_2}\ln\frac{y_{N_2,3}\cancel{p_3}}{\cancel{p_1}} - m_2 R_{CO_2}\ln\frac{y_{CO_2,3}\cancel{p_3}}{\cancel{p_2}}$$
$$= -\overline{R}(n_{N_2}\ln y_{N_2,3} + n_{CO_2}\ln y_{CO_2,3}) = 0.352 \text{ kJ/K}$$

TEST Analysis

Launch the PG/PG mixing, closed-process TESTcalc. Select N2 as gas-A, enter x_A1=1 (pure nitrogen), and evaluate State-1 from the given p1, T1 and Vol1. For State-2, select CO2 as gas-B, enter x_A2=0 (pure CO2), and evaluate the state from known p2, T2, and Vol2. Evaluate State-3 partially with Vol3=Vol1+Vol2, m3=m1+m2, and x3=m1/(m1+m2). On the process panel, load State-1 and State-2 as the composite beginning states and State-3 as the final state. Enter Q=W=0 and click Super-Calculate. Obtain T3 and p3 from State-3, and S_gen from the process panel to confirm the manual results.

What-If scenario

Make p2 an unknown. Enter m2 $= 1.108$ kg and the new value for T2. Click Super-Calculate. The final pressure and temperature are calculated as 356 kPa and 396 deg-C respectively. The entropy generation increases to 0.369 kJ/K.

Discussion

Temperature and pressure remaining constant, entropy is generated in this process due to mixing of the two species. This fundamental mechanism of entropy generation due to irreversible mixing of dissimilar substances is called **mixing friction**, one of the several components of thermodynamic friction discussed in Chapter 2. When a temperature difference is also present between the chambers, two mechanisms, mixing friction and thermal friction, combine to enhance entropy generation. The relative importance of mixing can be established from the marginal increase in entropy generation when the temperature difference is introduced.

EXAMPLE 11-14 Isentropic Expansion of Ideal Gas Mixture

A gas mixture with 70% CO_2 and 30% H_2 by volume expands isentropically in a nozzle (Fig. 11.25) from 300 kPa, 500 K, 5 m/s to an exit pressure of 100 kPa. Determine (a) the exit temperature and (b) exit velocity. **What-if scenario:** (c) What would the answers be if the mixture were equimolar? Use the IG mixture model.

SOLUTION

Use the isentropic condition to evaluate the exit state of the mixture.

Assumptions

The nozzle is steady, adiabatic, and internally reversible.

Analysis

For an ideal gas mixture, volume fraction and mole fraction are equal. With the mixture composition unaffected, $y_{CO_2,1} = y_{CO_2,2} = y_{CO_2} = 0.7$ and $y_{H_2,1} = y_{H_2,2} = y_{H_2} = 0.3$.

Simplifying the energy and entropy equations, Eqs. (4.2) and (4.3), for the nozzle, an open-steady device, we obtain $j_1 = j_2$ or $V_2 = \sqrt{2000(h_1 - h_2) + V_1^2}$; and $s_1 = s_2$ or $\Delta s = 0$. For an ideal gas mixture, $h = h(T)$. Hence the exit temperature is the only additional property that is necessary to find the exit velocity. Having evaluated the component mole fractions already, it is advantageous to use the molar representation of entropy, as $\Delta s = 0$ also implies $\Delta \bar{s} = 0$. Noting that the molar composition remains unchanged during expansion and $y_{CO_2} + y_{H_2} = 1$, Equation (11.114) yields

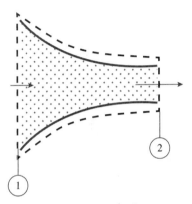

FIGURE 11.25 Schematic for Ex. 11-14.

$$\sum_k y_k \left(\bar{s}_k^o(T_2) - \bar{s}_k^o(T_1) - \bar{R} \ln \frac{p_2 y_{k,2}}{p_1 y_{k,1}} \right) = 0$$

$$\Rightarrow \quad y_{CO_2} \left(\bar{s}_{CO_2}^o(T_2) - \bar{s}_{CO_2}^o(T_1) - \bar{R} \ln \frac{p_2}{p_1} \right) + y_{H_2} \left(\bar{s}_{H_2}^o(T_2) - \bar{s}_{H_2}^o(T_1) - \bar{R} \ln \frac{p_2}{p_1} \right) = 0$$

$$\Rightarrow \quad y_{CO_2} \bar{s}_{CO_2}^o(T_2) + y_{H_2} \bar{s}_{H_2}^o(T_2)$$

$$= y_{CO_2} \bar{s}_{CO_2}^o(T_1) + y_{H_2} \bar{s}_{H_2}^o(T_1) + \bar{R} \ln \frac{p_2}{p_1}$$

The right-hand side can be evaluated using the ideal gas tables for CO_2 and H_2 (Tables D-6 and D-8).

$$y_{CO_2} \bar{s}_{CO_2}^o(T_1) + y_{H_2} \bar{s}_{H_2}^o(T_1) + \bar{R} \ln \frac{p_2}{p_1}$$

$$= (0.7)(234.8) + (0.3)(145.63) + (8.314) \ln \frac{100}{300} = 198.9 \, \frac{kJ}{kg \cdot K}$$

The left-hand side is now evaluated at several temperatures using the trial and error method. The temperature thus obtained is $T_1 = 390$ K, which produces

$$(0.7)(224.2) + (0.3)(138.3) = 198.4 \frac{kJ}{kg \cdot K}$$

The mixture molar mass can be found from the mole fractions as

$$\overline{M} = \sum_k y_k \overline{M}_k = (0.7)(44) + (0.3)(2) = 31.4 \frac{kg}{kmol}$$

The enthalpy difference between the inlet and exit states can be found from Eq. (11.108) using the ideal gas tables.

$$\Delta \overline{h} = \sum_k y_k [\, \overline{h}_k(T_2) - \overline{h}_k(T_1)\,] = (0.7)(3596 - 8314)$$

$$+ (0.3)(2666 - 5882) = -4267.4 \frac{kJ}{kmol}$$

The exit velocity now can be calculated.

$$V_2 = \sqrt{(1000 \text{ J/kJ})(2)(h_1 - h_2) + V_1^2} = \sqrt{-2000 \Delta h + V_1^2}$$

$$= \sqrt{-2000 \Delta \overline{h}/\overline{M} + V_1^2} = \sqrt{-2000(-4267.4)/31.4 + 5^2} = 521 \frac{m}{s}$$

TEST Analysis

Launch the n-IG single-flow TESTcalc. Compose the mixture by selecting CO2 and H2 and entering their amounts in percent by volume. Evaluate State-1 from the given properties: p1, T1, and Vel1. For State-2, make Vel2 an unknown and enter p2, j2=j1, and s2=s1. Click Calculate to produce T2 = 395 K and Vel2 = 509 m/s, which are reasonably close to the manually calculated results.

What-If scenario

Change the composition by entering a new value (50%) for each component and click Super-Calculate. The new answers are 388 K and 592 m/s respectively.

Discussion

The nozzle exit velocity increases when the proportion of hydrogen in the mixture is increased, all other parameters remaining the same. Using the PG mixture model, it can be shown that the exit velocity is proportional to the square root of c_p for negligible inlet kinetic energy. An increase in the level of hydrogen, therefore, boosts the mixture c_p, explaining the increase in the exit velocity. The effect on the specific heat ratio, however, is less pronounced. Therefore, the exit temperature does not change as significantly as the exit velocity.

11.4.5 Real Gas Mixture Model

When a gas mixture does not satisfy the ideal gas criteria (Sec. 3.8), it must be treated as a **real gas (RG) mixture**, with each component modeled as a **real gas**. In Sec. 11.3 we have developed formulas for evaluating properties of a pure real gas. Presence of the compressibility factor, which is a function of temperature and pressure, however, introduces some complications in extending the pure gas formulas to a RG mixture.

The real gas equation of state, like its ideal gas counterpart, can be expressed in the molar form:

$$pv = ZRT; \quad \Rightarrow \quad p\frac{V}{m} = ZRT; \quad \Rightarrow \quad pV = mZ\frac{\overline{R}}{\overline{M}}T = \frac{m}{\overline{M}}Z\overline{R}T = nZ\overline{R}T$$

Applying it to component k (Dalton model) and then summing over all components produces the following real gas counterpart of Dalton's law of partial pressure:

$$\sum_k p_k V = \sum_k n_k Z_k \overline{R} T; \tag{11.120}$$

$$\Rightarrow \quad \sum_k p_k = \frac{\overline{R}T}{\Psi}\sum_k n_k Z_k = \frac{p}{nZ}\sum_k n_k Z_k = \frac{p}{Z}\sum_k y_k Z_k; \quad \text{[kPa]} \qquad \textbf{(11.121)}$$

If Dalton's law, which has been experimentally found to be approximately satisfied by a real gas mixture, is assumed to be applicable to the RG mixture model, we can define mixture compressibility factor as:

$$Z \equiv \sum_k y_k Z_k \quad \text{where} \quad Z_k = Z_k(p_r, T_r) \qquad \textbf{(11.122)}$$

where Z_k must be evaluated at the reduced pressure and temperature of the mixture. Application of Amagat's law results in exactly the same formula for mixture compressibility, except $Z_k = Z_k(v_r, T_r)$ must be evaluated at the pseudo-reduced volume and temperature of the mixture. Unfortunately, the mixture compressibilities obtained by the two methods do not agree with each other, pointing out the fundamental problem in using the real gas equation of state for a mixture.

One approach to overcome this difficulty is to treat the mixture as a single pseudo-pure real gas with pseudo-critical properties given by **Kay's rule**.

$$p_{cr} \equiv \sum_k y_k p_{k,cr}; \quad \text{[kPa]} \quad \text{and} \quad T_{cr} \equiv \sum_k y_k T_{k,cr}; \quad \text{[K]} \qquad \textbf{(11.123)}$$

The effective critical pressure and temperature for a mixture can be found from the mixture composition and critical properties of its components. The compressibility and departure charts developed for a pure real gas can now be used without any modifications for real gas mixtures. The simplicity of Kay's rule outweighs the error, around 10% over a wide range of temperatures and pressures, it introduces.

EXAMPLE 11-15   Mixing of Two Real Gases

Two separate streams of nitrogen and oxygen, each at 9 MPa and 200 K and each having a flow rate of 1 kg/s, steadily enter (Fig. 11.26) an adiabatic mixing chamber. The atmospheric temperature is 300 K. If the pressure loss in mixing can be neglected, use TEST analysis to determine (a) the exit temperature, and (b) rate of exergy destruction in the mixing chamber's universe. **What-if scenario:** (c) What would the answers be if the oxygen stream entered at 500 K instead?

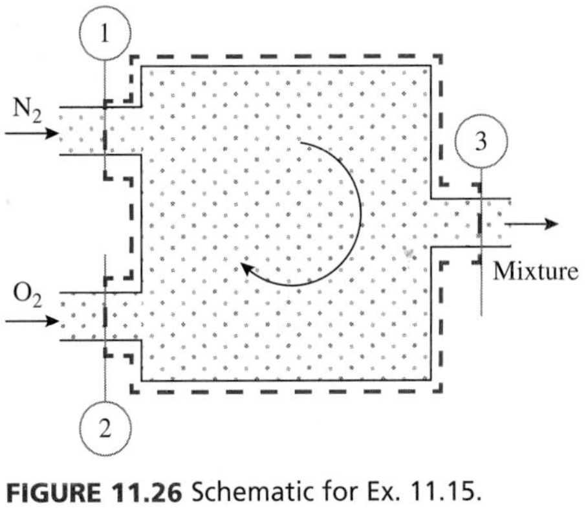

SOLUTION

Using the RG/RG multi-flow, mixing, TESTcalc, perform an energy and entropy analyses of the open-steady mixing chamber.

FIGURE 11.26 Schematic for Ex. 11.15.

Assumptions

The RG mixture model with Kay's rule can be applied. Equilibrium exists at the inlets, State-1 and State-2, and exit, State-3. The exit pressure is same as the inlet pressures. Negligible changes in ke and pe.

TEST Analysis

Launch the RG/RG multi-flow, mixing TESTcalc. Select N2 as Gas-A, enter x_ A1 = 1 (pure nitrogen), and evaluate State-1 from the given p1, T1, and mdot1. For State-2, select O2 as Gas-B, enter x_A2 = 0 (pure O_2), and evaluate the state from p2 = p1, T2 = T1, and mdot2. Evaluate State-3, partially with mdot3 = mdot1 + mdot2, p3 = p1, and x3 = mdot1/(mdot1 + mdot2). Kay's rule is automatically applied to any gas mixtures by this TESTcalc. On the device panel, load State-1 and State-2 as i1 and i2 states, and State-3 as the e1-State. To specify a single exit leave e2-State as Null (blocked). Enter T_B = 300 K, Qdot = Wdot_ext = 0, and click Super-Calculate. From State-3, obtain T3 as 200 K and from the device panel, obtain Sdot_gen = 0.382 kW/K. The system being adiabatic, Sdot_gen, univ = Sdot_gen. Therefore, the rate of exergy destruction can be calculated as (300)(0.382) = 114.6 kW.

What-If scenario

Change T2 to 500 K, press the Enter key to register the change and click Super-Calculate. The new answers are calculated as 332.7 K and 300*0.62549 = 188 kW respectively.

Discussion

With no difference in pressure or temperature, the entropy generation must be due to mixing of two dissimilar species (mixing friction). Unlike the ideal (or perfect) gas mixture model, the real gas mixture model treats the mixture as a pseudo-pure real gas even for entropy calculations. Information about mixture composition is represented by the mole fraction weighted critical properties.

11.5 CLOSURE

In this chapter we developed the theoretical foundation on which all material models are built. It began with a generalized development of the thermodynamic relations, including the *Tds* equations introduced in Chapter 3 and used throughout this book. This is followed by the development of Maxwell relations, Clapeyron equation, and Clapeyron–Clausius equation. A mnemonic diagram called the thermodynamic square is constructed to graphically represent various thermodynamic relations, including Maxwell equations, developed in this chapter. General relations for internal energy, enthalpy, entropy, and specific heats are derived for simply compressible substances. Using some of these relations, it is shown that one of the main assumptions on which the IG and PG models are rooted—internal energy being a function of temperature only—is a consequence of the ideal gas equation of state. New properties, namely Joule–Thomson coefficient, volume expansivity, and two types of compressibility, are introduced. The general relations are applied to further develop the RG model, stated in Chapter 3. Finally, three mixture models—the perfect gas, ideal gas, and real gas mixture models—are developed. With these new models, mixing is studied and mixing friction is established as a fundamental mechanism for generating entropy.

PROBLEMS

SECTION 11-1: THERMODYNAMIC RELATIONS

11-1-1 Using the Maxwell's relation and the equation of state, determine a relation for the partial of s with respect to v at constant T for the IG model. Verify the relation using TEST at 100 kPa and 300 K.

11-1-2 Derive the relation for the slope of the $n = $ constant lines on a T-p diagram for a gas that obeys the van der Waals equation of state.

11-1-3 Derive the relation for the volume expansivity (β) and the isothermal compressibility (κ_T) for (a) an ideal gas and (b) a gas whose equation of state is $p(v - b) = RT$.

11-1-4 Estimate the volume expansivity (β) and the isothermal compressibility (κ_T) of refrigerant-134a at 200 kPa and 30°C.

11-1-5 Estimate the Joule–Thomson coefficient (μ_J) of nitrogen at (a) 200 psia, 500°R and (b) 2000 psia, 400°R.

11-1-6 For $\beta \geq 0$, prove that at every point of a single-phase region of an h-s diagram, the slope of a constant-pressure ($p = $ constant) line is greater than the slope of a constant temperature ($T = $ constant) line, but less than the slope of a constant-volume ($v = $ constant) line.

11-1-7 Starting with the relation $dh = Tds + vdp$, show that the slope of a constant-pressure line in a h-s diagram (a) is constant in the saturation region, and (b) increases with temperature in the superheated region.

11-1-8 Derive relations for (a) Δu, (b) Δh, and (c) Δs of a gas that obeys the equation of state $(p + a/v^2)v = RT$ for an isothermal process.

11-1-9 Show that $c_v = -T(\partial v/\partial T)_s(\partial p/\partial T)_v$, and $c_p = T(\partial p/\partial T)_s(\partial v/\partial T)_p$.

11-1-10 Steam is throttled from 4.5 MPa and 400°C to 3.5 MPa. Estimate the temperature change (ΔT) of the steam during this process and the average Joule–Thomson coefficient (μ_J).

11-1-11 Consider an infinitesimal reversible adiabatic compression or expansion process. By taking $s = s(P,v)$ and using the Maxwell relations, show that for this process $pv^k = $ constant, where k is the isentropic expansion exponent, defined as $k = (v/p)(\partial p/\partial v)_s$. Also, show that the isentropic expansion exponent k reduces to the specific heat ratio (c_p/c_v) for an ideal gas.

11-1-12 Consider a mixture of two gases A and B. Show that when the mass fraction x_A and x_B are known, the mole-fraction can be determined from $y_A = M_B/M_A(1/x_A - 1)M_B$ and $y_B = 1 - y_A$ where M_A and M_B are the molar masses of A and B.

11-1-13 Nitrogen gas at 400 K and 300 kPa behaves as an ideal gas. Estimate the c_p and c_v of nitrogen at this state.

11-1-14 A system contains oxygen (ideal gas) at 400 K and 100 kPa. As a result of some disturbance, the conditions of the gas change to 404 K and 98 kPa. (a) Estimate the change in the specific volume (Δv) of the gas using the ideal-gas relation and using Taylor's theorem. (b) Determine the exact answer using the IG equation of state.

11-1-15 Estimate the specific-heat difference ($c_p - c_v$) for liquid water at 20 MPa and 60°C.

11-1-16 Estimate the specific-heat difference ($c_p - c_v$) for liquid water at 1000 psia and 150°F.

11-1-17 Plot the Joule–Thomson coefficient (μ_J) for nitrogen over the pressure range of 100 psia to 1500 psia at the enthalpy values of 100 Btu/lbm, 175 Btu/lbm and 225 Btu/lbm. Discuss the results.

11-1-18 Determine the enthalpy change (Δh) and the entropy change (Δs) of nitrogen per unit mole as it undergoes a change of state from 225 K and 6 MPa to 320 K and 12 MPa, (a) by assuming ideal-gas behavior, and (b) by accounting for the deviation from ideal-gas behavior through the use of generalized charts.

11-1-19 Determine the enthalpy change (Δh) and the entropy change (Δs) of carbon dioxide per unit mass as it undergoes a change of state from 250 K and 7 MPa to 280 K and 12 MPa, (a) by assuming ideal-gas behavior, and (b) by accounting for the deviation from ideal-gas behavior.

11-1-20 Methane is compressed adiabatically by a steady-state flow compressor from 2 MPa and -10°C to 10 MPa and 110°C at a rate of 0.8 kg/s. Using the generalized charts, determine the required power input to the compressor.

11-1-21 Methane gas flows through a pipeline with a mass flow rate of 110 lb/s at a pressure of 183 atm and a temperature of 56°F. Determine the volumetric flow rate ($\dot{V}$), in ft^3/s, using (a) the ideal gas equation, (b) van der Waals equation and (c) compressibility chart.

11-1-22 Determine the specific volume (v) of water vapor at 10 MPa and 360°C, in m^3/kg, using (a) the steam tables, (b) compressibility chart and (c) ideal gas equation.

11-1-23 Consider refrigerant R-12 vapor at 160°F and 0.5 ft^3/lb. Estimate the pressure (p) at this state, in atm, using (a) the ideal gas equation, (b) van der Waals equation and (c) compressibility chart.

11-1-24 For the functions $x = x(y,w)$, $y = y(z,w)$, $z = z(x,w)$, demonstrate that $(\partial x/\partial y)_w \ (\partial y/\partial z)_w$ $(\partial z/\partial x)_w = 1$.

11-1-25 The following expressions for the equation of state and the specific heat (c_p) are obeyed by a certain gas: $v = RT/paT^2$ and $c_p = ABTC_p$ where a, A, B, C are constants. Obtain an expression for (a) the Joule–Thomson coefficient (μ_J) and (b) the specific heat (c_v).

11-1-26 The differential of pressure obtained from a certain equation of state is given by the following expression, determine the equation of state: $dp = \{2(v - b)/RT\} dv + \{(v - b)^2/RT^2\} dT$.

11-1-27 The differential of pressure obtained from a certain equation of state is given by the following expression, determine the equation of state: $dp = \{-RT/(v - b)^2\} dv + \{R/(v - b)\} dT$.

11-1-28 Derive the relation $c_p = -T(\partial^2 g/\partial T^2)_p$.

11-1-29 Prove that $(\partial \beta/\partial p)_T = -(\partial \kappa_T/\partial T)_p$.

11-1-30 At certain states, the p-v-T data for a particular gas can be represented as $Z = 1 - Ap/T^4$, where Z is the compressibility factor and A is a constant. Obtain an expression for the difference in specific heats ($c_p - c_v$) in terms of Z.

SECTION 11-2: SYSTEM ANALYSIS USING RG (REAL GAS) MODEL

11-2-1 Determine the change in (a) enthalpy (Δh) and (b) entropy (Δs) of nitrogen as it undergoes a change of state from 200 K and 6 MPa to 300 K and 10 MPa by treating nitrogen as a perfect gas. *What-if scenario:* What would the change in enthalpy be if nitrogen were modeled using (c) the ideal gas, or (d) real gas model?

11-2-2 Calculate Δv, Δu, Δh, and Δs for the following change of state of superheated steam: State-1: $p_1 = 2$ MPa, saturated vapor; State-2: $p_2 = 33$ kPa, 400°C. Compare the following models: (a) PC model, (b) PG model, (c) IG model, (d) RG model (L-K) and (e) RG model (N-O). Your answers should be in a tabular form—one table for manual results and another for corresponding TEST results.

11-2-3 Methane is isothermally and reversibly compressed by a piston-cylinder device from 1 MPa, 100°C to 4 MPa. Using the Lee–Kesler RG model, calculate (a) the work done (w_B) and (b) heat transfer per unit mass (q). (c) *What-if scenario:* What would the work done be if the process were isentropic?

11-2-4 A cylindrical tank contains 4.0 kg of carbon monoxide at −45°C has an inner diameter of 0.2 m and a length of 1 m. Using the RG model (L-K charts), determine (a) the pressure exerted by the gas. (b) *What-if scenario:* What would the pressure exerted by the gas be if the IG model were used instead?

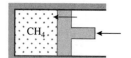

11-2-5 Methane is adiabatically compressed by a piston-cylinder device from 1 MPa, 100°C to 4 MPa. Calculate (a) the work done per unit mass (w_B). Assume the adiabatic efficiency to be 90%. Use the real gas model. (b) *What-if scenario:* What would the work done per unit mass be if the compressed gas were ethane instead?

11-2-6 Propane is compressed isothermally by a piston-cylinder device from 1.5 MPa, 90°C to 4 MPa. Using the Nelson–Obert charts, determine (include sign) (a) the work done (w_B) and (b) the heat transfer per unit mass of propane (q). (c) *What-if scenario:* What would be the work done if the PC model were used?

11-2-7 Methane is isothermally compressed by a piston-cylinder device from 1 MPa, 100°C to 4 MPa. Calculate (a) the entropy generation (s_{gen}) and (b) the irreversibility associated with the process if the ambient temperature is 25°C. Use the real gas model.

11-2-8 A piston-cylinder device contains 2 kg of H_2 and 14 kg of O_2 at 150 K and 5000 kPa. Heat is then transferred until the mixture expands at constant pressure (why does the pressure remains constant?) until the temperature rises to 200 K. Determine (a) the heat transfer (Q) by treating the mixture as perfect gas. (b) *What-if scenario:* What would the conclusion be if the device contained 3 kg of H_2 and 12 kg of O_2 instead?

11-2-9 A piston-cylinder device contains 1 lbm of O_2 and 9 lbm of N_2 at 300°R and 900 psia. The gas mixture is now heated at constant pressure to 400°R. Determine (a) the heat transfer (Q) during the expansion process by treating the mixture as a perfect gas mixture. (b) *What-if scenario:* What would the heat transfer be if the real gas mixture model (with Kay's rule) were used?

11-2-10 An insulated piston-cylinder device contains 0.1 kg of N_2 and 0.2 kg of CO_2 at 300 K and 100 kPa. The gas mixture is now compressed isentropically to a pressure of 1000 kPa. Determine (a) the final temperature (T_2) by treating the mixture as a perfect gas. (b) *What-if scenario:* What would the final temperature be if the ideal gas model were used?

11-2-11 A rigid tank contains 3 m^3 of argon at $-100°C$ and 1 MPa. Heat is transferred until the temperature rises to 0°C. Determine (a) the mass of argon, (b) the final pressure (p_2) and (c) heat transferred (Q). Use the real gas (L-K) model. (d) *What-if scenario:* What would the mass be if the tank contained 1 m^3 of argon?

11-2-12 A 0.5 m^3 well-insulated rigid tank contains oxygen at 200 K and 9 MPa. A paddle wheel placed in the tank is turned on, and the temperature of the oxygen rises to 240 K. Determine (a) the final pressure in the tank and (b) the paddle-wheel work (W) done during the process. Use the RG Model (L-K).

11-2-13 A closed, rigid, insulated vessel having a volume of 0.15 m^3 contains oxygen initially at 10 MPa and 280 K. The oxygen is stirred by a paddle wheel until pressure increases to 15 MPa. Stirring ceases and the gas attains a final equilibrium state. Using the RG model (N-O), determine (a) the final temperature (T_2), (b) work done (W) during the process and (c) the amount of availability destroyed in the process. Let $T_0 = 280$ K.

11-2-14 Steam is throttled from 10 MPa, 400°C to 3 MPa. If the ambient temperature is 25°C, determine (a) the change in temperature (ΔT) and (b) the irreversibility for a flow rate of 1 kg/s. Use the real gas (L-K) model. (c) *What-if scenario:* What would the answer in (b) be if steam were throttled from 10 MPa, 400°C to 5 MPa?

11-2-15 Oxygen is throttled from 10 MPa, 400 K to 2 MPa. Using the RG model (L-K), determine (a) the change in temperature (ΔT). (b) *What-if scenario:* What would the change in temperature be if steam were throttled from 15 MPa?

11-2-16 Nitrogen gas enters a turbine at 7 MPa, 500 K, 100 m/s and leaves at 1 MPa, 300 K, 150 m/s at a flow rate ($\dot{m}$) of 2 kg/s. Heat is being lost to the surroundings at 25°C at a rate of 100 kW. Determine (a) the power output ($\dot{W}_{ext}$) and (b) irreversibility. Use the real gas (L-K) model. (c) *What-if scenario:* What would the power output be if the mass flow rate were 1 kg/s?

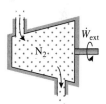

11-2-17 Nitrogen gas enters a turbine operating at steady state at 10 MPa, 26°C with a mass flow rate ($\dot{m}$) of 1 kg/s and exits at 4 MPa, $-28°C$. Using the RG model (N-O) and ignoring the heat transfer with the surrounding, determine (a) the work developed ($\dot{W}_{ext}$). (b) *What-if scenario:* What would the work developed be if the mass flow rate were 0.5 kg/s?

11-2-18 Methane at 9.3 MPa, 300 K enters the turbine operating at steady state at a mass flow rate ($\dot{m}$) of 1 kg/s, expands adiabatically through a 6:1 pressure ratio, and exits at 225 K. KE and PE effects are negligible. Using the RG model (L-K), determine (a) the power developed ($\dot{W}_{ext}$) and (b) the entropy produced ($\dot{S}_{gen}$). (c) *What-if scenario:* What would the power developed be if PG model were used?

11-2-19 Argon gas enters a turbine operating at steady state at 10 MPa, 51°C with a mass flow rate ($\dot{m}$) of 1 kg/s and expands adiabatically to 4 MPa, $-35°C$ with no change in KE or PE. Using the RG model (N-O), determine (a) the work developed ($\dot{W}_{ext}$) and (b) the entropy generated ($\dot{S}_{gen}$). (c) *What-if scenario:* What would the work developed be if IG model were used?

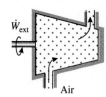

11-2-20 Air (79% N_2 and 21% O_2 by volume) is compressed isothermally at 500 K from 4 MPa to 8 MPa in a steady-flow compressor at a rate of 5 kg/s. Assuming no irreversibilities, determine the power input to the compressor. Treat air as a mixture of (a) perfect gases, (b) ideal gases and (c) real gases (L-K).

11-2-21 Methane is compressed adiabatically by a steady flow compressor from 3 MPa and $-15°C$ to 10 MPa and 100°C at a rate of 0.9 kg/s. Determine the power input ($\dot{W}_{ext}$) to the compressor. Use the real gas model (L-K).

11-2-22 Carbon dioxide enters an adiabatic nozzle at 10 MPa, 450 K with a low velocity and leaves at 3 MPa, 350 K. Using the RG model (N-O), determine the exit velocity (V_2).

11-2-23 Oxygen enters a nozzle operating at steady state at 6 MPa, 300 K, 1 m/s and expands isentropically to 3 MPa. Using the RG model (L-K), determine (a) the exit temperature (T_2) and (b) the exit velocity (V_2).

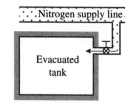

11-2-24 An adiabatic 1 m^3 rigid tank is initially evacuated. It is filled to a pressure of 10 MPa from a supply line that carries nitrogen at 275 K and 10 MPa. Determine (a) the final temperature (T_2) and (b) the mass in the tank. Use the real gas model (L-K). (c) *What-if scenario:* What would the mass in the tank be if the tank temperature were maintained at 300 K, the surrounding's temperature?

11-2-25 One kmol of argon at 320 K is initially confined to one side of a rigid, insulated container divided into equal volumes of 0.2 m^3 by a partition. The other side is initially evacuated. The partition is removed and argon expands to fill the entire container. Using the RG model (L-K), determine (a) the final temperature (T_2) of argon. (b) *What-if scenario:* What would the final temperature be if PG model were used?

SECTION 11-3: MIXTURE MODELS (PG/PG, IG/IG, AND RG/RG MODELS)

11-3-1 A gas mixture contains 5 kg of N_2 and 10 kg of O_2. Determine (a) the average molar mass and (b) gas constant. (c) *What-if scenario:* What would the average molar mass be if the gas mixture contained 10 kg of N_2 and 5 kg of O_2?

11-3-2 A tank contains 3 kmol N_2 and 7 kmol of CO_2 gases at 25°C, 10 MPa. Based on the ideal gas equation of state, determine (a) the average molar mass and (b) volume ($\dot{V}$) of the tank.

11-3-3 A gas mixture consists of 9 kmol H_2 and 2 kmol of N_2. Determine (a) the mass of each gas and (b) the apparent gas constant of the mixture. (c) *What-if scenario:* What would the answer in (b) be if hydrogen were replaced by oxygen?

11-3-4 A rigid tank contains 4 kmol O_2 and 5 kmol of CO_2 gases at 18°C, 100 kPa. Determine the volume ($\dot{V}$) of the tank.

11-3-5 A mixture of CO_2 and water vapor is at 100 kPa, 200°C. As the mixture is cooled at a constant pressure, water vapor begins to condense when the temperature reaches 70°C. Determine (a) the mole fraction and (b) the mass fraction of CO_2 in the mixture.

11-3-6 A 0.4 m^3 rigid tank contains 0.4 kg N_2 and 0.7 kg of O_2 gases at 350 K. Determine (a) the partial pressure of each gas and (b) the total pressure (p) of the mixture. (c) *What-if scenario:* What would the total pressure be if the volume of the rigid tank were 0.6 m^3?

11-3-7 A 1 kmol mixture of CO_2 and C_2H_6 (ethane) occupies a volume of 0.2 m^3 at a temperature of 410 K. The mole fraction of CO_2 is 0.3. Using the RG model (Kay's rule and L-K charts), determine (a) the mixture pressure. (b) *What-if scenario:* What would the mixture pressure be if IG model were used?

11-3-8 A mixture consisting of 0.18 kmol of methane and 0.274 kmol of butane occupies a volume of 0.3 m^3 at a temperature of 240°C. Using the IG model, determine (a) the pressure exerted by the mixture. (b) *What-if scenario:* What would the answer in (a) be if RG model (Kay's rule and L-K charts) were used?

11-3-9 Determine (a) the mass of 1 m^3 air (N_2: 79% and O_2: 21% by volume) at 10 MPa and 160 K, assuming air as a perfect gas mixture. (b) *What-if scenario:* What would the answer in (a) be if real gas mixture model (Kay's rule and L-K chart) were used?

11-3-10 An insulated rigid tank is divided into two compartments by a partition. One compartment contains 8 kg of oxygen gas at 42°C and 100 kPa, and the other compartment contains 4 kg of nitrogen gas at 20°C and 180 kPa. The partition is then removed and the two gases are allowed to mix. Determine (a) the mixture temperature and (b) the mixture pressure after equilibrium has been reached.

11-3-11 An insulated rigid tank is divided into two compartments by a partition. One compartment contains 4 kmol of O_2, and the other compartment contains 5 kmol of CO_2. Both gases are initially at 25°C and 150 kPa. The partition is then removed and the two gases are allowed to mix. Assuming the surroundings are at 25°C and both gases behave as ideal gases, determine (a) the entropy change (ΔS) and (b) the exergy destruction (I) associated with this process. (c) *What-if scenario:* What would the entropy change be if there were 2 kmol of O_2 instead?

11-3-12 An insulated rigid tank is divided into two compartments by a partition. One compartment contains 0.2 kmol of CO_2 at 26°C, 180 kPa and the other compartment contains 3 kmol of H_2 gas at 37°C, 340 kPa. The partition is then removed and the two gases are allowed to mix. Determine (a) the mixture temperature and (b) the mixture pressure after equilibrium has been established.

11-3-13 A rigid insulated tank is divided into two compartments by a membrane. One compartment contains 0.3 kmol of CO_2 at 25°C and 100 kPa, and the other compartment contains 4 kmol of H_2 gas at 40°C and 300 kPa. The membrane is then punctured and the two gases are allowed to mix. Determine (a) the mixture temperature and (b) the mixture pressure after equilibrium has been reached. (c) *What-if scenario:* What would the mixture temperature be if there were 2 kmol of H_2 instead?

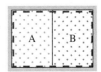

11-3-14 Two rigid, insulated tanks are interconnected by a valve. Initially 0.79 kmol of nitrogen at 200 kPa and 255 K fills one tank. The other tank contains 0.21 kmol of oxygen at 100 kPa and 300 K. The valve is opened and the gases are allowed to mix until a final equilibrium state is attained. During this process, there are no heat or work interactions between the tank contents and the surroundings. Determine (a) the final temperature of the mixture, (b) the final pressure of the mixture and (c) the amount of entropy produced (S_{gen}) in the mixing process.

11-3-15 A 1.1 m^3 rigid tank is divided into two equal compartments by a partition. One compartment contains Ne at 22°C and 120 kPa, and the other compartment contains Ar at 50°C and 200 kPa. The partition is then removed, and the two gases are allowed to mix. Heat is lost to the surrounding air at 20°C during this process in the amount of 18 kJ. Determine (a) the final mixture temperature and (b) the final mixture pressure. (c) *What-if scenario:* What would the final mixture temperature be if heat lost were in the amount of 20 kJ?

11-3-16 A rigid tank that contains 3 kg of N_2 at 25°C and 250 kPa is connected to another rigid tank that contains 2 kg of O_2 at 25°C and 450 kPa. The valve connecting the two tanks is opened, and the two gases are allowed to mix. If the final mixture temperature is 25°C, determine (a) the volume ($\dot{V}$) of each tank and (b) the final mixture pressure (p_m). Use the IG Model.

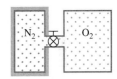

11-3-17 An insulated rigid tank that contains 1 kg of CO_2 at 100 kPa and 25°C and is connected to another insulated rigid tank that contains 1 kg of H_2 at 200 kPa and 500°C. The valve connecting the two tanks is opened, and the two gases are allowed to mix. Determine (a) the final mixture pressure, (b) temperature, (c) entropy generated (S_{gen}) during mixing, and (d) reversible work (W_{rev}) of mixing if the outside temperature is 25°C. Use the IG mixture model. (e) *What-if scenario:* How would the answer in (d) change if both gases were CO_2?

11-3-18 N_2 at 100 kPa, 30°C with a flow rate of 100 m^3/min is mixed with CO_2 at 200°C, 100 kPa with a flow rate of 50 m^3/min. Determine (a) the final temperature (T_2) and (b) rate of generation of entropy ($\dot{S}_{gen}$). Assume the gases to behave as a perfect gas mixture. (c) *What-if scenario:* What would the final temperature be if the ideal gas mixture model were used?

11-3-19 An equimolar mixture of oxygen and nitrogen enters a compressor operating at steady state at 10 bar, 220 K with a mass flow rate ($\dot{m}$) of 1 kg/s. The mixture exits the compressor at 60 bar, 400 K with no significant change in KE or PE. The heat transfer from the compressor can be neglected. Using the RG model (Kay's rule), determine (a) the required power ($\dot{W}_{ext}$) and (b) the rate of entropy production ($\dot{S}_{gen}$). (c) *What-if scenario:* What would the required power be if the ideal gas mixture model were used?

11-3-20 A mixture of 0.5 kg of carbon dioxide and 0.3 kg of nitrogen is compressed from 100 kPa, 300 K to 300 kPa in a polytropic process for which $n = 1.25$. Determine (a) the final temperature (T_2), (b) the work (W), (c) the heat transfer (Q) and (d) the change in entropy (ΔS) of the mixture.

11-3-21 Helium at 200 kPa, 20°C is heated by mixing it with argon at 200 kPa, 500°C in an adiabatic chamber. Helium enters the chamber at 2 kg/s and argon at 0.5 kg/s. If the mixture leaves at 200 kPa, determine (a) the temperature (T_2) at the exit and (b) the rate of entropy generation ($\dot{S}_{gen}$) due to mixing.

11-3-22 Repeat Problem 11-3-22 with argon entering the chamber at the same temperature as helium (all other conditions remaining the same). Explain the change in entropy generation rate ($\dot{S}_{gen}$).

11-3-23 Repeat Problem 11-3-23 with argon replaced by neon, entering the chamber at the same temperature as helium (all other conditions remaining the same). Explain the change in entropy generation rate ($\dot{S}_{gen}$).

11-3-24 Repeat Problem 11-3-26 with the hot gas argon replaced by neon (all other conditions remaining the same). Explain the change in entropy generation rate ($\dot{S}_{gen}$).

11-3-25 Hydrogen is mixed with oxygen in an adiabatic mixing chamber at 100 kPa, 25°C. The flow rate of hydrogen is 2 kmol/s and that of oxygen is 1 kmol/s. If the mixture leaves the chamber at 100 kPa, determine (a) the temperature at the exit (T_2) and (b) the rate of entropy generation ($\dot{S}_{gen}$) in the device. Assume no chemical reaction and use the PG mixture model.

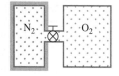

11-3-26 An insulated rigid tank that contains 1 kg of H_2 at 25°C and 100 kPa is connected to another insulated rigid tank that contains 1 kg of He at 25°C and 100 kPa. The valve connecting the two tanks is opened, and the two gases are allowed to mix adiabatically. Determine (a) the entropy generated (s_{gen}) during mixing per unit mass of the mixture. (b) Now replace the content of the second tank with different gases of same mass and for each gas determine the entropy generated per unit mass of the total mixture (2 kg). Plot a bar chart of entropy generation per unit mass against different gases on the x-axis. Use IG mixture model.

11-3-27 An insulated rigid tank that contains 1 kg of O_2 at 25°C and 500 kPa is connected to another insulated rigid tank that contains 1 kg of O_2 at 25°C and 1000 kPa. The valve connecting the two tanks is opened, and the two gases are allowed to mix adiabatically. Determine (a) the entropy generated (s_{gen}) during mixing per unit mass of the mixture. (b) Now repeat the calculations by varying the pressure in the second tank from a range of 10 kPa through 10 MPa and for each pressure determine the entropy generated per unit mass of the total mixture (2 kg). Plot the entropy generation per unit mass (y-axis) against the initial pressure ratio. Use IG mixture model.

11-3-28 An insulated rigid tank that contains 1 kg of CO_2 at 300 K and 500 kPa is connected to another insulated rigid tank that contains 1 kg of CO_2 at 400 K and 500 kPa. The valve connecting the two tanks is opened, and the two gases are allowed to mix adiabatically. Determine (a) the entropy generated (s_{gen}) during mixing per unit mass of the mixture. (b) Now repeat the calculations by varying the temperature in the second tank from a range of 100 K through 2000 K. Plot the entropy generation per unit mass (y-axis) against the initial temperature ratio. Use IG mixture model.

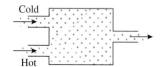

11-3-29 Carbon-dioxide at 100 kPa, 25°C enters an adiabatic mixing chamber with a mass flow rate of 1 kg/s is mixed with hydrogen entering at 100 kPa and 25°C. Plot the entropy generated per unit mass of the mixture (s_{gen}) as a function of mass fraction of hydrogen in the mixture (vary x from 0 to 1). (b) Repeat the plot with entropy generation expressed on the basis of unit mole (1 kmol) of the flow and mass fraction replaced with mole fraction.

11-3-30 Repeat Problem 11-3-30 with a completely different pair of gases. Can you come up with a generalized mixing criterion that maximizes entropy generation per unit mass or mole?

11-3-31 Argon at 100 kPa, 600 K enters an adiabatic mixing chamber with a mass flow rate of 1 kg/s is mixed with with nitrogen entering at 100 kPa, 600 K and 1 kg/s. (a) Determine the entropy generated per unit mass of the mixture (s_{gen}). (b) Vary the inlet temperature of nitrogen (from 200 K to 2000 K) and plot (s_{gen}) as a function of temperature ratio of the incoming streams.

11-3-32 Repeat Problem 11-3-32 with a completely different pair of gases. Can you come up with a generalized mixing curve where data from different pairs fall on the same line as far as entropy generation is concerned?

11-3-33 Argon at 1000 kPa, 300 K enters an adiabatic mixing chamber with a mass flow rate of 1 kg/s is mixed with nitrogen entering at 1000 kPa, 300 K and 1 kg/s. (a) Determine the entropy generated per unit mass of the mixture (s_{gen}). (b) Vary the mixing pressure (from 10 kPa through 10 MPa) and plot (s_{gen}) as a function of chamber pressure.

11-3-34 Repeat Problem 11-3-34 with a completely different pair of gases. Can you come up with a generalized mixing curve where data from different pairs fall on the same line as far as effect of pressure on entropy generation is concerned?

12 | PSYCHROMETRY

The study of air and water vapor mixture, commonly known as moist air, is called **psychrometrics**, which has application in heating, ventilation, and air conditioning (HVAC). Both components of moist air—dry air and water vapor, with the latter having relatively low partial pressure—behave like ideal gases. Also, the temperature variation in air conditioning applications is rather mild. Therefore, the PG mixture model, developed in Chapter 11 appears to be an ideal candidate for modeling moist air. However, the possibility that water vapor may condense, or vapor may be added through evaporation or injection of steam, means that the moist air composition may change in a psychrometric device or process. The PG mixture model, therefore, is modified leading to the development of the moist air (MA) model. By the *state postulate*, three properties are required to define the thermodynamic state of a gas mixture. To represent a state of moist air on a two-dimensional plot, special psychrometric charts are developed—each corresponding to a given mixture pressure (the third property). After the development of the MA model, mass and energy balance equations are customized for psychrometric systems and processes. Air conditioning applications—simple heating, simple cooling, heating with humidification, cooling with dehumidification, evaporative cooling, and cooling towers—are successively analyzed. As in other chapters, animations are used to illustrate different psychrometric processes, and TEST solutions are used to verify manual results and carry out "what-if" studies.

12.1 THE MOIST AIR MODEL

Atmospheric air is principally a mixture of nitrogen and oxygen with a small amount of water vapor, also called **moisture**, and some minor species. It is the presence of the water vapor, in various degrees, that gives air everyday adjectives such as dry, damp, moist, muggy, humid, sultry, etc. In psychrometric studies, we use the term **moist air**, or simply **air**, to describe atmospheric air and **dry air** for air with absolutely no moisture. In air conditioning applications, the composition of dry air—approximately 79% N_2 and 21% O_2 by volume—does not change and the temperature seldom goes outside the range of $-10°C$ to $50°C$; therefore, dry air can be modeled as a pseudo-pure perfect gas with $c_{pa} = 1$ kJ/kg·K, $\overline{M}_a = \sum_k y_k \overline{M}_k = (0.79)(28) + (0.21)(32) \cong 29$ kg/kmol and $R_a = \overline{R}/\overline{M}_a = 8.314/29 = 0.287$ kJ/kg·K (with an error of less than 1%).

The partial pressure of water vapor is generally quite low—less than 15 kPa in most situations (Fig. 12.1)—so that the reduced pressure is less than 0.001, well within the ideal gas criterion of Eq. (3.90). Moreover, temperature variation in psychrometric applications is moderate to small, allowing specific heats to be assumed constant. Therefore, water vapor in moist air, treated as a Dalton mixture component (Sec. 11.4.3), can be considered a perfect gas with $c_{pv} = 1.87$ kJ/kg·K, (average value in the range -10 to $50°C$), $\overline{M}_v = 18$ kg/kmol, and $R_v = 0.462$ kJ/kg·K. However, the PG mixture model, developed in Chapter 11, cannot be directly applied to moist air due to the presence of water vapor, bringing about the possibility of condensation or evaporation. We begin the development of the **moist air model**, which can be considered an enhanced PG mixture model that takes into account phase transformation of water vapor.

FIGURE 12.1 Moist air behaves like a perfect gas mixture of dry air and water vapor (see Anim. 11.C.*DaltonLawIG*).

12.1.1 Model Assumptions

(i) Moist air is a mixture of two perfect gases, water vapor and dry air. The dry air is assumed to be a pseudo-pure gas with $\overline{M}_a = 29 \dfrac{\text{kg}}{\text{kmol}}$ and $c_{pa} = 1$ kJ/kg·K.

(ii) When moist air is in equilibrium with liquid water or solid ice, the equilibrium between water vapor and the condensed phase (liquid water or ice) is not affected by the presence of dry air.

(iii) Constituents of dry air do not dissolve in the condensed phase of water.

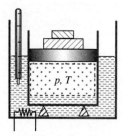

FIGURE 12.2 Moist air in this piston-cylinder device is saturated at pressure p and temperature T (see Anim. 12.A.*TConstHumidification*).

The significance of the second assumption is that it provides a simple means to determine the amount of water vapor in air. Suppose the air and liquid water trapped in the piston-cylinder device of Figure 12.2 are maintained at a constant pressure p (by placing the right amount of weight on the piston) and temperature T (by maintaining the bath at a fixed temperature). As time passes, some water may evaporate or some water vapor may condense (no other species can migrate between phases due to the third assumption). After sufficient time, the system eventually comes to thermodynamic equilibrium (Sec. 1.2) and from that point there are no further changes in the composition or properties of the moist air or the liquid water. When this equilibrium is achieved, the moist air is said to be **saturated** with the water vapor content at its peak for the given pressure and temperature.

By the second assumption, the presence of dry air in saturated air can be completely ignored, leaving water vapor at a partial pressure of $p_{v,max}$ in thermodynamic equilibrium with its liquid phase at the given temperature T. Therefore, the water vapor must be saturated vapor with its state along the g-line in a T-s diagram, as shown in Figure 12.3. Following the nomenclature developed in Chapter 3, the equilibrium pressure of water vapor in saturated air is given by $p_{v,max} = p_g = p_{sat@T}$ (see Anim. 12.A.*saturatedAir*). For instance, if water and air are allowed to equilibrate at 25°C, the equilibrium pressure can be obtained from the saturation table as $p_g = p_{sat@25°C} = 3.17$ kPa, which must also be the partial pressure p_v of water vapor in the saturated air. It is important to observe that this pressure is independent of the total pressure of the system and is only a function of the temperature. If the total pressure of air is 100 kPa, from Dalton's law of partial pressure [Eq. (11.93)] we can deduce the dry air pressure as $p_a = 100 - 3.17 = 96.83$ kPa. Similarly, it can be shown that when moist air at 100 kPa is in equilibrium with ice at −10°C, $p_v = p_g = p_{sat@-10°C} = 0.26$ kPa and $p_a = 99.75$ kPa.

Before we can describe the state of moist air in a general manner, a few new properties and processes must be introduced.

12.1.2 Saturation Processes

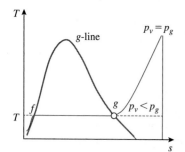

FIGURE 12.3 State of the water vapor in saturated (State-g) and unsaturated (dashed line) moist air (see Anim. 12.A.*saturatedAir*).

Moist air is not always saturated. At a given temperature T, the maximum possible partial pressure of water vapor is $p_{v,max} = p_g = p_{sat@T}$. Therefore, the condition for air to be **unsaturated** is simply $p_v < p_g$. Any state in the dashed line of Figure 12.3 satisfies this condition for a given temperature T (see Anim. 12.A.*moistAir*). The dashed line in the superheated region of the T-s diagram, so the state of vapor in unsaturated air is always superheated. If p_v and the temperature of the moist air are known, the state can be located on this diagram.

A process that leads moist air into saturation is called a **saturation process** and there are several such processes. In an **isothermal saturation process** (see Anim. 12.A.*TConstHumidification*), the temperature and total pressure of air are held constant as moisture is added by bringing the air in contact with liquid water or ice. Total pressure remaining constant, addition of water vapor increases the mole fraction of water vapor in the mixture, hence p_v, according to Eq. (11.96). As p_v continues to increase, the state of vapor, starting from State-1 in Figure 12.4, moves towards the saturated state, State-2, along the constant-temperature (horizontal) line. At State-2, the partial pressure of water vapor, $p_{v2} = p_g = p_{sat@T}$, must be a maximum as no further addition of moisture is possible.

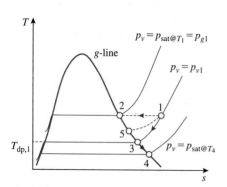

FIGURE 12.4 States of moist air in a T-s diagram in constant temperature (1–2), constant pressure (1–3), and adiabatic saturation (1–5) processes (see animations starting with 12.A.*pConstCooling*).

When air at State-1 is cooled at a constant total pressure (see Anim. 12.A.*pConstCooling*), the composition of moist air is not initially affected. The total pressure and vapor mole fraction remaining constant, the partial pressure $p_{v1} = y_{v1}p_1$ must remain constant, and the vapor state follows the constant-pressure line in Figure 12.4, from State-1 towards saturation at State-3. At the saturated state, State-3, the temperature can be obtained as the saturation temperature for the known pressure p_{v1} from the pressure saturation table of water (Table B-1). The temperature at which air is saturated by **constant-pressure cooling** is called the **dew point temperature** T_{dp}. It should be noted that the dew point temperature $T_3 = T_{sat@p_{v1}} = T_{dp,1}$ is a thermodynamic property of air at State-1, as is the saturation pressure $p_{v2} = p_{sat@T_1} = p_{g1}$. Obviously, the dew point temperature and saturation pressure remain constant during a constant-pressure cooling until saturation is achieved. Any further cooling below the dew point temperature, for example to T_4 in Figure 12.4, moves the saturated state down the g-line and the vapor pressure must decrease in accordance with $p_v = p_g = p_{sat@T_4}$ since $T_4 < T_3$. The only way this can be accomplished is through condensation that reduces the mole fraction of vapor in air.

Saturation can also be brought about through a **constant-volume** process, which is less common. In **adiabatic saturation**, air is allowed to pick up moisture in an adiabatic channel (Fig. 12.5). The temperature at the end of adiabatic saturation is lower than the inlet temperature because the air supplies the *latent heat* (Sec. 5.1.4B) of evaporation. After sufficient time, the system becomes steady (strictly speaking, some water must be supplied to replenish the evaporated water from the tray), the state of water on the tray stabilizes and its temperature is assumed to be the same as the saturated air at State-5. As a result, with the incoming air being warmer than water, none of the latent heat is supplied by the water. The exit temperature of the saturated air is known as the **adiabatic saturation temperature**. As the air picks up moisture, the vapor pressure increases as shown by the process 1–5 in Figure 12.4 (and in Anim 12.A.*adiabaticSaturation*). We will analyze adiabatic saturation in a later section.

FIGURE 12.5 Air flowing over the long liquid tray is saturated adiabatically (see Anim. 12.A.*adiabaticSaturation*). As y_v increases, so does $p_v = y_v p$ (see Fig. 12.4).

12.1.3 Absolute and Relative Humidity

In Chapter 11, we learned that the most logical way to describe the proportion of a particular component in a mixture is through its mass or mole fraction. Unfortunately, neither is quite suitable for moist air, as the total mass of moist air may change due to condensation or evaporation. Rather, the amount of dry air in the mixture is usually invariant in most psychrometric applications. A new property, called **absolute** or **specific humidity**, ω, is defined as the ratio of the mass of moisture m_v to the mass of dry air m_a in a system of moist air, so that the total mass $m = m_a + m_v$ can be expressed in terms of ω. In the case of a flow of moist air, the corresponding mass flow rates must be used in the definition of ω so that the total mass flow rate $\dot{m} = \dot{m}_a + \dot{m}_v$ can be similarly expressed:

$$\omega \equiv \frac{m_v}{m_a} = \frac{\dot{m}_v}{\dot{m}_a}; \quad \left[\frac{\text{kg H}_2\text{O}}{\text{kg d.a.}}\right] \tag{12.1}$$

$$\text{Total mass:} \quad m = m_a + m_v = m_a(1 + \omega); \quad [\text{kg}] \tag{12.2}$$

$$\text{Total mass flow rate:} \quad \dot{m} = \dot{m}_a + \dot{m}_v = \dot{m}_a(1 + \omega); \quad \left[\frac{\text{kg}}{\text{s}}\right] \tag{12.3}$$

Although absolute humidity is a ratio, notice how the unit is used for clarity. When specifying a mass, we will use kg H$_2$O or kg d.a. (dry air) whenever appropriate. In a vapor air mixture the mass of dry air remains constant, therefore, ω is proportional to the mass of water vapor. Using the Dalton model (Sec 11.4.3) for an ideal gas mixture, ω can be related to the vapor pressure p_v as follows:

$$\omega \equiv \frac{m_v}{m_a} = \frac{p_v \overline{V} \overline{M}_v/(\overline{R}T)}{p_a \overline{V} \overline{M}_a/(\overline{R}T)} = \frac{\overline{M}_v}{\overline{M}_a} \frac{p_v}{p_a} = 0.622 \frac{p_v}{p - p_v}; \quad \left[\frac{\text{kg H}_2\text{O}}{\text{kg d.a.}}\right] \tag{12.4}$$

Note that the relationship above is not linear between ω and p_v. It can be deduced from this relation that specific humidity ω has a minimum value of zero for dry air, and it reaches a maximum for saturated air where p_v reaches its peak value: $p_v = p_{v,\text{max}} = p_g = p_{\text{sat@}T}$. Even at its peak, ω is a small fraction since $p_g << p$. For instance, at 100 kPa, 25°C, $p_g = p_{\text{sat@25°C}} = 3.17$ kPa; therefore, the maximum value of ω can be calculated using $p_{v,\text{max}} = p_g$ in Eq. (12.4), producing $\omega_{\text{max}} = 0.020$ kg H$_2$O/kg d.a.

Although specific humidity turns out to be a very useful property in the mass and energy analysis of psychrometric systems, human comfort is more closely linked to a relative measure of humidity. **Relative humidity** is defined as the ratio of the mole fraction of water vapor in air to the maximum possible mole fraction of vapor the air can hold at the same temperature. Using the IG (ideal gas) equation of state for a component, Eq. (11.92), the relative humidity can be expressed as the ratio of vapor pressure to the saturation vapor pressure at the temperature of the moist air:

$$\phi \equiv \frac{n_v}{n_{v,\text{max}}} = \frac{y_v}{y_{v,\text{max}}} = \frac{p_v/(\overline{R}T)}{p_g/(\overline{R}T)} = \frac{p_v}{p_g} = \frac{p_v}{p_{\text{sat@}T}} \tag{12.5}$$

The minimum and maximum limits of relative humidity are 0 and 100% respectively. From Eq. (12.5), the vapor pressure p_v can be evaluated from a knowledge of ϕ and T and, if the total pressure is given, ω can also be obtained from Eq. (12.4). Conversely, if ω, p, and T are known, ϕ can be obtained from Eqs. (12.4) and (12.5).

Dry bulb thermometer gives the current air temperature.

Wick is dipped in water.

Thermometers are swung around handle.

When swung, water evaporates from the wick, cooling the wet-bulb thermometer. Drier is the comparative of dry air results in lower temperature.

FIGURE 12.6 A sling psychrometer is swung around the handle until the DBT and WBT reach their equilibrium values.

12.1.4 Dry- and Wet-Bulb Temperatures

The **dry-bulb temperature** of moist air is simply its thermodynamic temperature T measured by a simple liquid-in-bulb thermometer. When the bulb of the thermometer is kept moistened by a wick soaked in water, the indicated temperature decreases due to evaporation of water, if the surrounding air is not saturated. The steady state temperature recorded is the **wet-bulb temperature** and is represented by the symbol T_{wb}. An instrument with dry-bulb and wet-bulb thermometers mounted together is called a **psychrometer**. To reach local equilibrium faster, the thermometers of a **sling psychrometer** are whirled around in air by gripping a handle (Fig. 12.6). Unlike the adiabatic saturation temperature, the wet-bulb temperature T_{wb} is not a property, but a path function whose value is influenced by the rate of heat and mass transfer at the wick. It will be shown that the difference between the easy-to-measure T_{wb} and the adiabatic saturation temperature is small, making T_{wb} a very useful pseudo-property of moist air. In this textbook, we will treat T_{wb} as just another thermodynamic property, as is the dry-bulb temperature T of moist air.

12.1.5 Moist Air (MA) TESTcalcs

TEST offers psychrometric TESTcalcs for system and flow state evaluation involving moist air, and for analyzing closed processes and open-steady devices, including cooling towers. Moist air state TESTcalcs have the same look and feel as other state TESTcalcs. The working fluid selector offers several moist gases in addition to moist air and pure H_2O. When H_2O is selected, the model switches to the PC model and properties, such as quality, are activated. Saturation properties of H_2O can be calculated, without leaving the MA TESTcalcs, by using $x = 0$ (saturated liquid) or $x = 1$ (saturated vapor). When moist air, or any other moist gas is selected, the mass refers to dry air (gas) mass and the specific properties are based on unit mass of dry air (gas), a standard practice in psychrometry. The MA TESTcalcs can handle moist gases over a wide range of temperatures and pressures—much beyond the capabilities of standard psychrometric charts. Calculated states can be displayed on a psychrometric plot, which resembles a psychrometric chart, rather than the standard T-s diagram. Constant-property lines can be drawn through calculated states and the plot can be dragged and zoomed upon as in any other thermodynamic plot.

The process panel in the closed-process TESTcalc, and the device panel in the open-steady TESTcalc, are quite similar to the corresponding panels used in the generic process and device TESTcalcs. However, the mass equation differs for moist air as separate equations are needed for dry air and H_2O conservation.

b-state=State-1

30°C
100 kPa
60% R.H.

f-state=State-2

5°C
100 kPa
100% R.H.

FIGURE 12.7 Schematic for 12-1 (see Anim. 12.A.*pConstCooling*).

EXAMPLE 12-1 **Properties of Moist Air**

1 kg of moist air at 60% R.H., 30°C is cooled, at a constant pressure of 100 kPa, to a final temperature of 5°C. For the initial state, determine (a) the specific humidity, (b) dew point temperature, and (c) amount of moisture in the air. Also determine (d) the amount of condensate collected during the cooling process.

SOLUTION

Use the definitions introduced in this section to determine the moist air properties, at the beginning and final states, from the given data.

Assumptions

The moist air model can be applied. Air is in equilibrium at the principal states, State-1 and State-2 (Fig. 12.7).

Analysis

State-1 (given p_1, T_1, ϕ_1, m_1):

$$p_{v1} = \phi_1 p_{g1} = \phi_1 p_{sat@T_1} = \phi_1 p_{sat@30°C} = (0.6)(4.246) = 2.55 \text{ kPa};$$

$$p_{a1} = p_1 - p_{v1} = 100 - 2.55 = 97.45 \text{ kPa};$$

$$\text{Therefore,} \quad \omega_1 = (0.622)\frac{p_{v1}}{p_{a1}} = (0.622)\frac{(2.55)}{(97.45)} = 0.0163 \; \frac{\text{kg H}_2\text{O}}{\text{kg d.a.}}$$

The dry air and moisture mass can be calculated by using the definition of ω:

$$m_1 = m_{a1} + m_{v1} = m_{a1}(1 + \omega_1); \quad \Rightarrow \quad m_{a1} = \frac{m_1}{(1 + \omega_1)} = \frac{1}{1.0163} = 0.984 \text{ kg d.a.}$$

$$\text{Therefore,} \quad m_{v1} = \omega_1 m_{a1} = (0.0163)(0.984) = 0.0160 \text{ kg}$$

The dew point temperature can be obtained by looking up $T_{\text{sat}@p_{v1}}$ from the saturation pressure table (Table B-1):

$$T_{\text{dp},1} = T_{\text{sat}@p_{v1}} = T_{\text{sat}@2.55 \text{ kPa}} = 21.4°\text{C}$$

State-2 (given $p_2 = p_1, T_2, \phi_2 = 100\%, m_2 = m_1$):

$$p_{v2} = \phi_2 p_{g2} = p_{\text{sat}@T_2} = p_{\text{sat}@5°\text{C}} = 0.872 \text{ kPa};$$

$$p_{a2} = p_2 - p_{v2} = 100 - 0.872 = 99.128 \text{ kPa};$$

$$\omega_2 = (0.622)\frac{p_{v2}}{p_{a2}} = (0.622)\frac{(0.872)}{(99.128)} = 0.0055 \; \frac{\text{kg H}_2\text{O}}{\text{kg d.a.}}$$

$$\text{Therefore,} \quad m_{v2} = \omega_2 m_{a2} = (0.0055)(0.982) = 0.0054 \text{ kg}$$

The amount of condensate must be:

$$m_{v1} - m_{v2} = 0.0160 - 0.0054 = 0.0106 \text{ kg}$$

TEST Solution

Launch the MA system-state TESTcalc and select moist air as the working fluid. Evaluate State-1 partially by entering known values of p1, T1, and RH1. Now guess values for m1 (dry air) and check to see if m1+m_v1=1 kg. By adjusting m1 on a guess-and-check basis, obtain m1=0.984 kg. Calculate State-2 by entering p2=p1, T2, RH2=100%, and m2=m1 (dry air amount remains unchanged). Now that all the properties of the two states are determined, evaluate the expression '= m_v1−m_v2' as 0.0106 kg in the I/O panel.

Discussion

The volume of the system is fully occupied by moist air at State-1. At the final state, the liquid volume is only about 11 mL compared to the moist air volume (m2*v2) of 790 L; therefore the volume occupied by a condensate is generally neglected.

12.1.6 More properties of Moist Air

Specific Volume:

All specific properties of moist air are expressed in terms of a unit mass of dry air, which remains unaffected by condensation or evaporation of water. The specific volume v, defined as volume of moist air per unit mass of dry air, can be expressed in terms of partial pressure of air using the IG equation of state:

$$v \equiv \frac{V}{m_a} = \frac{V}{p_a V \overline{M}_a/(\overline{R}T)} = \frac{\overline{R}T}{p_a \overline{M}_a} = \frac{R_a T}{p_a}; \quad \left[\frac{\text{m}^3}{\text{kg d.a.}}\right] \qquad \textbf{(12.6)}$$

In a given system, the volume occupied by dry air, water vapor, or moist air must be the same and must be equal to the volume of the system V. The density of moist air, $\rho \equiv m/V = (m_a + m_v)/V = (1 + \omega)/v$, can no longer be expressed as the inverse of specific volume.

When the working fluid in the MA TESTcalcs is switched from moist air to H_2O, the symbol and unit displayed for the specific properties stay the same. Rolling the cursor over a specific property widget brings up a short explanation on the message panel regarding the subtle difference between the units used in two material models (MA and PC models).

Enthalpy:

For a perfect gas mixture of water vapor and dry air, the specific enthalpy, defined in terms of unit mass of dry air, must contain contributions from both the constituents. At the typically low partial pressure p_v, the vapor in moist air can be treated as a perfect gas (Sec. 3.8) and $h_v = h_v(p, T) = h_v(T)$ (Fig. 12.8). Given the possibility of a phase change, the enthalpy of water vapor must use the same datum as liquid water or ice. At a given temperature T, the value of $h_v(T)$ can be obtained from the saturation temperature table of water, which assigns the same datum to liquid and vapor, as $h_v(T) \cong h_g(T)$. The enthalpy of saturated liquid and saturated vapor at 0°C are 0 (reference value) and 2501.3 kJ/kg, respectively. Therefore, h_v and h_a can be approximately expressed by application of the PG model as:

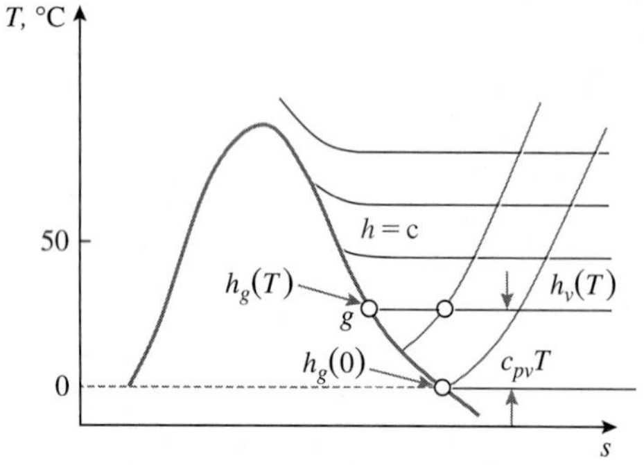

FIGURE 12.8 Graphical interpretation of enthalpy as expressed by Eq. (12.7). For $T < 50°C$, $h_v(T) \cong h_g(T) \cong h_g(0) + c_{pv}T$.

$$h_v \cong h_g(T) \cong h_g(0°C) + c_{pv}(T - 0) = 2501.3 + 1.87T; \quad \left[\frac{kJ}{kg}\right] \tag{12.7}$$

and

$$h_a = c_{pa}(T - 0) \cong T; \quad \left[\frac{kJ}{kg}\right] \tag{12.8}$$

where T in Celsius, is the **dry-bulb temperature**. Adding the contribution of h_a and h_v, the moist air enthalpy h per unit mass of dry air can be obtained:

$$h \equiv \frac{H}{m_a} = \frac{H_a}{m_a} + \frac{H_v}{m_a} = h_a + \frac{m_v}{m_a}\frac{H_v}{m_v} = h_a + \omega h_v = h_a + \omega h_{g@T}; \quad \left[\frac{kJ}{kg\ d.a.}\right] \tag{12.9}$$

It should be stressed that, unlike any other material model, the moist air enthalpy is based on unit mass of dry air and has a unit of kJ/kg d.a.

Internal Energy:

With the enthalpy expression evaluated, the specific internal energy can be obtained from the enthalpy definition and using the ideal gas equation of state for a component, Eq. (11.92), as:

$$u \equiv \frac{U}{m_a} = \frac{H - p\Psi}{m_a} = h - pv \cong h - R_a T \quad \left[\frac{kJ}{kg\ d.a.}\right] \tag{12.10}$$

where h can be evaluated from Eq. (12.9) and v is approximated with the help of Eq. (12.6).

Entropy:

Although an entropy analysis is outside the scope of this chapter, the PG mixture model (Sec. 11.4.3) can be used to derive an expression for moist air entropy. The contribution from vapor can be expressed following a similar logic as advanced for enthalpy:

$$s_v(p_v, T) = s_v(p_g, T) - [s_v(p_g, T) - s_v(p_v, T)]$$

$$= s_g(T) - R_v \ln\frac{p_v}{p_g} = s_g(T) - R_v \ln\phi; \quad \left[\frac{kJ}{kg\cdot K}\right] \tag{12.11}$$

Moist air entropy, therefore, can be expressed by combining the contributions from dry air and water vapor:

$$s \equiv \frac{S}{m_a} = \frac{S_a}{m_a} + \frac{S_v}{m_a} = s_a + \frac{m_v}{m_a m_v}S_v = s_a + \omega s_v;$$

$$\Rightarrow \quad \Delta s = \Delta s_a + \Delta(\omega s_v);$$

$$\Rightarrow \quad \Delta s = c_{pa} \ln \frac{T_2}{T_1} - R_a \ln \frac{p_{a2}}{p_{a1}} + \omega_2[s_g(T_2) - R_v \ln \phi_2]$$

$$- \omega_1[s_g(T_1) - R_v \ln \phi_1]; \left[\frac{kJ}{kg\ d.a. \cdot K}\right] \qquad (12.12)$$

We emphasize that all the specific properties (u, v, h, and s) are based on unit mass of dry air in the MA model, but they are based on per-unit total mass in all other models.

EXAMPLE 12-2 Properties of Moist Air

1 m^3 of moist air at 50% R.H., 25°C, and 100 kPa is compressed to saturation in a constant temperature process. Determine the final (a) volume and (b) pressure at which air becomes saturated and (c) the mass of water vapor in air at the end of the process.

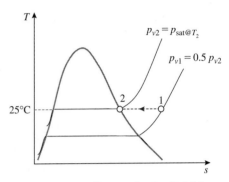

SOLUTION

Use the definitions introduced in this section to determine the moist air properties, at the initial and final states, from the given data.

Assumptions

The MA model can be applied. Air is in equilibrium at both the beginning and final states, State-1 and State-2, of the process (Fig. 12.9). The volume of condensate, if any, is negligible compared to that of moist air.

FIGURE 12.9 *T-s* diagram for Ex. 12.2 (also see Anim. 12.A.*moistAir*).

Analysis

Use Eqs. (12.1)–(12.6) to evaluate the following states:

State-1 (given p_1, T_1, ϕ_1, V_1):

$$p_{v1} = \phi_1 p_{g1} = \phi_1 p_{sat@T_1} = \phi_1 p_{sat@25°C} = (0.5)(3.169) = 1.585 \text{ kPa};$$

$$\omega_1 = (0.622)\frac{p_{v1}}{p_1 - p_{v1}} = (0.622)\frac{(1.585)}{(98.415)} = 0.0100 \frac{\text{kg H}_2\text{O}}{\text{kg d.a.}};$$

$$m_{v1} = \omega_1 m_{a1} = \omega_1 \frac{V_1}{v_1} = \frac{\omega_1 V_1 p_{a1}}{R_a T_1} = \frac{(0.0100)(1)(98.414)}{(0.287)(298)} = 0.0115 \text{ kg}$$

The mass of water vapor does not change during the process, at the end of which $m_{v2} = m_{v1} = 0.0115$ kg. The mass of air also remains unchanged; therefore, ω must remain constant.

State-2 (given $T_2 = T_1$, $\phi_2 = 100\%$, $\omega_2 = \omega_1$, $m_{a2} = m_{a1}$):

$$p_{v2} = \phi_2 p_{g2} = p_{sat@25°C} = 2p_{v1};$$

$$\omega_2 = (0.622)\frac{p_{v2}}{p_{a2}}; \quad \Rightarrow \quad p_{a2} = (0.622)\frac{p_{v2}}{\omega_2} = 2(0.622)\frac{p_{v1}}{\omega_1} = 2p_{a1}$$

Therefore,

$$p_2 = p_{a2} + p_{v2} = 2(p_{a1} + p_{v1}) = 2p_1 = 200 \text{ kPa};$$

$$V_2 = m_{a2}v_2 = m_{a1}\frac{R_a T_2}{p_{a2}} = m_{a1}\frac{R_a T_1}{2p_{a1}} = \frac{m_{a1}v_1}{2} = \frac{V_1}{2} = 0.5 \text{ m}^3$$

TEST Analysis

Launch the MA system-state TESTcalc. Select moist air and evaluate State-1 by entering the known values of p1, T1, RH1, and Vol1. Calculate State-2 by setting p2 as unknown, and entering T2=T1, RH2=100%, omega2=omega1, and m2=m1 (dry air mass). p2 and Vol2 are evaluated as part of State-2 and the manual solution is verified.

Discussion

During isothermal compression of air, the temperature remaining constant, vapor pressure reaches its peak when it is equal to the saturation vapor pressure, $p_{g1} = p_{sat@T_1}$, at the temperature of the air. Any further increase in the total pressure p will trigger condensation to reduce the vapor mole fraction y_v, so that $p_v = y_v p = p_g = $ constant.

12.2 MASS AND ENERGY BALANCE EQUATIONS

In this section, we will develop customized mass and energy balance equations, for a generic open-steady device and a generic closed process to serve as templates for the practical systems encountered in the rest of the chapter. Due to the fact that water is often added or separated from moist air in psychrometric processes, separate mass equations for dry air and H_2O are necessary.

12.2.1 Open-Steady Device

Consider a generic open psychrometric system, as shown in Figure 12.10 (and Anim. 12.B.*balanceEquations*), operating at steady state. For simplicity, we will assume that the device has only one inlet, State-1, and one exit, State-2, for moist air. Keeping in mind the generality of applications to be discussed, water injection at State-3, steam injection at State-4, condensate removal at State-5, cooling and heating coils, are all incorporated in this generic psychrometric system. Using the subscript a for dry air, v for water vapor, and w for externally injected or rejected liquid water or steam, the steady-state mass balance, Eq. (2.6), can be separately written for dry air and H_2O as:

$$\text{Dry Air:} \quad \dot{m}_{a1} = \dot{m}_{a2} = \dot{m}_a; \quad \left[\frac{\text{kg d.a.}}{\text{s}}\right] \tag{12.13}$$

$$H_2O: \quad \dot{m}_{v1} + \dot{m}_{w3} + \dot{m}_{w4} = \dot{m}_{v2} + \dot{m}_{w5};$$

$$\Rightarrow \quad \dot{m}_{a1}\omega_1 + \dot{m}_{w3} + \dot{m}_{w4} = \dot{m}_{a2}\omega_2 + \dot{m}_{w5};$$

$$\Rightarrow \quad \dot{m}_a(\omega_2 - \omega_1) = \dot{m}_{w3} + \dot{m}_{w4} - \dot{m}_{w5}; \quad \left[\frac{\text{kg } H_2O}{\text{s}}\right] \tag{12.14}$$

The energy equation, Eq. (2.11), can also be simplified by neglecting changes in ke and pe, assuming that there is no external work transfer, i.e., $\dot{W}_{ext} = 0$, and summing up contributions from dry air and moisture to represent energy transport by moist air:

$$0 = (\dot{m}_{a1}h_{a1} + \dot{m}_{v1}h_{v1}) + \dot{m}_{w3}h_{w3} + \dot{m}_{w4}h_{w4} - (\dot{m}_{a2}h_{a2} + \dot{m}_{v2}h_{v2}) - \dot{m}_{w5}h_{w5} + \dot{Q};$$

$$\Rightarrow \quad \dot{m}_a(h_{a1} + \omega_1 h_{v1}) + \dot{m}_{w3}h_{f3} + \dot{m}_{w4}h_{g4} + \dot{Q} = \dot{m}_a(h_{a2} + \omega_2 h_{v2}) + \dot{m}_{w5}h_{f5}; \quad [\text{kW}]$$

$$\text{Energy:} \quad \dot{m}_a h_1 + \dot{m}_{w3}h_{f@T_3} + \dot{m}_{w4}h_{g@T_4} + \dot{Q} = \dot{m}_a h_2 + \dot{m}_{w5}h_{f@T_5}; \quad [\text{kW}] \tag{12.15}$$

$$\text{where, } h \text{ is specific entahlpy of moist air:} \quad h = h_a + \omega h_v; \quad \left[\frac{\text{kJ}}{\text{kg d.a.}}\right]$$

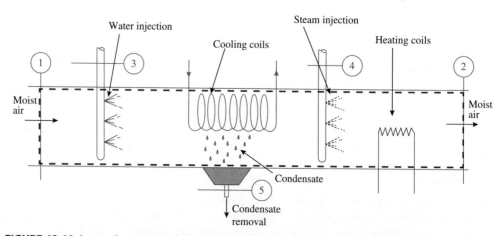

FIGURE 12.10 A generic open-steady psychrometric system used for derivation of mass and energy equations (see Anim. 12.B.*balanceEquations*).

Note that the moist air enthalpy $h = h_a + \omega h_v$ from Eq. (12.9) is substituted at State-1 and State-2. Enthalpies for liquid water at State-3 and State-5 are approximated as that of saturated liquid water at T_3 and T_5, respectively, as prescribed by the compressed liquid sub-model (Sec. 3.4.8). Injected steam is usually at the saturated vapor state, resulting in the substitution $h_{w4} = h_{g4} = h_{g@T_4}$; otherwise, superheated tables must be used to determine h_{w4}. Also, heat transfer from the heating and cooling coils are algebraically added to produce the net heat addition rate $\dot{Q}$.

EXAMPLE 12-3   **Mass and Energy Balance for an Open-Steady Device**

Moist air enters a duct at 100 kPa, 35°C, and 30% R.H. with a volumetric flow rate of 100 m³/min. The mixture is cooled and both the condensate and mixture leave at 10°C. For steady state operation, determine (a) the rate of heat transfer in tons and (b) the mass flow rate of the condensate in kg/min. *What-if scenario:* (c) What would the answer in part (b) be if the relative humidity at the inlet was 95%?

SOLUTION

Perform a mass and energy balance on the open-steady device enclosed within the red boundary in Figure 12.11.

Assumptions

The MA model is applicable. The exit state, State-2, must be saturated to allow condensation. The condensate leaves as saturated liquid. Negligible changes in ke and pe and LTE at all the principal states, State-1, State-2, and State-3.

Analysis

Evaluate the inlet and exit states, as best as possible, from the given data.

State-1 (given p_1, T_1, ϕ_1, $\dot{V}_1$):

$$p_{v1} = \phi_1 p_{g1} = \phi_1 p_{sat@35°C} = (0.3)(5.65) = 1.695 \text{ kPa};$$

$$p_{a1} = p_1 - p_{v1} = 100 - 1.695 = 98.31 \text{ kPa};$$

$$\omega_1 = (0.622)\frac{p_{v1}}{p_{a1}} = (0.622)\frac{(1.695)}{(98.31)} = 0.0107 \frac{\text{kg H}_2\text{O}}{\text{kg d.a.}}$$

$$\dot{m}_{a1} = \frac{\dot{V}_1}{v_1} = \frac{\dot{V}_1 p_{a1}}{R_a T_1} = \frac{(100)(98.31)}{(60)(0.287)(273 + 35)} = 1.853 \frac{\text{kg d.a.}}{\text{s}};$$

$$\dot{m}_{v1} = \omega_1 \dot{m}_{a1} = (0.0107)(1.853) = 0.01983 \text{ kg/s};$$

$$h_1 = h_{a1} + \omega_1 h_{g@35°C} = (1)(35) + (0.0107)(2565.3) = 62.45 \text{ kJ/kg}$$

State-2 (given $p_2 = p_1$, T_2, $\phi_2 = 100\%$, $\dot{m}_{a2} = \dot{m}_{a1}$):

$$p_{v2} = \phi_2 p_{g2} = p_{sat@10°C} = 1.228 \text{ kPa};$$

$$p_{a2} = p_2 - p_{v2} = 100 - 1.228 = 98.772 \text{ kPa};$$

$$\omega_2 = (0.622)\frac{p_{v2}}{p_{a2}} = (0.622)\frac{(1.228)}{(98.772)} = 0.0077 \frac{\text{kg H}_2\text{O}}{\text{kg d.a.}};$$

$$\dot{m}_{v2} = \omega_2 \dot{m}_{a2} = (0.0077)(1.853) = 0.01432 \text{ kg/s};$$

$$h_2 = h_{a2} + \omega_2 h_{g@10°C} = (1)(10) + (0.0077)(2519.8) = 29.4 \text{ kJ/kg}$$

State-3 (given $T_3 = T_2$):

$$h_3 = h_{f@10°C} = 42 \text{ kJ/kg}$$

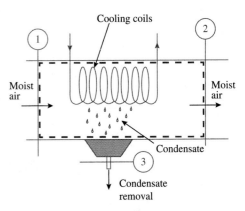

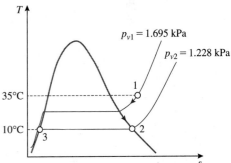

FIGURE 12.11 System schematic and *T-s* diagram for moisture in Ex. 12.3 (see Anim. 12.B.*coolingDehumidify*).

A mass balance for H_2O yields:

$$\dot{m}_{w3} = \dot{m}_{v1} - \dot{m}_{v2} = 0.0198 - 0.0143 = 0.0055 \text{ kg/s} = 0.033 \text{ kg/min}$$

The energy balance equation, Eq. (12.15), yields:

$$\dot{Q} = \dot{m}_a(h_2 - h_1) + \dot{m}_{w3}h_3 = (1.853)(29.4 - 62.5) + (0.0055)(42)$$
$$= -61.1 \text{ kW} = -17.4 \text{ ton}$$

TEST Analysis

Launch the open-steady psychrometry TESTcalc located in the systems, open, steady, specific, Psychrometry branch. Select moist air as the working fluid. Evaluate State-1 from the known values of p1, T1, RH1, and Voldot1, and State-2 (partially) from p2=p1, T2, and RH2=100%. For State-3, select H2O as the working fluid, enter T3=T2, x3=0 (saturated liquid), and click Calculate. In the device panel, load State-1 as the i1-State, States-2 and -3 as the e1- and e2-states, respectively; enter Wdot_ext=0 and click Super-Calculate. Obtain condensate mass from State-3 and Qdot from the device panel, verifying the manual results.

What-if scenario

Change RH1 to 95%, press Enter, and click Super-Calculate. The new heat transfer is calculated as –48.1 ton.

Discussion

The negative sign of heat transfer signifies heat loss in accordance with the WinHip sign convention (Sec. 0.7.2). The cooling rate is its absolute value. Notice the tremendous increase in the cooling rate when the air to be conditioned is more humid. The extra amount is necessary to absorb the additional vapor's latent heat of condensation (Sec. 5.1.4B) that must be removed in order to achieve the desired exit conditions.

12.2.2 Closed Process

Consider a closed system containing moist air at State-1. The system goes through a process in which heat and boundary work are transferred, and the final state consists of moist air at State-2, and possibly, liquid condensate at State-3 (Fig. 12.12). For simplicity, we will neglect any change in ke and pe and assume that the volume of the condensate is negligible compared to the total volume of the system.

For a closed system, the mass of dry air and water (vapor and liquid together) remain constant, and the energy balance equation for the process can be simplified starting with Eq. (5.2).

$$\text{Dry Air:} \quad m_{a1} = m_{a2} = m_a; \quad \text{[kg d.a.]} \tag{12.16}$$

$$H_2O: \quad m_{v1} = m_{v2} + m_{w3};$$

$$\Rightarrow \quad m_{a1}\omega_1 = m_{a2}\omega_2 + m_{w3} \quad \text{or,} \quad m_a(\omega_1 - \omega_2) = m_{w3}; \quad \text{[kg } H_2O] \tag{12.17}$$

$$\text{Energy:} \quad \Delta U = Q - W_B;$$

$$\Rightarrow \quad \Delta U = Q - W_B;$$

$$\Rightarrow \quad Q = (U_2 + U_{w2}) - U_1 + W_B = m_a(u_2 - u_1) + m_{w3}h_3 + W_B;$$

$$\Rightarrow \quad Q = m_a(h_2 - h_1) - R_a(T_2 - T_1) + m_{w3}h_{f2} + W_B; \quad \text{[kJ]} \tag{12.18}$$

In this derivation, the condensate is approximated as saturated liquid at the final temperature of the system, and Eq. (12.10) is inserted to express the internal energy in terms of enthalpy of moist air.

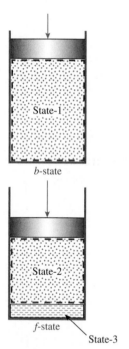

FIGURE 12.12 Moist air undergoing a closed process (see Anim. 12.A.*pConstCooling*).

State-1

b-state

State-2

f-state

State-3

EXAMPLE 12-4   Energy Analysis of a Closed Process

1 m³ of moist air at 50% R.H., 25°C, and 100 kPa is compressed to saturation in a constant temperature process (Ex. 12-2). Determine (a) the work and (b) heat transfer. *What-if scenario:* (c) How would the answer in part (b) change if the final volume were one tenth the original volume?

SOLUTION

Perform an energy balance for the closed process.

Assumptions

The system is uniform and is at equilibrium at the beginning state, State-1, and final state, State-2, of the process. Neglect changes in ke or pe.

Analysis

If the temperature remains constant, the enthalpy of dry air, $h_a = h_a(T)$, remains invariant. As long as there is no condensation, ω also remains constant; therefore, the enthalpy of moist air, $h = h_a + \omega h_v$ as expressed in Eq. (12.9), must remain constant, that is, $h_1 = h_2$. Substituting this result in Eq. (12.18), we obtain $Q = W_B$. Using Eq. (5.10), the work can be evaluated for the PG mixture as:

$$W_{B,\text{isothermal}} = mRT \ln \frac{v_f}{v_b} = p_1 V_1 \ln \frac{v_2}{v_1} = (100)(1) \ln \frac{1}{2} = -69.31 \text{ kJ}$$

Therefore, the heat transfer is also -69.31 kJ.

TEST Analysis

Launch the closed-process psychrometric TESTcalc located in the systems, closed, process, specific, Psychrometry branch. Calculate State-1 and -2, as explained in Example 12-2 (see TEST code in *TEST > TEST-codes*). In the process panel, load these states as the *b*- and *f*-state, respectively. Enter W as '= 100*ln(0.5) kJ' and click Calculate to verify the manual solution.

What-if scenario

Evaluate State-3 as the final moist air state with T3=T1, RH3=100%, Vol3=Vol1/10, and m3=m1 (dry air mass). Evaluate the condensate state, State-4, with T4=T1 and x4=0. In the process panel, load State-1 as the bA-State and States-2 and 3 as fA and fB states, and enter W as 100*ln(0.1) = −230 kJ. Calculate to obtain Qdot as −252 kJ. Note that if the presence of condensate is neglected in the energy balance, the answer changes only marginally to −253.5 kJ.

Discussion

Temperature remaining constant during isothermal compression of air, the vapor pressure reaches a maximum and remains constant once the air is saturated. As the saturated mixture is compressed further, condensation must occur in order to avoid an increase in vapor pressure. In analyzing closed processes involving condensation, it is a standard practice to neglect the volume occupied by the condensate.

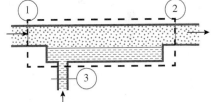

12.3 ADIABATIC SATURATION AND WET-BULB TEMPERATURE

Adiabatic saturation offers a simple way to determine the humidity of moist air. Examine the system shown in Figure 12.13, where a steady stream of moist air enters with an unknown ω_1 at the inlet state, State-1, and passes over a long insulated channel that contains a pool of liquid water. The chamber is long enough to ensure that the air picks up sufficient amount of moisture to become the saturated State-2 at the exit. The latent heat of vaporization must be supplied from within the adiabatic system and the exit temperature T_2, called the **adiabatic saturation temperature**, is less than T_1, provided State-1 is not saturated to start with. To keep the system at steady state, makeup water is supplied at State-3. By convention, T_3 is assumed to be equal to T_2.

To analyze the system, we neglect any changes in ke and pe and impose the restriction that there is no heat or external work transfer. The mass and energy balance equations, developed for a general purpose open system in Sec. 12.2, can be customized for the adiabatic saturation process as follows:

$$\text{Dry Air:} \quad \dot{m}_{a1} = \dot{m}_{a2} = \dot{m}_a; \quad [\text{kg d.a./s}] \tag{12.19}$$

$$\text{H}_2\text{O:} \quad \dot{m}_{v1} = \dot{m}_{v2} + \dot{m}_{w3}; \quad \Rightarrow \quad \dot{m}_{a1}\omega_1 = \dot{m}_{a2}\omega_2 + \dot{m}_{w3};$$

$$\Rightarrow \quad \dot{m}_a(\omega_2 - \omega_1) = \dot{m}_{w3}; \quad [\text{kg H}_2\text{O/s}] \tag{12.20}$$

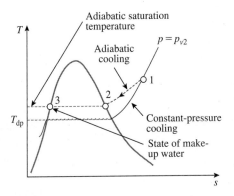

FIGURE 12.13 By measuring temperature at State-1 and State-2, the absolute humidity at State-1 can be determined from this simple saturation experiment (see Anim. 12.A.*adiabaticSaturation*).

Energy: $\quad 0 = \dot{m}_{a1}h_1 - \dot{m}_{a2}h_2 + \dot{m}_{w3}h_{w3}; \quad \Rightarrow \quad \dot{m}_a(h_2 - h_1) = \dot{m}_{w3}h_{f3};$ [kW] **(12.21)**

Substituting the mass equation for H_2O in the energy equation, and using Eq. (12.9) to expand the moist air enthalpy, we obtain:

$$\dot{m}_a(h_2 - h_1) = \dot{m}_a(\omega_2 - \omega_1)h_{f3};$$
$$\Rightarrow \quad (h_{a2} + \omega_2 h_{g2}) - (h_{a1} + \omega_1 h_{g1}) = \omega_2 h_{f2} - \omega_1 h_{f2};$$
$$\Rightarrow \quad \omega_2(h_{g2} - h_{f2}) = c_{pa}(T_1 - T_2) + \omega_1(h_{g1} - h_{f2}); \quad \left[\frac{kJ}{kg}\right] \qquad \textbf{(12.22)}$$

The exit air being saturated, Eqs. (12.4) and (12.22) can be combined to yield ω_1 and ω_2 as:

$$\omega_2 = 0.622\frac{p_{g@T_2}}{p - p_{g@T_2}};$$
$$\omega_1 = \frac{\omega_2(h_{g@T_2} - h_{f@T_2}) + c_{pa}(T_2 - T_1)}{(h_{g@T_1} - h_{f@T_2})} \qquad \textbf{(12.23)}$$

Equation (12.23) provides a simple, indirect way to determine ω_1 from the measurement of dry-bulb temperature at the inlet and exit of an adiabatic saturator.

Going back to the definition of wet-bulb temperature (Sec. 12.1.4), air around the wick is not necessarily in equilibrium with the water in the wick; however, in local systems at the air water interface around the wick, the situation is similar to the adiabatic saturation process. As a result, at atmospheric pressure, the **wet-bulb temperature** T_{wb} can be used in place of T_2 in Eq. (12.23) without much sacrifice in accuracy.

EXAMPLE 12-5 **Dry- and Wet-Bulb Temperature**

The dry- and wet-bulb temperatures of air in a room at 101 kPa are measured with a sling psychrometer to be 30°C and 15°C, respectively. Determine (a) the specific humidity, (b) relative humidity, and (c) enthalpy of the air in the room.

SOLUTION

Use Eq. (12.23) and other property relations to determine the desired properties.

Assumptions

Air is in equilibrium at State-1 with the wet-bulb temperature equal to the adiabatic saturation temperature of the adiabatically saturated state, State-2.

Analysis

The specific humidity at State-2 can be evaluated from Eq. (12.23):

$$\omega_2 = 0.622\frac{p_{g@15°C}}{p - p_{g@15°C}} = \frac{(0.622)(1.705)}{(101 - 1.705)} = 0.0107 \frac{kg\ H_2O}{kg\ d.a.};$$

$$\text{Therefore,} \quad \omega_1 = \frac{\omega_2 h_{fg@15°C} + c_{pa}(T_2 - T_1)}{(h_{g@30°C} - h_{f@15°C})}$$

$$= \frac{(0.0107)(2465.9) + (1.005)(15 - 30)}{(2556.3 - 63)} = 0.0045 \frac{kg\ H_2O}{kg\ d.a.}$$

To evaluate the relative humidity, evaluate the vapor pressure at State-1:

$$\omega_1 = 0.622\frac{p_{v1}}{p - p_{v1}}; \quad \Rightarrow \quad p_{v1} = \frac{p\omega_1}{\omega_1 + 0.622} = \frac{(101)(0.0045)}{0.0045 + 0.622} = 0.725\ kPa;$$

$$\text{Therefore,} \quad \phi_1 = \frac{p_{v1}}{p_{sat@30°C}} = \frac{0.725}{4.246} = 17.07\%$$

and $\quad h_1 = h_{a1} + \omega_1 h_{g@30°C} = (1.005)(30) + (0.0045)(2556.3) = 41.65$ kJ/kg d.a.

TEST Analysis

Launch the MA system-state TESTcalc. Select moist air, input **p1**, **T1**, and **T_wb1**, then click Calculate. The manual results are verified.

Discussion

Evaluate $h_2 = h_{a2} + \omega_2 h_{g@15°C} = 42.135$ kJ/kg d.a. Notice the small difference between h_1 and h_2. For approximate analysis, it is a common practice to assume enthalpy remains constant during an adiabatic saturation process.

12.4 PSYCHROMETRIC CHART

Moist air is a mixture of two perfect gases—dry air and water vapor—with the possibility of a change in composition due to the addition or removal of moisture. Specification of the composition requires an additional property, such as specific humidity, relative humidity, or vapor pressure. The *state postulate* will be shown (Chapter 14) to require three independent thermodynamic properties to specify an equilibrium for a binary mixture. In psychrometric applications, the total pressure often remains constant throughout a system. Therefore, for a given total pressure, an equilibrium state of moist air can be represented on a two-dimensional thermodynamic diagram with two independent properties for a given total pressure, usually 1 atm as axes. A psychrometric chart is a customized thermodynamic diagram where states and processes can be graphically displayed, properties can be quickly read without lengthy calculations, and regions of desired states can be marked in the analysis and design of air conditioning devices.

The psychrometric chart for a total pressure of 1 atm (101.325 kPa) can be found in Table F-1 of the appendix (also from the Property Tables module of TEST) with its basic features illustrated in Figure 12.14 (and Anim. 12.A.*psychrometricChart*). The abscissa is the dry-bulb temperature T in celsius within the range 0°C–50°C. The specific humidity ω, within the range 0–30 grams moisture per kilogram of dry air, appears on the ordinate, which is shifted to cross the abscissa, the dry-bulb temperature, at $T = 50$°C. The ordinate can also be calibrated for vapor pressure since vapor pressure p_v is a function of ω, as expressed by Eq. (12.4).

The boundary curve on the left hand side is called the saturation line, which corresponds to a relative humidity of 100%, that is, $\phi = 1$. The resemblance of its shape to the familiar saturation curve of a phase diagram (Fig. 3.26, Sec. 3.4.2) can be understood from the fact that in both these plots the axes are identical—p against T in the phase diagram and p_v against T in the psychrometric chart. Also, the $\phi = 1$ line represents the locus of states with $p_v = p_g = p_{sat@T}$, or saturated vapor states. Vapor pressure being zero along the abscissa, the $\phi = 0$ line must also be along the abscissa. The shapes of other $\phi = c$ ($0 < c < 1$) lines follow the interpolated shape between these two extreme constant-ϕ lines.

The constant-volume lines follow the linear relation $T = c_1 - c_2 p_v$, as can be deduced from Eq. (12.6) by substituting $p_a = p - p_v$, where c_1 and c_2 are constants. This explains the negative slope of the $v =$ constant lines. The equation for the constant-enthalpy lines can be expressed, in terms of constants c_3 through c_6 from Eqs. (12.7)–(12.9), as $c_3\omega + c_4 T + c_5\omega T = c_6$. The non-linear term in this equation is quite small ($c_5\omega \ll c_4$), so the constant-enthalpy lines are quite similar—linear with negative slope—to the constant-specific volume lines (Fig. 12.14). As discussed in Example 12-5, enthalpy remains approximately constant during adiabatic saturation; therefore the constant-wet-bulb temperature lines are almost parallel to the constant-enthalpy lines, though slightly steeper. Lines of constant vapor pressure are parallel to the abscissa, so the lines of constant dew point temperature must also be parallel. These are not shown in a psychrometric chart as T_{dp}, but can be obtained from the dry-bulb temperature at the saturation state on the $p_v = c$ line (click the Specific humidity/dew point radio-button in Anim. 12.A.*psychrometricChart*).

For any given state, the dry-bulb temperature can be obtained from the abscissa, the absolute humidity from the ordinate, relative humidity from the $\phi = c$ lines, wet-bulb temperature from the $T_{wb} = c$ lines, specific volume from the $v = c$ lines, enthalpy from the $h = c$ lines, and vapor pressure from the ordinate (which represents both ω and p_v). The dew point, wet-bulb, and adiabatic saturation temperatures can be obtained from the

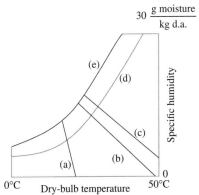

(a) $v =$ constant
(b) $T_{wb} =$ constant
(c) $h =$ constant
(d) $\phi =$ constant
(e) Saturation line, $\phi = 100\%$

FIGURE 12.14 Schematic for a psychrometric chart (see Anim. 12.A.*psychrometricChart*).

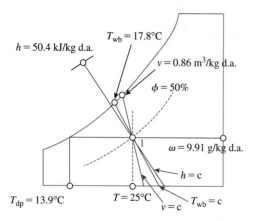

FIGURE 12.15 Obtaining a moist air state from the psychrometric chart at 1 atm (see Anim. 12.B.*moistAirStates*).

abscissa by following a vertical line drawn from a saturated state on the saturation line ($\phi = 100\%$), with the saturation achieved by following the $p_v = $ c line, $T_{wb} = $ c line, or $h = $ c line, respectively, from a given state (Fig. 12.15). If State-1 in Figure 12.15 is saturated, the dry-bulb temperature, the dew point temperature, and the wet-bulb temperature must all be identical.

Psychrometric charts are used exclusively in air conditioning, where most processes can be easily described. For instance, the moisture amount in air cannot change in any process represented by a horizontal line. An increase or decrease in ω suggests condensation or moisture addition since the amount of dry air, generally, remains constant. The value of the psychrometric chart will be appreciated when we discuss different air-conditioning processes.

EXAMPLE 12-6   Use of the Psychrometric Chart

Use psychrometric chart at 1 atm to answer parts (a) through (d) of Example 12-1.

SOLUTION

Repeat the solution of Example 12-1 by evaluating properties from the psychrometric chart of Table F-1.

Analysis

State-1 (given T_1, ϕ_1, m_1):

At 30°C and 60% R.H. we obtain: $\omega_1 = 16 \dfrac{\text{g of H}_2\text{O}}{\text{kg of d.a.}}$

The dew point temperature can be obtained by tracing the horizontal line from State-1 until it meets the saturation curve. The temperature at that point is $T_{dp,1} = 21°C$.

The moisture mass can be evaluated, as in Example 12-1:

$$m_{a1} = \frac{m_1}{(1 + \omega_1)} = \frac{1}{1.016} = 0.983 \text{ kg d.a.;}$$

$$m_{v1} = \omega_1 m_{a1} = (0.016)(0.984) = 0.016 \text{ kg;}$$

State-2 (given T_2, $\phi_2 = 100\%$):

$$\omega_2 = 0.55 \frac{\text{g H}_2\text{O}}{\text{kg d.a.}};$$

$$m_{v2} = \omega_2 m_{a2} = \omega_2 m_{a1} = (0.0055)(0.983) = 0.0054 \text{ kg}$$

Therefore, the amount of condensate is:

$$m_{v1} - m_{v2} = 0.0157 - 0.0054 = 0.0103 \text{ kg}$$

With the boundary work evaluated at constant pressure, the energy equation, Eq. (12.18), simplifies to

$$Q = m_a(h_2 - h_1) - R_a(T_2 - T_1) + m_{w3}h_{f@T_2} + pm_a(v_2 - v_1)$$

$$= (0.983)(18.5 - 71) - (0.287)(5 - 30)$$

$$+ (0.013)(125.8) + (100)(0.983)(0.795 - 0.88) = -51.2 \text{ kJ}$$

TEST Analysis

Launch the psychrometric process TESTcalc. Evaluate State-1 and State-2 as described in Example 12-1 (see TEST-code in *TEST > TEST-codes*). In the I/O panel, evaluate '=mdot1*(omega1−omega2)' to obtain the amount of condensate.

Discussion

Comparing the lengthy solution of Example 12-1 to the current solution, the advantage offered by the psychrometric chart for a manual solution becomes apparent. The sacrifice in accuracy is well within acceptable limits for most applications.

12.5 AIR-CONDITIONING PROCESSES

The human body can be looked upon as a heat engine that converts heat released from metabolism of food into useful work. To satisfy the second law, waste heat must be continuously rejected to the surroundings we live in. Assuming a diet of 3000 kcal (12.552 MJ) per day with a 40% thermal efficiency, a simple calculation yields a heat rejection rate of 87 W. Experiments reveal that the actual waste heat for an average male adult is about 87 W while sleeping, 115 W while resting or doing light work, and 440 W during heavy physical work.

It turns out that our comfort is directly related to how efficiently we can reject the waste heat. The body maintains a constant temperature of about 37°C, for an average person. Driven by the temperature difference between the body and the surroundings, heat transfer increases in a cold environment. Too much heat loss makes us feel cold and uncomfortable. In response, the body cuts down the blood circulation near the skin. We wear insulating barriers (jackets, blankets, etc.), reduce the exposed surface area by huddling up, or increase the heat release through exercise. A better engineering solution is to condition the air to a more comfortable state in the comfort zone shown in Figure 12.16.

The problem is just the opposite in a hot environment, where the heat transfer is reduced due to a smaller temperature difference. To enhance the heat transfer through latent heat of evaporation, the body increases perspiration. We turn on the fan to increase the convective heat transfer. Even if the surroundings temperature exceeds the body temperature, evaporative cooling is possible as long as the saturation vapor pressure at the body temperature is higher than the vapor pressure in surrounding air. For example, at 40°C and 50% relative humidity, the vapor pressure is $p_v = (0.5)p_{g@40°C} = 3.71$ kPa, which is lower than the saturation pressure $p_{g@37°C} = 6.27$ kPa, the vapor pressure in the immediate surroundings of the sweat droplet (by the assumption of LTE). The difference between the vapor pressures makes it possible for evaporative cooling to continue. If the humidity of the air is also high the perspiratory cooling also suffers.

Another factor important to our comfort is radiative heat transfer. A fireplace can warm us up even if the intervening air is much cooler than our body. Solar radiation during the day and radiative losses to the sky at night can significantly affect the heat transfer and, hence, our comfort.

In summary, human comfort depends on three external factors—the dry-bulb temperature, relative humidity, and air motion. As shown in Figure 12.16, most people feel comfortable when the temperature and humidity are within the ranges 22–27°C and 40–60%. The motion of air is also important, as it replaces the hot and humid air that builds up around the body with fresh air, enhancing both the convective and perspiratory heat losses. Air speed of about 15 m/min is considered comfortable.

Controlling the temperature and relative humidity of air is a major goal of air conditioning processes. Modern air-conditioning systems can heat, cool, humidify, dehumidify, filter particulates, clean, and even deodorize the air. We will analyze major air-conditioning processes and combine them to achieve any desired condition of air.

12.5.1 Simple Heating or Cooling

When the dry-bulb temperature of air is changed without affecting the amount of moisture and specific humidity ω, the resulting process can be depicted by a horizontal line on the psychrometric chart. In **simple heating** (Fig. 12.17 and Anim. 12.B.*simpleHeating*), the temperature is raised by

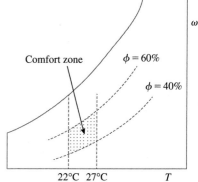

FIGURE 12.16 Comfort zones may vary with the individual, but most find the moist air states in the shaded area comfortable.

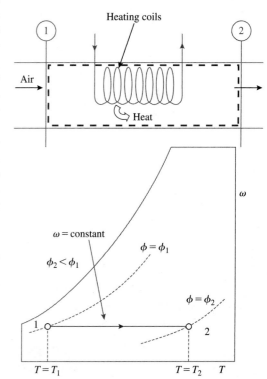

FIGURE 12.17 During simple heating or cooling, absolute humidity remains constant, but relative humidity changes (see Anim. 12.B.*simpleHeating*).

circulating steam, hot gases, or passing electric current through a resistor in a duct through which air is flowing. Such simple-heating systems are primarily employed in domestic applications. It is clear from the psychrometric plot in Figure 12.17, that although ω remains constant, the relative humidity decreases during simple heating. If the air is very cold to start with, simple heating results in relatively dry air even if the initial state is saturated.

In **simple cooling** (see Anim. 12.B.*simpleCooling*), cold water (brine) or refrigerant is circulated, absorbing heat from the air flowing through the duct without causing any condensation. As a result, the dry-bulb temperature decreases and relative humidity increases. The process on the psychrometric diagram is the reverse of the simple heating process shown in Figure 12.17. Only the inlet and exit states are required to be in thermodynamic equilibrium in a simple heating or cooling process. If the coolant has a temperature below the dew point temperature, condensation may occur locally around the cooling coil; however, before the exit state is reached, the moisture is recaptured by the air in its constant pursuit of equilibrium, resulting in no net condensation between the inlet and exit.

To analyze a heating or cooling system, the system boundary is carefully drawn to exclude the heating or cooling element so that there is only heat transfer and a single-flow of moist air across the boundary. Assuming steady-state operation, and neglecting any changes in ke and pe, the mass and energy equations, Eqs. (12.13)–(12.15), for an open-steady device can be simplified as follows:

$$\text{Dry Air:}\quad \dot{m}_{a1} = \dot{m}_{a2} = \dot{m}_a; \quad \left[\frac{\text{kg d.a.}}{\text{s}}\right] \tag{12.24}$$

$$\text{H}_2\text{O:}\quad \dot{m}_{v1} = \dot{m}_{v2}; \quad \left[\frac{\text{kg H}_2\text{O}}{\text{s}}\right], \quad \Rightarrow \quad \omega_1 = \omega_2 \tag{12.25}$$

$$\text{Energy}\quad \dot{Q} = \dot{m}_a(h_2 - h_1); \quad [\text{kW}] \tag{12.26}$$

The mass equation resulting in $\omega = c$ provides an important link between the inlet and exit state. If the two states can be evaluated, the required heat transfer can be obtained from the energy equation.

12.5.2 Heating with Humidification

Humidification is the process of adding moisture to relatively dry air. Generally there are two ways in which this can be accomplished—injection of steam or spraying water into the air stream. Steam injection not only raises the dry-bulb temperature, but also increases the absolute humidity (Fig. 12.18). Water spraying, on the other hand, results in the cooling of the air stream because the latent heat of evaporation is supplied by the air. The mass and energy equations, Eqs. (12.13)–(12.15), can be customized for humidification with steam or liquid water as follows:

$$\text{Dry Air:}\quad \dot{m}_{a1} = \dot{m}_{a2} = \dot{m}_a; \quad \left[\frac{\text{kg d.a.}}{\text{s}}\right] \tag{12.27}$$

$$\text{H}_2\text{O:}\quad \dot{m}_{v1} + \dot{m}_{w3} = \dot{m}_{v2}; \quad \Rightarrow \quad \dot{m}_a(\omega_2 - \omega_1) = \dot{m}_{w3}; \quad \left[\frac{\text{kg H}_2\text{O}}{\text{s}}\right] \tag{12.28}$$

$$\text{Energy}\quad \dot{Q} = \dot{m}_a(h_2 - h_1) - \dot{m}_{w3}h_{w3}; \quad [\text{kW}] \tag{12.29}$$

where h_{w3}, the enthalpy of the injected H_2O, can be obtained as $h_{w3} = h_{f3} = h_{f@T_3}$ for liquid water and $h_{w3} = h_{g3} = h_{g@T_3}$ for saturated steam. A combination of heating and water or steam spraying can be used to increase the temperature and humidity to any desired condition. The use of steam would reduce the heat requirement, as illustrated in the following example.

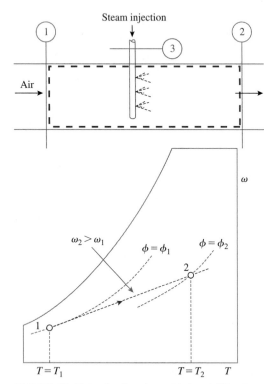

Steam injection

$\omega_2 > \omega_1$

$\phi = \phi_1$ $\phi = \phi_2$

$T = T_1$ $T = T_2$ T

FIGURE 12.18 During heating with humidification, the dry-bulb temperature and the specific humidity increase (see Anim. 12.B.*steamInjection*).

EXAMPLE 12-7 **Heating with Humidification**

An air-conditioning system is to take outdoor air at 5°C, 30% R. H. at a steady rate of 50 m³/min and to condition it to 25°C and 50% R.H. The air is first heated in a heating section, then humidified by spraying liquid water at 25°C. Assuming a constant pressure of 1 atm everywhere in the system, determine (a) the mass flow rate of water in kg/min and (b) the heating rate. ***What-if scenario:*** (c) What would the answers be if saturated steam at 100 kPa was used for humidification?

SOLUTION

Perform a mass and energy balance on the open-steady device enclosed within the red boundary of Figure 12.19.

Assumptions

The MA model is applicable. Equilibrium at the inlet states, State-1 and State-3, and exit state, State-2. Negligible changes in ke and pe.

Analysis

Use the psychrometric chart to evaluate States 1 and 2 and obtain the enthalpy of the water from the saturation temperature table of water, Table B-2.

State-1 (given T_1, ϕ_1, $\dot{V}_1$):

At 5°C and 30% R.H., obtain $\omega_1 = 1.6 \dfrac{\text{g H}_2\text{O}}{\text{kg d.a.}}$;

$$v_1 = 0.79 \frac{\text{m}^3}{\text{kg d.a.}}; \quad \dot{m}_{a1} = \frac{\dot{V}_1}{v_1} = \frac{(50/60)}{0.79}$$

$$= 1.055 \frac{\text{kg d.a.}}{\text{s}}; \quad h_1 = 9 \frac{\text{kJ}}{\text{kg d.a.}};$$

State-2 (given $T_2 = 25°C$, $\phi_2 = 50\%$):

$$\omega_2 = 10 \frac{\text{g H}_2\text{O}}{\text{kg d.a.}}; \quad h_2 = 51 \frac{\text{kJ}}{\text{kg d.a.}}$$

Therefore, the mass flow rate of water is:

$$\dot{m}_{w3} = \dot{m}_a(\omega_2 - \omega_1) = (1.055)(0.01 - 0.0016)(60) = 0.532 \text{ kg/min}$$

The heating rate is given by the energy equation as:

$$\dot{Q} = \dot{m}_a(h_2 - h_1) - \dot{m}_{w3}h_{f@T_3} = 1.055(51 - 9) - \frac{0.532}{60}(104.9) = 43.4 \text{ kW}$$

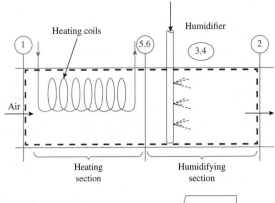

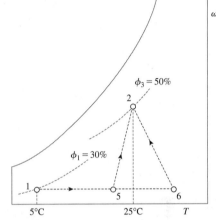

FIGURE 12.19 Schematic for Ex. 12.7. Note that State-3 need not be evaluated in this problem (see Anim. 12.B.*heatingHumidify*).

TEST Analysis

Launch the open-steady psychrometric TESTcalc. Evaluate State-1 and State-2 for moist air from the given properties (see TEST-code in *TEST > TEST-codes*). For State-3, select H2O as the working fluid and enter T3 and x3=0 (saturated liquid). In the device panel, load States 1 and 3 as the i1 and i2 states and State-2 as the e1-state. Enter Wdot_ext=0 and click Super-Calculate. The heat transfer is evaluated as Qdot = 42.4 kW. In State-3, mdot_3 is found as 0.525 kg/min.

What-if scenario

Evaluate State-4 by selecting saturated steam as the working fluid, then enter x4=1 and p4=100 kPa. In the device panel, select Device-B and load States 1 and 4 as the inlet states and State-2 as the exit state. Enter Wdot_ext=0 and click Calculate. The new answers are calculated as 19.9 kW and 0.525 kg/min.

Discussion

The amount of water necessary to raise the moisture level remains the same, regardless of the method employed. However, the heat transfer required reduces drastically if steam is injected instead of liquid water.

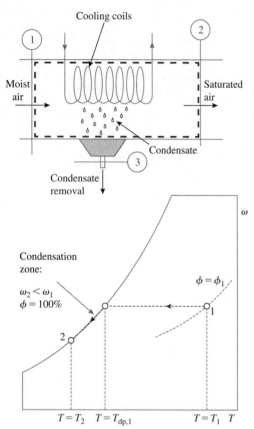

FIGURE 12.20 Cooling with dehumidification
(see Anim. 12.B.*coolingDehumidify*).

12.5.3 Cooling with Dehumidification

A schematic of the cooling and dehumidification process is shown in Figure 12.20 (and Anim. 12.B.*coolingDehumidify*). As air passes through the cooling coil, its temperature drops due to simple cooling. If the air is not saturated to start with, the specific humidity initially remains constant during the constant-pressure simple cooling process. However, as shown in the psychrometric diagram of Figure 12.20, the cooled air eventually becomes saturated when the dry-bulb temperature drops down to the dew point temperature. Any further cooling results in dehumidification, removal of moisture from air through condensation. As shown in Figure 12.20, the air remains saturated and follows the 100% relative humidity line during the rest of the process.

The condensed water, assumed to be at the temperature of the exit state of air by convention, leaves through a separate exit. The cool air that leaves the device is saturated, but its specific humidity is lowered significantly. To achieve a desired relative humidity and temperature, simple heating can be added at the end of the dehumidification process. Often the cool and saturated air is routed directly to the air-conditioned space, where it mixes with existing air creating the desired state. Such mixing will be analyzed later.

The mass and energy equations, Eqs. (12.13)–(12.15), can be customized for the cooling and dehumidification process as follows:

$$\text{Dry Air:} \quad \dot{m}_{a1} = \dot{m}_{a2} = \dot{m}_a; \quad \left[\frac{\text{kg d.a.}}{\text{s}}\right] \tag{12.30}$$

$$\text{H}_2\text{O:} \quad \dot{m}_{v1} = \dot{m}_{v2} + \dot{m}_{w3}; \quad \Rightarrow \quad \dot{m}_a(\omega_1 - \omega_2) = \dot{m}_{w3}; \quad \left[\frac{\text{kg H}_2\text{O}}{\text{s}}\right] \tag{12.31}$$

$$\text{Energy} \quad \dot{Q} = \dot{m}_a(h_2 - h_1) + \dot{m}_{w3}h_{w3}; \quad [\text{kW}] \tag{12.32}$$

where h_{w3}, the enthalpy of the condensate, can be obtained from $h_{w3} = h_{f@T_2}$ by assuming the condensate to be saturated liquid water at the exit temperature of air.

EXAMPLE 12-8   Cooling with Dehumidification

An air-conditioning system is to take outdoor air at 30°C, 90% R.H. at a steady rate of 20 m³/min and to condition it to 15°C with the condensate also leaving at 15°C. Assuming a constant pressure of 1 atm everywhere in the system, determine (a) the heat transfer rate in kJ/min and (b) the moisture removal rate in kg/min. **What-if scenario:** (c) What would the answers be if the outdoor air had a R.H. of 50% instead?

SOLUTION

Perform a mass and energy balance on the open-steady device enclosed within the red boundary of Figure 12.20.

Assumptions

The moist air model is applicable. Equilibrium at the inlet state, State-1, and exit states, State-2 for moist air (Fig. 12.21) and State-3 for the condensate. Negligible changes in ke and pe.

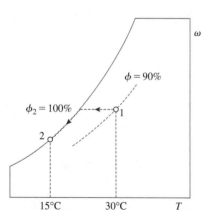

FIGURE 12.21 Cooling with
dehumidification for Ex. 12-8.

Analysis

Use the psychrometric chart to evaluate States 1 and 2 and obtain the enthalpy of the water from the temperature saturation table of water.

State-1 (given $T_1 = 30°C$, $\phi_1 = 90\%$, $\dot{V}_1$):

$$\omega_1 = 24.5\frac{\text{g H}_2\text{O}}{\text{kg d.a.}}; \quad v_1 = 0.893\frac{\text{m}^3}{\text{kg d.a.}};$$

$$\dot{m}_{a1} = \frac{\dot{V}_1}{v_1} = \frac{(20/60)}{0.893} = 0.37\frac{\text{kg d.a.}}{\text{s}}; \quad h_1 = 93\frac{\text{kJ}}{\text{kg d.a.}}$$

State-2 (given $T_2 = 15°C$, $\phi_2 = 100\%$):

$$\omega_2 = 10.7 \frac{\text{g } H_2O}{\text{kg d.a.}}; \quad h_2 = 42.5 \frac{\text{kJ}}{\text{kg d.a.}}$$

Therefore, the flow rate of condensate is:

$$\dot{m}_{w3} = \dot{m}_a(\omega_1 - \omega_2) = (0.37)(0.0245 - 0.0107)(60) = 0.309 \text{ kg/min}$$

The heat transfer rate is given by the energy equation as:

$$\dot{Q} = \dot{m}_a(h_2 - h_1) + \dot{m}_{w3}h_{f@T_2} = (0.37)(42.5 - 93) + \frac{0.309}{60}(63) = -18.5 \text{ kW}$$

TEST Analysis

Launch the open-steady psychrometric TESTcalc. Evaluate State-1 and State-2 for moist air from the given properties (see TEST-code in *TEST > TEST-codes*). For State-3, select condensed water as the working fluid and enter T3=T2 and x3=0. In the device panel, load State-1 as the i1-state, and States 2 and 3 as the e1 and e2 states, respectively. Enter Wdot_ext=0 and click Super-Calculate. The heat transfer rate is evaluated as Qdot = −18.2 kJ. In State-3, mdot_3 is calculated as 0.307 kg/min.

What-if scenario

Change RH1 to 50% and click Super-Calculate. The new answers are calculated as −8.27 kW and 0.06 kg/min.

Discussion

The removal of moisture in this example imposes more penalty in terms of power consumption (and cost of operation) than sensible cooling alone.

12.5.4 Evaporative Cooling

When water evaporates in air, the enthalpy or latent heat of evaporation h_{fg} must be supplied by some agent. In the absence of any external energy transfer, the surrounding air and the leftover water are the only sources to supply this heat, which they do in their quest for equilibrium (more on this in Chapter 14) at the expense of a drop in temperature. This phenomenon is known as **evaporative cooling**. In our daily life, we encounter evaporative cooling when we get out of a swimming pool, cool water by keeping it in an earthen pitcher (water seeping through the pores evaporates outside), or enjoy a breeze coming over a lake.

The driving force of evaporative cooling is the difference between the vapor pressure in the immediate vicinity of the water droplets and that in the surrounding air. Suppose a droplet is initially at the same temperature T of the surrounding air. Then the vapor pressure near the surface of the droplet can be assumed to be $p_{g@T}$ by applying the LTE hypothesis to the local air next to the water surface. If the air away from the droplet has a relative humidity of ϕ, the difference in vapor pressure, the driving force of evaporation, is $p_{g@T} - \phi p_{g@T}$. As the droplet cools down, this driving force weakens and finally stops when the air around becomes saturated at the wet-bulb temperature.

Clearly, evaporative cooling cannot work if the air is already saturated. However, in a dry climate, evaporative cooling can be used to substantially cool and humidify air without the need of a refrigeration cycle to supply the coolant. In a porus jug or **swamp cooler** (Fig. 12.22), this principle is utilized by circulating water over a porous screen through which air is drawn from outside. Makeup water is supplied to keep the cooler at a steady state.

The mass and energy equations, Eqs. (12.13)–(12.15), can be simplified for evaporative cooling by neglecting the enthalpy supplied by the makeup water.

$$\text{Dry Air:} \quad \dot{m}_{a1} = \dot{m}_{a2} = \dot{m}_a; \quad \left[\frac{\text{kg d.a.}}{\text{s}}\right] \quad (12.33)$$

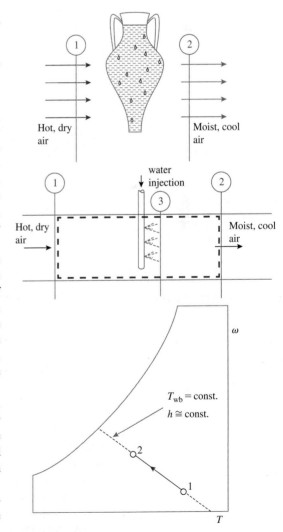

FIGURE 12.22 Water in a porous jug, or the evaporative cooler, work on the same principle (see Anim. 12.B.*evapCooling*).

$$H_2O: \quad \dot{m}_{v2} = \dot{m}_{v1} + \dot{m}_{w3}; \quad \Rightarrow \quad \dot{m}_a(\omega_2 - \omega_1) = \dot{m}_{w3}; \quad \left[\frac{kg\ H_2O}{s}\right] \qquad (12.34)$$

$$Energy: \quad \dot{Q}^0 = \dot{m}_a(h_2 - h_1) + \cancel{\dot{m}_{w3}h_{w3}}^{\ 0}; \quad [kW], \quad \Rightarrow \quad h_1 \cong h_2; \quad \left[\frac{kJ}{kg\ d.a.}\right] \qquad (12.35)$$

From the psychrometric chart, we can verify that the constant-wet-bulb temperature lines are almost parallel to the constant-enthalpy lines. Therefore, Eq. (12.35) implies that $T_{wb,1} \cong T_{wb,2}$ (Fig. 12.22). At a given state of air, the minimum temperature that can be achieved by evaporative cooling must be the wet-bulb temperature.

EXAMPLE 12-9   Evaporative Cooling

A swamp cooler (see Anim. 12.B.*swampCooler*) takes outdoor air at 100 kPa, 30°C, and 10% R.H. at a steady rate of 20 m³/min. At the exit, the relative humidity is 90%. Determine (a) the exit temperature and (b) the rate of water consumption. **What-if scenario:** (c) What would the answers be if the outdoor air had a R.H. of 50%?

SOLUTION

Use the energy equation to determine the exit state. Perform a mass balance to obtain the flow rate of makeup water.

Assumptions

The MA model is applicable. Equilibrium at the inlet and exit states, State-1 and State-2. Negligible changes in ke and pe. The pressure remains constant throughout the system.

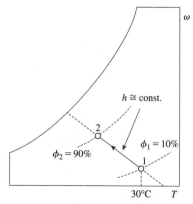

FIGURE 12.23 Psychrometric plot for Ex. 12-9.

Analysis

Use the psychrometric chart to evaluate State-1. To find the wet-bulb temperature, follow the constant T_{wb} line to the saturation curve, where the dry-bulb and wet-bulb temperatures are the same. At State-2 (Fig. 12.23), the wet-bulb temperature and the relative humidity are known.

State-1 (given $T_1 = 30°C$, $\phi_1 = 10\%$, $\dot{V}_1$):

$$\omega_1 = 2.6\frac{g\ H_2O}{kg\ d.a.}; \quad T_{wb,1} = 13.2°C; \quad v_1 = 0.862\frac{m^3}{kg\ d.a.};$$

$$\dot{m}_{a1} = \frac{\dot{V}_1}{v_1} = \frac{(20/60)}{0.862} = 0.387\frac{kg\ d.a.}{s}$$

State-2 (given $T_{wb,2} = T_{wb,1}$, $\phi_2 = 90\%$):

$$T_2 = 14.2°C; \quad \omega_2 = 9.1\frac{g\ H_2O}{kg\ d.a.}$$

Therefore, the flow rate of makeup water is:

$$\dot{m}_{w3} = \dot{m}_a(\omega_2 - \omega_1) = (0.387)(0.0091 - 0.0026)(60) = 0.151\ kg/min$$

TEST Solution

Launch the open-steady psychrometric TESTcalc. Evaluate State-1 and State-2 for moist air from the given properties (see TEST-code in *TEST > TEST-codes*). For State-3, select condensed water as the working fluid and enter T3=T1 and x3=0. In the device panel, load State-1 as the i1-state and states 2 and 3 as the e1 and e2 states, respectively. Enter Wdot_ext=0 and click Calculate and then Super-Calculate to update all panels. The exit temperature and water consumption rate are found in states 2 and 3 as 13.97 deg-C and 0.147 kg/min. The heat transfer, which should ideally be zero, can be found in the device panel as −0.31 kW. This slight error results from our assumption that the wet-bulb temperature remains constant, which is only approximately true.

What-if scenario

Change RH1 to 50% and click Super-Calculate. The new answers are calculated as 23.1 deg-C and 0.063 kg/min.

Discussion

From the what-if study, it can be seen that the effectiveness of a swamp cooler decreases rapidly with an increase in the relative humidity of air. In a swamp-cooler equipped room, recirculation of indoor air is inadvisable.

12.5.5 Adiabatic Mixing

Mixing two air streams is quite common in many air-conditioning applications, where the conditioned air is mixed with fresh air according to building codes, which are different for hospitals, large buildings, or process plants. The two streams are merged in a mixing section, as shown in Figure 12.24.

To simplify the analysis, heat transfer, external work transfer, and any change in ke or pe are neglected. The pressure of the two inlet streams must be equal to prevent the back flow of one stream into another. If the pressure drop, due to friction, can be neglected, the exit pressure must be the same as the inlet pressure. Additionally, let us assume that no condensation occurs during mixing. The mass and energy equations, Eqs. (12.13)–(12.15), for the adiabatic mixing process can be written:

$$\text{Dry Air:} \quad \dot{m}_{a1} + \dot{m}_{a2} = \dot{m}_{a3}; \quad \left[\frac{\text{kg d.a.}}{\text{s}}\right] \qquad \textbf{(12.36)}$$

$$\text{H}_2\text{O:} \quad \dot{m}_{v1} + \dot{m}_{v2} = \dot{m}_{v3}; \quad \Rightarrow \quad \dot{m}_{a1}\omega_1 + \dot{m}_{a2}\omega_2 = \dot{m}_{a3}\omega_3; \quad \left[\frac{\text{kg H}_2\text{O}}{\text{s}}\right] \qquad \textbf{(12.37)}$$

$$\text{Energy:} \quad \dot{m}_{a1}h_1 + \dot{m}_{a2}h_2 = \dot{m}_{a3}h_3; \quad [\text{kW}] \qquad \textbf{(12.38)}$$

Substituting $\dot{m}_{a3}$ from the dry air mass equation into the vapor mass and energy equations, we obtain:

$$\frac{\dot{m}_{a1}}{\dot{m}_{a2}} = \frac{\omega_2 - \omega_3}{\omega_3 - \omega_1} = \frac{h_2 - h_3}{h_3 - h_1} \qquad \textbf{(12.39)}$$

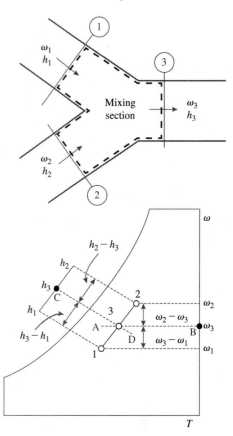

FIGURE 12.24 The relation between the exit state and inlet states of moist air during adiabatic mixing can be interpreted geometrically (see Anim. 12.B.*adiabaticMixing*).

To locate the exit state, State-3, on the psychrometric diagram for a given pair of inlet states, State-1 and State-2, Eq. (12.39) lends itself to a simple geometrical interpretation. On the ω axis of Figure 12.24, ω_3 can be located by realizing that the ratio of the segments $\omega_2 - \omega_3$ and $\omega_3 - \omega_1$, according to Eq. (12.39), is simply the ratio of the dry air mass flow rates $\dot{m}_{a1}$ and $\dot{m}_{a2}$. State-3 must, therefore, lie on the horizontal line AB. In a similar manner, line CD can be drawn by locating h_3 on the h axis. The intersection between the two lines, which satisfies both conditions, $h = h_3$ and $\omega = \omega_3$, must be State-3.

Having excluded any possibility of condensation, Eq. (12.39) is valid until State-3 approaches the saturation curve $\phi = 1$ in the psychrometric diagram, so that line 1–2 becomes a tangent. If either stream was located any closer to the saturation curve, condensation would result. Note that as long as the States 1, 2, and 3 satisfy $\phi \leq 1$, Eq. (12.39) would apply even if part of the line 1–2–3 intersected the saturation curve. Any condensate generated during mixing would evaporate to produce a condensate free exit state, as dictated by State-3.

EXAMPLE 12-10   Adiabatic Mixing of Moist Air Streams

A saturated stream of air, at 30°C with a volume flow rate of 50 m³/min, mixes with a stream of cooled air consisting of a flow rate of 25 m³/min at 15°C and 30% R.H. Assuming the total pressure to remain constant at 1 atm, determine (a) the temperature and (b) the relative humidity at the exit. *What-if scenario:* (c) What would the volume flow rate of the cooler stream be if the desired relative humidity at the exit was 65%?

SOLUTION

Use a mass and energy balance for the mixing device to determine the exit state.

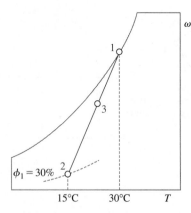

FIGURE 12.25 Psychrometric plot for Ex. 12-10.

Assumptions

The moist air model is applicable. Equilibrium at the inlet states, State-1 and State-2, and the exit state, State-3 (Fig. 12.25). Negligible changes in ke and pe.

Analysis

Use the psychrometric chart, or the MA flow-state TESTcalc, and Eq. (12.39) to evaluate the three states.

State-1 (given $T_1 = 30°C$, $\phi_1 = 100\%$, $\dot{V}_1$):

$$\omega_1 = 27.3 \frac{\text{g H}_2\text{O}}{\text{kg d.a.}}; \quad h_1 = 100 \frac{\text{kJ}}{\text{kg d.a.}}; \quad v_1 = 0.896 \frac{\text{m}^3}{\text{kg d.a.}};$$

$$\dot{m}_{a1} = \frac{\dot{V}_1}{v_1} = \frac{(50/60)}{0.896} = 0.93 \frac{\text{kg d.a.}}{\text{s}}; \quad \dot{m}_{v1} = \omega_1 \dot{m}_{a1} = 25.4 \frac{\text{g}}{\text{s}}$$

State-2 (given $T_2 = 15°C$, $\phi_2 = 30\%$, $\dot{V}_2$):

$$\omega_2 = 3.2 \frac{\text{g H}_2\text{O}}{\text{kg d.a.}}; \quad h_2 = 23.1 \frac{\text{kJ}}{\text{kg d.a.}}; \quad v_2 = 0.82 \frac{\text{m}^3}{\text{kg d.a.}};$$

$$\dot{m}_{a2} = \frac{\dot{V}_2}{v_2} = \frac{(25/60)}{0.82} = 0.508 \frac{\text{kg}}{\text{s}}; \quad \dot{m}_{v2} = \omega_2 \dot{m}_{a2} = 1.63 \frac{\text{g}}{\text{s}}$$

State-3: $\dot{m}_{a3} = \dot{m}_{a1} + \dot{m}_{a2} = 0.93 + 0.508 = 1.438 \dfrac{\text{kg d.a.}}{\text{s}};$

$$\omega_3 = \frac{\dot{m}_{v3}}{\dot{m}_{a3}} = \frac{\dot{m}_{v1} + \dot{m}_{v2}}{\dot{m}_{a3}} = \frac{25.4 + 1.63}{1.438} = 18.8 \frac{\text{g H}_2\text{O}}{\text{kg d.a.}};$$

$$h_3 = \frac{\dot{m}_{a1}h_1 + \dot{m}_{a2}h_2}{\dot{m}_{a3}} = \frac{(0.93)(100) + (0.508)(23.1)}{1.438} = 72.8 \frac{\text{kJ}}{\text{kg d.a.}};$$

$$\Rightarrow \quad T_3 = 25°C; \quad \phi_3 = 95\%$$

TEST Analysis

Launch the open-steady psychrometric TESTcalc. Evaluate State-1 and State-2 for moist air from the given properties (see TEST-code in *TEST > TEST-codes*). For State-3, enter p3=p1 and mdot3=mdot1+mdot2. Instead of using the energy balance expressions for h3, we can use the device panel to perform the energy analyses. In the device panel, load State-1 and State-2 as the i1 and i2 states, respectively, and the partially evaluated State-3 as the e1-state. Enter Wdot_ext=0, Qdot=0, and click Calculate and then Super-Calculate. The exit temperature and relative humidity are updated in State-3 as 24.8 deg-C and 94.0%.

What-if scenario

Change Voldot2 to 200 m³/min, press the Enter key, and click Super-Calculate. RH3 is calculated as 60%. Guess Voldot3 again and recalculate RH3. Repeat until RH3 is close to 65%. The required volume flow rate is about 155 m³/min. The exit temperature at this flow rate is 18.5 deg-C.

Discussion

When the exit state is equidistant from the inlet states in a mixing flow, enthalpy and specific humidity assume average values at the exit; however, the same is not true for other properties.

12.5.6 Wet Cooling Tower

Waste heat is generated by power plants and all other heat engines, air-conditioning systems, refrigeration systems, and many process industries. For small systems, such as the automobile engines or household refrigerator, the surrounding air serves as the heat sink. Large plants are generally located close to a water reservoir—lake, river, ocean, etc.—so that the waste heat can be rejected to circulating water. In arid climates, or in environments where the rise in water temperature causes significant thermal pollution, **cooling towers** provide an excellent environmentally friendly alternative

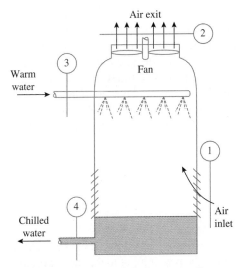

FIGURE 12.26 A forced-draft counter flow cooling tower (see Anim. 12.B.*coolingTower*).

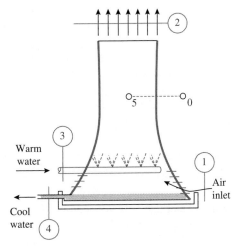

FIGURE 12.27 A natural-draft cooling tower (see Anim. 9.A.*coolingTowerNatural*).

by recirculating the warm water after chilling it through evaporative cooling (Fig. 9.1 and Anim. 12.B.*coolingTower*). The waste heat absorbed by water is ultimately rejected to the surrounding air. Cooling towers are also used simply to provide chilled water to meet various industry needs.

To accomplish evaporative cooling, warm water is sprayed into relatively dry air in the cooling tower. Depending on how the air is circulated, there are two major types of cooling towers. In a **forced**, or **induced-draft**, cooling tower (Fig. 12.26) air is drawn into the tower at the bottom and forced out at the top by a fan. Warm water is sprayed downward in a counter-flow configuration or sideways, creating a cross-flow configuration. A fine spray maximizes the surface area of contact between air and liquid water, reducing the size of the tower. If we track a warm falling droplet (at State-3 to State-4 in Fig. 12.26), its temperature decreases not only due to heat exchange with the relatively cool air it encounters, but also from evaporative cooling as part of the droplet evaporates. The air entering at the ambient condition, State-1, encounters increasingly warmer water and leaves at State-2, with a much higher temperature and humidity, at the top. The chilled water at State-4 accumulates at the bottom and is pumped back to the plant after make-up water is added to maintain steady state.

A **natural draft cooling tower** (Fig. 12.27) is tall—about 100 m high—and has a distinctive hyperbolic profile. Designed to take advantage of natural buoyancy-induced drafts to circulate air, such a cooling tower, does not require any external power for its operation. To understand the principle of buoyancy-induced flow, consider the ratio of density between warm and humid air inside the tower to the air outside at the same altitude. Assuming pressure variation to be hydrostatic, unaffected by the gentle motion of air, $p_5 = p_0$ in Figure 12.27. Therefore, using the IG equation of state, $\rho_5/\rho_0 = (\overline{M}_5 T_0)/(\overline{M}_0 T_5)$. Due to the presence of a higher mole fraction of water vapor in the gas mixture, humid air has a smaller molar mass than that of relatively drier air, that is, $\overline{M}_5 < \overline{M}_0$. Combined with the fact that $T_5 > T_0$, the warm and humid air is lighter than outside air. The lighter air rises in the tower, pushed upward by the buoyancy effect which lasts through the entire height. The rising air is replaced by fresh outside air entering through inlets at the bottom. The taller the tower, the longer the upward push lasts—this is known as the **chimney effect**. A cooling tower, therefore, is simply a huge chimney and its distinctive hyperbolic profile is based on structural considerations and has little to do with thermodynamics.

The mass and energy equations, Eqs. (12.13)–(12.15), can be simplified for the open-steady cooling tower, shown in Figure 12.27, as follows:

$$\text{Dry Air: } \dot{m}_{a1} = \dot{m}_{a2} = \dot{m}_a; \quad \left[\frac{\text{kg d.a.}}{\text{s}}\right]$$

$$\text{H}_2\text{O: } \dot{m}_{v1} + \dot{m}_{w3} = \dot{m}_{v2} + \dot{m}_{w4}; \tag{12.40}$$

$$\Rightarrow \dot{m}_a(\omega_2 - \omega_1) = \dot{m}_{w3} - \dot{m}_{w4} = \dot{m}_{\text{makeup}}; \quad \left[\frac{\text{kg H}_2\text{O}}{\text{s}}\right] \tag{12.41}$$

$$\text{Energy: } \dot{m}_a h_1 + \dot{m}_{w3} h_{f3} = \dot{m}_a h_2 + \dot{m}_{w4} h_{f4}; \quad [\text{kW}] \tag{12.42}$$

In addition to the standard simplifications, the power consumed by the fan is neglected in the derivation of the energy equation for a forced-draft cooling tower.

A variation of a cooling tower is the **spray pond**, where the warm water is sprayed into the air and is cooled convectively and evaporatively as it falls back into the pond. Although lower in cost compared to a cooling tower, a spray pond requires about 25 to 50 times the area of a cooling tower. Water loss due to air drift and water pollution from dust and dirt are other problems associated with a spray pond. In a **cooling pond**, which is merely an artificial lake, warm water is simply dumped into the pond. A very large surface area, about 20 times the area of a spray pond, is required to cool the water convectively.

EXAMPLE 12-11 | Wet Cooling Tower

Warm water leaves the condenser of a power plant and enters a wet cooling tower at 35°C with a mass flow rate of 120 kg/s. The water is cooled to 20°C by the counter-flow air which enters at 15°C, 50% R.H., and leaves at 30°C, 90% R.H. Neglecting the power input to the fan, determine the volume flow rate of air at (a) the inlet, (b) the exit, and (c) the mass flow rate of the makeup water. *What-if scenario:* (d) What would the answer in part (a) be if the air left the cooling tower at saturated condition?

SOLUTION

Perform a mass and energy balance for the cooling tower, an open-steady device.

Assumptions

The moist air model is applicable to air. Equilibrium at all inlet and exit states shown in Figure 12.28. The SL model is applicable to liquid water. Negligible changes in ke and pe. The pressure remains constant at 1 atm.

Analysis

Use the psychrometric chart and the saturation temperature table of water, or the MA and PC flow-state TESTcalcs, to determine the states, properties.

State-1 (Moist air; given $T_1 = 15°C$, $\phi_1 = 50\%$):

$$\omega_1 = 5.3 \frac{\text{g H}_2\text{O}}{\text{kg d.a.}}; \quad h_1 = 28.5 \frac{\text{kJ}}{\text{kg d.a.}}; \quad v_1 = 0.823 \frac{\text{m}^3}{\text{kg d.a.}}$$

State-2 (Moist air; given $T_2 = 30°C$, $\phi_2 = 90\%$):

$$\omega_2 = 24.5 \frac{\text{g H}_2\text{O}}{\text{kg d.a.}}; \quad h_2 = 93 \frac{\text{kJ}}{\text{kg d.a.}}; \quad v_2 = 0.893 \frac{\text{m}^3}{\text{kg d.a.}}$$

State-3 (Liquid water; given $T_3 = 35°C$, $x_3 = 0$, $\dot{m}_3 = 120$ kg/s):

$$h_3 = h_{f@T_3} = 147 \frac{\text{kJ}}{\text{kg d.a.}}$$

State-4 (Liquid Water; given $T_4 = 20°C$, $x_4 = 0$):

$$h_4 = h_{f@T_4} = 84 \frac{\text{kJ}}{\text{kg}}$$

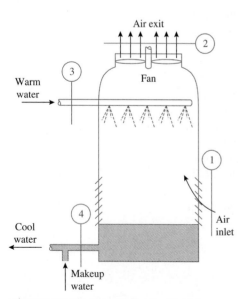

FIGURE 12.28 Schematic for Ex. 12-11 (see Anim. 12.B.*coolingTower*).

Manipulation of Eqs. (12.40)–(12.42) leads to:

$$\dot{m}_{a1} = \dot{m}_{a2} = \dot{m}_a = \frac{\dot{m}_3(h_3 - h_4)}{(h_2 - h_1) - h_4(\omega_2 - \omega_1)};$$

$$\Rightarrow \dot{m}_a = \frac{(120)(147 - 84)}{(93 - 28.5) - (84)(0.0245 - 0.0053)} = 120 \frac{\text{kg}}{\text{s}}$$

Therefore,

$$\dot{V}_1 = \dot{m}_a v_1 = (120)(0.823) = 99 \text{ m}^3/\text{s};$$

$$\dot{V}_2 = \dot{m}_a v_2 = (120)(0.893) = 107 \text{ m}^3/\text{s}; \quad \text{and,}$$

$$\dot{m}_{\text{makeup}} = \dot{m}_{w3} - \dot{m}_{w4} = \dot{m}_a(\omega_2 - \omega_1) = \frac{(120)(24.5 - 5.3)}{1000} = 2.3 \frac{\text{kg}}{\text{s}}$$

TEST Analysis

Launch the open-steady psychrometric TESTcalc. Evaluate State-1 and State-2 for moist air and State-3 and State-4 for saturated liquid water from the given properties (See TEST-code in *TEST > TEST-codes*). In the device panel, select Cooling Tower radio button, then load States 1 and 3 as the i1 and i2 states, and States 2 and 4 as e1 and e2 states. Enter Wdot_ext=0, Qdot=0, click Calculate and then click Super-Calculate. Voldot1 and Voldot2 are calculated as part of the states. Calculate the flow rate of the makeup water in the I/O panel using the expression '=mdot1*(omega2−omega1)'. The manual results are reproduced.

What-if scenario

Change RH2 to 100%, press the Enter key, and click Super-Calculate. Variable mdot1 is updated to 108.4 kg/s.

Discussion

It can be seen from the "what-if" study that by maximizing the relative humidity of the exiting air the mass flow rate of air can be reduced, lowering the power requirement for the fan. To ensure saturation at the exit, the tower must be very tall to allow the liquid droplets sufficient residence time for evaporation.

12.6 CLOSURE

We studied psychrometry and its application to air conditioning in this chapter. It began with a new material model for moist air by extending the PG mixture model developed in Chapter 11. New properties were introduced to describe psychrometric states, and they were expressed in terms of readily measurable properties. Mass and energy balance equations were customized for a generic open-steady device and a generic closed process. The psychrometric chart was introduced as an alternative to look-up tables for moist air properties. Frequently encountered air-conditioning processes—simple heating, simple cooling, heating and humidification, cooling and dehumidification, and evaporative cooling—were analyzed, ending with an analysis of a wet cooling tower. Several psychrometric TESTcalcs are introduced and used to verify manual solution and study what-if scenarios.

PROBLEMS

SECTION 12-1: MOIST AIR (MA) MODEL

12-1-1 Consider 100 m^3 of moist air at 100 kPa, 35°C and 80% R.H. Calculate (a) the humidity ratio (ω), (b) dew point, and (c) mass of vapor. (d) *What-if scenario:* What would the dew point be if the moist air was at 60% R.H.?

12-1-2 A tank contains 15 kg of dry air and 0.2 kg of water vapor at 22°C, 100 kPa total pressure. Determine (a) the specific humidity (ω), (b) relative humidity (ϕ) and (c) the volume (V) of the tank.

12-1-3 A room contains air at 18°C and 95 kPa with a relative humidity of 80%. Determine (a) the partial pressure of dry air, (b) the specific humidity (ω) of the air, and (c) the enthalpy per unit mass of dry air. (d) *What-if scenario:* What would the answers be if the pressure was 85 kPa?

12-1-4 A tank with a volume 20 m^3 contains a mixture of carbon dioxide and water vapor at 18°C and 100 kPa with a relative humidity (ϕ) of 80%. Determine (a) the partial pressure of dry CO_2, (b) the specific humidity (ω) of the mixture, (c) the dew point, (d) the amount of vapor, and (e) the mass of CO_2. (f) *What-if scenario:* What would the amount of vapor be if the mixture was composed of dry air and water vapor?

12-1-5 A room contains air at 65°F and 14.2 psia with a relative humidity (ϕ) of 75%. Determine (a) the partial pressure of dry air, (b) the specific humidity (ω) of the air, and (c) the enthalpy per unit mass (h) of dry air.

12-1-6 A 10 m^3 tank contains saturated air at 25°C, 101 kPa. Determine (a) the specific humidity (ω), (b) mass of dry air and (c) the enthalpy of the dry air, per unit mass of the dry air.

12-1-7 Determine (a) the masses of the dry air and (b) the water vapor, contained in a 200 m^3 room at 96 kPa, 22°C and 65% relative humidity (ϕ).

12-1-8 A house contains air at 22°C and 55% relative humidity (ϕ). (a) Will any moisture condense on the inner surfaces of the windows when the temperature of the windows drop to 10°C? (b) *What-if scenario:* Will any moisture condense on the windows at 13°C? (Answer should be 1 if Yes and 2 if No.)

12-1-9 Air in a room has a dry-bulb temperature of 24°C and a wet-bulb temperature of 18°C. Assuming a pressure of 100 kPa, determine (a) the specific humidity (ω), (b) the relative humidity (ϕ), and (c) the dew-point temperature. (d) *What-if scenario:* What would the relative humidity be if the wet-bulb temperature was 15°C?

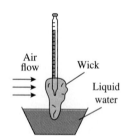

Air flow — Wick — Liquid water

12-1-10 Air in a room is at 1 atm, 29°C and 55 percent relative humidity (ϕ). Determine (a) the specific humidity (ω), (b) the enthalpy (h), (c) The wet-bulb temperature, (d) the dew-point temperature, and (e) the specific volume (v) of the air.

12-1-11 Air in a room has a pressure of 1 atm, a dry-bulb temperature of 26°C, and a wet-bulb temperature of 16°C. Determine (a) the specific humidity (ω), (b) the enthalpy (h), (c) the relative humidity (ϕ), (d) the dew-point temperature, and (e) specific volume (v) of the air.

SECTION 12-2: OPEN SYSTEMS ANALYSIS

12-2-1 Moist air at 12°C and 80% R.H. enters a duct at a rate of 150 m^3/min. The mixture is heated until it exits at 35°C. The pressure remains constant at 100 kPa. Determine (a) the rate of heat transfer ($\dot{Q}$) and (b) the relative humidity (ϕ) at the exit. (c) *What-if scenario:* What would the rate of heat transfer be if the mixture was heated until it exits at 50°C?

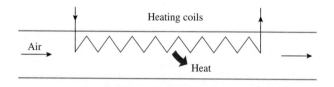

Heating coils

Air → Heat

12-2-2   Air enters a heating section, at 100 kPa, 9°C, 45% relative humidity, at rate of 10 m³/min, and it leaves at 22°C. Determine (a) the rate of heat transfer ($\dot{Q}$) in the heating section and (b) the relative humidity (ϕ) at the exit. (c) *What-if scenario:* What would the relative humidity at the exit be if the relative humidity of the mixture changed to 75%?

12-2-3   A heating section consists of a 30 cm diameter duct, which houses a 6 kW electric resistance heater. Air enters the heating section at 1 atm, 15°C, and 33% relative humidity (ϕ) at velocity 8.5 m/s. Determine (a) the exit temperature (T_2), (b) the exit relative humidity (ϕ_2) of the air, and (c) the exit velocity (V_2). (d) *What-if scenario:* What would the answers be if the inlet velocity was 5 m/s?

12-2-4   A heating section consists of a 10 inch diameter duct which houses a 8 kW electric resistance heater. Air enters the heating section at 14.7 psia, 40°F and 35% relative humidity (ϕ) at a velocity of 21 ft/s. Determine (a) the exit temperature (T_2), (b) the exit relative humidity (ϕ_2) of the air, and (c) the exit velocity (V_2).

12-2-5   Air enters a 30 cm diameter cooling section at 1 atm, 35°C, and 40% relative humidity (ϕ) at 35 m/s. Heat is removed from the air at a rate of 1400 kJ/min. Determine (a) the exit temperature (T_2), (b) the exit relative humidity (ϕ_2) and (c) the exit velocity (V_2). (d) *What-if scenario:* What would the answers be if the heat removal rate was 1000 kJ/min?

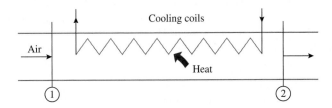

12-2-6   Air at 1 atm, 13°C, and 50% relative humidity (ϕ) is first heated to 18°C in a heating section, then humidified by introducing saturated water vapor at 1 atm. Air leaves the humidifying section at 22°C and 60% of relative humidity (ϕ). Determine (a) the amount of steam added to the air in kg H_2O/kg dry air and (b) the amount of heat transfer (q) to the air in the heating section in kJ/kg dry air.

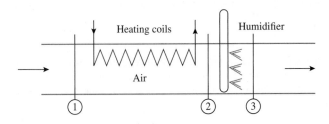

12-2-7   An air conditioning system that operates at a total pressure of 1 atm, consists of a heating section and humidifier which supplies wet steam (saturated water vapor) at 1 atm. Air enters the heating section at 15°C and 75% relative humidity (ϕ) at 60 m³/min, and it leaves the humidifying section at 24°C and 55% relative humidity (ϕ_2). Determine (a) the temperature and relative humidity (ϕ) of air when it leaves the heat section, (b) the rate of heat transfer ($\dot{Q}$) in the heating section, and (c) the rate at which water is added to the air in the humidifying section.

12-2-8   Moist air, at 40°C and 90% R.H., enters a dehumidifier at a rate of 300 m³/min. The condensate and the saturated air exit at 10°C through separate exits. The pressure remains constant at 100 kPa. Determine (a) the mass flow rate ($\dot{m}$) of dry air, (b) the water removal rate and (c) the required refrigeration capacity in tons. (d) *What-if scenario:* What would the required refrigeration be if the moist air entered the dehumidifier at a rate of 200 m³/min?

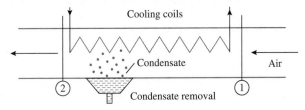

12-2-9 Air enters a window air conditioner at 1 atm, 36°C, and 75% relative humidity (ϕ) at a rate of 12 m³/min. It leaves as saturated air at 18°C. Part of the moisture in the air, which condenses during the process, is also removed at 18°C. Determine (a) the rate of heat ($\dot{Q}$) and (b) moisture removal from the air. (c) *What-if scenario:* What would the rate of heat removal be if moist air entered the dehumidifier at 95 kPa instead of 1 atm?

12-2-10 An air conditioning system is to take in air at 1 atm, 32°C, 65% relative humidity and deliver it at 22°C, 40% relative humidity. Air flows first over the cooling coils, where it is cooled and dehumidified, then over the resistance heating wires, where it is heated to the desired temperature. Assuming that the condensate is removed from the cooling section at 7°C, determine (a) the temperature (T) of air before it enters the heating section, (b) the amount of heat removed in the cooling section, and (c) the amount of heat transferred (Q) in the heating section, both in kJ/kg dry air.

12-2-11 Air enters a 20 cm diameter cooling section, at 1 atm, 32°C and 65% relative humidity, at 60 m/min. The air is cooled by passing it over a cooling coil, through which cold water flows. The water experiences a temperature rise of 10°C. Air leaves the cooling section saturated at 18°C. Determine (a) the rate of heat transfer ($\dot{Q}$), (b) the mass flow rate ($\dot{m}$) of the water, and (c) the exit velocity (V_2) of the airstreams. (d) *What-if scenario:* What would the rate of heat transfer be if moist air entered the dehumidifier at 95 kPa instead of 1 atm?

12-2-12 Air enters a 1.5 ft diameter cooling section, at 14.4 psia, 100°F and 70% relative humidity, at 500 ft/min. The air is cooled by passing it over a cooling coil, through which cold water flows. The water experiences a temperature rise of 15°F. Air leaves the cooling section saturated at 72°F. Determine (a) the rate of heat transfer ($\dot{Q}$), (b) the mass flow rate ($\dot{m}$) of the water, and (c) the exit velocity (V_2) of the airstreams.

12-2-13 A saturated stream of carbon dioxide enters a dehumidifier with a flow rate of 100 m³/min at 30°C, 100 kPa. The mixture is cooled to 10°C by circulating cold water before being electrically heated back to 30°C. Determine (a) the rate of water removal in kg/min, (b) the cooling load, (c) the heating load, and (d) the relative humidity (ϕ) at the exit.

12-2-14 Repeat Problem 12-2-13 assuming the gas mixture is composed of dry air and water vapor.

12-2-15 Moist air with dry and wet bulb temperatures of 20°C and 9°C, respectively, enters a steam-spray humidifier at rate of 100 kg of dry air/min. Saturated water vapor at 110°C is injected at 1 kg/min. The pressure remains constant at 100 kPa. Determine (a) the inlet R.H. and (b) exit R.H. (c) *What-if scenario:* What would the exit R.H. be if the saturated water vapor was injected at 0.5 kg/min?

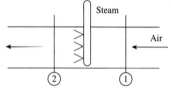

12-2-16 Air, at 110°F and 10% R.H., enters an evaporative cooler at a flow rate of 5500 ft³/min. The air leaves at 70°F. Determine (a) the mass flow rate ($\dot{m}$) of water and (b) the exit R.H. Assume the pressure (1 atm) and wet-bulb temperature to remain constant along the flow.

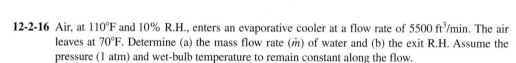
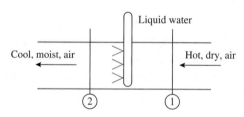

12-2-17 Air enters an evaporative cooler at 1 atm, 38°C and 15% relative humidity, at a rate of 5 m³/min and it leaves with a relative humidity of 80%. Determine (a) the exit temperature (T_2) of the air; (b) the required rate of water supply to the evaporative cooler. (c) *What-if scenario:* What would the required rate of water supply be if the moist air entered the evaporative cooler at a rate of 10 m³/min instead of 5 m³/min?

12-2-18 Air enters an evaporative cooler at 1 atm, 34°C and 20% relative humidity, at a rate of 8 m³/min and it leaves at 21°C. Determine (a) the the final relative humidity (ϕ) and (b) the amount of water added to the air. (c) *What-if scenario:* What would the answers be if air entered the evaporative cooler at 93 kPa?

12-2-19 Air, at 1 atm, 13°C and 55% relative humidity (ϕ), is first heated to 28°C in a heating section, then is passed through an evaporative cooler, where its temperature drops to 22°C. Determine (a) the exit relative humidity (ϕ) and (b) the amount of water added to the air in kg H_2O/kg dry air.

12-2-20 Determine the adiabatic saturation temperature of air at 100 kPa, 30°C and 50% relative humidity.

12-2-21 A 150 m³/min stream of air at 30°C and 65% R.H. is mixed with a 50 m³/min stream of air at 5°C and 90% R.H. in an adiabatic mixing chamber. Determine the R.H. at the exit. Assume pressure to be 1 atm.

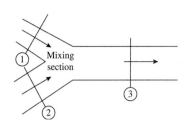

12-2-22 Two airstreams are mixed steadily and adiabatically. The first stream enters at 28°C and 35% relative humidity, at rate of 15 m³/min, while the second stream enters at 10°C and 90% of relative humidity at rate of 20 m³/min. Assuming that the mixing process occurs at a pressure of 1 atm, determine (a) the specific humidity (ω), (b) relative humidity (ϕ), (c) the dry-bulb temperature, and (d) the volume flow rate ($\dot{V}$) of the mixture. (e) *What-if scenario:* What would the answers be if the total mixing-chamber pressure was 92 kPa?

12-2-23 During an air conditioning process, 50 m³/min of conditioned air at 15°C and 33% relative humidity is mixed adiabatically with 10 m³/min of outside air at 32°C and 80% relative humidity at pressure of 1 atm. Determine (a) the temperature (T), (b) the specific humidity (ω), and (c) the relative humidity (ϕ) of the mixture. (d) *What-if scenario:* What would the temperature of the mixture be if the volumes were 40 m³/min and 25 m³/min respectively?

12-2-24 During an air conditioning process, 800 ft³/min of conditioned air at 60°F and 40% relative humidity is mixed adiabatically with 200 ft³/min of outside air at 84°F and 85% relative humidity at pressure of 14.7 psia. Determine (a) the temperature (T), (b) the specific humidity (ω), and (c) the relative humidity (ϕ) of the mixture.

12-2-25 A stream of warm air, with a dry-bulb temperature of 38°C and wet-bulb of 30°C, is mixed adiabatically with a stream of saturated cool air at 15°C. The dry mass flow rates of the warm and cool air streams are 8 kg/s and 6 kg/s, respectively. Assuming a total pressure of 1 atm, determine (a) the temperature (T), (b) the specific humidity (ω) and (c) relative humidity (ϕ) of the mixture.

12-2-26 Cooling water leaves the condenser of a power plant and enters a wet cooling tower at 35°C at a rate ($\dot{m}$) of 100 kg/s. The water is cooled to 22°C in the cooling tower by air which enters the tower at 100 kPa, 20°C, 60% R.H. and leaves saturated at 30°C. Neglecting the power input to the fan, determine (a) the volume flow rate ($\dot{V}$) of air into the cooling tower and (b) the mass flow rate ($\dot{m}$) of the required makeup water.

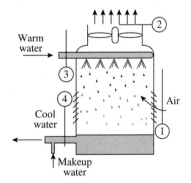

12-2-27 The cooling water from the condenser of a power plant enters a wet cooling tower at 45°C at a rate ($\dot{m}$) of 40 kg/s. The water is cooled to 22°C in the cooling tower by air which enters the tower at 1 atm, 20°C, 60% relative humidity (ϕ) and leaves saturated at 30°C. Neglecting the power input to the fan, determine (a) the volume flow rate ($\dot{V}$) of the air into the cooling tower and (b) the mass flow rate ($\dot{m}$) of the required makeup water. (c) *What-if scenario:* What would the mass flow rate in (b) be if mass flow rate of water was 60 kg/s instead of 40 kg/s?

12-2-28 The cooling water from the condenser of a power plant enters a wet cooling tower at 105°F at a rate of 90 lbm/s. The water is cooled to 85°F in the cooling tower by air which enters the tower at 1 atm, 73°F, 50% relative humidity and leaves saturated at 90°F. Neglecting the power input to the fan, determine (a) the volume flow rate ($\dot{V}$) of the air into the cooling tower and (b) the mass flow rate ($\dot{m}$) of the required makeup water.

12-2-29 A wet cooling tower is to cool 50 kg/s of water from 38°C to 24°C. Atmospheric air enters the tower at 1 atm with dry and wet-bulb temperatures of 20°C and 15°C, respectively, and leaves at 32°C with a relative humidity of 85%. Determine (a) the volume flow rate ($\dot{V}$) of air into the cooling tower and (b) the mass flow rate ($\dot{m}$) of the required makeup water.

12-2-30 A wet cooling tower is to cool 100 kg/s of cooling water from 45°C to 24°C at a location where the atmospheric pressure is 94 kPa. Atmospheric air enters the tower at 18°C and 65% relative humidity and leaves saturated at 32°C. Neglecting the power input to the fan, determine (a) the volume flow rate ($\dot{V}$) of air into the cooling tower and (b) the mass flow rate ($\dot{m}$) of the required makeup water.

SECTION 12-3: CLOSED SYSTEMS ANALYSIS

12-3-1 Consider 100 m³ of moist air at 100 kPa, 35°C and 80% R.H. Calculate (a) the amount of water vapor condensed if the mixture is cooled to 5°C in a constant pressure process. Also calculate (b) the heat transfer (Q). (c) **What-if scenario:** What would the heat transfer be if the mixture was cooled to 15°C in a constant pressure process?

12-3-2 Calculate (a) the amount of water vapor condensed if the mixture in previous example is cooled to 5°C in a constant volume process. Also calculate (b) the heat transfer (Q). (c) **What-if scenario:** What would the heat transfer be if the mixture was cooled to 15°C in a constant volume process?

12-3-3 A tank of volume 10 m³ contains dry air and water vapor mixture at 40°C and 100 kPa at a relative humidity of 90%. The tank is cools down to 10°C by transferring heat to the surroundings. Determine (a) the amount of water condensed and (b) the heat transfer (Q). (c) **What-if scenario:** What would the heat transfer be if the initial pressure was 1000 kPa?

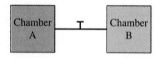

12-3-4 A 50 m³ insulated chamber, containing air at 40°C, 100 kPa and R.H. of 20%, is connected to another 50 m³ insulated chamber containing air at 20°C, 100 kPa and R.H. of 100%. The valve is opened and the system is allowed to reach thermal equilibrium. Determine (a) the final pressure (p_2), (b) temperature (T_2), and (c) humidity.

12-3-5 A 50 m³ insulated chamber, containing air at 5°C, 100 kPa and R.H. of 100%, is connected to another 50 m³ insulated chamber containing air at 22°C, 100 kPa and R.H. of 100%. The valve is opened and the system is allowed to reach thermal equilibrium. Will there be condensation? (Answer 1 if Yes and 2 if No.)

13 | COMBUSTION

This chapter introduces chemical reactions in general terms and combustion in particular. It begins with the meaning of a chemical reaction in mole and mass basis. The mixture models developed in Chapter 11 are extended for evaluation of states of reactants and products by taking into account possible changes in chemical composition. Using these modified mixture models, mass, energy, entropy, and exergy equations are customized for combustion systems. Open steady systems and closed processes are analyzed with application to internal combustion engines and gas turbines. Conversion of heat into work, through heat engines and fuel cells, and combustion efficiencies for various applications are also discussed. As in other chapters, animations are used to illustrate reacting systems; and TEST solutions are used to verify manual results and carrying out what-if studies. An Interactive is also introduced for simulating a combustion chamber.

13.1 COMBUSTION REACTION

In any chemical reaction, the molecules of the participating **reactants** are changed through rearrangement of electronic bonds to form **products** (see Anim. 13.B.*reaction*). Combustion is a special class of exothermic reaction where the **fuel**, typically a mixture of hydrocarbons, is rapidly oxidized by the **oxidizer**, typically the atmospheric air, forming combustion products, which are a mixture of leftover reactants, oxidizer, and many newly formed chemical species.

To illustrate the meaning of a combustion reaction, let us consider the familiar oxidation reaction of hydrogen and explore the various interpretations (Fig. 13.1) that can be attached to this simple reaction:

$$2H_2 + O_2 \rightarrow 2H_2O \qquad (13.1)$$

In such a reaction, called an **overall reaction**, hydrogen and oxygen constitute the *reactants*— hydrogen is the *fuel* and oxygen is the *oxidizer*—and water constitutes the *products* with the arrow in Eq. (13.1) pointing towards the products of the reaction. Although the end product of this reaction is water, many intermediate products are formed during the actual chemical reaction. However, such detailed consideration is not always necessary and an overall reaction connecting initial reactants with final products with a directed arrow is sufficient in most analysis. Phase composition of a component is an important consideration. If the products of a reaction are cooled to temperature below its dew point temperature—the *dew point temperature* is the saturation temperature at the partial pressure of water vapor in the products mixture—the water in the products condenses. To identify a known phase in a reaction, a suffix such as (*l*) or (*s*) is added to a component symbol. Liquid water, for instance, is identified by the symbol $H_2O(l)$ and solid carbon by the symbol $C(s)$.

A properly expressed reaction also carries important quantitative information. In Eq. (13.1), for example, 2 molecules of hydrogen combine with one molecule of oxygen to form 2 molecules of water. These proportions can be immediately scaled up to mole—the *mole* is the count of molecules in a system (in the unit of kmol)—and the reaction can be reinterpreted (Fig. 13.1 and Anim. 13.A.*stoichiometry*) as 2 kmol of hydrogen reacting with 1 kmol of oxygen forming 2 kmol of water. Because a reaction rearranges molecular bonds, mole is not conserved in a reaction, so the total amount of mole before and after the reaction can be different. Thus, in Eq. (13.1), 3 kmol of reactants produces only 2 kmol of products. Atoms, on the other hand, are conserved—the number of atoms of a particular element must be the same before and after a reaction. To ensure this, a reaction is **balanced** by evaluating the coefficients that appear before the components—2, 1, and 2 in this reaction—which are called **stoichiometric**

	$2H_2$	$+ O_2$	$\rightarrow 2H_2O$
Molecules	2	1	2
Dozen	2	1	2
mol	2	1	2
kmol	2	1	2

FIGURE 13.1 Meaning of a reaction in terms of number of molecules or mole involved (see Anim. 13.B.*stoichiometry*).

$$\underbrace{H_2}\ +\ \underbrace{0.5O_2}\ \rightarrow\ \underbrace{H_2O}$$

	H_2	$0.5O_2$	H_2O
Mole (v) (kmol)	1	0.5	1
Mass (η) (kg)	1	8	9

FIGURE 13.2 Meaning of a reaction in mole and mass basis (see Anim. 13.B.*stoichiometry*).

Component	$\overline{M}$ (kg/kmol)
H_2	2
C	12
N_2	28
O_2	32
CO_2	44
H_2O	18
Air	29
C_nH_m	$12n+m$

FIGURE 13.3 Molar masses of some common combustion species.

coefficients of a reaction. There can be as many *atom balance equations* as the number of elements—2 in this case. This process of balancing a reaction is illustrated in Ex. 13-1.

In an overall combustion reaction, it is customary to *normalize* a reaction on the basis of unit mole of fuel. When Eq. (13.1) is normalized in terms of 1 kmol of fuel (hydrogen), the coefficients can be interpreted as follows (Fig. 13.2):

$$\underbrace{H_2}_{1\text{ kmol}}\ +\ \underbrace{0.5O_2}_{0.5\text{ kmol}}\ \rightarrow\ \underbrace{H_2O}_{1\text{ kmol}}\ ; \tag{13.2}$$

Extending this concept, a generic overall reaction among any number of reactants forming any number of products can be written as:

$$\sum_r \nu_r W_r \rightarrow \sum_p \nu_p W_p \tag{13.3}$$

W_r and ν_r (ν is pronounced as *nu*) represent the rth reactant and its stoichiometric coefficient; and W_p and ν_p represent the pth product and its stoichiometric coefficient, respectively. By applying our normalizing convention, $\nu_F = 1$, where F stands for the fuel. For every other component, ν_k ($k = r$ or p) has the unit of kmol of k per-kmol of fuel.

A reaction can also be expressed on a mass basis by converting mole to mass, then normalizing the coefficients on the basis of unit fuel mass. In Eq. (13.1), $\overline{M}_{H_2}$ kg of hydrogen reacts with $0.5\overline{M}_{O_2}$ kg of oxygen, producing $\overline{M}_{H_2O}$ kg of water. Substituting molar masses from Figure 13.3, this translates to 1 kg of hydrogen reacting with 8 kg of oxygen producing 9 kg of water:

$$\underbrace{H_2}_{2/2=1\text{ kg}}\ +\ \underbrace{0.5O_2}_{(0.5)(32)/2=8\text{ kg}}\ \rightarrow\ \underbrace{H_2O}_{18/2=9\text{ kg}}\ ; \tag{13.4}$$

Similarly, we can convert the generic reaction, Eq. (13.3), into mass basis as follows:

$$\sum_r \eta_r W_r \rightarrow \sum_p \eta_p W_p \tag{13.5}$$

where

$$\eta_r = \frac{\nu_r \overline{M}_r}{\overline{M}_F};\ \left[\frac{\text{kg of } r}{\text{kg Fuel}}\right]\quad \text{and}\quad \eta_p = \frac{\nu_p \overline{M}_p}{\overline{M}_F};\ \left[\frac{\text{kg of } p}{\text{kg Fuel}}\right] \tag{13.6}$$

Here, η (pronounced *eta*) represents mass-based stoichiometric coefficient, the mass of a reactant consumed or product produced per unit mass of fuel, with $\eta_F = 1$. Conservation of mass requires that $\sum_r \eta_r = \sum_p \eta_p$. Note that the mass-based coefficient η_k can be converted to mole-based coefficient ν_k ($k = r$ or p) and vice versa by using the relation $\eta_k \overline{M}_F = \nu_k \overline{M}_k$ from Eq. (13.6). Molar masses of some frequently occurring combustion species are listed in Table G-1.

Although both mole and mass-based stoichiometric coefficients are ratios, ν_k and η_k have units, as two different species, k (which represents a reactant or a product component) and F (the fuel), are involved in this ratio. Only $\nu_F = 1$ and $\eta_F = 1$ are truly non-dimensional.

13.1.1 Combustion TESTcalcs

TEST offers a number of combustion TESTcalcs for balancing reactions, analyzing open-steady combustion chambers, and closed combustion processes. They are located in the open-steady and closed-process sub-branches under the specific branch in the TESTcalcs module. Combustion TESTcalcs are divided into two categories—premixed and non-premixed—based on whether the fuel and oxidizer are mixed or separated before they are introduced into a reaction chamber. Based on how the products mixture is modeled, each class of combustion TESTcalcs is further sub-divided into IG mixture and PG mixture TESTcalcs.

In addition to the state, device (or process), and I/O panel, there is a reaction panel in each combustion TESTcalc. In the reaction panel of all non-premixed TESTcalcs, fuel, oxidizer, and products are displayed in three separate blocks. To set up a reaction, we first select a fuel, an oxidizer, and product components from a pull-down menu provided in each block. Air is selected as the default oxidizer and a few common product components are already included in the products block. Some standard reactions can be obtained by simply selecting an action from the action menu. For instance, selecting 'Theoretical Air' produces the stoichiometric complete reaction between a selected fuel and air. To balance a reaction with excess or deficient air, enter 'lambda' (percent theoretical air used) or 'phi' (equivalence ratio), then select an appropriate action—'Excess Air' or 'Deficient Air.' Non-standard reactions can be balanced by specifying known amounts of a few components, then selecting 'Balance Reaction' from the action menu. A reaction can be toggled between mole and mass basis or SI and English units by selecting the appropriate radio button.

The state panel is similar to that of an IG or PG TESTcalc, except every new state must be associated with fuel, oxidizer, or products in the state panel. The analysis panel is identical to the corresponding panel in the generic open-steady or closed-process TESTcalcs. More operational details about the combustion TESTcalcs will be provided as we use those in examples in this chapter.

Another state TESTcalc that can be quite useful for property evaluation during manual problem solving is the n-IG system-state TESTcalc, which makes it very easy to evaluate partial enthalpy and entropy of components in a mixture. In this TESTcalc, we create a gas mixture by selecting a component, entering a value in kmol, kg, or percent, and adding it to the mixture. The mixture state (that includes the partial properties of its components) is updated as more components are added.

Finally, the combustion Interactives, located in Interactives, open-steady, specific branch, can be used by advanced users to simulate a combustion chamber and perform a parametric study involving incomplete reactions.

EXAMPLE 13-1   **Balancing a Reaction**

Determine (a) the molar stoichiometric coefficients and (b) the mass-based stoichiometric coefficients, for a reaction between acetylene (C_2H_2) and pure oxygen producing only carbon dioxide and water.

SOLUTION

Set up and solve the atom balance equations for balancing the reaction (Fig. 13.4).

Analysis

The overall reaction is expressed in terms of 1 kmol of fuel and unknown coefficients of all other components:

$$C_2H_2 + aO_2 \rightarrow bCO_2 + cH_2O$$

The three atom balance equations are:

$$\text{C: } 2 = b; \quad \text{H: } 2 = 2c; \quad \text{O: } 2a = 2b + c;$$

Solving, $b = 2$; $c = 1$; and $a = 2.5$, and the balanced equation can be written:

$$C_2H_2 + 2.5O_2 \rightarrow 2CO_2 + H_2O$$

Therefore, $\nu_F = 1$, $\nu_{O_2} = 2.5$, $\nu_{CO_2} = 2$, and $\nu_{H_2O} = 1$, each having the unit kmol/(kmol of fuel).

The molar mass of C_2H_2 is $\overline{M}_F = 2 \times 12 + 2 \times 1 = 26$ kg/kmol. Those of other compounds can be similarly calculated (Fig. 13.3 and Table G-1). The mass-based coefficients can now be obtained from Eq. (13.6) in the unit of kg/(kg fuel):

$$\eta_F = \frac{\nu_F \overline{M}_F}{\overline{M}_F} = 1; \quad \eta_{O_2} = \frac{\nu_{O_2} \overline{M}_{O_2}}{\overline{M}_F} = \frac{2.5 \times 32}{26} = 3.08;$$

$$\eta_{CO_2} = \frac{\nu_{CO_2} \overline{M}_{CO_2}}{\overline{M}_F} = \frac{2 \times 44}{26} = 3.38;$$

$$\eta_{H_2O} = \frac{\nu_{H_2O} \overline{M}_{H_2O}}{\overline{M}_F} = \frac{1 \times 18}{26} = 0.69$$

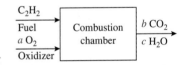

FIGURE 13.4 Schematic for Ex. 13-1.

Fuel	Proven energy reserves in ZJ (end of 2009)
Coal	19.8
Oil	8.1
Gas	8.1

FIGURE 13.5 Proven energy reserves [1 ZJ (zettajoule) = 10^{15} MJ] in 2009 (see Anim. 13.A.*fuelStats*).

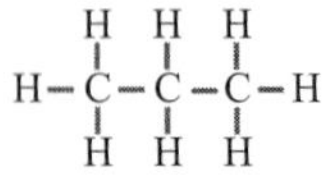

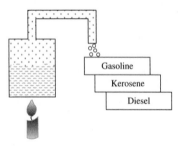

Propane: Saturated chain.

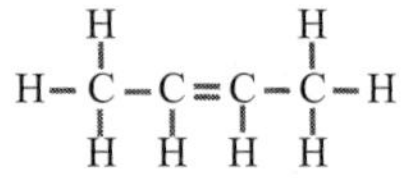

Butene: Unsaturated chain.

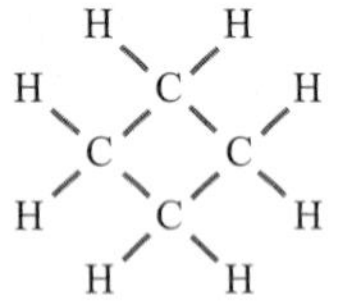

Cyclobutane: Ring, Saturated

FIGURE 13.6 Molecular structure of hydrocarbon fuels (see Anim. 13.A.*fuelClassified*).

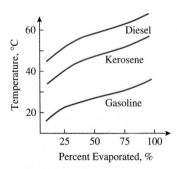

FIGURE 13.7 Liquid hydrocarbon fuels are obtained by distilling crude oil (in Anim. 13.A.*distillation*).

TEST Analysis

Launch the PG (or IG) non-premixed open-steady combustion TESTcalc. Select Acetylene as the fuel, and in the fuel block and execute Theoretical Combustion with oxygen from the action menu to obtain the balanced reaction on a mole basis. To convert it into mass basis, click on the Mass radio-button and run Normalize from the action menu.

Discussion

Atoms are preserved by a balanced reaction while molecules are not. While the total mole decreases by 0.5 kmol in this reaction, the total mass of the reactants and products are equal, 4.08 kg each.

13.1.2 Fuels

We began a discussion of fuels in Sec. 7.10, in connection with automobile fuels. Here we take a closer look at hydrocarbon fuels, the chief source of energy in the world (Fig. 13.5). The major elements of hydrocarbon fuels are carbon, hydrogen, and to a lesser extent, sulfur. Most of the energy release comes from the oxidation of carbon and hydrogen; however, sulfur can be a significant polluting source that is associated with acid rain and corrosion.

Hydrocarbon fuels (Table 13-1) can exist in any of the three phases—solid, liquid and gas. They come in two types of structures—**chain** and **ring** structures (Fig. 13.6), and each of these types can be either **saturated** or **unsaturated**, depending upon whether all the carbon atoms are joined by single bonds or not. Some hydrocarbons have the same formula but different molecular structures; they are called **isomers**. Octane (C_8H_{18}), for instance, has several isomers (Sec. 7.4.3). This aromatic family contains a ring structure, indicated by the prefix "cyclo-", with various degrees of saturation. Alcohols (suffix "-anol") are also used in different fuel blends, their structure is derived from the paraffin family by replacing one of the hydrogen atoms with an OH radical. For instance ethanol, C_2H_5OH, is derived from ethane, and methanol from methane.

Most liquid fuels—gasoline, kerosene, diesel fuel, and fuel oil, etc.—are mixtures of hydrocarbons obtained from crude oil through distillation (Fig. 13.7) and cracking processes. Their differences are brought out by the distillation curves shown in Figure 13.8, which plots the amount of vapor that is recovered through condensation against the temperature to which a fuel is slowly heated—the lower a curve, the more volatile is the fuel. Each fuel has a variety of grades depending on its location on the distillation curve. Fuels derived from gaseous hydrocarbons, such as (liquefied petroleum gas), do not contain any sulfur and are considered relatively cleaner.

For modeling purposes, **gasoline** is represented by octane, C_8H_{18} and **diesel** by dodecane $C_{12}H_{26}$. Similarly, **natural gas**, which contains several different hydrocarbons, is modeled by methane CH_4. **Coal** is the most common solid fuel, whose composition varies depending on its source. The composition of coal in terms of the major elements present on a mass basis is called the **ultimate analysis**.

When all the combustible elements in a fuel—carbon, hydrogen, and sulfur—are fully oxidized to carbon dioxide, water, and sulfur dioxide, combustion is said to be **complete** (Fig. 13.9). Using this terminology, the reaction in Ex. 13-1 could be specified as a *complete* reaction. It is redundant to mention a list of products for a complete reaction.

Gaseous fuels can be modeled by the PG or IG model, while the solid and liquid fuels are modeled by the SL model, as long as the reference properties used in these models include the

FIGURE 13.8 Distillation curves for a few hydrocarbon fuels.

Temperature, °C	

Table 13-1 A partial list of hydrocarbon families

Family	Formula	Suffix	Example	Structure
Paraffin	C_nH_{2n+2}	-ane	Methane: CH_4	Chain
Olefin	C_nH_{2n}	-ene	Propene: C_3H_6	Chain
Diolefin	C_nH_{2n-2}	-diene	Butadiene: C_4H_6	Chain
Alcohol	$C_nH_{2n+1}OH$	-ol	Methanol: CH_3OH	Chain
Aromatic	C_nH_{2n-6}	-benzene	Benzene: C_6H_6	Ring

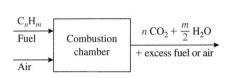

FIGURE 13.9 Combustion is complete when all combustible components are fully oxidized (see Anim. 13.B.*theoretical Reaction*).

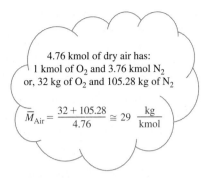

FIGURE 13.10 One kmol of O_2 is accompanied by 3.76 kmol of nitrogen in 4.76 kmol of air.

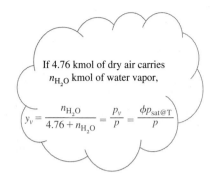

FIGURE 13.11 The Dalton model for gas mixture can be used to find n_{H_2O} in moist air.

possibility of chemical changes. The complexity associated with the change of phase of a condensed fuel is somewhat simplified by neglecting the volume occupied by its condensed phase in a mixture of condensed fuel and air.

13.1.3 Air

Air is the most common oxidizer, although pure oxygen is sometimes used in specialized applications such as rocket engines, acetylene torches for welding, etc.

As discussed in the last chapter, dry air can be looked upon as an ideal gas (IG) mixture of oxygen and nitrogen with mole fractions $y_{O_2} = 0.21$ and $y_{N_2} = 0.79$. A kmol of air can be represented by the formula $(0.21O_2 + 0.79N_2)$, resulting in an effective molar mass (Sec. 11.4.1) of approximately $\overline{M}_{Air} = 29$ kg/kmol (a more precise value of 28.97 kg/kmol is obtained when the presence of heavier components, such as carbondioxide and argon, is factored in). The oxygen in air is the primary chemical reactant. Air (Fig. 13.10) is often expressed in terms of unit mole of oxygen: $O_2 + (0.79/0.21)N_2$ or $O_2 + 3.76N_2$ represents the composition of 4.76 kmol of air.

The presence of water vapor in air can be taken into account (Fig. 13.11) by modifying the formula of air to $O_2 + 3.76N_2 + n_{H_2O}H_2O$, where n_{H_2O} can be related to the relative humidity of moist air at pressure p and temperature T by applying Eqs. (11.96) and (12.5):

$$\frac{n_{H_2O}}{n_{Air} + n_{H_2O}} = \frac{p_v}{p}; \quad \Rightarrow \quad \frac{n_{H_2O}}{4.76 + n_{H_2O}} = \frac{\phi p_{sat@T}}{p}; \tag{13.7}$$

$$\Rightarrow \quad n_{H_2O} = 4.76 \frac{\phi p_{sat@T}}{p - \phi p_{sat@T}}; \quad [\text{kmol}] \tag{13.8}$$

Even for saturated air at atmospheric temperature, n_{H_2O} is rather small (for instance, we can calculate $n_{H_2O} = 0.15$ for saturated air at 1 atm and 25°C); therefore, air will be represented by dry air in combustion analysis, unless mentioned otherwise.

When air supplied to a reaction carries just enough oxygen for *complete* combustion, the reaction is said to be *stoichiometric*, or **theoretical**, (Fig. 13.12 and Anim. 13.B.*theoreticalReaction*) and the amount of air supplied as oxidizer is called **theoretical air**. In a theoretical reaction, all the combustible elements are fully oxidized and there is no left over fuel or oxygen. The theoretical amount of air, however, does not guarantee a theoretical reaction as the reaction may not be *complete*.

To illustrate how the theoretical air is calculated, let us revisit Eq. (13.2) with the oxygen replaced by air as the oxidizer. Recall that each kmol of O_2 in air is accompanied by 3.76 kmol of N_2. The theoretical reaction can be written:

$$\underbrace{H_2}_{1 \text{ kmol}} + \underbrace{0.5(O_2 + 3.76N_2)}_{(0.5)(1 + 3.76) = 2.38 \text{ kmol}} \rightarrow H_2O + 1.88N_2 \tag{13.9}$$

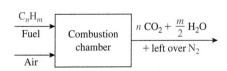

FIGURE 13.12 A reaction is theoretical if it is *complete* and there is no leftover fuel or oxygen (see Anim. 13.B.*airFuelRatio*).

The theoretical amount of air per unit mole of fuel, ν_{Air} in kmol per kmol of fuel, is called the **molar theoretical air-fuel ratio** $(\overline{AF})_{th}$. For the above reaction, $(\overline{AF})_{th} = \nu_{Air} = 2.38$ kmol/(kmol of fuel). The bar is used on the symbol for molar air-fuel ratio to be consistent with our convention of representing any mole-based specific property. On a unit fuel mass basis, the reaction can be reinterpreted:

$$\underbrace{H_2}_{2/2=1 \text{ kg}} + \underbrace{0.5(O_2 + 3.76N_2)}_{(0.5)(4.76)(29)/2 = 34.51 \text{ kg}} \rightarrow H_2O + 1.88N_2 \qquad (13.10)$$

The theoretical mass of air per-unit mass of fuel, η_{Air}, is the mass-based **theoretical air-fuel ratio** $(AF)_{th}$, which is calculated as 34.51 kg/kg fuel for this reaction. A slightly different value for $(AF)_{th}$ results if the mass of air is calculated by summing the mass of oxygen and nitrogen:

$$\eta_{O_2} + \eta_{N_2} = (0.5)(32)/2 + (0.5)(3.76)(28)/2 = 34.32 \text{ kg/kg fuel}.$$

This discrepancy is due to assigning a molar mass of 29 kg/kmol to air. We will stick to the practice of obtaining mass of air from $\nu_{Air}\overline{M}_{Air}$ rather than adding the masses of its components, oxygen and nitrogen.

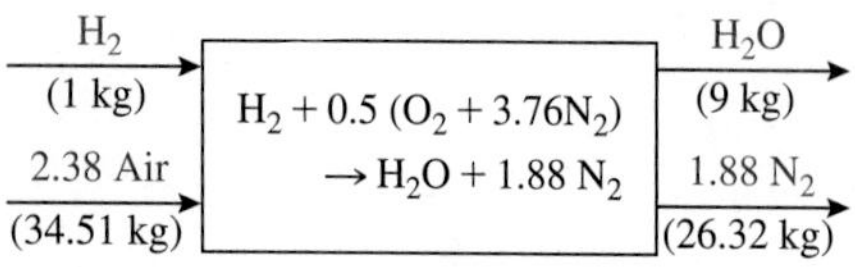

Air-fuel ratio can also be expressed on a mass basis as the minimum mass of air necessary for complete combustion of unit mass of fuel. For a generic reaction, the two kinds of air-fuel ratios, $\overline{AF}$ and AF, are defined and related (Fig. 13.13) to each other as follows:

$$\overline{AF} \equiv \frac{n_{Air}}{n_F} = \frac{\dot{n}_{Air}}{\dot{n}_F} = \frac{\nu_{Air}}{\nu_F(=1)} = \nu_{Air} \quad \left[\frac{\text{kmol Air}}{\text{kmol Fuel}}\right] \qquad (13.11)$$

$$AF \equiv \frac{m_{Air}}{m_F} = \frac{\dot{m}_{Air}}{\dot{m}_F} = \frac{\eta_{Air}}{\eta_F(=1)} = \eta_{Air} = \frac{\nu_{Air}\overline{M}_{Air}}{\nu_F\overline{M}_F} \quad \left[\frac{\text{kg Air}}{\text{kg Fuel}}\right] \qquad (13.12)$$

$$\text{Therefore,} \quad AF = \eta_{Air} = \frac{\nu_{Air}\overline{M}_{Air}}{\overline{M}_F} = \overline{AF}\left(\frac{\overline{M}_{Air}}{\overline{M}_F}\right) \quad \left[\frac{\text{kg Air}}{\text{kg Fuel}}\right] \qquad (13.13)$$

FIGURE 13.13 In this theoretical reaction, AF = 34.51 and $\overline{AF}$ is 2.38. (see Anim. 13.B.*airFuelRatio*).

A reaction can be based on fixed amounts of fuel and air (known n's or m's) or on fixed flow rates of fuel and air (known $\dot{n}$'s or $\dot{m}$'s) into a combustion chamber. In either case, the air-fuel ratio can be obtained from a balanced reaction using Eqs. (13.11)–(13.13).

Air is not always supplied in *theoretical* proportion to fuel in a given reaction. Relative amounts of air in a reaction can be expressed in several ways. A fuel-air mixture is said to be **lean**, **theoretical**, or **rich**, depending on whether the air-fuel ratio is greater than, equal to, or less than the theoretical air-fuel ratio. A quantitative way to express the relative amount of air is through fraction, or **percent of theoretical air**, defined as:

$$\lambda \equiv \frac{AF}{(AF)_{th}} = \left[\overline{AF}\left(\frac{\overline{M}_{Air}}{\overline{M}_F}\right)\right] \bigg/ \left[\overline{AF}\left(\frac{\overline{M}_{Air}}{\overline{M}_F}\right)\right]_{th} = \frac{\overline{AF}}{(\overline{AF})_{th}} \qquad (13.14)$$

A 200% theoretical air, or $\lambda = 2$, in the oxidation of hydrogen means that $2 \times 2.38 = 4.76$ kmol of air per kmol of fuel, or $2 \times 34.51 = 69.0$ kg of air per kg of hydrogen, is used in the actual reaction. Sometimes the actual amount of air is expressed through **percent excess** or **percent deficient** air relative to the theoretical amount. They can be related to the percent theoretical air by:

$$\lambda = 1 + (\% \text{ excess air})/100; \quad [\text{lean combustion: } \lambda > 1] \qquad (13.15)$$

$$\lambda = 1 - (\% \text{ deficient air})/100; \quad [\text{rich combustion: } \lambda < 1] \qquad (13.16)$$

Another way to describe an actual fuel-air mixture is through the **equivalence ratio**, ϕ, which is defined as the ratio of actual fuel-air ratio to the theoretical fuel-air ratio, and is simply the reciprocal of the *percent theoretical air*:

$$\phi \equiv \frac{1}{\lambda} \qquad (13.17)$$

A rich mixture, therefore, is characterized by $\phi > 1$ and a lean mixture by $\phi < 1$, with $\phi = \lambda = 1$ representing a *theoretical* mixture (see Anim. 13.B.*airFuelRatio*).

EXAMPLE 13-2   **Air-Fuel Ratio**

Determine the mass-based air-fuel ratio for the complete combustion of methane if the air supplied is: (a) theoretical amount, (b) 150% theoretical air. *What-if scenario:* What would the answer in part (a) be if the fuel was (c) octane or (d) dodecane?

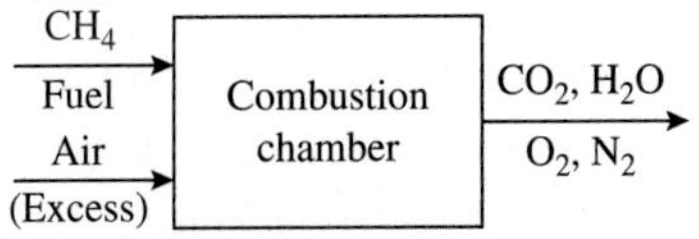

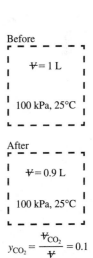

CH_4
Fuel $\longrightarrow$ | Combustion chamber | $\longrightarrow$ CO_2, H_2O
Air $\longrightarrow$ | | O_2, N_2
(Excess)

FIGURE 13.14 Schematic for Ex. 13-2.

SOLUTION

Balance the theoretical and actual reaction.

Assumptions

In air, each kmol of oxygen is accompanied by 3.76 kmol of nitrogen, which does not participate in the reaction. Air has a molar mass of 29 kg/kmol.

Analysis

The theoretical reaction is expressed in terms of 1 kmol of fuel and unknown mole of air and products components:

$$CH_4 + a(O_2 + 3.76N_2) \rightarrow bCO_2 + cH_2O + dN_2$$

The four atom balance equations are:

$$C: 1 = b; \quad H: 4 = 2c; \quad O: 2a = 2b + c; \quad N: 7.52a = 2d$$

Solving, the theoretical reaction can be written:

$$CH_4 + 2(O_2 + 3.76N_2) \rightarrow CO_2 + 2H_2O + 7.52N_2 \qquad \textbf{(13.18)}$$

The mass-based air-fuel ratio can now be obtained from Eq. (13.12):

$$(AF)_{th} = \frac{m_{Air}}{m_F} = \frac{\nu_{Air}\overline{M}_{Air}}{\nu_F \overline{M}_F} = \frac{(2 \times 4.76)(29)}{(1)(16)} = 17.26 \frac{\text{kg of air}}{\text{kg of fuel}}$$

For 150% theoretical air (which can be described by $\lambda = 1.5$ or 50% excess air), the air-fuel ratio can be obtained from Eq. (13.14) as:

$$AF = \lambda(AF)_{th} = 1.5(AF)_{th} = 1.5(17.26) = 25.89 \frac{\text{kg of air}}{\text{kg of fuel}}$$

TEST Analysis

Launch the IG non-premixed open-steady combustion TESTcalc. In the reaction panel, select methane in the fuel block and execute Theoretical Combustion with Air from the action menu to display the balanced reaction on the basis of 1 kmol of fuel. To obtain the air-fuel ratio on a mass basis, click on the Mass radio button, then execute Normalize Reaction from the action menu. The mass of air is displayed as 17.2 kg, the desired air-fuel ratio. Now enter Lambda = 1.5 and execute Excess/Deficient Air from the action menu. The new air-fuel ratio is calculated as 25.8. To separate air into its component, execute Air → O2, N2 from the action menu.

What-if scenario

In the fuel block, deselect methane by clicking its checkbox and select octane from the fuel menu. From the action menu, select Theoretical Air and the mass of air is displayed as 15.1 kg. Similarly, for dodecane, calculate the air-fuel ratio as 15.01 kg of air/kg of fuel.

Discussion

The air-fuel ratio is probably the most critical parameter in controlling combustion. Due to the difficulty associated with measuring the fuel and air flow rates, an indirect approach is to balance a reaction based on the analysis of combustion products, which is discussed next.

Before

$\Psi = 1$ L

100 kPa, 25°C

After

$\Psi = 0.9$ L

100 kPa, 25°C

$$y_{CO_2} = \frac{\Psi_{CO_2}}{\Psi} = 0.1$$

FIGURE 13.15 In an Orsat analyzer, the mole fraction of CO_2 can be found by comparing the volume of combustion gases, before and after removing CO_2 through absorption by a suitable chemical.

13.1.4 Combustion Products

When combustion of a hydrocarbon fuel is *complete,* the products contain CO_2, H_2O, N_2, and either O_2 or unburned fuel, depending on whether excess ($\lambda > 1$) or deficient ($\lambda < 1$) air is used in the reaction. In a *theoretical* reaction ($\lambda = 1$) there is no leftover oxygen or fuel in the products. Complete combustion, however, is rare for various reasons—insufficient time for reaction, poor mixing between fuel and air, incorrect air-fuel ratio, and heat loss are a few examples. But, as will be discussed in the next chapter (see Anim. 14.D.*equilibriumReaction*), the overriding reason for incomplete combustion is the quest for equilibrium by a reacting system. Dictated by the second law of thermodynamics, numerous species are formed and consumed by a large number of simultaneous reactions during combustion, resulting in a huge number of components (some in trace amounts) in the combustion products.

Products of incomplete combustion can be many, but a few are quite prominent. Carbon monoxide, unburned fuel, soot and other particulates, which are definite health hazards, are present in significant amounts in the products of rich combustion ($\lambda < 1$). On the other hand, various oxides of nitrogen—collectively known as NOX, a major source of atmospheric pollution (smog)—are formed in lean combustion ($\lambda > 1$). Measurements by devices such as an the Orsat analyzer (see Fig. 13.15), gas chromatograph, infrared analyzer, and flame ionization detector are used to determine the volume fractions (and mole fractions) of the gaseous products. The analysis of the data, known as **dry products analysis**, is reported on the basis of 100 kmol of dry products (after H_2O has been removed). As will be shown in the next example, the actual overall reaction can be obtained through atom balance equations if the dry products analysis is given (see Anim. 13.B.*productsAnalysis*).

Once a reaction is balanced, the actual air-fuel ratio and λ, the percent air used in the reaction, can be deduced by comparing the actual reaction with the corresponding theoretical one. The vapor pressure of water, dew point temperature, and relative humidity of the products can be obtained from the following relations (derived in Chapter 12):

$$p_v = \frac{\nu_{H_2O}}{\sum_p \nu_p} p; \quad [\text{kPa}], \quad T_{dp} = T_{\text{sat}@p_v}; \quad [°C], \quad p_v = \phi p_{\text{sat}@T}; \quad [\text{kPa}] \qquad \textbf{(13.19)}$$

In these relations, p and T are the pressure and temperature of the products mixture. If the products temperature falls below the dew point temperature T_{dp}, condensation begins (Sec. 12.5.3). The condensate becomes acidic by absorbing combustion gases and can harm the exhaust system.

EXAMPLE 13-3 **Fuel and Products Analysis**

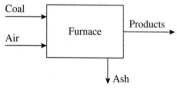

FIGURE 13.16 Schematic for Ex. 13-3.

Coal from a certain mine has the following ultimate analysis on a dry basis in percent by mass: Sulfur, 1.0%; Hydrogen, 5.8%; Carbon, 80.0%; Oxygen, 10.0%; Ash: 3.2%. The coal is burned in a furnace (Fig. 13.16) with an unknown amount of air. Analysis of the products on a dry basis produces the following distribution in percent by volume: CO_2, 13.09%; CO, 1.46%; SO_2, 0.07%; O_2, 5.0%; N_2, 80.38%. Calculate air-fuel ratio on a mass basis.

SOLUTION

Balance the reaction based on 100 kg of fuel.

Assumptions

In air, each kmol of oxygen is accompanied by 3.76 kmol of nitrogen, which does not participate in the reaction. Air has a molar mass of 29 kg/kmol. Ash is assumed to be a non-participant agent.

Analysis

100 kg of coal has the following molar composition:

$$\left(\frac{80}{\overline{M}_C}\right)C + \left(\frac{5.8}{\overline{M}_{H_2}}\right)H_2 + \left(\frac{10}{\overline{M}_{O_2}}\right)O_2 + \left(\frac{1}{\overline{M}_S}\right)S$$

$$= 6.67C + 2.90H_2 + 0.31O_2 + 0.03S$$

In an ideal gas mixture, the volume fraction and mole fraction of a component are identical. Therefore, a volumetric analysis of the products is the same as a molar analysis. The reaction can be expressed in terms of 3 unknown coefficients a, b, and c:

$$[6.67C + 2.90H_2 + 0.31O_2 + 0.03S] + a(O_2 + 3.76N_2)$$
$$\rightarrow b(13.09CO_2 + 1.46CO + 0.07SO_2 + 5.00O_2 + 80.38N_2) + cH_2O$$

Three of the four atom balance equations are:

$$C: 6.67 = (13.09 + 1.46)b; \quad H: 5.8 = 2c; \quad N: 7.52a = 160.76b$$

Solving: $a = 9.795$, $b = 0.458$, and $c = 2.9$. The extra equations, arising from oxygen and sulfur balance, can be used to check the solution. The mass of air used can be calculated as $a(4.76)(29) = 1352$ kg per 100 kg of coal.

Therefore, $AF = 13.52$ kg of air/kg fuel.

TEST Analysis

The combustion TESTcalc cannot be used to balance this reaction as it cannot handle partial description of both fuel and products compositions. For a pure fuel, the Balance Reaction command can be used after entering the dry products composition based on 100 kmol of products. Also, a manually balanced reaction can be entered in the reaction panel for further analysis by running the command Read As Is from the action menu.

Discussion

When released from the smoke stacks of coal burning power plants, sulfur dioxide reacts with rain water forming sulfuric acid (H_2SO_4), which has been strongly linked with acid rain downwind of industrial areas.

EXAMPLE 13-4 Products Analysis

Methane, CH_4, is burned (Fig. 13.17) with moist air at 100 kPa, 25°C, and 50% relative humidity. The molar analysis of the products on a dry basis is CO_2, 8.70%; CO, 0.46%; O_2, 4.80%; N_2, 86.04%. Calculate (a) the molar air-fuel ratio, (b) the percent theoretical air used, (c) the equivalence ratio, and (d) the dew point temperature of the products if the pressure is 100 kPa.

FIGURE 13.17 Schematic for Ex. 13-4 (see Anim. 13.B.*productsAnalysis*).

SOLUTION

Balance the reaction based on the given dry products analysis and humidity of the air.

Assumptions

In air, each kmol of oxygen is accompanied by 3.76 kmol of nitrogen, which does not participate in the reaction. Air has a molar mass of 29 kg/kmol. Air and the products mixture can be modeled by the IG mixture model.

Analysis

The overall reaction can be written on the basis of 100 kmol of dry products, leaving all other (four) coefficients unknown:

$$aCH_4 + b(O_2 + 3.76N_2) + cH_2O \rightarrow 8.7CO_2 + 0.46CO + 4.8O_2 + 86.04N_2 + dH_2O$$

The four atom balance equations are:

$$C: a = 8.7 + 0.46; \quad H: 4a + 2c = 2d;$$
$$O: 2b + c = 27.46 + d; \quad N: 7.52b = 172.08$$

However, only three of these are independent, resulting in $a = 9.16$, $b = 22.88$, and $d - c = 18.32$. The missing link is supplied by Eq. (13.7), relating vapor mole fraction in moist air with its partial pressure:

$$\frac{n_{H_2O}}{n_{Air} + n_{H_2O}} = \frac{p_v}{p}; \quad \Rightarrow \quad \frac{c}{4.76b + c} = \frac{\phi p_{sat@25°C}}{p};$$

$$\Rightarrow \quad \frac{c}{108.91 + c} = \frac{(0.5)(3.16)}{100}; \quad \Rightarrow \quad c = 1.75$$

The balanced equation after normalizing (dividing by a) becomes:

$$CH_4 + 2.50(O_2 + 3.76N_2) + 0.19H_2O$$

$$\rightarrow 0.95CO_2 + 0.05CO + 0.52O_2 + 9.39N_2 + 2.19H_2O \qquad \text{(13.20)}$$

The air-fuel ratio only involves dry air and fuel.

Therefore, $\overline{AF} = (2.5)(4.76) = 11.9 \dfrac{\text{kmol of air}}{\text{kmol of fuel}}$

The theoretical reaction between methane and dry air, derived in Ex. 13-2, is:

$$CH_4 + 2(O_2 + 3.76N_2) \rightarrow CO_2 + 2H_2O + 7.52N_2 \qquad \text{(13.21)}$$

Comparing the theoretical reaction, Eq. (13.21), with the actual reaction, Eq. (13.20), the percent theoretical air can be obtained as $\lambda = 2.5/2.0 = 125\%$. The equivalence ratio is obtained from its definition as $\phi = 1/\lambda = 1/1.25 = 0.8$:

Using Eq. (13.19):

$$\frac{\nu_{H_2O}}{\sum\limits_{p} \nu_p} = \frac{p_v}{p}; \quad \Rightarrow \quad p_v = (100)\frac{2.19}{13.1} = 16.72 \text{ kPa};$$

Therefore, $T_{dp} = T_{sat@16.72\,kPa} = 56.2°C$

TEST Analysis

With dry air as the oxidizer, the reaction could be balanced with the help of the IG non-premixed combustion TESTcalc. Select methane in the fuel block, populate the products block with CO2, CO, O2, and N2. Execute Balance Reaction from the action menu and the reaction is balanced. Execute Normalize to express the reaction in terms of 1 kmol of fuel. For the theoretical reaction, after selecting the fuel, all that is necessary is to execute Theoretical Reaction from the action menu. The amount of water vapor in the reactants or the dew point temperature must be calculated manually.

Discussion

The presence of moisture in the air does not greatly affect the dew point of the products. With completely dry air, the vapor pressure can be recalculated as 15.49 kPa, yielding a dew point temperature of 54.62°C. This example illustrates that most of the moisture in the products comes from the oxidation of hydrogen, not from the moisture present in the air.

13.2 SYSTEM ANALYSIS

The governing balance equations—mass, energy, and entropy equations, derived in Chapter 2, for a generic, open, unsteady system—have been applied without modifications to a variety of applications in the subsequent chapters. Simplified forms of these equations were achieved by incorporating specific system characteristics—steady vs. unsteady, closed vs. open, etc.—without affecting the basic framework. Using the same approach, we will customize the governing equations for a reacting system and use them to analyze practical combustion systems, primarily open combustion chambers operating at steady state and combustion processes occurring in closed rigid chambers.

13.3 OPEN-STEADY DEVICE

Fuel and oxidizer enter separately, or as a mixture, through a single port in a combustion chamber, which is basically a mixing chamber with some sort of ignition device. The products generally leave through a single exit port as a mixture, which is often modeled as an ideal gas mixture. With the reaction represented by the generic reaction $\sum\limits_{r} \nu_r W_r \rightarrow \sum\limits_{p} \nu_p W_p$, both the reactant and product mixtures can be split into equivalent flows of their components as shown in Figure 13.18, without losing any generality. If the mixture flowing through a port can be treated as an ideal gas mixture, application of the Dalton model will require each component to flow at the temperature of the mixture with a pressure equal to the partial pressure of the component at the port (Sec. 11.4.4). With this equivalence, the summations over inlets and exits of the transport terms of the governing equations can be replaced by summations over the components.

The mass equation at steady state, Eq. (2.5), can be expressed:

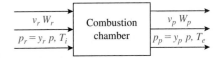

$$\sum_i \dot{m}_i = \sum_e \dot{m}_e; \quad \Rightarrow \quad \sum_r \dot{m}_r = \sum_p \dot{m}_p; \quad \left[\frac{\text{kg}}{\text{s}}\right] \tag{13.22}$$

$$\Rightarrow \quad \sum_r \left(\frac{\dot{m}_r}{\dot{m}_F}\right) = \sum_p \left(\frac{\dot{m}_p}{\dot{m}_F}\right); \quad \Rightarrow \quad \sum_r \eta_r = \sum_p \eta_p; \quad \left[\frac{\text{kg}}{\text{kg Fuel}}\right] \tag{13.23}$$

FIGURE 13.18 Using the Dalton model, gaseous reactants and products can be split into components flowing at their respective partial pressures for mixtures of ideal gases (click on any component formula in Anim. 13.B.*productsPropertiesIG*).

where $\dot{m}_k = \dot{m}_F \eta_k$ relates the mass flow rate of a component with the mass flow rate of the fuel and the mass-based stoichiometric coefficient in the overall combustion reaction as expressed by Eq. (13.6). The steady-state mass equation can be converted into molar form:

$$\sum_i \dot{m}_i = \sum_e \dot{m}_e; \quad \Rightarrow \quad \sum_r \dot{n}_r \overline{M}_r = \sum_p \dot{n}_p \overline{M}_p; \quad \left[\frac{\text{kg}}{\text{s}}\right] \tag{13.24}$$

$$\Rightarrow \quad \sum_r \left(\frac{\dot{n}_r}{\dot{n}_F}\right) \overline{M}_r = \sum_p \left(\frac{\dot{n}_p}{\dot{n}_F}\right) \overline{M}_p; \quad \Rightarrow \quad \sum_r \nu_r \overline{M}_r = \sum_p \nu_p \overline{M}_p; \quad \left[\frac{\text{kg}}{\text{kmol Fuel}}\right] \tag{13.25}$$

where $\dot{n}_k = \dot{n}_F \nu_k$ relates mole flow rate of component k with the stoichiometric coefficient and the mole flow rate of the fuel. The molar form of the mass equation makes it clear that the mole flow rate is not conserved, that is, $\sum_i \dot{n}_i$ at the combustion chamber inlet need not be equal to $\sum_e \dot{n}_e$ at the exit (see Anim. 13.B.*massBalanceEqn*).

To customize the energy equation, Eq. (2.8), we set the unsteady term to zero, neglect ke and pe, and express the equation on the basis of unit mass of fuel (see Anim. 13.B.*energyBalanceEqn*).

$$\text{Energy (Mass Based):} \quad 0 = \sum_r \dot{m}_r h_r - \sum_p \dot{m}_p h_p + \dot{Q} - \dot{W}_{\text{ext}}; \quad [\text{kW}]$$

$$\Rightarrow \quad 0 = \dot{m}_F \sum_r \left(\frac{\dot{m}_r}{\dot{m}_F}\right) h_r - \dot{m}_F \sum_p \left(\frac{\dot{m}_p}{\dot{m}_F}\right) h_p + \dot{Q} - \dot{W}_{\text{ext}}; \quad [\text{kW}]$$

$$\Rightarrow \quad 0 = \underbrace{\sum_r \eta_r h_r}_{h_R} - \underbrace{\sum_p \eta_p h_p}_{h_P} + \frac{\dot{Q}}{\dot{m}_F} - \frac{\dot{W}_{\text{ext}}}{\dot{m}_F}; \quad \left[\frac{\text{kJ}}{\text{kg Fuel}}\right]$$

$$\Rightarrow \quad 0 = h_R - h_P + \frac{\dot{Q}}{\dot{m}_F} - \frac{\dot{W}_{\text{ext}}}{\dot{m}_F}; \quad \left[\frac{\text{kJ}}{\text{kg Fuel}}\right] \tag{13.26}$$

$$\text{where} \quad h_R \equiv \sum_r \eta_r h_r; \quad \text{and} \quad h_P \equiv \sum_p \eta_p h_p; \quad \left[\frac{\text{kJ}}{\text{kg Fuel}}\right]$$

The molar form of the energy equation can be obtained by substituting expressions for η_r and η_p from Eq. (13.6) in the above equation, and using Eq. (1.29) to relate molar specific properties with the corresponding mass based specific properties: $\overline{h}_k = \overline{M}_k h_k$ and $\dot{m}_F = \dot{n}_F \overline{M}_F$:

$$0 = \overline{h}_R - \overline{h}_P + \frac{\dot{Q}}{\dot{n}_F} - \frac{\dot{W}_{\text{ext}}}{\dot{n}_F}; \quad \left[\frac{\text{kJ}}{\text{kmol Fuel}}\right] \tag{13.27}$$

$$\text{where} \quad \overline{h}_R \equiv \sum_r \nu_r \overline{h}_r; \quad \text{and} \quad \overline{h}_P \equiv \sum_p \nu_p \overline{h}_p; \quad \left[\frac{\text{kJ}}{\text{kmol Fuel}}\right]$$

Recall that $\overline{h}_k = \overline{h}_k(p_k, T)$, representing the specific molar enthalpy of a particular reactants or products component, $\overline{h}_r$ or $\overline{h}_p$ in Eq. (13.27), is called the *partial molar specific enthalpy* of component k (Sec. 11.4.3). The reactants and products can be modeled by the IG mixture model, $\bar{h}_k(p_k,T) = \bar{h}_k(T)$, where $\bar{h}_k(T)$ is the *molar specific enthalpy* of the pure gas k (as listed in the ideal gas table). Therefore, in combustion analysis involving gas mixtures, it is customary to assume that the partial specific enthalpy of component k, $h_k(p_k, T)$, is equal to the specific enthalpy of the pure gas k, $h_k(T)$ at the mixture temperature T.

In a similar manner, the entropy equation can be expressed in the following mass and mole based forms (see Anim. 13.B.*entropyBalanceEqn*):

$$\text{Mass based:} \quad 0 = s_R - s_P + \frac{\dot{Q}}{\dot{m}_F T_B} + \frac{\dot{S}_{gen}}{\dot{m}_F}; \quad \left[\frac{kJ}{(kg\ Fuel) \cdot K}\right] \qquad (13.28)$$

$$\text{Mole based:} \quad 0 = \bar{s}_R - \bar{s}_P + \frac{\dot{Q}}{\dot{n}_F T_B} + \frac{\dot{S}_{gen}}{\dot{n}_F}; \quad \left[\frac{kJ}{(kmol\ Fuel) \cdot K}\right] \qquad (13.29)$$

where $\quad s_R \equiv \sum_r \eta_r s_r(p_r,T_i) \quad$ and $\quad s_P \equiv \sum_p \eta_p s_p(p_p,T_e); \quad \left[\frac{kJ}{(kg\ Fuel) \cdot K}\right],$

$$\bar{s}_R \equiv \sum_r \nu_r \bar{s}_r(p_r,T_i) \quad \text{and} \quad \bar{s}_P \equiv \sum_p \nu_p \bar{s}_p(p_p,T_e); \quad \left[\frac{kJ}{(kmol\ Fuel) \cdot K}\right]$$

In these expressions, $\bar{s}_k(p_k, T)$ is the *partial molar specific entropy* (Sec. 11.4.3) of component k and is greater than $\bar{s}_k(p, T)$, the *molar specific enthalpy* of the pure gas at pressure p and temperature T since $p_k < p$. Likewise, $s_k(p_k, T)$ is the *partial specific entropy* of component k and is greater than the *specific enthalpy* $s_k(p, T)$ of the pure gas k. Browse Anims. 13.B.*componentPropertiesIG*, *partialPropertiesIG*, and *productsPropertiesIG* for a visual examination of pure, partial, and mixture properties.

Before these equations can be applied to a reacting system, a procedure must be established to evaluate partial specific enthalpy and entropy of a component in a mixture, taking into account the changes in chemical composition likely to occur in a chemical reaction.

13.3.1 Enthalpy of Formation

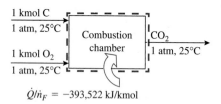

$\dot{Q}/\dot{n}_F = -393{,}522$ kJ/kmol

FIGURE 13.19 The negative sign of heat transfer (following the WinHip sign convention) indicates an exothermic reaction (see Anim 13.B.*formation Enthalpy*).

Consider the open-steady combustion chamber shown in Figure 13.19. Solid carbon and gaseous oxygen enter the chamber at 1 atm (about 0.1 MPa or 1 bar) and 25°C with their relative flow rates adjusted to *theoretical* ratio. Combustion is assumed to be *complete* and carbon dioxide is the only product. Heat transfer between the chamber and the surroundings ensures that the gaseous carbon dioxide, formed in the reaction, leaves the exit at 25°C. Assuming no pressure loss in the chamber, the exit pressure is same as that at the inlet, 1 atm. The overall reaction can be written:

$$\underbrace{C}_{1\ kg} + \underbrace{O_2}_{32/12\,=\,2.67\ kg} \rightarrow \underbrace{CO_2}_{44/12\,=\,3.67\ kg} \qquad (13.30)$$

where the mass-based stoichiometric coefficient η_k is indicated under each component. With no external work transfer, the heat transfer per kmol of fuel (carbon) is given by Eq. (13.27):

$$\frac{\dot{Q}}{\dot{n}_F} = \bar{h}_P - \bar{h}_R = \bar{h}_{CO_2} - \bar{h}_C - \bar{h}_{O_2}; \quad \left[\frac{kJ}{kmol\ Fuel}\right] \qquad (13.31)$$

We know from experience that $\dot{Q}$ must be leaving the chamber (and negative according to the WinHip sign convention) to keep the product at the same temperature as the reactants. However, evaluation of enthalpies with the material models developed in Chapter 3 runs into a contradiction. These models were developed in Sec. 3.7 with the assumption of fixed chemical composition. The Enthalpy of each substance was set to an arbitrary reference value (mostly 0) at the standard reference state (see Anim. 3.D.*propertiesIG*). Given that both the reactants and products in Eq. (13.30) are at the standard state, a simple application of the SL and PG model with the reference enthalpy of 0 to Eq. (13.31) will produce $Q = 0$, which is clearly inconsistent with the heat release (indicated by a large negative value) we expect in an exothermic reaction. In the mixing analysis of Sec. 11.4.4 this was not an issue because, in the absence of chemical reaction, the mass of each component was conserved and the reference enthalpies was always cancelled out in the energy equation. But in the context of a reacting system, reference properties must be standardized to account for transformation and creation of molecules.

To define a universal datum, that takes into account the possibility of changing chemical composition, we realize that chemical reactions preclude transformation of one element to another. Therefore, without losing generality, we can arbitrarily assign a zero value of enthalpy to all stable elements at the standard state of 1 atm and 25°C. Enthalpy of any pure substance at the standard state can then be obtained through an energy balance that takes into account the heat release in an exothermic reaction and heat absorption in an endothermic reaction. Known as the **enthalpy of formation**, the enthalpy per-unit mass or unit-mole of a substance at the reference state is represented by the symbol $h_{f,k}^o$ or $\bar{h}_{f,k}^o$, where the superscript "o" is used to represent the standard atmospheric pressure (1 atm). Thus, stable forms of hydrogen (H_2), nitrogen (N_2), oxygen (O_2), and carbon (C) have a zero enthalpy of formation. Enthalpy of formation of a compound can be related to heat transfer by an energy balance, such as Eq. (13.31). In fact, if the heat transfer in Eq. (13.31) could be precisely measured, $\dot{Q}/\dot{n}_F$ would be found as $-393{,}522$ kJ/kmol. Substituting this, and enthalpies of stable elements in Eq. (13.31), we obtain:

$$\bar{h}_{f,CO_2}^o = -393{,}522 + \bar{h}_{f,C}^{o\,0} + \bar{h}_{f,O_2}^{o\,0} = -393{,}522\ \frac{kJ}{kmol} \qquad \textbf{(13.32)}$$

This is the value listed for CO_2, along with formation enthalpies of many other substances, in Table G-1.

Enthalpy formulas from Chapter 3 can now be generalized by using the formation enthalpy at the standard state (Sec. 3.7) as the reference value (Fig. 13.20 and Anims. 13.B.*componentPropertiesSL* and *componentPropertiesIG*):

$$\bar{h}(p, T) = \bar{h}_f^o + [\bar{h}(p, T) - \bar{h}(p_0, T_0)]^{MM};\quad \left[\frac{kJ}{kmol}\right] \qquad \textbf{(13.33)}$$

In this definition of enthalpy of a pure substance, p_0 and T_0 represent the standard conditions; and superscript MM stands for the particular material model (SL, IG, or PG) used to evaluate the difference in enthalpy, called the **sensible enthalpy** (as opposed to latent), from the model-specific formulas or charts developed in Chapter 3. Note how Eq. (13.33) separates the chemical contribution from the thermal (sensible) contribution towards enthalpy of a substance.

In a mixture, the contribution from component k is called the *partial molar enthalpy* (Sec. 11.4.3) and for the IG model, Eq. (13.33), can be modified see Anim. 13.B.*partialPropertiesIG*:

$$\bar{h}_k^{IG}(p_k, T) = \bar{h}_k^{IG}(T) = \bar{h}_{f,k}^o + [\bar{h}_k(T) - \bar{h}_k(T_0)]^{IG} = \bar{h}_{f,k}^o + \Delta\bar{h}_k^{IG};\quad \left[\frac{kJ}{kmol\ of\ k}\right] \textbf{(13.34)}$$

where $\Delta\bar{h}_k^{IG}$, the sensible enthalpy change between temperature T and the standard temperature T_0, can be obtained from the IG tables (Table D's). For the PG mixture model, Eq. (13.34) simplifies further when Eq. (3.69) is applied to evaluate the sensible change in enthalpy:

$$\bar{h}_k^{PG}(p_k, T) = \bar{h}_k^{PG}(T) = \bar{h}_{f,k}^o + \Delta\bar{h}_k^{PG} = \bar{h}_{f,k}^o + \bar{c}_{p,k}(T - T_0);\quad \left[\frac{kJ}{kmol\ of\ k}\right] \textbf{(13.35)}$$

in which, $\bar{c}_{p,k}$, the molar specific heat of component k, is a material property that can be obtained from Table C ($\bar{c}_{p,k} = \bar{M}_k c_{p,k}$). For a solid or a liquid (mostly fuels or liquid water), substitution of Eq. (3.21) from the SL model produces:

$$\bar{h}_k^{SL}(p_k, T) = \bar{h}_k^{SL}(T) = \bar{h}_{f,k}^o + \Delta\bar{h}_k^{SL} = \bar{h}_{f,k}^o + \bar{c}_{p,k}(T - T_0);\quad \left[\frac{kJ}{kmol\ of\ k}\right] \textbf{(13.36)}$$

In using Eq. (3.21) for sensible enthalpy change, we assume the contribution from the term $v\Delta p$ to be negligible compared to the other terms, given the small value of v for a solid or a liquid (see Anim. 13.B.*componentPropertiesSL*). While evaluating the products' or reactants' enthalpies, in Eq. (13.27), liquid or solid fuels and condensed water in the products must be modeled by the SL model. In the SL model, and both PG and IG mixture models, partial specific enthalpy is a function of mixture temperature alone and is independent of partial or total pressure. Partial molar specific enthalpy, therefore, is same as the molar specific enthalpy in IG, PG, and SL models, $\bar{h}_k^{IG,PG,SL}(p_k, T) = \bar{h}_k^{IG,PG,SL}(T)$. The superscript "o" in $\bar{h}_{f,k}^o$ is retained as a reminder that the formation enthalpies are evaluated at the standard atmospheric pressure.

You may notice two entries for H_2O in Table G-1—one for liquid water, the equilibrium phase of H_2O at the standard state, and one for H_2O as an ideal gas at the standard temperature.

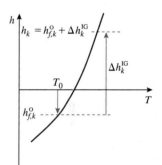

FIGURE 13.20 Specific enthalpy is the sum of formation enthalpy and sensible enthalpy evaluated from a particular material model such as the IG, PG, or SL model (see Anim. 13.B.*component PropertiesIG*).

The difference between the two can be verified as the enthalpy of vaporization (Sec. 5.1.4B), h_{fg} at T_0 (not at p_0 because enthalpy of an ideal gas is a function of temperature, not pressure).

13.3.2 Energy Analysis

Having developed a methodology to determine the enthalpy of a component in a reacting mixture, we are ready to begin a comprehensive energy analysis of reacting open-steady systems.

EXAMPLE 13-5 **Energy Analysis of a Combustion Chamber**

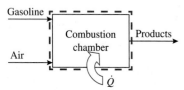

FIGURE 13.21 Schematic for Ex. 13-5. $\dot{Q}$ is negative following WinHip sign convention (see Sec. 0.7.2).

Gasoline enters a combustion chamber (Fig. 13.21), at 1 atm, 298 K with a rate of 0.08 kg/min, where it burns steadily and completely with 50% excess air that enters the chamber at 1 atm and 200 K. If the exit temperature of the combustion products is 600 K, determine (a) the mass flow rate of air and (b) the rate of heat transfer. Use the IG mixture model for gases and assume no pressure loss. **What-if scenario:** (c) What is the answer in part (b) if the excess air used is increased to 100%? Represent gasoline by liquid octane $C_8H_{18}(l)$.

SOLUTION

Perform a mass and energy balance on the open-steady device.

Assumptions

Each kmol of oxygen is accompanied by 3.76 kmol of nitrogen, which does not participate in the reaction. Air has a molar mass of 29 kg/kmol. The IG mixture model for gas mixtures and the SL model for liquid is applicable. No changes in ke and pe.

Analysis

The theoretical reaction is expressed in terms of 1 kmol of fuel and unknown amounts of air and products:

$$C_8H_{18}(l) + a(O_2 + 3.76N_2) \rightarrow bCO_2 + cH_2O + dN_2$$

The four atom balance equations produce $a = 12.5$, $b = 8$, $c = 9$, and $d = 47$; and the balanced reaction can be written:

$$C_8H_{18}(l) + 12.5(O_2 + 3.76N_2) \rightarrow 8CO_2 + 9H_2O + 47N_2$$

For 50% excess air, $\lambda = 1.5$, and the actual reaction can be set up as:

$$C_8H_{18}(l) + 18.75(O_2 + 3.76N_2) \rightarrow 8CO_2 + 9H_2O + 70.5N_2 + 6.25O_2$$

The mass-based air-fuel ratio can now be obtained from Eq. (13.13):

$$AF = \overline{AF}\left(\frac{\overline{M}_{Air}}{\overline{M}_F}\right) = (18.75)(4.76)\frac{29}{8 \times 12 + 18 \times 1} = 22.70 \; \frac{\text{kg of air}}{\text{kg of fuel}}$$

Therefore, the mass flow rate of air is:

$$\dot{m}_{Air} = (\dot{m}_F)(AF) = (0.08)(22.7) = 1.82 \; \frac{\text{kg of air}}{\text{min}}$$

From Eq. (13.27), the rate of heat transfer can be expressed:

$$\frac{\dot{Q}}{\dot{n}_F} = \sum_p \nu_p \bar{h}_p - \sum_r \nu_r \bar{h}_r + \frac{\dot{W}_{ext}^{\nearrow 0}}{\dot{n}_F} = \bar{h}_P - \bar{h}_R; \tag{13.37}$$

Enthalpy of the reactants (fuel at 298 K and air at 200 K) and products (at 600 K) can be evaluated using data from Table G, C and D (see Anim. 13.B.*completeSolution*). At 100 kPa, 600 K water is a superheated vapor; therefore, the formation enthalpy of $H_2O(g)$ is used in calculating the products enthalpy. The n-IG state TESTcalcs can also be used to obtain component and mixture properties.

Species	ν_k	$\bar{h}^o_{f,k}$	$\Delta\bar{h}^{IG}_k$ (kJ/kmol)	$\bar{h}_k$
$C_8H_{18}(l)$	1	−249,910	0	−249,910
O_2	18.75	0	−2,868	−2,868
N_2	70.5	0	−2,857	−2,857
			$\bar{h}_R = \sum_r \nu_r \bar{h}_r = -505,103$ kJ	
CO_2	8	−393,520	12,906	−380,614
$H_2O(g)$	9	−241,820	10,499	−231,321
N_2	70.5	0	8,894	8,894
O_2	6.25	0	9,245	9,245
			$\bar{h}_P = \sum_p \nu_p \bar{h}_p = -4,441,993$ kJ	

The mole flow rate of fuel:

$$\dot{n}_F = \frac{\dot{m}_F}{\bar{M}_F} = \frac{0.08}{(60)(114)} = 1.1696 \times 10^{-5} \frac{\text{kmol}}{\text{s}}$$

Therefore, the heat transfer rate in kW can be obtained from Eq. (13.37):

$$\dot{Q} = (1.1696 \times 10^{-5})[-4,441,993 - (-505,103)] = -46.05 \text{ kW}$$

TEST Analysis

Launch the open-steady non-premixed IG combustion TESTcalc. In the reaction panel, select octane(L) in the fuel block. Enter Lambda=1.5 and execute Excess/Deficient Air from the action menu to balance the reaction on the basis of 1 kmol of fuel. Convert the reaction to mass basis by clicking the Mass radio button and selecting Normalize from the action menu. Adjust the fuel mass to 0.08/60 = 0.001333 kg by inputting 0.00133 in the Scaling Factor box and selecting Multiply from the action menu. Now switch to the state panel and evaluate States 1, 2, and 3 for fuel, oxidizer (air), and products, respectively. The mass flow rates are automatically imported from the reaction panel. In the device panel, import the fuel and oxidizer states as i1 and i2 states and the products state as the e-state. Enter Wdot_ext=0 and click Calculate to produces Qdot = −45.8 kW, which is very close to the manual solution. As an alternative to the combustion TESTcalc, you could try the combustion Interactive (Interactives>Open Steady>Specifc branch) and see how much easier it is to solve the problem through its intuitive graphical interface.

What-if scenario

In the reaction panel, change Lambda to 2 (100% excess air) and run Excess Air from the action menu. Adjust the fuel mass, as before, and click Super-Calculate to produce the new heat transfer rate as −41.79 kW.

Discussion

An increase in the supply of excess air (from 50% to 100%) reduces the magnitude of heat transfer significantly. The additional air acts as a dead load that must also be heated to the exit temperature. In Figure 13.22, $\bar{h}_R$ and $\bar{h}_P$ are qualitatively plotted against temperature. $\dot{Q}/\dot{n}_F$ can be interpreted in this diagram as the vertical distance between the exit and inlet state.

The energy analysis used in Example 13-5 can be generalized for any *complete* reaction. In the mass-based energy equation, Eq. (13.26), the IG mixture model for enthalpy can be introduced to express the energy transfer:

$$\frac{\dot{Q}}{\dot{m}_F} - \frac{\dot{W}_{ext}}{\dot{m}_F} = h_P - h_R = \sum_p \eta_p(h^o_{f,p} + \Delta h^{IG}_p) - \sum_r \eta_r(h^o_{f,r} + \Delta h^{IG}_r)$$

$$= \underbrace{\sum_p \eta_p h^o_{f,p}}_{h_P(T_0)} - \underbrace{\sum_r \eta_r h^o_{f,r}}_{h_R(T_0)} + \sum_p \eta_p \Delta h^{IG}_p - \sum_r \eta_r \Delta h^{IG}_r$$

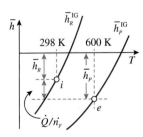

FIGURE 13.22 The energy equation, Eq. (13.37), in Ex. 13-5 interpreted on a $\bar{h}$-T diagram (click energy balance in Anim. 13.B.*completeSolution*).

$$\Rightarrow \quad \frac{\dot{Q}}{\dot{m}_F} - \frac{\dot{W}_{\text{ext}}}{\dot{m}_F} = \underbrace{h_P(T_0) - h_R(T_0)}_{\substack{\Delta h_C^\circ:\text{ Enthalpy of}\\\text{Combustion}}} + \underbrace{\sum_p \eta_p \Delta h_p^{\text{IG}} - \sum_r \eta_r \Delta h_r^{\text{IG}}}_{\text{Sensible Enthalpy Change}}; \quad \left[\frac{\text{kJ}}{\text{kg Fuel}}\right] \quad \textbf{(13.38)}$$

Note that while h_P and h_R are evaluated at the products and reactants temperatures, $h_P(T_0)$ and $h_R(T_0)$ are evaluated at standard atmospheric conditions. Dividing both sides by $\overline{M}_F$, and using Eq. (13.6), the molar form can be obtained. Alternatively, Eq. (13.27) can be used for this purpose:

$$\frac{\dot{Q}}{\dot{n}_F} - \frac{\dot{W}_{\text{ext}}}{\dot{n}_F} = \underbrace{\sum_p \nu_p \overline{h}_{f,p}^\circ}_{\overline{h}_P(T_0)} - \underbrace{\sum_r \nu_r \overline{h}_{f,r}^\circ}_{\overline{h}_R(T_0)} + \sum_p \nu_p \Delta\overline{h}_p^{\text{IG}} - \sum_r \nu_r \Delta\overline{h}_r^{\text{IG}}$$

$$\Rightarrow \quad \frac{\dot{Q}}{\dot{n}_F} - \frac{\dot{W}_{\text{ext}}}{\dot{n}_F} = \underbrace{\overline{h}_P(T_0) - \overline{h}_R(T_0)}_{\Delta\overline{h}_C^\circ = \Delta h_C \overline{M}_F} + \underbrace{\sum_p \nu_p \Delta\overline{h}_p^{\text{IG}} - \sum_r \nu_r \Delta\overline{h}_r^{\text{IG}}}_{\text{Molar Sensible Enthalpy Change}}; \quad \left[\frac{\text{kJ}}{\text{kmol Fuel}}\right] \quad \textbf{(13.39)}$$

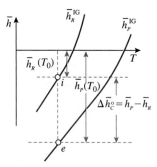

FIGURE 13.23 Heat of combustion interpreted graphically [see Eq. (13.39) and Anim. 13.B.*heatingValues*].

In these equations, the left-hand side is the net energy transfer per unit mass (or mole) of fuel into the reactor and must be negative for the exothermic combustion reaction. On the right-hand side, the change in enthalpies of the products and reactants are separated into two groups. The first group, Δh_C° (or its molar counterpart, $\Delta\overline{h}_C^\circ = \Delta h_C^\circ \overline{M}_F$), is a constant for a given reaction and is called the **enthalpy of combustion** for a complete reaction. The second group can be identified as the change in *sensible enthalpy,* which depends on the products and reactants temperatures. If the reactants and products are at their standard states, there is no contribution from the sensible component in Eq. (13.38). The *enthalpy of combustion,* therefore, can be interpreted as the heat transfer per unit mass (or mole) of fuel when the reactants and products of *complete* combustion are at standard conditions and external work transfer is absent. In an exothermic reaction heat is released; therefore, the enthalpy of combustion, $\Delta h_C^\circ = h_P(T_0) - h_R(T_0) = \dot{Q}/\dot{n}_F$, must be negative to be consistent with the WinHip sign convention of heat transfer. Likewise, the enthalpy of vaporization (Sec. 5.1.4B) is positive because heat is absorbed. The molar enthalpy of combustion $\Delta\overline{h}_C^\circ = \overline{M}\Delta h_C^\circ$ is graphically interpreted in Figure 13.23 by applying Eq. (13.39) to a combustion chamber with the reactants and products at their standard states.

A more general term for the enthalpy of combustion is the **heat of reaction**, or heat of combustion, defined as the difference between h_P and h_R (the vertical distance in Fig. 13.23) at a given T, that is $\Delta h_T^\circ(T) \equiv h_P(T) - h_R(T)$. The heat of reaction depends on the temperature, and its value at the standard temperature is the heat of combustion, $\Delta h_C^\circ = \Delta h_T^\circ(T_0)$. Perhaps the most commonly used term in this regard is the **heating value** of a fuel, which is the absolute value $|\Delta h_C^\circ|$ of its enthalpy of combustion. It is the heat released per unit mass of fuel during complete combustion at standard conditions. Although pressure has no effect on the heating value, the superscript "o" is consistently used to indicate standard pressure before and after the reaction. The use of excess or deficient air cannot change the heating value as long as the reaction is complete [we can deduce this from Eq. (13.38)]. However, the heating value is sensitive to the phase composition of water in the products; if water is assumed to be in the liquid phase, the use of $h_{f,\,H_2O(l)}^\circ$ for condensed water results in what is known as the **higher heating value (HHV)** of the fuel (see Anim. 13.B.*heatingValues*). In most situations the products contain water as superheated vapor and the use of $h_{f,\,H_2O(g)}^\circ$ results in the **lower heating value (LHV)** of the fuel. The difference between the two can be shown from Eq. (13.38):

$$\text{HHV} - \text{LHV} = \eta_{H_2O} h_{fg,H_2O}; \quad \left[\frac{\text{kJ}}{\text{kg Fuel}}\right] \quad \textbf{(13.40)}$$

Or in molar terms:

$$\overline{\text{HHV}} - \overline{\text{LHV}} = \nu_{H_2O} \overline{h}_{fg,H_2O}; \quad \left[\frac{\text{kJ}}{\text{kmol Fuel}}\right] \quad \textbf{(13.41)}$$

Heating values for several fuels are listed in Table G-2 and further explored in Anim. 13.B.*heatingValues*.

The large changes in temperature that typically accompany combustion justify the use of IG mixture model for air and the products. However, the simplicity of the PG and SL models makes them attractive whenever accuracy can be traded for mathematical insight. With the additional assumption of $\bar{c}_{p,k} = \bar{c}_p$, uniform specific heat across all components, the PG (and, when necessary, SL) model can be used to simplify Eq. (13.38) for a combustion chamber with inlet temperature T_i and exit temperature T_e:

$$\frac{\dot{Q}}{\dot{m}_F} - \frac{\dot{W}_{ext}}{\dot{m}_F} = \Delta h_C^o + \sum_p \eta_p \Delta h_p^{PG/SL} - \sum_r \eta_r \Delta h_r^{PG/SL}$$

$$= \Delta h_C^o + c_p(T_e - T_0)\sum_p \eta_p - c_p(T_i - T_0)\sum_r \eta_r$$

$$= \Delta h_C^o + c_p(T_e - T_i)\sum_r \eta_r; \quad \left[\frac{kJ}{kg\ Fuel}\right]$$

$$\Rightarrow \quad \frac{\dot{Q}}{\dot{m}_F} - \frac{\dot{W}_{ext}}{\dot{m}_F} = \Delta h_C^o + c_p(T_e - T_i)(1 + AF); \quad \left[\frac{kJ}{kg\ Fuel}\right] \quad \textbf{(13.42)}$$

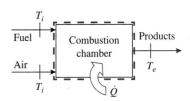

FIGURE 13.24 Combustion chamber used for derivation of Eq. (13.42).

Formulas for sensible enthalpy are identical for the PG and SL model, so condensed phases can be included as part of the reactants and products without affecting the final expression. To illustrate the usefulness of this expression, apply it to estimate the heat transfer calculated in Ex. 13-5. Substituting $c_p = 1.005$ kJ/kg·K (air, PG model), $T_e = 600$ K, $T_i = 200$ K, AF = 22.7 kg, and $\Delta h_C^o = -LLV = -44,430$ kJ/kg (Table G-2) yields $\dot{Q} = -46.5$ kW. This is off by only about 1% from the more sophisticated analysis of Ex. 13-5. Moreover, Eq. (13.42) gives us an analytical insight into the energy balance in a steady-state reactor, revealing how the net energy transfer depends upon the heating value, air-fuel ratio, specific heat, and the difference in temperature between the inlet and exit. To show that the enthalpy of combustion does not depend on air-fuel ratio, all we need to do is substitute $T_i = T_e = T_0$ and $\dot{W}_{ext} = 0$ in this equation.

As indicated by Eqs. (13.38) and (13.42), the energy released in combustion goes partly into raising the sensible enthalpy of the products while the rest is transferred out as heat and/or external work. If the energy transfer through heat and external work are reduced, the sensible enthalpy of the products and the exit temperature must increase. In the limiting case of an adiabatic combustion chamber, the entire amount of energy released must go into raising the sensible enthalpy so that the resulting temperature of the products reaches a peak, known as the **adiabatic flame temperature** T_{af}. To evaluate T_{af}, we simplify the energy equation, Eq. (13.38), by substituting $\dot{Q} = \dot{W}_{ext} = 0$:

$$h_P = h_R; \quad \left[\frac{kJ}{kg\ Fuel}\right]; \quad \text{or,} \quad \bar{h}_P = \bar{h}_R; \quad \left[\frac{kJ}{kmol\ Fuel}\right] \quad \textbf{(13.43)}$$

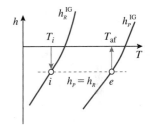

FIGURE 13.25 Adiabatic flame temperature graphically interpreted (see also Anim. 13.B. *adiabaticFlameTemp*).

For a given reaction, trends of h_P and h_R are plotted as functions of the inlet and exit temperatures in Figure 13.25. Starting with a known inlet temperature, h_P, h_R, and T_{af} can be graphically related as shown in this figure. It can be seen from this plot that T_{af} is the maximum temperature that can be obtained for a given inlet temperature T_i (fixing State-i) and a given reaction (that fixes the h_R curve).

The immense difference between T_i and T_{af} makes the IG mixture model more suitable than the PG mixture model for evaluating T_{af}. From the given states of the reactants, $\bar{h}_R$ can be evaluated using the procedure illustrated in Ex. 13-5. h_p on the other hand, depends on the products temperature, T_{af}, the desired unknown. A trial and error procedure, therefore, must be followed, in which $\bar{h}_P$ is calculated from successive guesses of T_{af} (by following the procedure used in Ex. 13-5) until it is approximately equal to $\bar{h}_R$.

Setting $\dot{Q}$ and $\dot{W}_{ext}$ to zero in Eq. (13.42), the PG model yields the following simplified expression for the adiabatic flame temperature:

$$T_{af}^{PG} = T_e = T_i + \frac{(-\Delta h_C^o)}{c_p(1 + AF)} = T_i + \frac{LHV}{c_p(1 + AF)}; \quad [K] \quad \textbf{(13.44)}$$

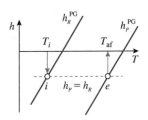

FIGURE 13.26 The PG model generally overestimates the value of T_{af} as enthalpy varies linearly with temperature (see Anim. 13.B. *adiabaticTempIGvsPG*).

Although this solution overly predicts T_{af} (due to constant c_p assumption), it clearly establishes the dependence of T_{af} on other parameters for a given reaction (Fig. 13.26 and Anim. 13.B. *adaiabaticTempParam*). The lower heating value is used for the calculation of T_{af} as there is no possibility of water condensation at such a high temperature. The dependence of T_{af} on the air-fuel ratio suggests that the peak temperature of combustion can be directly controlled by adjusting the amount of excess air in a reaction.

EXAMPLE 13-6 Adiabatic Flame Temperature

Methane enters an adiabatic combustion chamber at 1 atm, and 25°C and is burned with air that enters at the same conditions. Determine the adiabatic flame temperature for (a) the theoretical reaction, (b) complete combustion with 200% air and (c) incomplete combustion with 90% theoretical air producing some CO (but no leftover fuel). Use the PG mixture model with a constant c_p of 1 kJ/kg·K. **What-if scenario:** (d) What would the answers be if the IG mixture model was used instead?

SOLUTION

Obtain the perfect gas solution using Eq. (13.44) and use TESTcalc for the IG solution.

Assumptions

Each kmol of oxygen is accompanied by 3.76 kmol of nitrogen, which does not participate in the reaction. Air has a molar mass of 29 kg/kmol. The IG mixture model for gas mixtures and the SL model for liquid are applicable. No changes in ke and pe.

Analysis

The theoretical reaction between methane and air can be obtained through atom balance equations:

$$\underbrace{CH_4}_{16/16\,=\,1\,kg} + \underbrace{2(O_2 + 3.76N_2)}_{2(4.76)(29)/16\,=\,17.26\,kg} \rightarrow CO_2 + 2H_2O + 7.52N_2$$

From the fuel and air mass calculated above, $(AF)_{th} = 17.26$ kg/kg fuel. Using formation enthalpies from Table G-1, the enthalpy of combustion can be calculated:

$$\Delta \bar{h}_C^o = \sum_p \nu_p \bar{h}_{f,p}^o - \sum_r \nu_r \bar{h}_{f,r}^o = \bar{h}_{f,CO_2}^o + 2\bar{h}_{f,H_2O}^o - \bar{h}_{f,CH_4}^o$$

$$= -393{,}520 + 2(-241{,}820) - (-74{,}850) = -802{,}310 \text{ kJ/kmol}$$

Therefore, $\Delta h_C^o = \Delta \bar{h}_C^o / \bar{M}_{CH_4} = -802{,}310/16 = -50{,}144$ kJ/kg
Equation (13.44) yields:

$$T_{af}^{PG} = T_i + \frac{(-\Delta h_C^o)}{c_p(1 + AF)} = 25 + \frac{50{,}144}{(1)(1 + 17.26)} = 2771°C$$

To obtain the IG solution using TEST, launch the IG open-steady non-premixed combustion TESTcalc. Select methane in the fuel block and Theoretical Air from the action menu to obtain the complete reaction. In the state panel, evaluate the fuel (State-1) and air (State-2) states from the known pressure and temperature. For the products (State-3), leave temperature as an unknown. In the device panel, import the three states to the two inlet and one exit ports, enter Qdot and Wdot_ext as zero, and click Super-Calculate. Go back to State-3 to find T3, the adiabatic flame temperature, which is calculated as 2056 deg-C. Alternatively, the combustion Interactive, which is much easier to use, can be used to generate this solution.

(b) With 200% ($\lambda = 2$) theoretical air, $AF = 2 \times (AF)_{th} = 34.51$ kg/kg fuel. The enthalpy of combustion remains unchanged as long as combustion is complete:

Therefore, $T_{af}^{PG} = T_i + \frac{(-\Delta h_C^o)}{c_p(1 + AF)} = 25 + \frac{50{,}144}{(1)(1 + 34.51)} = 1437°C$

To obtain the IG solution with the combustion TESTcalc, change Lambda to 2, select Excess/Deficient Air from the action menu, and click Super-Calculate. All calculations are updated with T3 calculated as 1209 deg-C.

(c) With 90% theoretical air $AF = 0.9 \times (AF)_{th} = 15.53$ kg/kg fuel and the reaction with the addition of CO in the products can be balanced (Ex. 13-2):

$$\underbrace{CH_4}_{16/16\,=\,1\,kg} + \underbrace{1.8(O_2 + 3.76N_2)}_{1.8(4.76)(29)/16\,=\,15.53\,kg} \rightarrow 0.6CO_2 + 0.4CO + 2H_2O + 6.77N_2$$

Since the stoichiometric coefficients of participating components are different, the enthalpy of combustion has to be recalculated:

$$\Delta \bar{h}_C^o = \sum_p \nu_p \bar{h}_{f,p}^o - \sum_r \nu_r \bar{h}_{f,r}^o = 0.6\bar{h}_{f,CO_2}^o + 0.4\bar{h}_{f,CO}^o + 2\bar{h}_{f,H_2O}^o - \bar{h}_{f,CH_4}^o$$

$$= (0.6)(-393,520) + (0.4)(-47,540) + 2(-241,820) - (-74,850)$$

$$= -663,918 \text{ kJ/kmol}$$

Therefore,

$$\Delta h_C^o = \Delta \bar{h}_C^o / \bar{M}_{CH_4} = -663,918/16 = -41,495 \text{ kJ/kg}$$

and $\quad T_{af}^{PG} = T_i + \dfrac{(-\Delta h_C^o)}{c_p(1 + AF)} = 25 + \dfrac{41,495}{(1)(1 + 15.53)} = 2,535°C$

For the IG solution, set up the reaction by selecting 1 kmol of methane in the fuel block, 8.57 kmol of Air (90% of theoretical amount) as the oxidizer, and unknown amounts of CO2, H2O, N2, and CO as products. Run Balance Reaction from the action menu. With the reaction balanced, work on the state and device panels as described in part (a). Click Super-Calculate to produce T3 as 1944 deg-C.

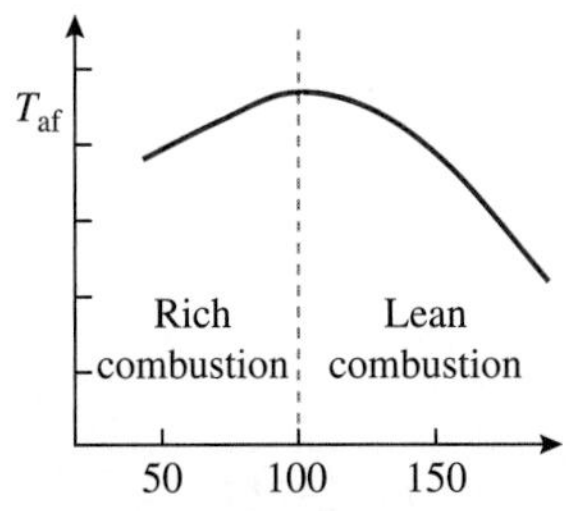

Discussion

The PG model, as expected, overpredicts the adiabatic flame temperature in all three situations when compared to the more accurate IG model. Despite this limitation, the trends are correctly reproduced by it. Both models show that T_{af} reaches a maximum for the theoretical reaction and decreases for both lean and rich combustion. By repeating the study over a wide range of air fuel ratios, this trend can be firmly established for any given reaction (Fig. 13.27).

FIGURE 13.27 T_{af} reaches a peak for the theoretical reaction. Use the combustion chamber simulator (TEST. Interactives) to reproduce this trend (switch to Known Heat Transfer tab, and follow steps 1 through 10.

13.3.3 Entropy Analysis

The difficulty faced with the reference value of enthalpy for a reacting substance does not arise for entropy. This is because the **third law** of thermodynamics, also known as the **Nernst-Simon postulate**, establishes the concept of absolute entropy. Originating from the studies of chemical reactions at low temperatures, the third law asserts that the entropy of a pure crystalline substance is zero at absolute zero temperature (0 K), thus providing a natural datum for entropy for all substances, reacting or non-reacting.

Absolute entropy:

$\bar{s}_k(p_0, T_0)$ of a substance at the standard state, p_0, T_0, can be obtained from precise measurement of energy transfers and specific heat data between absolute zero and the standard state by integrating the first Tds equation, Eq. (3.7), at constant pressure:

$$\bar{s}_k(p_0, T_0) - \bar{s}_k(p_0, 0)^0 = \int_{p_0,0}^{p_0,T_0} d\bar{s} = \int_{p_0,0}^{p_0,T_0} \frac{d\bar{h}}{T} = \int_{p_0,0}^{p_0,T_0} \frac{\bar{c}_p dT}{T}$$

$$\Rightarrow \quad \bar{s}_k(p_0, T_0) = \int_{p_0,0}^{p_0,T_0} \frac{\bar{c}_p dT}{T}; \quad \left[\frac{\text{kJ}}{(\text{kmol of } k) \cdot \text{K}} \right] \quad \textbf{(13.45)}$$

A shorter symbol such as $\bar{s}_k^o(T_0)$, or simply $\bar{s}_k^o$, is used in place of $\bar{s}_k(p_0, T_0)$, with the superscript indicating standard pressure (see Anim. 13.B.*componentPropertiesIG*). It is the third law that asserts the value of $\bar{s}_k^o(0)$ to zero in deriving Eq. (13.45). Also, as $T \geq 0$ so does $\bar{c}_p$, which keeps the integral from exploding. An alternative for evaluating this integral is to use computational chemistry to analyze molecular data. Table G-1 lists $\bar{s}_k^o(T_0)$ along with $h_{f,k}^o$ for many common combustion species. Note that these values are universal, independent of the material model. For a given state at p, T, the absolute entropy can be obtained from:

$$\bar{s}_k(p,T) = \bar{s}_k(p_0, T_0) + [\bar{s}_k(p,T) - \bar{s}_k(p_0, T_0)]; \quad \left[\frac{\text{kJ}}{(\text{kmol of } k) \cdot \text{K}} \right]$$

$$\Rightarrow \quad \bar{s}_k(p,T) = \underbrace{\bar{s}_k^o(T_0)}_{\substack{\text{Absolute entropy at standard state.}}} + \underbrace{[\bar{s}_k(p,T) - \bar{s}_k(p_0,T_0)]^{\text{MM}}}_{\substack{\text{Contributions from temperature and pressure} \\ \text{change obtained using a particular material model.}}} \quad \textbf{(13.46)}$$

where the absolute entropy at the standard state is obtained from Table G-1 and the change in entropy inside the bracketed term is evaluated by using a suitable material model—the IG or PG model for gases and the SL model for solids and liquids. In the IG model, the pressure contribution can be separated from the temperature contribution using Eq. (3.62):

$$\bar{s}_k^{\text{IG}}(p,T) = \underbrace{\bar{s}_k^o(T_0) + [\bar{s}_k(p_0,T) - \bar{s}_k(p_0,T_0)]^{\text{IG}}}_{\bar{s}_k^o(T)\text{: Temperature contribution}} + \underbrace{[\bar{s}_k(p,T) - \bar{s}_k(p_0,T)]^{\text{IG}}}_{\text{Pressure contribution}}$$

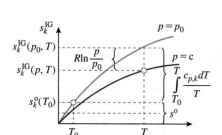

$$\Rightarrow \quad \bar{s}_k^{\text{IG}}(p,T) = \bar{s}_k^o(T) - \bar{R} \ln \frac{p}{p_0}; \quad \left[\frac{\text{kJ}}{(\text{kmol of } k) \cdot \text{K}} \right] \quad \textbf{(13.47)}$$

Equation (13.47) is essentially the same (except for a factor of $\overline{M}_k$) as Eq. (3.62), developed for a pure ideal gas. Different terms in this equation are graphically represented on the $T\text{-}\bar{s}$ diagram of Figure 13.28. To simplify the evaluation of entropy using the IG model, $\bar{s}_k^o(T)$ is listed as a function of T for various species in the ideal gas tables (Tables D-4–D-15). The pressure correction is added to $\bar{s}_k^o(T)$ to obtain specific entropy of a pure gas k at p, T from Eq. (13.47). For the PG model, the use of tables can be avoided by directly evaluating the temperature's contribution (Sec. 3.5.3):

$$\bar{s}_k^{\text{PG}}(p,T) = \bar{s}_k^o(T_0) + \bar{c}_{p,k} \ln \frac{T}{T_0} - \bar{R} \ln \frac{p}{p_0}; \quad \left[\frac{\text{kJ}}{(\text{kmol of } k) \cdot \text{K}} \right] \quad \textbf{(13.48)}$$

FIGURE 13.28 Graphical explanation of Eq. (13.47). See Anim. 13.B.*component PropertiesIG.*

In the SL model (Sec. 3.3), there is no contribution from pressure and $\bar{c}_{v,k} = \bar{c}_{p,k}$, simplifying the entropy expression to:

$$\bar{s}_k^{\text{SL}}(p,T) = \bar{s}_k^o(T_0) + \bar{c}_{p,k} \ln \frac{T}{T_0}; \quad \left[\frac{\text{kJ}}{(\text{kmol of } k) \cdot \text{K}} \right] \quad \textbf{(13.49)}$$

Equations (13.48) and (13.49) are also same (except for a factor of $\overline{M}_k$) as the expressions derived in Chapter 3 for the PG and SL models, Eqs. (3.55) and (3.18).

In a gas mixture at p, T, the pressure of component k is the partial pressure $p_k = y_k p$, where y_k is the mole fraction of component k. The contribution of entropy from component k to the mixture entropy is called the *partial molar specific entropy* (Sec. 11.4.3), which can be expressed for the IG model as follows:

$$\bar{s}_k^{\text{IG}}(p_k,T) = \bar{s}_k^o(T) - \bar{R} \ln \frac{p_k}{p_0} = \bar{s}_k^o(T) - \bar{R} \ln \frac{y_k p}{p_0}; \quad \left[\frac{\text{kJ}}{(\text{kmol of } k) \cdot \text{K}} \right] \quad \textbf{(13.50)}$$

where $\bar{s}_k^o(T)$ can be obtained from the ideal gas table for the species k at temperature T. In a similar manner, the *partial molar specific entropy* for the PG and SL model can be expressed:

$$\bar{s}_k^{\text{PG}}(p_k,T) = \bar{s}_k^o(T_0) + \bar{c}_{p,k} \ln \frac{T}{T_0} - \bar{R} \ln \frac{y_k p}{p_0}; \quad \left[\frac{\text{kJ}}{(\text{kmol of } k) \cdot \text{K}} \right] \quad \textbf{(13.51)}$$

$$\bar{s}_k^{\text{SL}}(p_k,T) = \bar{s}_k^{\text{SL}}(p,T) = \bar{s}_k^o(T_0) + \bar{c}_{p,k} \ln \frac{T}{T_0}; \quad \left[\frac{\text{kJ}}{(\text{kmol of } k) \cdot \text{K}} \right] \quad \textbf{(13.52)}$$

The partial specific properties of components in an ideal or perfect gas mixture can be evaluated by using the n-IG or n-PG state TESTcalcs. The following example illustrates entropy calculations using the IG mixture model.

EXAMPLE 13-7 **Entropy Analysis of Steady State Combustor**

In Ex. 13-5, determine the rate of entropy generation in the system's universe, assuming heat is transferred from the combustion chamber to a reservoir at 500 K. Use the IG mixture model.

What-if scenario: (d) What would the answers be if the combustion chamber pressure was 50 atm instead?

SOLUTION

Perform an entropy balance on the open-steady device.

Assumptions

The same as in Ex. 13-5.

Analysis

The rate of entropy generation in the system's universe within the red boundary in Figure 13.29 is given by the entropy balanced equation, Eq. (13.29):

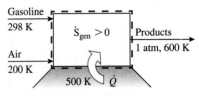

FIGURE 13.29 Schematic for Ex. 13-7 (click Entropy balance in Anim. 13.B.*completeSolution*).

$$\dot{S}_{\text{gen,univ}} = \dot{n}_F \left[\sum_p \nu_p \bar{s}_p - \sum_r \nu_r \bar{s}_r \right] - \frac{\dot{Q}}{T_B} = \dot{n}_F [\bar{s}_P - \bar{s}_R] - \frac{\dot{Q}}{T_B}$$

Entropy of the reactants (fuel at 298 K and air at 200 K) and products (at 600 K) can be evaluated using data from Table C and the IG tables (Table D). Alternatively, use the n-IG state TESTcalc for partial properties of gases in a mixture.

Species k	ν_k	y_k	T_k	$\bar{s}_k^o$	$-\bar{R} \ln (p_k/p_0)$ $p_k = y_k p$ kJ/kmol·K	$\bar{s}_k(p_k, T_k)$
$C_8H_{18}(l)$	1	1	298	360.57	0	360.57
O_2	18.75	0.21	200	193.480	12.97	206.45
N_2	70.5	0.79	200	179.98	1.96	181.94
				$\bar{s}_R = \sum_r \nu_r \bar{s}_r = 17{,}021$ kJ/kmol·K		
CO_2	8	0.085	600	243.2	20.49	263.69
$H_2O(g)$	9	0.096	600	212.9	19.48	232.38
N_2	70.5	0.752	600	212.1	2.37	214.47
O_2	6.25	0.067	600	226.3	22.47	248.77
				$\bar{s}_P = \sum_p \nu_p \bar{s}_p = 20{,}876$ kJ/kmol·K		

The mole flow rate of fuel and the heat transfer rate have been already determined in Ex 13-5. Therefore,

$$\dot{S}_{\text{gen,univ}} = \dot{n}_F [\bar{s}_P - \bar{s}_R] - \dot{Q}/T_B$$
$$= (1.1696 \times 10^{-5})[20{,}876 - 17{,}058] - \frac{(-46.05)}{500}$$
$$= 0.137 \text{ kW/K}$$

TEST Analysis

In the device panel of the TEST solution of Ex. 13-5, enter T_B as 500 K. Click the Calculate button to obtain Sdot_gen as 0.136 kW/K. To verify different parts of this solution, you may evaluate expressions, such as '= mdot1*s1+mdot2*s2' and '= mdot3*sdot3', after setting the reaction on the basis of 1 kmol of fuel and clicking Super-Calculate.

What-if scenario

In the state panel, change p1 to 50 atm and click Super-Calculate. The heat transfer rate remains unchanged (enthalpy is a function of temperature alone) while Sdot_gen decreases very slightly to 0.134 kW/K, showing a weak pressure effect.

Discussion

Vigorous entropy generation during the reaction, as is evident from this example, can be attributed to irreversibilities associated with reaction or **chemical friction**. The accompanying exergy destruction can be considerable, raising the possibility of improving work output through reversible oxidation of fuel.

13.3.4 Exergy Analysis

In a reacting open-steady system, specific chemical exergy of a fuel ψ^{ch}, or **molar specific chemical exergy** $\overline{\psi}^{ch}$ is the maximum useful work that can be extracted from a flow of fuel at standard conditions per unit mass or unit mole in a reversible manner. The fuel is assumed to be already in thermal and mechanical equilibrium with the surroundings so that the extracted useful work is solely due to chemical reaction.

To simplify the derivation of an expression for the chemical exergy, consider an open-steady combustion chamber that is in thermal communication with only the atmospheric reservoir TER-0 (Fig. 13.30). We will continue to use the subscript 0 to represent the conditions of the surroundings. Although it is not necessary, we will make no distinction between the conditions of the surroundings and the standard state without much sacrifice of generality. Eliminating $\dot{Q}$ between Eqs. (13.27) and (13.29) and neglecting changes in kinetic and potential energies, we obtain:

$$\dot{W}_{\text{ext}} = \underbrace{\dot{n}_F \left[\sum_r \nu_r(\overline{h}_r - T_0\overline{s}_r) - \sum_p \nu_p(\overline{h}_p - T_0\overline{s}_p) \right]}_{\dot{W}_{\text{rev}}} - \underbrace{T_0\dot{S}_{\text{gen,univ}}}_{\dot{I}} ; \quad [\text{kW}] \quad \textbf{(13.53)}$$

For an open system, atmospheric work being absent, all of the external work transfer is useful, that is, $\dot{W}_{\text{ext}} = \dot{W}_u$. Also, comparing Eq. (13.53) with Eq. (6.16), we can interpret the last term on the right hand side as the rate of exergy destruction and the term in the bracket as the reversible work—the maximum useful work when thermodynamic friction is completely eliminated.

Expressions for $\dot{I}$ and $\dot{W}_{\text{rev}}$ can be expanded by using the entropy balance equation, Eq. (13.29) and introducing Eq. (13.33) for component enthalpies:

$$\dot{I} = T_0\dot{S}_{\text{gen,univ}} = \dot{n}_F T_0 \left[\sum_p \nu_p\overline{s}_p - \sum_r \nu_r\overline{s}_r \right]; \quad [\text{kW}] \quad \textbf{(13.54)}$$

$$\dot{W}_{\text{rev}} = \dot{W}_u + \dot{I} = \dot{n}_F \left[\sum_r \nu_r(\overline{h}_{f,r}^{\circ} + \Delta\overline{h}_r - T_0\overline{s}_r) - \sum_p \nu_p(\overline{h}_{f,p}^{\circ} + \Delta\overline{h}_p - T_0\overline{s}_p) \right]; \quad [\text{kW}]$$

$$\textbf{(13.55)}$$

where the procedure for calculating the partial specific enthalpy and entropy of a component, using the IG or PG mixture model, has already been developed. As expected, $\dot{W}_{\text{rev}}$, the maximum useful work produced by the combustion device, depends solely on the states of the reactants (inlet) and products (exit) (Fig. 13.31 and Anim. 13.B.*reversibleWork*). Now suppose the

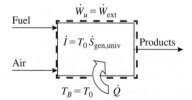

FIGURE 13.30 A steady work-producing combustion device which exchanges heat with the surrounding atmosphere at p_0 and T_0.

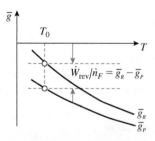

FIGURE 13.31 Graphical interpretation of Eq. (13.56). (See Anim. 13.B.*reversibleWork*).

reactants are in thermal and mechanical equilibrium (but not chemical) with the surroundings at p_0, T_0. If the products are also brought to thermo-mechanical equilibrium with the surroundings, the reversible work of Eq. (13.55) is solely due to chemical reaction and reaches a peak given by:

$$\frac{\dot{W}_{\text{rev}}}{\dot{n}_F} = \sum_r \nu_r [\bar{h}_r(T_0) - T_0 \bar{s}_r(p_r, T_0)] - \sum_p \nu_p [\bar{h}_p(T_0) - T_0 \bar{s}_p(p_p, T_0)]$$

$$= \underbrace{\sum_r \nu_r \bar{g}_r(p_r, T_0)}_{\bar{g}_R(T_0)} - \underbrace{\sum_p \nu_p \bar{g}_p(p_p, T_0)}_{\bar{g}_P(T_0)}; \quad \left[\frac{\text{kJ}}{\text{kmol Fuel}}\right]$$

$$\Rightarrow \quad \frac{\dot{W}_{\text{rev}}}{\dot{n}_F} = -[\bar{g}_P(T_0) - \bar{g}_R(T_0)] = -\Delta \bar{g}_C^{\circ}(T_0); \quad \left[\frac{\text{kJ}}{\text{kmol Fuel}}\right] \qquad \textbf{(13.56)}$$

where the combination property $\bar{g}_k(p_k, T_0)$ can be expressed:

$$\bar{g}_k(p_k, T_0) \equiv \bar{h}_k(T_0) - T_0 \bar{s}_k(p_k, T_0); \quad \left[\frac{\text{kJ}}{\text{kmol of } k}\right] \qquad \textbf{(13.57)}$$

$\bar{g}_k(p_k, T_0)$ is an important partial molar property (see Anim. 13.B.*partialPropertiesIG*) known as the **partial molar specific Gibbs function** of component k at its partial pressure p_k, mixture pressure p_0, and mixture temperature T_0. (We will discuss this function, also known as the chemical potential of component k, more thoroughly in Chapter 14.) The superscript in $\Delta \bar{g}_C^{\circ}(T_0)$ symbolizes that both the reactants and products mixture are at standard atmospheric pressure. The reversible work output can be shown to decrease if T_0 increases, as shown in Figure 13.31.

The reversible work of Eq. (13.56) can also be interpreted as the chemical flow exergy at the inlet of a chemically reacting system because the exit state at p_0 and T_0 is like the dead state of the products mixture. To be precise, the combustion products thrown away into the atmosphere still have some potential to produce useful work if the irreversibilities associated with mixing of the products can be exploited (Sec. 11.4.4). By convention, the dead state for a combustion mixture is assumed to have the following distribution (very close to atmospheric component distribution) of mole fractions: $y_{O_2,0} = 0.20248$, $y_{CO_2,0} = 0.02471$, and $y_{H_2O,0} = 0.00032$. The **molar specific chemical exergy** $\bar{\psi}^{\text{ch}}$ of a fuel is defined (Fig. 13.32 and Anim. 13.B.*chemicalExergy*) as the maximum useful work that can be extracted when the fuel at the standard state (p_0, T_0) burns completely with atmospheric air, also at standard state, and the products leave the system at the dead state, that is, at p_0 and T_0 with the standard distribution of components.

For a mixture at p_0 and T_0, the partial molar specific entropy of a component k in the products can be expressed in terms of the corresponding property of the pure component k using Eq. (13.50) as $\bar{s}_k(p_k, T_0) = \bar{s}_k^{\circ}(p_0, T_0) - \bar{R} \ln y_k$. Substituting this expression in Eq. (13.56) and setting the reactants and products at their respective dead states, the chemical exergy of a fuel can be expressed:

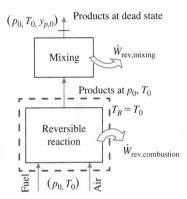

FIGURE 13.32 Chemical exergy is the total useful work that can be extracted as products of the reversible reaction are brought to dead state. (See Anim. 13.B.*chemicalExergy*).

$$\bar{\psi}^{\text{ch}} \equiv \frac{\dot{W}_{\text{rev,max}}}{\dot{n}_F} = \left[\underbrace{\sum_r \nu_r \bar{h}_r - \sum_p \nu_p \bar{h}_p}_{h_R(T_0,p_0) - h_P(T_0,p_0)}\right] - T_0 \left[\underbrace{\sum_r \nu_r \bar{s}_r - \sum_p \nu_p \bar{s}_p}_{\bar{s}_R(T_0,p_0) - \bar{s}_P(T_0,p_0)}\right]$$

$$+ \bar{R} T_0 \ln \left[\frac{\prod_r (y_{r,0})^{\nu_r}}{\prod_p (y_{p,0})^{\nu_p}}\right]; \quad \left[\frac{\text{kJ}}{\text{kmol Fuel}}\right] \qquad \textbf{(13.58)}$$

in which the first two bracketed terms on the RHS are evaluated at the dead state; and the last term contains products of terms arising out of the pressure contribution terms such as $\nu_k \bar{R} \ln y_{k,0} = \bar{R} \ln y_{k,0}^{\nu_k}$. The above equation can be simplified by realizing that the fuel (in liquid or solid phase) is not part of the gaseous reactants mixture, and non-participants components, such as nitrogen, make no contribution to any of the three terms.

As an example, let us develop an expression for the chemical exergy for a generic hydrocarbon, C_aH_b. First, set up the theoretical reaction between the fuel and oxygen in air:

$$C_aH_b + \left(a + \frac{b}{4}\right)O_2 \rightarrow aCO_2 + \frac{b}{2}H_2O(g)$$

Now, simplify Eq. (13.58) to produce:

$$\bar{\psi}^{ch} = \left[\bar{h}_F + \left(a + \frac{b}{4}\right)\bar{h}_{O_2} - a\bar{h}_{CO_2} - \frac{b}{2}\bar{h}_{H_2O(g)}\right]_{(p_0,T_0)}$$

$$- T_0\left[\bar{s}_F + \left(a + \frac{b}{4}\right)\bar{s}_{O_2} - a\bar{s}_{CO_2} - \frac{b}{2}\bar{s}_{H_2O(g)}\right]_{(p_0,T_0)} + \bar{R}T_0\ln\left[\frac{(y_{O_2,0})^{a+b/4}}{(y_{CO_2,0})^a(y_{H_2O,0})^{b/2}}\right]$$

$$= \left[\bar{h}_R - \bar{h}_P\right]_{(p_0,T_0)} - T_0\left[\bar{s}_R - \bar{s}_P\right]_{(p_0,T_0)} + \bar{R}T_0\ln\left[\frac{(y_{O_2,0})^{a+b/4}}{(y_{CO_2,0})^a(y_{H_2O,0})^{b/2}}\right]$$

$$= \left[\bar{g}_R - \bar{g}_P\right]_{(p_0,T_0)} + \bar{R}T_0\ln\left[\frac{(y_{O_2,0})^{a+b/4}}{(y_{CO_2,0})^a(y_{H_2O,0})^{b/2}}\right]; \quad \left[\frac{kJ}{kmol\ of\ Fuel}\right] \qquad \textbf{(13.59)}$$

where $g \equiv h - Ts$ is substituted at the last step. The first term, decrease of the Gibbs function, can be identified with the reversible work of combustion; the second term stems from the work extracted while mixing the products in a reversible manner resulting in the dead state. This is shown conceptually in Figure 13.32. As illustrated in example 13-8, the last term is not very significant, and reversible work of combustion can be used as an acceptable estimate of the chemical exergy of a fuel.

EXAMPLE 13-8 Chemical Exergy and Heating Values

Determine (a) the HHV, (b) the LHV, and (c) the chemical exergy of ethane (C_2H_6) in kJ/kg. Assume methane and the atmosphere to be in the standard conditions of 1 atm and 25°C. (d) Discuss how these values compare with the reversible work for isothermal combustion at the standard state.

SOLUTION
Obtain the heating values from an energy balance and exergy from Eq. (13.58).

Assumptions
Air has the following composition at the dead state: $y_{O_2,0} = 0.20248$, $y_{CO_2,0} = 0.02471$, and $y_{H_2O,0} = 0.00032$. No changes in ke and pe.

Analysis
The presence of neutral components, such as nitrogen, cannot affect either the heating value or the exergy, and the theoretical reaction between methane and air can be represented by:

$$C_2H_6 + 3.5O_2 \rightarrow 2CO_2 + 3H_2O(g)$$

The enthalpy and entropy of the reactants and products, at standard state, can be evaluated using data from Table G-1 (or we can use the n-IG system-state TESTcalc to obtain hbar_k and s0_k*MM_k) as follows:

Species	ν_k	$\bar{h}^o_{f,k}$ (kJ/kmol)	$\bar{s}^o_k$ (kJ/kmol·K)
C_2H_6	1	−84,680	229.49
O_2	3.5	0	205.03

$$\bar{h}_R(T_0) = \sum_r \nu_r\bar{h}^o_{f,p} = -84,680\ kJ/kg; \quad \bar{s}^o_R(T_0) = \sum_r \nu_r\bar{s}^o_r(T_0) = 947.1\ kJ/kmol\cdot K$$

Species	ν_k	$\overline{h}^o_{f,k}$ (kJ/kmol)	$\overline{s}^o_k$ (kJ/kmol · K)
CO_2	2	−393,520	213.69
$H_2O(g)$	3	−241,820	188.72

$$\overline{h}_P(T_0) = \sum_p \nu_p \overline{h}^o_{f,p} = -1512,500 \text{ kJ}; \quad \overline{s}^o_P(T_0) = \sum_r \nu_p \overline{s}^o_p(T_0) = 993.5 \text{ kJ/kmol} \cdot \text{K}$$

From Eq. (13.39): $\overline{\text{LHV}} = \overline{h}_R(T_0) - \overline{h}_P(T_0) = -84.680 - (-1512.500) = 1427.8 \dfrac{\text{MJ}}{\text{kmol Fuel}}$

Also, from Eq. (13.41):

$$\overline{\text{HHV}} = \overline{\text{LHV}} + \nu_{H_2O}\overline{h}_{fg,H_2O} = 1427.8 + 3 \times 18 \times 2.442 = 1559.7 \dfrac{\text{MJ}}{\text{kmol Fuel}}$$

where $\overline{h}_{fg,H_2O} = \overline{M}_{H_2O}h_{fg,H_2O}$ is obtained from the steam table at a saturation temperature of 25°C. The chemical exergy of fuel can be obtained from Eq. (13.58) as:

$$\overline{\psi}^{ch} = \underbrace{1427.8 - 298\frac{(-46.4)}{1000}}_{1441.6} + \underbrace{\frac{8.314(298)}{1000}\ln\left[\frac{(0.20248)^{3.5}}{(0.02471)^2(0.00032)^3}\right]}_{64.3}$$

$$= 1506 \dfrac{\text{MJ}}{\text{kmol of } C_2H_6} \tag{13.60}$$

Dividing by the molar mass of ethane, 30 kg/kmol, the desired answers can be obtained as:

$$\text{LHV} = 47.6 \frac{\text{MJ}}{\text{kg}}; \quad \text{LHV} = 52.0 \frac{\text{MJ}}{\text{kg}}; \quad \overline{\psi}^{ch} = 50.2 \frac{\text{MJ}}{\text{kg}}$$

TEST Analysis

Launch the IG non-premixed open-steady combustion TESTcalc. Choose the Mass radio button. In the reaction panel, select ethane in the fuel block. Execute Theoretical Combustion with oxygen from the action menu to set up the theoretical reaction between ethane and pure oxygen on the basis of 1 kg of fuel. The water in the products is assumed to be in gaseous state by default. Evaluate the fuel, oxidizer, and products' states, at 1 atm and 25 deg-C, as States 1, 2 and 3. In the device panel, import the fuel and oxidizer states as i1 and i2 states and the products state as the e-state. Enter Wdot_ext = 0 and click Calculate to produce Qdot = −47.5 MW, which must be the LHV of the fuel. To calculate the HHV, change H2O to H2O(L) in the reaction panel. Choose Balance Reaction from the action menu, then click Super-Calculate. The device panel updates Qdot as –51.9 MW, meaning HHV = 51.9 MJ/kg. The contribution of the first two terms to chemical exergy can be calculated in the I/O panel as (mdot11*g1+mdot2*g2−mdot3*g3)=49.0 MJ kg. The last term must be manually obtained. You can repeat the solution with air as the oxidizer without any change in the answers.

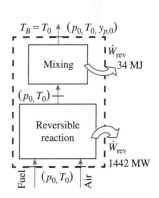

FIGURE 13.33 Interpretation of Eq. (13.60). Fuel exergy can be approximated as the reversible work produced by isothermal combustion at p_0, T_0 for a fuel flow rate of 1 kmol/s.

Discussion

A fuel's chemical exergy usually falls somewhere between its highest and lowest heating values, as in this case. Notice the small contribution from the mixing term; about 98% of the fuel exergy is extracted by the reversible work with the products achieving thermo-mechanical equilibrium with the dead state. The rest, about 2%, can be recouped as the products' composition equilibrates with the standardized dead state.

EXAMPLE 13-9   **Exergy Analysis of a Steady State Combustor**

Hydrogen gas enters a steady-flow combustion chamber at 1 atm, 25°C and is mixed with the theoretical amount of oxygen, which also enters at 1 atm and 25°C. For each kmol of hydrogen, determine (a) the rate of exergy destruction and (b) reversible work for the following

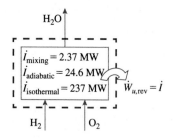

FIGURE 13.34 The reversible work is a maximum for isothermal combustion in Ex. 13-9.

cases: Case-I: Adiabatic mixing with no chemical reaction; Case-II: Theoretical isothermal reaction at 25°C. Case-III: Theoretical adiabatic combustion. Neglect pressure loss in the combustion chamber. Use TEST.

SOLUTION

Perform an energy and entropy balance on the open-steady reaction chamber using TESTcalcs.

Assumptions

The IG mixture model is applicable. No change in ke or pe. The analysis is based on a hydrogen flow rate of 1 kmol/s. Neglect the contribution of mixing to the exergy expression.

TEST Analysis

The theoretical reaction between hydrogen and oxygen is:

$$H_2 + 0.5O_2 \rightarrow H_2O$$

CASE-I:

Launch the IG non-premixed open-steady combustion TESTcalc. In the reaction panel, select H2 in the fuel and execute Theoretical Combustion with oxygen from the action menu to obtain the balanced reaction. Evaluate the fuel, oxidizer, and products' states in the state panel—State-1 for hydrogen with p1=1 atm and T1=25 deg-C, State-2 for oxygen with p2=p1 and T2=T1, and State-3 for products with p3=p1 (partially calculated state). In the device panel, import these states as i1, i2, and e-states. Enter **Qdot=0 and Wdot_ext=0**, and click Super-Calculate. As expected, the products temperature is calculated as 25 deg-C. The rate of entropy generation can be found in the device panel as 7.94 kW/K (or kJ/kmol · K). For a surrounding temperature of 298 K, the rate of exergy destruction can be calculated as 298*7.94 = 2366 kW, which must also be the reversible work (since $\dot{W}_u = \dot{W}_{ext} = 0$). In the I/O panel, evaluate 'm3*g3−m1*g1−m2*g2', which also yields $\dot{W}_{rev} = -2.37$ MW (or MJ/kmol of fuel).

CASE-II:

Launch the same combustion TESTcalc. Select H2 (1 kmol), O2, and H2O(L) in the fuel, oxidizer, and products blocks. Execute Balance Reaction from the action menu to obtain the balanced reaction on the basis of 1 kmol of H2.

   Evaluate the states as in Case-I, except for State-3 for which T3 is also known ('=T1'). Set up the device panel as before, enter **Wdot_ext=0**, and click Super-Calculate. Qdot and Sdot_gen are calculated as −285.82 MW and 795.2 kW/K. The rate of irreversibility is 298* 795.2/1000 = 237 MW, which is also be the reversible work since there is no useful work transfer in this device. Evaluations of the change in Gibbs function, as in case-I, produce the rate of irreversibilities as 237 MW.

CASE-III:

Launch the same combustion TESTcalc again. Obtain the reaction as in Case I. Water is assumed to be in gaseous form at the adiabatic flame temperature of the exit.

   Set up the three states and the device panel as in Case-I and click Super-Calculate. T3 and Sdot_gen are calculated as 4991 K and 82.7 kW/K. The rate of irreversibility, therefore, is 298* 82.7/1000 = 24.6 MW, which must also be the reversible work. This can be directly evaluated in the I/O panel from =m3*(h3−298*s3)−m1*(h1−298*s1)−m2*(h2−298*s2) as 24.6 MW. Note that the Gibbs function cannot be used here as the system is not isothermal.

Discussion

There is no actual work transfer in these three situations; therefore, the exergy destroyed is equal to the reversible work [Eq. (6.16)]. Without any chemical reaction or heat transfer, the exergy destruction is due only to mixing friction, and the reversible work is the lowest for pure mixing. Adiabatic combustion raises the exergy destruction rate by almost an order of magnitude through chemical friction. Another order of magnitude increase occurs with isothermal combustion because of the thermal friction associated with heat transfer to the surroundings.

   The above example raises two distinct possibilities for extracting useful work from a fuel; (1) Isothermal combustion at the ambient temperature T_0 with simultaneous work production, and (2) raise the temperature of the products then run a heat engine between the products and the ambient atmosphere. Two practical means of achieving this are discussed next.

13.3.5 Isothermal Combustion—Fuel Cells

A **fuel cell** is an electrochemical device in which useful electrical work is extracted from the fuel through isothermal chemical reaction without direct contact between the fuel and oxidizer. Used by NASA to produce power and water in spacecraft for decades, hydrogen-oxygen fuel cells are recently receiving increasing interest from automakers for commercialization of fuel cell-powered cars.

A fuel cell works much like a battery, except it operates at steady state and does not need recharging. There are several types of fuel cell technologies, based on the nature of electrolyte used. Figure 13.35 illustrates a hydrogen-oxygen alkaline fuel cell (AFC), in which the flows of fuel and oxidizer are separated by two porous electrodes—the anode and the cathode. The space between them is filled with a suitable electrolyte, such as a solution of potassium hydroxide in water. Hydrogen, diffusing through the anode, reacts with the hydroxyl ion coming from the cathode side. Free electrons are released as a result according to:

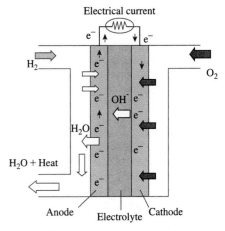

FIGURE 13.35 Hydrogen-oxygen alkaline fuel cell (AFC) (see Anim. 13.B.*fuelCell*).

$$\underbrace{H_2 + 2OH^-}_{1\ kmol} \rightarrow 2H_2O + \underbrace{2e^-}_{2\ kmol}$$

Electrons released by this reaction travel through the external circuit, and the water diffuses through the electrolyte into the cathode, where it reacts with the incoming oxygen and produce hydroxyl ions:

$$0.5O_2 + 2H_2O + 2e^- \rightarrow 2OH^- + H_2O$$

Finally, the hydroxyl ions diffuse through the electrolyte to the anode surface, completing the loop. For every kmol of hydrogen, 0.5 kmol of oxygen is consumed, producing water according to the overall reaction, $H_2 + 0.5O_2 \rightarrow H_2O$. Although it looks the same as the exothermic combustion reaction discussed earlier, this is an isothermal reaction producing useful electrical work without the need for a heat engine.

The mass, energy, entropy, and exergy analysis, developed in earlier sections for an open-steady reacting device, can also be applied to a fuel cell. When the reactants and products are both at the pressure and temperature of the surroundings p_0, T_0, the reversible work reaches a maximum, given by $-\Delta \bar{g}_C^o(T_0)$, as expressed by Eq. (13.56). If the cell operates at a higher temperature $T > T_0$, the reversible work from the cell alone can be derived (see Anim. 14.D.*revWorkAtConstTp*):

$$\frac{\dot{W}_{rev}}{\dot{n}_F} = \underbrace{\sum_r \nu_r \bar{g}_r(p_r, T)}_{\bar{g}_R(T)} - \underbrace{\sum_p \nu_p \bar{g}_p(p_p, T)}_{\bar{g}_P(T)};$$

$$\Rightarrow \frac{\dot{W}_{rev}}{\dot{n}_F} = -[\bar{g}_P(T) - \bar{g}_R(T)] = -\Delta \bar{g}_C(T, p_0) = -\Delta \bar{g}_C^o(T); \quad \left[\frac{kJ}{kmol\ Fuel}\right] \quad \textbf{(13.61)}$$

The superscript o in $\Delta \bar{g}_C^o(T)$ indicates that pressure remains constant at the standard atmospheric pressure. The reversible work, given by Eq. (13.61), can be seen to decrease (Fig. 13.31) as the operating temperature is raised. This is an important consideration as fuel cells with solid oxides, such as the electrolyte (SOFC), operate at much higher temperature. Additional work can be theoretically obtained from a heat engine running between the cell at T and the surroundings at T_0 utilizing the waste heat. The maximum possible work output, however, is given by $-\Delta \bar{g}_C(p_0, T_0)$ at the lowest possible temperature of the fuel cell, and is used as an upper limit of any fuel cell output.

An energy balance, using Eq. (13.27), produces the net energy transfer by a cell temperature of T:

$$\frac{(\dot{W}_{ext} - \dot{Q})}{\dot{n}_F} = \sum_r \nu_r \bar{h}_r(T) - \sum_p \nu_p \bar{h}_p(T) = -\Delta \bar{h}_C^o(T); \quad \left[\frac{kJ}{kmol\ Fuel}\right] \quad \textbf{(13.62)}$$

DID YOU KNOW?

- The capital cost (in 2004) of a fuel cell was $4000 per kW vs. $100 per kW for an automobile engine.

Equation (13.62) can be subtracted from Eq. (13.61) to produce the heat transfer for a reversible fuel cell:

$$\frac{\dot{Q}_{rev}}{\dot{n}_F} = \Delta\bar{h}^o_C(T) - \Delta\bar{g}^o_C(T) = T\Delta\bar{s}^o_C(T); \quad \left[\frac{kJ}{kmol\ Fuel}\right] \tag{13.63}$$

where the change in entropy is calculated in a similar manner as the change in Gibbs function. This equation for heat transfer is consistent with the general expression for reversible heat transfer derived in Sec. 4.1.3.

Another important parameter of a fuel cell is its open circuit voltage, which can be obtained by equating the ideal electrical work (voltage times current) with the reversible work from Eq. (13.61). The ideal electrical work can be expressed in terms of charge carried by particles—electrons in this case—flowing across a potential V_i:

$$\frac{\dot{W}_i}{\dot{n}_F} = \underbrace{(V_i)}_{\substack{\text{Ideal} \\ \text{voltage}}} \underbrace{(\nu_e N)}_{\substack{\text{Number of} \\ \text{electrons}}} \underbrace{(e)}_{\substack{\text{Charge of} \\ \text{electron}}}; \quad \left[\frac{kJ}{kmol\ Fuel}\right] \tag{13.64}$$

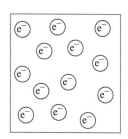

FIGURE 13.36 Amount of electrons, just like molecules, can be counted in mole (in the unit of kmol).

where ν_e is the mole of electrons (Fig. 13.36) freed per kmol of fuel, N is the Avogadro number (which converts mole into actual number), and e is the charge of an electron in coulombs. Substituting the constants N and e in Eq. (13.64), and using the reversible limit of work from Eq. (13.61), allow us to obtain the ideal voltage that can be generated by a fuel cell at a given temperature:

$$V_i = \frac{-\Delta\bar{g}^o_C(T)}{96,487\nu_e}; \quad [V] \tag{13.65}$$

In this expression, V_i is in volts, $\Delta\bar{g}^o_C(T)$ is in kJ/kmol, and ν_e is obtained from the anode or cathode reaction. For the hydrogen-oxygen fuel cell $\nu_e = 2$.

13.3.6 Adiabatic Combustion—Power Plants

Although fuel cells seem to provide an elegant solution for extracting useful work from a fuel, there are many practical issues—production, delivery, storage, and safety of hydrogen; durability of precious-metal catalysts used in the electrolyte; and high capital cost—that still impede large scale utilization of this technology.

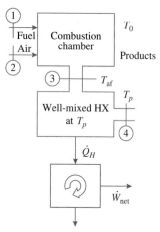

FIGURE 13.37 Schematic for deriving Eq. (13.66).

Conventionally, heat released from the combustion of fossil fuels is converted to work by placing a heat engine between the products of combustion and the atmosphere. An ideal set up is shown in Figure 13.37. The products leaving at the adiabatic flame temperature, T_{af}, cool down in a well-mixed heat exchanger to a temperature T_p, while supplying $\dot{Q}_H$ to the heat engine. To maximize the work output, an increase in $\dot{Q}_H$ may seem like a good idea, but it is accompanied by a decrease in T_p causing a decrease in the efficiency of the heat engine. To find the optimum value of T_p, the engine is replaced by a reversible Carnot engine in Figure 13.37, operating between T_p and the ambient temperature T_0. Simplified expressions for the heat transfer and work produced can be obtained by using the PG mixture model with uniform specific heat for all components in the products. With reference to the schematic in Figure 13.37, $\dot{Q}_H$ and $\dot{W}_{net}$ can be expressed as follows:

$$\frac{\dot{Q}_H}{\dot{m}_F} = \sum_p \eta_p(h_{p3} - h_{p4}) = \sum_p \eta_p c_{p,p}(T_{p3} - T_{p4}); \quad \left[\frac{kJ}{kg\ Fuel}\right]$$
$$= c_p(T_3 - T_4)\sum_p \eta_p = c_p(T_{af} - T_p)(1 + AF) \tag{13.66}$$

$$\frac{\dot{W}_{net}}{\dot{m}_F} = \frac{\dot{Q}_H}{\dot{m}_F}(1 - T_C/T_H) = c_p(T_{af} - T_p)(1 - T_0/T_p)(1 + AF); \quad \left[\frac{kJ}{kg\ Fuel}\right] \tag{13.67}$$

As can be seen from these expressions, the heat transfer increases with a decrease in T_p, but the Carnot efficiency suffers, reducing the net work. An optimum value of T_p can be found by differentiating the expression for $\dot{W}_{net}$ with respect to T_p and equating it to zero. The resulting expression for the optimum T_p and the maximum work can be shown to be:

$$T_{p,\text{opt}} = \sqrt{T_0 T_{\text{af}}}; \quad [\text{K}] \qquad \textbf{(13.68)}$$

and

$$\frac{\dot{W}_{\text{net,max}}}{\dot{m}_F} = c_p(1 + \text{AF})T_{\text{af}}\left[1 - \sqrt{T_0/T_{\text{af}}}\right]^2; \quad \left[\frac{\text{kJ}}{\text{kg Fuel}}\right] \qquad \textbf{(13.69)}$$

It can be easily verified that the maximum work from adiabatic combustion, as given by Eq. (13.69), is considerably smaller than the reversible work of isothermal combustion. Methane, for instance, has a $-\Delta \bar{g}_C^\circ(T_0) = 80$ MJ per kmol at 1 atm and 25°C (Use premixed IG combustion TESTcalc to set up theoretical reaction of methane with air and calculate reactants and products states at 100 kPa, 298 K. In the I/O panel calculate m2*g2-m1*g1 = -79.89 kJ/kmol)), meaning that the reversible work of isothermal combustion is 50 MJ/kg of methane. The theoretical air-fuel ratio and T_{af} have been evaluated in Ex. 13-6. Using these values, Eq. (13.69) yields $T_{p,\text{opt}}$ as 952 K and $\dot{W}_{\text{net,max}}/\dot{m}_F$ as 26.2 MJ/kg of methane. The work output, therefore, is about half as much as the reversible isothermal work.

TEST can be used for a parametric study to evaluate the optimum temperature and heat transfer. Set up the theoretical reaction for methane and air with the IG non-premixed combustion TESTcalc and evaluate State-3 as the adiabatic products (Ex. 13-7). Now evaluate State-4 for products with no pressure loss (p4=p3) and leave temperature T4 as an unknown. Set up Device-B in the device panel with State-3 as i1-State and State-4 as *e*-state, and set Wdot_ext=0. For a given T4, Qdot can be found by clicking Super-Calculate and the maximum work output can be evaluated using the Carnot efficiency, based on T4 and the ambient temperature. A parametric study yields the value of the optimum temperature as 950 K (Fig. 13.38) and the maximum work output as 24.5 MJ/kg fuel, which confirms the reasonable accuracy from the simpler PG model.

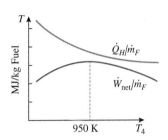

FIGURE 13.38 Result of a parametric study to evaluate the optimum value of T_p in Fig. 13.37 using the IG model.

EXAMPLE 13-10 | **Analysis of a Fuel Cell**

A hydrogen–oxygen fuel cell (Fig. 13.39) operates steadily at 1 atm, 298 K with a fuel consumption rate of 1 kg/h. Determine (a) the maximum possible electrical power in kW, (b) the accompanying heat transfer rate, and (c) the ideal electrical potential. ***What-if scenario:*** (d) What would the answer in part (a) be if the operating temperature increased to 500 K? Use TEST.

SOLUTION

Set up the overall reaction between hydrogen and oxygen and perform an energy and entropy balance to evaluate the change in enthalpy and Gibbs function.

Assumptions

IG mixture model is applicable. No changes in ke and pe. No pressure loss.

TEST Analysis

Launch the open-steady non-premixed IG combustion TESTcalc and click the Mass radio button. In the fuel block, select H2, then enter its amount as 1/3600 = 0.0002778 kg. Select O2 as the oxidizer and H2O(L) as the products (you have to de-select the default components first). Execute Balance Reaction from the action menu.

Switch to the state panel and evaluate State-1 as the fuel state with T1=298 K and p1=1 atm. Similarly, evaluate State-2 for the oxidizer with T2=T1 and p2=p1, and State-3 for products with T3=T1 and p3=p1. Determine the decrease in the Gibbs function by evaluating the expression '=m1*g1+m2*g2−m3*g3' in the I/O panel as 32.7 kW. This must be the maximum electrical power that can be produced by a reversible fuel cell according to Eq.(13.61). The reversible heat transfer, similarly, can be evaluated as '=298*(m3*s3−m1*s1−m2*s2)' = −6.71 kW.

In order to calculate the ideal potential using Eq. (13.65), the decrease in Gibbs function per unit mole of fuel must first be obtained. To convert the reaction to a molar basis, select the Mole radio button in the reaction panel and Normalize from the action menu, then click

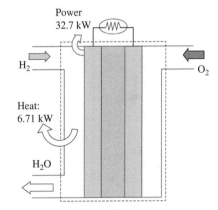

Power
32.7 kW

H₂ → ← O₂

Heat:
6.71 kW

H₂O

FIGURE 13.39 A reversible fuel cell operating at 298 K—schematic for Ex. 13-10.

Super-Calculate. In the I/O panel, evaluate the decrease in the Gibbs function as 237,138 kJ/kmol of H2; therefore, V_i = 237,138/(96,487*2) = 1.23 volts.

What-if scenario

At 500 K, the water in the products cannot be in the liquid phase. Reset the reaction with H2O in the products. To do so, deselect H2O(L) in the products block and select H2O. Choose Balance Reaction from the action menu. Now change T1 to **500 K** in the state panel and click Super-Calculate. Recalculate the reversible work transfer as –30.12 kW, the reversible heat transfer as –2.03 kW, and the ideal voltage as 1.135 volts.

Manual Analysis

Follow the systematic procedure laid out in Exs. 13-5 and 13-7 for calculating component enthalpy $\bar{h}_k(T)$ and partial entropy $\bar{s}_k(p_k,T)$. The partial molar Gibbs function can be calculated in a separate column, from $\bar{g}_k(T) \equiv \bar{h}_k(T) - T\bar{s}_k(p_k,T)$. We can use the n-IG state TESTcalc to verify the partial properties. Finally, apply Eq. (13.61) to obtain the reversible work.

Discussion

An actual fuel cell cannot be expected to be reversible. Nevertheless, the ideal limits for work, heat, and voltage are used to characterize a cell. A negative value for heat transfer suggests the possibility of space heating, by a fuel cell, as a by-product.

13.4 CLOSED PROCESS

When fuel and oxidizer are ignited in a closed chamber, the closed process executed by the system involves combustion. Customized balance equations for a generic closed process, in which a closed system transitions from a well defined beginning state to a final state, were derived in Sec. 5.1.1. To further customize these equations for a reacting process, the internal energy u_k of component k in a mixture must take into account the chemical contribution. Leveraging expressions developed for $\bar{h}_k$ earlier in Sec. 13.3.1, the partial internal energy of a component k in a mixture of ideal gases (see Anim. 13.B.*componentPropertiesIG*) can be expressed:

$$U_k^{IG}(p_k,T) = H_k^{IG}(p_k,T) - p_k\mathcal{V} = H_k^{IG}(T) - n_k\bar{R}T; \quad [\text{kJ}]$$

$$\Rightarrow \quad \bar{u}_k^{IG}(T) = \frac{U_k^{IG}(T)}{n_k} = \bar{h}_k^{IG}(T) - \bar{R}T; \quad \left[\frac{\text{kJ}}{\text{kmol of } k}\right] \tag{13.70}$$

Therefore, $u_k^{IG}(T) = \dfrac{\bar{u}_k^{IG}(T)}{\overline{M}_k} = h_k^{IG}(T) - \left(\dfrac{\bar{R}}{\overline{M}_k}\right)T;$

$$\Rightarrow \quad u_k^{IG}(T) = h_k^{IG}(T) - R_kT; \quad \left[\frac{\text{kJ}}{\text{kg of } k}\right] \tag{13.71}$$

For a liquid or solid component (condensed fuel, for instance) of a mixture, $\bar{u}_k^{SL} = \bar{h}_k^{SL}$, since the volume occupied by the condensed phase can be considered negligible. Also, the partial pressure of a condensed fuel vapor in the mixture is so small that $p_k = 0$ for a solid or liquid fuel, that is, the entire amount of fuel is assumed to be in the condensed phase.

Realizing that the beginning and final states (*b*- and *f*- states in Fig. 13.40) of the process are composed of reactants and products, and neglecting changes in KE and PE, the energy equation for a closed process, Eq. (5.2), can be modified as:

Energy (Mole based): $\quad Q - W_{\text{ext}} = \Delta U = \sum_p n_p\bar{u}_p - \sum_r n_r\bar{u}_r; \quad [\text{kJ}]$

$$\Rightarrow \quad \frac{Q}{n_F} - \frac{W_{\text{ext}}}{n_F} = \sum_p \nu_p(\bar{h}_p - \bar{R}T) - \sum_r \nu_r(\bar{h}_r - \bar{R}T)$$

$$= \underbrace{\sum_p \nu_p\bar{h}_p}_{\bar{h}_P} - \underbrace{\sum_r \nu_r\bar{h}_r}_{\bar{h}_R} - \bar{R}T_f\sum_p \nu_p + \bar{R}T_b\sum_r \nu_r;$$

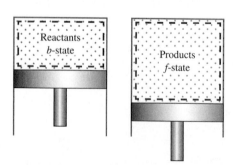

FIGURE 13.40 The energy equation, Eq. (13.72) can be applied to any reacting closed process, such as the one shown here (see Anim. 13.C.*ClEngineComb*).

Reactants·
b-state

Products·
f-state

$$\Rightarrow \quad \frac{Q}{n_F} - \frac{W_{\text{ext}}}{n_F} = \bar{h}_P - \bar{h}_R - \bar{R}\underbrace{\left[T_f \sum_p \nu_p - T_b \sum_r \nu_r\right]}_{\text{Gaseous components only}}; \quad \left[\frac{kJ}{\text{kmol Fuel}}\right] \quad (13.72)$$

As explained previously, only the gaseous components should be considered for the summation terms. It is left as an exercise to derive the mass-based counterpart of the energy equation. Notice that the heat transfer, per unit mole of fuel, in a constant-volume process can be different from a constant-pressure process (the open-steady systems discussed earlier are constant-pressure systems) if there is a change in the total number of mole in a reaction. Heating values determined from constant-volume experiments, therefore, are different from the heating values based on enthalpies by an amount,

$$\bar{R}T_0\left(\sum_p \nu_p - \sum_r \nu_r\right)$$

The discrepancy, however, is within experimental error in most situations. The entropy equation is relatively straightforward and can be derived from Eq. 5.4:

Entropy: $\quad S_{\text{gen}} = \Delta S - \dfrac{Q}{T_B} = \sum_p \nu_p \bar{s}_p(p_p, T_f) - \sum_r \nu_r \bar{s}_r(p_r, T_b) - \dfrac{Q}{T_B}; \quad \left[\dfrac{kJ}{K}\right]$ **(13.73)**

The evaluation of partial molar specific enthalpy and entropy in a mixture has been already discussed in Secs. 13.3.1 and 13.3.3 and is illustrated in the following analysis of a constant-volume reacting process.

EXAMPLE 13-11   Analysis of a Reacting Closed Process

A rigid chamber contains 0.1 kg of liquid octane and 2.3 kg of air at 1 atm, 200 K. The mixture is ignited and heat is transferred until the final temperature reaches 600 K. Assuming complete combustion, determine (a) the final pressure and (b) the rate of heat transfer. ***What-if scenario:*** (c) What would the final temperature and pressure be if the chamber was adiabatic?

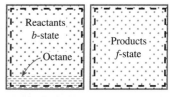

FIGURE 13.41 Schematic for Ex. 13-11. Volume occupied by the liquid fuel is ignored in the analysis (see Anim. 13.C.*SIEngineComb*).

SOLUTION

Perform a mass, energy, and entropy balance for the closed process shown in Figure 13.41.

Assumptions

Liquid octane occupies negligible volume compared to air. Use IG mixture model for reactants and products. No change in KE or PE. Vapor pressure of octane is negligible.

Analysis

For complete combustion, the reaction can be expressed in terms of unknown stoichiometric coefficients and balanced with the help of atom balance equations (Ex. 13-1) to give:

$$C_8H_{18}(l) + 19(O_2 + 3.76N_2) \rightarrow 8CO_2 + 9H_2O + 71.44N_2 + 6.5O_2$$

In setting up this reaction, the presence of octane vapor in the reactants is ignored. Applying $p\forall = n\bar{R}T$, the IG equation of state, to the reactants and products mixture, the final pressure can be evaluated:

$$\frac{p_f\forall_f}{p_b\forall_b} = \frac{n_f\bar{R}T_f}{n_b\bar{R}T_b}; \quad \Rightarrow \quad p_f = p_b\frac{T_f\sum\limits_p \nu_p}{T_b\sum\limits_r \nu_r} = (1 \text{ atm})\frac{(600)94.94}{(200)90.44} = 3.15 \text{ atm}$$

The assumption $\forall_f = \forall_b$, used in the above derivation, results from neglecting the volume occupied by the liquid. Enthalpies of the components at 200 K and 600 K have been already evaluated in Ex. 13-5, except for the fuel. In the following calculations, the contribution from sensible enthalpy of liquid fuel is obtained from the SL model with the specific heat of octane read from Table A-1.

$$\bar{h}_R = \nu_F \bar{h}_F + \nu_{O_2} \bar{h}_{O_2} + \nu_{N_2} \bar{h}_{N_2} = (\bar{h}^o_{f,F} + \Delta\bar{h}^{SL}) + \nu_{O_2}\Delta\bar{h}^{IG}_{O_2} + \nu_{N_2}\Delta\bar{h}^{IG}_{N_2}$$

$$= -249,910 + 254.2(200 - 298) + 19(-2868) + 71.44(-2857)$$

$$= -533,418 \text{ kJ/(kmol of Fuel)}$$

At 600 K, the products can be treated as an IG mixture with:

$$\bar{h}_P = \nu_{CO_2}(\bar{h}^o_f + \Delta\bar{h}^{IG})_{CO_2} + \nu_{H_2O(g)}(\bar{h}^o_f + \Delta\bar{h}^{IG})_{H_2O(g)}$$

$$+ \nu_{N_2}\Delta\bar{h}^{IG}_{N_2} + \nu_{O_2}\Delta\bar{h}^{IG}_{O_2} = 8(-380,614) + 9(-231,321)$$

$$+ 71.44(8,894) + 6.5(9,245) = -4,431,321 \text{ kJ/(kmol of Fuel)}$$

With the chamber volume remaining constant, there cannot be any boundary work and the heat transfer can be evaluated from Eq. (13.72) as:

$$\frac{Q}{n_F} = \bar{h}_P - \bar{h}_R - \bar{R}\left[T_f\sum_p\nu_p - T_b\sum_r\nu_r\right]$$

$$= (-4,431,321) - (-533,418) - 8.314[600(94.94) - 200(90.44)]$$

$$= -4,221,118 \text{ kJ/(kmol of Fuel)}$$

The mole of fuel used is $n_F = 0.1/114 = 0.000877$ kmol. Therefore:

$$Q = -4,221,118(0.000877) = 3702 \text{ kJ}$$

TEST Analysis

Launch the IG premixed closed-process combustion TESTcalc and select the Mass radio button. In the reaction panel, select octane(L) and Air in the fuel block, then enter their amounts in kg. In the products block, select the product components and execute Balance Reaction from the action menu. (An alternative is to balance a reaction in a non-premixed TESTcalc, manually enter the balanced reaction in a premixed reaction panel, then select Read As Is from the action menu.)

Switch to the state panel. Evaluate State-1 for the reactants from the given T1 and p1 and State-2 for the products from No: T2=600K and Vol2=Vol1. The final pressure is calculated as p2 = 3.15 atm. In the process panel, import the reactants and products as b- and f-states, enter W_ext=0, and click Calculate to obtain Q as –3696 kJ.

What-if scenario

In the state panel, make T2 an unknown, enter Q = 0 in the process panel, then click Super-Calculate. The new answers are p2 = 11.3 atm and T2 = 2148 K.

Discussion

The temperature calculated in the "what-if" study is the adiabatic flame temperature for constant-volume combustion, which is substantially higher than the corresponding constant-pressure adiabatic flame temperature (Ex. 13-6).

13.5 COMBUSTION EFFICIENCIES

Fossil fuels supply the required input to a great many thermal systems including furnaces, steam generators, gas turbines, burners, combustion chambers, and fuel cells. The desired output varies, and, depending on the application, combustion efficiency can assume a variety of meanings.

In the combustion chamber of a gas turbine, for example, the objective is to raise the temperature of the products to as high a temperature as the turbine blades can withstand. The typical air-fuel ratio in a gas turbine far exceeds the theoretical air-fuel ratio to moderate the products temperature. The fuel requirement, per unit mass of air, reaches a minimum when combustion is complete and there is no heat loss from the combustion chamber. To make up for incomplete combustion and heat loss, extra fuel is burned in a gas turbine combustor. The combustion efficiency is defined to reflect this:

$$\eta_{\text{comb,gas turbine}} \equiv \frac{(AF)_{\text{ideal}}}{(AF)_{\text{actual}}} \qquad (13.74)$$

where both the ideal and actual configuration produce the same exit temperature.

In a steam generator (Fig. 13.42), the objective is the transfer of heat into the steam, so the efficiency is defined as:

$$\eta_{\text{steam generator}} \equiv \frac{\dot{m}_{H_2O}\Delta h_{H_2O}}{\dot{m}_F(HHV)} \qquad (13.75)$$

where the numerator is the increase in flow energy and the denominator is the maximum possible heat released by the fuel.

The overall efficiency of a power plant, however, is defined as the ratio of the net work to the maximum possible heat released by the fuel:

$$\eta_{\text{overall}} \equiv \frac{\dot{W}_{\text{net}}}{\dot{m}_F(HHV)} = \frac{\dot{W}_{\text{net}}}{\dot{m}_{H_2O}\Delta h_{H_2O}} \frac{\dot{m}_{H_2O}\Delta h_{H_2O}}{\dot{m}_F(HHV)} = \eta_{\text{th}}\eta_{\text{steam generator}} \qquad (13.76)$$

The overall efficiency of a gas turbine or an internal combustion engine can be defined the same way.

Like a steam generator, all furnaces have a desired heat output. The efficiency of a furnace can be defined as the ratio of the heat output to the energy input; the latter, by convention, being the higher heating value. By directly measuring the heat output and the fuel flow rate, the furnace efficiency can be determined. Also, the efficiency can be indirectly determined from an analysis of the combustion products:

$$\eta_{\text{furnace}} \equiv \frac{\dot{Q}_{\text{used}}}{\dot{m}_F(HHV)} = \frac{h_R - h_P - \dot{Q}_L/\dot{m}_F}{HHV} = \frac{\bar{h}_R - \bar{h}_P - \dot{Q}_L/\dot{n}_F}{\overline{HHV}} \qquad (13.77)$$

where the numerator is obtained from an energy balance using Eq. (13.27) on the furnace shown in Figure 13.43.

A more logical choice of overall efficiency for a work producing device should be its exergetic efficiency [Eq. (6.19)], which compares the net work produced with the maximum possible useful work—the reversible work—that can be produced as the reactants convert to products. As already discussed, the maximum limit of this reversible work is the chemical exergy of the fuel, the useful work released when the products equilibrate with the dead state thermally, mechanically, and chemically. The values of chemical exergy, generally, lie somewhere between LHV and HHV. As a result, the overall exergetic efficiency is only slightly greater than the overall first-law efficiency based on HHV.

For a fuel cell, the exergetic efficiency, defined as the ratio of the actual work output to the reversible work, is a better choice than the thermal efficiency because there is no actual heat input. However, an energetic efficiency is still defined as the ratio of the reversible work output to the enthalpy change between the reactants and products:

$$\eta_{\text{fuel cell}} \equiv \frac{\dot{W}_{\text{rev}}}{\dot{n}_F(h_R - h_P)} = \frac{(g_R - g_P)}{(h_R - h_P)} = \frac{\Delta g_R}{\Delta h_R} \qquad (13.78)$$

There are two important issues that should be kept in mind with the above definition. First, the numerator is not the actual work output, but the ideal output based on a 100% exergetic efficiency. Also, the denominator is not the heat input, but the net flow energy input per unit mass of fuel, as can be seen from Eq. (13.62). In Ex. 13-10, the efficiency of the fuel cell can be calculated as 83% at 298 K and 89.8% at 500 K. While the fuel cell involves direct conversion of chemical energy into work, a heat engine converts heat to work. Therefore, comparing fuel cell efficiency with Carnot efficiency can be misleading.

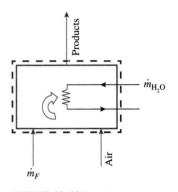

FIGURE 13.42 In a steam generator the desired output is the heat transferred to the steam.

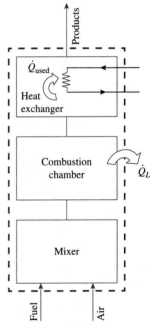

FIGURE 13.43 A gas-fired furnace schematic. $\dot{Q}_L$ is the heat lost and $\dot{Q}_{\text{used}}$ is the useful heat transfer (absolute values).

EXAMPLE 13-12   Efficiency of a Gas Turbine Combustor

The combustion chamber of a gas turbine uses diesel, which can be modeled as liquid dodecane ($C_{12}H_{26}$). During testing, the following data were obtained; Fuel temperature, 40°C; Air temperature, 375 K; Air velocity, 100 m/s; Products temperature, 1000 K; Products velocity, 150 m/s.

The fuel-air ratio was measured as 0.019 kg fuel/kg air. Determine the combustion efficiency. Use TEST.

SOLUTION

Using excess air as a parameter, determine the fuel-air ratio that produces zero heat and work transfer while satisfying the given inlet and exit conditions. Compare it with the actual ratio given.

Assumptions

IG mixture model for gaseous reactants and products. Changes in ke and pe are non-negligible.

TEST Analysis

Launch the open-steady non-premixed IG combustion TESTcalc and select the Mass radio button. Select Dodecane(L) and enter Lambda as 2.5. From the action menu execute Excess/Deficient Air to display the balanced reaction, on the basis of 1 kg of fuel. Calculate State-1 for fuel with T1=40 deg-C and Vel1=0. Similarly, evaluate State-2 for Air from the known T2 and Vel2, and State-3 for products for the given values of T3 and Vel3. In the device panel, import these states and enter Wdot_ext=0. Calculate Qdot as –16,879 kW. The negative value according to the WinHip sign convention signifies that heat must be rejected to achieve the exit temperature.

Increase the value of Lambda to 5 then balance the reaction by selecting Excess/Deficient Air from the action menu. Click Super-Calculate to produce Qdot as 8,675 kW—a positive quantity. A few more trials yield Lambda as 4.15 for adiabatic (Qdot close to zero) combustion. The ideal fuel-air ratio can be obtained from the balanced equation as 1/62.3 = 0.016 kg fuel/kg air; therefore, the combustion efficiency is 0.016/0.019 = 84.5%.

Manual Analysis

The problem asks for a TEST solution, but we will also outline a manual solution. Balance the complete reaction in terms of a single unknown stoichiometric coefficient ν_{O_2}. The energy equation for the open-steady adiabatic chamber, Eq. (13.43), can be extended to include the kinetic energy terms, producing $\bar{j}_P = \bar{j}_R$. Expanding:

$$\sum_r \nu_r \left(\bar{h}_{f,r}^o + \Delta\bar{h}_r^{IG} + \frac{\overline{M}_r V_r^2}{2000} \right) = \sum_p \nu_p \left(\bar{h}_{f,p}^o + \Delta\bar{h}_p^{IG} + \frac{\overline{M}_p V_p^2}{2000} \right)$$

where the velocity of the fuel can be neglected, velocity of oxygen and nitrogen is the same as the air velocity, and the velocity of all the products is same as the exit velocity. Knowing the temperature of the reactants and products, the enthalpy of each component can be evaluated using the procedure described in Ex. 13-5. The equation then yields the only unknown; ν_{O_2}. The resulting balanced reaction yields the ideal fuel-air ratio, from which the combustion efficiency can be obtained.

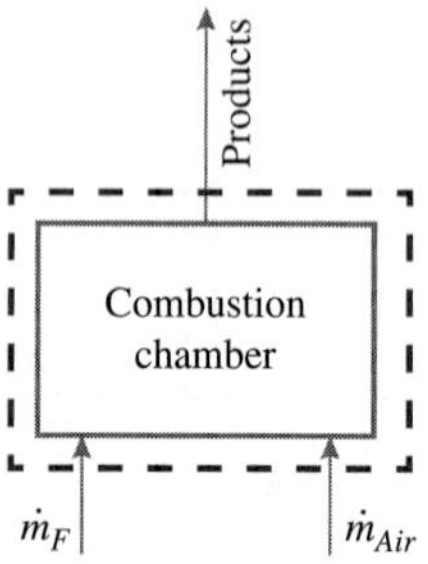

Discussion

The inlet and exit conditions of a combustor, as well as the mass flow rate of air, are usually specified during the design of the overall gas turbine cycle. If the combustor efficiency is known, the analysis presented in this example can be used to predict the fuel injection rate necessary to produce the desired turbine inlet conditions.

FIGURE 13.44 In the energy analysis of Ex. 13-12, the kinetic energy of air and products are not neglected.

13.6 CLOSURE

In this chapter we analyzed reacting systems in general and combustion systems in particular. Fundamental combustion concepts such as balancing a reaction, complete and theoretical reactions, indirect evaluation of air-fuel ratio through product analysis, equivalent ratio, etc., were introduced first. The governing mass, energy, and entropy equations, developed in Chapter 2, were customized for reacting systems, particularly for open and steady combustion chambers and closed processes involving combustion. The IG, PG and SL models, developed in Chapter 3, were expanded to take into account the formation enthalpy of a species. The third law of thermodynamics was briefly discussed in connection with entropy evaluation. Energy analysis of reacting systems led to the definitions of the various heating values of fuels and adiabatic flame temperature. An entropy analysis gave rise to the concept of reversible work, leading to an exergy analysis. It was shown that the fuel exergy was somewhere between its LHV and HHV. Two methods of extracting work from a reacting system were further analyzed; one led to a discussion of fuel cells, and the other established an optimum temperature for maximum power output in a thermal power plant. Reacting closed processes in rigid chambers were analyzed as an extension of concepts introduced in Chapter 5. Finally, various application-dependent combustion efficiencies were introduced. Several combustion TESTcalcs were used in this chapter to verify manual results and performing parametric studies.

PROBLEMS

SECTION 13-1: BALANCING THEORETICAL AND ACTUAL REACTIONS

13-1-1 Methane is burned with the theoretical amount of air during a combustion process. Assuming complete combustion, determine the air-fuel ratio on (a) mass basis, (b) mole basis, and (c) volume basis.

13-1-2 Acetylene (C_2H_2) is burned with the stoichiometric amount of air during a combustion process. Assuming complete combustion, determine (a) the air-fuel ratio on a mass basis and (b) the air-fuel ratio on a mole basis. (c) *What-if scenario:* What would the air to fuel ratio be if propene (C_3H_6) was burned instead of acetylene?

13-1-3 For a theoretical (stoichiometric) hydrogen–air reaction at 200 kPa, find (a) the fuel-to-air mass ratio, (b) the mass of fuel per unit mass of reactants, and (c) the partial pressure of water vapor in the products. Assume the products' temperature is high enough for all of the H_2O to be in vapor form. (d) *What-if scenario:* What minimum temperature, of the products, is required to ensure water stays as vapor (in the products).

13-1-4 One kmol of octane (C_8H_{18}) is burned with air that contains 21 kmol of O_2. Assuming the products contain only CO_2, H_2O, O_2 and N_2, determine the (a) mole of each gas in the products and (b) the mass-based air-fuel ratio for this combustion process.

13-1-5 Ethane (C_2H_6) is burned with 30% excess air during a combustion process. Assuming complete combustion and a total pressure of 100 kPa, determine (a) the mass-based air-fuel ratio and (b) the dew point temperature of the products. (c) *What-if scenario:* What would the air to fuel ratio be if methane was burned instead of ethane?

13-1-6 One kmol of ethane (C_2H_6) is burned with an unknown amount of air. If the combustion is assumed to be complete, and there are 2 kmol of free O_2 in the products, determine (a) the percent theoretical air used, (b) the excess air used (in percent), and (c) the equivalence ratio.

13-1-7 Determine the mass-based air-fuel ratio for hydrogen (H_2) burning with (a) 50% excess air and (b) 50% deficient air.

13-1-8 Determine the air-fuel ratio on a mass basis for the complete combustion of octane, C_8H_{18}, with (a) the theoretical amount of air. (b) *What-if scenario:* What would the theoretical amount of air be if complete combustion occurred in 120% theoretical air (20% excess air)?

13-1-9 Octane, C_8H_{18}, is burned with 150% theoretical air. Determine the (a) molar analysis of the products of combustion and (b) the dew point of the products if the pressure is 0.1 MPa.

13-1-10 Octane, C_8H_{18}, is burned with theoretical amount of air at 500 kPa. Determine (a) the air fuel ratio on a mole basis and (b) the air fuel ratio on a mass basis. (c) If the products are cooled at a constant pressure of 500 kPa, at what temperature will dew start to form? (d) Suppose the products are cooled below the dew point temperature so that all the water from the products is removed (neglect any vapor which may still be present at a very low amount). What is the volume fraction of CO_2 in the dry products? If you have a smog test report of your gasoline burning car at hand, compare the CO_2 amount (in percent).

13-1-11 Methane, CH_4, is burned with dry air. The molar analysis of the products on a dry basis is CO_2, 9.7%; CO, 0.5%; O_2, 2.95%; and N_2, 86.85%. Determine (a) the air fuel ratio, on both a molar and a mass basis, and (b) the percent theoretical air.

13-1-12 A natural gas has the following molar analysis: CH_4, 81.62%; C_2H_6, 4.41%; C_3H_8, 1.85%; C_4H_{10}, 1.62%, N_2, 10.50%. The gas is burned with dry air, giving products having a molar analysis on a dry basis: CO_2, 8.8%; CO, 0.2%; O_2, 6%; N_2, 85 %. Determine the air fuel ratio on a mole basis.

13-1-13 One kmol of ethane (C_2H_6) is burned with an unknown amount of air during a combustion process. An analysis of the combustion products reveals that the combustion is complete and there are 3 kmol of free O_2 in the products. Determine (a) the mass-based air fuel ratio and (b) the percentage of theoretical air (λ) used during the process.

13-1-14 Coal from Kentucky has the following analysis on a dry basis, percent by mass: S, 0.8%; H_2, 5.4%; C, 80.1%; O_2, 9.5%; N_2, 1.2%; Ash, 3% . This coal is burned with 40% excess air. Determine the (a) air-fuel ratio on a mass basis. (b) *What-if scenario:* What is the air to fuel ratio if the coal is burned with the theoretical amount of air?

13-1-15 Coal with a mass analysis of 79% carbon, 5% sulfur and 17% noncombustible ash burns completely with 110% of theoretical air. Determine the amount of SO_2 produced, in kg per kg of coal.

13-1-16 Diesel (dodecane) is burned with air at an air-fuel ratio of 30 kg of air/kg fuel. Determine the percent of theoretical air (λ) used.

13-1-17 Octane (C_8H_{18} in gaseous form) is burned with dry air. The volumetric analysis of the products on a dry basis is: 8.86% CO_2, 0.66% CO, 7.51% O_2, and 82.97% N_2. Determine (a) the mass-based air-fuel ratio and (b) the percentage of stoichiometric air used. If the initial pressure and temperature of the fuel air mixture are 100 kPa, 25°C, determine (c) the final pressure (p_2), if the combustion chamber is an insulated closed tank and the amount of fuel is 1 kmol.

13-1-18 Analysis of the dry exhaust products from a burner which uses natural gas and air reads 5% O_2 and 9% CO_2. Find the excess air used in this burner.

13-1-19 One kmol of ethane (C_2H_6) is burned with an unknown amount of air. If the combustion is complete, and there are 2 kmol of free O_2 in the products, determine (a) the percent of theoretical air (λ) used, (b) the excess air used (in percent), and (c) the equivalence ratio (φ).

13-1-20 Producer gas from bituminous coal has the following analysis on a mole basis: CH_4, 3%; H, 14%; N_2, 50.9%; O_2, 0.6%; CO, 27%; CO_2, 4.5%. It is burned with 30% excess air. Determine the air-fuel ratio on mass basis.

13-1-21 A fuel mixture with a molar analysis of 70% CH_4, 20% CO, 5% O_2 and 5% N_2 burns completely with 130% of theoretical air. Determine (a) the air-fuel ratio on a mass and (b) mole basis.

13-1-22 Octane (C_8H_{18}) is burned with dry air. The volumetric analysis of the products on a dry basis is: 10.02% CO_2, 0.88% CO, 5.62% O_2 and 83.48% N_2. Determine (a) the mass-based air-fuel ratio and (b) the percentage of theoretical air (λ) used.

13-1-23 A fuel mixture with a molar analysis of 65% CH_4, 25% C_2H_6, 10% N_2 is supplied to a furnace where it burns completely with 120% of theoretical air. Determine (a) the air-fuel ratio on a mass basis. (b) *What-if scenario:* What is the air to fuel ratio if the fuel has 25% of C_2H_4 instead of C_2H_6?

13-1-24 A dry analysis of products from the combustion of coal yields: 9.7% CO_2, 0.5% CO, 2.95% O_2, and the rest N_2 by volume. (a) Determine the percent of theoretical air (λ) used in the reaction and (b) the equivalent ratio (φ).

13-1-25 Carbon is burned with dry air. The volumetric analysis of the products produces 10.06% CO_2, 0.42% CO, and 10.69% O_2 (and the rest N_2). Determine (a) the air-fuel ratio on a mass basis, (b) the percentage of theoretical air (λ) used in the reaction, and (b) the equivalent ratio (φ).

13-1-26 Octane (C_8H_{18}) is burned with the theoretical amount of air at a pressure of 500 kPa. Determine (a) the air fuel ratio on a mole basis and (b) the air fuel ratio on a mass basis. (c) If the products are cooled at a constant pressure of 100 kPa, at what temperature will dew start to form? (d) Suppose the products are cooled way below the dew point temperature so that all the water from the products is removed (neglect any vapor which may still be present at very low amount). What is the volume fraction of CO_2 in the dry products? (e) If you have a smog test report of your gasoline burning car at hand, compare the CO_2 level of your analysis with the actual smog test result and discuss your finding.

13-1-27 A coal sample has a mass analysis of 76.39% carbon, 4.2% hydrogen (H_2), 5.32% oxygen (O_2), 1.63% nitrogen (N_2), 1.5% sulfur and the rest is ash. For complete combustion with 120% of the theoretical air, determine the air-fuel ratio on a mass basis.

13-1-28 A fuel mixture with the molar analysis 75% CH_4, 15% CO, 5% O_2 and 5% N_2 burns completely with 10% excess air in a reactor operating at steady state. If the mole flow rate of fuel is 0.2 kmol/h, determine the mole flow rate of air in kmol/h.

SECTION 13-2: OPEN SYSTEMS ANALYSIS

13-2-1 (a) Explain why the enthalpy ($\Delta \bar{h}_C^o$) of formation of CO_2 in MJ/kmol is the same as the heating value of carbon in MJ/kmol. (b) Express these quantities in the unit of MJ/kg.

13-2-2 In a combustion chamber, propane (C_3H_8) is burned at a rate of 5 kg/h and air enters at a rate of 140 kg/h. Determine (a) the percent of excess air used, if the reactants enter at 25°C, and also determine (b) the adiabatic flame temperature (T_{af}). Assume the products to be a perfect gas mixture with $c_p = 1.005$ kJ/kg · K.

13-2-3 Calculate the enthalpy of combustion of methane (CH_4) at (a) 298 K and (b) 500 K.

13-2-4 Calculate the enthalpy of combustion of propane (C_3H_8) at 25°C, on a kilogram basis.

13-2-5 In an adiabatic combustion chamber, methane (CH_4) is burned at a rate of 1 kg/s with 1 kg/s of oxygen. Both the fuel and oxidizer enter the chamber separately at 100 kPa and 300 K. Determine (a) the temperature of the products and (b) the rate of entropy generation ($\dot{S}_{gen}$).

13-2-6 Hydrogen (H_2) at 10°C is burned with 30% excess air, that is also at 10°C, during an adiabatic steady flow combustion process. Assuming complete combustion, (a) determine the exit temperature of the product gases using the PG mixture model with a uniform value of $c_p = 1.005$ kJ/kg · K. (b) *What-if scenario:* What is the exit temperature if the IG mixture model is used?

13-2-7 Methane gas, at 350 K and 1 atm, enters a combustion chamber, where it is mixed with air entering at 550 K and 1 atm. The products of combustion exit at 1500 K and 1 atm with the product analysis given in Problem 13-1-11. For an operation at steady state, neglecting KE an PE, determine (a) the rate of heat transfer ($\dot{Q}$) from the combustion chamber in kJ per kmol of fuel. Use the IG mixture model. (b) *What-if scenario:* What is the rate of heat transfer if the PG mixture model is used?

13-2-8 Diesel fuel is burned with 25% excess air in a steady-state combustor. Both fuel and air enter at 77°F. The products leave at 800°R. Assuming complete combustion, determine (a) the mass flow rate ($\dot{m}$) necessary to supply a heating rate of 2000 Btu/s and (b) the change in entropy (ΔS) per lbm of fuel. Assume the pressure remains constant at 14.7 psia.

13-2-9 Gasoline enters a combustion chamber, at 1 atm, 298 K and a rate of 0.09 kg/min, where it burns steadily and completely, with 70% excess air that enters the chamber at 1 atm, 200 K. If the exit temperature of the combustion products is 800 K, determine (a) the mass flow rate ($\dot{m}$) of air and (b) the rate of heat transfer ($\dot{Q}$). Use the IG mixture model for gases. Represent gasoline with liquid octane. (c) *What-if scenario:* What is the rate of heat transfer if the excess air used is only 20%?

13-2-10 Gaseous ethane (C_2H_6) and 300% excess oxygen, both at 25°C, 100 kPa, react in a steady-flow reaction chamber. The products exit at 3000 K. Determine the amount of heat transfer (q) per kg of ethane.

13-2-11 Determine the adiabatic flame temperature (T_{af}) for octane (C_8H_{18}) burning with 20% excess air. The inlet conditions are 100 kPa and 298 K. Use the IG model.

13-2-12 Benzene gas (C_6H_6) at 25°C is burned during a steady flow combustion process, with 90% theoretical air that enters the combustion chamber at 25°C. All the hydrogen in the fuel burns to H_2O, but part of the carbon burns to CO. If the products leave at 1500 K, determine (a) the mole fraction of the CO in the products (in percent) and (b) the heat transfer ($\overline{Q}$) from the combustion chamber during this process.

13-2-13 Methane (CH_4) enters a furnace at 100 kPa and 300 K. It is burned with 25% excess air that also enters at 300 K and 100 kPa. Assuming the exhaust temperature to be 500 K and the heat transfer load to be 20 kW, determine (a) the fuel consumption rate in kg/h. (b) *What-if scenario:* What is the fuel consumption rate if the theoretical amount of air is used? Assume complete combustion and use the IG model for gas mixtures.

13-2-14 Methane (CH_4) enters a steady flow adiabatic combustion chamber at 100 kPa and 25°C. It is burned with 100% excess air that also enters at 25°C and 100 kPa. Assuming complete combustion, determine (a) the temperature (T) of the products, (b) the entropy generation ($\dot{S}_{gen}$), and (c) the reversible work. Assume that $T_0 = 298$ K and the products leave the combustion chamber at 100 kPa.

13-2-15 Liquid octane (C_8H_{18}) enters the combustion chamber of a gas turbine steadily at 100 kPa, 25°C and it is burned with air that enters the combustion chamber at the same state. Disregarding any changes in KE and PE, determine the adiabatic flame temperature (T_{af}) for (a) complete combustion with 100% theoretical air and (b) complete combustion with 200% theoretical air. (c) *What-if scenario:* What is the answer in part (a) for incomplete combustion where some CO is produced with 80% theoretical air?

13-2-16 Methane gas, at 450 K and 100 kPa, enters a combustion chamber, where it burns steady and completely with the theoretical amount of air entering at 500 K and 100 kPa. The products of combustion exit at 1900 K and 100 kPa. For operation at steady state, neglecting the KE and PE, determine the rate of heat transfer ($\dot{Q}$) from the combustion chamber in kJ per kmol of fuel.

13-2-17 Ethene (C_2H_4), at 25°C and 1 atm, is burned with 300% excess air at 25°C and 1 atm. Assume that this reaction takes place reversibly at 25°C and that the products leave at 25°C and 1 atm. (a) Determine the reversible work (w_{rev}) for this process per kg of fuel. Use the IG mixture model. (b) *What-if scenario:* What is the reversible work if the PG mixture model is used?

13-2-18 Consider the same combustion process as in Problem 13-2-17, but let it take place adiabatically. Assume that each constituent in the product is at 1 atm and at the adiabatic flame temperature (T_{af}). The temperature of the surroundings is 25°C. Determine the irreversibility of the adiabatic combustion process per kg of fuel.

13-2-19 Ethene (C_2H_4), at 25°C and 1 atm, is burned with 300% excess air at 25°C and 1 atm. Assume that this reaction takes place adiabatically at 25°C and that the products leave at 25°C and 1 atm. Determine the irreversibility of the process per kg of fuel. Use the PG mixture model. Assume c_p to be uniform at 1.005 kJ/kg · K.

13-2-20 Liquid propane (C_3H_8) enters a combustion chamber, at 25°C, 100 kPa with a rate of 0.5 kg/min, where it is mixed and burned with 150% theoretical air which enters at 10°C. Only 90% of C is converted to CO_2, the rest to CO, while the hydrogen burns completely into H_2O. The products leave at 1000 K and 100 kPa. Determine (a) the mass flow rate $\dot{m}$ of air, (b) the rate of heat transfer ($\dot{Q}$), and (c) the rate of entropy generation ($\dot{S}_{gen}$), assuming the surroundings to be at 300 K.

13-2-21 Liquid octane enters an internal combustion engine operating at steady state with a mass flow rate ($\dot{m}$) of 0.0018 kg/s and is mixed with the theoretical amount of air. The fuel and air enter the engine at 25°C, 100 kPa. The mixture burns completely and the products leave the engine at 600°C. The engine develops a power output of 30 horsepower. Neglecting KE and PE, determine (a) the rate of heat transfer ($\dot{Q}$) from the engine. (b) *What-if scenario:* What is the rate of heat transfer if the PG model is used?

13-2-22 Octane (C_8H_{18}), at 25°C and 100 kPa, enters a well insulated reactor and reacts with air, entering at the same temperature and pressure. For steady state operation, and negligible effects of KE and PE, determine the temperature of the combustion products for the complete combustion with (a) the theoretical amount of air and (b) 300% theoretical air.

13-2-23 Ethylene (C_2H_4) enters an adiabatic combustion chamber at 25°C, 1 atm and is burned with 40% excess air that enters at 25°C, 1 atm. The combustion is complete and the products leave the combustion chamber at 1 atm. Assuming $T_0 = 25$°C, determine (a) the temperature (T) of the products, (b) the entropy generation ($\dot{S}_{gen}$), and (c) the irreversibility.

13-2-24 Calculate the enthalpy (Δh_C^o) of combustion of gaseous methane, in kJ per kg of fuel, (a) at 25°C, 100 kPa with water vapor in the products, (b) at 25°C, 100 kPa with liquid water in the products. (c) *What-if scenario:* What is the enthalpy in part (a) at 850 K, 1 atm?

13-2-25 Ethylene burns with 50% excess air, both entering at 25°C, 100 kPa. The product exits at the same temperature and pressure. Determine (a) the lower heating value, (b) the higher heating value and (c) the heat transfer (Q), per kmol of fuel, without assuming complete vapor or complete liquid in the product.

13-2-26 Propane burns with a theoretical amount of air, both entering at 25°C, 100 kPa. The products exits at the same temperature and pressure. Determine (a) the lower heating value (LHV) and (b) the higher heating value (HHV). (c) *What-if scenario:* What is the lower heating value if liquid propane is used?

13-2-27 Octane gas (C_8H_{18}) at 25°C is burned steadily with 50% excess air at 25°C, 100 kPa, and 40% relative humidity. Assuming combustion is complete, and the products leave the combustion chamber at 800 K, determine (a) the heat transfer ($\dot{Q}$) for this process. (b) *What-if scenario:* What is the heat transfer if hexane (C_6H_{14}) is used?

13-2-28 Liquid octane (C_8H_{18}) enters an internal combustion engine operating at steady state with a mass flow rate ($\dot{m}$) of 0.002 kg/s and is mixed with 10% excess air. Fuel and air enter the engine at 25°C, 1 atm and the products leave the engine at 500 K. The engine develops a power output ($\dot{W}_{ext}$) of 40 kW. (a) Assuming complete combustion, and neglecting KE and PE effects, determine the rate of heat transfer ($\dot{Q}$) from the engine. Use the PG mixture model. (b) *What-if scenario:* What is the rate of heat transfer if the IG mixture model is used?

13-2-29 Liquid octane (C_8H_{18}) enters an internal combustion engine operating at steady state with a mass flow rate ($\dot{m}$) of 0.002 kg/s and is mixed with 20% excess air. Fuel and air enter the engine at 25°C, 1 atm and the products leave the engine at 500 K. The engine develops a power output of 35 kW. Assuming complete combustion, and neglecting KE and PE effects, determine the rate of heat transfer ($\dot{Q}$) from the engine. Use the IG mixture model.

13-2-30 Repeat Problem 13-2-29 for 10% excess air. *What-if scenario:* What is the answer if the reactor pressure is 50 atm?

13-2-31 Repeat Problem 13-2-29 for 10% deficient air.

13-2-32 Octane (C_8H_{18}), at 25°C, 100 kPa, enters a combustion chamber and reacts with 100% theoretical air entering at the same conditions. Determine the adiabatoc flame temperature, assuming the complete combustion. Use the combustion Interactive linked from the left margin. (c) *What-if scenario:* How will the answer change if the air flow is increased to 120% theoretical air?

SECTION 13-3: CLOSED SYSTEMS ANALYSIS

13-3-1 A rigid tank contains a mixture of 1 lbm of methane gas and 5 lbm of O_2 at 77°F and 25 psia. Upon ignition, the contents of the tank burn completely. If the final temperature is 1500°R, determine (a) the final pressure (p_2) in the tank and (b) the amount of heat transfer (Q) during the process.

13-3-2 A 10 m^3 insulated, rigid tank contains a mixture of 1 kmol of octane (liquid) and the theoretical amount of air at 25°C. The contents are ignited and the mixture burns completely. Determine the final temperature (T_2) and pressure (p_2). Assume the products to be (a) a perfect gas mixture with a c_p of 1.005 kJ/kg · K and (b) an ideal gas mixture.

13-3-3 A mixture of 1 kmol of gaseous methane and 2 kmol of oxygen, initially at 298 K and 100 kPa, burns completely in a closed, rigid container. Heat transfer occurs until the products are cooled to 1000 K. If the reactants and the products each form ideal gaseous mixtures, determine (a) the amount of heat transfer (Q) and (b) the final pressure (p_2). (c) *What-if scenario:* What is the final pressure if the heat transfer occurs until the products are cooled to 800 K?

13-3-4 A constant-volume tank contains 1 kmol of methane (CH_4) gas and 3 kmol of O_2 at 25°C and 100 kPa. The contents of the tank are ignited and the methane gas burns completely. If the final temperature is 700°C, determine (a) the final pressure (p_2) in the tank and (b) the heat transfer (Q) during this process.

13-3-5 Ethane (C_2H_6) is burned with 200% theoretical air at 500 kPa. Assuming complete combustion at constant pressure, determine (a) the air-fuel ratio and (b) the dew point temperature of the products.

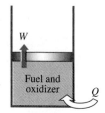

13-3-6 A closed combustion chamber is designed so that it maintains a constant pressure (p) of 120 kPa during the combustion process. The combustion chamber has an initial volume (V_1) of 0.6 m^3 and contains a stoichiometric mixture of octane (C_8H_{18}) gas and air at 25°C. The mixture is ignited and the product gases are observed to be at 900 K at the end of the combustion process. Assuming complete combustion, and treating both the reactants and the products as ideal gases, determine (a) the heat transfer (Q) from the combustion chamber during this process. (b) *What-if scenario:* What is the heat transfer if the initial volume is 0.8 m^3?

13-3-7 A constant-volume tank contains a mixture of 1 kmol of benzene (C_6H_6) gas and 20% excess air at 25°C and 1 atm. The contents of the tank are ignited and all hydrogen in the fuel burns to H_2O, but only 93% of the carbon burns to CO_2; the remaining 7% forms CO. If the final temperature (T_2) in the tank is 1000 K, determine (a) the heat transfer (Q) from the combustion chamber during this process. Use the IG mixture model. (b) *What-if scenario:* What is the heat transfer if the PG mixture model if used?

13-3-8 A constant-volume tank contains a mixture of 150 g of methane (CH_4) gas and 750 g air at 25°C, 150 kPa. The contents of the tank now ignited and part of the methane gas burns completely until all the oxygen is depleted. If the final temperature in the tank is 1100 K, determine (a) the final pressure (p_2) in the tank and (b) the heat transfer (Q) during this process.

13-3-9 One kmol of gaseous ethene (C_2H_4) and 4 kmol of oxygen at 25°C react in a constant volume bomb. Heat is transferred until the products are cooled to 800 K. Determine (a) the amount of heat transfer (Q) from the system. (b) *What-if scenario:* What is the amount of heat transfer if ethene reacts with 3 kmol of oxygen?

13-3-10 An adiabatic constant-volume tank contains a mixture of 1 kmol of hydrogen (H_2) gas and the stoichiometric amount of air at 25°C and 1 atm. The contents of the tank are ignited. Assuming complete combustion, determine (a) the final temperature (T_2) and (b) pressure (p_2) in the tank. Use the PG mixture model. (c) *What-if scenario:* What will the final temperature and pressure be if the IG mixture model is used?

13-3-11 Consider the same combustion process as in Problem 13-2-17, but assume that the reactants consists of a mixture at 1 atm, 25°C and that the product also consist of a mixture at 1 atm, 25°C. Determine the heat transfer per unit mole of fuel.

14

EQUILIBRIUM

Classical thermodynamics is based on the concept of equilibrium. We have discussed this important concept in varying degrees in Chapters 1, 2, 3, and 11. In this chapter, we fully explore the criteria for equilibrium, exploiting the third postulate of the second law of thermodynamics, introduced in Chapter 2. After establishing the general criterion, called the Gibbs state postulate, mechanical and thermal equilibrium are studied for single-phase systems, in which we analyze equilibrium distribution of properties in large non-uniform systems. In multi-phase systems that allow phase transformation, we study how individual components of a mixture distribute themselves among multiple phases after equilibrium is achieved. Finally, we apply the principle of equilibrium to reacting systems. Even when sufficient time is allowed, a theoretical mixture of reactants may achieve an equilibrium state that is very different from the prediction of a complete overall reaction. A general methodology will be developed to obtain the equilibrium distribution of products in reacting systems. Just as thermal equilibrium removes any temperature imbalance, mechanical equilibrium eliminates force (pressure) imbalance, and chemical equilibrium (as will be shown) eliminates the imbalance of a new property, called the *chemical potential*, among different phases or species. The powerful chemical equilibrium TESTcalc, which has similar capabilities as the industry benchmark NASA CEA, is be used for verification of manual results and analyzing complex equilibrium problems involving multiple species for which a manual solution is prohibitively time consuming. The Interactive for simulating a combustion chamber, introduced in Chapter 13, is used for parametric studies involving equilibrium temperature and emissions.

14.1 CRITERIA FOR EQUILIBRIUM

Entropy Maximum Principle:

The third postulate of the second law (Sec. 2.1.3) states that entropy can only be generated, not destroyed. Applied to an isolated system (Fig. 14.1), the entropy balance equation simplifies to $\frac{dS}{dt} = \dot{S}_{gen} \geq 0$. This means the entropy S of an isolated system can never decrease. As an isolated system changes spontaneously, due to the presence of internal mechanical, thermal, or chemical imbalances, only those changes are allowed by the second law, for which entropy is generated. With no possibility of an entropy transfer through mass or heat transfer and entropy destruction forbidden by the second law, the entropy of the system increases monotonically. This is illustrated in Figure 14.2, where the system's entropy is shown to continually increase as the system evolves from its initial non-equilibrium state (with many internal imbalances). The mechanisms of entropy generation, collectively called the *thermodynamic friction,* include mechanical or viscous friction driven by velocity gradient; thermal friction driven by temperature gradient; mixing friction driven by concentration gradient; and chemical friction, studied in Chapter 13, for which the driving force will be established to be the gradient of a new property called *chemical potential.*

We know from experience that all these internal imbalances gradually subside as the strength of the driving forces gradually decreases. This is accompanied by a gradual reduction in the intensity of entropy generation. After sufficient time, there are no driving forces left to cause any change, and the system achieves internal mechanical, thermal, and chemical balance and said to be in **thermodynamic equilibrium**. The temperature, pressure, and *chemical potential* of all components (different phase or chemical species) become uniform at equilibrium (see Anim. 3.A.*equilibrium*). With the entropy generation

FIGURE 14.1
An isolated system (see Anim. 14.A. *isolatedSystemEx*) spontaneously marches towards equilibrium, as thermodynamic friction accompanied by entropy generation gets rid of all internal imbalances.

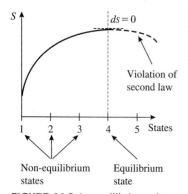

FIGURE 14.2 At equilibrium, the entropy of an isolated system reaches a maximum (see Anim. 14.A.*isolatedSystem*).

coming to a standstill, the entropy of the isolated system reaches a maximum at equilibrium, as shown in Figure 14.2 (and Anim. 14.A.*isolatedSystem*). Known as the **entropy maximum principle**, the condition for equilibrium in an isolated system can be stated (Fig. 14.2):

$$dS = 0; \quad \left[\frac{kJ}{K}\right] \tag{14.1}$$

and

$$\dot{S}_{gen} = 0; \quad \left[\frac{kW}{K}\right] \quad \text{since} \quad \frac{dS}{dt} = \dot{S}_{gen}; \quad \Rightarrow \quad dS^0 = \dot{S}_{gen}\, dt \tag{14.2}$$

where Eq. (14.1) directly expresses the entropy maximum principle, and the second equation is deduced as a corollary for an isolated system.

EXAMPLE 14-1   Entropy Maximum Principle

A 2 kg copper block at 100°C is brought in thermal contact with a 5 kg aluminum block at 50°C. Treating the combined system as an isolated system, determine the final temperature (a) by assuming that at equilibrium the temperatures of the two blocks are equal and (b) by using the entropy maximum principle, without assuming the final temperatures to be equal.

SOLUTION

Perform an energy and entropy analysis of the closed process executed by the combined non-uniform system within the red boundary of Figure 14.3.

Assumptions

The SL model is applicable to the subsystems and a composite state made of two local states in LTE can describe the global beginning and final states. No changes in KE or PE.

Analysis

The copper block is sub-system A and the aluminum block is sub-system B. From Table A-1, or any SL TESTcalc, we obtain $c_{vA} = 0.386$ kJ/kg · K and $c_{vB} = 0.90$ kJ/kg · K. Designating the composite beginning state (bA and bB states) by States 1 and 2 and the composite final state (fA and fB states) by States 3 and 4 (Fig. 14.3), the energy equation for the closed process, Eq. (5.2), can be simplified:

$$\Delta U = \cancel{Q}^0 - \cancel{W}^0; \quad \Rightarrow \quad \Delta(U_A + U_B) = 0; \quad \Rightarrow \quad \Delta U_A = -\Delta U_B;$$
$$\Rightarrow \quad m_A c_{vA}(T_{fA} - T_{bA}) = m_B c_{vB}(T_{bB} - T_{fB});$$
$$\Rightarrow \quad m_A c_{vA}(T_3 - T_1) = m_B c_{vB}(T_2 - T_4)$$

Substituting the known variables, the energy equation reduces to:

$$T_3 = 2256 - 5.83 T_4 \quad [\text{K}] \tag{14.3}$$

Given (and also known from our experience) that $T_4 = T_3 = T_f$, Eq. (14.3) produces T_f as 330.3 K or 57.3°C.

(b) The entropy of the combined system, which is isolated, reaches a maximum at equilibrium. To locate the maximum, one approach is to plot the entropy of the combined system against all possible final states. The entropy of the composite system can be expressed in terms of T_3 and T_4 using the SL model:

$$S_f - S_b = (S_{fA} + S_{fB}) - (S_{bA} + S_{bB});$$
$$\Rightarrow \quad \Delta S = (S_{fA} - S_{bA}) + (S_{fB} - S_{bB}) = m_A c_{vA} \ln\frac{T_{fA}}{T_{bA}} + m_B c_{vB} \ln\frac{T_{fB}}{T_{bB}};$$
$$\Rightarrow \quad S_f = (S_b - m_A c_{vA} \ln T_1 - m_B c_{vB} \ln T_2) + m_A c_{vA} \ln T_3 + m_B c_{vB} \ln T_4$$

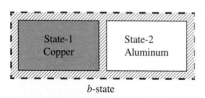

b-state

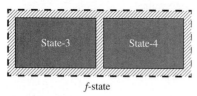

f-state

FIGURE 14.3
Schematic for Ex. 14-1 (see Anim. 14.A.*thermalEquilibrium*).

Representing the terms inside the parentheses by a constant c, and substituting the constraint from the energy equation, Eq. (14.3), the final entropy can be expressed as a function of T_4 alone:

$$S_f = c + 0.772 \ln (2256 - 5.83T_4) + 4.5 \ln T_4 \quad [\text{kJ/K}] \qquad \textbf{(14.4)}$$

$(S_f - c)$, evaluated from this expression, is plotted against T_4 (which represents the possible final temperature of sub-system B) in Figure 14.4. To obtain the precise temperature at which the entropy peaks, set the derivative of S_f, with respect to T_4, to zero and solve the resulting equation to show that $T_4 = 57.3°C$ at the maximum. This result is consistent with the maximum exhibited by the graph in Figure 14.4.

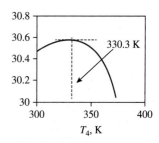

FIGURE 14.4

Entropy $(S_f - c)$ of the combined system (in kJ/K) as a function of T_4 in Ex. 14-1.

TEST Analysis

Launch the SL/SL non-uniform, non-mixing, closed-process TESTcalc. Evaluate State-1 and State-2 with the given information (mass and temperature) and leave the temperatures unknown for State-3 and State-4. For each state, make sure that the working substance is correctly selected (there are two working substance selectors, the left one can be used for sub-system A and the other one for sub-system B). In the process panel, set up the adiabatic process by selecting the composite anchor states and entering process variables Q=W_ext=0. Now guess a value of T3, say, 75 deg-C, and click Super-Calculate. The entropy change of the system, Delta_S, is calculated as 0.0058 kJ/K. Change T3 to a new value and Super-Calculate to update the value of Delta_S. It can be verified that when T3 is 57.3 deg-C, entropy change reaches a maximum, and T4 becomes equal to T3. See TEST-code (*TEST > TEST-codes*) for further details.

Discussion

If the two subsystems consist of two gases separated by a sliding piston, instead of two solid blocks, the entropy maximum principle leads to equality of pressure as well (see Anim. 14.A. *mechEquilibrium*). While the equality of temperature or pressure among subsystems at equilibrium is almost trivial, our experience falls far short in predicting more challenging equilibrium conditions when there is a phase or chemical transformation involved.

Maximization of Reversible Work Principle:

Another approach is to understand that useful work can be extracted—exploiting the internal mechanical, thermal, and chemical imbalances in a system—by building innovating reversible devices (see Anim. 14.A.*revWorkMax*). For instance, if there is a temperature difference inside a system, we could install a Carnot heat engine and produce work until the temperature becomes uniform throughout the system. When a system reaches a thermodynamic equilibrium, however, with all the driving forces for any spontaneous internal changes completely gone, no useful work can be extracted from within the system. This is not to say that the system is at its dead state, since it may not be in equilibrium with its surroundings (a condition for exergy to be zero). Reversible work is the maximum limit of the work (Sec. 6.1.1) that can be extracted from a system; therefore, setting reversible work to zero becomes an alternative criterion for equilibrium. For an isolated system, with no external work transfer, we can write $\dot{W}_{rev} = \dot{I} = T_0 \dot{S}_{gen} = 0$, from Eq. (6.16). This leads to the same equation as Eq. (14.2). As it turns out, $\dot{W}_{rev} = 0$ is a general criteria of equilibrium, from which all other criterion of equilibrium can be derived.

There are many practical situations, where the system is not isolated, but its temperature remains constant as the system marches towards equilibrium. Let us consider a large isothermal system, such as a natural gas reservoir, and deduce the criterion for equilibrium using the fact that reversible work that can be extracted from the system must be zero at equilibrium. To deduce a quantitative criterion in terms of state properties, assume a conceptual device (Fig. 14.5) extracts the reversible work in a pseudo-steady manner while the working fluid gets rid of the internal imbalances during its quest for equilibrium. The boundary of the system is drawn internally, as we are interested in the equilibrium of the internal system rather than the system's universe. With heat transfer, if any, taking place across the boundary at T, the energy and entropy equations, Eqs. (4.2) and (4.3), for the device can be written:

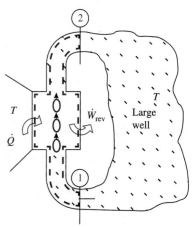

FIGURE 14.5 When an isothermal system reaches equilibrium, the combination property $g + ke + pe$ is minimized (see Anim. 14.A.*revWorkMax*).

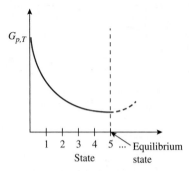

FIGURE 14.6 A system going through spontaneous isobaric and isothermal processes achieves equilibrium when $G_{p,T}$ is minimized.

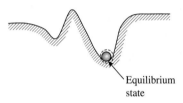

FIGURE 14.7 When an isothermal mechanical system achieves equilibrium, mechanical energy $ke + pe$ is minimized.

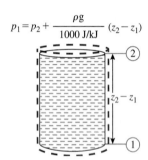

FIGURE 14.8 Hydrostatic pressure variation in an isothermal incompressible liquid.

$$0 = -\dot{m}\Delta j + \dot{Q} - \dot{W}_{ext}; \quad [\text{kW}] \quad 0 = -\dot{m}\Delta s + \frac{\dot{Q}}{T} + \dot{S}_{gen}; \quad \left[\frac{\text{kW}}{\text{K}}\right] \quad \textbf{(14.5)}$$

where Δj and Δs represent changes in j and s between two arbitrary local states, State-2 and State-1. Eliminating $\dot{Q}$ between the two equations and substituting $j \equiv h + ke + pe$ and $g \equiv h - Ts$ (specific Gibbs function), the external work can be expressed:

$$\dot{W}_{ext} = -\dot{m}\Delta(j - Ts) - T\dot{S}_{gen} = -\dot{m}(\Delta g + \Delta ke + \Delta pe) - T\dot{S}_{gen}; \quad [\text{kW}] \quad \textbf{(14.6)}$$

The last term in this equation is the internal irreversibility, which quantifies the lost work due to thermodynamic friction. Introducing the definition of reversible work—reversible work is the maximum work output when thermodynamic friction is completely eliminated without changing the anchor states (Section 6.1.1)—we rearrange Eq. (14.6) to obtain:

$$\frac{\dot{W}_{rev}}{\dot{m}} \equiv \dot{W}_{ext} + T\dot{S}_{gen} = -\Delta(g + ke + pe); \quad \left[\frac{\text{kW}}{(\text{kg/s})} = \frac{\text{kJ}}{\text{kg}}\right] \quad \textbf{(14.7)}$$

A drop in the combination property $g + ke + pe$, therefore, is the reversible work output for a given pair of inlet and exit states, provided the temperature remains unchanged. As a system marches towards equilibrium, the reversible work output $\dot{W}_{rev}/\dot{m}$, and $g + ke + pe$, monotonically decreases (Fig. 14.6), there is no more driving force left for changes that can be exploited to produce work, when the system is in equilibrium, and $g + ke + pe$ is minimized so that $d(\dot{W}_{rev}/\dot{m}) = d(g + ke + pe) = 0$ at equilibrium. This means that between any two local states, within the system, the combination property $g + ke + pe$ must be equal when all the internal differences have been eliminated.

Several interesting limiting situations can be derived from this conclusion. In motion, involving solids under atmospheric conditions, the use of the SL model yields $g = h - Ts = g(T)$. In isothermal systems involving solid objects (Fig. 14.7), therefore, the mechanical energy of the system $ke + pe$ is minimized at equilibrium—this is a well known theorem in mechanics and we are somewhat familiar with it in our daily life (just watch a pendulum come to equilibrium after you give it a push). If the kinetic energy is not significant, Eq. (14.7) for isothermal systems simplifies to:

$$\Delta g_T + \Delta(pe) = 0; \quad \text{or}, \quad dg_T + d(pe) = 0; \quad \left[\frac{\text{kJ}}{\text{kg}}\right] \quad \textbf{(14.8)}$$

Subscript T in the above equation underscores the fact that the system is isothermal at T. For a liquid in hydrostatic equilibrium, the equilibrium condition leads to a familiar formula (Eq. 1.10) for hydrostatic pressure variation (Fig. 14.8) as follows:

$$\Delta g_T + \Delta(pe) = 0; \quad \Rightarrow \quad \Delta(h_T - Ts) + \Delta(pe) = 0;$$

$$\Rightarrow \quad \Delta(h - Ts)_T + \Delta(pe) = 0; \quad \Rightarrow \quad \Delta(u - Ts + pv)_T + \Delta(pe) = 0;$$

$$\Rightarrow \quad v\Delta p + \Delta\left(\frac{gz}{1000 \text{ J/kJ}}\right) = 0; \quad \Rightarrow \quad \Delta p = -\Delta\left(\frac{\rho gz}{1000 \text{ J/kJ}}\right)$$

$$\Rightarrow \quad p_2 - p_1 = \frac{\rho g}{(1000 \text{ J/kJ})}(z_1 - z_2); \quad [\text{kPa}] \quad \textbf{(14.9)}$$

For small systems, the variation in potential energy is negligible and pressure can also be considered uniform throughout the system. The equilibrium criterion simplifies further to what is known as the **minimization of Gibbs function principle**, expressed:

$$\Delta g_{p,T} = 0, dg_{p,T} = 0; \quad \left[\frac{\text{kJ}}{\text{kg}}\right], \quad \Delta G_{p,T} = 0, \quad \text{and} \quad dG_{p,T} = 0; \quad [\text{kJ}] \quad \textbf{(14.10)}$$

Equation (14.10) means that out of all possible states, the one with the minimum value of Gibbs function represents the equilibrium state. Subscripts p and T emphasize that pressure and temperature must remain unchanged as the system transitions to equilibrium. It is left as an exercise to show that the last two of these relations can also be directly obtained by analyzing the series of spontaneous processes executed by a closed local system, sketched in Figure 14.5, at a constant pressure and temperature.

Although we arrived at the minimization of Gibbs function principle using phenomeno-logical arguments, it can be rigorously derived from the entropy maximum principle. Think about an irreversible, differential process, involving a closed system at constant temperature and pressure, that allows heat and boundary work (only $pd\Psi$ type) transfer, dQ and dW_B. If the pressure remains constant, $dW_B = pd\Psi$ is the actual work transfer, even for an irreversible process, and the energy and entropy equations for a differential process, Eqs. (5.29) and (5.31), can be written:

$$dQ = dU + pd\Psi; \text{ and } TdS = dQ + TdS_{\text{gen,int}}; \quad [\text{kJ}] \qquad (14.11)$$

The differential of the Gibbs function at constant p and T can be manipulated using these relations:

$$dG_{p,T} = d(H - TS) = d(U + p\Psi - TS)$$
$$= dU + pd\Psi - TdS = dQ - (dQ + TdS_{\text{gen,int}})$$
$$\Rightarrow \quad dG_{p,T} = -TdS_{\text{gen,int}}; \quad [\text{kJ}] \qquad (14.12)$$

The second law dictates that $dS_{\text{gen,int}} \geq 0$, and therefore, at constant p and T, $dG_{p,T} \leq 0$. As the system marches toward equilibrium, the Gibbs function decreases due to entropy generation. At equilibrium, when the entropy reaches a maximum and entropy generation ceases, substitution of $dS_{\text{gen,int}} = 0$ from the entropy maximum principle, Eq. (14.1), leads to the Gibbs function minimization principle of Eq. (14.10).

Any form of Eq. (14.10) is known as the **general equation for equilibrium** at constant pressure and temperature. As simple as it looks, this is a powerful relation that leads us to the equilibrium composition of a multi-component, multi-phase system, with or without the possi-bility of chemical reaction.

Chemical Potential:

Given the importance of the Gibbs function, let us review formulas developed in earlier chapters for evaluating G of a mixture. Recognizing that G is an extensive property of the mixture, we can use Eqs. (11.115) and (11.116) to express:

$$G = \sum_k \bar{g}_k n_k = G(p, T, n_i); \quad [\text{kJ}] \qquad (14.13)$$

where

$$\bar{g}_k \equiv \left(\frac{\partial G}{\partial n_k} \right)_{p,T,n_{i,i \neq k}} = \bar{g}_k(p, T, y_{i,i \neq k}); \quad \left[\frac{\text{kJ}}{\text{kmol}} \right] \qquad (14.14)$$

In this expression, $\bar{g}_k$ is the partial molar Gibbs function, which measures the change in the Gibbs function of the mixture as a unit mole of a component k, at constant pressure and temperature, is introduced into the mixture (see Anim. 11.C.*partialMolarProperty*). Gibbs recognized its similarity to electric and gravitational potential and called $\bar{g}_k$ the **chemical potential** of component k (see Anim. 14.B.*chemicalPotential*). As we will see shortly, the chemical potential is responsible for the movement of chemicals from one substance to another; just as electric potential drives a charge in an electric field (which is not allowed in a simple system), a difference in temperatures drives heat transfer or a pressure difference drives a flow. Chemical potential is generally symbolized by μ, but to be consistent with all other molar specific properties, in this book we will use the bar on top the symbol:

$$\bar{\mu}_k(p, T, y_i) \equiv \bar{g}_k(p, T, y_i) = \bar{M}_k g_k(p, T, y_i); \quad \left[\frac{\text{kJ}}{\text{kmol}} \right] \qquad (14.15)$$

We will use the terms *chemical potential* and *molar specific Gibbs function* interchange-ably. However, while explaining the movement of a species, the term chemical potential is preferred. Just as temperature, an intensive property, determines the direction of heat transfer for a system to reach thermal equilibrium, we will soon deduce that the chemical potential of a component in a mixture, also an intensive property, decides the direction of migration for that component during diffusion, phase transformation, or even chemical reactions to reach chemical equilibrium (see Anim. 14.A.*chemEquilibrium*). In fact, the chemical potential is such a fundamental thermodynamic property in its own right, that a new unit called the Gibb (1 G = 1 kJ/kmol) has been proposed for its measurement.

For a pure substance (a single-component mixture), Eq. (14.14) simplifies to $\bar{\mu} = \bar{g} = \bar{h} - T\bar{s}$; that is, the molar specific Gibbs function is also the chemical potential of a pure substance (see Anim. 14.C.*chemPotentialPure*). The chemical potential for an isothermal liquid, for example, can be further simplified by applying the SL model $\bar{\mu} = c_1 + c_2 p$, where c_1 and c_2 are constants. In a hydrostatic equilibrium, which was shown by Eq. (14.9) to be a special case of chemical equilibrium, the chemical potential not change along the horizontal direction, and its change in the vertical direction is compensated by a change in pe so that $g + $ pe is minimized. A mass flow driven by a pressure difference can be thought of as being driven by a difference in chemical potential.

14.2 EQUILIBRIUM OF GAS MIXTURES

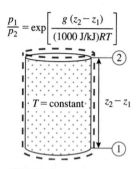

$$\frac{p_1}{p_2} = \exp\left[\frac{g\,(z_2 - z_1)}{(1000 \text{ J/kJ})RT}\right]$$

FIGURE 14.9 Pressure distribution in a gas column at a constant temperature T.

Let us consider the isothermal gas column shown, in Figure 14.9. Applying the equilibrium criterion, Eq. (14.8), between the two states, State-1 and State-2 at the bottom and top, we obtain:

$$0 = \Delta\bar{h}^0 - T\Delta s + \Delta(\text{pe}); \quad \Rightarrow \quad \Delta\left(\frac{gz}{1000 \text{ J/kJ}}\right) = T\left[c_p \ln\frac{T_2}{T_1}^{\,1} - R\ln\frac{p_2}{p_1}\right]; \quad \left[\frac{\text{kJ}}{\text{kg}}\right]$$

$$\Rightarrow \quad \frac{p_1}{p_2} = \exp\left[\frac{g(z_2 - z_1)}{(1000 \text{ J/kJ})RT}\right] \quad \text{(14.16)}$$

For an ideal gas mixture, Dalton's model allows us to apply the above result to each component k of the mixture, with p_1 and p_2 replaced with the partial pressures p_{k1} and p_{k2} at the corresponding states, and the gas constant R replaced with R_k. This is illustrated in Example 14-2.

EXAMPLE 14-2 Equilibrium of an Ideal Gas Mixture

A natural gas reservoir contains a mixture of methane and nitrogen at 298 K. At the top of the well, the mixture pressure is 100 kPa and the molar composition is found to be 50% methane and 50% nitrogen. Determine (a) the pressure and (b) the molar composition at the bottom, 3 km below the surface. Assume uniform temperature and ideal gas behavior.

SOLUTION

Determine the partial pressure of each component at the bottom of the well, using Eq. (14.16), from which the molar composition and total pressure can be deduced.

Assumptions

Uniform temperature throughout the well. The IG mixture model is applicable for the gas mixture.

Analysis

From Table C-1, or any IG TESTcalc, we obtain R_{CH_4} as 0.512 kJ/kg·K and R_{N_2} as 0.297 kJ/kg·K. Representing the state of the gas at the bottom and top by State-1 and State-2, respectively, the component partial pressures can be evaluated at the two states as:

$$p_{CH_4,2} = y_{CH_4,2}\,p_2 = (0.5)(100) = 50 \text{ kPa};$$

$$p_{N_2,2} = p_2 - p_{CH_4,2} = 50 \text{ kPa}$$

Using Eq. (14.16),

$$p_{CH_4,1} = p_{CH_4,2}\exp\left[\frac{g(z_2 - z_1)}{(1000 \text{ J/kJ})R_{CH_4}T}\right] = (50)\exp\left[\frac{(9.81)(3000)}{(1000 \text{ J/kJ})(0.512)(298)}\right]$$

$$= 60.6 \text{ kPa}$$

Similarly,

$$p_{N_4,1} = (50) \exp\left[\frac{(9.81)(3000)}{(1000 \text{ J/kJ})(0.297)(298)}\right]$$

$$= 69.7 \text{ kPa}$$

Therefore, the total pressure at the bottom, State-1, is:

$$p_1 = p_{CH_4,1} + p_{N_2,1} = 60.6 + 69.7 = 130.3 \text{ kPa}$$

And the molar composition is: $y_{CH_4,1} = p_{CH_4,1}/p_1 = 60.6/130.3 = 0.465$;

$$y_{N_4,1} = 1 - p_{CH_4,1} = 1 - 0.465 = 0.535$$

TEST Analysis

Launch the n-IG system-state TESTcalc. Add 1 kmol of methane as the only component. Set up State-2 with p2=50 kPa, z2=3000 m, and T2=298 K. Evaluate State-1, with T1=T2, z1=0, and g1=g2+9.81*z2/1000, yielding p1 as 60.5 kPa. Similarly, find the pressure of nitrogen by repeating the solution for pure nitrogen. Given that the two gases exist independently, the evaluated pressures are partial pressures from which the mole fractions can be evaluated.

Discussion

Methane and nitrogen coexist in Dalton's model without any special intermolecular attraction or repulsion. An ideal solution can be similarly treated, with each component being in equilibrium independently. For example, under equilibrium condition, the concentration of salt can be expected to vary with depth in the oceans. However, the time required to reach equilibrium through diffusion is much longer than the time of mixing due to ocean currents; therefore, oceans have a remarkably constant concentration of salt at all depths.

Separating the components of a mixture has been a longstanding engineering problem. Imagine how great it would be to inexpensively separate oxygen from air, or drinking water from sea water. The equilibrium criteria we have developed can be used to explore the minimum costs of such separation. One method of separation is to use special membranes that allow preferential passage of a particular component while blocking others. The minimum work required to separate a mixture can be evaluated if we first calculate the work of reversible mixing, the opposite (or reverse) process of separation.

For an ideal gas mixture, application of the Dalton model simplifies the expression for the partial molar Gibbs function as $\bar{g}_k(p, T, y_i) = \bar{g}_k(p_k, T)$. For a given p and T, the temperature and pressure contributions to $\bar{g}_k$ (or $\bar{\mu}_k$) can be separated (see Anim. 14.B.*GibbsFunctionIG*) to give:

$$\bar{g}_k(p_k, T) \equiv \bar{h}_k(T) - T\bar{s}_k(p_k, T) = [\bar{h}_k(T) - T\bar{s}_k^{\circ}(T)] + \bar{R}T \ln\frac{p_k}{p_0}$$

$$= \bar{g}_k^{\circ}(T) + \bar{R}T \ln\frac{p_k}{p_0} = \bar{g}_k^{\circ}(T) + \bar{R}T \ln\frac{p}{p_0} + \bar{R}T \ln\frac{p_k}{p}$$

$$\Rightarrow \quad \bar{g}_k(p_k, T) = \bar{\mu}_k(p_k, T) = \bar{g}_k(p, T) + \bar{R}T \ln\frac{p_k}{p}; \quad \left[\frac{\text{kJ}}{\text{kmol of } k}\right] \qquad \textbf{(14.17)}$$

Note that $\bar{g}_k^{\circ}(T)$ is the Gibbs function (chemical potential) of the pure component k at standard pressure p_0 and temperature T, while $\bar{g}_k(p, T)$ is the Gibbs function (chemical potential) at the total pressure p and temperature T. In this derivation, we have substituted Eq. (13.47) for the partial molar entropy. For an ideal gas mixture, application of the Dalton model (Sec. 11.4.3) produces $p_k = y_k p$, and Eq. (14.17) can be further simplified (click Pure vs Partial in Anim. 14.B.*GibbsFunctionIG*) as:

$$\bar{g}_k(p_k, T) = \bar{\mu}_k(p_k, T) = \bar{g}_k(p, T) - \bar{R}T \ln(1/y_k); \quad \left[\frac{\text{kJ}}{\text{kmol of } k}\right] \qquad \textbf{(14.18)}$$

The first term in Eq. (14.18) is the chemical potential of a pure component evaluated at the mixture conditions p and T, while the second term is a correction term that depends logarithmically

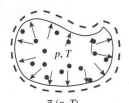

FIGURE 14.10 The same amount (mole) of red molecules at the same pressure and temperature has a lower partial molar Gibbs function in the mixture than in the pure system, as dictated by Eq. (14.18).

on the mole fraction y_k of the component in the mixture (Fig. 14.10). The lower the mole fraction of a species in a mixture, the lower its chemical potential. Using Eq. (14.18) and the definition of mole fraction, $y_k = n_k/n$, we can express the total Gibbs function of a mixture, $G(p, T)$, as:

$$G(p, T) = \sum_k n_k \bar{g}_k(p_k, T) = \sum_k n_k \left[\bar{g}_k(p, T) - \bar{R}T \ln (1/y_k) \right]$$

$$\Rightarrow \quad G(p, T) = \sum_k n_k \bar{g}_k(p, T) - n\bar{R}T \sum_k y_k \ln (1/y_k); \quad [kJ] \qquad \textbf{(14.19)}$$

Therefore, we expect the total Gibbs function of a mixture to be less than the weighted sum of the chemical potentials of the pure components at the same pressure and temperature, or: $G(p, T) < \sum_k n_k \bar{g}_k(p, T)$.

Now consider steady-state mixing among different gases (Fig. 14.11a) at a constant temperature and pressure. Heat transfer is allowed so that the temperature does not change and frictional loss is assumed negligible, thus the pressure at each inlet, and the total pressure at the exit, are all equal. Neglecting changes in ke and pe, the energy and entropy equations, Eqs. (11.118) and (11.119), for the isothermal, isobaric mixing system can be written in molar terms:

Energy: $\quad 0 = \sum_{k(\text{inlet})} \dot{n}_k \bar{h}_k(T) - \sum_{k(\text{exit})} \dot{n}_k \bar{h}_k(T) + \dot{Q} - \dot{W}_{\text{ext}}; \quad [kW]$

Entropy: $\quad 0 = \sum_{k(\text{inlet})} \dot{n}_k \bar{s}_k(p, T) - \sum_{k(\text{exit})} \dot{n}_k \bar{s}_k(p_k, T) + \dfrac{\dot{Q}}{T} + \dot{S}_{\text{gen,univ}}; \quad \left[\dfrac{kW}{K}\right]$

$\qquad\qquad\qquad\qquad\qquad\qquad\qquad\qquad\qquad\qquad\qquad\qquad\qquad\qquad\qquad\qquad$ **(14.20)**

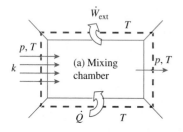

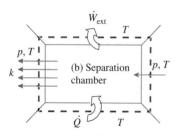

FIGURE 14.11 While a mixing chamber has the potential to produce useful work, external work input is a must in a separation chamber.

The summation at the exit involves partial molar properties $\bar{h}_k$ and $\bar{s}_k$ of each component. However, to enter the chamber through separate pipes, each component must have the same pressure as the total pressure p at the exit. In the absence of any chemical reaction, $\dot{n}_k$ for each component k remains unchanged between the inlet and exit. Eliminating $\dot{Q}$ between the energy and entropy equations, and substituting $\dot{I} = T_0 \dot{S}_{\text{gen,univ}} = T\dot{S}_{\text{gen}}$ for an isothermal system, we obtain:

$$\dot{W}_{\text{rev}} \equiv \dot{W}_{\text{ext}} + \dot{I} = \dot{W}_{\text{ext}} + T\dot{S}_{\text{gen}} = -\left[\sum_{k(\text{exit})} \dot{n}_k \bar{g}_k(p_k, T) - \sum_{k(\text{inlets})} \dot{n}_k \bar{g}_k(p, T) \right]; \quad [kW] \quad \textbf{(14.21)}$$

Introducing the ideal gas assumption, we substitute Eq. (14.18) for the partial molar Gibbs function and divide the entire equation by the total mole flow rate $\dot{n} = \sum_{k(\text{exits})} \dot{n}_k = \sum_{k(\text{inlets})} \dot{n}_k$ (this holds true only in the absence of chemical reaction) to obtain (see Anim. 14.B.*mixingAnd Separation*):

$$\dfrac{\dot{W}_{\text{rev,mixing}}}{\dot{n}} = -\bar{R}T \sum_{k(\text{exits})} y_k \ln \left(\dfrac{p_k}{p}\right) = \bar{R}T \sum_{k(\text{exits})} y_k \ln \left(\dfrac{1}{y_k}\right); \quad \left[\dfrac{kJ}{kmol}\right] \qquad \textbf{(14.22)}$$

Mixing is generally accomplished in chambers without any useful work transfer. However, y_k and $\ln\left(\dfrac{1}{y_k}\right)$ are always positive ($0 < y_k < 1$) and hence $\dot{W}_{\text{rev,mixing}}$ is also positive, indicating that useful work can be extracted during mixing without any change in the end results. With no useful work transfer, the entire amount of $\dot{W}_{\text{rev}}$ is wasted through exergy destruction due to mixing friction.

Equation (14.22) can also be used to formulate a criterion for the direction of a spontaneous flow of a given component k during mixing. A **semi-permeable membrane** allows migration of a particular component through the membrane. Now imagine a smart membrane, as shown in Figure 14.12, which produces useful power as a pure component k migrates towards a mixture to the left. The maximum work output that is thermodynamically possible—the reversible work—is obtained from Eq. (14.21):

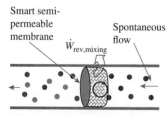

Smart semi-permeable membrane $\dot{W}_{\text{rev,mixing}}$ — Spontaneous flow

Mixture of hydrogen and carbon monoxide — Hydrogen

FIGURE 14.12 The spontaneous flow of hydrogen into the mixture is exploited by this concept heat engine.

$$\dfrac{\dot{W}_{\text{rev,mixing}}}{\dot{n}_k} = -\Delta \bar{g}_k = \bar{g}_{k,i} - \bar{g}_{k,e} = \bar{g}_k(p, T) - \bar{g}_k(p_k, T); \quad \left[\dfrac{kJ}{kmol \text{ of } k}\right] \qquad \textbf{(14.23)}$$

As long as $\bar{g}_k(p, T) > \bar{g}_k(p_k, T)$, that is, the chemical potential of the pure species k is greater than the chemical potential of species k in the mixture, useful work can be obtained by exploiting the spontaneous migration of component k. If an ordinary membrane is used, the entire amount of reversible work will be wasted as exergy destruction; however, the spontaneous migration will still be driven by the difference in the chemical potential of species k on the two sides of the membrane. Component k will flow in the direction of decreasing chemical potential. When chemical equilibrium is achieved, there is no driving force left and the criterion for equilibrium can be obtained from Eq. (14.23) by setting $\dot{W}_{rev,mixing}$ to zero.

Equilibrium Criterion:

$$\bar{g}_k(p, T) = \bar{g}_k(p_k, T); \quad \left[\frac{kJ}{kmol\ of\ k}\right] \quad \textbf{(14.24)}$$

In other words, the migration of component k will come to a halt when the partial Gibbs function (or chemical potential) of component k is equalized on the two sides of the membrane.

The equilibrium criterion for an ideal gas mixture can be simplified by applying Eq. (14.17) to express $\bar{g}_k(p_k, T)$ as a function of p, p_k, and T. It can be seen from this relation that a component will migrate from its reservoir to a mixture, as long as $p > p_k$, reaching equilibrium when $p = p_k$.

Going back to the separation device (Fig. 14.11b), the reversible work of separation must be the negative of the reversible work of mixing, as the two processes are mutually reversible. Therefore:

$$\dot{W}_{rev,separation} = -\dot{W}_{rev,mixing} = -\dot{n}_k\bar{R}T\sum_{k(inlets)} y_k \ln(1/y_k); \quad [kW] \quad \textbf{(14.25)}$$

This equation can be used to obtain the minimum work necessary for separation of a gas mixture as illustrated Example 14-3.

EXAMPLE 14-3 **Equilibrium Across a Membrane**

Hydrogen is produced (Fig. 14.13), at 100 kPa and 298 K with a rate of 1 kg/s, from a mixture of hydrogen and methane containing 10% hydrogen and 90% methane by volume. If the device works by raising the pressure of the mixture on one side of a semi-permeable membrane, determine (a) the minimum power consumption and (b) the minimum mixture pressure necessary to produce hydrogen.

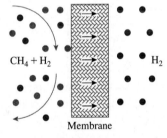

FIGURE 14.13 Schematic for Ex. 14-3 (click Separation in Anim. 14.B.*mixingAndSeparation*).

SOLUTION

Determine the reversible power of separation from Eq. (14.23) and the mixture pressure by applying the equilibrium criterion, Eq. (14.24).

Assumptions

The IG mixture model for the gas mixture, and the IG model for pure hydrogen, are applicable. Changes in ke and pe are negligible. The mixture composition is not affected by the transfer of hydrogen.

Analysis

Substitute the expression for the partial Gibbs function in Eq. (14.23) to obtain:

$$\dot{W}_{rev,separation}/\dot{n}_{H_2} = -\dot{W}_{rev,mixing}/\dot{n}_{H_2}$$

$$= \bar{g}_{H_2}(p_{H_2}, T) - \bar{g}_{H_2}(p, T)$$

$$= \left[\bar{g}_{H_2}(p, T) + \bar{R}T \ln\frac{p_{H_2}}{p}\right] - \bar{g}_{H_2}(p, T)$$

$$= \bar{R}T \ln y_{H_2}$$

For an ideal gas mixture, the volume fraction is equal to mole fraction [Eq. (11.96)], producing $y_{H_2} = 0.1$. Therefore, the minimum power consumption, the reversible power, is given by:

$$\dot{W}_{rev} = \dot{n}_{H_2}\bar{R}T \ln y_{H_2} = \frac{\dot{m}_{H_2}}{M_{H_2}}\bar{R}T \ln y_{H_2}$$

$$= \frac{(1)(8.314)(298)}{2} \ln(0.1) = -2{,}852\ kW$$

At equilibrium, the partial pressure of hydrogen must be equal on the two sides of the separating membrane. Therefore, if p_{H_2} on the mixture side is 100 kPa, the total mixture pressure is $p_{H_2}/y_{H_2} = 100/0.1 = 1,000$ kPa, which is the minimum pressure of the mixture before hydrogen starts separating.

TEST Analysis

Launch the n-IG flow-state TESTcalc. Create a mixture with 90% CH4 and 10% H2 by volume. Evaluate State-1 with p1=100 kPa and T1=298 K. The partial specific Gibbs function of H2, $g_{H_2}(p_{H_2}, T)$, is calculated (displayed in the mixture panel) as −22,164 kJ/kg. For pure H2 (add zero amount of CH4 to the mixture to eliminate CH4) at 100 kPa and 298 K, calculate the specific Gibbs function $g_{H_2}(p, T)$, as −19,335 kJ/kg. For a flow rate of 1 kg/s, the power required can now be calculated as the difference between the two Gibbs functions, 2829 kW. The partial specific Gibbs function of hydrogen in the mixture must be increased to 19,335 kJ/kg (at which point the chemical potential of hydrogen on the two sides become equal) before hydrogen starts separating. Increase p1 and calculate the state to see how $g_{H_2}(p_{H_2}, T)$ changes with total pressure. For p1 = 1000 kPa, g_k, reported on the mixture panel, equals −19,335 kJ/kg, verifying the manual result.

Discussion

Gas mixtures produced in industrial processes, such as coal gasification, often contain hydrogen. Given its clean burning potential and application to fuel cells, hydrogen production through separation is attracting renewed interest from the research community.

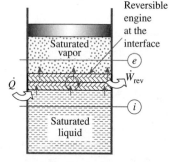

FIGURE 14.14 Reversible engine that exploits the driving force for phase transformation. The power output is zero at equilibrium when $\bar{g}_f = \bar{g}_g$ (see Anim. 14.C.*vaporAndLiquid*).

14.3 PHASE EQUILIBRIUM

When different phases of a substance are not in equilibrium, transfer of mass occurs across the interface. For instance, we are accustomed to seeing ice cubes melt when added to a glass of water at room temperature and water vapor condensing around the glass.

To seek a specific criterion for equilibrium among different phases, first consider a spontaneous transfer of mass between the liquid and vapor phases of a pure substance like water. Suppose a mixture of liquid water and vapor, kept in a piston-cylinder device, is subjected to a given pressure and temperature. Heat and boundary work transfer are allowed so that the pressure and temperature remain constant as the system seeks equilibrium. During its transition to a new equilibrium, it is possible for some of the vapor to condense or some of the liquid to evaporate, depending on which phase is favored by the new equilibrium. A concept heat engine, shown in Figure 14.14, exploits the energy transported by the mass, say, from the liquid phase to the vapor phase, to extract reversible work. By treating the two phases as different components (at the same total pressure and temperature), we can use Eq. (14.23) to obtain the reversible work as:

$$\frac{\dot{W}_{rev}}{\dot{n}} = -\Delta\bar{g} = \bar{g}_i - \bar{g}_e = \bar{g}_f(p, T) - \bar{g}_g(p, T); \quad \left[\frac{kJ}{kmol}\right] \quad \textbf{(14.26)}$$

The equation is simplified as there are no partial properties involved for a pure substance (water). Subscripts f and g refer to saturated liquid and vapor phase, respectively, and the mass transfer in Figure 14.14 is assumed to be due to the evaporation of liquid. A positive reversible work implies a spontaneous process. Therefore, mass transfer from the liquid to vapor phase takes place as long as $\bar{g}_f > \bar{g}_g$; that is, the chemical potential of the liquid phase is greater than that of the vapor phase. Conversely, if $\bar{g}_g > \bar{g}_f$, positive work can be obtained only if $\dot{n}$, the mole flow rate, is negative, meaning spontaneous condensation will result if the chemical potential of the saturated vapor phase is higher than that of the saturated liquid. At equilibrium, there is no driving force left for any net transfer. Equating the reversible power from the concept engine to zero, we obtain the *equilibrium criterion for phase equilibrium*:

$$\bar{g}_f(p, T) = \bar{g}_g(p, T); \quad \left[\frac{kJ}{kmol}\right] \quad \textbf{(14.27)}$$

That is, transfer of mass between the liquid and vapor phases of a pure substance will stop, and a phase equilibrium will be established when the chemical potential (molar specific Gibbs function) of the working substance on the two phases become equal. This can be quickly verified from any saturation table. To do so, launch the system-state PC TESTcalc and select any fluid, for example, R-134a (Fig. 14.15). Evaluate State-1 with T1=0 deg-C and x1=0; and State-2 with T2=T1 and x2=1. In the I/O panel, evaluate g1 as '=h1−T1*s1'=3.856 kJ/kg and g2 as '=h2−T2*s2' =3.857 kJ/kg. A two-phase mixture can stay in equilibrium, without any net transfer of mass between the two phases, only when the Gibbs functions (chemical potentials) are equal in the two phases.

As another example of mass transfer between phases driven by an imbalance of chemical potential, consider what happens when an ice cube is dropped in warmer water. We can use the SL model (Sec. 3.3) to show that $g = h - Ts = c_1 + c_pT(1 - \ln T)$ (where c_1 is a constant), which means the chemical potential of a solid or liquid decreases with an increase in temperature (Fig. 14.16 and Anim. 14.C.*chemPotentialPure*). We know that under standard atmospheric pressure the two phases coexist in equilibrium at 273 K. Therefore, the equilibrium condition of Eq. (14.26) will dictate that the chemical potential of the two phases must be equal at 273 K, explaining why the chemical potential curves in Figure 14.16 intersect at 273 K. Moreover, c_p of liquid water being higher than that of ice, the curve for the liquid water can be expected to be steeper (dashed line in Fig. 14.16). Ice must have higher chemical potential than warmer water ($T > 273$ K), which explains the transfer of mass from the solid to the liquid phase, that is, **melting**. At the interface between the two phases, the temperature difference will drive a heat transfer from water to ice, but the relatively higher chemical potential of ice will drive the mass transfer in the opposite direction until thermal (equality of temperature) and chemical equilibrium (equality of chemical potential) are established or all the ice melts. Now suppose the ice block is so cold to start with that its temperature is still less than 273 K, even after the entire amount of liquid water has been brought down to 273 K. Continued heat transfer (driven by temperature difference) will try to cool the water to below 273 K. The inversion of the curves in Figure 14.16 suggests that the higher chemical potential of liquid water will cause transfer of mass from liquid to ice; this is known as **solidification**. The latent heat of solidification will prevent the water temperature from dropping further until all the water freezes or equilibrium is reached.

As a thermodynamic property, the Gibbs function (chemical potential) for a pure substance can be expressed in terms of pressure and temperature. Therefore, Eq. (14.27) provides a relation between p and T whenever two phases of a pure substance are in equilibrium. Equation (14.27) dictates a constant temperature during phase transformation at a given pressure, as long as equilibrium is assumed to prevail. Even during the boiling of water in an open pot, local thermodynamic equilibrium will ensure that the temperature of boiling is controlled by the pressure to which the water vapor mixture at the surface is subjected, which is atmospheric pressure (nitrogen and oxygen very close to the surface are totally displaced by water vapor during boiling). Equation (14.27) also suggests that either pressure or temperature can be independently varied in a saturated mixture at equilibrium, thus establishing the physics behind the assumption we made in Chapter 3, that $p_{sat} = f(T_{sat})$, and vice versa.

In a multi-component, multi-phase mixture, the conditions for phase equilibrium can be generalized for a given component k by placing an imaginary semi-permeable membrane at the interface that allows transfer of component k only. Assuming each component acts independently, the reversible work from the migration of each species from one phase to another must be zero at equilibrium. Therefore, Eq. (14.27) can be extended for each component for every pair of phases, implying that the chemical potential of a given component must be equal across all phases in the mixture.

A multi-phase mixture is definitely not an ideal gas mixture, so Eq. (14.14), for the functional dependence of the chemical potential must be further generalized to account for the phases that may be present in equilibrium. Representing individual phases by the number of primes used in the superscript (phase-I with one prime, phase-II with two primes) and the total number of components by C, we can generalize the equilibrium condition, Eq. (14.24), for component k in phases I and II as:

$$\bar{g}'_k(p, T, y'_1, \ldots, y'_{C-1}, y''_1, \ldots, y''_{C-1})$$

$$= \bar{g}''_k(p, T, y''_1, \ldots, y''_{C-1}, y'_1, \ldots, y'_{C-1}); \quad \left[\frac{kJ}{kmol \text{ of } k}\right] \tag{14.28}$$

Here, $\bar{g}_k$ is the partial molar specific Gibbs function interchangeably used for the *chemical potential* $\bar{\mu}_k(p, T, y'_1, \ldots, y'_{C-1}, y''_1, \ldots, y''_{C-1})$ of a component in a given phase and C is the total number

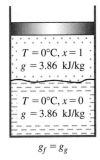

$g_f = g_g$

FIGURE 14.15 Eq. (14.27) interpreted for a saturated liquid-vapor mixture of refrigerant R-134a.

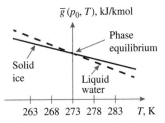

FIGURE 14.16 At standard atmospheric pressure, the chemical potential curves for ice and water intersect at 273.15 K, establishing phase equilibrium (see Anim. 14.C.*iceAndWater*).

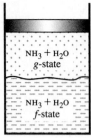

$g_f, NH_3 = g_g, NH_3$
$g_f, H_2O = g_g, H_2O$

FIGURE 14.17 Eq. (14.28) interpreted for ammonia water mixture.

of components in the system (see Anim. 14.B.*chemicalPotential*). The sum of all mole fractions in a given phase is being unity, only $C - 1$ mole fractions can be independent in a particular phase.

Equation (14.28) is known as the **equation of phase equilibrium**. As an example of the application of the equation of phase equilibrium, consider an ammonia–water solution (Fig. 14.17). The partial pressure of ammonia and water in the vapor mixture will be dictated by Eq. (14.28), which requires the chemical potential of each component—ammonia and water—to be equal in each phase, liquid and vapor. Just as a temperature difference drives a heat flow, a difference in $\bar{g}_k$ (or $\bar{\mu}_k$) between two phases will drive the flow of species k until the chemical potentials are equalized. Recall that we reached a similar conclusion while analyzing diffusion of a species through a membrane in Sec. 14.2.

EXAMPLE 14-4   Triple Point

Prove that three phases of water—solid, liquid, and vapor—can exist in equilibrium at a unique pair of values of pressure and temperature.

SOLUTION

Use Eq. (14.28) for a pure substance (a single component) to show that a unique solution exists for a given p and T.

Analysis

Equation (14.28), for a pure substance with $C = 1$, reduces to:

$$\bar{g}_k'(p, T) = \bar{g}_k''(p, T) = \bar{g}_k'''(p, T).$$

These are two independent equations involving two unknowns, p and T; therefore, a unique solution for p and T can be found at which all three phases can coexist.

Discussion

While a mixture of two phases of a pure substance can be in equilibrium over a range of temperatures (we have one equation relating p and T which makes them dependent on each other), there is only a single temperature and pressure, called the **triple point**, where a mixture of three different phases can be in equilibrium. When we see ice melting into water, the chemical potential of ice must be greater than that of water, driving the phase transformation until $\bar{g}_k'(p, T) = \bar{g}_k''(p, T)$ equilibrium is reached. This implies that the pressure and temperature of ice-water at equilibrium must be functions of each other.

EXAMPLE 14-5   Equilibrium of Moist Air with Liquid Water

Determine the pressure of water vapor at equilibrium with liquid water, at 35°C, when (a) H_2O is the only component and (b) when the gas phase is composed of water vapor and air at a mixture pressure of 500 kPa.

SOLUTION

For pure water, use the saturation table. For part (b), use the phase equilibrium equation for H_2O.

Assumptions

The vapor mixture can be modeled as an IG mixture. Air is not soluble in liquid water.

Analysis

For the liquid-vapor mixture involving pure H_2O, the pressure can be obtained from the saturation temperature table for water, or any PC TESTcalc (with **T1=35 deg-C, x1=0**), as:

$$p_{H_2O,sat@35°C} = 5.63 \text{ kPa}$$

It can be verified that the specific Gibbs function for liquid water and saturated water vapor at 5.63 kPa are approximately equal, as dictated by the equilibrium criterion, Eq. (14.28).

As the total pressure is increased to 500 kPa in the presence of air, we apply Eq. (14.28) for the component H_2O, which is present in both the phases (liquid phase represented by a single prime), to obtain:

$$\bar{g}_{H_2O}(p_{H_2O}, T) = \bar{g}'_{H_2O}(p, T)$$

$$\Rightarrow \quad \bar{h}_{H_2O}(p_{H_2O}, T) - T\bar{s}(p_{H_2O}, T) = \bar{h}'_{H_2O}(p, T) - T\bar{s}'(p, T)$$

$$\Rightarrow \quad \bar{h}_{H_2O}(p_{H_2O}, T) - \bar{h}'_{H_2O}(p, T) = T[\bar{s}(p_{H_2O}, T) - \bar{s}'(p, T)] \tag{14.29}$$

In this equation, the prime as a superscript represents the liquid phase, while no superscript is used to represent the vapor phase. The liquid water is subjected to the mixture pressure $p = 500$ kPa, while the water vapor in the mixture is at the partial pressure p_{H_2O}. Applying the compressed liquid sub-model (Sec. 3.4.8), the enthalpy and entropy of the compressed liquid can be expressed in terms of the known saturation properties. Another simplification results from the fact that water vapor in the mixture can be treated as an ideal gas, and the enthalpy of an ideal gas is a function of temperature alone. Therefore, we can write:

Vapor phase: $\bar{h}_{H_2O}(p_{H_2O}, T) \cong \bar{h}_{g,H_2O@T}$;

Liquid phase: $\bar{h}'_{H_2O}(p, T) \cong \bar{h}_{f,H_2O@T} + \bar{v}_{f,H_2O@T}(p - p_{sat@T})$;

Liquid phase: $\bar{s}'(p, T) \cong \bar{s}_{f,H_2O@T}$ $\tag{14.30}$

The IG model can be used to relate entropy $\bar{s}(p_{H_2O}, T)$ with $\bar{s}(p_{sat@T}, T) = \bar{s}_{g,H_2O@T}$ through a pressure correction (see Anim. 3.D.*propertiesIG*) producing:

$$\bar{s}(p_{H_2O}, T) \cong \bar{s}(p_{sat@T}, T) - \bar{R}\ln\frac{p_{H_2O}}{p_{sat@T}} = \bar{s}_{g,H_2O@T} - \bar{R}\ln\frac{p_{H_2O}}{p_{sat@T}} \tag{14.31}$$

Substituting Eqs. (14.31) and (14.30) in Eq. (14.29) and recognizing (from Eq. 11.30) that $\bar{h}_{g,H_2O@T} - \bar{h}_{f,H_2O@T} = T(\bar{s}_{g,H_2O@T} - \bar{s}_{f,H_2O@T})$, we obtain:

$$\ln\frac{p_{H_2O}}{p_{sat@T}} = \frac{\bar{v}_{f,H_2O@T}(p - p_{sat@T})}{\bar{R}T}$$

$$\Rightarrow \quad p_{H_2O} = p_{sat@T}\exp\left[\frac{\overline{M}_{H_2O}v_{f,H_2O@T}(p - p_{sat@T})}{\bar{R}T}\right]$$

$$= (5.63)\exp\left[\frac{(18)(0.001)(500 - 5.63)}{(8.314)(273 + 35)}\right] = 5.65 \text{ kPa}$$

Discussion

Expressed as a percentage, the departure of p_{H_2O} from $p_{sat@T}$ is less than 0.4%. At 100 kPa, this discrepancy is be shown to decrease even further. The equilibrium between the liquid and vapor phases of a PC fluid is not appreciably affected by the presence of dry air. Therefore, the use of $p_v = p_{sat@T}$ for saturated air in Chapter 12 is justifiable.

Phase Rule

Let us return to the general mixture of C components, each having P phases. For a mixture that is modeled as an IG mixture, we see that intensive properties, such as partial molar specific properties, can be expressed as functions of partial pressure of the component ($p_k = y_kp$) and temperature. However, for a general mixture, independent variables must include mole fractions of all components present in all possible phases in the mixture—the set of independent variables $[p, T, y', \ldots, y'_{C-1}, y'', \ldots, y''_{C-1}, \ldots]$ totals $2 + P(C - 1)$ under the most general scenario of each component being present in each of the P phases. Out of C components, only $C - 1$ mole fractions can be independently varied as $\sum_k y_k = 1$ in each phase. Given that each component will produce $(P - 1)$ phase equilibrium relations, in accordance with Eq. (14.28), the total number of independent equations reduces to $C(P - 1)$.

The difference between the total number of variables and equations is the number of independent variables:

$$q = 2 + P(C - 1) - C(P - 1) = C - P + 2 \tag{14.32}$$

Known as the **Gibbs, phase rule**, this equation determines the number of intensive properties $(q(q \geq 0))$ that can be independently varied at equilibrium. For a pure substance $C = 1$ in a single phase $(P = 1)$, $q = 1 - 1 + 2 = 2$; therefore, only two properties can be independently varied at equilibrium. This was referred to as the **state postulate** in Sec 3.1.2. The existence of a unique state of a pure substance (triple point in Example 14-4) where three different phases can be in equilibrium can be deduced by simply substituting $C = 1$ and $q = 0$ (p and T are fixed).

As a more complex illustration of the phase rule, consider the different phase compositions that are possible in the triple-point counterpart of a **binary mixture**, a mixture of two components, such as ammonia and water. Substitution of $C = 2$ and $q = 0$ in Eq. (14.32) produces $P = 4$, which means that a maximum of four phases can coexist at a unique state defined by a unique pressure, temperature, and molar composition in each phase. Alternatively, three phases can be in equilibrium $(q = 2 - 3 + 2 = 1)$ allowing only one property to be independently varied. The rest of the properties can be expressed as functions of that independent property, for example, pressure, yielding $T = T(p)$, $y_{NH_3} = y_{NH_3}(p)$, $y'_{NH_3} = y'_{NH_3}(p)$, and $y''_{NH_3} = y''_{NH_3}(p)$, etc. With only two phases in equilibrium, there will be two independent properties $(q = 2 - 2 + 2 = 2)$, yielding $T = T(p, y'_{NH_3})$ and $y_{NH_3} = y_{NH_3}(p, y'_{NH_3})$. Finally, in a single-phase binary mixture, pressure, temperature, and one mole fraction, say, y'_{NH_3} can be arbitrarily varied since $q = 2 - 1 + 2 = 3$.

A thermodynamic diagram of a binary mixture in the $T - y_k$ coordinates is called a **phase diagram**, which is sketched in Figure 14.18 for an ammonia–water mixture at a total pressure of 1 atm, with only liquid and vapor as possible phases. In the liquid solution at equilibrium, three $(q = 2 - 1 + 2 = 3)$ independent properties are required to define a state. With the total pressure given for the entire diagram, a state can be located by the two properties—temperature T and mole fraction y'_{NH_3} of ammonia in the liquid solution. Similarly, for a superheated vapor mixture $(q = 3)$, a state can be located by T and y_{NH_3}. The phase diagram in Figure 14.18 takes advantage of the fact that y_{NH_3} and y'_{NH_3} can be represented on a single axis.

In a saturated mixture of the two phases, the composition of liquid and vapor doesn't need to be identical. With the number of independent properties reduced to two $(q = 2 - 2 + 2 = 2)$, the saturation temperature T_{sat} can be expressed as a function of total pressure and one more property, y'_{NH_3} or y_{NH_3}. For a given total pressure, T_{sat} can be experimentally determined as a function of y'_{NH_3} or y_{NH_3} and plotted in the phase diagram (the lower and upper red curve respectively) by using the abscissa to represent both y'_{NH_3} and y_{NH_3}. Suppose a liquid ammonia–water solution, originally at State-1 $(y'_{NH_3} = y'_{NH_3,1})$, is heated at constant pressure. The temperature will increase without any change in y'_{NH_3} $(y'_{NH_3,2} = y'_{NH_3,1})$ until the solution is saturated at State-2. At that point, evaporation of ammonia will begin; however, the vapor mixture produced can have a much different composition, in accordance with $y_{NH_3} = y_{NH_3}(p, y'_{NH_3})$, resulting in a saturated vapor state, State-3, at the constant saturation temperature $T_2 = T_{sat}(p, y'_{NH_3}) = T_{sat}(p, y_{NH_3}) = T_3$. A saturated liquid-vapor mixture at State-4, therefore, contains liquid at State-2 and vapor at State-3 $(T_4 = T_2 = T_3)$, much like wet steam contains saturated liquid and saturated vapor of water at f- and g-states. If the temperature of the saturated vapor mixture is further increased, the superheated state, State-5, is reached and the mixture's composition remains fixed $(y_{NH_3,3} = y_{NH_3,4})$.

To determine the composition, the simplest approach is to model the vapor phase as an ideal gas mixture and the liquid phase as an **ideal solution**. We are already familiar with the Dalton model (Sec. 11.4.3) for an ideal gas mixture, where the components are assumed to coexist independently, sharing the same volume and temperature, but each component at its own partial pressure:

$$p_k = y_k p; \quad \text{and,} \quad p = \sum p_k; \quad [\text{kPa}] \tag{14.33}$$

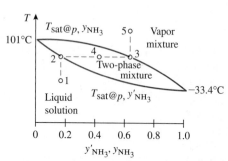

FIGURE 14.18 Phase diagram for two-phase mixture of ammonia and water at 1 atm. The prime superscript symbolizes the liquid phase.

For a single-component mixture of liquid and vapor, the pressure is simply the saturation vapor pressure at the given temperature, or, $p = p_{sat@T}$. **Raoult's model**, the counterpart of Dalton's law of partial pressure for an ideal solution, relates p_k, the partial vapor pressure in the vapor phase, to the mole fraction of component k in the solution (Fig. 14.19):

$$p_k = y'_k p_{k,\text{sat}@T}; \quad \text{where} \quad p = \sum p_k = \sum y'_k p_{k,\text{sat}@T}; \quad [\text{kPa}] \qquad \textbf{(14.34)}$$

Therefore, if the molar composition of a liquid mixture is known, the composition in the vapor mixture can be determined from Eqs. (14.33) and (14.34).

Chemical engineering handbooks carry the **Cox chart**, which relates the vapor pressure of several pure liquids with temperature. Any PC TESTcalc in TEST can be used to obtain a reasonably accurate value of vapor pressure at a given temperature (enter temperature and any value for the quality). Dissimilar substances do not behave like ideal solutions and, sometimes, do not even mix, a familiar example is oil and water. Completely immiscible liquid mixtures are still miscible in the vapor phase. Raoult's law, Eq. (14.34), can still be applied by setting $y'_k = 1$, as the liquids maintain separate phases.

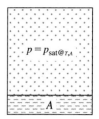

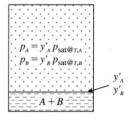

FIGURE 14.19 Raoult's law relates the partial pressure of a component in the vapor phase with its mole fraction in the liquid phase.

EXAMPLE 14-6   Equilibrium in Binary Mixtures of Two Phases

In absorption refrigeration systems, liquid-vapor equilibrium mixtures of ammonia and water are frequently used. Consider such an equilibrium mixture at 20°C. If the molar composition of the liquid phase is 50% NH_3 and 50% H_2O, determine (a) the mixture pressure and (b) the molar composition of the vapor phase.

SOLUTION

Apply Raoult's law to obtain the partial pressure of each component in the vapor mixture.

Assumptions

The solution is an ideal solution so that Raoult's model can be applied. The vapor phase can be modeled by the IG mixture model.

Analysis

From the PC system-state TESTcalc, obtain $p_{NH_3,\text{sat}@20°C} = 857$ kPa and $p_{H_2O,\text{sat}@20°C} = 2.34$ kPa. Using Eq. (14.34), the partial pressures of the components in the vapor phase are given as:

$$p_{NH_3} = y'_{NH_3} p_{NH_3,\text{sat}@20°C} = (0.5)(857) = 428.5 \text{ kPa};$$

$$p_{H_2O} = y'_{H_2O} p_{H_2O,\text{sat}@20°C} = (0.5)(2.34) = 1.17 \text{ kPa}$$

Therefore, the mixture pressure is:

$$p = p_{NH_3} + p_{H_2O} = 429.67 \text{ kPa}$$

The molar composition can be obtained from Eq. (14.33):

$$y_{NH_3} = p_{NH_3}/p = 428.5/429.67 = 0.997;$$

$$y_{H_2O} = 1 - y_{NH_3} = 0.003$$

Discussion

In the phase diagram of Figure 14.20, State-1 and State-2 represent the liquid and vapor states of the initial equilibrium mixture. If the temperature of the mixture is increased (at a constant total pressure), the phase diagram requires both the liquid and vapor states to migrate the left (towards States 3 and 4, respectively), indicating a lowered mole fraction of ammonia in both the liquid and vapor mixtures.

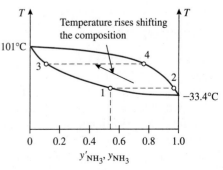

FIGURE 14.20 Phase diagram for Ex. 14-6.

14.3.1 Osmotic Pressure and Desalination

To further explore the equilibrium of a liquid vapor mixture in a binary system, consider the arrangement shown in Figure 14.21, where a liquid mixture of components A and B with mole fractions y'_A and $y'_B (= 1 - y'_A)$, is separated from the pure liquid

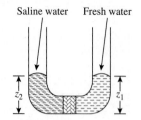

FIGURE 14.21 Fresh water and saline water are separated by a semi-permeable membrane that allows the transfer of water only. The osmotic pressure makes z_2 greater than z_1.

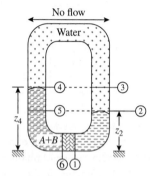

FIGURE 14.22 Osmotic pressure is the hydrostatic pressure difference $\Delta p = p_6 - p_1$ at equilibrium.

of A by a membrane that is permeable to A only. Both columns of liquid, which are initially the same height ($z_1 = z_2$), are exposed to the atmosphere, as shown in the figure. As a specific example, we can have pure water (component A) on the right side and saline water (a solution of A and B) on the left, with the membrane permeable to water only. We would like to answer a simple question: how would the column heights, z_1 and z_2 in Figure 14.21, adjust as the system moves toward equilibrium? Obviously, if a pure fluid is on both sides, mechanical equilibrium (hydrostatic balance) will require the two heights to be equal.

In many respects, the situation is similar to the equilibrium of a gas mixture and pure gas discussed in Sec. 14.2. In fact, it can be shown that at equilibrium, the partial molar Gibbs function $\bar{g}_k$ (or chemical potential) of the permeable component k must be same on each side of the membrane. However, expressing $\bar{g}_k$ for a solution component is not an task and involves new concepts such as fugacity, affinity, etc.

As an alternative approach, consider a similar arrangement where we look at the equilibrium of vapor and liquid phases together by connecting the left and right tube, as in Figure 14.22, allowing transfer of the permeable component through the vapor phase as well. For simplicity, we assume that component B (salt, in our example) does not evaporate—a reasonable assumption for many types of solutions. At equilibrium, the permeable component A must stop migrating through the membrane, as well as through the vapor pathway. In fact, the final equilibrium would be the same regardless of the pathway. With no flow anywhere the pressure variation must be hydrostatic. Using Raoult's law, p_4 and p_2 can be expressed:

$$p_4 = p_{A,\text{sat}@T} y'_A \quad \text{and} \quad p_2 = p_{A,\text{sat}@T}$$

where y'_A is the mole fraction of component A (water) in the solution (saline water). Eq. (14.16) can also be used to relate p_3 with $\Delta z = z_3 - z_2$:

$$p_3 = p_2 \exp\left[\frac{-g(z_3 - z_2)}{(1000 \text{ J/kJ})RT}\right] = p_{A,\text{sat}@T} \exp\left[\frac{-g\Delta z \bar{M}_A}{(1000 \text{ J/kJ})\bar{R}T}\right]; \quad [\text{kPa}]$$

Hydrostatic pressure variation in the vapor phase requires $p_3 = p_4$ (horizontal points). Equating the expressions for p_3 and p_4, we obtain:

$$\Delta z = \frac{(1000 \text{ J/kJ})\bar{R}T}{g\bar{M}_A} \ln\frac{1}{y'_A}; \quad [\text{m}] \tag{14.35}$$

The pressures on the two sides of the membrane can be related using hydrostatic formulas:

$$p_6 = p_4 + \frac{z_4 \rho_{A+B} g}{(1000 \text{ N/kN})}$$

$$= p_{A,\text{sat}@T} y'_A + \frac{\Delta z \rho_{A+B} g}{(1000 \text{ N/kN})} + \frac{z_5 \rho_{A+B} g}{(1000 \text{ N/kN})}; \quad [\text{kPa}]$$

and

$$p_1 = p_2 + \frac{z_2 \rho_A g}{(1000 \text{ N/kN})} = p_{A,\text{sat}@T} + \frac{z_2 \rho_A g}{(1000 \text{ N/kN})}; \quad [\text{kPa}]$$

Therefore, the **osmotic pressure**, defined as the pressure difference between the two sides of the membrane at equilibrium, is given by

$$\Delta p = p_6 - p_1$$

$$= \frac{\Delta z \rho_{A+B} g}{(1000 \text{ N/kN})} - p_{A,\text{sat}@T}(1 - y'_A) + \frac{z_2(\rho_{A+B} - \rho_A)g}{(1000 \text{ N/kN})}; \quad [\text{kPa}]$$

Generally the vapor pressure $p_{A,\text{sat}@T}$ and the difference in liquid density $\rho_{A+B} - \rho_A$ are quite small, so the last two terms can be neglected. Substituting the expression derived for Δz from Eq. (14.35), we obtain:

$$\Delta p \cong \frac{\Delta z \rho_{A+B} g}{(1000 \text{ N/kN})} = \frac{\rho_{A+B} \bar{R}T}{\bar{M}_A} \ln\frac{1}{y'_A}; \quad [\text{kPa}] \tag{14.36}$$

Osmotic pressure, therefore, is expected to increase linearly with temperature and logarithmically with salinity (as salinity increases, y'_A decreases). Also, the reversible work necessary to produce a flow of fresh water against this osmotic pressure difference—a process that is known as **reverse osmosis**—can be obtained from Eq. (4.10) as:

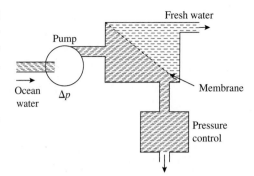

FIGURE 14.23 Schematic of a simple desalination system (see Anim. 14.B. *reverseOsmosis*).

$$\dot{W}_{rev} = -\dot{m}_{A+B} \int_i^e v_{A+B} dp = -\frac{\dot{m}_{A+B} \Delta p}{\rho_{A+B}} = -\frac{\dot{m}_{A+B} \bar{R} T}{\bar{M}_A} \ln \frac{1}{y'_A}; \quad [kW] \quad \textbf{(14.37)}$$

This expression gives the minimum pumping power requirement (see Anim. 14.B.*reverseOsmosis*) for the reverse osmosis process. The negative sign signifies work input in accordance with the WinHip sign convention. A simple desalination device that uses this principle is depicted in Figure 14.23. The saline water is pressurized against a membrane only permeable to water. An increase in salinity raises the osmotic pressure; therefore, a fresh supply of saline water must be used. If useful work is not extracted from the pressurized saline water exiting the device, the actual work of reverse osmosis will far exceed the reversible work.

The minimum work requirement, predicted by Eq. (14.37), is also the negative of the maximum power that can be extracted by exploiting the driving force for mixing between fresh water and saline water. When fresh water from a river meets ocean water, in principle reversible salination can be exploited to produce a considerable amount of useful work. Practical difficulties, however, prevent implementation of a salination power plant, such as the conceptual device sketched in Figure 14.24. Nature seems to be a step ahead of engineers in this regard. Exploiting the osmotic pressure difference between the fresh water in the soil and the fluid inside the root, trees have been pumping water to their leaves hundreds of feet above the ground level since the prehistoric time.

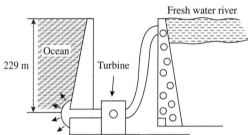

FIGURE 14.24 The osmotic head, 228 m calculated in Ex. 14-7, can be utilized by a concept turbine to produce useful work when a river meets the ocean.

EXAMPLE 14-7 Desalination through Reverse Osmosis

(a) Determine the minimum useful work required for desalination per unit mass of ocean water at 10°C and 1 atm through reverse osmosis. (b) Compare it with the energy required for desalination by distillation. (c) Also calculate the osmotic head. Use the following data: ocean water has a vapor pressure of 1.207 kPa at 10°C and a density of 1024 kg/m^3.

SOLUTION

Use Eq. (14.37) to obtain the energy required for reverse osmosis. For distillation use an energy balance in an open-steady device.

Assumptions

Ocean water behaves as an ideal solution. Vapor pressure of the salt or any of its ions is zero. There is no recovery of energy during distillation. Enthalpy of ocean water can be approximated by that of pure water.

Analysis

The mole fraction of pure water in the ocean can be obtained from Raoult's law, which relates the vapor pressure on the ocean water's surface with the mole fraction of water in the saline water of the ocean:

$$p_{v,ocean} = p_{v,fresh} y'_{H_2O} = p_{sat@10°C} y'_{H_2O}$$

$$\Rightarrow \quad y'_{H_2O} = \frac{p_{v,ocean}}{p_{sat@10°C}} = \frac{1.207}{1.228} = 0.983$$

The reversible work, the minimum useful work required, per unit mass of ocean water is now obtained from Eq. (14.37):

$$\frac{\dot{W}_{rev}}{\dot{m}_{ocean}} = -\frac{\bar{R}T}{\bar{M}_A} \ln \frac{1}{y'_A} = -\frac{(8.314)(283)}{(18)} \ln \frac{1}{0.983} = -2.25 \frac{kJ}{kg}$$

To calculate the energy required to distill ocean water, a steady state energy balance produces:

$$\frac{\dot{Q} - \dot{W}_{ext}}{\dot{m}_{ocean}} = j_e - j_i \cong h_{g@100\,kPa} - h_{f@10°C} = 2675 - 42 = 2633 \frac{kJ}{kg}$$

The osmotic pressure can be calculated from Eq. (14.36):

$$\Delta p \cong \frac{\rho_{A+B}\overline{R}T}{\overline{M}_A} \ln \frac{1}{y'_A} = \frac{\rho_{A+B}\dot{W}_{rev}}{\dot{m}_{sal.water}} = \frac{\rho_{A+B}g\Delta z}{(1000\ Pa/kPa)}\ [kPa]$$

$$\Rightarrow \quad \Delta z = \frac{\Delta p}{\rho_{A+B}g/1000} = \frac{1000}{g}\frac{\dot{W}_{rev}}{\dot{m}_{sal.water}} = \frac{1000}{9.81}(2.25) = 229\ m$$

Discussion

The comparison presented above, is somewhat flawed due to the fact that the best case scenario for reverse osmosis has been compared with the worst case scenario for distillation. The energy required for distillation can be supplied from a heat source at a low temperature (just above 100°C) to improve the exergetic efficiency and much of the waste heat can be recovered through regeneration. The work required for the reverse osmosis, on the other hand, can be significantly higher by using a realistic value for exergetic efficiency. Despite these practical considerations, desalination by reverse osmosis is far more energy efficient than distillation.

14.4 CHEMICAL EQUILIBRIUM

In the last chapter, we analyzed reacting systems in a comprehensive manner. However, there was one important restriction: the products composition of a reaction was always specified; either the reaction was assumed to be *complete* or an experimentally determined *dry analysis* of the products was supplied. Equilibrium thermodynamics provides a purely theoretical way in which the products composition can be predicted by assuming thermodynamic equilibrium, which is a reasonable assumption, particularly, for fast exothermic combustion reactions.

Let us begin the discussion with a simple reaction. Suppose we have a steady-flow reactor (Fig. 14.25) into which a mixture of hydrogen and oxygen enters steadily in *theoretical* proportions and the products leave at the same temperature and pressure as the reactants. The familiar *theoretical* reaction (Sec. 13.1.3) can be written:

$$H_2 + 0.5O_2 \rightarrow H_2O \tag{14.38}$$

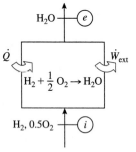

FIGURE 14.25 A theoretical mixture at the inlet does not guarantee a theoretical reaction (see Anim. 13.B.*theoretical Reaction*).

The reactants being in theoretical proportions, we can jump to the conclusion that the final products contain only water, regardless of the temperature or pressure in the reaction chamber. This is highly unlikely, as a theoretical mixture never guarantees a theoretical reaction.

To understand what really happens, let us briefly explore the rudimentary mechanism of the above reaction. As the hydrogen and oxygen molecules collide, only a fraction of the collisions are energetic enough to break the electronic bonds, in the hydrogen and oxygen molecules, to liberate the atoms, which then recombine to form water molecules. Such collisions can also occur in the reverse direction. Collisions among sufficiently energetic water molecules produce hydrogen and oxygen molecules; however, the energy required for the **forward reaction** can be quite different from that in the backward reaction. The backward reaction is commonly known as the **inverse reaction**. In fact, the threshold energy for a collision, called the **activation energy**, is much higher for the inverse reaction, making dissociation of water less likely. But, as the temperature increases, so does the kinetic energy of the water molecules, and the likelihood of overcoming the activation barrier sharply increases.

Generally speaking, any reaction is accompanied by the corresponding inverse reaction. To include this possibility, the theoretical reaction between hydrogen and oxygen is more precisely expressed with the forward and backward arrows as:

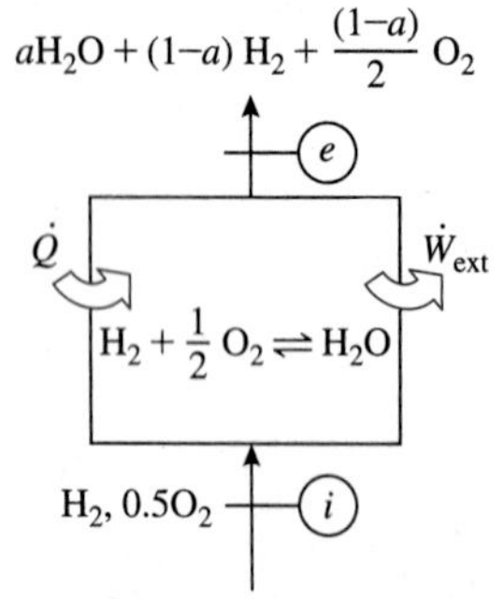

$$H_2 + 0.5O_2 \leftrightarrows H_2O \tag{14.39}$$

FIGURE 14.26 The overall reaction connects the inlet with exit, while stoichiometric reactions describe the actual chemistry inside the reactor (see Anim. 14.D.*overallVsElementary*).

Such a balanced bi-directional reaction indicated by the reversible arrow is known as an **elementary step**, or a **stoichiometric reaction** (Fig. 14.26). For a given reactants mixture, there can be a large number of elementary steps leading to the final products (see Anim. 14.D.*overallVsElementary*).

Now suppose Eq. (14.39) is the only elementary step involving H_2, O_2, and H_2O and we are interested in the equilibrium products composition of a theoretical mixture of hydrogen and oxygen kept at a given pressure and temperature. Also assume that only the elementary step shown above is possible. In that case, the forward reaction will initially be dominant; however, as the proportion of water molecules increases, the inverse reaction will eventually kick in, as the probability of sufficiently energetic collisions among water molecules increases. Eventually, a dynamic equilibrium is reached, where the rate of reactions in the two directions equalize, resulting in a stable composition of the mixture comprised of H_2, O_2, and H_2O at a given temperature and pressure. The mixture is then said to be in **chemical equilibrium** (see Anim. 14.D.*equilibriumReaction*). The goal of an equilibrium analysis is to evaluate the composition of the products at equilibrium.

An **overall** reaction lists the inlet (or initial) composition on the LHS and the exit (final) equilibrium composition on the RHS, as shown in Eq. (14.40) for the oxidation of hydrogen. By restricting the reaction mechanism to a single elementary step, Eq. (14.39), we have a products mixture that can contain a maximum of three components. The unknown coefficients, in this case, can be reduced to a single unknown a, using the atom balance equations:

$$H_2 + 0.5O_2 \rightarrow aH_2O + \underbrace{b}_{(1-a)} H_2 + \underbrace{c}_{(1-a)/2} O_2 \qquad \textbf{(14.40)}$$

In reality, far more than a single elementary step is involved as a reacting system seeks equilibrium. Therefore, the number of components and, hence, the number of unknowns in the overall reaction is much higher.

Consider a steady-flow reactor with inlet and exit compositions described by the general overall reaction:

$$\sum_r \nu_r W_r \rightarrow \sum_p \nu_p W_p \qquad \textbf{(14.41)}$$

Here, ν_r represents the known stoichiometric coefficient of the reactants r and ν_p represents the unknown coefficients of the products component p. The reactor is allowed to communicate with a TER at T so that the temperature remains constant at T throughout the reactor. Also, pressure loss due to friction is assumed to be negligible so that p remains constant. In other words, our goal is to evaluate the equilibrium composition of a reacting mixture at a given p and T. It may appear that this is a very restrictive condition; after all, we expect the pressure and temperature to rise when a fuel-air mixture is ignited in a closed chamber. However, once we develop a procedure for finding the equilibrium composition at a known pressure and temperature, we can use iterative techniques (or other direct approaches) to evaluate an unknown temperature or pressure.

Using the procedure we developed in Sec. 14.1, let us obtain an expression for the reversible work that can be extracted from this reactor as the reactants seek chemical equilibrium. By substituting $\dot{Q}$ from the entropy equation, Eq. (13.29), into the energy equation, Eq. (13.27), and setting $\dot{S}_{gen} = 0$ (no irreversibilities to maximize the useful work), we obtain:

$$\frac{\dot{W}_{rev}}{\dot{n}_F} = \sum_r \nu_r[\bar{h}_r(p_r, T) - T\bar{s}_r(p_r, T)] - \sum_e \nu_p[\bar{h}_p(p_p, T) - T\bar{s}_p(p_p, T)]$$

$$\Rightarrow \quad \frac{\dot{W}_{rev}}{\dot{n}_F} = \underbrace{\sum_r \nu_r \bar{g}_r(p_r, T)}_{\bar{g}_R} - \underbrace{\sum_e \nu_p \bar{g}_p(p_p, T)}_{\bar{g}_P}$$

$$\Rightarrow \quad \frac{\dot{W}_{rev}}{\dot{n}_F} = -[\bar{g}_P(p, T) - \bar{g}_R(p, T)] = -\Delta\bar{g}(p, T) = -\Delta\bar{g}_{p,T}; \quad \left[\frac{kJ}{kmol}\right] \qquad \textbf{(14.42)}$$

Although developed with a combustion reaction in mind, this expression for reversible work holds true for any overall reaction with $\dot{n}_F$ representing the mole flow rate of a principal reactant or as a proportionality constant so that $\dot{n}_k = \nu_k \dot{n}_F$. As long as $\bar{g}_R > \bar{g}_P$, useful work can be extracted from the reactor. Different possible compositions of the products will yield different values of $\bar{g}_P$. With p, T, and $\bar{g}_R$ remaining constant for a given reactants mixture, the products composition controls the reversible work—the smaller the products Gibbs function, the higher is the reversible work output. When the work extracted reaches a maximum, $d\dot{W}_{rev} = 0$, there are no more driving forces left to exploit and the products state reaches thermodynamic equilibrium (see Anim. 14.D.*revWorkAndGmin*). The products composition that produces the smallest value of $\bar{g}_P$ (Fig. 14.27) must be the **equilibrium composition** at the exit (see Anim. 14.D.*reactionEquilibriumEqn*).

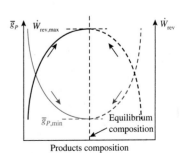

FIGURE 14.27 As the Gibbs function reaches a minimum, no more work can be extracted ($d\dot{W}_{rev} = 0$), signaling chemical equilibrium.

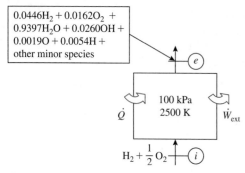

FIGURE 14.28 Equilibrium composition at the exit state at 100 kPa, 2500 K. Use the combustion chamber simulator (TEST. Interactives) to reproduce this reaction (select H2 as the fuel, pure oxygen as the oxidizer, select desired species in the products, set the temperature, and click Calculate).

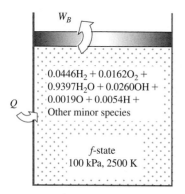

FIGURE 14.29 The same equilibrium composition, as in Fig. 14.27, is achieved in an isobaric, isothermal process at 100 kPa, 2500 K from a reactant mixture of H_2 + $0.5O_2$.

The conclusion that the minimum value of $\bar{g}_P$ signals chemical equilibrium at a given pressure and temperature is quite general. If we started with a closed system undergoing an isobaric, isothermal process, and sought the condition for maximum reversible work, we would end up with the same conclusion: at equilibrium $\bar{g}_P$ is minimized. The equilibrium composition would be identical, regardless of how the system arrived at equilibrium—through a steady state reactor or a constant-pressure, constant-temperature process executed by the reactants (Figs. 14.28 and 14.29).

A direct approach for computational evaluation of equilibrium composition is to seek a product mixture consistent with atom balance constraints, which minimizes the specific Gibbs function of the product mixture. Given the large number of components and their distribution that have to be considered, such an approach is not suitable for hand calculation. But the advantage of this approach is that there is no need to keep track of the large number of simultaneous elementary steps that take place in an equilibrium reaction. The equilibrium TESTcalcs use this approach for calculating the equilibrium composition.

In an alternative development, consider a reacting mixture passing through a steady-flow reactor. Out of all possible reactions occurring, we focus on a particular *elementary step*:

$$\sum_r \nu_r W_r \leftrightarrows \sum_p \nu_p W_p \tag{14.43}$$

This step affects only a subset of all the components present in the mixture. Let us also assume that this is the only reaction that is possible. If this reaction were to go to completion, in the forward direction, the resulting reversible work would be:

$$\frac{\dot{W}_{\mathrm{rev}}}{\dot{n}_F} = \underbrace{\sum_r \nu_r \bar{g}_r(p_r, T)}_{\bar{g}_R(p,T)} - \underbrace{\sum_p \nu_p \bar{g}_p(p_p, T)}_{\bar{g}_P(p,T)} = -\Delta \bar{g}_{p,T}; \quad \left[\frac{\mathrm{kJ}}{\mathrm{kmol}}\right] \tag{14.44}$$

Of course, fuel in this equation means an arbitrarily selected reactant of the elementary step, Eq. (14.43). If the reaction does not go to completion, as is likely to be the case, the reversible work will be only a fraction of, but proportional to, $\Delta \bar{g}_{p,T}$, depending on the degree of completion. As long as $\bar{g}_R > \bar{g}_P$, even an infinitesimal forward progress will produce useful work. Now imagine the reactants passing through a large sequence of reaction chambers, each producing a small amount $d\dot{W}_{\mathrm{rev}} = \dot{n}_F(\bar{g}_R - \bar{g}_P)d\varepsilon$ of reversible work, where $d\varepsilon$ is a very small number representing an infinitesimal degree of completion of the reaction. Contributions from the individual reactors add up to $\dot{W}_{\mathrm{rev}}$. As the mixture approaches equilibrium, there is no chemical, thermal, or mechanical driving force to exploit at the last chamber. Therefore, $d\dot{W}_{\mathrm{rev}} = \dot{n}_F(\bar{g}_R - \bar{g}_P)d\varepsilon = 0$ becomes the condition of equilibrium. For the given elementary step, the equilibrium condition reduces to:

$$\bar{g}_R = \bar{g}_P, \quad \text{or,} \quad \sum_r \nu_r \bar{g}_r(p_r, T) = \sum_r \nu_p \bar{g}_p(p_p, T); \quad \left[\frac{\mathrm{kJ}}{\mathrm{kmol}}\right] \tag{14.45}$$

The molar specific Gibbs functions, $\bar{g}_R$ and $\bar{g}_P$, for a given elementary step are based on unit mole of one principal reactant and can be interpreted as the mole averaged chemical potential of the reactants and products sides. A positive $d\dot{W}_{\mathrm{rev}}$ implies a spontaneous direction of internal changes. Therefore, an elementary step moves towards the direction of decreasing chemical potential as long as $\bar{g}_R > \bar{g}_P$, reaching equilibrium when $\bar{g}_R = \bar{g}_P$.

Of course, there are numerous elementary steps that constitute an overall reaction. Contribution from each elementary step must be added to obtain $d\dot{W}_{\mathrm{rev}}$, which must be zero at equilibrium:

$$\frac{d\dot{W}_{\mathrm{rev}}}{\dot{n}_F} = 0 = (\bar{g}_R - \bar{g}_P)_1 d\varepsilon_1 + (\bar{g}_R - \bar{g}_P)_2 d\varepsilon_2 + \ldots; \quad \left[\frac{\mathrm{kJ}}{\mathrm{kmol}}\right] \tag{14.46}$$

In this equation, each group on the RHS represents a different elementary step, with different degrees of completion represented by the differentials $d\varepsilon_1$, $d\varepsilon_2$, etc. Given that each elementary step is independent, so must be its degree of completion. The right hand side can only sum up to zero when each term in parentheses is independently zero. Therefore, the equilibrium criterion $\bar{g}_R = \bar{g}_P$ must be obeyed by each elementary step (Fig. 14.30 and see Anim. 14.D.*reactionEquilibriumEqn*). This is known as the **equation of reaction equilibrium**, which conforms to the general equation, Eq. (14.10).

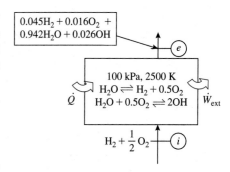

FIGURE 14.30 The equilibrium composition at the exit is based on only two simultaneous stoichiometric reactions. For each reaction, $\bar{g}_R = \bar{g}_P$ if the mixture at the exit is at equilibrium.

14.4.1 Equilibrium TESTcalcs

TEST offers several TESTcalcs for solving equilibrium problems. The n-IG state TESTcalcs are useful for evaluating properties of gas mixtures, including partial properties such as the chemical potential, partial molar enthalpy or entropy, etc. For setting up the reactants side of a combustion reaction with excess or deficient air, the reaction panel of the non-premixed combustion TESTcalcs (Sec. 13.1.1) can be used.

There are three types of powerful equilibrium TESTcalcs: state TESTcalcs, open-steady TESTcalcs, and closed-process TESTcalcs. The IGE (ideal gas equilibrium) state TESTcalcs, located in the states, system states branch, can be used to evaluate system and flow states of an ideal gas mixture, where the equilibrium composition is evaluated as part of the equilibrium state. In addition to the standard state and I/O panels, the IGE state TESTcalcs offer a composition panel. To compose the reactants mixture, select the components and, for each component, enter its mass or mole. For the products, select the desired components (all eligible products are highlighted). In the state panel, select a radio button to choose products or reactants, enter temperature and pressure, and click Calculate to evaluate the state. If the Products radio button is selected, the products' composition is evaluated and displayed in the composition panel. Based on the products' composition (treated as an IG mixture), all state properties are evaluated.

The open-steady and closed-process chemical equilibrium TESTcalcs, located in the systems, open-steady, specific, combustion & equilibrium and systems, closed-process, specific, combustion & equilibrium branches, build upon the state TESTcalcs by adding a reaction panel and an analysis panel for device or process analysis. Specifying a reactants mixture is simplified through the reaction panel (Sec. 13.1.1). Once a reaction is set up, fuel, oxygen, and nitrogen moles are automatically imported into the composition panel, where the products' components are selected, and reactants' and products' states are evaluated, as in the IGE state TESTcalcs. The computed states can be imported to the device or process panel to perform energy and entropy analysis for a device or a process. For a parametric study, you can change any variable (add a new component in the products for example) and click Super-Calculate to update the analysis.

The combustion chamber simulation Interactive, linked from the Interactives module, can be used for visually calculating a products' composition, after a fuel and an equivalence ratio is selected. It is capable of visual parametric studies.

EXAMPLE 14-8   Minimization of Gibbs Function

0.5 kmol of hydrogen and 0.5 kmol of oxygen are kept in a container maintained at 1 atm and 2500 K. Which of the following component sets is more likely to be the equilibrium composition? (a) Set A: H_2, O_2, H_2O; (b) Set B: H_2, O_2, H_2O, OH; (c) Set C: H_2, O_2, H_2O, OH, O; (d) Set D: H_2, O_2, H_2O, OH, O, H. Use the IGE system-state TESTcalc.

SOLUTION

Evaluate the Gibbs function of each mixture. The one with the minimum value of the function is the most likely candidate for the equilibrium mixture.

Assumptions

The Ideal gas mixture model is applicable.

TEST Analysis

Launch the IGE system-state TESTcalc. In the composition panel, enter the moles of the reactants (select the checkbox, type in the value in a yellow field, and click on a blank cell to register the input). Then select the desired set of products components. In the state panel, select State-1, mark the products' radio-button, enter p1=100 kPa and T1=2500 K, and calculate the state. The mixture composition can now be found in the composition panel and the Gibbs function g1 in the state panel.

> Composition A: The products mixture is found to have g1 $= -34149$ kJ/kg and the composition is calculated as $0.0050\ H_2 + 0.2525\ O_2 + 0.4949\ H_2O$.
>
> Composition B: The products mixture is found to have g1 $= -34190$ kJ/kg and the composition is calculated as $0.0050\ H_2 + 0.2441\ O_2 + 0.4781\ H_2O + 0.0337\ OH$.
>
> Composition C: The products mixture is found to have g1 $= -34199$ kJ/kg and the composition is calculated as $0.0050\ H_2 + 0.2410\ O_2 + 0.4781\ H_2O + 0.0337\ OH + 0.0062\ O$.
>
> Composition D: The products mixture is found to have g1 $= -34200$ kJ/kg and the composition is calculated as $0.0050\ H_2 + 0.2414\ O_2 + 0.4774\ H_2O + 0.0337\ OH + 0.0062\ O + 0.0015\ H$.

Discussion

Composition D, with its lowest value of Gibbs function, is the most likely candidate for equilibrium composition. In reality, there are many more components, some in trace amounts, that are present at equilibrium. The Gibbs function of the mixture will continue to decline as more components are included. Note that the TESTcalc produces Gibbs function per unit mass of the mixture, while $\bar{g}_P$, in our terminology, is the Gibbs function of the products mixture per unit mole of the principal reactant (generally, a fuel).

14.4.2 Equilibrium Composition

Direct minimization of the Gibbs function, as used by the equilibrium TESTcalc, is a *constrained minimization* problem that must be solved by an iterative approach. For a manual solution, the *equations of reaction equilibrium* provide a simpler alternative as long as the number of simultaneous elementary steps is restricted to a small number. A manual solution, of even the simplest of equilibriums, provides an insight that cannot be obtained from a computer solution, which should be used only for verification of manual results and studies of complex systems.

To explain the procedure for using the equations of reaction equilibrium, let us limit this discussion to the equilibrium of IG mixtures. In that case, the IG Gibbs function (see Anim. 14.B.*GibbsFunctionIG*) can be substituted from Eq. (14.17) in the equation of reaction equilibrium, Eq. (14.45), producing:

$$\sum_r \nu_r \left[\bar{g}_r^\circ(p_0, T) + \bar{R}T \ln \frac{p_r}{p_0} \right] = \sum_p \nu_p \left[\bar{g}_p^\circ(p_0, T) + \bar{R}T \ln \frac{p_p}{p_0} \right]; \quad \left[\frac{kJ}{kmol} \right]$$

$$\Rightarrow \quad \underbrace{\sum_p \nu_p \bar{g}_p^\circ(p_0, T) - \sum_r \nu_r \bar{g}_r^\circ(p_0, T)}_{\Delta \bar{g}_T^\circ} = -\bar{R}T \left[\sum_p \ln \left(\frac{p_p}{p_0} \right)^{\nu_p} - \sum_r \ln \left(\frac{p_r}{p_0} \right)^{\nu_r} \right]; \quad \left[\frac{kJ}{kmol} \right]$$

$$\tag{14.47}$$

In this equation, $\Delta \bar{g}_T^\circ$ is the change in Gibbs function at temperature T and standard pressure p_0, and can be evaluated for any given elementary step by following the procedure developed in Sec. 13.3.4. Equation (14.47) can be manipulated to relate the partial pressures of the components in the products and reactants with $\Delta \bar{g}_T^\circ$ to produce:

$$-\frac{\Delta \bar{g}_T^\circ}{\bar{R}T} = \ln \frac{\prod_p p_p^{\nu_p}}{\prod_r p_r^{\nu_r}} (p_0)^{\sum_r \nu_r - \sum_p \nu_p} = \ln \frac{\prod_p y_p^{\nu_p}}{\prod_r y_r^{\nu_r}} \left(\frac{p}{p_0} \right)^{\sum_p \nu_p - \sum_r \nu_r} \tag{14.48}$$

The LHS of this equation establishes the equation to be a function of temperature alone. Expressed in logarithmic form, it defines the **equilibrium constant** K as:

$$\ln K \equiv -\frac{\Delta \bar{g}_T^\circ}{\bar{R}T} \quad \text{or} \quad K \equiv \exp\left(-\frac{\Delta \bar{g}_T^\circ}{\bar{R}T} \right) \tag{14.49}$$

where

$$K \equiv \frac{\prod\limits_p y_p^{\nu_p}}{\prod\limits_r y_r^{\nu_r}} \left(\frac{p}{p_0}\right)^{\sum_p \nu_p - \sum_r \nu_r} \qquad (14.50)$$

Equation (14.50) is known as the **chemical equilibrium equation**, which relates the mole fractions of different components in the products mixture with the equilibrium constant K of an elementary step. Although pressure p appears prominently on the RHS of Eq. (14.50), it must be stressed that K is a thermodynamic function of the elementary step and depends on temperature only. Any effect of changing the total pressure p at a given temperature T is completely absorbed by appropriate changes in the mole fractions y_p's of the products components, so that K is unaffected (see Anim. 14.D.*equilibriumConstant*).

It is customary to evaluate $\ln K$ or $\log_{10} K$ for a given elementary step from Eq. (14.49) and list it against temperature. Such a listing is provided for several reactions in Table G-3. The IGE state TESTcalcs can also be used to obtain $\ln K$ for an elementary step by setting it up in the composition panel. The reactants are specified by their moles (stoichiometric coefficients) and the products' components, of the elementary step, are selected by checking the appropriate checkboxes. In the state panel, the temperature and pressure are entered and Calculate produces the balanced reaction in the composition panel along with the equilibrium constant. By recalculating the state with a new temperature, the value of $\ln K$ can be updated.

As a specific application of the equilibrium equation, suppose a chamber contains gases A, B, C, D, E, F, and G in chemical equilibrium at a given pressure and temperature. Further, suppose the following elementary steps are the only reactions possible:

$$(1) \quad \nu_{A1}A + \nu_B B \rightarrow \nu_C C + \nu_D D \qquad (14.51)$$

$$(2) \quad \nu_{A2}A + \nu_E E \rightarrow \nu_F F + \nu_G G \qquad (14.52)$$

In that case, the equilibrium constants for these reactions provide two independent equations relating the mole fractions of the seven components:

$$K_1 = \frac{y_C^{\nu_C} y_D^{\nu_D}}{y_A^{\nu_{A1}} y_B^{\nu_B}} \left(\frac{p}{p_0}\right)^{\nu_C + \nu_D - \nu_{A1} - \nu_B} \quad \text{and} \quad K_2 = \frac{y_F^{\nu_F} y_G^{\nu_G}}{y_A^{\nu_{A2}} y_E^{\nu_E}} \left(\frac{p}{p_0}\right)^{\nu_F + \nu_G - \nu_{A2} - \nu_E} \qquad (14.53)$$

Coupled with the atom balance equations, these equations provide the necessary closure to solve for the stoichiometric coefficients of the products in the overall reaction. Let us go over a few examples to illustrate this procedure.

EXAMPLE 14-9   **Equilibrium Constant**

Evaluate $\ln K$ at 298 K for the following elementary steps: (a) $H_2 + 0.5O_2 \rightarrow H_2O$ and (b) $2H_2O \rightarrow 2H_2 + O_2$. Assume water to be in the vapor phase. **What-if scenario:** What would the answer in part (a) be if the temperature was (c) 1000 K or (d) 2500 K?

SOLUTION

Evaluate $\Delta \bar{g}_T^{\circ}$ for an elementary step using properties from the ideal gas tables or the n-IG state TESTcalc, then obtain $\ln K$ from Eq. (14.49).

Assumptions

The IG mixture model is applicable.

Analysis

$\Delta \bar{g}_T^{\circ}$ can be expressed in terms of enthalpy and entropy as:

$$\Delta \bar{g}_T^{\circ} = \Delta[\bar{h}(p_0, T) - T\bar{s}(p_0, T)] = \Delta\bar{h}(T) - T\Delta\bar{s}^{\circ}(T)$$

Using the procedure developed in the last chapter, we evaluate $\Delta\bar{h}(T)$ and $\Delta\bar{s}^{\circ}(T)$ for the elementary step $H_2 + 0.5O_2 \rightarrow H_2O$.

$$\Delta\bar{h}(T) = \bar{h}_P(T) - \bar{h}_R(T) = \sum_p \nu_p \bar{h}_p(T) - \sum_r \nu_r \bar{h}_r(T)$$

$$= \nu_{H_2O}\bar{h}^\circ_{f,H_2O} - \nu_{H_2}\bar{h}^\circ_{f,H_2} - \nu_{O_2}\bar{h}^\circ_{f,O_2}$$

$$= (1)(-241,820) - (1)(0) - (0.5)(0) = -241,820 \text{ kJ/kmol}$$

$$\Delta\bar{s}^\circ(T) = \bar{s}^\circ_P(T) - \bar{s}^\circ_R(T) = \sum_p \nu_p \bar{s}^\circ_p(T) - \sum_r \nu_r \bar{s}^\circ_r(T)$$

$$= \nu_{H_2O}\bar{s}^\circ_{H_2O@298 \text{ K}} - \nu_{H_2}\bar{s}^\circ_{H_2@298 \text{ K}} - \nu_{O_2}\bar{s}^\circ_{O_2@298 \text{ K}}$$

$$= (1)(188.7) - (1)(130.7) - (0.5)(205.0)$$

$$= -44.5 \text{ kJ/kmol} \cdot \text{K}$$

Combining,

$$\Delta\bar{g}^\circ_T = \Delta\bar{h}(T) - T\Delta\bar{s}^\circ(T)$$

$$= -241,820 - (298)(-44.5) = -228,559 \text{ kJ/kmol}$$

Therefore,

$$\ln K \equiv -\frac{\Delta\bar{g}^\circ_T}{\bar{R}T} = -\frac{-228,559}{(8.314)(298)} = 92.25$$

The second reaction, $2H_2O \rightarrow 2H_2 + O_2$, represented here by the suffix (b), is simply the inverse of the first reaction multiplied by 2. Therefore,

$$\Delta\bar{g}^\circ_{T(b)} = -(2)(-228,559) = (2)(228,559) \text{ kJ/kmol}$$

$$\text{and} \quad \ln K_{(b)} \equiv -\frac{\Delta\bar{g}^\circ_{T(b)}}{\bar{R}T} = -\frac{(2)(228,559)}{(8.314)(298)} = -(2)(92.25) = -184.5$$

TEST Analysis

Launch the IGE system-state TESTcalc. In the composition panel, select the reactants, H2 and O2 for the first reaction and enter their stoichiometric coefficients, 1 and 0.5, in kmol (to register the change you must click on an empty cell). Select H2O in the products block. In the state panel, enter p1=100 kPa and T1=298 K, select the Products radio button, and click Calculate. Obtain ln K from the composition panel as 92.26.

What-if scenario

Change T1 to 1000 K and re-calculate State-1. Evaluate ln K as 23.16. At 2500 K ln K can be computed as 5.13.

Discussion

The huge value of the equilibrium constant $e^{92.26}$ at 298 K, indicates that the forward reaction for the elementary step $H_2 + 0.5O_2 \rightarrow H_2O$ dominates at room temperature; and hydrogen and oxygen are almost nonexistent at equilibrium (that is why water is so stable at atmospheric conditions). As the temperature is increased, there is a drastic drop in the value of K. At 2500 K, therefore, we will expect some hydrogen and oxygen to be present in the mixture as the inverse reaction strengthens. We will verify that in the next example.

EXAMPLE 14-10 Equilibrium Composition

One kmol of hydrogen and one kmol of oxygen react to produce an equilibrium mixture of water, hydrogen, and oxygen at 2500 K. (a) Determine the equilibrium molar composition if the mixture pressure is 1 atm. *What-if scenario:* What would the mole fraction of water in the mixture be if (b) the pressure increased to 10 atm without changing the temperature and if (c) the temperature was reduced to 1000 K at 1 atm?

SOLUTION

The elementary step $H_2 + 0.5O_2 \leftrightarrows H_2O$ is not available in Table G-3, but the inverse reaction is listed with $\ln K = -5.13$ (interpolated) at 2500 K. For the chosen elementary step, therefore, $\ln K = 5.13$. This has also been verified in the previous example. Using this equilibrium constant, the unknown stoichiometric coefficients of an overall reaction can be evaluated.

Assumptions

The IG mixture model is applicable for the equilibrium mixture.

Analysis

The overall reaction, with the initial mixture on the LHS and the equilibrium mixture on the RHS, can be written in terms of three unknown coefficients:

$$H_2 + O_2 \rightarrow aH_2 + bO_2 + cH_2O$$

Using the atom balance equations, b and c can be expressed in terms of a:

$$H_2 + O_2 \rightarrow aH_2 + \frac{1 + a}{2} O_2 + (1 - a)H_2O$$

The total mole in the equilibrium mixture is $\sum_p \nu_p = \dfrac{(3 + a)}{2}$; therefore, the mole fractions of the components in the equilibrium mixture are:

$$y_{H_2} = \frac{2a}{3 + a}; \quad y_{O_2} = \frac{1 + a}{3 + a}; \quad \text{and} \quad y_{H_2O} = \frac{2(1 - a)}{3 + a}$$

Substituting these expressions in the equilibrium reaction equation, Eq. (14.50), we obtain:

$$K = \frac{y_{H_2O}}{y_{H_2} y_{O_2}^{0.5}} \left(\frac{p}{p_0}\right)^{1-1-0.5}$$

$$= \frac{2(1 - a)}{3 + a} \left(\frac{2a}{3 + a}\right)^{-1} \left[\frac{1 + a}{3 + a}\right]^{-0.5} \left(\frac{p}{p_0}\right)^{-0.5}$$

(14.54)

In the previous example, $\ln K$ at 2500 K was calculated as 5.13. At $p = 1$ atm ($p = p_0$), the equilibrium relation reduces to:

$$e^{5.13} = \frac{2(1 - a)}{3 + a} \left(\frac{2a}{3 + a}\right)^{-1} \left(\frac{1 + a}{3 + a}\right)^{-0.5}$$

$$\Rightarrow \quad 170.91 = \frac{1 - a}{a} \sqrt{\frac{3 + a}{1 + a}}$$

Iteration with a calculator produces $a = 0.01$.

The molar composition can now be evaluated as:

$$y_{H_2} = 0.0067; \quad y_{H_2O} = 0.6578; \quad y_{O_2} = 0.3356$$

TEST Analysis

Launch the IGE system-state TESTcalc. In the composition panel, select H2 and O2 as the reactants and enter their moles. Also select H2, H2O, and O2 as the possible products components. Switch to the state panel and evaluate State-1 for the products (make sure that the Products radio button is checked) at p1=100 kPa and T1=2500 K. In the I/O panel, the mole fractions can be found as 0.0067, 0.6578, and 0.3356, respectively, verifying the manual Solution.

What-if scenario

(b) Change p1 to 10 atm and click Calculate to obtain $y_{H_2O} = 0.664$. (c) Change p1 to 1 atm and T1 to 1000 K. Click Calculate to obtain $y_{H_2O} = 0.667$.

Discussion

At very high temperatures, water starts dissociating into hydrogen and oxygen and the elementary step shifts to the left. On the other hand, as pressure is increased, the reaction shifts to the right, producing more water. A shift to the right accompanies a decrease in mole (from 1.5 to 1), mitigating the rise in pressure. A shift in equilibrium due to an external factor, generally, tends to offset the effect of the external factor. This is a well-known principle called the **Le Chatelier principle**.

14.4.3 Significance of Equilibrium Constant

Keeping in mind that the goal of chemical equilibrium analysis is to predict the equilibrium state of a reacting mixture, the following observations summarize several important aspects of application of the chemical equilibrium equation.

1. It is important to distinguish an overall reaction from the numerous simultaneous elementary steps. The overall reaction is an input/output type expression, connecting the inlet and exit compositions of a steady-flow reactor or the initial and final compositions of a reacting process. The elementary steps, on the other hand, represent the detailed reaction mechanism on which the equilibrium equations are based. While the partial pressure of a component can be related to the stoichiometric coefficients of the products of the overall reaction, the exponents in an equilibrium equation come from the corresponding elementary step.
2. There can be many simultaneous elementary steps in an equilibrium mixture and not all equilibrium equations obtained from these reactions are independent. If the total number of components in the equilibrium mixture is C, and the number of independent equations from atom balance of the overall equation is A, then only $C - A$ equations of equilibrium are necessary for closure. Any combinations of $C - A$ independent elementary steps will produce the same equilibrium mixture.
3. The equilibrium constant K is associated with a particular form of the elementary step. If a given step is inverted—$H_2 + 0.5O_2 \rightarrow H_2O$ vs. $H_2O \rightarrow H_2 + 0.5O_2$, for instance—products and reactants exchange their meanings, resulting in a change of sign for $\Delta \bar{g}_T^o$. Therefore, from Eq. (14.49), it can be deduced that $\ln K^* = -\ln K$, where K^* is the equilibrium constant for the inverse reaction. Similarly, if a reaction is multiplied by a constant α—$H_2 + 0.5O_2 \rightarrow H_2O$ vs. $2H_2 + O_2 \rightarrow 2H_2O$, for instance—$\Delta \bar{g}_T^o$ changes by a factor of α and so must $\ln K$. This can be verified from the results of Example 14-9.
4. The equilibrium constant K, for a given stoichiometric reaction, is not really a constant but a function of temperature. To explore the functional relationship between K and temperature, let us differentiate Eq. (14.49) with respect to T:

$$\frac{d}{dT}[\ln K] = -\frac{d}{dT}\left[\frac{\Delta \bar{g}_T^o}{\bar{R}T}\right] = \frac{\Delta \bar{g}_T^o}{\bar{R}T^2} - \frac{1}{\bar{R}T}\frac{d(\Delta \bar{g}_T^o)}{dT}; \quad \left[\frac{1}{K}\right]$$

$$= \frac{\Delta \bar{h}_T^o - T\Delta \bar{s}_T^o}{\bar{R}T^2} - \frac{1}{\bar{R}T}\frac{d(\Delta \bar{h}_T^o)}{dT} + \frac{1}{\bar{R}T}\frac{d(T\Delta \bar{s}_T^o)}{dT} \quad \textbf{(14.55)}$$

$$= \frac{\Delta \bar{h}_T^o}{\bar{R}T^2} - \frac{1}{\bar{R}T}\frac{d(\Delta \bar{h}_T^o)}{dT} + \frac{1}{\bar{R}}\frac{d(\Delta \bar{s}_T^o)}{dT}$$

This relation can be further simplified by expanding the last term and recognizing that the *Tds* relation for a species k at a constant pressure p_0 yields $Td[\bar{s}_k^o(T)] = d[\bar{h}_k^o(T)]$:

$$\frac{d(\Delta \bar{s}_T^o)}{dT} = \frac{d}{dT}\left[\sum_p \nu_p \bar{s}_p^o(T) - \sum_r \nu_r \bar{s}_r^o(T)\right]; \quad \left[\frac{kJ}{kmol \cdot K^2}\right]$$

$$= \sum_p \nu_p \frac{d}{dT}[\bar{s}_p^o(T)] - \sum_r \nu_r \frac{d}{dT}[\bar{s}_r^o(T)]$$

$$= \sum_p \nu_p \frac{1}{T}\frac{d}{dT}[\bar{h}_p^o(T)] - \sum_r \nu_r \frac{1}{T}\frac{d}{dT}[\bar{h}_r^o(T)] = \frac{1}{T}\frac{d}{dT}[\Delta \bar{h}_T^o]$$

Substituting this expression into Eq. (14.55), we obtain:

$$\frac{d}{dT}[\ln K] = \frac{\Delta \bar{h}_T^o}{\bar{R}T^2}; \quad \left[\frac{1}{K}\right] \quad \textbf{(14.56)}$$

This is known as the **van't Hoff equation**. It shows that for an exothermic reaction, K must decrease with temperature since $\Delta \bar{h}_T^o$ is negative. For a combustion reaction, this equation can be further simplified by recognizing that $\Delta \bar{h}_T^\circ$ remains approximately constant over a wide range of temperatures (Fig. 14.31), so that $\Delta \bar{h}_T^\circ \cong \Delta \bar{h}_C^\circ$. The enthalpy of combustion being a constant for a given reaction, the van't Hoff equation can be integrated to yield a simpler relation between K and T:

$$\ln \frac{K_2}{K_1} \cong \frac{-\Delta \bar{h}_C^\circ}{\bar{R}} \left(\frac{1}{T_2} - \frac{1}{T_1} \right)$$ (14.57)

where K_1 and K_2 are equilibrium constants at T_1 and T_2, respectively. Not only does this relation show that $\ln K$ varies linearly with $1/T$, but it also presents a novel way for evaluating enthalpy of combustion using equilibrium composition data or equilibrium constants from enthalpy data.

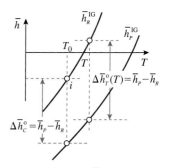

FIGURE 14.31 $\Delta \bar{h}_T^\circ$ does not vary significantly with temperature for most combustion reactions (see Anim. 13.B.*heatingValues*).

5. The larger the value of K for a given elementary step, the more is its degree of completion. This is evident from the equilibrium equation, Eq. (14.49). In Example 14-9, for example, K at 298 K for the reaction $H_2 + 0.5O_2 \rightarrow H_2O$ was calculated as $e^{92.25} = 1.158 \times 10^{40}$. According to Eq. (14.54), a large value of K implies that $y_{H_2} y_{O_2}^{0.5} \cong 0$, that is, at least one of the reactants is almost totally consumed, signaling completion of the reaction in the forward direction. As a rule of thumb, a reaction is assumed to proceed to completion when $K > 1000$. Conversely, $K < 0.001$ signifies that the reaction does not proceed at all and can be ignored in the calculations of equilibrium composition.

6. The degree of completeness of an elementary step and the equilibrium composition can be affected by overall pressure, even though the equilibrium constant is independent of the total pressure. The Le Chatelier principle, as discussed in Ex 14-10, states that an increase in pressure shifts the reaction in the direction of decreasing mole. Ammonia production by the reaction $N_2 + 3H_2 \rightarrow 2NH_3$, for example, will increase if the pressure is increased.

7. Inert gases may not participate in a reaction directly, but their presence can affect the equilibrium composition indirectly without affecting the equilibrium constants. Presence of an inert gas reduces the mole fraction of each component, thereby affecting other terms in the equilibrium equation.

8. Equilibrium calculations are completely silent on the time needed to achieve equilibrium. According to the fifth observation, the reaction $H_2 + 0.5O_2 \rightarrow H_2O$ goes to completion at 298 K, leaving no traces of hydrogen or oxygen. But we know from our experience that hydrogen and oxygen can coexist at 298 K. This is possible due to the sluggish nature of reaction kinetics—the fraction of molecules with high enough kinetic energy, 298 K, to overcome the activation energy of the reaction is almost negligible. However, the reaction rate increases exponentially with temperature, and the likelihood of equilibrium at the typical temperatures of combustion is quite high. Emissions and flame temperature in combustion, therefore, can be calculated with good degree of accuracy through equilibrium calculations.

To illustrate how these observations can be helpful in understanding the characteristics of a chemical equilibrium, consider the production of plasma—a mixture of ionized gases—through the **dissociation reaction** $O_2 \rightarrow 2O$, followed by the **ionization reaction** $O \rightarrow O^+ + e^-$. Such reactions generally occur only at very high temperature. At 2600 K, for instance, the equilibrium constant for the dissociation reaction can be read from Table G-3 as only 0.0005, barely enough for dissociation to begin. Ionization requires even higher temperatures. Electrons released in the ionization reaction can be treated as an ideal gas like the ionized oxygen. In both the dissociation and ionization reactions of oxygen, the mole doubles in the forward direction. From the Le Chatelier principle, we expect both of these reactions to shift to the right at lower pressures. This explains why ionization is more pronounced at very high temperatures and low pressures. The procedure for quantitative determination of the equilibrium composition remains unaltered: set up the overall equations with unknown coefficients; use atom balance equations to reduce the number of unknowns; and, finally, use an appropriate number of independent equations of equilibrium for closure. Chemical properties of electron gas are obtained from application of statistical thermodynamics. To reinforce these steps, let us go through some more examples.

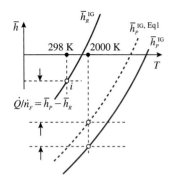

FIGURE 14.33 Heat released in combustion is reduced if the products are assumed to be in equilibrium (Ex. 14-13). You can verify this numerically by using the combustion chamber simulator (TEST. Interactives). Select a fuel and obtain the heat release for a products temperature of 2000 K for Complete and Equilibrium reactions.

Substituting these expressions in the equilibrium equation, Eq. (14.50), we obtain:

$$K = \frac{y_{NO}}{y_{N_2}^{0.5} y_{O_2}^{0.5}} \left(\frac{p}{p_0}\right)^{1-0.5-0.5} = \left(\frac{1-2a}{12.9}\right)\left(\frac{a+8.9}{12.9}\right)^{-0.5}\left(\frac{a}{12.9}\right)^{-0.5};$$

$$\Rightarrow \quad e^{-3.931} = \frac{1-2a}{\sqrt{a(a+8.9)}}; \quad \text{Solving,} \quad a = 0.4792$$

The overall reaction, therefore, is evaluated as:

$$CH_4 + 2.5(O_2 + 3.76N_2) \rightarrow CO_2 + 2H_2O + 0.4792O_2 + 9.3792N_2 + 0.0416NO$$

From the products composition, the mole fraction of NO can be found:

$$y_{NO} = \frac{\nu_{NO}}{\sum_p \nu_p} = \frac{0.0416}{12.9} = 0.003225 \text{ or } 3225 \text{ ppm}$$

The heat transfer can be evaluated from the energy equation, Eq. (13.27), for the overall reaction:

$$\frac{\dot{Q}}{\dot{n}_F} = \sum_p \nu_p \bar{h}_p - \sum_r \nu_r \bar{h}_r = \bar{h}_P - \bar{h}_R$$

where $\bar{h}_P$ and $\bar{h}_R$ can be evaluated using the procedure described in Example 13-5 (Fig. 14.33):

$$\bar{h}_P = \bar{h}_{CO_2} + 2\bar{h}_{H_2O} + 0.479\bar{h}_{O_2} + 9.379\bar{h}_{N_2} + 0.0416\bar{h}_{NO}$$

$$= -302,082 + 2(-169,137) + 0.479(59,199) + 9.379(56,141) + 0.0416(148,150)$$

$$= -79,290 \text{ kJ/kmol};$$

$$\bar{h}_R = \bar{h}_{CH_4} + 2.5\bar{h}_{O_2} + 9.4\bar{h}_{N_2} = \bar{h}_{CH_4}$$

$$= -74,876 \text{ kJ/kmol}$$

Therefore, $\dot{Q}/\dot{n}_F = \bar{h}_P - \bar{h}_R = -79,290 - (-74,876) = -4,414 \text{ kJ/kmol}$

TEST Analysis

Launch the open-steady chemical equilibrium TESTcalc from the Specific, Combustion and Equilibrium branch. In the reaction panel, choose CH4 in the fuel block, enter lambda = 1.25, and select Excess/Deficient Air from the action menu. The reaction is balanced for a complete reaction. In the composition panel, the reactants—1 kmol of CH4, 9.4 kmol of N2, and 2.5 kmol of O2—are automatically populated. Add NO to the default list of products (scroll through the list to find NO and select the checkbox). Evaluate the reactants state, State-1, with p1=1 MPa and T1=298 K, and the products state, State-2, with p2=p1 and T2=2000 K (select the Products radio-button). The mole fraction of NO is displayed in the composition (or I/O) panel as 0.00328. To determine the heat transfer, enter mdot1=361 kg/s and mdot2=mdot1 (total amount of products for burning 1 kmol of fuel). In the device panel, load the inlet and exit states and enter Wdot_ext=0 to obtain Qdot as −3617 kW, which is somewhat different from the manual solution due to slight difference between the products composition evaluated by the TESTcalc.

What-if scenario

Include CO in the products mixture and click Super-Calculate. A very small amount of CO ($y = 0.00017$) is reported in the products mixture. There seems to be no appreciable effect on the amount of NO.

Discussion

If the overall reaction was assumed to be complete (no NO in the products), the exit enthalpy at the exit can be shown (by using TEST) to be $\bar{h}_E = \bar{h}_{CO_2} + 2\bar{h}_{H_2O} + 0.5\bar{h}_{O_2} + 9.4\bar{h}_{N_2} = -84,432 \text{ kJ/kmol}$. Therefore, as shown in Figure 14.33, heat transfer calculated on the assumption of complete combustion is larger in magnitude. If all likely species are included in the equilibrium calculations, the enthalpy of the products (shown by the dotted lines in Fig. 14.33) may significantly deviate from the corresponding complete combustion results.

For adiabatic combustion, the exit temperature based on equilibrium composition at the exit is called the **equilibrium flame temperature**. As shown in Figure 14.34, the equilibrium

FIGURE 14.34 The adiabatic flame temperature for complete reaction and equilibrium reaction are graphically compared (see Anim. 13.B.*adiabatic TempParam*). To calculate the adiabatic flame temperature using the combustion chamber simulator (TEST. Interactives), switch to Known Heat Transfer tab, select a fuel, and click Calculate. For a complete reaction, select the Complete button.

flame temperature is always smaller than the adiabatic flame temperature for complete combustion (Sec. 13.3.2). The actual flame temperature is likely to be slightly less than the equilibrium temperature due to radiative losses from the flame.

EXAMPLE 14-14   **Equilibrium Flame Temperature**

Methane and air enter an adiabatic combustion chamber at 1 atm, 298 K. Do a parametric study of how the adiabatic flame temperature T_{af} depends on the equivalence ratio ϕ. Vary ϕ over a range 0.5 through 1.5. Neglect pressure loss and change in ke or pe. Use TESTcalcs with (a) the PG mixture model (complete combustion), (b) IG mixture model (complete combustion), and (c) IGE mixture model (equilibrium combustion) with the following common species in the combustion products: CO_2, CO, H_2O, OH, O_2, N_2, NO, NO_2.

TEST Analysis

The h-T diagram of Figure 14.34 graphically predicts $T_{eql} < T_{af}^{IG} < T_{af}^{PG}$. To verify this numerically, launch the PG open-steady, non-premixed combustion TESTcalc. On separate tabs, open the corresponding IG combustion TESTcalc and the open-steady equilibrium TESTcalc.

In the reaction panel of the PG combustion TESTcalc, select Methane in the fuel block and Theoretical Air from the action menu. The balanced reaction is displayed. In the state panel, evaluate State-1 and State-2 as the fuel and oxidizer state from the known pressure and temperature. Evaluate State-3 for the products partially by entering p1 only. In the device panel, import the fuel, oxidizer, and products' states, enter **Qdot=Wdot_ext=0**, and click Super-Calculate. Obtain the adiabatic flame temperature for the theoretical reaction T3 from State-3. Alternatively, you can use the energy equation to enter j3 as =(mdot1*j1+mdot2*j2)/(mdot1+mdot2), which will produce a Qdot of 0 in the I/O panel.

The Super-Calculate operation generates the TEST-code in the I/O panel. Copy the TEST code and switch to the IG TESTcalc tab in your browser. Set up the reaction, paste the TEST code onto the I/O panel, and click the Load button. All the states are updated and the adiabatic flame temperature can be picked up as T3 in State-3. In the reaction panel of the equilibrium TESTcalc, set up the complete reaction as already described. In the composition panel, with the reactants already populated, select the eight species in the products block. In the state panel, enter p1 and T1, select the Reactants radio button, and calculate the state. Select State-2, enter **p2=p1** and **h2=h1**, select the Products radio button, and Calculate. The adiabatic flame temperature is calculated as T2 along with the equilibrium composition of the products mixture.

Copy the temperatures from the three different models (PG, IG, and IGE) into a spreadsheet. Now enter an equivalence ratio (phi), say, 0.8 in the reaction panel of the PG combustion TESTcalc and select Excess/Deficient Air from the action menu. Super-Calculate to update T3. Repeat the same procedure with the IG TESTcalc and the equilibrium TESTcalc. The results are compiled in a spreadsheet and plotted in Figure 14.35.

For equilibrium computations, you can also use the combustion Interactive located in the TEST.Interactives module. Once the Interactive is launched, go to the Known Heat Transfer tab and select the fuel and the desired products. On the Parametric Study box, enter the lower and upper limits of the equivalence ratio ϕ, and the number of steps. Click Analyze and the equilibrium temperature variation will be plotted in real time as the computations are performed in the Graphical Results tab (you have to select View Results by Run).

Discussion

The diabatic flame temperature, predicted by the PG model, is consistently much higher than other models. For the theoretical mixture, the predictions from the three models are 2792 K, 2329 K, and 2231 K, respectively. The equilibrium flame temperature is the smallest of the three as can be expected from a graphical interpretation (Fig. 14.34). It is also closest to the actual

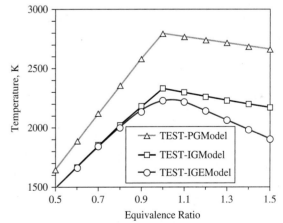

FIGURE 14.35 Adiabatic flame temperature for methane and air as a function of equivalence ratio. Prediction from the equilibrium computation is the most accurate. Use the combustion chamber simulator (TEST. Interactives) to reproduce the IG and IGE model trends (Switch to Known Heat Transfer tab, select CH4 as the fuel, and follow steps 1 through 10.

flame temperature. As the equivalence ratio is increased (rich burning), incomplete combustion becomes increasingly more important, which is captured by the IGE model. For lean combustion ($\phi < 1$), the flame temperature rapidly decreases (Fig. 14.35), and the discrepancy between the IG and IGE models diminishes, confirming that equilibrium computation assumes important role only at high temperatures. If pure oxygen is used as the oxidizer, a much higher flame temperature will result and the discrepancy between the results of complete combustion and equilibrium combustion will be much larger.

14.5 CLOSURE

We discussed thermodynamic equilibrium in various chapters of this book, but it was here where we did an in-depth study of this important topic as applied to phase and chemical equilibrium. The entropy maximum principle of equilibrium was developed first, leading to a Gibbs function minimization principle for systems with fixed pressure and temperature. An equivalent criterion for equilibrium was developed, based on the fact that at equilibrium no useful work can be extracted from a system. The equilibrium criterion was applied to study the equilibrium of a pure gas separated from a gas mixture by a semi-permeable membrane and equilibrium among different phases. A new property called the chemical potential is introduced to explain mass transfer between phases or through membranes. Osmotic pressure and desalination using reverse osmosis were discussed. The theory of equilibrium was then extended to chemical equilibrium. The equilibrium criterion was derived for simultaneous reactions, and equilibrium composition was manually evaluated using equilibrium constants. Powerful chemical equilibrium TESTcalcs were introduced to analyze complex reacting systems and to perform parametric studies of equilibrium composition and equilibrium flame temperature.

PROBLEMS

SECTION 14-1: EQUILIBRIUM CONCEPT

14-1-1 A 2 kg copper block at 100°C is brought in thermal contact with a 5 kg copper block at 50°C. Treating the combined system as an isolated system, show that at equilibrium the entropy (S) of the combined system reaches a maxima.

14-1-2 An isolated cylindrical container is divided into two chambers separated by a pinned piston. The left chamber contains 0.1 kg of H_2 at 200 kPa, 20°C and the right chamber contains 0.2 kg of He at 100 kPa, 20°C. After the pin is removed, the piston oscillates and eventually comes to equilibrium. (a) Show that the entropy (S) of the combined isolated system reaches a maximum when the pressure (p) on two sides are equal. Assume the piston is thermally conductive and use the PG model. (b) Can equilibrium be reached if friction is assumed to be absent?

14-1-3 Since the beginning of the atomic age, it has been a practice to store uranium as gaseous uranium hexafluoride (UF_6, molar mass: 352 kg/kmol) in abandoned oil wells. The gas at the top of a 2 km deep well was sampled and found to be pure UF_6 at 100 kPa, 300 K. Assuming equilibrium throughout the well, isothermal condition, and ideal gas behavior, determine the pressure (p) at the bottom of the well.

14-1-4 A mixture of hydrogen and uranium hexafluoride (UF_6, molar mass: 352 kg/kmol) is stored in an abandoned oil well (in a spherical chamber radioactive uranium can become critical leading to nuclear meltdown). After a long time and after thermodynamic equilibrium is achieved, a sampling at the top reveals that the mixture consists of a 50-50 (by volume) mixture at 100 kPa and 300 K. Determine the (a) total pressure (p) and (b) molar composition 3 km below the surface. Assume isothermal conditions and ideal gas behavior.

14-1-5 Hydrogen is produced, at 100 kPa and 298 K at a rate of 1 kg/s, from a mixture of hydrogen and methane containing 20% hydrogen and 80% methane by volume. If the device works by raising the pressure of the mixture on one side of a semipermeable membrane, determine (a) the minimum power consumption ($\dot{W}$) and (b) the minimum mixture pressure necessary to produce hydrogen.

14-1-6 An ideal gas mixture contains 9 kmol of argon and 1 kmol of helium at a total pressure of 100 kPa and a temperature of 25°C. Determine the chemical potential ($\overline{\mu}_k$) of argon and helium in the mixture.

14-1-7 Determine the chemical potential (molar specific Gibbs function ($\overline{g}_k$)) of pure oxygen at (a) 100 kPa, 300 K; (b) 1000 kPa, 300 K; and (c) 100 kPa, 3000 K. Verify your answers using the n-IG system-state TESTcalc.

14-1-8 A membrane, permeable to oxygen, separates pure oxygen at 100 kPa, 300 K from an ideal gas mixture of oxygen, nitrogen, and hydrogen of equal volume fractions at 300K. What is the minimum pressure (p) of the mixture at which oxygen starts separating from the mixture?

14-1-9 A two-phase, liquid-vapor mixture of R-134a is in equilibrium at 22°C. Show that the specific Gibbs functions of the saturated liquid and saturated vapor are equal.

14-1-10 Determine the chemical potential (molar specific Gibbs function) of pure hydrogen at (a) 100 kPa, 300 K; (b) 1000 kPa, 300 K; and (c) 100 kPa, 3000 K. Verify your answers using the n-IG system-state TESTcalc.

14-1-11 Use the n-IG system-state TESTcalc to (a) determine the chemical potential ($\overline{\mu}_k$) of hydrogen in an equimolar mixture of hydrogen and carbon dioxide at 100 kPa and 300 K. Plot how the chemical potential ($\overline{\mu}_k$) of hydrogen varies with pressure (in a range of 50 kPa through 10000 kPa) as the temperature is held constant at (b) 300 K and (c) 1000 K.

14-1-12 Oxygen is produced at 100 kPa and 298 K at a rate of 1 kg/s, from air (21% oxygen and 79% nitrogen by volume). If the device works by raising the pressure of the mixture on one side of a semipermeable membrane, determine (a) the minimum power consumption ($\dot{W}$) and (b) the minimum mixture pressure (p) necessary to produce oxygen.

14-1-13 Hydrogen is produced at 100 kPa and 298 K, from a mixture of hydrogen and methane containing 10% hydrogen and 90% methane by volume. If the device has an exergetic efficiency of 20%, (a) determine the power consumption ($\dot{W}$) in kW used to produce hydrogen at a rate of 1 kg/s. (b) **What-if scenario:** What is the power consumption if the temperature is 500 K instead?

SECTION 14-2: PHASE EQUILIBRIUM

14-2-1 What is the maximum number of phases that can stay in equilibrium when the system has: (a) one component, (b) two components, (c) four components.

14-2-2 Consider a liquid-vapor mixture of ammonia and water in equilibrium at 30°C. If the molar composition of the liquid phase is 55% NH_3 and 45% H_2O, determine the composition of the vapor phase of this mixture.

14-2-3 Determine the vapor pressure adjacent to the surface of a lake at 15°C, (a) assuming air does not affect the equilibrium between the water vapor and the liquid water, (b) and taking into account the presence of air, which affects the chemical potential (Gibbs function).

14-2-4 For a mixture of saturated vapor and saturated liquid of water at 200°C, use tabulated properties or the PC system-state TESTcalc to show that the specific Gibbs functions of the two phases are equal.

14-2-5 For a mixture of saturated vapor and saturated liquid of R-134a at 100 kPa, use tabulated properties or the PC system-state TESTcalc to show that the chemical potentials of the two phases are equal.

14-2-6 Show that an equilibrium mixture of saturated vapor and saturated liquid of water at 100°C satisfies the criterion for phase equilibrium.

14-2-7 Consider a liquid-vapor mixture of ammonia and water in equilibrium at 15°C. If the molar composition of the liquid phase is 50% NH_3 and 50% H_2O, determine the composition of the vapor phase of this mixture.

14-2-8 A two-phase mixture of ammonia and water is in equilibrium at 40°C. If the molar composition of the vapor phase is 98% NH_3 and 2% H_2O, determine the composition of the liquid phase of the this mixture.

14-2-9 Repeat 14-1-9 for water at 100°C.

14-2-10 Consider a glass of water in a room at 20°C and 100 kPa. If the relative humidity in the room is 100%, and the water and air are in thermal and phase equilibrium, determine (a) the mole fraction of the water vapor in the air and (b) the mole fraction of air in the water.

14-2-11 Water is sprayed into air at 75°F and 14.3 psia and the falling water droplets are collected in a container on the floor. Determine (a) the mass and (b) mole fractions of air dissolved in the water.

14-2-12 A mixture of 1 kmol of H_2 and 1 kmol of Ar is heated in a reaction chamber at a constant pressure of 1 atm until 15% of H_2 dissociates into monatomic hydrogen (H). Determine the final temperature (T_2) of the mixture.

14-2-13 Carbon monoxide at 300 K, 1 atm reacts with theoretical amount of air at 300 K and 1 atm in a chamber. An equilibrium mixture of CO_2, CO, O_2 and N_2 exits the chamber at 1 atm. Determine the composition and temperature (T) of the exiting mixture.

14-2-14 Derive an expression for estimating the pressure (p) at which graphite and diamond exist in equilibrium at 300 K and 100 kPa, in terms of the specific Gibbs function.

14-2-15 In a closed chamber at 30°C, 100 kPa, liquid water is in equilibrium with water vapor and dry air. Assuming air does not dissolve in water, determine (a) the partial pressure of water vapor (a) using equality of Gibbs function (chemical potential) and (b) Rault's law.

14-2-16 Consider a two-phase, liquid-vapor NH_3-H_2O system in equilibrium at 30°C. The mole fraction of ammonia in the liquid phase is 80%. Determine the pressure (p) in kPa and the mole fraction of ammonia in the vapor phase. Use Rault's law.

14-2-17 Consider a two-phase, liquid-vapor NH_3-H_2O system in equilibrium at 40°C, 150 kPa. Determine the mole fractions of ammonia in the liquid and vapor phases. Use Rault's law.

14-2-18 In a closed chamber at 10°C and 100 kPa, liquid water is in equilibrium with water vapor and dry air. Assuming air does not dissolve in water, determine (a) the partial pressure of water vapor using equality of Gibbs function (chemical potential). Compare your result with saturation vapor pressure. (c) *What-if scenario:* What would be the answer in part (a) if the pressure in the chamber was 1 MPa?

14-2-19 Fresh water is to be extracted at a rate of 100 L/s from brackish water at 15°C with a salinity of 0.05% (by mass). Determine (a) the mole fraction of water in the brackish water, (b) the minimum power (in kW) required, and (c) the minimum pressure to which the brackish water must be pumped if the fresh water is to be obtained by reverse osmosis using semipermeable membrane.

14-2-20 Fresh water is extracted from brackish water at 15°C with a salinity of 0.1% (by mass). Determine (a) the minimum work required to separate 1 kg of brackish water completely into pure water and pure salts and (b) the minimum work required to obtain 1 kg of fresh water.

14-2-21 In Problem 14-2-19 change the salinity from 0.05% through 5% and plot how the minimum power requirement varies with the salinity of the brackish water.

14-2-22 A desalination plant produces fresh water from seawater at 10°C with a salinity of 3.1% (by mass) through reverse osmosis at a rate of 1.5 m^3/s while consuming 9 MW of power. The amount of fresh water is negligible compared to the seawater used. Determine the exergetic efficiency of the plant.

14-2-23 A river discharges fresh water at 20°C at a rate of 500,000 m^3/s into an ocean at the same temperature with a salinity of 3.5% (by mass). Determine the amount of power that can be generated if the river water mixes with the ocean water in a reversible manner.

14-2-24 A desalination plant produces fresh water through reverse osmosis at a rate of 1m^3/s, consuming 6 MW of power. The plant has an exergetic efficiency of 20%. Determine the power (W) that can be produced from reversible mixing of the produced fresh water with seawater.

SECTION 14-3: CHEMICAL EQUILIBRIUM

14-3-1 For the reaction $A + B \leftrightarrows C + D$, $\Delta \bar{g}°$ is calculated to be 0 at 500 K. Starting with a mixture of 1 kmol of A and 1 kmol of B, (a) evaluate what percent of the mixture is converted to products at equilibrium at 500 kPa, 500 K. (b) What would be the answer if the pressure (p) was 1000 kPa?

14-3-2 For an elementary step $A + B \leftrightarrows C + D$, $\Delta \bar{g}°$ is tabulated as 22.819 at 298 K, 0 at 1000 K, and −229.724 MJ/kmol at 3000 K. Starting with a mixture of 1 kmol of A and 1 kmol of B, evaluate what percent of the mixture is converted to products at equilibrium at (a) 298 K, (b) 1000 K, and (c) 3000 K.

14-3-3 Evaluate $\Delta \bar{g}°$ at 298 K for the reaction $H_2 + (1/2)O_2 \leftrightarrows H_2O$ using (a) formation enthalpy ($\bar{h}_f°$) and entropy ($\bar{s}°$) values; and (b) using formation Gibbs function from Table G.1.

14-3-4 Evaluate (a) $\Delta \bar{h}°$, (b) $\Delta \bar{g}°$, and (c) ln$K$ at 298 K for the reaction $CO + (1/2)O_2 \leftrightarrows CO_2$ at 1 atm.

14-3-5 A mixture of CO_2, CO, and O_2 is in equilibrium at a specified temperature and pressure. If the pressure is tripled: (a) Will the equilibrium constant K change? (b) Will the moles of CO_2, CO, and O_2 change? How?

14-3-6 Suppose the equilibrium constant of the dissociation reaction $H_2 = 2H$ at 2000 K and 1 atm is K_1. Express the equilibrium constants of the following reactions at 2000 K in terms of K_1: (a) $H_2 \leftrightarrows 2H$ at 4 atm, (b) $2H \leftrightarrows H_2$ at 1 atm, (c) $2H_2 \leftrightarrows 4H$ at 1 atm, (d) $H_2 + 2N_2 \leftrightarrows 2H + 2N_2$ at 2 atm, and (e) $6H \leftrightarrows 3H_2$ at 4 atm.

14-3-7 A mixture of NO, O_2, and N_2 is in equilibrium at a specified temperature and pressure. If the pressure is doubled: (a) Will the equilibrium constant K change? (b) Will the moles of NO, O_2 and, N_2, change? How?

14-3-8 Suppose the equilibrium constant of the reaction $CO + (1/2)O_2 \leftrightarrows CO_2$ at 1500 K and 1 atm is K_1. Express the equilibrium constant of the following reactions at 1500 K in terms of K_1. (a) $CO + (1/2)O_2 \leftrightarrows CO_2$ at 2 atm, (b) $CO_2 \leftrightarrows CO + (1/2)O_2$ at 1 atm, (c) $CO + O_2 \leftrightarrows CO_2 + (1/2)O_2$ at 1 atm, (d) $CO + 2O_2 + 6N_2 \leftrightarrows CO_2 + 1.5O_2 + 6N_2$ at 5 atm, and (e) $2CO + O_2 \leftrightarrows 2CO_2$.

14-3-9 For the dissociation of nitrogen tetraoxide, according to the reaction, $N_2O_4 \leftrightarrows 2NO_2$, express the degree of dissociation at equilibrium in terms of the initial and equilibrium volumes.

14-3-10 For the dissociation of nitrogen tetraoxide described in Problem 14-3-9, there is a 77.7% increase in volume when equilibrium is reached at 50°C, 125 kPa. Determine the value of the equilibrium constant (K).

14-3-11 A chamber contains a mixture of CO_2, CO and, O_2 is in equilibrium at a specified temperature and pressure. How will (a) increasing the temperature (T) at constant pressure and (b) increasing the pressure (p) at constant temperature affect the mole of CO?

14-3-12 Determine the change in the Gibbs function $\Delta \bar{g}°$ at 25°C, in kJ/kmol, for the reaction $CH_4(g) + 2O_2 \leftrightarrows CO_2 + 2H_2O(g)$ using enthalpy of formation and absolute entropy data.

14-3-13 Calculate the equilibrium constant (K) for $CO_2 \leftrightarrows CO + (1/2)O_2$ at (a) 500 K and (b) 1000 K.

14-3-14 One kmol of CO_2 is heated at a constant pressure of 100 kPa to 3000 K. (a) Calculate the equilibrium composition of CO_2. Use the IGE system-state TESTcalc to verify your answers. (b) Use the n-IG system-state TESTcalc to calculate the Gibbs function of the mixture $(1 − x)CO_2 + xCO + (x/2)O_2$ at 100 kPa, 3000 K with x varying from 0 to 1. Show that for $x = 0.438$, the Gibbs function reaches a minimum.

14-3-15 Calculate the equilibrium constant (K) for the water-gas reaction $CO + H_2O(g) \leftrightarrows CO_2 + H_2$ at (a) 298 K and (b) 1000 K.

14-3-16 Determine the equilibrium constant (K) for the reaction $H_2 + (1/2)O_2 \leftrightarrows H_2O$ at (a) 298 K and (b) 2500 K.

14-3-17 Determine the equilibrium constant (K) for the reaction $CH_4 + 2O_2 \leftrightarrows CO_2 + 2H_2O$ at 25°C.

14-3-18 Determine the equilibrium constant (K) for the dissociation process $CO_2 \leftrightarrows CO + (1/2)O_2$ at (a) 298 K and (b) 2000 K.

14-3-19 For the chemical reaction $CO_2 + H_2 \leftrightarrows CO + H_2O$, the equilibrium value of the degree of reaction (forward completion fraction) at 1200 K is 0.56. Determine the equilibrium constant and the change in the Gibbs function.

14-3-20 Determine the temperature at which 2% of diatomic oxygen (O_2) dissociates into monatomic oxygen (O) at a pressure of 3 atm.

14-3-21 Oxygen (O_2) is heated to 3000 K at a constant pressure of 5 atm. Determine the percentage of O_2 that will dissociate into O during this process.

14-3-22 $H_2O(g)$ dissociates into an equilibrium mixture at 3000 K and 100 kPa. Assume the equilibrium mixture consists of $H_2(g)$, H_2O, O_2, and OH, determine the equilibrium composition of (a) H_2O and (b) OH (in kmols).

14-3-23 Nitrogen (N_2) is heated to 3000 K at a constant pressure of 5 atm. (a) Determine the percentage of N_2 that will dissociate into N during this process. *What-if scenario:* What would the answer be if the conditions were (b) 5 atm, 5000 K, and (c) 0.1 atm, 5000 K?

14-3-24 Carbon dioxide (CO_2) is heated to 3200 K at a constant pressure of 2 atm. Determine the percentage of CO_2 that will dissociate into CO and O_2 during the process.

14-3-25 Consider the dissociation of 1 kmol of A_2 through the elementary step $A_2 \leftrightarrows 2A$. If x stands for the degree of dissociation (fraction of A_2 by mole that is dissociated), the overall reaction can be represented by $A_2 \leftrightarrows (1-x)A_2 + 2xA$. At 500 K and 100 kPa, the equilibrium constant is calculated as 1. (a) Determine x. (b) What is the value of x if the pressure were reduced to 0.01 kPa. (c) *What-if scenario:* Plot how the degree of dissociation changes as the pressure is increased from 0.01 MPa to 1 MPa.

14-3-26 One kmol of carbon dioxide (CO_2) is heated from 300 K to 4000 K at a constant pressure of 1 atm. Use the IGE system-state TESTcalc, set up the reactants as 1 kmol of CO_2, and select CO_2, CO, and O_2 as the possible products. Now calculate a series of equilibrium products' states (in the state panel, select the Products radio-button) for several temperatures ranging from 300 K through 3000 K. Plot how the fraction of CO_2 dissociated changes with temperature.

14-3-27 A mixture of 1 kmol of CO and 2 kmol of O_2 is heated to 2000 K at a pressure of 2 atm. Determine the equilibrium composition of (a) O_2 and (b) CO_2 (in kmols) assuming the mixture consists of CO_2, CO and O_2.

14-3-28 A mixture of 1 kmol of CO_2, 1/2 kmol of O_2, and 1/2 kmol of N_2 is heated to 2900 K at a pressure of 1 atm. Determine the equilibrium composition of (a) CO_2, (b) O_2 assuming the mixture consists of CO_2, CO, O_2, and N_2.

14-3-29 One kmol of CO reacts with 1 kmol of O_2 to form an equilibrium mixture of CO_2, CO, and O_2 at 2800 K. Determine the equilibrium composition of CO_2 at (a) 1 atm and (b) 5 atm.

14-3-30 One kmol of carbon monoxide (CO), reacts with the theoretical amount of air to form an equilibrium mixture of CO_2, CO, O_2, and N_2 at 2200 K and 1 atm. Determine the equilibrium composition of (a) CO and (b) CO_2.

14-3-31 A mixture of 2 kmol of N_2, 1 kmol of O_2, and 0.1 kmol of Ar is heated to 2500 K at a constant pressure of 10 atm. Assuming the equilibrium mixture consists of N_2, O_2, Ar, and NO, determine the equilibrium composition of (a) O_2 and (b) N_2 (in kmols).

14-3-32 A chamber initially contains a gaseous mixture of 4 kmol of CO_2, 8 kmol of CO, and 2 kmol of H_2. Assume the equilibrium mixture formed consists of CO_2, CO, H_2O, H_2 and O_2 at 2600 K and 100 kPa. Determine the equilibrium composition of (a) CO_2 and (b) CO (in kmols).

14-3-33 Butane (C_4H_{10}) burns inside a vessel with 50% excess air to form an equilibrium mixture at 1400 K and 1 MPa. The equilibrium mixture is composed of CO_2, O_2, N_2, $H_2O(g)$, NO_2, and NO. Determine the balanced reaction equation. Use $K_p = 8.4 \times 10^{-10}$ for this reaction.

14-3-34 Methane (CH_4) reacts with 125% of theoretical air inside a chamber to form an equilibrium mixture consists of CO_2, CO, $H_2O(g)$, H_2, and N_2 at 1200 K, 100 kPa. Determine the equilibrium composition of (a) CO_2 and (b) H_2O (in kmols).

14-3-35 Octane (C_8H_{18}) reacts with 110% of theoretical air inside a chamber to form an equilibrium mixture consists of CO_2, CO, $H_2O(g)$, H_2, and N_2 at 1650 K, 100 kPa. Determine the composition of the equilibrium mixture.

14-3-36 Starting with n_0 kmol of NH_3, which dissociates according to $NH_3 \leftrightarrows (1/2)N_2 + (3/2)H_2$, evaluate an expression for K, in terms of the degree of dissociation ε and pressure p.

14-3-37 Determine the mole fraction of sodium that ionizes according to the reaction $Na \leftrightarrows Na^+ + e^-$ at 2300 K and 0.5 atm (use $K = 0.688$ for this reaction).

14-3-38 Determine the percent ionization of cesium that ionizes according to the reaction $Ce \leftrightarrows Ce^+ + e^-$ at 2000 K and 1 atm (use $K = 15.63$ for this reaction).

14-3-39 Determine (a) the pressure if the ionization of Ar is 80% complete according to the reaction $Ar \leftrightarrows Ar^+ + e^-$ at 10,000 K (use $K = 4.2 \times 10^{-4}$ for this reaction). (b) *What-if scenario:* What is the pressure if the ionization of Ar is 50% complete?

14-3-40 One kmol of H_2O is heated to 3000 K at a pressure of 1 atm. Determine the equilibrium composition of H_2O (in kmols) assuming that only H_2O, OH, O_2, and H_2 are present.

14-3-41 A mixture of 2 kmol of CO_2 and 1 kmol of O_2 is heated to 3200 K at a pressure of 5 atm. Determine the equilibrium composition of (a) CO and (b) CO_2 (in kmols), assuming that only CO_2, CO, O_2, and O are present.

14-3-42 Air (21% O_2, 79% N_2) is heated to 2800 K at a pressure of 1 atm. Determine the equilibrium composition, assuming that only O_2, N_2, O, and NO are present. Can the presence of NO in the equilibrium mixture be neglected?

14-3-43 The equilibrium constant for the dissociation of N_2O_4 is 0.664 and 0.141 at 318 K and 298 K, respectively. Calculate the average heat of reaction within this temperature range.

14-3-44 At an average temperature of 2000 K, the slope of the graph of log K against $1/T$ for the dissociation of water vapor (into hydrogen and oxygen), is found to be -13000. (a) Determine the heat of dissociation. (b) Is this an exothermic reaction?

14-3-45 Potassium is ionized according to the equation $K \leftrightarrows K^+ + e^-$. The values of the equilibrium constants at 3000 K and 3500 K are measured to be 8.33×10^{-6} and 1.33×10^{-4}, respectively. Determine the average heat of reaction in MJ/kmol.

14-3-46 The equilibrium constant for the reaction $SO_3 \leftrightarrows SO_2 + O$ has the following values:

T	800 K	900 K	1000 K	1105 K
K	0.0319	0.153	0.540	1.59

Determine the average heat of dissociation using graphical method.

14-3-47 1 kmol of carbon, at 25°C, 0.1 MPa, reacts with 2.5 kmol of oxygen at 25°C, 0.1 MPa forming an equilibrium mixture of CO_2, CO, and O_2 at 3000 K, 0.1 MPa. Determine (a) the amount of CO_2 present in the products mixture and (b) heat transfer for the process.

14-3-48 For the reaction in Problem 14-3-47, plot how heat transfer varies with equivalence ratio varying from 0.1 to 1, all other parameters remaining unchanged.

SECTION 14-4: CHEMICAL EQUILIBRIUM AND ENERGY BALANCE

14-4-1 An equimolar mixture of carbon monoxide and water vapor, at 1 atm and 298 K, enters a reactor operating at steady state. The equilibrium mixture, composed of CO_2, CO, $H_2O(g)$, and H_2, leaves at 2000 K. Determine (a) equilibrium composition of CO_2 in the mixture and (b) the heat transfer (Q) between the reactor and surroundings in kJ per kmol of CO entering the reactor.

14-4-2 Carbon dioxide gas, at 1 atm, 298 K, enters a reactor operating at steady state. If an equilibrium mixture of CO_2, CO, and O_2 exits at 2800 K, 1 atm, determine (a) the composition of the CO_2 in the products and (b) the heat transfer to the surroundings, per unit mass of carbon dioxide.

14-4-3 Hydrogen is heated in an open-steady device at 100 kPa from 300 K to 3000 K at a rate of 0.5 kg/min. Determine the rate of heat transfer ($\dot{Q}$) in kW, assuming (a) no dissociation takes place; (b) dissociation takes place.

14-4-4 An equimolar mixture of CO_2, O_2, and N_2 enters a reactor operating at steady state. An equilibrium mixture of CO_2, O_2, N_2, NO, and CO exits at 3500 K and 600 kPa. Determine the (a) mole of CO_2 (in kmols) in the mixture and (b) heat transfer (q) to the surroundings per unit mass of CO_2. Assume the surroundings are 300 K and 100 kPa.

14-4-5 Hydrogen (H_2) is heated, during a steady-flow process at 1 atm, from 298 K to 3000 K at a rate of 0.6 kg/min. Determine the rate of heat transfer ($\dot{Q}$) needed during this process, assuming (a) some H_2 dissociates into H and (b) no dissociation takes place.

14-4-6 Carbon, at 300 K, 100 kPa, enters a chamber and reacts with oxygen entering at the same mole rate at 400 K, 100 kPa. An equilibrium mixture consisting of CO_2, CO, and O_2 exits at 3000 K, 100 kPa. Determine the heat transfer to the surroundings in kJ/kmol of carbon.

14-4-7 Steam enters a heat exchanger operating at steady state. An equilibrium mixture of H_2O, H_2, O_2, H, and OH exits at 2500 K and 100 kPa. Determine the (a) equilibrium composition of H_2O (in kmols) and (b) heat transfer (q) to the surroundings per unit mass of steam. Assume the surroundings are 300 K and 100 kPa.

14-4-8 Propane gas (C_3H_8), at 300 K and 100 kPa, enters a combustion chamber operating at steady state and reacts with 100% of excess air entering at 350 K and 100 kPa. An equilibrium mixture of CO_2, CO, $H_2O(g)$, H_2, and N_2 exits at 2000 K and 100 kPa. Determine (a) the heat transfer (Q) to the surroundings, in kJ/kmol of propane. (b) *What-if scenario:* What is the heat transfer if 200% excess air enters the chamber instead?

14-4-9 Methane gas (CH_4), at 300 K and 100 kPa, enters combustion chamber operating at steady state and reacts with 50% of excess air entering at 400 K and 100 kPa. An equilibrium mixture of CO_2, CO, $H_2O(g)$, H_2, and N_2 exits at 2000 K and 100 kPa. Determine (a) the heat transfer (Q) to the surroundings, in kJ/kmol of methane. (b) *What-if scenario:* What is the heat transfer if 100% excess air enters the chamber instead?

14-4-10 CO_2 gas, at 300 K and 400 kPa, enters a steady state heat exchanger. An equilibrium mixture of CO_2, O_2, and CO exits at 2700 K and 350 kPa. Determine the (a) composition of CO_2 in the exiting mixture and (b) heat transfer (Q) to the gas, in kJ/kmol of CO_2.

14-4-11 Carbon monoxide, at 20°C and 100 kPa, enters a combustion chamber and burns with 25% excess air entering at the same temperature and pressure. An equilibrium mixture of CO_2, CO, O_2, and N_2 exits at 1200°C and 100 kPa. Determine (a) the heat transfer (Q) to the surroundings in kJ/kmol of CO. (b) *What-if scenario:* What is the heat transfer to the surroundings if 85% excess air enters the chamber instead?

14-4-12 Methane (CH_4), at 25°C and 1 atm, enters a well-insulated reactor and reacts with air entering at the same conditions. For steady-state operation, negligible effects of ke and pe, and negligible pressure loss, plot the temperature of the combustion products against the equivalence ratio ranging from 0.5 through 3.0. (a) Assume complete combustion with the PG mixture model. (b) Assume complete combustion with the IG mixture model. (c) Assume equilibrium combustion with the IGE mixture model and having CO_2, H_2O, N_2, O_2, NO, OH, CO, O, H_2, and H in the products mixture. For part (c), also plot the mole fraction of CO and NO in the products against the equivalence ratio.

14-4-13 Repeat 14-4-12 if the reactants enter the chamber at 800°C and 1 atm.

14-4-14 Repeat 14-4-12 if the reactants enter the chamber at 600°C and 10 atm.

14-4-15 Octane (C_8H_{18}), at 25°C, 1 atm, enters a combustion chamber and reacts with 111% theoretical air entering at the same conditions. If the products exiting at 1800 K contains CO, CO_2, CN, H, H_2, H_2O, HCN, O, O_2, O_3, OH, N, N_2, N_2O, NO, and NO_2, determine (a) the molar fraction of NO in the products, in ppm, and (b) the heat transfer rate ($\dot{Q}$) for a fuel mass flow rate ($\dot{m}$) of 1 kg/s. (c) *What-if scenario:* What would the molar fraction of NO be if the chamber was adiabatic? (d) What is the equilibrium flame temperature?

14-4-16 Octane (C_8H_{18}), at 25°C and 100 kPa, enters a combustion chamber and reacts with 100% theoretical air entering at the same conditions. Determine the equilibrium flame temperature, assuming the products contain CO, CO_2, CN, H, H_2, H_2O, HCN, O, O_2, O_3, OH, N, N_2, N_2O, NO, and NO_2. *What-if scenario:* How would the temperature change if the chamber pressure was increased to 1 MPa?

14-4-17 In Problem 13-2-10, an equilibrium mixture consisting of CO_2, CO, H_2, H_2O and O_2 exits at 3000 K, 100 kPa. Determine (a) the molar composition of CO_2 at equilirium and (b) the amount of heat transfer per kg of fuel. (c) *What-if scenario:* What is the heat transfer if there is no CO present in the products mixture?

15 GAS DYNAMICS

In a way, this chapter is an extension of Chapter 4, which is devoted to the study of open-steady systems. The objective here is to re-examine the high-speed flow of gases, called **compressible flow**, through variable-area ducts such as nozzles and diffusers. An understanding of such flows is important for designing rockets, jet engines, and turbo machineries like compressors, turbines, etc. A non-dimensional extrinsic property called the Mach number is shown to control the nature of high-speed flows. Modeling the working fluid as a perfect gas, the mass, energy, entropy, and momentum equations for a one-dimensional flow are recast, introducing several new properties of an extended flow state—Mach number, total pressure, total temperature, and critical area. Isentropic relations between thermodynamic and total properties are tabulated as functions of the flow Mach number, and a solution procedure for isentropic flows is developed based on an isentropic table. The phenomenon of a normal shock wave is analyzed, leading to the development of the normal shock table, which relates the flow states before and after a normal shock. The effect of back pressure on a converging and converging-diverging nozzle is studied with the help of the isentropic and shock tables. Finally, nozzles and diffusers are revisited for a more refined analysis. A dedicated gas dynamics TESTcalc, which customizes the PG single-flow open-steady TESTcalc for high-speed flows, is used to verify manual solutions and pursue "what-if" scenarios.

15.1 ONE-DIMENSIONAL FLOW

Analysis of a flow through a confined passage, such as a nozzle or a diffuser, can be considerably simplified by assuming that the flow states are uniform across a flow; all properties, including flow velocity, remain invariant across the flow at any given cross-section. Because properties can only vary along the axial direction, this assumption is known as the **one-dimensional flow assumption**, and has been implicitly assumed (Sec. 0.5) at the inlet and exit ports of all open devices (Fig. 15.1). In this section, we will define several new properties of a flow state, which will prove useful in the mass, energy, entropy, and momentum analysis of one-dimensional, high-speed flows.

The energy and entropy equations for a single-flow, open-steady system, summarized in Sec. 4.1, are also applicable to a one-dimensional, steady flow through a confined passage, the open system enclosed by the red internal boundary of Figure 15.1. Denoting the inlet and exit flow states through generic symbols, i and e, respectively, and recognizing that there is no external work transfer, the energy and entropy equations can be rewritten as follows:

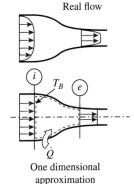

FIGURE 15.1 All flow states along the flow become uniform when a flow is assumed to be a one-dimensional flow.

Energy:

$$j_e = j_i + \frac{\dot{Q}}{\dot{m}}; \quad \left[\frac{kJ}{kg}\right] \tag{15.1}$$

Entropy:

$$s_e = s_i + \frac{\dot{Q}}{\dot{m}T_B} + \frac{\dot{S}_{\text{gen,int}}}{\dot{m}} \quad \left[\frac{kJ}{kg \cdot K}\right] \tag{15.2}$$

If there is no heat transfer, these equations simplify to $j_e = j_i$ and $s_e \geq s_i$ because $\dot{S}_{\text{gen,int}} \geq 0$. That is, the specific flow energy remains constant while specific entropy must increase or, at best, remain constant in the total absence of thermodynamic friction. The internal entropy generation is mostly due to viscous and thermal friction (Sec. 2.1.3). If the system is assumed to be internally reversible (Sec. 4.1.3) in other words the flow is thermodynamically frictionless, the entropy equation simplifies to $s_e = s_i$. Thus, an adiabatic and frictionless flow is known as an **isentropic flow** characterized by a constant entropy s

and a constant flow energy j. If the density of the working fluid can be regarded as a constant, as in a liquid modeled by the SL model, an isentropic flow can be shown to produce Bernoulli's equation (Sec. 4.2.1 and Anim. 4.A.*bernoulliEquation*). In a constant-area isentropic duct, there is no change in pressure or temperature (as dictated by the isentropic relation), and the flow state does not change along the flow (see Anim. 15.A.*constantAreaDuct*). In a variable area duct, however, variation of gas density has a tremendous impact and Bernouli's equation can no longer be used.

15.1.1 Static, Stagnation and Total Properties

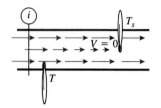

FIGURE 15.2 Thermometer A reads the stagnation temperature and thermometer B reads the static temperature of the flow at State-*i*. (see Anim. 15.A.*stagnation VsTotal*).

One of the ways to generate a high-speed flow of a gas is to store it at a high pressure inside a large chamber, called the **stagnation chamber**, and let it expand through a nozzle into a duct (see Anim. 15.A.*staticVsTotal*). Conversely, any high-speed flow can be imagined to have started from a stagnation chamber. In an isentropic flow, the flow state. When a one-dimensional adiabatic flow accelerates from zero velocity in the stagnation chamber to its current velocity or is brought to rest by an obstruction (Fig. 15.2), the state of the flow, including its *thermodynamic properties* (pressure, temperature, etc.), changes as it accelerates or decelerates. Between any two-flow states, the specific flow energy remains unchanged, according to Eq. (15.1). A flow state is called a **stagnation state**, and its properties **stagnation properties**, when the flow velocity approaches zero (as inside an originating chamber or at the surface of the thermometer in Fig. 15.2). Designating the stagnation state with the subscript s and neglecting any change in potential energy, the properties of a local flow state can be related to the corresponding stagnation properties by the energy equation:

$$j = j_s; \quad \Rightarrow \quad h + \text{ke} = h_s + \text{ke}_s^{\,0}; \quad \Rightarrow \quad h_s = h + \frac{V^2}{2(1000 \text{ J/kJ})}; \quad \left[\frac{\text{kJ}}{\text{kg}}\right] \quad (15.3)$$

where h_s is called the **stagnation enthalpy**. It is a *thermodynamic property* describing the equilibrium inside the stagnation chamber. However, if we take the literal definition of h_s, the sum of h, a thermodynamic property of the local flow state, and ke, an *extrinsic property*, it can be considered a new *extrinsic property* of the flow. Just like the flow energy j, h_s remains constant along an adiabatic, one-dimensional flow. Modeling the working fluid as a perfect gas (the PG model is discussed in Sec. 3.5.3), Eq. (15.3) can be rearranged to relate the temperature at the stagnation state with the local temperature:

$$h_s - h = c_p(T_s - T) = \text{ke}; \quad \left[\frac{\text{kJ}}{\text{kg}}\right] \quad \Rightarrow \quad T_s = T + \frac{V^2}{2(1000 \text{ J/kJ})c_p}; \quad [\text{K}] \quad (15.4)$$

Like h_s, the stagnation temperature, T_s, also have two identities: it can be regarded as a thermodynamic property of the stagnation chamber or an extrinsic property of the local flow state. The core thermodynamic state of a flow state is called the **static state** and the thermodynamic properties of a static state, p, T, v, u, h, s, etc., are called **static properties** in a high-speed flow (see Anim. 15.A.*staticVsTotal*). An observer has to move with the flow (inside the local system) to measure its static properties. While the static pressure can be measured by mounting a probe at the wall (see Anim. 15.A.*staticVsTotal*), the static temperature cannot be directly measured as any thermometer inserted into the flow will tend to measure the stagnation temperature. It will be soon established that the velocity of sound depends on the static temperature and their relation can be exploited to measure the static temperature of a high speed flow.

When an adiabatic flow is brought to stagnation in a frictionless (isentropic) manner, the resulting stagnation state is called the **isentropic stagnation state**, or **total state**, and its properties are called **total properties**, designated with the subscript t. The simplified energy equations, Eqs. (15.3) and (15.4), are still applicable regardless of presence or absence of friction as long as the flow is adiabatic. Therefore:

$$h_t = h_s; \quad \left[\frac{\text{kJ}}{\text{kg}}\right] \quad \text{and,} \quad T_t = T_s; \quad [\text{K}] \quad (15.5)$$

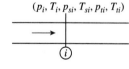

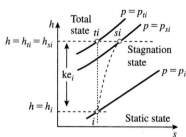

FIGURE 15.3 Stagnation and total (isentropic stagnation) state for the static (thermodynamic) state, State-*i* (see Anim. 15.A.*stagnationVsTotal*).

Figure 15.3 illustrates the relations among static, stagnation, and total properties with the help of a h-s diagram for a given flow state, State-*i*. In this diagram, States *i*, *si*, and *ti* represent the static state, stagnation state, and isentropic stagnation state. Applying Eq. (15.2) between the static and stagnation states and recognizing that $\dot{S}_{\text{gen}} > 0$, it can show that $s_s > s_t$. This is why State-*si* is to the right of State-*ti* in Figure 15.3. To clarify the distinction among the three states shown in Figure 15.3, we can write:

Stagnation State: $V_{si} = 0;$ $\left[\dfrac{m}{s}\right],$ $h_{si} = h_i + ke_i;$ $\left[\dfrac{kJ}{kg}\right],$ $s_{si} \geq s_i;$ $\left[\dfrac{kJ}{kg \cdot K}\right]$ **(15.6)**

Total State: $V_{ti} = 0;$ $\left[\dfrac{m}{s}\right],$ $h_{ti} = h_i + ke_i = h_{si};$ $\left[\dfrac{kJ}{kg}\right],$ $s_{ti} = s_i;$ $\left[\dfrac{kJ}{kg \cdot K}\right]$ **(15.7)**

While the static (thermodynamic) state of State-*i* is defined by the static properties p and T, State-*si* is defined by h_{si} and s_{si}, and State-*ti* by h_{ti} and s_{ti}. As mentioned before, the same property h_{ti} can be considered a thermodynamic property, when treated as a property of the total state, or as an extrinsic property, when treated as part of the flow-state *i*. Although an extended flow state bundles the static and total properties of a flow into a single state, it is convenient to visualize total properties as part of a separate total state in a thermodynamic plot such as the *h-s* diagram of Fig. 15.3.

To summarize the concepts introduced so far, the core thermodynamic properties (see Anim. 3.A.*extendedStates*) of a flow state, p, T, h, s, etc., are labeled as the *static properties* of the flow in gas dynamics. The stagnation and total properties, such as p_s, T_s, h_s, s_s, p_t, T_t, h_t, s_t, etc., of a flow are extrinsic properties that depend on the ke of the flow. However, when the stagnation and total states are treated as separate equilibriums, the stagnation and total properties assume the role of thermodynamic properties defining those states, which are system states with no ke. The total state can be considered an ideal stagnation state, with the stagnation brought about in a frictionless manner so that entropy remains constant.

We have already established that $T_s = T_t$ in Eq. (15.5), therefore, for the flow state State-*i* of Figure 15.3, T is $T_{si} = T_{ti}$. Also, according to Eq. (15.3), the vertical separation between the static and stagnation, or total enthalpy, is equal to the specific kinetic energy ke_i of the flow; or, $ke_i = (h_{ti} - h_i) = (h_{si} - h_i)$. To satisfy the two conditions $h_{si} = h_{ti}$ and $s_{si} \geq s_{ti}$, the State-*si* must be to the right of State-*ti* in Figure 15.3. Recalling the relative positions of constant-pressure lines for an ideal gas (Fig. 3.56), it can be concluded that the stagnation pressure p_{si} must be lower than the total pressure p_{ti}. If the frictional losses decrease, State-*si* will move closer to State-*ti* and the difference between p_{ti} and p_{si} will diminish. The stagnation temperature measured by the thermometer in Figure 15.2 is same as the total temperature, but the stagnation pressure measured by the pitot tube in Figure 15.4 is expected to be lower than the total pressure due to frictional losses.

The relations between static and total properties can be formalized by employing the PG model property relations. Equations (15.3), (15.4), and the PG isentropic relations, Eq. (3.71), can be used to relate p_t, ρ_t, and T_t to the corresponding static properties, p, ρ, and T:

$$h_t = h + \frac{V^2}{2000}; \quad \left[\frac{kJ}{kg}\right] \quad \text{and,} \quad T_t = T + \frac{V^2}{2000c_p}; \quad [K] \quad \textbf{(15.8)}$$

$$\frac{p_t}{p} = \left(\frac{\rho_t}{\rho}\right)^k = \left(\frac{T_t}{T}\right)^{k/(k-1)} \quad \textbf{(15.9)}$$

In applying the isentropic relations, the static and total properties are treated as properties of two separate equilibrium states that are isentropic to each other.

Isentropic flows are reversible and bringing a flow to rest isentropically is equivalent to its reverse, creating an isentropic flow from rest, as shown in Figure 15.5. If an isentropic flow originates from a reservoir (Fig. 15.5), the properties inside the reservoir can be regarded as the total properties of any resulting isentropic flow; $p_t = p_r$ and $T_t = T_r$ for any flow state. Measuring total properties is, therefore, equivalent to measuring the properties inside the stagnation chamber, where the flow originates.

15.1.2 The Gas Dynamics TESTcalc

TEST offers a dedicated gas dynamics TESTcalc, located in the system, open-steady, specific, gas dynamics branch. Like any other open-steady TESTcalc, the gas dynamics TESTcalc also has a flow-state panel, a device panel, and an I/O panel. Additionally, it has a table panel where different gas dynamics tables can be readily accessed for a large number of gases (not just air). Beside the standard flow-state properties, the state panel also includes the newly defined total properties and two other properties—the Mach number and the critical area—as part of

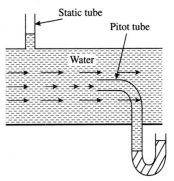

FIGURE 15.4 The pitot tube reads the stagnation pressure of the flow, which may be smaller than the total pressure (see Anim. 15.A.*staticVsTotal*).

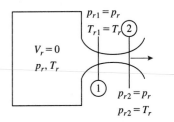

FIGURE 15.5 The reservoir properties are the same as the total properties of any flow state along the isentropic flow.

the extended flow state. The table panel, consisting of the isentropic table, normal shock table, oblique shock table, and Prandtl-Meyer table, are dynamically updated as a new gas is selected in the state panel.

The operation of this TESTcalc is similar to any other open-steady TESTcalc (sec 4.1.1): States are calculated fully or partially from the given properties; calculated states are loaded in the device panel; device variables are entered; and clicking the Calculate button evaluates the missing properties or device variables. When two different solutions—one subsonic and one supersonic—are possible for the same input, only one solution is displayed and the Mach number of the alternative solution is posted on the message panel. Then the alternative solution can be obtained by using this alternative Mach number in place of some other independent property (perhaps the flow area). The flow structure Interactive, located in the Interactives>Open-steady>Specific branch, can be used for advanced study of flow structure outside a converging-diverging nozzle.

EXAMPLE 15-1 Stagnation and Total Properties

Steam flows through a passage between two turbine blades at 500 kPa, 500 K, and 250 m/s. Determine the total pressure and temperature using (a) the PC model and (b) the PG model with $k = 1.38$. **What-if scenario:** (c) What would the answers be if the velocity was 25 m/s?

SOLUTION

Determine the total state for steam using the PC and PG models.

Assumptions

One-dimensional flow.

Analysis

Let State-1 represent the flow state in the passage and State-2 the total (isentropic stagnation) state.

(a) For the PC model, we can use the PC flow-state TESTcalc or the manual approach (Sec. 3.4) to evaluate properties.

State-1 (given p_1, T_1, V_1): $h_1 = 2912 \dfrac{\text{kJ}}{\text{kg}}$; $s_1 = 7.1728 \dfrac{\text{kJ}}{\text{kg} \cdot \text{K}}$;

State-2 (given $j_2 = j_1$, $s_2 = s_1$, $V_2 = 0$):

$$h_2 = h_1 + \text{ke}_1 = h_1 + \frac{V_1^2}{2(1000 \text{ J/kJ})} = 2912 + \frac{250^2}{2000} = 2943 \frac{\text{kJ}}{\text{kg}}$$

From the known enthalpy and entropy, the state can be completely determined using the manual approach (Sec. 3.4) or using a PC state TESTcalc, producing:

$$p_{t1} = p_2 = 575 \text{ kPa}; \quad \text{and} \quad T_{t1} = T_2 = 516 \text{ K}$$

(b) For the PG model, we can obtain c_p from the given value of the specific heat ratio by using Eq. (3.46):

$$c_p = \frac{kR}{k-1} = \frac{k}{k-1}\frac{\bar{R}}{M} = \left(\frac{1.38}{1.38-1}\right)\left(\frac{8.314}{18}\right) = 1.677 \frac{\text{kJ}}{\text{kg} \cdot \text{K}}$$

Now the total temperature can be evaluated from Eq. (15.8):

$$T_{t1} = T_1 + \frac{V_1^2}{2(1000 \text{ J/kJ})c_p} = 500 + \frac{250^2}{2000(1.677)} = 519 \text{ K}$$

Also, the total pressure can be obtained from the isentropic relation, Eq. (15.9):

$$p_{t1} = p_1\left(\frac{T_{t1}}{T_1}\right)^{k/(k-1)} = 500\left(\frac{519}{500}\right)^{1.38/(1.38-1)} = 571 \text{ kPa}$$

TEST Analysis

Launch the PC flow-state TESTcalc and select H2O as the working fluid. Evaluate State-1 for the given values of p1, T1, and Vel1. Evaluate State-2 with j2 = j1 (energy equation), s2 = s1 (entropy equation), and Vel2 = 0, to obtain p2 and T2 as the desired total properties verifying the manual results. For the PG model, launch the gas dynamics TESTcalc on a separate browser tab, select Custom Gas from the menu, and enter k1 and MM1 (molar mass). Evaluate State-1 with the given values of p1, T1, and Vel1. Total properties p_t1 and T_t1 are evaluated as extrinsic properties of State-1, verifying the manual results.

What-if scenario

In the TEST solution, change Vel1 to the new value and click Super-Calculate. The new answers are as follows: PC Model, p2 = 500.7 kPa and T2 = 500.2 K; PG Model, p_t1 = 500.7 kPa and T_t1 = 500.2 K.

Discussion

The PG model seems to be reasonable accurate for superheated steam, under the given conditions. As previously stated, the total properties can be treated either as thermodynamic properties of a real or imaginary isentropic state or as extrinsic properties of the local (static) state. The latter interpretation is more convenient for gas dynamics analyses, as an extended state can encapsulate all relevant properties. Comparing the state panels of the gas dynamics TESTcalc and PG flow-state TESTcalc can be quite useful at this point. The static and total states in the *h-s* or *T-s* diagram of Figure 15.6 must be represented by two separate states in the PG state TESTcalc, while in the gas dynamics TESTcalc a single state can bundle all the properties of the two states in a single extended flow state.

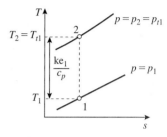

FIGURE 15.6 Thermodynamic plots for Ex. 15-1. The *T-s* diagram is valid only for the PG model (see Anim. 15.A.*staticVsTotal*).

15.2 ISENTROPIC FLOW OF A PERFECT GAS

A steady flow through any open device is isentropic when it is adiabatic and internally reversible. Suppose State-*i* and State-*e* represent two flow states in an isentropic flow through a confined passage. The energy equation reduces to $j_i = j_e$ and the entropy equation produces the isentropic condition $s_i = s_e$. That is, the specific flow energy and entropy remain constant along the flow. As simple as these equations are, it is more convenient to cast them in terms of total properties in gas dynamics.

With the potential energy pe neglected for gas or vapor flows, the energy equation, Eq. (15.1), for an adiabatic flow can be cast in terms of stagnation enthalpy, which is same as the total enthalpy [see Eq. (15.5)]:

$$j_i = j_e; \quad \Rightarrow \quad h_{ti} = h_{te}; \quad \left[\frac{\text{kJ}}{\text{kg}}\right] \qquad \textbf{(15.10)}$$

Like flow energy, the total enthalpy h_t remains constant along the flow. If the gas can be modeled as a perfect gas, the total temperature T_t can be shown to remain constant:

$$h_{ti} - h_{te} = 0 \quad \Rightarrow \quad c_p(T_{ti} - T_{te}) = 0; \quad \left[\frac{\text{kJ}}{\text{kg}}\right]; \quad \Rightarrow \quad T_{ti} = T_{te}; \quad [\text{K}] \qquad \textbf{(15.11)}$$

The flow does not have to be internally reversible for Eqs. (15.10) or (15.11) to apply; therefore, even in the presence of friction (non-isentropic flow), the total temperature remains constant in an adiabatic, one-dimensional flow.

The isentropic relation for a perfect gas can be applied between the inlet and exit states of an isentropic flow:

$$\frac{p_{te}}{p_{ti}} = \frac{p_{te}}{p_e} \frac{p_i}{p_{ti}} \frac{p_e}{p_i} = \left(\frac{T_{te}}{T_e}\right)^{k/(k-1)} \left(\frac{T_i}{T_{ti}}\right)^{k/(k-1)} \left(\frac{T_e}{T_i}\right)^{k/(k-1)} = 1 \qquad \textbf{(15.12)}$$

The entropy balance equation for an isentropic flow can be recast in terms of total pressure as:

$$p_{ti} = p_{te}; \quad [\text{kPa}] \qquad \textbf{(15.13)}$$

DID YOU KNOW?

• NASA experimental aircraft X-43a set the world record for speed by an air-breathing aircraft (not a rocket) in 2004. The needle nosed scramjet engine reached a speed of Mach 7, or about 5000 miles per hour.

That is, the total pressure remains constant in an isentropic flow of a perfect gas. If an isentropic flow originates from a reservoir, then the energy and entropy equations imply that the total temperature and total pressure along the flow are not only invariant but are also equal to the temperature and pressure in the reservoir (see Anim. 15.B.*isentropicFlow*).

EXAMPLE 15-2   **Isentropic Discharge from a Tank**

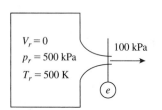

$V_r = 0$
$p_r = 500$ kPa
$T_r = 500$ K

100 kPa

e

FIGURE 15.7 For a large reservoir, the flow through the nozzle can be considered steady (Ex. 15-2).

Air is discharged from a large reservoir (Fig. 15.7) into the atmosphere through a nozzle. The reservoir conditions are $p_r = 500$ kPa and $T_r = 500$ K, and the atmospheric pressure is 100 kPa. Assuming the flow to be isentropic and the exit pressure to be the same as the outside pressure, determine (a) the exit velocity of the air. ***What-if scenario:*** (b) What would the answer be if the reservoir temperature was 1000 K?

SOLUTION

Use isentropic relation to obtain the exit temperature and Eq. (15.8) to obtain the exit velocity.

Assumptions

One-dimensional flow. Velocity in the reservoir is negligible and the flow is isentropic, so that $p_r = p_t$ and $T_r = T_t$. The pressure at the exit is equal to the outside pressure.

Analysis

From Table C-1, or the gas dynamics TESTcalc, we obtain $k = 1.4$ and $c_p = 1.005$ kJ/kg · K for air. In a one-dimensional isentropic flow of a perfect gas, the total temperature and total pressure remain constant along the flow. Therefore:

$$T_{te} = T_r = 500 \text{ K}; \quad \text{and} \quad p_{te} = p_r = 500 \text{ kPa}$$

The relation between the total and static properties, Eq. (15.9), yields:

$$\frac{T_{te}}{T_e} = \left(\frac{p_{te}}{p_e}\right)^{(k-1)/k} = \left(\frac{500}{100}\right)^{(1.4-1)/1.4} = 1.584;$$

$$\Rightarrow \quad T_e = T_{te}/1.584 = 500/1.584 = 315.7 \text{ K}$$

From Eq. (15.8):

$$\frac{V_e^2}{2(1000 \text{ J/kJ})c_p} = T_{te} - T_e;$$

$$\Rightarrow \quad V_e = \sqrt{2(1000)(1.005)(500 - 315.7)} = 608.7 \text{ m/s}$$

TEST Analysis

Launch the gas dynamics TESTcalc and select air as the working fluid. Evaluate State-1 as the reservoir state for the given p1, T1, and Vel1 = 0. For the exit, State-2, enter p2, T_t2 = T1 and p_t2 = p1 to obtain Vel1 = 608.5 m/s.

What-if scenario

Change T1 to 1000 K and click Super-Calculate to update the solution. The new exit velocity is calculated as 860.6 m/s. By increasing p1 and updating the solution, the exit velocity can be shown to increase with the chamber pressure as well.

Discussion

Soon we will develop a simpler approach for analyzing isentropic flows and revisit this problem.

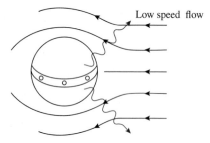

Low speed flow

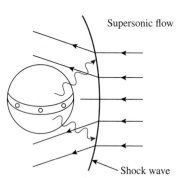

Supersonic flow

Shock wave

Information from the baseball cannot travel upstream of the shock wave.

FIGURE 15.8 Streamlines showing how air moves around a sphere. For $M > 1$, the flow has to adjust suddenly. The resulting discontinuity is known as a shock wave (see Anim. 15.B.*machNumber*).

15.3 MACH NUMBER

When a baseball travels through air, the air in front adjusts to the ball by quickly flowing around the ball to make room for it. This can be demonstrated in a simple wind tunnel experiment in which a ball is held stationary in a flow of air. The flow around the ball is visualized using smoke trails, as sketched in Figure 15.8. Air, even before reaching the ball, seems to react to the shape of the obstruction ahead as the velocity field starts changing accordingly. This is possible because the presence and shape of an obstruction is telegraphed upstream by small pressure disturbances emitted from the surface of the object. These weak pressure disturbances are the same as **sound waves**, which propagate under the same principle.

Naturally, the **speed of sound**, c, plays an important role in a high-speed flow and must be regarded as a legitimate property of the flow. If the flow velocity is small compared to c, the working fluid has ample time to adjust. However, at speeds faster than c, the flow is not aware of any downstream objects. The non-dimensional number obtained by dividing the flow velocity V with the local speed of sound c is known as the **Mach number**, yet another property of the flow:

$$M \equiv \frac{V}{c} \qquad \textbf{(15.14)}$$

A flow is called **supersonic** when $M > 1$, **subsonic** when $M < 1$, and **sonic** when $M = 1$. The term **hypersonic** (highly supersonic) is used for $M > 5$, which is a subset of the supersonic regime. The term **transonic** is used to indicate the transition regime: $0.8 < M < 1.2$.

Going back to the flow around the baseball, when air approaches the baseball at supersonic speed, the flow undergoes discrete changes through a discontinuity (Fig. 15.8) called a **shock wave**, which abruptly turns the flow into a subsonic flow, and thereafter the flow can negotiate an obstruction as before, guided by the sound waves. Propagation of sound waves, therefore, plays an important role in high-speed flows.

To analyze a sound wave, consider an infinitesimal disturbance, a compression wave propagating at a constant velocity c in a stationary gas (Fig. 15.9), which is created by a differential velocity dV of a piston. Although sound is a series of weak, alternating compression and rarefaction waves, analysis of a compression wave is adequate for our purpose. As the wave moves ahead of the piston, the fluid on its left assumes the steady velocity dV of the piston (to satify mass conservation). Other properties also undergo differential changes on the disturbed (left) side. However, the undisturbed fluid upstream of the wave retains its original state, unaware of the approaching compression wave.

To set up the governing equations for the wave, we have two choices for selecting the system boundary. If the system is attached to the cylinder of Figure 15.9, the problem becomes unsteady (the system image changes with time as the wave propagates) and difficult to analyze. On the other hand, by attaching the system to the wave itself, as shown in Figure 15.10, the system becomes steady; with the undisturbed fluid entering the system with a velocity c from the right and leaving with a velocity $c - dV$ on the left. Note that the high velocity at the inlet and exit only affects the extrinsic properties; thermodynamic properties are not affected by how the system is chosen.

The mass equation for this open-steady system of Figure 15.10 can be written as:

$$\dot{m}_i = \dot{m}_e; \quad \Rightarrow \quad \rho A c = (\rho + d\rho)A(c - dV); \quad \left[\frac{\text{kg}}{\text{s}}\right]$$

Simplifying, and neglecting the second-order term $d\rho dV$, we get:

$$c(d\rho) = \rho(dV); \quad \left[\frac{\text{kg}}{\text{m}^2 \cdot \text{s}}\right] \qquad \textbf{(15.15)}$$

The lateral boundaries of the system are parallel to the flow, so the internal pressure distribution cannot create any thrust (Sec. 8.7.2). Pressure at the inlet and exit planes create the only force on the internal system, so the momentum equation, Eq. (8.25), in the direction of the flow, reduces to:

$$0 = \frac{1}{(1000 \text{ N/kN})}[\dot{m}c - \dot{m}(c - dV)] + pA - (p + dp)A; \quad [\text{kN}]$$

$$\Rightarrow \quad \frac{\dot{m}dV}{(1000 \text{ N/kN})} = A dp; \quad [\text{N}]$$

$$\Rightarrow \quad \frac{\rho c dV}{(1000 \text{ N/kN})} = dp; \quad \left[\frac{\text{kN}}{\text{m}^2} = \text{kPa}\right] \qquad \textbf{(15.16)}$$

Substituting dV from Eq. (15.16) into Eq. (15.15), we obtain:

$$c^2 = \frac{(1000 \text{ N/kN})dp}{d\rho}; \quad \left[\frac{\text{m}^2}{\text{s}^2}\right] \qquad \textbf{(15.17)}$$

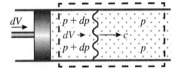

FIGURE 15.9 The system around the sound wave fixed to the laboratory coordinates is an unsteady system (see Anim. 15.B.*speedOfSound*).

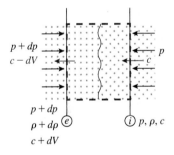

FIGURE 15.10 The system around the sound wave, when fixed to the wave, renders the system steady.

In this equation, the ratio of two infinitisimal quantities $dp/d\rho$ has to be a partial derivative, as p, a thermodynamic property, must be a function of two independent properties (Sec. 3.1.2) for a pure substance (the working gas). A sound wave propagates so rapidly that there is not enough time for any significant heat transfer. Also, because the wave is a very weak disturbance with only differential changes of properties, the irreversibilities are negligible. It is reasonable to assume a sound wave is isentropic, as the flow (with respect to the wave) satisfies both the adabatic and frictionless (reversible) conditions. With entropy remaining constant, Eq. (15.17) can be rewritten, producing the following expression for the velocity of sound:

$$c = \sqrt{(1000 \text{ N/kN})\left(\frac{\partial p}{\partial \rho}\right)_s}; \quad \left[\frac{\text{m}}{\text{s}}\right] \tag{15.18}$$

In an incompressible medium (think SL model), the density does not change with pressure; therefore, the partial derivative of pressure, with respect to density in Eq. (15.18) causes c to become infinity. This explains why the velocity of sound is much higher through water (water is not truly incompressible) than through air.

For a perfect gas the isentropic relation, Eq. (3.71), yields $p = A\rho^k$, where A is a constant. Differentiating this expression with respect to ρ and substituting $p = \rho RT$, Eq. (15.18) simplifies to:

$$c = \sqrt{(1000 \text{ N/kN})kRT}; \quad \left[\frac{\text{m}}{\text{s}}\right] \tag{15.19}$$

As a function of thermodynamic property T and gas properties k and R, c must also be considered a *thermodynamic* property. The Mach number, on the other hand, depends on the velocity and is an *extrinsic* property (Sec. 1.6).

EXAMPLE 15-3 | Velocity of Sound

Steam flows through a duct with a velocity of 300 m/s at 100 kPa and 200°C. Determine the velocity of sound using (a) the PG model and (b) PC model for steam.

SOLUTION
Use Eq. (15.19) for the PG model and Eq. (15.18) for the PC model to evaluate the flow Mach number.

Assumptions
One-dimensional flow.

Analysis
From Table C-1, we obtain the material properties of H_2O as $k = 1.327$ and $R = 0.462$ kJ/kg·K. Equation (15.19) produces:

$$c = \sqrt{(1000 \text{ N/kN})(1.327)(0.462)(473)} = 538.5 \text{ m/s}$$

For the PC model, the partial derivative of Eq. (15.18) must be evaluated first. Representing the given conditions by State-1 and its differential neighbor—an isentropic state with, say, a one percent variation in pressure—by State-2 (Fig. 15.11), use the PC flow state TESTcalc, or the manual procedure described in Chapter 3, to obtain:

State-1 (given p_1, T_1): $\rho_1 = 0.46036 \dfrac{\text{kg}}{\text{m}^3}$; $s_1 = 7.8342 \dfrac{\text{kJ}}{\text{kg} \cdot \text{K}}$

State-2 (given $s_2 = s_1, p_2 = 1.01p_1$): $\rho_2 = 0.46381 \dfrac{\text{kg}}{\text{m}^3}$

Therefore,

$$\left(\frac{\partial p}{\partial \rho}\right)_s \cong \left(\frac{\Delta p}{\Delta \rho}\right)_s = \frac{101 - 100}{0.46381 - 0.46036} = 290 \frac{\text{kPa} \cdot \text{m}^3}{\text{kg}}$$

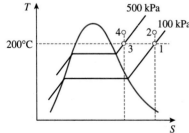

FIGURE 15.11 Perfect gas assumption works for superheated steam at State-1, but not as well at State-3.

and

$$c = \sqrt{\left(1000 \, \frac{N}{kN}\right)\left(\frac{\partial p}{\partial \rho}\right)_s} = \sqrt{(1000)(290) \, \frac{Pa}{kPa} \, \frac{kPa \cdot m^3}{kg}} = 538.5 \text{ m/s}$$

TEST Analysis

Launch the gas dynamics TESTcalc. Select H2O as the working fluid and evaluate State-1 with T1 = 200 deg-C. The velocity of sound c1 is calculated as 538.6 m/s. To use the PC model, launch the PC flow state TESTcalc. After evaluating State-1 with the given pressure and temperature, evaluate State-2 with p2 = 1.01*p1, and s2 = s1. In the I/O panel, evaluate the expression '=sqrt(1000*(p2 − p1)/(rho2 − rho1))' as 538.4 m/s.

Discussion

Instead of varying pressure by 1%, any other property could be used to select a neighboring isentropic state. Under the given conditions, superheated steam behaves like a perfect gas, thereby producing almost identical results for the PG and PC models. As a state gets closer to saturation, the PG model results will not be as accurate. This can be verified by calculating the speed of sound at State-3 at 500 kPa (Fig. 15.11). Eq. (15.19) should be used only when the PG model assumptions (Sec. 3.5.2) are appropriate.

15.4 SHAPE OF AN ISENTROPIC DUCT

In an isentropic flow through a confined passage, the flow must be adiabatic and reversible, that is, frictionless. While the adiabatic condition is achieved through good insulation, how can we ensure that the flow is frictionless?

Anticipating that an optimal nozzle shape may minimize viscous friction, let us explore how the flow area, a property of the flow state, varies along the isentropic flow through the nozzle of Example 15-2. Flow states State-1 through State-5 mark several intermediate stations of known static pressure from the inlet to the exit of the nozzle sketched in Figure 15.12. Using the solution procedure of Example 15-2, temperature and velocity at each station can be calculated from the ratio of the static pressure at a station to the constant total pressure of the flow. Additionally, the density of the gas can be obtained from the ideal gas equation of state $\rho = p/(RT)$. The steady-state mass equation can be used to relate the local area to the exit area through $\rho A V = \rho_e A_e V_e$. Finally, the Mach number can be obtained from Eq. (15.14) after the local temperature and velocity are found.

The results, tabulated below, can be verified using the gas dynamics TESTcalc. To do so, set State-1 as the state in the reservoir with Vel1 = 0, p1 = 500 kPa, and T1 = 500 K. For each subsequent state, State-i (where, i = 2 through 5), enter pi, p_ti = p1 and T_ti = T1, then obtain Ai/A5 by evaluating the expression '=rho5*Vel5/(rho$i$*Vel$i$)' in the I/O panel or, alternatively, setting A5 to be 1 m^2 and using mdot$i$ = mdot5 for each state.

Table 15-1 displays several interesting features of the isentropic flow of a perfect gas through a variable-area duct:

1. The pressure, temperature, and gas density monotonically decrease along the flow from the inlet to the exit. A monotonic decrease in density can be interpreted as an expansion of the gas along the entire length of the nozzle.
2. The velocity increases monotonically in an isentropic nozzle, while the local velocity of sound (not shown in Table 15-1) must decrease due to a decrease in temperature. Both these changes contribute to an increase of the Mach number along the flow.
3. The area ratio exhibits the most intriguing feature—it decreases, reaches a minimum at $M = 1$ at State-3, and then increases again. Obviously, we have no direct control over how the properties change along the flow, except for the flow area. However, to maintain an isentropic flow, the flow area must follow a strict **converging-diverging** pattern, as computed in the table for a specific inlet conditions and exit pressure.

It may appear counterintuitive that the flow velocity increases in the diverging section, but considering the drastic drop of density, an increase in area must be combined with an increase

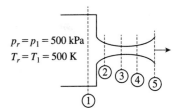

$p_r = p_1 = 500$ kPa
$T_r = T_1 = 500$ K

FIGURE 15.12 States at these five stations are displayed in Table 15-1. You can also use the flow structure Interactive to calculate the flow states (by changing the location of the probe).

Table 15-1 Change in properties of air along the isentropic flow in the nozzle of Ex. 15-2 (see Fig. 15.12).

State	p	T	V	ρ	M	A/A_e
	kPa	K	m/s	kg/m^3		
1	500	500	0	3.48	0	∞
2	400	469	249	2.97	0.57	0.91
3	264	417	409	2.21	1.00	0.74
4	200	385	481	1.81	1.22	0.77
5	100	316	609	1.10	1.71	1.00

in velocity to maintain the steady flow (ρAV is constant). Station-3, where the area reaches a minimum in a variable-area duct, is called the **throat** of the duct, which plays a critical role in high-speed flows.

To generalize the conclusions reached in this numerical case study, let us apply the governing balance equations between two neighboring flow states along the flow. The energy equation, $h_t = $ constant, can be expressed in a differential form:

$$d(h_t) = 0; \quad \Rightarrow \quad dh + d(\text{ke}) = 0; \quad \left[\frac{\text{kJ}}{\text{kg}}\right]$$

$$\Rightarrow \quad [Td\cancel{s}^0 + vdp] + d\left(\frac{V^2}{2(1000 \text{ J/kJ})}\right) = 0; \quad \Rightarrow \quad \frac{dp}{\rho} + \frac{VdV}{(1000 \text{ J/kJ})} = 0; \quad \left[\frac{\text{kJ}}{\text{kg}}\right]$$

$$(15.20)$$

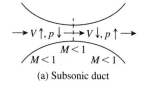

→$V\uparrow, p\downarrow$—→$V\downarrow, p\uparrow$→
$M<1$
$M<1 \qquad M<1$
(a) Subsonic duct

where the second Tds relationship (Eq. 11.8) for isentropic flow ($s = $ constant) is used to simplify dh. With ρ and V always positive, Eq. (15.20) implies that an increase in velocity (positive dV) along the duct must accompany a drop in pressure (negative dp) and vice versa, regardless of the Mach number (subsonic or supersonic duct) or the location along the nozzle. For a nozzle whose purpose is to accelerate a flow, the pressure must decrease monotonically and the reverse must be true for a diffuser (Fig. 15.13). This provides the rationale behind our choice of monotonically decreasing pressure (Table 15-1) along the flow direction in the nozzle of Figure 15.12.

→$V\uparrow, p\downarrow$—→$V\uparrow, p\downarrow$→
$M=1$
$M<1 \qquad M>1$
(b) Supersonic nozzle

The mass equation, $\rho AV = $ constant, can similarly be expressed by first taking the logarithm, then evaluating the differential on both sides of the equation:

$$\frac{d\rho}{\rho} + \frac{dA}{A} + \frac{dV}{V} = 0 \qquad (15.21)$$

Eliminating dV from Eqs. (15.21) and (15.20), and realizing that entropy remains constant during these changes along the flow, we obtain:

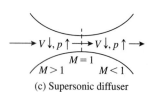

→$V\downarrow, p\uparrow$—→$V\downarrow, p\uparrow$→
$M=1$
$M>1 \qquad M<1$
(c) Supersonic diffuser

FIGURE 15.13 Consequences of Eq. (15.20) for a converging-diverging isentropic duct.

$$\frac{dA}{A} = (1000 \text{ J/kJ})\frac{dp}{\rho V^2}\left[1 - V^2\left(\frac{\partial \rho}{\partial p}\right)_s\right]; \quad \Rightarrow \quad \frac{dA}{A} = (1000 \text{ J/kJ})\frac{dp}{\rho V^2}[1 - M^2]$$

$$(15.22)$$

Substitution of dp from Eq. (15.20) leads to:

$$\frac{dA}{A} = -\frac{dV}{V}[1 - M^2] \qquad (15.23)$$

This relationship between dA and dV, coupled with the already established relation between dV and dp, provides significant insight into the shape of a variable-area isentropic duct.

According to Eq. (15.23), a positive dV implies a negative dA, that is, a converging shape as long as $M < 1$, a diverging shape (positive dA) for $M > 1$, and $dA = 0$ (throat) for $M = 1$. In a subsonic flow through a converging-diverging nozzle, velocity must increase in the converging section and decrease in the diverging section. Therefore, it must reach a maximum at the throat.

By definition, a nozzle (Fig. 15.14) must accelerate a flow, that is, dV is positive along the flow direction. And the pressure must monotonically decrease along the flow because a positive dV, according to Eq. (15.20), must accompany a negative dp. The converging-diverging shape alone cannot guarantee a supersonic flow in a nozzle or a sonic flow at the throat; the pressure at the exit must be sufficiently low. Conversely, a reduction of pressure at the exit cannot guarantee a supersonic flow unless a nozzle is converging-diverging in shape.

It follows from the argument above, the highest Mach number that can be achieved for a converging nozzle is one, and it must occur at the throat ($dA = 0$) of the nozzle. Any further decrease in the pressure outside the nozzle cannot affect the flow inside the nozzle because the throat cuts off propagation of all information upstream (Fig. 15.15). At that point, the flow is said to be **choked**.

In a diffuser, whose purpose is to increase pressure, a positive dp must accompany a monotonic decrease in velocity (negative dV). Following similar arguments, as in the case of an isentropic nozzle, we can establish that the supersonic section must be converging, the subsonic section diverging, and the transition $M = 1$ must take place at the throat of an isentropic diffuser.

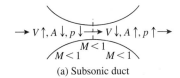
(a) Subsonic duct

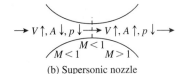
(b) Supersonic nozzle

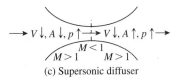
(c) Supersonic diffuser

FIGURE 15.14 Consequences of Eq. (15.23) for a converging-diverging duct. Pressure variation can be predicted by using Fig. 15.13. Upward and downward arrows indicate rise and fall of a property.

15.5 ISENTROPIC TABLE FOR PERFECT GASES

The Mach number, M, plays a vital role in the analysis of isentropic flow of perfect gases. As is shown below, using the PG model, various property ratios at a given flow state can be expressed as functions of M only, allowing these ratios to be tabulated against M in what is known as the isentropic table.

The ratio of total and static temperatures from Eq. (15.8) can be expressed in terms of M by using Eqs. (15.19), (15.14), and (3.65):

$$\frac{T_t}{T} = 1 + \frac{V^2}{2(1000\ \text{J/kJ})c_p T} = 1 + \frac{kR}{2}\frac{1}{c_p}\left(\frac{V^2}{(1000\ \text{N/kN})kRT}\right) = 1 + \frac{kR}{2}\frac{(k-1)}{kR}\left(\frac{V^2}{c^2}\right);$$

$$\Rightarrow \quad \frac{T_t}{T} = 1 + \frac{k-1}{2}M^2 \tag{15.24}$$

Substituting the temperature ratio into Eq. (15.9), we obtain corresponding ratios for pressure and density:

$$\frac{p_t}{p} = \left(\frac{T_t}{T}\right)^{k/(k-1)} = \left(1 + \frac{k-1}{2}M^2\right)^{k/(k-1)} \tag{15.25}$$

and

$$\frac{\rho_t}{\rho} = \left(\frac{T_t}{T}\right)^{1/(k-1)} = \left(1 + \frac{k-1}{2}M^2\right)^{1/(k-1)} \tag{15.26}$$

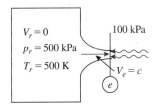

FIGURE 15.15 The flow is choked at the throat because the exit velocity equals the velocity of sound, the speed at which pressure information propagates upstream.

To express the mass flow rate in terms of total properties, a reference state, called the **critical state** is defined at the location where $M = 1$ occurs along an isentropic flow. In the case of a converging-diverging duct, we have already established that $M = 1$ where $dA = 0$—at the throat. A distinction should be made between the critical state and the flow state at a throat. In a subsonic flow, it is possible for a throat to have $M < 1$, in which case the critical state can be thought of as the state inside an imaginary throat with $M = 1$. To emphasize this difference, we will use the subscripts * and th for the critical and throat properties (see Anim. 15.B.*isentropicFlow*).

The mass equation, applied between a flow state anywhere along the flow and the critical state, yields:

$$\rho A V = \rho_* A_* V_*;\quad \left[\frac{\text{kg}}{\text{s}}\right] \quad \Rightarrow \quad \frac{A}{A_*} = \frac{\rho_*}{\rho}\frac{V_*}{V}\frac{c}{c_*}\frac{c_*}{c} = \frac{\rho_*}{\rho}\frac{M_*}{M}\frac{\sqrt{kRT_*}}{\sqrt{kRT}}$$

$$\Rightarrow \quad \frac{A}{A_*} = \frac{\rho_*}{\rho_t}\frac{\rho_t}{\rho}\frac{1}{M}\sqrt{\frac{T_*}{T_t}\frac{T_t}{T}} = \frac{1}{M}\frac{\rho_t}{\rho}\left(\frac{\rho_t}{\rho_*}\right)^{-1}\left(\frac{T_t}{T}\right)^{1/2}\left(\frac{T_t}{T_*}\right)^{-1/2}$$

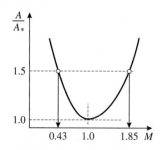

FIGURE 15.16 Area ratio according to Eq. (15.27). A must be larger than A_* for all M's except $M = 1$.

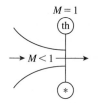

(a) Converging choked duct

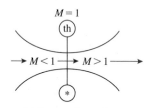

(b) Supersonic nozzle

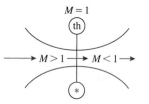

(c) Supersonic diffuser

FIGURE 15.17 If choking occurs in an isentropic flow, the flow state at the throat is also the critical state.

Substituting the property ratios from Eqs. (15.26) and (15.24) and realizing that, by definition, $M_* = 1$, we obtain:

$$\frac{A}{A_*} = \frac{1}{M}\left[\left(\frac{2}{k+1}\right)\left(1 + \frac{k-1}{2}M^2\right)\right]^{(k+1)/(2k-2)} \quad (15.27)$$

The area ratio from this expression, plotted for air ($k = 1.4$) in Figure 15.16 as a function of the Mach number, reveals some interesting features about this result. As expected, $M = 1$ corresponds to the critical location where $A/A_* = 1$. For a converging or a converging-diverging passage, we have already established that if flow velocity becomes sonic, it must occur at the throat, or $A_{\text{th}} = A_*$ (Fig. 15.17). Figure 15.16, obtained from Eq. (15.27), reconfirms the shape of a supersonic isentropic nozzle, numerically established earlier, where A_* can be interpreted as the minimum area necessary to pass the flow. If the flow is subsonic everywhere in the same duct (which can happen if the back pressure is not sufficiently low), the critical state becomes a hypothetical state, as shown in Figure 15.18. With $A_{\text{th}} > A_*$, A_{th}/A_* can be obtained from Figure 15.16 (or an isentropic table) if $M_{\text{th}}(< 1)$ is known. Another interesting feature of Figure 15.16 is that two different Mach numbers—one belonging to the subsonic branch and another to the supersonic branch—correspond to the same value of the area ratio A/A_*. Out of the two possible solutions for a given A/A_*, the correct solution can be deduced if the local shape (whether dA is positive or negative) is known. Like total and stagnation properties, **critical properties** of a flow (A_*, ρ_*, etc.) that describe the critical state—can also be treated as additional properties of a local flow state.

A simple expression for the mass flow rate through a variable-area duct can be derived by using critical properties:

$$\dot{m} = \rho_* A_* V_* = \rho_t \frac{\rho_*}{\rho_t}A_* c_* = \frac{p_t}{RT_t}\frac{\rho_*}{\rho_t}A_* c_* = \frac{A_* p_t}{R\sqrt{T_t}}\frac{\rho_*}{\rho_t}\sqrt{(1000 \text{ J/kJ})kR\frac{T_*}{T_t}}; \quad \left[\frac{\text{kg}}{\text{s}}\right]$$

Substituting static-to-total property ratios from Eqs. (15.26) and (15.24), we obtain:

$$\Rightarrow \quad \dot{m} = A_* p_t \sqrt{\frac{(1000 \text{ J/kJ})k}{RT_t}}\left(\frac{2}{k+1}\right)^{(k+1)/[2(k-1)]}; \quad \left[\frac{\text{kg}}{\text{s}}\right] \quad (15.28)$$

The critical area, therefore, is a system (extensive) property as it depends on the mass flow rate. It appears counterintuitive that the mass flow rate for a given gas is a function of total properties and the critical area alone. After all, the difference between the reservoir pressure ($p_r = p_t$) and the outside pressure (**back pressure** p_b) is driving the flow, and $\dot{m}$ can be expected to depend on the pressure difference $p_r - p_b$. When the flow reaches sonic speed at the throat, the mass flow rate becomes insensitive to any further decrease in p_b as information about the back pressure, traveling at the local speed of sound, gets blocked at the throat, where $M = 1$, and cannot propagate upstream (Fig. 15.15). The mass flow, therefore, reaches a maximum and the flow is said to be **choked** for the given reservoir conditions and throat area ($A_* = A_{\text{th}}$).

For a subsonic flow, the mass flow rate must depend on the back pressure p_b. With $p_t = p_r$ and $T_t = T_r$ dependent only on the reservoir condition, how does Eq. (15.28) involve the back pressure? To answer that question, note that in Eq. (15.28) A_* is no longer equal to A_{th}, but is related to the throat Mach number ($M_{\text{th}} < 1$) through Eq. (15.27), which in turn will be shown to depend on the back pressure p_b. We will thoroughly examine the effect of back pressure on the mass flow rate in the next section.

We have already seen how the energy and entropy equations can be alternatively expressed as $T_t = $ constant and $p_t = $ constant for the isentropic flow of a perfect gas. The mass flow rates from Eq. (15.28), at two different flow states, can be equated, producing:

$$A_{*i} = A_{*e}; \quad [\text{m}^2] \quad (15.29)$$

The mass equation reduces to a simple statement; the critical area remains constant along an isentropic flow through a variable-area duct (Figure 15.19). If the area ratio of two flow states, State-1 and State-2, is known, Eqs. (15.29) and (15.27) can be used to determine the Mach number at State-2 from that at State-1. For that, we first obtain A_2/A_{*2} from:

$$\frac{A_2}{A_{*2}} = \frac{A_2}{A_1}\frac{A_1}{A_{*1}}\frac{A_{*1}}{A_{*2}} = \frac{A_2}{A_1}\left(\frac{A}{A_*}\right)_1 = \frac{A_2}{A_1}\left(\frac{A}{A_*}\right)_{@M_1} \quad (15.30)$$

From Figure 15.16 it is evident that the Mach number is a double-valued function of A/A_*. Two possible solutions for M_2 can be obtained by solving Eq. (15.27) for the area ratio given by Eq. (15.30).

The isentropic relations—Eqs. (15.25), (15.24), (15.26), and (15.27)—are used to generate an **isentropic table**. Table H-1 lists p/p_t, T/T_t, ρ/ρ_t, and A/A_* as functions of the local Mach number M for air. It is dynamically created by the gas dynamics TESTcalc for a wide selection of gases in its table panel. Any of the four variables can be entered to obtain the other three. If A/A_* is entered, one of the branches—subsonic and supersonic—must be specified through a radio button. Use of an isentropic table greatly simplifies isentropic flow calculations.

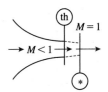

(a) Converging subsonic duct

EXAMPLE 15-4   Use of an Isentropic Table

In Example 15-2, assume $p_r = 150$ kPa and $T_r = 500$ K. If the exit area is 10 cm^2, determine (a) the exit Mach number, (b) exit velocity, (c) critical pressure, and (d) critical area. Use the isentropic table.

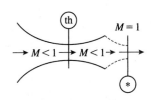

(b) Converging-diverging subsonic duct

FIGURE 15.18 The critical state is a hypothetical state for a subsonic flow in a converging or a converging-diverging isentropic duct.

SOLUTION

Total temperature remaining constant along an isentropic flow, the ratio of static to total pressure can be found at the exit, and the exit Mach number can be obtained from the isentropic table for air (Table H-1).

Assumptions

Same as in Example 15-2.

Analysis

The exit pressure being equal to the back pressure, the static-to-total pressure ratio at the exit, State-2, is:

$$\left(\frac{p}{p_t}\right)_2 = \frac{p_2}{p_r} = \frac{p_b}{p_r} = \frac{100}{150} = 0.667$$

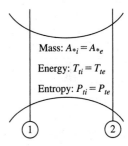

Mass: $A_{*i} = A_{*e}$

Energy: $T_{ti} = T_{te}$

Entropy: $P_{ti} = P_{te}$

FIGURE 15.19 Mass, energy, and entropy balance equation for an isentropic flow of a perfect gas (see Anim. 15.B.*isentropicFlow*).

From Table H-1 (or the table panel of the gas dynamics TESTcalc), the corresponding Mach number and the temperature ratio can be interpolated as:

$$M_2 = M_{@p/p_t = 0.667} = 0.784; \quad \left(\frac{T}{T_t}\right)_e = \left(\frac{T}{T_t}\right)_{@M_2} = 0.891$$

Using $T_{t2} = T_r$:

$$V_2 = M_2 c_2 = M_2 \sqrt{(1000 \text{ N/kN})kRT_2} = M_e \sqrt{(1000)kR\frac{T_2}{T_{t2}}\frac{T_{t2}}{T_r}T_r}$$

$$= (0.784)\sqrt{(1000)(1.4)(0.287)(0.891)(500)} = 331.5 \text{ m/s}$$

To calculate the critical pressure, find the pressure ratio corresponding to $M = 1$ from Table H-1 or the TESTcalc:

$$\frac{p_*}{p_t} = \left(\frac{p}{p_t}\right)_{@M=1} = 0.5283;$$

$$\Rightarrow \quad p_* = 0.5283 p_t = 0.5283(150) = 79.24 \text{ kPa}$$

Similarly, the critical exit area can be calculated with the help of Table H-1:

$$\frac{A_2}{A_*} = \left(\frac{A}{A_*}\right)_{@M=0.784} = 1.045;$$

$$\Rightarrow \quad A_* = A_2/1.045 = 10/1.045 = 9.57 \text{ cm}^2$$

TEST Analysis

Launch the gas dynamics TESTcalc and select Air. Evaluate States 1 and 2, as in Example 15-2, except p1=150 kPa. To evaluate the critical state, calculate State-3 with Mach3=1,T_t3=T1, p_t3=p1, and A3=Astar2. All critical properties, including pressure, are displayed as part of State-3 and the manual results are verified. See TEST-code (posted in *TEST* > TEST-codes) for details.

DID YOU KNOW?

• Critical flow offers a simple, yet elegant, way to produce precise flow rates, which can be used to calibrate gas flow meters.

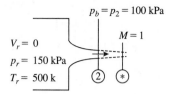

$p_b = p_2 = 100$ kPa

$V_r = 0$
$p_r = 150$ kPa
$T_r = 500$ k

$M = 1$

FIGURE 15.20 The critical area is smaller than the throat (exit) area in Ex. 15-4.

Discussion

If the back pressure is reduced to 79.25 kPa, the flow will be choked at the exit regardless of the exit area. On the other hand, if only the exit area is reduced, the exit Mach number will remain unchanged and only the mass flow rate will change, according to Eq. (15.28). The critical state does not coincide with the nozzle exit, as shown in Figure 15.20. However, if the nozzle is pinched in the middle to create an area of 9.89 cm², the imaginary critical state will then shift to that pinched location, which becomes the new throat of the reshaped converging-diverging duct. How do you think the flow will adjust if the pinched area is reduced below the critical area?

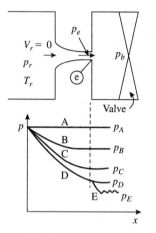

$V_r = 0$
p_r
T_r

p_e
p_b

Valve

p
A — p_A
B — p_B
C — p_C
D — p_D
E — p_E

x

FIGURE 15.21 The back pressure is the only parameter that is changed in this experiment (see Anim. 15.B.*ConvNozzleBackP*).

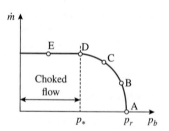

$\dot{m}$

E | D
C
Choked flow
B
A

p_* p_r p_b

FIGURE 15.22 Mass flow rate through a converging nozzle as the back pressure is varied in the set up described in Fig. 15.21.

15.6 EFFECT OF BACK PRESSURE: CONVERGING NOZZLE

The pressure in the immediate surroundings of a jet is called the **back pressure**. A jet ejecting into the atmospheric air has a back pressure of the surrounding atmosphere, which can change drastically with altitude. The exit pressure, on the other hand, is a property of the exit state, the last state within the duct carrying the flow before it emerges from the duct as a jet. Unless proven otherwise, we will assume that there is no discontinuity at the exit, and the pressure is equal to the back pressure.

To thoroughly explore the effect of back pressure on a flow, consider a converging nozzle, shown in Figure 15.21, connected between a reservoir and a chamber whose pressure can be adjusted by a valve. In this setup, the reservoir pressure p_r, temperature T_r, and exit area A_e are fixed, and only the back pressure p_b can be changed to any desired value. For isentropic flow of a perfect gas in the nozzle, the mass, energy, and entropy equations dictate that $A_* = $ constant ($A_* = A_{th}$ only if the flow is choked at the throat), $T_t = T_r$, and $p_t = p_r$.

Several cases, labeled A through E for different values of p_b, are considered in Figure 15.21 and the corresponding mass flow rates are plotted in Figure 15.22. For $p_b = p_t = p_r$ (case A), there is no driving force dp anywhere along the duct; therefore, $dV = 0$. With $V_r = 0$, the gas must be stationary everywhere, and $\dot{m} = 0$. Any reduction in the back pressure will create a flow according to Eq. (15.20).

First examine cases B and C, where the flow is assumed to be subsonic throughout the nozzle. In these cases, information about the back pressure, which travels at sonic speed, is propagated upstream all the way to the reservoir, and the flow adjusts to the back pressure to ensure that $p_e = p_b$. Recognizing that $(p/p_t)_e = p_b/p_r$, the exit Mach number M_e and the area ratio $(A/A_*)_e$ can be obtained from an isentropic table such as Table H-1.

Because the flow is subsonic, it can be deduced from Figure 15.16 that $A/A_* < 1$ throughout the converging nozzle so that $A_* < A_e$. As $p_e = p_b$ is reduced while $p_t = p_r$ remains constant, M_e must increase according to Eq. (15.25). An increase in M_e along the subsonic branch of Figure 15.16 requires A_e/A_* to decrease; therefore, for the given nozzle with a fixed exit (throat) area, the critical area A_* must increase (click different back pressure buttons in Anim. 15.B.*convNozzleBackP*). This explains the increase in the mass flow rate in accordance with Eq. (15.28).

For a given subsonic Mach number at the exit, A_e/A_* can be found in the isentropic table. At any other cross section, identified by the area ratio A/A_e, the local Mach number can be obtained from the isentropic table by selecting the subsonic solution for $A/A_* = (A/A_e)/(A_e/A_*)$. Once the local Mach number is found, static pressure or any other local property can be calculated from M, p_t, and T_t following the procedure outlined in Example 15-4. Figure 15.21 plots the variation of pressure, along the flow from p_r to p_b, in a qualitative manner. Similar plots can be made for other properties, including the local Mach number.

As p_b continues reduce, a point is reached when M_e reaches one and the flow is choked at the exit. From Table H-1, p/p_t at $M = 1$ can be found as 0.528 for air. This means if the back pressure is 52.8% of the reservoir pressure, the flow of air through a converging nozzle is choked at the exit, irrespective of the exit area. In such a special case, represented by case D, the exit state itself is the critical state and $p_e = p_b = p_*$. Flow properties at any cross-section can still be calculated, as in the case of subsonic flow outlined above. With the critical area reaching its highest value of A_e (Fig. 15.16), the mass flow rate must also reach a maximum, as predicted by Eq. (15.28) (p_t and T_t remain constant and A_* is maximized) and shown in Fig. 15.22. At that point communication between outside and inside is completely cut off because the exit velocity is equal to the velocity of sound.

Any further decrease in p_b below p_* cannot affect the flow inside the nozzle and case E ($p_b < p_*$) becomes identical to case D in all respects, including the variation of pressure inside the nozzle (Fig. 15.21). However, outside the nozzle, the adjustment from the exit pressure, $p_e = p_*$, to the sub-critical back pressure, $p_b < p_*$, takes place through a series of expansion waves (case E), a discussion of which is outside the scope of this book Open the flow structure simulation Interactive, select Fixed Geometry button, and vary the back pressure to numerically explore its effect on the flow structure. An important fact is established by case E: the back pressure can be smaller than the exit pressure in a choked flow through a converging nozzle.

The exit of a convergent nozzle must be shaped like the throat of a convergent-divergent nozzle so that $dA = 0$; otherwise, at $M_e = 1$, Eq. (15.23) will not be satisfied and the flow will cease to be isentropic. That is why the exit of an isentropic converging nozzle is also known as its throat.

EXAMPLE 15-5	**Flow Through a Converging Nozzle**

Air is discharged from a reservoir, at 500 kPa and 500 K, through an isentropic converging nozzle with an exit area of 10 cm^2. Determine the mass flow rate for a back pressure of (a) 0 kPa, (b) 100 kPa, (c) 200 kPa, and (d) 300 kPa.

SOLUTION

Determine if the flow is choked for a given back pressure and use Eq. (15.28) to determine the mass flow rate.

Assumption

Isentropic, one-dimensional flow of a perfect gas.

Analysis

From the isentropic table of air, Table H-1, or the table panel of the gas dynamics TESTcalc with air selected as the working fluid, the critical pressure ratio p_*/p_t for $M = 1$ can be obtained as 0.5283 so that:

$$p_* = \frac{p_*}{p_t}p_t = \left(\frac{p}{p_t}\right)_{@M=1} p_r = (0.5283)(500) = 264.15 \text{ kPa}$$

For back pressures less than the critical pressure 264.15 kPa—0 kPa, 100 kPa, and 200 kPa—the flow is choked.

For choked flow: $T_{th} = T_* = \left(\frac{T_*}{T_t}\right)T_t = \left(\frac{T}{T_t}\right)_{@M=1} T_t = (0.8333)(500) = 416.7 \text{ K.}$

The mass flow rate for the choked flow can be directly calculated:

$$\dot{m} = \rho_* A_* V_* = \rho_* A_* c_* = \frac{p_*}{RT_*} A_{th} \sqrt{(1000 \text{ N/kN})kRT_*}$$

$$= \frac{(264.15)(0.001)\sqrt{1000(1.4)}}{\sqrt{(0.287)(416.7)}}$$

$$= 0.904 \frac{\text{kg}}{\text{s}}$$

The same answer can be obtained by application of Eq. (15.28).

For a back pressure of 300 kPa, the flow is subsonic and the back pressure must be equal to the exit pressure. For $p_{th}/p_t = 300/500 = 0.6$, the isentropic table produces $M_{th} = 0.886$ and $T_{th} = (0.864)(500) = 432$ K. Therefore:

$$\dot{m} = \rho_{th} A_{th} V_{th} = \frac{p_{th}}{RT_{th}} A_{th} M_{th} \sqrt{(1000 \text{ N/kN})kRT_{th}}$$

$$= \frac{(300)(0.001)(0.886)\sqrt{1000(1.4)}}{\sqrt{(0.287)(432.1)}}$$

$$= 0.893 \frac{\text{kg}}{\text{s}}$$

TEST Analysis

Launch the gas dynamics TESTcalc. Select Air as the working fluid. Evaluate State-1 with p1 = 500 kPa, T1 = 500 K, and Vel1 = 0 as the reservoir (total) state. For the choked throat state, evaluate State-2, with M2 = 1, T_t2 = T1, p_t2 = p1, and A2 = 10 cm2. For the non-choked state, State-3, use p3, T_t3, p_t3, and A3 = A2. Mass flow rates are calculated as part of the complete flow states. For more details, see TEST-code (posted in *TEST* > TEST-codes module).

Discussion

By gradually increasing p3, and recalculating State-3, mdot3 can be plotted as a function of the back pressure, as shown in Figure 15.23. A similar analysis can be used to determine the exit area required to create a desired mass flow rate for a given total pressure and temperature. We could use Eq. (15.28) to obtain the mass flow rates, since p_t and T_t are known from the reservoir conditions and A_* can be obtained from Table H-1 for a given Mach number.

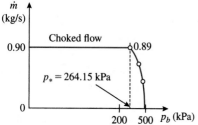

FIGURE 15.23 Mass flow rate as a function of back pressure in Ex. 15-5.

15.7 EFFECT OF BACK PRESSURE: CONVERGING-DIVERGING NOZZLE

As we did with the converging nozzle, let us examine several cases in which the back pressure is gradually reduced, this time, for the converging-diverging nozzle shown in Figure 15.24.

Case A for $p_b = p_t = p_r$ is still a trivial case with no possibilities of a flow. As the back pressure is slightly lowered to p_B (case B), a subsonic flow is created throughout the duct. Dictated by Eq. (15.23), the velocity must increase in the converging section and decrease in the diverging section, reaching a maximum at the throat (Sec. 15.4). Also, according to Eq. (15.22), pressure must decrease in the first section while it must increase in the latter section. Thus, the converging section acts like a nozzle and the diverging section like a diffuser. Subsonic analysis of the converging nozzle also applies to this case. With $p_e = p_b$, the exit Mach number M_e and the area ratio A_e/A_* can be obtained from the isentropic table. At any other location, say, the throat, M can be evaluated from the area ratio $A_{th}/A_* = (A_{th}/A_e)(A_e/A_*)$ using the subsonic branch of the two possible solutions (Fig. 15.16). Realizing that $p_t = p_r$ and $T_t = T_r$ for an isentropic flow, all other properties can be evaluated from M, p_t, and T_t.

When the back pressure is reduced to p_C in Figure 15.24, the exit Mach number is still subsonic, well below one, while the flow just reaches the critical level at the throat ($M = 1$). Consequently, $A_* = A_{th}$ or $A_e/A_* = A_e/A_{th}$, which produces two solutions for the Mach number at the exit according to the isentropic table (or Fig. 15.16)—one subsonic, corresponding to $p_e/p_{te} = p_C/p_r$ (case C in Fig. 15.24) and another supersonic for $p_e/p_{te} = p_G/p_r$ (case G). While, for both cases, the pressure decreases in the same manner in the converging section, for $p_b = p_C$ (case C), the pressure begins to increase (thus, velocity decreases and becomes subsonic downstream of the throat) in the diverging section, whereas for $p_b = p_G$ (case G), the pressure continues to decrease (thereby, accelerating the flow into the supersonic territory).

Based on the exit-to-throat area ratio, known simply as the **area ratio** of a converging-diverging duct, M_C and M_G can be determined from Eq. (15.27) and p_C and p_G can be obtained from Eq. (15.25) or the isentropic table. Flow states at any desired location for case C can be obtained using the same procedure outlined for case B. In case G, the flow is supersonic in the diverging section; therefore, for a given $A/A_* = A/A_{th}$, the supersonic solution is chosen from the isentropic table. Once the Mach number is found, the rest of the properties can be obtained from the fact that T_t and p_t remain invariant from the reservoir through the exit. In the converging section, the flow state solutions for both cases, C and G, are identical because downstream information cannot travel upstream of the throat, which is choked. Cases D, E, F, and H will be discussed after we introduce the normal shock wave in the next section.

The mass flow rate, plotted in Figure 15.25 against the back pressure p_b, behaves in a similar manner as in the case of a converging nozzle. After the flow is choked at the throat (case C), any further reduction of the back pressure (cases C through G) fails to affect the flow in the converging section. Therefore, the mass flow rate remains constant even though the flow in the diverging section undergoes major changes.

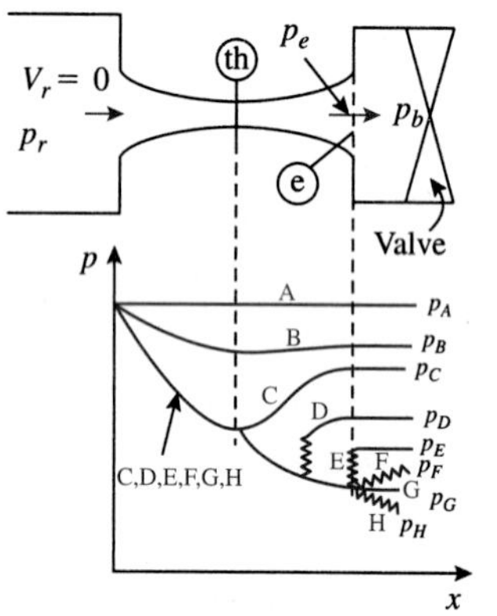

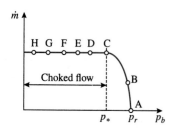

FIGURE 15.24 Effect of back pressure on a flow through a converging-diverging nozzle (see Anim. 15.B.*convDivNozzleBackP*).

FIGURE 15.25 Mass flow rate for different cases of Fig. 15.24.

EXAMPLE 15-6 **Flow through a Converging-Diverging Nozzle**

Air enters a converging-diverging nozzle (Fig. 15.26) with a velocity of 50 m/s at 400 kPa, 400 K. The nozzle has an exit-to-throat area ratio of 2. Determine (a) the maximum back pressure that will still choke the flow and (b) the design back pressure for isentropic supersonic flow.

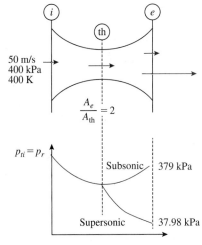

FIGURE 15.26 Schematic for Ex. 15-6.

SOLUTION

Evaluate the total pressure from the inlet conditions. The maximum pressure to choke the flow is the back pressure that corresponds to curve C in Figure 15.24. Curve G corresponds to the supersonic design condition. Obtain the design Mach number from the area ratio of the nozzle.

Assumption

Isentropic, one-dimensional flow of a perfect gas.

Analysis

The inlet Mach number is:

$$M_i = \frac{V_i}{\sqrt{(1000 \text{ N/kN})kRT_i}} = \frac{50}{\sqrt{(1000)(1.4)(0.287)(400)}} = 0.125$$

Using the isentropic table, the total pressure can be obtained:

$$p_t = p_{ti} = p_i \frac{p_{ti}}{p_i} = p_i \left[\left(\frac{p}{p_t}\right)_{@M_i = 0.125} \right]^{-1} = 400(0.9892)^{-1} = 404.4 \text{ kPa}$$

When the nozzle is choked—$M = 1$ occurs at the throat—the throat area becomes the critical area. Additionally, the flow must be subsonic and decelerating in the diverging section to maximize the exit pressure. The subsonic branch of the isentropic table (or the gas dynamics TESTcalc) for $A_e/A_* = 2$ yields:

$$M_e = 0.306; \quad \text{and} \quad p_e = p_b = p_t \left(\frac{p}{p_t}\right)_{@M_e = 0.306} = (404.4)(0.937) = 379 \text{ kPa}$$

The same area ratio produces the following supersonic isentropic solution:

$$M_e = 2.197; \quad \text{and} \quad p_e = p_b = p_t \left(\frac{p}{p_t}\right)_{@M_e = 2.197} = (404.4)(0.0939) = 37.98 \text{ kPa}$$

TEST Analysis

Launch the gas dynamics TESTcalc and select air as the working fluid. Evaluate State-1 as the inlet state from the given properties p1, T1, and Vel1, and State-2 as the critical throat state with M2 = 1, T_t2 = T_t1, p_t2 = p_t1, and an arbitrarily chosen A2 = 1 m². For the exit state, State-3, enter A3 = 2*A2, p_t3 = p_t1, T_t3 = T_t1, and Astar3 = A2. When you click the Calculate button the subsonic solution is displayed. At the same time, the Mach number for the alternative supersonic solution is displayed in the message panel as 2.197. Evaluate State-4 with M4 = 2.197, A4 = 2*A2, p_t4 = p_t1, and T_t4 = T_t1. The manual solution is now verified.

Discussion

The supersonic exit Mach number is called the design Mach number of a converging-diverging nozzle. It is a function of the exit-to-throat area ratio for a given gas.

EXAMPLE 15-7 **Thrust Calculation for a Rocket**

A rocket nozzle with an exit area of 0.1 m² is designed to work supersonically with a chamber pressure of 2.5 MPa and a chamber temperature of 1500 K at sea level, where the ambient pressure is 101 kPa. (a) Determine the thrust at sea level. Assume the chamber conditions remain steady. Model the exhaust gas as a perfect gas with $k = 1.3$ and $R = 0.42$ kJ/kg · K. *What-if scenario:* (b) What would the thrust be if the rocket was in space where the pressure can be assumed to be 0 kPa?

SOLUTION

Calculate the exit velocity and then use Eq. (8.27), derived in Chapter 8, to evaluate the thrust.

Assumption

Isentropic one-dimensional flow of a perfect gas.

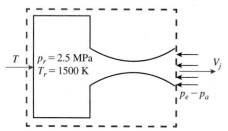

FIGURE 15.27 Schematic for Ex. 15-7 (see Anim. 15.B.*rocketThrust*). The thrust *T* can be interpreted as the force necessary to keep the rocket stationary.

Analysis

The thrust equation, Eq. (8.27), can be simplified for the rocket shown in Figure 15.27 by eliminating the inlet transport term.

$$T = \frac{\dot{m}}{(1000 \text{ N/kN})}(V_j - V_a^0) + A_j(p_j - p_a) = A_e\left[\frac{\rho_e V_e^2}{1000} + (p_e - p_a)\right]$$

The pressure thrust in this equation comes from the exit pressure, which is not always equal to the back pressure. The nozzle is designed for sea level; therefore, the exit pressure must be same as the back pressure, 101 kPa, at sea level for isentropic expansion. As a result, $p_e/p_t = 101/2500 = 0.0404$. Solving Eqs. (15.25) and (15.24), we obtain $M_e = 2.7$ and $T_e = 715$ K.

$$V_e = M_e\sqrt{(1000 \text{ N/kN})kRT_e} = 2.7\sqrt{(1000)(1.3)(0.42)(715)} = 1690 \text{ m/s};$$

$$\rho_e = \frac{p_e}{RT_e} = \frac{101}{(0.42)(715)} = 0.336 \text{ kg/m}^3;$$

At sea level:

$$T = (0.1)\left[\frac{(0.336)1690^2}{(1000)} + (101 - 101)\right] = 96.0 \text{ kN}$$

Because the flow at the exit is supersonic, the exit state is not affected by a reduction in back pressure as the rocket passes through the atmosphere into space. The thrust, however, varies with altitude as p_a decreases.

TEST Analysis

Launch the gas dynamics TESTcalc, select Custom working gas, and enter the given values of R and k. Evaluate the chamber state, State-1, from the given properties p1, T1, and Vel1 = 0. Evaluate the isentropic exit state, State-2, from p2 = 101 kPa, p_t2 = p1, T_t2 = T1, and A2 = 0.1m². In the I/O panel, calculate the thrust by evaluating the expression = mdot2*Vel2/1000 as 96.02 kN.

What-if scenario

The exit velocity remains unchanged in space as the supersonic flow adjusts to the zero back pressure *outside* the nozzle. However, the pressure component of the thrust increases, producing:

$$T = (0.1)\left[\frac{(0.336)1690^2}{1000} + (101 - 0)\right] = 106.1 \text{ kN}$$

Discussion

The pressure of the exit stream, which remains unchanged at 101 kPa as long as the back pressure is less than 101 kPa, augments the thrust in space by 0.1*(101 − 0) = 10.1 kN. At sea level, the same pressure, 101 kPa, acts all around the system boundary, resulting in no net pressure contribution to the thrust. (see Anim. 15.B.*rocketThrust*)

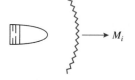

FIGURE 15.28 The projectile traveling at a supersonic speed drives a shock wave in front. The flow becomes steady when the coordinates are attached to the shock (see Anim. 15.B.*movingShocks*).

15.7.1 Normal Shock

We have already introduced the concept of a shock wave (Fig. 15.8), which was necessary for a supersonic flow to negotiate an obstruction. In the case of a projectile traveling at supersonic speed in stationary air (Fig. 15.28), a shock wave moving at the speed of the projectile makes abrupt changes in the stationary air, unaware of the projectile, to quickly adjust to make room for the projectile to pass. If an observers sits on the projectile, the situation becomes identical to that described in Fig. 15.8. In a converging-diverging nozzle, stationary or **standing shock waves** are sometime created so that the flow can satisfy boundary conditions (going from a given reservoir pressure to a given back pressure) for supersonic flows. Shock waves produced under different scenarios are summarized in Anim. 15.B.*standingShocks*.

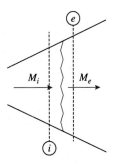

FIGURE 15.29 A stationary normal shock in the diverging section of a supersonic nozzle (see Anim. 15.B.*nozzleShock*).

To begin a shock wave analysis, go back to the case studies of Figure 15.24. In a flow through a converging-diverging nozzle, the back pressure has to be just right ($p_b = p_C$ or $p_b = p_G$ in Fig. 15.24) for the flow to be isentropic while being choked ($M = 1$) at the throat. The isentropic condition also requires that the flow in the diverging passage must be subsonic if $p_b = p_C$ and supersonic if $p_b = p_G$. If the back pressure is somewhere in between, say, $p_b = p_D$ (case D), the flow is supersonic part of the way before it abruptly transitions to a subsonic flow by passing through a standing shock wave (Fig. 15.29) somewhere in the diverging section. A shock wave is called a **normal shock wave** if it is perpendicular to the incoming flow, as in the one-dimensional flow in the nozzle. Across the shock, properties change abruptly (typical thickness of a shock is about 0.25 micrometer) with a step increase in pressure, temperature, and density, and a step decrease in the Mach number and velocity. The flow being subsonic after the shock, the diverging section acts as a diffuser, where an increase in area ensures a reduction of velocity and increase in pressure (Sec. 15.4).

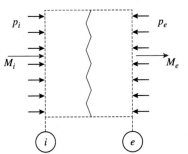

FIGURE 15.30 The system anchored around the normal shock wave is steady even for a moving shock (see Anim. 15.B.*normalShock*).

The severity of the step changes in properties across a shock depends on the Mach number. The **shock strength**, measured by the Mach number of the flow just upstream of the stationary shock, increases if the shock is located further along the diverging section. The particular location where the shock wave occurs is a function of the back pressure; the lower the back pressure, the closer is the normal shock to the exit, as shown in Figure 15.24. Conversely, the higher the back pressure, the closer the shock moves towards the throat until it disappears when the shock strength is one.

Although the shock waves associated with supersonic objects as in Figure 15.28 is curved, the central part of a detached shock can be treated as a normal shock. Now look at a normal shock wave, as shown in Figure 15.30. The system drawn around the shock wave is attached to the wave so that it is steady, regardless of whether the shock is stationary, as in the nozzle (Fig. 15.29), or moving, as in the case of the projectile (Fig. 15.28). Let States i and e represent the flow states of the one-dimensional flow just before and after the shock. Just as isentropic relations connect properties of two isentropic states, our goal is to develop normal shock relations to relate the two flow states across a normal shock. In the case of a moving shock in a still atmosphere, the upstream air approaches the shock with the velocity of the shock wave in the coordinates attached to the shock. Note that the thermodynamic properties p, T, ρ, etc., maintain their atmospheric value, regardless of whether the shock is standing or moving. However, the extrinsic properties, such as p_t or T_t, assume quite different values based on the relative velocity of the flow, with respect to the shock. Given the extremely small thickness of a shock, A_i and A_e can be assumed to be equal even when a shock is situated in a diverging section.

The mass, energy, entropy, and momentum equations for the system sketched in Figure 15.30 can be simplified as follows:

$$\text{Mass:} \quad \dot{m}_i = \dot{m}_e = \dot{m}; \quad \left[\frac{\text{kg}}{\text{s}}\right] \quad \Rightarrow \quad \rho_i V_i = \rho_e V_e; \quad \left[\frac{\text{kg}}{\text{m}^2 \cdot \text{s}}\right] \tag{15.31}$$

$$\text{Energy:} \quad 0 = \dot{m}(j_i - j_e) + \dot{Q}^0; \quad \Rightarrow \quad h_i + \text{ke}_i = h_e + \text{ke}_e; \quad \left[\frac{\text{kJ}}{\text{kg}}\right] \tag{15.32}$$

$$\text{Entropy:} \quad 0 = \dot{m}(s_i - s_e) + \frac{\dot{Q}^0}{T_B} + \dot{S}_{\text{gen}}; \quad \left[\frac{\text{kW}}{\text{K}}\right] \quad \Rightarrow \quad s_i + \frac{\dot{S}_{\text{gen}}}{\dot{m}} = s_e; \quad \left[\frac{\text{kJ}}{\text{kg} \cdot \text{K}}\right] \tag{15.33}$$

$$\text{Momentum:} \quad 0 = \frac{\dot{m}(V_i - V_e)}{(1000 \text{ N/kN})} + p_i A_i - p_e A_e; \quad [\text{kN}]$$

$$\Rightarrow \quad p_i + \frac{\rho_i V_i^2}{(1000 \text{ J/kJ})} = p_e + \frac{\rho_e V_e^2}{(1000 \text{ J/kJ})}; \quad [\text{kPa}] \tag{15.34}$$

In deriving the energy equation, the flow is assumed to be adiabatic. This is because temperature variation in a normal shock is assumed to occur in the flow direction; and the inlet and exit states are sufficiently removed from the shock so that the temperature gradients at those locations are negligible. In a variable area duct—a ramjet for instance—the variable pressure on the internal surface creates a net force in the direction of the flow, which is accounted for by an additional term, thrust T, in the momentum equation (Sec. 8.7.2). The distance between the inlet and exit states in Figure 15.30 is so small that the contribution to the axial force from the wall pressure is negligible compared to the forces at the inlet and exit planes, and there is no need for a thrust term in the momentum equation for the normal shock.

To obtain a solution for the exit state from the set of equations, assume that the gas can be modeled as a perfect gas so that the property relations $p = \rho RT$, $dh = c_p dT$, and $c^2 = (1000 \text{ J/kJ})kRT$ are applicable. Although a shock is a source of significant amount of irreversibilities, the energy equation, Eq. (15.32), is identical to that for an isentropic flow, Eq. (15.10). Therefore, $T_{ti} = T_{te}$, the total temperature does not change across a shock. Substituting Eq. (15.24) on both sides of this energy equation results in:

$$\frac{T_e}{T_i} = 1 + \frac{1 + M_i^2(k-1)/2}{1 + M_e^2(k-1)/2} \tag{15.35}$$

The momentum equation, after a little manipulation, similarly yields the static pressure ratio as a function of Mach numbers:

$$p_i\left[1 + \frac{V_i^2}{(1000 \text{ J/kJ})RT_i}\right] = p_e\left[1 + \frac{V_e^2}{(1000 \text{ J/kJ})RT_e}\right]; \quad [\text{kN}],$$

$$\Rightarrow \quad \frac{p_e}{p_i} = \frac{1 + kM_i^2}{1 + kM_e^2} \tag{15.36}$$

The density ratio can also be expressed in terms of M_i and M_e by applying the ideal gas equation of state to Eqs. (15.35) and (15.36).

However, M_e still remains undetermined. To express it in terms of inlet properties, the mass equation can be manipulated to yield:

$$\rho_i V_i = \rho_e V_e; \quad \Rightarrow \quad \frac{p_i}{RT_i}M_i\sqrt{(1000 \text{ J/kJ})kRT_i} = \frac{p_e}{RT_e}M_e\sqrt{(1000 \text{ J/kJ})kRT_e}; \quad \left[\frac{\text{kg}}{\text{m}^2 \cdot \text{s}}\right]$$

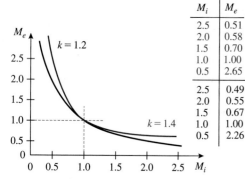

M_i	M_e
2.5	0.51
2.0	0.58
1.5	0.70
1.0	1.00
0.5	2.65
2.5	0.49
2.0	0.55
1.5	0.67
1.0	1.00
0.5	2.26

FIGURE 15.31 Exit Mach number as a function of inlet Mach number of a normal shock in air.

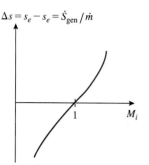

FIGURE 15.32 Δs from Eq. (15.38) plotted as a function of inlet Mach number.

Substituting the expressions for temperature and pressure ratios from Eqs. (15.35) and (15.36) into this mass equation, M_e can be expressed in terms of M_i:

$$M_e^2 = \frac{M_i^2 + 2/(k-1)}{2kM_i^2/(k-1) - 1} \tag{15.37}$$

To explore the nature of this relation, M_e is plotted as a function of M_i for two different values of k in Figure 15.31 (see Anim. 15.B.*normalShock*). The plot shows that for a supersonic inlet ($M_i > 1$), the flow after the shock must be subsonic ($M_e < 1$). From the property ratios derived above, $M_i > M_e$ implies an increase in pressure and temperature across the shock wave. That is why a normal shock is also called a **compression shock**. However, Figure 15.31 also reveals the possibility of an **expansion shock**, the inverse of a compression shock for $M_i < 1$, where a subsonic flow abruptly turns into supersonic flow with an accompanying decrease in pressure and temperature.

To investigate the feasibility of an expansion shock, express the change in entropy between the flow states across the shock wave using the PG model formula, Eq. (3.70):

$$\Delta s = s_e - s_i = c_p \ln\frac{T_e}{T_i} - R\ln\frac{p_e}{p_i}; \quad \left[\frac{\text{kJ}}{\text{kg} \cdot \text{K}}\right] \tag{15.38}$$

The entropy balance equation for the shock, Eq. (15.33), can be recast in terms of Δs as $\Delta s = \dot{S}_{gen}/\dot{m} \geq 0$. With T_e/T_i and p_e/p_i already expressed in terms of M_i and M_e, and M_e expressed in terms of M_i, Δs can also be expressed as a function of M_i. The functional relationship is qualitatively plotted in Figure 15.32. It can be seen from this plot that only for $M_i > 1$, Δs is non-negative, that is, $\dot{S}_{gen} \geq 0$. We would expect significant irreversibilities to be present in a shock wave due to thermal friction caused by heat transfer across a steep temperature gradient (Fig. 15.33). An expansion wave with $M_i < 1$, on the other hand, produces a negative value for Δs, that is, $\dot{S}_{gen} < 0$, violating the second law of thermodynamics, which forbids entropy destruction. An expansion shock wave, therefore, is impossible. Figures 15.31 and 15.32 also establish that the higher the inlet Mach number M_i, the lower the exit Mach number M_e and the higher the entropy generation rate in the shock wave. The degree of irreversibility of a compression shock, therefore, increases with its strength (M_i).

By manipulating Eq. (15.38), it can be shown that the total pressure decreases across the shock. It is left as an exercise to express the total pressure ratio across a normal shock wave:

$$\frac{p_{te}}{p_{ti}} = \frac{M_i}{M_e}\left[\frac{1 + M_e^2(k-1)/2}{1 + M_i^2(k-1)/2}\right]^{(k+1)/(2k-2)} \quad \textbf{(15.39)}$$

For a given k and M_i, the total pressure can be shown from Eq. (15.39) to drop across a shock, unlike the total temperature which remains constant. The drop in total pressure can be attributed to the irreversibilities (caused by thermodynamic friction) generated by the shock; the higher the strength of the shock, the greater the loss of total pressure. Another useful relation results by applying the mass flow rate expression of Eq. (15.28) to both sides of the normal shock, yielding:

$$\frac{A_{*e}}{A_{*i}} = \frac{p_{ti}}{p_{te}} \quad \textbf{(15.40)}$$

The critical area after a shock increases (Fig. 15.34) in inverse proportion to the total pressure, which decreases according to Eq. (15.39). As an interesting application of this equation, consider a stationary shock wave in a duct connecting two converging-diverging nozzles with supersonic flows, as shown in Figure 15.35. To handle the mass flow passing through the first nozzle, the second throat must be larger than the first one because $p_{te} < p_{ti}$.

A normal shock table lists different property ratios derived in this section as a function of M_i for a given gas. Table H-2 is the normal shock table for air. The table panel of the gas dynamics TESTcalc uniquely combines the isentropic table and normal shock table into a single table. The table is activated only after a state is partially calculated after selecting a gas and clicking the Calculate button. In the table panel, enter any variable to calculate the rest. If an alternative solution exists (say, A/A_* is known), an appropriate message is displayed in the message panel. The static to total property ratios for the incoming flow are also displayed in the same table.

15.7.2 Normal Shock in a Nozzle

Having analyzed a normal shock wave, the task of exploring the back pressure's effect on the flow through a converging-diverging nozzle can be completed. In case D of Figure 15.24, the flow accelerates isentropically to supersonic speed until it undergoes a normal shock transition. The flow, after the shock in the diverging duct, becomes subsonic and resumes a diffuser-like subsonic isentropic flow through the diverging section. As in a diffuser, the pressure increases, reaching p_D at the exit. For a given p_D, there is only one location where the strength of the shock is just right to satisfy the pressure boundary condition.

If the location of the shock in the diverging section is specified in terms of the area ratio A/A_{th}, M_i (the Mach number at the shock location) can be obtained from $A/A_{th} = A/A_*$ by using the supersonic branch of the isentropic table. With a known shock strength, the normal shock table can be used to obtain p_{te} and A_{*e} from the known values of the corresponding properties at State-i: $p_{ti} = p_r$ and $A_{*i} = A_{th}$. Also, from the energy equation, we know that $T_{te} = T_{ti}$. With the total pressure, total temperature, and critical area determined after the shock, the subsonic isentropic flow downstream can be analyzed, as was outlined for case B in Sec. 15.7.

As the back pressure is further reduced, the shock moves towards the exit, its strength increasing as M_i increases along the diverging section (Fig. 15.36) until the shock stands at the exit, but still inside the nozzle, as in case F of Figure 15.24. In that case, the flow is isentropic everywhere in the nozzle, except a very thin region right before the exit. Any further reduction in the back pressure causes the shock to open up and become slanted at an angle to the flow, while still anchored at the edge of the nozzle exit. Such shocks are known as oblique shocks whose angle with the flow keeps decreasing until they completely disappear as the flow turns isentropic for the design condition of $p_b = p_G$. For $p_b < p_G$ (case H), expansion waves (Fig. 15.24)

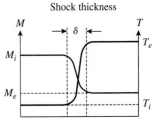

FIGURE 15.33 Sharp temperature gradient across a shock is a strong driving force for heat transfer and the accompanying thermal friction and entropy generation.

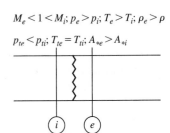

$M_e < 1 < M_i; p_e > p_i; T_e > T_i; \rho_e > \rho$

$p_{te} < p_{ti}; T_{te} = T_{ti}; A_{*e} > A_{*i}$

FIGURE 15.34 The normal shock table lists property ratios across the shock as a function of inlet Mach number (see Anim. 15.B.*normalShock*).

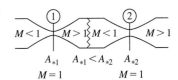

FIGURE 15.35 The second throat must be larger than the first one according to Eq. (15.40).

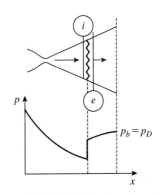

FIGURE 15.36 The location of the normal shock in the diverging section of a supersonic nozzle depends on the back pressure (see Anim. 15.B.*nozzleShock*).

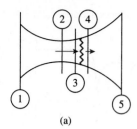

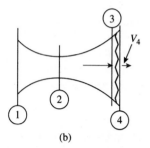

FIGURE 15.37 Schematic for Ex. 15-8, showing the flow states used in the analysis: (a) normal shock in the diverging section, (b) normal shock at the exit.

enable the flow to expand further outside the nozzle without having any effect on the flow inside. Analysis of oblique shock and expansion waves are outside the scope of this book, although the table panel of the gas dynamics TESTcalc offers the relevant tables for such advanced analysis. To explore the effect of back pressure through a numerical experiment, launch the flow structure simulation Interactive and with the default settings of Fixed Geometry Nozzle, change the back pressure (by using the $+$ or $-$ button) and observe how it affects the solution.

EXAMPLE 15-8   Flow through a Converging-Diverging Nozzle

In Example 15-6, determine the back pressure for which a normal shock stands (a) in the diverging section at an area 50% larger than the throat area and (b) at the exit.

SOLUTION

Do an isentropic analysis before and after the shock wave with the normal shock table relating properties on the two sides of the shock.

Assumption

One-dimensional flow of a perfect gas. Isentropic flow everywhere except across the shock wave.

Analysis

Represent the various states along the nozzle by State-1 through State-5, with State-2 being the throat (critical state), as shown in Figure 15.37(a). The flow is isentropic until State-3; therefore, M_3 can be obtained from the isentropic table (or the table panel of the gas dynamics TESTcalc). From $A_3/A_* = A_3/A_2 = 1.5$, we obtain $M_3 = 1.854$. Therefore:

$$p_3 = \left(\frac{p}{p_t}\right)_{@M_3} p_{t3} = (0.1602)p_{t1} = (0.1602)(404.4) = 64.78 \text{ kPa}$$

Using the normal shock table (or the gas dynamics TESTcalc), we can obtain State-4:

$$p_{t4} = \frac{p_{t4}}{p_{t3}}p_{t3} = \left(\frac{p_{te}}{p_{ti}}\right)_{@M_3} p_{t3} = (0.7884)(404.64) = 318.83 \text{ kPa};$$

and,

$$A_{*4} = \frac{A_{*4}}{A_{*3}}A_{*3} = \left(\frac{A_{*e}}{A_{*i}}\right)_{@M_3} A_2 = (1.268)A_2$$

The flow being isentropic after the shock, the critical area and total pressure remain constant downstream of State-4. Therefore:

$$\left(\frac{A}{A_*}\right)_{@M_5} = \frac{A_5}{A_{*5}} = \frac{A_5}{A_{*4}} = \frac{2A_2}{1.268A_2} = 1.577, \text{ producing } M_5 = 0.404$$

and, $\dfrac{p_5}{p_{t5}} = \left(\dfrac{p}{p_t}\right)_{M_5} = 0.894; \quad \Rightarrow \quad p_5 = (0.894)(318.83) = 285.03 \text{ kPa}$

When the shock is at the exit, the states are renumbered, as shown in Figure 15.37(b), where p_4 is the desired unknown. Realizing that the flow is isentropic up to the shock location, properties at State-3 can be obtained from the isentropic table:

$$\left(\frac{A}{A_*}\right)_{@M_3} = \frac{A_3}{A_{*3}} = \frac{A_3}{A_2} = 2; \quad \Rightarrow \quad M_3 = 2.20; \quad p_3 = \left(\frac{p}{p_t}\right)_{@M_3} p_{t3}$$
$$= (0.0939)(404.4) = 37.97 \text{ kPa}$$

From the shock table, we can obtain the static pressure ratio for $M_i = 2.20$ to find:

$$p_4 = \frac{p_4}{p_3}p_3 = \left(\frac{p_e}{p_i}\right)_{@M_3} p_3 = (5.466)(37.97) = 207.5 \text{ kPa}$$

TEST Analysis

Launch the gas dynamics TESTcalc and select air as the working fluid. Evaluate the five states, as described in the TEST-code (see *TEST* > TEST-codes), to verify the manual results.

Discussion

The mass flow rate can be calculated if the throat area is known. The presence of a shock wave cannot change the mass flow rate which is a function of chamber pressure, chamber temperature, and throat area. The flow is choked at the throat; therefore, what happens in the diverging section cannot affect any property upstream of the throat.

EXAMPLE 15-9 **Meteorite Entry**

A meteorite enters the earth's atmosphere at Mach 20 with an attached shock wave, as shown in Figure 15.38. Determine the stagnation (a) pressure and (b) temperature the meteorite is subjected to. The local ambient conditions are 1 kPa and 200 K. Assume the air behaves as a perfect gas with $k = 1.4$ (no dissociation).

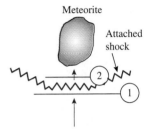

Meteorite
Attached shock

FIGURE 15.38 Schematic for Ex. 15-9.

SOLUTION

Use the isentropic and normal shock tables to determine the total properties after the shock.

Assumption

Close to the meteorite, the shock can be assumed to be normal to the direction of the flow. Flow is isentropic before and after the shock, and air can be assumed to be a perfect gas.

Analysis

Obtain the stagnation properties at State-1 before the shock by using the table panel of the gas dynamics TESTcalc and entering M1 = 20 (roll the pointer over a variable to view a more accurate value on the message panel):

$$p_{t1} = \left[\left(\frac{p}{p_t}\right)_{@M_1}\right]^{-1} p_1 = \left(\frac{1}{2.0788 \times 10^{-7}}\right)\left(\frac{1}{1000}\right) = 4{,}810 \text{ MPa};$$

$$T_{t1} = \left[\left(\frac{T}{T_t}\right)_{@M_1}\right]^{-1} T_1 = \left(\frac{1}{0.012355}\right)(200) = 16{,}188 \text{ K}$$

The total temperature remains unchanged after the shock, but the total pressure decreases as dictated by the shock strength:

$$p_{t2} = \frac{p_{t2}}{p_{t1}} p_{t1} = \left(\frac{p_{te}}{p_{ti}}\right)_{@M_3} p_{t1} = (1.0714 \times 10^{-4})(4{,}810) = 0.515 \text{ MPa}$$

The flow after the shock is isentropic; therefore, the stagnation pressure must be the same as the total pressure:

$$p_{s2} = p_{t2} = 0.515 \text{ MPa}; \quad T_{s2} = T_{t2} = T_{t1} = 16{,}188 \text{ K}$$

TEST Analysis

Launch the gas dynamics TESTcalc and select air as the working fluid. Evaluate the states, as described in the TEST-code (see *TEST* > TEST-codes), to verify the manual results.

Discussion

The high values of stagnation pressure and temperature explain why most meteorites disintegrate upon entering the earth's atmosphere. At such extreme conditions, dissociation of oxygen and nitrogen is very likely, and would alter the gas properties, so the results presented in this example should be considered approximate.

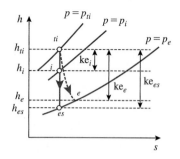

FIGURE 15.39 Actual and isentropic nozzle on a *h-s* diagram (see Anim. 15.B.*nozzleEfficiency*).

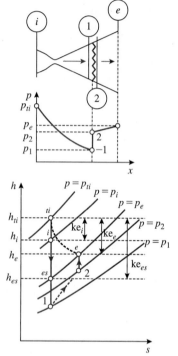

FIGURE 15.40 A normal shock in the diverging section rendering the supersonic nozzle non-isentropic (see Anim. 15.B.*nozzleShock*).

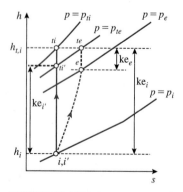

FIGURE 15.41 Various states used to define the diffuser efficiency on a *h-s* diagram (see Anim. 15.B. *diffuserEfficiency*).

15.8 NOZZLE AND DIFFUSER COEFFICIENTS

In analyzing nozzles and diffusers in Sec. 4.2.2, we used isentropic flow through these devices as the basis for defining device efficiencies. Having gained new insight into isentropic flow through converging-diverging passages, review these devices and refine some of the concepts developed in Chapter 4.

We begin with the isentropic efficiency of a nozzle, which was defined by Eq. (4.22) as the ratio of the exit kinetic energy of an actual nozzle to that of an isentropic nozzle operating under the same inlet conditions and exit pressure. In terms of total properties, the nozzle efficiency can be redefined (see Anim. 15.B.*nozzleEfficiency*) using Eqs. (15.3) and (15.5) as:

$$\eta_{\text{nozzle}} \equiv \frac{\text{ke}_e}{\text{ke}_{es}} = \frac{j_i - h_e}{j_i - h_{es}} = \frac{h_{ti} - h_e}{h_{ti} - h_{es}} \tag{15.41}$$

In this equation, the subscript *es* symbolizes the isentropic exit state and *t* represents total properties. By definition (Sec. 4.2.2), the isentropic exit state is at the same pressure as the actual exit state. The difference between the above definition and Eq. (4.22) is that by replacing h_i with h_{ti}, we no longer have to neglect the inlet kinetic energy.

Different enthalpies appearing in Eq. (15.41) are shown in the *h-s* diagram of Figure 15.39 for a nozzle. In this diagram, static (thermodynamic) states *i, e,* and *es* represent the inlet, actual exit, and the isentropic exit states. The isentropic stagnation state (composed of all total properties), for any state in the isentropic nozzle, is represented by State-*ti*. The exact location of State-*ti* depends on the inlet kinetic energy; and State-*e* depends on the irreversibilities in the nozzle. The actual and ideal (isentropic) exit kinetic energies—ke$_e$ and ke$_{es}$, as well as the inlet kinetic energy ke$_i$—are graphically interpreted as the difference between total enthalpy h_{ti} at the inlet and static (thermodynamic) enthalpy at the exit, isentropic exit, and inlet states, respectively. As the irreversibilities increase, the exit state will move further to the right along the $p = p_e$ line, reducing ke$_e$ and, hence, the nozzle efficiency. Inclusion of ke$_i$ in the definition increases both the numerator and denominator and, therefore, does not affect the nozzle efficiency significantly.

Irreversibilities in a nozzle are largely due to the viscous friction in the boundary layer next to the wall. That is why the nozzle efficiency, which typically ranges from 90% to 99%, is sensitive to the nozzle shape. By carefully shaping the contour, irreversibilities in the boundary layer can be considerably reduced, improving the nozzle efficiency. Obviously, the presence of an irreversible shock will drastically reduce the nozzle efficiency. In an otherwise isentropic nozzle, the presence of a shock is indicated by State-1 and State-2 in Figure 15.40. The increase in pressure is accompanied by an increase in entropy, moving the exit state to the right of the isentropic state *es*. However, the divergent section after the shock acts as a diffuser and the exit pressure becomes much higher than the design pressure (see Anim. 15.B.*nozzleShock*), drastically reducing the exit kinetic energy, and the nozzle efficiency.

Two other coefficients, the velocity coefficient C_V and the discharge coefficient C_D, are also used to compare actual nozzle performance with the corresponding isentropic nozzle. The velocity coefficient is defined as the ratio of actual to isentropic exit velocity and can be related to the nozzle efficiency:

$$C_V \equiv \frac{V_e}{V_{es}} = \sqrt{\frac{V_e^2}{V_{es}^2}} = \sqrt{\frac{\text{ke}_e}{\text{ke}_{es}}} = \sqrt{\eta_{\text{nozzle}}} \tag{15.42}$$

The discharge coefficient C_D, on the other hand, compares the mass flow rates through an actual nozzle to that in an isentropic nozzle:

$$C_D \equiv \frac{\dot{m}_e}{\dot{m}_{es}} \tag{15.43}$$

The isentropic mass flow rate is calculated following the procedure described in Section 15.5. For subsonic flows, the back pressure is assumed to prevail at the exit. However, if the nozzle is choked, regardless of the back pressure, the flow is assumed to be isentropic until the exit and $\dot{m}_{es}$ is calculated from Eq. (15.28).

The diffuser efficiency, defined by Eq. (4.26), can also be refined by not neglecting the exit kinetic energy, which can have a significant contribution towards the exit flow energy. In the accompanying *h-s* diagram (Fig. 15.41), State-*i* represents the inlet state and State-*ti* represents the isentropic stagnation (total) state so that ke_i is simply the vertical distance between the two states. Likewise, State-e (the actual exit static state) and State-te (the corresponding isentropic stagnation or total state) at the exit are separated by ke_e. Now consider State-i', an ideal inlet state which is at the same static (thermodynamic) state as State-i, but with a different inlet kinetic energy $\text{ke}_{i'}$, such that the corresponding total state, State-ti', is isentropic to State-i or State-i' and has the same total pressure (and velocity) as the actual exit state so that $p_{ti'} = p_{te}$. State-i', therefore, carries the minimum kinetic energy to produce the same exit total pressure as the actual diffuser. The diffuser efficiency, which compares the ideal inlet kinetic energy to the actual inlet kinetic energy to produce the same total exit pressure (see Anim. 15.B.*diffuserEfficiency*), can be expressed:

$$\eta_{\text{diffuser}} \equiv \frac{\text{ke}_{i'}}{\text{ke}_i} = \frac{j_{ti'} - h_i}{j_i - h_i} = \frac{h_{ti'} - h_i}{h_{ti} - h_i} \tag{15.44}$$

For an isentropic diffuser, all the stagnation states in Figure 15.41 converge to State-ti and the diffuser efficiency becomes 100%, as expected. Typical diffuser efficiency ranges from 90% to close to 100%. In addition to viscous friction, formation of a shock wave can adversely affect the diffuser efficiency.

The loss of stagnation pressure in a diffuser is quantified by the **pressure recovery factor** defined as:

$$F_p \equiv \frac{p_{te}}{p_{ti}} \tag{15.45}$$

For a perfect gas, it is left as an exercise to show that the diffuser efficiency can be related to the pressure recovery factor and the inlet Mach number through:

$$\eta_{\text{diffuser}} = \frac{[1 + M_i^2(k - 1)/2](F_p)^{(k-1)/k} - 1}{M_i^2(k - 1)/2} \tag{15.46}$$

There cannot be any loss of stagnation pressure in an isentropic flow; therefore, a pressure recovery factor of one implies a diffuser efficiency of 100% and vice versa. This can be verified from Eq. (15.46).

One drawback in the way the diffuser efficiency is defined is that a straight duct carrying an isentropic flow has a diffuser efficiency of 100%, even though such a device is completely useless in raising the static pressure, the purpose of a diffuser. A pressure recovery factor of unity is also misleading for the same reason. That is why a separate parameter, called the **pressure rise coefficient** C_p (not to be confused with c_p, the specific heat at constant pressure) is used to measure the effectiveness of a diffuser. It is defined as the ratio of the pressure rise in the actual diffuser to that in an isentropic diffuser where the exit state is the isentropic stagnation (total) state. Using the same symbols as in Figure 15.41, C_p can be expressed:

$$C_p = \frac{p_e - p_i}{p_{ti} - p_i} \tag{15.47}$$

For a straight isentropic duct, C_p is zero because the inlet and exit states are identical. For most diffusers, the value of C_p is less than 0.8 even if the efficiency is close to 100%. While the diffuser efficiency measures the degree of irreversibilities, the pressure recovery factor indicates the effectiveness of the diffuser in converting kinetic energy into pressure component pv of the flow energy (recall $j = u + pv + \text{ke} + \text{pe}$).

EXAMPLE 15-10 **Non-isentropic Nozzle Analysis**

Carbondioxide, at 250 kPa, 600 K, enters a 90% efficient adiabatic converging nozzle with a velocity of 50 m/s. Determine the exit (a) velocity, (b) temperature, (c) velocity coefficient, and (d) discharge coefficient if the exit area is 10 cm^2 and the back pressure is 100 kPa.

SOLUTION

Determine if the nozzle is choked and, accordingly, obtain the isentropic solution for the exit state, then use the nozzle efficiency to find the actual exit conditions.

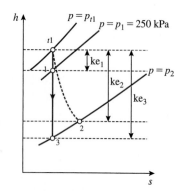

FIGURE 15.42 *h-s diagram* showing different principal states of the nozzle of Ex. 15-10.

Assumptions

One-dimensional, adiabatic flow of a perfect gas.

Analysis

Different principal states of the nozzle are shown in Figure 15.42. State-1 is the inlet flow state with its total properties represented by a separate (total) state, State-$t1$. State-2 is the actual exit state. And State-3 is the isentropic exit state at the pressure of the exit (p_2). From the PG table (Table C-1), or the gas dynamics TESTcalc, obtain the gas properties for CO_2.

State-1 (given p_1, T_1, V_1):

$$M_1 = \frac{V_1}{\sqrt{(1000 \text{ N/kN})kRT_1}} = \frac{50}{\sqrt{(1000)(1.289)(0.189)(600)}} = 0.131$$

Using the isentropic table or (gas dynamics TESTcalc),

$$T_{t1} = 601.5 \text{ K}; \quad p_{t1} = 252.8 \text{ kPa}$$

State-2, -3 (given A_2, p_2): First check if the back pressure is low enough to choke the nozzle. To do so, assume $M_2 = 1$, determine the expected exit pressure, and compare it with the actual back pressure, 100 kPa.

Using the energy equation, $T_{t1} = T_{t2}$, which is also valid for a non-isentropic nozzle, T_2 can be obtained with the help of the isentropic table:

$$\frac{T_2}{T_{t2}} = \left(\frac{T}{T_t}\right)_{M_2=1} = 0.874; \quad \Rightarrow \quad T_2 = 0.874(T_{t1}) = (0.874)(601.5) = 525.7 \text{ K}$$

However, to evaluate p_2, we must evaluate the isentropic exit state, State-3, first and use the fact that $p_2 = p_3$. Toward that goal, T_3 is first determined from the known isentropic efficiency:

$$\eta_{\text{nozzle}} = \frac{h_{t1} - h_2}{h_{t1} - h_3} = \frac{T_{t1} - T_2}{T_{t1} - T_3}; \quad \Rightarrow \quad T_3 = T_{t1} - (T_{t1} - T_2)/0.9 = 517.5 \text{ K}$$

State-3 is isentropic to State-1; therefore, $p_{t3} = p_{t1} = 252.8$ kPa. The energy equation also produces $T_{t3} = T_{t1} = 601.5$ K. The isentropic exit state now can be determined as follows:

$$\frac{T_3}{T_{t3}} = \left(\frac{T}{T_t}\right)_{M_3} = 0.86; \quad \Rightarrow \quad M_3 = 1.063;$$

$$p_3 = \frac{p_3}{p_{t3}}p_{t3} = \left(\frac{p}{p_t}\right)_{M_3}p_{t3} = (0.51)(252.8) = 128.9 \text{ kPa}$$

$$V_3 = M_3\sqrt{(1000 \text{ N/kN})kRT_3} = 377.0 \text{ m/s}$$

The actual and isentropic exit states share the same pressure; therefore, $p_2 = 128.9$ kPa for an exit Mach number of unity.

Because the back pressure, 100 kPa, is less than the calculated exit pressure p_2, the nozzle is choked, and State-2 is the actual exit state. From the known exit temperature, the exit velocity now can be calculated as:

$$T_2 = 525.7 \text{ K}; \quad V_2 = M_2\sqrt{(1000 \text{ N/kN})kRT_2} = 357.7 \text{ m/s}$$

The velocity coefficient can be obtained from Eq. (15.42) or directly from the values of V_2 and V_3 as:

$$C_V \equiv \frac{V_2}{V_3} = \sqrt{\eta_{\text{nozzle}}} = 0.949$$

To obtain the discharge coefficient, the mass flow rates at State-2 and -3 are calculated first:

$$\dot{m}_2 = \rho_2 A_2 V_2 = \frac{p_2}{RT_2}A_2V_2 = \frac{(128.9)(0.001)(357.7)}{(0.189)(525.7)} = 0.464 \frac{\text{kg}}{\text{s}}$$

$$\dot{m}_3 = \rho_3 A_3 V_3 = \frac{p_3}{RT_3}A_2V_3 = \frac{(128.9)(0.001)(377.0)}{(0.189)(517.5)} = 0.497 \frac{\text{kg}}{\text{s}}$$

Therefore,

$$C_D = \frac{\dot{m}_2}{\dot{m}_3} = \frac{0.464}{0.497} = 0.933$$

TEST Solution

Launch the gas dynamics TESTcalcs and select Air as the working fluid. Calculate States 1, 2, and 3 as described in the TEST-code (see *TEST* > TEST-codes). State-2 is fully evaluated only after State-3 is found. The velocity coefficient and discharge coefficients are calculated in the I/O panel to confirm the manual results.

Discussion

If the nozzle is not choked, p_3 will be found to be less than the back pressure, 100 kPa. In that case, we have to reject $M_3 = 1$, and repeat the isentropic calculations with $p_3 = 100$ kPa. Note that the definition of isentropic efficiency requires the pressure of the isentropic exit state to be the same as the actual exit pressure. Although the actual nozzle is convergent, the hypothetical isentropic nozzle may be convergent-divergent type as in this case with $M_3 > 1$.

EXAMPLE 15-11   Non-isentropic Diffuser Analysis

Air, at 100 kPa, 300 K, enters a 90% efficient adiabatic diffuser with a velocity of 300 m/s. If the exit velocity is 50 m/s, determine (a) the exit pressure, (b) pressure recovery factor, (c) pressure rise coefficient, and (d) exit-to-inlet area ratio. *What-if scenario:* (e) What would the answers be if the exit velocity was 75 m/s?

SOLUTION

Different principal states of the diffuser are shown in Figure 15.43. States 1 and 2 are the actual inlet and exit states, with t1 and t2 denoting the corresponding total states. State-3 is the ideal inlet state (with reduced kinetic energy) such that the corresponding total state has the same total pressure at the exit state, that is, $p_{t3} = p_{t2}$. The diffuser efficiency is the ratio of ke_3 to ke_1, which relates T_{t3}, T_{t1}, and T_1.

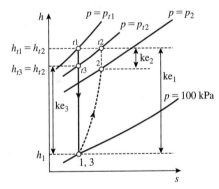

FIGURE 15.43 *h-s* diagram for Ex. 15-11.

Assumptions

One-dimensional, adiabatic flow of a perfect gas.

Analysis

For the inlet state, State-1:

$$M_1 = \frac{V_1}{\sqrt{(1000 \text{ N/kN})kRT_1}} = \frac{300}{\sqrt{(1000)(1.4)(0.287)(300)}} = 0.864$$

Using the isentropic table (or the gas dynamics TESTcalc):

$$T_{t1} = T_1\left(\frac{T_1}{T_{t1}}\right)^{-1} = T_1\left[\left(\frac{T}{T_t}\right)_{@M_1}\right]^{-1} = (300)(0.870)^{-1} = 344.8 \text{ K};$$

$$p_{t1} = p_1\left(\frac{p_1}{p_{t1}}\right)^{-1} = p_1\left[\left(\frac{p}{p_t}\right)_{@M_1}\right]^{-1} = (100)(0.6144)^{-1} = 162.76 \text{ kPa}$$

Irrespective of the presence of friction, the energy equation, Eq. (15.10), for the diffuser can be written as $T_{t1} = T_{t2}$, where State-2 is the actual exit state. With the exit velocity given, the static temperature at the exit can be obtained from Eq. (15.8) as:

$$T_2 = T_{t2} - ke_2 = T_{t2} - \frac{V^2}{2(1000 \text{ J/kJ})c_p} = 344.8 - \frac{50^2}{2000(1.005)} = 343.56 \text{ K}$$

The Mach number and the static-to-total pressure ratio at the exit now can be calculated:

$$M_2 = \frac{V_2}{\sqrt{(1000 \text{ N/kN})kRT_2}} = 0.135; \quad \Rightarrow \quad \frac{p_2}{p_{t2}} = \left(\frac{p}{p_t}\right)_{@M_2} = 0.987$$

Because the flow is not isentropic in the diffuser, $p_{t2} < p_{t1}$. To obtain $p_{t2} = p_{t3}$, we first use the diffuser efficiency to get T_{t3}:

$$\eta_{\text{diffuser}} = \frac{ke_3}{ke_1} = \frac{h_{t3} - h_3}{h_{t1} - h_1} = \frac{T_{t3} - T_1}{T_{t1} - T_1}$$

$$\Rightarrow \quad T_{t3} = 300 + 0.9(344.8 - 300) = 340.3 \text{ K}$$

The total pressure p_{t2} can be calculated from η_{diffuser} and p_{t1} using Eq. (15.46). A more fundamental approach is to use the isentropic relation, Eq. (3.71), between State-3 and State-$t3$ to obtain:

$$p_{t2} = p_{t3} = \frac{p_{t3}}{p_3}p_3 = p_3\left(\frac{T_{t3}}{T_3}\right)^{k/(k-1)} = p_1\left(\frac{T_{t3}}{T_1}\right)^{k/(k-1)} = 155.5 \text{ kPa};$$

Therefore,

$$p_2 = \frac{p_2}{p_{t2}}p_{t2} = \left(\frac{p}{p_t}\right)_{@M_2 = 0.135} p_{t2} = (0.987)(155.5) = 153.5 \text{ kPa}$$

The pressure recovery factor and pressure rise coefficient can now be calculated:

$$F_p \equiv \frac{p_{t2}}{p_{t1}} = \frac{155.5}{162.76} = 0.955 \quad \text{and} \quad C_p = \frac{p_2 - p_1}{p_{t1} - p_1} = \frac{153.5 - 100}{162.76 - 100} = 0.85$$

Due to frictional losses in the diffuser, the critical area between the inlet and exit must be different. The mass flow rate expression of Eq. (15.28), when applied to the inlet and exit, yields:

$$A_{*1}p_{t1} = A_{*2}p_{t2}; \quad \Rightarrow \quad \frac{A_{*1}}{A_{*2}} = \frac{p_{t2}}{p_{t1}} = F_p = 0.955; \qquad \textbf{(15.48)}$$

Therefore, the area ratio between the exit and inlet can be calculated by using the isentropic table as:

$$\frac{A_2}{A_1} = \frac{A_2}{A_{*2}}\frac{A_{*2}}{A_{*1}}\frac{A_{*1}}{A_1} = \left(\frac{A}{A_*}\right)_{@M_2}\left(\frac{A_*}{A}\right)_{@M_1}\frac{A_{*2}}{A_{*1}} = (4.35)\left(\frac{1}{1.017}\right)\left(\frac{1}{0.955}\right) = 4.48 \quad \textbf{(15.49)}$$

TEST Solution

Launch the gas dynamics TESTcalcs and select Air as the working fluid. Calculate States 1, 2, and 3 as described in the TEST-code (see *TEST* > TEST-codes). Although A1 is assumed to be 1 m^2, the answers can be verified to be independent of the value of A1.

What-if scenario

Change Vel2 to 75 m/s and click Super-Calculate. The new p2 is found to be 151.1 kPa. While the pressure recovery factor remains unchanged, the pressure rise factor is calculated in the I/O panel as 0.814 and the area ratio as A2/A1 = 3.02.

Discussion

The exit velocity can be seen to have a significant effect on the shape and pressure rise factor of a diffuser. Therefore, the exit velocity should not be neglected in a diffuser analysis, as in Example 4-6.

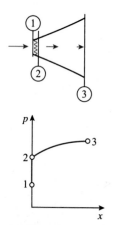

FIGURE 15.44 Schematic for Ex. 15-12.

EXAMPLE 15-12　Supersonic Diffuser with Normal Shock

Air at 10 kPa approaches a supersonic diffuser at Mach 3.0. A normal shock stands at the inlet, as shown in Figure 15.44. If $A_e/A_i = 3$, determine (a) the exit Mach number, (b) the exit pressure, and (c) the pressure recovery factor. Assume the flow is isentropic after the shock.

SOLUTION

The three principal states of the flow, shown in Figure 15.44, are redrawn in the h-s diagram of Figure 15.45, along with the corresponding total states. Evaluate the flow states, State-1 and State-2, before and after the normal shock. While there is a drop in total pressure from State-1 to State-2 ($p_{t2} < p_{t1}$), caused by the normal shock, the flow is isentropic thereafter, so that $p_{t2} = p_{t3}$. Do an isentropic analysis for the flow between State-2 and State-3 where the diverging section works as a subsonic diffuser.

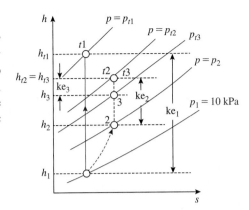

FIGURE 15.45 h-s diagram for Ex. 15-12.

Assumptions

One-dimensional, isentropic flow of a perfect gas after the shock.

Analysis

The total pressure at the inlet, State-1, can be found from the isentropic table:

$$h_{t1} = h_{t2}; \quad p_{t1} = p_1\left(\frac{p_1}{p_{t1}}\right)^{-1} = p_1\left[\left(\frac{p}{p_t}\right)_{@M_1=3}\right]^{-1} = 10(0.02722)^{-1} = 367.38 \text{ kPa}$$

Using the shock table or the gas dynamics TESTcalc, State-2 can be obtained:

$$p_{t2} = p_{t1}\frac{p_{t2}}{p_{t1}} = p_{t1}\left(\frac{p_{te}}{p_{ti}}\right)_{@M_1=3} (367.38)(0.3282) = 120.57 \text{ kPa}; \quad M_2 = 0.475$$

The flow is isentropic downstream of State-2. Using the mass equation for an isentropic flow between State-2 and -3, $A_{*3} = A_{*2}$, we obtain:

$$\left(\frac{A}{A_*}\right)_{@M_3} = \frac{A_3}{A_{*3}} = \frac{A_3}{A_{*2}} = \frac{A_3}{A_2}\frac{A_2}{A_{*2}} = \frac{A_3}{A_1}\frac{A_2}{A_{*2}} = (3)\left(\frac{A}{A_*}\right)_{@M_1 = 0.475} = 3(1.39) = 4.17$$

Using the isentropic table we get:

$$M_3 = 0.14; \quad \text{and} \quad p_3 = p_{t3}\frac{p_3}{p_{t3}} = p_{t2}\left(\frac{p}{p_t}\right)_{@M_3} = (120.57)(0.986) = 118.9 \text{ kPa}$$

Therefore, $F_p \equiv \dfrac{p_{t3}}{p_{t1}} = \dfrac{p_{t2}}{p_{t1}} = \dfrac{120.57}{367.38} = 0.328$

TEST Solution

Launch the gas dynamics TESTcalcs and select Air as the working fluid. Calculate State-1 partially from p1, M1, and A1=1 m^2 (assumed). Obtain M2 and p_t2/p_t1 from the shock table in the table panel. Calculate State-2 with M2, p_t2 = 0.328*p_t1, and A2 = A1. Finally, evaluate State-3 with A3 = 3*A2, p_t3 = p_t2 (entropy equation), and Astar3 = Astar2 (mass equation). See TEST-code (posted in *TEST* > TEST-codes) for details.

Discussion

With the pressure recovery factor evaluated, the isentropic efficiency can be calculated from Eq. (15.46). Obviously, the presence of the shock is the main reason behind the poor performance of this diffuser.

15.9 CLOSURE

In this chapter we refined and customized the steady-state analysis of open systems (introduced in Chapter 4), for high-speed flows of gases through variable area ducts—mostly nozzles and diffusers. Several new extrinsic properties of a flow state, the stagnation and total properties of a flow, the Mach number, and critical area were introduced. Application of the PG model led to simplified formulation of the mass, energy, and entropy balance equations for a one-dimensional gas flow. For an isentropic one-dimensional flow of a perfect gas, the total temperature, total pressure, and critical area were shown to remain constant along the flow. The non-dimensional

Mach number was shown to control the flow behavior in converging and converging-diverging nozzles and diffusers. Solution procedures for isentropic flows were developed using the isentropic relations or the isentropic table. The effect of back pressure on a converging and converging-diverging nozzle was studied, leading to the concept of a choked flow and the phenomenon of normal shock. Properties before and after a normal shock were related by subjecting a shock wave to mass, energy, entropy, and momentum analysis. Finally gas flow through nozzles and diffusers, introduced in Chapter 4, was revisited by applying gas dynamics fundamentals.

PROBLEMS

SECTION 15-1: STATIC AND TOTAL PROPERTIES

15-1-1   Air flows at Mach 0.2 through a circular duct that has internal diameter of 50 cm. The total pressure of the flow is 500 kPa and the total temperature is 200°C. (a) Calculate the mass flow rate ($\dot{m}$) through the channel. (b) *What-if scenario:* What would the mass flow rate be if the gas was helium?

15-1-2   Determine (a) the velocity (V) of sound, (b) the Mach number, (c) the total temperature, (T_t) and (d) the total pressure (p_t) of air that is flowing at 50 kPa, 250 K and 500 m/s. (e) *What-if scenario:* What would the Mach number be if the gas was helium?

15-1-3   A subsonic airplane is flying at an altitude of 3500 m, where the atmospheric conditions are 72 kPa and 260 K. A pitot tube measures the difference between the static and total pressures as 10 kPa. Determine (a) the speed of the airplane and (b) the flight Mach number.

15-1-4   Saturated steam, at 200 kPa, is flowing with a velocity of 500 m/s. A pitot tube brings the flow to stagnation and the stagnation pressure is measured as 350 kPa. Determine (a) the total pressure (p_t), (b) the total temperature (T_t), and (c) the stagnation temperature. Use the PC model.

15-1-5   Air flowing, at 100 kPa and 298 K, is brought to rest, and the stagnation pressure and temperature are measured as 130 kPa and 329 K, respectively. Determine (a) the flow velocity and (b) the total pressure (p_t) of the flow.

15-1-6   Determine (a) the back pressure necessary for a normal shock to appear at the exit of a converging-diverging nozzle, with an exit to throat area ratio of 2, if the reservoir conditions are 1 MPa and 850 K. (b) *What-if scenario:* What would the back pressure be if the working gas was helium?

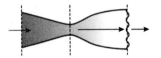

15-1-7   Carbon dioxide enters an adiabatic nozzle, at 1200 K with a velocity of 100 m/s, and leaves at 500 K. Determine the Mach number (a) at the inlet and (b) at the exit of the nozzle.

15-1-8   Air enters an adiabatic nozzle, at 1200 K with a velocity of 100 m/s, and leaves at 500 K. Determine (a) the Mach number at the inlet, (b) the velocity (V_2), and (c) the Mach number at the exit of the nozzle.

15-1-9   Determine the velocity of sound in air (a) at a temperature of 273 K and (b) at 1000 K.

15-1-10   Determine the velocity of sound in steam at 800 kPa and 350°C. Assume a (a) 1% variation of temperature across the wave and using the PC model, (b) assume steam to behave as a perfect gas. Use the PC flow-state TESTcalc to evaluate steam properties.

15-1-11   Determine the velocity of sound in steam at 800 kPa and 350°C assuming (a) 1% variation of pressure across the wave and using the PC model, (b) steam to behave as a perfect gas. Use the PC flow-state TESTcalc to evaluate steam properties.

15-1-12   Refrigerant R-134a flows at 100 m/s at the exit of a nozzle where the pressure and temperature are 500 kPa and 30°C, respectively. Determine the Mach number of the flow state. Use the PC model for R-134a and assume a 1% variation in temperature across a sound wave.

15-1-13   A needle nose projectile, traveling at a speed of $M = 2$, passes 250 m above the observer. Determine (a) the projectile's velocity (V) and (b) how far beyond the observer the projectile will first be heard. Assume the static temperature as 15°C.

15-1-14   Air enters a diffuser with a velocity of 180 m/s and an inlet temperature of 303 K. Determine (a) the velocity of sound and (b) the flow Mach number at the diffuser inlet. (c) *What-if scenario:* What would the flow Mach number at the diffuser inlet be if the gas was nitrogen?

15-1-15   Air flows through a device such that the total pressure is 700 kPa. The total temperature is 300°C and the velocity is 500 m/s. Determine (a) the static pressure (p_s) and (b) the static temperature (T_s) of air at this state. (c) *What-if scenario:* What would the static temperature be if the gas was oxygen?

15-1-16   Air leaves a compressor in a pipe with a total temperature and pressure of 180°C, 350 kPa and a velocity of 150 m/s. The pipe has a cross sectional area of 0.02 m². Determine (a) the static pressure (p_s), (b) the static temperature (T_s), and (c) the mass flow rate ($\dot{m}$). (d) *What-if scenario:* What would the mass flow rate be if the gas was hydrogen?

15-1-17 Determine the total temperature for the following substances flowing through a duct: (a) helium at 180 kPa, 40°C, and 250 m/s; (b) nitrogen at 180 kPa, 40°C, and 250 m/s; and (c) steam at 3 MPa, 300°C, and 400 m/s.

15-1-18 In Problem 15-1-17 determine the total pressure in each case.

15-1-19 Steam, at 1.2 MPa and 450°C, flows through a pipe with a velocity of 300 m/s. Treating the super-heated steam as a perfect gas, determine (a) the velocity of sound and (b) the flow Mach number for the steam. (c) *What-if scenario:* What would the flow Mach number for the steam be if the velocity of steam through the pipe was 480 m/s?

15-1-20 Steam, at 250°C and quality 95%, is flowing through a duct with a velocity of 250 m/s. Determine the total properties, (a) temperature (T_t), (b) pressure (p_t), (c) quality and (d) density. Use the PC model for steam. (e) *What-if scenario:* What would the temperature be if the steam was treated as a perfect gas?

15-1-21 Air flows in a duct at a pressure of 200 kPa with a velocity of 200 m/s. The temperature of the air is 300 K. Determine (a) the total pressure (p_t) and (b) the total temperature (T_t).

15-1-22 Determine (a) the total temperature (T_s) and (b) the total pressure (p_t) of air that is flowing at 40 kPa, −25°C and 400 m/s.

15-1-23 Nitrogen is discharged from a large reservoir, at 500 K and 150 kPa, through an adiabatic nozzle. At the exit, the pressure is 100 kPa, the area is 10 cm², and the Mach number is 0.7. Determine (a) the exit temperature (T_2), (b) the mass flow rate $(\dot{m})$, and (c) the total pressure (p_2) at the exit. (d) Is the flow through the nozzle isentropic?

15-1-24 Repeat Problem 15-1-23 with carbon dioxide as the working fluid.

15-1-25 An aircraft is cruising with a velocity of 1000 km/h and at an altitude of 10 km, where the static temperature and pressure are −50°C and 26.5 kPa. Determine (a) the Mach number of the air-craft. Also determine (b) the pressure (p) and (c) temperature (T) of the air brought to rest isentropically by a diffuser. (d) *What-if scenario:* How would the answer in part (c) change if the diffuser was adiabatic but irreversible?

15-1-26 An aircraft is cruising with a velocity of 1000 km/h and at an altitude of 10 km, where the static temperature and pressure are −50°C and 26.5 kPa. Determine (a) the Mach number of the air-craft. Also determine (b) the pressure (p) and (c) temperature (T) of the air brought to rest isentropically by a diffuser. (d) *What-if scenario:* How would the answer in part (c) change if the diffuser was adiabatic but irreversible?

15-1-27 Steam, was at 450°C and 1 MPa, is flowing through a nozzle with a velocity of 300 m/s. Determine (a) the total temperature (T_t) and (b) pressure (p_t) of the flow. Use the PG model for steam. (c) *What-if scenario:* What would the total temperature and pressure of the flow be if the PC model was used?

SECTION 15-2: ISENTROPIC FLOWS

15-2-1 Air, at 1 MPa and 550°C, enters a converging nozzle with a throat area 50 cm². The air enters with a velocity of 100 m/s. Determine the mass flow rate for a back pressure of (a) 0.7 MPa, (b) 0.4 MPa, and (c) 0.2 MPa.

15-2-2 Air enters a converging-diverging nozzle at 700 K and 1000 kPa with negligible velocity. The exit Mach number is 2 and the throat area is 20 cm². Assuming steady isentropic flow, determine (a) the throat velocity, (b) mass flow rate $(\dot{m})$, and (c) the exit area.

15-2-3 Stationary nitrogen, at 0.9 MPa and 450 K, is accelerated isentropically to a Mach number of 0.6. Determine (a) the temperature (T) and the pressure (p) of nitrogen after acceleration.

15-2-4 Air is expanded in an isentropic nozzle from 1.0 MPa and 800 K to an exhaust pressure of 100 kPa. If air enters the nozzle with a velocity of 80 m/s, determine the exhaust velocity.

15-2-5 Helium enters a converging-diverging nozzle at a pressure of 800 kPa, 700 K and, 100 m/s. Determine (a) the lowest temperature and (b) the lowest pressure that can be obtained at the throat of the nozzle. (c) *What-if scenario:* What would the lowest pressure in (b) be if the working gas was air?

15-2-6 Nitrogen flows steadily through a variable-area duct with a mass flow rate of 3 kg/s. It enters the duct at 1200 kPa and 250°C with a low velocity and expands to a pressure of 300 kPa. The duct is designed so that the flow can be approximated as isentropic. Determine (a) the Mach number, (b) the flow area, and (c) the velocity at the exit and the throat.

15-2-7 A convergent nozzle has an exit area of 4 cm². Air enters the nozzle with a total pressure of 1200 kPa and a total temperature of 400 K. Assuming isentropic flow, determine the mass flow rate ($\dot{m}$) for a back pressure of (a) 900 kPa, (b) 634 kPa, and (c) 400 kPa.

15-2-8 Air enters a nozzle at 3000 kPa, 400 K and a velocity of 180 m/s. Assuming isentropic flow, determine (a) the temperature (T), (b) the pressure (p) of the air at a location where the air velocity equals the velocity of sound, and (c) the ratio of area at this location to the entrance area. (d) *What-if scenario:* What would the temperature in (a) be if air entered the nozzle at 200 m/s?

15-2-9 An ideal gas with $k = 1.5$ is flowing through a nozzle so that the Mach number is 3 where the flow area is 30 cm². Assuming the flow to be isentropic, determine (a) the flow area at the location where the Mach number is 1.4.

15-2-10 Air enters a converging-diverging nozzle of a supersonic wind tunnel at 1000 kPa and 35°C with a low velocity. The flow area of the test section is equal to the exit area of the nozzle, which is 0.5 m². Determine (a) the pressure (p), temperature (T), (b) velocity (V), and (c) mass flow rate ($\dot{m}$) in the test section for a $M = 2$.

15-2-11 Compressed air is discharged through a converging nozzle, as shown in the accompanying figure. The conditions in the tank are 1 MPa and 500 K, while the outside pressure is 100 kPa. The inlet area of the nozzle is 100 cm² and the exit area is 35 cm². Determine (a) the exit velocity, (b) the exit temperature, and (c) the force of the air on the nozzle. Assume the conditions inside to remain unchanged during the discharge.

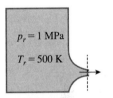

15-2-12 For the converging-diverging nozzle, shown in the accompanying figure, (a) find the maximum back pressure below which the flow is choked and (b) the mass flow rate for a choked nozzle. The throat area is 10 cm² and the exit area is 40 cm².

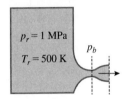

15-2-13 A converging-diverging nozzle with an exit area of 35 cm² and a throat area of 10 cm² is attached to a reservoir which contains air at 700 kPa and 20°C absolute. Determine (a) the two exit pressures that result in $M = 1$ at the throat for an isentropic flow and (b) the associated exit temperatures and (c) velocities.

15-2-14 A rocket motor is fired on a test stand. Hot exhaust gases leave the exit with a velocity of 700 m/s at a mass flow rate of 10 kg/s. The exit area is 0.01 m² and the exit pressure is 50 kPa. For an ambient pressure of 100 kPa, determine the rocket motor thrust that is transmitted to the stand. Assume steady state and one-dimensional flow.

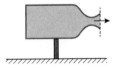

15-2-15 A rocket nozzle has an exit-to-throat area ratio of 4.0 and a throat area of 100 cm². The exhaust gases are generated in a combustion chamber with stagnation pressure equal to 4 MPa and a stagnation temperature equal to 2000 K. Assume the working fluid behaves as a perfect gas with $k = 1.3$ and molar mass = 20 kg/kmol. Determine (a) the rockets exhaust velocity and (b) the mass flow rate. Assume isentropic steady flow.

15-2-16 In Problem 15-2-15, determine the thrust if the outside pressure is 0 kPa.

15-2-17 On a test stand, a nozzle operates isentropically with a chamber pressure of 2 MPa and chamber temperature of 2500 K. If the products of combustion are assumed to behave as a perfect gas, with $k = 1.3$ and molar mass = 20 kg/kmol, determine the rocket motor thrust that is transmitted to the stand. Assume the nozzle exit area to be 0.01 m² and the back pressure to be 50 kPa.

15-2-18 Nitrogen enters a duct with varying flow area at 500 K, 100 kPa and Mach number 0.4. Assuming a steady isentropic flow, determine (a) the Mach number, (b) the pressure (p), and (c) the temperature (T) at a location where the flow area has been reduced by 30%. (d) *What-if scenario:* What would the pressure be for a location where the flow area was reduced by 20%?

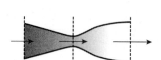

15-2-19 Products of combustion enter the nozzle of a gas turbine at the design conditions of 420 kPa, 1200 K, and 200 m/s, and they exit at a pressure of 290 kPa at a rate of 3 kg/s. $k = 1.34$ and $c_p = 1.16$ kJ/kg · k are the combustion products. Assuming an isentropic flow, determine (a) whether the nozzle is converging or converging-diverging, (b) the exit velocity and (c) the exit area.

15-2-20 Combustion products enter a nozzle with total temperature of 700 K and total pressure of 200 kPa. For a back pressure of 60 kPa and a nozzle efficiency of 90%, determine (a) the exit pressure and (b) the exit velocity. Assume the combustion gases have the properties of air, with $k = 1.33$.

15-2-21 A converging-diverging nozzle has an exit area to throat area ratio of 1.8. Air enters the nozzle with a total pressure of 1100 kPa and a total temperature of 400 K. The throat area is 5 cm². If the velocity at the throat is sonic, and the diverging section acts as a nozzle, determine (a) the mass flow rate, (b) the exit pressure and temperature, (c) the exit Mach number, and (d) the exit velocity. (e) *What-if scenario:* What would the exit pressure and temperature be if the velocity at the throat was sonic, and the diverging section acted as a diffuser?

15-2-22 A jetliner is flying at 275 m/s at a high altitude where the atmospheric pressure and temperature are 50 kPa and 250 K. Air is first decelerated in a diffuser, before it is compressed by an isentropic compressor with a compression ratio of 10. Determine (a) the pressure at the compressor inlet and (b) the compressor work per unit mass of air.

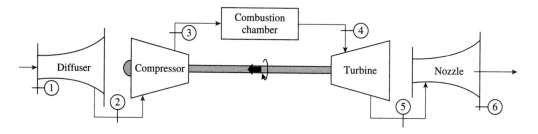

15-2-23 Carbon dioxide flows steadily through a variable-area duct with a mass flow rate of 2 kg/s. It enters the duct at 1400 kPa and 250°C with negligible velocity and expands to a pressure of 200 kPa isentropically. Determine (a) the exit Mach number, (b) flow area, and (c) the critical flow area. (d) *What-if scenario:* What would the critical flow area be if the mass flow rate of carbon dioxide was 3 kg/s?

15-2-24 A converging-diverging nozzle has a throat area of 100 mm² and an exit area of 160 mm². The inlet flow is helium at a total pressure of 1 MPa and total temperature of 375 K. Determine the back pressure that will give sonic conditions at the throat, but subsonic everywhere else.

15-2-25 Oxygen flows at Mach 0.5 in a channel with a cross-sectional area of 0.16 m². The temperature and pressure are 800 K and 800 kPa. (a) Calculate the mass flow rate through the channel. The cross-sectional area is now reduced to 0.15 m². Determine the (b) Mach number and (c) flow velocity at the reduced area. Assume the flow to be isentropic.

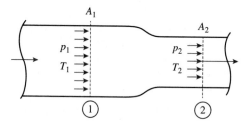

15-2-26 Air flows at Mach 0.5 in a channel with a cross-sectional area of 0.16 m². The temperature and pressure are 800 K and 800 kPa. (a) Calculate the mass flow rate ($\dot{m}$) through the channel. (b) What should the cross-sectional area be reduced to before the mass flow rate is affected? Assume the flow to be isentropic.

15-2-27 Air enters a diffuser with a velocity of 200 m/s, a static pressure of 80 kPa, and a temperature of 268 K. The velocity when leaving the diffuser is 50 m/s and the static pressure at the diffuser exit is 90 kPa. Determine (a) the static temperature (T_s) at the diffuser exit, (b) the total pressure (p_t) at the the exit, and (c) the diffuser efficiency (%).

15-2-28 Helium enters a variable-area duct at 400 K, 100 kPa, and $M = 0.3$. Assuming steady isentropic flow, determine (a) the Mach number at a location where the area is 20% smaller. (b) *What-if scenario:* What would the Mach number be if the area was 30% smaller?

15-2-29 A converging-diverging nozzle has an area ratio of 3:1 and a throat area of 50 cm². The nozzle is supplied from a tank containing helium at 100 kPa and 270 K. Find (a) the maximum mass flow rate possible through the nozzle and (b) the range of back pressures over which the mass flow can be attained. (c) *What-if scenario:* What would the answers be if hydrogen was the working fluid?

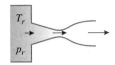

15-2-30 A symmetric converging diverging duct, with an area ratio of 2, is placed in a wind tunnel where it encounters a 300 m/s flow of air at 50 kPa, 300 K. Determine the bypass ratio (diverted flow/incoming flow).

15-2-31 In Problem 15-2-30, determine the air speed below which flow bypass is not necessary.

SECTION 15-3: NORMAL SHOCK WAVES AND OBLIQUE SHOCK WAVES

15-3-1 An airstream, with a velocity of 600 m/s, a pressure of 50 kPa, and a temperature of 250 K, undergoes a normal shock. Determine (a) the velocity (V) and (b) the pressure (p) at the exit. (c) Also, determine the loss of total pressure due to the shock. (d) *What-if scenario:* What would answers be if the inlet velocity was 1200 m/s instead?

15-3-2 Air enters a normal shock at 30 kPa, 220 K and 700 m/s. Determine (a) the total pressure and Mach number upstream of the shock, (b) pressure, temperature, velocity, Mach number and total pressure downstream of the shock.

15-3-3 A meteorite at a Mach number of 25 is entering the earth's outer atmosphere where the pressure and temperature are 1 kPa and 200 K respectively. Determine (a) the velocity (V) of the meteorite with respect to the ground, (b) the relative velocity of the meteorite with respect to the immediate surroundings, and (c) the temperature (T) to which the meteorite is subjected.

15-3-4 Air enters a converging-diverging nozzle at 700 K and 1000 kPa with negligible velocity. The exit Mach number is designed to be 2 and the throat area is 20 cm². Now suppose the air experiences a normal shock at the exit plane. Determine (a) the total pressure (p) after the shock, (b) mass flow rate ($\dot{m}$), and (c) the exit velocity.

15-3-5 Air enters a converging-diverging nozzle of a supersonic wind tunnel at 400 K and 1 MPa with a low velocity. If a normal shock wave occurs at the exit plane of the nozzle at Mach equal to 2, determine (a) the pressure (p), the temperature (T), (b) the Mach number, (c) the velocity (V), and (d) the total pressure after the shock wave. (e) *What-if scenario:* What would the Mach number be if a normal shock wave occurred at the nozzle's exit plane at Mach equal to 1.5?

15-3-6 Air enters a converging-diverging nozzle with a low velocity at 1.5 MPa and 120°C. If the exit area of the nozzle is 3 times the throat area, determine (a) the back pressure to produce a normal shock at the exit plane of the nozzle. (b) *What-if scenario:* What would the back pressure be for a normal shock to occur at a location where the cross-sectional area was twice the throat area?

15-3-7 Carbon dioxide enters a converging-diverging nozzle at 900 kPa and 900 K with a negligible velocity. The flow is steady and isentropic. For a exit Mach number of 2.5 and a throat area of 25 cm², determine (a) the exit area, (b) the mass flow rate ($\dot{m}$) and (c) the exit velocity.

15-3-8 Air enters a converging-diverging nozzle at 700 K and 1000 kPa with a velocity of 75 m/s. The exit Mach number is designed to be 2 and the throat area is 20 cm². Suppose the air experiences a normal shock at the exit plane. Determine (a) the mass flow rate ($\dot{m}$) and (b) the nozzle efficiency (%). (c) *What-if scenario:* What would the mass flow rate and nozzle efficiency be if the inlet kinetic energy was neglected (%)?

15-3-9 Consider the convergent-divergent nozzle in Problem 15-2-21 where the diverging section acts as a supersonic nozzle. Assume that a normal shock stands in the exit plane of the nozzle. Determine (a) the static pressure, temperature, and (b) total pressure just downstream of the normal shock.

15-3-10 Consider the convergent-divergent nozzle in Problem 15-3-9. Assume that there is a normal shock wave standing at the point where $M = 1.5$. Determine (a) the exit plane pressure, (b) temperature, and (c) Mach number.

15-3-11 Air flowing steadily in a nozzle experiences a normal shock at a Mach number of 2.5. If the pressure and temperature of the air are 60 kPa and 273 K, upstream of the shock, determine (a) the pressure (p), temperature (T), (b) velocity (V), (c) Mach number, and (d) the total pressure downstream of the shock. (e) *What-if scenario:* What would the velocity be for helium undergoing a normal shock under the same conditions?

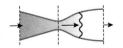

15-3-12 An airstream at Mach 2.0, with a pressure of 100 kPa and a temperature of 300 K, enters a diverging channel with a area ratio of 3 between the exit and inlet. Determine the back pressure necessary to produce a normal shock in the channel at an area equal to twice the inlet area.

15-3-13 A supersonic flow of air at $M = 3.0$ is to be slowed down via a normal shock in a diverging channel with an exit to inlet area ratio of 2. At the exit $M = 0.5$. Find the ratio of exit to inlet pressure.

15-3-14 Air approaches a diffuser with a pressure of 20 kPa at a Mach number of 2. A normal shock occurs at the inlet of the channel, as shown in the accompanying figure. For an exit to inlet area ratio of 3, find (a) the loss of total pressure and (b) the Mach number at the exit. (c) *What-if scenario:* What would the answers be if the shock occurred at the exit?

15-3-15 In Problem 15-3-14, determine the diffuser efficiency if the shock is (a) at the inlet and (b) at the exit (%).

15-3-16 A rocket nozzle has an exit-to-throat area ratio of 4. The exhaust gases are generated in a combustion chamber with stagnation pressure equal to 4 MPa and a stagnation temperature equal to 2000 K. Assume the working fluid behaves as a perfect gas with $k = 1.3$ and molar mass $= 20$ kg/kmol. Determine (a) the rocket exhaust velocity and (b) the mass flow rate. Assume isentropic steady flow, except at the nozzle exit where a normal shock is located as shown in the accompanying figure.

15-3-17 In Problem 15-3-16, (a) determine the thrust if the outside pressure is 0 kPa. (b) *What-if scenario:* What would the thrust be if the normal shock did not exist?

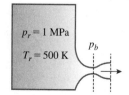

15-3-18 For the converging-diverging nozzle, shown in the accompanying figure, (a) find the range of back pressures for which a normal shock appears in the diverging section and (b) the mass flow rate ($\dot{m}$) when the shock is present. The throat area is 10 cm² and the exit area is 40 cm².

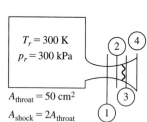

15-3-19 For the converging-diverging nozzle, shown in the accompanying figure, find the exit Mach number.

15-3-20 A symmetric, converging diverging duct, with an area ratio of 2, is placed in a wind tunnel where it encounters a 700 m/s flow of air at 50 kPa and 300 K. Determine the bypass ratio (diverted flow/incoming flow).

APPENDIX A

Table A-1 SL Model: Material Properties of Common Solids and Liquids

Properties of Common Solids

Substance	Density kg/m³ ρ	Sp. Heat kJ/kg·K $c_p = c_v$	Mol. Mass kg/kmol $\overline{M}$
Metals			
Aluminum	2,700	0.902	26.98
Bronze (76% Cu, 2% Zn, 2% Al)	8,280	0.400	50.12
Brass, yellow (65% Cu, 35% Zn)	8,310	0.400	64.29
Copper	8,900	0.386	63.55
Iron	7,840	0.450	55.85
Lead	11,310	0.128	207.20
Magnesium	1,730	1.000	24.31
Nickel	8,890	0.440	58.69
Silver	10,470	0.235	107.87
Steel, mild	7,830	0.500	55.71
Tungsten	19,400	0.130	183.85
Nonmetals			
Asphalt	2,110	0.920	10.91
Brick, common	1,922	0.790	59.49
Brick, fireclay (500°C)	2,300	0.960	78.96
Concrete	2,300	0.653	270.10
Clay	1,000	0.920	258.16
Diamond	2,420	0.616	12.01
Glass, window	2,700	0.800	60.08
Glass, pyrex	2,230	0.840	62.86
Graphite	2,500	0.711	12.01
Granite	2,700	1.017	62.44
Gypsum or plaster board	800	1.090	172.17
Ice (0°C)	921	2.110	18.02
Limestone	1,650	0.909	100.09
Marble	2,600	0.880	100.09
Plywood (Douglas Fir)	545	1.210	162.14
Rubber (soft)	1,100	1.840	68.12
Rubber (hard)	1,150	2.009	68.12
Sand	1,520	0.800	60.08
Stone	1,500	0.800	66.42
Woods, hard (Maple, Oak, etc.)	721	1.260	162.14

Properties of Common Liquids

Substance	Temp. °C T	Density kg/m³ ρ	Sp. Heat kJ/kg·K $c_p = c_v$	Mol. Mass kg/kmol $\overline{M}$
Ammonia	25.0	602	4.80	17.03
Argon	−185.6	1,394	1.14	39.95
Benzene	20.0	879	1.72	78.12
Brine (20% sodium cloride by mass)	20.0	1,150	3.11	26.10
n-Butane	−0.5	601	2.31	58.12
Carbon Dioxide	0	298	0.59	44.01
Ethanol	25.0	783	2.46	46.07
Ethyl Alcohol	20.0	789	2.84	46.07
Ethylene Glycol	20.0	1,109	2.84	62.07
Glycerine	20.0	1,261	2.32	92.09
Helium	−268.9	146	22.80	4.00
Hydrogen	−252.8	71	10.00	2.02
Isobutane	−11.7	594	2.28	58.12
Kerosene	20.0	820	2.00	170.34
Mercury	25.0	13,560	0.14	200.59
Methane		423	3.49	16.04
	−161.5	423	3.49	16.04
	−100.0	301	5.79	16.04
Methanol	25.0	787	2.55	32.04
Nitrogen		809	2.06	28.01
	−195.8	809	2.06	28.01
	−160.0	596	2.97	28.01
Octane	20.0	703	2.10	114.23
Oil (light)	25.0	910	1.80	114.23
Oxygen	−183.0	1,141	1.71	32.00
Petroleum	20.0	640	2.00	114.23
Propane		529	2.53	44.09
	−42.1	581	2.25	44.09
	0	529	2.53	44.09
	50.0	449	3.13	44.09
Refrigerant-134a		1,294	1.34	102.03
	−50.0	1,443	1.23	102.03
	−26.1	1,374	1.27	102.03

(continued)

Properties of Common Solids

Substance	Density kg/m³	Sp. Heat kJ/kg·K	Mol. Mass kg/kmol
	ρ	$c_p = c_v$	$\overline{M}$
Nonmetals			
Woods, soft (Fir, Pine, etc.)	513	1.380	162.14

Properties of Common Liquids

Substance	Temp. °C	Density kg/m³	Sp. Heat kJ/kg·K	Mol. Mass kg/kmol
	T	ρ	$c_p = c_v$	$\overline{M}$
	0	1,294	1.34	102.03
	25.0	1,206	1.42	102.03
Water		997	4.184	18.02
	0	1,000	4.23	18.02
	25.0	997	4.18	18.02
	50.0	988	4.18	18.02
	75.0	975	4.19	18.02
	100.0	958	4.22	18.02

Note: The unit kJ/kg·K is equivalent to kJ/kg·°C.

Values in red are estimates only.

Table A-2 Material Properties of Elements in Periodic Table (SL model)

Properties of Elements

Name	Symbol	$\overline{M}$ kg/kmol	$T_{melting}$ °C	$T_{boiling}$ °C	ρ kg/m³	$c_p = c_v$ kJ/kg·K	λ W/m·K	Δh_{fusion} kJ/kg	Δh_{evap} kJ/kg
Aluminum	Al	26.98	660.40	2467.00	2700.00	0.897	237.0000	397.0	10896.0
Antimony (Stibium)	Sb	121.75	630.70	1750.00	6697.00	0.207	24.4000	163.0	–
Argon	Ar	39.95	−189.20	−185.70	1.78	0.521	0.0177	28.0	161.0
Arsenic	As	74.92	817.00	613.00	5727.00	0.329	50.0000	326.0	1703.0
Barium	Ba	137.33	725.00	1640.00	3510.00	0.204	18.4000	51.8	1019.0
Beryllium	Be	9.01	1278.00	2970.00	1850.00	1.825	201.0000	877.0	–
Bismuth	Bi	208.98	271.30	1560.00	9780.00	0.122	7.9200	54.1	723.0
Boron	B	10.81	2079.00	2250.00	2340.00	1.026	27.4000	4644.0	44403.0
Bromine(l)	Br(l)	79.90	−7.20	58.78	3102.80	0.474	0.1220	66.2	188.0
Bromine(g)	Br2(g)	159.81	–	–	7.59	0.226	0.0048	–	–
Cadmium	Cd	112.41	320.90	765.00	8650.00	0.232	96.9000	55.1	888.0
Calcium	Ca	40.08	839.00	1484.00	1550.00	0.647	201.0000	213.0	3.8
Carbon	C	12.01	5530.00	5530.00	2146.00	0.709	1.5900	8709.0	–
Cesium	Cs	132.91	28.40	669.30	1930.00	0.242	35.9000	15.8	394.0
Cesium(l)	Cs(l)	132.91	–	–	1843.00	–	19.7000	–	–
Chlorine	Cl	35.45	−100.98	−34.60	3.20	0.479	0.0089	90.3	288.0
Chlorine(l)	Cl(l)	35.45	–	–	3.20	–	0.1340	–	–
Chromium	Cr	52.00	1857.00	2672.00	7150.00	0.449	93.9000	404.0	6.6

Properties of Elements

Name	Symbol	$\overline{M}$	$T_{melting}$	$T_{boiling}$	ρ	$c_p = c_v$	λ	Δh_{fusion}	Δh_{evap}
		kg/kmol	°C	°C	kg/m³	kJ/kg·K	W/m·K	kJ/kg	kJ/kg
Cobalt	Co	51.93	1495.00	2870.00	8900.00	0.421	100.0000	312.0	6390.0
Copper (Cuprum)	Cu	63.55	1083.40	2567.00	8960.00	0.385	401.0000	209.0	4700.0
Fluorine	F	19.00	−219.60	−188.10	1.70	0.824	0.0279	13.4	174.0
Gold (Aurum)	Au	196.97	1064.40	3080.00	19300.00	0.129	318.0000	63.7	1645.0
Helium	He	4.00	−272.20	−268.90	0.18	5.193	0.1520	2.1	20.7
Hydrogen	H	1.01	−259.10	−252.90	0.09	14.304	0.1815	59.5	446.0
Indium	In	114.82	156.60	2080.00	7310.00	0.233	81.8000	28.6	2019.0
Iodine	I	126.91	113.50	184.40	4933.00	0.145	0.4490	61.0	164.0
Iodine(l)	I(l)	126.91	–	–	4927.28	–	0.1160	–	–
Iridium	Ir	192.22	2410.00	4130.00	22650.00	0.131	147.0000	214.0	3185.0
Iron (Ferrum)	Fe	55.85	1535.00	2750.00	7860.00	0.449	80.4000	247.0	6260.0
Krypton	Kr	83.80	−156.60	−152.30	3.75	0.248	0.0095	16.3	108.0
Lanthanum	La	138.91	920.00	3454.00	6162.00	0.195	13.4000	44.6	2980.0
Lead (Plumbum)	Pb	207.20	327.50	1740.00	11340.00	0.129	35.3000	23.0	866.0
Lithium	Li	6.94	180.50	1342.00	534.00	3.582	84.8000	432.0	21340.0
Magnesium	Mg	24.31	648.80	1090.00	1738.00	1.023	156.0000	349.0	5240.0
Manganese	Mn	54.94	1244.00	1962.00	7210.00	0.479	7.8000	235.0	4110.0
Mercury (Hydrargyrum)	Hg	200.59	−38.87	356.60	13600.00	0.140	8.3000	11.4	295.0
Molybdenum	Mo	95.94	2617.00	4612.00	10280.00	0.251	138.0000	391.0	–
Neon	Ne	20.12	−248.70	−246.00	0.90	1.030	0.0493	16.9	85.0
Nickel	Ni	58.69	1453.00	2732.00	8908.00	0.444	90.9000	298.0	6310.0
Nitrogen	N	14.01	−209.90	−195.80	1.25	1.040	0.0260	25.4	199.0
Oxygen	O	16.00	218.40	−182.96	1.43	0.918	0.0267	13.8	213.0
Phosphorus, white	P	30.97	44.10	280.00	1823.00	0.769	0.2360	21.3	400.0
Platinum	Pt	195.08	1772.00	3827.00	21450.00	0.133	71.6000	114.0	2610.0
Plutonium	Pu	244.00	641.00	3232.00	19816.00	-	6.7000	11.6	1410.0
Potassium (Kalium)	K	39.10	63.25	759.90	890.00	0.757	102.5000	59.3	2050.0
Radon	Rn	222.00	−71.00	−61.80	0.01	0.094	0.0036	12.3	83.0
Rubidium	Rb	85.47	38.90	686.00	1532.00	0.363	58.2000	25.6	810.0
Selenium	Se	78.96	217.00	684.90	4810.00	0.321	0.5200	84.7	1209.0
Silicon	Si	28.09	1410.00	2355.00	2330.00	0.705	149.0000	1788.0	14050.0
Silver (Argentum)	Ag	107.87	961.90	2212.00	–	0.235	429.0000	105.0	2323.0

(continued)

Properties of Elements

Name	Symbol	$\overline{M}$	$T_{melting}$	$T_{boiling}$	ρ	$c_p = c_v$	λ	Δh_{fusion}	Δh_{evap}
		kg/kmol	°C	°C	kg/m³	kJ/kg·K	W/m·K	kJ/kg	kJ/kg
Sodium (Natrium)	Na	22.99	97.80	882.90	968.00	1.228	142.0000	113.0	4260.0
Strontium	Sr	87.62	769.00	1384.00	2640.00	0.301	35.4000	84.8	–
Sulfur	S	32.06	112.80	444.70	2080.00	0.710	0.2000	53.6	1404.0
Thallium	Tl	204.38	303.50	1457.00	11850.00	0.129	46.1000	20.3	806.0
Tin (Stannum)	Sn	118.71	232.00	2270.00	7265.00	0.228	66.8000	59.2	2500.0
Titanium	Ti	47.88	1660.00	3287.00	4506.00	0.523	21.9000	296.0	8790.0
Tungsten (Wolfram)	W	183.85	3410.00	5660.00	19250.00	0.132	173.0000	285.0	4500.0
Uranium	U	238.03	1132.00	3818.00	19100.00	0.116	27.5000	38.4	1950.0
Vanadium	V	50.94	1890.00	3380.00	6000.00	0.489	30.7000	422.0	8870.0
Xenon	Xe	131.29	−111.90	−107.10	5.89	0.158	0.0057	13.8	96.0
Zinc	Zn	65.39	419.60	907.00	7140.00	0.388	116.0000	112.0	1770.0
Zirconium	Zr	91.22	1852.00	4377.00	6.52	0.278	22.7000	230.0	6400.0

λ: Thermal conductivity

Table B-1   PC Model: Pressure Saturation Table, H_2O

Saturation Pressure Table (PC Model) of H_2O

Press.	Sat. Temp.	Spec. Volume		Spec. Int. Energy		Spec. Enthalpy		Spec. Entropy	
		m³/kg		kJ/kg		kJ/kg		kJ/kg·K	
kPa/Mpa	°C	Sat. liquid	Sat. vapor	Sat. liquid	Sat. vapor	Sat. liquid	Sat. vapor	Sat. liquid	Sat. vapor
p, kPa	$T_{sat@p}$	v_f	v_g	u_f	u_g	h_f	h_g	s_f	s_g
0.6113	0.01	0.001000	206.14	0	2375.3	0	2501.4	0	9.1562
1.0	6.98	0.001000	129.21	29.30	2385.0	29.30	2514.2	0.1059	8.9237
1.5	13.03	0.001001	87.98	54.71	2393.3	54.71	2525.3	0.1957	8.8279
2.0	17.50	0.001001	67.00	73.48	2399.5	73.48	2533.5	0.2607	8.7237
2.5	21.08	0.001002	54.25	88.48	2404.4	88.49	2540.0	0.3120	8.6432
3.0	24.08	0.001003	45.67	101.04	2408.5	101.05	2545.5	0.3545	8.5776
4.0	28.96	0.001004	34.80	121.45	2415.2	121.46	2554.4	0.4226	8.4746
5.0	32.88	0.001005	28.19	137.81	2420.5	137.82	2561.5	0.4764	8.3951
7.5	40.29	0.001008	19.24	168.78	2430.5	168.79	2574.8	0.5764	8.2515
10	45.81	0.001010	14.67	191.82	2437.9	191.83	2584.7	0.6493	8.1502

Saturation Pressure Table (PC Model) of H$_2$O

Press.	Sat. Temp.	Spec. Volume		Spec. Int. Energy		Spec. Enthalpy		Spec. Entropy	
		m^3/kg		kJ/kg		kJ/kg		kJ/kg·K	
kPa/Mpa	°C	Sat. liquid	Sat. vapor	Sat. liquid	Sat. vapor	Sat. liquid	Sat. vapor	Sat. liquid	Sat. vapor
p, kPa	$T_{sat@p}$	v_f	v_g	u_f	u_g	h_f	h_g	s_f	s_g
15	53.97	0.001014	10.02	225.92	2448.7	225.94	2599.1	0.7549	8.0085
20	60.06	0.001017	7.649	251.38	2456.7	251.40	2609.7	0.8320	7.9085
25	64.97	0.001020	6.204	271.90	2463.1	271.93	2618.2	0.8931	7.8314
30	69.10	0.001022	5.229	289.20	2468.4	289.23	2625.3	0.9439	7.7686
40	75.87	0.001027	3.993	317.53	2477.0	317.58	2636.8	1.0259	7.6700
50	81.33	0.001030	3.240	340.44	2483.9	340.49	2645.9	1.0910	7.5939
75	91.78	0.001037	2.217	384.31	2496.7	384.39	2663.0	1.2130	7.4564
p, MPa	$T_{sat@p}$	v_f	v_g	u_f	u_g	h_f	h_g	s_f	s_g
0.1	99.63	0.001043	1.6940	417.36	2506.1	417.46	2675.5	1.3026	7.3594
0.125	105.99	0.001048	1.3749	444.19	2513.5	444.32	2685.4	1.3740	7.2844
0.150	111.37	0.001053	1.1593	466.94	2519.7	467.11	2693.6	1.4336	7.2233
0.175	116.06	0.001057	1.0036	486.80	2524.9	486.99	2700.6	1.4849	7.1717
0.200	120.23	0.001061	0.8857	504.49	2529.5	504.70	2706.7	1.5301	7.1271
0.225	124.00	0.001064	0.7933	520.47	2533.6	520.72	2712.1	1.5706	7.0878
0.250	127.44	0.001067	0.7187	535.10	2537.2	535.37	2716.9	1.6072	7.0527
0.275	130.60	0.00107	0.6573	548.59	2540.5	548.89	2721.3	1.6408	7.0209
0.300	133.55	0.001073	0.6058	561.15	2543.6	561.47	2725.3	1.6718	6.9919
0.325	136.30	0.001076	0.5620	572.90	2546.4	573.25	2729.0	1.7006	6.9652
0.350	138.88	0.001079	0.5243	583.95	2548.9	584.33	2732.4	1.7275	6.9405
0.375	141.32	0.001081	0.4914	594.40	2551.3	594.81	2735.6	1.7528	6.9175
0.40	143.63	0.001084	0.4625	604.31	2553.6	604.74	2738.6	1.7766	6.8959
0.45	147.93	0.001088	0.4140	622.77	2557.6	623.25	3743.9	1.8207	6.8565
0.50	151.86	0.001093	0.3749	639.68	2561.2	640.23	2748.7	1.8607	6.8213
0.55	155.48	0.001097	0.3427	655.32	2564.5	665.93	2753.0	1.8973	6.7893
0.60	158.85	0.001101	0.3157	669.90	2567.4	670.56	2756.8	1.9312	6.7600
0.65	162.01	0.001104	0.2927	683.56	2570.1	684.28	2760.3	1.9627	6.7331
0.70	164.97	0.001108	0.2729	696.44	2572.5	697.22	2763.5	1.9922	6.7080
0.75	167.78	0.001112	0.2556	708.64	2574.7	709.47	2766.4	2.0200	6.6847
0.80	170.43	0.001115	0.2404	720.22	2576.8	721.11	2769.1	2.0462	6.6628
0.85	172.96	0.001118	0.2270	731.27	2578.7	732.22	2771.6	2.0710	6.6421
0.90	175.38	0.001121	0.2150	741.83	2580.5	742.83	2773.9	2.0946	6.6226
0.95	177.69	0.001124	0.2042	751.95	2582.1	753.02	2776.1	2.1172	6.6041
1.00	179.91	0.001127	0.19444	761.68	2583.6	762.81	2778.1	2.1387	6.5865

(continued)

Saturation Pressure Table (PC Model) of H₂O

Press.	Sat. Temp.	Spec. Volume		Spec. Int. Energy		Spec. Enthalpy		Spec. Entropy	
		m³/kg		kJ/kg		kJ/kg		kJ/kg·K	
kPa/Mpa	°C	Sat. liquid	Sat. vapor	Sat. liquid	Sat. vapor	Sat. liquid	Sat. vapor	Sat. liquid	Sat. vapor
p, kPa	$T_{sat@p}$	v_f	v_g	u_f	u_g	h_f	h_g	s_f	s_g
1.10	184.09	0.001133	0.17753	780.09	2586.4	781.34	2871.7	2.1792	6.5536
1.20	187.99	0.001139	0.16333	797.29	2588.8	798.65	2784.8	2.2166	6.5233
1.30	191.64	0.001144	0.15125	813.44	2591.0	814.93	2787.6	2.2515	6.4953
1.40	195.07	0.001149	0.14084	828.70	2592.8	830.30	2790.0	2.2842	6.4693
1.50	198.32	0.001154	0.13177	843.16	2594.5	844.89	2792.2	2.3150	6.4448
1.75	205.76	0.001166	0.11349	876.46	2597.8	878.50	2796.4	2.3851	6.3896
2.00	212.42	0.001177	0.09963	906.44	2600.3	908.79	2799.5	2.4474	6.3409
2.25	218.45	0.001189	0.08875	933.83	2602.0	936.49	2801.7	2.5035	6.2972
2.50	223.99	0.001197	0.07998	959.11	2603.1	962.11	2803.1	2.5547	6.2575
3.00	233.90	0.001217	0.06668	1004.78	2604.1	1008.42	2804.2	2.6457	6.1869
3.50	242.60	0.001235	0.05707	1045.43	2603.7	1049.75	2803.4	2.7253	6.1253
4	250.40	0.001252	0.04978	1082.31	2602.3	1087.31	2801.4	2.7964	6.0701
5	263.99	0.001286	0.03944	1147.81	2597.1	1154.23	2794.3	2.9202	5.9734
6	275.64	0.001319	0.03244	1205.44	2589.7	1213.35	2784.3	3.0267	5.8892
7	285.88	0.001351	0.02737	1257.55	2580.5	1267.00	2772.1	3.1211	5.8133
8	295.06	0.001384	0.02352	1305.57	2569.8	1316.64	2758.0	3.2068	5.7432
9	303.40	0.001418	0.02048	1350.51	2557.8	1363.26	2742.1	3.2858	5.6722
10	311.06	0.001452	0.018026	1393.04	2544.4	1407.56	2724.7	3.3596	5.6141
11	318.15	0.001489	0.015987	1433.70	2529.8	1450.10	2705.6	3.4295	5.5527
12	324.75	0.001527	0.014263	1473.00	2513.7	1491.30	2684.9	3.4962	5.4924
13	330.93	0.001567	0.012780	1511.10	2496.1	1531.50	2662.2	3.5606	5.4323
14	336.75	0.001611	0.011485	1548.60	2476.8	1571.10	2637.6	3.6232	5.3717
15	342.24	0.001658	0.010337	1585.60	2455.5	1610.50	2610.5	3.6848	5.3098
16	347.44	0.001711	0.009306	1622.70	2431.7	1650.10	2580.6	3.7461	5.2455
17	352.37	0.001770	0.008364	1660.20	2405.0	1690.30	2547.2	3.8079	5.1777
18	357.06	0.001840	0.007489	1698.90	2374.3	1732.00	2509.1	3.8715	5.1044
19	361.54	0.001924	0.006657	1739.90	2338.1	1776.50	2464.5	3.9388	5.0228
20	365.81	0.002036	0.005834	1785.60	2293.0	1826.30	2409.7	4.0139	4.9269
21	369.89	0.002207	0.004952	1842.10	2230.6	1888.40	2334.6	4.1075	4.8013
22	373.80	0.002742	0.003568	1961.90	2087.1	2022.20	2165.6	4.3110	4.5327
22.09	374.14	0.003155	0.003155	2029.60	2029.6	2099.30	2099.3	4.4298	4.4298

Table B-2   PC Model: Temperature Saturation Table, H_2O

		Saturation Temperature Table (PC Model) of H_2O							
Temp.	Sat. Press.	Spec. Volume		Spec. Int. Energy		Spec. Enthalpy		Spec. Entropy	
		m^3/kg		kJ/kg		kJ/kg		kJ/kg·K	
°C	kPa/MPa	Sat. liquid	Sat. vapor	Sat. liquid	Sat. vapor	Sat. liquid	Sat. vapor	Sat. liquid	Sat. vapor
T	$p_{sat@T}$ kPa	v_f	v_g	u_f	u_g	h_f	h_g	s_f	s_g
0.01	0.6113	0.001000	206.140	0.00	2375.3	0	2501.4	0	9.1562
5	0.8721	0.001000	147.120	20.97	2382.3	20.98	2510.6	0.0761	9.0257
10	1.2276	0.001000	106.380	42.00	2389.2	42.01	2519.8	0.1510	8.9008
15	1.7051	0.001001	77.930	62.99	2396.1	62.99	2528.9	0.2245	8.7814
20	2.3390	0.001002	57.790	83.95	2402.9	83.96	2538.1	0.2966	8.6672
25	3.1690	0.001003	43.360	104.88	2409.8	104.89	2547.2	0.3674	8.5580
30	4.2460	0.001004	32.890	125.78	2416.6	125.79	2556.3	0.4369	8.4533
35	5.6280	0.001006	25.220	146.67	2423.4	146.68	2565.3	0.5053	8.3531
40	7.3840	0.001008	19.520	167.56	2430.1	167.57	2574.3	0.5725	8.2570
45	9.5930	0.001010	15.260	188.44	2436.8	188.45	2583.2	0.6387	8.1648
50	12.3490	0.001012	12.030	209.32	2443.5	209.33	2592.1	0.7038	8.0763
55	15.7580	0.001015	9.568	230.21	2450.1	230.23	2600.9	0.7679	7.9913
60	19.9400	0.001017	7.671	251.11	2456.6	251.13	2609.6	0.8312	7.9096
65	25.0300	0.001020	6.197	272.02	2463.1	272.06	2618.3	0.8935	7.8310
70	31.1900	0.001023	5.042	292.95	2469.6	292.98	2626.8	0.9549	7.7553
75	38.5800	0.001026	4.131	313.90	2475.9	313.93	2643.7	1.0155	7.6824
80	47.3900	0.001029	3.407	334.86	2482.2	334.91	2635.3	1.0753	7.6122
85	57.8300	0.001033	2.828	355.84	2488.4	335.90	2651.9	1.1343	7.5445
90	70.1400	0.001036	2.361	376.85	2494.5	376.92	2660.1	1.1925	7.4791
95	84.5500	0.001040	1.982	397.88	2500.6	397.96	2668.1	1.2500	7.4159
T	$p_{sat@T}$ MPa	v_f	v_g	u_f	u_g	h_f	h_g	s_f	s_g
100	0.10135	0.001044	1.6729	418.94	2506.5	419.04	2676.1	1.3069	7.3549
105	0.12082	0.001048	1.4194	440.02	2512.4	440.15	2683.8	1.3630	7.2958
110	0.14327	0.001052	1.2102	461.14	2518.1	461.30	2691.5	1.4185	7.2387
115	0.16906	0.001056	1.0366	482.30	2523.7	482.48	2699.0	1.4734	7.1833
120	0.19853	0.001060	0.8919	503.50	2529.3	503.71	2706.3	1.5276	7.1296
125	0.2321	0.001065	0.7706	524.74	2534.6	524.99	2713.5	1.5813	7.0775
130	0.2701	0.001070	0.6685	546.02	2539.9	546.31	2720.5	1.6344	7.0269
135	0.3130	0.001075	0.5822	567.35	2545.0	567.69	2727.3	1.6870	6.9777
140	0.3613	0.001080	0.5089	588.74	2550.0	589.13	2733.9	1.7391	6.9299
145	0.4154	0.001085	0.4463	610.18	2554.9	610.63	2740.3	1.7907	6.8833

(continued)

Saturation Temperature Table (PC Model) of H₂O

Temp.	Sat. Press.	Spec. Volume		Spec. Int. Energy		Spec. Enthalpy		Spec. Entropy	
		m³/kg		kJ/kg		kJ/kg		kJ/kg·K	
°C	kPa/MPa	Sat. liquid	Sat. vapor	Sat. liquid	Sat. vapor	Sat. liquid	Sat. vapor	Sat. liquid	Sat. vapor
T	$p_{sat@T}$ MPa	v_f	v_g	u_f	u_g	h_f	h_g	s_f	s_g
150	0.4758	0.001091	0.3928	631.68	2559.5	632.20	2746.5	1.8418	6.8379
155	0.5431	0.001096	0.3468	653.24	2564.1	653.84	2752.4	1.8925	6.7935
160	0.6178	0.001102	0.3017	674.87	2568.4	675.55	2758.1	1.9427	6.7502
165	0.7005	0.001108	0.2727	696.56	2572.5	697.34	2763.5	1.9925	6.7078
170	0.7917	0.001114	0.2428	718.33	2576.5	719.21	2768.7	2.0419	6.6663
175	0.8920	0.001121	0.2168	740.17	2580.2	741.17	2773.6	2.0909	6.6256
180	1.0021	0.001127	0.19405	762.09	2583.7	763.22	2778.2	2.1396	6.5857
185	1.1227	0.001134	0.17409	784.10	2587.0	785.37	2782.4	2.1879	6.5465
190	1.2544	0.001141	0.15654	806.19	2590.0	807.62	2786.4	2.2359	6.5079
195	1.3978	0.001149	0.14105	828.37	2592.8	829.98	2790.0	2.2835	6.4698
200	1.5538	0.001157	0.12736	850.65	2595.3	852.45	2793.2	2.3309	6.4323
205	1.7230	0.001164	0.11521	873.04	2597.5	875.04	2796.0	2.3780	6.3952
210	1.9062	0.001173	0.10441	985.53	2599.5	897.76	2798.5	2.4248	6.3585
215	2.1040	0.001181	0.09479	918.14	2601.1	920.62	2800.5	2.4714	6.3221
220	2.3180	0.001190	0.08619	940.87	2602.4	943.62	2802.1	2.5178	6.2861
225	2.5480	0.001199	0.07849	963.73	2603.3	966.78	2803.3	2.5639	6.2503
230	2.7950	0.001209	0.07158	986.74	2603.9	990.12	2804.0	2.6099	6.2146
235	3.0600	0.001219	0.06537	1009.89	2604.1	1013.62	2804.2	2.6558	6.1791
240	3.3440	0.001229	0.05976	1033.21	2604.0	1037.32	2803.8	2.7015	6.1437
245	3.6480	0.001240	0.05471	1056.71	2603.3	1061.23	2803.0	2.7472	6.1083
250	3.9730	0.001251	0.05013	1080.39	2602.4	1085.36	2801.5	2.7927	6.0730
255	4.3190	0.001263	0.04598	1104.28	2600.9	1109.73	2799.5	2.8383	6.0375
260	4.6880	0.001276	0.04221	1128.39	2599.0	1134.37	2796.9	2.8838	6.0019
265	5.0810	0.001289	0.03877	1152.74	2596.0	1159.28	2793.6	2.9294	5.9662
270	5.4990	0.001302	0.03564	1177.36	2593.7	1184.51	2789.7	2.9751	5.9301
275	5.9420	0.001317	0.03279	1202.25	2590.2	1210.07	2785.0	3.0208	5.8938
280	6.4120	0.001332	0.03017	1227.46	2586.1	1235.99	2779.6	3.0668	5.8571
285	6.9090	0.001348	0.02777	1253.00	2581.4	1262.31	2773.3	3.1130	5.8199
290	7.4360	0.001366	0.02557	1278.92	2576.0	1289.07	2766.2	3.1594	5.7821
295	7.9930	0.001384	0.02354	1305.20	2569.9	1316.30	2758.1	3.2062	5.7437
300	8.5810	0.001404	0.02167	1332.00	2563.0	1344.00	2749.0	3.2534	5.7045
305	9.2020	0.001250	0.019948	1359.30	2555.2	1372.40	2738.7	3.3010	5.6643
310	9.8560	0.001447	0.018350	1387.10	2546.4	1401.30	2727.3	3.3493	5.6230

Saturation Temperature Table (PC Model) of H$_2$O

Temp.	Sat. Press.	Spec. Volume		Spec. Int. Energy		Spec. Enthalpy		Spec. Entropy	
		m^3/kg		kJ/kg		kJ/kg		kJ/kg·K	
°C	kPa/MPa	Sat. liquid	Sat. vapor	Sat. liquid	Sat. vapor	Sat. liquid	Sat. vapor	Sat. liquid	Sat. vapor
T	$P_{sat@T}$ MPa	v_f	v_g	u_f	u_g	h_f	h_g	s_f	s_g
315	10.5470	0.001472	0.016867	1415.50	2536.6	1431.00	2714.5	3.3982	5.5804
320	11.2740	0.001499	0.015488	1444.60	2525.5	1461.50	2700.1	3.4480	5.5362
330	12.8450	0.001561	0.012996	1505.30	2498.9	1525.30	2665.9	3.5507	5.4417
340	14.5860	0.001638	0.010797	1570.30	2464.6	1594.20	2622.0	3.6594	5.3357
350	16.5130	0.001740	0.008813	1641.90	2418.4	1670.60	2563.9	3.7777	5.2112
360	18.6510	0.001893	0.006945	1725.20	2351.5	1760.50	2481.0	3.9147	5.0526
370	21.0300	0.002213	0.004925	1844.00	2228.5	1890.50	2332.1	4.1106	4.7971
374.14	22.0900	0.003155	0.003155	2029.60	2029.6	2099.30	2099.3	4.4298	4.4298

Table B-3   PC Model: Superheated Vapor Table, H$_2$O

Superheated Table (PC Model), H$_2$O

°C	m^3/kg	kJ/kg	kJ/kg	kJ/kg·K	m^3/kg	kJ/kg	kJ/kg	kJ/kg·K	m^3/kg	kJ/kg	kJ/kg	kJ/kg·K
T	$p = 0.01$ MPa ($T_{sat} = 45.81$°C)				$p = 0.05$ MPa ($T_{sat} = 81.33$°C)				$p = 0.10$ MPa ($T_{sat} = 99.63$°C)			
	v	u	h	s	v	u	h	s	v	u	h	s
Sat.	14.674	2437.9	2584.7	8.1502	3.240	2483.9	2645.9	7.5939	1.6940	2506.1	2675.5	7.3594
50	14.869	2443.9	2592.6	8.1749	–	–	–	–	–	–	–	–
100	17.196	2515.5	2687.5	8.4479	3.418	2511.6	2682.5	7.6947	1.6958	2506.7	2676.2	7.3614
150	19.512	2587.9	2783.0	8.6882	3.889	2585.6	2780.1	7.9401	1.9364	2582.8	2776.4	7.6143
200	21.825	2661.3	2879.5	8.9038	4.356	2659.9	2877.7	8.1580	2.1720	2658.1	2875.3	7.8343
250	24.136	2736.0	2977.3	9.1002	4.820	2735.0	2976.0	8.3556	2.4060	2733.7	2974.3	8.0333
300	26.445	2812.1	3076.5	9.2813	5.284	2811.3	3075.5	8.5373	2.6390	2810.4	3074.3	8.2158
400	31.063	2968.9	3279.6	9.6077	6.209	2968.5	3278.9	8.8642	3.1030	2967.9	3278.2	8.5435
500	35.679	3132.3	3489.1	9.8978	7.134	3132.0	3488.7	9.1546	3.5650	3131.6	3488.1	8.8342
600	40.295	3302.5	3705.4	10.1608	8.057	3302.2	3705.1	9.4178	4.0280	3301.9	3704.4	9.0976
700	44.911	3479.6	3928.7	10.4028	8.981	3479.4	3928.5	9.6599	4.4900	3479.2	3928.2	9.3398
800	49.526	3663.8	4159.0	10.6281	9.904	3663.6	4158.9	9.8852	4.9520	3663.5	4158.6	9.5652
900	54.141	3855.0	4396.4	10.8396	10.828	3854.9	4396.3	10.0967	5.4140	3854.8	4396.1	9.7767
1000	58.757	4053.0	4640.6	11.0393	11.751	4052.9	4640.5	10.2964	5.8750	4052.8	4640.3	9.9764
1100	63.372	4257.5	4891.2	11.2287	12.674	4257.4	4891.1	10.4859	6.3370	4257.3	4891.0	10.1659
1200	67.987	4467.9	5147.8	11.4091	13.597	4467.8	5147.7	10.6662	6.7990	4467.7	5147.6	10.3463
1300	72.602	4683.7	5409.7	11.5811	14.521	4683.6	5409.6	10.8382	7.2600	4683.5	5409.5	10.5183
T	$p = 0.20$ MPa ($T_{sat} = 120.23$°C)				$p = 0.30$ MPa ($T_{sat} = 133.55$°C)				$p = 0.40$ MPa ($T_{sat} = 143.63$°C)			
	v	u	h	s	v	u	h	s	v	u	h	s
Sat.	0.8857	2529.5	2706.7	7.1272	0.6058	2543.6	2725.3	6.9919	0.4625	2553.6	2738.6	6.8959
150	0.9596	2576.9	2768.8	7.2795	0.6339	2570.8	2761.0	7.0778	0.4708	2564.5	2752.8	6.9299

Superheated Table (PC Model), H₂O

°C	m³/kg	kJ/kg	kJ/kg	kJ/kg·K	m³/kg	kJ/kg	kJ/kg	kJ/kg·K	m³/kg	kJ/kg	kJ/kg	kJ/kg·K
T	$p = 0.20$ MPa ($T_{sat} = 120.23°C$)				$p = 0.30$ MPa ($T_{sat} = 133.55°C$)				$p = 0.40$ MPa ($T_{sat} = 143.63°C$)			
	v	u	h	s	v	u	h	s	v	u	h	s
200	1.0803	2654.4	2870.5	7.5066	0.7163	2650.7	2865.6	7.3115	0.5342	2646.8	2860.5	7.1706
250	1.1988	2731.2	2971.0	7.7086	0.7964	2728.7	2967.6	7.5166	0.5951	2726.1	2964.2	7.3789
300	1.3162	2808.6	3071.8	7.8926	0.8753	2806.7	3069.3	7.7022	0.6548	2804.8	3066.8	7.5662
400	1.5493	2966.7	3276.6	8.2218	1.0315	2965.6	3275.0	8.0330	0.7726	2964.4	3273.4	7.8985
500	1.7814	3130.8	3487.1	8.5133	1.1867	3130.0	3486.0	8.3251	0.8893	3129.2	3484.9	8.1913
600	2.0130	3301.4	3704.0	8.7770	1.3414	3300.8	3703.2	8.5892	1.0055	3300.2	3702.4	8.4558
700	2.2440	3478.8	3927.6	9.0194	1.4957	3478.4	3927.1	8.8319	1.1215	3477.9	3926.5	8.6987
800	2.4750	3663.1	4158.2	9.2449	1.6499	3662.9	4157.8	9.0576	1.2372	3662.4	4157.3	8.9244
900	2.7050	3854.5	4395.8	9.4566	1.8041	3854.2	4395.4	9.2692	1.3529	3853.9	4395.1	9.1362
1000	2.9370	4052.5	4640.0	9.6563	1.9581	4052.3	4639.7	9.4690	1.4685	4052.0	4939.4	9.3360
1100	3.1680	4257.0	4890.7	9.8458	2.1121	4256.8	4890.4	9.6585	1.5840	4256.5	4890.2	9.5256
1200	3.3990	4467.5	5147.5	10.0262	2.2661	4467.2	5147.1	9.8389	1.6996	4467.0	5146.8	9.7060
1300	3.6300	4683.2	5409.3	10.1982	2.4201	4683.0	5409.0	10.0110	1.8151	4682.8	5408.8	9.8780
T	$p = 0.50$ MPa ($T_{sat} = 151.86°C$)				$p = 0.60$ MPa ($T_{sat} = 158.85°C$)				$p = 0.80$ MPa ($T_{sat} = 170.43°C$)			
	v	u	h	s	v	u	h	s	v	u	h	s
Sat.	0.3749	2561.2	2748.7	6.8213	0.3175	2567.4	2756.8	6.7600	0.2404	2576.8	2769.1	6.6628
200	0.4249	2642.9	2855.4	7.0592	0.3520	2638.9	2850.1	6.9665	0.2608	2630.6	2839.3	6.8158
250	0.4744	2723.5	2960.7	7.2709	0.3938	2720.9	2957.2	7.1816	0.2931	2715.5	2950.0	7.0384
300	0.5226	2802.9	3064.2	7.4599	0.4344	2801.0	3061.6	7.3724	0.3241	2797.2	3056.5	7.2328
350	0.5701	2882.6	3167.7	7.6329	0.4742	2881.2	3165.7	7.5464	0.3544	2878.2	3161.7	7.4089
400	0.6173	2963.2	3271.9	7.7938	0.5137	2962.1	3270.3	7.7079	0.3843	2959.7	3267.1	7.5716
500	0.7109	3128.4	3483.9	8.0873	0.5920	3127.6	3482.8	8.0021	0.4433	3126.0	3480.6	7.8673
600	0.8041	3299.6	3701.7	7.3522	0.6697	3299.1	3700.9	8.2674	0.5018	3297.9	3699.4	8.1333
700	0.8969	3477.5	3925.9	8.5952	0.7472	3477.0	3925.3	8.5107	0.5601	3476.2	3924.2	8.3770
800	0.9896	3662.1	4156.9	8.8211	0.8245	3661.8	4156.6	8.7367	0.6181	3661.1	4155.6	8.6033
900	1.0822	3853.6	4394.7	9.0329	0.9017	3853.4	4394.4	8.9486	0.6761	3852.8	4393.7	8.8153
1000	1.1747	4051.8	4639.1	9.2328	0.9788	4051.5	4638.8	9.1485	0.7340	4051.0	4638.2	9.0153
1100	1.2672	4256.3	4889.9	9.4224	1.0559	4256.1	4889.6	9.3381	0.7919	4255.6	4889.1	9.2050
1200	1.3956	4466.8	5146.6	9.6029	1.1330	4466.5	5146.3	9.5185	0.8497	4466.1	5145.9	9.3855
1300	1.4521	4682.5	5408.6	9.7749	1.2101	4682.3	5408.3	9.6906	0.9076	4681.8	5407.9	9.5575
T	$p = 1.00$ MPa ($T_{sat} = 179.91°C$)				$p = 1.20$ MPa ($T_{sat} = 187.99°C$)				$p = 1.40$ MPa ($T_{sat} = 195.07°C$)			
	v	u	h	s	v	u	h	s	v	u	h	s
Sat.	0.19444	2583.6	2778.1	6.5865	0.16333	2588.8	2784.4	6.5233	0.14084	2592.8	2790.0	6.4693
200	0.20600	2621.9	2827.9	6.6940	0.16930	2612.8	2815.9	6.5898	0.14302	2603.1	2803.3	6.4975
250	0.23270	2709.9	2942.6	6.9247	0.19234	2704.2	2935.0	6.8294	0.16350	2698.3	2927.2	6.7467
300	0.25790	2793.2	3051.2	7.1229	0.21380	2789.2	3045.8	7.0317	0.18228	2785.2	3040.4	6.9534
350	0.28250	2875.2	3157.7	7.3011	0.23450	2872.2	3153.6	7.2121	0.20030	2869.2	3149.5	7.1360
400	0.30660	2957.3	3263.9	7.4651	0.25480	2954.9	3260.7	7.3774	0.21780	2952.5	3257.5	7.3026
500	0.35410	3124.4	3478.5	7.7622	0.29460	3122.8	3476.3	7.6759	0.25210	3121.1	3474.1	7.6027
600	0.40110	3296.8	3697.9	8.0290	0.33390	3295.6	3696.3	7.9435	0.28600	3294.4	3694.8	7.8710
700	0.44780	3475.3	3923.1	9.2731	0.37290	3474.4	3922.0	8.1881	0.31950	3473.6	3920.8	8.1160

Superheated Table (PC Model), H_2O

°C	m³/kg	kJ/kg	kJ/kg	kJ/kg·K	m³/kg	kJ/kg	kJ/kg	kJ/kg·K	m³/kg	kJ/kg	kJ/kg	kJ/kg·K
T	$p = 1.00$ MPa ($T_{sat} = 179.91$°C)				$p = 1.20$ MPa ($T_{sat} = 187.99$°C)				$p = 1.40$ MPa ($T_{sat} = 195.07$°C)			
	v	u	h	s	v	u	h	s	v	u	h	s
800	0.49430	3660.4	4154.7	8.4996	0.41180	3659.7	4153.8	8.4148	0.35280	3659.0	4153.0	8.3431
900	0.54070	3852.2	4392.9	8.7118	0.45050	3851.6	4392.2	8.6272	0.38610	3851.1	4391.5	8.5556
1000	0.58810	4050.5	4637.6	8.9119	0.48920	4050.0	4637.0	8.8274	0.41920	4049.5	4636.4	8.7559
1100	0.63350	4255.1	4888.6	9.1017	0.52780	4254.6	4888.0	9.0172	0.45240	4254.1	4887.5	8.9457
1200	0.67980	4465.6	5145.4	9.2822	0.56650	4465.1	5144.9	9.1977	0.48550	4464.7	5144.4	9.1262
1300	0.72610	4681.3	5407.4	9.4543	0.60510	4680.9	5407.0	9.3698	0.51860	4680.4	5406.5	9.2984

°C	m³/kg	kJ/kg	kJ/kg	kJ/kg·K	m³/kg	kJ/kg	kJ/kg	kJ/kg·K	m³/kg	kJ/kg	kJ/kg	kJ/kg·K
T	$p = 1.60$ MPa ($T_{sat} = 201.41$°C)				$p = 1.80$ MPa ($T_{sat} = 207.15$°C)				$p = 2.00$ MPa ($T_{sat} = 212.42$°C)			
	v	u	h	s	v	u	h	s	v	u	h	s
Sat.	0.12380	2596.0	2794.0	6.4218	0.11042	2598.4	2791.1	6.3794	0.09963	2600.3	2799.5	6.3409
225	0.13287	2644.7	2857.3	6.5518	0.11673	2636.6	2846.7	6.4808	0.10377	2628.3	2835.8	6.4147
250	0.14184	2692.3	2919.2	6.6732	0.12497	2686.0	2911.0	6.6066	0.11144	2679.6	2902.5	6.5453
300	0.15862	2781.1	3034.8	6.8844	0.14021	2776.9	3029.2	6.8226	0.12547	2772.6	3023.5	6.7664
350	0.17456	2866.1	3145.4	7.0694	0.15457	2863.0	3141.2	7.0100	0.13857	2859.8	3137.0	6.9563
400	0.19005	2950.1	3254.2	7.2374	0.16847	2947.7	3250.9	7.1794	0.15120	2945.2	3247.6	7.1271
500	0.22030	3119.5	3472.0	7.5390	0.19550	3117.9	3469.8	7.4825	0.17568	3116.2	3467.6	7.4317
600	0.25000	3293.3	3693.2	7.8080	0.22200	3292.1	3691.7	7.7523	0.19960	3290.9	3690.1	7.7024
700	0.27940	3472.7	3919.7	8.0535	0.24820	3471.8	3918.5	7.9983	0.22320	3470.9	3917.4	7.9487
800	0.30860	3658.3	4152.1	8.2808	0.27420	3657.6	4151.2	8.2258	0.24670	3657.0	4150.3	8.1765
900	0.33770	3850.5	4390.8	8.4935	0.30010	3849.9	4390.1	8.4386	0.27000	3849.3	4389.4	8.3895
1000	0.36680	4049.0	4635.8	8.6938	0.32600	4048.5	4635.2	8.6391	0.29330	4048.0	4634.6	8.5901
1100	0.39580	4253.7	4887.0	8.8837	0.35180	4253.2	4886.4	8.8290	0.31660	4252.7	4885.9	8.7800
1200	0.42480	4464.2	5143.9	9.0643	0.37760	4467.7	5143.4	9.0096	0.33980	4463.3	5142.9	8.9607
1300	0.45380	4679.9	5406.0	9.2364	0.40340	4679.5	5405.6	9.1818	0.36310	4679.0	5405.1	9.1329

°C	m³/kg	kJ/kg	kJ/kg	kJ/kg·K	m³/kg	kJ/kg	kJ/kg	kJ/kg·K	m³/kg	kJ/kg	kJ/kg	kJ/kg·K
T	$p = 2.50$ MPa ($T_{sat} = 223.99$°C)				$p = 3.00$ MPa ($T_{sat} = 233.90$°C)				$p = 3.50$ MPa ($T_{sat} = 242.60$°C)			
	v	u	h	s	v	u	h	s	v	u	h	s
Sat.	0.07998	2603.1	2803.1	6.2575	0.06668	2604.1	2804.2	6.1869	0.05070	2603.7	2803.4	6.1253
225	0.08027	2605.6	2806.3	6.2639	–	–	–	–	–	–	–	–
250	0.08700	2662.6	2880.1	6.4085	0.07058	2644.0	2855.8	6.2872	0.05872	2623.7	2829.2	6.1749
300	0.09890	2761.6	3008.8	6.6438	0.08114	2750.1	2993.5	6.5390	0.06842	2738.0	2977.5	6.4461
350	0.10976	2851.9	3126.3	6.8403	0.09053	2843.7	3115.3	6.7428	0.07678	2835.3	3104.0	6.6579
400	0.12010	2939.1	3239.3	7.0148	0.99360	2932.8	3230.9	6.9212	0.84530	2926.4	3222.3	6.8405
450	0.13014	3025.5	3350.8	7.1746	0.10787	3020.4	3344.0	7.0834	0.09196	3015.3	3337.2	7.0052
500	0.13993	3112.1	3462.1	7.3234	0.11619	3108.0	3456.5	7.2338	0.09918	3103.0	3450.9	7.1572
600	0.15930	3288.0	3686.3	7.5960	0.13243	3285.0	3682.3	7.5085	0.11324	3282.1	3678.4	7.4339
700	0.17832	3468.7	3914.5	7.8435	0.14838	3466.5	3911.7	7.7571	0.13699	3464.3	3908.8	7.6837
800	0.19716	3655.3	4148.2	8.0720	0.16414	3653.5	4145.9	7.9862	0.14056	3651.8	4143.7	7.9134
900	0.21590	3847.9	4387.6	8.2853	0.17980	3846.5	4385.9	8.1999	0.15402	3845.0	4384.1	8.1276
1000	0.23460	4046.7	4633.1	8.4861	0.19541	4045.4	4631.6	8.4009	0.16743	4044.1	4630.1	8.3288
1100	0.25320	4251.5	4884.6	8.6762	0.21098	4250.3	4883.3	8.5912	0.18080	4249.2	4881.9	8.5192
1200	0.27180	4462.1	5141.7	8.8569	0.22652	4460.9	5140.5	8.7720	0.19415	4459.8	5139.3	8.7000
1300	0.29050	4677.8	5404.0	9.0291	0.24206	4676.6	5402.8	8.9442	0.20749	4675.5	5401.7	8.8723

(continued)

Superheated Table (PC Model), H₂O

°C	m³/kg	kJ/kg	kJ/kg	kJ/kg·K	m³/kg	kJ/kg	kJ/kg	kJ/kg·K	m³/kg	kJ/kg	kJ/kg	kJ/kg·K
T	$p = 2.50$ MPa ($T_{sat} = 223.99$°C)				$p = 3.00$ MPa ($T_{sat} = 233.90$°C)				$p = 3.50$ MPa ($T_{sat} = 242.60$°C)			
	v	u	h	s	v	u	h	s	v	u	h	s
Sat.	0.04978	2602.3	2801.4	6.0701	0.04406	2600.1	2798.3	6.0198	0.03944	2597.1	2794.3	5.9734
275	0.05457	2667.9	2886.2	6.2285	0.04730	2650.3	2863.2	6.1401	0.04141	2631.3	2838.3	6.0544
300	0.05884	2725.3	2960.7	6.3615	0.05135	2712.0	2943.1	6.2828	0.04532	2698.0	2924.5	6.2084
350	0.06645	2826.7	3092.5	6.5821	0.05840	2817.8	3080.6	6.5131	0.05194	2808.4	3068.4	6.4493
400	0.07341	2919.9	3213.6	6.7690	0.06475	2913.3	3204.7	6.7047	0.05781	2906.6	3195.7	6.6459
450	0.08002	3010.2	3330.3	6.9363	0.07074	3005.0	3323.3	6.8746	0.06330	2999.7	3316.2	6.8186
500	0.08643	3099.5	3445.3	7.0901	0.07651	3095.3	3439.6	7.0301	0.06857	3091.0	3433.8	6.9759
600	0.09885	3279.1	3674.4	7.3688	0.08765	3276.0	3670.5	7.3110	0.07869	3273.0	3666.5	7.2589
700	0.11095	3462.1	3905.9	7.6198	0.09847	3459.9	3903.0	7.5631	0.08849	3457.6	3900.1	7.5122
800	0.12287	3650.0	4141.5	7.8502	0.10911	3648.3	4339.3	7.7942	0.09811	3646.6	4137.1	7.7440
900	0.13469	3843.6	4382.3	8.0647	0.11965	3842.2	4380.6	8.0910	0.10762	3840.7	4378.8	7.9593
1000	0.14645	4042.4	4628.7	8.2662	0.13013	4041.6	4627.2	8.2108	0.11707	4040.4	4625.7	8.1612
1100	0.15817	4248.0	4880.6	8.4567	0.14056	4246.8	4879.3	8.4015	0.12648	4245.6	4878.0	8.3520
1200	0.16987	4458.6	5138.1	8.6376	0.15098	4457.5	5136.9	8.5825	0.13587	4456.3	5135.7	8.5331
1300	0.18156	4674.3	5400.5	8.8100	0.16139	4673.1	5399.4	8.7549	0.14526	4672.0	5398.2	8.7055
T	$p = 6.0$ MPa ($T_{sat} = 275.64$°C)				$p = 7.0$ MPa ($T_{sat} = 285.88$°C)				$p = 8.0$ MPa ($T_{sat} = 295.06$°C)			
	v	u	h	s	v	u	h	s	v	u	h	s
Sat.	0.03244	2589.7	2784.3	5.8892	0.02737	2580.5	2772.1	5.8133	0.02352	2569.8	2758.0	5.7432
300	0.03616	2667.2	2884.2	6.0674	0.02947	2632.2	2838.4	5.9305	0.02426	2590.9	2785.0	5.7903
350	0.04223	2789.6	3043.0	6.3335	0.03524	2769.4	3016.0	6.2283	0.02995	2747.7	2987.3	6.1301
400	0.04739	2892.9	3177.2	6.5408	0.03993	2878.6	3158.1	6.4478	0.03432	2863.8	3138.3	6.3634
450	0.05214	2988.9	3301.8	6.7193	0.04416	2978.0	3287.1	6.6327	0.03817	2966.7	3272.0	6.5551
500	0.05665	3082.2	3422.2	6.8803	0.04814	3073.4	3410.3	6.7975	0.04175	3064.3	3398.3	6.7240
550	0.06101	3174.6	3540.6	7.0288	0.05195	3167.2	3530.9	6.9486	0.04516	3159.8	3521.0	6.8778
600	0.06525	3266.9	3658.4	7.1677	0.05565	3260.7	3650.3	7.0894	0.04845	3254.4	3642.0	7.0206
700	0.07352	3453.1	3894.2	7.4234	0.06283	3448.5	3888.3	7.3476	0.05481	3443.9	3882.4	7.2812
800	0.08160	3643.1	4132.7	7.6566	0.06981	3639.5	4128.2	7.5822	0.06097	3636.0	4123.8	7.5193
900	0.08958	3837.8	4375.3	7.8727	0.07669	3835.0	4371.8	7.7991	0.06702	3832.1	1368.3	7.7351
1000	0.09749	4037.8	4622.7	8.0751	0.08350	4035.3	4619.8	8.0020	0.07301	4032.8	4616.9	7.9384
1100	0.10536	4243.3	4875.4	8.2661	0.09270	4240.9	4872.8	8.1933	0.07896	4238.6	4870.3	8.1300
1200	0.11321	4454.0	5133.3	8.4474	0.09703	4451.7	4530.9	8.3747	0.08489	4449.5	5128.5	8.3115
1300	0.12106	4669.6	5396.0	8.6199	0.10377	4667.3	5393.7	8.5475	0.09080	4665.0	5391.5	8.4842
T	$p = 9.0$ MPa ($T_{sat} = 303.40$°C)				$p = 10.0$ MPa ($T_{sat} = 311.06$°C)				$p = 12.5$ MPa ($T_{sat} = 327.89$°C)			
	v	u	h	s	v	u	h	s	v	u	h	s
Sat.	0.02048	2557.8	2742.1	5.6772	0.018026	2554.4	2724.7	5.6141	0.013495	2505.1	2673.8	5.4624
325	0.02327	2646.6	2856.0	5.8712	0.019861	2610.4	2809.1	5.7568	–	–	–	–
350	0.02580	2724.4	2956.6	6.0361	0.02242	2699.2	2923.4	5.9443	0.016126	2624.6	2826.2	5.7118
400	0.02993	2848.4	3117.8	6.2854	0.02641	2832.4	3096.5	6.2120	0.02000	2789.3	3039.3	6.0417
450	0.03350	2855.2	3256.6	6.4844	0.02975	2943.4	3240.9	6.4190	0.02299	2812.5	3199.8	6.2719
500	0.03677	3055.2	3386.1	6.6576	0.03279	3045.8	3373.7	6.5966	0.02560	3021.7	3341.8	6.4618
550	0.03987	3152.2	3511.0	6.8142	0.03564	3144.6	3500.4	6.7561	0.02801	3125.0	3475.2	6.6290

Superheated Table (PC Model), H₂O

°C	m³/kg	kJ/kg	kJ/kg	kJ/kg·K	m³/kg	kJ/kg	kJ/kg	kJ/kg·K	m³/kg	kJ/kg	kJ/kg	kJ/kg·K
T	$p = 9.0$ MPa ($T_{sat} = 303.40°C$)				$p = 10.0$ MPa ($T_{sat} = 311.06°C$)				$p = 12.5$ MPa ($T_{sat} = 327.89°C$)			
	v	u	h	s	v	u	h	s	v	u	h	s
600	0.04285	3248.1	3633.7	6.9589	0.03837	3241.7	3625.3	6.9029	0.03029	3225.4	3604.0	6.7810
650	0.04574	3343.6	3755.3	7.0943	0.04101	3338.2	3748.2	7.0398	0.03248	3324.4	3730.4	6.9218
700	0.04857	3439.3	3876.5	7.2221	0.04358	3434.7	3870.5	7.1687	0.03460	3422.9	3855.3	7.0536
800	0.05409	3632.5	4119.3	7.4596	0.04859	3628.9	4114.8	7.4077	0.03869	3620.0	4103.9	7.2965
900	0.05950	3829.3	4364.8	7.6783	0.05349	3826.3	4361.2	7.6272	0.04267	3812.1	4352.5	7.5182
1000	0.06485	4030.3	4614.0	7.8821	0.05832	4027.8	4611.0	7.8315	0.04658	4216.0	4606.8	7.7237
1100	0.07016	4236.3	4867.7	8.0740	0.06312	4234.0	4865.1	8.0237	0.05045	4228.2	4858.8	7.9165
1200	0.07544	4447.2	5126.2	8.2556	0.06789	4444.9	5123.8	8.2055	0.05430	4439.3	5118.0	8.0937
1300	0.08072	4662.7	5389.2	8.4284	0.07264	4460.5	5387.0	8.3783	0.05813	4654.8	5381.4	8.2717
T	$p = 15.0$ MPa ($T_{sat} = 342.24°C$)				$p = 17.5$ MPa ($T_{sat} = 354.75°C$)				$p = 20.0$ MPa ($T_{sat} = 365.81°C$)			
	v	u	h	s	v	u	h	s	v	u	h	s
Sat.	0.010337	2455.5	2610.5	5.3098	0.007920	2390.2	2528.8	5.1419	0.005834	2293.0	2409.7	4.9269
350	0.011470	2520.4	2692.4	5.4421	–	–	–	–	–	–	–	–
400	0.015649	2740.7	2975.5	5.8811	0.012447	2685.0	2902.9	5.7213	0.009942	2619.3	2818.1	5.5540
450	0.018445	2879.5	3156.2	6.1404	0.015174	2844.2	3109.7	6.0184	0.012695	2806.2	3060.1	5.9017
500	0.02080	2996.6	3308.6	6.3443	0.017358	2970.3	3274.1	6.2383	0.014768	2942.9	3238.2	6.1401
550	0.02293	3104.7	3448.6	6.5199	0.019288	3083.9	3421.4	6.4230	0.016555	3062.4	3393.5	6.3348
600	0.02491	3208.6	3582.3	6.6776	0.02106	3191.5	3560.1	6.5866	0.018178	3174.0	3537.6	6.5048
650	0.02680	3310.3	3712.3	6.8224	0.02274	3296.0	3693.9	6.7357	0.019693	3281.4	3675.3	6.6582
700	0.02861	3410.9	3840.1	6.9572	0.02434	3398.7	3824.6	6.8736	0.02113	3386.4	3809.0	6.7993
800	0.03210	3610.9	4092.4	7.2040	0.02738	3601.8	4081.1	7.1244	0.02385	3592.7	4069.7	7.0544
900	0.03546	3811.9	4343.8	7.4279	0.03031	3804.7	4335.1	7.3507	0.02645	3797.5	4326.4	7.2830
100	0.03875	4015.4	4596.6	7.6348	0.03316	4009.3	4589.5	7.5589	0.02897	4003.1	4582.5	7.4925
1100	0.04200	4222.6	4852.6	7.8283	0.03597	4216.9	4846.4	7.7531	0.03145	4211.3	4840.2	7.6874
1200	0.04523	4433.8	5112.3	8.0108	0.03876	4428.3	5106.6	7.9360	0.03391	4422.8	5101.0	7.8707
1300	0.04845	4649.1	5376.0	8.1840	0.04154	4643.5	5370.5	8.1093	0.03636	4638.0	5365.1	8.0442
T	$p = 25.0$ MPa				$p = 30.0$ MPa				$p = 35.0$ MPa			
	v	u	h	s	v	u	h	s	v	u	h	s
375	0.0019731	1798.7	1848.0	4.0320	0.0017892	1737.8	1791.5	3.9305	0.0017003	1702.9	1762.4	3.8722
400	0.006004	2430.1	2580.2	5.1418	0.002790	2067.4	2151.1	4.4728	0.002100	1914.1	1987.6	4.2126
425	0.007881	2609.2	2806.3	5.4723	0.005303	2455.1	2614.2	5.1504	0.003428	2253.4	2373.4	4.7747
450	0.009162	2720.7	2949.7	5.6744	0.006735	2619.3	2821.4	5.4424	0.004961	2498.7	2672.4	5.1962
500	0.011123	2884.3	3162.4	5.9592	0.008678	2820.7	3081.1	5.7905	0.006927	2751.9	2994.4	5.6282
550	0.012724	3017.5	3335.6	6.1765	0.010168	2970.3	3275.4	6.0342	0.008345	2921.0	3213.0	5.9026
600	0.014137	3137.6	3491.4	6.3602	0.011446	3100.5	3443.9	6.2331	0.009527	3062.0	3395.5	6.1179
650	0.015433	3251.6	3637.4	6.5229	0.012596	3221.0	3598.9	6.4058	0.010575	3189.8	3559.9	6.3010
700	0.016646	3361.3	3777.5	6.6707	0.013661	3335.8	3745.9	6.5606	0.011533	3309.8	3713.5	6.4631
800	0.018912	3574.3	4047.1	6.9345	0.015623	3555.5	4024.2	6.8332	0.013278	3536.7	4001.5	6.7450
900	0.021045	3783.0	4309.1	7.1680	0.017448	3768.5	4291.9	7.0718	0.014883	3754.0	4274.9	6.9386
1000	0.02310	3990.9	4568.5	7.3802	0.019196	3978.8	4554.7	7.2867	0.016410	3966.7	4541.1	7.2064
1100	0.02512	4200.2	4828.2	7.5765	0.020903	4189.2	4816.3	7.4845	0.017895	4178.3	4804.6	7.4037

(continued)

Superheated Table (PC Model), H₂O

°C	m³/kg	kJ/kg	kJ/kg	kJ/kg·K	m³/kg	kJ/kg	kJ/kg	kJ/kg·K	m³/kg	kJ/kg	kJ/kg	kJ/kg·K
T	$p = 25.0$ MPa				$p = 30.0$ MPa				$p = 35.0$ MPa			
	v	u	h	s	v	u	h	s	v	u	h	s
1200	0.02711	4412.0	5089.9	7.7605	0.022589	4401.3	5079.0	7.6692	0.019360	4390.7	5068.3	7.5910
1300	0.02910	4626.9	5354.4	7.9342	0.024266	4616.0	5344.0	7.8432	0.020815	4605.1	5333.6	7.7653
T	$p = 40.0$ MPa				$p = 50.0$ MPa				$p = 60.0$ MPa			
	v	u	h	s	v	u	h	s	v	u	h	s
375	0.0016407	1677.1	1742.8	3.8290	0.0015594	1638.6	1716.6	3.7639	0.0015028	1609.4	1699.5	3.7141
400	0.0019077	1854.6	1930.9	4.1135	0.0017309	1788.1	1874.6	4.0031	0.0016335	1745.4	1843.4	3.9318
425	0.002532	2096.9	2198.1	4.5029	0.002007	1959.7	2060.0	4.2734	0.0018165	1892.7	2001.7	4.1626
450	0.003693	2365.1	2512.8	4.9459	0.002486	2159.6	2284.0	4.5884	0.002085	2053.9	2179.0	4.4121
500	0.005622	2678.4	2903.3	5.4700	0.003892	2525.5	2720.1	5.1726	0.002956	2390.6	2567.9	4.9321
550	0.006984	2869.7	3149.1	5.7785	0.005118	2763.6	3019.5	5.5485	0.003956	2658.8	2896.2	5.3441
600	0.008094	3022.6	3346.4	6.0144	0.006112	2942.0	3247.6	5.8178	0.004834	2861.1	3151.2	5.6452
650	0.009063	3158.0	3520.6	6.2054	0.006966	3093.5	3441.8	6.0342	0.005595	3028.8	3364.5	5.8829
700	0.009941	3283.6	3681.2	6.3750	0.007727	3230.5	3616.8	6.2189	0.006272	3177.2	3553.5	6.0824
800	0.011523	3517.8	3978.7	6.6662	0.009076	3479.8	3933.6	6.5290	0.007459	3441.5	3889.1	6.4109
900	0.012962	3739.4	4257.9	6.9150	0.010283	3710.3	4224.4	6.7882	0.008505	3681.0	4191.5	6.6805
1000	0.014324	3954.6	4527.6	7.1356	0.011411	3930.5	4501.1	7.0146	0.009480	3906.4	4475.2	6.9127
1100	0.015642	4167.4	4793.1	7.3364	0.012496	4145.7	4770.5	7.2184	0.010409	4124.1	4748.6	7.1195
1200	0.016940	4380.1	5057.7	7.5224	0.013561	4359.1	5037.2	7.4058	0.011317	4338.2	5017.2	7.3083
1300	0.018229	4594.3	5323.5	7.6969	0.014616	4572.8	5303.6	7.5805	0.012215	4551.4	5284.3	7.4837

Table B-4   PC Model: Compressed Liquid Table, H₂O

Compressed Liquid Table (PC Model), H₂O

°C	m³/kg	kJ/kg	kJ/kg	kJ/kg·K	m³/kg	kJ/kg	kJ/kg	kJ/kg·K	m³/kg	kJ/kg	kJ/kg	kJ/kg·K
T	$p = 5$ MPa ($T_{sat} = 263.99$°C)				$p = 10$ MPa ($T_{sat} = 311.06$°C)				$p = 15$ MPa ($T_{sat} = 342.24$°C)			
	v	u	h	s	v	u	h	s	v	u	h	s
Sat.	0.0012859	1147.8	1154.2	2.9202	0.0014524	1393.0	1407.6	3.3596	0.0016581	1585.60	1610.5	3.6848
0	0.0009977	0	5.0	0.0001	0.0009952	0.1	10.0	0.0002	0.0009928	0.15	15.1	0.0004
20	0.0009995	83.7	88.7	0.2956	0.0009972	83.4	93.3	0.2945	0.0009950	83.06	98.0	0.2934
30	0.0010056	167.0	172.0	0.5705	0.0010034	166.4	176.4	0.5686	0.0010013	165.76	180.8	0.5666
60	0.0010149	250.2	255.3	0.8285	0.0010127	249.4	259.5	0.8258	0.0010105	248.51	263.7	0.8232
80	0.0010268	333.7	338.9	1.0720	0.0010245	332.6	342.8	1.0688	0.0010222	331.48	346.8	1.0656
100	0.0010410	417.5	422.7	1.3030	0.0010385	416.1	426.5	1.2992	0.0010361	414.74	430.3	1.2955
120	0.0010576	501.8	507.1	1.5233	0.0010549	500.1	510.6	1.5189	0.0010522	498.40	514.2	1.5145
140	0.0010768	586.8	592.2	1.7343	0.0010737	584.7	595.4	1.7292	0.0010707	582.66	598.7	1.7242
160	0.0010988	672.6	678.1	1.9375	0.0010953	670.1	681.1	1.9317	0.0010918	667.71	684.1	1.9260
180	0.0011240	759.6	765.3	2.1341	0.0011199	756.7	767.8	2.1275	0.0011159	753.76	770.5	2.1210
200	0.0011530	848.1	853.9	2.3255	0.0011480	844.5	856.0	2.3178	0.0011433	841.00	858.2	2.3104
220	0.0011866	938.4	944.4	2.5128	0.0011805	934.1	945.9	2.5039	0.0011748	929.90	947.5	2.4953
240	0.0012264	1031.4	1037.5	2.6979	0.0012187	1026.0	1038.1	2.6872	0.0012114	1020.80	1039.0	2.6771

Compressed Liquid Table (PC Model), H₂O

°C	m³/kg	kJ/kg	kJ/kg	kJ/kg·K	m³/kg	kJ/kg	kJ/kg	kJ/kg·K	m³/kg	kJ/kg	kJ/kg	kJ/kg·K
T	$p = 5$ MPa ($T_{sat} = 263.99$°C)				$p = 10$ MPa ($T_{sat} = 311.06$°C)				$p = 15$ MPa ($T_{sat} = 342.24$°C)			
	v	*u*	*h*	*s*	*v*	*u*	*h*	*s*	*v*	*u*	*h*	*s*
260	0.0012749	1127.9	1134.3	2.8830	0.0012645	1121.1	1133.7	2.8699	0.0012550	1114.60	1133.4	2.8576
280	–	–	–	–	0.0013216	1220.9	1234.1	3.0548	0.0013084	1212.50	1232.1	3.0393
300	–	–	–	–	0.0013972	1328.4	1342.3	3.2469	0.0013770	1316.60	1337.3	3.2260
320	–	–	–	–	–	–	–	–	0.0014724	1431.10	1453.2	3.4247
340	–	–	–	–	–	–	–	–	0.0016311	1567.50	1591.9	3.6546
T	$p = 20$ MPa ($T_{sat} = 365.81$°C)				$p = 30$ MPa				$p = 50$ MPa			
	v	*u*	*h*	*s*	*v*	*u*	*h*	*s*	*v*	*u*	*h*	*s*
Sat.	0.0020360	1785.6	1826.3	4.0139	–	–	–	–	–	–	–	–
0	0.0009904	0.2	20.0	0.0004	0.0009856	0.3	29.8	0.0001	0.0009766	0.20	49.0	0.0014
20	0.0009928	82.8	102.6	0.2923	0.0009886	82.2	111.8	0.2899	0.0009804	81.00	130.0	0.2848
40	0.0009992	165.2	185.2	0.5646	0.0009951	164.0	193.9	0.5607	0.0009872	161.86	211.2	0.5527
60	0.0010084	247.7	267.9	0.8206	0.0010042	246.1	276.2	0.8154	0.0009962	242.98	292.8	0.8052
80	0.0010199	330.4	350.8	1.0624	0.0010156	328.3	358.8	1.0561	0.0010073	324.34	374.7	1.0440
100	0.0010337	413.4	434.1	1.2917	0.0010290	410.8	441.7	1.2844	0.0010201	405.88	456.9	1.2703
120	0.0010496	496.8	517.8	1.5102	0.0010445	493.6	524.9	1.5018	0.0010348	487.65	539.4	1.4857
140	0.0010678	580.7	602.0	1.7193	0.0010621	576.9	608.8	1.7098	0.0010515	569.77	622.4	1.6915
160	0.0010885	665.4	687.1	1.9204	0.0010821	660.8	693.3	1.9096	0.0010703	652.41	705.9	1.8891
180	0.0011120	751.0	773.2	2.1147	0.0011047	745.6	778.7	2.1024	0.0010912	735.69	790.3	2.0794
200	0.0011388	837.7	860.5	2.3031	0.0011302	831.4	865.3	2.2893	0.0011146	819.70	875.5	2.2634
220	0.0011695	925.9	949.3	2.4870	0.0011590	918.3	953.1	2.4711	0.0011408	904.70	961.7	2.4419
240	0.0012046	1016.0	1040.0	2.6674	0.0011920	1006.9	1042.6	2.6490	0.0011702	990.70	1049.2	2.6158
260	0.0012462	1108.6	1133.5	2.8459	0.0012303	1097.4	1134.3	2.8243	0.0012034	1078.10	1138.2	2.7860
280	0.0012965	1204.7	1230.6	3.0248	0.0012755	1190.7	1229.0	2.9986	0.0012415	1167.20	1229.3	2.9537
300	0.0013596	1306.1	1333.3	3.2071	0.0013307	1287.9	1327.8	3.1741	0.0012860	1258.70	1323.0	3.1200
320	0.0014437	1415.7	1444.6	3.3979	0.0013997	1390.7	1432.7	3.3539	0.0013388	1353.30	1420.2	3.2868
340	0.0015684	1539.7	1571.0	3.6075	0.0014920	1501.7	1546.5	3.5426	0.0014032	1452.00	1522.1	3.4557
360	0.0018226	1702.8	1739.3	3.8772	0.0016265	1626.6	1675.4	3.7494	0.0014838	1556.00	1630.2	3.6291
380	–	–	–	–	0.0018691	1781.4	1837.5	4.0012	0.0015884	1667.20	1746.6	3.8101

Table B-5   PC Model: Temperature Saturation Table, Ice-Water Mixture

Saturation Ice-Water Vapor Table (PC Model)

Temp.	Sat. Press.	Spec. Volume		Spec. Int. Energy		Spec. Enthalpy		Spec. Entropy	
		m³/kg		kJ/kg		kJ/kg		kJ/kg·K	
°C	kPa	Sat. ice	Sat. vapor	Sat. ice	Sat. vapor	Sat. ice	Sat. vapor	Sat. ice	Sat. vapor
T	$p_{sat@T}$	$v_i \times 10^3$	v_g	u_i	u_g	h_i	h_g	s_i	s_g
0.01	0.6113	1.0908	206.1	−333.40	2375.3	−333.40	2501.4	−1.221	9.156
0	0.6108	1.0908	206.3	−333.43	2375.3	−333.43	2501.3	−1.221	9.157
−2	0.5176	1.0904	241.7	−337.62	2372.6	−337.62	2497.7	−1.237	9.219

(continued)

Saturation Ice-Water Vapor Table (PC Model)

Temp.	Sat. Press.	Spec. Volume		Spec. Int. Energy		Spec. Enthalpy		Spec. Entropy	
		m³/kg		kJ/kg		kJ/kg		kJ/kg·K	
°C	kPa	Sat. ice	Sat. vapor	Sat. ice	Sat. vapor	Sat. ice	Sat. vapor	Sat. ice	Sat. vapor
T	$p_{sat@T}$	$v_i \times 10^3$	v_g	u_i	u_g	h_i	h_g	s_i	s_g
−4	0.4375	1.0901	283.8	−341.78	2369.8	−341.78	2494.0	−1.253	9.283
−6	0.3689	1.0898	334.2	−345.91	2367.0	−345.91	2490.3	−1.268	9.348
−8	0.3102	1.0894	394.4	−350.02	2364.2	−350.02	2486.6	−1.284	9.414
−10	0.2602	1.0891	466.7	−354.09	2361.4	−354.09	2482.9	−1.299	9.481
−12	0.2176	1.0888	553.7	−358.14	2358.7	−358.14	2479.2	−1.315	9.550
−14	0.1815	1.0884	658.8	−362.15	2355.9	−362.15	2475.5	−1.331	9.619
−16	0.1510	1.0881	786.0	−366.14	2353.1	−366.14	2471.8	−1.346	9.690
−18	0.1252	1.0878	940.5	−370.10	2350.3	−370.10	2468.1	−1.362	9.762
−20	0.1035	1.0874	1128.6	−374.03	2347.5	−374.03	2464.3	−1.377	9.835
−22	0.0853	1.0871	1358.4	−377.93	2344.7	−377.93	2460.6	−1.393	9.909
−24	0.0701	1.0868	1640.1	−381.80	2342.0	−381.80	2456.9	−1.408	9.985
−26	0.0574	1.0864	1986.4	−385.64	2339.2	−385.64	2453.2	−1.424	10.062
−28	0.0469	1.0861	2413.7	−389.45	2336.4	−389.45	2449.5	−1.439	10.141
−30	0.0381	1.0858	2943.0	−393.23	2333.6	−393.23	2445.8	−1.455	10.221
−32	0.0309	1.0854	3600.0	−396.98	2330.8	−396.98	2442.1	−1.471	10.303
−34	0.0250	1.0851	4419.0	−400.71	2328.0	−400.71	2438.4	−1.486	10.386
−36	0.0201	1.0848	5444.0	−404.40	2325.2	−404.40	2434.7	−1.501	10.470
−38	0.0161	1.0844	6731.0	−408.06	2322.4	−408.06	2430.9	−1.517	10.556
−40	0.0129	1.0841	8354.0	−411.70	2319.6	−411.70	2427.2	−1.532	10.644

Table B-6 PC Model: Temperature Saturation Table, R-134a

Saturation Temperature Table (PC Model) of R-134a

Temp.	Sat. Press.	Spec. Volume		Spec. Int. Energy		Spec. Enthalpy		Spec. Entropy	
		m³/kg		kJ/kg		kJ/kg		kJ/kg·K	
°C	MPa	Sat. liquid	Sat. vapor	Sat. liquid	Sat. vapor	Sat. liquid	Sat. vapor	Sat. liquid	Sat. vapor
T	$p_{sat@T}$	v_f	v_g	u_f	u_g	h_f	h_g	s_f	s_g
−24	0.11160	0.0007296	0.1728	19.21	213.57	19.29	232.85	0.0798	0.9370
−22	0.12192	0.0007328	0.1590	21.68	214.70	21.77	234.08	0.0897	0.9351
−20	0.13299	0.0007361	0.1464	24.17	215.84	24.26	235.31	0.0996	0.9332

Saturation Temperature Table (PC Model) of R-134a

Temp.	Sat. Press.	Spec. Volume		Spec. Int. Energy		Spec. Enthalpy		Spec. Entropy	
		m³/kg		kJ/kg		kJ/kg		kJ/kg·K	
°C	MPa	Sat. liquid	Sat. vapor	Sat. liquid	Sat. vapor	Sat. liquid	Sat. vapor	Sat. liquid	Sat. vapor
T	$p_{sat@T}$	v_f	v_g	u_f	u_g	h_f	h_g	s_f	s_g
−18	0.14483	0.0007395	0.1350	26.67	216.97	26.77	236.53	0.1094	0.9315
−16	0.15748	0.0007428	0.1247	29.18	218.10	29.30	237.74	0.1192	0.9298
−12	0.18540	0.0007498	0.1068	34.25	220.36	34.39	240.15	0.1388	0.9267
−8	0.21704	0.0007569	0.0919	39.38	222.60	39.54	242.54	0.1583	0.9239
−4	0.25274	0.0007644	0.0794	44.56	224.84	44.75	244.90	0.1777	0.9213
0	0.29282	0.0007721	0.0689	49.79	227.06	50.02	247.23	0.1970	0.9190
4	0.33765	0.0007801	0.0600	55.08	229.27	55.35	249.53	0.2162	0.9169
8	0.38756	0.0007884	0.0525	60.43	231.46	60.73	251.80	0.2354	0.9150
12	0.44294	0.0007971	0.0460	65.83	233.63	66.18	254.03	0.2545	0.9132
16	0.50416	0.0008062	0.0405	71.29	235.78	71.69	256.22	0.2735	0.9116
20	0.57160	0.0008157	0.0358	76.80	237.91	77.26	258.35	0.2924	0.9102
24	0.64566	0.0008257	0.0317	82.37	240.01	82.90	260.45	0.3113	0.9089
26	0.68530	0.0008309	0.0298	85.18	241.05	85.75	261.48	0.3208	0.9082
28	0.72675	0.0008362	0.0281	88.00	242.08	88.61	262.50	0.3302	0.9076
30	0.77006	0.0008417	0.0265	90.84	243.10	91.49	263.50	0.3396	0.9070
32	0.81528	0.0008473	0.0250	93.70	244.12	94.39	264.48	0.3490	0.9064
34	0.86247	0.0008530	0.0236	96.58	245.12	97.31	265.45	0.3584	0.9058
36	0.91168	0.0008590	0.0223	99.47	246.11	100.25	266.40	0.3678	0.9053
38	0.96298	0.0008651	0.0210	102.38	247.09	103.21	267.33	0.3772	0.9047
40	1.01640	0.0008714	0.0199	105.30	248.06	106.19	268.24	0.3866	0.9041
42	1.07200	0.0008780	0.0188	108.25	249.02	109.19	269.14	0.3960	0.9035
44	1.12990	0.0008847	0.0177	111.22	249.96	112.22	270.01	0.4054	0.9030
48	1.25260	0.0008989	0.0159	117.22	251.79	118.35	271.68	0.4243	0.9017
52	1.38510	0.0009142	0.0142	123.31	253.55	124.58	273.24	0.4432	0.9004
56	1.52780	0.0009308	0.0127	129.51	255.23	130.93	274.68	0.4622	0.8990
60	1.68130	0.0009488	0.0114	135.82	256.81	137.42	275.99	0.4814	0.8973
70	2.11620	0.0010027	0.0086	152.22	260.15	154.34	278.43	0.5302	0.8918
80	2.63240	0.0010766	0.0064	169.88	262.14	172.71	279.12	0.5814	0.8827
90	3.24350	0.0011949	0.0046	189.82	261.34	193.69	276.32	0.6380	0.8655
100	3.97420	0.0015443	0.0027	218.60	248.49	224.74	259.13	0.7196	0.8117

Table B-7 PC Model: Superheated Vapor Table, R-134a

Superheated Table (PC Model), R-134a											
°C	m³/kg	kJ/kg	kJ/kg	kJ/kg·K	m³/kg	kJ/kg	kJ/kg	kJ/kg·K	m³/kg	kJ/kg	kJ/kg	kJ/kg·K
	$p = 0.06$ MPa ($T_{sat} = -37.07°C$)				$p = 0.10$ MPa ($T_{sat} = -26.43°C$)				$p = 0.14$ MPa ($T_{sat} = -18.80°C$)			
T	v	u	h	s	v	u	h	s	v	u	h	s
Sat.	0.31003	206.12	224.72	0.9520	0.19170	212.18	231.35	0.9395	0.13945	216.52	236.04	0.9322
−20	0.33536	217.86	237.98	1.0062	0.19770	216.77	236.54	0.9602	0.14549	223.03	243.40	0.9606
−10	0.34992	224.97	245.96	1.0371	0.20686	224.01	244.70	0.9918	0.15219	230.55	251.86	0.9922
0	0.36433	232.24	254.10	1.0675	0.21587	231.41	252.99	1.0227	0.15875	238.32	260.43	1.0230
10	0.37861	239.69	262.41	1.0973	0.22473	238.96	261.43	1.0531	0.16520	246.01	269.13	1.0532
20	0.39279	247.32	270.89	1.1267	0.23349	246.67	270.02	1.0829	0.17155	253.96	277.97	1.0828
30	0.40688	255.12	279.53	1.1557	0.24216	254.54	278.76	1.1122	0.17783	262.06	286.96	1.1120
40	0.42091	263.10	288.35	1.1844	0.25076	262.58	287.66	1.1411	0.18404	270.32	296.09	1.1407
50	0.43487	271.25	297.34	1.2126	0.59300	270.79	296.72	1.1696	0.19020	278.74	305.37	1.1690
60	0.44879	279.58	306.51	1.2405	0.26779	279.16	305.94	1.1977	0.19633	287.32	314.80	1.1969
70	0.46266	288.08	315.84	1.2681	0.27623	287.70	315.32	1.2254	0.20241	296.06	324.39	1.2244
80	0.47650	296.75	325.34	1.2954	0.28464	296.40	324.87	1.2528	0.20846	304.95	334.14	1.2516
90	0.49031	305.58	335.00	1.3224	0.29302	305.27	344.57	1.2799	0.21449	314.01	344.04	1.2785
T	$p = 0.18$ MPa ($T_{sat} = -12.73°C$)				$p = 0.20$ MPa ($T_{sat} = -10.09°C$)				$p = 0.24$ MPa ($T_{sat} = -5.37°C$)			
	v	u	h	s	v	u	h	s	v	u	h	s
Sat.	0.10983	219.94	239.71	0.9273	0.09933	221.43	241.30	0.9253	0.08343	224.07	244.09	0.9222
−10	0.11135	222.02	242.06	0.9362	0.09938	221.50	241.38	0.9256	–	–	–	–
0	0.11678	229.67	250.69	0.9684	0.10438	229.23	250.10	0.9582	0.08574	228.31	248.89	0.9399
10	0.12207	237.44	259.41	0.9998	0.10922	237.05	258.89	0.9898	0.08993	236.26	257.84	0.9721
20	0.12723	245.33	268.23	1.0304	0.11394	244.99	267.78	1.0206	0.09339	244.30	266.85	1.0034
30	0.13230	253.36	277.17	1.0604	0.11856	253.06	276.77	1.0508	0.09794	252.45	275.95	1.0339
40	0.13730	261.53	286.24	1.0898	0.12311	261.26	285.88	1.0804	0.10181	260.72	285.16	1.0637
50	0.14222	269.85	295.45	1.1187	0.12758	269.61	295.12	1.1094	0.10562	269.12	294.47	1.0930
60	0.14710	278.31	304.79	1.1472	0.13201	278.10	304.50	1.1380	0.10937	277.67	303.91	1.1218
70	0.15193	286.93	314.28	1.1753	0.13639	286.74	314.02	1.1661	0.11307	286.35	313.49	1.1501
80	0.15672	295.71	323.92	1.2030	0.14073	295.53	323.68	1.1939	0.11674	295.18	323.19	1.1780
90	0.16148	304.63	333.70	1.2303	0.14504	304.47	333.48	1.2212	0.12037	304.15	333.04	1.2055
100	0.16622	313.72	343.69	1.2573	0.14932	313.57	343.43	1.2483	0.12398	313.27	343.03	1.2326
T	$p = 0.28$ MPa ($T_{sat} = -1.23°C$)				$p = 0.32$ MPa ($T_{sat} = 2.48°C$)				$p = 0.40$ MPa ($T_{sat} = 8.93°C$)			
	v	u	h	s	v	u	h	s	v	u	h	s
Sat.	0.07193	226.38	246.52	0.9197	0.06322	228.43	248.66	0.9177	0.05089	231.97	252.32	0.9145
0	0.07240	227.37	247.64	0.9238	–	–	–	–	–	–	–	–
10	0.07613	235.44	256.76	0.9566	0.06576	234.61	255.65	0.9427	0.05119	232.87	253.35	0.9182
20	0.07972	243.59	265.91	0.9883	0.06901	242.87	264.95	0.9749	0.05397	241.37	262.96	0.9515
30	0.08320	251.83	275.12	1.0192	0.07214	251.19	274.28	1.0062	0.05662	249.89	272.54	0.9837
40	0.08660	260.17	284.42	1.0494	0.07518	259.61	283.67	1.0367	0.05917	258.47	282.14	1.0148
50	0.08992	268.64	293.81	1.0789	0.07815	268.14	293.15	1.0665	0.06164	267.13	291.79	1.0452
60	0.09319	277.23	303.32	1.1079	0.08106	276.79	302.72	1.0957	0.06405	275.89	301.51	1.0748
70	0.09641	285.96	312.95	1.1364	0.08392	285.56	312.41	1.1243	0.06641	284.75	311.32	1.1038
80	0.09960	294.82	322.71	1.1644	0.08674	294.46	322.22	1.1525	0.06873	293.73	321.23	1.1322

Superheated Table (PC Model), R-134a

°C	m³/kg	kJ/kg	kJ/kg	kJ/kg·K	m³/kg	kJ/kg	kJ/kg	kJ/kg·K	m³/kg	kJ/kg	kJ/kg	kJ/kg·K
T	$p = 0.28$ MPa ($T_{sat} = -1.23$°C)				$p = 0.32$ MPa ($T_{sat} = 2.48$°C)				$p = 0.40$ MPa ($T_{sat} = 8.93$°C)			
	v	*u*	*h*	*s*	*v*	*u*	*h*	*s*	*v*	*u*	*h*	*s*
90	0.10275	303.83	332.60	1.1920	0.08953	303.50	332.15	1.1802	0.07102	302.84	331.25	1.1602
100	0.10587	312.98	342.62	1.2193	0.09229	312.68	342.21	1.1076	0.07327	312.07	341.38	1.1878
110	0.10897	322.27	352.78	1.2461	0.09503	322.00	352.40	1.2345	0.07550	321.44	351.64	1.2149
120	0.11205	331.71	363.08	1.2727	0.09774	331.45	362.73	1.2611	0.07771	330.94	362.03	1.2417
130	–	–	–	–	–	–	–	–	0.07991	340.58	372.54	1.2681
140	–	–	–	–	–	–	–	–	0.08208	350.35	383.18	1.2941
T	$p = 0.50$ MPa ($T_{sat} = 15.74$°C)				$p = 0.60$ MPa ($T_{sat} = 21.58$°C)				$p = 0.70$ MPa ($T_{sat} = 26.72$°C)			
	v	*u*	*h*	*s*	*v*	*u*	*h*	*s*	*v*	*u*	*h*	*s*
Sat.	0.04086	253.64	256.07	0.9117	0.03408	238.74	259.19	0.9097	0.02918	241.42	261.85	0.9080
20	0.04188	239.40	260.34	0.9264	–	–	–	–	–	–	–	–
30	0.04416	248.20	270.28	0.9597	0.03581	246.41	267.89	0.9388	0.02979	244.51	265.37	0.9197
40	0.04633	256.99	280.16	0.9918	0.03774	255.45	278.09	0.9719	0.03157	253.83	275.93	0.9539
50	0.04842	265.83	290.04	1.0229	0.03958	264.48	288.23	1.0037	0.03324	263.08	286.35	1.9867
60	0.05043	274.73	299.95	1.0531	0.04134	273.54	298.35	1.0346	0.03482	272.31	296.69	1.0182
70	0.05240	283.72	309.92	1.0825	0.04304	282.66	308.48	1.0645	0.03634	281.57	307.01	1.0487
80	0.05432	292.80	319.96	1.1114	0.04469	291.86	318.67	1.0938	0.03781	290.88	317.35	1.0784
90	0.05620	302.00	330.10	1.1397	0.04631	301.14	328.93	1.1225	0.03924	300.27	327.74	1.1074
100	0.05805	311.31	340.33	1.1675	0.04790	310.53	339.27	1.1505	0.04064	309.74	338.19	1.1358
110	0.05988	320.74	350.68	1.1949	0.04946	320.03	349.70	1.1781	0.04201	319.31	348.71	1.1637
120	0.06168	330.30	361.14	1.2218	0.05099	329.64	360.24	1.2053	0.04335	328.98	359.33	1.1910
130	0.06347	339.98	371.72	1.2484	0.05251	339.38	370.88	1.2320	0.04468	338.76	370.04	1.2179
140	0.06524	349.79	382.42	1.2746	0.05402	349.23	381.64	1.2584	0.04599	348.66	380.86	1.2444
150	–	–	–	–	0.05550	359.21	392.52	1.2844	0.04729	358.68	391.79	1.2706
160	–	–	–	–	0.05698	369.32	403.51	1.3100	0.04857	368.82	402.82	1.2963
T	$p = 0.80$ MPa ($T_{sat} = 31.33$°C)				$p = 0.90$ MPa ($T_{sat} = 35.53$°C)				$p = 1.00$ MPa ($T_{sat} = 39.33$°C)			
	v	*u*	*h*	*s*	*v*	*u*	*h*	*s*	*v*	*u*	*h*	*s*
Sat.	0.02547	243.78	264.15	0.9066	0.02255	245.88	266.18	0.9054	0.02020	247.77	267.97	0.9043
40	0.02691	252.13	273.66	0.9374	0.02325	250.32	271.25	0.9217	0.02029	248.39	268.68	0.9066
50	0.02846	261.62	284.39	0.9711	0.02472	260.09	282.34	0.9566	0.02171	258.48	280.19	0.9428
60	0.02992	271.04	294.98	1.0034	0.02609	269.72	239.21	0.9897	0.02301	268.35	291.36	0.9768
70	0.03131	280.45	305.50	1.0345	0.02738	279.30	303.94	1.0214	0.02423	278.11	302.34	1.0093
80	0.03264	289.89	316.00	1.0647	0.02861	288.87	314.62	1.0521	0.02538	287.82	313.20	1.0405
90	0.03393	299.37	326.52	1.0940	0.02980	298.46	325.28	1.0819	0.02649	297.53	324.01	1.0707
100	0.03519	308.93	337.08	1.1227	0.03095	308.11	335.96	1.1109	0.02755	307.27	334.82	1.1000
110	0.03642	318.57	347.71	1.1508	0.03207	317.82	346.68	1.1392	0.02858	317.06	345.65	1.1286
120	0.03762	328.31	358.40	1.1784	0.03316	327.62	357.47	1.1670	0.02959	326.93	356.52	1.1567
130	0.03881	338.14	369.19	1.2055	0.03423	337.52	368.33	1.1943	0.03058	336.88	367.46	1.1841
140	0.03997	348.09	380.07	1.2321	0.03529	347.51	379.27	1.2211	0.03154	346.92	378.46	1.2111
150	0.04113	358.15	391.05	1.2584	0.03633	357.61	390.31	1.2475	0.03250	357.06	389.56	1.2376
160	0.04227	368.32	402.14	1.2843	0.03736	367.82	401.44	1.2735	0.03344	367.31	400.74	1.2638
170	0.04340	378.61	413.33	1.3098	0.03838	378.14	412.68	1.2992	0.03436	377.66	412.02	1.2895
180	0.04452	389.02	424.63	1.3351	0.03939	388.57	424.02	1.3245	0.03528	388.12	423.40	1.3149

(continued)

Superheated Table (PC Model), R-134a

°C	m³/kg	kJ/kg	kJ/kg	kJ/kg·K	m³/kg	kJ/kg	kJ/kg	kJ/kg·K	m³/kg	kJ/kg	kJ/kg	kJ/kg·K
T	$p=1.20$ MPa ($T_{sat}=46.32$°C)				$p=1.40$ MPa ($T_{sat}=52.43$°C)				$p=1.60$ MPa ($T_{sat}=57.92$°C)			
	v	u	h	s	v	u	h	s	v	u	h	s
Sat.	0.01663	251.03	270.99	0.9023	0.01405	253.74	273.40	0.9003	0.01208	256.00	275.33	0.8982
50	0.01712	254.98	275.52	0.9164	–	–	–	–	–	–	–	–
60	0.01835	265.42	287.44	0.9527	0.01495	262.17	283.10	0.9297	0.01233	258.48	278.20	0.9069
70	0.01947	275.59	298.96	0.9868	0.01603	272.87	295.31	0.9658	0.01340	269.89	291.33	0.9457
80	0.02051	285.62	310.24	1.0192	0.01701	283.29	307.10	0.9997	0.01435	280.78	303.74	0.9813
90	0.02150	295.59	321.39	1.0503	0.01792	293.55	318.63	1.0319	0.01521	291.39	315.72	1.0148
100	0.02244	305.54	332.47	1.0804	0.01878	303.73	330.02	1.0628	0.01601	301.84	327.46	1.0467
110	0.02335	315.50	343.52	1.1096	0.01960	313.88	341.32	1.0927	0.01677	312.20	339.04	1.0773
120	0.02423	325.51	354.58	1.1381	0.02039	324.05	352.59	1.1218	0.01750	322.53	350.53	1.1069
130	0.02508	335.58	365.68	1.1600	0.02115	334.25	363.86	1.1501	0.01820	332.87	361.99	1.1357
140	0.02592	345.73	376.83	1.1933	0.02189	344.50	375.15	1.1777	0.01887	343.24	373.44	1.1638
150	0.02674	355.95	388.04	1.2201	0.02262	354.82	386.49	1.2048	0.01953	353.66	384.91	1.1912
160	0.02754	366.27	399.33	1.2465	0.02333	365.22	397.89	1.2315	0.02017	364.15	396.43	1.2181
180	0.02912	387.21	422.16	1.2980	0.02472	386.29	420.90	1.2834	0.02142	385.35	419.62	1.2704
190	–	–	–	–	0.02541	396.96	432.53	1.3088	0.02203	396.08	431.33	1.2960
190	–	–	–	–	0.02541	396.96	432.53	1.3088	0.02203	396.08	431.33	1.2960
200	–	–	–	–	0.02608	407.73	444.24	1.3338	0.02263	406.90	443.11	1.3212

Table B-8   PC Model: Temperature Saturation Table, R-22

Saturation Temperature Table (PC Model) of R-22

Temp.	Sat. Press.	Spec. Volume		Spec. Int. Energy		Spec. Enthalpy		Spec. Entropy	
		m³/kg		kJ/kg		kJ/kg		kJ/kg·K	
°C	kPa	Sat. liquid	Sat. vapor	Sat. liquid	Sat. vapor	Sat. liquid	Sat. vapor	Sat. liquid	Sat. vapor
T	$p_{sat@T}$	v_f	v_g	u_f	u_g	h_f	h_g	s_f	s_g
−60	0.3749	0.0006833	0.53700	−21.57	203.67	−21.55	223.81	−0.0964	1.0547
−50	0.6451	0.0006966	0.32390	−10.89	207.70	−10.85	228.60	−0.0474	1.0256
−45	0.8290	0.0007037	0.25640	−5.50	209.70	−5.44	230.95	−0.0235	1.0126
−40	1.0522	0.0007109	0.20520	−0.07	211.68	0	233.27	0	1.0005
−36	1.2627	0.0007169	0.17300	4.29	213.25	4.38	235.09	0.0186	0.9914
−32	1.5049	0.0007231	0.14680	8.68	214.80	8.79	236.89	0.0369	0.9828
−30	1.6389	0.0007262	0.13550	10.88	215.58	11.00	237.78	0.0460	0.9787
−28	1.7819	0.0007294	0.12520	13.09	216.34	13.22	238.66	0.0551	0.9746
−26	1.9345	0.0007327	0.11590	15.31	217.11	15.45	239.53	0.0641	0.9707
−22	2.2698	0.0007393	0.09970	19.76	218.62	19.92	241.24	0.0819	0.9631
−20	2.4534	0.0007427	0.09260	21.99	219.37	22.17	242.09	0.0908	0.9595
−18	2.6482	0.0007462	0.08610	24.23	220.11	24.43	242.92	0.0996	0.9559
−16	2.8547	0.0007497	0.08020	26.48	220.85	26.69	243.74	0.1084	0.9525

Saturation Temperature Table (PC Model) of R-22

Temp.	Sat. Press.	Spec. Volume		Spec. Int. Energy		Spec. Enthalpy		Spec. Entropy	
		m³/kg		kJ/kg		kJ/kg		kJ/kg·K	
°C	kPa	Sat. liquid	Sat. vapor	Sat. liquid	Sat. vapor	Sat. liquid	Sat. vapor	Sat. liquid	Sat. vapor
T	$p_{sat@T}$	v_f	v_g	u_f	u_g	h_f	h_g	s_f	s_g
−14	3.0733	0.0007533	0.07480	28.73	221.58	28.97	244.56	0.1171	0.9490
−12	3.3044	0.0007569	0.06980	31.00	222.30	31.25	245.36	0.1258	0.9457
−10	3.5485	0.0007606	0.06520	33.27	223.02	33.54	246.15	0.1345	0.9424
−8	3.8062	0.0007644	0.06100	35.54	223.13	35.83	246.93	0.1431	0.9392
−6	4.0777	0.0007683	0.05710	37.83	224.43	38.14	247.70	0.1517	0.9361
−4	4.3638	0.0007722	0.05350	40.12	225.13	40.46	248.45	0.1602	0.9330
−2	4.6647	0.0007762	0.05010	42.42	225.82	42.78	249.20	0.1688	0.9300
0	4.9811	0.0007803	0.04700	44.73	226.50	45.12	249.92	0.1773	0.9271
2	5.3133	0.0007844	0.04420	47.04	227.17	47.46	250.64	0.1857	0.9241
4	5.6619	0.0007887	0.04150	49.37	227.83	49.82	251.34	0.1941	0.9213
6	6.0275	0.0007930	0.03910	51.71	228.48	52.18	252.03	0.2025	0.9184
8	6.4105	0.0007974	0.03680	54.05	229.13	54.56	252.70	0.2109	0.9157
10	6.8113	0.0008020	0.03460	56.40	229.76	56.95	253.35	0.2193	0.9129
12	7.2307	0.0008066	0.03260	58.77	230.38	59.35	253.99	0.2276	0.9102
16	8.1268	0.0008162	0.02910	63.53	231.59	64.19	255.21	0.2442	0.9048
20	9.1030	0.0008263	0.02590	68.33	232.76	69.09	256.37	0.2607	0.8996
24	10.1640	0.0008369	0.02320	73.19	233.87	74.04	257.44	0.2772	0.8944
28	11.3130	0.0008480	0.02080	78.09	234.92	79.05	258.43	0.2936	0.8893
32	12.5560	0.0008599	0.01860	83.06	235.91	84.14	259.32	0.3101	0.8842
36	13.8970	0.0008724	0.01680	88.08	236.83	89.29	260.11	0.3265	0.8790
40	15.3410	0.0008858	0.01510	93.18	237.66	94.53	260.79	0.3429	0.8738
45	17.2980	0.0009039	0.01320	99.65	238.59	101.21	261.46	0.3635	0.8672
50	19.4330	0.0009238	0.01160	106.26	239.34	108.06	261.90	0.3842	0.8603
60	24.2810	0.0009705	0.00890	120.00	240.24	122.35	261.96	0.4264	0.8455

Table B-9   PC Model: Superheated Vapor Table, R-22

Superheated Table (PC Model), R-22

°C	m³/kg	kJ/kg	kJ/kg·K	m³/kg	kJ/kg	kJ/kg·K
T	$p = 0.04$ MPa ($T_{sat} = -58.86$°C)			$p = 0.06$ MPa ($T_{sat} = -51.40$°C)		
	v	h	s	v	h	s
Sat.	0.50559	224.36	1.0512	0.34656	227.93	1.0294
−55	0.51532	226.53	1.0612	–	–	–

(continued)

			Superheated Table (PC Model), R-22			
°C	**m³/kg**	**kJ/kg**	**kJ/kg·K**	**m³/kg**	**kJ/kg**	**kJ/kg·K**
T	$p = 0.04$ MPa ($T_{sat} = -58.86°C$)			$p = 0.06$ MPa ($T_{sat} = -51.40°C$)		
	v	**h**	**s**	**v**	**h**	**s**
−50	0.52787	229.38	1.0741	0.34895	228.74	1.0330
−45	0.54037	232.24	1.0868	0.35747	231.65	1.0459
−40	0.55284	235.13	1.0993	0.36594	234.58	1.0586
−35	0.56526	238.05	1.1117	0.37437	237.52	1.0711
−30	0.57766	240.99	1.1239	0.38277	240.49	1.0835
−25	0.59002	243.95	1.1360	0.39114	243.49	1.0956
−20	0.60236	246.95	1.1479	0.39948	246.51	1.1077
−15	0.61468	249.97	1.1597	0.40779	249.55	1.1196
−10	0.62697	253.01	1.1714	0.41608	252.62	1.1314
−5	0.63925	256.09	1.1830	0.42436	255.71	1.1430
0	0.65151	259.19	1.1944	0.43261	258.83	1.1545
T	$p = 0.08$ MPa ($T_{sat} = -45.73°C$)			$p = 0.10$ MPa ($T_{sat} = -41.09°C$)		
	v	**h**	**s**	**v**	**h**	**s**
Sat.	0.26503	230.61	1.0144	0.21518	232.77	1.0031
−45	0.26597	231.04	1.0163	–	–	–
−40	0.27245	234.01	1.0291	0.21633	233.42	1.0059
−35	0.27890	236.99	1.0418	0.22158	236.44	1.0187
−30	0.28530	239.99	1.0543	0.22679	239.48	1.0313
−25	0.29167	243.02	1.0666	0.23197	242.54	1.0438
−20	0.29801	246.06	1.0788	0.23712	245.61	1.0560
−15	0.30433	249.13	1.0908	0.24224	248.70	1.0681
−10	0.31062	252.22	1.1026	0.24734	251.82	1.0801
−5	0.31690	255.34	1.1143	0.25241	254.95	1.0919
0	0.32315	258.47	1.1259	0.25747	258.11	1.1035
5	0.32939	261.64	1.1374	0.26251	261.29	1.1151
10	0.33561	264.83	1.1488	0.26753	264.50	1.1265
T	$p = 0.15$ MPa ($T_{sat} = -32.08°C$)			$p = 0.20$ MPa ($T_{sat} = -25.18°C$)		
	v	**h**	**s**	**v**	**h**	**s**
Sat.	0.14721	236.86	0.9830	0.11232	239.88	0.9691
−30	0.14872	238.16	0.9883	–	–	–
−25	0.15232	241.30	1.0011	0.11242	240.00	0.9696
−20	0.15588	244.45	1.0137	0.11520	243.23	0.9825
−15	0.15941	247.61	1.0260	0.11795	246.47	0.9952
−10	0.16292	250.78	1.0382	0.12067	249.72	1.0076
−5	0.16640	253.98	1.0502	0.12336	252.97	1.0199
10	0.16987	257.18	1.0621	0.12603	256.23	1.0310
5	0.17331	260.41	1.0738	0.12868	259.51	1.0438
10	0.17674	263.66	1.0854	0.13132	262.81	1.0555
15	0.18015	266.93	1.0968	0.13393	266.12	1.0671
20	0.18355	270.22	1.1081	0.13653	269.44	1.0786
25	0.18693	273.53	1.1193	0.13912	272.79	1.0899

°C	m³/kg	kJ/kg	kJ/kg·K	m³/kg	kJ/kg	kJ/kg·K
	Superheated Table (PC Model), R-22					

Superheated Table (PC Model), R-22

°C	m³/kg	kJ/kg	kJ/kg·K	m³/kg	kJ/kg	kJ/kg·K
T	$p = 0.25$ MPa ($T_{sat} = -19.51°C$)			$p = 0.30$ MPa ($T_{sat} = -14.66°C$)		
	v	h	s	v	h	s
Sat.	0.09097	242.29	0.9586	0.07651	244.29	0.9502
−15	0.09303	245.29	0.9703	–	–	–
−10	0.09528	248.61	0.9831	0.07833	247.46	0.9623
−5	0.09751	251.93	0.9956	0.08025	250.86	0.9751
0	0.09971	255.26	1.0078	0.08214	254.25	0.9876
5	0.10189	258.59	1.0199	0.08400	257.64	0.9999
10	0.10405	261.93	1.0318	0.08585	261.04	1.0120
15	0.10619	265.29	1.0436	0.08767	264.44	1.0239
20	0.10831	268.66	1.0552	0.08949	267.85	1.0357
25	0.11043	272.04	1.0666	0.09128	271.28	1.0472
30	0.11253	275.44	1.0779	0.09307	274.72	1.0587
35	0.11461	278.86	1.0891	0.09484	278.17	1.0700
40	0.11669	282.30	1.1002	0.09660	281.64	1.0811
T	$p = 0.35$ MPa ($T_{sat} = -10.39°C$)			$p = 0.40$ MPa ($T_{sat} = -6.56°C$)		
	v	h	s	v	h	s
Sat.	0.06605	246.00	0.9431	0.05812	247.48	0.9370
−10	0.06619	246.27	0.9441	–	–	–
−5	0.06789	249.75	0.9572	0.05860	248.60	0.9411
0	0.06956	253.21	0.9700	0.06011	252.14	0.9542
5	0.07121	256.67	0.9825	0.06160	225.66	0.9670
10	0.07284	260.12	0.9948	0.06306	259.18	0.9795
15	0.07444	263.57	1.0069	0.06450	262.69	0.9918
20	0.07603	267.03	1.0188	0.06592	266.19	1.0039
25	0.07760	270.50	1.0305	0.06733	269.71	1.0158
30	0.07916	273.97	1.0421	0.06872	273.22	1.0274
35	0.08070	227.46	1.0535	0.07010	276.75	1.0390
40	0.08224	280.97	1.0648	0.07146	280.28	1.0504
45	0.08376	284.48	1.0759	0.01282	283.83	1.0616
T	$p = 0.45$ MPa ($T_{sat} = -3.08°C$)			$p = 0.50$ MPa ($T_{sat} = 0.12°C$)		
	v	h	s	v	h	s
Sat.	0.05189	248.80	0.9316	0.04686	249.97	0.9269
0	0.05275	251.03	0.9399	–	–	–
5	0.05411	254.63	0.9529	0.0481	253.57	0.9399
10	0.05545	258.21	0.9657	0.04934	257.22	0.9530
15	0.05676	261.78	0.9782	0.05056	260.85	0.9657
20	0.05805	265.34	0.9904	0.05175	264.47	0.9781
25	0.05933	268.90	1.0025	0.05293	268.07	0.9903
30	0.06059	272.46	1.0143	0.05409	271.68	1.0023
35	0.06184	276.02	1.0259	0.05523	275.28	1.0141
40	0.06308	279.59	1.0374	0.05636	278.89	1.0257
45	0.06430	283.17	1.0488	0.05748	282.50	1.0371
50	0.06552	286.76	1.0600	0.05859	286.12	1.0484
55	0.06672	290.36	1.0710	0.05969	289.75	1.0595

°C	m³/kg	kJ/kg	kJ/kg·K	m³/kg	kJ/kg	kJ/kg·K
			Superheated Table (PC Model), R-22			

T	$p = 0.55$ MPa ($T_{sat} = 3.08°C$)			$p = 0.60$ MPa ($T_{sat} = 5.85°C$)		
	v	h	s	v	h	s
Sat.	0.04271	251.02	0.9226	0.03923	251.98	0.9186
5	0.04317	252.46	0.9278	–	–	–
10	0.04433	256.20	0.9411	0.04015	255.14	0.9299
15	0.04547	259.90	0.9540	0.04122	258.91	0.9431
20	0.04658	263.57	0.9667	0.04227	262.65	0.9560
25	0.04768	267.23	0.9790	0.04330	266.37	0.9685
30	0.04875	270.88	0.9912	0.04431	270.07	0.9808
35	0.04982	274.53	1.0031	0.04530	273.76	0.9929
40	0.05086	278.17	1.0148	0.04628	277.45	1.0048
45	0.05190	281.82	1.0264	0.04724	281.13	1.0164
50	0.05293	285.47	1.0378	0.04820	284.82	1.0279
55	0.05394	289.13	1.0490	0.04914	288.51	1.0393
60	0.05495	292.80	1.0601	0.05008	292.20	1.0504

T	$p = 0.70$ MPa ($T_{sat} = 10.91°C$)			$p = 0.80$ MPa ($T_{sat} = 15.45°C$)		
	v	h	s	v	h	s
Sat.	0.03371	253.64	0.9117	0.02953	255.05	0.9056
15	0.03451	256.86	0.9229	–	–	–
20	0.03547	260.75	0.9363	0.03033	258.74	0.9182
25	0.03639	264.59	0.9493	0.03118	262.70	0.9315
30	0.03730	268.4	0.9619	0.03202	266.66	0.9448
35	0.03819	272.19	0.9743	0.03283	270.54	0.9574
40	0.03906	275.96	0.9865	0.03363	274.42	0.9700
45	0.03992	279.72	0.9984	0.03440	278.26	0.9821
50	0.04076	283.48	1.0101	0.03517	282.10	0.9941
55	0.04160	287.23	1.0216	0.03592	285.92	1.0058
60	0.04242	290.99	1.0330	0.03667	289.74	1.0174
65	0.04324	294.75	1.0442	0.03741	293.56	1.0287
70	0.04405	298.51	1.0552	0.03814	297.38	1.0400

T	$p = 0.90$ MPa ($T_{sat} = 19.59°C$)			$p = 1.00$ MPa ($T_{sat} = 23.40°C$)		
	v	h	s	v	h	s
Sat.	0.02623	256.25	0.9001	0.02358	257.28	0.8952
20	0.02630	256.59	0.9013	–	–	–
30	0.02789	264.83	0.9289	0.02457	262.91	0.9139
40	0.02939	272.82	0.9549	0.02598	271.17	0.9407
50	0.03082	280.68	0.9795	0.02732	279.22	0.9660
60	0.03219	288.46	1.0033	0.02860	287.15	0.9902
70	0.03353	296.21	1.0262	0.02984	295.03	1.0135
80	0.03483	303.96	1.0484	0.03104	302.88	1.0361
90	0.03611	311.73	1.0701	0.03221	310.74	1.0580
100	0.03736	319.53	1.0913	0.03337	318.61	1.0794
110	0.03860	327.37	1.1120	0.03450	326.52	1.1003
120	0.03982	335.26	1.1323	0.03562	334.46	1.1207

Superheated Table (PC Model), R-22

°C	m³/kg	kJ/kg	kJ/kg·K	m³/kg	kJ/kg	kJ/kg·K
T	$p = 0.90$ MPa ($T_{sat} = 19.59°C$)			$p = 1.00$ MPa ($T_{sat} = 23.40°C$)		
	v	h	s	v	h	s
130	0.04103	343.21	1.1523	0.03672	342.46	1.1408
140	0.04223	351.22	1.1719	0.03781	350.51	1.1605
150	0.04342	359.29	1.1912	0.03889	358.63	1.1790
T	$p = 1.20$ MPa ($T_{sat} = 30.25°C$)			$p = 1.40$ MPa ($T_{sat} = 36.29°C$)		
	v	h	s	v	h	s
Sat.	0.01955	258.94	0.8864	0.01662	260.16	0.8786
40	0.02083	267.62	0.9146	0.01708	263.70	0.8900
50	0.02204	276.14	0.9413	0.01823	272.81	0.9186
60	0.02319	284.43	0.9666	0.01929	281.53	0.9452
70	0.02428	292.58	0.9907	0.02029	290.01	0.9703
80	0.02534	300.66	1.0139	0.02125	298.34	0.9942
90	0.02636	308.70	1.0363	0.02217	306.60	1.0172
100	0.02736	316.73	1.0582	0.02306	314.80	1.0395
110	0.02834	324.78	1.0794	0.02393	323.00	1.0612
120	0.02930	332.85	1.1002	0.02478	331.19	1.0823
130	0.03024	340.95	1.1205	0.02562	339.41	1.1029
140	0.03118	349.09	1.1405	0.02644	347.65	1.1231
150	0.03210	357.29	1.1601	0.02725	355.94	1.1429
160	0.03301	365.54	1.1793	0.02805	364.26	1.1624
170	0.03392	373.84	1.1983	0.02884	372.64	1.1815
T	$p = 1.60$ MPa ($T_{sat} = 41.73°C$)			$p = 1.80$ MPa ($T_{sat} = 46.69°C$)		
	v	h	s	v	h	s
Sat.	0.01440	261.04	0.8715	0.01265	261.64	0.8649
50	0.01533	269.18	0.8971	0.01301	265.14	0.8758
60	0.01634	278.43	0.9252	0.01401	275.09	0.9061
70	0.01728	287.30	0.9515	0.01492	284.43	0.9337
80	0.01817	295.93	0.9762	0.01576	293.40	0.9595
90	0.01901	304.42	0.9999	0.01655	302.16	0.9839
100	0.01983	312.82	1.0228	0.01731	310.77	1.0073
110	0.02062	321.17	1.0448	0.01804	319.30	1.0299
120	0.02139	329.51	1.0663	0.01874	327.78	1.0517
130	0.02214	337.84	1.0872	0.01943	336.24	1.0730
140	0.02288	346.19	1.1077	0.02011	344.70	1.0937
150	0.02361	354.56	1.1277	0.02077	353.17	1.1139
160	0.02432	362.97	1.1473	0.02142	361.66	1.1338
170	0.02503	371.42	1.1666	0.02207	370.19	1.1532
T	$p = 2.00$ MPa ($T_{sat} = 51.26°C$)			$p = 2.40$ MPa ($T_{sat} = 59.46°C$)		
	v	h	s	v	h	s
Sat.	0.01124	261.98	0.8586	0.00907	261.99	0.8463
60	0.01212	271.43	0.8873	0.00913	262.68	0.8484
70	0.01300	281.36	0.9167	0.01006	274.43	0.8831
80	0.01381	290.74	0.9436	0.01085	284.93	0.9133

(continued)

°C	m³/kg	kJ/kg	kJ/kg·K	m³/kg	kJ/kg	kJ/kg·K
	Superheated Table (PC Model), R-22					
T	**p = 2.00 MPa (T_{sat} = 51.26°C)**			**p = 2.40 MPa (T_{sat} = 59.46°C)**		
	v	**h**	**s**	**v**	**h**	**s**
90	0.01457	299.80	0.9689	0.01156	294.75	0.9407
100	0.01528	308.65	0.9929	0.01222	304.18	0.9663
110	0.01596	317.37	1.0160	0.01284	313.35	0.9906
120	0.01663	326.01	1.0383	0.01343	322.35	1.0137
130	0.01727	334.61	1.0598	0.01400	331.25	1.0361
140	0.01789	343.19	1.0808	0.01456	340.08	1.0577
150	0.01850	351.76	1.1013	0.01509	348.87	1.0787
160	0.01910	360.34	1.1214	0.01562	357.64	1.0992
170	0.01969	368.95	1.1410	0.01613	366.41	1.1192
180	0.02027	377.58	1.1603	0.01663	375.20	1.1388

Table B-10 PC Model: Temperature Saturation Table, R-12

Temp.	Sat. Press.	Spec. Volume		Spec. Int. Energy		Spec. Enthalpy		Spec. Entropy	
		m³/kg		kJ/kg		kJ/kg		kJ/kg·K	
°C	kPa	Sat. liquid	Sat. vapor	Sat. liquid	Sat. vapor	Sat. liquid	Sat. vapor	Sat. liquid	Sat. vapor
T	$p_{sat@T}$	v_f	v_g	u_f	u_g	h_f	h_g	s_f	s_g
−90	2.8	0.000608	4.41555	−43.29	133.91	−43.28	146.46	−0.2086	0.8273
−80	6.2	0.000617	2.13835	−34.73	137.82	−34.72	151.02	−0.1631	0.7984
−70	12.3	0.000627	1.12728	−26.14	141.81	−26.13	155.64	−0.1198	0.7749
−60	22.6	0.000637	0.63791	−17.50	145.86	−17.49	160.29	−0.0783	0.7557
−50	39.1	0.000648	0.38310	−8.80	149.95	−8.78	164.95	−0.0384	0.7401
−45	50.4	0.000654	0.30268	−4.43	152.01	−4.40	167.28	−0.019	0.7334
−40	64.2	0.000659	0.24191	−0.04	154.07	0	169.59	0	0.7274
−35	80.7	0.000666	0.19540	4.37	156.13	4.42	171.90	0.0187	0.7219
−30	100.4	0.000672	0.15937	8.79	158.19	8.86	174.20	0.0371	0.7170
−29.8	101.3	0.000672	0.15803	8.98	158.28	9.05	174.29	0.0379	0.7168
−25	123.7	0.000679	0.13117	13.24	160.25	13.33	176.48	0.0552	0.7126
−20	150.9	0.000685	0.10885	17.71	162.31	17.82	178.74	0.0731	0.7087
−15	182.6	0.000693	0.09102	22.20	164.35	22.33	180.97	0.0906	0.7051
−10	219.1	0.000700	0.07665	26.72	166.39	26.87	183.19	0.1080	0.7019
−5	261.0	0.000708	0.06496	31.26	168.42	31.45	185.37	0.1251	0.6991
0	308.6	0.000716	0.05539	35.83	170.44	36.05	187.53	0.1420	0.6965
5	362.6	0.000724	0.04749	40.43	172.44	40.69	189.65	0.1587	0.6942

Saturation Temperature Table (PC Model) of R-12

Temp.	Sat. Press.	Spec. Volume		Spec. Int. Energy		Spec. Enthalpy		Spec. Entropy	
		m³/kg		kJ/kg		kJ/kg		kJ/kg·K	
°C	kPa	Sat. liquid	Sat. vapor	Sat. liquid	Sat. vapor	Sat. liquid	Sat. vapor	Sat. liquid	Sat. vapor
T	$p_{sat@T}$	v_f	v_g	u_f	u_g	h_f	h_g	s_f	s_g
10	423.3	0.000733	0.04091	45.06	174.42	45.37	191.74	0.1752	0.6921
15	491.4	0.000743	0.03541	49.73	176.38	50.10	193.78	0.1915	0.6902
20	567.3	0.000752	0.03078	54.45	178.32	54.87	195.78	0.2078	0.6884
25	651.6	0.000763	0.02685	59.21	180.23	59.70	197.73	0.2239	0.6868
30	744.9	0.000774	0.02351	64.02	182.11	64.59	199.62	0.2399	0.6853
35	847.7	0.000786	0.02064	68.88	183.95	69.55	201.45	0.2559	0.6839
40	960.7	0.000798	0.01817	73.82	185.74	74.59	203.20	0.2718	0.6825
45	1084.3	0.000811	0.01603	78.83	187.49	79.71	204.87	0.2877	0.6811
50	1219.3	0.000826	0.01417	83.93	189.17	84.94	206.45	0.3037	0.6797
55	1366.3	0.000841	0.01254	89.12	190.78	90.27	207.92	0.3197	0.6782
60	1525.9	0.000858	0.01111	94.43	192.31	95.74	209.26	0.3358	0.6765
65	1698.8	0.000877	0.00985	99.87	193.73	101.36	210.46	0.3521	0.6747
70	1885.8	0.000897	0.00873	105.46	195.03	107.15	211.48	0.3686	0.6726
75	2087.5	0.000920	0.00772	111.23	196.17	113.15	212.29	0.3854	0.6702
80	2304.6	0.000946	0.00682	117.21	197.11	119.39	212.83	0.4027	0.6672
85	2538.0	0.000976	0.00600	123.45	197.80	125.93	213.04	0.4204	0.6636
90	2788.5	0.001012	0.00526	130.02	198.14	132.84	212.80	0.4389	0.6590
95	3056.9	0.001056	0.00456	137.01	197.99	140.23	211.94	0.4583	0.6531
100	3344.1	0.001113	0.00390	144.59	197.07	148.31	210.12	0.4793	0.6449
105	3650.9	0.001197	0.00324	153.15	194.73	157.52	206.57	0.5028	0.6325
110	3978.5	0.001364	0.00246	164.12	188.20	169.55	197.99	0.5333	0.6076
112	4116.8	0.001792	0.00179	176.06	176.06	183.43	183.43	0.5689	0.5689

▮ Table B-11   PC Model: Superheated Vapor Table, R-12

Superheated Table (PC Model), R-12

°C	m³/kg	kJ/kg	kJ/kg·K	m³/kg	kJ/kg	kJ/kg·K	m³/kg	kJ/kg	kJ/kg·K
	$p = 0.025$ MPa ($T_{sat} = -58.26$°C)			$p = 0.05$ MPa ($T_{sat} = -45.18$°C)			$p = 0.10$ MPa ($T_{sat} = -30.10$°C)		
T	v	h	s	v	h	s	v	h	s
Sat.	0.58130	161.10	0.7527	0.30515	167.19	0.7336	0.15999	174.15	0.7171
−30	0.66179	176.19	0.8187	0.32738	175.55	0.7691	0.16006	174.21	0.7174
−20	0.69001	181.74	0.8410	0.34186	181.17	0.7917	0.16770	179.99	0.7406
−10	0.71811	187.40	0.8630	0.35623	186.89	0.8139	0.17522	185.84	0.7633
0	0.74613	193.17	0.8844	0.37051	192.70	0.8356	0.18265	191.77	0.7854
10	0.77409	199.03	0.9055	0.38472	198.61	0.8568	0.18999	197.77	0.8070

(continued)

Superheated Table (PC Model), R-12

°C	m³/kg	kJ/kg	kJ/kg·K	m³/kg	kJ/kg	kJ/kg·K	m³/kg	kJ/kg	kJ/kg·K
T	$p = 0.025$ MPa ($T_{sat} = -58.26$°C)			$p = 0.05$ MPa ($T_{sat} = -45.18$°C)			$p = 0.10$ MPa ($T_{sat} = -30.10$°C)		
	v	h	s	v	h	s	v	h	s
20	0.80198	204.99	0.9262	0.39886	204.62	0.8776	0.19728	203.85	0.8281
30	0.82982	211.05	0.9465	0.41296	210.71	0.8981	0.20451	210.02	0.8488
40	0.85762	217.20	0.9665	0.42701	216.89	0.9181	0.21169	216.26	0.8691
50	0.88538	223.45	0.9861	0.44103	223.16	0.9378	0.21884	222.58	0.8889
60	0.91312	229.77	1.0054	0.45502	229.51	0.9572	0.22596	228.98	0.9084
70	0.94083	236.19	1.0244	0.46898	235.95	0.9762	0.23305	235.46	0.9276
80	0.96852	242.68	1.0430	0.48292	242.46	0.9949	0.24011	242.01	0.9464
90	0.99618	249.26	1.0614	0.49684	249.05	1.0133	0.24716	248.63	0.9649
100	1.02384	255.91	1.0795	0.51074	255.71	1.0314	0.25419	255.32	0.9831
110	1.05148	262.63	1.0972	0.52463	262.45	1.0493	0.26121	262.08	1.0009
120	1.07910	269.43	1.1148	0.53851	269.26	1.0668	0.26821	268.91	1.0185

°C	m³/kg	kJ/kg	kJ/kg·K	m³/kg	kJ/kg	kJ/kg·K	m³/kg	kJ/kg	kJ/kg·K
T	$p = 0.20$ MPa ($T_{sat} = -12.53$°C)			$p = 0.30$ MPa ($T_{sat} = -0.86$°C)			$p = 0.40$ MPa ($T_{sat} = 8.15$°C)		
	v	h	s	v	h	s	v	h	s
Sat.	0.08354	182.07	0.7035	0.05690	187.16	0.6969	0.04321	190.97	0.6928
0	0.08861	189.80	0.7325	0.05715	187.72	0.6989	–	–	–
10	0.09255	196.02	0.7548	0.05998	194.17	0.7222	0.04363	192.21	0.6972
20	0.09642	202.28	0.7766	0.06273	200.64	0.7446	0.04584	198.91	0.7204
30	0.10023	208.60	0.7978	0.06542	207.12	0.7663	0.04797	205.58	0.7428
40	0.10399	214.97	0.8184	0.06805	213.64	0.7875	0.05005	212.25	0.7645
50	0.10771	221.41	0.8387	0.07064	220.19	0.8081	0.05207	218.94	0.7855
60	0.11140	227.90	0.8585	0.07319	226.79	0.8282	0.05406	225.65	0.8060
70	0.15506	234.46	0.8779	0.07571	233.44	0.8479	0.05601	232.40	0.8259
80	0.11869	241.09	0.8969	0.07820	240.15	0.8671	0.05794	239.19	0.8454
90	0.12230	247.77	0.9156	0.08067	246.90	0.8860	0.05985	246.02	0.8645
100	0.12590	254.53	0.9339	0.08313	253.72	0.9045	0.06173	252.89	0.8831
110	0.12948	261.34	0.9519	0.08557	260.58	0.9226	0.06360	259.81	0.9015
120	0.13305	268.21	0.9696	0.08799	267.50	0.9405	0.06546	266.79	0.9194
130	0.13661	275.15	0.9870	0.09041	274.48	0.9580	0.06730	273.81	0.9370
140	0.14016	282.14	1.0042	0.09281	281.51	0.9752	0.06913	280.88	0.9544
150	0.14370	289.19	1.0210	0.09520	288.59	0.9922	0.07095	287.99	0.9714

°C	m³/kg	kJ/kg	kJ/kg·K	m³/kg	kJ/kg	kJ/kg·K	m³/kg	kJ/kg	kJ/kg·K
T	$p = 0.50$ MPa ($T_{sat} = 15.60$°C)			$p = 0.75$ MPa ($T_{sat} = 30.26$°C)			$p = 1.00$ MPa ($T_{sat} = 41.64$°C)		
	v	h	s	v	h	s	v	h	s
Sat.	0.03482	194.03	0.6899	0.02335	199.72	0.6852	0.01744	203.76	0.6820
30	0.03746	203.96	0.7235	–	–	–	–	–	–
40	0.03921	210.81	0.7457	0.02467	206.91	0.7086	–	–	–
50	0.04091	217.64	0.7672	0.02595	214.18	0.7314	0.01837	210.32	0.7026
60	0.04257	224.48	0.7881	0.02718	221.37	0.7533	0.01941	217.97	0.7259
70	0.04418	231.33	0.8083	0.02837	228.52	0.7745	0.02040	225.49	0.7481
80	0.04577	238.21	0.8281	0.02952	235.65	0.7949	0.02134	232.91	0.7695
90	0.04734	245.11	0.8473	0.03064	242.76	0.8148	0.02225	240.28	0.7900
100	0.04889	252.05	0.8662	0.03174	249.89	0.8342	0.02313	247.61	0.8100
110	0.05041	259.03	0.8847	0.03282	257.03	0.8530	0.02399	254.93	0.8293

Superheated Table (PC Model), R-12

°C	m³/kg	kJ/kg	kJ/kg·K	m³/kg	kJ/kg	kJ/kg·K	m³/kg	kJ/kg	kJ/kg·K
T	$p = 0.50$ MPa ($T_{sat} = 15.60$°C)			$p = 0.75$ MPa ($T_{sat} = 30.26$°C)			$p = 1.00$ MPa ($T_{sat} = 41.64$°C)		
	v	h	s	v	h	s	v	h	s
120	0.05193	266.06	0.9028	0.03388	264.19	0.8715	0.02483	262.25	0.8482
130	0.05343	273.12	0.9205	0.03493	271.38	0.8895	0.02566	269.57	0.8665
140	0.05492	280.23	0.9379	0.03596	278.59	0.9072	0.02647	276.90	0.8845
150	0.05640	287.39	0.9550	0.03699	285.84	0.9246	0.02728	284.26	0.9021
160	0.05788	294.59	0.9718	0.03801	293.13	0.9416	0.02807	291.63	0.9193
170	0.05934	301.83	0.9884	0.03902	300.45	0.9583	0.02885	299.04	0.9362
180	0.06080	309.12	1.0046	0.04002	307.81	0.9747	0.02963	306.47	0.9528
T	$p = 1.50$ MPa ($T_{sat} = 59.22$°C)			$p = 2.00$ MPa ($T_{sat} = 72.88$°C)			$p = 4.00$ MPa ($T_{sat} = 110.32$°C)		
	v	h	s	v	h	s	v	h	s
Sat.	0.01132	209.06	0.6768	0.00813	211.97	0.6713	0.00239	196.90	0.6046
80	0.01305	226.73	0.7284	0.00870	219.02	0.6914	–	–	–
90	0.01377	234.77	0.7508	0.00941	228.23	0.7171	–	–	–
100	0.01446	242.65	0.7722	0.01003	236.94	0.7408	–	–	–
110	0.01512	250.41	0.7928	0.01061	245.34	0.7630	–	–	–
120	0.01575	258.10	0.8126	0.01116	253.53	0.7841	0.00375	225.18	0.6777
130	0.01636	265.74	0.8318	0.01168	261.58	0.8043	0.00433	238.69	0.7116
140	0.01696	273.35	0.8504	0.01217	269.53	0.8238	0.00478	249.93	0.7392
150	0.01754	280.94	0.8686	0.01265	277.41	0.8426	0.00517	260.12	0.7636
160	0.01811	288.52	0.8863	0.01312	285.24	0.8609	0.00552	269.71	0.7860
170	0.01867	296.11	0.9036	0.01357	293.04	0.8787	0.00585	278.90	0.8069
180	0.01922	303.70	0.9205	0.01401	300.82	0.8961	0.00615	287.82	0.8269
190	0.01977	311.31	0.9371	0.01445	308.59	0.9131	0.00643	296.55	0.8459
200	0.02203	318.93	0.9534	0.01488	316.36	0.9297	0.00671	305.14	0.8642
210	0.02084	326.58	0.9694	0.01530	324.14	0.9459	0.00697	313.61	0.8820
220	0.02137	334.24	0.9851	0.01572	331.92	0.9619	0.00723	322.01	0.8992

Table B-12 PC Model: Temperature Saturation Table, Ammonia (NH₃)

Saturation Temperature Table (PC Model) of Ammonia (NH₃)

Temp.	Sat. Press.	Spec. Volume		Spec. Int. Energy		Spec. Enthalpy		Spec. Entropy	
		m³/kg		kJ/kg		kJ/kg		kJ/kg·K	
°C	kPa	Sat. liquid	Sat. vapor	Sat. liquid	Sat. vapor	Sat. liquid	Sat. vapor	Sat. liquid	Sat. vapor
T	$p_{sat@T}$	v_f	v_g	u_f	u_g	h_f	h_g	s_f	s_g
−50	40.9	0.001424	2.62700	−43.82	1265.2	−43.76	1372.6	−0.1916	6.1554
−45	54.5	0.001437	2.00632	−22.01	1271.4	−21.94	1380.8	−0.0950	6.0534
−40	71.7	0.001450	1.55256	−0.10	1277.4	0	1388.8	0	5.9567
−35	93.2	0.001463	1.21613	21.93	1283.3	22.06	1396.5	0.0935	5.8650
−30	119.5	0.001476	0.96339	44.08	1288.9	44.26	1404.0	0.1856	5.7778

(continued)

Saturation Temperature Table (PC Model) of Ammonia (NH₃)

Temp.	Sat. Press.	Spec. Volume		Spec. Int. Energy		Spec. Enthalpy		Spec. Entropy	
		m³/kg		kJ/kg		kJ/kg		kJ/kg·K	
°C	kPa	Sat. liquid	Sat. vapor	Sat. liquid	Sat. vapor	Sat. liquid	Sat. vapor	Sat. liquid	Sat. vapor
T	$p_{sat@T}$	v_f	v_g	u_f	u_g	h_f	h_g	s_f	s_g
−25	151.6	0.001490	0.77119	66.36	1294.3	66.58	1411.2	0.2763	5.6947
−20	190.2	0.001504	0.62334	88.76	1299.5	89.05	1418.0	0.3657	5.6155
−15	236.3	0.001519	0.50838	111.30	1304.5	111.66	1424.6	0.4538	5.5397
−10	290.9	0.001534	0.41808	133.96	1309.2	134.41	1430.8	0.5408	5.4673
−5	354.9	0.001550	0.34648	156.76	1313.7	157.31	1436.7	0.6266	5.3977
0	429.6	0.001566	0.28920	179.69	1318.0	180.36	1442.2	0.7114	5.3309
5	515.9	0.001583	0.24299	202.77	1322.0	203.58	1447.3	0.7951	5.2666
10	615.2	0.001600	0.20541	225.99	1325.7	226.97	1452.0	0.8779	5.2045
15	728.6	0.001619	0.17462	249.36	1329.1	250.54	1456.3	0.9598	5.1444
20	857.5	0.001638	0.14922	272.89	1332.2	274.30	1460.2	1.0408	5.0860
25	1003.2	0.001658	0.12813	296.59	1335.0	298.25	1463.5	1.1210	5.0293
30	1167.0	0.001680	0.11049	320.46	1337.4	322.42	1466.3	1.2005	4.9738
35	1350.4	0.001702	0.09567	344.50	1339.4	346.80	1468.6	1.2792	4.9196
40	1554.9	0.001725	0.08313	368.74	1341.0	371.43	1470.2	1.3574	4.8662
45	1782.0	0.001750	0.07248	393.19	1342.1	396.31	1471.2	1.4350	4.8136
50	2033.1	0.001777	0.06337	417.87	1342.7	421.48	1471.5	1.5121	4.7614
55	2310.1	0.001804	0.05555	442.79	1342.7	446.96	1471.0	1.5888	4.7095
60	2614.4	0.001834	0.04880	467.99	1342.1	472.79	1469.7	1.6652	4.6577
65	2947.8	0.001866	0.04296	493.51	1340.9	499.01	1467.5	1.7415	4.6057
70	3312.0	0.001900	0.03787	519.39	1338.9	525.69	1464.4	1.8178	4.5533
75	3709.0	0.001937	0.03341	545.70	1336.1	552.88	1460.1	1.8943	4.5001
80	4140.5	0.001978	0.02951	572.50	1332.4	580.69	1454.6	1.9712	4.4458
85	4608.6	0.002022	0.02606	599.90	1327.7	609.21	1447.8	2.0488	4.3901
90	5115.3	0.002071	0.02300	627.99	1321.7	638.59	1439.4	2.1273	4.3325
95	5662.9	0.002126	0.02028	656.95	1314.4	668.99	1429.2	2.2073	4.2723
100	6253.7	0.002188	0.01784	686.96	1305.3	700.64	1416.9	2.2893	4.2088
105	6890.4	0.002261	0.01564	718.30	1294.2	733.87	1402.0	2.3740	4.1407
110	7575.7	0.002347	0.01363	751.37	1280.5	769.15	1383.7	2.4625	4.0665
115	8313.3	0.002452	0.01178	786.82	1263.1	807.21	1361.0	2.5566	3.9833
120	9107.2	0.002589	0.01003	825.77	1240.3	849.36	1331.7	2.6593	3.8861
125	9963.5	0.002783	0.00833	870.69	1208.4	898.42	1291.4	2.7775	3.7645
130	10891.6	0.003122	0.00649	929.29	1156.2	963.29	1227.0	2.9326	3.5866
132.3	11333.2	0.004255	0.00426	1037.62	1037.6	1085.85	1085.9	3.2316	3.2316

Table B-13   PC Model: Superheated Vapor Table, Ammonia (NH₃)

°C	m³/kg	kJ/kg	kJ/kg·K	m³/kg	kJ/kg	kJ/kg·K	m³/kg	kJ/kg	kJ/kg·K
	Superheated Table (PC Model), Ammonia (NH₃)								
T	$p = 0.050$ MPa ($T_{sat} = -46.53°C$)			$p = 0.075$ MPa ($T_{sat} = -39.16°C$)			$p = 0.100$ MPa ($T_{sat} = -33.60°C$)		
	v	h	s	v	h	s	v	h	s
Sat.	2.17521	1378.3	6.0839	1.48922	1390.1	5.9411	1.13806	1398.7	5.8401
−30	2.34484	1413.4	6.2333	1.55321	1410.1	6.0247	1.15727	1406.7	5.8734
−20	2.44631	1434.6	6.3187	1.62221	1431.7	6.1120	1.21007	1428.8	5.9626
−10	2.54711	1455.7	6.4006	1.69050	1453.3	6.1954	1.26213	1450.8	6.0477
0	2.64736	1476.9	6.4795	1.75823	1474.8	6.2756	1.31362	1472.6	6.1291
10	2.74716	1498.1	6.5556	1.82551	1496.2	6.3527	1.36465	1494.4	6.2073
20	2.84661	1519.3	6.6293	1.89243	1517.7	6.4272	1.41532	1516.1	6.2826
30	2.94578	1540.6	6.7008	1.95906	1539.2	6.4993	1.46569	1537.7	6.3553
40	3.04472	1562.0	6.7703	2.02547	1560.7	6.5693	1.15182	1559.5	6.4258
50	3.14348	1583.5	3.8379	2.09168	1582.4	6.6373	1.56577	1581.2	6.4943
60	3.24209	1605.1	6.9038	2.15775	1604.1	6.7036	1.61557	1603.1	6.5609
70	3.34058	1626.9	6.9682	2.22369	1626.0	6.7683	1.66525	1625.1	6.6258
80	3.43897	1648.8	7.0312	2.28954	1648.0	6.8315	1.71482	1647.1	6.6892
100	3.63551	1693.2	7.1533	2.42099	1692.4	6.5939	1.81373	1691.7	6.8120
120	3.83183	1738.2	7.2708	2.55221	1737.5	7.0716	1.91240	1736.9	6.9300
140	4.02797	1783.9	7.3842	2.68326	1783.4	7.1853	2.01091	1782.8	7.0439
160	4.22398	1830.4	7.4941	2.81418	1829.9	7.2953	2.10927	1829.4	7.1540
180	4.41988	1877.7	7.6008	2.94499	1877.2	7.4021	2.20754	1876.8	7.2609
200	4.61570	1925.7	7.7045	3.07571	1925.3	7.5059	2.30571	1924.9	7.3648
T	$p = 0.125$ MPa ($T_{sat} = -29.07°C$)			$p = 0.150$ MPa ($T_{sat} = -25.22°C$)			$p = 0.200$ MPa ($T_{sat} = -18.86°C$)		
	v	h	s	v	h	s	v	h	s
Sat.	0.92365	1405.4	5.7620	0.77870	1410.9	5.6983	0.59460	1419.6	5.5979
−20	0.96271	1425.9	5.8446	0.79774	1422.9	5.7465	–	–	–
−10	1.00506	1448.3	5.9314	0.83364	1445.7	5.8349	0.61926	1440.6	5.6791
0	1.04682	1470.5	6.0141	0.86892	1468.3	5.9189	0.64648	1463.8	5.7659
10	1.08811	1492.5	6.0933	0.90373	1490.6	5.9992	0.67319	1486.8	5.8484
20	1.12903	1514.4	6.1694	0.93815	1512.8	6.0761	0.69951	1509.4	5.9270
30	1.16964	1536.3	6.2428	0.97227	1534.8	6.1502	0.72553	1531.9	6.0025
40	1.21003	1558.2	6.3138	1.00615	1556.9	6.2217	0.75129	1554.3	6.0751
50	1.25022	1580.1	6.3827	1.03984	1578.9	6.2910	0.77685	1576.6	6.1453
60	1.29026	1602.1	6.4496	1.07338	1601.0	6.3583	0.80226	1598.9	6.2133
70	1.33017	1624.1	6.5149	1.10678	1623.2	6.4238	0.82754	1621.3	6.2794
80	1.36998	1646.3	6.5785	1.14009	1645.4	6.4877	0.85271	1643.7	6.3437
100	1.44937	1691.0	6.7017	1.20646	1690.2	6.6112	0.90282	1688.8	6.4679
120	1.52852	1736.3	6.8199	1.27259	1735.6	6.7297	0.95268	1734.4	6.5869
140	1.60749	1782.2	6.9339	1.33855	1781.7	6.8439	1.00237	1780.6	6.7015
160	1.68633	1828.9	7.0443	1.40437	1828.4	6.9544	1.05192	1827.4	6.8123
180	1.76507	1876.3	7.1513	1.47009	1875.9	7.0615	1.10136	1875.0	6.9196
200	1.84371	1924.5	7.2553	1.53572	1924.1	7.1656	1.15072	1923.3	7.0239
220	1.92229	1973.4	7.3566	1.60127	1973.1	7.2670	1.20000	1972.4	7.1255

(continued)

Superheated Table (PC Model), Ammonia (NH₃)

°C	m³/kg	kJ/kg	kJ/kg·K	m³/kg	kJ/kg	kJ/kg·K	m³/kg	kJ/kg	kJ/kg·K
T	$p = 0.25$ MPa ($T_{sat} = -13.66$°C)			$p = 0.30$ MPa ($T_{sat} = -9.24$°C)			$p = 0.35$ MPa ($T_{sat} = -5.36$°C)		
	v	h	s	v	h	s	v	h	s
Sat.	0.48213	1426.3	5.5201	0.40607	1431.7	5.4565	0.35108	1436.6	5.4026
0	0.51293	1459.3	5.6441	0.42383	1454.7	5.5420	0.36011	1449.9	5.4532
10	0.53481	1482.9	5.7288	0.44251	1478.9	5.6290	0.37654	1474.9	5.5427
20	0.55629	1506.0	5.8093	0.46077	1502.6	5.7113	0.39251	1499.1	5.6270
30	0.57745	1529.0	5.8861	0.47870	1525.9	5.7896	0.40814	1522.9	5.7068
40	0.59835	1551.7	5.9599	0.49636	1549.0	5.8645	0.42350	1546.3	8.7828
50	0.61904	1574.3	6.0309	0.51382	1571.9	5.9365	0.43865	1569.5	5.8557
60	0.63958	1596.8	6.0997	0.53111	1594.7	6.0060	0.45362	1592.6	5.9259
70	0.65998	1619.4	6.1663	0.54827	1617.5	6.0732	0.46846	1615.5	5.9938
80	0.68028	1641.9	6.2312	0.56532	1640.2	6.1385	0.48319	1638.4	6.0596
100	0.72063	1687.3	6.3561	0.59916	1685.8	6.2642	0.51240	1684.3	6.1860
120	0.76073	1733.1	6.4756	0.63276	1731.8	6.3842	0.54135	1730.5	6.3066
140	0.80065	1779.4	6.5906	0.66618	1778.3	6.4996	0.57012	1777.2	6.4223
160	0.84044	1826.4	6.7016	0.69946	1825.4	6.6109	0.59876	1824.4	6.5340
180	0.88012	1874.1	6.8093	0.73263	1873.2	6.7188	0.62728	1872.3	6.6421
200	0.91972	1922.5	6.9138	0.76572	1921.7	6.8235	0.65571	1920.9	6.7470
220	0.95923	1971.6	7.0155	0.79872	1970.9	6.9254	0.68407	1970.2	6.8491
240	0.99868	2021.5	7.1147	0.83167	2020.9	7.0247	0.71237	2020.3	9.9486
260	1.03808	2072.2	7.2115	0.86455	2071.6	7.1217	0.74060	2071.0	7.0456
T	$p = 0.40$ MPa ($T_{sat} = -1.89$°C)			$p = 0.50$ MPa ($T_{sat} = 4.13$°C)			$p = 0.60$ MPa ($T_{sat} = 9.28$°C)		
	v	h	s	v	h	s	v	h	s
Sat.	0.30942	1440.2	5.3559	0.25035	1446.5	5.2776	0.21038	1451.4	5.2133
10	0.32701	1470.7	5.4663	0.25757	1462.3	5.3340	0.21115	1453.4	5.2205
20	0.34129	1495.6	5.5525	0.26949	1488.3	5.4244	0.22154	1480.8	5.3156
30	0.35520	1519.8	5.6338	0.28103	1513.5	5.5090	0.23152	1507.1	5.4037
40	0.36884	1543.6	5.7111	0.29227	1538.1	5.5889	0.24118	1532.5	5.4862
50	0.38226	1567.1	5.7850	0.30328	1562.3	5.6647	0.25059	1557.3	5.5641
60	0.39550	1590.4	5.8560	0.31410	1586.1	5.7373	0.25981	1581.6	5.6383
70	0.40860	1613.6	5.9244	0.32478	1609.6	5.8070	0.26888	1605.7	5.7094
80	0.42160	1636.7	5.9907	0.33535	1633.1	5.8744	0.27783	1629.5	5.7778
100	0.44732	1682.8	6.1179	0.35621	1679.8	6.0031	0.29545	1676.8	5.9081
120	0.47279	1729.2	6.2390	0.37681	1726.6	6.1253	0.31281	1724.0	6.0314
140	0.49808	1776.0	6.3552	0.39722	1773.8	6.2422	0.32997	1771.5	6.1491
160	0.52323	1823.4	6.4671	0.41748	1821.4	6.3548	0.34699	1819.4	6.2623
180	0.54827	1871.4	6.5755	0.43764	1869.6	6.4636	0.36389	1867.8	6.3717
200	0.57321	1920.1	6.6806	0.45771	1918.5	6.5691	0.38071	1916.9	6.4776
220	0.59809	1969.5	6.7828	0.47770	1968.1	6.6717	0.39745	1966.6	6.5806
240	0.62289	2019.6	6.8825	0.49763	2018.3	6.7717	0.41412	2017.1	6.6808
260	0.64764	2070.5	6.9797	0.51749	2069.3	6.8692	0.43073	2068.2	6.7786
280	0.67234	2122.1	7.0747	0.53731	2121.1	6.9644	0.44729	2120.1	6.8741

Superheated Table (PC Model), Ammonia (NH₃)

°C	m³/kg	kJ/kg	kJ/kg·K	m³/kg	kJ/kg	kJ/kg·K	m³/kg	kJ/kg	kJ/kg·K
T	$p = 0.70$ MPa ($T_{sat} = 13.80$°C)			$p = 0.80$ MPa ($T_{sat} = 17.85$°C)			$p = 0.90$ MPa ($T_{sat} = 21.52$°C)		
	v	h	s	v	h	s	v	h	s
Sat.	0.18148	1455.3	5.1586	0.15958	1458.6	5.1110	0.14239	1461.2	5.0686
20	0.18721	1473.0	5.2196	0.16138	1464.9	5.1328	–	–	–
30	0.19610	1500.4	5.3115	0.16947	1493.5	5.2287	0.14872	1486.5	5.1530
40	0.20464	1526.7	5.3968	0.17720	1520.8	5.3171	0.15582	1514.7	5.2447
50	0.21293	1552.2	5.4770	0.18465	1547.0	5.3996	0.16263	1541.7	5.3296
60	0.22101	1577.1	5.5529	0.19189	1572.5	5.4774	0.16922	1567.9	5.4093
70	0.22894	1601.6	5.6254	0.19896	1597.5	5.5513	0.17563	1593.3	5.4847
80	0.23674	1625.8	5.6949	0.20590	1622.1	5.6219	0.18191	1618.4	5.5565
100	0.25205	1673.7	5.8268	0.21949	1670.6	5.7555	0.19416	1667.5	5.6919
120	0.26709	1721.4	5.9512	0.32380	1718.7	5.8811	0.20612	1716.1	5.8187
140	0.28193	1769.2	6.0698	0.24590	1766.9	6.0006	0.21787	1764.5	5.9389
160	0.29663	1817.3	6.1837	0.25886	1815.3	6.1150	0.22948	1813.2	6.0541
180	0.31121	1866.0	6.2935	0.27170	1864.2	6.2254	0.24097	1862.4	6.1649
200	0.32570	1915.3	6.3999	0.28445	1913.6	6.3322	0.25236	1912.0	6.2721
220	0.34012	1965.2	6.5032	0.29712	1963.7	6.4358	0.26368	1962.3	6.3762
240	0.35447	2015.8	6.6037	0.30973	2014.5	6.5367	0.27493	2013.2	6.4774
260	0.36876	2067.1	6.7018	0.32228	2065.9	6.6350	0.28612	2064.8	6.5760
280	0.38299	2119.1	6.7975	0.33477	2118.0	6.7310	0.29726	2117.0	6.6722
300	0.39718	2171.8	6.8911	0.34722	2170.9	6.8242	0.30835	2170.0	6.7662
T	$p = 1.00$ MPa ($T_{sat} = 24.90$°C)			$p = 1.20$ MPa ($T_{sat} = 30.94$°C)			$p = 1.40$ MPa ($T_{sat} = 36.26$°C)		
	v	h	s	v	h	s	v	h	s
Sat.	0.12852	1463.4	5.0304	0.10751	1466.8	4.9635	0.09231	1469.0	4.9060
30	0.13206	1479.1	5.0826	–	–	–	–	–	–
40	0.13868	1508.5	5.1778	0.11287	1495.4	5.0564	0.09432	1481.6	4.9463
50	0.14499	1536.3	5.2654	0.11846	1525.1	5.1497	0.09943	1513.4	5.0462
60	0.15106	1563.1	5.3471	0.12378	1553.3	5.2357	0.10423	1543.1	5.1370
70	0.15695	1589.1	5.4240	0.12890	1580.5	5.3159	0.10882	1571.5	5.2209
80	0.16270	1614.6	5.4971	0.13387	1606.8	5.3916	0.11324	1598.8	5.2994
100	0.17389	1664.3	5.6342	0.14347	1658.0	5.5325	0.12172	1651.4	5.4443
120	0.18477	1713.4	5.7622	0.15275	1708.0	5.6631	0.12986	1702.5	5.5775
140	0.19545	1762.2	5.8834	0.16181	1757.5	5.7860	0.13777	1752.8	5.7023
160	0.20597	1811.2	5.9992	0.17071	1807.1	5.9031	0.14552	1802.9	5.8208
180	0.21638	1860.5	6.1105	0.17950	1856.9	6.0156	0.15315	1853.2	5.9343
200	0.22669	1910.4	6.2182	0.18819	1907.1	6.1241	0.16068	1903.8	6.0437
220	0.23693	1960.8	6.3226	0.19680	1957.9	6.2292	0.16813	1955.0	6.1495
240	0.24710	2011.9	6.4241	0.20534	2009.3	6.3313	0.17551	2006.7	6.2523
260	0.25720	2063.6	6.5229	0.21382	2061.3	6.4308	0.18283	2059.0	6.3523
280	0.26726	2116.0	6.6194	0.22225	2114.0	6.5278	0.19010	2111.9	6.4498
300	0.27726	2169.1	6.7137	0.23063	2167.3	6.6225	0.19732	2165.5	6.5450
320	0.28723	2222.9	6.8059	0.23897	2221.3	6.7151	0.20450	2219.8	6.6380

(continued)

Superheated Table (PC Model), Ammonia (NH₃)

°C	m³/kg	kJ/kg	kJ/kg·K	m³/kg	kJ/kg	kJ/kg·K	m³/kg	kJ/kg	kJ/kg·K
T	$p = 1.60$ MPa ($T_{sat} = 41.03°C$)			$p = 1.80$ MPa ($T_{sat} = 45.38°C$)			$p = 2.00$ MPa ($T_{sat} = 49.37°C$)		
	v	h	s	v	h	s	v	h	s
Sat.	0.08079	1470.5	4.8553	0.07174·	1471.3	4.8096	0.06444	1471.5	4.7680
50	0.08506	1501.0	4.9510	0.07381	1487.9	4.8614	0.06471	1473.9	4.7754
60	0.08951	1532.5	5.0472	0.07801	1521.4	4.9637	0.06875	1509.8	4.8848
70	0.09372	1562.3	5.1351	0.08193	1552.7	5.0561	0.07246	1542.7	4.9821
80	0.09774	1590.7	5.2167	0.08565	1582.2	5.1410	0.07595	1573.5	5.0707
100	0.10539	1644.8	5.3659	0.09267	1638.0	5.2948	0.08248	1631.1	5.2294
120	0.11268	1696.9	5.5018	0.09931	1691.2	5.4337	0.08861	1685.5	5.3714
140	0.11974	1748.0	5.6286	0.10570	1743.1	5.5624	0.09447	1738.2	5.5022
160	0.12662	1798.7	5.7485	0.11192	1794.5	5.6838	0.10016	1790.2	5.6251
180	0.13339	1849.5	5.8631	0.11801	1845.7	5.7995	0.10571	1842.0	5.7420
200	0.14005	1900.5	5.9734	0.12400	1897.2	5.9107	0.11116	1893.9	5.8540
220	0.14663	1952.0	6.0800	0.12990	1949.1	6.0180	0.11652	1946.1	5.9621
240	0.15314	2004.1	6.1834	0.13574	2001.4	6.1211	0.12182	1998.8	6.0668
260	0.15959	2056.7	6.2834	0.14152	2054.3	6.2232	0.12705	2052.0	6.1685
280	0.16599	2109.9	6.3819	0.14724	2107.8	6.3217	0.13224	2105.8	6.2675
300	0.17234	2163.7	6.4775	0.15291	2161.9	6.4178	0.13737	2160.1	6.3641
320	0.17865	2218.2	6.5710	0.15854	2216.7	6.5116	0.14246	2215.1	6.4583
340	0.18492	2273.3	6.6624	0.16414	2272.0	6.6034	0.14751	2270.7	6.5505
360	0.19115	2329.1	6.7519	0.16969	2328.0	6.6932	0.15253	2326.8	6.6406
T	$p = 5.00$ MPa ($T_{sat} = 88.9°C$)			$p = 10.00$ MPa ($T_{sat} = 125.20°C$)			$p = 20.00$ MPa		
	v	h	s	v	h	s	v	h	s
Sat.	0.02365	1441.4	4.3454	0.00826	1289.4	3.7587	–	–	–
100	0.02636	1501.5	4.5091	–	–	–	–	–	–
120	0.03024	1586.3	4.7306	–	–	–	–	–	–
140	0.03350	1657.3	4.9068	0.01195	1461.3	4.1839	0.00251	918.9	2.7630
160	0.03643	1721.7	5.0591	0.01461	1578.3	4.4610	0.00323	1097.2	3.1838
180	0.03916	1782.7	5.1968	0.01666	1667.2	4.6617	0.00490	1329.7	3.7087
200	0.04174	1841.8	5.3245	0.01842	1744.5	4.8287	0.00653	1497.7	4.0721
220	0.04422	1900.0	5.4450	0.02001	1816.0	4.9767	0.00782	1618.7	4.3228
240	0.04662	1957.9	5.5600	0.02150	1884.2	5.1123	0.00891	1718.6	4.5214
260	0.04895	2015.6	5.6704	0.02290	1950.6	5.2392	0.00988	1807.6	4.6916
280	0.05123	2073.6	5.7771	0.02424	2015.9	5.3596	0.01077	1890.5	4.8442
300	0.05346	2131.8	5.8805	0.02552	2080.7	5.4746	0.01159	1969.6	4.9847
320	0.05565	2190.3	5.9809	0.02676	2145.2	5.5852	0.01237	2046.3	5.1164
340	0.05779	2249.2	6.0786	0.02796	2209.6	5.6921	0.01312	2121.6	5.2412
360	0.05990	2308.6	6.1738	0.02913	2274.1	5.7955	0.01382	2195.8	5.3603
380	0.06198	2368.4	6.2668	0.03026	2338.7	5.8960	0.01450	2269.4	5.4748
400	0.06403	2428.6	6.3576	0.03137	2403.5	5.9937	0.01516	2342.6	5.5851
420	0.06606	2489.3	6.4464	0.03245	2468.5	6.0888	0.01579	2415.4	5.6917
440	0.06806	2550.4	6.5334	0.03351	2533.7	6.1815	0.01641	2488.1	5.7950

Table B-14   PC Model: Temperature Saturation Table, Nitrogen (N₂)

Saturation Temperature Table (PC Model) of Nitrogen (N₂)

Temp.	Sat. Press.	Spec. Volume		Spec. Int. Energy		Spec. Enthalpy		Spec. Entropy	
		m³/kg		kJ/kg		kJ/kg		kJ/kg·K	
K	kPa	Sat. liquid	Sat. vapor	Sat. liquid	Sat. vapor	Sat. liquid	Sat. vapor	Sat. liquid	Sat. vapor
T	$p_{sat@T}$	v_f	v_g	u_f	u_g	h_f	h_g	s_f	s_g
63.1	12.5	0.001150	1.48189	−150.92	45.94	−150.91	64.48	2.4234	5.8343
65	17.4	0.001160	1.09347	−147.19	47.17	−147.17	66.21	2.4816	5.7645
70	38.6	0.001191	0.52632	−137.13	50.40	−137.09	70.70	2.6307	5.5991
75	76.1	0.001223	0.28174	−127.04	53.43	−126.95	74.87	2.7700	5.4609
77.3	101.3	0.001240	0.21639	−122.27	54.76	−122.15	76.69	2.8326	5.4033
80	137.0	0.001259	0.16375	−116.86	56.20	−116.69	78.63	2.9014	5.3429
85	229.1	0.001299	0.10148	−106.55	58.65	−106.25	81.90	3.0266	5.2401
90	360.8	0.001343	0.06611	−96.06	60.70	−95.58	84.55	3.1466	5.1480
95	541.1	0.001393	0.04476	−85.35	62.25	−84.59	86.47	3.2627	5.0634
100	779.2	0.001452	0.03120	−74.33	63.17	−73.20	87.48	3.3761	4.9829
105	1084.6	0.001522	0.02218	−62.89	63.29	−61.24	87.35	3.4883	4.9034
110	1467.6	0.001610	0.01595	−50.81	62.31	−48.45	85.71	3.6017	4.8213
115	1939.3	0.001729	0.01144	−37.66	59.70	−34.31	81.88	3.7204	4.7307
120	2513.0	0.001915	0.00799	−22.42	54.21	−17.61	74.30	3.8536	4.6195
125	3208.0	0.002355	0.00490	−0.83	39.90	6.73	55.60	4.0399	4.4309
126.2	3397.8	0.003194	0.00319	18.94	18.94	29.79	29.79	4.2193	4.2193

Table B-15   PC Model: Superheated Vapor Table, Nitrogen (N₂)

Superheated Table (PC Model), Nitrogen (N₂)

K	m³/kg	kJ/kg	kJ/kg	kJ/kg·K	m³/kg	kJ/kg	kJ/kg	kJ/kg·K	m³/kg	kJ/kg	kJ/kg	kJ/kg·K
T	$p = 0.10$ MPa ($T_{sat} = 77.24$ K)				$p = 0.20$ MPa ($T_{sat} = 83.62$ K)				$p = 0.50$ MPa ($T_{sat} = 93.98$ K)			
	v	u	h	s	v	u	h	s	v	u	h	s
Sat.	0.21903	54.707	76.61	5.4059	0.11520	58.010	81.05	5.2673	0.04834	61.980	86.15	5.0802
100	0.29103	72.837	101.94	5.6944	0.14252	71.736	100.24	5.4775	0.05306	67.930	94.46	5.1660
120	0.35208	87.942	123.15	5.8878	0.17397	87.136	121.93	5.6753	0.06701	84.615	118.12	5.3821
140	0.41253	102.947	144.20	6.0501	0.20476	102.328	143.28	5.8399	0.08007	100.405	140.44	5.5541
160	0.47263	117.907	165.17	6.1901	0.23519	117.402	164.44	5.9812	0.09272	115.860	162.22	5.6996
180	0.53254	132.836	186.09	6.3132	0.26542	132.406	185.49	6.1052	0.10515	131.125	183.70	5.8261
200	0.59231	147.739	206.97	6.4232	0.29551	147.378	206.48	6.2157	0.11744	146.280	205.00	5.9383
220	0.65199	162.631	227.83	6.5227	0.32552	162.306	227.41	6.3155	0.12964	161.360	226.18	6.0392

(continued)

Superheated Table (PC Model), Nitrogen (N₂)

K	m³/kg	kJ/kg	kJ/kg	kJ/kg·K	m³/kg	kJ/kg	kJ/kg	kJ/kg·K	m³/kg	kJ/kg	kJ/kg	kJ/kg·K
T	$p=0.10$ MPa ($T_{sat}=77.24$ K)				$p=0.20$ MPa ($T_{sat}=83.62$ K)				$p=0.50$ MPa ($T_{sat}=93.98$ K)			
	v	*u*	*h*	*s*	*v*	*u*	*h*	*s*	*v*	*u*	*h*	*s*
240	0.71161	177.509	248.67	6.6133	0.35546	177.228	248.32	6.4064	0.14177	176.385	247.27	6.1310
260	0.77118	192.392	269.51	6.6967	0.38535	192.140	269.21	6.4900	0.15385	191.385	268.31	6.2152
280	0.83072	207.258	290.33	6.7739	0.41520	207.040	290.08	6.5674	0.16590	206.360	289.31	6.2930
300	0.89023	222.137	311.16	6.8457	0.44503	221.934	310.94	6.6393	0.17792	221.320	310.28	6.3653
350	1.03891	259.349	363.24	7.0063	0.51952	259.186	363.09	6.8001	0.20788	258.690	362.63	6.5267
400	1.18752	296.658	415.41	7.1456	0.59392	296.526	415.31	6.9396	0.23777	296.105	414.99	6.6666
450	1.33607	334.163	467.77	7.2690	0.66827	334.046	467.70	7.0630	0.26759	333.695	467.49	6.7902
500	1.48458	371.952	520.41	7.3799	0.74258	371.854	520.37	7.1740	0.29739	371.545	520.24	6.9014
600	1.78154	448.786	626.94	7.5741	0.89114	448.712	626.94	7.3682	0.35691	448.475	626.93	7.0959
700	2.07845	527.735	735.58	7.7415	1.03965	527.680	735.61	7.5357	0.41637	527.495	735.68	7.2635
800	2.37532	609.068	846.60	7.8897	1.18812	609.016	846.64	7.6839	0.47581	608.875	846.78	7.4118
900	2.67217	692.793	960.01	8.0232	1.33657	692.756	960.07	7.8175	0.53522	692.630	960.24	7.5454
1000	2.96900	778.780	1075.68	8.1451	1.48501	778.748	1075.75	7.9393	0.59462	778.650	1075.96	7.6673
T	$p=0.60$ MPa ($T_{sat}=96.37$ K)				$p=0.80$ MPa ($T_{sat}=100.38$ K)				$p=1.00$ MPa ($T_{sat}=103.73$ K)			
	v	*u*	*h*	*s*	*v*	*u*	*h*	*s*	*v*	*u*	*h*	*s*
Sat.	0.04046	62.574	86.85	5.0411	0.03038	63.216	87.52	4.9768	0.02416	63.350	87.51	4.9237
120	0.05510	83.730	116.79	5.3204	0.04017	81.884	114.02	5.2191	0.03117	79.910	111.08	5.1357
140	0.06620	99.750	139.47	5.4953	0.04886	98.412	137.50	5.4002	0.03845	97.020	135.47	5.3239
160	0.07689	115.336	161.47	5.6422	0.05710	114.270	159.95	5.5501	0.04522	113.200	158.42	5.4772
180	0.08734	130.696	183.10	5.7696	0.06509	129.818	181.89	5.6793	0.05173	128.940	180.67	5.6082
200	0.09766	145.904	204.50	5.8823	0.07293	145.166	203.51	5.7933	0.05809	144.430	202.52	5.7234
220	0.10788	161.032	225.76	5.9837	0.08067	160.404	224.94	5.8954	0.06436	159.750	224.11	5.8263
240	0.11803	176.102	246.92	6.0757	0.08835	175.550	246.23	5.9880	0.07055	174.980	245.53	5.9194
260	0.12813	191.132	268.01	6.1601	0.09599	190.628	267.42	6.0728	0.07670	190.130	266.83	6.0047
280	0.13820	206.130	289.05	6.2381	0.10358	205.676	288.54	6.1511	0.08281	205.230	288.04	6.0833
300	0.14824	221.116	310.06	6.3105	0.11115	220.700	309.62	6.2238	0.08889	220.290	309.18	6.1562
350	0.17326	258.524	362.48	6.4722	0.12998	258.186	362.17	6.3858	0.10401	257.860	361.87	6.3187
400	0.19819	295.976	414.89	6.6121	0.14873	295.696	414.68	6.5260	0.11905	295.420	414.47	6.4591
450	0.22308	333.572	467.42	6.7359	0.16743	333.336	467.28	6.6500	0.13404	333.110	467.15	6.5832
500	0.24792	371.448	520.20	6.8471	0.18609	371.248	520.12	6.7613	0.14899	371.050	520.04	6.6947
600	0.29755	448.400	626.93	7.0416	0.22335	448.250	626.93	6.9560	0.17883	448.090	626.92	6.8895
700	0.34712	527.428	735.70	7.2093	0.26056	527.312	735.76	7.1237	0.20862	527.190	735.81	7.0573
800	0.39666	608.824	846.82	7.3576	0.29773	608.726	846.91	7.2721	0.23837	608.630	847.00	7.2057
900	0.44618	692.592	960.30	7.4912	0.33488	692.516	960.42	7.4058	0.26810	692.440	960.54	7.3394
1000	0.49568	778.612	1076.02	7.6131	0.37202	778.544	1076.16	7.5277	0.29782	778.480	1076.30	7.4614

Table B-16   PC Model: Temperature Saturation Table, Propane (C_3H_8)

Saturation Temperature Table (PC Model) of Propane (C_3H_8)

Temp.	Sat. Press.	Spec. Volume		Spec. Int. Energy		Spec. Enthalpy		Spec. Entropy	
		m^3/kg		kJ/kg		kJ/kg		kJ/kg·K	
K	kPa	Sat. liquid	Sat. vapor	Sat. liquid	Sat. vapor	Sat. liquid	Sat. vapor	Sat. liquid	Sat. vapor
T	$p_{sat@T}$	v_f	v_g	u_f	u_g	h_f	h_g	s_f	s_g
85.5	1.65×10^{-7}	0.001364	9.570×10^7	−495.72	50.78	−495.72	66.90	1.8813	8.4639
100	2.52×10^{-5}	0.001392	7.470×10^6	−467.82	61.30	−467.82	80.15	2.1826	7.6624
120	2.95×10^{-3}	0.001432	7.660×10^3	−428.94	77.05	−428.94	99.67	2.5370	6.9421
140	7.87×10^{-2}	0.001475	3.354×10^2	−389.60	94.10	−389.60	120.49	2.8402	6.4837
160	8.47×10^{-1}	0.001521	3.558×10^1	−349.68	112.31	−349.68	142.44	3.1067	6.1824
180	5.0	0.00157	6.69013	−308.95	131.56	−308.95	165.35	3.3465	5.9814
200	20.1	0.001624	1.84938	−267.18	151.68	−267.14	188.92	3.5665	5.8468
220	60.5	0.001684	0.66753	−224.06	172.47	−223.95	212.83	3.7719	5.7573
231.1	101.3	0.00172	0.41342	−199.50	184.18	−199.33	226.07	3.8808	5.7219
240	147.9	0.001752	0.29076	−179.27	193.72	−179.01	236.71	3.9668	5.6989
260	310.5	0.001831	0.14475	−132.45	215.19	−131.88	260.14	4.1542	5.6619
280	581.5	0.001925	0.07919	−83.17	236.52	−82.05	282.57	4.3369	5.6391
300	997.4	0.002043	0.04618	−30.84	257.10	−28.81	303.16	4.5176	5.6242
320	1598.4	0.0022	0.02793	25.38	275.78	28.90	320.42	4.6996	5.6106
340	2430.6	0.002433	0.01693	87.18	290.00	93.09	331.15	4.8883	5.5884
360	3554.0	0.002894	0.00949	160.40	291.07	170.69	324.79	5.1013	5.5293
369.9	4247.7	0.004535	0.00454	240.34	240.34	259.61	259.61	5.3376	5.3376

Table B-17   PC Model: Superheated Vapor Table, Propane (C_3H_8)

Superheated Table (PC Model), Propane (C_3H_8)

K	m^3/kg	kJ/kg	kJ/kg·K	m^3/kg	kJ/kg	kJ/kg·K	m^3/kg	kJ/kg	kJ/kg·K
T	p = 50 kPa (T_{sat} = 216.22 K)			p = 100 kPa (T_{sat} = 230.77 K)			p = 200 kPa (T_{sat} = 247.72 K)		
	v	h	s	v	h	s	v	h	s
Sat.	0.79617	208.31	5.7715	0.41851	225.72	5.7227	0.21917	245.83	5.6826
240	0.89026	241.87	5.9187	0.43749	239.29	5.7804	–	–	–
260	0.96824	271.64	6.0378	0.47787	269.57	5.9016	0.23242	265.25	5.7591
280	1.04557	302.97	6.1539	0.51755	301.27	6.0190	0.25338	297.76	5.8795
300	1.12246	335.96	6.2677	0.55677	334.53	6.1337	0.27382	331.61	5.9962
320	1.19904	370.68	6.3797	0.59566	369.45	6.2463	0.29391	366.96	6.1103
340	1.27540	407.16	6.4902	0.63433	406.09	6.3574	0.31375	403.93	6.2223
360	1.35160	445.41	6.5995	0.67283	444.47	6.4670	0.33341	442.57	6.3327
380	1.42767	485.45	6.7077	0.71120	484.61	6.5755	0.35294	482.91	6.4418

(continued)

				Superheated Table (PC Model), Propane (C$_3$H$_8$)					
K	m^3/kg	kJ/kg	kJ/kg·K	m^3/kg	kJ/kg	kJ/kg·K	m^3/kg	kJ/kg	kJ/kg·K
T	*p* = 50 kPa (*T*$_{sat}$ = 216.22 K)			*p* = 100 kPa (*T*$_{sat}$ = 230.77 K)			*p* = 200 kPa (*T*$_{sat}$ = 247.72 K)		
	v	*h*	*s*	*v*	*h*	*s*	*v*	*h*	*s*
400	1.50365	527.25	6.8149	0.74947	526.49	6.6829	0.37237	524.96	6.5496
420	1.57956	570.79	6.9211	0.78767	570.10	6.7893	0.39172	568.72	6.6563
440	1.65540	616.06	7.0264	0.82581	615.43	6.8947	0.41101	614.16	6.7620
460	1.73119	663.01	7.1307	0.86389	662.43	6.9991	0.43024	661.27	6.8667
480	1.80694	711.61	7.2341	0.90194	711.08	7.1027	0.44943	710.00	6.9704
500	1.88266	761.83	7.3366	0.93995	761.33	7.2052	0.46859	760.34	7.0731
520	1.95835	813.62	7.4382	0.97793	813.16	7.3068	0.48772	812.23	7.1749
540	2.03402	866.95	7.5388	1.01589	866.52	7.4075	0.50683	865.65	7.2757
560	2.10966	921.77	7.6385	1.05382	921.37	7.5073	0.52591	920.56	7.3755
580	2.18529	978.05	7.7372	1.09174	977.67	7.6060	0.54497	976.91	7.4744
600	2.26090	1035.74	7.8350	1.12964	1035.38	7.7039	0.56402	1034.67	7.5722
T	*p* = 300 kPa (*T*$_{sat}$ = 258.99 K)			*p* = 400 kPa (*T*$_{sat}$ = 267.69 K)			*p* = 500 kPa (*T*$_{sat}$ = 274.89 K)		
	v	*h*	*s*	*v*	*h*	*s*	*v*	*h*	*s*
Sat.	0.14956	258.97	5.6634	0.11371	268.92	5.6518	0.09172	276.97	5.6439
280	0.16516	294.08	5.7937	0.12092	290.21	5.7296	0.09424	286.11	5.6768
300	0.17940	328.57	5.9127	0.13211	325.42	5.8510	0.10367	322.14	5.8011
320	0.19326	364.39	6.0282	0.14288	361.75	5.9682	0.11261	359.03	5.9202
340	0.20684	401.71	6.1413	0.15336	399.45	6.0825	0.12124	397.14	6.0356
360	0.22024	440.63	6.2525	0.16363	438.66	6.1945	0.12965	436.65	6.1485
380	0.23350	481.19	6.3622	0.17377	479.44	6.3047	0.13791	477.67	6.2594
400	0.24665	523.42	6.4705	0.18379	521.86	6.4135	0.14606	520.28	6.3687
420	0.25973	567.32	6.5775	0.19372	565.91	6.5210	0.15411	564.49	6.4765
440	0.27273	612.89	6.6835	0.20359	611.6	6.6272	0.16210	610.31	6.5831
460	0.28569	660.10	6.7884	0.21341	658.92	6.7324	0.17003	657.74	6.6885
480	0.29860	708.92	6.8923	0.22318	707.84	6.8365	0.17792	706.75	6.7928
500	0.31147	759.34	6.9952	0.23291	758.33	6.9395	0.18577	757.32	6.8960
520	0.32432	811.30	7.0971	0.24262	810.37	7.0415	0.19359	809.43	6.9982
540	0.33714	864.78	7.1980	0.25229	863.91	7.1426	0.20139	863.04	7.0993
560	0.34994	919.75	7.2979	0.26195	918.93	7.2426	0.20916	918.12	7.1994
580	0.36271	976.15	7.3969	0.27159	975.39	7.3417	0.21691	974.62	7.2986
600	0.37548	1033.95	7.4949	0.28121	1033.24	7.4397	0.22465	1032.52	7.3967
T	*p* = 600 kPa (*T*$_{sat}$ = 281.08 K)			*p* = 700 kPa (*T*$_{sat}$ = 286.56 K)			*p* = 800 kPa (*T*$_{sat}$ = 291.48 K)		
	v	*h*	*s*	*v*	*h*	*s*	*v*	*h*	*s*
Sat.	0.07680	283.74	5.6381	0.06598	289.58	5.6336	0.05776	294.69	5.6299
300	0.08463	318.71	5.7585	0.07097	315.12	5.7207	0.06066	311.33	5.6862
320	0.09239	356.22	5.8795	0.07791	353.32	5.8440	0.06701	350.31	5.8120
340	0.09980	394.77	5.9963	0.08446	392.33	5.9622	0.07294	389.84	5.9318
360	0.10698	434.60	6.1102	0.09077	432.51	6.0770	0.07860	430.39	6.0477
380	0.11400	475.88	6.2217	0.09691	474.05	6.1893	0.08408	472.20	6.1607
400	0.12090	518.68	6.3315	0.10292	517.06	6.2996	0.08943	515.43	6.2715
420	0.12770	563.05	6.4397	0.10884	561.60	6.4082	0.09468	560.14	6.3806

Superheated Table (PC Model), Propane (C₃H₈)

K	m³/kg	kJ/kg	kJ/kg·K	m³/kg	kJ/kg	kJ/kg·K	m³/kg	kJ/kg	kJ/kg·K
	p = 600 kPa (T_{sat} = 281.08 K)			p = 700 kPa (T_{sat} = 286.56 K)			p = 800 kPa (T_{sat} = 291.48 K)		
T	v	h	s	v	h	s	v	h	s
440	0.13444	609.01	6.5466	0.11468	607.69	6.5154	0.09985	606.37	6.4881
460	0.14112	656.54	6.6522	0.12046	655.35	6.6213	0.10497	654.14	6.5943
480	0.14775	705.65	6.7567	0.12620	704.55	6.7260	0.11004	703.45	6.6692
500	0.15435	756.31	6.8601	0.13190	755.30	6.8296	0.11507	754.28	6.8029
520	0.16091	808.49	6.9625	0.13757	807.55	6.9321	0.12006	806.61	6.9055
540	0.16745	862.17	7.0637	0.14321	861.29	7.0335	0.12503	860.41	7.0071
560	0.17397	917.30	7.1640	0.14883	916.48	7.1338	0.12998	915.66	7.1075
580	0.18046	973.86	7.2632	0.15443	973.09	7.2331	0.13490	972.33	7.2069
600	0.18694	1031.80	7.3614	0.16001	1031.08	7.3314	0.13981	1030.37	7.3053
T	p = 1000 kPa (T_{sat} = 300.10 K)			p = 2000 kPa (T_{sat} = 330.42 K)			p = 4000 kPa (T_{sat} = 366.53 K)		
	v	h	s	v	h	s	v	h	s
Sat.	0.04606	303.26	5.6241	0.02157	327.17	5.6011	0.00715	308.67	5.4748
340	0.05675	384.63	5.8787	0.02353	352.08	5.6754	–	–	–
360	0.06153	426.00	5.9969	0.02695	400.50	5.8139	–	–	–
380	0.06610	468.41	6.1115	0.02990	447.30	5.9404	0.01064	383.27	5.6754
400	0.07053	512.09	6.2236	0.03259	494.04	6.0602	0.01312	447.49	5.8402
420	0.07485	557.16	6.3335	0.03511	541.37	6.1757	0.01500	503.81	5.9777
440	0.07910	603.69	6.4417	0.03753	589.66	6.2880	0.01663	557.89	6.1034
460	0.08328	651.71	6.5484	0.03986	639.09	6.3979	0.01811	611.45	6.2225
480	0.08741	701.22	6.6538	0.04213	689.76	6.5057	0.01949	665.27	6.3370
500	0.09150	752.23	6.7579	0.04435	741.75	6.6118	0.02080	719.74	6.4482
520	0.09555	804.71	6.8608	0.04653	795.07	6.7163	0.02206	775.09	6.5567
540	0.09958	858.65	6.9626	0.04869	849.73	6.8195	0.02328	831.45	6.6631
560	0.10358	914.02	7.0632	0.05081	905.74	6.9213	0.02447	888.91	6.7676
580	0.10757	970.79	7.1628	0.05291	963.07	7.0219	0.02564	947.50	6.8703
600	0.11153	1028.93	7.2614	0.05499	1021.71	7.1213	0.02678	1007.24	6.9716
T	p = 10,000 kPa			p = 20,000 kPa			p = 50,000 kPa		
	v	h	s	v	h	s	v	h	s
380	0.002618	207.08	5.1520	0.002290	194.83	5.0560	0.001991	207.83	4.9236
400	0.002999	278.45	5.3350	0.002437	254.18	5.2082	0.002056	261.94	5.0623
420	0.003608	356.92	5.5263	0.002609	315.64	5.3581	0.002124	317.54	5.1979
440	0.004454	437.64	5.7141	0.002809	378.97	5.5054	0.002196	374.58	5.3306
460	0.005371	513.72	5.8833	0.003037	443.82	5.6495	0.002272	433.00	5.4604
480	0.006225	583.73	6.0323	0.003291	509.79	5.7899	0.002351	492.72	5.5875
500	0.006996	649.74	6.1670	0.003568	576.48	5.9260	0.002432	553.69	5.7119
520	0.007700	713.66	6.2924	0.003861	643.58	6.0576	0.002517	615.85	5.8338
540	0.008353	776.65	6.4113	0.004164	710.85	6.1846	0.002604	679.14	5.9532
560	0.008968	839.42	6.5254	0.004470	778.21	6.3070	0.002693	743.52	6.0703
580	0.009553	902.39	6.6359	0.004775	845.66	6.4254	0.002784	808.94	6.1851
600	0.010115	965.84	6.7434	0.005076	913.27	6.5400	0.002876	875.37	6.2977

| Table C-1 | PG Model: Material Properties of Common Perfect Gases |

Material Properties of Perfect Gases (PG Model) (at 298 K)

Gas	Formula	Molar Mass kg/kmol $\overline{M}$	Gas Constant kJ/kg·K R	Spec. Heat at Const. Press. kJ/kg·K c_p	Spec. Heat at Const. Vol. kJ/kg·K c_v	Spec. Heat Ratio k
Air	–	28.97	0.2870	1.0050	0.7180	1.400
Argon	Ar	39.95	0.2081	0.5203	0.3122	1.667
Butane	C_4H_{10}	58.12	0.1433	1.7164	1.5734	1.091
Carbon Dioxide	CO_2	44.01	0.1889	0.8460	0.6570	1.289
Carbon Monoxide	CO	28.01	0.2968	1.0400	0.7440	1.400
Ethane	C_2H_6	30.07	0.2765	1.7662	1.4897	1.186
Ethylene	C_2H_4	28.05	0.2964	1.5482	1.2518	1.237
Helium	He	4.00	2.0769	5.1926	3.1156	1.667
Hydrogen	H_2	2.02	4.1240	14.3070	10.1830	1.405
Methane	CH_4	16.04	0.5182	2.2537	1.7354	1.299
Neon	Ne	20.18	0.4119	1.0299	0.6179	1.667
Nitrogen	N_2	28.01	0.2968	1.0390	0.7430	1.400
Octane	C_8H_{18}	114.23	0.0729	1.7113	1.6385	1.044
Oxygen	O_2	32.00	0.2598	0.9180	0.6580	1.395
Propane	C_3H_8	44.10	0.1885	1.6794	1.4909	1.126
Steam	H_2O	18.02	0.4614	1.8677	1.4063	1.327

Note: The unit kJ/kg·K is equivalent to kJ/kg·°C

| Table D-1 | IG Model: Polynomial Relation for Specific Heat for Common Gases |

Polynomial Relation for Ideal Gas Specific Heat as a Function of Temperature

Substance	Formula	$\overline{c}_p = \overline{M}c_p = a + bT + cT^2 + dT^3$ (c_p in kJ/(kg K), T in K)				Temperature Range, K	% error	
		a	b	c	d		Max.	Avg.
Nitrogen	N_2	28.900	-0.1571×10^{-2}	0.8081×10^{-5}	-2.873×10^{-9}	273–1800	0.59	0.34
Oxygen	O_2	25.480	1.5200×10^{-2}	-0.7155×10^{-5}	1.312×10^{-9}	273–1800	1.19	0.28
Air	-	28.110	0.1967×10^{-2}	0.4802×10^{-5}	-1.966×10^{-9}	273–1800	0.72	0.33
Hydrogen	H_2	29.110	-0.1916×10^{-2}	0.4003×10^{-5}	-0.870×10^{-9}	273–1800	1.01	0.26
Carbon Monoxide	CO	28.160	0.1675×10^{-2}	0.5372×10^{-5}	-2.222×10^{-9}	273–1800	0.89	0.37
Carbon Dioxide	CO_2	22.260	5.9810×10^{-2}	-3.5010×10^{-5}	7.469×10^{-9}	273–1800	0.67	0.22
Water Vapor	H_2O	32.240	0.1923×10^{-2}	1.0550×10^{-5}	-3.595×10^{-9}	273–1800	0.53	0.24
Nitric Oxide	NO	29.340	-0.0940×10^{-2}	0.9747×10^{-5}	-4.187×10^{-9}	273–1500	0.97	0.36
Nitrous Oxide	N_2O	24.110	5.8632×10^{-2}	-3.5620×10^{-5}	10.580×10^{-9}	273–1500	0.59	0.26
Nitrogen Dioxide	NO_2	22.900	5.7150×10^{-2}	-3.5200×10^{-5}	7.870×10^{-9}	273–1500	0.46	0.18
Ammonia	NH_3	27.568	2.5630×10^{-2}	0.9907×10^{-5}	-6.691×10^{-9}	273–1500	0.91	0.36
Sulfur	S_2	27.210	2.2180×10^{-2}	-1.6280×10^{-5}	3.986×10^{-9}	273–1800	0.99	0.38
Sulfur Dioxide	SO_2	25.780	5.7950×10^{-2}	-3.8120×10^{-5}	8.612×10^{-9}	273–1800	0.45	0.24
Sulfur Trioxide	SO_3	16.400	14.5800×10^{-2}	-11.2000×10^{-5}	32.420×10^{-9}	273–1300	0.29	0.13

Polynomial Relation for Ideal Gas Specific Heat as a Function of Temperature

Substance	Formula	$\bar{c}_p = \overline{M} c_p = a + bT + cT^2 + dT^3$ (c_p in kJ/(kg K), T in K)				Temperature Range, K	% error	
		a	b	c	d		Max.	Avg.
Acetylene	C_2H_2	21.800	9.2143×10^{-2}	-6.5270×10^{-5}	18.210×10^{-9}	273–1500	1.46	0.59
Benzene	C_6H_6	−36.220	48.4750×10^{-2}	-31.5700×10^{-5}	77.620×10^{-9}	273–1500	0.34	0.20
Methanol	CH_4O	19.000	9.1520×10^{-2}	-1.2200×10^{-5}	-8.039×10^{-9}	273–1000	0.18	0.08
Ethanol	C_2H_6O	19.900	20.9600×10^{-2}	-10.3800×10^{-5}	20.050×10^{-9}	273–1500	0.40	0.22
Hydrogen Chloride	HCl	30.330	-0.7620×10^{-2}	1.3270×10^{-5}	-4.338×10^{-9}	273–1500	0.22	0.08
Methane	CH_4O	19.890	5.0240×10^{-2}	1.2690×10^{-5}	-11.010×10^{-9}	273–1500	1.33	0.57
Ethane	C_2H_6	6.900	17.2700×10^{-2}	-6.4060×10^{-5}	7.285×10^{-9}	273–1500	0.83	0.28
Propane	C_3H_8	−4.040	30.4800×10^{-2}	-15.7200×10^{-5}	31.740×10^{-9}	273–1500	0.40	0.12
n-Butane	C_4H_{10}	3.960	37.1500×10^{-2}	-18.3400×10^{-5}	35.000×10^{-9}	273–1500	0.54	0.24
i-Butane	C_4H_{10}	−7.913	41.6000×10^{-2}	-23.0100×10^{-5}	49.910×10^{-9}	273–1500	0.25	0.13
n-Pentane	C_5H_{12}	6.774	45.4300×10^{-2}	-22.4600×10^{-5}	42.290×10^{-9}	273–1500	0.56	0.21
n-Hexane	C_6H_{14}	6.938	55.2200×10^{-2}	-28.6500×10^{-5}	57.690×10^{-9}	273–1500	0.72	0.20
Ethylene	C_2H_4	3.950	15.6400×10^{-2}	-8.3440×10^{-5}	17.670×10^{-9}	273–1500	0.54	0.13
Propylene	C_3H_6	3.150	23.8300×10^{-2}	-12.1800×10^{-5}	24.620×10^{-9}	273–1500	0.73	0.17

Table D-2   IG Model: Specific Heat Values for Six Common Gases

Temperature T	c_p	c_v	k	c_p	c_v	k	c_p	c_v	k
K	kJ/kg·K	kJ/kg·K		kJ/kg·K	kJ/kg·K		kJ/kg·K	kJ/kg·K	
	Air			Carbon Dioxide, CO_2			Carbon Monoxide, CO		
250	1.003	0.716	1.401	0.791	0.602	1.314	1.039	0.743	1.400
300	1.005	0.718	1.400	0.846	0.657	1.288	1.040	0.744	1.399
350	1.008	0.721	1.398	0.895	0.706	1.268	1.043	0.746	1.398
400	1.013	0.726	1.395	0.939	0.750	1.252	1.047	0.751	1.395
450	1.020	0.733	1.391	0.978	0.790	1.239	1.054	0.757	1.392
500	1.029	0.742	1.387	1.014	0.825	1.229	1.063	0.767	1.387
550	1.040	0.753	1.381	1.046	0.857	1.220	1.075	0.778	1.382
600	1.051	0.764	1.376	1.075	0.886	1.213	1.087	0.790	1.376
650	1.063	0.776	1.370	1.102	0.913	1.207	1.100	0.803	1.370
700	1.075	0.788	1.364	1.126	0.937	1.202	1.113	0.816	1.364
750	1.087	0.800	1.359	1.148	0.959	1.197	1.126	0.829	1.358
800	1.099	0.812	1.354	1.169	0.980	1.193	1.139	0.842	1.353
900	1.121	0.834	1.344	1.204	1.015	1.186	1.163	0.866	1.343
1000	1.142	0.855	1.336	1.234	1.045	1.181	1.185	0.888	1.335
	Hydrogen, H_2			Nitrogen, N_2			Oxygen, O_2		
250	14.051	9.927	1.416	1.039	0.742	1.400	0.913	0.653	1.398
300	14.307	10.183	1.405	1.039	0.743	1.400	0.918	0.658	1.395
350	14.427	10.302	1.400	1.041	0.744	1.399	0.928	0.668	1.389
400	14.476	10.352	1.398	1.044	0.747	1.397	0.941	0.681	1.382
450	14.501	10.377	1.398	1.049	0.752	1.395	0.956	0.696	1.373

(continued)

Temperature T	c_p	c_v	k	c_p	c_v	k	c_p	c_v	k
K	kJ/kg·K	kJ/kg·K		kJ/kg·K	kJ/kg·K		kJ/kg·K	kJ/kg·K	
	Hydrogen, H$_2$			Nitrogen, N$_2$			Oxygen, O$_2$		
500	14.513	10.389	1.397	1.056	0.759	1.391	0.972	0.712	1.365
550	14.530	10.405	1.396	1.065	0.768	1.387	0.988	0.728	1.358
600	14.546	10.422	1.396	1.075	0.778	1.382	1.003	0.743	1.350
650	14.571	10.447	1.395	1.086	0.789	1.376	1.017	0.758	1.343
700	14.604	10.480	1.394	1.098	0.801	1.371	1.031	0.771	1.337
750	14.645	10.521	1.392	1.110	0.813	1.365	1.043	0.783	1.332
800	14.695	10.570	1.390	1.121	0.825	1.360	1.054	0.794	1.327
900	14.822	10.698	1.385	1.145	0.849	1.349	1.074	0.814	1.319
1000	14.983	10.859	1.380	1.167	0.870	1.341	1.090	0.830	1.313

Table D-3   IG Model: Ideal Gas Properties of Air

Ideal-Gas Properties of Air (In IG TESTcalcs, select Air*)

$$\overline{h}^{\,o}_{f,298} = 0 \text{ kJ/kmol}; \quad \overline{M} = 28.97 \text{ kg/kmol}$$

T	h	p_r	u	v_r	s^o
K	kJ/kg		kJ/kg		kJ/kg·K
200	199.97	0.3363	142.56	1707.0	1.29559
210	209.97	0.3987	149.69	1512.0	1.34444
220	219.97	0.4690	156.82	1346.0	1.39105
230	230.02	0.5477	164.00	1205.0	1.43557
240	240.02	0.6355	171.13	1084.0	1.47824
250	250.05	0.7329	178.28	979.0	1.51917
260	260.09	0.8405	185.45	887.8	1.55848
270	270.11	0.9590	192.60	808.0	1.59634
280	280.13	1.0889	199.75	738.0	1.63279
285	285.14	1.1584	203.33	706.1	1.65055
290	290.16	1.2311	206.91	676.1	1.66802
295	295.17	1.3068	210.49	647.9	1.68515
300	300.19	1.3860	214.07	621.2	1.70203
305	305.22	1.4686	217.67	596.0	1.71865
310	310.24	1.5546	221.25	572.3	1.73498
315	315.27	1.6442	224.85	549.8	1.75106
320	320.29	1.7375	228.42	528.6	1.76690
325	325.31	1.8345	232.02	508.4	1.78249
330	330.34	1.9352	235.61	489.4	1.79783
340	340.42	2.149	242.82	454.1	1.82790

Ideal-Gas Properties of Air (In IG TESTcalcs, select Air*)

$$\overline{h}_{f,298}^{o} = 0 \text{ kJ/kmol}; \quad \overline{M} = 28.97 \text{ kg/kmol}$$

T	h	p_r	u	v_r	s^o
K	kJ/kg		kJ/kg		kJ/kg·K
350	350.49	2.379	250.02	422.2	1.85708
360	360.58	2.626	257.24	393.4	1.88543
370	370.67	2.892	264.46	367.2	1.91313
380	380.77	3.176	271.69	343.4	1.94001
390	390.88	3.481	278.93	321.5	1.96633
400	400.98	3.806	286.16	301.6	1.99194
410	411.12	4.153	293.43	283.3	2.01699
420	421.26	4.522	300.69	266.6	2.04142
430	431.43	4.915	307.99	251.1	2.06533
440	441.61	5.332	315.30	236.8	2.08870
450	451.80	5.775	322.62	223.6	2.11161
460	462.02	6.245	329.97	211.4	2.13407
470	472.24	6.742	337.32	200.1	2.15604
480	482.49	7.268	344.70	189.5	2.17760
490	492.74	7.824	352.08	179.7	2.19876
500	503.02	8.411	359.49	170.6	2.21952
510	513.32	9.031	366.92	162.1	2.23993
520	523.63	9.684	374.36	154.1	2.25997
530	533.98	10.37	381.84	146.7	2.27967
540	544.35	11.10	389.34	139.7	2.29906
550	555.74	11.86	396.86	133.1	2.31809
560	565.17	12.66	404.42	127.0	2.33685
570	575.59	13.50	411.97	121.2	2.35531
580	586.04	14.38	419.55	115.7	2.37348
590	596.52	15.31	427.15	110.6	2.39140
600	607.02	16.28	434.78	105.8	2.40902
610	617.53	17.30	442.42	101.2	2.42644
620	628.07	18.36	450.09	96.92	2.44356
630	638.63	19.84	457.78	92.84	2.46048
640	649.22	20.64	465.50	88.99	2.47716
650	659.84	21.86	473.25	85.34	2.49364
660	670.47	23.13	481.01	81.9	2.50985
670	681.14	24.46	488.81	78.61	2.52589
680	691.82	25.85	496.62	75.50	2.54175
690	702.52	27.29	504.45	72.56	2.55731

(continued)

Ideal-Gas Properties of Air (In IG TESTcalcs, select Air*)

$\overline{h}_{f,298}^{\circ} = 0$ kJ/kmol; $\overline{M} = 28.97$ kg/kmol

T	h	p_r	u	v_r	s^o
K	kJ/kg		kJ/kg		kJ/kg·K
700	713.27	28.80	512.33	69.76	2.57277
710	724.04	30.38	520.23	67.07	2.58810
720	734.82	32.02	528.14	64.53	2.60319
730	745.62	33.72	536.07	62.13	2.61803
740	756.44	35.50	544.02	59.82	2.63280
750	767.29	37.35	551.99	57.63	2.64737
760	778.18	39.27	560.01	55.54	2.66176
780	800.03	43.35	576.12	51.64	2.69013
800	821.95	47.75	592.30	48.08	2.71787
820	843.98	52.59	608.59	44.84	2.74504
840	866.08	57.60	624.95	41.85	2.77170
860	888.27	63.09	641.40	39.12	2.79783
880	910.56	68.98	657.95	36.61	2.82344
900	932.93	75.29	674.58	34.31	2.84856
920	955.38	82.05	691.28	32.18	2.87324
940	977.92	89.28	708.08	30.22	2.89748
960	1000.55	97.00	725.02	28.40	2.92128
980	1023.25	105.2	741.98	26.73	2.94468
1000	1046.04	114.0	758.94	25.17	2.96770
1020	1068.89	123.4	776.10	23.72	2.99034
1040	1091.85	133.3	793.36	23.29	3.01260
1060	1114.86	143.9	810.62	21.14	3.03449
1080	1137.89	155.2	827.88	19.98	3.05608
1100	1161.07	167.1	845.33	18.896	3.07732
1120	1184.28	179.7	862.79	17.886	3.09825
1140	1207.57	193.1	880.35	16.946	3.11883
1160	1230.92	207.2	897.91	16.064	3.13916
1180	1254.34	222.2	915.57	15.241	3.15916
1200	1277.79	238.0	933.33	14.470	3.17888
1220	1301.31	254.7	951.09	13.747	3.19834
1240	1324.93	272.3	968.95	13.069	3.21751
1260	1348.55	290.8	986.90	12.435	3.23638
1280	1372.24	310.4	1004.76	11.835	3.25510

Ideal-Gas Properties of Air (In IG TESTcalcs, select Air*)

$\overline{h}_{f,298}^{o} = 0$ kJ/kmol; $\overline{M} = 28.97$ kg/kmol

T	h	p_r	u	v_r	s^o
K	kJ/kg		kJ/kg		kJ/kg·K
1300	1395.97	330.9	1022.82	11.275	3.27345
1320	1419.76	352.5	1040.88	10.747	3.29160
1340	1443.60	375.3	1058.94	10.247	3.30959
1360	1467.49	399.1	1077.10	9.780	3.32724
1380	1491.44	424.2	1095.26	9.337	3.34474
1400	1515.42	450.5	1113.52	8.919	3.36200
1420	1539.44	478.0	1131.77	8.526	3.37901
1440	1563.51	506.9	1150.13	8.153	3.39586
1460	1587.63	537.1	1168.49	7.801	3.41247
1480	1611.79	568.8	1186.95	7.468	3.42892
1500	1635.97	601.9	1205.41	7.152	3.44516
1520	1660.23	636.5	1223.87	6.854	3.46120
1540	1684.51	672.8	1242.43	6.569	3.47712
1560	1708.82	710.5	1260.99	6.301	3.49276
1580	1733.17	750.0	1279.65	6.046	3.50829
1600	1757.57	791.2	1298.30	5.804	3.52364
1620	1782.00	834.1	1316.96	5.574	3.53879
1640	1806.46	878.9	1335.72	5.355	3.55381
1660	1830.96	925.6	1354.48	5.147	3.56867
1680	1855.50	974.2	1373.24	4.949	3.58335
1700	1880.1	1025	1392.7	4.761	3.5979
1750	1941.6	1161	1439.8	4.328	3.6336
1800	2003.3	1310	1487.2	3.994	3.6684
1850	2065.3	1475	1534.9	3.601	3.7023
1900	2127.4	1655	1582.6	3.295	3.7354
1950	2189.7	1852	1630.6	3.022	3.7677
2000	2252.1	2068	1678.7	2.776	3.7994
2050	2314.6	2303	1726.8	2.555	3.8303
2100	2377.7	2559	1775.3	2.356	3.8605
2150	2440.3	2837	1823.8	2.175	3.8901
2200	2503.2	3138	1872.4	2.012	3.9191
2250	2566.4	3464	1921.3	1.864	3.9474

Table D-4   IG Model: Ideal Gas Properties of Nitrogen (N₂)

Ideal-Gas (IG Model) Properties of Nitrogen (N₂)

$\overline{h}^{\circ}_{f,298} = 0$ kJ/kmol; $\overline{M} = 28.013$ kg/kmol

T	$\overline{h}$	$\overline{u}$	$\overline{s}^{\circ}$	T	$\overline{h}$	$\overline{u}$	$\overline{s}^{\circ}$
K	kJ/kmol	kJ/kmol	kJ/kmol·K	K	kJ/kmol	kJ/kmol	kJ/kmol·K
0	−8,669	−8,669	0	600	8,894	3,906	212.066
220	−2,278	−4,107	182.639	610	9,195	4,123	212.564
230	−1,986	−3,898	183.938	620	9,497	4,342	213.055
240	−1,694	−3,689	185.180	630	9,799	4,561	213.541
250	−1,403	−3,482	186.370	640	10,103	4,782	214.018
260	−1,111	−3,273	187.514	650	10,406	5,002	214.489
270	−820	−3,065	188.614	660	10,711	5,224	214.954
280	−528	−2,856	189.673	670	11,016	5,446	215.413
290	−237	−2,648	190.695	680	11,322	5,668	215.866
298	0	−2,478	191.502	690	11,628	5,891	216.314
300	54	−2,440	191.682	700	11,935	6,115	216.756
310	345	−2,232	192.638	710	12,243	6,340	217.192
320	637	−2,023	193.562	720	12,551	6,565	217.624
330	928	−1,816	194.459	730	12,860	6,791	218.059
340	1,219	−1,608	195.328	740	13,170	7,018	218.472
350	1,511	−1,399	196.173	750	13,480	7,245	218.889
360	1,802	−1,191	196.995	760	13,791	7,472	219.301
370	2,094	−982	197.794	770	14,103	7,701	219.709
380	2,386	−773	198.572	780	14,416	7,931	220.113
390	2,678	−564	199.331	790	14,729	8,161	220.512
400	2,971	−355	200.071	800	15,045	8,394	220.907
410	3,263	−146	200.794	810	15,358	8,624	221.298
420	3,556	64	201.499	820	15,673	8,856	221.684
430	3,849	274	202.189	830	15,989	9,088	222.067
440	4,142	484	202.863	840	16,305	9,321	222.447
450	4,436	695	203.523	850	16,623	9,556	222.822
460	4,730	906	204.170	860	16,941	9,791	223.194
470	5,024	1,116	204.803	870	17,259	10,026	223.562
480	5,319	1,328	205.424	880	17,579	10,263	223.927
490	5,616	1,542	206.033	890	17,899	10,500	224.288
500	5,912	1,755	206.630	900	18,221	10,738	224.647
510	6,207	1,967	207.216	910	18,541	10,975	225.002
520	6,503	2,180	207.792	920	18,863	11,214	225.353

Ideal-Gas (IG Model) Properties of Nitrogen (N₂)

$\overline{h}_{f,298}^{\circ} = 0$ kJ/kmol; $\overline{M} = 28.013$ kg/kmol

T	$\overline{h}$	$\overline{u}$	$\overline{s}^{\circ}$	T	$\overline{h}$	$\overline{u}$	$\overline{s}^{\circ}$
K	kJ/kmol	kJ/kmol	kJ/kmol·K	K	kJ/kmol	kJ/kmol	kJ/kmol·K
530	6,800	2,394	208.358	930	19,185	11,453	225.701
540	7,097	2,607	208.914	940	19,509	11,694	226.047
550	7,395	2,822	209.461	950	19,832	11,934	226.389
560	7,694	3,038	209.999	960	20,157	12,176	226.728
570	7,993	3,254	210.528	970	20,482	12,417	227.064
580	8,293	3,471	211.049	980	20,807	12,659	227.398
590	8,593	3,688	211.562	990	21,134	12,903	227.728
1000	21,460	13,146	228.057	1760	47,558	32,925	247.396
1020	22,115	13,635	228.706	1780	48,269	33,470	247.798
1040	22,773	14,126	229.344	1800	48,982	34,017	248.195
1060	23,432	14,619	229.973	1820	49,694	34,563	248.589
1080	24,093	15,114	230.591	1840	50,406	35,108	248.979
1100	24,757	15,612	231.199	1860	51,121	35,657	249.365
1120	25,423	16,111	231.799	1880	51,835	36,205	249.748
1140	26,091	16,613	232.391	1900	52,551	36,754	250.128
1160	26,761	17,117	232.973	1920	53,267	37,304	250.502
1180	27,435	17,624	233.549	1940	53,985	37,856	250.874
1200	28,108	18,131	234.115	1960	54,712	38,417	251.242
1220	28,783	18,640	234.673	1980	55,421	38,959	251.607
1240	29,460	19,151	235.223	2000	56,141	39,513	251.969
1260	30,138	19,662	235.766	2050	57,943	40,899	252.858
1280	30,819	20,177	236.302	2100	59,748	42,289	253.726
1300	31,501	20,693	236.831	2150	61,557	43,682	254.578
1320	32,184	21,210	237.353	2200	63,371	45,080	255.412
1340	32,870	21,729	237.867	2250	65,187	46,481	256.227
1360	33,558	22,251	238.376	2300	67,007	47,885	257.027
1380	34,246	22,773	238.878	2350	68,827	49,289	257.810
1400	34,936	23,296	239.375	2400	70,651	50,697	258.580
1420	35,626	23,820	239.865	2450	72,480	52,111	259.332
1440	36,319	24,347	240.350	2500	74,312	53,527	260.073
1460	37,013	24,875	240.827	2550	76,145	54,944	260.799
1480	37,708	25,403	241.301	2600	77,981	56,365	261.512
1500	38,404	25,933	241.768	2650	79,819	57,787	262.213
1520	39,102	26,465	242.228	2700	81,659	59,211	262.902

(continued)

Ideal-Gas (IG Model) Properties of Nitrogen (N$_2$)

$\overline{h}^{\circ}_{f,298} = 0$ kJ/kmol; $\overline{M} = 28.013$ kg/kmol

T	$\overline{h}$	$\overline{u}$	$\overline{s}^{\circ}$	T	$\overline{h}$	$\overline{u}$	$\overline{s}^{\circ}$
K	kJ/kmol	kJ/kmol	kJ/kmol·K	K	kJ/kmol	kJ/kmol	kJ/kmol·K
1540	39,801	26,997	242.685	2750	83,502	60,639	263.577
1560	40,499	27,529	243.137	2800	85,345	62,066	264.241
1580	41,200	28,064	243.585	2850	87,190	63,495	264.895
1600	41,902	28,600	244.028	2900	89,036	64,925	265.538
1620	42,606	29,137	244.464	2950	90,887	66,361	266.170
1640	43,311	29,676	244.896	3000	92,738	67,796	266.793
1660	44,017	30,216	245.324	3050	94,591	69,233	267.404
1680	44,724	30,756	245.747	3100	96,446	70,673	268.007
1700	45,430	31,296	246.166	3150	98,303	72,114	268.601
1720	46,138	31,838	246.580	3200	100,161	73,556	269.186
1740	46,847	32,381	246.990	3250	102,021	75,001	269.763

Table D-5   IG Model: Ideal Gas Properties of Oxygen (O$_2$)

Ideal-Gas (IG Model) Properties of Oxygen (O$_2$)

$\overline{h}^{\circ}_{f,298} = 0$ kJ/kmol; $\overline{M} = 32$ kg/kmol

T	$\overline{h}$	$\overline{u}$	$\overline{s}^{\circ}$	T	$\overline{h}$	$\overline{u}$	$\overline{s}^{\circ}$
K	kJ/kmol	kJ/kmol	kJ/kmol·K	K	kJ/kmol	kJ/kmol	kJ/kmol·K
0	0	0	0	600	17,929	12,940	226.346
220	6,404	4,575	196.171	610	18,250	13,178	226.877
230	6,694	4,782	197.461	620	18,572	13,417	227.400
240	6,984	4,989	198.696	630	18,895	13,657	227.918
250	7,275	5,197	199.885	640	19,219	13,898	228.429
260	7,566	5,405	201.027	650	19,544	14,140	228.932
270	7,858	5,613	202.128	660	19,870	14,383	229.430
280	8,150	5,822	203.191	670	20,197	14,626	229.920
290	8,443	6,032	204.218	680	20,524	14,871	230.405
298	8,682	6,203	205.033	690	20,854	15,116	230.885
300	8,736	6,242	205.213	700	21,184	15,364	231.358
310	9,030	6,453	206.177	710	21,514	15,611	231.827
320	9,325	6,664	207.112	720	21,845	15,859	232.291
330	9,620	6,877	208.020	730	22,177	16,107	232.748
340	9,916	7,090	208.904	740	22,510	16,357	233.201
350	10,213	7,303	209.765	750	22,844	16,607	233.649
360	10,511	7,518	210.604	760	23,178	16,859	234.091

Ideal-Gas (IG Model) Properties of Oxygen (O₂)

$\overline{h}^\circ_{f,298} = 0 \text{ kJ/kmol}; \quad \overline{M} = 32 \text{ kg/kmol}$

T	$\overline{h}$	$\overline{u}$	$\overline{s}°$	T	$\overline{h}$	$\overline{u}$	$\overline{s}°$
K	kJ/kmol	kJ/kmol	kJ/kmol·K	K	kJ/kmol	kJ/kmol	kJ/kmol·K
370	10,809	7,733	211.423	770	23,513	17,111	234.528
380	11,109	7,949	212.222	780	23,850	17,364	234.960
390	11,409	8,166	213.002	790	24,186	17,618	235.387
400	11,711	8,384	213.765	800	24,523	17,872	235.810
410	12,012	8,603	214.510	810	24,861	18,126	236.230
420	12,314	8,822	215.241	820	25,199	18,382	236.644
430	12,618	9,043	215.955	830	25,537	18,637	237.055
440	12,923	9,264	216.656	840	25,877	18,893	237.462
450	13,228	9,487	217.342	850	26,218	19,150	237.864
460	13,525	9,710	218.016	860	26,559	19,408	238.264
470	13,842	9,935	218.676	870	26,899	19,666	238.660
480	14,151	10,160	219.326	880	27,242	19,925	239.051
490	14,460	10,386	219.963	890	27,584	20,185	239.439
500	14,770	10,614	220.589	900	27,928	20,445	239.823
510	15,082	10,842	221.206	910	28,272	20,706	240.203
520	15,395	11,071	221.812	920	28,616	20,967	240.580
530	15,708	11,301	222.409	930	28,960	21,228	240.953
540	16,022	11,533	222.997	940	29,306	21,491	241.323
550	16,338	11,765	223.576	950	29,652	21,754	241.689
560	16,654	11,998	224.146	960	29,999	22,017	242.052
570	16,971	12,232	224.708	970	30,345	22,280	242.411
580	17,290	12,467	225.262	980	30,692	22,544	242.768
590	17,609	12,703	225.808	990	31,041	22,809	242.120
1040	32,789	24,142	244.844	1800	60,371	45,405	264.701
1060	33,490	24,677	245.513	1820	61,118	45,986	265.113
1080	34,194	25,214	246.171	1840	61,866	46,568	265.521
1100	34,899	25,753	246.818	1860	62,616	47,151	265.925
1120	35,606	26,294	247.454	1880	63,365	47,734	266.326
1140	36,314	26,836	248.081	1900	64,116	48,319	266.722
1160	37,023	27,379	248.698	1920	64,868	48,904	267.115
1180	37,734	27,923	249.307	1940	65,620	49,490	267.505
1200	38,447	28,469	249.906	1960	66,374	50,078	267.891
1220	39,162	29,018	250.497	1980	67,127	50,665	268.275

(continued)

Ideal-Gas (IG Model) Properties of Oxygen (O$_2$)

$\overline{h}^{\,\circ}_{f,298} = 0$ kJ/kmol; $\overline{M} = 32$ kg/kmol

T	$\overline{h}$	$\overline{u}$	$\overline{s}^{\circ}$	T	$\overline{h}$	$\overline{u}$	$\overline{s}^{\circ}$
K	kJ/kmol	kJ/kmol	kJ/kmol·K	K	kJ/kmol	kJ/kmol	kJ/kmol·K
1240	39,877	29,568	251.079	2000	67,881	51,253	268.655
1260	40,594	30,118	251.653	2050	69,772	52,727	269.588
1280	41,312	30,670	252.219	2100	71,668	54,208	270.504
1300	42,033	31,224	252.776	2150	73,573	55,697	271.399
1320	42,753	31,778	253.325	2200	75,484	57,192	272.278
1340	43,475	32,334	253.868	2250	77,397	58,690	273.136
1360	44,198	32,891	254.404	2300	79,316	60,193	273.891
1380	44,923	33,449	254.932	2350	81,243	61,704	274.809
1400	45,648	34,008	255.454	2400	83,174	63,219	275.625
1420	46,374	34,567	255.968	2450	85,112	64,742	276.424
1440	47,102	35,129	256.475	2500	87,057	66,271	277.207
1460	47,831	35,692	256.978	2550	89,004	67,802	277.979
1480	48,561	36,256	257.474	2600	90,956	69,339	278.738
1500	49,292	36,821	257.965	2650	92,916	70,883	279.485
1520	50,024	37,387	258.450	2700	94,881	72,433	280.219
1540	50,756	37,952	258.928	2750	96,852	73,987	280.942
1560	51,490	38,520	259.402	2800	98,826	75,546	281.654
1580	52,224	39,088	259.870	2850	100,808	77,112	282.357
1600	52,961	39,658	260.333	2900	102,793	78,682	283.048
1620	53,696	40,227	260.791	2950	104,785	80,258	283.728
1640	54,434	40,799	261.242	3000	106,780	81,837	284.399
1660	55,172	41,370	261.690	3050	108,778	83,419	285.060
1680	55,912	41,944	262.132	3100	110,784	85,009	285.713
1700	56,652	42,517	262.571	3150	112,795	86,601	286.355
1720	57,394	43,093	263.005	3200	114,809	88,203	286.989
1740	58,136	43,669	263.435	3250	116,827	89,804	287.614

Table D-6   IG Model: Ideal Gas Properties of Carbon Dioxide (CO$_2$)

Ideal-Gas (IG Model) Properties of Carbon Dioxide (CO$_2$)

$\overline{h}^{\,\circ}_{f,298} = -393,522$ kJ/kmol; $\overline{M} = 44.01$ kg/kmol

T	$\overline{h}$	$\overline{u}$	$\overline{s}^{\circ}$	T	$\overline{h}$	$\overline{u}$	$\overline{s}^{\circ}$
K	kJ/kmol	kJ/kmol	kJ/kmol·K	K	kJ/kmol	kJ/kmol	kJ/kmol·K
0	0	0	0	600	22,280	17,291	243.299
220	6,601	4,772	202.966	610	22,754	17,683	243.983
230	6,938	5,026	204.464	620	23,231	18,076	244.758

Ideal-Gas (IG Model) Properties of Carbon Dioxide (CO$_2$)

$\overline{h}^o_{f,298} = -393{,}522$ kJ/kmol; $\overline{M} = 44.01$ kg/kmol

T	$\overline{h}$	$\overline{u}$	$\overline{s}^o$	T	$\overline{h}$	$\overline{u}$	$\overline{s}^o$
K	kJ/kmol	kJ/kmol	kJ/kmol·K	K	kJ/kmol	kJ/kmol	kJ/kmol·K
240	7,280	5,285	205.920	630	23,709	18,471	245.524
250	7,627	5,548	207.337	640	24,190	18,869	246.282
260	7,979	5,817	208.717	650	24,674	19,270	247.032
270	8,335	6,091	210.062	660	25,160	19,672	247.773
280	8,697	6,369	211.376	670	25,648	20,078	248.507
290	9,063	6,651	212.660	680	26,138	20,484	249.233
298	9,364	6,885	213.685	690	26,631	20,894	249.952
300	9,431	6,939	213.915	700	27,125	21,305	250.663
310	9,807	7,230	215.146	710	27,622	21,719	251.368
320	10,186	7,526	216.351	720	28,121	22,134	252.065
330	10,570	7,826	217.534	730	28,622	22,522	252.755
340	10,959	8,131	218.694	740	29,124	22,972	253.439
350	11,351	8,439	219.831	750	29,629	23,393	254.117
360	11,748	8,752	220.948	760	29,135	23,817	254.787
370	12,148	9,068	222.044	770	30,644	24,242	255.452
380	12,552	9,392	223.122	780	31,154	24,669	256.110
390	12,960	9,718	224.182	790	31,165	25,097	256.762
400	13,372	10,046	225.225	800	32,179	25,527	257.408
410	13,787	10,378	226.250	810	32,694	25,959	258.048
420	14,206	10,714	227.258	820	33,212	26,394	258.682
430	14,328	11,053	228.252	830	33,730	26,829	259.311
440	15,054	11,393	229.230	840	34,251	27,267	259.934
450	15,483	11,742	230.194	850	34,773	27,706	260.551
460	15,916	12,091	231.144	860	35,296	28,125	261.164
470	16,351	12,444	232.080	870	35,821	28,588	261.770
480	16,791	12,800	233.004	880	36,347	29,031	262.371
490	17,232	13,158	233.916	890	36,876	29,476	262.968
500	17,678	13,521	234.814	900	37,405	29,922	263.559
510	18,126	13,885	235.700	910	37,935	30,369	264.416
520	18,576	14,253	236.575	920	38,467	30,818	264.728
530	19,029	14,622	237.439	930	39,000	31,268	265.304
540	19,485	14,996	238.292	940	39,535	31,719	265.877
550	19,945	15,372	239.135	950	40,070	32,171	266.444
560	20,407	15,751	239.962	960	40,607	32,625	267.007

(continued)

Ideal-Gas (IG Model) Properties of Carbon Dioxide (CO₂)

$\overline{h}_{f,298}^{\circ} = -393{,}522$ kJ/kmol; $\overline{M} = 44.01$ kg/kmol

T	$\overline{h}$	$\overline{u}$	$\overline{s}^{\circ}$	T	$\overline{h}$	$\overline{u}$	$\overline{s}^{\circ}$
K	kJ/kmol	kJ/kmol	kJ/kmol·K	K	kJ/kmol	kJ/kmol	kJ/kmol·K
570	20,870	16,131	240.789	970	41,145	33,081	267.566
580	21,337	16,515	241.602	980	41,685	33,537	268.119
590	21,807	16,902	232.405	990	42,226	33,995	268.670
1000	42,796	34,455	269.215	1760	86,420	71,787	301.543
1020	43,859	35,378	270.293	1780	87,612	72,812	302.217
1040	44,953	36,306	271.354	1800	88,806	73,840	302.884
1060	46,051	37,238	272.400	1820	90,000	74,868	303.544
1080	47,153	38,174	273.430	1840	91,196	75,897	304.198
1100	48,258	39,112	274.445	1860	92,394	76,929	304.845
1120	49,369	40,057	275.444	1880	93,593	77,962	305.487
1140	50,484	41,006	276.430	1900	94,793	78,996	306.122
1160	51,602	41,957	277.403	1920	95,995	80,031	306.751
1180	52,424	42,913	278.361	1940	97,197	81,067	307.374
1200	53,848	43,871	297.307	1960	98,401	82,105	307.992
1220	54,977	44,834	280.238	1980	99,606	83,144	308.604
1240	56,108	45,799	281.158	2000	100,804	84,185	309.210
1260	57,244	46,768	282.066	2050	103,835	86,791	310.701
1280	58,381	47,739	282.962	2100	106,864	89,404	312.160
1300	59,522	48,713	283.847	2150	109,898	92,023	313.589
1320	60,666	49,691	284.722	2200	112,939	94,648	314.988
1340	61,813	50,672	285.586	2250	115,984	97,577	316.356
1360	62,963	51,656	286.000	2300	119,035	99,912	317.695
1380	64,116	52,643	439.000	2350	122,091	102,552	319.011
1400	65,271	53,631	287.283	2400	125,152	105,197	320.302
1420	66,427	54,621	288.106	2450	128,219	107,849	321.566
1440	67,586	55,614	288.934	2500	131,290	110,504	322.808
1460	68,748	56,609	289.743	2550	134,368	113,166	324.026
1480	66,911	57,606	290.542	2600	137,449	115,831	325.222
1500	71,078	58,606	291.333	2650	140,533	118,500	326.396
1520	72,246	59,609	292.114	2700	143,620	121,172	327.549
1540	73,417	60,613	292.888	2750	146,713	123,849	328.684
1560	74,590	61,620	293.654	2800	149,808	134,528	329.800
1580	76,767	62,630	295.161	2850	152,908	129,212	330.896
1600	76,944	63,741	295.901	2900	156,009	131,898	331.975
1620	78,123	64,653	296.632	2950	159,117	134,589	333.037

Ideal-Gas (IG Model) Properties of Carbon Dioxide (CO$_2$)

$\overline{h}_{f,298}^{\circ} = -393,522$ kJ/kmol; $\overline{M} = 44.01$ kg/kmol

T	$\overline{h}$	$\overline{u}$	$\overline{s}^{\circ}$	T	$\overline{h}$	$\overline{u}$	$\overline{s}^{\circ}$
K	kJ/kmol	kJ/kmol	kJ/kmol·K	K	kJ/kmol	kJ/kmol	kJ/kmol·K
1640	79,303	65,668	297.356	3000	162,226	137,283	334.084
1660	80,486	66,592	298.072	3050	165,341	139,982	335.114
1680	81,670	67,702	298.781	3100	168,456	142,682	336.126
1700	82,856	68,721	299.482	3150	171,576	145,385	337.124
1720	84,043	69,742	300.177	3200	174,695	148,089	338.106
1740	85,231	70,764	300.863	3250	177,822	150,801	339.069

Table D-7   IG Model: Ideal Gas Properties of Carbon Monoxide (CO)

Ideal-Gas (IG Model) Properties of Carbon Monoxide (CO)

$\overline{h}_{f,298}^{\circ} = -110,527$ kJ/kmol; $\overline{M} = 28.01$ kg/kmol

T	$\overline{h}$	$\overline{u}$	$\overline{s}^{\circ}$	T	$\overline{h}$	$\overline{u}$	$\overline{s}^{\circ}$
K	kJ/kmol	kJ/kmol	kJ/kmol·K	K	kJ/kmol	kJ/kmol	kJ/kmol·K
0	0	0	0	600	17,611	12,622	218.204
220	6,391	4,562	188.683	610	17,915	12,843	218.708
230	6,683	4,771	189.980	620	18,221	13,066	219.205
240	6,975	4,979	191.221	630	18,527	13,289	219.695
250	7,266	5,188	192.411	640	18,833	13,512	220.179
260	7,558	5,396	193.554	650	19,141	13,736	220.656
270	7,849	5,604	194.654	660	19,449	13,962	221.127
280	8,140	5,812	195.713	670	19,758	14,187	221.592
290	8,432	6,020	196.735	680	20,068	14,414	222.052
298	8,669	6,190	197.543	690	20,378	14,641	222.505
300	8,723	6,229	197.723	700	20,690	14,870	222.953
310	9,014	6,437	198.678	710	21,002	15,099	223.396
320	9,306	6,645	199.603	720	21,315	15,328	223.833
330	9,597	6,854	200.500	730	21,628	15,558	224.265
340	9,889	7,062	201.371	740	21,943	15,789	224.692
350	10,181	7,271	202.217	750	22,258	16,022	225.115
360	10,473	7,480	203.040	760	22,573	16,255	225.533
370	10,765	7,689	203.842	770	22,890	16,488	225.947
380	11,058	7,899	204.622	780	23,208	16,723	226.357
390	11,351	8,108	205.383	790	23,526	16,957	226.762
400	11,644	8,319	206.125	800	23,844	17,193	227.162
410	11,938	8,529	206.850	810	24,164	17,429	227.559

(continued)

Ideal-Gas (IG Model) Properties of Carbon Monoxide (CO)

$\overline{h}^{\circ}_{f,298} = -110,527$ kJ/kmol; $\overline{M} = 28.01$ kg/kmol

T	$\overline{h}$	$\overline{u}$	$\overline{s}^{\circ}$	T	$\overline{h}$	$\overline{u}$	$\overline{s}^{\circ}$
K	kJ/kmol	kJ/kmol	kJ/kmol·K	K	kJ/kmol	kJ/kmol	kJ/kmol·K
420	12,232	8,740	207.549	820	24,483	17,665	227.952
430	12,526	8,951	208.252	830	24,803	17,902	228.339
440	12,821	9,163	208.929	840	25,124	18,140	228.724
450	13,116	9,375	209.593	850	25,446	18,379	229.106
460	13,412	9,587	210.243	860	25,768	18,617	229.482
470	13,708	9,800	210.880	870	26,091	18,858	229.856
480	14,005	10,014	211.504	880	26,415	19,099	230.227
490	14,302	10,228	212.117	890	26,740	19,341	230.593
500	14,600	10,443	212.719	900	27,066	19,583	230.957
510	14,898	10,658	213.310	910	27,392	19,826	231.317
520	15,197	10,874	213.890	920	27,719	20,070	231.674
530	15,497	11,090	214.460	930	28,046	20,314	232.028
540	15,797	11,307	215.020	940	28,375	20,559	232.379
550	16,097	11,524	215.572	950	28,703	20,805	232.727
560	16,399	11,743	216.115	960	29,033	21,051	233.072
570	16,701	11,961	216.649	970	29,362	21,298	233.413
580	17,003	12,181	217.175	980	29,693	21,545	233.752
590	17,307	12,401	217.693	990	30,024	21,793	234.088
1,000	30,355	22,041	234.421	1,760	56,756	42,123	253.991
1,020	31,020	22,540	235.079	1,780	57,473	42,673	254.398
1,040	31,688	23,041	235.728	1,800	58,191	43,225	254.797
1,060	32,357	23,544	236.364	1,820	58,910	43,778	255.194
1,080	33,029	24,049	236.992	1,840	59,629	44,331	255.587
1,100	33,702	24,557	237.609	1,860	60,351	44,886	255.976
1,120	34,377	25,065	238.217	1,880	61,072	45,441	256.361
1,140	35,054	25,575	238.817	1,900	61,794	45,997	256.743
1,160	35,733	26,088	239.407	1,920	62,516	46,552	257.122
1,180	36,406	26,602	239.989	1,940	63,238	47,108	257.497
1,200	37,095	27,118	240.663	1,960	63,961	47,665	257.868
1,220	37,780	27,637	241.128	1,980	64,684	48,221	258.236
1,240	38,466	28,426	241.686	2,000	65,408	48,780	258.600
1,260	39,154	28,678	242.236	2,050	67,224	50,179	259.494
1,280	39,844	29,021	242.780	2,100	69,044	51,584	260.370
1,300	40,534	29,725	243.316	2,150	70,864	52,988	261.226
1,320	41,226	30,251	243.844	2,200	72,688	54,396	262.065

Ideal-Gas (IG Model) Properties of Carbon Monoxide (CO)

$\bar{h}_{f,298}^{o} = -110{,}527$ kJ/kmol; $\overline{M} = 28.01$ kg/kmol

T	$\bar{h}$	$\bar{u}$	$\bar{s}^{o}$	T	$\bar{h}$	$\bar{u}$	$\bar{s}^{o}$
K	kJ/kmol	kJ/kmol	kJ/kmol·K	K	kJ/kmol	kJ/kmol	kJ/kmol·K
1,340	41,919	30,778	244.366	2,250	74,516	55,809	262.887
1,360	42,613	31,306	244.880	2,300	76,345	57,222	263.692
1,380	43,309	31,836	245.388	2,350	78,178	58,649	264.480
1,400	44,027	32,367	245.889	2,400	80,015	60,060	265.253
1,420	44,707	32,900	246.385	2,450	81,852	61,482	266.012
1,440	45,408	33,434	246.876	2,500	83,692	62,906	266.755
1,460	46,110	33,971	247.360	2,550	85,537	64,335	267.485
1,480	46,813	34,508	247.839	2,600	87,383	65,766	268.202
1,500	47,517	35,046	248.312	2,650	89,230	67,197	268.905
1,520	48,222	35,046	248.312	2,700	91,077	68,628	269.596
1,540	48,928	35,584	248.240	2,750	92,930	70,066	270.285
1,560	49,635	36,665	249.695	2,800	94,784	71,504	270.943
1,580	50,344	37,207	250.147	2,850	96,639	72,945	271.602
1,600	51,053	37,750	250.592	2,900	98,495	74,383	272.249
1,620	51,763	38,293	251.033	2,950	100,352	75,825	272.884
1,640	52,472	38,837	251.470	3,000	102,210	77,267	273.508
1,660	53,184	39,382	251.901	3,050	104,073	78,715	274.123
1,680	53,895	39,927	252.329	3,100	105,939	80,164	274.730
1,700	54,609	40,474	252.751	3,150	107,802	81,612	275.326
1,720	55,323	41,023	253.169	3,200	109,667	83,061	275.914
1,740	56,039	41,572	253.582	3,250	111,534	84,513	276.494

Table D-8 IG Model: Ideal Gas Properties of Hydrogen (H$_2$)

Ideal-Gas (IG Model) Properties of Hydrogen (H$_2$)

$\bar{h}_{f,298}^{o} = 0$ kJ/kmol; $\overline{M} = 2.016$ kg/kmol

T	$\bar{h}$	$\bar{u}$	$\bar{s}^{o}$	T	$\bar{h}$	$\bar{u}$	$\bar{s}^{o}$
K	kJ/kmol	kJ/kmol	kJ/kmol·K	K	kJ/kmol	kJ/kmol	kJ/kmol·K
0	0	0	0	1440	42,808	30,835	177.410
260	7,370	5,209	126.636	1480	44,091	31,786	178.291
270	7,657	5,412	127.719	1520	45,384	32,746	179.513
280	7,945	5,617	128.765	1560	46,683	33,713	179.995
290	8,233	5,822	129.775	1600	47,990	34,687	180.820
298	8,468	5,989	130.574	1640	49,303	35,668	181.632
300	8,522	6,027	130.754	1680	50,622	36,654	182.428

(continued)

Ideal-Gas (IG Model) Properties of Hydrogen (H$_2$)

$\bar{h}^o_{f,298} = 0$ kJ/kmol; $\overline{M} = 2.016$ kg/kmol

T	$\bar{h}$	$\bar{u}$	$\bar{s}^o$	T	$\bar{h}$	$\bar{u}$	$\bar{s}^o$
K	kJ/kmol	kJ/kmol	kJ/kmol·K	K	kJ/kmol	kJ/kmol	kJ/kmol·K
320	9,100	6,440	132.621	1720	51,947	37,646	183.208
340	9,680	6,853	134.378	1760	53,279	38,645	183.973
360	10,262	7,268	136.039	1800	54,618	39,652	184.724
380	10,843	7,684	137.612	1840	55,962	40,663	185.463
400	11,426	8,100	139.106	1880	57,311	41,680	186.190
420	12,010	8,518	140.529	1920	58,668	42,705	186.904
440	12,594	8,936	141.888	1960	60,031	43,735	187.607
460	13,179	9,355	143.187	2000	61,400	44,771	188.297
480	13,764	9,773	144.432	2050	63,119	46,074	189.148
500	14,350	10,193	145.628	2100	64,847	47,386	189.979
520	14,935	10,611	146.775	2150	66,584	48,708	190.796
560	16,107	11,451	148.945	2200	68,328	50,037	191.598
600	17,280	12,291	150.968	2250	70,080	51,373	193.385
640	18,453	13,133	152.863	2300	71,839	52,716	193.159
680	19,630	13,976	154.645	2350	73,608	54,069	193.921
720	20,087	14,821	156.328	2400	75,383	55,429	194.669
760	21,988	15,669	157.923	2450	77,168	56,798	195.403
800	23,171	16,520	159.440	2500	78,960	58,175	196.125
840	24,359	17,375	160.891	2550	80,775	59,554	196.837
880	25,551	18,235	162.277	2600	82,558	60,941	197.539
920	26,747	19,098	163.607	2650	84,368	62,335	198.229
960	27,948	19,966	164.884	2700	86,186	63,737	198.907
1000	29,154	20,839	166.114	2750	88,008	65,144	199.575
1040	30,364	21,717	167.300	2800	89,838	66,558	200.234
1080	31,580	22,601	168.449	2850	91,671	67,976	200.885
1120	32,802	23,490	169.560	2900	93,512	69,401	201.527
1160	34,028	24,384	170.636	2950	95,358	70,831	202.157
1200	35,262	25,284	171.682	3000	97,211	72,268	202.778
1240	36,502	26,192	172.698	3050	99,065	73,707	203.391
1280	37,749	27,106	173.687	3100	100,926	75,152	203.995
1320	39,002	28,027	174.652	3150	102,793	76,604	204.592
1360	40,263	28,955	175.593	3200	104,667	78,061	205.181
1400	41,530	29,889	176.510	3250	106,545	79,523	205.765

Table D-9   IG Model: Ideal Gas Properties of Water Vapor (H$_2$O)

Ideal-Gas (IG Model) Properties of Water Vapor (H$_2$O)

$\overline{h}^{\circ}_{f,298} = -241{,}826$ kJ/kmol; $\overline{M} = 18.015$ kg/kmol

T	$\overline{h}$	$\overline{u}$	$\overline{s}^{\circ}$	T	$\overline{h}$	$\overline{u}$	$\overline{s}^{\circ}$
K	kJ/kmol	kJ/kmol	kJ/kmol·K	K	kJ/kmol	kJ/kmol	kJ/kmol·K
0	0	0	0	600	20,402	15,413	212.920
220	7,295	5,466	178.576	610	20,765	15,693	213.529
230	7,628	5,715	180.054	620	21,130	15,975	214.122
240	7,961	5,965	181.471	630	21,495	16,257	214.707
250	8,294	6,215	182.831	640	21,862	16,541	215.285
260	8,627	6,466	184.139	650	22,230	16,826	215.856
270	8,961	6,716	185.399	660	22,600	17,112	216.419
280	9,296	6,968	186.616	670	22,970	17,399	216.976
290	9,631	7,219	187.791	680	23,342	17,688	217.527
298	9,904	7,425	188.720	690	23,714	17,978	218.071
300	9,966	7,472	188.928	700	24,088	18,268	218.610
310	10,302	7,725	190.030	710	24,464	18,561	219.142
320	10,639	7,978	191.098	720	24,840	18,854	219.668
330	10,976	8,232	192.136	730	25,218	19,148	220.189
340	11,314	8,487	193.144	740	25,597	19,444	220.707
350	11,652	8,742	194.125	750	25,977	19,741	221.215
360	11,992	8,998	195.081	760	26,358	20,039	221.720
370	12,331	9,255	196.012	770	26,741	20,339	222.221
380	12,672	9,513	196.920	780	27,125	20,639	222.717
390	13,014	9,771	197.807	790	27,510	20,941	223.207
400	13,356	10,030	198.673	800	27,896	21,245	223.693
410	13,699	10,290	199.521	810	28,284	21,549	224.174
420	14,043	10,551	200.350	820	28,672	21,855	224.651
430	14,388	10,813	201.160	830	29,062	22,162	225.123
440	14,734	11,075	201.955	840	29,454	22,470	225.592
450	15,080	11,339	202.734	850	29,846	22,779	226.057
460	15,428	11,603	203.497	860	30,240	23,090	226.517
470	15,777	11,869	204.247	870	30,635	23,402	226.973
480	16,126	12,135	204.982	880	31,032	23,715	227.426
490	16,477	12,403	205.705	890	31,429	24,029	227.875
500	16,828	12,671	206.413	900	31,828	24,345	228.321
510	17,181	12,940	207.112	910	32,228	24,662	228.763

(continued)

Ideal-Gas (IG Model) Properties of Water Vapor (H₂O)

$\overline{h}^o_{f,298} = -241{,}826$ kJ/kmol; $\overline{M} = 18.015$ kg/kmol

T	$\overline{h}$	$\overline{u}$	$\overline{s}^o$	T	$\overline{h}$	$\overline{u}$	$\overline{s}^o$
K	kJ/kmol	kJ/kmol	kJ/kmol·K	K	kJ/kmol	kJ/kmol	kJ/kmol·K
520	17,534	13,211	207.799	920	32,629	24,980	229.202
530	17,889	13,482	208.475	930	33,032	25,300	229.637
540	18,245	13,755	209.139	940	33,436	25,621	230.070
550	18,601	14,028	209.795	950	33,841	25,943	230.499
560	18,959	14,303	210.440	960	34,247	26,265	230.924
570	19,318	14,579	211.075	970	34,653	26,588	231.347
580	19,678	14,856	211.702	980	35,061	26,913	231.767
590	20,039	15,134	212.320	990	35,472	27,240	232.184
1000	35,882	27,568	232.597	1760	70,535	55,902	258.151
1020	36,709	28,228	233.415	1780	71,523	56,723	258.708
1040	37,542	28,895	234.223	1800	72,513	57,547	259.262
1060	38,380	29,567	235.020	1820	73,507	58,375	259.811
1080	39,223	30,243	235.806	1840	74,506	59,207	260.357
1100	40,071	30,925	236.584	1860	75,506	60,042	260.898
1120	40,923	31,611	237.352	1880	76,511	60,880	261.436
1140	41,780	32,301	238.110	1900	77,517	61,720	261.969
1160	42,642	32,997	238.859	1920	78,527	62,564	262.497
1180	43,509	33,698	239.600	1940	79,540	63,411	263.022
1200	44,380	34,403	240.333	1960	80,555	64,259	26.542
1220	45,256	35,112	241.057	1980	81,573	65,111	264.059
1240	46,137	35,827	241.773	2000	82,593	65,965	264.571
1260	47,022	36,546	242.482	2050	85,156	68,111	265.838
1280	47,912	37,270	243.183	2100	87,753	70,275	267.081
1300	48,807	38,000	243.877	2150	90,330	72,454	268.301
1320	49,707	38,732	244.564	2200	92,940	74,649	269.500
1340	50,612	39,470	245.243	2250	95,562	76,855	270.679
1360	51,521	40,213	245.915	2300	98,199	79,076	271.839
1380	52,434	40,960	246.582	2350	100,846	81,308	272.978
1400	53,351	41,711	247.241	2400	103,508	83,553	274.098
1420	54,273	42,466	247.895	2450	106,183	85,811	275.201
1440	55,198	43,226	248.543	2500	108,868	88,082	276.286
1460	56,128	43,989	249.185	2550	111,565	90,364	277.354
1480	57,062	44,756	249.820	2600	114,283	92,656	278.407
1500	57,999	45,528	250.450	2650	116,991	94,958	279.441
1520	58,942	46,304	251.074	2700	119,717	97,269	280.462

Ideal-Gas (IG Model) Properties of Water Vapor (H₂O)

$\overline{h}_{f,298}^{o} = -241,826$ kJ/kmol; $\overline{M} = 18.015$ kg/kmol

T	$\overline{h}$	$\overline{u}$	$\overline{s}^{\circ}$	T	$\overline{h}$	$\overline{u}$	$\overline{s}^{\circ}$
K	kJ/kmol	kJ/kmol	kJ/kmol·K	K	kJ/kmol	kJ/kmol	kJ/kmol·K
1540	59,888	47,084	251.693	2750	122,453	99,588	281.464
1560	60,838	47,868	252.305	2800	125,198	101,917	282.453
1580	61,792	48,655	252.912	2850	127,952	104,256	283.429
1600	62,748	49,445	253.513	2900	130,717	106,605	284.390
1620	63,709	50,240	254.111	2950	133,486	108,959	285.338
1640	64,675	51,039	254.703	3000	136,264	111,321	286.273
1660	65,643	51,841	255.290	3050	139,051	113,692	287.194
1680	66,614	52,646	255.873	3100	141,846	116,072	288.102
1700	67,589	53,455	256.450	3150	144,648	118,458	288.999
1720	68,567	54,267	257.022	3200	147,457	120,851	289.884
1740	69,550	55,083	257.589	3250	150,272	123,250	290.756

Table D-10   IG Model: Ideal Gas Properties of Nitrogen Dioxide (NO₂)

Ideal-Gas (IG Model) Properties of Nitrogen Dioxide (NO₂)

$\overline{h}_{f,298}^{o} = 33,100$ kJ/kmol; $\overline{M} = 46.005$ kg/kmol

T	$\overline{h}$	$\overline{u}$	$\overline{s}^{\circ}$	T	$\overline{h}$	$\overline{u}$	$\overline{s}^{\circ}$
K	kJ/kmol	kJ/kmol	kJ/kmol·K	K	kJ/kmol	kJ/kmol	kJ/kmol·K
0	−10,186	−10,186	0	1800	76,008	61,043	325.861
100	−6,861	−7,692	202.563	1900	91,624	75,827	328.898
200	−3,495	−5,158	225.852	2000	87,259	70,631	331.788
298	0	−2,478	240.034	2200	98,578	80,287	337.182
300	68	−2,426	240.263	2400	109,948	89,994	342.128
400	3,927	601	251.342	2600	121,358	99,742	346.695
500	8,099	3,942	260.638	2800	132,800	109,521	350.934
600	12,555	7,567	268.755	3000	144,267	119,325	354.890
700	17,250	11,430	275.988	3200	155,756	129,151	358.597
800	22,138	15,487	282.513	3400	167,262	138,994	362.085
900	27,180	19,697	288.450	3600	178,783	148,853	365.378
1000	32,344	24,030	293.889	3800	190,316	158,723	368.495
1100	37,606	28,461	298.904	4000	201,860	168,604	371.456
1200	42,946	32,969	303.551	4400	224,973	188,391	376.963
1300	48,351	37,543	307.876	4800	248,114	208,207	381.997

(continued)

Ideal-Gas (IG Model) Properties of Nitrogen Dioxide (NO₂)

$\overline{h}^{o}_{f,298} = 33,100$ kJ/kmol; $\overline{M} = 46.005$ kg/kmol

T	$\overline{h}$	$\overline{u}$	$\overline{s}^{o}$	T	$\overline{h}$	$\overline{u}$	$\overline{s}^{o}$
K	kJ/kmol	kJ/kmol	kJ/kmol·K	K	kJ/kmol	kJ/kmol	kJ/kmol·K
1400	53,808	42,168	311.920	5200	271,276	228,043	386.632
1500	59,309	46,838	315.715	5600	294,455	247,897	390.926
1600	64,846	51,544	319.289	6000	317,648	267,764	394.926
1700	70,414	56,280	322.664	–	–	–	–

Table D-11 IG Model: Ideal Gas Properties of Nitrogen Monoxide (NO)

Ideal-Gas (IG Model) Properties of Nitric Oxide (NO)

$\overline{h}^{o}_{f,298} = 90,291$ kJ/kmol; $\overline{M} = 30.006$ kg/kmol

T	$\overline{h}$	$\overline{u}$	$\overline{s}^{o}$	T	$\overline{h}$	$\overline{u}$	$\overline{s}^{o}$
K	kJ/kmol	kJ/kmol	kJ/kmol·K	K	kJ/kmol	kJ/kmol	kJ/kmol·K
0	−9,192	−9,192	0	1800	50,557	35,592	269.282
100	−6,073	−6,904	177.031	1900	54,201	38,404	271.252
200	−2,951	−4,614	198.747	2000	57,859	41,231	273.128
298	0	−2,478	210.759	2200	65,212	46,921	276.632
300	55	−2,439	210.943	2400	72,606	52,652	279.849
400	3,040	−286	219.529	2600	80,034	58,418	282.822
500	6,059	1,902	226.263	2800	87,491	64,212	285.585
600	9,144	4,156	231.886	3000	94,973	70,031	288.165
700	12,308	6,488	236.762	3200	102,477	75,872	290.587
800	15,548	8,897	241.088	3400	110,000	81,732	292.867
900	18,858	11,375	244.985	3600	117,541	87,611	295.022
1000	22,229	13,915	248.536	3800	125,099	93,506	297.065
1100	25,653	16,508	251.799	4000	132,671	99,415	299.007
1200	29,120	19,143	254.816	4400	147,857	111,275	302.626
1300	32,626	21,818	257.621	4800	163,094	123,187	305.940
1400	36,164	24,524	260.243	5200	178,377	135,144	308.998
1500	39,729	27,258	262.703	5600	193,703	147,145	311.838
1600	43,319	30,017	265.019	6000	209,070	159,186	314.488
1700	46,929	32,795	267.208	–	–	–	–

Table D-12   IG Model: Ideal Gas Properties of Hydroxyl Radical (OH)

Ideal-Gas (IG Model) Properties of Hydroxyl Radical (OH)

$\overline{h}_{f,298}^{\circ} = 38,987$ kJ/kmol; $\overline{M} = 17.007$ kg/kmol

T	$\overline{h}$	$\overline{u}$	$\overline{s}^{\circ}$	T	$\overline{h}$	$\overline{u}$	$\overline{s}^{\circ}$
K	kJ/kmol	kJ/kmol	kJ/kmol·K	K	kJ/kmol	kJ/kmol	kJ/kmol·K
0	0	0	0	2400	77,015	57,061	248.628
298	9,188	6,709	183.594	2450	78,801	58,431	249.364
300	9,244	6,749	183.779	2500	80,592	59,806	250.088
500	15,181	11,024	198.955	2550	82,388	61,186	250.799
1000	30,123	21,809	219.624	2600	84,189	62,572	251.499
1500	46,046	33,575	232.506	2650	85,995	63,962	252.187
1600	49,358	36,055	234.642	2700	87,806	65,358	252.864
1700	52,706	38,571	236.672	2750	89,622	66,757	253.530
1800	56,089	41,123	238.606	2800	91,442	68,162	254.186
1900	59,505	43,708	240.453	2850	93,266	69,570	254.832
2000	62,952	46,323	242.221	2900	95,095	70,983	255.468
2050	64,687	47,642	243.077	2950	96,927	72,400	256.094
2100	66,428	48,968	243.917	3000	98,763	73,820	256.712
2150	68,177	50,301	244.740	3100	102,447	76,673	257.919
2200	69,932	51,641	245.547	3200	106,145	79,539	259.093
2250	71,694	52,987	246.338	3300	109,855	82,418	260.235
2300	73,462	54,339	247.116	3400	113,578	85,309	261.347
2350	75,236	55,697	247.879	3500	117,312	88,212	262.429

Table D-13   IG Model: Ideal Gas Properties of Oxygen Ion (O)

Ideal-Gas (IG Model) Properties of Oxygen Ion (O)

$\overline{h}_{f,298}^{\circ} = 249,170$ kJ/kmol; $\overline{M} = 16$ kg/kmol

T	$\overline{h}$	$\overline{u}$	$\overline{s}^{\circ}$	T	$\overline{h}$	$\overline{u}$	$\overline{s}^{\circ}$
K	kJ/kmol	kJ/kmol	kJ/kmol·K	K	kJ/kmol	kJ/kmol	kJ/kmol·K
0	0	0	0	2400	50,894	30,940	204.932
298	6,852	4,373	160.944	2450	51,936	31,566	205.362
300	6,892	4,398	161.079	2500	52,979	32,193	205.783
500	11,197	7,040	172.088	2550	54,021	32,820	206.196
1000	21,713	13,398	186.678	2600	55,064	33,447	206.601
1500	32,150	19,679	195.143	2650	56,108	34,075	206.999
1600	34,234	20,931	196.488	2700	57,152	34,703	207.389
1700	36,317	22,183	197.751	2750	58,196	35,332	207.772

(continued)

Ideal-Gas (IG Model) Properties of Oxygen Ion (O)

$\overline{h}^o_{f,298} = 249{,}170$ kJ/kmol; $\overline{M} = 16$ kg/kmol

T	$\overline{h}$	$\overline{u}$	$\overline{s}^o$	T	$\overline{h}$	$\overline{u}$	$\overline{s}^o$
K	kJ/kmol	kJ/kmol	kJ/kmol·K	K	kJ/kmol	kJ/kmol	kJ/kmol·K
1800	38,400	23,434	198.941	2800	59,241	35,961	208.148
1900	40,482	24,685	200.067	2850	60,286	36,590	208.518
2000	42,564	25,935	201.135	2900	61,332	37,220	208.882
2050	43,605	26,560	201.649	2950	62,378	37,851	209.240
2100	44,646	27,186	202.151	3000	63,425	38,482	209.592
2150	45,687	27,811	202.641	3100	65,520	39,746	210.279
2200	46,728	28,436	203.119	3200	67,619	41,013	210.945
2250	47,769	29,062	203.588	3300	69,720	42,283	211.592
2300	48,811	29,688	204.045	3400	71,824	43,556	212.220
2350	49,852	30,314	204.493	3500	73,932	44,832	212.831

Table D-14   IG Model: Ideal Gas Properties of Nitrogen Ion (N)

Ideal-Gas (IG Model) Properties of Nitrogen Ion (N)

$\overline{h}^o_{f,298} = 472{,}680$ kJ/kmol; $\overline{M} = 14.007$ kg/kmol

T	$\overline{h}$	$\overline{u}$	$\overline{s}^o$	T	$\overline{h}$	$\overline{u}$	$\overline{s}^o$
K	kJ/kmol	kJ/kmol	kJ/kmol·K	K	kJ/kmol	kJ/kmol	kJ/kmol·K
0	−6,197	−6,197	0	1800	31,218	16,253	190.672
100	−4,119	−4,950	130.593	1900	33,296	17,499	191.796
200	−2,040	−3,703	154.001	2000	35,375	18,747	192.863
298	0	−2,478	153.300	2200	39,534	21,243	194.845
300	38	−2,456	153.429	2400	43,695	23,741	196.655
400	2,117	−1,209	159.409	2600	47,860	26,244	198.322
500	4,196	39	164.047	2800	52,033	28,754	199.868
600	6,274	1,286	167.837	3000	56,218	31,276	201.311
700	8,353	2,533	171.041	3200	60,420	33,815	202.667
800	10,431	3,780	173.816	3400	64,646	36,378	203.948
900	12,510	5,027	176.265	3600	68,902	38,972	205.164
1000	14,589	6,275	178.455	3800	73,194	41,601	206.325
1100	16,667	7,522	180.436	4000	77,532	44,276	207.437
1200	18,746	8,769	182.244	4400	86,367	49,785	209.542
1300	20,825	10,017	183.908	4800	95,457	55,550	211.519
1400	22,903	11,263	185.448	5200	104,843	61,610	213.397
1500	24,982	12,511	186.883	5600	114,550	67,992	215.195
1600	27,060	13,758	188.224	6000	124,590	74,706	216.926
1700	29,139	15,005	189.484	–	–	–	–

Table D-15   IG Model: Ideal Gas Properties of Hydrogen Ion (H)

Ideal-Gas (IG Model) Properties of Hydrogen Ion (H)

$\overline{h}^{\circ}_{f,298} = 217{,}999$ kJ/kmol; $\overline{M} = 1.008$ kg/kmol

T	$\overline{h}$	$\overline{u}$	$\overline{s}^{\circ}$	T	$\overline{h}$	$\overline{u}$	$\overline{s}^{\circ}$
K	kJ/kmol	kJ/kmol	kJ/kmol·K	K	kJ/kmol	kJ/kmol	kJ/kmol·K
0	−6,197	−6,197	0	1800	31,218	16,253	152.089
100	−4,119	−4,950	92.009	1900	33,296	17,499	153.212
200	−2,040	−3,703	106.417	2000	35,375	18,747	154.279
298	0	−2,478	114.716	2200	39,532	21,241	156.260
300	38	−2,456	114.845	2400	43,689	23,735	158.069
400	2,117	−1,209	120.825	2600	47,847	26,231	159.732
500	4,196	39	125.463	2800	52,004	28,725	161.273
600	6,274	1,286	129.253	3000	56,161	31,219	162.707
700	8,353	2,533	132.457	3200	60,318	33,713	164.048
800	10,431	3,780	135.233	3400	64,475	36,207	165.308
900	12,510	5,027	137.681	3600	68,633	38,703	166.497
1000	14,589	6,275	139.871	3800	72,790	41,197	167.620
1100	16,667	7,522	141.852	4000	76,947	43,691	168.687
1200	18,746	8,769	143.661	4400	85,261	48,679	170.668
1300	20,825	10,017	145.324	4800	93,476	53,569	172.476
1400	22,903	11,263	146.865	5200	101,890	58,657	174.140
1500	24,982	12,511	148.299	5600	110,205	63,647	175.681
1600	27,060	13,758	149.640	6000	118,519	68,635	177.114
1700	29,139	15,005	150.900	–	–	–	–

Table E-1   RG-Model: Critical Properties of Phase-Change (PC) Fluids

Material Properties Used in the Real Gas (RG) Model

Substance	Formula	Molar Mass	Critical-State Properties		
			Temperature	Pressure	Volume
		kg/kmol	K	MPa	m³/kmol
		$\overline{M}$	T_{cr}	p_{cr}	$\overline{v}_{cr}$
Air	–	28.97	132.5	3.770	0.0883
Ammonia	NH_3	17.03	405.5	11.280	0.0724
Argon	Ar	39.95	151.0	4.860	0.0749
Benzene	C_6H_6	78.12	562.0	4.920	0.2603
Bromine	Br_2	159.81	584.0	10.340	0.1355
n-Butane	C_4H_{10}	58.12	425.2	3.800	0.2547
Carbon Dioxide	CO_2	44.01	304.2	7.390	0.0943
Carbon Monoxide	CO	28.01	133.0	3.500	0.0930
Carbon Tetrachloride	CCl_4	153.82	556.4	4.560	0.2759

Material Properties Used in the Real Gas (RG) Model

Substance	Formula	Molar Mass kg/kmol $\overline{M}$	Critical-State Properties Temperature K T_{cr}	Critical-State Properties Pressure MPa p_{cr}	Critical-State Properties Volume m³/kmol $\overline{v}_{cr}$
Chlorine	Cl_2	70.91	417.0	7.710	0.1242
Chloroform	$CHCl_3$	119.38	536.6	5.470	0.2403
Dichlorodifluoromethane (R-12)	CCl_2F_2	120.91	384.7	4.010	0.2179
Dichlorofluoromethane (R-21)	$CHCl_2F$	102.92	451.7	5.170	0.1973
Ethane	C_2H_6	30.07	305.5	4.480	0.1480
Ethyl Alcohol	C_2H_5OH	46.07	516.0	6.380	0.1673
Ethylene	C_2H_4	28.05	282.4	5.120	0.1242
Helium	He	4.00	5.3	0.230	0.0578
n-Hexane	C_6H_{14}	86.18	507.9	3.030	0.3677
Hydrogen	H_2	2.02	33.3	1.300	0.0649
Krypton	Kr	83.80	209.4	5.500	0.0924
Methane	CH_4	16.04	191.1	4.640	0.0993
Methyl Alcohol	CH_3OH	32.04	513.2	7.950	0.1180
Methyl Chloride	CH_3Cl	50.49	416.3	6.680	0.1430
Neon	Ne	20.18	44.5	2.730	0.0417
Nitrogen	N_2	28.01	126.2	3.390	0.0899
Nitrous Oxide	N_2O	44.01	309.7	7.270	0.0961
Oxygen	O_2	32.00	154.8	5.080	0.0780
Propane	C_3H_8	44.10	370.0	4.260	0.1998
Propylene	C_3H_6	42.08	365.0	4.620	0.1810
Sulfur Dioxide	SO_2	64.06	430.7	7.880	0.1217
Tetrafluoroethane (R-134a)	CF_3CH_2F	102.03	374.3	4.067	0.1847
Trichlorofluoromethane (R-11)	CCl_3F	137.37	471.2	4.380	0.2478
Water	H_2O	18.02	647.3	22.090	0.0568
Xenon	Xe	131.30	289.8	5.880	0.1186

Table E-8   Constants Used in BWR Equation of State

Constants in Benedict-Webb-Rubin (BWR) Equation of State

p in atm; T in K
$\overline{v}$ in L/mol
$\overline{R} = 0.08206$ L·atm/mol·K

$$p = \frac{\overline{R}T}{\overline{v}} + \left(B_0\overline{R}T - A_0 - \frac{C_0}{T^2}\right)\frac{1}{\overline{v}^2} + \frac{b\overline{R}T - a}{\overline{v}^3} + \frac{a\alpha}{\overline{v}^6} + \frac{c}{\overline{v}^3T^2}\left(1 + \frac{\gamma}{\overline{v}^2}\right)e^{-\gamma/\overline{v}^2}$$

Gas	Formula	a	A_0	b	B_0	c	C_0	α	γ
Nitrogen	N_2	2.54	106.73	0.002328	0.04074	7.379×10^4	8.164×10^5	1.272×10^{-4}	0.0053
Methane	CH_4	5.00	187.91	0.003380	0.04260	2.578×10^5	2.286×10^6	1.244×10^{-4}	0.0060
Carbon Monoxide	CO	3.71	135.87	0.002623	0.05450	1.054×10^5	8.673×10^5	1.350×10^{-4}	0.0060
Carbon Dioxide	CO_2	13.86	277.30	0.007210	0.04991	1.511×10^6	1.404×10^7	8.470×10^{-5}	0.0054
n-Butane	C_4H_{10}	190.68	1021.60	0.039998	0.12436	3.205×10^7	1.006×10^8	1.101×10^{-3}	0.0340

Table G-1   Reactions: Enthalpy of Formation Table

Molar Specific Enthalpy of Formation, Gibbs Function of Formation, and Absolute Entropy at 25°C, 1 atm

Substance	Formula (Phase)	$\overline{M}$ kg/kmol	$\overline{h}_f^\circ$ kJ/kmol	$\overline{g}_f^\circ$ kJ/kmol	$\overline{s}^\circ$ kJ/kmol·K
Carbon	C(s)	12.01	0	0	5.74
Hydrogen	H_2(g)	2.02	0	0	130.68
Nitrogen	N_2(g)	28.01	0	0	191.61
Oxygen	O_2(g)	32.00	0	0	205.04
Carbon Monoxide	CO(g)	28.01	−110,530	−137,150	197.65
Carbon Dioxide	CO_2(g)	44.00	−393,520	−394,360	213.80
Water Vapor	H_2O(g)	18.02	−241,820	−228,590	188.83
Water	H_2O(l)	18.02	−285,820	−237,180	69.92
Hydrogen Peroxide	H_2O_2(g)	34.02	−136,310	−105,600	232.63
Ammonia	NH_3(g)	17.03	−46,190	−16,590	192.33
Methane	CH_4(g)	16.04	−74,850	−50,790	186.16
Acetylene	C_2H_2(g)	26.04	226,730	209,170	200.85
Ethylene	C_2H_4(g)	28.05	52,280	68,120	219.83
Ethane	C_2H_6(g)	30.07	−84,680	−32,890	229.49
Propylene	C_3H_6(g)	42.05	20,410	62,720	266.94
Propane	C_3H_8(g)	44.10	−103,850	−23,490	269.91
n-Butane	C_4H_{10}(g)	58.12	−126,150	−15,710	310.12
n-Octane(l)	C_8H_{18}(l)	114.23	−249,950	6,610	360.79
n-Octane(g)	C_8H_{18}(g)	114.23	−208,450	16,530	466.73
n-Dodecane	$C_{12}H_26$(g)	170.22	−291,010	50,150	622.83
Benzene	C_6H_6(g)	78.11	82,930	129,660	269.20
Methyl Alcohol	CH_3OH(g)	32.04	−200,670	−162,000	239.70
Methyl Alcohol	CH_3OH(l)	32.04	−238,660	−166,360	126.80
Ethyl Alcohol	C_2H_5OH(g)	46.07	−235,310	−168,570	282.59
Ethyl Alcohol	C_2H_5OH(l)	46.07	−277,690	−174,890	160.70
Oxygen (atomic)	O(g)	16.00	249,190	231,770	161.06
Hydrogen (atomic)	H(g)	1.01	218,000	203,290	114.72
Nitrogen (atomic)	N(g)	14.01	472,650	455,510	153.30
Hydroxyl (radical)	OH(g)	17.01	39,460	34,290	183.70

Table G-2   Reactions: Heating Values of Common Fuels

Properties of Some Common Fuels and Hydrocarbons

Fuel (Phase)	Formula	Molar Mass kg/kmol $\overline{M}$	Density[1] kg/L ρ	Enthalpy of Vaporization[2] kJ/kg Δh_v	Specific Heat[1] kJ/kg·K c_p	Higher Heating Value[3] kJ/kg HHV	Lower Heating Value[4] kJ/kg LHV
Carbon(s)	C	12.01	2.000	–	0.708	32,800	32,800
Hydrogen(g)	H_2	2.02	–	–	14.40	141,800	120,000

(continued)

Properties of Some Common Fuels and Hydrocarbons

Fuel (Phase)	Formula	Molar Mass kg/kmol $\overline{M}$	Density[1] kg/L ρ	Enthalpy of Vaporization[2] kJ/kg Δh_v	Specific Heat[1] kJ/kg·K c_p	Higher Heating Value[3] kJ/kg HHV	Lower Heating Value[4] kJ/kg LHV
Carbon Monoxide(g)	CO	28.01	–	–	1.05	10,100	10,100
Methane(g)	CH_4	16.04	–	509	2.20	55,530	50,050
Methanol(l)	CH_4O	32.04	0.790	1168	2.53	22,660	19,920
Acetylene(g)	C_2H_2	26.04	–	–	1.69	49,970	48,280
Ethane(g)	C_2H_6	30.07	–	172	1.75	51,900	47,520
Ethanol(l)	C_2H_6O	46.07	0.790	919	2.44	29,670	26,810
Propane(l)	C_3H_8	44.10	0.500	420	2.77	50,330	46,340
Butane(l)	C_4H_{10}	58.12	0.579	362	2.42	49,150	45,370
1-Pentene(l)	C_5H_{10}	70.13	0.641	363	2.20	47,760	44,630
Isopentane(l)	C_5H_{12}	72.15	0.626	–	2.32	48,570	44,910
Benzene(l)	C_6H_6	78.11	0.877	433	1.72	41,800	40,100
Hexene(l)	C_6H_{12}	84.16	0.673	392	1.84	47,500	44,400
Hexane(l)	C_6H_{14}	86.18	0.660	366	2.27	48,310	44,740
Toluene(l)	C_7H_8	92.14	0.867	412	1.71	42,400	40,500
Heptane(l)	C_7H_{16}	100.20	0.684	365	2.24	48,100	44,600
Octane(l)	C_8H_{18}	114.23	0.703	363	2.23	47,890	44,430
Decane(l)	$C_{10}H_{22}$	142.29	0.730	361	2.21	47,640	44,240
Gasoline(l)	$CnH_{1.87n}$	100 - 110	0.72 - 0.78	350	2.4	47,300	44,000
Light Diesel(l)	$CnH_{1.8n}$	170	0.78 - 0.84	270	2.2	46,100	43,200
Heavy Diesel(l)	$CnH_{1.7n}$	200	0.82 - 0.88	230	1.9	45,500	42,800
Natural Gas(g)	$CnH_{3.8n}N0.1n$	18	–	–	2.0	50,000	45,000

1: 1 atm and 20°C.
2: At 25°C for liquid fuels, and at saturation temperature at 1 atm for gaseous fuels.
3: H_2O in liquid phase in products.
4: H_2O in vapor phase in products.

Table G-3 Reactions: Equilibrium Constants Table

Natural log of equilibrium constant, $\ln(K)$, for the reaction $\nu_A A + \nu_B B \rightleftharpoons \nu_C C + \nu_D D$

$$\text{where,} \quad K \equiv \frac{p_C^{\nu_C} p_D^{\nu_D}}{p_A^{\nu_A} p_B^{\nu_B}} = \frac{Y_C^{\nu_C} Y_D^{\nu_D}}{Y_A^{\nu_A} Y_B^{\nu_B}} \left(\frac{p}{p_0}\right)^{\nu_C + \nu_D - \nu_{A1} - \nu_B}$$

Temp. K	$H_2 \rightleftharpoons 2H$	$O_2 \rightleftharpoons 2O$	$N_2 \rightleftharpoons 2N$	$H_2O \rightleftharpoons H_2 + (1/2)O_2$	$H_2O \rightleftharpoons OH + (1/2)H_2$	$CO_2 \rightleftharpoons CO + (1/2)O_2$	$(1/2)N_2 + (1/2)O_2 \rightleftharpoons NO$	$CO_2 + H_2 \rightleftharpoons CO + H_2O$
298	−164.005	−186.975	−367.480	−92.208	−106.208	−103.762	−35.052	−11.554
500	−92.827	−105.630	−213.372	−52.691	−60.281	−57.616	−20.295	−4.925
1000	−39.803	−45.150	−99.127	−23.163	−26.034	−23.529	−9.388	−0.366
1200	−30.874	−35.005	−80.011	−18.182	−20.283	−17.871	−7.569	0.311
1400	−24.463	−27.742	−66.329	−14.609	−16.099	−13.842	−6.270	0.767

Natural log of equilibrium constant, ln (K), for the reaction $\nu_A A + \nu_B B \rightleftharpoons \nu_C C + \nu_D D$

$$\text{where, } K \equiv \frac{p_C^{\nu_C} \, p_D^{\nu_D}}{p_A^{\nu_A} \, p_B^{\nu_B}} = \frac{Y_C^{\nu_C} \, Y_D^{\nu_D}}{Y_A^{\nu_A} \, Y_B^{\nu_B}} \left(\frac{p}{p_0}\right)^{\nu_C + \nu_D - \nu_{A1} - \nu_B}$$

Temp. K	$H_2 \rightleftharpoons 2H$	$O_2 \rightleftharpoons 2O$	$N_2 \rightleftharpoons 2N$	$H_2O \rightleftharpoons H_2 + (1/2)O_2$	$H_2O \rightleftharpoons OH + (1/2)H_2$	$CO_2 \rightleftharpoons CO + (1/2)O_2$	$(1/2)N_2 + (1/2)O_2 \rightleftharpoons NO$	$CO_2 + H_2 \rightleftharpoons CO + H_2O$
1600	−19.637	−22.285	−56.055	−11.921	−13.066	−10.830	−5.294	1.091
1800	−15.866	−18.030	−48.051	−9.826	−10.657	−8.497	−4.536	1.328
2000	−12.840	−14.622	−41.645	−8.145	−8.728	−6.635	−3.931	1.510
2200	−10.353	−11.827	−36.391	−6.768	−7.148	−5.120	−3.433	1.648
2400	−8.276	−9.497	−32.011	−5.619	−5.832	−3.860	−3.019	1.759
2600	−6.517	−7.521	−28.304	−4.648	−4.719	−2.801	−2.671	1.847
2800	−5.002	−5.826	−25.117	−3.812	−3.763	−1.894	−2.372	1.918
3000	−3.685	−4.357	−22.359	−3.086	−2.937	−1.111	−2.114	1.976
3200	−2.534	−3.072	−19.937	−2.451	−2.212	−0.429	−1.888	2.022
3400	−1.516	−1.935	−17.800	−1.891	−1.576	0.169	−1.690	2.061
3600	−0.609	−0.926	−15.898	−1.392	−1.088	0.701	−1.513	−
3800	0.202	−0.019	−14.199	−0.945	−0.501	1.176	−1.356	−
4000	0.934	0.796	−12.660	−0.542	−0.044	1.599	−1.216	−
4500	2.486	2.513	−9.414	0.312	0.920	2.490	−0.921	−
5000	3.725	3.895	−6.807	0.996	1.689	3.197	−0.686	−
5500	4.743	5.023	−4.666	1.560	2.318	3.771	−0.497	−
6000	5.590	5.963	−2.865	2.032	2.843	4.245	−0.341	−

Table H-1   High Speed Flows: Isentropic Table for Air ($k = 1.4$)

Isentropic Table for Air (Perfect Gas with $k = 1.4$)

M	$\dfrac{A}{A_*}$	$\dfrac{p}{p_t}$	$\dfrac{\rho}{\rho_t}$	$\dfrac{T}{T_t}$
0	∞	1.00000	1.00000	1.00000
0.10	5.8218	0.99303	0.99502	0.99800
0.20	2.9635	0.97250	0.98027	0.99206
0.30	2.0351	0.93947	0.95638	0.98232
0.40	1.5901	0.89562	0.92428	0.96899
0.50	1.3398	0.84302	0.88517	0.95238
0.60	1.1882	0.78400	0.84045	0.93284
0.70	1.0944	0.72092	0.79158	0.91075
0.80	1.0382	0.65602	0.74000	0.88652
0.90	1.0089	0.59126	0.68704	0.86058
1.00	1.0000	0.52828	0.63394	0.83333
1.10	1.0079	0.46835	0.58169	0.80515

(continued)

	Isentropic Table for Air (Perfect Gas with $k = 1.4$)			
M	$\dfrac{A}{A_*}$	$\dfrac{p}{p_t}$	$\dfrac{\rho}{\rho_t}$	$\dfrac{T}{T_t}$
1.20	1.0304	0.41238	0.53114	0.77640
1.30	1.0663	0.36092	0.48291	0.74738
1.40	1.1149	0.31424	0.43742	0.71839
1.50	1.1762	0.27240	0.39498	0.68965
1.60	1.2502	0.23527	0.35573	0.66138
1.70	1.3376	0.20259	0.31969	0.63372
1.80	1.4390	0.17404	0.28682	0.60680
1.90	1.5552	0.14924	0.25699	0.58072
2.00	1.6875	0.12780	0.23005	0.55556
2.10	1.8369	0.10935	0.20580	0.53135
2.20	2.0050	0.09352	0.18405	0.50813
2.30	2.1931	0.07997	0.16458	0.48591
2.40	2.4031	0.06840	0.14720	0.46468
2.50	2.6367	0.05853	0.13169	0.44444
2.60	2.8960	0.05012	0.11787	0.42517
2.70	3.1830	0.04295	0.10557	0.40684
2.80	3.5001	0.03685	0.09462	0.38941
2.90	3.8498	0.03165	0.08489	0.37286
3.00	4.2346	0.02722	0.07623	0.35714
3.50	6.7896	0.01311	0.04523	0.28986
4.00	10.7190	0.00658	0.02766	0.23810
4.50	16.5620	0.00346	0.01745	0.19802
5.00	25.0000	0.00189	0.01134	0.16667
6.00	53.1800	0.000633	0.00519	0.12195
7.00	104.1430	0.000242	0.00261	0.09259
8.00	190.1090	0.000102	0.00141	0.07246
9.00	327.1890	0.0000474	0.000815	0.05814
10.00	535.9380	0.0000236	0.000495	0.04762
∞	∞	0	0	0

Table H-2 High Speed Flows: Normal Shock Table for Air ($k = 1.4$)

		Normal Shock Table for Air (Perfect Gas with $k = 1.4$)					
M_i	M_e	$\dfrac{p_e}{p_i}$	$\dfrac{\rho_e}{\rho_i}$	$\dfrac{T_e}{T_i}$	$\dfrac{p_{t,e}}{p_{t,i}}$	$\dfrac{A_{*,e}}{A_{*,i}}$	$\dfrac{V_e}{V_i}$
1.00	1.00000	1.0000	1.0000	1.0000	1.00000	1.00000	1.0000
1.10	0.91177	1.2450	1.1691	1.0649	0.99892	1.00107	0.8554

Normal Shock Table for Air (Perfect Gas with $k = 1.4$)

M_i	M_e	$\dfrac{p_e}{p_i}$	$\dfrac{\rho_e}{\rho_i}$	$\dfrac{T_e}{T_i}$	$\dfrac{p_{t,e}}{p_{t,i}}$	$\dfrac{A_{*,e}}{A_{*,i}}$	$\dfrac{V_e}{V_i}$
1.20	0.84217	1.5133	1.3416	1.1280	0.99280	1.00725	0.7453
1.30	0.78596	1.8050	1.5157	1.1909	0.97935	1.02106	0.6597
1.40	0.73971	2.1200	1.6896	1.2547	0.95819	1.04364	0.5918
1.50	0.70109	2.4583	1.8621	1.3202	0.92978	1.07553	0.5370
1.60	0.66844	2.8200	2.0317	1.3880	0.89520	1.11709	0.4921
1.70	0.64055	3.2050	2.1977	1.4583	0.85573	1.16864	0.4550
1.80	0.61650	3.6133	2.3592	1.5316	0.81268	1.23054	0.4238
1.90	0.59562	4.0450	2.5157	1.6079	0.76735	1.30325	0.3974
2.00	0.57735	4.5000	2.6666	1.6875	0.72088	1.38732	0.3749
2.10	0.56128	4.9784	2.8119	1.7704	0.67422	1.48338	0.3556
2.20	0.54706	5.4800	2.9512	1.8569	0.62812	1.59221	0.3388
2.30	0.53441	6.0050	3.0846	1.9468	0.58331	1.71466	0.3241
2.40	0.52312	6.5533	3.2119	2.0403	0.54015	1.85170	0.3113
2.50	0.51299	7.1250	3.3333	2.1375	0.49902	2.00438	0.2999
2.60	0.50387	7.7200	3.4489	2.2383	0.46012	2.17387	0.2899
2.70	0.49563	8.3383	3.5590	2.3429	0.42359	2.36144	0.2809
2.80	0.48817	8.9800	3.6635	2.4512	0.38946	2.56846	0.2729
2.90	0.48138	9.6450	3.7629	2.5632	0.35773	2.79639	0.2657
3.00	0.47519	10.333	3.8571	2.6790	0.32834	3.04681	0.2592
4.00	0.43496	18.500	4.5714	4.0469	0.13876	7.21309	0.2187
5.00	0.41523	29.000	5.0000	5.8000	0.06172	16.22510	0.1999
10.00	0.38757	116.50	5.7143	20.388	0.00304	329.56100	0.1749
∞	0.37796	∞	6.000	∞	0	∞	0.1667

Table H-3   High Speed Flows Atmospheric Property Variation

Properties of Atmospheric Air as Functions of Altitude

Altitude	Temp.	Pressure	Gravity	Speed of Sound	Density	Viscosity	Thermal Conductivity
m	°C	kPa	m/s²	m/s	kg/m³	kg/m·s	W/m·K
z	T	p	g	c	ρ	μ	λ
0	15.00	101.33	9.807	340.3	1.225	1.789×10^{-5}	0.0253
200	13.70	98.95	9.806	339.5	1.202	1.783×10^{-5}	0.0252
400	12.40	96.61	9.805	338.8	1.179	1.777×10^{-5}	0.0252
600	11.10	94.32	9.805	338.0	1.156	1.771×10^{-5}	0.0251
800	9.80	92.08	9.804	337.2	1.134	1.764×10^{-5}	0.0250

(continued)

Properties of Atmospheric Air as Functions of Altitude

Altitude	Temp.	Pressure	Gravity	Speed of Sound	Density	Viscosity	Thermal Conductivity
m	°C	kPa	m/s²	m/s	kg/m³	kg/m·s	W/m·K
z	T	p	g	c	ρ	μ	λ
1000	8.50	89.88	9.804	336.4	1.112	1.758×10^{-5}	0.0249
1200	7.20	87.72	9.803	335.7	1.090	1.752×10^{-5}	0.0248
1400	5.90	85.60	9.802	334.9	1.069	1.745×10^{-5}	0.0247
1600	4.60	83.53	9.802	334.1	1.048	1.739×10^{-5}	0.0245
1800	3.30	81.49	9.801	333.3	1.027	1.732×10^{-5}	0.0244
2000	2.00	79.50	9.800	332.5	1.007	1.726×10^{-5}	0.0243
2200	0.70	77.55	9.800	331.7	0.987	1.720×10^{-5}	0.0242
2400	−0.59	75.63	9.799	331.0	0.967	1.713×10^{-5}	0.0241
2600	−1.89	73.76	9.799	330.2	0.947	1.707×10^{-5}	0.0240
2800	−3.19	71.92	9.798	329.4	0.928	1.700×10^{-5}	0.0239
3000	−4.49	70.12	9.797	328.6	0.909	1.694×10^{-5}	0.0238
3200	−5.79	68.36	9.797	327.8	0.891	1.687×10^{-5}	0.0237
3400	−7.09	66.36	9.796	327.0	0.872	1.681×10^{-5}	0.0236
3600	−8.39	64.94	9.796	326.2	0.854	1.674×10^{-5}	0.0235
3800	−9.69	63.28	9.795	325.4	0.837	1.668×10^{-5}	0.0234
4000	−10.98	61.66	9.794	324.6	0.819	1.661×10^{-5}	0.0233
4200	−12.30	60.07	9.794	323.8	0.802	1.655×10^{-5}	0.0232
4400	−13.60	58.52	9.793	323.0	0.785	1.648×10^{-5}	0.0231
4600	−14.90	57.00	9.793	322.2	0.769	1.642×10^{-5}	0.0230
4800	−16.20	55.51	9.792	321.4	0.752	1.635×10^{-5}	0.0229
5000	−17.50	54.05	9.791	320.5	0.736	1.628×10^{-5}	0.0228
5200	−18.80	52.62	9.791	319.7	0.721	1.622×10^{-5}	0.0227
5400	−20.10	51.23	9.790	318.9	0.705	1.615×10^{-5}	0.0226
5600	−21.40	49.86	9.789	318.1	0.690	1.608×10^{-5}	0.0224
5800	−22.70	48.52	9.785	317.3	0.675	1.602×10^{-5}	0.0223
6000	−24.00	47.22	9.788	316.5	0.660	1.595×10^{-5}	0.0222
6200	−25.30	45.94	9.788	315.6	0.646	1.588×10^{-5}	0.0221
6400	−26.60	44.69	9.787	314.8	0.631	1.582×10^{-5}	0.0220
6600	−27.90	43.47	9.786	314.0	0.617	1.575×10^{-5}	0.0219
6800	−29.20	42.27	9.785	313.1	0.604	1.568×10^{-5}	0.0218
7000	−30.50	41.11	9.785	312.3	0.590	1.561×10^{-5}	0.0217
8000	−36.90	35.65	9.782	308.1	0.526	1.527×10^{-5}	0.0212
9000	−43.40	30.80	9.779	303.8	0.467	1.493×10^{-5}	0.0206

Properties of Atmospheric Air as Functions of Altitude

Altitude	Temp.	Pressure	Gravity	Speed of Sound	Density	Viscosity	Thermal Conductivity
m	°C	kPa	m/s^2	m/s	kg/m^3	kg/m·s	W/m·K
z	T	p	g	c	ρ	μ	λ
10000	−49.90	26.50	9.776	299.5	0.414	1.458×10^{-5}	0.0201
12000	−56.50	19.40	9.770	295.1	0.312	1.422×10^{-5}	0.0195
14000	−56.50	14.17	9.764	295.1	0.228	1.422×10^{-5}	0.0195
16000	−56.50	10.53	9.758	295.1	0.166	1.422×10^{-5}	0.0195
18000	−56.50	7.57	9.751	295.1	0.122	1.422×10^{-5}	0.0195

APPENDIX B

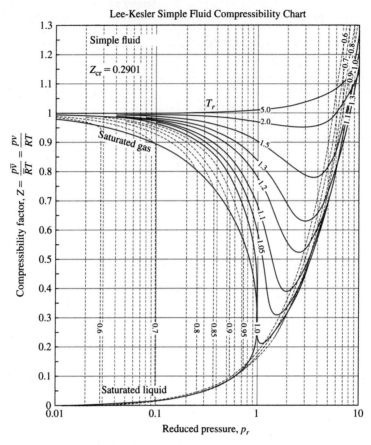

FIGURE E.2 RG Model: Lee-Kesler Compressibility Chart

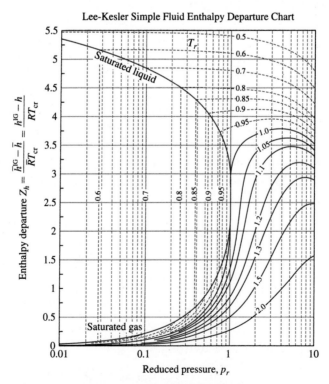

FIGURE E.3 RG Model: Lee-Kesler Enthalpy Departure Chart

Lee-Kesler Simple Fluid Entropy Departure Chart

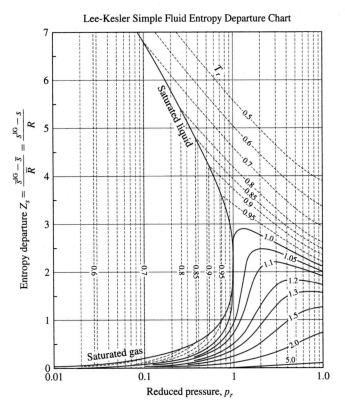

FIGURE E.4 RG Model: Lee-Kesler Entropy Departure Chart

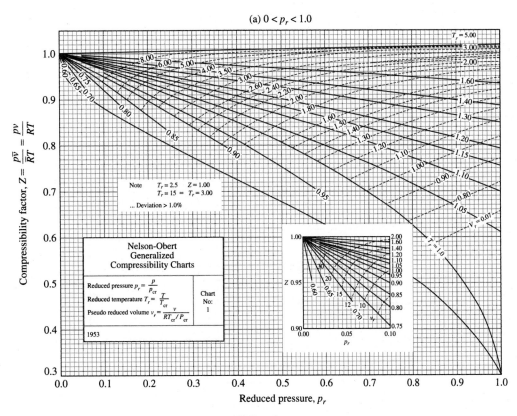

FIGURE E.5 RG Model: Nelson-Obert Compressibility Charts

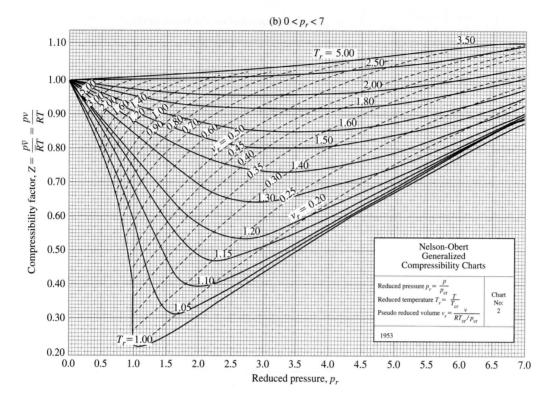

(b) $0 < p_r < 7$

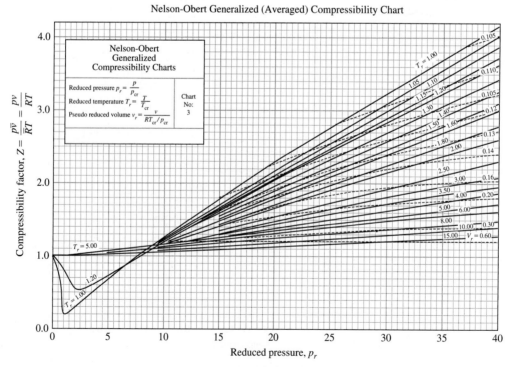

Nelson-Obert Generalized (Averaged) Compressibility Chart

FIGURE E.5 (continued)

Nelson-Obert Generalized (Averaged) Enthalpy Departure Chart

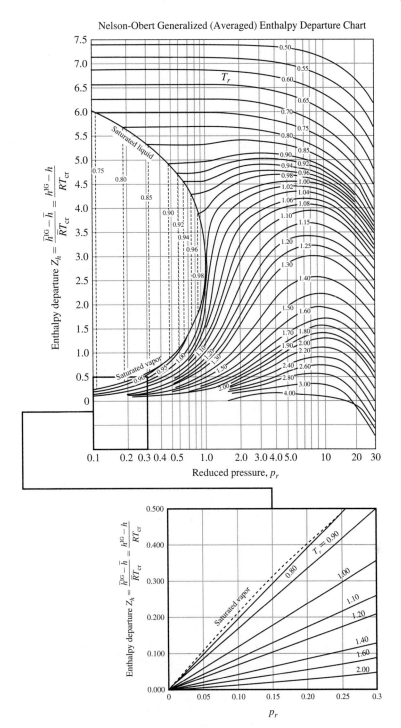

FIGURE E.6 RG Model: Nelson-Obert Enthalpy Departure Chart

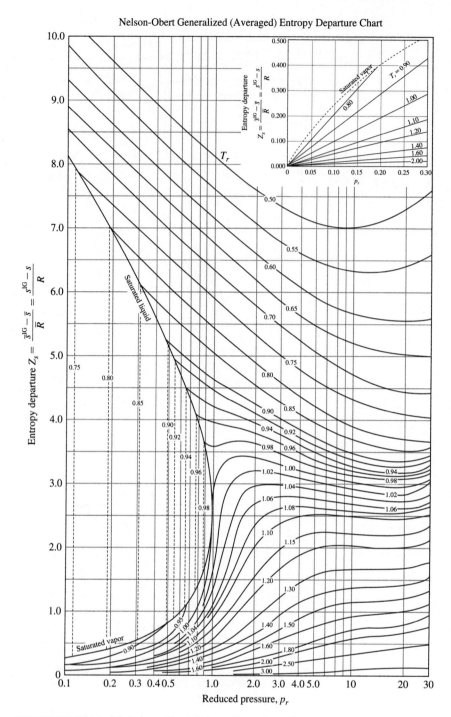

FIGURE E.7 RG Model: Nelson-Obert Entropy Departure Chart

Psychrometric Chart for Air at 101.325 kPa

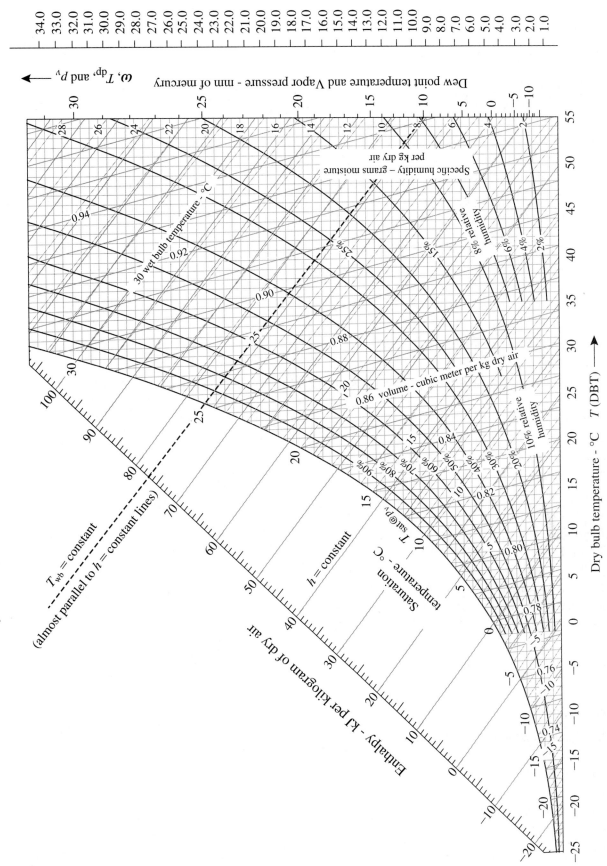

FIGURE F.1 Moist Air (MA) Model: Psychrometric Chart (SI units).

FIGURE F.1 Moist Air (MA) Model: Psychrometric Chart for 1 atm (SI units).

ANSWERS TO KEY PROBLEMS

Chapter 0

0-1-1: (a) 0.01 kN, (b) 0.02 kN

0-1-12: (a) 101 kPa, (b) 102 kPa

0-2-6: (a) -1, (b) -1

0-3-2: (a) 0.49 m^3/kg, (b) 0.098 m^3/s

0-4-1: (a) 981 kJ, (b) 49.05 kW

0-4-12: 418 m

0-4-22: (a) -12.81 kJ, (b) -12.51 kJ

0-4-39: (a) 60 kJ, (b) -300 kJ, (c) 170 kJ

0-6-2: (a) 19.62 kW, (b) 1.4 g/s

0-6-5: (a) 2923 kJ, (b) 0.221 kg

Chapter 1

1-1-13: 495 m

1-1-31: (a) 0.8 kJ/kg, (b) 19.63 kg/s, (c) 15.7 kW

1-1-40: (a) 3 MW, (b) 4.05 kW

1-1-47: (a) 1, (b) 0, (c) 0, (d) 0, (e) 0, (f) 1, (g)1

1-2-2: (a) -20.93 kJ, (b) 0.047%

1-2-20: (a) 127°C

1-2-26: (a) 20838 kW, (b) 0.0955 kW, (c) 18721.6 kW, (d) 18721.7 kW, (e) 20838 kW, (f) 2116.4 kW

Chapter 2

2-1-2: (a) 0.1 kg/s, (b) 11 kg

2-2-2: 10.37 kW

2-2-3: (a) 0 kW, (b) 0

2-2-19: (a) 0.49 kW, (b) 0.50 kW

2-2-23: (a) 0, (b) 0.1 kJ/kg

2-2-44: 270.7 kW

2-3-4: (a) 0.004 kW/K, (b) 0.0067 kW/K

2-3-9: (a) -2 kW, (b) 0.004 kW/K, (c) 0.0067 kW/K

2-3-18: (a) 550 m/s, (b) 6.149 kJ/kg · K

2-4-3: (a) -1 kW, (b) -0.003663 kW/K, (c) 0.003663 kW/K

2-5-5: 132.275 tons/h

2-5-8: (a) 45.72 kg/min, (b) 82.35%

2-5-15: (a) 2000 kW, (b) 4.667 kW/K

2-5-27: (a) 0.3 kW, (b) 11

Chapter 3

3-1-1: (a) 1.014 kJ/kg · K, (b) 1.102 kJ/kg · K

3-2-1: (a) 0.903 kJ/kg, (b) 0 kJ/kg · K

3-2-4: 111.73 kJ/K

3-3-2: (a) 311°C, (b) 0, (c) 100°C

3-3-28: (a) 101 kPa, (b) 198.5 kPa, (c) 216.2 kPa

3-3-49: (a) 198.3°C, (b) -108 kJ

3-4-1: (a) 146 kg, (b) 7.3 kg, (c) 7.0 kg

3-4-13: (a) 502.6 kJ/kg, (b) 541.3 kJ/kg,

3-5-1: (a) 0.00205 m^3/kg, (b) 0.0021 m^3/kg

3-5-7: (a) 15.26%, (b) 1%

Chapter 4

4-1-2: (a) 98%, (b) 0.1%

4-1-13: (a) 2591 kW, (b) 0.06 kg/s

4-1-16: (a) 299.6 K, (b) 29.96 m/s

4-1-24: 30.3 min

4-1-29: (a) 5.9 cm^2, (b) 359°C, (c) 0.175 kW/K

4-1-45: (a) 146.63 kg/s, (b) 56.6°C, (c) 4.39 kW/K

4-1-53: (a) 26.58 kW, (b) 2.09

4-1-78: (a) 187.7 kW, (b) 0.05 kW/K, (c) 191.0 kW

4-1-93: (a) 7.53 kg/s, (b) 15.07 kg/s

4-1-94: (a) 99.6°C, (b) 5.55 kg/s

4-2-2: 0.66 kg/s

4-3-1: (a) 0.2541, (b) 0.887 kW/K, (c) 0.215, (d) 1.005 kW/K

Chapter 5

5-1-4: (a) 1710 kJ, (b) 2.984 kJ/K, (c) 1.637 kJ/K

5-1-18: (a) 0.46 kg, (b) 68.75 kJ

5-1-31: (a) 477°C, (b) 680 kJ

5-1-40: (a) −173.88 kJ, (b) −173.88 kJ

5-1-44: (a) −12.83 kJ, (b) −13.20 kJ

5-2-3: (a) 51.7°C, (b) 13.4 kPa, (c) 0.08 kJ/K

5-2-12: (a) 454.5 kPa, (b) −44.98 kJ, (c) 0.24 kJ/K

5-3-4: (a) 200°C, (b) 0.097 kJ/K

5-3-3: (a) 20.11°C, (b) 0.170 kJ/K

5-4-3: (a) 13.49 kg, (b) −665.4kJ, (c) 2.7 kJ/K

Chapter 6

6-1-1: (a) 4.905 kW, (b) 4.905 kW

6-1-3: (a) 8 kW, (b) 2.5 kW

6-2-1: (a) 0.77 kW, (b) 14.2 kW

6-2-4: (a) 3.6 kW, (b) 2.69 kW

6-1-7: (a) −13.33 kJ, (b) 166.66 kJ

6-2-16: (a) 305.9 kPa, (b) 134.2°C, (c) 253.6 kJ

6-2-20: 270 kJ

6-3-1: (a) 7.534 kg/s, (b) 1.85 kW, (c) 37%

6-3-3: (a) 67.4%

6-3-11: (a) 5.79 MW, (b) 86.35%, (c) 790 kW, (d) 844 kW

6-3-15: (a) 2.98 kW, (b) 2.98 kW

Chapter 7

7-1-2: (a) 1.33 kJ, (b) 1333 kPa, (c) 60 kg/h

7-2-2: (a) 70%, (b) 0.30 kJ

7-3-5: (a) 56.5%, (b) 710.3 kPa, (c) 51%

7-4-3: (a) 2691 K, (b) 55.9 %, (c) 1201 kPa, (d) 48.9 %

7-5-2: (a) 68%, (b) 0.527 MPa

Chapter 8

8-1-2: (a) 520 K, (b) 40.8%, (c) 42.7%

8-1-4: (a) 48.23 %, (b) 1135.8 kW, (c) 58.9 %

8-1-14: (a) 39.15%, (b) 277.8 (c) 46.65%

8-2-3: 6.5 kN

8-2-18: (a) 842 kPa, (b) 1328 m/s, (c) 1467.6 m/s

8-3-1: (b) 48.2%, (c) 57.7%

Chapter 9

9-1-1: (a) 45.7 %, (b) 37.9 %, (c) 1317 kJ/kg

9-1-4: (a) 27 %, (b) 28.6 %

9-1-18: (a) 1.29 MPa, (b) 43.9%, (c) 44.3%

9-2-1: (a) 2902 kW, (b) 684 kW, (c) 719 kW, (d) 748.6 kW

9-2-9: (a) 1.6 kg/s, (b) 8.3 MW, (c) 64.7%, (d) 65.6%

9-3-2: (a) 0, (b) 39.52%

Chapter 10

10-1-2: (a) 9.70 kW, (b) 1.42 kW, (c) 6.83

10-1-7: (a) 1.165 tons, (b) 1.36 kW, (c) 3.01, (d) 3.33

10-1-26: (a) 167.5 kW, (b) 2.52, (c) 139.6 kW, (c.2) 2.45

10-1-28: (a) 0.2215, (b) 29.9 tons, (c) 3.49, (d) 3.18

10-1-35: (a) 0.075 kg/s, (b) 3.15, (c) 3.05

10-2-2: (a) 3.24 kW, (b) 1.57 kW, (c) 2.06, (d) 1.12

10-2-15: (a) 0.12 kW, (b) 0.069 kW, (c) 1.69

10-3-11: (a) −0.00441 kW, (b) 12.97 kW, (c) 6.15

10-4-2: (a) 2.83, (b) 7.65 ton, (c) 2.704 kW, (c.2) 1.107 kW

10-4-3: (b) 39.15%, (c) 2.83

Chapter 11

11-2-3: (a) −211.5 kJ, (b) −226.7 kJ/kg, (c) −249.0 kJ

11-2-6: (a) −120.96 kJ/kg, (b) −193.02 kJ/kg, (c) −42.71 kJ/kg

11-2-17: (a) 48.9 kW, (b) 24.4 kW

11-3-2: (a) 39.2 kg/kmol, (b) 2.47 m^3

11-3-7: (a) 12.1 MPa, (b) 17 MPa

11-3-13: (a) 38.57°C, (b) 264.2 kPa, (c) 35.7°C

11-3-26: (a) 1.98 kJ/kg · K

Chapter 12

12-1-2: (a) 0.01333, (b) 79.35%, (c) 12.98m^3

12-1-7: (a) 222.60 kg, (b) 2.52 kg,

12-2-1: (a) 4242.8 kJ/min, (b) 19.85%, (c) 7009.1 kJ/min

12-2-6: (a) 0.0053, (b) 5.067 kJ/kg d.a.

12-2-13: (a) 4.057 kg/min, (b) 67.1 tons, (c) 97.8 kW, (d) 28.9%

12-2-19: (a) 45.65 %, (b) 0.00239 kg H_2O/ kg dry air.

12-2-27: (a) 58.01 m^3/s, (b) 1.27 kg/s, (c) 1.91 kg/s

12-3-1: (a) 2.58 kg, (b) −8973 kJ, (c) −6733 kJ

Chapter 13

13-1-2: (a) 13.2 kg air/kg fuel, (b) 11.9 kmol air/kmol fuel, (c) 14.7 kg air/kg fuel

13-1-7: (a) 3.57, (b) 1.19

13-1-13: (a) 29.8 kg air/kg fuel, (b) 186%

13-1-24: (a) 126%, (b) 0.791

13-2-2: (a) 78.6 %, (b) 1836 K

13-2-3: (a) 55530 kJ/kg, (b) 53720 kJ/kg

13-2-13: (a) 1.60 kg/h, (b) 1.57 kg/h

13-2-19: (a) 22488 kJ/kg

13-2-28: (a) −41.32 kW, (b) −41.1 kW

13-3-2: (a) 3866.4 K, (a.2) 205.8 MPa, (b) 2911.28 K, (b.2) 154.9 MPa

13-3-8: (a) 553.41 kPa, (b) −1336.91 kJ

Chapter 14

14-1-7: (a) −61.569 MJ/kmol, (b) −55.826 MJ/kmol, (c) −755.695 MJ/kmol

14-1-11: (a) −40.965 MJ/kmol

14-3-1: (a) 50

14-3-5: (a) 1, (b) 1, (b.2) 2, (b.3) 2.

14-3-14: 0.5615 kmol

14-3-24: 50.6%

14-3-44: (a) 248,890 kJ/kmol, (b) No

14-4-5: (a) 618.62 kW, (b) 440.22 kW

Chapter 15

15-1-1: (a) 61.55 kg/s, (b) 24.89 kg/s

15-1-6: (a) 513.4 kPa, (b) 476.6 kPa

15-1-11: (a) 609 m/s, (b) 618 m/s

15-1-27: (a) 474°C, (b) 1.14 MPa

15-2-1: (a) 6.77 kg/s, (b) 7.17 kg/s, (c) 7.17 kg/s

15-2-7: (a) 0.857 kg/s, (b) 0.969 kg/s, (c) 0.969 kg/s

15-2-14: 6.5 kN

15-2-20: (a) 100.3 KPa, (b) 478 m/s

15-2-22: (a) 81.69 KPa, (b) 268 kJ/kg

15-2-24: (a) 884.93 kPa

15-3-12: 296 kPa

INDEX

UNIT CONVERSION (ENGLISH TO SI) AND SOME IMPORTANT CONSTANTS

(A more comprehensive resource is the Converter TESTcalc or the Android app Engineering Unit Converter)

Length/Velocity:

$1\,\text{ft} = 0.3048\,\text{m}$	$1\,\text{in} = 25.4\,\text{mm}$	$1\,\text{mm} = 1 \times 10^{-3}\,\text{m}$
$1\,\text{mile} = 1.61\,\text{km}$	$1\,\text{mile (nautical)} = 1.85\,\text{km}$	
$1\,\text{mile/hour} = 0.447\,\text{m/s}$	$1\,\text{km/hour} = 0.2777\,\text{m/s}$	

Volume/Flow Rate:

$1\,\text{ft}^3 = 0.02832\,\text{m}^3$	$1\,\text{gal} = 3.785\,\text{L}$	$1\,\text{L} = 1 \times 10^{-3}\,\text{m}^3$
$1\,\text{in}^3 = 16.387\,\text{mL}$	$1\,\text{mL} = 1 \times 10^{-3}\,\text{L}$	$1\,\text{quart} = 0.9464\,\text{L}$
$1\,\text{ounce} = 29.574\,\text{mL}$	$1\,\text{pint} = 0.473\,\text{L}$	
$1\,\text{ft}^3/\text{min (cfm)} = 4.72 \times 10^{-4}\,\text{m}^3/\text{s}$	$1\,\text{gal/hour} = 1.0514 \times 10^{-6}\,\text{m}^3/\text{s}$	

Mass:

$1\,\text{slug} = 14.594\,\text{kg}$	$1\,\text{lbm} = 0.4536\,\text{kg}$	$1\,\text{g} = 1 \times 10^{-3}\,\text{kg}$
$1\,\text{ounce} = 28.35\,\text{g}$		
$1\,\text{Ton (long)} = 1016\,\text{kg}$	$1\,\text{Ton (short)} = 2000\,\text{lbm} = 907.1847\,\text{kg}$	

Force:

$1\,\text{lbf} = 4.448\,\text{N}$	$1\,\text{kgf} = 9.81\,\text{N}$	$1\,\text{N} = 1 \times 10^{-3}\,\text{kN}$
$1\,\text{Ounce} - \text{Force} = 0.278\,\text{N}$	$1\,\text{Dyne} = 10\,\text{mN}$	$1\,\text{KIP} = 4.44.2\,\text{kN}$

Pressure:

$1\,\text{psi} = 6.895\,\text{kPa}$	$1\,\text{bar} = 100\,\text{kPa}$	$1\,\text{inch of Hg} = 3.374\,\text{kPa}$
$1\,\text{inch of water} = 0.2486\,\text{kPa}$	$1\,\text{Torr} = 1\,\text{mm of Hg}$	$1\,\text{mm of Hg} = 0.1333\,\text{kPa}$

Energy/Power:

$1\,\text{Btu} = 1.055\,\text{kJ}$	$1\,\text{MJ} = 1 \times 10^3\,\text{kJ}$	$1\,\text{J} = 1 \times 10^{-3}\,\text{kJ}$
$1\,\text{erg} = 1\,\text{mJ}$	$1\,\text{Therm} = 105.5\,\text{MJ}$	$1\,\text{kWh} = 3.6\,\text{MJ}$
$1\,\text{Cal (food)} = 4.187\,\text{kJ}$	$1\,\text{calorie} = 4.187\,\text{J}$	$1\,\text{kilo} - \text{calorie} = 4.187\,\text{kJ}$
$1\,\text{Horsepower} - \text{hour} = 2.6845\,\text{MJ}$	$1\,\text{ft} - \text{lbf} = 1.3558\,\text{J}$	
$1\,\text{Btu/ft}^3 = 0.0373\,\text{MJ/m}^3$	$1\,\text{Btu/lbm} = 2.3258\,\text{kJ/kg}$	$1\,\text{cal/g} = 4.187\,\text{kJ/kg}$
$1\,\text{Btu/hour} = 0.2931\,\text{W}$	$1\,\text{Horsepower} = 0.7457\,\text{kW}$	
$1\,\text{Ton of Refrigeration} = 3.517\,\text{kW}$		

Temperature:

$T(\text{K}) = T(°\text{C}) + 273$	$\Delta T(\text{K}) = \Delta T(°\text{C})$	$T(°\text{F}) = 1.8T(°\text{C}) + 32$
$\Delta T(°\text{F}) = 1.8\Delta T(°\text{C})$	$T(\text{R}) = 1.8T(\text{K})$	$\Delta T(\text{R}) = 1.8\Delta T(\text{K})$

Specific Heat:

$1\,\text{J/(g} \cdot °\text{C)} = 1\,\text{kJ/(kg} \cdot °\text{C)} = 1\,\text{kJ/(kg} \cdot \text{K)}$

$1\,\text{Btu/(lbm} \cdot °\text{F)} = 4.187\,\text{kJ/(kg} \cdot \text{K)}$

$1\,\text{Btu/(lbm} \cdot \text{R)} = 4.187\,\text{kJ/(kg} \cdot \text{K)}$

$1\,\text{Btu/(1bmol} \cdot \text{R)} = 4.187\,\text{kJ/(kmol} \cdot \text{K)}$

Thermal Conductivity:

$1\,\text{W/(m} \cdot °\text{C)} = 1\,\text{W/(m} \cdot \text{K)} = 1 \times 10^{-3}\,\text{kW/(m} \cdot \text{K)}$

$1\,\text{Btu/(ft} \cdot \text{hr} \cdot °\text{F)} = 1.7307\,\text{W/(m} \cdot \text{K)}$

Absolute and Kinematic Viscosity:

$1\,\text{Pa.}s = 1\,\text{N.}s/\text{m}^2 = 1\,\text{kg/(s} \cdot \text{m)}$ (absolute)

$1\,\text{poise} = 0.1\,\text{Pa.s}$ $1\,\text{lbm/(ft} \cdot \text{s)} = 1.4882\,\text{Pa.s}$ (absolute)

$1\,\text{stoke} = 100\,\text{m}^2/\text{s}$ (kinematic)